Applications of Computational Tools in Drug Design and Development

S. N. Koteswara Rao G.
Rajasekhar Reddy Alavala
Editors

Applications of Computational Tools in Drug Design and Development

Springer

Editors
S. N. Koteswara Rao G.
Shobhaben Pratapbhai Patel School of
Pharmacy and Technology Management
SVKM's NMIMS Deemed to be University
Mumbai, Maharashtra, India

Rajasekhar Reddy Alavala
Shobhaben Pratapbhai Patel School of
Pharmacy and Technology Management
SVKM's NMIMS Deemed to be University
Mumbai, Maharashtra, India

ISBN 978-981-96-4153-6 ISBN 978-981-96-4154-3 (eBook)
https://doi.org/10.1007/978-981-96-4154-3

This Springer imprint is published by the registered company Springer Nature Singapore Pte Ltd.
The registered company address is: 152 Beach Road, #21-01/04 Gateway East, Singapore 189721, Singapore

Dedicated to all the researchers, faculty, and students
May this book serve as a valuable resource in your journey toward transforming computational insights into therapeutic breakthroughs

Preface

The world of drug discovery and development is a complex, multifaceted field that addresses the pressing challenges in disease treatment. As scientific understanding advances and technology progresses, the integration of computational tools has become increasingly vital, offering new dimensions of insight and precision to researchers. Initially focusing on drug discovery, the computational tools have now amalgamated every aspect of drug discovery and development, including drug discovery, drug delivery, biopharmaceutics, and pharmacokinetics, as well as pharmacology and toxicology. Not many books touch on all of these aspects. Offering a broad perspective, this book explores the computational tools utilized in every phase of drug discovery and development.

Chapters 1–10 of the book delves into applications of computational tools for drug delivery. This part covers application of computational tools for preformulation studies as well as use of design of experiments (DoE) software for optimizing the formulation parameters. Another emerging area of focus is the delivery of drugs to the lungs. The Chap. 7 focuses about computational fluid dynamics and is used to simulate the flow of drug aerosol in the respiratory track and its application in the design and optimization of aerosol-based delivery systems. Likewise, the use of simulation to analyze and optimize key parameters in the pharmaceutical die filling process is discussed in this part.

Chapters 11–19 of the book explores the computational tools for predicting biopharmaceutical as well as pharmacokinetic parameters. Detailed in this part are computational tools for the prediction of solubility, advances in computational prediction of absorption as well as permeation, and forecasting protein binding using computational tools. Also included in this part is a detailed explanation about visualizing and understanding the interactions between drug molecules and excipients using simulations and how this information is used for the selection of best excipient under the given conditions.

Chapters 20–27 focuses on specific therapeutic areas. Through the mechanistic understanding of diseases such as Alzheimer's, diabetes, atherosclerosis, and cancer, this section highlights how computational tools such as QSAR, pharmacophore mapping, docking, and homology modeling as well as network analysis are being used for design of better drugs. The design of drugs against epilepsy and hypertension is also included in this part. The final chapters explore drug discovery and computational strategies in the context of multidrug-resistant tuberculosis and the network pharmacology approach to uncover the pharmacological mechanisms of natural products. Drug repurposing, highlighted during COVID, although practiced much before it, in the treatment of cancer is also discussed in this part.

This book offers a holistic view of multiple computational tools used at various stages of drug discovery and development. It will be a valuable resource to researchers and students working in this field.

Urmila J. Joshi
Retd. Principal and Head of Department
of Pharmaceutical Chemistry
Principal K. M. Kundnani College of Pharmacy
Mumbai, Maharashtra, India

Foreword

The book *Applications of Computational Tools in Drug Design and Development* is a pleasant blend of computational methods employed in both drug discovery and drug development. It covers how computational methods can be used at all stages in the drug discovery and development process. There are not many volumes that have been published to date that comprehensively cover how *in silico* methods are used in drug discovery and development, and therefore, this volume will be welcomed by all those working in drug discovery and development.

Understanding the mechanism of action of a drug is central to the development and design of new drugs. A variety of experimental methods is used to unravel the mechanism of action of drugs. The use of computational tools like network analysis (STRING, Cytoscape, DAVID, etc.) and molecular modeling tools like docking, QSAR, receptor-based pharmacophore mapping, etc. to understand the pathological mechanisms of anti-Alzheimer agents and drugs used for the treatment of epilepsy and atherosclerosis has been illustrated in Chaps. 20, 22, and 24.

Chemoinformatics, which combines concepts and methods in chemistry, computer science, and information technology, is used to streamline the process of identifying and developing new leads. The tools of chemoinformatics for storage, analysis, visualization, and interpretation of chemical data, and their significance in drug discovery and development are covered in Chap. 9.

Drug likeness aims to predict how promising a molecule is to be developed as a drug. Drug likeness as embodied by Lipinski's rule-of-five has been a guiding principle to identify molecules that are most likely to succeed as novel therapeutic agents. Tools to assess drug likeness have been described in Chap. 19 titled "Recent Advances in Drug-Likeness Screening by Using the Software and Online Tools."

In silico approaches have revolutionized drug discovery by decreasing the time to identify hits, optimize leads, and develop a clinical candidate. The use of *in silico* methods like molecular docking, molecular dynamics (MD) simulations, QSAR, and protein structure determination (homology modeling) to identify new targets and to decipher drug receptor interactions coupled with the prediction of ADME attributes have been exemplified in Chap. 21 in the design of novel antidiabetic agents, and in Chap. 25 in the design of novel antihypertensive agents.

The safety aspect of a drug is an integral part of the drug design process, and the accurate prediction of toxicity helps mitigate the risk associated with the appearance of life-threatening effects during preclinical or clinical studies. Toxicity databases such as ToxCast, ACToR (Aggregated Computational Toxicology Resource), and PubChem can be queried to assess the potential toxicity of known or structurally related substances. Additionally, toxicogenomics tools like GeneGo (MetaCore) and pathway analysis tools, along with in silico toxicity prediction platforms such as ADMET Predictor, the Tox21 collaboration, and DEREK Nexus, can be used to evaluate compound toxicity across multiple biological endpoints. Some toxicity prediction tools are covered in Chap. 16.

Repurposing an existing therapeutic for a new medical condition has several key advantages, such as reduced time to bring the drug to market, since the safety profile of the drug is already established. Viral-based therapies like gendicine, oncorine, and imlygic are examples of therapies that have been repurposed for cancer treatment. These examples and the initiatives by the National Center for Advancing Translational Sciences (NCATS) are covered in Chap. 23 titled "Modern Computational Intelligence Based Drug Repurposing for Cancer."

The appearance of drug resistance across various therapeutic segments is posing serious threats to public health in terms of failure of treatment, development of chronic infections, escalation in the cost of treatment, and narrowing the options available for treatment. The appearance of MDR, XDR, and TDR-TB is creating havoc worldwide in the management of this disease. Chapter 26 on "Drug Discovery and Computational Strategies in Multi-Drug-Resistant Tuberculosis" attempts an overview of various computational strategies, like structure-based and ligand-based, that have been used in the development of new drugs for the treatment of MDR-TB.

Natural products, with their diversity of chemical structures, have been the founding structures on which many therapeutic agents have been designed. The vast untapped biodiversity found in marine ecosystems, tropical rainforests, and the animal world holds the promise of discovery of novel compounds with unique therapeutic attributes. Chapter 27 titled "The Network Pharmacology Approach to Uncover the Pharmacological Mechanism of Natural Products" expounds a multidisciplinary approach based on systems biology and polypharmacology.

After the target has been identified and validated, hits recognized, and leads optimized, preclinical studies that evaluate the solubility, safety, efficacy, drug binding, and pharmacokinetics of lead compounds in laboratory and animal models prior to human testing is conducted. All these aspects are covered in Chaps. 11, 12, 13, and 15. A pivotal aspect of preclinical drug development is the assessment of solubility of the drug candidate. This is essential for formulation design and satisfactory pharmacokinetics. The shake flask method is the gold standard for measurement of solubility; however, quantum mechanical calculations based on DFT, molecular dynamics simulations, QSPR, and machine learning models offer cheaper and accurate ways to model solvation energy. These computational tools are covered in Chap. 11 titled "Computational Tools for Solubility Prediction." Along with solubility, another fundamental theme in preclinical studies is the prediction of absorption, distribution, metabolism, and excretion (ADME) of the drug. Understanding absorption is essential for predicting a drug's bioavailability, its dosing regimen, and potential effectiveness. Simcyp, GastroPlus, SwissADME, and QikProp are some of the finest tools used for prediction of ADME attributes; these could have been covered in depth along with the other methods described in Chap. 12 titled "Advances in the Computational Prediction of Absorption Prediction of Pharmaceuticals." The amount of drug available for interaction with the target is dictated by the fraction bound/unbound (free) to plasma proteins (e.g., albumin, a1-acid glycoprotein, lipoproteins, etc.) and is a key parameter that is studied in the preclinical phase of drug development. Computational tools used to study ADME, described above, have modules to assess the fraction of drugs bound to plasma proteins. Case studies of how computational methods have been used to optimize protein binding and hasten the drug development process are elucidated in Chap. 13 titled "Protein Binding Prediction by Computational Methods." Biopermeability studies are linked with the prediction of drug absorption and are used to understand how well a drug will cross biological membranes to enter systemic circulation. Poor permeability may necessitate modifications in the chemical structure or alterations in the formulation development. The use of advanced molecular dynamics simulations to predict the bioavailability of pharmaceuticals and its relationship to experimental data has been discussed in Chap. 15.

One of the important stages in drug development is formulation design. In this phase, the focus is on developing a stable, effective, and safe formulation that can be used in human clinical trials. Decisions must be made about the dosage form, whether tablet or injection or any other, and the method of delivery. Computational methods for design of the formulation and mode of delivery—oral, ophthalmic, parenteral, or inhalation—are covered in Chaps. 1, 2, 8, and 10. There are several challenges in the delivery of drugs to the lungs. Factors such as the design of the device, formulation aspects like particle size, dose control, and patient competence all dictate how efficiently the drug is delivered to the lungs. Computational fluid dynamics (CFD) has played a significant role in the aerospace industry and is now making a deep impact in the design and optimization of inhalers, in formulation design (Chaps. 3 and 7), and in the die filling process during tablet compression (Chap. 6). Exhaustive simulations can be conducted to get insights into airflow patterns, particle behavior, and sites of drug deposition within the respiratory system.

The underlying relationships between the dependent and independent variable in a process can be understood by Design of Experiments (DOE). How DOE can be used to optimize the effectiveness of drug delivery systems has been well explained in Chap. 5 titled "Design of Experiments: Understanding Optimization."

A significant percentage of the excipients in pharmaceutical formulations are polymers—natural, synthetic, biodegradable, or mucoadhesive. Computational methods (Chap. 18) are used to understand how a formulation chemist can select the "right" polymer to avoid drug-polymer interactions that could limit the stability, release, or bioavailability of the drug.

All drug discovery programs begin with target identification and target validation. Discussion of this facet of drug discovery, which is missing in this volume, may be covered in a future edition of this book.

In essence, this volume gives a comprehensive insight into the numerous computational methods and tools that are used in various stages of drug discovery and development. This volume would benefit both the novice and experienced pharmaceutical researchers and scientists, clinical researchers, medical professionals, graduate students in a drug discovery program, and the regulatory agencies and professionals.

Evans Coutinho
Dean Research
St John Technical and Educational Campus
Palghar, Maharashtra, India

Acknowledgments

We express our deepest gratitude to **Mr. Fumiko Yamaguchi**, Senior Editor, Books, Biomedical and Life Sciences, at Springer Nature, for his invaluable support in accepting our book proposal and providing exceptional guidance throughout the development of this book. His expertise and encouragement have been instrumental in bringing this work to fruition. We would also like to extend our sincere thanks to **Ms. Kavitha Jayakumar**, Production Editor (Books), for her dedicated assistance in overseeing the production process of this book.

We extend our heartfelt thanks to the **Dean, SPPSPTM** and **all the authorities of SVKM's NMIMS**, Mumbai, for providing the necessary facilities and fostering an environment of encouragement and academic excellence. Their unwavering support has greatly contributed to the successful completion of this project.

We are profoundly grateful to all the **contributing authors** whose dedication, expertise, and commitment to advancing knowledge have enriched this book. Their collective efforts have made this endeavor possible, and we deeply value their contributions.

Finally, we acknowledge the support and encouragement of our families, colleagues, and friends, whose belief in this project has been a source of motivation throughout this journey.

Regards
Dr. S. N. Koteswara Rao G.
Dr. Rajasekhar Reddy Alavala

Contents Volume-II

Contents Volume-I

Chapter 13
Protein Binding Prediction by Computational Methods

Ritam Mondal, Deepika Paliwal, Aman Thakur, Rakesh Sahu, and Roja Rani Budha

Contents

R. Mondal · D. Paliwal (✉) · R. Sahu
School of Medical & Allied Science, Galgotias University, Greater Noida, Uttar Pradesh, India

A. Thakur
DCO, Government of Rajasthan, Bharatpur, Rajasthan, India

R. R. Budha
H.K. College of Pharmacy, Oshiwara, Jogeshwari (W), Mumbai, Maharashtra, India

S. N. K. Rao G., R. R. Alavala (eds.), *Applications of Computational Tools in Drug Design and Development*, https://doi.org/10.1007/978-981-96-4154-3_13

Abstract A key component of drug design and discovery is protein-ligand binding prediction, which helps to create new therapeutic approaches with higher specificity and effectiveness. In this endeavour, computational approaches have proven to be invaluable tools, providing insights into the sophisticated chemical interactions among proteins and ligands. This chapter presents a thorough review of the methods, techniques, and applications of the computational algorithms used to predict protein binding. An overview of the importance of protein-ligand interaction in drug development and the difficulties involved in experimentally determining binding affinities opens the chapter. The chapter covers about number of methods, including solvation effects, protein flexibility, and scoring function optimisation, that may be used to improve the precision and dependability of protein binding predictions, and also demonstrating the usefulness of computational techniques in predicting the bind modes, affinities ranking, structure-activity relationship, examples and case studies. Additionally, the chapter examines contemporary developments in the area, such as the incorporation of massive amounts of data for large-scale virtual screening, quantum physics/molecular mechanics (QM/MM) modelling, and multi-scale modelling methodologies. This chapter emphasizes about how crucial computational techniques are hastening the drug development process by making it easier to rationally designed ligands with the necessary binding characteristics.

Keywords Computational technique · Protein-ligand binding molecular docking · Molecular dynamics simulation · Quantum mechanics

13.1 Introduction

Protein binding prediction is essential for comprehending molecular interactions, clarifying biological processes, and developing efficacious therapeutic treatments, it is highly valued in many areas of medical research and drug development. Drug-target binding relies heavily on protein-ligand interactions, which also affect the safety, effectiveness, and selectivity of pharmaceuticals. Estimating protein binding affinities helps with lead compound optimisation, pharmacokinetic property prediction, and identification of possible therapeutic options [1]. Protein binding prediction-based virtual screening techniques makes it possible to quickly and affordably screen huge chemical libraries, which helps with hit discovery and lead optimisation in drug development pipelines. Deciphering the molecular mechanisms behind biological processes, including signal transmission, enzyme catalysis, and gene regulation, requires an understanding of protein binding interactions. Understanding cellular signalling pathways, protein complexes, and regulatory networks through thoughts of protein-protein and protein-ligand interactions sheds information on disease processes and putative treatment targets. By aiding in the prediction, modelling, and refining of protein structures, protein binding prediction advances structural biology [2]. Rational drug design is aided by the determination of binding sites on proteins, characterization to their binding modes, which permits the synthesis of ligands with enhanced binding affinity. Anticipating the reaction between proteins -ligands aids in clarifying structure-activity relationships (SAR), directing the creation of ligands with appropriate pharmacological characteristics, and mitigating off-target effects. By forecasting individual patient reactions to certain medications based on genetic differences, protein expression patterns, and pharmacogenomic data, protein binding prediction might assist personalised medicine techniques [3]. Comprehending molecular protein binding interactions facilitates the development of customised treatments aimed at certain disease indicators, patient segments, or hereditary markers. In the study of protein structure-function correlations, binding of ligand dynamics, and interactions between protein and ligand under various physiological circumstances, protein binding prediction is a useful tool in biomedical research. The research and development of proteins with improved interaction specificity, enzyme activity, or therapeutic value is made easier by computational approaches for protein binding prediction, which advances the field of biotechnology and medicine [4]. Nearly every biological activity depends on interactions between proteins. Understanding the binding affinity as well as the model of protein with protein complexes is important to comprehending biological role of protein with protein interactions. Although slight, transient changes in protein with protein reactions can have important operational implications during signal transduction, for the cell or metabolic regulation. A greater range of protein–protein complex compositions have been determined recently. When paired with template-based or docking approaches, it is able possible to create crystalline representations to several possible protein with protein complex and to generate fresh protein with protein interactions. Accurate affinity binding forecasting is becoming more important in determining the functional importance of predicted or hypothesised protein-protein complexes. Several AI tools, starting with force-field or Many techniques may be found to establish the relative or exact binding affinity for complexes, such as scoring of individual protein with protein complexes, group oriented techniques, and accurate simulations of binding free energy [5]. With the establishment of molecular modelling models in the years 1960 and 1970, computational approaches for analysing protein binding became available. Researchers started using docking algorithms in the 1980s to forecast how tiny compounds will attach to protein targets and how strongly they would bind. Protein-ligand interactions began to be studied in the 1990s using quantum mechanics techniques including density functional theory, or and quantum mechanics (QM/MM) computations. In the 1990s, the use of molecular dynamics simulations to examine protein with ligand interactions grew in popularity. In the twenty-first century, protein with ligand interactions have been predicted via the application of algorithms for learning (ML) and computational intelligence (AI) approaches [6].

13.1.1 Role of Computational Techniques in Protein Binding Prediction

In contemporary medical studies and drug development, computational techniques play a critical role in protein binding prediction. These approaches make use of modelling tools, simulations, and computer algorithms to clarify the molecular interactions that occur among proteins compare with molecules, including acid with nucleic acid, amino acid chain, tiny ligand, other proteins. Quantitative information on the intensity of interactions among proteins and ligands may be obtained by estimating the binding affinities between them using computational techniques [7]. Computational approaches assist in the rational design of pharmaceutical medicines with desired pharmacological features by modelling ligand-protein complexes and computing binding free energies. This helps prioritise lead compounds and forecast their binding potency. In order to make it easier to identify possible binding pockets and important residues that contribute to ligand recognition, molecular docking algorithms estimate the position and mode of binding of tiny molecules within protein binding sites. Through the use of computational docking simulations, ligand-protein interactions may be investigated at the atomic level, which aids in the thoughtful creation of ligands with the best possible binding geometry and specificity [8]. Molecular dynamics simulations offer valuable insights into the stability, adaptability, and binding kinetics of protein with ligand complexes by capturing their dynamic habit and conformational correction over time. Protein-ligand interactions are influenced by solvent interactions, environmental variables, and binding and unbinding processes, all of which are clarified by MD simulations [9]. Massive collections of protein-ligand complexes are analysed using computer methods and machine learning techniques in order to identify significant characteristics and create prediction models for affinity for binding, specificity, and selectivity. Random forests, vector Support machines (SVM), and deep neural networks are examples of machine-learning predictive structures that are useful for lead optimisation, virtual screening, and medication repurposing. Docking of molecules, MD simulations, machine learning, and other computational approaches are used in hybrid methods to improve the precision, dependability of protein interaction predictions.

To create new structural models, support binding theories, and forecast the impact of mutations or modifications to conformation on protein with ligand interactions, integrative modelling techniques integrate experimental data with computer simulations. High-throughput screening of huge chemical libraries is made possible by computational approaches, which also expedite the process of finding new drugs by ranking the most likely candidates for validation in experiments. Computational approaches make it easier to explore chemical space, optimise leads, and find new therapeutic targets by cutting down on the time and expense involved in experimental screening [10].

13.2 Molecular Docking

13.2.1 The Fundamentals of Molecular Docking

The docking of molecules is a computer a technique that forecast between the direction, intensity for the small molecule to a target protein's receptor. It is used in fields of structural biology, bio informatics and drug development. Both ligand and the protein receptor must ready for docking. To do this, water molecules must be eliminated, hydrogen atoms must be added, and the structures must be optimised to guarantee that they are in the right conformations for docking. Scoring functions are used by docking algorithms to assess a ligand's fitness within a binding site [11]. Typically, these functions take into account things like electrostatic interactions, hydrogen bonding, steric conflicts, breakdown energy, and changes in entropy following binding. To investigate structural environment of the ligand inside protein attachment, docking methods use search techniques. Lamarckian Genetic Algorithm (LGA), and Genetic algorithms are examples of common search algorithms. For docking to successful, the site of bind of protein must verified. Binding site can be found using computational techniques like binding site algorithmic forecasting or experimental techniques like NMR spectroscopy and X-ray crystallography. Certain situations may include changes in conformational in both the protein-ligand during interaction. Flexibility is explained by flexible docking techniques, which permit the protein-ligand to modify their simultaneously conformations throughout the docking process [11]. Validating docking results requires comparing experimental data with expected binding energies and modes. Validation aids in evaluating the docking predictions' correctness and dependability. The docking findings are analysed using visualisation techniques, such as calculating binding energies, visualising the binding modes, and identifying significant interactions that occur between the ligand & the protein [12, 13]. The steps of protein-ligand interaction are shown in Fig. 13.1.

13.2.2 Docking Programs

The prediction and analysis of interactions among proteins and various other atoms such as ligands or other proteins or peptides are done computationally using protein binding algorithms. These programs model interactions between proteins or between proteins and ligands using a variety of techniques. These are a few well-known binding protein algorithms that are utilised in drug development and computational biology [14]. The molecular docking classification are shown in Fig. 13.2.

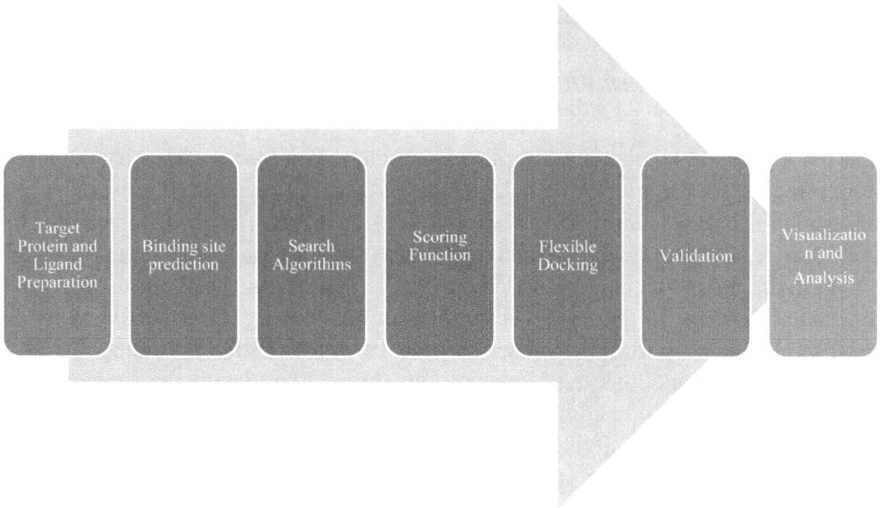

Fig. 13.1 Steps of protein-ligand interaction

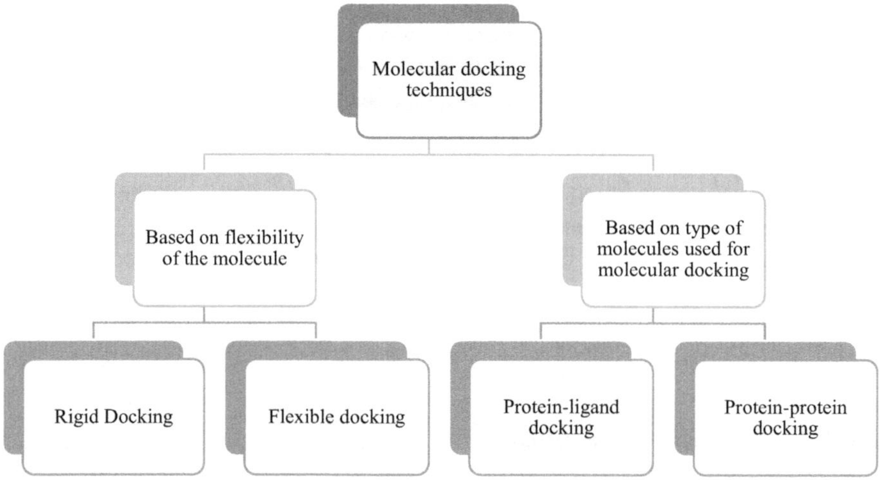

Fig. 13.2 Molecular docking techniques classification

13.2.2.1 Based on Flexibility of the Molecule, They Are of Two Types

- Rigid Docking- In structural biology, rigid docking is a technique which compute that to predicts the three-dimensional configuration of molecules—typically proteins or protein-ligand complexes—when they connect. The molecules in this approach are regarded as hard bodies, which means that throughout the docking process, their structures cannot alter.

Rigid docking seeks to anticipate the most energetically advantageous spatial configuration of the molecules, known as the "docking pose," taking into account a number of variables including hydrogen bonding, the forces of electrostatic interactions and van der Waals. Understanding a binding affinity of molecules is important for both drug development and biological process comprehension, and these predictions can shed light on these relationships.

In rigid docking algorithms, the most advantageous binding mode is usually found by sifting through a huge number of potential molecule orientations and conformations. Computational methods including geometric matching, simulations using Monte Carlo, and molecular dynamics simulations are frequently used in this procedure.

Rigid docking has drawbacks even if it offers insightful information about molecular interactions. For instance, it might not correctly represent the conformational changes that many biological systems experience during binding. Flexible docking techniques have been developed to overcome this restriction by permitting a certain amount of structural movement in the molecules during the docking process. Rigid docking is still a popular and helpful method, nonetheless, especially for preliminary screening and investigation of protein-ligand interactions [15].

- Flexible Docking: In the context of structural biology and drug development, flexible docking—also referred to as flexible the ligand connecting or induced-fit docking—is a technique which compute that predicts determine a ligand—usually a small molecule—will bind to a receptor—typically a protein—while permitting flexibility between the ligand and the protein target.

Flexible docking considers the conformational adaptability of the molecules, in contrast to rigid docking, which presumes that ligand and receptor retain stable configurations during the docking process. This adaptability is crucial because, upon interacting to one another in several biological interactions, the ligand nor the receptor may experience conformational changes. The interaction's specificity and binding affinity may be strongly impacted by these modifications [16].

Generally, flexible docking techniques consist of two primary steps:

- Conformational sampling: This technique looks for energetically favourable conformations that are expected to arise during the binding process by examining the conformational space between the ligand and the receptor. Numerous methods, including systematic search algorithms, and MD simulation, can be used to carry out this sampling [17].
- Docking and scoring: Following conformational sampling, a series of variables, with hydrogen bond, desolvation effects, electrostatic reaction, and van der Waals forces are used to score various ligand-receptor complexes in order to assess their binding affinity. The most energetically advantageous binding mode is then predicted by the docking algorithm, which takes the effectibility of ligand and protein into consideration [18].

When examining protein-ligand interactions wherein conformational changes are important, such in enzyme-substrate interaction or receptor-ligand recognition,

Fig. 13.3 Steps of protein-ligand docking

flexible docking techniques are very helpful. Flexible docking techniques take flexibility into account, which can lead to more precise estimates of binding affinities as well as the identification of new drug candidates with higher potencies and selectivity. However, because conformational sampling is more difficult in flexible docking than in rigid docking, it requires more computing power [19].

13.2.2.2 Based on Type of Molecules Used for Molecular Docking, They Are of Two Types

- Protein-ligand docking: To predict the chemical reactions between ligands, or small molecules, and protein targets in drug development, a computer approach known as protein-ligand docking is employed. By forecasting their binding affinities and modes inside the protein's active region, it mimics the binding process and aids in the identification of putative therapeutic candidates [20]. The steps of protein-ligand docking are shown in Fig. 13.3.
- Protein-protein docking: A computer technique called protein with protein docking is found that forecast three-dimensional form to complexes made up of at least two protein molecules. By simulating the process of protein binding, it sheds light on the relationships and functional functions that proteins play in biological processes. By locating possible targets for therapeutic intervention, this approach can help in drug development and is essential for comprehending protein-protein interactions [21]. Popular docking programs used in prediction of the binding protein interactions, Docking Algorithms, Scoring Functions are shown in Tables 13.1, 13.2, and 13.3.

13.3 Molecular Dynamics Simulation

13.3.1 Fundamental of Molecular Dynamic

Computational techniques called simulations of molecular dynamics are working to look over the motions and interaction of atoms and molecules across time. MD simulations provide important insights into how protein ligand complexes behave dynamically, including changes in conformation, binding kinetics, and interaction thermodynamics, when it comes to forecasting protein binding [22]. The relationships among atoms and molecules are described in MD simulations using

Table 13.1 Popular docking programs used in prediction of the binding protein interactions

Name of programs	Working	Founder	website
Autodock	A popular molecular docking programme for predicting small molecule ligand binding affinities and modes to protein targets is called AutoDock.	Scripps Research Institute	https://autodock.scripps.edu/
Autodock Vina	Agradient-based optimisation approach is used by AutoDock Vina, an enhanced version of AutoDock, to provide quicker and more precise docking predictions.	Dr. Oleg Trott	https://vina.scripps.edu/
Glide	Schrödinger created the molecular docking programme Glide, which effectively searches the conformational environment of ligands inside protein binding sites using a hierarchical docking technique.	Schrödinger	https://newsite.schrodinger.com/platform/products/glide/
RosettaLigand	A part of the Rosetta software package, RosettaLigand combines sampling and scoring routines to provide customisable ligand docking and scoring.	Andrew Leaver-Fay	https://meilerlab.org/rosetta-ligand/
HADDOCK	A versatile docking programme called HADDOCK (High Ambiguity Driven biomolecularDOCKing) predicts protein-protein and protein-ligand complexes by fusing computational techniques with experimental data.	Arnau Navarro	https://wenmr.science.uu.nl/
HEX	Using a spherical polar Fourier correlation approach, Hex is a quick and effective molecular docking programme that finds the ideal binding orientation for protein-protein or ligand-protein complexes.	Curtis Lary	https://hex.loria.fr/
GOLD	A dynamic algorithm-based docking programme called GOLD (Genetic Optimisation for Ligand Docking) optimises ligand conformations and interactions with proteins to forecast binding poses and affinities.	GOLD	https://www.ccdc.cam.ac.uk/solutions/software/gold
FlexX	FlexX is a docking molecular programme that predicts protein-ligand binding modes by exploring ligand conformational space using a flexible ligand docking technique.	Christine Ju	http://biophys.umontreal.ca/nrg/resources.html

(continued)

Table 13.1 (continued)

Name of programs	Working	Founder	website
DOCK	A geometric matching technique is used by the molecular docking programme DOCK to find comparable binding configurations among ligands and binding locations for proteins.	Swssdock	http://www.swissdock.ch/
PLANTS	A evolutionary algorithm-based molecular docking programme called PLANTS (Protein-Ligand Ant System) uses solvation effects and protein flexibility to predict the affinities and modalities of protein-ligand binding.	GitHub	https://github.com/discoverdata/parallel-PLANTS

Table 13.2 Docking algorithms

Name of algorithms	Working	website
Bspred	A programme called BSpred uses neural networks to detrmine binding protein sites according to amino acid sequences. The system was thoroughly trained using sequence-based information, such as the hydrophobicity levels of amino acids, secondary structure estimation, and protein sequence profiles.	https://zhanggroup.org/BSpred/
PPI Affinity	PPI-Affinity is a programme that use SVM (support vector machine) predictions of affinity for binding to produce and rank variants of a specific structure in addition to screening drafts of protein with protein and protein-amino acids complexes.	https://protdcal.zmb.uni-due.de/PPIAffinity/
InFOLD	An all-in-one interface for 3D modelling, domain prediction, intrinsic disorder prediction, tertiary structure prediction, and 3D model quality evaluation estimation of residues bound by proteins and ligands.	https://www.creative-proteomics.com/services/molecular-docking-service.html
Biskit	Biskit integrates third-party apps into automated processes.	https://bio.tools/patchdock
Phyre and phyre2	*Ab initio*, multi-templates, 3D modelling, remote template detection, and alignment.	http://www.sbg.bio.ic.ac.uk/phyre2/html/page.cgi?id=index/
ROBETTA	Ginzu domain prediction and *ab initio* fragment assembly using Rosetta homology modelling.	https://robetta.bakerlab.org/
AWSEM-suite	Molecular dynamics simulation using optimised folding landscapes that are directed by templates and enhanced by coevolution.	https://bio.tools/awsem-suite

mathematical models known as force fields. Parameters for bound interactions (stretching of bond, bending of angle etc.) and non-bonded interactions (electrostatic forces, interaction of van der waal etc.) are included to force fields. A variety of force fields with differing accuracies and applicability's have been created, such

Table 13.3 Scoring functions

Search functions	Working
Local search algorithm	These algorithms search for energetically favourable binding poses by examining the conformational space surrounding a particular ligand or protein structure. Tabu Search, Simulated Annealing, and Metropolis Monte Carlo are a few examples.
Global optimization algorithm	These algorithms investigate a wider range of conformations in an attempt to find the protein-ligand complex's global minimal energy conformation. Particle Swarm Optimisation, Ant Colony Optimisation, and Genetic Algorithms are a few examples.
Gradient based optimization	These techniques optimise binding energy of complexes of protein and ligand repeatedly by utilizing gradient information. Gradient descent, conjugate gradient, and Newton's method are a few examples.
Hybrid search algorithm	These methods improve the precision and efficiency of protein binding predictions by combining many search techniques. Memetic algorithms, hybrid genetic algorithms, and hybrid Monte Carlo methods are a few examples.
Interactive search algorithm	To increase prediction accuracy, these algorithms iteratively enhance the interaction postures and energetic of protein-ligand complexes. Some instances are: Iterative Stochastic Elimination, Iterative Docking, and Iterated Local Search.

as AMBER, CHARMM, and GROMOS [23]. The protein-ligand complex's initial conformation, usually found in docking studies or experimental structures, serves as the foundation of MD model. In order for neutralise system's charge, counterions may be injected after the system has been dissolved in a box containing water molecules. The locations and speeds of atoms are propagated across time by MD simulations using numerical of Newton's equations of motion. Algorithms such as the leapfrog integration technique and the Varlet algorithm with a predetermined time step are used to carry out this integration. In order to prevent edge effects and simulate an infinite system, periodic boundary conditions, or PBCs, are used. This guarantees that molecules cannot "feel" the simulation box's edges and enables the modelling of bulk-like behaviour. To regulate the system's temperature and pressure, respectively, MD simulations frequently use thermostats and baristas. While baristas, such as the Parrinello-Rahman barista, maintain constant pressure, common thermostat include the Berendsen and Nose-Hoover models. The length of MD simulations can change according on how long the biological process under study takes. While lengthier simulation (microseconds to milliseconds) would be required to reveal broader conformational changes, shorter models (nanoseconds to microseconds) may detect local movements and binding events. On MD trajectories, a variety of analysis may be carried out to gain understand of protein with ligand interaction. Applying methods like MM/PBSA or MM/GBSA, these studies compute the terms represent square mean root fluctuations and root mean square deviation, formation of H bond of free binding energy. Timescale constraints may be addressed and uncommon occurrences, such as ligand binding and releasing processes, can be investigated by using advanced molecular dynamics (MD)

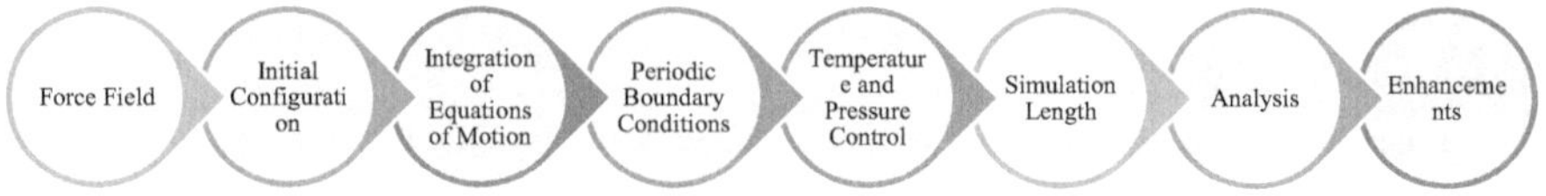

Fig. 13.4 Steps of molecular dynamics simulation

techniques, such as proper sampling approaches (such as sampling by umbrella, Meta dynamics), increase MD [24]. The steps of molecular dynamics simulation are shown in Fig. 13.4.

13.3.2 Concept of Force Field in Molecular Dynamic Simulations

Force fields are essential for docking molecules and simulations of molecular dynamics in predicting protein binding. In order to comprehend the interactions among ligands and proteins (small molecules or similar proteins) and to forecast the affinity for binding and mechanism of binding, these computational techniques are frequently employed in structural biology and drug development. Molecular dynamics (MD) computations use force fields to modulate the dynamic behaviour of ligand and protein complexes all the time. MD model investigate conformational domain of protein-ligand complexes and offer perspectives on binding kinetics, changes in conformation, and bound free energies by imposing of Newton's equations. In virtual screening of structure-based, force fields are used for forecast how well small molecule ligands will bind to protein targets. by virtual screening, possible lead compounds are identified for additional experimental confirmation by analysing the interactions among ligands and proteins binding sites by force -field simulations [25]. In subject of protein and ligand binding prediction, a number of force fields are often employed. Commonly employed force fields in MD simulations are mentioned in Table 13.4.

13.3.3 Various Analysis Tools Employed in the MD Simulations

In order to make inferences about binding interactions that are relevant to prediction to protein binding, analysis of data produced by the simulations must also be done. These represent a few of most important analytical methods, instruments for binding prediction of protein.

Table 13.4 Commonly employed force fields in MD simulations

Name of force field	Uses	Website
CHARMM (Chemistry at HARvard Macromolecular Mechanics)	Protein-ligand binding studies, among other biomolecular simulations, frequently employ CHARMM force fields. A prominent force field to nucleic acids and protein is CHARMM36.	https://www.charmm.org/
AMBER (Assisted Model Building to Energy Refinement)	Parameters of tiny molecules, proteins, nucleic acids are included within the AMBER force field suite. Protein-ligand complex molecular dynamics simulations make full use of it.	https://ambermd.org/AmberTools.php
OPLS (Optimized Potential for Liquid Simulations)	Protein properties have been added to OPLS force fields, which were first created for tiny chemical molecules. In research on protein-ligand interaction, the force fields OPLS-AA (All Atom) and OPLS-UA (United Atom) are frequently utilised.	http://www.swissdock.ch/
GROSMOS (GROningen Molecular Simulation)	Biomolecule molecular dynamics simulations commonly employ GROMOS force fields. GROMOS force fields were first created for tiny organic molecules, but parameters for proteins have since been added.	https://www.gromacs.org/
GAFF (Generalized Amber Force Field)	Force fields like GAFF, which are made especially for tiny organic molecules, are commonly employed in molecular docking research where ligands and proteins are subjected to separate force fields.	
Consensus Force Fields	To increase the precision of protein-ligand binding predictions, some research employ consensus techniques that include averaging or combining various force fields.	[26]

13.3.3.1 Compute the Affinity to Bind

Main studies carried out following the simulations is the determination of the protein with ligand figures binding affinity. Molecule of surface region of mechanics/Poisson-Boltzmann (MM/PBSA), mechanics/generalized Molecular Born area of molecules (MM/GBSA), and free energy calculations are a few techniques that may be used for this. By cracking down the binding free energy to contributions from several energetic components, including solvation effects, electrostatic interactions and van der Waals conversations these approaches calculate the binding free energy [27].

13.3.3.2 Fingerprint Interaction

The interactions among the ligand and protein are described both quantitatively and qualitatively by interaction fingerprints. These fingerprints usually assess the prevalence, strength of interactions all the over the simulation trajectory and classify them into distinct categories (e.g., hydrophobic interactions, salt bridges, and hydrogen bonds [28].

13.3.3.3 Analysis of Site of Binding

Examining binding site of protein can reveal important residues that contribute to interaction as well as structural elements that influence the binding's specificity and affinity [29]. Properties of binding pockets may be analysed and they can be identified using tools like SiteMap, POVME, and CASTp.

13.3.3.4 Analysis of MD Simulation

This type of analysis looks at the dynamic and model characteristics of the protein with ligand complex as it changes over time. Conformational modifications, flexibility, and dominating structural motifs in the complex can be found using methods like principal component analysis (PCA), clustering analysis, deviation of root-mean-square (RMSD), and fluctuation of root-mean-square (RMSF) [30].

13.3.3.5 Energy Decomposition Analysis

These techniques break down the overall binding free energy into the contributions of each individual interaction between a protein and a ligand. These techniques may be used to determine important binding factors and offer conclusions about the relative relevance of various interaction types [31].

13.3.3.6 Docking Analysis and Virtual Screening

Virtual screening looks for possible binders for an intended protein by going through huge libraries of chemicals. Evaluating ligands' docking poses, classifying ligands according to docking scores or binding energy, and eliminating false positives are all part of the analysis of docking findings [32].

13.3.3.7 Machine Learning Techniques and Statistical Analysis

Prediction models for protein-ligand interaction may be constructed by utilising machine learning techniques to evaluate simulation data, identify pertinent characteristics, and extract information. Significant links and correlations between molecular characteristics and binding affinity may also be found using statistical analytic techniques including regression analysis, correlation analysis, and hypothesis testing [33].

13.3.3.8 Visualisation Tools Such as PyMOL and Visual Molecular Dynamics (VMD)

These tools are frequently used to visualise molecular trajectories, structures, and interaction patterns. These technologies make it easier to communicate findings and analyse outcome of simulation. By employing these analytical instruments and methodologies, scientists may get a more profound comprehension of the interactions between proteins and ligands, pinpoint crucial binding factors, and direct the logical development of innovative treatments and inhibitors [34]. Name of some Analysis tools are mentioned in Table 13.5.

13.3.4 Limitation and Challenges of MD Simulations

While protein-ligand binding interactions may be effectively studied using molecular dynamics (MD) simulations, there are a several of fail and difficulties that scientists must be used. The following are some of the main drawbacks, difficulties with MD simulations of protein binding prediction:

13.3.4.1 Force Field Accuracy

To represent the relationships of atoms and molecules, MD simulations utilise force fields. The intricate energy environment of protein-ligand interaction may not always be adequately represented by force fields, though, as they are dependent on empirical values and approximations. Binding affinity and binding modes can be incorrectly predicted due to force field imperfections [35].

13.3.4.2 Simulation Time Scale

It might be difficult to capture the time scales at which protein-ligand interaction events often take place in MD simulations. Although Molecular Dynamics simulations are useful tool of understanding the dynamics to attaching conversations, their scope is usually restricted to time ranges of nanoseconds to microseconds, while binding events might take place on time scales of milliseconds to seconds. Due to this restriction, transitory binding events and a partial sample of the conformational domain may be overlooked.

Table 13.5 Name of some analysis tools

Name	Working	Founder	Website
Pymol	PyMOL is a biochemical visualisation programme that lets users see and examine ligands, protein-ligand complexes, and protein structures in three dimensions. It offers resources for calculating distances, modelling molecular surfaces, and examining interactions like hydrophobic and hydrogen bonding.	Warren Lyford DeLano. Dr. DeLano	https://pymol.org/
UCSF Chimera	A molecular modelling tool called UCSF Chimaera makes it possible to interactively see and analyse molecular structures, such as those of nucleic acids, proteins, and tiny molecules. It offers instruments for computing molecular surface and electrostatic potentials, as well as for measuring lengths, angles, and torsional angles.	Zheng Yang	https://www.cgl.ucsf.edu/chimera/download.html/
MOE (Molecular Operating Environment)	With capabilities for protein-ligand docking, molecule-by-molecule modelling, and structure-based drug design, MOE is a complete molecular modelling software suite. It provides modules for determining binding energies, examining protein-ligand interactions, and displaying the structures and characteristics of molecules.	The Chemical Computing Group	https://www.chemcomp.com/Products.htm
DS Visualizer (Discovery Studio Visualizer)	One of the tools provided by the Discovery Studio software package for molecular visualisation and analysis is called DS Visualizer. It has tools for creating 2D and 3D structural representations, estimating binding energies, and visualising protein-ligand interactions.	Dassault Systemes BIOVIA	https://www.3ds.com/products/biovia/discovery-studio/visualization/
LigPlot	A programme called LigPlot+ is used to create schematic diagrams of interactions between proteins and ligands. Hydrogen chains, hydrophobic relationships, and other non-covalent interactions are found by analysing protein-ligand complexes and are then visually represented in 2D diagrams.	Tariq	https://www.ebi.ac.uk/thornton-srv/insoftware/LigPlus/

(continued)

Table 13.5 (continued)

Name	Working	Founder	Website
Autodock Tools	A graphical interface of user called Autodock Tools is used to make and analyse protein-ligand dock simulations that are run using AutoDock or AutoDock Vina. It enables users to compute binding energies, examine binding postures, and visualise docking findings.	Dr. Oleg Trott	https://autodock.scripps.edu/
PyRx	With an intuitive interface, PyRx is a virtual screening programme that incorporates AutoDock, which, AutoDock Vina, and additional docking programmes. It offers resources for virtual screening, lead optimisation, and the visualisation and analysis of protein-ligand interactions.	Sarkis Dallakian	https://pyrx.sourceforge.io/

13.3.4.3 Size of the System

Because of the many atoms and intricate nature of the interactions involved, simulating large protein-ligand complexes, membrane protein systems, or systems involving interactions between proteins can be computationally costly. Because of this, scientists would have to use coarse-grained simulations or simplified models, which might not be able to represent all relevant details of binding process [36].

13.3.4.4 Selection of Initial Structures

The quality of the starting structures utilised as the simulations' beginning points determines how accurate MD simulations will be. It can be difficult to get precise protein-ligand intricate structures from molecular docking poses or experimental data, and mistakes in the first structures might spread throughout the simulation [37].

13.3.4.5 Handling of Solvent Effects

Although solvent molecules are essential in mediating interactions between proteins and ligands, precisely simulating solvent effects in molecular dynamics simulations may be difficult. While expressive solvent models need huge simulation boxes and extensive simulation periods to sample solvent fluctuations sufficiently, implicit solvent models trade accuracy for computational speed [38].

13.3.4.6 Protein Flexibility

Upon interacting to a ligand, proteins, which are naturally flexible molecules, undergo conformational changes. To effectively anticipate binding affinities and modes, protein flexibility must be captured in MD simulations. However, in order to sample pertinent structural states, modelling protein flexibility increases simulation complexity and necessitates longer simulation timeframes [39].

13.3.4.7 Variability in Free Energy Estimates

MD simulations are frequently utilized to predict binding affinities through the utilization of free energy computations, including MM/GBSA or MM/PBSA. However, there are several inherent errors and approximations in these computations, such as the precision of the fundamental force fields, sample convergence, and end-state selection [40]. Limitation and Challenges of MD simulations in Protein binding prediction are mentioned in Fig. 13.5.

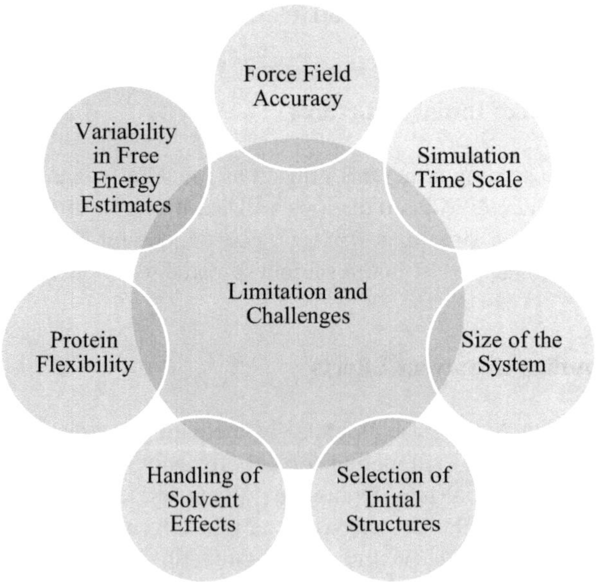

Fig. 13.5 Limitation and challenges of MD simulations in protein binding prediction

13.4 Hybrid Approaches for the Prediction of the Protein Binding

A hybrid model is a strategy that combines many computational techniques or algorithms to increase the precision and dependability of predictions in the setting of protein binding prediction. To get beyond the drawbacks of individual approaches, these hybrid models take use of the advantages of many approaches and combine them in a way that works well together. This is an example of how a hybrid model for protein binding prediction may be built. Hybrid Approaches of protein binding prediction are mentioned in Fig. 13.6.

In this chapter I briefly discussed two most common approach which are mostly uses in Hybrid approach of protein binding prediction. They are—(a) Integration of docking and molecular dynamics and (b) Physics-Based and Machine Learning Approaches.

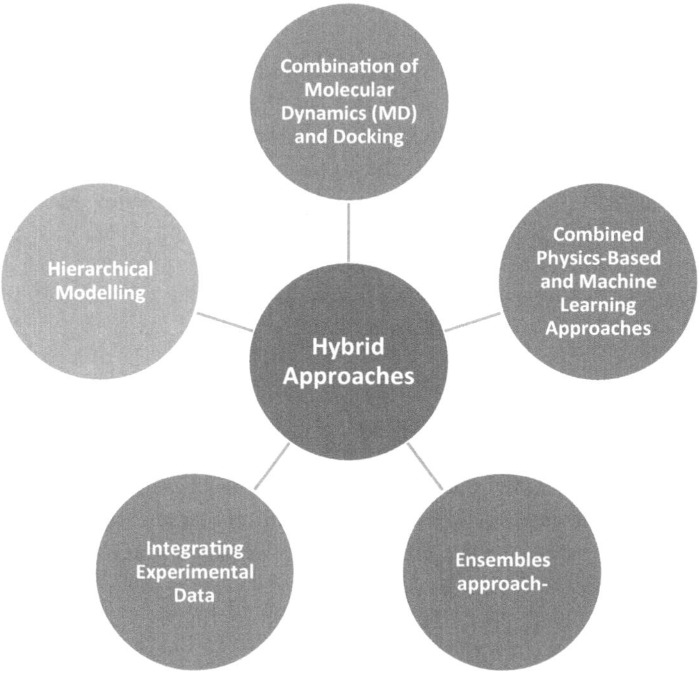

Fig. 13.6 Hybrid approaches of protein binding prediction

13.4.1 Integration of Docking and Molecular Dynamics

One effective method for more consistently and accurately thought protein-ligand binding reactions is the combination of docking of molecules and molecular dynamic (MD) simulations. In order to give idea interaction modalities, dynamics, energy of protein-ligand complexes, this combination approach makes use of the advantages of both methodologies. The making of predicting protein binding by combination of docking and Monte Carlo simulations is as follows:

13.4.1.1 First Docking of Molecular Screening

For forecast the bind positions and affinities to tiny molecules or ligands inside the protein binding site, molecular docking is utilised as a first-pass screening technique. Docking algorithms examine the ligand's conformational space and provide a binding posture score based on the protein's complimentary interactions. The outcomes of docking serve as a foundation for additional characterisation and refining [41].

13.4.1.2 Enhancement Using Molecular Dynamics Models

MD simulations are used to further define and characterise docking positions that are identified as possible binders. The dynamic behaviour and suppleness of the complex of protein-ligand over time may be investigated using MD simulations. MD simulations shed light on the kinetics, flexibility, and stability of the binding interactions by sample the space of conformations and recording dynamic variations. The impact of ions, solvent molecules, and external variables on the binding procedure may also be seen by MD simulations [42].

13.4.1.3 Evaluation and Verification

The structural and dynamic characteristics of the protein-ligand complex, such as deviation of root-mean-square (RMSD), fluctuation of root-mean-square (RMSF), and interacting signatures, are evaluated by the analysis of MD simulation trajectories. Using techniques like surface area of molecular mechanics/generalized Born (MM/GBSA) or molecular mechanics/Poisson-Boltzmann surface area (MM/PBSA), binding affinities and energies may be determined from MD simulations. By contrasting simulation findings with actual data—such as bonding constants, mutagenesis research, and structural details from spectroscopy of X-ray crystallography—the projected binding mechanisms and affinities may be validated [43].

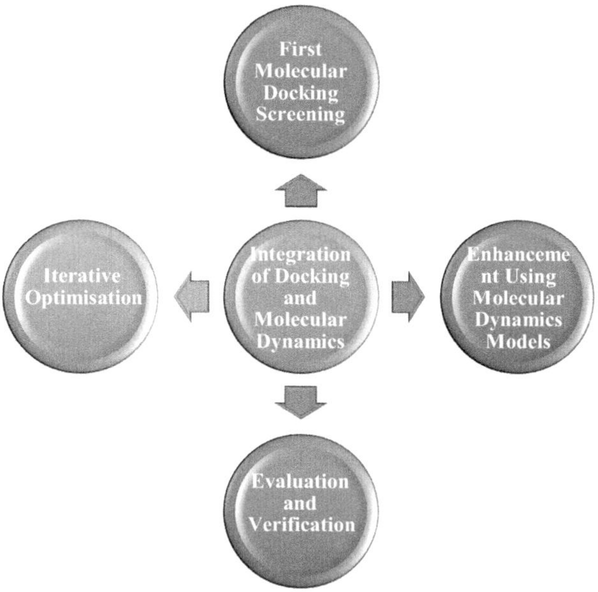

Fig. 13.7 Integration of docking and molecular dynamics

13.4.1.4 Iterative Optimisation

Through repeated cycles of MD simulations and analysis, docking poses are improved and verified. This is made possible by the combination of docked and MD simulations. It is possible to increase the precision and dependability of the anticipated protein-ligand binding reactions by iteratively optimising the binding attitudes and parameters [44]. Integration of Docking and Molecular Dynamics are mentioned in Fig. 13.7.

13.4.2 Combination of Machine Learning and Physics-Based Model

For protein-ligand binding interaction prediction, integrating machine learning (ML) with models based on physics provides a potent strategy that capitalises on the advantages of both approaches. Here's how to make the most of the combination:

13.4.2.1 Obtaining Features from Models Based on Physics

Protein-ligand complex structural, vibrant, and dynamic features are described by copious quantities of data produced by physics-based models like molecular docking or molecular dynamics simulations. These physics-based simulations may be utilised to extract pertinent information using machine learning methods. These traits may comprise geometric qualities of binding sites, kinetic contribution from force fields, dynamic characteristics determined from simulation trajectories, and descriptors of protein-ligand interactions [45].

13.4.2.2 Training Learning Models

Neural network of Convolutional (CNNs) and neural recurrent of network networks (RNNs), two types of deep learning models, as well as a variety of algorithms like machine of support vector (SVM), random forests, and boosting of recurrent, can all be trained using the extracted features as input. Usually, known complexes of protein-ligand with experimentally confirmed binding affinities or additional relevant features make up the training data. Based on training data, machine learning models discover how to link the input characteristics with the intended output (binding affinity, for example) [46].

13.4.2.3 Model Training for Machine Learning

The collected features may be used as training data for a range of models of deep learning, including networks of neural Convolutional (CNNs) and recurrent neural network networks (RNNs), two types of deep learning models, and support vector machines (SVMs), as well as random forests and boosting of gradient methods. The training data typically consists of known protein-ligand complexes with experimentally verified binding affinities or other pertinent properties. Machine learning models learn to connect what is expected (binding affinity, for example) with the input features based on training data [47].

13.4.2.4 Interpretability and Perspectives

By illuminating the fundamental factors that influence protein-ligand binding interactions, machine learning models can offer insights into the underlying connections between molecular characteristics and binding affinity. Methods that can assist in identifying significant characteristics and comprehending the molecular underpinnings of binding affinity prediction include feature significance evaluation, SHAP (Shapley Additive ex Planations) principles, and attention processes in neural networks [48].

13.4.2.5 Iterative Enhancement

Iterative improvement is made possible by the coupling of learning of machine and physics-based models, since predictions from the former can inform the latter's creation of more focused and effective physics-based simulations. On the other hand, knowledge gathered from physics-based simulations may guide the thoughts of machine learning models and the selection of important features [49].

13.4.3 Limitation and Challenges of Hybrid Approach for the Protein Binding Prediction

Although hybrid techniques have great potential to predict protein-ligand binding interactions, they also have a number of drawbacks and difficulties. These include combining machine learning (ML) algorithms with physics-based models. The following are some significant drawbacks and difficulties with hybrid techniques to protein binding prediction:

13.4.3.1 Quantity and Quality of Data

High-quality and varied datasets are needed for training, validating, and testing machine learning models. However, the number and quality of research findings on protein-ligand conversations, such as adsorption affinities, are frequently restricted.

In the field, it is still difficult to get complete datasets covering an extensive spectrum of amino acid targets, ligands that and binding affinities [50].

13.4.3.2 Engineering Features

For hybrid techniques, efficient feature engineering is essential since it entails obtaining useful features from molecular descriptors as well as physics-based simulations.

It takes domain knowledge and careful thought to design meaningful features that represent the pertinent components of protein-ligand interactions, including geometric qualities, energetic investments, and dynamic characteristics [51].

13.4.3.3 Interpretability of the Model

Understanding the underlying links between input data and output predictions—particularly in the setting of protein-ligand binding—requires that hybrid models be interpretable.

While ML models—especially deep learning architectures—may be difficult to comprehend because of their complexity and opacity, physics-based models provide characteristics that can be understood from basic principles [52].

13.4.3.4 Generalisation and Overfitting

One typical issue with ML-based techniques is overfitting, when an algorithm achieve well on data training but badly with unknown structure. Strong regularisation methods, cross-validation plans, and meticulous assessment on separate test sets are necessary to guarantee that hybrid models generalise to novel protein-ligand complexes [53].

13.4.3.5 Heterogeneous Data Source Integration

Integrating diverse data sources, such as structural details from protein-ligand complexes, quantitative readings, and molecular descriptors, is a common step in hybrid techniques. Technical difficulties arise in data preparation, extracting features, and model training when dealing with and integrating heterogeneous data kinds, formats, and sizes [54].

13.4.3.6 Computing Capabilities

Hybrid techniques could need a lot of processing power to execute physics-based simulations and train intricate machine learning models. Large-scale virtual screening and hybrid methodologies for chemical space exploration can be computationally demanding and may necessitate high-performance computer resources [55].

13.4.3.7 Benchmarking and Validation

Evaluation criteria, validation methodologies, and standardised datasets are needed to compare against current methods and assess how well hybrid systems work. Due to variations in methodology and presumptions, comparing the accuracy of predictions of combination models with purely ML or physics-based techniques can be difficult [56].

13.4.3.8 Dynamics and Complexity of Biology

The processes of protein-ligand binding reactions are intrinsically dynamic and complicated, impacted by a multitude of variables such as solvent effects, allosteric modulation, and protein flexibility. It is still very difficult to fully capture the intricacy and dynamism of protein-ligand relationships in hybrid models, necessitating developments in ML and physics-based modelling methodologies [57]. Limitation and Challenges of Hybrid Approaches in protein binding prediction are mentioned in Fig. 13.8.

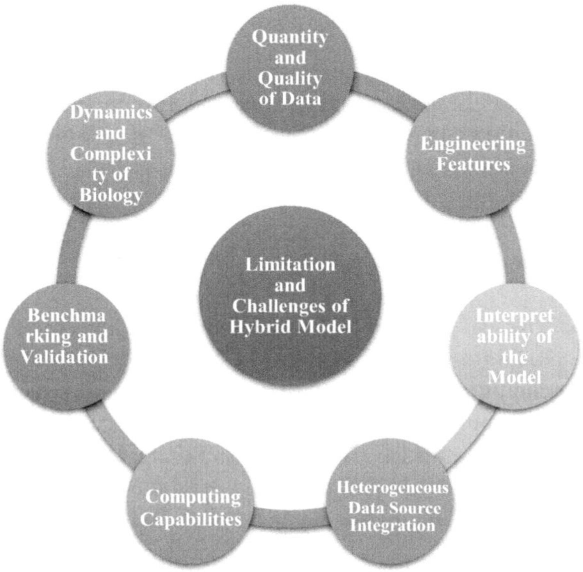

Fig. 13.8 Limitation and challenges of hybrid approaches in protein binding prediction

13.5 Challenges and Future Perspectives of the Computational Approaches of the Protein Binding Prediction

13.5.1 Overcoming Bottlenecks in Computation

Combining algorithmic advances, parallel computing power, and optimisation techniques is necessary to break through computational bottlenecks in protein-ligand binding interaction prediction. The following strategies aim to alleviate computational bottlenecks:

13.5.1.1 Parallel Processing and Dispersed Computing

To speed up simulations and calculations, make use of parallel computing architectures like distributed computing clusters, multi-core CPUs, and graphics processing units (GPUs). To effectively divide the computing effort among several processing units, use data parallelism and parallel algorithms. Use software frameworks and libraries designed for parallel processing, including TensorFlow, CUDA, MPI, OpenMP, or MPI, to fully use the computational capacity of contemporary hardware architectures [58].

13.5.1.2 Hardware Acceleration

To speed up computationally demanding activities, make use of specialised increase the hardware like GPUs, gate arrays of field-programmable (FPGAs) Quantum chemistry computations and molecular dynamics simulations that are GPU-accelerated can run much faster and have fewer computational bottlenecks [59].

13.5.1.3 Optimisation via Algorithms

Create and use effective computational techniques and algorithms to a variety of applications, including docking molecules, MD simulations, and energies calculations. Reducing computing complexity of simulations without sacrificing tolerable accuracy can be accomplished by using approximation techniques, coarse-graining approaches, or simpler models. To increase sampling efficiency and concentrate computing resources on areas of interest, use machine learning-guided sampling, improved sampling strategies, or adaptive sampling approaches [60].

13.5.1.4 Workflow Automation and Optimisation

To cut down on manual involvement and boost productivity, simplify and automate computational operations utilising workflow management platforms, scripting languages, or process orchestration tools. Optimise simulation and computational software settings, parameters, and configurations for best results on certain hardware platforms. To guarantee the resilience and dependability of simulations, make use of checkpointing, restarting capabilities, and handling errors techniques [61].

13.5.1.5 Online Resources with Cloud Computing

Utilise on-demand resources and cloud computing platforms to have access to scalable and reasonably priced computing infrastructures for protein-ligand binding prediction. Utilise cloud-based services to build and expand computational workflows dynamically in response to workload demands. Examples of these services include Microsoft Azure, Google Cloud Platform (GCP), and Amazon Web Services (AWS) [62].

13.5.1.6 Compression & Reduction of Data

Employ data reduction, compression as well, or complexity reduction strategies to reduce the overhead associated with storing and transferring huge datasets produced by simulations. To minimise the size of simulated output files and make data analysis and storage easier, use feature selection techniques, data collection methods, or

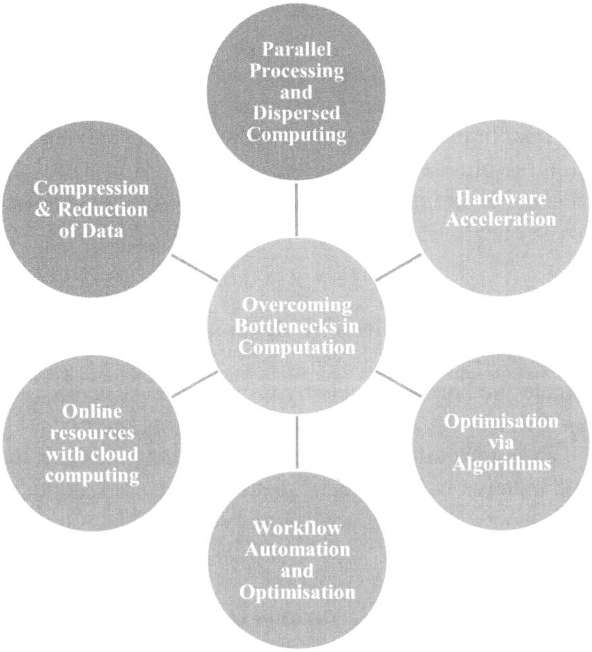

Fig. 13.9 Overcoming bottlenecks in computation of protein binding prediction

lossless or lossy data compression algorithms [63]. Overcoming Bottlenecks in Computation of protein binding prediction are mentioned in Fig. 13.9.

13.5.2 Increasing Precision and Predictive Capability

Increasing the precision and prediction ability of protein-ligand binding interactions necessitates a multimodal strategy that incorporates developments in data analysis, experimental design, and computational methodologies. The following techniques can be used to improve precision and forecasting power:

13.5.2.1 Superior Experimental Data

To train and validate prediction models, get high-quality experimental data, including as binding affinities, structure of crystals, and biophysical measurements. To guarantee the generalizability of prediction models, make use of different datasets that encompass a broad variety of protein goals, ligands, and binding modalities [64].

13.5.2.2 Molecular Models and Enhanced Force Fields

Achieve greater precision in force fields and molecule models utilised in docking investigations and molecular dynamics simulations by optimising parameters, adding quantum mechanical effects, and taking solvent polarisation into consideration. To more precisely capture the subtleties of protein-ligand interactions, create force fields that are advanced and customised to certain protein targets or receptor classes [65].

13.5.2.3 Data-Driven Methods and Machine Learning

Create prediction models by utilising the structure and sequencing data of ligands and proteins by utilising machine learning methods.

Examine cutting-edge machine learning methods to extract intricate correlations and patterns from protein-ligand binding data, including deep learning, supervised neural networks with graphs, and reinforcement training [65].

13.5.2.4 Combining Various Data Sources

Incorporate a variety of data sources, including as structural, sequence, chemical, and biophysical data, to enhance the feature space and raise the prediction models' accuracy.

To develop hybrid models that take use of the advantages of both methods and produce predictions that are more trustworthy, combine computational forecasts with experimental data [66].

13.5.2.5 Interpretability and Explainability of the Model

Create comprehensible machine learning models which shedding energy on the fundamental principles guiding the interactions to proteins with ligands. Make forecasts more visible and comprehensible by using strategies like explainable AI, attention mechanisms, and feature significance analysis to clarify the contributions of particular features [67].

13.5.2.6 Benchmarking and Validation

In order to evaluate the accuracy, durability, and generalizability of predictive models, thoroughly validate them using separate test sets, cross-validation methods, and external benchmarks.

To determine how well predictive models perform in comparison to current methods, benchmark them against cutting-edge techniques and standardised datasets [68].

13.5.2.7 Combining Computational and Experimental Methods

To confirm computational predictions and corroborate findings, combine them with experimental validation approaches including binding assays, structural biology techniques, and mutagenesis research.

Make use of feedback loops that exist between experimental data and computer predictions to iteratively enhance and improve prediction models [69].

13.5.2.8 Community Involvement with Open Science

Encourage cooperation and information exchange among scientists by taking part in cooperative research projects, data-sharing programmes, and open-source software creation initiatives. Assist in the distribution of instruments, techniques, and information for forecasting protein-ligand binding interactions by making contributions to community resources, data bases, and benchmarking platforms [70]. Increasing Precision and Predictive Capability in protein binding prediction are mentioned in Fig. 13.10.

13.5.3 Combining Experimental Methods with Integration

Modelling protein-ligand binding interactions more accurately and consistently requires integration with experimental approaches. Here's how experimental and computational approaches may be combined to predict protein binding:

13.5.3.1 Verification and Adjustment

To validate and calibrate computer models, use experimental data such as binding affinities evaluated by methods such as isotherm calorimetry (ITC) and fluorescence tests. To determine precision and dependability of the estimates and pinpoint areas for development, contrast computational forecasts with experimental data [71].

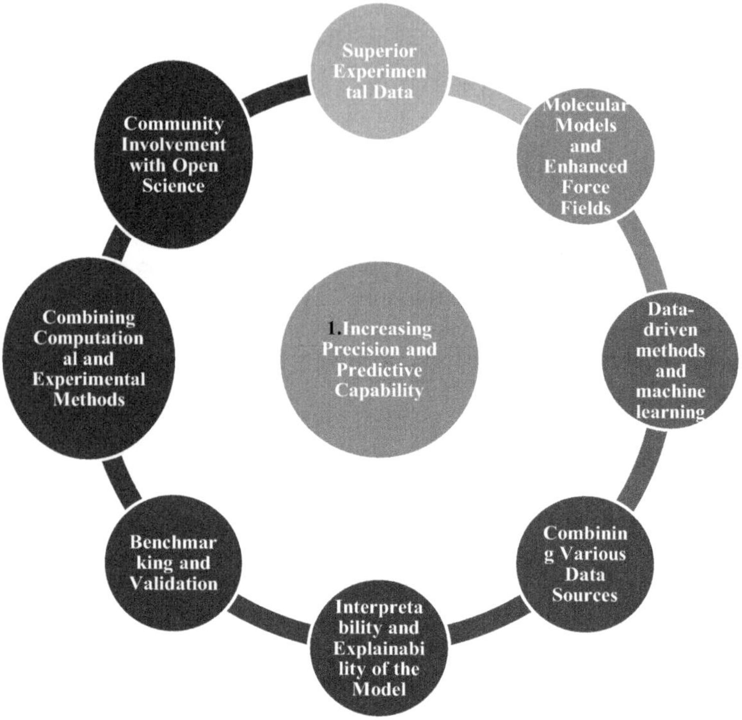

Fig. 13.10 Increasing precision and predictive capability in protein binding prediction

13.5.3.2 Determining the Structure

To ascertain the excellent quality structures of protein-ligand complexes, employ experimental structural biology methods as spectroscopy using nuclear magnetic resonance cryo-electron microscopy, and X-ray crystallography. Use experimental structures as validation datasets to evaluate the precision of docking predictions, or as a starting point for molecular docking simulations [72].

13.5.3.3 Identification of the Binding Site

To find and define binding sites on proteins, combine computational techniques with experimental strategies like binding tests or site-directed mutagenesis. Make experimental validations of your computationally predicted potential binding sites according to protein structure and physical properties [73].

13.5.3.4 Prediction of Binding Mode

Predict the ligand binding modalities inside protein binding sites by combining experimental structural data with computational docking simulations.

To verify the correctness of anticipated binding modes, verify docking predictions utilising experimental techniques such structural biology approaches or site-directed mutagenesis in conjunction with functional experiments [74].

13.5.3.5 Clarification of Mechanism

To understand dynamic conduct and conformational shifts of ligand-protein complexes across time, use computer molecular dynamics simulations. Use experimental methods, like single-molecule energy transfer using fluorescence resonance energy (smFRET) or hydrogen-deuterium interchange mass spectrometry (HDX-MS), to validate computer simulations of binding processes and dynamics [75].

13.5.3.6 Drug Development and Enhancement

Utilise computational techniques in conjunction with fragment-based screening (FBS) or empirical high-throughput screening (HTS) to identify and prioritise new ligands that exhibit enhanced affinity for binding and specificity.

Utilise computational structure-based drug development methods to direct the creation and improvement of small-molecule inhibitors, such as pharmacophore modelling or virtual screening [76].

13.5.3.7 The Modelling of the Qsar

To create model that are predictive of ligand binding affinity, combine measurements from experiments of ligand attributes, such as physicochemical characteristics or biological activities, with computational QSAR modelling. To make sure QSAR models are accurate and dependable in predicting ligand binding interactions, validate them with experimental data [77].

13.5.3.8 Loop of Iterative Feedback

Create an iterative cycle of feedback whereby experimental data is used to test computational models and computational predictions are improved upon, and vice versa. To continuously increase prediction accuracy, use experimental data to inform the setting of parameters of computational models, the validation of predictions, and the selection of input characteristics [78].

13.5.4 New Developments in Trends and Technologies

The study of protein with ligand binding interactions at being shaped by number of new developments in technology and trends. These developments might lead to improvements in structural biology and drug discovery in terms of precision, effectiveness, and application. The following technologies and trends are noteworthy:

13.5.4.1 Artificial Intelligence (AI) and Deep Learning

With learning of deep techniques such as neural networks of convolutional (CNN), recurrent relationships of neural networks containing (RNN), and neural network including graph (GNN), protein-ligand binding interactions are becoming more and more frequent to predict. Large-scale datasets with intricate interactions may be captured by AI-driven methods, which can also be used to develop representations of ligand and protein structures and produce predictions that are easier to understand [79].

13.5.4.2 Generative Ligand Design Models

To create new ligand structures with desirable binding characteristics, generative models are being employed, such as VAEs, or variational autoencoders, and adversarial generative networks (GANs). These models can produce a wider range of varied and synthetically plausible ligand possibilities for drug development, as well as more efficiently explore chemical space [80].

13.5.4.3 Hybrid Methods

Hybrid methodologies that use many computing techniques, such quantum mechanics calculations, learning of machine, docking of molecule, and dynamics simulations of molecular, can combine the advantages of each methodology to increase prediction accuracy. Complete modelling of protein-ligand interactions, including calculation of binding affinity and prediction of binding posture, is made possible by hybrid techniques [81].

13.5.4.4 Methods Based on Graphs

An increasing number of people are using graph-based visualisations of macromolecules and protein molecules, in which atoms and bonds are shown as edges and nodes in a graph. More precise and scalable predictions may be made by using graph neural networks (GNNs) alongside additional graph-based techniques to capture the three-dimensional structure and biochemical environment of molecules and proteins [82].

13.5.4.5 Interpretability and Explainable AI

Creating explainable AI techniques that offer comprehensible forecasts and insights into the variables influencing protein-ligand binding interactions is becoming more and more important. Methods like saliency maps, feature significance analysis, and attention processes help make sound choices in drug design and improve comprehension of model predictions [83].

13.5.4.6 High-Volume Experimental Tests

Rapid empirical verification of computational predictions is made possible by developments in high-throughput testing (HTS) technologies, including mass spectrometry-based assays, microfluidics-based assays, and label-free detection techniques. Large-scale experimental data are made available by HTS assays, which may be training, validating models for prediction of protein with ligand binding relationships [84].

13.5.4.7 Combination with Methods of Structural Biology

Validating computational models and facilitating structural characterisation of protein-ligand complexes are made easier by combining computational predictions with empirical structural biology methods including microscopy by cryo-electron (cryo-EM), NMR, and crystallography of X-ray [85]. Combining Experimental Methods with Integration in protein binding prediction are mentioned in Fig. 13.11.

13.5.5 Regulatory and Ethical Considerations

Although medication development and molecular biology studies heavily rely on protein binding prediction, it's crucial to consider the moral and legal implications of such efforts. Some important ethical and legal considerations are as follows:

13.5.5.1 Data Security and Privacy

In many cases, biological systems depend on interactions between protein and their attachment partners, which can affect human health and disease. Researchers must adhere to data privacy rules and use caution while handling sensitive biological data

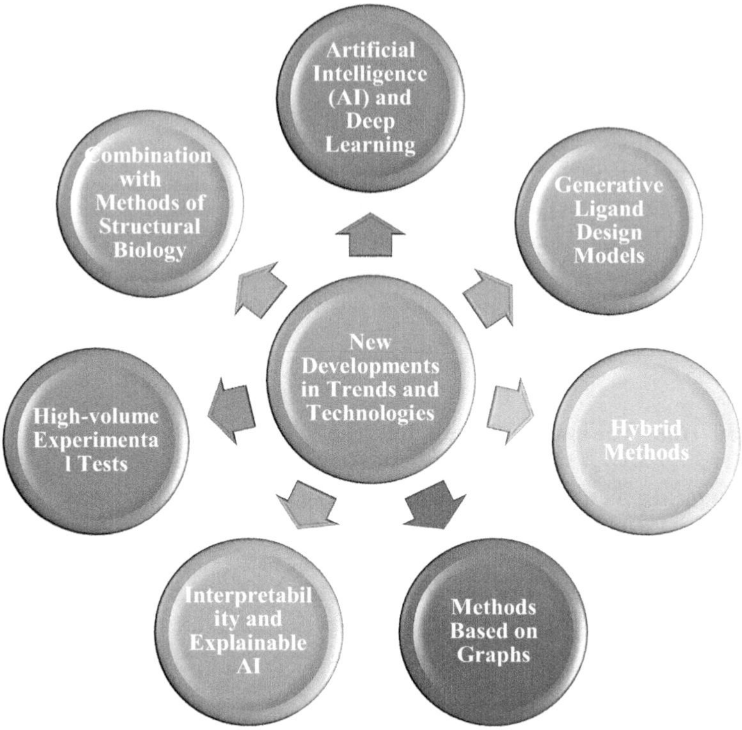

Fig. 13.11 Combining experimental methods with integration in protein binding prediction

in order to protect confidential and personally identifiable information from misuse or unauthorised access [86].

13.5.5.2 Aware Consent and Ethical Data Use

Researchers are required to get consent in writing from participants and follow ethical rules for the use of personal information while performing studies with human subjects or gathering biological samples from humans.

A voluntary participation policy, participant risk reduction, and institutional review boards' (IRBs') or ethics committees' approval are all important ethical issues [87].

13.5.5.3 Animal Welfare

Animal models may be used in some protein-ligand binding investigations for preclinical or experimental drug candidate testing. When it comes to the handling and utilisation of laboratory animals, researchers have to go by ethical criteria. These guidelines include minimising pain and discomfort, guaranteeing humane treatment, and adhering to established procedures for animal testing [88].

13.5.5.4 Openness and Repeatability

Improving the truthfulness as well as dependability of results from studies in protein binding prediction requires transparency and repeatability.To help others duplicate and confirm their findings, researchers should include detailed explanations of their computational techniques, data sources, and experimental procedures [89].

13.5.5.5 Data Sharing and Intellectual Property

When using protein binding predictions to produce innovative results, methods, or medication candidates, intellectual property issues may come up. Researchers should support data sharing and cooperation among the scientific community while still abiding by intellectual property rules and regulations that regulate patents, copyrights, and licencing agreements [90].

13.5.5.6 Adherence to Regulations

Agencies like The United States Food and Drug Administration (FDA) in the US and the European Medicines Agency (EMA) in the European Union are in charge of regulating the prediction of protein binding for drug development purposes. When creating therapeutic candidates using predictions of protein-ligand interaction, researchers have to abide by regulations for preclinical testing, risk evaluations, and clinical trials [91].

13.5.5.7 Fairness and Bias

Predictive models must be free from prejudice and discrimination, especially when it comes to sensitive characteristics like colour, race, sexual orientation, or socio-economic position. This is one of the ethical issues. In order to ensure justice and equity in protein binding predictions, researchers should evaluate and reduce prejudices in data used for training, algorithm design, and decision-making procedures [92].

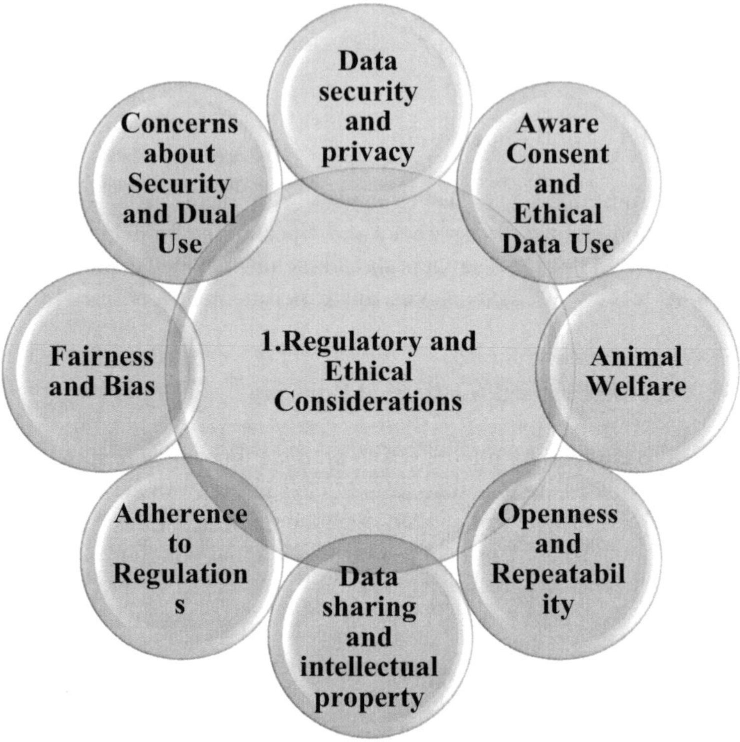

Fig. 13.12 Regulatory and ethical considerations in protein binding predictions

13.5.5.8 Concerns about Security and Dual Use

Predictions about protein-ligand interaction may have dual-use consequences, with possible uses in biodefense and medicinal development. Concerns with security pertaining to the improper use of prediction models or the creation of bioactive compounds with undesirable or unexpected effects should be known to researchers [93]. Regulatory and Ethical Considerations in protein binding predictions are mentioned in Fig. 13.12.

13.6 Case Studies

Baskaran et al., Using conventional medication such as donepezil, molecular docking with rivastigmine and chlorzoxazone has been carried out with on these *in-silico* studies. The searching of the information show that each of the ligands docked effectively with the AChE protein's crysal structure at the active region of 4EY6 and 4EY7; however, 4EY5 exhibits a higher number of hydrophobic contacts with the ligands.Out of the three protein architectures, 4EY7 has the highest binding affinity when it comes to ligand interactions. It was possible for chlorzoxazone to bind into the AChE (4EY5, 4EY6, and 4EY7 active sites). As a result, insilico research on chlorzoxazone and disease like are done, and chlorzoxazone may work well in treating Alzheimer's disease. In this research paper the author used molecular docking as computational technique as protein binding [94].

Dege et al., the synthesised structure's characterisation results indicate that van der Walls contact forces and H-bonds provide molecular stability. The stability and H-bond formation of the molecule are facilitated by the nitro groups. By using MEP calculations, the attack areas of the molecule that are electrophilic and nucleophilic have been identified. These groups additionally contributed significantly to the docking poses' ligand-protein interactions. The author employed molecular docking as a computational method for protein binding in their study [95]. Here in the QSAR computation technique SAR has been described. SAR of piperidine 1-yl derivative is given in Fig. 13.13.

Namasivayam et al., The human ABCA7 homology model that was produced and utilised in the docking process with the PSO@AutoDock particle swarm optimisation (PSO) tool.The swarm size was adjusted to 60 unique particles and the default settings of the varCPSO-ls algorithm, were set to 6.05. The algorithm's remaining parameters were set to their default settings. Using 50,000 assessment steps as the termination criteria, a total of 50 separate docking computations were carried out. Visual inspection was used to examine the docked postures acquired from docking experiments for the chosen compounds, and potential binding poses were chosen.

The first stage involved creating a model of homology of human ABCA7 utilising the Protein Preparation Wizard, a tool in Schrödingers software. Three stages were included in the protein preparation process: using the Epik module, a tool that can anticipate the protein structure's protonation state at a normal pH of 7.4, to (i) assign bond ordering; (ii) add the absence of hydrogen atoms; and (iii) generate het states. The H-bond assignment and optimisation utilising the PROPKA module implemented in Schrödinger 2020–3 come next. Ultimately, the amino acid structureis reduced by employing a (OPLS3) force to establish and restrict the heavy atoms to a maximum RMSD value of 0.30. The Generation of Receiver Grid panel of Glide was used to create the grid files needed for the docking process. Grid points were generated to enclose a box that covered the whole transmembrane area, much as PSO@AutoDock. Following grid setup, the chosen compounds were docked utilising Glide XP (increased precision) docking onto the transporter grids that were constructed. The parameters for the docking method were set to their default values. Ten separate docking computations were carried out. When creating the structure, the top rank docked poses of the 10 chosen compound from PSO@AutoDock were

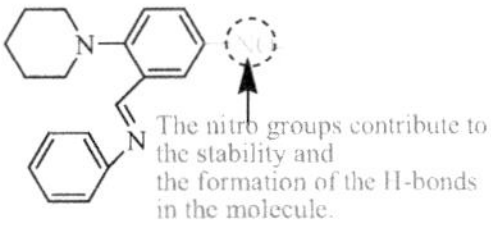

Fig. 13.13 SAR of piperidine 1-yl derivative

taken into account. The pharmacophore characteristics in this module were computed using a consensus pharmacophore function automatically. With the use to the action, the feature are clustered into probable model that are more preserved than threshold and tolerance distances of 30% and 1.5 Å, respectively [96].

Mukherjee et al., they were to look at possible medications that target Keap1 to treat AD. A particular inhibition of Keap1, an adverse regulator of Nrf2, can assist in preventing amyloid-beta-induced neuronal damage. Random Forest regressor was the most effective model that we implemented after comparing popular machine learning methods to create a model that can be used to calculate the inhibition activity (pIC50) against Keap1. Five chemical compounds have been discovered. Keap1 is protein which found in PDB. The author docked molecule by computational technique [97].

Liang et al., they have thoroughly examined every ACP identification technique that has been developed to far in this work and have introduced ACPredStackL, a novel predictor. They nearly completely covered all of the benchmark datasets that are now utilised by predictors when we conducted a thorough benchmarking study of the ACP prediction tools that are currently available on several datasets. More than 7000 peptides that are found naturally have been linked to human physiology. Clinical studies are now being conducted to test many peptide therapies. Under this situation, high-throughput computing techniques is very necessary to extract promising peptides from sequences that may have medicinal applications. Unfortunately, the majority of computational techniques are designed to work with a certain class of peptides, meaning they are limited to identifying unique peptides within that class. We aim to create a more thorough and dependable model. In future that will be able to forecast many distinct peptide types at once [98].

Yadalam et al., the activities and signalling pathways of rheumatoid arthritis and periodontitis are mediated by a number of differently expressed genes, which further contribute to the development and progression of both conditions. Our created integrated network revealed several typical connections between the two illnesses. Furthermore, our network offers IFI44Las, a shared therapeutic target on molecular docking and virtual screening, which is inhibitable by vemurafenib. To validate findings, further cell line research is necessary. These results add to our understanding of both of these inflammatory illnesses and may facilitate further research on medication development. Molecular docking, a computer method for protein binding, was employed by the author in this study [99].

Mustafa et al., A thorough phylogenetic study of GDF11, protein with ligand, and protein with protein docked has shown that TGF-β members have the potential to be used as pharmaceutical inhibitors of IL-6 and TNF-α receptor proteins. Three tetrapeptides (SAGP against IL-6, AFDP and AGPC against TNF-α) that were synthesised from GDF11 and had strong docking scores met all the requirements to be considered potential drugs.

The proinflammatory cytokines TNF-α and IL-6 control the inflammatory reactions and signals during infections. When these cytokines are dysregulated, autoimmunity and a lack of self-tolerance develop as a result of the conditions. Several methods were used in the current investigation to investigate the inhibitory nature of

IL-6 reaction (a) and bound patterns (b) of SAGP and IL-6. and interaction patterns (b) of TETV with IL-6. Figure 9S shows the GSAG's reactions (a) and binding patterns (b) with IL-6. The computational method of protein binding employed in this study was molecular docking [100].

Sharma et al., revealed that ARL15, a new susceptibility gene for RA, codes to a tiny GTP-binding protein whose action be unclear carried out on a case-control cohort of northern India. ARL15 may prove to be a viable therapeutic receptor for RA, given the group's extensive functional and structural research' encouraging early discoveries about the protein's possible involvement in the pathogenesis of RA. Lead discovery CADD methods cannot be effectively carried out in the absence of an empirically established crystal structure. Thus, two distinct approaches were used to determine models of the 3D structure of ARL15: an *ab initio* model prediction method and a template-based threading approach. The ab initio models showed a slightly higher overall packaging quality and orientation compared to the other models, and they fared marginally better in the quality evaluation. Disorganised N-terminal helix, it seems. These findings were verified, which wes run for a brief period of time (10 ns) for all of the projected models. Subsequently, longer simulations were run for the two best models predicted by each approach, Qmod5 (ab initio) and Tmod2 (threading).Qmod5 was selected for the putative GTP-binding site prediction of ARL15 since the ab initio method appeared to produce higher quality tertiary structures for ARL15. This was accomplished by mapping the resolved binding site of GTP for a close homolog (ARF6) on the anticipated structure of ARL15 through different alignment. Thus, the excellent, revised projected model for the tertiary structure of ARL15 might be useful in identifying possible agonists or inhibitors using CADD methods. Such as HTVS, and direct and support work on experimental structural characterisation. The author employed molecular docking as a computational method for protein binding in their study [101].

Qamar et al., A rigid docking investigation was carried out using PADI4 receptor-filtered natural and synthetic drugs. The three structure, hydroxyanthroquinone (−7.40 kcal/mol), vallinic acid (−5.90 kcal/mol), and thymol (−5.97 kcal/mol), which exhibit very strong binding affinity with the active site of PADI4 receptor, were screened out by taking into account the several confirmation. Because they stop the disease's progression by blocking our PADI4 receptor's ability to citrullinate histone and protein residues that cause pathogenesis, these phytochemicals are both affordable and useful as a preventative measure against this illness. PADI4 is protein which found in PDB. The author docked molecule by computational technique [102].

The findings of Kumar et al. work suggest that repurposed antiviral medications may be able to bind in the SARS-CoV-2 major protease active site in a highly specific manner that is akin to the crystal-bound -ketoamide Mpro structure. Analysis validated the stability and structural flexibility of the evaluated drugs—lopinavir-ritonavir, raltegravir, and tipranavir—in enzyme active site, where they demonstrated the strongest binding. Since every medication found in this study complies with established pharmacokinetic standards and has been given FDA approval for use in treating the corresponding illnesses in humans, it might be possible to move directly to clinical trials in accordance with established protocols and receive FDA approval for new indications, which would expedite the development of SARS-CoV-2 therapeutics. Here used MD simulation as computational technique [103].

The protein's active site is being occupied by the -ketoamide 13b, which is also forming multiple bonds of H and hydronic interactions that involve the residues of the binding pocket's other substrate-binding residues. This is shown by the crystal structure. Using drug repurposing virtual screening techniques, we have tested over 75 antivirals in nature, anticancer, and anti-malarial medications in the current work in order to identify possible therapeutic compounds. The majority of the tested pharmacological compounds interact with the SARS-CoV-2 Mpro proteins and share a binding pocket with comparable interacting amino acid residues, according to molecular docking studies. Strong bond interactions are seen between the SARS-CoV-2 bound ligand (improved-ketoamide) and surrounding amino acids within the range of 4 Å at several subsites where it accepts H bonds via the structural amides on Gly143, Cys145, and Ser144 in addition to forming H bonds with the active site His41. This interaction to the amino acid and the ligand demonstrates a potent suppression of viral protease. Molecular docking, a computer method for protein binding, was employed by the author in this study [104].

Ahmad et al., they took most potent molecules of Ganoderma luci-dum extract, which include GAA, D, and F stand for ganoderic acid, f, and ganoderic acid., Ganoderenic Acid D, and Ganoderenic Acid B, which shows the binding affinity of −3.0, −3.2, −3.3, −3.4, and −3.0 kcal/mol, respectively. They performed further studies like molecular dynamics simulation and MMGBSA; they concluded that GAA is the most promising Ganoderma lucidium extract. The computational method of protein binding employed in this study was molecular docking [105].

Pande et al., Through the use of molecular docking simulations, they have been able to gather important information on possible interactions with therapeutically significant small molecules. By means of these modelling, we have investigated the natural substances' binding affinities and mechanisms with the pathogenic protein, including Withaniasomnifera (ashwagandha), Gotu Kola (Centellaasiatica), Asiatic Acid, and Curcumin (turmeric). Our study is important because it adds to our knowledge of the molecular causes of disease like Alzheimer's. They had set the stage for the creation of innovative treatment determination by analysing the structural subtleties of PHFs and spotting possible interactions with bioactive substances. These results suggest that focused therapies may be able to lessen the devasting consequences of Alzheimer's. Here used molecular simulation technique as computational technique for protein technique [106].

13.7 Conclusion

To sum up, the use of computational techniques to predict protein binding has great potential for the fields of drug development and bioinformatics. By employing several approaches, such molecular docking, machine learning, and MD simulations, scientists can get a deeper understanding of the molecular interactions that occur between ligands and proteins. By predicting binding affinities, locating binding locations, and investigating binding processes, these techniques aid in the development and improvement of new medicinal medicines.But even with these advances in computer methods, precise protein-ligand binding prediction remains a difficulty. Obstacles still exist in the form of issues like appropriately describing molecular interactions, accounting for solvent effects, and protein flexibility. Furthermore, experiment is important to guarantee usefulness and dependability to computational prediction. In the future, multidisciplinary cooperation among biologists, chemists, and computer scientists will be crucial to improving and practicality of protein binding predictions techniques. We can open up new avenues for drug development and discovery by combining data from experiments with mathematical models and improving algorithms. This will ultimately lead to a deeper comprehension of interactions between protein and ligand and their function in biological processes.

References

1. Cao L, Coventry B, Goreshnik I, Huang B, Sheffler W, Park JS, Jude KM, Marković I, Kadam RU, Verschueren KHG, Verstraete K, Walsh STR, Bennett N, Phal A, Yang A, Kozodoy L, DeWitt M, Picton L, Miller L, et al. Design of protein-binding proteins from the target structure alone. Nature. 2022;605(7910):551–60. https://doi.org/10.1038/s41586-022-04654-9.
2. Hiranuma N, Park H, Baek M, et al. Improved protein structure refinement guided by deep learning based accuracy estimation. Nat Commun. 2021;12:1340. https://doi.org/10.1038/s41467-021-21511-x.
3. Michael D, Bankoti N, Biswas A, Sekar K. Aspects of protein structure, function, and dynamics in rational drug designing. In: Singh SK, editor. Innovations and implementations of computer aided drug discovery strategies in rational drug design. Singapore: Springer; 2021. https://doi.org/10.1007/978-981-15-8936-2_3.
4. Wang M, Cang Z, Wei GW. A topology-based network tree for the prediction of protein–protein binding affinity changes following mutation. Nat Mach Intell. 2020;2:116–23. https://doi.org/10.1038/s42256-020-0149-6.
5. Siebenmorgen T, Zacharias M. Computational prediction of protein–protein binding affinities. Wiley Interdiscip Rev Comput Mol Sci. 2020;10(3) Blackwell Publishing Inc https://doi.org/10.1002/wcms.1448.
6. Sliwoski G, Kothiwale S, Meiler J, Lowe EW. Computational methods in drug discovery. Pharmacol Rev. 2014;66(1):334–95. https://doi.org/10.1124/pr.112.007336.
7. Hu L, Wang X, Huang YA, Hu P, You ZH. A survey on computational models for predicting protein-protein interactions. Brief Bioinform. 2021;22(5) Oxford University Press https://doi.org/10.1093/bib/bbab036.
8. Zhao J, Cao Y, Zhang L. Exploring the computational methods for protein-ligand binding site prediction. Comput Struct Biotechnol J. 2020;18:417–26. https://doi.org/10.1016/J.CSBJ.2020.02.008.
9. Espinosa YR, Alvarez HA, Howard EI, Carlevaro CM. Molecular dynamics simulation of the heart type fatty acid binding protein in a crystal environment. J Biomol Struct Dyn. 2021;39(10):3459–68. https://doi.org/10.1080/07391102.2020.1773315.
10. Rube HT, Rastogi C, Feng S, et al. Prediction of protein–ligand binding affinity from sequencing data with interpretable machine learning. Nat Biotechnol. 2022;40:1520–7. https://doi.org/10.1038/s41587-022-01307-0.

11. Tripathi MK, Ahmad S, Tyagi R, Dahiya V, Yadav MK. Fundamentals of molecular modeling in drug design. In: Computer Aided Drug Design (CADD): from ligand-based methods to structure-based approaches; 2022. p. 125–55. https://doi.org/10.1016/B978-0-323-90608-1.00001-0.
12. Agarwal S, Mehrotra R. An overview of molecular docking. JSM Chem. 2016;4(2):1024. https://www.researchgate.net/publication/303897563
13. Gschwend DA, Good AC, Kuntz ID. Molecular docking towards drug discovery. J Mol Recognit. 1996;9:175–86.
14. Vemula D, Jayasurya P, Sushmitha V, Kumar YN, Bhandari V. CADD, AI and ML in drug discovery: a comprehensive review. Eur J Pharm Sci. 2023;181. Elsevier B.V. https://doi.org/10.1016/j.ejps.2022.106324.
15. Fan J, Fu A, Zhang L. Progress in molecular docking. Quant Biol. 2019;7:83–9. https://doi.org/10.1007/s40484-019-0172-y.
16. Andrusier N, Mashiach E, Nussinov R, Wolfson HJ. Principles of flexible protein-protein docking. Proteins. 2008;73(2):271–89. https://doi.org/10.1002/prot.22170.
17. Hatfield M, Lovas S. Conformational sampling techniques. Curr Pharm Des. 2014;20(20):3303–13. https://doi.org/10.2174/13816128113199990603.
18. Kitchen DB, Decornez H, Furr JR, Bajorath J. Docking and scoring in virtual screening for drug discovery: methods and applications. Nat Rev Drug Discov. 2004;3(11):935–49. https://doi.org/10.1038/nrd1549.
19. Bonvin AM. Flexible protein–protein docking. Curr Opin Struct Biol. 2006;16(2):194–200. https://doi.org/10.1016/J.SBI.2006.02.002.
20. Yang C, Chen EA, Zhang Y. Protein–ligand docking in the machine-learning era. Molecules. 2022;27(14) MDPI https://doi.org/10.3390/molecules27144568.
21. Sunny S, Jayaraj PB. Protein–protein docking: past, present, and future. Protein J. 2022;41:1–26. https://doi.org/10.1007/s10930-021-10031-8.
22. Yao H, Liu J, Xu M, Ji J, Dai Q, You Z. Discussion on molecular dynamics (MD) simulations of the asphalt materials. Adv Colloid Interf Sci. 2022;299:102565. https://doi.org/10.1016/J.CIS.2021.102565.
23. Liu H, Fu H, Chipot C, Shao X, Cai W. Accuracy of alternate nonpolarizable force fields for the determination of protein-ligand binding affinities dominated by cation-πInteractions. J Chem Theory Comput. 2021;17(7):3908–15. https://doi.org/10.1021/acs.jctc.1c00219.
24. Wan S, Bhati AP, Zasada SJ, Coveney PV. Rapid, accurate, precise and reproducible ligand-protein binding free energy prediction: binding free energy prediction. Interface Focus. 2020;10(6) Royal Society Publishing https://doi.org/10.1098/rsfs.2020.0007.
25. González MA. Force fields and molecular dynamics simulations. École Thématique de La Société Française de La Neutronique. 2011;12:169–200. https://doi.org/10.1051/sfn/201112009.
26. Blanes-Mira C, Fernández-Aguado P, de Andrés-López J, Fernández-Carvajal A, Ferrer-Montiel A, Fernández-Ballester G. Comprehensive survey of consensus docking for high-throughput virtual screening. Molecules. 2023;28(1) MDPI https://doi.org/10.3390/molecules28010175.
27. da Silva AD, Bitencourt-Ferreira G, de Azevedo WF. Taba: a tool to analyze the binding affinity. J Comput Chem. 2020;41(1):69–73. https://doi.org/10.1002/jcc.26048.
28. Wang DD, Chan MT, Yan H. Structure-based protein–ligand interaction fingerprints for binding affinity prediction. Comput Struct Biotechnol J. 2021;19:6291–300. Elsevier B.V. https://doi.org/10.1016/j.csbj.2021.11.018.
29. Clark JJ, Orban ZJ, Carlson HA. Predicting binding sites from unbound versus bound protein structures. Sci Rep. 2020;10:15856. https://doi.org/10.1038/s41598-020-72906-7.
30. Badar MS, Shamsi S, Ahmed J, Alam A. Molecular dynamics simulations: concept, methods, and applications; 2022. p. 131–51. https://doi.org/10.1007/978-3-030-94651-7_7.

31. Nunes-Alves A, Ormersbach F, Wade RC. Prediction of the drug-target binding kinetics for flexible proteins by comparative binding energy analysis. J Chem Inf Model. 2021;61(7):3708–21. https://doi.org/10.1021/acs.jcim.1c00639.
32. Sahayarayan JJ, Rajan KS, Vidhyavathi R, Nachiappan M, Prabhu D, Alfarraj S, Arokiyaraj S, Daniel AN. In-silico protein-ligand docking studies against the estrogen protein of breast cancer using pharmacophore based virtual screening approaches. Saudi J Biol Sci. 2021;28(1):400–7. https://doi.org/10.1016/j.sjbs.2020.10.023.
33. Yuan Y, Chang S, Zhang Z, Li Z, Li S, Xie P, Yau WP, Lin H, Cai W, Zhang Y, Xiang X. A novel strategy for prediction of human plasma protein binding using machine learning techniques. Chemom Intell Lab Syst. 2020;199:103962. https://doi.org/10.1016/J.CHEMOLAB.2020.103962.
34. Fitzkee N. A beginner's guide to molecular visualization using *PyMOL*. n.d.. http://sourceforge.net/projects/pymol/files/Legacy/.
35. Fu X, Wu Z, Wang W, Xie T, Keten S, Gomez-Bombarelli R, Jaakkola T. Forces are not enough: benchmark and critical evaluation for machine learning force fields with molecular simulations. 2022. http://arxiv.org/abs/2210.07237
36. Shukla R, Tripathi T. Molecular dynamics simulation in drug discovery: opportunities and challenges. In: Singh SK, editor. Innovations and implementations of computer aided drug discovery strategies in rational drug design. Singapore: Springer; 2021. https://doi.org/10.1007/978-981-15-8936-2_12.
37. Gomes PSFC, Gomes DEB, Bernardi RC. Protein structure prediction in the era of AI: challenges and limitations when applying to in silico force spectroscopy. Front Bioinform. 2022;2 https://doi.org/10.3389/fbinf.2022.983306.
38. Khabiri M, Minofar B, Brezovský J, Damborský J, Ettrich R. Interaction of organic solvents with protein structures at protein-solvent interface. J Mol Model. 2013;19(11):4701–11. https://doi.org/10.1007/s00894-012-1507-z.
39. Sanavia T, Birolo G, Montanucci L, Turina P, Capriotti E, Fariselli P. Limitations and challenges in protein stability prediction upon genome variations: towards future applications in precision medicine. Comput Struct Biotechnol J. 2020;18:1968–79. Elsevier B.V. https://doi.org/10.1016/j.csbj.2020.07.011.
40. Gapsys, V., Michielssens, S., Peters, J. H, de Groot, B. L., & Leonov, H. (2015). Calculation of binding free energies. Methods Mol Biol (Clifton, N.J.), 1215, 173–209. https://doi.org/10.1007/978-1-4939-1465-4_9.
41. Grasso G, di Gregorio A, Mavkov B, Piga D, Labate GFD, Danani A, Deriu MA. Fragmented blind docking: a novel protein–ligand binding prediction protocol. J Biomol Struct Dyn. 2022;40(24):13472–81. https://doi.org/10.1080/07391102.2021.1988709.
42. Lazim R, Suh D, Choi S. Advances in molecular dynamics simulations and enhanced sampling methods for the study of protein systems. Int J Mol Sci. 2020;21(17):1–20. MDPI AG. https://doi.org/10.3390/ijms21176339.
43. Mei S, Li F, Leier A, Marquez-Lago TT, Giam K, Croft NP, Akutsu T, Ian Smith A, Li J, Rossjohn J, Purcell AW, Song J. A comprehensive review and performance evaluation of bioinformatics tools for HLA class I peptide-binding prediction. Brief Bioinform. 2019;21(4):1119–35. Oxford University Press. https://doi.org/10.1093/bib/bbz051.
44. Ochoa R, Cossio P, Fox T. Protocol for iterative optimization of modified peptides bound to protein targets. J Comput Aided Mol Des. 2022;36:825–35. https://doi.org/10.1007/s10822-022-00482-1.
45. Guedes IA, Barreto AMS, Marinho D, et al. New machine learning and physics-based scoring functions for drug discovery. Sci Rep. 2021;11:3198. https://doi.org/10.1038/s41598-021-82410-1.
46. Ahmed A, Mam B, Sowdhamini R. DEELIG: a deep learning approach to predict protein-ligand binding affinity. Bioinform Biol Insights. 2021;15 https://doi.org/10.1177/11779322211030364.

47. Rezaei MA, Li Y, Wu D, Li X, Li C. Deep learning in drug design: protein-ligand binding affinity prediction. IEEE/ACM Trans Comput Biol Bioinform. 2022;19(1):407–17. https://doi.org/10.1109/TCBB.2020.3046945.
48. Wang R, Jin J, Zou Q, Nakai K, Wei L. Predicting protein-peptide binding residues via interpretable deep learning. Bioinformatics. 2022;38(13):3351–60. https://doi.org/10.1093/bioinformatics/btac352.
49. McGuffin LJ, Edmunds NS, Genc AG, Alharbi SMA, Salehe BR, Adiyaman R. Prediction of protein structures, functions and interactions using the IntFOLD7, MultiFOLD and ModFOLDdock servers. Nucleic Acids Res. 2023;51(W1):W274–80. https://doi.org/10.1093/nar/gkad297.
50. Zeng Y, Gong M, Lin M, Gao D, Zhang Y. A review about transcription factor binding sites prediction based on deep learning. IEEE Access. 2020;8:219256–74. https://doi.org/10.1109/ACCESS.2020.3042903.
51. Narayanan H, Seidler T, Luna MF, Sokolov M, Morbidelli M, Butté A. Hybrid models for the simulation and prediction of chromatographic processes for protein capture. J Chromatogr A. 2021;1650:462248. https://doi.org/10.1016/J.CHROMA.2021.462248.
52. Yu TH, Su BH, Battalora LC, Liu S, Tseng YJ. Ensemble modeling with machine learning and deep learning to provide interpretable generalized rules for classifying CNS drugs with high prediction power. Brief Bioinform. 2022;23(1) https://doi.org/10.1093/bib/bbab377.
53. Domingo L, Djukic M, Johnson C, et al. Binding affinity predictions with hybrid quantum-classical convolutional neural networks. Sci Rep. 2023;13:17951. https://doi.org/10.1038/s41598-023-45269-y.
54. Ye Q, Hsieh CY, Yang Z, et al. A unified drug–target interaction prediction framework based on knowledge graph and recommendation system. Nat Commun. 2021;12:6775. https://doi.org/10.1038/s41467-021-27137-3.
55. Lin Z, Zou J, Liu S, Peng C, Li Z, Wan X, Fang D, Yin J, Gobbo G, Chen Y, Ma J, Wen S, Zhang P, Yang M. A cloud computing platform for scalable relative and absolute binding free energy predictions: new opportunities and challenges for drug discovery. J Chem Inf Model. 2021;61(6):2720–32. https://doi.org/10.1021/acs.jcim.0c01329.
56. Hahn DF, Bayly CI, Boby ML, Bruce Macdonald HE, Chodera JD, Gapsys V, Mey ASJS, Mobley DL, Benito LP, Schindler CEM, Tresadern G, Warren GL. Best practices for constructing, preparing, and evaluating protein-ligand binding affinity benchmarks [article v1.0]. Living J Comput Mol Sci. 2022;4(1) https://doi.org/10.33011/livecoms.4.1.1497.
57. Chance MR, Farquhar ER, Yang S, Lodowski DT, Kiselar J. Protein Footprinting: auxiliary engine to power the structural biology revolution. J Mol Biol. 2020;432(9):2973–84. https://doi.org/10.1016/j.jmb.2020.02.011.
58. Masoudi-Sobhanzadeh Y, Jafari B, Parvizpour S, Pourseif MM, Omidi Y. A novel multi-objective metaheuristic algorithm for protein-peptide docking and benchmarking on the LEADS-PEP dataset. Comput Biol Med. 2021;138:104896. https://doi.org/10.1016/J.COMPBIOMED.2021.104896.
59. Sadybekov AV, Katritch V. Computational approaches streamlining drug discovery. Nature. 2023;616:673–85. https://doi.org/10.1038/s41586-023-05905-z.
60. Pucci F, Schwersensky M, Rooman M. Artificial intelligence challenges for predicting the impact of mutations on protein stability. Curr Opin Struct Biol. 2022;72:161–8. https://doi.org/10.1016/J.SBI.2021.11.001.
61. Sauer S, Matter H, Hessler G, Grebner C. Optimizing interactions to protein binding sites by integrating docking-scoring strategies into generative AI methods. Front Chem. 2022;10 https://doi.org/10.3389/fchem.2022.1012507.
62. Wang Y, Wu S, Duan Y, Huang Y. A point cloud-based deep learning strategy for protein-ligand binding affinity prediction. Brief Bioinform. 2022;23(1) https://doi.org/10.1093/bib/bbab474.

63. Lim S, Lu Y, Cho CY, Sung I, Kim J, Kim Y, Park S, Kim S. A review on compound-protein interaction prediction methods: data, format, representation and model. Comput Struct Biotechnol J. 2021;19:1541–56. Elsevier B.V. https://doi.org/10.1016/j.csbj.2021.03.004.
64. Pei Q, Wu L, Zhu J, Xia Y, Xie S, Qin T, Liu H, Liu TY, Yan R. Breaking the barriers of data scarcity in drug–target affinity prediction. Brief Bioinform. 2023;24(6) https://doi.org/10.1093/bib/bbad386.
65. Zhao L, Ciallella HL, Aleksunes LM, Zhu H. Advancing computer-aided drug discovery (CADD) by big data and data-driven machine learning modeling. Drug Discov Today. 2020;25(9):1624–38. https://doi.org/10.1016/J.DRUDIS.2020.07.005.
66. Adam G, Rampášek L, Safikhani Z, et al. Machine learning approaches to drug response prediction: challenges and recent progress. NPJ Precis Onc. 2020;4:19. https://doi.org/10.1038/s41698-020-0122-1.
67. Monteiro NRC, Simões CJV, Ávila HV, et al. Explainable deep drug–target representations for binding affinity prediction. BMC Bioinform. 2022;23:237. https://doi.org/10.1186/s12859-022-04767-y.
68. Horlacher M, Cantini G, Hesse J, Schinke P, Goedert N, Londhe S, Moyon L, Marsico A. A systematic benchmark of machine learning methods for protein-RNA interaction prediction. Brief Bioinform. 2023;24(5) Oxford University Press https://doi.org/10.1093/bib/bbad307.
69. Seffernick JT, Lindert S. Hybrid methods for combined experimental and computational determination of protein structure. J Chem Phys. 2020;153(24) American Institute of Physics Inc https://doi.org/10.1063/5.0026025.
70. Sapoval N, Aghazadeh A, Nute MG, et al. Current progress and open challenges for applying deep learning across the biosciences. Nat Commun. 2022;13:1728. https://doi.org/10.1038/s41467-022-29268-7.
71. Yan J, Zhu M. A review about RNA–protein-binding sites prediction based on deep learning. IEEE Access. 2020;8:150929–44. https://doi.org/10.1109/ACCESS.2020.3014996.
72. Jisna VA, Jayaraj PB. Protein structure prediction: conventional and deep learning perspectives. Protein J. 2021;40:522–44. https://doi.org/10.1007/s10930-021-10003-y.
73. Liao J, Wang Q, Wu F, Huang Z. In silico methods for identification of potential active sites of therapeutic targets. Molecules. 2022;27(20) MDPI https://doi.org/10.3390/molecules27207103.
74. Zhou Y, Jiang Y, Chen SJ. RNA–ligand molecular docking: advances and challenges. Wiley Interdiscip Rev Comput Mol Sci. 2022;12(3) John Wiley and Sons Inc https://doi.org/10.1002/wcms.1571.
75. Gutnik D, Evseev P, Miroshnikov K, Shneider M. Using AlphaFold predictions in viral research. Curr Issues Mol Biol. 2023;45(4):3705–32. MDPI. https://doi.org/10.3390/cimb45040240.
76. Howard M, Hill J, Galluppi G, McLean M. Plasma protein binding in drug discovery and development. Comb Chem High Throughput Screen. 2010;13(2):170–87. https://doi.org/10.2174/138620710790596745.
77. Peter SC, Dhanjal JK, Malik V, Radhakrishnan N, Jayakanthan M, Sundar D, Sundar D. Quantitative structure-activity relationship (QSAR): modeling approaches to biological applications. In: Encyclopedia of bioinformatics and computational biology: ABC of bioinformatics, vols. 1–3. Elsevier; 2018. p. 661–76. https://doi.org/10.1016/B978-0-12-809633-8.20197-0.
78. Soleymani F, Paquet E, Viktor H, Michalowski W, Spinello D. Protein–protein interaction prediction with deep learning: a comprehensive review. Computat Struct Biotechnol J. 2022;20:5316–41. Elsevier B.V. https://doi.org/10.1016/j.csbj.2022.08.070.
79. Dhakal A, McKay C, Tanner JJ, Cheng J. Artificial intelligence in the prediction of protein-ligand interactions: recent advances and future directions. Brief Bioinform. 2022;23(1) Oxford University Press https://doi.org/10.1093/bib/bbab476.
80. Li Y, Pei J, Lai L. Structure-based: de novo drug design using 3D deep generative models. Chem Sci. 2021;12(41):13664–75. https://doi.org/10.1039/d1sc04444c.

81. Rangwala H, Karypis G. Introduction to protein structure prediction: methods and algorithms. Wiley, 2010.
82. Blinov ML, Yang J, Faeder JR, Hlavacek WS. Graph theory for rule-based modeling of biochemical networks. InTransactions on Computational Systems Biology VII 2006 Jan 1 (pp. 89–106). Berlin, Heidelberg: Springer Berlin Heidelberg.
83. Medina-Ortiz D, Khalifeh A, Anvari-Kazemabad H, Davari MD. Interpretable and explainable predictive machine learning models for data-driven protein engineering. n.d. https://doi.org/10.1101/2024.02.18.580860.
84. Howard ML, Hill JJ, Galluppi GR, Mclean MA. Plasma protein binding in drug discovery and development. Comb Chem High Throughput Scree. 2010;13:170–87.
85. Cerofolini L, Fragai M, Ravera E, Diebolder CA, Renault L, Calderone V. Integrative approaches in structural biology: a more complete picture from the combination of individual techniques. Biomolecules. 2019;9(8) MDPI AG https://doi.org/10.3390/biom9080370.
86. Thapa C, Camtepe S. Precision health data: requirements, challenges and existing techniques for data security and privacy. Comput Biol Med. 2021;129:104130. https://doi.org/10.1016/J.COMPBIOMED.2020.104130.
87. Müller EJ, Munn BR, Redinbaugh MJ, Lizier J, Breakspear M, Saalmann YB, Shine JM. The non-specific matrix thalamus facilitates the cortical information processing modes relevant for conscious awareness. Cell Rep. 2023;42(8) https://doi.org/10.1016/j.celrep.2023.112844.
88. Mellor DJ, Beausoleil NJ, Littlewood KE, McLean AN, McGreevy PD, Jones B, Wilkins C. The 2020 five domains model: including human–animal interactions in assessments of animal welfare. Animals. 2020;10(10):1–24. MDPI AG. https://doi.org/10.3390/ani10101870.
89. Corti C, Cobanaj M, Dee EC, Criscitiello C, Tolaney SM, Celi LA, Curigliano G. Artificial intelligence in cancer research and precision medicine: applications, limitations and priorities to drive transformation in the delivery of equitable and unbiased care. Cancer Treat Rev. 2023;112:102498. https://doi.org/10.1016/J.CTRV.2022.102498.
90. Jiménez-Luna J, Grisoni F, Weskamp N, Schneider G. Artificial intelligence in drug discovery: recent advances and future perspectives. Expert Opin Drug Discov. 2021;16(9):949–59. https://doi.org/10.1080/17460441.2021.1909567.
91. Pike A, Williamson B, Harlfinger S, Martin S, McGinnity DF. Optimising proteolysis-targeting chimeras (PROTACs) for oral drug delivery: a drug metabolism and pharmacokinetics perspective. Drug Discov Today. 2020;25(10):1793–800. https://doi.org/10.1016/J.DRUDIS.2020.07.013.
92. Pancotti C, Benevenuta S, Birolo G, Alberini V, Repetto V, Sanavia T, Capriotti E, Fariselli P. Predicting protein stability changes upon single-point mutation: a thorough comparison of the available tools on a new dataset. Brief Bioinform. 2022;23(2) https://doi.org/10.1093/bib/bbab555.
93. Li J, Zhao H, Zheng L, An W. Advances in synthetic biology and biosafety governance. Front Bioeng. Biotechnol. 2021;9. Frontiers Media S.A https://doi.org/10.3389/fbioe.2021.598087.
94. Baskaran KP, Arumugam A, Kandasamy R, Alagarsamy S. Insilico method for prediction of maximum binding affinity and ligand – protein interaction studies on alzheimer's disease. Int J Res GRANTHAALAYAH. 2020;8(11):362–70. https://doi.org/10.29121/granthaalayah.v8.i11.2020.2472.
95. Şahin S, Dege N. A newly synthesized small molecule: the evaluation against Alzheimer's disease by in silico drug design and computational structure analysis methods. J Mol Struct. 2021;1236 https://doi.org/10.1016/j.molstruc.2021.130337.
96. Namasivayam V, Stefan K, Pahnke J, Stefan SM. Binding mode analysis of ABCA7 for the prediction of novel Alzheimer's disease therapeutics. Comput Struct Biotechnol J. 2021;19:6490–504. https://doi.org/10.1016/J.CSBJ.2021.11.035.
97. Mukherjee A, Gopalakrishnan AV. The mechanistic insights of the antioxidant Keap1-Nrf2 pathway in oncogenesis: a deadly scenario. Med Oncol. 2023;40:248. https://doi.org/10.1007/s12032-023-02124-.

98. Liang X, Li F, Chen J, Li J, Wu H, Li S, Song J, Liu Q. Large-scale comparative review and assessment of computational methods for anti-cancer peptide identification. Brief Bioinform. 2021;22(4) Oxford University Press https://doi.org/10.1093/bib/bbaa312.
99. Yadalam PK, Sivasankari T, Rengaraj S, Mugri MH, Sayed M, Khan SS, Kamil MA, Bhandi S, Raj AT, Patil S, Heboyan A. Gene interaction network analysis reveals IFI44L as a drug target in rheumatoid arthritis and periodontitis. Molecules. 2022;27(9) https://doi.org/10.3390/molecules27092749.
100. Mustafa G, Mahrosh HS, Arif R. In silico characterization of growth differentiation factors as inhibitors of TNF-alpha and IL-6 in immune-mediated inflammatory disease rheumatoid arthritis. Biomed Res Int. 2021;2021 https://doi.org/10.1155/2021/5538535.
101. Sharma A, Saini M, Kundu S, Thelma BK. Computational insight into the three-dimensional structure of ADP ribosylation factor like protein 15, a novel susceptibility gene for rheumatoid arthritis. J Biomol Struct Dyn. 2022;40(10):4626–41. https://doi.org/10.1080/07391102.2020.1860826.
102. Garg P, Srivastava P, Qamar T. A computational study of PADI4 receptor for identifying novel drug target against rheumatoid arthritis a multifaceted review journal in the field of pharmacy a computational study of PADI4 receptor for identifying novel drug target against rheumatoid arthritis. Syst Rev Pharm. n.d.;13 https://doi.org/10.2210/pdb.
103. Kumar Y, Singh H, Patel CN. In silico prediction of potential inhibitors for the main protease of SARS-CoV-2 using molecular docking and dynamics simulation based drug-repurposing. J Infect Public Health. 2020;13(9):1210–23. https://doi.org/10.1016/j.jiph.2020.06.016.
104. Zhang L, Lin D, Sun X, Curth U, Drosten C, Sauerhering L, et al. Crystal structure of SARS-CoV-2 main protease provides a basis for design of improved -ketoamide inhibitors. Science. 2020;368:409–12. https://doi.org/10.1126/science.abb3405.
105. Ahmad F. Ganoderic acid A targeting leucine-rich repeat kinase 2 involved in Parkinson's disease–a computational study. Aging Med. 2023;6(3):272–80. https://doi.org/10.1002/agm2.12235.
106. Mukherjee SU. Computational Analysis for Dementia Drug Target Identification and Optimization. Corrosion and Protection. 2023; 51(2).

Chapter 14
Physiologically Based Pharmacokinetic (PBPK) Modeling and Pharmacokinetic Pharmacodynamic (PKPD) Modelling

Siddhi Varnekar, Mehak Sohal, Nayan Tikale, Ruchitha Reddy S., Anushree Bose, Shrushti Pawar, and Prashant Kurkute

Contents

S. Varnekar · M. Sohal · N. Tikale · Ruchitha Reddy S. · A. Bose · S. Pawar · P. Kurkute (✉)
Shobhaben Pratapbhai Patel School of Pharmacy & Technology Management, SVKM's NMIMS, Mumbai, Maharashtra, India
e-mail: prashant.kurkute@nmims.edu

S. N. K. Rao G., R. R. Alavala (eds.), *Applications of Computational Tools in Drug Design and Development*, https://doi.org/10.1007/978-981-96-4154-3_14

Abstract Physiologically Based Pharmacokinetic (PBPK) and Pharmacokinetic Pharmacodynamic (PKPD) modeling are two crucial branches of clinical pharmacology that study the effects of drugs on the body over time. PBPK modeling involves the application of mathematical models to simulate drug absorption, distribution, metabolism, and excretion, as well as the interaction between drug properties and the body. It is used to assess drug safety, dosing parameters, and side effects. Pharmacokinetics and pharmacodynamics (PK-PD) modeling is a statistical model that studies the interrelationship of both pharmacokinetics and pharmacodynamics of drugs to assess their dosage regime. PK-PD modeling offers a quantitative depiction of the drug's absorption and elimination processes within the body, while PD modeling evaluates the effects of drugs on the body, considering the drug's mechanism of action and primary biological mechanisms that restrict its rate of action. The relationship between drug exposure and response can be measured using PK and PD models. These models can also be employed to assess the impact of drug-specific, delivery system-specific, physiological, and pathological system-specific factors on this relationship. It is currently common practice for pharmaceutical companies to integrate the PBPK modeling strategy into the dossiers that they submit to regulatory bodies. In conclusion, PBPK and PKPD modeling are essential tools in drug development, risk assessment, and personalized medicine. They help predict drug absorption, distribution, metabolism, and excretion, as well as the effects of various factors on drug interactions. Future directions for PBPK and PKPD modeling include advances in modeling techniques and the integration of these models for comprehensive drug effect predictions.

Keywords Physiologically Based Pharmacokinetic (PBPK) · Pharmacokinetic Pharmacodynamic (PKPD) · Risk assessment · Personalized medicine · Drug-drug interaction

14.1 Introduction

Pharmacokinetics and pharmacodynamics are two main important branches of clinical pharmacology. Pharmacokinetics is the study of the drugs in the body with time course for their action, distribution, metabolism, and elimination (ADME) to assess their safety, dosing parameters, and side effects. Pharmacodynamics is defined as the study of the relationship between the drug's concentration and their physiological effect with the course of time. The various mathematical models were developed to simulate the pharmacokinetics and pharmacodynamics of the drug molecule. The equation of mathematical models to simulate ADME was put forward for the first time by Teorell in 1937. Initially their application was limited due to the inability to solve such complicated equation, but with the advancement of computational power, they are now an indispensable and integral part of drug discovery and development.

Physiologically based pharmacokinetic (PBPK) modeling is the application of mathematical models to simulate the various pharmacokinetics behaviors of the drugs and study the effect of various factors like intrinsic factors (organ dysfunction, sex, race, age, and mutation), and extrinsic factors such as drug-drug interaction due to induction or inhibition metabolic enzymes, alone or in combination with other drugs to assess their safety and dosing parameters.

Pharmacokinetics and pharmacodynamics (PK-PD) modeling is a statistical model to study the interrelationship of both pharmacokinetics and pharmacodynamics of drugs to assess their dosage regime. More specifically, pharmacokinetic (PK) modeling offers a quantitative depiction of the drug's absorption and elimination processes within the body. PD modeling evaluates the effects of drugs on the body, considering the drug's mechanism of action and the primary biological mechanisms that restrict its rate of action, over a period. The relationship between drug exposure and response can be measured using PK and PD models. These models can also be employed to assess the impact of drug-specific, delivery system-specific, physiological, and pathological system-specific factors on this relationship. Drug-specific attributes, such as the strength of binding to receptors and the rate at which the drug is eliminated from the body, provide insight into how a drug interacts with the body. The parameters specific to the drug delivery system represent the carrier's clearance, release rate, and internalization rate. The system-specific physiological parameters encompass physiological variables such as blood flow, cell longevity, enzyme expression, and transporter expression.

Over the past decades, various software platforms have been developed to understand the PBPK modeling. It is currently common practice for pharmaceutical companies to integrate the PBPK modeling strategy into the dossiers that they submit to regulatory bodies. Both the United States Food and Drug Administration and the European Medicines Agency have recently updated guidance documents for the utilization of PBPK modeling on the assessment of the drug-drug interaction (DDI) potential of medications that are currently in the process of development. Through the FDA Modernization Act of 1997 and the FDA "effectiveness" guidance of 1998, the US Food and Drug Administration (FDA) made it permissible to use exposure-response data in conjunction with a single pivotal clinical study as adequate proof of efficacy. The FDA has specifically recognised the value of population modelling as a useful tool for transferring knowledge from adult indications to paediatric ones. Pharmacometrics analyses of 198 submissions were surveyed; of these, 126 (64%) contributed to choices about medication approval, and 133 (67%) to decisions about labelling [1].

14.2 Fundamentals of PBPK Modeling

Physiologically based pharmacokinetic (PBPK) modeling is a mathematical approach that simulates the absorption, distribution, metabolism, and excretion (ADME) of drugs in the body. PBPK models are built by integrating drug-specific data with independent prior knowledge of the physiology and biology of the organism. These models consider different organs and tissues, each characterized by specific blood-flow rates, volumes, tissue-partition coefficients, and permeability. The model predicts the concentration-time profiles of drugs in various compartments, providing a quantitative characterization of drug exposure in the respective compartments [2].

The structure of a PBPK model consists of two main parts:

- An anatomical "backbone" that contains species-specific physiological parameters
- A drug-specific part that includes the individual drug's ADME properties.

The model parameters can be either drug-dependent (e.g., binding to blood, tissue-to-plasma distribution coefficient, tissue permeability-surface area product, enzymatic activity) or drug-independent (e.g., blood flows, tissue volumes, tissue composition) [3].

The development of a PBPK model involves several steps:

1. Acquisition of drug-dependent parameters: This step requires a bottom-up approach that integrates a large number of drug-specific data, parameters on species physiology, and a good understanding of all active processes affecting the pharmacokinetic properties of a drug.
2. Model parameterization: This step involves determining the anatomical and physiological parameters of the respective species, such as rat, dog, human, etc. These parameters include body weight, height, organ volume, blood flow rates, tissue composition, and other relevant factors.
3. Model verification and validation: This step ensures the consistency and accuracy of the model by comparing the predicted results with experimental and clinical data [4].

PBPK models have various applications in drug development, including:

1. Prediction of drug concentration-time profiles: PBPK models can estimate the pharmacokinetic (PK) profile of a compound based on its preclinical ADME data.
2. Assessment of exposure in a target organ: PBPK models can assess the exposure in a target organ after the administration of a drug by taking into account the rate of absorption and disposition in that organ, as well as metabolism within that organ if it is applicable.
3. Evaluation of different dose schedules and routes: Based on the PK data generated from one dose schedule, the PBPK model can be used to evaluate the PK profile of different dose schedules and/or dose routes.
4. Prediction of PK profile in different ethnic populations and age and disease stages: Based on the PK data from one ethnic population, the PBPK model can be used to predict the PK profile in different ethnic populations as well as populations of various age and disease stages [5].

PBPK Modeling: Applications of Software Tools

PBPK models can be used to guide research and development of new drugs, predict drug pharmacokinetics in paediatric populations, or predict drug interactions [6, 7]. These models can also estimate the pharmacokinetic (PK) profile of a compound based on its preclinical ADME data and assess the exposure in a target organ after the administration of a drug [6, 7].

There are different software packages available for PBPK modelling, ranging from general mathematical and engineering modelling software to specialized PBPK modelling software [6, 7]. General modelling software packages, such as MATLAB (Simbiology is a MATLAB tool used to model, simulate, and analyse biological systems including physiologically-based pharmacokinetics) Model Maker, Berkeley Madonna, and acslX, provides a programming language for the model code, numerical solutions for the ordinary differential equations that defines the system being modelled, and a graphical output of the simulation results [6, 7].

Specialized PBPK modelling software packages, such as PK-Sim, Simcyp, and GastroPlus, provide less flexibility in model development but require less mathematical and modelling experience [2, 3]. These software tools either provide the user with a click-and-drag assembly of the model structure or an already built model and can simulate particular PK-relevant processes or constitute a generic whole-body PBPK model [6, 7].

PBPK modelling and simulation currently receives extensive attention during drug discovery and development and in submissions for regulatory filing and reviews. It is a unique tool that can be used to estimate the pharmacokinetic (PK) profile of a compound based on its preclinical ADME data and can be used to assess the exposure in a target organ after the administration of a drug [6, 7].

Limitations of PBPK Modelling

The limitations of PBPK modelling in drug development include the estimation of unknown parameter values, the integration of PBPK models in the regulatory framework, and the use of PBPK models in assessment of trial design, paediatric formulation, and toxicology [2, 3, 8].

The estimation of unknown parameter values is a significant challenge in PBPK model building. To improve PBPK modelling, several regulatory agencies such as the EMA and FDA have published guidelines on reporting of modelling and simulation to enable assessment by regulators. These guidelines recommend that all the model parameters, including assumptions, system-dependent parameters, compound-specific parameters, and models, should be presented clearly and justified whenever required. An analysis plan should be built along with sensitivity analysis during both development and application. Also, parameter identification can be performed for optimising data [3, 8].

The integration of PBPK models in the regulatory framework is still limited. To improve PBPK models for integration in the regulatory framework, several challenges need to be addressed, such as the estimation of unknown parameter values, the validation of PBPK models, and the communication of PBPK modelling results to regulators [3, 8].

The use of PBPK models in assessment of trial design, paediatric formulation, and toxicology is still limited. PBPK models can be used to estimate the pharmacokinetic (PK) profile of a compound based on its preclinical ADME data and can be used to assess the exposure in a target organ after the administration of a drug. However, the use of PBPK models in assessment of trial design, paediatric formulation, and toxicology is still not widespread [3, 8].

Therefore, to improve PBPK modelling, it is essential to address the challenges of estimating unknown parameter values, integrating PBPK models in the regulatory framework, and expanding the use of PBPK models in various aspects of drug development, such as trial design, paediatric formulation, and toxicology [2, 3, 8].

14.3 Building PBPK Models

The concept of Physiologically Based Pharmacokinetic (PBPK) modeling has a long history, dating back to the introduction of multi compartmental models by Teorell in 1937. Over time, PBPK modeling has evolved, with early reports by Bischoff contributing significantly to its development. Initially, the complexity and data requirements of PBPK models limited their widespread use in the pharmaceutical industry. However, advancements in predicting key drug-specific parameters, like tissue to plasma partition coefficients, from in vitro and in silico data have made PBPK modeling more accessible [9].

In recent years, there has been a substantial increase in publications on PBPK modeling, indicating its growing popularity across the scientific community. Commercial platforms integrating PBPK methodologies, such as Simcyp, Gastro Plus, and PK-Sim, have further fueled interest in PBPK modeling within the pharmaceutical industry. This methodology is now extensively applied in drug discovery and development, leveraging preclinical ADME data for modeling. Notably, pharmaceutical companies are incorporating PBPK modeling in regulatory submissions, with regulatory agencies like the FDA and EMA endorsing its use. The future of

PBPK modeling looks promising, with its continued integration in drug development processes and regulatory practices [9]. Different PK model types: Classical Vs PBPK models are represented Table 14.1. Figure 14.1 represented building blocks in PBPK modelling.

Table 14.1 Different PK model types: classical Vs PBPK models [9]

S no.	Classical PK models	PBPK models
1.	Typically has a central compartment representing plasma, linked to one or two peripheral compartments via rate constants I parameterized using known physiology, with a larger number of compartments corresponding to different organs or tissues in the body	Parameterized using known physiology, with a larger number of compartments corresponding to different organs or tissues in the body
2.	Model parameters do not generally have any physiological meaning but can be transformed to provide more interpretable PK descriptors I provides a quantitative mechanistic framework by which scaled drug-specific parameters can be used to predict the plasma and tissue concentration-time profiles of new drugs	Provides a quantitative mechanistic framework by which scaled drug-specific parameters can be used to predict the plasma and tissue concentration-time profiles of new drugs
3.	Offers a concise and standard representation of both preclinical and clinical experimental results	Can be used to extrapolate a dose in healthy volunteers to one in a disease population if the relevant physiological properties of the target population are available
4.	Cannot accommodate all prior information on both the drug and the physiology, restricting the ability to predict PK for a similar drug or to extrapolate the PK to different physiological conditions	Provides estimates of common PK parameters, like clearance, volume of distribution, and effective half-life
5.	Used to rank compounds for further investigation or to link PK parameters to physicochemical, in vitro, or structural properties in preclinical settings.	Used to inform initial dose selection or personalize dosage in subgroups of patients in clinical settings
6.	Built using compartments or "building blocks"	Built using a similar mathematical framework as classical PK models
7.	Provides estimates of common PK parameters, like clearance, volume of distribution, and effective half-life	Connected by flow rates that parallel the circulating blood system
8.	Used to describe the relationship between the plasma or relevant tissue concentration of the drug and time	Provides a more physiologically relevant representation of drug behaviour within the human body
9.	Predictive capabilities: Assess drug PK in diverse populations for first-in-human dosing, study optimization, formulation evaluation, and drug-drug interaction predictions	Innovative simulator components: Organ specific models
10.	Predicts drug behaviour within the human body, aiding in various stages of drug development	Suitable for various drug types, including small molecules, biologics, ADCs, bispecific, and oligonucleotides

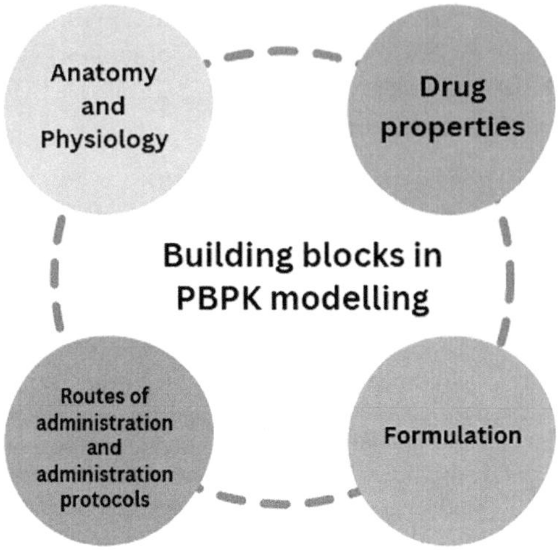

Fig. 14.1 Building blocks in PBPK modelling

14.3.1 Anatomy and Physiology

Understanding the physiology of the organism under study is crucial for developing Physiologically Based Pharmacokinetic (PBPK) models, as it involves incorporating relevant organs with their anatomical and physiological characteristics. Each organ is represented based on properties like volume, tissue composition, and perfusion blood flow rates. Additionally, compartments within organs, such as plasma, interstitium, cellular space, and red blood cells, can be further divided to account for specific processes like enzymatic reactions within cells. Blood flow rates vary among organs and play a crucial role in distributing substances throughout the body.

To enhance model accuracy, detailed descriptions of certain organs like the brain, liver, kidney, and lungs, or processes like intestinal absorption, can be included. PBPK models typically feature arterial and venous blood compartments connecting various organ compartments. However, in clinical practice, blood samples are often taken from peripheral veins, leading to discrepancies between measured concentrations and model predictions. To address this, some models include an additional organ representing tissues drained by peripheral veins.

For orally administered drugs, various factors including physiological, compound-related, and formulation-related aspects influence parameters like solubility, dissolution rate, and intestinal permeability. The gastrointestinal (GI) tract can be segmented into regions like the stomach, small intestine, and large intestine, each with specific roles. These segments can further be divided into zones representing different tissues, allowing for a realistic description of absorption processes. Drug absorption in the intestine can occur via transcellular or Para cellular routes, with factors like pH affecting solubility and permeability. Charged molecules, influenced by pH, may have different permeability compared to neutral forms, necessitating consideration in the model [4].

14.3.2 Drug Properties

Compound-specific parameters are vital inputs for PBPK models and encompass physico-chemical characteristics like molecular weight (MW), lipophilicity, solubility, and pKa values, typically determinable through in vitro methods. Lipophilicity, along with MW, influences membrane permeability, with adjustments made for halogen atoms to better represent molecular size. Solubility, especially crucial for oral drugs, affects absorption in the gastrointestinal tract, while pKa values inform on pH-dependent solubility changes. Pharmacokinetic parameters, like fraction unbound, are compound-specific and organism-dependent. These parameters, combined with physiological inputs, facilitate the calculation of more complex parameters crucial for PBPK modeling, such as tissue-plasma partition coefficients. Fraction unbound, reflecting the proportion of drug not bound to plasma proteins, significantly affects drug distribution and clearance. Experimental measurement, concentration estimation, or prediction from drug lipophilicity can determine fraction unbound. These compound-specific parameters play key roles in quantifying passive processes like membrane permeation and organ/plasma partitioning. Conversely, parameters for active processes, such as enzymatic activity (Vmax), require alternative methods, such as in vitro determination or fitting from in vivo data, followed by conversion to enzymatic rate or related parameters [3].

14.3.3 Formulation

Creating oral dosage forms for new drugs is a common strategy in pharmaceutical development, and mechanistic models that accurately simulate in vivo bioavailability can aid in their design and development. In vitro dissolution profiles in bio relevant media like Fasted State Simulated Intestinal Fluid (FaSSIF) and Fed State Simulated Intestinal Fluid (FeSSIF) have been established to mimic the disintegration and dissolution behavior of orally administered forms. These profiles, alongside drug pharmacokinetic (PK) profiles, can establish an in vitro-in vivo correlation (IVIVC). This correlation allows estimation of in vivo PK profiles from in vitro dissolution profiles of different formulations, facilitating optimization of dosage forms with fewer animal experiments or clinical trials.

Alternatively, dissolution functions can be incorporated into gastrointestinal (GI) compartments of PBPK models to represent disintegration and dissolution processes in the GI tract. Literature describes combining such dissolution profiles into GI physiological models and integrating them with whole-body PBPK models [4].

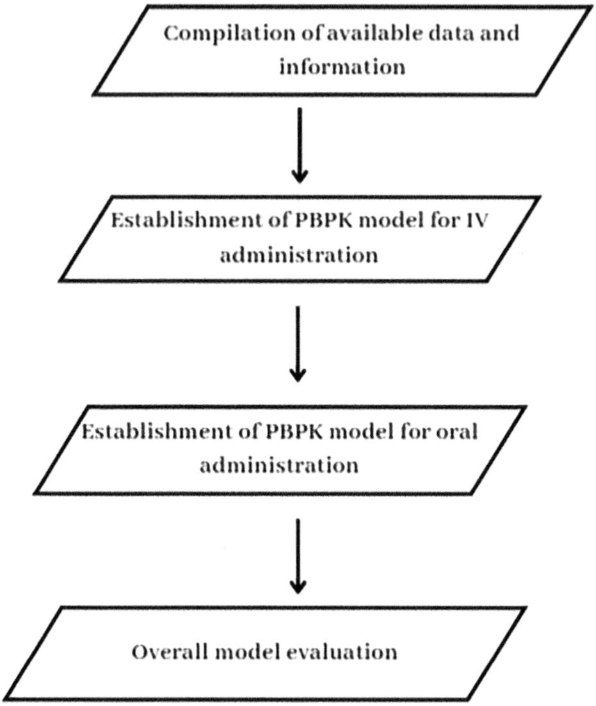

Fig. 14.2 Building a Physiologically-Based Pharmacokinetic (PBPK) model

14.3.4 Routes of Administration and Administration Protocols

The administration protocol encompasses the route of administration, dose amount, and dosing schedule for multiple administrations. Typically, the administration protocol used in model simulations matches that of the study being evaluated. Various administration routes, such as intravenous (IV) and oral (po), as well as alternative routes like inhalation, subcutaneous, intramuscular, topical, or ocular administration, are considered in PBPK models. Each route affects drug absorption and distribution in the body differently. For routes other than iv, the PBPK model should also incorporate mechanistic considerations related to local drug release and absorption from the administration site. For example, intramuscular administration involves describing progressive drug release and diffusion from the injection site [4]. Figure 14.2 showed building a Physiologically-Based Pharmacokinetic (PBPK) Model.

How to Build PBPK Model

1. Compilation of available data and information

As an initial step in model development, all available information regarding the drug's Absorption, Distribution, Metabolism, and Excretion (ADME) properties is collected. This encompasses drug-specific parameters, utilized as input parameters, and the characterization of the organism or population. Default physiological parameters are only adjusted if there is a mechanistic rationale, particularly in the case of special populations. Physicochemical properties of the drug are vital parameters informing drug distribution. Additionally, any further ADME information, such as clearance processes, transporters, or specific binding partners, is relevant for constructing physiologically plausible models. The level of detail in representing these processes and the criteria for model evaluation vary based on the model's aims, objectives, and applications, as well as the availability of data and information.

Experimental pharmacokinetic (PK) data are crucial for identifying unknown or uncertain parameters and validating the model. Commonly used data in PBPK model building include time-concentration profiles of the drug in plasma (across different preclinical species and/or humans, various administration routes, dosages, or application schemes), time-concentration profiles in relevant tissues, percentages of drug excreted via urine and feces, mass balance data, and metabolite data (time-concentration profiles and excretion data).

2. Establishment of the PBPK model for IV administration

The initial step involves selecting a distribution model, also known as organ/plasma partition coefficients calculation method, to characterize the drug's distribution behavior. This model should consistently describe the drug distribution across species and administration protocols. By comparing model simulations with in vivo pharmacokinetic (PK) measurements following intravenous (iv) administration, different distribution models can be assessed and clearance parameters estimated. These parameters are derived from experimental in vivo data, particularly plasma concentration-time profiles, and mass balance data may supplement information on different elimination pathways. Metabolism rates can also be estimated using In Vitro-In Vivo Extrapolation (IVIVE) approaches.

Surrogate compound parameters, such as lipophilicity, are adjusted slightly to improve agreement with experimental IV data. There are no strict guidelines on the extent of adjustment, as it depends on the initial lipophilicity measure used. Despite a PBPK model potentially containing numerous ordinary differential equations, the number of independent model parameters for a new compound is typically small, usually fewer than 5 per compound, due to the incorporation of extensive prior physiological information.

Similar to distribution models, compound-specific parameters are generally kept constant across different species and administration protocols. However, if adjustments to distribution models and parameters fail to adequately describe IV data, it suggests that relevant processes may be missing from the model. In such cases, additional processes like active transport or binding partners can be incorporated based on existing knowledge or testable hypotheses. Including as much experimental data as possible, such as Km and Kd values and enzyme or binding partner abundance, is advisable when introducing additional processes. Notably, data from different doses, demonstrating dose non-linearity, are necessary to identify Michaelis-Menten parameters for transporters or clearances.

3. Establishment of the PBPK model for oral administration

After establishing an intravenous (iv) model, the next step involves creating a model for oral (po) administration or other extra-vascular routes. During this stage, parameters influencing distribution or metabolism/excretion remain unchanged, and only parameters affecting oral absorption are adjusted. Common parameters estimated during po model development include intestinal permeability and parameters related to meal events. Additionally, formulation-related parameters describing drug release and solubility may be adjusted, particularly if aqueous solubility is used as the starting value instead of FaSSIF and FeSSIF solubility. Subsequently, additional absorption-related processes like enterohepatic recycling (EHC) may be incorporated if relevant to the drug.

This stepwise approach emphasizes that during oral absorption model establishment, information on distribution and clearance processes from IV data should be utilized to identify relevant absorption parameters. However, in some cases where the IV model includes processes pertinent to oral absorption, such as uptake transporters, it may be preferable to perform parameter identification using IV and po data simultaneously. Although IV data are valuable for informing distribution and clearance processes independently of drug absorption, depending on modeling objectives and required precision, a PBPK model may be developed solely using po data. In the absence of human IV data, scaling a human po model from an animal model established using both IV and po data is an option. However, without IV data, model uncertainty may be higher, particularly depending on the Biopharmaceutics Classification System (BCS) classification of the drug and involvement of transporters or gut-wall metabolism.

4. Overall model evaluation

After completing model development, it is essential to evaluate the quality of the model to ensure it meets its intended purpose. The evaluation criteria depend on the project's goals, and several tests or criteria can be employed:

(i) *Agreement with experimental data:* Typically, the model is assessed by visually comparing simulated and experimental concentration-time profiles, focusing on absolute concentrations and dynamic profile shapes. Error functions like root-mean-square deviation (RMSD), area under the curve (AUC) error, or concordance correlation coefficient can also be used. Key pharmacokinetic (PK) parameters such as Cmax, tmax, AUC, and $t^{1/2}$ can be compared between simulated and experimental data. It is important to acknowledge uncertainty in both model simulations and experimental data. External validation data can also be used to assess model consistency beyond the data used for model building.
(ii) *Consistency across different doses, species, special populations, and compounds:* Evaluating the model's performance across various dose levels provides insight into potential structural shortcomings. Drug-dependent parameters and calculation methods for distribution and cellular permeability should remain consistent across species. Any deviations should be explained,

such as species-specific processes. If PK data for special populations are available, the model can be evaluated by adjusting physiological parameters accordingly. Additionally, the model can be assessed for consistency across compounds with similar ADME properties.

(iii) *Sensitivity analysis and best- and worst-case scenarios*: Conducting sensitivity analysis for relevant parameters helps assess uncertainties in model results. This analysis can focus on uncertain parameters for active processes or all parameters in the model to identify the most sensitive ones for specific model outputs. Best- and worst-case scenarios can simulate the effects of extreme parameter values within experimental and physiological uncertainties. Assessing model alternatives based on underlying mechanisms, helps evaluate uncertainties. Results from sensitivity analysis and scenarios aid in determining the robustness of modeling conclusions.

A hierarchical Representation of Physiological and Drug Parameters in a Physiologically-Based Pharmacokinetic (PBPK) Model [10]

A hierarchical representation of physiological and drug parameters in a Physiologically Based Pharmacokinetic (PBPK) model involves organizing these parameters into a structured framework based on their relationships and dependencies. This hierarchical structure helps to understand the complexity of the model and how different parameters interact with each other to influence drug behavior within the body.

At the highest level of the hierarchy, there are overarching categories that encompass various physiological and drug-related parameters. These categories may include:

(i) Physiological Parameters:

This category encompasses parameters related to the anatomy and physiology of the organism, such as organ volumes, blood flow rates, tissue composition, and metabolic enzyme activities. These parameters provide the foundation for modeling how drugs are distributed, metabolized, and eliminated within the body.

(ii) Drug-Specific Parameters:

This category includes parameters that describe the physicochemical properties of the drug, such as molecular weight, lipophilicity, solubility, and pKa values. These parameters influence how drugs interact with biological systems and affect their absorption, distribution, metabolism, and excretion.

Within each overarching category, there are subcategories representing specific types of parameters. For example:

(a) Physiological Parameters:

- Organ-specific parameters: Parameters specific to individual organs, such as organ volume, blood flow rate, and tissue composition.
- Systemic parameters: Parameters that describe overall systemic physiology, such as cardiac output and total body water volume.

(b) Drug-Specific Parameters:

- Physicochemical properties: Parameters that describe the physical and chemical characteristics of the drug, such as molecular weight, lipophilicity, and solubility.
- Pharmacokinetic parameters: Parameters that describe the drug's disposition within the body, such as clearance rate, volume of distribution, and absorption rate constant.

Each parameter within these subcategories may have further hierarchical levels, representing more detailed aspects or variations. For instance, organ-specific parameters may include subcategories for different compartments within an organ (e.g., plasma, interstitial fluid, intracellular space), while pharmacokinetic parameters may include subcategories for different routes of administration (e.g., oral, intravenous, inhalation).

By organizing physiological and drug parameters hierarchically, PBPK modelers can systematically define and incorporate these parameters into their models, ensuring that all relevant factors are considered and accurately represented. This hierarchical representation facilitates model development, validation, and interpretation, ultimately enhancing the predictive capabilities of PBPK models in drug development and risk assessment. Figure 14.3 showed Hierarchical representation of the main physiological and drug parameters in a PBPK model [10].

14.4 Applications of PBPK Modeling

PBPK models simulate human anatomy and physiology using interconnected compartments representing organs and tissues connected by blood circulation. These mathematical models predict drug pharmacokinetics, including absorption, distribution, metabolism, and excretion. Various commercial platforms offer tools to build PBPK models, some with preset models and others allowing custom creations [13].

14.4.1 Predicting Drug Absorption, Distribution, Metabolism, and Excretion

- Predicting drug absorption

It is possible to expedite and enhance drug development by using PBPK modeling, a potent technique that predicts oral drug absorption and offers trustworthy pharmacokinetic insights across a variety of demographic and ethnic groups. It is commonly used to predict oral drug absorption by mechanistically simulating a compound's pharmacokinetics under various situations, such as fed or fasting, and

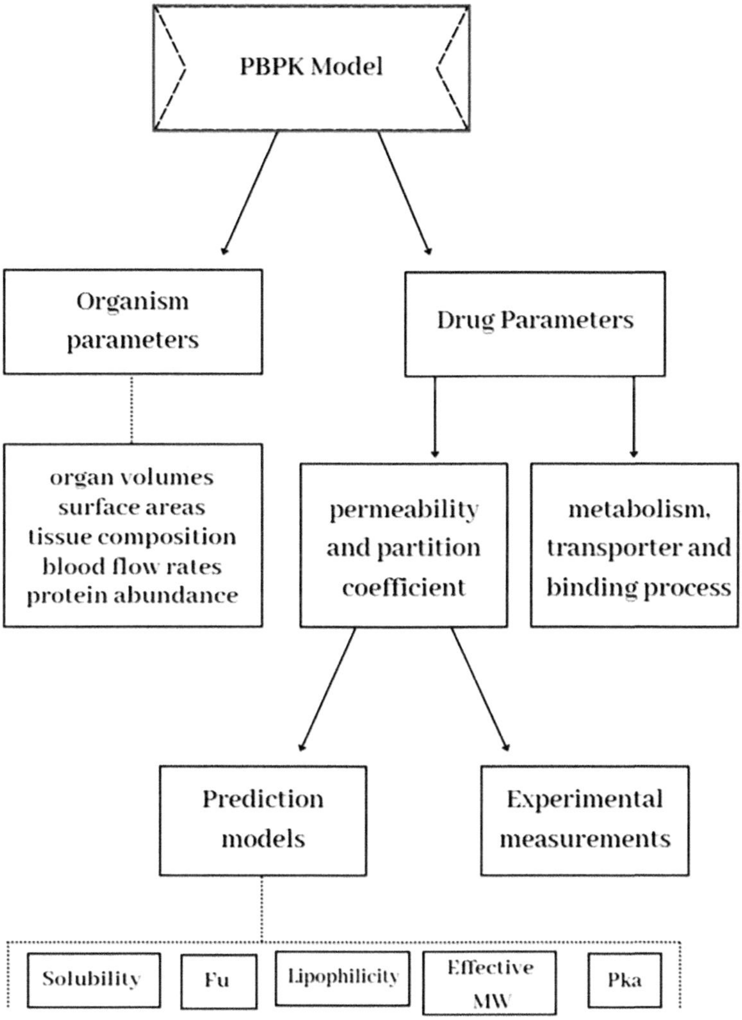

Fig. 14.3 Hierarchical representation of the main physiological and drug parameters in a PBPK model [10]

integrating drug properties like permeability and solubility. With the use of this modeling approach, drug interactions in the gastrointestinal system can be predicted while accounting for factors such as pH fluctuations, gastric emptying rates, bile secretion, and changes in blood flow brought on by food consumption. Changes in luminal metabolism, interactions between food and drugs, and transporter inhibition also influence drug absorption. Inhibiting drug-metabolizing enzymes or efflux transporters in the gastrointestinal tract can enhance the rate of drug absorption [11]. By placing all available in vitro data within an in vivo environment, the use of oral PBPK models built with restricted drug discovery information can aid in the evaluation of the oral absorption of possible therapeutic candidates [12]. Figure 14.4 showed effect of food on pharmacokinetic profile of drug [13].

Changes in drug pharmacokinetics are minor and fall within typical variations seen after oral administration. Positive food effects increase systemic drug exposure, while negative food effects decrease it. PBPK models are used in drug discovery throughout the phases of clinical trials. The effect of food in oral drug absorption can be investigated through pharmacokinetic studies. Drug-specific parameters like molecular weight, log P, pKa, particle size, aqueous solubility, and metabolic clearance rate are used as inputs for the models [11, 12].

- Distribution

One particularly useful tool for predicting drug distribution is physiologically based pharmacokinetic (PBPK) modelling, which allows for the prediction of drug concentrations in a variety of body tissues and organs. Physiologically Based Pharmacokinetic (PBPK) modeling stands out as a valuable asset in forecasting drug metabolism, facilitating the calculation of drug clearance rates and the delineation of metabolic routes. This approach is particularly helpful for some populations, such as youngsters, expecting mothers, and individuals with hepatic or renal problems, where individual differences might significantly affect medication disposal [14]. The application of PBPK (Physiologically Based Pharmacokinetic) modeling in distribution involves using these models to predict the dispersion of chemicals, such as drugs or toxic materials like manganese, throughout the human body. Specifically, as research by Ramoju et al. demonstrates, manganese levels in human brain tissue can be measured using PBPK models. In order to examine the relationship between manganese-induced neurotoxicity and occupational exposure in humans, this study used PBPK modeling with categorical regression [15]. Figure 14.5 representation of flow of drug distribution in PBPK models along with organ compartments and arrows representing blood flow [16].

- Metabolism

One of the most important tools for predicting drug metabolism is physiologically based pharmacokinetic (PBPK) modelling, which allows for the evaluation of drug clearance rates and metabolic pathways. When predicting the impact of food on drug metabolism, PBPK modeling incorporates physiological changes that occur

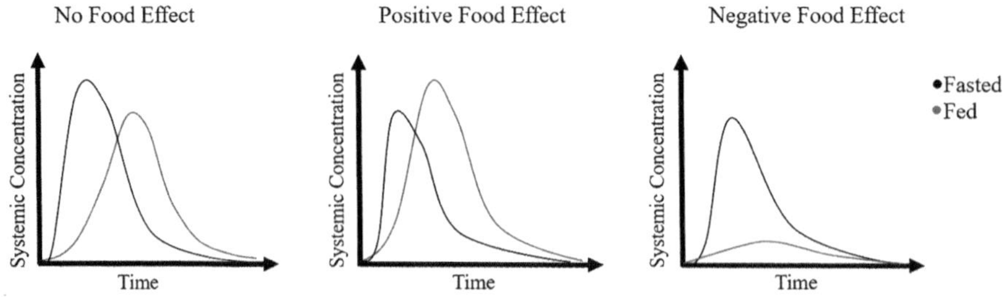

Fig. 14.4 Effect of food on pharmacokinetic profile of drug [13]

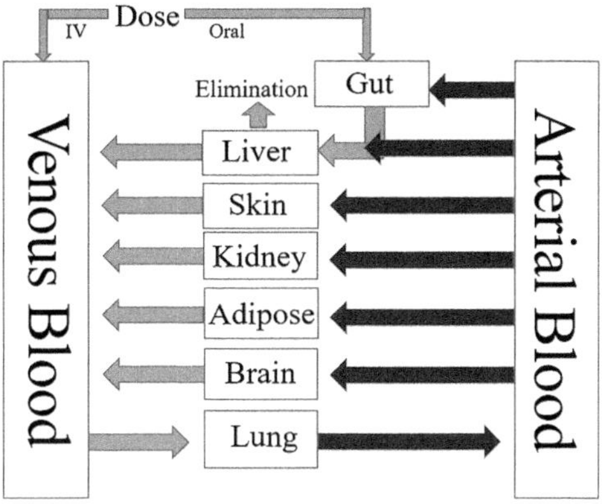

Fig. 14.5 Schematic representation of flow of drug distribution in PBPK models along with organ compartments and arrows representing blood flow [16]

during digestion, such as changes in gastrointestinal pH, stomach emptying rates, and splanchnic blood flow. It is an essential tool for understanding the metabolism of various substances, ranging from drugs to toxic substances like manganese. Through the integration of biochemical processes, enzyme kinetics, and physiological variables, it provides a mechanistic understanding of these substances metabolism [15]. Using GastroPlus™ to forecast the metabolism of CYP3A substrates in the human gut is one assessment and case study illustrating the use of PBPK modeling in metabolism [18]. The purpose of the study was to show how well PBPK modeling predicts human intestinal metabolism of CYP3A substrates. According to the study, the intestinal metabolism of CYP3A substrates could be predicted with high accuracy using the PBPK model, and the predictions agreed well with the data that was observed. This work demonstrates how PBPK modeling can be used to forecast a drug's intestinal metabolism, which is a crucial aspect of the drug's overall pharmacokinetics. Another illustration is the application of PBPK modeling to forecast dextromethorphan (DXM) metabolism according to a person's CYP2D6 genotype [9]. The results of the study showed that PBPK modeling can be used to anticipate certain metabolic characteristics and to tailor dosage regimens to the genetic profiles of individuals. The study also emphasized the significance of PBPK models' repeatability and accessibility as well as the requirement for reliable real-world data to support them.

- Excretion

An essential tool for evaluating how medication properties affect excretion is PBPK modeling. PBPK models predict the distribution, metabolism, and excretion of pharmaceuticals across many tissues and organs by combining physiological and drug-specific data. The increasing utilisation of PBPK modelling in the domains of drug development and safety assessment highlights its importance. Example of the application of PBPK modeling in excretion studies is the evaluation and case study using GastroPlus™ to predict human intestinal metabolism of CYP3A substrates [17]. This study demonstrated the utility of PBPK modeling in predicting the intestinal metabolism of CYP3A substrates in humans, showcasing the accuracy and applicability of PBPK models in understanding drug excretion pathways. Two-compartment PBPK model was used to predict the pharmacokinetics of a compound in humans. The excretion of the compound was modeled through urine via the kidney. The PBPK model was able to accurately predict the pharmacokinetics of the compound, including the excretion pathway, in humans [19].

14.4.2 Inter-species Extrapolation

When extrapolating across species, PBPK modelling is incredibly helpful as it provides a mechanistic understanding of how medications act in different animals. This method expedites and improves the drug development process by enabling the efficient planning of clinical trials and accurate forecasting of pharmacokinetic patterns across various groups. PBPK modeling proves essential for cross-species extrapolation in drug development and toxicology. It enables the forecasting of drug or toxicant levels in different tissues and organs across a range of species, including humans. By integrating species-specific physiological and biochemical data, PBPK models offer a mechanistic insight into variations in drug or toxicant handling and metabolism among species [15]. PK modeling predicts a drug's effectiveness and safety in humans using data from animal studies. This method bridges preclinical research with human outcomes, informing dosing strategies and evaluating the drug's risks and benefits in clinical scenarios. The pharmacokinetic modeling is used to assess oxytetracycline and gold nanoparticles in different species, providing insights into the potential effects and safety profiles of these compounds across species [20].

14.4.3 Impact of Drug Properties on PK

Drug concentrations in different tissues and organs can be determined with the use of PBPK modeling, a crucial technique that eventually improves and speeds up the drug development process by anticipating the manner in which pharmacological characteristics affect pharmacokinetics (PK). Here the example highlights the influence of drug properties on PK modeling is the development of PBPK models for metoprolol. This model was designed based on different CYP2D6 genotypes using in vitro data. By integrating drug-specific parameters and accounting for the variability in CYP2D6 genotypes, the PBPK models successfully predicted the pharmacokinetics of metoprolol across various genotypes [4].

14.4.4 Simulation of Food Effects and Special Populations

Utilising PBPK (Physiologically Based Pharmacokinetic) modelling is crucial when attempting to replicate the impacts of food and unique demographics. PBPK models predict how food affects drug absorption, distribution, metabolism, and elimination by combining physiological and biochemical data. This kind of strategy helps improve medication research by providing insight into the effects of food on therapeutic performance in a variety of settings, including specialised populations such as children, the elderly, and people with liver or kidney diseases. PBPK modeling provides insights into the ways in which drug distribution is influenced by age, gender, pregnancy, health problems, and genetic features within special populations. For example, PBPK models can anticipate the effects of pregnancy on drug handling and fetal exposure by accounting for changes in organ size, blood circulation, and the function of drug-metabolizing enzymes. Similarly, PBPK models can account for changes in organ function and drug metabolism with age, predicting the impact of aging on drug distribution and clearance [15].

Drug-Drug Interaction

In order to quantitatively anticipate pharmacokinetic (PK)-based medication interactions, physiologically-based pharmacokinetic (PBPK) modelling techniques are becoming more and more popular as an alternative to clinical investigations. To anticipate drug-drug interactions (DDIs), PBPK modelling combines important in vitro data with species-specific characteristics and the physicochemical properties of a novel chemical entity. Additionally, the DDI potential for untested scenarios can be predicted by evaluating different doses and dosing regimens using the PBPK technique [16].

When a molecule has been demonstrated to either inhibit or stimulate any of the enzymes or transporters involved in drug elimination and/or distribution, either in vitro or in vivo, interaction parameters are necessary for the creation of the PBPK model. Therefore, the compound may be a "perpetrator" of DDIs. when the offender modifies its own metabolism (auto-induction, auto-inhibition, or both). When a drug is verified by clinical DDI research, it is possible to evaluate both the fm (in terms of its function as a substrate) and the interaction parameters in the model (in terms of its function as a perpetrator) [17].

14.5 Software Available

At various phases of drug discovery and development, PK and PD modeling and simulation are made easier by a number of commercial and open-source software programs. MATLAB ® (MathWorks, Natick, MA, USA), SAAM II, acslXtreme ®, ADAPT II, SAS ®, S-PLUS ® (Seattle, WA, USA), R, Xpose, NONMEM ®, Monolix, WinNonMix ®, WinBUGS, POPT, WinPOPT, WinNonLin ®, Trial

Simulator ®, Drug Model Explorer ®, PK-Sim ®, GastroPlus TM, Cloe ® Predict, and Simcyp ® are a non-exhaustive list of such software [18].

In both academia and industry, computer-aided modeling and simulation are being used more often to study the ADME of medications. Comprising of physiologically based pharmacokinetic (PBPK) modeling, these techniques are used in programs like GastroPlus, SimCYP, and PKSim. Utilizing these programs, ADME behavior may be modeled by taking into account how pathology and physiology affect drug pharmacokinetics. However, in order to utilize the tools properly, one must comprehend both the underlying scientific ideas and the software [19].

14.5.1 Simcyp

A platform and database for "bottom-up" mechanistic modeling and simulation of the processes of oral absorption, tissue distribution, metabolism, and excretion of drugs and drug candidates in healthy and disease populations is the Simcyp ® population-based absorption, distribution, metabolism, and excretion simulator. It integrates experimental data from in vitro enzyme and cellular systems—which are often generated during preclinical drug discovery and development—as well as pertinent physicochemical properties of compounds and dosage forms with demographic, physiological, and genetic data on various patient populations. The Simcyp Simulator's mechanistic approach enables the simulation of intricate drug interactions, parent drug and metabolite profiles, time- and dose-dependent phenomena like auto-induction and auto-inhibition, as well as complex outcomes related to absorption, distribution, metabolism, and excretion. The simulator offers the option to choose between two simulation modes: "PK Parameters" and "PK Profiles." "PK Parameters" provide straightforward "steady-state" solutions that are typically sufficient at the early phases of drug discovery and development because they eliminate time- and, in certain situations, concentration-dependent phenomena.

In a virtual population that ideally replicates the range of inter-subject variability found in real patient populations, users "administer" their substances. These virtual populations can be handled similarly to compound files and are stored in files that resemble them. The data needed to create the populations can be found on several screens under the headings "Tissue Composition," "GI Tract," "Demographic," "Liver," "Kidney," and "Tissue Blood Flow Rates." Up to four compounds can be administered simultaneously via a flexible study design interface, which also enables the simulation of parent drug profiles and metabolite profiles. The basic features of Simcyp are provided by the Simulator engine, which also houses a potent real-time differential equation solver and the suite of algorithms that make up the simulations [18].

14.5.2 PK-Sim

PK-Sim® is an innovative 'whole-body' physiology-based pharmacokinetic simulation software. It allows the description of the fate of xenobiotics, particularly drugs, in the mammalian body on the basis of a mathematical representation of the determinant physical and physiological processes. Its modular architecture is tailored to serve the different needs of a drug research and development company from the early research and lead optimization phase up to preclinical and clinical development stages, despite its undoubted value, it has so far not been widely established as a standard tool in the drug R&D process [20].

The key to a realistic simulation of a substance's fate in the body is how the organs are represented in a PBPK model. Four mammalian species—the mouse, rat, dog, and human—can be PK-simulated using PK-Sim®. The pertinent physiological data for each species has been gathered from a variety of literature sources. One can estimate the prospective behaviour in specific sub-populations (children, old, high or low metabolizers, etc.) by simply modifying the physiological data in a simulation to match the sub-populations' physiological characteristics [20].

The PBPK model for small molecules has separate blood pool compartments and 15 organs or tissues. Adipose tissue, the brain, the bone, the gonads, the heart, the kidneys, the large and small intestines, the liver, the lung, the muscle, the pancreas, the skin, the spleen, and the stomach are among the organs and tissues that are specifically portrayed. Each organ is made up of smaller sections that correspond to the interstitial space, cellular space, plasma, and blood cells, which collectively make up the vascular space [21] (Fig. 14.6).

14.5.3 GastroPlus

GastroPlus is a software simulation tool with a mechanistic foundation that can mimic several aspects of drug absorption in people and animals, including intravenous, oral, oral cavity, ophthalmic, inhalation, dermal, subcutaneous, and intramuscular absorption, as well as biopharmaceutics, pharmacokinetics, and pharmacodynamics. To assist you in making quicker and better-informed project decisions, this seamlessly integrated platform blends a user-friendly interface with potent science. To make it easier for businesses to license just the capabilities they require in each area, GastroPlus has been divided into modules.

GastroPlus® is software for modeling and the conducting of simulations of properties of drugs or chemicals including absorption, release rate, bioavailability, pharmacokinetics, and pharmacodynamics; computer software which allows for prediction of drug-to-drug interactions; computer software for the prediction of drug effects on animals and virtual patient populations; computer software which allows researchers to adjust pharmacodynamic models to observed data and to then use the resulting fitted models to predict pharmacodynamic changes due to changes

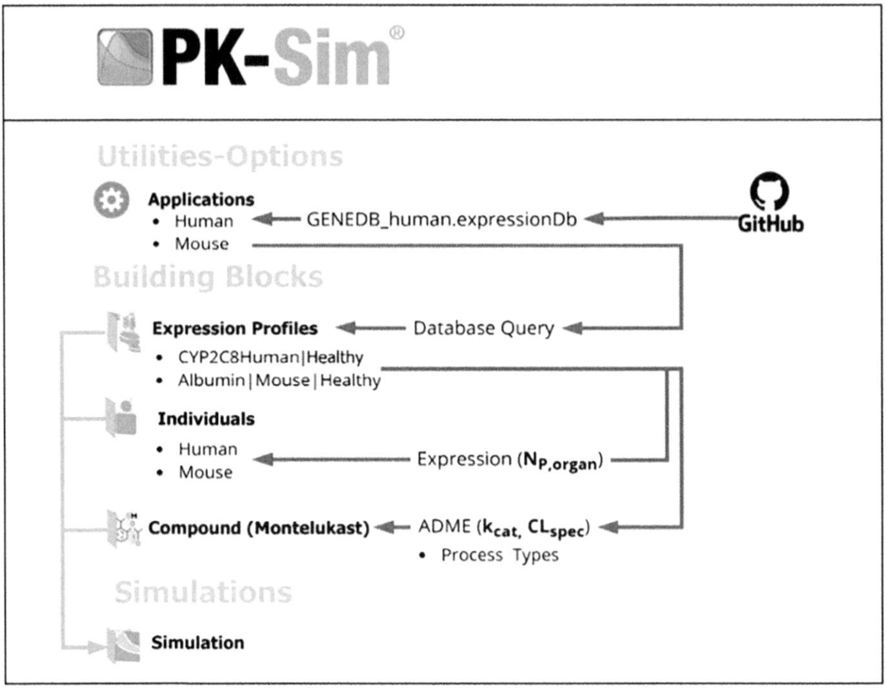

Fig. 14.6 PK-Sim workflow [22]

in dose, dosage form, and dosing regimen of a drug or chemical; computer software which allows for the generation of in vitro-in vivo correlations; computer software which allows for the prediction of absorption and systemic distribution/elimination for large molecules [23].

14.6 Challenges and Limitations

Although useful, PBPK models have drawbacks and restrictions. One drawback is that these models might not take inter-individual heterogeneity into account, instead describing the typical person with a particular condition [28]. The absence of supporting data on formulation design, manufacturing procedures, and API characteristics—all of which have a substantial impact on the in vivo solubility and bioavailability of medications—is another important drawback of PBPK models [29]. Accurate PBPK modeling requires biorelevant inputs like as dissolution profiles, permeability, and solubility, as well as adequate validation for the intended use. This presents another difficulty. Additionally, the paucity of in vivo PK datasets for model building and validation may provide a challenge for PBPK models, necessitating a thorough approach that takes into account all available data [29]. Despite these difficulties, pharmacokinetic drug prediction using PBPK modeling is a significant tool for risk assessment and medication development since it provides a mechanistic approach based on physiological and anatomical features [28].

14.6.1 Residual Simplification in the PBPK Model

The concept of residual simplification in PBPK models refers to the process of reducing the complexity of these models while maintaining accuracy in predicting drug behavior. This method entails making the model more comprehensible by simplifying its structure or lowering the number of parameters without sacrificing the model's accuracy [30]. In order to expedite parameter estimation, increase model efficiency, and improve the model's capacity to predict pharmacokinetic features across species and for population PK studies, researchers are trying to simplify PBPK models [30]. The objective is to develop mechanistic models that distinguish between factors particular to the medication and the system, enabling precise forecasting even in the lack of comprehensive tissue concentration information [30]. Proper lumping and balanced truncation are examples of residual simplification techniques that are used to simplify PBPK models without compromising their predictive ability [31].

14.6.2 Balancing Complexity and Accuracy

Balancing complexity and accuracy in PBPK modeling is a critical aspect of utilizing these models effectively. The difficulty is in keeping the model's accuracy in forecasting drug behavior across species and for population pharmacokinetic (PK) studies while streamlining its structure and cutting down on its number of parameters [30]. This balance is essential for enabling parameter estimation, increasing model effectiveness, and strengthening the model's capacity to forecast physiologically significant PK parameters devoid of tissue concentration information [30]. Researchers can create a model that is both interpretable and predictive by keeping complexity levels that capture key physiological processes without adding needless details. This will enable population PK analysis and reliable extrapolation of PK properties from animals to humans [30]. In order to effectively describe the underlying biological processes and provide precise drug behavior predictions, the model must be balanced between being sufficiently reduced to be useful and efficient and maintaining the requisite complexity.

14.6.3 Uncertainty and Sensitivity Analysis

Sensitivity analysis and uncertainty are essential elements in the creation and verification of physiologically-based pharmacokinetic (PBPK) models. Through these investigations, the uncertainty surrounding the model's predictions is measured, and the parameters that most significantly affect the model's results are determined.

14.6.4 Uncertainty Analysis

In PBPK modeling, uncertainty analysis refers to assessing the level of uncertainty in the model's predictions as a result of the intrinsic variability in the input parameters. This is especially crucial because experimental data and empirical estimation, on which PBPK models are predicated, are prone to errors and limits. The level of uncertainty surrounding the model's predictions has the potential to substantially impact both the validity of the model's validation and the degree of confidence in its forecasts [31, 32].

Sensitivity Analysis
Sensitivity analysis is a technique used in PBPK modeling to determine how modifications to the input parameters impact the model's results. Researchers can concentrate on the factors that are most important to the accuracy of the model by using this analysis to help find the parameters that have the biggest effects on the model's predictions. Numerous techniques are available for performing sensitivity analysis, such as algorithmic techniques like tempering MCMC and Monte Carlo simulations [32, 33].

Importance of Uncertainty and Sensitivity Analysis
For a number of reasons, sensitivity analysis and uncertainty must be integrated into PBPK modeling.

Model Validation
An essential component of model validation is determining how reliable the model's predictions are, and uncertainty analysis assists I n this regard. Sensitivity analysis aids in this process by pointing out the variables that have the biggest impact on the model's results, enabling focused enhancements to the model's accuracy [32].

Parameter Fixing
To increase the computational effectiveness of the model calibration process, non-influential parameters in the model can be found via sensitivity analysis and fixed [32].

Model Simplification
Sensitivity analysis can help simplify complex PBPK models by pointing out the most important factors, which will make the models easier to handle and more effective [32].

Modeling for Regulatory Purposes
For regulatory purposes, where choices about drug development and approval depend heavily on the accuracy and dependability of the model's predictions, the application of uncertainty and]sensitivity analysis in PBPK modeling is very significant [33].

14.7 Introduction to PK-PD Modeling

Pharmacokinetics and pharmacokinetics (PK-PD) modeling is statistical model that combine both pharmacokinetics and pharmacodynamics to describe response of drug and dose over time [29]. PK is how body process drug in terms of absorption, distribution, metabolism and excretion while PD describes how drug affects the body. By merging these two concepts, PK-PD modeling which helps to predict efficacy of drug dosing and guiding experimental trails in drug development [35].

PK-PD modeling plays an essential role in optimizing drug delivery systems, predicting dosing strategies, and enhancing the drug efficacy and safety profiles. It predicts toxicity profiles, analyzes drug-drug interactions. Additionally, PK-PD modeling allows for the quantification of the relationship between drug exposure and response, considering various parameters like drug clearance, release rates, and physiological values such as blood flow and enzyme expression [35].

In PK-PD modeling, the concentration-response relationship is a major aspect that describes how drug concentration influences the pharmacodynamic response. This relationship is significant in understanding how different drug concentrations lead to specific effects in the body, various PK-PD models will help you to understand this relationship [36].

14.8 Mechanistic PK-PD Modeling

14.8.1 Simple Direct Effect Models

The majority of pharmacokinetic/pharmacodynamic models utilize the sigmoidal Emax model, which is derived from the Hill equation and classical receptor occupancy theory, to represent the connection between drug concentration and its effects. Receptor theory serves as the foundation for two main purposes: (a) predicting the link between drug concentration and its effects in living organisms, and (b) understanding the kinetics of target association and dissociation, which influence the temporal delay known as hysteresis in development of the drug's effect. This approach inherently differentiates between drug-specific and biological-system specific factors, enabling precise extrapolation from in vitro to in vivo and across different species. These methods play a key role in translational drug research [37].

The Hill equation predicts that the effects of drugs (E) are directly proportional to the activation of receptors (i.e., linear transduction). It presupposes that the concentration of pharmaceuticals in the plasma is in rapid equilibrium with the site where the action occurs. This equation reflects a key relationship in pharmacodynamics as follows:

$$E = Eo \pm \frac{E_{\max} \times C_p}{EC_{50} + C_p}$$

The equation, referred to as the E_{max} model, explains the link between concentration and effect by considering a baseline effect (E_0), the greatest effect achievable (E_{max}), and the drug concentration that produces half of the maximum effect (EC_{50}). The specified variables or conditions. The relationship between effect and log-concentration can be easily understood by examining a graph. In this plot, the plateau at high concentrations represents E_{max}, whereas EC_{50} corresponds to the drug concentration at which $E = 0.5 * E_{max}$ [38].

The Hill equation is a widely used pharmacological model that explains the correlation between drug concentration and the subsequent physiological response. The equation combines the intermediary processes from receptor association to receptor activation-response relationship into a single equation. However, it does not explicitly differentiate between drug-specific parameters (such as receptor affinity and intrinsic activity) and system-specific parameters (such as receptor density and the relationship between receptor activation and pharmacological effect). This limitation indicates that the Hill equation is not easily able to forecast the correlation between concentration and effect in various tissues or under varied stimulus-response circumstances. The Hill equation assumes the presence of equilibrium circumstances, which may not always hold true. The inherent descriptive character of this restricts its use for making predictions beyond the available data. To get more precise predictions, a model based on the mechanisms involved, which distinguishes between drug-specific and system-specific parameters, is required [39].

14.8.2 Biophase Distribution Model

The term "biophase" was introduced by Furchgott to refer to the specific location where a drug exerts its effects. Plasma drug concentrations are quantified using a suitable pharmacokinetic model, and the rate of change of drug concentrations in the biophase (C_e) is defined as the first-order distribution rate constant (k_{eo}):

$$\frac{dC_e}{dt} = keo \times Cp - keo \times Ce$$

Where, k_{eo} is distribution rate constant, C_p represent the blood concentration and C_e represent the effect side i.e., biophase concentration.

Initially, distinct rate constants for the processes of production and loss were suggested. However, for the sake of being able to determine and distinguish these processes, they are frequently assumed to be equal and represented by the term k_{eo}. The quantity of drug entering and exiting this compartment is considered to be insignificant, and so does not affect the pharmacokinetics of the drug. The biophase model is specifically applicable only for explaining delayed responses caused by drugs distribution. Due to its initial use in modeling delayed drug reactions, this approach frequently gets incorrectly applied to systems where the rate-limiting phase is not connected to drug distribution. This leads to inaccurate fitting of this model and represents the produces values that are not feasible in the context of the problem being studied [38].

14.8.3 Indirect Response Models

Indirect response models are a valuable type of models that involve reversible interactions between drugs and receptors, which in return affect the formation or loss of biomarker response variables. These models have been employed to examine the pharmacodynamics of various drugs effects, and their mathematical characteristics have been thoroughly described. The four fundamental models consist of the inhibition of production (Model I) or dissipation (Model II) of response, as well as the stimulation of production (Model III) or dissipation of response (Model IV). These models are described by the following differential equations [38]:

Model I: $$\frac{dR}{Dt} = K_{in}\left(1 - \frac{I_{max} \times C_p}{IC_{50} + C_p}\right) - k_{out} \times R.$$

Model II: $$\frac{dR}{Dt} = K_{in} - k_{out}\left(1 - \frac{I_{max} \times C_p}{IC_{50} + C_p}\right) \times R.$$

Model III: $$\frac{dR}{Dt} = K_{in}\left(1 - \frac{S_{max} \times C_p}{SC_{50} + C_p}\right) - k_{out} \times R.$$

Model IV: $$\frac{dR}{Dt} = K_{in} - k_{out}\left(1 - \frac{S_{max} \times C_p}{SC_{50} + C_p}\right) \times R.$$

Where,

k_{in} = zero-order production rate constant,
k_{out} = first-order elimination rate constant,

I_{max} and S_{max} = maximum fractional factors of inhibition,
IC_{50} and SC_{50} = EC_{50}.

14.8.4 Signal Transduction Models

Significant delays in the observed pharmacodynamic response might arise due to several time-dependent processes that take place between the binding of the medication to the receptor and the final pharmacological reaction. A transit compartment technique can be used to explain the delay between the concentration of a drug and the observed effects due to time-dependent signal transduction. The rate of change of the initial transit compartment (M1) can be described by the following differential equation, assuming rapid receptor binding [38].

$$\frac{dM_1}{dt} = \frac{1}{\tau}\left(\frac{E_{max} \times C_p}{EC_{50} + C_p} - M_1\right)$$

wherein the,

E_{max} = the drug–receptor interaction, and
τ = mean transit time

14.8.5 Parameter Estimation

Parameter estimation in PK/PD modeling involves obtaining the optimal values of the model parameters that most accurately correspond to the observed data. Typically, non-linear regression techniques are employed to compare simulations from the model with experimental data and optimize the model parameters to reduce the difference between them.

For example, a pharmacokinetic model with a single compartment, where the dose follows a first-order pattern and the clearance is linear. The parameters of this model consist of three variables: the absorption rate constant (ka), the elimination rate constant (ke), and the volume of distribution (V). One method to determine these parameters based on data involves simulating the model using initial estimates for the parameters, then consequently adjusting the parameters to minimize the difference between the anticipated and observed concentrations.

A frequently used approach to estimate parameters is maximum a posteriori (MAP) Bayesian estimation. This method combines prior knowledge of parameter values with observed data to derive a posterior distribution for the parameters. The MAP estimate corresponds to the mode of the posterior distribution, indicating the parameter value that is most likely given the data and prior knowledge.

An alternative method involves use non-linear regression to estimate fixed effects. This involves comparing simulations generated by the model with experimental data and modifying the model parameters to minimize the disparity between the two. PK/PD models may incorporate random effects to consider the variation across individuals in the parameters, in addition to calculating fixed effects. Random effects can be evaluated using mixed effects modeling techniques, which entail estimating both the fixed effects and the variance components of the random effects. This methodology is illustrated in a research project that assesses the accuracy of parameter estimates for sigmoid E_{max} models in exposure-response relationships. When doing parameter estimate in PK/PD modeling, it is crucial to take into account the data's quality and quantity, along with the model's assumptions and restrictions. Cross-validation and predictive tests are employed to assess the model's performance and verify its ability to accurately represent the data [40].

14.8.6 Model Validation

Quantitative time course experimental data sets are essential for compare with computer simulations in order to construct a PBPK/PD model. The objective and aim is for computer simulations to align perfectly with experimental data, as previously said. The data sets utilized for constructing the model are known as working sets or training sets. In order to validate the model, it is essential to utilize data sets that are distinct from the working sets to compare against the simulated outputs of the model. For example, let's consider an example where a PBPK/PD model was created utilizing PK and PD datasets obtained from the administration of two different dose levels of medication A via intravenous route, as reported by laboratory X. The validation of this physiologically-based pharmacokinetic/pharmacodynamic (PBPK/PD) model for medication A should be conducted using distinct datasets, including pharmacokinetic (PK) and pharmacodynamic (PD) data obtained from oral and dermal administration of drug A, as reported by laboratories Y and Z, respectively. Typically, the PBPK/PD model becomes more robust when a greater number of diverse data sets are included in the validation process. Once a physiologically-based pharmacokinetic/pharmacodynamic (PBPK/PD) model has been verified and confirmed to be accurate, it can be utilized for the purpose of making predictions. Currently, the predictive potential of the state-of-the-art is restricted to the specific compound for which the PBPK/PD model is created, or at least to similar compounds with the same mechanism of action.

14.8.7 Case Studies Illustrating PKPD Relationships

Penicillin: A study was conducted to assess the bactericidal efficacy of piperacillin against Escherichia coli in immunocompromised infected rats. The authors employed a PK/PD model, namely a modified sigmoidal E_{max} model that has been previously constructed to evaluate this activity. This model was found to be appropriate to evaluate the bacterial decrease profile over time to demonstrate the bactericidal impact of piperacillin. This study conducted a comparison between the parameters acquired from in vitro and in vivo models. The obtained results emphasize the significance of conducting both in vitro and in vivo models to evaluate the activity of a drug. The in vitro model provides insights into the elements that can impact the pharmacological effect of the drug in the in vivo model. The in vitro pharmacokinetic-pharmacodynamic characteristics for PIP against E. coli are calculated as follows: a generation rate constant of 1.30 ± 0.10 h-1, a maximum killing effect of 3.11 ± 0.27 h-1, and a concentration required to achieve 50% of the maximal effect of 5.44 ± 0.03 lg ml-1. The in vivo pharmacokinetic-pharmacodynamic characteristics for PIP against E. coli in immunocompromised infected rats are calculated as follows: a generation rate constant of 0.76 ± 0.20 h-1, a maximum killing effect of 1.38 ± 0.20 h-1, and a concentration required to achieve 50% of the maximal effect of 1.31 ± 0.27 lg ml-1. Comparing the in vitro and in vivo parameters is complex and requires careful consideration of the inherent variations between the models being used. The primary utilization of the PK-PD model assessed is to compare the potency and efficacy of various antimicrobial agents under identical conditions.

The study emphasizes the significance of taking into account the disparities in bacterial behaviour between laboratory conditions (in vitro) and living organisms (in vivo) when simulating the impact of antibiotics on killing bacteria [41].

14.9 Integration of PBPK and PKPD Models

- PBPK Model: Physiologically based pharmacokinetics (PBPK) is a specialized field within pharmacokinetics that integrates the physiology, anatomy, and biochemistry of the body with computer simulation to model the behaviour of chemicals of interest [42]. This model is based on physiology and considers factors like organ function, blood flow, and tissue composition. It predicts how a drug moves through the body, gets absorbed, distributed, metabolized, and excreted.
- PK/PD Model: This model describes the relationship between drug concentration in the body (pharmacokinetics) and the resulting effects on the body (pharmacodynamics). It helps understand how a drug dose translates into a therapeutic response.

Physiologically based pharmacokinetic (PBPK) modeling offers a methodical framework for combining the drug's concentration time profile with accurate physiological and biological processes in the body. This modeling approach has an advantage over traditional compartmental modeling approaches because it has the potential to extrapolate and investigate conditions that have not been studied in pharmacokinetic (PK) studies. This can provide valuable information and speed up the drug development process [43].

Since the main focus in drug development is the effectiveness and safety of the drug, the use of PBPK linked pharmacodynamic (PD) models can be highly beneficial. These models provide a framework for studying the impact of variability in physiological, biochemical, and formulation factors on the drug's response. This is particularly useful in situations where clinical studies have not been conducted or are not feasible [44]. One significant benefit of PBPK connected PD models is the capability to establish a connection between the drug concentration at the likely site of action and its toxicological and/or therapeutic effect. It is especially important when the amount of medicine in the blood is not an accurate representation of the amount of drug at the specific location where it has its effect,as demonstrated in a recent study with rosuvastatin, a cholesterol lowering drug which is a substrate of the OATP1B1 influx transporter [45].

14.9.1 Combining PBPK and PKPD Models for Comprehensive Drug Effect Predictions

14.9.1.1 Physiologically Based Pharmacokinetic/Pharmacodynamic Models to Evaluate the Interaction Between Nifedipine and Apatinib

The study examined the possibility of a drug interaction between nifedipine, a widely used prescription for hypertension, and apatinib, a substance utilized in the treatment of some types of cancer. They accomplished this by constructing a computational model (PBPK/PD) capable of forecasting the concentration of nifedipine in the circulatory system and its efficacy in reducing blood pressure.

The researchers initially verified the validity of their model by assuring its precise representation of real-world data regarding the effects of nifedipine. Subsequently, the model was employed to simulate the impact of apatinib on the action of nifedipine [5]. The findings demonstrated that apatinib significantly enhances the systemic concentration of nifedipine. The occurrence is due to the interference of apatinib with an enzyme that typically facilitates the degradation of nifedipine. Nevertheless, the augmentation in nifedipine was insufficient to have a substantial effect on its ability to reduce blood pressure.

In short, apatinib influences the metabolism of nifedipine in the body, although this does not significantly alter nifedipine's effectiveness in regulating blood pressure. These findings indicate that the concurrent use of nifedipine and apatinib is likely to be safe without requiring any dosage adjustments for nifedipine [46].

14.9.1.2 Development of Physiologically Based Pharmacokinetic/Pharmacodynamic Model for Indomethacin Disposition in Pregnancy

This study created a pregnancy physiological-based pharmacokinetic/pharmacodynamic (PBPK/PD) model for indomethacin to explain the changes in pharmacokinetics between pregnancy and non-pregnancy [47]. To define indomethacin PK in pregnant women and compare it to non-pregnant patients, a whole-body PBPK model with pregnancy-related physiological alterations was created. Literature data on maternal physiological and biological changes was added into the structural PBPK model for non-pregnant PK data. In pregnant women, indomethacin AUC, Cmax, and Cave were projected to change. The model-simulated PK profiles matched the data. For indomethacin Cave, the predicted mean ratio (non-pregnant: T2) was 1.6 and the observed value was 1.59. Additionally, the projected steady-state apparent clearance (CL/Fss) ratio was 0.46, almost identical to the observed (0.42). Sensitivity study revealed that CYP2C9 and, to a lesser extent, UGT2B7 activity most affects indomethacin disposition during pregnancy. The PBPK model, which combines physiological information, in vitro, and in vivo data, predicted indomethacin distribution during T2. The PBPK/PD model advised greater indomethacin dose during pregnancy [47].

14.9.1.3 Physiologically Based Pharmacokinetic/Pharmacodynamic Modeling to Predict the Impact of CYP2C9 Genetic Polymorphisms, Co-medication and Formulation on the Pharmacokinetics and Pharmacodynamics of Flurbiprofen

This study shows that translational PBPK/PD modeling and simulation can mechanistically describe flurbiprofen absorption and anticipate the effects of formulation and CYP2C9 genetic variation. The biopharmaceutic analysis began with adequately planned bio relevant in vitro tests of several flurbiprofen formulations, followed by data analysis and extrapolation to in vivo using a translational framework. Our extensive PBPK/PD investigations revealed how genotype and dissolution rate affect PK/PD. These findings led us to suggest clinically meaningful exposure metrics and dosage adjustments. Our PBPK model also predicted gene dose-dependent drug–drug interactions, demonstrating its robustness. The current PBPK/PD model could be used in biopharmaceutical applications, dose optimization rationale in healthy populations with genetic variances, and PK extrapolations to patients or special populations like rheumatoid arthritis and paediatrics [48].

Genetic differences and formulation in vivo performance appear to be key factors of pharmacological efficacy and safety variability, making medication development difficult. This study uses the translational PBPK/PD strategy to bridge the gap between in vitro–in silico–in vivo and make accurate and robust clinical predictions for target populations and genotypes, enabling personalized therapy.

14.9.2 Examples of Drug Efficacy and Safety Assessments

14.9.2.1 A PBPK/PD Model of Histone Deacetylase (HDAC) Inhibitor Vorinostat for Paediatric and Adult Patients Dosing

A novel PBPK/PD model of vorinostat has been created, allowing for the accurate prediction of the concentration-time profiles of vorinostat in adults and the effects of vorinostat on its primary pharmacodynamic biomarkers, specifically the inhibition of HDAC.

Furthermore, there is a significant reduction in thrombocytes that is limited by the dosage. The PBPK/PD model was utilized to propose paediatric dosages for individuals aged 0 to 17 years and potentially more efficacious vorinostat administration schedules for adults [49]. Additional investigation is required to identify the enzymes responsible for vorinostat metabolism, understand the pharmacokinetics of vorinostat in paediatric patients under the age of 5, and establish the relationship between pharmacokinetics and pharmacodynamics in paediatric patients. These findings demonstrate the use of a mechanistic modeling technique that could assist researchers in designing clinical studies and perhaps enhancing the dosage schedules of both newly created and already existing medications.

14.9.2.2 Use of Physiologically Based Pharmacokinetic/Pharmacodynamic (PBPK/PD) Modeling for Exploring Risk Assessment Approaches of Chemical Mixtures

The primary advantage of PBPK/PD modeling in risk assessment, specifically for chemical mixtures, lies in its capacity for Predictive Toxicology by accurately predicting the distribution of substances in tissues at both the pharmacokinetic and pharmacodynamic levels. Combining PBPK/PD modeling with statistical methods and experiments on chemical mixtures holds tremendous promise for risk assessment. The researchers propose integrating computer simulations with targeted toxicology experiments to create a more realistic approach to complex mixtures. Their goal is to develop a predictive toxicology strategy for mixtures, ultimately leading to innovative risk assessment methods.

They have utilized PBPK/PD modeling to examine the toxicological interactions of a chemical mixture consisting of seven commonly found groundwater contaminants (arsenic, benzene, chloroform, chromium, lead, phenol, TCE) [50]. The researchers sought to determine the specific components inside this intricate mixture that are responsible for toxicological interactions by examining its impact on different hazardous endpoints. The study emphasized the significance of comprehending the toxicology of chemical combinations for thorough risk evaluation in environmental and occupational contexts.

14.9.3 Prediction of the Pharmacokinetics, Pharmacodynamics, and Efficacy of a Monoclonal Antibody, Using a Physiologically Based Pharmacokinetic FcRn

Monoclonal antibodies (mAbs) interact with FcRn receptors, affecting their distribution in the body. Traditional PBPK models describe mAb distribution but lack the ability to predict treatment response [51]. Integrating PBPK with a PD model can predict both drug concentration variations and treatment response. This is especially valuable because the PD model can consider tissue compartments, crucial for simulating interactions with membrane-bound receptors.

The PBPK/PD model accurately forecasted the pharmacokinetics and effectiveness of efalizumab, a monoclonal antibody. The model integrated crucial elements such as target-mediated drug distribution, binding interactions, and catabolism. The simulations conducted using the model provided an appropriate representation of the clinical data pertaining to efalizumab concentrations and efficacy markers. The study showcased the model's capacity to anticipate variations in PASI scores, suggesting its effectiveness in treating psoriasis [52]. In general, the PBPK/PD model demonstrated potential in forecasting the pharmacokinetic and pharmacodynamic characteristics of monoclonal antibodies such as efalizumab.

14.10 Future Perspective

14.10.1 Advances in Modeling Techniques

14.10.1.1 Integration of PBPK and AI for Nano Medicine Research

Conventional physiologically-based pharmacokinetic (PBPK) models face challenges in accurately forecasting the behaviour of nanoparticles (NPs) in the body since they heavily depend on animal data. This innovative method employs artificial intelligence (AI) to forecast crucial kinetic characteristics associated with nanoparticle (NP) administration, including tumour formation, eliminating the requirement for significant animal experimentation. By integrating AI-generated predictions into the PBPK framework, this strategy allows for precise assessments of tumour transport efficiency. This not only speeds up the progress of Nano medicine development, but it also provides the opportunity to understand and analyse the behaviour mechanisms of nanoparticles, offering vital insights. A thorough understanding of the mechanisms at play is crucial for improving the development and utilization of nanoparticle (NP) techniques, leading to the production of more effective Nano medicines in the long run. This novel strategy has the potential to completely transform Nano medicine research by providing a quicker, more ethical, and scientifically informed way to produce life-saving treatments [53].

14.10.1.2 PBPK Modeling on Organs on Chips

PBPK models are being explored for use with Organ-on-a-Chip (OoC) systems in drug discovery. A 2017 study by Tsamandouras et al. combined gut- and liver-on-a-chip with a PK model to investigate drug transport. This integration shows promise for improved drug development [53]. PBPK models and Organ-on-a-Chip (OoC) systems are a new tool for drug development. Shinha et al. (2020) [54] showed this by using an OoC system with a PBPK model to predict how medications might interact. This approach has promise for creating lab-grown chips that mimic how drugs interact in the body [55]. The ultimate objective of in vitro PBPK/PKPD research with OoC is to develop complementing experimental and computational techniques that can effectively extrapolate in vitro findings to the in vivo situation [56].

14.10.1.3 PBPK Modeling for Nano Medicine Design and Gene Therapy

Recent developments in physiologically-based pharmacokinetic (PBPK) models involve utilizing population pharmacokinetic (PopPK) methodologies to enhance model accuracy, the influence of these models on regulatory sciences, and their use in forecasting drug pharmacokinetic/pharmacodynamics (PK/PD) characteristics in both healthy individuals and patients. These developments have facilitated the shift from being a mere academic interest to being recognized and approved by regulatory authorities. This has been achieved by broadening the traditional paradigm to include a concept of 'predict-learn-confirm-apply'. The growing utilization of PBPK models in regulatory submissions, conferences organized by regulatory agencies to deliberate on optimal methodologies, and the inclusion of PBPK simulations on medicinal labels have collectively contributed to the transformation of PBPK from an academic curiosity to a standard practice in the industry [57].

14.10.1.4 PBPK Modeling in Advancement of Nano Particle Delivery

Physiologically based pharmacokinetic (PBPK) modeling has advanced to address the complex interactions between nanoparticles and biological systems, providing a mechanistic understanding of influential factors and sources of pharmacokinetic (PK) variability. This tool has been widely accepted by pharmaceutical industry and regulatory agencies for predicting drug exposure, efficacy, and toxicity. PBPK modeling, when combined with pharmacodynamic (PD) models, can predict the effects of nanoparticles on target tissues. Challenges in studying nanoparticle disposition include their structural and functional complexity, as well as the need to quantify phagocytic cell levels in tissues. Despite these challenges, PBPK modeling is expected to play a significant role in characterizing and predicting the disposition and biological effects of nanoparticles as the field continues to evolve [58].

14.10.1.5 Adapting Physiologically-Based Pharmacokinetic Models for Machine Learning Applications

PBPK modeling research boomed between 2012 and 2018, focusing on cancer drugs and drug interactions. Basic physiologically based pharmacokinetic (PBPK) models necessitate numerous molecule-specific inputs from in vitro experiments, posing challenges to the efficiency and accuracy of these models in early drug discovery, prompting the adoption of machine learning (ML) models. This approach is increasingly used for therapeutic drug monitoring, especially for medications with a narrow effective dose range [59].

14.10.2 PBPK Model in Personalized Medicine

Personalized medicine aims to deliver the "right drug at the right dose" by taking into account differences between people (inter-person variability), which is one reason why some treatments don't work for certain groups of patients [60]. As personalized medicine moves forward, physiologically-based pharmacokinetic (PBPK) modeling is an important tool for analysing complex clinical situations. It uses physiological information and physicochemical data to simulate different physiological states and predict how pharmacokinetic responses will be distributed.

PBPK modelling is useful in personalized medicine for two reasons:

1. It helps figure out which subpopulations are most important in a distribution of pharmacokinetic responses for a given drug formulation; and
2. It figures out the formulation design space needed to get the desired drug plasma concentration profile.

The initial application utilizes the predictive powers of PBPK modeling and its capacity to adjust physiological functions. The use of parameters to represent individual populations allows for a more comprehensive knowledge of how different segments of the pharmacokinetic response distribution can be categorized based on subpopulation effects. The second application employs a reverse methodology by utilizing a deconvolution technique to ascertain the specific formulation required to align with a targeted plasma concentration profile for a particular population [61].

Advancements in computing technology, along with the urgent requirement to rapidly and effectively analyse the pharmacokinetics and absorption of drugs, have led to the emergence of PBPK modelling as a potent simulation tool for guiding pharmaceutical development. PBPK models utilize data from both in vitro studies and in vivo preclinical and clinical data, in contrast to standard pharmacokinetic models that rely on a "top down" approach and "bottom up" approaches that advocate modeling solely on a virtual basis. The model parameters are obtained via computational estimations, laboratory trials, and preclinical data. These parameters can be categorized into two main groups: drug-dependent and system-dependent. Drug-dependent parameters for various therapies can be divided into two categories: intrinsic compound properties (such as molecular weight, diffusion coefficient, solubility at different gastrointestinal pH levels, and ionization) and formulation factors (including dosage form and composition, particle size and shape, excipients, tablet dimensions, powder compression force, coating composition, and thickness) [62].

PBPK modeling involves a highly complex process that seeks to bridge the gap in understanding between the limited experimental data available on customized physiology and the specific action or mechanism of a drug. When applied to the entire population, PBPK models take into account the physiology of different patient subgroups or special populations, which are characterized by comparable intrinsic and/or extrinsic characteristics. These models are used to forecast drug exposure [63]. Subpopulation modeling can be accomplished by employing an iterative and integrative approach using the adopt, adapt, develop, assess, amend, and deliver $(AAD)^2$ computational technique, which was presented within the framework of systems engineering [64]. The success of PBPK modeling in personalized medicine depends greatly on accurately identifying patient-specific factors that account for the reported pharmacokinetic parameters in individuals or specific groups of individuals [65].

References

1. Mould DR, Upton RN. Basic concepts in population modeling, simulation, and model-based drug development. CPT Pharmacometrics Syst Pharmacol. 2012;1(9):e6. https://doi.org/10.1038/psp.2012.4.
2. Monine M, Norris D, Wang Y, Nestorov I. A physiologically-based pharmacokinetic model to describe antisense oligonucleotide distribution after intrathecal administration. J Pharmacokinet Pharmacodyn. 2021;48(5):639–54. https://doi.org/10.1007/s10928-021-09761-0.
3. Zhuang X, Lu C. PBPK modeling and simulation in drug research and development. Acta Pharm Sin B. 2016;6(5):430–40. https://doi.org/10.1016/j.apsb.2016.04.004.
4. Kuepfer L, Niederalt C, Wendl T, Schlender JF, Willmann S, Lippert J, Block M, Eissing T, Teutonico D. Applied concepts in PBPK modeling: how to build a PBPK/PD model. CPT Pharmacometrics Syst Pharmacol. 2016;5(10):516–31. https://doi.org/10.1002/psp4.12134.
5. Liu XI, Momper JD, Rakhmanina NY, Green DJ, Burckart GJ, Cressey TR, Mirochnick M, Best BM, van den Anker JN, Dallmann A. Physiologically based pharmacokinetic modeling framework to predict neonatal pharmacokinetics of transplacentally acquired emtricitabine, dolutegravir, and raltegravir. Clin Pharmacokinet. 2021;60(6):795–809. https://doi.org/10.1007/s40262-020-00977-w.
6. Clewell HJ III, Reddy MB, Lave T, Andersen ME. Physiologically based pharmacokinetic modeling. In: Preclinical development handbook: ADME and biopharmaceutical properties. Hoboken: Wiley-Interscience; 2008. p. 1167–227.
7. Peters SA. Physiologically-based pharmacokinetic modeling and simulations: principles, methods, and applications in the pharmaceutical industry. Hoboken: Wiley; 2012.
8. Deepika D, Kumar V. The role of "Physiologically Based Pharmacokinetic Model (PBPK)" New Approach Methodology (NAM) in pharmaceuticals and environmental chemical risk assessment. Int J Environ Res Public Health. 2023;20(4):3473. https://doi.org/10.3390/ijerph20043473.

9. Jones H, Rowland-Yeo K. Basic concepts in physiologically based pharmacokinetic modeling in drug discovery and development. CPT Pharmacomet Syst Pharmacol. 2013;2(8):e63. https://doi.org/10.1038/psp.2013.41.
10. Kuepfer L, Niederalt C, Wendl T, et al. Applied Concepts in PBPK Modeling: How to Build a PBPK/PD Model. CPT Pharmacometrics Syst Pharmacol. 2016;5(10):516–531.
11. Cheng L, Wong H. Food effects on oral drug absorption: application of physiologically-based pharmacokinetic modeling as a predictive tool. Pharmaceutics. 2020;12(7):672. https://doi.org/10.3390/pharmaceutics12070672.
12. Lin L, Wong H. Predicting oral drug absorption: mini review on physiologically-based pharmacokinetic models. Pharmaceutics. 2017;9(4):41. https://doi.org/10.3390/pharmaceutics9040041.
13. Kuepfer L, Niederalt C, Wendl T, Schlender JF, Willmann S, Lippert J, et al. Applied concepts in PBPK modeling: how to build a PBPK/PD model. CPT: pharmacometrics & systems pharmacology. 2016;5(10):516–31. https://ascpt.onlinelibrary.wiley.com/doi/10.1002/psp4.12134
14. Sager JE, Jingjing Y, Ragueneau-Majlessi I, Isoherranen N. Physiologically Based Pharmacokinetic (PBPK) modeling and simulation approaches: a systematic review of published models, applications, and model verification. Drug Metab Dispos. 2015;43(11):1823–37. https://doi.org/10.1124/dmd.115.065920.
15. Ramoju SP, Mattison DR, Milton B, McGough D, Shilnikova N, Clewell HJ, Yoon M, Taylor MD, Krewski D, Andersen ME. The application of PBPK models in estimating human brain tissue manganese concentrations. Neuro Toxicol. 2017;58:226–37. ISSN 0161-813X, https://doi.org/10.1016/j.neuro.2016.12.001. Siebinga H, de Wit-van der Veen BJ, Beijnen JH, Stokkel MPM, Dorlo TPC, Huitema ADR, Hendrikx JJMA. A physiologically based pharmacokinetic (PBPK) model to describe organ distribution of 68Ga-DOTATATE in patients without neuroendocrine tumors. EJNMMI Res. 2021;11(1):73. https://doi.org/10.1186/s13550-021-00821-7
16. Bolleddula J, Ke A, Yang H, Prakash C. PBPK modeling to predict drug-drug interactions of ivosidenib as a perpetrator in cancer patients and qualification of the Simcyp platform for CYP3A4 induction. CPT Pharmacometrics Syst Pharmacol. 2021;10(6):577–88. https://doi.org/10.1002/psp4.12619.
17. Ezuruike U, Zhang M, Pansari A, De Sousa M, Mendes XP, Neuhoff S, Gardner I. Guide to development of compound files for PBPK modeling in the Simcyp population-based simulator. CPT Pharmacometrics Syst Pharmacol. 2022;11(7):805–21. https://doi.org/10.1002/psp4.12791.
18. Jamei M, Marciniak S, Feng K, Barnett A, Tucker G, Rostami-Hodjegan A. The Simcyp® population-based ADME simulator. Expert Opin Drug Metab Toxicol. 2009;5(2):211–23. https://doi.org/10.1517/17425250802691074.
19. Morningstar-Kywi N, Morris DN, Romero RM, Haworth IS. Teaching of drug disposition using physiologically based pharmacokinetic modeling software: GastroPlus as an educational tool. Adv Physiol Educ. 2023;47(4):718–25. https://doi.org/10.1152/advan.00051.2023.
20. Willmann S, Lippert J, Sevestre M, Solodenko J, Fois F, Schmitt W. PK-Sim®: a physiologically based pharmacokinetic 'whole-body' model. https://doi.org/10.1016/S1478-5382(03)02342-4.
21. Niederalt C, Kuepfer L, Solodenko J, et al. A generic whole body physiologically based pharmacokinetic model for therapeutic proteins in PK-Sim. J Pharmacokinet Pharmacodyn. 2018;45:235–57. https://doi.org/10.1007/s10928-017-9559-4.
22. Quick Guide. PK Sim documentation. Open Systems Pharmacology. https://docs.open--systems-pharmacology.org/working-with-pk-sim/pk-sim-documentation/pk-sim-quick-guide
23. GastroPlus software. SimulationsPlus. https://www.simulations-plus.com/software/gastroplus/
24. Heikkinen AT, Baneyx G, Caruso A, Parrott N. Application of PBPK modeling to predict human intestinal metabolism of CYP3A substrates – An evaluation and case study using GastroPlus™. Eur J Pharm Sci. 2012;47(2):375–86. https://doi.org/10.1016/j.ejps.2012.06.013. ISSN 0928-0987

25. Jan G, Janosch B, Matthias K. Physiologically based pharmacokinetic (PBPK) modeling of the role of CYP2D6 polymorphism for metabolic phenotyping with dextromethorphan. Front Pharmacol. 2022;13 https://doi.org/10.3389/fphar.2022.1029073. ISSN=1663-9812
26. Lin Z, Jaberi-Douraki M, He C, Jin S, Yang RSH, Fisher JW, Riviere JE. Performance assessment and translation of physiologically based pharmacokinetic models from acslX to Berkeley Madonna, MATLAB, and R language: oxytetracycline and gold nanoparticles as case examples. Toxicol Sci. 2017;158(1):23–35. https://doi.org/10.1093/toxsci/kfx070.
27. Prieto Garcia L, Lundahl A, Ahlström C, Vildhede A, Lennernäs H, Sjögren E. Does the choice of applied physiologically-based pharmacokinetics platform matter? A case study on simvastatin disposition and drug–drug interaction. CPT Pharmacometr Syst Pharmacol. 2022;11(9):1194–209.
28. Zhuang X, Lu C. PBPK modeling and simulation in drug research and development. Acta Pharm Sinica B. 2016;6(5):430–40.
29. Wu F. Challenges and opportunities when using oral PBPK to support risk assessment and biowaiver in regulatory submissions. https://www.fda.gov/media/166598/download
30. Yau E, Olivares-Morales A, Ogungbenro K, Aarons L, Gertz M. Investigation of simplified physiologically-based pharmacokinetic models in rat and human. CPT Pharmacometr Syst Pharmacol. 2023;12(3):333–45. https://doi.org/10.1002/psp4.12911.
31. Snowden TJ, van der Graaf PH, Tindall MJ. Model reduction in mathematical pharmacology. J Pharmacokinet Pharmacodyn. 2018;45:537–55. https://doi.org/10.1007/s10928-018-9584-y.
32. Fàbrega F, Nadal M, Schuhmacher M, Domingo JL, Kumar V. Influence of the uncertainty in the validation of PBPK models: A case-study for PFOS and PFOA. Regul Toxicol Pharmacol. 2016;77:230–9. https://doi.org/10.1016/j.yrtph.2016.03.009.
33. Uncertainty, Variability, and Sensitivity Analyses in Toxicological Modeling, Nan-Hung Hsieh. https://nanhung.rbind.io/slide/191002_dpr
34. Pereira LC, Fátima MA, Santos VV, Brandão CM, Alves IA, Azeredo FJ. Pharmacokinetic/pharmacodynamic modeling and application in antibacterial and antifungal pharmacotherapy: a narrative review. Antibiotics. 2022;11(8):986. https://doi.org/10.3390/antibiotics11080986.
35. Tuntland T, Ethell B, Kosaka T, Blasco F, Zang RX, Jain M, Gould T, Hoffmaster K. Implementation of pharmacokinetic and pharmacodynamic strategies in early research phases of drug discovery and development at Novartis Institute of Biomedical Research. Front Pharmacol. 2014;5:174. https://doi.org/10.3389/fphar.2014.00174.
36. Derendorf H, Meibohm B. Modeling of pharmacokinetic/pharmacodynamic (PK/PD) relationships: concepts and perspectives. Pharm Res. 1999;16(2):176–85. https://doi.org/10.1023/a:1011907920641.
37. Danhof M, de Jongh J, De Lange EC, Della Pasqua O, Ploeger BA, Voskuyl RA. Mechanism-based pharmacokinetic-pharmacodynamic modeling: biophase distribution, receptor theory, and dynamical systems analysis. Annu Rev Pharmacol Toxicol. 2007;47(1):357–400.
38. Felmlee MA, Morris ME, Mager DE. Mechanism-based pharmacodynamic modeling. Comput Toxicol. 2012;I:583–600.
39. de Lange EC. PKPD aspects of brain drug delivery in a translational perspective. In: Drug delivery to the brain: physiological concepts, methodologies and approaches. New York: Springer New York; 2013. p. 233–68.
40. Choe S, Lee D. Parameter estimation for sigmoid Emax models in exposure-response relationship. Transl Clin Pharmacol. 2017;25(2):74.
41. De Araujo BV, Diniz A, Palma EC, Buffé C, Costa TD. PK-PD modeling of β-lactam antibiotics: in vitro or in vivo models? J Antibiot. 2011;64(6):439–46.
42. van de Waterbeemd H, Gifford E. ADMET in silico modelling: towards prediction paradise? Nat Rev Drug Discov. 2003;2(3):192–204. https://doi.org/10.1038/nrd1032.
43. Chen Y, Jin JY, Mukadam S, Malhi V, Kenny JR. Application of IVIVE and PBPK modeling in prospective prediction of clinical pharmacokinetics: strategy and approach during the drug discovery phase with four case studies. Biopharm Drug Dispos. 2012;33(2):85–98. https://doi.org/10.1002/bdd.1769.

44. Rostami-Hodjegan A. Response to "the link between pharmacodynamics and physiologically based pharmacokinetic models". Clin Pharmacol Ther. 2013;93(2):152. https://doi.org/10.1038/clpt.2012.216.
45. Rose RH, Neuhoff S, Abduljalil K, Chetty M, Rostami-Hodjegan A, Jamei M. Application of a physiologically based pharmacokinetic model to predict OATP1B1-related variability in pharmacodynamics of rosuvastatin. CPT Pharmacometr Syst Pharmacol. 2014;3(7):e124. https://doi.org/10.1038/psp.2014.24.
46. Liu H, Yu Y, Liu L, Wang C, Guo N, Wang X, Xiang X, Han B. Application of physiologically-based pharmacokinetic/pharmacodynamic models to evaluate the interaction between nifedipine and apatinib. Front Pharmacol. 2022;13:970539. https://doi.org/10.3389/fphar.2022.970539.
47. Alqahtani S, Kaddoumi A. Development of physiologically based pharmacokinetic/pharmacodynamic model for indomethacin disposition in pregnancy. PLoS One. 2015;10(10):e0139762. https://doi.org/10.1371/journal.pone.0139762.
48. Loisios-Konstantinidis I, Cristofoletti R, Jamei M, Turner D, Dressman J. Physiologically based pharmacokinetic/pharmacodynamic modeling to predict the impact of CYP2C9 genetic polymorphisms, co-medication and formulation on the pharmacokinetics and pharmacodynamics of Flurbiprofen. Pharmaceutics. 2020;12(11):1049. https://doi.org/10.3390/pharmaceutics12111049.
49. Moj D, Britz H, Burhenne J, Stewart CF, Egerer G, Haefeli WE, Lehr T. A physiologically based pharmacokinetic and pharmacodynamic (PBPK/PD) model of the histone deacetylase (HDAC) inhibitor vorinostat for pediatric and adult patients and its application for dose specification. Cancer Chemother Pharmacol. 2017;80(5):1013–26. https://doi.org/10.1007/s00280-017-3447-x.
50. Yang RSH, El-Masri HA, Thomas RS, Constan AA, Tessari JD. The application of physiologically based pharmacokinetic/pharmacodynamic (PBPK/PD) modeling for exploring risk assessment approaches of chemical mixtures. Toxicol Lett. 1995;79(1–3, 193):–200. https://doi.org/10.1016/0378-4274(95)03370-Z. ISSN 0378-4274
51. Shah DK, Betts AM. Towards a platform PBPK model to characterize the plasma and tissue disposition of monoclonal antibodies in preclinical species and human. J Pharmacokinet Pharmacodyn. 2012;39(1):67–86. https://doi.org/10.1007/s10928-011-9232-2.
52. Chetty M, Li L, Rose R, Machavaram K, Jamei M, Rostami-Hodjegan A, Gardner I. Prediction of the pharmacokinetics, pharmacodynamics, and efficacy of a monoclonal antibody, using a physiologically based pharmacokinetic FcRn model. Front Immunol. 2015;5:670. https://doi.org/10.3389/fimmu.2014.00670.
53. Chou W-C, Chen Q, Yuan L, Cheng Y-H, He C, Monteiro-Riviere NA, Riviere JE, Lin Z. An artificial intelligence-assisted physiologically-based pharmacokinetic model to predict nanoparticle delivery to tumors in mice. J Control Release. 2023;361:53–63. https://doi.org/10.1016/j.jconrel.2023.07.040. ISSN 0168-3659
54. Tsamandouras N, Chen WLK, Edington CD, Stokes CL, Griffith LG, Cirit M. Integrated gut and liver microphysiological systems for quantitative in vitro pharmacokinetic studies. AAPS J. 2017;19(5):1499–512. https://doi.org/10.1208/s12248-017-0122-4.
55. Shinha K, Nihei W, Ono T, Nakazato R, Kimura H. A pharmacokinetic-pharmacodynamic model based on multi-organ-on-a-chip for drug-drug interaction studies. Biomicrofluidics. 2020;14(4):044108. https://doi.org/10.1063/5.0011545.
56. Przekwas A, Somayaji M. Computational pharmacokinetic modeling of organ-on-chip devices and microphysiological systems; 2020. https://doi.org/10.1016/B978-0-12-817202-5.00011-5.
57. Jamei M. Recent advances in development and application of Physiologically-Based Pharmacokinetic (PBPK) models: a transition from academic curiosity to regulatory acceptance. Curr Pharmacol Rep. 2016;2(3):161–9. https://doi.org/10.1007/s40495-016-0059-9.
58. Yuan D, He H, Wu Y, Fan J, Cao Y. Physiologically based pharmacokinetic modeling of nanoparticles. J Pharm Sci. 2019;108(1):58–72. https://doi.org/10.1016/j.xphs.2018.10.037.

59. Perry C, Davis G, Conner TM, Zhang T. Utilization of physiologically based pharmacokinetic modeling in clinical pharmacology and therapeutics: an overview. Curr Pharmacol Rep. 2020;6(3):71–84. https://doi.org/10.1007/s40495-020-00212-x.
60. Lesko LJ, Schmidt S. Individualization of drug therapy: history, present state, and opportunities for the future. Clin Pharmacol Ther. 2012;92(4):458–66. https://doi.org/10.1038/clpt.2012.113.
61. Hartmanshenn C, Scherholz M, Androulakis IP. Physiologically-based pharmacokinetic models: approaches for enabling personalized medicine. J Pharmacokinet Pharmacodyn. 2016;43(5):481–504. https://doi.org/10.1007/s10928-016-9492-y.
62. Jamei M, Turner D, Yang J, Neuhoff S, Polak S, Rostami-Hodjegan A, Tucker G. Population-based mechanistic prediction of oral drug absorption. AAPS J. 2009;11(2):225–37. https://doi.org/10.1208/s12248-009-9099-y.
63. Zhao P, Rowland M, Huang SM. Best practice in the use of physiologically based pharmacokinetic modeling and simulation to address clinical pharmacology regulatory questions. Clin Pharmacol Ther. 2012;92(1):17–20. https://doi.org/10.1038/clpt.2012.68.
64. Androulakis IP. Systems engineering meets quantitative systems pharmacology: from low-level targets to engaging the host defenses. Wiley Interdiscip Rev Syst Biol Med. 2015;7(3):101–12. https://doi.org/10.1002/wsbm.1294.
65. Price PS, Conolly RB, Chaisson CF, Gross EA, Young JS, Mathis ET, Tedder DR. Modeling interindividual variation in physiological factors used in PBPK models of humans. Crit Rev Toxicol. 2003;33(5):469–503.

Chapter 15
Prediction of Bio-permeability of Pharmaceuticals with Advanced Dynamic Simulation Studies

Mushtaque A. S. Shaikh and Tamanna K. Khan

Contents

Abstract Understanding and accurately estimating bio-permeability, which is one of the main factors when evaluating clinical efficacy of drugs, is critical to the drug development process. This chapter describes, in particular, how computer-based dynamic simulation studies are used for forecasting the bio-permeability of drugs, addressing the issues of prediction techniques. For example, we commence with a reasoned overview of traditional methodologies and their shortcomings, prior to wading into the depictions of Monte Carlo, Molecular Dynamics simulations, and

---

M. A. S. Shaikh (✉)
Department of Pharmaceutical Chemistry, Vivekanand Education Society's College of Pharmacy (Autonomous), Mumbai, Maharashtra, India
e-mail: mushtaque.shaikh@ves.ac.in

T. K. Khan
Department of Chemistry, GN Khalsa College of Art Science and Commerce (Autonomous), Mumbai, Maharashtra, India

S. N. K. Rao G., R. R. Alavala (eds.), *Applications of Computational Tools in Drug Design and Development*, https://doi.org/10.1007/978-981-96-4154-3_15

others state of the art techniques. A good understanding concerning the analysis is provided in the methodology section that describes in detail how the selection of drugs was done, what data collection methods were utilized, and the challenges of the development of the simulation. This chapter centres on the use of dynamic simulation studies. Results and discussions support the forecast on bio-permeability, describing some drugs and allowing comparative research. Attention is also paid to the relationship between the simulation and experimental data, emphasizing the benefits, limitations, and unique expertise that this novel technique brings to the table. It addresses the actual problem of practical usage of the simulation outcomes in evaluating the effects of a drug development. The transformational potential impact on processes within the pharmaceutical business is depicted vividly, including from the very early issue identification to drug candidate optimization. Future prospects and obstacles include ethical dilemmas, the necessity of careful decisions, and the development of forecasting science.

The discussion of both the bio-permeability prediction and applicability of the advance dynamic simulations is systematically presented in this chapter providing the pathway for future pharmaceutical development. Aside from enhancing knowledge about certain aspects of the drug development process, it offers opportunities to enhance the drug design process that entails the employment of advances predictive modeling techniques.

Keywords Biopermeability · Lipid bilayer · Drug discovery · Formulation development · ADMET · Molecular dynamics · Steered molecular dynamics · Metadynamics · Umbrella sampling · Solvation energy · Potential of mean force · Artificial intelligence and machine learning

15.1 Introduction

15.1.1 Definitions and Importance of Bio-permeability in Pharmaceuticals

In pharmaceutical and medicinal sciences, the permeation refers to "*the ability of drug substances to transport across gastrointestinal membrane and also crucial for proper drug distribution to pharmacological target organs and cells, and is therefore frequently utilized in drug discovery and development*" [1]. More specifically, penetration usually entails piercing a variety of physiological barriers, including epidermal layers, cellular membranes, and the blood-brain barrier. One should be aware by now that there are a number of mechanisms via which permeation occurs, including transcellular and paracellular transport, endocytosis and exocytosis, passive diffusion, facilitated diffusion, active transport, and others.

15.1.2 Impact on Drug Development and Efficacy

Likewise, the evaluation of the effect of this property on drug delivery systems is critical and contributes to the drug exposure and effectiveness evaluation because it enables the investigation and associations of the following aspects: drug transport interactions, drug regimens, drug administration techniques, drug metabolism, drug equivalence studies, disease management, and biopharmaceutical classification. Moreover, bio-permeability affects pharmacokinetic activities such as pharmacokinetic studies, product formulation, dosage regulation, clinical trials, regulatory submission, and drug effectiveness. Due to this bio-permeability, its contribution to pharmaceutical development is very high.

A few parameters will be revisited on what affects drug penetration before we go to this chapter, which includes size, lipophilicity, ionization, pH, efflux pumps, and the barrier being breached. Some considerations about these mechanisms and variables are necessary for the design of practical and efficient drug delivery systems as well as for achieving better clinical outcomes.

15.1.3 Challenges, Techniques and Factors Affecting Bio-permeability

There are a number of techniques for forecasting and for calculating the permeation and permeability of drug compounds through the particle membrane [2]. However, what motivates the development of this methodology is the fact that all systems (membranes) structurally and functionally complicated hence difficult to isolate, and replicate their role as the biological barriers, there is a need to describe them computationally. Biologic barriers such as the GI tract, BBB, and skin have organized architectures but their permeation capabilities are varied. Nevertheless, biological systems are complex and barrier function is still not very well understood thus making it difficult to predict such interactions between drugs and biological barriers and how quickly they penetrate through them. In addition, hundred percent bioavailability is hard to assess as evaluation of the effect of transporter mediated permeation on bioavailability requires understanding of transporter kinetics and substrate specificity which is not very easy to include in experimental designs. In the same vein, this picture also allows to estimate how different formulations and routes of administration affect bio-permeability. It is also important to emphasize on the need of thorough experimental validation of the models in order to ensure experimental models are developed that are both accurate and reliable. Thus, addressing these challenges in bio-permeability prediction requires interdisciplinary approaches integrating pharmacology, physiology, chemistry, computational modeling, and data science. Advances in computational modeling techniques, high-throughput screening methods, and translational research efforts are essential for improving the accuracy and reliability of bio-permeability prediction in pharmaceutical development.

There are multiple computational methodologies applied to study drug permeation which aid in revealing the underlying mechanics. These include molecular modeling (docking, pharmacophore mapping, Computational Fluid Dynamics (CFD) Modeling) and simulation techniques to quantitative structure-activity relationship (QSAR) modeling which includes Machine Learning and Artificial Intelligence (AI) Approaches as well as pharmacokinetic modelling. An example of such simulation techniques is Molecular Dynamics (MD) simulation, a technique that models the movements and interactions between atoms or molecules over time in order to gain insight into molecular behaviour of biological systems and pharmaceuticals. The drug-protein interactions mediating drug permeation properties, as well as the drug permeation over biological barriers such as membranes can be studied by MD simulations. It is important to state in advance that no single technique will be able to capture the entire process of permeation. The computational techniques can thus be combined to establish the sequence of processes in drug development, explain the science behind drug design/formulation and offer comprehensive understanding of mechanisms of drug permeation. In general, if computational methods are interpolated into an iterative process of experimental validation enhanced by computational predictivity, the success rate of drug discovery and development could be optimised."

15.1.4 Mechanisms and Models of Biopermeability Studies

When microscopic molecules diffuse from the bulk solvent to the membrane's surface, surface binding and penetration take place. Hydrophobic or electrostatic interactions can start this process. The size, shape, physicochemical properties, membrane composition, and environmental factors like pH, temperature, and ionic strength all affect how tiny molecules interact with one another. Seelig et al. provided an explanation of the mechanisms underlying peptide adsorption and insertion into membranes in one of the investigations [3]. Additionally, the characteristics of the tiny molecule affect such an affinity.

The process of permeation, which was previously thought to adhere to specific models, is the subsequent stage. A process with three stages which involves binding to the water-lipid interface and insertion into the membrane core, passive diffusion across the membrane core, and unbinding at the opposite water-lipid interface. This is known as a homogeneous solubility-diffusion model. The lipid bilayer, however, is divided into four components by the inhomogeneous solubility-diffusion model, each of which represents a leaflet [1] referred as the low head group density, [2] the high head group density [3], the high tail density, and [4] the low tail density area [4]. It has been demonstrated that small molecules' molecular weight, shape, hydrophobicity/lipophilicity, and net charge play an essential role in their penetration [5].

Permeant Counting Studies is the third model which computes passive membrane permeability by impartial simulations that only count permeant transport events. Since the accuracy of these methods depends on the quantity of permeation events that are recorded during simulation, they work best for permeants with high permeability. Such studies are pretty straight forward as no external bias is required but are limited due to sampling concern and are time consuming [6].

15.1.5 Molecular Dynamics Simulations and Its Prerequisites for Biopermeability Studies

15.1.5.1 General Principle

The permeation is not an abrupt process, in almost all instances some equilibrium can be presented. Either concentration of permeant or intervention by some permeating agents causes this transport. Since this process is dynamic, molecular dynamics can assist the researcher in observing the process over a period of time. By numerically resolving Newton's equations of motion, molecular dynamics (MD) simulations offer in-depth understanding of the kinetic and structural characteristics of intricate molecular systems at the atomic level.

The starting positions and velocities of atoms in an MD simulation are determined by the system of interest's molecular structure. Force fields are used to describe the interactions between atoms and are an approximation of the potential energy surface that governs molecular motion. These force fields take into consideration a variety of interatomic forces, including non-bonded interactions (e.g., van der Waals and electrostatic forces) and bond stretching and angle bending.

As the simulation commences, in discrete time steps, the atoms' locations and velocities are updated according to their forces and accelerations. MD simulations are able to trace the paths of individual atoms and molecules, providing insights into phenomena including protein folding, ligand binding, membrane penetration, and chemical reactions. This is achieved by continually computing these updates across millions of time steps.

The dynamic behaviour of materials and biomolecules can be better understood by using MD simulations, which can explain conformational changes, stability, flexibility, and interactions with solvent molecules and other biomolecules. Moreover, MD simulations can be used to determine how ligand binding and environmental variables (such as pH, pressure, and temperature) affect molecular structure and function, which helps with drug development, protein engineering, and material science studies.

Overall, MD simulation is an effective technique for investigating complicated molecular systems, providing atomistic-level insights that supplement actual discoveries while also influencing the rational design of novel treatments, materials, and biological systems. With this foundation in MD, it is clear that it can assist a researcher in simulating the movement of a single small or large molecule across the membrane, identifying the sequence of events, the factors influencing, and calculating the change in free energy throughout this process. This change in free energy can be used to compare permeability changes within the same molecule or across a

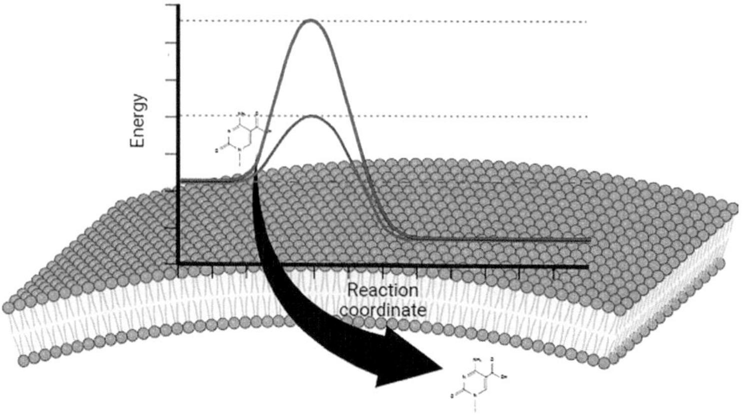

Fig. 15.1 Schematic diagram for molecule crossing membrane as a reaction and change in energy. (Created in BioRender)

group of molecules. The process of permeation can be viewed as a chemical reaction where two entities reacting with each other, interacting mutually for some time and then producing the product at the end. Hence a reaction coordinate diagram (Fig. 15.1) can be plotted to calculate the energy of activation.

As depicted by membrane model in Fig. 15.1, the membranes at most of the lipid bilayer membrane made up of different fatty acid or such hydrophobic molecules. Hence, for more realistic simulation these systems are solvated in suitable water model, ionized to mimic right pH, with added ions to mimic isotonicity.

Figure 15.2 shows the classification of scenarios and MD based techniques simulating the permeation of biomolecules through different membrane types under various schemes.

We have already spoken about energetics of processes. One of the most popular method for permeation energetics is Potential of mean force (PMF) with MD simulations. In molecular dynamics (MD) simulations, the PMF method is a computational tool that determine free energy profile of a reaction coordinate or a particular pathway. It is useful for studying various complex processes like molecule penetration across membranes, ligand binding to receptors and chemical reactions. This method attempts to map out the free energy landscape of a system by sampling configurations along a given reaction coordinate. A reaction coordinate is a collective variable that describes the progress of some particular phenomenon (e.g., crossing a membrane by a molecule or a conformational change in a protein).

The system can be subjected to advanced sampling methods such as umbrella sampling and metadynamics to bias the sampling of configurations along the reaction coordinate and thus allowing study of phase space regions that are energetically unfavourable in traditional MD simulations. The free energy profile along the reaction coordinate is reconstructed from the sampled configurations using statistical analysis methods as the weighted histogram analysis method (WHAM) or the Bennett acceptance ratio (BAR) methodology. The mean force potential refers to

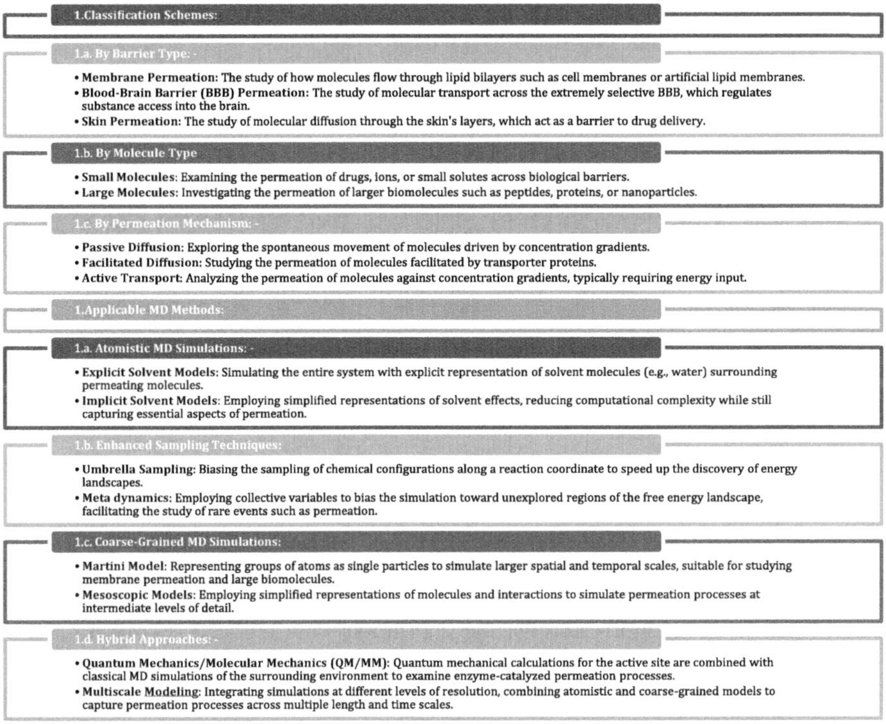

Fig. 15.2 Classification and applicable MD methods in permeation studies

the free energy surface along the supplied coordinate. If the system of interest is in a solvent, the PMF considers its effects. PMF calculations can show the energetics of drug or solute penetration across membranes, providing information about membrane selectivity and transport pathways. This will be discussed in more details later in this chapter.

Let us assume a canonical ensemble, however the ideas are equally applicable to alternative ensembles. Assume one can calculate the probability distribution function P(x) for a degree of freedom, i.e. a coordinate x. It can be presented as

$$P(x) = \delta\left[x - x(q)\right] \tag{15.1}$$

In this case, x is a specific coordinate value, and x(q) is the coordinate expressed as a function of all of the system's instantaneous particle coordinates. The angle brackets represent an ensemble average (which requires integrating across all q coordinates) with a weight function, such as a Boltzmann factor. The averaged value is a Dirac delta function, which can be thought of as a histogram reflecting the frequency of occurrence of x values in the statistical ensemble, separated into very small bins. Writing this average down in, the canonical ensemble yields the following result:

$$P(x)=\frac{\int dq\delta\left[x-x(q)\right]\exp\left(-\frac{E(q)}{kBT}\right)}{\int d\exp\left(-\frac{E(q)}{kBT}\right)}=Q(x)/Q \tag{15.2}$$

Here, Q is the partition function; Q(x) can also be conceived of as a partition function for a system bound to lie on a surface defined by the equation

$$x(q)=x \tag{15.3}$$

This relates to free energy. Using the general force formula F = −kBTlnQ and a similar equation for Q(x), we can define the "Landau free energy" as follows.

$$F(x)=F-kBT\ln P(x) \tag{15.4}$$

Because F is only a constant in this equation and the change of F(x) with x is the most interesting, the F term is sometimes left out.

Other than the restriction to a particular value of x, the standard conditions for equilibrium are met. For example, the coordinate x could be a chemical reaction coordinate or the total magnetization of an Ising system. The pair distribution function for a simple liquid is written P(x), and the potential mean force is the log of this. A free energy barrier between two stable, or metastable, system states is frequently calculable as a function of x. This type of function is often of interest; in the classical theory of nucleation, for instance, the free energy is determined as a function of the developing nucleus's radius.

Further manipulations can show that the mechanical force ensemble average, $\langle -dE/dx\rangle x\rangle$, is equal to the amount −dF(x)/dx, where the subscript indicates that the ensemble average is evaluated at the given value of x. The phrase "potential of mean force" originates from this.

To simulate a penetration over a bilayer membrane, a few basic requirements must be met, including choosing a membrane model, the medication to be evaluated, a solvent, and counterions [7]. Periodic boundary conditions are usually added to the simulation box to model a continuous system with a finite number of particles N in three dimensions. The simulation box's dimensions change based on the information that the MD trajectory is expected to produce. However, orthorhombic forms have been reported often, with sizes ranging from about 4–8 nm for width and depth and 7–11 nm for height [8, 9]. Now that we've mentioned the boundary conditions, it's only natural to bring up ensembles. Macroscopic boundary conditions in a simulation can be addressed in a variety of ways. Typically, the temperature T and particle count N stay constant. Three scenarios may exist here. NlAT, NlpT, and NlγT, where l is perpendicular box size, A is lateral area, p is bulk pressure, and γ is surface tension, which can be maintained constant while there are numerous ways to monitor volume or pressure.

15.1.5.2 Methodology

To accurately mimic a biological membrane, the membrane model must always be surrounded by a significant amount of water, whether it is positioned in the centre or at one end of the container. The biological system in the microenvironment is considered solvated, so even water molecules must be simulated using normal methods. Water molecules are frequently modelled in molecular dynamics (MD) simulations using a variety of models designed to reflect the main properties of water's structure and movement while minimizing computing expense. One prominent method is Transferable Intermolecular Potential with 3 Points (TIP3P), which depicts water molecules as three-point charges (one oxygen and two hydrogen atoms) connected by harmonic links and angles. Aside from physical presence, TIP3P reproduces many of the attributes of liquid water, including density, dielectric constant, and radial distribution functions. A recent TIP4P water model incorporates an additional fake site to mimic the lone pair of electrons on the oxygen atom. This enables TIP4P to better capture the hydrogen bonding features of water. On the other hand, a simpler model is Simple Point Charge (SPC), which represents water as three-point charges (one oxygen and two hydrogen atoms) organized linearly. Other well-known but slightly more expensive models in terms of computational cost include SPC/E (extended) and TIP5P. The exact research question, computational resources, and required level of accuracy all influence the model selection. The following is a general approach for performing the MD simulation [10]. However, before beginning a simulation technique, we must make decisions about the level of theory required, our grasp of the real process being replicated, the software to be used, and an appropriate force field for our system. Other considerations include the amount of time and computational power available. MD simulations are performed in fractional seconds (*fs*) iterations; hence a suitable number of iterations must be defined for a sufficiently long simulation (often nanoseconds and microseconds in current times). As a matter of fact some of the experts [11]. recommend that "*due to the increasing computers power, there is no reason to simulate any membrane for less than 150 ns of production time*", in addition 20 plus 100 ns equilibration of system while for estimation of diffusion coefficients about 1 μs must be preferred (Fig. 15.3).

15.1.5.3 Techniques and Tools

Picking the right technique is only the first step. Secondly, we must decide which molecular simulation program to utilize. There are presently many respectable substitutes that are becoming more and more accessible daily. The most well-known ones are CHARMM, GROMACS, Amber, NAMD, and LAMMPS. The good thing is that now a days most of them are freeware while a few needs a license. Also, versions are available to run different operating systems and hardware like normal workstations, super computer (parallel MPI) and GPU versions.

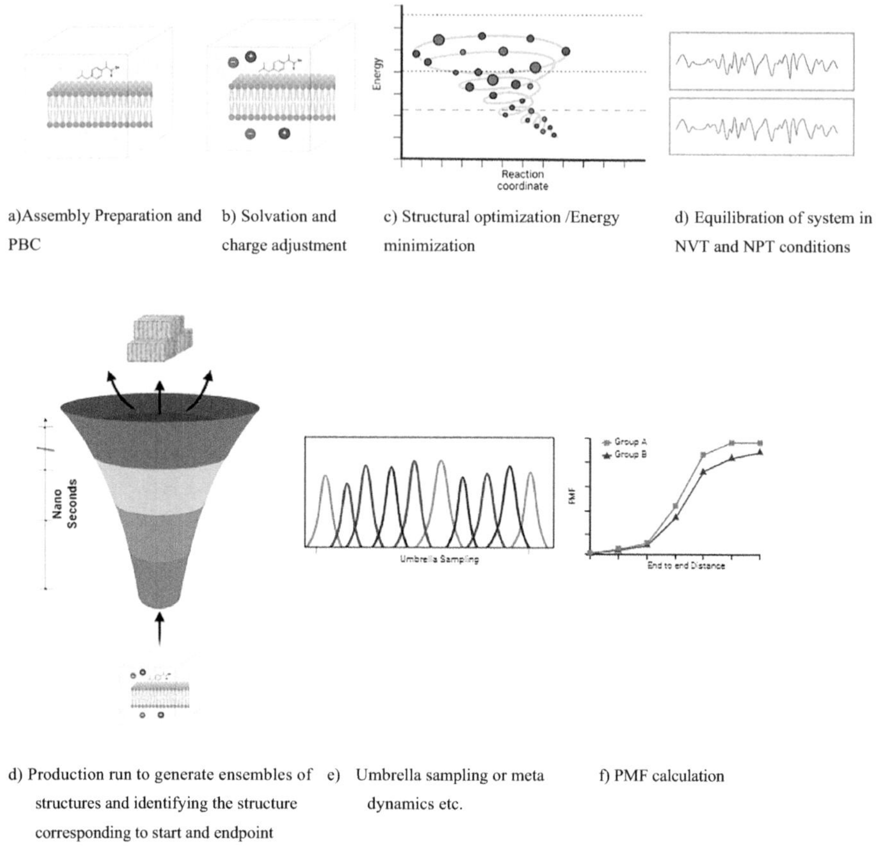

Fig. 15.3 Flowchart of representative MD protocol. (**a**) Assembly Preparation and PBC. (**b**) Solvation and charge adjustment. (**c**) Structural optimization /Energy minimization. (**d**) Equilibration of system in NVT and NPT conditions. (**e**) Production run to generate ensembles of structures and identifying the structure corresponding to start and endpoint. (**f**) Umbrella sampling or meta dynamics etc. (**g**) PMF calculation. Inspired by reference [10]

The final step is to pick the appropriate force field for our unique situation of medicines and bilayer membrane. The table below depicts many topological parameters accessible in force fields for lipid molecules particularly. The reader should keep in mind that these characteristics were produced as scientific groups expanded their grasp of the field and made minute observations. The effectiveness of including these parameters was compared to various properties such as surface area per lipid, NMR order parameters, and structure factors, including temperature dependence, scattering data, lipid lateral diffusion, membrane thickness, membrane dipole potential, and so on (Table 15.1).

It is also possible to use different combinations of forcefield and molecular simulation package, however it will be essential to check the compatibility between them.

Table 15.1 Parametrization for forcefield for lipid molecules [12]

Sr. no	Parameter	Force field	Topology description
1.	Non-bonded term	All	Force constants and measures for bonds, bond angles and torsions (k_b,k_θ and V_φ) and ((r_0,$\theta_0$andφ_0), Lennard-Jones parameters (ε and σ) and partial charges(q) can be computed from quantum mechanics. Here Lennard-Jones potential can be used to calculate the attractive (r^{-6}) and repulsive forces (r^{-12}). Electrostatic interactions are calculated from interaction of fixed-point charges from coulomb equation. Unfortunately, the effect of conformation over the point charges is still neglected.
2.	Dihedral angles and conformations		Based on the present convention of proteins, which defines staggered and eclipsed rotamers methodically, dihedral angles are specified [13]. The phosphate group has been found major contributor to the dynamics of lipid bilayers.
3.	GROMOS non-bonded parameters (45A3 [14], G53A [15] and G54A [16]	GROMOS	Modification in the non-boded terms and inclusion of Berger parameter set
4.	Berger parameter set [17]	Berger lipid FF	Specifically, the Ryckaert-Bellemans potential is used to characterize torsional rotations of the hydrocarbon chains of lipids.
5.	43 A1-S3 parameters set [18]	GROMOS	Data on thermodynamics for liquid alkanes The area per lipid prediction was improved using ab initio computations.
6.	G53A6	GROMOS	It improved the fluidity of 16:0/16:0PC (DSPC) lipid bilayer
7.	A wider range of parameters for dihedral angles and the Urey Bradley term for covalent angles [19]	CHARMM	Better description and fitting to experimental data
8.	C27r [20] and right atomic charges [21]	CHARMM	Alkane torsion parameters, computes area per lipid more accurately, thus simulates the liquid crystalline phase
9.	C36 parameter set [22]	CHARMM	In the NPT ensemble, partial atom charges and torsion potentials are reparametrized to appropriately mimic the bilayer properties of fully saturated and monounsaturated lipid molecules.
10.	C36p [23], C36c [24]	CHARMM	An update to simulate polyunsaturated [23] Cholesterol [24] and sphingolipids [25] bilayer in NPT ensemble

(continued)

Table 15.1 (continued)

Sr. no	Parameter	Force field	Topology description
11.	C36-UA [26]	CHARMM	Using a united atom (UA) description of hydrocarbons allows for more accurate predictions of surface area per lipid, chain order, and other physicochemical parameters while incurring less computing cost.
12.	Modified C36 parameters	Slipids FF [27–29]	Applicable to a variety of lipid bilayers and admixtures, with replicating area per lipid, NMR order parameters, and structure factors, including temperature dependency.
13.	LIPID11 [30]	AMBER	Foremost set of parameters for lipids in AMBER force field
14.	LIPID14 [31]	AMBER	Possibility of different heads and tails, leading to comparable area per lipid, volume per lipid, bilayer thickness, NMR order parameters, scattering data, and lipid lateral diffusion.
15.	All general lipid parameters [32]	OPLS-AA	For DPPC in particular, led to good agreement with experimental data, yet there is scope for development
16.	Polarization	CHARMM Drude polarizable FF [33]	Dielectric constant of water, parametrization of molecules in water phase may affect the energies, partitioning, and significantly change the partial electrostatic interactions between the atomic charges [34, 35].
17.	coarse-grained (CG) description	MARTINI FF [36]	Individual atoms in lipids are grouped into pseudo-particles, yielding a coarse-grained (CG) representation of a bilayer. A general 3-site lipid model binds clusters of 3–5 heavy atoms into a single CG site [37]

15.1.5.4 Materials and Resources

Molecular Dynamics simulation can be executed in four discrete steps viz.; System preparation, Minimization, Equilibration, and Production. System preparation refers to prepare a molecular assembly with different components, setting a simulation box around, solvation and charge neutralization. The initial assembly may be comprised of Ligands which could be small molecules that bind to proteins, receptors, or enzymes, influencing their activity or function. Examples of small molecules include drug molecules, natural ligands, and substrate analogs. At times one needs to consider Cofactors which may include non-protein molecules or ions that are required for the catalytic activity of enzymes or the structural stability of proteins. Examples include metal ions (e.g., Mg2+, Zn2+), coenzymes (e.g., NAD+, FAD), and prosthetic groups (e.g., heme). Often, Inhibitors which bind to enzymes or receptors and inhibit their activity can also be considered. Now diverting our attention to the most important component of the system the lipid bilayer. A broad class of chemicals known as lipids is essential to cellular membranes, energy storage, and signalling. Since they have both hydrophilic (which attracts water) and

1.Phospholipids:

- **Phosphatidylcholine (PC)**: One of the most prevalent phospholipids in biological membranes is phosphatidylcholine (PC), which is made up of a phosphorylcholine head group and a glycerol backbone connected to two fatty acid chains. Cell membrane outer leaflets are often modeled using PC.
- **Phosphatidylethanolamine (PE)**: Like PC, except instead of choline, it has an ethanolamine head group. PE is involved in membrane fusion and curvature and is widely distributed in the inner leaflet of cell membranes.
- **Phosphatidylserine (PS)**: The serine head group of phosphatidylserine (PS) is primarily located in the inner leaflet of cell membranes. PS is involved in blood coagulation, apoptosis, and cell signaling.
- **Phosphatidylglycerol (PG)**: Glycerol head group and glycerol backbone make up phosphatidylglycerol (PG). Bacterial membranes include PG, which is important in the production of the cell envelope.
- **Cardiolipin (CL)**: A dimeric phospholipid that is mostly present in the inner membrane of the mitochondria. The structure and operation of the mitochondria, especially in the areas of energy metabolism and apoptosis, depend on CL.

2.Cholesterol:

- **Cholesterol:** A sterol molecule featuring a hydroxyl group and a stiff ring shape. Animal cell membranes require cholesterol to regulate the permeability and fluidity of the membrane. It also acts as a precursor for bile acids and steroid hormones.

3.Glycolipids:

- **Glycosphingolipids:** These are lipids with a carbohydrate moiety attached to a sphingosine or ceramide backbone. Glycosphingolipids are mostly located in the outer leaflet of cell membranes and have a role in signaling and cell-to-cell recognition.
- **Gangliosides**: A class of glycosphingolipid with residues of sialic acid. Particularly prevalent in nerve cell membranes, gangliosides are involved in cell adhesion, signaling, and neuronal growth.

4.Other Lipids:

- **Sphingomyelin**:Phosphorylcholine head group and sphingosine backbone make up sphingomyelin. The myelin sheath of nerve cells contains a large amount of sphingomyelin, which is important in membrane construction and signal transduction.
- **Lipid Rafts**: Compact, movable regions rich in sphingolipids and cholesterol found in cell membranes. Signal transduction, membrane trafficking, and membrane organization are all facilitated by lipid rafts.
- **Lipid Bilayers**: Phospholipid assemblages organized in two dimensions as a bilayer structure. Cell membranes are made of lipid bilayers, which also act as a barrier against solute penetration.

Fig. 15.4 Classification of lipids employed in MD simulations for permeation studies

hydrophobic (which repels water) sections, they are amphiphilic molecules. Lipids play a crucial role in maintaining the structural integrity of cell membranes and controlling a range of cellular functions. The presentation in Fig. 15.4 is a general classification of representative of lipids in bilayer. The Table 15.2 illustrates applicability of these membrane lipids in MD Simulation and Fig. 15.5 details the structural features of these classes of compounds.

Table 15.2 Some illustration of permeation studies through molecular dynamics in literature

Permeation study	Type of permeation	Cell line	Membrane lipid abbreviation	Membrane lipid name
General cell membrane	Passive	HeLa	DOPC	Dioleoylphosphatidylcholine
General cell membrane	Passive	CHO	DPPC	Dipalmitoylphosphatidylcholine
Renal	Passive	MDCK	POPC	Palmitoyloleoylphosphatidylcholine
Intestinal	Passive	Caco-2	DMPC	Dimyristoylphosphatidylcholine
General cell membrane	Passive	HEK293	DOPS	Dioleoylphosphatidylserine
General cell membrane	Passive	Jurkat	DPPS	Dipalmitoylphosphatidylserine
Pulmonary	Passive	A549	POPS	Palmitoyloleoylphosphatidylserine
General cell membrane	Active	HeLa	DOPA	Dioleoylphosphatidic acid
General cell membrane	Active	CHO	DPPA	Dipalmitoylphosphatidic acid
Renal	Active	MDCK	POPA	Palmitoyloleoylphosphatidic acid
Intestinal	Passive	Caco-2	DOPG	Dioleoylphosphatidylglycerol
General cell membrane	Passive	HEK293	DPPG	Dipalmitoylphosphatidylglycerol
General cell membrane	Passive	Jurkat	POPG	Palmitoyloleoylphosphatidylglycerol
Pulmonary	Active	A549	DOPE	Dioleoylphosphatidylethanolamine
General cell membrane	Active	HeLa	DPPE	Dipalmitoylphosphatidylethanolamine
General cell membrane	Active	CHO	POPE	Palmitoyloleoylphosphatidylethanolamine
Renal	Active	MDCK	SM	Sphingomyelin
Intestinal	Passive	Caco-2	Chol	Cholesterol
Neural (e.g. BBB)	Active	HEK293	GM1	Monosialotetrahexosylganglioside

Albeit the choice of drug depends on the problem at the hands of the computational chemists, at times it will be required to select molecules from available set. Making the right drug choices is essential to ensure the accuracy and applicability of research design. Comprehensive insights into the parameters influencing biopermeability can be obtained using MD simulations by picking medications based on these criteria and guaranteeing variety in chemical structures.

Following criteria can help a computational chemist,

- **Known Permeability Data:** Here it could be two scenarios, the first being Where there is no knowledge of prior experimentation for permeation of our desired class of compound. Since a lot has been already done in this field of science at times, new scenarios do pop up. This is in the sense that lot more has been

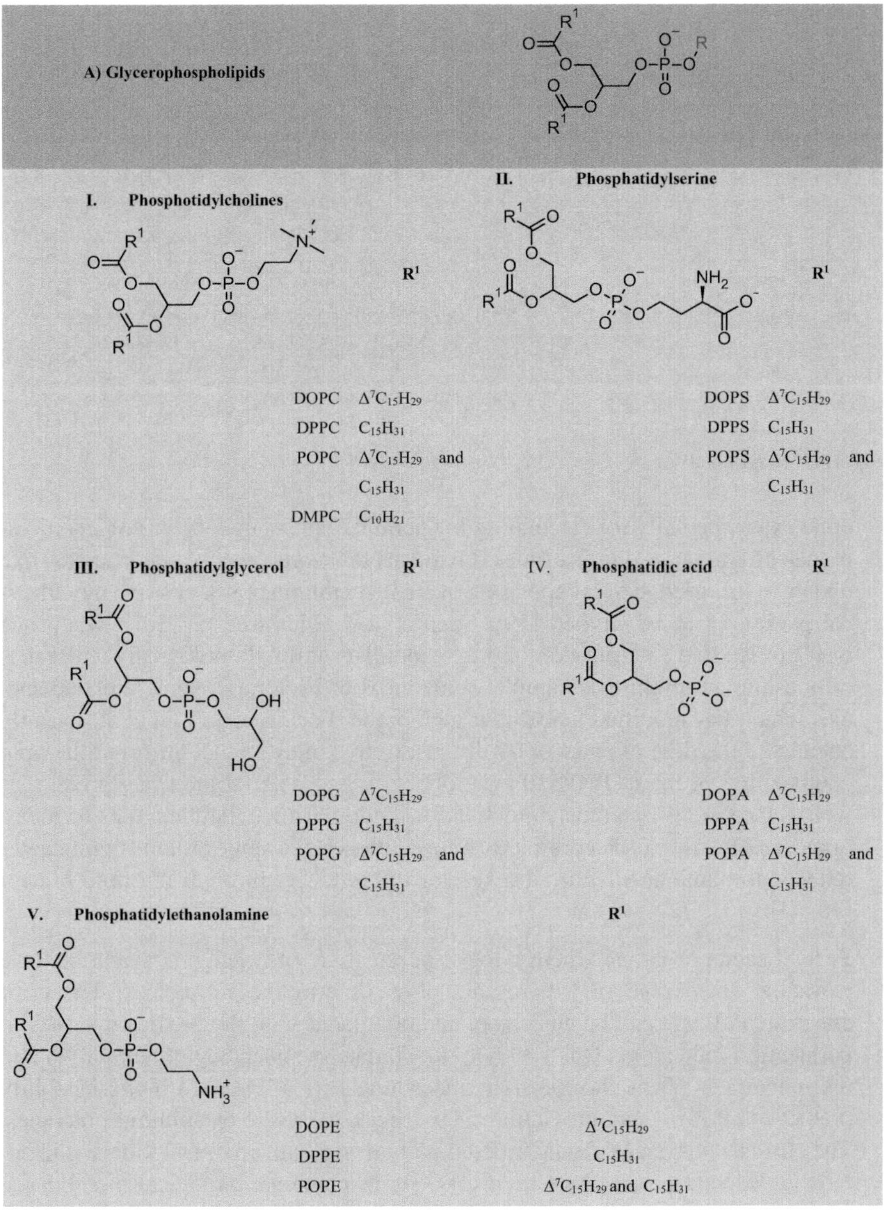

Fig. 15.5 Structural information of some the lipids in bilayer

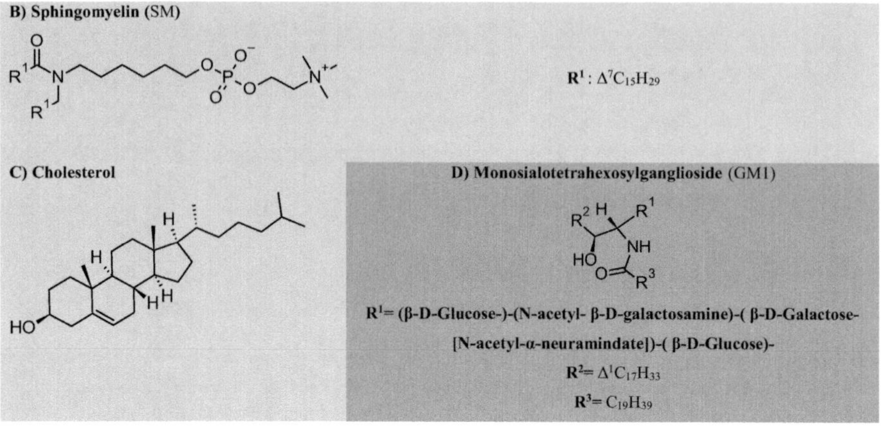

Fig. 15.5 (continued)

done experimentally for calculating the permeation through the biological membranes of Human origin. At times it happens that some new scenario arises, like update in knowledge of composition of some membrane [38]. or some hypothetical parameter to be studied [39]. Such a case holds true typically, when one wants to study the permission of drug candidates through biological membranes of non-human origin for example, some virus or bacteria or some mutated species which has modified its cellular membrane. For instance, Liu et al. recently calculated the time of entry of 79 drugs into five bilayers that mirrored the lipid composition of the COVID-19 envelope with and without the spike protein, as well as the Golgi apparatus, mitochondria, endoplasmic reticulum, plasma membrane, and lysosome. According to these estimates, the spike protein significantly reduces the amount of time it takes for drugs to get through the lipid bilayer [40, 41].

- Second scenario is one where a lot of permeation knowledge is available for a particular membrane or a particular class of druglike substances. The prior experimental data can be correlated and compared with the predicted values of permeation indicators. This will help to validate the accuracy of simulation and the outcomes. When choosing pharmaceuticals for MD-based biopermeability prediction, it is essential to define clear criteria to ensure meaningful outcomes. The chosen substances should be exemplary drugs with pharmacokinetic characteristics depend critically on their capacity to permeate biological membranes, such as the blood-brain barrier or the gastrointestinal system. The MD simulation results should ideally be validated by some experimental permeability data (e.g., from PAMPA, in vivo research, or Caco-2 experiments). Such benchmarking support the computational predictions.
- **Therapeutic Importance:** Select pharmaceuticals that are either in clinical use or under development, especially those where biopermeability is a known bottleneck in their therapeutic efficacy or bioavailability.

- **Log P (Partition Coefficient):** A wide range of lipophilicity values should be covered, as Log P is a key factor influencing membrane permeability. Pharmaceuticals with low, moderate, and high Log P values should be included to understand how MD simulations predict permeability across varying physicochemical properties.
- **Size and Molecular Weight:** Molecular size affects permeability, with smaller molecules generally having higher membrane permeability. Therefore, including small, medium, and large-sized drugs helps assess MD's ability to handle different molecular sizes.
- **Solubility:** Drugs can be picked depending on their spectrum of water solubilities (highly soluble, poorly soluble), as solubility determines lipid bilayer interaction and membrane penetration.
- **Chemical Stability:** Pharmaceuticals with varying degrees of stability in physiological conditions (pH, enzymatic activity) may offer insights into MD's capacity to predict permeability under different environmental stresses.
- The diversity of chemical structures is crucial for ensuring reduction in advent of chance errors and false conclusions through MD simulation and thus enhancing applicable of the results. When studying mechanistic or validating, or benchmarking for a new concept, method, software, or quantitating some studies Structural diversity also enables the simulation to include a wide range of membrane permeability and interaction with biological membranes. Here's how structural diversity comes into play:
- **Functional Groups:** One must include pharmaceuticals (Drugs or New chemical entities or biochemicals) with varied functional groups, such as amines, carboxylates, hydroxyls, and halogens. Functional groups have a substantial impact on permeability via influencing interactions with the lipid bilayer, such as hydrogen bonding and ionic interactions.
- **Polarity and Charge:** As the membrane is hydrophobic in nature, the polarity is going to be an important factor affecting permeation. Hence compounds with varying degrees of polarity and charge (neutral, zwitterionic, cationic, anionic) should be selected. Charged or polar molecules interact differently with membrane lipids compared to neutral, hydrophobic compounds. For instance, MD simulations can explore how charged drugs permeate through protein transporters or ion channels, while neutral compounds may passively diffuse across the membrane. Such few endeavours will be discussed in this chapter.
- **Rigid vs. Flexible Structures:** Drugs have varied interactions with membranes based on their rigid or flexible structures. Flexible molecules can adopt conformations that increase permeability, whereas rigid molecules may encounter obstacles. Compounds from this range are chosen to evaluate MD's ability to handle conformational dynamics and their impact on permeability.
- **Molecular Topology:** Diverse molecular shapes (linear, branched, cyclic, aromatic) should be represented in the study. Different topologies would influence the way the molecules navigate the lipid bilayer, with cyclic and planar molecules having different interaction modes compared to branched or flexible drugs.

Sizes, configurations and conformations of molecules may also affect the topologies and hence permeation of molecule.
- **Presence of Stereoisomers:** Enantiomers or stereoisomers of the same drug can exhibit vastly different permeability profiles, as stereochemistry can influence interactions with biological membranes. Including stereoisomers helps explore the stereoselective nature of membrane permeation.
- **Prodrug Inclusion:** Since prodrugs are designed to enhance permeability or target specific tissues, including prodrugs in the study allows for an examination of how MD simulations can predict enhanced permeability through structural modifications.

In this context, one good example is publication by Milanett et al. [42] who tested on 23 neutral structurally diverse, small molecular weight, with experimental data *i.e.* around 277 Caco-2 cell permeability values obtained from a number of literature sources referring to different experimental conditions.

In MD simulations, lipid molecules are often modeled using atomistic or coarse-grained representations, allowing researchers to study lipid dynamics, membrane properties, and lipid-protein interactions in detail. Lipid models are essential components of membrane simulations, enabling the investigation of membrane permeation processes, lipid-protein interactions, and membrane-mediated phenomena in biological systems.

The aforementioned Atomistic MD Simulations provides the most detailed representation of the system by explicitly modelling every atom in it. Here, detailed definitions for every chemical bond, bond angle, and dihedral angle are available in the form of a background database referred as Force field. As a result, atomic-level simulations of molecular dynamics and interactions offer delicate understanding of molecular structures, conformational changes, and interactions. For set research objectives, atomic simulations can be used to study small- to medium-sized molecules like proteins, nucleic acids, and small ligands. They can also be used to investigate particular molecular mechanisms like ligand binding, protein folding, and enzyme catalysis. In todays times A CPU can handle systems with 50,000 atoms while with GPU one can simulate larger systems to million atoms. Though the hardware capabilities are increasing day by day, but certainly there is a limit to it.

Alternatively, using Coarse-Grained MD In simulations, the user can combine several atoms into a single bead or interaction site, therefore decreasing the system's level of detail. Molecular interactions are usually represented at a mesoscopic or macroscopic level using coarse-grained models. Therefore, while compromising on atomic-level information, coarse-grained simulations offer a simplified depiction of molecular systems, including collective behaviors and large-scale motions. Hence the obvious application is for studying large biomolecular assemblies such as membranes, protein complexes, and viruses. In particular cases where exploration of long-timescale processes such as membrane fusion, protein aggregation, and self-assembly are being made. Also, for investigating system properties such as phase transitions, membrane permeability, and lipid bilayer dynamics.

In summary, because of the fewer degrees of freedom and simpler interactions, coarse-grained simulations are computationally more efficient than atomistic simulations.

The selection of this option is determined by the system size, research objectives, and computational resources (hardware and software). In practice, scientists often combine coarse-grained and atomistic simulations to leverage the strengths of each approach and gain a comprehensive understanding of complex molecular systems. It would be interesting for the reader to click on the following website to access an INSANE online tool for creating coarse grain systems: https://mad.ibcp.fr/tutorial/system.html [43].

15.1.5.5 Process and Data Generation

The next phase of the simulation is the energy minimization approach, which is just one level of structural optimization leading to a potential energy minimum without taking kinetic energy into account. Therefore, additional study is needed in order to conduct the MD simulation and optimize the system. The process minimizes steric collisions and helps determine the initial velocities of the atoms.

The initial velocities (vi) at the temperature of interest are selected from a Maxwell-Boltzmann distribution.

$$P(vi) = \sqrt{\frac{m_i}{2\pi k_b T}} \exp\left(-\frac{m_i v_i^2}{2k_B T}\right) \tag{15.5}$$

where P(vi) is the likelihood that, at temperature T, an atom *i* with mass m will have velocity vi. To stop the simulation box from drifting, the velocities must be changed until the system's total momentum is zero.

The equilibration phase, the first step of an MD experiment, can begin after the key elements of our experiment have been determined. In order to bring the system to equilibrium, researchers must first restore it to its initial configuration. This requires monitoring several parameters until they stabilize. The parameters that must be taken into account are significantly influenced by the sort of thermodynamical ensemble that was employed in the simulation.

It is typically advised to finish an equilibration run at constant temperature and volume before moving on to the production phase due to stability concerns. That is why an NVT equilibration run is usually the initial step. Therefore, a quick simulation with constant temperature (T) and volume (V) can be run while the pressure (P) fluctuates. Often, a simulation with a constant pressure (NPT ensemble) is used during the second equilibration step. This suggests that P and T will stay constant as the volume equilibrates. Changes in temperature, pressure, volume, kinetic, potential, and total energy should be permitted, but they shouldn't drastically alter the original structure. To determine whether the system has successfully attained equilibrium, it is essential to calculate the coordinates' Root Mean Square Deviation, or RMSD. The system is prepared for production when the RMSD is roughly constant.

The production phase technique often entails a simulation run in an ensemble, such NPT or NVT. Since constant pressure reactions are frequently conducted in the lab, the NPT ensemble is usually selected since it most closely resembles a genuine experiment. The final step involves creating a trajectory that depicts the mobility of the molecule, which can be used to see how its structure and properties changed throughout the course of the simulation and in real time [11, 44].

15.1.5.6 Validation

To validate the MD run and the trajectory's physical reliability, a number of factors need to be examined at after the simulation is complete. Two conceptual sections can be discussed in this analysis. The first step is to identify technical problems for which the trajectory is examined thoroughly. It is simple to identify aberrant atomic interactions or ruptured bonds by visual inspection. The behavior of the membrane phase is another important characteristic that is essential to make sure the membrane simulation exhibits the proper phase behavior matching the specified simulation settings. The escape of lipids from hydrophobic to hydrophilic i.e. the water phase would obviously a concern to a computational chemist. Hence visual inspection is especially important at the start of the simulation (first 100 *ps*) to identify potential issues. This can result in optimal use of computational resources. Validating thermal equilibration is the next step wherein potential energy, temperature fluctuations, area per lipid are inspected graphically to judge the quality of simulation.

Typically, root mean square fluctuation (RMSF) and root mean square deviation (RMSD) are the two metrics most frequently used in the study of molecular dynamics (MD) simulations. The deviation of a coordinate set from a reference set is represented by RMSD. Since the starting position of the simulation serves as the molecular dynamics reference set, we can compute the RMSD to see how these coordinates vary from the starting point.

RMSF assesses the unique flexibility of each residue, or the amount that a particular residue moves during a simulation, as opposed to highlighting spatial disparities in time. The RMSF vs. residue number and RMSD vs. time charts were analysed after MD simulations.

The degree of structural compactness of proteins is gauged by the radius of gyration, or RoG. It is focused on the question of how closely regular secondary structures can fit into molecules' three-dimensional configurations. A decrease in protein structural compactness, or greater flexibility and less stability, is suggested by an increase in RoG levels.

The RMSF of the average positions of the atoms and the crystallographic isotropic temperature factor per atom (Beta factor, B) are related. The underlying premise of this connection is that average energy variations of order k_BT around the native X-ray structure correspond to average atomic position fluctuations. Unless there are size limitation issues (for small N) causing high pressure fluctuations in NVT or high volume fluctuations in NPT, NVT and NPT simulations run under the same conditions (temperature, N, and roughly the same density (or volume)) should yield the same results. This is because atomic fluctuations, and therefore the B-factor, are dependent on temperature.

Quantitating the resilience and reliability of MD simulations is the next task to be completed once the simulation has been validated. These are important factors to take into account in order to guarantee the validity and correctness of the simulation findings. A comparison of experimental data and are capable of accurately reproducing experimental observables such as structural properties, thermodynamic quantities, and dynamic behaviors can advocate for quality of simulation. Quality control measures, such as systematic testing of simulation protocols and reproducibility checks, are essential for ensuring the reliability of MD simulations. A significant members of scientific community tend to report results of multiple simulation juts like an experiment performed at benchtop and then calculating indicators of ruggedness and robustness.

15.1.5.7 Data Analysis and Interpretation

The main objective of the MD simulation, which was to identify many significant structural and dynamical properties of a single membrane, can now be accomplished by the researcher. The total average membrane area divided by the number of lipid molecules in a single leaflet yields the area per lipid molecule in a pure bilayer. The composition of the membrane, the degree of hydration, the ionic force, pH, temperature, and pressure all affect the area per lipid molecule. The ordering and other dynamic and structural characteristics of the membrane are represented by the area per lipid. It is often used to validate lipid force fields and is a reliable indicator of membrane phase.

The transverse structure of the membrane is described by its thickness. Hydrophobicity, one of the bilayer's fundamental characteristics, influences membrane performance. Integral proteins become less functional due to mismatches. The lipid composition of the membrane mostly determines its thickness. Acetate chains' length, headgroup, and degree of unsaturation all have an impact on their value.

Because of its ambiguity, membrane thickness can be defined in various ways for proper scientific understanding. For pure membranes, calculating the distances between phosphorous atoms is easy. The typical locations of phosphorus atoms in opposing membrane leaflets are indicated by this distance [45].

A membrane density profile is a time average of the extended properties of a system, perpendicular to the bilayer plane and along the z-direction, such as mass and electronic density, charge, scattering length, frequency, etc. Membrane density profiles are interesting because they can be obtained experimentally and via MD simulations.

For pure saturated lipids, another analytical method is deuterium order parameter, or SCD. This value gives information on how ordered an acyl chain is at the chosen segment and can be directly compared to the order parameter found in NMR research. The formula for SCD is:

$$S_{CD} = \left(\frac{1}{2}\right)\left\langle 2\cos^2\theta - 1\right\rangle \tag{15.6}$$

where θ is the angle formed by the bilayer normal and the carbon-deuterium bond. This is very helpful in simulating membranes with a force field composed of all of the atoms. For instance, all 16 groups of carbon atoms in the acyl chain are included in the index file for DPPC. The atoms in the acyl chain must be in the same sequence as the groups. This happens when the positions of atoms n + 1 and n−1 are used to determine the SCD of the n carbon. The SCD values for the initial (oxygen-bonded) and final (terminal methyl group) atoms in the chain are not determined by this method.

The absence of a distinct structural ordering indicates that the fluid nature of the membrane is the consequence of a balance between attractive intermolecular interaction and disorganized thermal motion. Membranes in fluid phase nevertheless display spatial organization in contrary to this. With this method, the interactions between the many membrane components that make up the overall structure are identified and described. Radial distribution functions (RDF) are commonly used in this analysis. At a distance of r or r + dr from particle B, it indicates the probability of finding particle A. A simple most 2D RDF is discussed below, addresses ordering on the membrane plane (xy-plane).

The RDF is calculated using the following equation: $A_{seg}(r)$, or the area of the annulus, with ratio of A_{total}, the area of the simulation box, and N_{total}, or the total count of A particles; and $N_{seg}(r)$, or the count of particle A in an annulus of radius r and width dr surrounding particle B.

$$RDF_{xy}(\mathrm{r}) = \frac{\left[\left(\frac{N_{seg}(r)}{A_{seg}(r)}\right]\right.}{\left(\frac{N_{total}}{A_{total}}\right)} \tag{15.7}$$

One intriguing dynamical characteristic is related to the lipids' rotating modes. Any intramolecular vector's so-called autocorrelation functions (ACF) can be examined to determine these modes. These functions quantify the system's ability to remember a specific configuration across time (vector direction). Thus, the slower the analyzed mode, the longer the configuration is maintained.

A key dynamic aspect is diffusion laterally. It measures how well the lipids can pass through the leaflet. Lipid lateral diffusion is described by computing the diffusion coefficient.

$$D(\tau) = \lim_{\tau \to \infty} \left(\frac{MSD(\tau)}{4\tau} \right) \tag{15.8}$$

The elapsed time (τ) and the number of lipids (MSD) can be used to compute it. MSD is the mean-square displacement of the centre of mass of the chosen lipids in the xy-plane averaged over various start time frames and number of lipids.

15.1.5.8 Modern Applications and Advancements in Molecular Dynamics for Permeation Studies

When using molecular dynamics simulation for permeation studies, one can compare the penetration of a drug through multiple membranes to address selectivity, the process of permeation through different biological membranes, the drug-membrane interaction, or the comparison of different drugs or analogue for permeation through similar membranes. According to studies on the effects of drug binding to the membrane, the drug may modify its bilayer area, membrane thickness, phospholipid chain order, and membrane curvature (especially if the binding is asymmetric). In certain situations, bilayer-bound chemicals can cause lipid phase separation as well as the creation or prevention of membrane holes and defects.

Pharmacological substances bonded to membranes have the ability to boost ion and molecular penetration via defects and channels. Drug substances may alter the lipids' phase separation characteristics, changing the way the membrane is packed. Many protein receptor are embedded in lipid bilayer and interact with biochemical and drug molecule. Proteins on either side of the membrane may bind to the drug. Such proteins are activated by binding this effector molecule, which opens the ion channel and initiates the signalling cascade. Once across the membrane, the medicinal molecule might attach itself to a cytoplasmic protein. W. Kopec et al. had assembled a nice list of significant case studies related to these simulations of tiny compounds with lipid bilayers. We will discuss a few typical cases that align with the goal of this chapter, which is to advance the use of MD in penetration studies. Please be aware that it might not be in the chronological order of events [46].

Anaesthetic substances that congregate within cellular membranes present an intriguing and fitting example. Many studies have been conducted on MD simulations of anaesthetic drugs with lipid membranes. All these research is based on the fundamental hypothesis that an anaesthetics' partitioning characteristics in the bilayer indicate its efficacy. Porasso et al. evaluated the likelihood of discovering a benzocaine molecule on either side of the membrane using thermodynamic data and MD modelling. Porasso et al. proposed that the zone of actuation for benzocaine will be located in the hydrocarbon area or at the end of the lipid head based on the free energy profile. This will depend on whether charged lipids (DPPS) are present in the leaflet. It will be interesting to describe how the Gibbs free energy equation was used to determine the ratio of rate constants indicated as the likelihood of discovering medicine at a given side of the membrane in order to achieve this knowledge.

$$\int_{-\infty}^{0} \exp\left[-\frac{\Delta G_1}{k_B T}\right] / \int_{0}^{\infty} \exp\left[-\frac{\Delta G_2}{k_B T}\right] = k_1 / k_2 \tag{15.9}$$

In this particular case, a ratio of 1.02 was obtained, meaning that the chance of discovering benzocaine in the leaflet was 50.3% in the presence of DPPS and 49.7% in its absence. The following equation was used to determine the free energy information partition calculation, and the result was 7044, which was extremely near to the experimental value for lidocaine.

$$K = \exp\left[-\frac{\Delta G_{\min}}{k_B T}\right] \tag{15.10}$$

They also noticed a slight rise in membrane problem, specifically in DPPS disorder. The researchers determined the static and dynamic physicochemical features of benzocaine, such as its dipole orientation, translational diffusion coefficient, and rotational relaxation time. Thus, the influence of both the membrane and the medication on each other's structure and properties were investigated [47].

Leading in the same direction when scientific communities were able to predict the penetration of drugs across membrane, it was also tried to compare the penetrability among the drug candidates, through permeability rankings or permeability predictor" $P_{\Delta G}$ calculated as follows

$$P_{\Delta G} = \frac{k_B T}{\Delta G_{\max} - \Delta G_{\min}} \tag{15.11}$$

The sequence of permeability for a few medications was determined by Orsi and colleagues as follows: atenolol < pindolol < (progesterone ~ alprenolol) < testosterone. The dual resolution approach was used to run this simulation, with atoms representing the drug molecules and coarse grain beads representing the water and lipid molecules in the system. This yields, in theory, simulations that are as accurate as all atom techniques and as efficient as coarse grain methods. 'Enhanced sampling' techniques, including 'Umbrella sampling,' are also used in these simulations to speed up ordinarily slow-moving events. One big simulation may be successfully divided into several smaller simulations using this strategy, and they can all be run in parallel, greatly enhancing computational efficiency [48].

One similar approach was followed by Milanetti et al. [42] who applied regression analysis to molecular dynamics to arrive at the best fit model as

$$P_{pred} = 3.06 * I_n + 0.04 I_{-} + 3096 \tag{15.12}$$

where I_n are derived from the probability values of identifying water molecules in the first and second hydration shells and give information about the compound's hydrophobic characteristics. I_- is correlated with the (negative) charges of the analysed substance and, consequently, with the dipole orientation of the water molecules surrounding it. The experimental Caco-2 permeability values and this P_{pred} indicator correlated very well.

Along with developments in the field of medicinal chemistry, the researchers were also able to conclude that charges on molecules will undoubtedly change how well drug candidates penetrate the body. Another well-known and extensive class of pharmacological molecules are the non-steroidal anti-inflammatory medicines (NSAIDs), which are weak carboxylic acids with aromatic rings. Therefore, they can exist in both uncharged and anionic forms in water soluble states (as well as in biological fluids). This feature makes this class of compounds intriguing for research on membrane penetration. Positively charged local anaesthetics and negatively charged NSAIDs affect the zwitterionic membrane characteristics in a comparable way. Due to the hydration of their charged groups, charged molecules are located closer to the water/membrane contact and produce the least amount of disruption to the bilayer structure [49–51].

W. Kopec et al. wisely collated and summarized the outcome of such endeavors and some common traits [46]. Also, it was figured that Drugs close to a membrane have a distinct net charge when their polar half contains functional groups with varying propensities to protonate [52]. Drug partitioning and the impact of pendant hydrophobic rings on lipid bilayer characteristics are impacted by it [53]. Conversely, chemicals like fusidic acid and sterols, which have flat ring-like structures, organize fluid lipid membranes and affect the profiles of lateral pressure. As demonstrated with prilocaine, slight modifications to the guest molecule, such as flatness or net charge, can reverse the effect. The head group-water interface is where molecules with short aromatic rings that are negatively charged are more prone to split than the head group-lipid tail contact. As with perillic acid, the negatively charged moiety of the guest molecule forms a connection with the positively charged choline head group [54] and salicylate [55]. A thinner bilayer, more area per lipid, and disordered acyl tails close to the lipid-water interface are the outcomes of the aromatic ring's disruption of the head group region's packing. Propofol, articaine, unprotonated lidocaine, positively charged phenothiazines, and other uncharged substances establish hydrogen bonds with the glycerol backbone. Like cholesterol, flat aromatic rings are submerged in the hydrophobic zone, which improves the arrangement of lipid tails.

In a related work, Yue et al. attempted to determine how a charged medication, such as propranalol, is transported, only to discover later that it neutralizes at the hydrophobic boundary of the POPC Bilayer. It has a significant impact on the free energy landscape of permeation and suggests that dynamic variations in the protonation state should be taken into account in further simulations of permeation [56].

At this point the researchers figured that high amounts of cholesterol are present in mammalian plasma membranes which significantly affect their physical characteristics. Eventually these moieties can alter the order and interaction patterns [57,

58]. This onwards simulations had been performed in a system as close as possible to the biological scenarios.

One such example is case of permeation studies of paracetamol, where various parameters were studied to figure plausible biological scenario, i.e. exact type of lipid in membrane, number of drug molecules etc., where they studied Hydrogen bonds and Electrostatic potential along with parameters mentioned earlier. Both of them can help in understanding the mechanism (particularly for ion channels). If the small molecule is making more of hydrogen bonds it will be difficult for it to pass through. If it is disturbing the hydrogen bonds between the lipid molecules it will permeate faster but can rupture membranes. Similarly, a rise in electrostatic potential can lead to deformation in lipid tail and enhanced interactions in the tail area. It may also indicate a rise in the amount of carbonyl heads on lipid molecules interacting with the aqueous phase.

As mentioned in the in the text earlier space about the use of potential mean force for the calculation of energetics permeation one such case study we mentioned over here. In this study Meng and Xu [10] subjected Doxorubicin, Ibuprofen and Atenolol to permeate through the DPPC membrane. The trajectories and energetics indicated that low permeability drugs would have a higher free energy barrier, while high permeability drugs would have a lower barrier. Drugs with hydrophobic groups would easily traverse the membrane, but those with hydrophilic groups would trail behind due to strong interactions with DPPC polar groups and surrounding water molecules.

For neuroactive drugs to be effective, the blood-brain barrier must be crossed. The BBB should not be crossed by drugs that target other parts of the body to prevent possible psychiatric side effects. Using the gromos53a6 FF and SPC models, computational predictions of these parameters were obtained for a sample of 12 small molecule compounds. The mean force potential of each molecule is found using molecular dynamics simulations using a 1,2-dioleoyl-sn-glycero-3--phosphocholine (PAMPA) bilayer. This setup of the system is often used as a basic BBB mimic. Furthermore, molecular dynamics trajectories are used to produce one-dimensional position-dependent diffusion coefficients. The effective permeability (P_{eff}) of each sample material is obtained by combining the diffusion coefficient with the free energy landscape. The logBB and logPS values obtained from experimentation are compared with the relative permeabilities. Here, PMF computations are also applied. However, the scientists came to the conclusion that a mixed-lipid system, a much longer equilibration time, and an improved force field were necessary for the practical scenario [59].

Transdermal drug administration through the skin is another key field of research in which Molecular Dynamic Simulation has been used. Lundborga and colleagues investigated the transdermal permeation of some drug agents (benzene, codeine, dimethyl sulfoxide, ethanol, naproxen, nicotine, testosterone, and water) as well as the permeability-enhancing effect of some penetration enhancers (Azone, DMSO, oleic acid, stearic acid, and water), and discovered that permeation enhancement is dependent on the permeating compound and the concentration of penetration enhancer inside the skin's barrier structure. Here, local diffusion coefficient calculations and the potential of mean force (PMF) were applied [60]. Recently, Wennberg et al. conducted comparable research at a larger scale, examining the skin penetration of caffeine and naproxen in the presence of many permeation enhancers, such as water, stearic acid, lauric acid, geraniol, and thymol. Numerous innovative aspects of the experiment were investigated, including the length of the simulation and updated indications, the location and count of permeation enhancers, and the use of a novel method. Solvation free energy was computed using the Accelerated Weight Histogram (AWH) method for alchemical free energy calculations. From the obtained potentials of mean force (PMFs) across the barrier structure of the skin, the permeability coefficient, K_P, and the permeation resistance, R, were estimated as follows.

$$\frac{1}{K_P} = (30 \pm 6) X \int_{z1}^{z2} \frac{e^{\beta \Delta G_{transfer}(z)}}{D(z)} dz \tag{15.13}$$

where D is the local diffusion coefficient across the skin's barrier structure (dz), β is the thermodynamic β, and $\Delta G_{transfer}$ is the free energy in relation to the solvation free energy. The friction metric tensor g(z) is estimated by the AWH simulation and utilized to calculate the diffusion coefficient D(z) = g − 1(z) across the barrier structure. Additionally, the partition coefficient of the solute into the skin barrier structure was inferred as follows from the data on solvation free energy and PMF.

$$\log K_{w-lip} = \frac{\Delta G_{rel\_water}}{RT \ln(10)} \tag{15.14}$$

where R is the gas constant, T is the temperature, and ΔG_{rel_water} is the Boltzmann-weighted free energy in the lipid system relative to water. Both in the presence and absence of permeation enhancers, the anticipated values of the indicators matched the experimental results [61].

Molecular dynamics simulation has been used by computational scientists on multiple occasions to support or refute particular phenomena. One such instance is a work by Bennett et al. that used molecular dynamic modelling to comprehend how antimicrobial peptides generate pores in lipid bilayers [62]. The researchers believed that prolonged interactions between the charged peptides and the polar interface can result in bilayer deformations, such as membrane rupture and the stabilization of a hydrophilic pore by the peptides. Here antimicrobial peptide CM15 (KWKLFKKIGAVLKVL-NH2) was studied to observe pore formation in a POPC bilayer. In this venture though Bennett couldn't succeed in proving that the CM15 peptide would causeway any pore formation in the biological membrane. However certain instances definitely point out that work in this direction is pending and something was missing which can be studied later opening a new dimension for other scientists.

On the similar lines, another phenomenon observed was transformation in the lipid bilayer while the penetrant traverses through the membrane. At times it may not break or form the pore through the membrane but may cause changing the arrangement which may lead to changes in membrane behaviour. The alignment of the tail portion gets perturbed and the thickness of the membrane reduces drastically, this is referred as Gel to fluid phase transition. Interestingly, the majority of cell membranes exist in the fluid phase, which maintains their integrity and allows them to function properly. For this reason, researching the interactions between tiny molecules and membranes is best done in this state. This phenomenon is also applicable to non-biological human made membranes like those in liposomes, miscellas and bilayer sheets, proving its importance to drug formulation scientists [63]. In such scenario it will be desired to learn about membrane changes in structure, dynamics, and stability while interacting with tiny molecules, including orientation and location.

Such studies typically entail the analysis of small molecule membrane interactions with parameters such as area per lipid (APL), membrane thickness, lipid tail order, diffusivity of a particle using the Brownian motion (Einstein-Smoluchowski relation or Green-Kubo relations), free energy changes studied over reaction coordinate, and permeation coefficient [64]. In the fluid phase, lipid tails have a low tail order and are very mobile. Lipid tail order increases but lipid tail mobility decreases in the gel phase. The fluid-to-gel transition temperature (T_{melt}), which shows how the liquid crystalline phase of plasma membranes stabilizes and how their biological activity is preserved at lower temperatures, is another crucial signal.

Among other consequences of the medication, a shift in lateral pressure, i.e. is, pressure parallel to the membrane surface along the bilayer normal may cause structural and functional alterations to the lipid membranes. It has been observed that certain drugs can change the hydration of membranes. Furthermore, it has been found that changes in the orientation of lipid headgroups, especially those connected to signalling, may affect the binding of peripheral membrane proteins. This means that these kinds of impacts need to be monitored closely when doing MD simulations [65].

Molecular dynamics can also be utilized to comprehend the mechanics of various drug penetration processes, such as the bile's role in medication absorption. Kabedev et al. used a mix of computational simulations and experimental studies to perform a thorough analysis of the danazol transport pathways to lipid bilayers. There are two potential mechanisms at work in this process. The micelle may deliver the danazol directly to the membrane if the drug molecule can be absorbed without passing through the aqueous phase. In the other method, the micelle acts as an elevator to carry the solubilized danazol into the membrane along with it because the colloidal structure is there. The laboratory experiments confirmed the elevator hypothesis by showing that the colloidal structures of the simulated intestinal fluid efficiently and rapidly integrated into the lipid bilayers. It was seen using the coarse-grain approach of molecular dynamics that the complete micelle fused and dispersed within the membrane at the same moment [66].

As was previously highlighted, Hossain et al. showed another intriguing use for MD in formulation development. The researchers used molecular dynamics (MD) to validate their hypothesis on drug delivery via electric field, which might affect the phospholipid bilayer structure and result in the formation of temporary nanoscale pores in it. They applied a controlled shock impulse to the neighbouring water molecules of the DPPC layer along the z-direction, using a MARTINI coarse grain force field to the dipalmitoyl phosphatidylcholine (DPPC) bilayer system. It should be noted that Hossain et al. found a decrease in water content in the hydrophobic region upon disruption, indicating that a pore had formed on the DPPC bilayer [67].

According to the chapter's theme, advanced applications encompass both the innovative utilization and the size at which the issues or solutions are investigated. Using high throughput coarse-grained (HTCG) simulations, Menichetti and his team estimated the PMF of drug-membrane partitioning for an unprecedented 511,427 small compounds. Bulk partitioning free energy and pK_a were computed for coarse-grained systems [68].

That suggests that it's critical to understand that chemicals that permeate the membrane system don't always travel in a straight line. The simulation of the entire system may or may not follow a straight path, even though the computational scientist might prefer to see the migration of a small molecule in one direction—that is, towards the end of the membrane. To recreate the same, Zhang et al. devised a highly parallelizable weighted ensemble (WE) path sampling method [69]. The WESTPA 2.0 software package's Python API was used in this study to execute WE simulations. Trajectories that achieved the target state of compartment A ($z = 3.0$ nm) were "recycled," meaning that a new trajectory was started from the initial state, in order to preserve non-equilibrium steady-state conditions. Two distinct approaches were used to partition a one-dimensional progress coordinate into bins: a minimal binning scheme and a manual, fixed binning scheme. Therefore, the slowest motion pertinent to the rare-event process of interest and the separation between the molecule's and the lipid bilayer's centres of mass were used to determine an appropriate progress coordinate for a WE simulation.

Following this trend, the community of computational chemists began to believe that the polarity, size, and most likely entropy of the simulation system are what drive molecules through the membrane at their own pace. A quick, easy, and dependable way to determine the permeability relative to other materials was required. Using directed molecular dynamics, Thai et al. calculate the BBB penetration by applying a push in a particular direction to the permeating molecules and tracking the nonequilibrium work, or W_{neq}, needed to permeate the membrane. A good correlation was discovered between the W_{neq} estimated for approximately 26 small-molecule compounds and the experimental values of logBB, logPS, and topological polar surface area (TPSA) [70]. The ligand was dragged in the direction of the bilayer in the z direction at a constant velocity of v, covering a distance of $\Delta zc = vt$. As a result, the compound experienced an elastic force of $F = k(\Delta z - vt)$ at a certain point during the pulling process. Based on information from atomic force microscopy tests, k was fixed to 600 kJ/(mol·nm2) and various pulling velocities (i.e., v = 0.5, 1.5, and 5.0 nm/ns) in this study. This information allows for the easy calculation of the work, W_{neq}, required to move the compound across the bilayer over the distance $Lz = \int dz$., where Lz is the bilayer's width in the z direction. The force, F, may be determined as a function of time and the compound's position, in the z direction, using the formula

$$W_{neq} = \int F\, dz \tag{15.15}$$

Up until now, we have only talked about small compounds, but there is a wealth of literature that also reports on investigations on larger molecules. The research done by Sugita and colleagues on comparatively bigger molecules including 100 cyclic peptides with various chemical structures is a pertinent example that is being examined here. Here, the researchers used the previously described guided molecular dynamics (steered MD) approach and improved sampling with the replica-exchange umbrella sampling (REUS) method to try and predict the membrane permeability of cyclic peptides using the inhomogeneous solubility-diffusion model. It was interestingly discovered that the peptides confronted an energy barrier in the middle of the lipid bilayer and were unable to cross the membrane, only remaining in the vicinity of the boundary. Another significant discovery supported the earlier theory that, in order to penetrate the membrane, they must adopt a structure that conceals the polar groups. It was discovered that peptides alter their shape and orientation in response to their environment [71]. This suggests further that interactions with the binding site of the target protein may be affected by such conformational changes [65].

This discussion exposes us to an entire field of research where scientists have tried to classify molecules' behaviour inside membranes according to where they are located within the membrane. The location could be anywhere in the following ranges: (a) immediately below the lipid headgroups; (b) immediately below the lipid headgroups; (c) at the location of the hydrophobic lipid tails; and (d) at the precise centre of the membrane core, in the gap between the bilayer's two leaves. This would be the area where the lipid headgroups and the bulk solvent phase interface. Dr. Berendsen's group from the University of Groningen created this classification [72] According to studies using MD modeling, the degree of hydration or the makeup of the lipids determine where the drug molecule is located. In a hydrated bilayer, for example, curcumin has been shown to locate below the water-membrane contact and to orient parallel to the bilayer normal [73].

Artificial intelligence needs to be explored if sophisticated computational chemistry approaches are being used. Recently, Dutta et al. used the permeation and molecular dynamics data to test the sure-independence screening and sparsifying operator (SISSO), an artificial intelligence method that blends symbolic regression with compressed sensing. This work was based on coarse-grained simulations of around 400,000 tiny molecules from a massive data set. The simulation was conducted under the assumptions of an inhomogeneous solubility–diffusion model over many asymptotic acidity regimes. It's interesting to note that a SISSO 1D model connected the PMFs of the neutral, protonated, and deprotonated species in water to their dissociation constants (apK$_a$, bpK$_a$), as well as the pH of the surrounding medium.

$$f^{1D} = C^{1D}0 + C^{1D}1(apK_a - bpK_a - 2\beta\Delta G_{W\to O1} \tag{15.16}$$

where the only two fitting factors are c1D0 and c1D1, the water/octanol partitioning coefficient is ΔGW → Ol, and the acidity is represented by apK$_a$ and bpK$_a$. This study therefore linked acidity and hydrophobicity in passive permeation [74]. Thus, with AI, enormous data from MD was translated to a simple and sensible equation.

Another example of recent computational progress is the application of machine learning to interpret and analyse MD data, whereas conventional ML models like support vector machines (SVMs) and k-nearest neighbours (KNN) are widely used. The use of advanced neural networks to comprehend passive permeation and lipophilicity has been extensively documented in the literature. A pertinent study that warrants note is that conducted by Dutta et al. with the aim of making use of the massive amounts of data produced by simulations [75]. CG Martini model based molecular dynamics simulations of 18 molecules in lipid and water bilayers were carried out in this extensive and detailed investigation in order to collect multidimensional time-series data of eight chemical characteristics. To estimate free energy from the time-series data, deep learning models were built using this data of molecular characteristics. These models included Long Short-Term Memory networks, attention mechanisms, and thick layers. With R^2 values of roughly 0.99 and 0.82 for the test set and an external validation set, respectively, the prediction errors were much smaller than those of the state-of-the-art data-driven approaches.

15.1.5.9 Limitations

Till this point the reader must have judged that though enough work has been done in this field of science, yet lot more can be and is to be done. The fundamental barrier in studying drug translocation through lipid bilayers is the limited time scale available for MD simulations with all atom precision. More the simulation length better would be sampling and then reliability of the calculations. Through the advent of new computational methods, fast hardware and GPU based simulation software in future will help.

Despite enormous automation like Automated Topology Builder server [76], CHARMM GUI [77] etc. for Force-Field Development and Small-Molecule Parameterization the expansion in knowledge of chemicals limits their utility.

The absence of explicit polarizability in force fields used for biomolecule simulations is another barrier that could affect the translocation and partitioning of small molecules into lipid bilayers. A method for calculating free energy profiles using simulations and partial charges in the potential set obtained from QM calculations carried out in a polarizable continuum was proposed by Jämbeck and Lyubartsev [34].

The experimental data is required not only to design the simulation approach, but also to validate it. Although vast datasets are available, the prediction will become more reliable as additional experimental data is collected. There is obviously opportunity for more data to help with the continued development of molecular dynamics techniques.

Although simulations have enhanced capabilities, they should not be used to replace experimental validation. Relying primarily on computational results can result in dangerous or ineffective drug candidates proceeding through development, posing potential threats to public health. Simulations sometimes rely on assumptions or training data, which may add bias and skew outcomes. Ethical use requires researchers to rigorously evaluate these limitations and guarantee that simulation results are well-founded and unbiased. Simulations frequently simplify complex biological environments (for example, mimicking membrane or solvent conditions), which may not accurately represent in vivo conditions, lowering predicted accuracy for real-world applications. However, they provide meaningful results to drug discovery experts and a rationale for the approach.

Because drug development ultimately affects human health, simulations should always be employed with patient safety as a top priority. This includes rigorous validation before implementing findings in clinical settings. Researchers should offer full explanations of simulation parameters, models, and data to allow others to duplicate their findings. This transparency is critical for validating discoveries and establishing trust among the scientific community. The simulation results must be consistent across multiple processing platforms and situations. This necessitates precise documentation and sharing of both raw data and code so that other researchers may verify the results and reduce the chance of errors or misinterpretation. To prevent drawing false results, researchers must be honest and clear about their models' limits, including any uncertainties or sources of mistake.

15.2 Conclusion

15.2.1 Summary of Findings

15.2.1.1 Key Insights from Dynamic Simulation Studies

Dynamic simulation studies have proven beneficial in understanding molecular behaviour, especially in the context of drug discovery. These investigations have produced important information about molecular interactions, energy landscapes, and conformational dynamics. Notably, simulations have improved our capacity to predict molecule permeability across lipid membranes, revealing extensive information about the structural and physicochemical factors that regulate these processes.

15.2.1.2 Contributions to Bio-permeability Prediction

Significant improvements in bio-permeability prediction accuracy have been realized by combining powerful machine learning models with dynamic simulations. Simulations have found crucial physical variables that affect permeability, such as electrostatic potential and hydrophobicity. Furthermore, the use of deep learning architectures has enabled more precise mapping of chemical structure to permeability, enhancing the field of predictive modelling in drug discovery. In this chapter we have learnt that molecules may either follow some path, at time may concentrate in membrane and some instances can affect the membrane structure in some or other ways.

15.3 Closing Remarks

15.3.1 Significance for the Pharmaceutical Industry

The findings of dynamic simulation studies have enormous implications for the pharmaceutical industry since they provide a computationally efficient method of screening medication candidates. By precisely predicting membrane permeability, these simulations decrease the need for extensive in vitro testing, accelerating medication development while lowering expenses. They also provide useful insights into drug-membrane interactions, which are important in evaluating medication efficacy and bioavailability. Some examples of lipid membrane formulation has been exemplified in the chapter also effect of permeation enhancer was highlighted.

15.3.2 Potential Impact on Future Drug Development

As computer power increases and machine learning techniques improve, dynamic simulations' significance on drug development will only expand. These innovations have the potential to transform high-throughput virtual screening, resulting in more efficient and tailored drug design. In the coming future faster computer, better software, empirical data development of algorithms like AI and machine learning will certainly impact the drug discovery and development. The ongoing combination of simulations and experimental validation will result in more biologically relevant predictions, paving the way for more tailored and successful medicines in the future.

References

1. Wang J, Skolnik S. Permeability diagnosis model in drug discovery: a diagnostic tool to identify the most influencing properties for gastrointestinal permeability. Curr Top Med Chem. 2013;13(11):1308–16.
2. Mtewa AG, Ngwira KJ, Lampiao F, Weisheit A, Tolo CU, Ogwang PE, et al., editors. Fundamental methods in drug permeability, pKa, LogP and LogDx determination. J Drug Res Dev. 2028;5(1): 1–6.s.
3. Seelig J. Thermodynamics of lipid-peptide interactions. Biochim Biophys Acta. 2004;1666(1–2):40–50.
4. Marrink S-J, Berendsen HJC. Simulation of water transport through a lipid membrane. J Phys Chem. 1994;98(15):4155–68.
5. Cocucci E, Kim JY, Bai Y, Pabla N. Role of passive diffusion, transporters, and membrane trafficking-mediated processes in cellular drug transport. Clin Pharmacol Ther. 2017;101(1):121–9.
6. Bernardi A, Bennett WFD, He S, Jones D, Kirshner D, Bennion BJ, et al. Advances in computational approaches for estimating passive permeability in drug discovery. Membranes (Basel). 2023;13(11):851.

7. Tieleman DP, Berendsen HJC. Molecular dynamics simulations of a fully hydrated dipalmitoylphosphatidylcholine bilayer with different macroscopic boundary conditions and parameters. J Chem Phys. 1996;105(11):4871–80.
8. Sajadi F, Rowley CN. Simulations of lipid bilayers using the CHARMM36 force field with the TIP3P-FB and TIP4P-FB water models. PeerJ. 2018;6:e5472.
9. Zhao L, Cao Z, Bian Y, Hu G, Wang J, Zhou Y. Molecular dynamics simulations of human antimicrobial peptide LL-37 in model POPC and POPG lipid bilayers. Int J Mol Sci. 2018;19(4):1186.
10. Meng F, Xu W. Drug permeability prediction using PMF method. J Mol Model. 2013;19(3):991–7.
11. Martinez-Seara H, Rog T. Molecular dynamics simulations of lipid bilayers: simple recipe of how to do it. Methods Mol Biol. 2013;924:407–29.
12. Lyubartsev AP, Rabinovich AL. Force field development for lipid membrane simulations. Biochimica et Biophys Acta. 2016;1858(10):2483–97.
13. Pezeshkian W, Khandelia H, Marsh D. Lipid configurations from molecular dynamics simulations. Biophys J. 2018;114(8):1895–907.
14. Chandrasekhar I, Kastenholz M, Lins RD, Oostenbrink C, Schuler LD, Tieleman DP, et al. A consistent potential energy parameter set for lipids: dipalmitoylphosphatidylcholine as a benchmark of the GROMOS96 45A3 force field. Eur Biophys J. 2003;32(1):67–77.
15. Oostenbrink C, Villa A, Mark AE, Van Gunsteren WF. A biomolecular force field based on the free enthalpy of hydration and solvation: The GROMOS force-field parameter sets 53A5 and 53A6. J Comput Chem. 2004;25(13):1656–76.
16. Poger D, Van Gunsteren WF, Mark AE. A new force field for simulating phosphatidylcholine bilayers. J Comput Chem. 2010;31(6):1117–25.
17. Berger O, Edholm O, Jähnig F. Molecular dynamics simulations of a fluid bilayer of dipalmitoylphosphatidylcholine at full hydration, constant pressure, and constant temperature. Biophys J. 1997;72(5):2002–13.
18. Chiu S-W, Pandit SA, Scott HL, Jakobsson E. An improved united atom force field for simulation of mixed lipid bilayers. J Phys Chem B. 2009;113(9):2748–63.
19. MacKerell AD Jr, Wiorkiewicz-Kuczera J, Karplus M. An all-atom empirical energy function for the simulation of nucleic acids. J Am Chem Soc. 1995;117(48):11946–75.
20. Klauda JB, Brooks BR, MacKerell AD, Venable RM, Pastor RW. An ab Initio study on the torsional surface of alkanes and its effect on molecular simulations of alkanes and a DPPC bilayer. J Phys Chem B. 2005;109(11):5300–11.
21. Sonne J, Jensen MØ, Hansen FY, Hemmingsen L, Peters GH. Reparameterization of all-atom dipalmitoylphosphatidylcholine lipid parameters enables simulation of fluid bilayers at zero tension. Biophys J. 2007;92(12):4157–67.
22. Klauda JB, Venable RM, Freites JA, O'Connor JW, Tobias DJ, Mondragon-Ramirez C, et al. Update of the CHARMM all-atom additive force field for lipids: validation on six lipid types. J Phys Chem B. 2010;114(23):7830–43.
23. Klauda JB, Monje V, Kim T, Im W. Improving the CHARMM force field for polyunsaturated fatty acid chains. J Phys Chem B. 2012;116(31):9424–31.
24. Lim JB, Rogaski B, Klauda JB. Update of the cholesterol force field parameters in CHARMM. J Phys Chem B. 2012;116(1):203–10.
25. Venable Richard M, Sodt Alexander J, Rogaski B, Rui H, Hatcher E, MacKerell Alexander D, et al. CHARMM all-atom additive force field for sphingomyelin: elucidation of hydrogen bonding and of positive curvature. Biophys J. 2014;107(1):134–45.
26. Lee S, Tran A, Allsopp M, Lim JB, Hénin J, Klauda JB. CHARMM36 united atom chain model for lipids and surfactants. J Phys Chem B. 2014;118(2):547–56.
27. Jämbeck JPM, Lyubartsev AP. Another piece of the membrane puzzle: extending slipids further. J Chem Theory Comput. 2013;9(1):774–84.
28. Jämbeck JPM, Lyubartsev AP. An extension and further validation of an all-atomistic force field for biological membranes. J Chem Theory Comput. 2012;8(8):2938–48.

29. Jämbeck JPM, Lyubartsev AP. Derivation and systematic validation of a refined all-atom force field for phosphatidylcholine lipids. J Phys Chem B. 2012;116(10):3164–79.
30. Skjevik ÅA, Madej BD, Walker RC, Teigen K. LIPID11: a modular framework for lipid simulations using amber. J Phys Chem B. 2012;116(36):11124–36.
31. Dickson CJ, Madej BD, Skjevik ÅA, Betz RM, Teigen K, Gould IR, et al. Lipid14: the amber lipid force field. J Chem Theory Comput. 2014;10(2):865–79.
32. Maciejewski A, Pasenkiewicz-Gierula M, Cramariuc O, Vattulainen I, Rog T. Refined OPLS all-atom force field for saturated phosphatidylcholine bilayers at full hydration. J Phys Chem B. 2014;118(17):4571–81.
33. Vanommeslaeghe K, MacKerell AD. CHARMM additive and polarizable force fields for biophysics and computer-aided drug design. Biochim Biophys Acta Gen Subj. 2015;1850(5):861–71.
34. Jämbeck JPM, Lyubartsev AP. Implicit inclusion of atomic polarization in modeling of partitioning between water and lipid bilayers. Phys Chem Chem Phys. 2013;15(13):4677–86.
35. Bauer BA, Patel S. Recent applications and developments of charge equilibration force fields for modeling dynamical charges in classical molecular dynamics simulations. Theor Chem Accounts. 2012;131(3):1153.
36. López CA, Sovova Z, van Eerden FJ, de Vries AH, Marrink SJ. Martini force field parameters for glycolipids. J Chem Theory Comput. 2013;9(3):1694–708.
37. Cooke IR, Deserno M. Solvent-free model for self-assembling fluid bilayer membranes: stabilization of the fluid phase based on broad attractive tail potentials. J Chem Phys. 2005;123(22):224710.
38. Rabinovich AL, Lyubartsev AP, Zhurkin DV. Unperturbed hydrocarbon chains and liquid phase bilayer lipid chains: a computer simulation study. Eur Biophys J. 2018;47(2):109–30.
39. Tse CH, Comer J, Wang Y, Chipot C. Link between membrane composition and permeability to drugs. J Chem Theory Comput. 2018;14(6):2895–909.
40. Gordon DE, Jang GM, Bouhaddou M, Xu J, Obernier K, White KM, et al. A SARS-CoV-2 protein interaction map reveals targets for drug repurposing. Nature. 2020;583(7816):459–68.
41. Liu C, Elvati P, Violi A. On drug-membrane permeability of antivirals for SARS-CoV-2. J Phys Chem Lett. 2021;12(5):1384–9.
42. Milanetti E, Raimondo D, Tramontano A. Prediction of the permeability of neutral drugs inferred from their solvation properties. Bioinformatics. 2016;32(8):1163–9.
43. Wassenaar TA, Ingólfsson HI, Böckmann RA, Tieleman DP, Marrink SJ. Computational lipidomics with insane: a versatile tool for generating custom membranes for molecular simulations. J Chem Theory Comput. 2015;11(5):2144–55.
44. Setting up a Molecular Dynamics Simulation. 2022 2022/7/25. Available from: https://www.compchems.com/setting-up-a-molecular-dynamics-simulation/
45. Nagle JF, Tristram-Nagle S. Structure of lipid bilayers. Biochim Biophys Acta. 2000;1469(3):159–95.
46. Kopec W, Telenius J, Khandelia H. Molecular dynamics simulations of the interactions of medicinal plant extracts and drugs with lipid bilayer membranes. FEBS J. 2013;280(12):2785–805.
47. Porasso RD, Bennett WF, Oliveira-Costa SD, Lopez Cascales JJ. Study of the benzocaine transfer from aqueous solution to the interior of a biological membrane. J Phys Chem B. 2009;113(29):9988–94.
48. Orsi M, Essex JW. Permeability of drugs and hormones through a lipid bilayer: insights from dual-resolution molecular dynamics. Soft Matter. 2010;6(16):3797–808.
49. Boggara MB, Krishnamoorti R. Partitioning of nonsteroidal antiinflammatory drugs in lipid membranes: a molecular dynamics simulation study. Biophys J. 2010;98(4):586–95.
50. Lichtenberger LM, Zhou Y, Jayaraman V, Doyen JR, O'Neil RG, Dial EJ, et al. Insight into NSAID-induced membrane alterations, pathogenesis and therapeutics: characterization of interaction of NSAIDs with phosphatidylcholine. Biochim Biophys Acta. 2012;1821(7):994–1002.

51. Song Y, Guallar V, Baker NA. Molecular dynamics simulations of salicylate effects on the micro- and mesoscopic properties of a dipalmitoylphosphatidylcholine bilayer. Biochemistry. 2005;44(41):13425–38.
52. Peters GH, Dahmen-Levison U, de Meijere K, Brezesinski G, Toxvaerd S, Möhwald H, et al. Influence of surface properties of mixed monolayers on lipolytic hydrolysis. Langmuir. 2000;16(6):2779–88.
53. Osterberg T, Svensson M, Lundahl P. Chromatographic retention of drug molecules on immobilised liposomes prepared from egg phospholipids and from chemically pure phospholipids. Eur J Pharm Sci. 2001;12(4):427–39.
54. Witzke S, Duelund L, Kongsted J, Petersen M, Mouritsen OG, Khandelia H. Inclusion of terpenoid plant extracts in lipid bilayers investigated by molecular dynamics simulations. J Phys Chem B. 2010;114(48):15825–31.
55. Khandelia H, Witzke S, Mouritsen OG. Interaction of salicylate and a terpenoid plant extract with model membranes: reconciling experiments and simulations. Biophys J. 2010;99(12):3887–94.
56. Yue Z, Li C, Voth G, Swanson J. Dynamic protonation dramatically affects the membrane permeability of drug-like molecules. J Am Chem Soc. 2019;141:13421–33.
57. Orlowski A, Grzybek M, Bunker A, Pasenkiewicz-Gierula M, Vattulainen I, Mannisto PT, et al. Strong preferences of dopamine and l-dopa towards lipid head group: importance of lipid composition and implication for neurotransmitter metabolism. J Neurochem. 2012;122(4):681–90.
58. Lin F, Wang R. Hemolytic mechanism of dioscin proposed by molecular dynamics simulations. J Mol Model. 2010;16(1):107–18.
59. Carpenter TS, Kirshner DA, Lau EY, Wong SE, Nilmeier JP, Lightstone FC. A method to predict blood-brain barrier permeability of drug-like compounds using molecular dynamics simulations. Biophys J. 2014;107(3):630–41.
60. Lundborg M, Wennberg CL, Narangifard A, Lindahl E, Norlén L. Predicting drug permeability through skin using molecular dynamics simulation. J Control Release. 2018;283:269–79.
61. Wennberg C, Lundborg M, Lindahl E, Norlén L. Understanding drug skin permeation enhancers using molecular dynamics simulations. J Chem Inf Model. 2023;63(15):4900–11.
62. Bennett WFD, Hong CK, Wang Y, Tieleman DP. Antimicrobial peptide simulations and the influence of force field on the free energy for pore formation in lipid bilayers. J Chem Theory Comput. 2016;12(9):4524–33.
63. Bunker A, Róg T. Mechanistic understanding from molecular dynamics simulation in pharmaceutical research 1: drug delivery. Front Mol Biosci. 2020;7:604770.
64. Martinotti C, Ruiz-Perez L, Deplazes E, Mancera RL. Molecular dynamics simulation of small molecules interacting with biological membranes. ChemPhysChem. 2020;21(14):1486–514.
65. Róg T, Girych M, Bunker A. Mechanistic understanding from molecular dynamics in pharmaceutical research 2: lipid membrane in drug design. Pharmaceuticals (Basel). 2021;14(10):1062.
66. Kabedev A, Hossain S, Hubert M, Larsson P, Bergström CAS. Molecular dynamics simulations reveal membrane interactions for poorly water-soluble drugs: impact of bile solubilization and drug aggregation. J Pharm Sci. 2021;110(1):176–85.
67. Hossain S, Abdelgawad A. Analysis of membrane permeability due to synergistic effect of controlled shock wave and electric field application. Electromagn Biol Med. 2020;39(1):20–9.
68. Menichetti R, Kanekal KH, Bereau T. Drug–membrane permeability across chemical space. ACS Central Sci. 2019;5(2):290–8.
69. Zhang S, Thompson JP, Xia J, Bogetti AT, York F, Skillman AG, et al. Mechanistic insights into passive membrane permeability of drug-like molecules from a weighted ensemble of trajectories. J Chem Inf Model. 2022;62(8):1891–904.
70. Thai NQ, Theodorakis PE, Li MS. Fast estimation of the blood–brain barrier permeability by pulling a ligand through a lipid membrane. J Chem Inf Model. 2020;60(6):3057–67.
71. Sugita M, Sugiyama S, Fujie T, Yoshikawa Y, Yanagisawa K, Ohue M, et al. Large-scale membrane permeability prediction of cyclic peptides crossing a lipid bilayer based on enhanced sampling molecular dynamics simulations. J Chem Inf Model. 2021;61(7):3681–95.

72. Marrink SJ, Berendsen HJC. Permeation process of small molecules across lipid membranes studied by molecular dynamics simulations. J Phys Chem. 1996;100(41):16729–38.
73. Alsop RJ, Dhaliwal A, Rheinstädter MC. Curcumin protects membranes through a carpet or insertion model depending on hydration. Langmuir. 2017;33(34):8516–24.
74. Dutta A, Vreeken J, Ghiringhelli LM, Bereau T. Data-driven equation for drug-membrane permeability across drugs and membranes. J Chem Phys. 2021;154(24):244114.
75. Dutta P, Jain D, Gupta R, Rai B. Deep learning models for the estimation of free energy of permeation of small molecules across lipid membranes. Digital Discov. 2023;2(1):189–201.
76. Malde AK, Zuo L, Breeze M, Stroet M, Poger D, Nair PC, et al. An Automated Force Field Topology Builder (ATB) and repository: version 1.0. J Chem Theory Comput. 2011;7(12):4026–37.
77. Jo S, Kim T, Iyer VG, Im W. CHARMM-GUI: a web-based graphical user interface for CHARMM. J Comput Chem. 2008;29(11):1859–65.

Chapter 16
Accomplishment of Toxicity Prediction by CADD Tools

Niranjan Kaushik and Satendra Kumar

Contents

N. Kaushik (✉)
Department of Pharmacy, School of Medical & Allied Sciences, Galgotias University, Greater Noida, Uttar Pradesh, India
e-mail: niranjan.kaushik@galgotiasuniversity.edu.in

S. Kumar
Department of Pharmacy, School of Medical & Allied Sciences, Galgotias University, Greater Noida, Uttar Pradesh, India

SRM Modinagar College of Pharmacy, SRM Institute of Science and Technology, Delhi NCR Campus, Ghaziabad, Uttar Pradesh, India

S. N. K. Rao G., R. R. Alavala (eds.), *Applications of Computational Tools in Drug Design and Development*, https://doi.org/10.1007/978-981-96-4154-3_16

Abstract In the pursuit of transforming drug discovery and guaranteeing the progress of safer, more potent therapeutic options, the incorporation of Computer-Aided Drug Design (CADD) tools has emerged as a fundamental strategy. This comprehensive overview underscores the multifaceted role of CADD in complementing toxicity prediction, refining drug design, and aligning therapeutic innovations with patient-centric considerations. From the foundational steps of target identification and validation, which anchor drug development in precise disease pathology and scientific evidence, to the nuanced optimization of drug selectivity and specificity, CADD stands at the forefront of modern pharmacology. It not only streamlines the path to discovery but also enhances the safety and efficacy profiles of candidate compounds through meticulous ADME prediction and toxicological assessments. The strategic employment of CADD in early toxicity prediction and risk mitigation underscores a proactive approach to drug safety, minimizing adverse effects and optimizing therapeutic outcomes. Furthermore, the competitive landscape of drug development, particularly In the context of intellectual property rights, highlights importance of securing patents to protect innovation and investment. This summarizes a comprehensive approach that not only ensures uniqueness but also enhances the profitability of launching new drugs, thereby boosting their commercial viability. Central to evolution of drug design is increasing incorporation of end-user needs and preferences, signifying a shift towards patient-centred development. By leveraging CADD to enhance dosage forms, administration routes, and regimen simplicity, drug developers are better positioned to improve patient compliance and therapeutic success. The early integration of patient-centered design principles marks a paradigm shift, guaranteeing that pills are not only scientifically innovative but also practically aligned with the lifestyles and needs of those they aim to serve.

Keywords Computer-aided drug design · Toxicity prediction · Precision medicine · Drug discovery · Selectivity · Specificity · Patient-centered design · Intellectual property · Therapeutic outcomes · Drug safety

16.1 Introduction

CADD is fundamental process that uses a different of computer technologies to discover and create a prospective drug development lead [1]. CADD drug discovery and advancing the growth of safer, more effective therapeutic solutions, the integration of Computational Chemistry, molecular modeling, molecular design, and rational drug design has become a cornerstone strategy. Its access to vast amounts of data does not ensure the production of usable predictive models. To anticipate therapeutic success and adverse effects, approaches must be developed that systematically handle high-volume, complex, and sparse data sources [2]. New experiments employing classical drug discovery and advance the development of safer, more effective therapeutic options, the integration of Computer-Aided Drug Design (CADD), artificial intelligence (AI), machine learning (ML), and breakthrough deep learning technologies has emerged as a fundamental strategy [3]. Discovery and innovative prospective therapeutics for treatment of neurodegenerative disorders are vital for increasing both longevity and quality of life, particularly in the older population. However, this endeavour is fraught with difficulty because of ineffectiveness of standard therapies, which frequently fail to prevent or reduce the development of neurodegeneration [4]. Neurodegeneration is a process that includes a variety of neuropathological disorders as well as cognitive impairments such as memory and learning deficiencies. Neurodegenerative diseases cause neuronal loss and contribute to brain aging, ultimately leading to death. Conditions affecting the brain, including Alzheimer's disease, Parkinson's disease, Huntington's disease, and amyotrophic lateral sclerosis, contribute to over a quarter of worldwide fatalities and disabilities [5]. Novel therapeutic interventions to address & escalate burden of NDs on individuals and society as a whole. CADD is a tool computer programs that use algorithms and simulations to predict the characteristics and behaviors of molecules [6]. These approaches for toxicity prediction are coarsely divided into two groups: drug discovery and enhance therapeutic efficacy, two primary methodologies are employed: ligand-based and structure-based approaches. In the ligand-based approach, data pertaining to the chemical structure and properties of known ligands are utilized to predict the toxicity and efficacy of potential drugs. On the other hand, structure-based systems trust on the three-D structure of target protein to guide the design and optimization of drug candidates. By leveraging computational chemistry and molecular modeling techniques, structure-based approaches offer insights into the interactions between drugs and their target proteins, facilitating the prediction of toxicity and refining drug design. Conversely, ligand-based approaches utilize diverse sets of data, such as chemical fingerprints QSAR models, to infer the toxicity and efficacy of analogues based on their similarity to known ligands or their activity profiles use the chemical properties of known toxic or non-toxic compounds to generate such predictions. The benefits of using CADD technologies in toxicity prediction are numerous [7]. For starters, they can examine large databases of compounds far faster than previous approaches. It is techniques can predict the toxicity of compounds that have yet to be synthesised, reducing the need on animal testing [8]. CADD has techniques provide essential insights into the mechanics of toxicology, aiding in the optimization of drug design. it has few therapies available for neurodegenerative disorders & many patients do not have access to really effective medications [9]. The scarcity of approved medications for neurodegenerative diseases (NDs) is compounded by the significant rate of principal compound failures during clinical trials. Since ofcomplexity of NDs and their varied pathogenesis, traditional drug discovery strategies struggle to find innovative treatment alternatives [10]. CADD & CAMD has caused a paradigm change in drug development in recent years. CADD, which employs computational methodologies, has appeared as appreciated tool for speeding up drug development process. It has been canned easily discover possible lead compounds and accelerate the development of novel therapeutics for a wide range of illnesses, including NDs, through leveraging computational control of computers. CADD offers several benefits compared to prior methods, such as the ability to swiftly accommodate extensive compound libraries in a fraction oftime. CADD uses a variety of computational techniques to find promising lead compounds, raising hopes for the creation of much-needed ND therapies. Drug development uses automation to screen large numbers of compounds in search of those that elicit the required biological reaction [11]. This technique has the benefit of requiring little compound design or prior expertise, and screening technologies for huge libraries have grown more efficient. However, conventional High-Throughput Screening (HTS) frequently produces numerous hit compounds. Develop content, a portion of which can evolve into lead compounds and, ultimately, groundbreaking therapies [12, 13]. The success rate of High-Throughput Screening (HTS) is frequently very low [14]. Because of its poor hit rate, High-Throughput Screening (HTS) is only used in research programmes that can screen enormous chemical libraries. However, within the last decade, CADD has reemerged as a valuable tool for significantly reducing the number of compounds needed for screening while maintaining the same level of lead compound discovery [15, 16]. CADD allows many molecules that are projected to be inactive to be skipped, while those that are predicted to be active to be prioritised. Number of mixtures tested lowers the expense and burden of a complete HTS screen while preserving lead discovery [17]. Figure 16.1 showed from the initial choice of targets to regulatory clearance, the process of drug discovery and development entails a comprehensive journey.

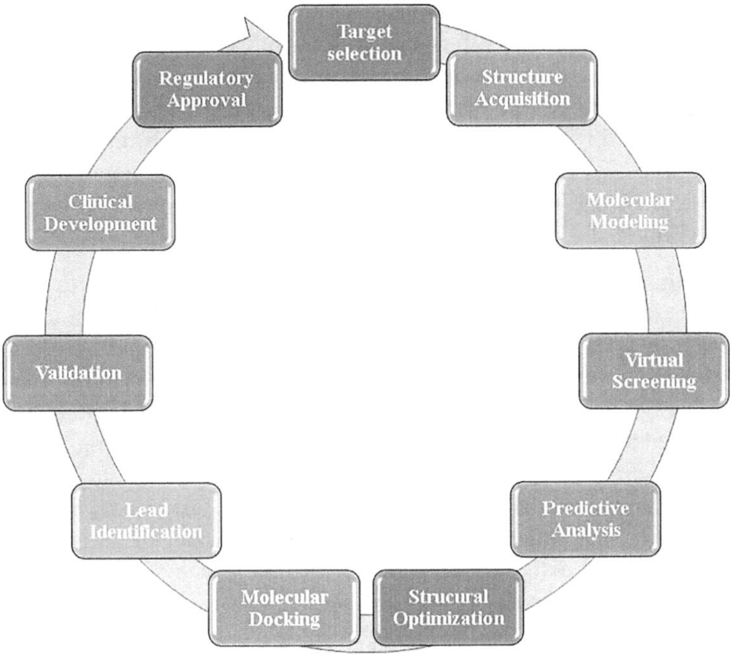

Fig. 16.1 From the initial choice of targets to regulatory clearance, the process of drug discovery and development entails a comprehensive journey

16.2 Advantages of Computer-Aided Drug Design (CADD)

The landscape of drug research has been transformed by the advent of computer-aided drug design (CADD) techniques. These tools assist researchers in identifying potential compounds for further exploration by employing computational methods to evaluate the toxicity and efficacy of new therapeutic candidates. Molecular docking represents one ofmethodologies offered by CADD, molecular dynamics simulations, and quantitative structure-activity relationship (QSAR) modelling. This is exceptionally effective at predicting toxicity of novel drug candidates, making them a valuable tool in drug development process. it has the advantage of accurately forecasting the toxicity of novel drug candidates, which can assist to minimise the number of animal trials necessary for drug development. CADD technologies assist researchers in finding Toxicities can be recognised early in the medicine development process, reducing time and expense. CADD technology may also be used to modify the structure of therapeutic candidates, lowering toxicity and resulting in safer, more effective pharmaceuticals.

Traditional high-throughput screening (HTS) assays require substantial time and financial investment for development and validation before they can be effectively utilized. In contrast, Computer-Aided Drug Design (CADD) procedures have minimal preparation time, enabling researchers to conduct CADD investigations concurrently while traditional HTS tests are being prepared. This simultaneous utilization of both technologies offers an additional advantage to CADD in drug development endeavours. For instance, scientists at Pharmacia (now part of Pfizer) utilized CADD methods to search for inhibitors of tyrosine phosphatase-1B, an enzyme associated with diabetes. Their virtual screening identified 365 compounds exhibiting effective inhibition, resulting in a hit rate of approximately 35%. Meanwhile, the same team conducted a conventional HTS against the same target showed in Fig. 16.2.

(a) **Low Hit Rate in Traditional Screening:** This process involves testing large libraries of compounds to identify those with desired biological activity, often aiming to find potential drug candidates. High-throughput screening (HTS) revolutionizes drug discovery and chemical biology by enabling rapid testing ofbiological or biochemical activity of a vast number of compounds. Utilizing automated systems, HTS runs experiments in parallel, allowing for the screening of hundreds to millions of chemicals against a given biological target. The statement mentions that out of 400,000 compounds tested, only 81 showed inhibition [18]. This underscores the vast scale of compound libraries commonly screened in drug discovery endeavors. Screening such a large number of compounds requires significant time, resources, and infrastructure. With only 81

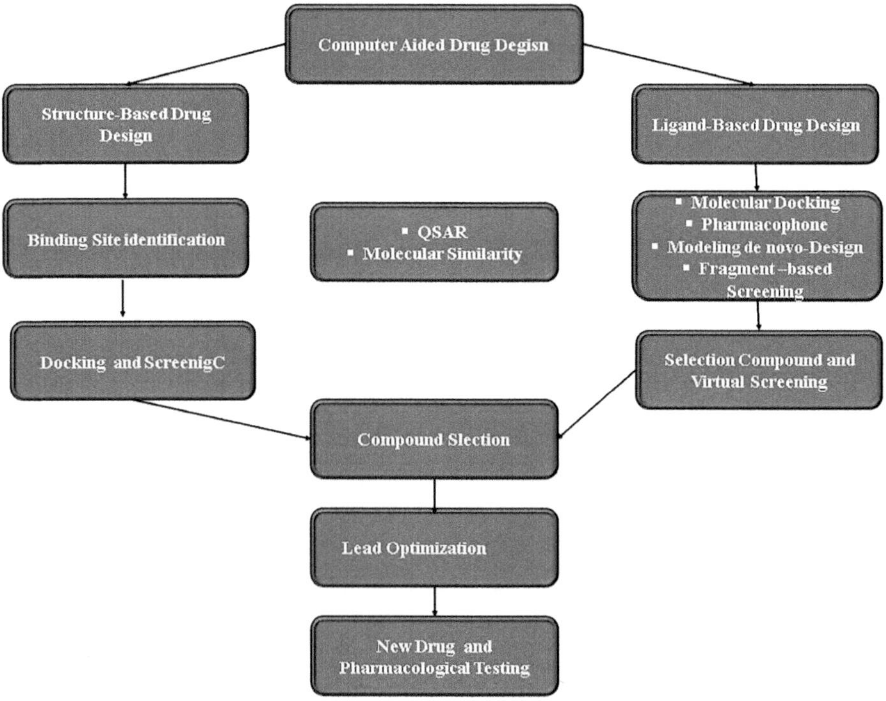

Fig. 16.2 Effectiveness of CADD in drug discovery

compounds showing inhibition out of 400,000 tested, the hit rate is calculated to be 0.021%. This low hit rate illustrates the challenge of identifying potential drug candidates amidst a vast sea of chemical compounds [19, 20].

(b) **Prevalence of Target**: Screening programs play a crucial role in public health by identifying individuals at risk of certain diseases or conditions before symptoms manifest. However, the effectiveness of screening trials container be hindered by factors such as the occurrence of target condition within the population. The prevalence oftarget condition influences hit rates in screening, particularly focusing on cases where the condition is rare within the population. The Effect of Target Prevalence on Hit Rates: The hit rate of a showing trial has defined as proportion true positives amongst all individuals tested. When the target condition is prevalent in the population, the likelihood of encountering individuals with the condition during screening is higher, leading to a higher hit rate. Conversely, in populations where the target condition is rare, the hit rate naturally decreases due to fewer positive cases being detected. This phenomenon can be illustrated using medical examples. For instance, consider a showing trial for a rare genetic disorder in a population where the occurrence of disorder is only 1 in 10,000 individuals. Even if the test has high sensitivity and specificity, the low prevalence means that only a small proportion of those tested will have the disorder, resulting in a low hit rate. Strategies to Address Low Prevalence: Addressing the challenge of low prevalence in screening requires careful consideration and tailored approaches. Several strategies can be employed to improve detection accuracy and maximize the utility of screening tests in populations with low prevalence of target condition

(c) **Sensitivity and Specificity:** Screening tests are crucial tools in healthcare for early detection and prevention of diseases. However, they often involve a trade-off between sensitivity and specificity, which can suggestively consequence their effectiveness. Sensitivity denotes a test's capacity to accurately detect individuals afflicted with the condition, while specificity pertains to its ability to appropriately identify Individuals lacking the ailment. A high-sensitivity test is excellent at finding true positives, which means It will accurately identify most individuals affected by the illness. However, It might also generate additional false positives, inaccurately labelling certain individuals as having the condition. Conversely, test with high specificity is adept at correctly identifying true negatives, meaning it accurately identifies most individuals without the condition. However, it may also produce more false negatives, failing to detect some individuals who actually have the condition [21]. The balance between sensitivity and specificity is crucial because it directly influences the test's ability to accurately classify individuals and determine their true disease status. If a test is overly sensitive, it may lead to unnecessary follow-up tests, treatments, and Excessive worry for individuals mistakenly identified as having the condition [18]. Conversely, if a test is overly specific, it may miss detecting individuals who have the condition, delaying diagnosis and potentially leading to adverse health. Several factors contribute to the sensitivity and specificity of a screening test, including the test's design, the population being screened, and the prevalence of condition. Moreover, the cutoff values used to define positivity or negativity can also impact sensitivity and specificity. Modifying these cutoff values can be beneficial optimize balance between sensitivity & specificity based on specific goals of screening program. For example, in breast cancer detection through mammography, a highly sensitive test is desirable to ensure early detection of tumors. However, this may result in a higher rate of false positives, leading to unnecessary biopsies and additional imaging tests. Conversely, a test with greater specificity could decrease the occurrence of false positives. But could miss some early-stage cancers, resulting in false negatives. Therefore, when designing and implementing screening programs, healthcare professionals must carefully consider trade-offs between sensitivity and specificity to maximize test's overall accuracy and effectiveness. Finding a balance that reduces both false positives and false negatives is crucial, all while achieving the primary objective of early disease detection and prevention. Table 16.1 [20].

16.2.1 Human Error

Human error remains a significant challenge in traditional screening methods, particularly those that heavily rely on human operators to interpret results or identify potential threats. This reliance introduces vulnerabilities that can compromise the effectiveness of security measures. Several factors contribute to human error. Prolonged periods of vigilance can lead to fatigue among human operators, diminishing their ability to detect threats accurately [22]. Fatigue may arise due to prolonged work shifts and repetitive tasks or inadequate rest periods between shifts. Fatigue can impair cognitive functions and decision-making capabilities, increasing the likelihood of errors in threat detection [23]. Human operators may become distracted by various factors, such as personal concerns, environmental stimuli, or competing tasks. Distraction can distract their care away from monitoring screens or evaluating potential threats, leading to missed detections or delayed responses. In high-pressure environments, distractions can be particularly detrimental, impairing situational awareness and compromising security. Cognitive biases can influence

Table 16.1 Molecular modelling techniques

Technique	Description
Molecular docking	Predicts preferred alignment of one molecule (ligand) when bound to target (receptor)
QSAR	Demonstrates the connection between biological activity and chemical structure
Molecular dynamics	Simulates the motion of atoms and molecules over time to study their behavior
Pharmacophore modelling	Identifies the essential features required for molecular recognition and interaction.

the judgment and decision-making of human operators during threat assessment. These biases may stalk from preconceived notions, stereotypes, or contextual factors that distort perception and evaluation. For example, Confirmation bias, where operators understand material in a way that checks their existing principles, can lead to overlooking or downplaying indicators of potential threats. Similarly, stereotype bias can influence assessments based on demographic or behavioral characteristics, potentially resulting in profiling or discrimination. Discrepancies in training and experience levels among human operators can impact their effectiveness in threat detection. Inadequate training may leave operators ill-equipped to recognize subtle cues or emerging threats, while inexperienced personnel may lack the proficiency needed to interpret complex situations accurately. Furthermore, turnover rates and staff shortages can compromise continuity in training efforts, exacerbating skill disparities and increasing the likelihood of errors [24]. Traditional screening methods often rely on human operators to interpret results or identify potential threats. Human error, including fatigue, distraction, or bias, can contribute to missed detections or false alarms, thereby reducing the hit rate.

16.2.2 Resource Allocation Constraints

Limited resources, such as time, manpower, or funding, impose constraints on the thoroughness and frequency of security screenings. These constraints can compromise the efficacy of screening efforts, as they may hinder the implementation of comprehensive or frequent screening protocols necessary to detect targets effectively [25]. For instance, staffing shortages or budget constraints may result in reduced personnel availability or operational hours, leading to gaps in coverage and increased vulnerability to security threats. Similarly, competing priorities for resource allocation, such as addressing operational demands or infrastructure maintenance, can divert attention and resources away from screening activities, undermining their effectiveness [26]. As threats evolve and resource limitations persist, traditional screening methods face significant challenges in maintaining effectiveness. Two key factors contributing to this are the evolving nature of threats and constraints in resource allocation. Continually innovate new techniques, substances, or strategies to circumvent existing screening methods [27]. This dynamic landscape poses a significant challenge to security measures, as traditional screening protocols may become outdated or insufficient to detect emerging threats effectively [28]. For example, advancements in concealment methods, such as sophisticated disguises or concealment materials, can render conventional screening technologies less effective in detecting prohibited items or dangerous substances. Similarly, the proliferation of novel weapons or tactics by hostile actors demands constant adaptation and enhancement of screening capabilities to mitigate evolving risks [29]. The comparison between these two approaches highlights the strengths and capabilities of CADD in drug discovery. While traditional HTS methods are effective in identifying lead compounds, CADD offers the advantage of rapidly screening large chemical libraries in silico, thereby accelerating drug discovery process & potentially leading to identification novel therapeutics with improved efficacy and selectivity. Virtual screening of compounds yielded 87 results for the Biogen Idec group, the best of which had the same structure as the lead molecule found by Eli Lilly's conventional High-Throughput Screening (HTS). This illustrates how Computer-Aided Drug Design (CADD) may achieve similar outcomes to more conventional approaches at a lower cost and labour. The same lead compound as a full-scale HTS might be produced in this situation using CADD, highlighting its potential for streamlining drug discovery processes [30]. Mankind has continually faced threats from various microorganisms such as bacteria, viruses, protozoa, fungi, etc., necessitating the development of drugs with minimal adverse effects and potent pharmacological activities. The process of inventing a new drug for a specific disease condition typically incurs substantial costs, estimated at around 2–3 billion dollars, and consumes considerable time, averaging between 12 to 15 years. To mitigate the duration and expenses associated with drug discovery, a rational drug design strategy has been employed for several decades [31]. Finding new drug candidates to treat unmet medical needs is the goal ofdifficult and resource-intensive process of drug development. It has traditionally taken a proportion of experimental effort; this could incur significant costs and consume valuable time. But with advances in computational techniques, especially at the pre-clinical stage of drug development, there's been a move towards using computational tools to speed up and improve the discovery process [32]. Computer-aided drug design (CADD) encompasses a suite of computational techniques employed to discover, design, and assess active compounds and potential drug candidates possessing specific biological attributes [33]. Complementing toxicity prediction with Computer-Aided Drug Design (CADD) tools represents a significant advancement in the pharmaceutical industry's quest to develop safer and more effective drugs [34]. This integration offers a sophisticated approach to undeerstanding the potential toxicological profiles of new chemical entities long before they enter costly clinical trials [35–38].

16.2.3 Revolutionizing Drug Discovery

Traditional drug discovery processes are time-consuming and resource-intensive. To address these challenges, pharmaceutical researchers have increasingly turned to computational approaches like CADD. By integrating principles from chemistry, biology, and computer science, CADD facilitates the identification and optimization of potential drug candidates [39]. CADD utilizes various computational techniques such as molecular modelling, molecular dynamics simulations, and structure-based drug design to simulate drug-receptor interactions [40]. These methods allow researchers to predict the binding affinity and selectivity of drug molecules towards target receptors, providing insights into their therapeutic effects and safety profiles. CADD tools stand at the forefront of drug discovery, offering a sophisticated platform for simulating the complex interactions between drug molecules and biological receptors [40]. This section introduces the fundamental concepts of CADD, including its historical development, the computational models used, and its significance in identifying and optimizing drug candidates [41].

(a) **Modeling Drug-Receptor Interaction:** CADD revolutionizes drug discovery by simulating and analysing the interactions between drug molecules and their target receptors. This section explores the various types of interactions that CADD models can simulate, including hydrogen bonding, hydrophobic effects, and electrostatic interactions, and elucidates how these simulations aid in predicting the affinity and specificity of drug molecules for their targets, crucial factors in designing effective drugs [40]. Hydrogen bonds are pivotal in molecular recognition and interactions between drugs and receptors. In computer-aided drug design (CADD), simulations can predict the strength and stability of interactions by modeling hydrogen bonding patterns between a drug molecule and its target receptor [41]. Hydrogen bonds form when a hydrogen atom, which is bonded covalently to an electronegative atom like oxygen or nitrogen, interacts with another electronegative atom. By examining the structure and energy dynamics of hydrogen bonds, computer-aided drug design (CADD) models can predict the probability of a drug molecule forming advantageous interactions with particular residues in the binding site of a receptor. Hydrophobic interactions arise from the tendency of nonpolar molecules or regions to aggregate in aqueous environments to minimize contact with water molecules. In drug-receptor interactions, hydrophobic effects play crucial role in stabilizing binding of lipophilic drug particles to hydrophobic pockets within the receptor. CADD models simulate the hydrophobicity profiles of drug molecules and receptors to identify complementary binding interfaces and predict the strength of hydrophobic interactions. Understanding the hydrophobic properties of drug molecules is essential for optimizing their binding affinity and selectivity for specific receptor targets [42]. Electrostatic interactions result from the attraction or repulsion between charged particles and significantly influence compulsory attraction and specificity drug-receptor connections. In CADD, electrostatic interactions between charged or polar groups in drug molecules and receptor residues are modeled to assess their contribution to binding energetics. Coulombic interactions between absolutely electric amino acid residues (e.g., arginine) & harmfully electric functional groups in drugs (e.g., carboxylate, phosphate) can enhance binding affinity, while electrostatic repulsions may hinder binding. By quantifying electrostatic contributions to binding energy [39].

(b) **Technological Advancements Driving CADD:** CADD relies on a suite advanced technologies & algorithms to simulate and analyze molecular interactions between drug candidates and target receptors [43]. These cutting-edge methodologies revolutionize drug discovery process by smoothing the identification and optimization of potential therapeutics with enhanced precision and efficiency [44]. Molecular docking is cornerstone performance in CADD that predicts the binding mode and affinity of small molecules to target proteins or receptors. By simulating the complementary fit between ligands and binding sites, docking algorithms rank & interactions with key residues. Recent advancements in docking algorithms, such as improved scoring functions and enhanced sampling strategies, have significantly enhanced the accuracy and reliability of docking predictions. Pharmacophore modeling identifies key structural features or chemical properties essential for ligand binding and biological activity [45]. By mapping common pharmacophoric elements from a set of active ligands, pharmacophore models guide the design and optimization of novel drug candidates with improved potency and selectivity [46]. Advances in pharmacophore modeling techniques, such as ensemble-based approaches and dynamic pharmacophore generation, enable the exploration of diverse chemical space and the identification of novel lead compounds. QM/MM simulations combine quantum mechanics (QM) and molecular mechanics (MM) methods to describe the electronic structure and energetics of biomolecular systems accurately [47]. By treating the active site region quantum mechanically and the surrounding environment classically, QM/MM simulations provide insights into chemical reactions and binding processes at atomic resolution. Recent developments in QM/MM methodologies, such as hybrid density functional theory (DFT) and enhanced sampling techniques, enable the exploration of complex enzyme mechanisms and the rational design of enzyme inhibitors. Advancements in computational power and machine learning have propelled CADD to new heights by enabling more accurate predictions and the exploration of previously intractable drug targets [48].

(c) **Impact on Drug Discovery and Development:** CADD has significantly influenced the pharmaceutical sector by providing a streamlined and economical route to drug discovery. Its direct contributions to the creation of novel therapeutics underscore its ability to expedite development timelines, improve drug effectiveness, and mitigate unwanted side effects [49]. CADD has transformed the pharmaceutical industry by providing more effectual and cost-effective way to drug discovery. Through the utilization of computational techniques and sophisticated algorithms, CADD empowers researchers to accelerate the discovery, refinement, and validation of potential therapeutic agents. CADD has directly contributed to growth of novel drugs, showcasing its role in reducing development times, enhancing drug efficacy, and minimizing adverse effects. CADD is a design of Tamiflu by facilitating the optimization of its active compound, oseltamivir carboxylate. Molecular docking and molecular dynamics simulations were employed to discover and enhance lead compounds that exhibit strong binding attraction for the neuraminidase enzyme, a crucial target complex in replication ofinfluenza virus. Tamiflu became one offirst-line treatments for influenza globally, demonstrating the significant impact of CADD in accelerating the development of antiviral therapies [50].

(d) **Identification and Validation of Target:** In pharmaceutical research and development, drug design stands as a pivotal endeavor, blending the complexities of biology with the precision of chemistry to create compounds capable of combating diseases [51]. The journey from conceptualization to clinical application is fraught with challenges and requires meticulous consideration at every step. This section delves into the fundamental considerations that guide the path of drug design, ensuring that the pursuit of therapeutic innovation is both effective and efficient [52].

(e) **Identification of Biological Targets:** The foundation of any drug design initiative lies in identification of life target, typically a gene related with an illness process. This initial step is paramount, as the selected target must be unequivocally linked to the pathological condition it aims to treat [53]. Target validation is essential, requiring rigorous demonstration that manipulating the target will alter the disease course in a beneficial manner [54]. The cornerstone of any drug design effort lies in identifying a precise life target, such as protein or gene, that plays a crucial role in disease. This target serves as the focal point for designing therapeutic interventions [51]. This step is crucial; the chosen target must be unequivocally linked to the pathological condition it aims to treat. Validation involves demonstrating that manipulating the target will alter the disease course in a beneficial manner drug design, identifying the right biological target is paramount. A biological target is typically a specific molecule, such as a protein or gene, that is involved in the disease process and is amenable to therapeutic modulation [51].

(f) **Identifying the target represents the first phase in process of medication design:** The initial stage of drug design starts with recognizing the target, laying the groundwork for discovering and developing novel therapeutic compounds [55]. This crucial phase involves recognizing a biological thing—such as a protein, enzyme, receptor that plays key role in a disease's pathology [53]. The rationale is that by modulating the activity or function of this target, one can influence the disease process in a way that leads to therapeutic benefits. Understanding why target identification is pivotal reveals the intricacies and challenges of drug discovery [56]. Target identification usually commences by comprehensively grasping the molecular mechanisms at play in the disease. This may involve studying the pathways, signalling cascades, and interactions between biomolecules that contribute to the disease phenotype [51].

(g) **Significance of identifying the target in drug design:** Identifying the target is a pivotal starting point in the drug discovery journey, serving as the cornerstone for eventual drug development. This process entails pinpointing particular biological targets—like proteins, enzymes, receptors, or nucleic acids—that hold significant sway in disease development. Underscoring its paramount importance, target identification profoundly shapes the efficiency and efficacy of subsequent drug discovery endeavors.

16.3 Lead Identification and Optimization

Once an approved target is recognized, the quest for a lead compound commences. This stage frequently utilizes high-throughput screening to assess the interactions between thousands of molecules and the target [57]. Upon identifying promising candidates, optimization is necessary to enhance their efficacy, selectivity, and pharmacokinetic properties. This iterative process refines these molecules into viable drug candidates that are potent, less toxic, and more drug-like in their behavior within the body [58]. His multifaceted process involves the systematic screening of vast compound libraries to identify promising candidates capable of interacting with validated molecular targets [59, 60]. Through high-throughput screening and subsequent hit confirmation, researchers sift through numerous compounds to isolate those demonstrating desirable activity against the target of interest [61]. Following hit identification, the iterative process of lead optimization ensues, wherein chemical modifications are made to enhance the potency, selectivity, and pharmacokinetic properties of lead compounds [62, 63]. Structure-activity relationship (SAR) studies play a critical role during this phase, guiding the refinement of chemical structures to optimize drug-like characteristics. Additionally, thorough assessment of pharmacokinetics and toxicity profiles ensures the selection of lead candidates with favorable drug properties and reduced risk of adverse effects [64].

16.3.1 Crucial Role of Target Identification in Disease Pathology

In the dynamic landscape of drug discovery and development, target identification emerges as a critical early phase, serving as the linchpin that anchors subsequent efforts towards therapeutic innovation. This pivotal step revolves around the meticulous selection of a biological molecule or pathway intricately entwined in the pathology of a disease [65–67]. Consider the quest to combat cancer, a multifaceted disease characterized by uncontrolled cell proliferation and tumor growth. Researchers embarking on the journey of target identification delve deep into the intricate molecular mechanisms underpinning cancer initiation, progression, and metastasis [67]. Amidst this complexity, they strive to pinpoint key players—proteins, enzymes, receptors, or segments of nucleic acid (DNA/RNA)—that orchestrate the aberrant cellular behaviors driving tumorigenesis. In oncogene addiction, cancer cells exhibit dependence on the continuous activation of specific oncogenic signaling pathways to sustain their survival and proliferation. Through meticulous molecular profiling and functional studies, researchers identify oncogenic driver mutations, such as those in the EGFR or BRAF genes, which serve as prime targets for therapeutic intervention [68]. The chosen target represents a pivotal nexus in the disease's mechanism, making it ripe for modulation through pharmacological means. Fortified with this knowledge, embark on a journey of drug design and development aimed at harnessing the therapeutic potential of these targets. Strategies may encompass minor molecule inhibitors, monoclonal antibodies, gene therapies, or immunotherapeutic agents, tailored to precisely disrupt or augment the action of target with exquisite specificity [69]. The ultimate goal of target identification transcends mere molecular elucidation—it is the quest for transformative therapies that tip the scales in favor of patients, offering hope for enhanced and enhanced quality of life [66].

16.3.2 Computational Identification of Off-Target Interactions

These progressions enable the discovery of fresh molecules targeting diverse therapeutic aims. Among these breakthroughs are computational approaches, which have become crucial weapons in drug discovery cache. The computational method for drug development is often categorized into two main paradigms: protein structure-based (target-centric) and ligand-based (ligand-centric) approaches. The former leverages our understanding of proteins' three-dimensional structure and the intricate nature of their binding sites [70]. Using this knowledge, scientists may create compounds that interact effectively with the target protein, maximising therapeutic efficacy while minimising off-target consequences. In contrast, ligand-based techniques concentrate on changes cutting-edge biological responses caused by Molecules with diverse chemical structures serve as ligands [71]. Pharmacophore modeling and quantitative structure-activity relationship (QSAR) analysis are employed to detect patterns in interactions between ligands and receptors, facilitating the creation of new drugs with enhanced pharmacologic characteristics. At the core of drug-receptor binding lies the intricate interplay of numerous thermodynamic factors [72]. The interaction relies on intermolecular recognition, which is dictated by a combination of geometric and physicochemical matching, along with steric considerations [73]. Computational tools are critical in clarifying these aspects, working in tandem with in-vitro & in-vivo experimental procedures to evaluate the possibility for intermolecular recognition [74, 75]. These principles serve as the foundation for developing unique therapeutic methods. Researchers get crucial insights into prospective therapeutic targets by meticulously examining the protein's three-D architecture and spatial arrangement its binding sites.

The ligand-centric approach, on the other hand, focuses on the various chemical structures of ligands and their interactions with living systemspharmacophore modeling and quantitative structure-activity relationship (QSAR) analysis enable scientists to discern recurring patterns in how ligands interact with receptors. By understanding these patterns, researchers can design molecules tailored to fit these interactions more effectively, resulting in the development of customized drugs with enhanced pharmacological properties [42, 76, 77]. At the heart of drug effectiveness and safety is first occasion of drug-receptor binding. This interaction, which is influenced by a quantity of thermodynamic parameters, dictates the drug's affinity to its target [42, 78, 79].

16.3.3 Exploring Protein Structure and Ligand-Centric Methods

Small molecule (SM) ligand libraries to target the three-dimensional (3D) pockets of protein structures is fundamental for identifying and refining leads in drug discovery. This method entails screening SM libraries against protein structures to evaluate their 3-D & chemical suitability [80]. Automated docking software such as Glide and AutoDock are widely utilized for calculating the compatibility between protein drug binding sites and chemical compounds. For example, in a study [81] computational process where natural products e-viniferin and meranzin were virtually docked on series of protein structures. Through this docking process, potential Protein targets such as PDE4, COX1, COX2, and PPARγ were recognized [82, 83] utilized docking to predict targets for a novel chemotype derived from combinatorial chemistry, identifying PLA2 as a top target [84].Furthermore, [83] docking procedures forecast potential targets relevant to toxicity for clinical candidates. Impressively, they achieved an 83% accuracy rate in predicting empirically known side effects and targets related with toxicity [46]. However, despite its utility, challenges accompany this approach. Computational docking is computationally intensive, leading to a high computational burden [72]. Docking method generates numerous potential targets, about of these may be false positives, complicating the identification of true targets. Additionally, linking predicted targets to toxicity can be complex, and verifying all predicted targets experimentally presents logistical difficulties [85, 86].

16.4 Computational Methods for Predicting Off-Target Interactions

Drug effectiveness and safety are fundamentally dependent on the initial event of drug-receptor binding. This interaction, mediated by a variety of thermodynamic parameters, defines the drug's affinity for its target. Among these parameters, intermolecular recognition appears as a key determinant, impacted by geometric and physicochemical complementarity, Additionally, steric constraints [87, 88]. To investigate the possibility for intermolecular recognition, use variety of computational methodologies as well as in-vitro & in-vivo experimental approaches. These tools shed light on the complicated interplay between pharmacological molecules and their biological targets, opening the path for rational design of therapeutic medicines with increased efficacy and lower toxicity [89, 90]. Swiss Target Prediction, CSNAP, Super Pred, DINES, STITCH, SEA LINCS, Poly pharma, Tar Pred. Other modelling components like Symmetry, Chemo-targets, SEA, and CLARITY are also used. AbbVie employs internal tools such as APEX, ANTELOPE, and 3Decision for its off-target safety assessment strategy. These methods represent common cheminformatics and target-based prediction approaches. However, cheminformatics methods are affected by the chemical diversity ofdatabase's active molecules [89, 90]. In this comparison, cheminformatics approaches are like a large net spread over the forest, quickly collecting primary and secondary targets based on their chemical makeup's closeness to known targets in a curated database. Despite its quickness, method may mistakenly capture non-targets that have a remarkable similarity to the desired targets, causing false positives. Targeted prediction algorithms, Alternatively, behave more like precision-guided arrows. They rely on precise protein structure mapping, which targets specific binding sites in a manner similar to homing in on certain branches within the forest. These arrows attempt to strike the bullseye of accurate off-target identification for new chemicals by assuming that structurally identical binding sites share similar molecular functions. However, like with pointing an arrow, a clear [91] (Fig. 16.3).

(a) **Disease Research:** This process involves a multifaceted approach encompassing the identification of disease-associated genes, elucidation of key cellular pathways, and deciphering the aberrant molecular mechanisms driving the disease phenotype [92]. Reflecting on Alzheimer's disease (AD) serves as a prime example. AD is a multifaceted neurodegenerative condition marked by a gradual decline in cognitive function and the deterioration of neurons. Scientists venturing into the realm of AD pathology delve into the complex molecular terrain of brain, with the goal of deciphering the molecular mechanisms that

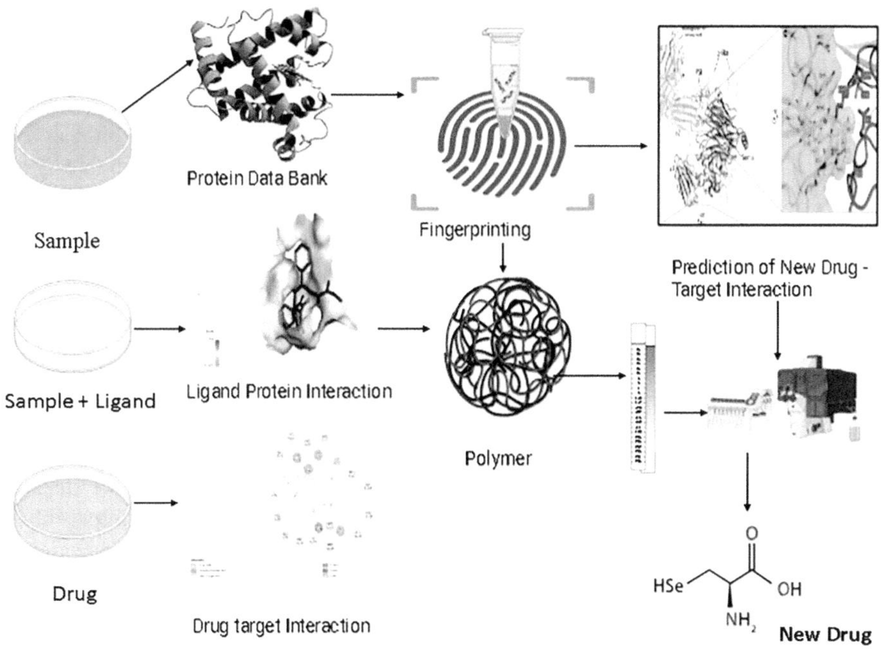

Fig. 16.3 Auto-docking as a computational tool for protein-ligand interaction analysis and detection of potential drug

trigger and advance the disease [93]. Alzheimer's disease (AD) investigation lies the quest to pinpoint genes implicated in susceptibility to and progression ofdisease. Through genome-wide association studies (GWAS) and functional genomics techniques, researchers have unveiled a plethora of genetic variations linked to AD risk. These findings have illuminated critical molecular actors such as amyloid precursor protein (APP), presenilin 1 (PSEN1), and apolipoprotein E (APOE). Concurrently, aberrant phosphorylation and aggregation of tau proteins disrupt neuronal microtubule function, contributing to neurofibrillary tangle formation and synaptic dysfunction [94]. Through meticulous experimentation and systems biology approaches, researchers aim to delineate the intricate interplay between these pathological processes, elucidating the molecular events that drive neurodegeneration in AD. Insights gleaned from preclinical models and clinical studies provide valuable clues about potential therapeutic targets and intervention strategies aimed at halting or slowing disease progression [95].

(b) **Genetic and Genomic Studies:** Methods like genome-wide association studies (GWAS), CRISPR-Cas9 gene editing, and RNA interference (RNAi) screening represent potent instruments in the pursuit of uncovering genes and pathways pivotal for disease development. Picture a scenario where a group of researchers is delving into the intricate molecular mechanisms driving neurodegenerative conditions like Alzheimer's disease (AD) [96]. Armed with GWAS data from large-scale population cohorts, researchers conduct comprehensive analyses to pinpoint genetic variants associated with increased disease susceptibility. Through meticulous scrutiny of genomic loci, they identify key genes implicated in AD pathophysiology, shedding light on novel therapeutic targets. Moreover, CRISPR-Cas9 gene editing emerges as a revolutionary technique, enabling precise manipulation ofgenome to elucidate the functional consequences of genetic alterations [97]. By selectively targeting candidate genes identified through GWAS or other omics approaches, researchers can engineer cellular models or animal models to assess the influence of gene perturbations on disease phenotypes. This approach facilitates the validation of potential therapeutic targets and elucidates their role in disease progression [98]. Furthermore, RNA interference (RNAi) screening represents a high-throughput approach to systematically interrogate gene function and pathway dependencies. By selectively silencing gene expression using RNAi libraries, researchers can uncover essential genes and signaling pathways that drive disease phenotypes. Integration of RNAi screening data with genomic datasets from GWAS or CRISPR-Cas9 studies provides a comprehensive understanding of disease-associated pathways and unveils druggable targets for therapeutic intervention. Through the synergistic application of these advanced techniques, researchers gain profound insights into the molecular landscape of disease, uncovering a myriad of potential therapeutic targets. By demonstrating how the modulation of specific genes or pathways can influence disease outcomes, these studies Clear the path for development of precision therapies tailored to the underlying genetic architecture of complex diseases [99] Table 16.2.

Table 16.2 The summary of KEGG release 12.0 (October 1999)

S.no	Database	Content
1.	Trail	The dataset includes 2706 entries consisting of pathway diagrams sourced from 143 diagrams crafted manually.
2.	Genetic factor	In total, there are 110,018 records distributed among 24 whole genomes and 12 incomplete genomes.
3.	Ligand	The compound segment includes 5645 entries, the enzyme segment consists of 3705 entries, and the reaction segment covers 5207 reactions.
4.	Auxiliary numbers	Content
5.	Ortholog group	61 tables
6.	Genome map	The dataset comprises 23 entire genomes and one incomplete genome
7.	Genetic map comparison	Comparison of complete genomes across 23 pairs.
8.	Representation of gene expression	Four collections of gene expression maps
9.	Inventory of genes present	Four groups of gene expression charts
10.	Molecular components	Eight compilations of data
11.	Medical ailment repository	Three collections or compilations

(c) **Bioinformatics and Computational Biology:** Integration of computational biology and molecular medicine has inaugurated a fresh phase in drug discovery. Here, bioinformatics tools assume a central role in pinpointing potential therapeutic targets with unparalleled accuracy. By harnessing the extensive genetic and molecular information amassed from initiatives like the Human Genome Project and comprehensive omics studies, these tools provide invaluable insights into the complex molecular terrain of diseases. Envision a team of researchers embarking on a mission to unveil new drug targets for combating cancer [100]. Harnessing bioinformatics algorithms and databases, they embark on a multifaceted analysis of genomic, transcriptomic, and proteomic data derived from patient samples and experimental models. Firstly, they employ gene expression profiling techniques to elucidate aberrant gene expression patterns characteristic of cancer cells. By comparing the transcriptomic profiles of tumor tissues with those of healthy counterparts, they identify genes that are dysregulated in cancer, thereby pinpointing potential drivers of tumorigenesis [101]. Furthermore, the researchers delve into protein-protein interaction networks, utilizing network analysis algorithms to decipher the intricate web of molecular interactions underlying cancer progression. By mapping out protein interaction networks and identifying hub proteins that play central roles in disease pathogenesis, they uncover promising candidates for targeted therapy. Moreover, pathway analysis tools enable researchers to dissect the complex signaling cascades and biological pathways dysregulated in cancer. By interrogating pathway databases and conducting enrichment analyses, they elucidate key signaling pathways perturbed in cancer cells, offering insights into potential nodes for therapeutic intervention. Through the integrative analysis of diverse omics datasets and the application of sophisticated bioinformatics algorithms, the research team identifies a cadre of candidate drug targets poised at the forefront of cancer therapy. These targets represent molecular vulnerabilities specific to cancer cells, offering the promise of precision medicine approaches tailored to the unique molecular signatures of individual tumors [101] Table 16.3.

(d) **Experimental Validation:** Following the identification of potential therapeutic targets through advanced genomic and screening techniques, the next crucial step in drug discovery involves rigorous experimental validation of their role in disease pathology. This validation process entails a comprehensive assessment using both in-vitro & in-vivo models to confirm that modulation oftarget yields the desired effect on the disease phenotype. Imagine a scenario where researchers have identified a novel protein implicated in the progression of pancreatic cancer, a deadly malignancy with limited treatment options [102]. Leveraging data from genomic studies and functional screens, the researchers hypothesize that targeting this protein could impede tumor growth and improve patient outcomes. However, before progressing to clinical development, it is imperative to validate this target experimentally. In the laboratory setting, in vitro studies Play a crucial role in elucidating the functional implication of identified target.

Table 16.3 The data content of GENES database entry

Field	Content	Links	Data source
Entrance	Identification code for a gene (gene accession number)	Aggregate the information from the VLinkDB repository	Genbank, primary database
Title	Gene identifiers and alternative nomenclature	Compound/Enzyme database, SWISS-PROT database, and PubMed database	GenBank
Description	Description of gene function	Aggregate the KEGG/ PATHWAY database	WISS-PROT, and KEGG.
Category	Classification of genes according to the KEGG pathways.	The primary databases and the NCBI Entrez database	Kegg.
Position	Genetic location on a chromosome	The primary databases and the NCBI Entrez repository	Computed
D-blinks	External hyperlinks		
Codon Usage	Codon usage		
A-aseq	The sequence of amino acids		
N-tseq	Sequence of nucleotides	See footnotea	The primary database functions as an extensive repository containing genetic sequence information.

Cultured cancer cells derived from pancreatic tumors are subjected to various assays to assess the impact of target modulation on critical cellular processes such as proliferation, apoptosis, and invasion. Through molecular techniques such as gene silencing or pharmacological inhibition, researchers can selectively manipulate the expression or activity oftarget and observe its effects on cancer cell behavior. Complementing in vitro experiments, in vivo studies using animal models provide invaluable insights into the physiological relevance and therapeutic potential oftarget. Mouse models engineered to develop pancreatic tumors are employed to evaluate the efficacy of targeting the identified protein in vivo. Treatment regimens involving candidate drugs or therapeutic interventions are administered, and tumor growth, metastasis, and overall survival are meticulously monitored and analyzed. By integrating data from both in-vitro & in-vivo studies, researchers can ascertain the validity ofidentified target as a bona fide therapeutic target for pancreatic cancer. Successful validation not only strengthens the rationale for further drug development efforts but also provides critical preclinical evidence to support the translational potential of targeting this protein in clinical settings [102].

(e) **Target Validation in Drug Development: Ensuring Therapeutic Impact Through Rigorous Scientific Evidence** The intricate interplay between human genetics and drug development unveils a compelling narrative of therapeutic innovation, where insights gleaned from genetic studies illuminate the path towards effective treatments for a myriad of diseases. By identifying alleles associated with disease risk or medical traits, researchers can discern potential drug targets, setting the stage for transformative therapeutic interventions [103]. Historical precedence lends credence to this approach, with numerous examples of gene-drug pairs underscoring the profound impact of human genetics on drug discovery and development. These examples span three distinct categories: prospective, retrospective, and repurposing [104]. Emergence of 5-alpha-reductase inhibitors serves as a notable example. Mutations in the SRD5A2 gene, identified in select instances of pseudo hermaphroditism, unexpectedly Cleared the path for discovery of finasteride as a remedy for benign prostatic hyperplasia and male pattern baldness [105]. In the retrospective category, several approved drugs were initially developed without direct human genetics data, only to have subsequent genetic studies identify their targets as pivotal players in disease pathogenesis. Examples include HMGCR inhibitors (statins) for hyperlipidemia and peroxisome proliferator-activated receptor-γ (PPARG) agonists for type 2 diabetes.

These examples underscore the symbiotic relationship between human genetics and drug development, where genetic discoveries inform the identification of druggable targets, validate therapeutic strategies, and facilitate the repurposing of existing drugs for new indications [106] (Table 16.4).

The understanding of biological function of causal gene and its associated variant is crucial. Particularly, discerning whether the variant leads to a gain or loss of function is essential. Investigations conducted in human tissues provide invaluable

Table 16.4 Recent advancements in CADD (Computer-Aided Drug Design) target validation in drug development from 2015 to 2024.

Year	Advancement
2015	Integration of structural biology techniques with CADD for more accurate target identification and validation
2016	The rise of cryo-electron microscopy (cryo-EM) as a potent method for revealing protein structures has significantly contributed to target validation efforts.
2017	Artificial intelligence (AI) and machine learning (ML) algorithms is bolstering predictions of protein-ligand interactions, thereby improving target validation.
2018	Advancements in virtual screening methodologies, such as fragment-based screening and structure-based virtual screening, leading to improved target validation
2019	Development of computational models incorporating pharmacokinetic and pharmacodynamic properties for comprehensive target validation
2020	Application of network pharmacology approaches to analyze complex interactions between drug targets and biological pathways, refining target validation strategies
2021	Utilization of deep learning models to predict protein structures and interactions, accelerating target validation processes
2022	Blending multi-omics information (genomics, transcriptomics, proteomics, metabolomics) with CADD enables comprehensive target validation.
2023	Advancements in protein-protein interaction prediction tools for assessing target engagement and downstream effects, enhancing target validation
2024	Adoption of cryo-electron tomography (cryo-ET) for studying dynamic protein complexes, providing deeper insights into target biology and validation

insights into properties specific alleles, while animal models Paved the way for playing a crucial role in elucidating the functionality ofgene itself. This gene, like the puzzle's centrepiece, has numerous causal variants, each contributing to the complicated picture of genetic influence on qualities. The genetic foundation of features is analogous to piecing together a jigsaw puzzle.

Consider the LDLR, PCSK9, and CFTR genes to be separate jigsaw pieces. Each gene reflects a unique component ofgenetic landscape, yet they all contribute to a common path, much to linking parts in a cohesive artwork. Polymorphisms in the LDLR, PCSK9, and HMGCR genes all influence cholesterol metabolism, indicating how different genes can have a similar biological effect [107]. Spread among communities like well-known landmarks, while some are rare, elusive jewels waiting. Furthermore, these variations can use a variety of abilities. Some may improve the gene's function, such as enchanted artefacts that offer particular powers (gain-of-function), while others may weaken or disrupt it, such as cursed relics that drain its strength (loss-of-function). This diversity adds levels of intricacy to the genetic jigsaw, making the environment all the more interesting. Despite this intricacy, the combined influence of these causal genes on moulding phenotypes is significant. Understanding the phenotypic effects of genetic variations needs calibration over a spectrum of dosages, much as constructing a jigsaw necessitates putting together its diverse parts to show the whole image [108–110]. Table 16.5.

Table 16.5 Recent advancements in gene-drug pair associations

Gene	Drug	Disease	Indication	Genetic approach comment
BRCA1	Olaparib (Lynparza)	Breast Cancer	HER2-negative metastatic breast cancer with germline BRCA mutation	Olaparib is a PARP inhibitor targeting cancer cells with defective DNA repair mechanisms, such as those with BRCA mutations. This exemplifies precision medicine in oncology [111].
CFTR	Ivacaftor (Kalydeco)	Cystic Fibrosis	Gating mutations in CFTR gene	Ivacaftor is a CFTR potentiator that enhances chloride transport, benefiting identified protein targets provided specific benefits for cystic fibrosis patients CFTR mutations. This represents a paradigm shift in treating the underlying cause ofdisease [87].
BCR-ABL	Imatinib (Gleevec)	Chronic Myeloid Leukemia	Philadelphia chromosome-positive CML	Imatinib, a tyrosine kinase inhibitor, precisely targets the BCR-ABL fusion protein, showcasing the effectiveness of targeted therapy in hematologic cancers.

(continued)

Table 16.5 (continued)

Gene	Drug	Disease	Indication	Genetic approach comment
PCSK9	Evolocumab (Repatha)	Hypercholesterolemia	Familial hypercholesterolemia	Evolocumab is a monoclonal antibody targeting PCSK9, which regulates LDL receptor degradation, leading to LDL cholesterol reduction and cardiovascular risk reduction. This highlights the potential of biologics in managing lipid disorders.
BRAF	Vemurafenib (Zelboraf)	Melanoma	BRAF V600E mutation-positive metastatic melanoma	Vemurafenib is a BRAF inhibitor designed to inhibit the oncogenic BRAF V600E mutation, showcasing the success of targeted therapy in melanoma treatment [49].

(f) **Allelic Variants: Disease Protection vs. Risk.** The rationale behind prioritizing loss-of-function alleles over gain-of-function ones stems from ease of developing drugs that hinder instead of stimulate protein targets. This principle is exemplified by PCSK9 variants Loss-of-function mutations provide protection against coronary heart disease. Disease, while gain-of-function mutations increase the risk. Furthermore, homozygous loss-of-function mutations, which completely knock out a gene, produce the maximum phenotypic effect on target modulation [112].

(g) **Genetic Trait Alignment with Treatment Target** The cases of type 1 diabetes and rheumatoid arthritis, In situations where the biological pathways responsible for disease progression may vary from those guiding symptomatic manifestations. To ensure successful drug development, it's imperative to precisely define the clinical indication and establish a robust connection between the biological pathways implicated in the genetic trait and those associated with the targeted treatment. Consider the example of a loss-of-function mutation in the amyloid precursor protein (APP) gene, known to confer protection against Alzheimer's disease and cognitive decline. If replicated, as indicated by a small follow-up study, finding grasps promise for pharmacological intervention targeting the APP gene or its associated pathways as a viable therapy for Alzheimer's disease.

(h) **GWAS Insights for Drug Targets in Complex Diseases:** Genome-wide association studies (GWASs) have emerged as potent tools for deciphering the genetic roots of complex diseases. However, harnessing this vast genetic data to pinpoint potential drug targets poses a challenge. Despite the majority of disease-linked alleles residing outside protein-coding regions, each allele merits meticulous scrutiny to assess its potential impact on protein structure or gene expression. An astounding 85% of alleles linked to complex diseases are situated outside protein-coding sequences. Nonetheless, it's vital to determine if These alleles are associated with variants that do influence protein structure, such as nonsynonymous or truncating Mutations that cause early termination codons. Identification of such associations warrants prompt follow-up with further functional studies in human cells and animal models

16.5 Challenges in Genetics-Based Target Validation

All genes in the human genome possess an allelic series to establish function-phenotype dose-response curves. Remarkably effective drugs have been developed without direct genetic evidence, underscoring that genetics alone may not always identify drug targets. For instance, although biological agents aimed at inhibiting tumor necrosis factor effectively treat rheumatoid arthritis, TNF was not initially identified as a drug target solely through genetics [113, 114]. Secondly, the complexity between genetic predisposition and disease pathogenesis should not be underestimated. Human genetics represents just the first step in drug development. Substantial investments in functional follow-up studies, spanning humans, animal models, and cellular models, are essential to fully harness the potential of human genetics in drug discovery. However, anchoring drug discovery solely in human genetics may slow down the process, especially when targeting poorly understood biology or unconventional disease mechanisms. Thirdly, disease-associated alleles identified by genome-wide association studies (GWASs) often exert a minimal effect on disease risk. Direct testing is necessary to ascertain whether pharmacological modulation of these targets yields clinically significant effects beyond those observed in genetics. For instance, a common polymorphism in HMGCR affecting LDL cholesterol levels highlights the nuanced relationship between genetic perturbation and pharmacological modulation. Human genetics may indicate potential targets, but not necessarily optimal levels of target modulation for therapeutic efficacy [114, 115].

16.5.1 CADD in Drug Discovery: Understanding Disease Biology

The Computer-Aided Drug Design (CADD) stage Plays a crucial role in drug discovery pipeline, offering a sophisticated blend of computational techniques to model, predict, and simulate potential drug-target interactions before any physical experiments are conducted. This stage is not only cost-effective and time-efficient but also critical for determining the direction and feasibility of a drug discovery project. Successful implementation of CADD relies heavily on a comprehensive understanding of disease biology and pathophysiology, highlighting its importance in initial phases of drug development.

16.5.2 Deep Understanding of Disease Biology

A profound understanding of disease biology is fundamental for the effective use of CADD. This understanding encompasses knowledge about the molecular mechanisms underlying a disease, the pathological processes involved, These processes. From normal physiological states. By comprehensively understanding the disease at the molecular level, discover appropriate targets for... therapeutic intervention. CADD tools can then be employed to explore the interactions between potential drug molecules and the biological target, predict the binding affinity, and simulate the molecular dynamics of these interactions over time.

16.5.3 Role of Pathophysiology in CADD

Pathophysiology, the investigation of aberrant physiological processes linked with disease or injury, assumes paramount importance during the computer-Aided Drug Design (CADD) stage. A comprehensive grasp of pathophysiological mechanisms empowers researchers to pinpoint pivotal junctures within these processes that hold potential for targeted drug intervention, thereby effectively altering the disease trajectory. For example, in instances where a disease pathogenesis is attributed to the dysregulation of a specific signaling pathway, CADD serves as a potent tool to discern and refine molecules capable of modulating this pathway [116].

16.5.4 Importance of Multidisciplinary Expertise

The effective application of CADD requires a multidisciplinary approach, integrating expertise in computational chemistry, molecular biology, pharmacology, and bioinformatics, among others. The collaborative efforts of scientists from these diverse fields enable the translation of complex disease biology and pathophysiological insights into actionable drug design strategies.

16.5.5 *Selectivity and Specificity: Importance of CADD Selectivity and Specificity*

The importance of selectivity and specificity in Computer-Aided Drug Design (CADD) cannot be overstated, as these characteristics are pivotal in the development of safe and effective pharmaceuticals. CADD methodologies leverage computational tools to forecast potential drug particles interact with biological targets, aiming to design compounds that fix to a precise target (specificity) without affecting other, unintended targets (selectivity). This focus is crucial for several reasons:

- Minimizing Side Effects
- Enhancing Drug Efficacy
- Avoiding Drug Resistance
- Facilitating Drug Approval and Market Success
- Enabling Personalized Medicine
- Implementing Selectivity and Specificity in CADD

16.5.6 *Optimizing Drug Design: The Critical Role of Selectivity and Specificity in Minimizing Off-Target Effects and Enhancing Safety*

The quest for the ideal drug molecule in the realm of Computer-Aided Drug Design (CADD) centres on two pivotal attributes: selectivity and specificity. These attributes are essential for... ensuring that a drug effectively binds to its intended biological target without interacting with other, unintended targets, thereby minimizing off-target effects and reducing the risk of side effects. The emphasis on selectivity and specificity in CADD underscores a comprehensive approach to drug design, aiming to optimize therapeutic outcomes while safeguarding patient safety.

(a) **Selectivity: The Key to Minimizing Off-Target Effects:** Selectivity refers to a drug's ability to preferentially bind to its intended target over other similar biological molecules. High selectivity is critical for preventing the drug from interacting with non-target proteins or enzymes that could lead to undesirable off-target effects. In the design phase, CADD utilizes various computational tools, such as molecular docking and pharmacophore modeling, to predict and enhance the selectivity of drug candidates. By carefully analyzing the structural and chemical assets of target and potential off-targets, CADD strategies can be employed to design particles that exhibit high affinity for target while avoiding unintended interactions.

(b) **Specificity: Ensuring Therapeutic Precision:** Specificity involves the drug's ability to elicit a desired therapeutic effect upon binding to its target. A highly specific drug not only binds to the correct target but also triggers the appropriate biological response, such as enzyme inhibition or receptor activation. This precision is paramount for achieving the intended therapeutic outcome without inducing side effects related to non-specific actions. CADD holds significant importance in enhancing specificity by enabling the detailed simulation of drug-target interactions, thereby facilitating the design of molecules that fit precisely into the target's active site and modulate its function as desired.

16.5.7 Precision in Therapy: The Crucial Role of Selectivity in Drug-Target Interactions

Electivity is a fundamental concept in drug design, referring to the molecule's ability to preferentially interact with its intended biological target over other similar targets within the body. This characteristic is crucial for the development of effective and safe therapeutic agents. High selectivity ensures that a drug exerts its desired effect on the disease-causing target without affecting similar, non-involved biological pathways or molecules, which could lead to adverse effects or toxicity.

16.5.8 Strategies for Enhancing Selectivity

(a) **Molecular Docking and Virtual Screening:** Computational methods can predict how well a compound fits into the binding site of its intended target versus other proteins. This screening process helps identify compounds with the best fit and, by extension, higher selectivity.
(b) **Structure-Based Drug Design (SBDD):** Utilizing detailed structural information about the target and related proteins, drug designers can craft molecules that interact with unique features oftarget site, minimizing interactions with similar sites on other proteins.
(c) **Pharmacophore Modeling:** This involves identifying the essential features that a molecule must possess to interact with the target. By focusing on these key features, compounds can be designed or selected for their ability to bind selectively to the intended target.
(d) **Use of High-Throughput Screening (HTS):** HTS can evaluate thousands to millions of Contribute to advancing the development of identifying those with selective interactions. This process often includes counter-screening against related targets to ensure selectivity.

16.5.9 Precision Targeting in Drug Development: The Critical Role of Selectivity in Molecular Interactions

Selectivity is a fundamental concept in drug design, referring to the molecule's ability to preferentially interact with its intended biological target over other similar targets within the body. This characteristic is crucial for the development of effective and safe therapeutic agents. High selectivity ensures that a drug exerts its desired effect on the disease-causing target without affecting similar, non-involved biological pathways or molecules, which could lead to adverse effects or toxicity.

16.5.10 Strategies for Enhancing Selectivity

(a) **Molecular Docking and Virtual Screening:** Computational methods can predict how well a compound fits into the binding site of its intended target versus other proteins. This screening process helps identify compounds with the best fit and, by extension, higher selectivity.
(b) **Structure-Based Drug Design (SBDD):** Utilizing detailed structural information about the target and related proteins, drug designers can craft molecules that interact with unique features oftarget site, minimizing interactions with similar sites on other proteins.
(c) **Pharmacophore Modeling:** This involves identifying the essential features that a molecule must possess to interact with the target. By focusing on these key features, compounds can be designed or selected for their ability to bind selectively to the intended target.
(d) **Use of High-Throughput Screening (HTS):** HTS can evaluate thousands to millions of compounds for their activity against a target, identifying those with selective interactions. This process often includes counter-screening against related targets to ensure selectivity.

16.5.11 Mastering Target Precision: The Key Role of Selectivity and Specificity in Drug Design Through Structural Insights and Molecular Interaction

Achieving high selectivity and specificity in drug design is a complex and nuanced challenge that hinges on a deep understanding oftarget's molecular structure and the interactions that govern binding affinity and activity. This involves a multidisciplinary approach, combining insights from structural biology, medicinal chemistry, pharmacology, and computational modeling to design drugs that interact precisely with their intended targets while minimizing interactions with other molecules.

Here's how detailed knowledge of target structure and molecular interactions facilitates the development of selective and specific drugs:

16.5.12 Empathetic Target Structure

The Empathetic Target Structure is a concept often used in psychology and communication studies to understand how individuals can effectively communicate with empathy.

(a) **3D Structure Analysis:** High-resolution structures of target, obtained through X-ray crystallography, NMR spectroscopy, or cryo-electron microscopy, offer comprehensive understanding of binding sites and crucial residues engaged in ligand interaction. Such insights are paramount for identifying the molecular features that a drug must have to bind selectively and specifically to the target.
(b) **Conformational Flexibility:** Many proteins exhibit conformational changes upon binding to ligands. Understanding these dynamic aspects of target structure lets for the design of drugs that can either stabilize a specific conformation of protein that is conducive to therapeutic effects or prevent conformational changes associated with disease progression.
(c) **Homology with Other Proteins:** Identifying structural similarities and differences between the target and other proteins within the same family or within the proteome at large is vital for designing medications that are selective for the target and do not cross-react with other proteins.

16.5.13 Analysing Molecular Interactions

Analysing molecular interactions is crucial biological processes, drug design, and various other scientific fields.

(a) **Binding Affinity:** The intensity of interaction between the drug and its target is a key determinant of drug's efficacy. Detailed knowledge of types of interactions (hydrogen bonds, hydrophobic interactions, van der Waals forces, etc.) that contribute to binding affinity enables design of molecules with optimized binding properties.
(b) **Activity and Functional Impact:** Understanding how binding translates into modulation oftarget's activity is essential. This requires insight into the mechanism of action at the molecular level, including how the drug affects the target's catalytic activity, signalling pathways, or structural integrity.

(c) **Off-Target Interactions:** A comprehensive analysis of potential off-target interactions is necessary to predict and minimize side effects. This involves comparing the target structure and binding site characteristics with those of other proteins to identify possible cross-reactivity.

16.5.14 Computational Modeling and Simulation

(a) **Molecular Docking:** Computational docking simulates the interaction between the drug and target, predicting the most likely binding modes and orientations. This helps in identifying key interactions that contribute to specificity and selectivity.
(b) **Molecular Dynamics Simulations:** These simulations provide insights into the dynamic behavior of drug-target complex over time, offering a deeper sympathetic of how conformational changes in protein or ligand affect binding and activity.
(c) **Quantitative Structure-Activity Relationship (QSAR) Models:** QSAR modeling uses statistical techniques to correlate structural or property descriptors of compounds with their biological activity. This can guide the modification of drug molecules to enhance selectivity and specificity Table 16.6.

16.6 Machine Learning

Physiologically speaking, learning is a complex cognitive process involving various mental activities aimed at retaining specific behaviours or information for future use. Interestingly, machines and algorithms can replicate this process to a certain extent. Consider the scenario of sorting emails into the spam folder. The server utilizes a query based on sender or message content, drawing from past user activity to determine the fate of each message. However, this process is imperfect, often leading to desired emails being mistakenly marked as spam. This highlights two key aspects of learning: leveraging past experiences or data to make decisions and discerning and discarding superstitions or false information. Building on these

Table 16.6 Computational chemistry methods

S. no	Method	Description
1	Quantum mechanics	Solves the Schrödinger equation to describe the electronic structure of molecules accurately
2	Molecular mechanics	Employs classical mechanics principles to model molecular systems based on force fields
3	Density functional theory	Approximates the quantum mechanical behavior of electrons in a molecule based on electron density

principles, algorithms mimicking knowledge can be developed. However, payable to their rudimentary nature, these algorithms heavily rely on extensive data to make significant decisions [117].

16.6.1 Machine Learning in Drug Discovery

The proliferation of data repositories such as PubChem and ChEMBL has revolutionized access to immense amounts of data. This development has paved way for data-driven discovery, where of these datasets enables the extraction and identification of patterns that yield valuable predictions. Machine learning techniques applied to various aspects of drug discovery. From virtual screening to target prediction, machine learning demonstrates its versatility in addressing diverse challenges within the field. However, it's significant to note that while machine learning holds immense promise, it often represents just the initial step towards more sophisticated models. Specifically, deep learning emulates the perception process observed in the human brain, employing multilayered neural networks to extract features after diverse. These neural networks are trained on various topologies to identify features across a spectrum of complexity, enabling the recognition of intricate patterns within data [118] Fig. 16.4.

16.6.2 Artificial Intelligence

In the ever-evolving landscape of medical science, artificial intelligence (AI) stands as a transformative force, revolutionizing diagnostics, treatment, and drug discovery through its unparalleled computational prowess. Stemming from the convergence of

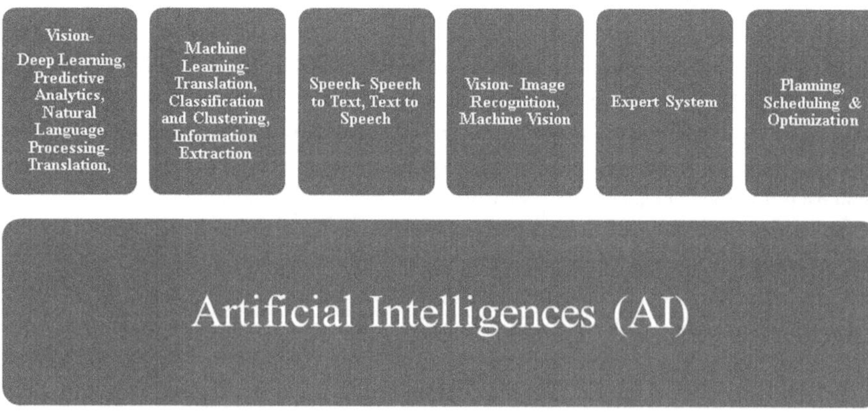

Fig. 16.4 Machine learning and artificial intelligence methods

Table 16.7 Analysing various machine learning techniques in drug discovery and development

Method	Description
Deep learning	Utilizes neural networks with multiple layers to learn complex patterns and relationships in data
Support vector machines	Classifies compounds based on their features by finding the optimal hyperplane that separates them
Random forest	Constructs an ensemble of decision trees to predict molecular properties or activities

information and computer science, AI endeavors to embed intelligence into machines and computers, allowing them to achieve tasks traditionally associated with human cognition [119]. The addition of AI into medical practice is evident in the proliferation of digital assistants such as Apple's Siri, Amazon's Alexa, and Microsoft's Cortana, which streamline data management and optimize scheduling in various settings [120]. However, the true potential of AI unfolds in healthcare environments, where it has made significant inroads in intensive care unit (ICU) management and surgical scheduling, enhancing efficiency Table 16.7.

Among the most promising applications of AI medical science lies in diagnosis and medical care, where data-driven paradigms have ushered in a new era of precision medicine [121]. Notably, AI has demonstrated remarkable capabilities in imaging analysis and recognition, raising the possibility of automating diagnostic processes and potentially surpassing human performance in the near future [122]. While the distinction between deep learning and AI methods remains subject to debate current advancements in AI draw inspiration from neuroscience to enhance model complexity and capabilities. Informatics studies underscore the importance of simplicity in predictive modelling, with the principle of Occam's razor guiding the development of AI algorithms for medical applications [121]. In medicinal chemistry, AI has revolutionized QSAR studies, augmenting their performance and predictive power (Klopman, 1984). This extends to drug design, molecular docking, and predictive toxicology, where AI algorithms optimize drug discovery pipelines with unprecedented speed and accuracy [123]. Notably, AI implementations for drug discovery often leverage metaheuristic algorithms such as tabu search, simulated annealing, genetic algorithms, and particle swarm optimization, facilitating heuristic optimization and accelerating the identification of lead compounds Table 16.8.

Table 16.8 Computer-Aided Drug Design (CADD), providing a diverse array of capabilities aimed at streamlining the drug discovery and development pipeline. Ranging from predicting molecular properties to facilitating protein-ligand docking and virtual screening, these tools harness cutting-edge machine learning algorithms and software infrastructures.

Table 16.8 Recent advancements in machine learning for CADD functional tools

Machine learning tool	Function	Software platform	Mechanism
DeepChem	Drug discovery and development	Python, TensorFlow, PyTorch	Neural networks for molecular property prediction
Chem prop	Compound property prediction	Python, PyTorch	Graph convolutional networks for molecular graphs
Auto Dock-GNN	Protein-ligand docking	AutoDock, TensorFlow	Graph neural networks for docking prediction
Molecular Transformer	Molecular representation learning	PyTorch	Transformer architecture for chemical space mapping
Ligand Net	Ligand-based virtual screening	Python, PyTorch	Siamese neural networks for molecular similarity
Molecular Graph Convolutions	Compound activity prediction	TensorFlow, PyTorch	Graph convolutional networks for bioactivity
Variational Autoencoder	Molecular generative models	TensorFlow, PyTorch	Latent space representation of chemical structures
Chem BERTa	Molecular property prediction	Python, Hugging Face	Transformer-based language model for chemistry

16.6.3 Integration in CADD

In the realm of drug discovery, the integration ADME and toxicological predictions within the framework of Computer-Aided Drug Design (CADD) revolutionizes the early stages of compound evaluation. This integration facilitates the identification of compounds with optimal pharmacokinetic and safety profiles, thus accelerating the drug development process while minimizing risks [111].

(a) **Virtual Screening:** Imagine a vast virtual library of chemical compounds awaiting scrutiny. Through sophisticated computational algorithms, CADD sifts through this library, discerning molecules with promising ADME properties and minimal toxicological risks. By excluding compounds with unfavorable characteristics, such as poor oral bioavailability or genotoxicity, virtual screening ensures that only the most promising candidates progress to the next stage of development [87].

(b) **Lead Optimization:** Picture a lead compound emerging from virtual screening—a potential therapeutic agent ripe for refinement. CADD steps in once more, guiding the modification of this led compound to enhance its pharmacokinetic properties and mitigate toxicity concerns. By analyzing predictive modeling outcomes, researchers strategically tweak molecular structures, optimizing drug candidates for improved ADME profiles

(c) [49].

Table 16.9 Recent advancements in tool integration for ADME virtual screening, lead optimization, and risk assessment

Tool integration method	Description
ADME virtual screening	Integration of computational tools for predicting ADME properties of drug candidates. Enables rapid screening of large compound libraries to identify molecules with favorable pharmacokinetic profiles.
Lead optimization	Integration of software and algorithms for refining lead compounds to enhance potency, selectivity, and pharmacokinetic properties. Combines structure-activity relationship (SAR) analysis, molecular modeling, and predictive modeling techniques to iteratively improve drug candidates.
Risk assessment and management system	Integration of tools for evaluating and dealing safety risks associated with drug candidates. Includes predictive toxicology models, adverse event prediction algorithms, and decision support systems to identify and mitigate potential safety concerns throughout the drug development process.

(d) **Risk Assessment:** Envision a crossroads in the drug development journey—a pivotal moment where decisions carry profound implications. Here, CADD serves as a beacon of insight, offering early warnings of potential safety issues through comprehensive risk assessment. By leveraging predictive models to anticipate toxicological risks, steering drug development efforts towards safer and more efficacious outcomes. Armed with this foresight, stakeholders navigate the intricate landscape of drug development with heightened confidence and prudence, ultimately advancing towards the realization of transformative therapies (Table 16.9).

16.6.4 Pharmacokinetic Optimization

Optimizing a drug's pharmacokinetic profile involves a careful balance between ensuring effective therapeutic levels site of act and minimizing the risk of accumulation and toxicity. This includes:

- **Predictive Modeling:** Using computer-aided drug design (CADD) and pharmacokinetic modeling to predict ADME properties early in the drug development process.
- **Dosage Form Design:** Developing controlled-release formulations that maintain therapeutic concentrations while minimizing peak levels that could lead to toxicity [124].
- **Personalized Medicine:** Tailoring drug choice and dosages based on individual patient characteristics, such as genetic makeup, age, liver and kidney function, to optimize safety and efficacy [60, 125].

16.7 Ensuring Patient Safety: The Crucial Role of Assessing and Optimizing Toxicological Properties in Drug Development

Assessing and optimizing a compound's toxicological properties is a critical aspect ofdrug development process, aimed at ensuring safety for human use. Toxicology studies are designed to identify potential adverse effects a compound may have on the body, providing essential data for evaluating the risk-benefit profile of a drug. The optimization of these properties involves modifying the chemical structure or formulation of a compound to minimize its toxicity while retaining therapeutic efficacy [126]. Here's an overview of how toxicological properties are assessed and optimized:

16.7.1 Toxicological Assessment

(a) **In Vitro Studies:** Early toxicology assessments often involve in-vitro tests using educated cells bacterial systems to evaluate cytotoxicity, genotoxicity (DNA damage), and mutagenicity (mutations). These investigations can rapidly pinpoint substances that may have adverse effects.
(b) **In Vivo Studies:** Compounds showing promise in vitro proceed to in vivo testing in animal models to study systemic toxicity, organ-specific toxicity (such as hepatotoxicity or cardiotoxicity), reproductive toxicity, carcinogenicity, and other long-term effects. These investigations can rapidly pinpoint compounds that may have adverse effects.
(c) **Mechanistic Studies:** Understanding the mechanism of action for both the therapeutic and toxic effects of a compound can help in identifying and mitigating risks. For instance, if a drug's toxic effect is mediated through a specific metabolic pathway, altering the compound to avoid that pathway can reduce toxicity.
(d) **ADME Studies:** Assessing how a drug is ADME can provide serious information for predicting toxicological outcomes. Accumulation of a drug or its metabolites may lead to toxicity, emphasizing the importance of these studies in safety assessment [127].

16.7.2 Optimization of Toxicological Properties

(a) **Structural Modifications:** Chemical modifications to the compound can reduce toxicity. For example, altering functional groups might decrease reactivity with off-target molecules or reduce the formation of toxic metabolites.
(b) **Formulation Changes:** The way a drug is formulated can influence its toxicological profile. Developing slow-release formulations or using carriers that tar-

get the drug to specific tissues can minimize exposure to non-target tissues and reduce side effects.

(c) **Prodrug Strategies:** A prodrug is an inactive precursor that is metabolized in the body to produce the active drug. Designing prodrugs can improve the selectivity of drug activation, reducing systemic toxicity.

(d) **Predictive Toxicology:** Leveraging computational models and databases of toxicological information can predict potential toxicities of new compounds, guiding early-stage modifications before extensive in vivo testing.

(e) **Biomarker Development:** Identifying biomarkers of toxicity can help in early detection and monitoring of adverse properties during clinical trials, allowing for timely intervention and dose adjustments.

16.7.3 Regulatory Considerations

Regulatory agencies require a comprehensive assessment of a drug's toxicological profile before approval for human use. The data generated from toxicology studies must meet stringent guidelines to ensure public safety. The optimization of toxicological properties is thus a critical component ofdrug development process, influencing regulatory decisions and the potential market success of a new therapeutic agent.

16.8 Revolutionizing Drug Safety: Leveraging CADD for Early Toxicity Prediction and Risk Mitigation in Compound Design

Assessing and optimizing a compound's toxicological properties is a critical aspect of drug development process, aimed at ensuring safety for human use. Toxicology studies are designed to identify potential adverse effects a compound may have on the body, providing essential data for evaluating the risk-benefit profile of a drug. The optimization of these properties involves modifying the chemical structure or formulation of a compound to minimize its toxicity while retaining therapeutic efficacy. Here's an overview of how toxicological properties are assessed and optimized.

16.8.1 Toxicological Assessment

(a) **In Vitro Studies:** Early toxicology assessments often involve In-vitro tests using cultivated cells or bacterial systems to evaluate cytotoxicity, genotoxicity (DNA damage), and mutagenicity (mutations). These studies can quickly recognize mixtures with potentially harmful effects.

(b) **In Vivo Studies:** Compounds showing promise in vitro proceed to in vivo testing in animal models to study systemic toxicity, organ-specific toxicity (such as hepatotoxicity or cardiotoxicity), reproductive toxicity, carcinogenicity, and other long-term effects. These studies offer insights into the compound's safety profile and identify safe dosage ranges.
(c) **Mechanistic Studies:** Understanding the mechanism of action for both the therapeutic and toxic effects of a compound can help in identifying and mitigating risks. For instance, if a drug's toxic effect is mediated through a specific metabolic pathway, altering the compound to avoid that pathway can reduce toxicity [128].
(d) **ADME Studies:** Assessing how a drug is absorbed, distributed, metabolized, and excreted can provide serious information for predicting toxicological outcomes. Accumulation of a drug or its metabolites may lead to toxicity, emphasizing the importance of these studies in safety assessment.

A. **Optimization of Toxicological Properties**

   (a) **Structural Modifications:** Chemical modifications to the compound can reduce toxicity. For example, altering functional groups might decrease reactivity with off-target molecules or reduce the formation of toxic metabolites.
   (b) **Formulation Changes:** The way a drug is formulated can influence its toxicological profile. Developing slow-release formulations or using carriers that target the drug to specific tissues can minimize exposure to non-target tissues and reduce side effects.
   (c) **Prodrug Strategies:** A prodrug is an inactive precursor that is metabolized in the body to produce the active drug. Designing prodrugs can improve the selectivity of drug activation, reducing systemic toxicity.
   (d) **Predictive Toxicology:** Leveraging computational models and databases of toxicological information can predict potential toxicities of new compounds, guiding early-stage modifications before extensive in vivo testing.
   (e) **Biomarker Development:** Identifying biomarkers of toxicity can help in early detection and monitoring of adverse effects during clinical trials, allowing for timely intervention and dose adjustments.

B. **Regulatory Considerations:** Regulatory agencies require a comprehensive assessment of a drug's toxicological profile before approval for human use. The data generated from toxicology studies must meet stringent guidelines to ensure public safety. The optimization of toxicological properties is thus a critical component of drug development process, influencing regulatory decisions and the potential market success of a new therapeutic agent.
C. **Revolutionizing Safety in Drug Development: Leveraging CADD for Early Toxicity Identification and Risk Mitigation:** Computer-Aided Drug Design (CADD) plays a pivotal role in modern drug discovery and development, especially in identifying potential toxicities early in the design process and modifying compounds to mitigate these risks. This proactive approach is crucial for

developing safer drugs and streamlining the drug development pipeline by reducing the likelihood of late-stage failures due to toxicity concerns. Here's how CADD integrates toxicity prediction and mitigation into the drug design process:

A. **Early Identification of Potential Toxicities.**

(a) **In Silico Toxicity Prediction:** CADD utilizes various computational models and databases to predict the toxicological profile of new compounds before they are synthesized and tested in biological systems. These models can assess a wide range of toxicities, including genotoxicity, hepatotoxicity, cardiotoxicity, and mutagenicity, based on the compound's structure and known toxicity mechanisms [129].

(b) **Structure-Activity Relationship (SAR) Analysis:** By analyzing the relationship between chemical structure and biological activity, including toxic effects, SAR models can help identify structural features associated with toxicity. This information guides the modification of molecular structures to reduce or eliminate toxicological risks [88].

(c) **Pharmacophore Modeling:** Identifying the pharmacophoric elements that contribute to both therapeutic activity and potential toxicity enables researchers to design molecules that retain the desired biological activity while minimizing harmful effects [130–132].

B. **Modifying Compounds to Mitigate Risks**

(a) **Chemical Modification:** Based on insights from in silico predictions and SAR analysis, chemists can modify functional groups, change stereochemistry, or alter the molecular scaffold of a compound to reduce its potential for toxicity while maintaining or enhancing its therapeutic activity [42, 133].

(b) **Prodrug Design:** In some cases, designing a prodrug—a compound that is metabolized in the body to produce the active drug—can improve the safety profile. Prodrugs can be designed to be activated specifically in target tissues, reducing systemic exposure and minimizing toxicity.

(c) **Targeted Delivery Systems:** CADD can also aid in the design of drug delivery systems that target the active compound to specific tissues or cells, thereby reducing off-target effects and associated toxicities [134].

C. **Integrating ADME Predictions**

(a) **Optimizing ADME Properties:** Good ADME properties are essential for ensuring that a drug reaches its target in the body at therapeutic concentrations without accumulating to toxic levels. CADD models can predict ADME characteristics early in the design process, allowing for the optimization of these properties alongside toxicity mitigation efforts.

(b) **Metabolite Prediction:** Understanding how a compound is metabolized can identify potential toxic metabolites. CADD tools can predict the metabolic pathways of new compounds, guiding structural modifications to avoid the formation of harmful metabolites.

D. **Collaborative and Iterative Process:** The process of identifying potential toxicities and modifying compounds is iterative and involves close collaboration between computational chemists, medicinal chemists, toxicologists, and pharmacologists. CADD provides a powerful set of tools that, when combined with experimental validation, can significantly enhance the safety profile of new drug candidates.
E. **Intellectual Property and Patentability:** Intellectual Property (IP) and patentability are crucial considerations in the field of Computer-Aided Drug Design (CADD), providing a framework for protecting the innovative work involved in discovery and development of new drugs. In the competitive landscape of pharmaceuticals, securing IP rights through patents not only incentivizes innovation but also offers a legal mechanism to prevent unauthorized use of proprietary technology, methods, and the compounds themselves [135]. Here's an overview of importance of IP and patentability in CADD, along with strategies for navigating these aspects.

16.9 Innovative Tools and Applications in Toxicity Prediction

The evaluation of the toxicity of chemicals is vital in recognising their potential harmful effects on humans, animals, plants, and the environment, hence serving as an important aspect of drug design and development. Since traditional in vivo animal tests are commonly employed, they encounter notable challenges concerning ethical considerations, financial implications, and time constraints [136]. In order to access these limitations, in silico toxicology presents a computational approach, utilizing techniques such as machine learning (ML), deep learning (DL), and transfer learning (TL) to effectively predict chemical toxicity. The use of these advanced techniques enhances traditional testing methods, facilitating prioritisation of chemicals, advising toxicity evaluations, and preventing late-stage failures in drug development processes [137].

Recent advancements have shown the effectiveness of graph-based deep learning models, which surpass traditional machine learning methods in predicting human organ-level toxicity endpoints, showing enhanced precision as well as dependability. TL proved its value in improving model performance in particular instances, such as skin sensitization, by effectively integrating a variety of data sources, including both in vivo and in vitro datasets. The implementation of those computational methods enhances the efficiency of identifying toxic compounds, which in turn diminishes the need for animal testing, decreases costs, and lessens environmental impacts [138].

Among the available tools, ProTox is distinguished as a web-based platform specifically crafted for predicting rodent oral toxicity. It employs a method of compound similarity analysis and identifies potentially harmful fragments to offer dependable predictions. ProTox integrates pharmacophore models to forecast potential toxicity targets linked to adverse drug reactions, providing strong performance metrics across a variety of compound classes. Furthermore, the development of multidimensional QSAR methodologies and virtual test kits has been undertaken to evaluate receptor-mediated toxicity and drug-drug interactions, highlighting the promising capabilities of in silico approaches in the initial phases of drug discovery [137].

In summary, computational toxicology presents itself as a swift, reliable, and economically viable option compared to conventional toxicity testing methods. By improving the efficiency and sustainability of drug development, it encourages innovation while addressing the ethical and financial challenges that often accompany traditional methods. The recent advancements play a significant role in enhancing the safety of drug design while simultaneously fostering sustainable practices within pharmaceutical research and ensuring environmental safety [136, 139].

Conflict of Interest NA

References

1. Duelen R, Corvelyn M, Tortorella I, Leonardi L, Chai YC, Sampaolesi M. Medicinal biotechnology for disease modeling, clinical therapy, and drug discovery and development. In: Introduction to biotech entrepreneurship: from idea to business: a European perspective. Springer International Publishing; 2019. p. 89–128.
2. Barua R, Das D, Biswas N. (1AD) revolutionizing drug discovery with artificial intelligence: enhancing efficiency, addressing ethical concerns, and overcoming limitations. https://services.igi-global.com/resolvedoi/resolve.aspx?doi=104018/979-8-3693-2238-3.ch004. pp 62–85.
3. Niazi SK, Mariam Z. Computer-aided drug design and drug discovery: a prospective analysis. Pharmaceuticals. 2023;17:22.
4. Cufer T, Kosty MP. ESMO/ASCO recommendations for a global curriculum in medical oncology edition 2023. ESMO Open. 2023;8:101631.
5. Sharma P, Sharma K, Nandave M. Computational approaches in drug discovery and design. In: Computational approaches in drug discovery, development and systems pharmacology. Elsevier; 2023. p. 53–93.
6. Rudrapal M, Khan J, editors. CADD and informatics in drug discovery. 2023. https://doi.org/10.1007/978-981-99-1316-9.
7. Yadav V, Reang J, Vinita, Tonk RK. Artificial intelligence and machine learning in drug discovery. Springer Nature Singapore; 2023. p. 205–30.
8. Shah A, Patel V, Jain M, Parmar G. Network pharmacology and systems biology in drug discovery. Springer Nature Singapore; 2023. p. 231–52.
9. Pawar B, Behera SK, Tekade M, Al-Shar'i N, Tekade RK. Computer-aided technologies in drug discovery and toxicity prediction. In: Essentials of pharmatoxicology in drug research: toxicity and toxicodynamics, vol. 1. Elsevier; 2023. p. 239–54.
10. Vemula D, Jayasurya P, Sushmitha V, Kumar YN, Bhandari V. CADD, AI and ML in drug discovery: a comprehensive review. Eur J Pharm Sci. 2023;181:106324.
11. Cox PB, Gupta R. Contemporary computational applications and tools in drug discovery. ACS Med Chem Lett. 2022;13:1016–29.
12. MacArron R, Banks MN, Bojanic D, et al. Impact of high-throughput screening in biomedical research. Nat Rev Drug Discov. 2011;10(3):188–95.

13. Putt KS, Du Y, Fu H, Zhang ZY. High-throughput screening strategies for space-based radiation countermeasure discovery. Life Sci Space Res (Amst). 2022;35:88–104.
14. van Vlijmen H, Ortholand JY, Li VMJ, de Vlieger JSB. The European Lead Factory: an updated HTS compound library for innovative drug discovery. Drug Discov Today. 2021;26:2406–13.
15. Genick CC, Wright SK. Biophysics: for HTS hit validation, chemical lead optimization, and beyond. Expert Opin Drug Discov. 2017;12:897–907.
16. Guido VCR, Oliva G, Andricopulo AD. Modern drug discovery technologies: opportunities and challenges in lead discovery. Comb Chem High Throughput Screen. 2011;14(10):830–9.
17. Kenny BA, Bushfield M, Parry-Smith DJ, Fogarty S, Treherne JM. The application of high-throughput screening to novel lead discovery. Prog Drug Res. 1998;51:245–69.
18. Walter LC, Covinsky KE. Cancer screening in elderly patients: a framework for individualized decision making. JAMA. 2001;285:2750–6.
19. Schwartz LM, Woloshin S, Fowler FJ, Welch HG. Enthusiasm for cancer screening in the United States. JAMA. 2004;291:71–8.
20. Bibbins-Domingo K, Grossman DC, Curry SJ, et al. Screening for colorectal cancer: US Preventive Services Task Force recommendation statement. JAMA. 2016;315:2564–75.
21. Smith RA, Andrews KS, Brooks D, Fedewa SA, Manassaram-Baptiste D, Saslow D, Wender RC. Cancer screening in the United States, 2019: a review of current American Cancer Society guidelines and current issues in cancer screening. CA Cancer J Clin. 2019;69:184–210.
22. Cha JS, Yu D. Objective measures of surgeon non-technical skills in surgery: a scoping review. Hum Factors. 2022;64:42–73.
23. Senders JW, Moray NP. Human error: cause, prediction, and reduction; 2020. https://doi.org/10.1201/9781003070375.
24. Endsley MR, Bolte B, Jones DG. Designing for situation awareness: an approach to user-centered design. Designing for situation awareness; 2003. https://doi.org/10.1201/9780203485088.
25. Brooks SK, Greenberg N. Non-deployment factors affecting psychological wellbeing in military personnel: literature review. J Ment Health. 2018;27:80–90.
26. Kang SC. Volunteer involvement and organizational performance: the use of volunteer officers in public safety. Public Perform Manag Rev. 2019;42:554–79.
27. Davies B, Challis D. Matching resources to needs in community care: an evaluated demonstration of a long-term care model. In: Matching resources to needs in community care: an evaluated demonstration of a long-term care model; 2018. p. 1–658.
28. Laufs J, Bowers K, Birks D, Johnson SD. Understanding the concept of 'demand' in policing: a scoping review and resulting implications for demand management. Polic Soc. 2021;31:1–24.
29. Brundage M, Avin S, Wang J, et al. Toward trustworthy AI development: mechanisms for supporting verifiable claims. 2020.
30. Clarke RV, Bowers K. Seven misconceptions of situational crime prevention. In: Handbook of crime prevention and community safety. Routledge; 2017. p. 109–42.
31. National Academies of Sciences E, National Academies of Sciences E Global health and the future role of the United States. 360.
32. National Academies of Sciences E Enhancing the resilience of the nation's electricity system. 156.
33. Simon HA, Herbert A, Laird J. The sciences of the artificial. 256.
34. Hunt K, Agarwal P, Zhuang J. Technology adoption for airport security: modeling public disclosure and secrecy in an attacker-defender game. Reliab Eng Syst Saf. 2021;207:107355.
35. Qin P, Fu Y, Zhang J, Geng S, Liu J, Zhao X. DRL-based resource allocation and trajectory planning for NOMA-enabled multi-UAV collaborative caching 6G network. IEEE Trans Veh Technol. 2024; https://doi.org/10.1109/TVT.2024.3357086.

36. Greenberg MR, Schneider D. Environmental health and the U.S. federal system: sustainably managing health hazards. In: Environmental health and the US Federal System: sustainably managing health hazards; 2019. p. 1–296.
37. Pathirana D. Towards better regulation of unmanned aerial vehicles in national airspace: a comparative analysis of selected national regulations. 2018.
38. Ransbotham S, Mitra S. The impact of immediate disclosure on attack diffusion and volume. Workshop on the Economics of Information Security; 2011. p. 1–12.
39. Kalyaanamoorthy S, Chen YPP. Structure-based drug design to augment hit discovery. Drug Discov Today. 2011;16:831–9.
40. Schneider G. Virtual screening: an endless staircase? Nat Rev Drug Discov. 2010;9(4):273–6.
41. Shoichet BK. Virtual screening of chemical libraries. Nature. 2004;432(7019):862–5.
42. Kitchen DB, Decornez H, Furr JR, Bajorath J. Docking and scoring in virtual screening for drug discovery: methods and applications. Nat Rev Drug Discov. 2004;3(11):935–49.
43. Lu X, Fang D, Ito S, Okamoto Y, Ovchinnikov V, Cui Q. QM/MM free energy simulations: recent progress and challenges. Mol Simul. 2016;42:1056–78.
44. Senn HM, Thiel W. QM/MM methods for biomolecular systems. Angew Chem Int Ed. 2009;48:1198–229.
45. Guedes IA, de Magalhães CS, Dardenne LE. Receptor-ligand molecular docking. Biophys Rev. 2014;6:75–87.
46. Trott O, Olson AJ. AutoDock Vina: improving the speed and accuracy of docking with a new scoring function, efficient optimization, and multithreading. J Comput Chem. 2010;31:455–61.
47. Koes DR, Camacho CJ. ZINCPharmer: pharmacophore search of the ZINC database. Nucleic Acids Res. 2012;40:W409–14.
48. Ehrlich HC, Rarey M. Systematic benchmark of substructure search in molecular graphs – From Ullmann to VF2. J Cheminform. 2012;4:1–17.
49. Lipinski CA, Lombardo F, Dominy BW, Feeney PJ. Experimental and computational approaches to estimate solubility and permeability in drug discovery and development settings. Adv Drug Deliv Rev. 2012;64:4–17.
50. Von Itzstein M, Wu WY, Kok GB, et al. Rational design of potent sialidase-based inhibitors of influenza virus replication. Nature. 1993;363(6428):418–23.
51. Drews J. Drug discovery: a historical perspective. Science (1979). 2000;287:1960–4.
52. Arrowsmith J, Miller P. Phase II and phase III attrition rates 2011–2012. Nat Rev Drug Discov. 2013;12:569–70.
53. DiMasi JA, Grabowski HG, Hansen RW. Innovation in the pharmaceutical industry: new estimates of R&D costs. J Health Econ. 2016;47:20–33.
54. Rossant J. Stem cells from the Mammalian blastocyst. Stem Cells. 2001;19:477–82.
55. Schenone M, Dančík V, Wagner BK, Clemons PA. Target identification and mechanism of action in chemical biology and drug discovery. Nat Chem Biol. 2013;9(4):232–40.
56. Hopkins AL, Groom CR. The druggable genome. Nat Rev Drug Discov. 2002;1(9):727–30.
57. Stanford SM, Diaz MA, Ardecky RJ, et al. Discovery of orally bioavailable purine-based inhibitors of the low-molecular-weight protein tyrosine phosphatase. J Med Chem. 2021;64:5645–53.
58. Coussens NP, Sittampalam GS, Guha R, et al. Assay guidance manual: quantitative biology and pharmacology in preclinical drug discovery. Clin Transl Sci. 2018;11:461–70.
59. Sittampalam GS, Kahl SD, Janzen WP. High-throughput screening: advances in assay technologies. Curr Opin Chem Biol. 1997;1:384–91.
60. Leeson PD, Springthorpe B. The influence of drug-like concepts on decision-making in medicinal chemistry. Nat Rev Drug Discov. 2007;6(11):881–90.
61. Mosedale M, Eaddy JS, Trask OJ, et al. miR-122 release in exosomes precedes overt tolvaptan-induced necrosis in a primary human hepatocyte micropatterned coculture model. Toxicol Sci. 2018;161:149–58.

62. Hughes JP, Rees SS, Kalindjian SB, Philpott KL. Principles of early drug discovery. Br J Pharmacol. 2011;162:1239–49.
63. Lovering F, Bikker J, Humblet C. Escape from flatland: Increasing saturation as an approach to improving clinical success. J Med Chem. 2009;52:6752–6.
64. Bleicher KH, Böhm HJ, Müller K, Alanine AI. Hit and lead generation: beyond high-throughput screening. Nat Rev Drug Discov. 2003;2(5):369–78.
65. Cell Press: Cell Press. https://www.cell.com/abstract/S0092-8674%2811%2900127-9?ref=https://githubhelp.com. Accessed 5 Apr 2024.
66. Weinstein IB, Joe A. Oncogene addiction. Cancer Res. 2008;68:3077–80.
67. Weinstein IB, Joe AK. Mechanisms of disease: oncogene addiction – a rationale for molecular targeting in cancer therapy. Nat Clin Pract Oncol. 2006;3:448–57.
68. Sawyers C. Targeted cancer therapy. Nature. 2004;432(7015):294–7.
69. Vogelstein B, Papadopoulos N, Velculescu VE, Zhou S, Diaz LA, Kinzler KW. Cancer genome landscapes. Science (1979). 2013;340:1546–58.
70. Schneider G, Fechner U. Computer-based de novo design of drug-like molecules. Nat Rev Drug Discov. 2005;4(8):649–63.
71. Bajorath J. Integration of virtual and high-throughput screening. Nat Rev Drug Discov. 2002;1(11):882–94.
72. Jones G, Willett P, Glen RC, Leach AR, Taylor R. Development and validation of a genetic algorithm for flexible docking. J Mol Biol. 1997;267:727–48.
73. Kuntz ID, Blaney JM, Oatley SJ, Langridge R, Ferrin TE. A geometric approach to macromolecule-ligand interactions. J Mol Biol. 1982;161:269–88.
74. Kubinyi H. Similarity and dissimilarity: a medicinal chemist's view. Perspect Drug Discov Des. 1998;9:225–52.
75. Muegge I, Mukherjee P. An overview of molecular fingerprint similarity search in virtual screening. Expert Opin Drug Discov. 2016;11:137–48.
76. Böhm HJ. The development of a simple empirical scoring function to estimate the binding constant for a protein-ligand complex of known three-dimensional structure. J Comput Aided Mol Des. 1994;8:243–56.
77. Kim S, Lim SW, Choi J. Drug discovery inspired by bioactive small molecules from nature. Anim Cells Syst (Seoul). 2022;26:254–65.
78. Seol MJ, Hwang SH, Han JW, Jang HW, Kim SY. Recent progress of halide perovskites applied to five senses sensors. ACS Appl Electron Mater. 2023;5:5261–77.
79. Kim HD, Choi H, Abekura F, Park JY, Yang WS, Yang SH, Kim CH. Naturally-occurring tyrosinase inhibitors classified by enzyme kinetics and copper chelation. Int J Mol Sci. 2023;24:8226.
80. Friesner RA, Banks JL, Murphy RB, et al. Glide: a new approach for rapid, accurate docking and scoring. 1. Method and assessment of docking accuracy. J Med Chem. 2004;47:1739–49.
81. Wachowicz M, Macedo J, Renso C, Ligtenberg A. The role of a multitier ontological framework in reasoning to discover meaningful patterns of sustainable mobility. In: Geographic data mining and knowledge discovery. 2nd ed. Boca Raton: CRC Press; 2009. p. 367–88.
82. Chen X, Ji ZL, Chen YZ. TTD: therapeutic target database. Nucleic Acids Res. 2002;30:412–5.
83. Kaur H, Kaur M, Bhattacharyya A, et al. Indian contribution toward biomedical research and development in COVID-19: a systematic review. Indian J Pharmacol. 2021;53:63–72.
84. Halgren TA, Murphy RB, Friesner RA, Beard HS, Frye LL, Pollard WT, Banks JL. Glide: a new approach for rapid, accurate docking and scoring. 2. Enrichment factors in database screening. J Med Chem. 2004;47:1750–9.
85. Koes DR, Baumgartner MP, Camacho CJ. Lessons learned in empirical scoring with smina from the CSAR 2011 benchmarking exercise. J Chem Inf Model. 2013;53:1893–904.
86. McGovern SL, Helfand BT, Feng B, Shoichet BK. A specific mechanism of nonspecific inhibition. J Med Chem. 2003;46:4265–72.

87. Baell JB, Holloway GA. New substructure filters for removal of pan assay interference compounds (PAINS) from screening libraries and for their exclusion in bioassays. J Med Chem. 2010;53:2719–40.
88. Fourches D, Muratov E, Tropsha A. Trust, but verify: on the importance of chemical structure curation in cheminformatics and QSAR modeling research. J Chem Inf Model. 2010;50:1189–204.
89. Wang Y, Xiao J, Suzek TO, Zhang J, Wang J, Bryant SH. PubChem: a public information system for analyzing bioactivities of small molecules. Nucleic Acids Res. 2009;37:W623–33.
90. Gaulton A, Hersey A, Nowotka ML, et al. The ChEMBL database in 2017. Nucleic Acids Res. 2017;45:D945–54.
91. Kuhn M, Szklarczyk D, Pletscher-Frankild S, Blicher TH, Von Mering C, Jensen LJ, Bork P. STITCH 4: integration of protein–chemical interactions with user data. Nucleic Acids Res. 2014;42:D401–7.
92. Hardy J, Allsop D. Amyloid deposition as the central event in the aetiology of Alzheimer's disease. Trends Pharmacol Sci. 1991;12:383–8.
93. Bertram L, Tanzi RE. Thirty years of Alzheimer's disease genetics: the implications of systematic meta-analyses. Nat Rev Neuroscie. 2008;9(10):768–78.
94. Holtzman DM, Morris JC, Goate AM. Alzheimer's disease: the challenge of the second century. Sci Transl Med. 2011;3:77sr1. https://doi.org/10.1126/SCITRANSLMED.3002369.
95. Lane CA, Hardy J, Schott JM. Alzheimer's disease. Eur J Neurol. 2018;25:59–70.
96. Lambert JC, Ibrahim-Verbaas CA, Harold D, et al. Meta-analysis of 74,046 individuals identifies 11 new susceptibility loci for Alzheimer's disease. Nat Genet. 2013;45(12):1452–8.
97. Mohr SE, Smith JA, Shamu CE, Neumüller RA, Perrimon N. RNAi screening comes of age: improved techniques and complementary approaches. Nat Rev Mol Cell Biol. 2014;15(9):591–600.
98. Doudna JA, Charpentier E. The new frontier of genome engineering with CRISPR-Cas9. Science (1979). 2014;346 https://doi.org/10.1126/SCIENCE.1258096.
99. Visscher PM, Wray NR, Zhang Q, Sklar P, Mccarthy MI, Brown MA, Yang J. 10 years of GWAS discovery: biology, function, and translation. Am J Hum Genet. 2017;101:5–22.
100. Ashburner M, Ball CA, Blake JA, et al. Gene Ontology: tool for the unification of biology. Nat Genet. 2000;25(1):25–9.
101. Huang DW, Sherman BT, Lempicki RA. Systematic and integrative analysis of large gene lists using DAVID bioinformatics resources. Nat Protoc. 2008;4(1):44–57.
102. Jiang H, Hegde S, Knolhoff BL, et al. Targeting focal adhesion kinase renders pancreatic cancers responsive to checkpoint immunotherapy. Nat Med. 2016;22(8):851–60.
103. Musunuru K. The hope and hype of CRISPR-Cas9 genome editing: a review. JAMA Cardiol. 2017;2:914–9.
104. Frayling TM, Timpson NJ, Weedon MN, et al. A common variant in the FTO gene is associated with body mass index and predisposes to childhood and adult obesity. Science (1979). 2007;316:889–94.
105. Hnisz D, Weintrau AS, Day DS, et al. Activation of proto-oncogenes by disruption of chromosome neighborhoods. Science (1979). 2016;351:1454–8.
106. Ripke S, Neale BM, Corvin A, et al. Biological insights from 108 schizophrenia-associated genetic loci. Nature. 2014;511(7510):421–7.
107. Schmelzer JWP, Tropin TV, Schick C. Effects of structural relaxation of glass-forming melts on the overall crystallization kinetics in cooling and heating. Entropy. 2023;25 https://doi.org/10.3390/E25111485.
108. Atienza-Mateo B, Serrano-Combarro A, Del-Val N, et al. OP0126 long-term follow up of Abatacept in usual interstitial pneumonia associated to rheumatoid arthritis. National multicenter study of 172 patients. Anne Rheum Dis. 2023;82(Suppl 1):82–3.
109. Meagher NS, White KK, Wilkens LR, Webb PM, Wu AH, Peres LC, Merritt MA. Abstract B113: associations between risk factors and epithelial ovarian cancer survival by racial

and ethnic group: an analysis from the ovarian cancer association consortium. Cancer Res. 2024;84:B113.
110. Deng ZH, Li Q, Wang DM, Gao SQ, Zeng JQ. Genetic polymorphisms of Y-STR haplotypes in unrelated male blood donors from the surname of Li, Wang and Zhang population in Shenzhen. Yi Chuan. 2007;29:1336–44.
111. Hughes JD, Blagg J, Price DA, et al. Physiochemical drug properties associated with in vivo toxicological outcomes. Bioorg Med Chem Lett. 2008;18:4872–5.
112. Stratford JK, Carnes MU, Willis C, et al. Identifying compounds to treat opiate use disorder by leveraging multi-omic data integration and multiple drug repurposing databases. medRxiv. 2024; https://doi.org/10.1101/2024.01.17.24301329.
113. Guadamuz JS, Pittell H, Pierre A, Ryals CA, Ho G, Calip GS. Multi-level factors underlying racial/ethnic inequities in clinical trial participation among patients with hematologic cancers: lessons for the development of diversity plans. Blood. 2023;142:2393.
114. Atienza-Mateo B, Serrano-Combarro A, Del-Val N, et al. OP0126 long-term follow up of Abatacept in usual interstitial pneumonia associated to rheumatoid arthritis. National multicenter study of 172 patients. Ann Rheum Dis. 2023;82:82–3.
115. Pozzi C, Vanet A, Francesconi V, Tagliazucchi L, Tassone G, Venturelli A, Spyrakis F, Mazzorana M, Costi MP, Tonelli M. Antitarget, anti-SARS-CoV-2 leads, drugs, and the drug discovery-genetics alliance perspective. J Med Chem. 2023;66:3664–702.
116. Kumar V, Abbas AK, Fausto N, Aster JC. Robbins and Cotran pathological basis of disease. 8th ed. Saunders Elsevier; 2010. Kumar, V. et al. (2010) Robbins and Cotran Pathological Basis of Disease. 8th edn. Saunders Elsevier, Elsevier Health Sciences. Elsevier Health Sciences 2:897–957
117. Schwartz CE, Aylsworth AS, Allanson J, et al. Personal journeys to and in human genetics and dysmorphology. Am J Med Genet A. 2024;194:e63514. https://doi.org/10.1002/AJMG.A.63514.
118. Riesenhuber M, Poggio T. Hierarchical models of object recognition in cortex. Nat Neurosci. 1999;2(11):1019–25.
119. Tzafestas SG. An introduction to robophilosophy cognition, intelligence, autonomy, consciousness, conscience, and ethics. In: An introduction to robophilosophy cognition, intelligence, autonomy, consciousness, conscience, and ethics. River Publishers; 2016. p. 1–309.
120. Patel VL, Arocha JF, Kaufman DR. Diagnostic reasoning and medical expertise. Psychol Learn Motivat. 1994;31:187–252.
121. Miller DD, Brown EW. Artificial intelligence in medical practice: the question to the answer? Am J Med. 2018;131:129–33.
122. Klopman G. Artificial intelligence approach to structure-activity studies. Computer automated structure evaluation of biological activity of organic molecules. J Am Chem Soc. 1984;106:7315–21.
123. Duch W, Swaminathan K, Meller J. Artificial intelligence approaches for rational drug design and discovery. Curr Pharm Des. 2007;13:1497–508.
124. Evans WE, McLeod HL. Pharmacogenomics — drug disposition, drug targets, and side effects. N Engl J Med. 2003;348:538–49.
125. Overholser BR, Sowinski KM. Principles of pharmacokinetics and pharmacodynamics. Remington: The Science and Practice of Pharmacy; 2021. p. 219–42.
126. Kapoor MS, Khan SA, Gupta SK, Choudhary R, Bodakhe SH. Animal models of erectile dysfunction. J Pharmacol Toxicol Methods. 2015;76:43–54.
127. Biomarkers in Toxicology – Google Books. https://books.google.co.in/books?hl=en&lr=&id=tAuIDwAAQBAJ&oi=fnd&pg=PP1&dq=Hayes,+A.+W.+(Ed.).+(2014).+Principles+and+Methods+of+Toxicology+(6th+ed.).+CRC+Press.&ots=7AEp1zwMJM&sig=cHOwMeaIsTzbRRHipFfMfM521dQ&redir_esc=y#v=onepage&q&f=false. Accessed 5 Apr 2024.
128. Assessment UENC for E (2009) Casarett and Doull's toxicology: The basic science of poisons.

129. Ciallella HL, Russo DP, Sharma S, Li Y, Sloter E, Sweet L, Huang H, Zhu H. Predicting prenatal developmental toxicity based on the combination of chemical structures and biological data. Environ Sci Technol. 2022;56:5984–98.
130. Schultz TW, Netzeva TI, Cronin MTD. Selection of data sets for QSARs: analyses of tetrahymena toxicity from aromatic compounds. SAR QSAR Environ Res. 2003;14:59–81.
131. Patrick Walters W, Stahl MT, Murcko MA. Virtual screening—an overview. Drug Discov Today. 1998;3:160–78.
132. Tropsha A. Best practices for QSAR model development, validation, and exploitation. Mol Inform. 2010;29:476–88.
133. Stella VJ, Nti-Addae KW. Prodrug strategies to overcome poor water solubility. Adv Drug Deliv Rev. 2007;59:677–94.
134. Torchilin VP. Recent advances with liposomes as pharmaceutical carriers. Nat Rev Drug Discov. 2005;4(2):145–60.
135. Langer R, Tirrell DA. Designing materials for biology and medicine. Nature. 2004;428(6982):487–92.
136. Hu Y, Ren Q, Liu X, Gao L, Xiao L, Yu W. In silico prediction of human organ toxicity via artificial intelligence methods. Chem Res Toxicol. 2023;36:1044–54.
137. Raies AB, Bajic VB. In silico toxicology: computational methods for the prediction of chemical toxicity. WIREs Comput Mol Sci. 2016;6:147–72.
138. Dearden JC. In silico prediction of drug toxicity. J Comput Aided Mol Des. 2003;17:119–27.
139. Vedani A, Dobler M, Lill MA. The challenge of predicting drug toxicity in silico. Basic Clin Pharmacol Toxicol. 2006;99:195–208.

Chapter 17
Metabolism Prediction: Identification of Potential Sites of Metabolism in Lead Molecules

Anjani Umarani Wunnava, Kit-Kay Mak, Shiming Zhang, Jia-Chern Pang, and Mallikarjuna Rao Pichika

Contents

A. U. Wunnava (✉)
Centre for Bioactive Molecules and Drug Delivery, Institute for Research Development and Innovation, International Medical University, Kuala Lumpur, Malaysia

K.-K. Mak
Centre for Bioactive Molecules and Drug Delivery, Institute for Research Development and Innovation, International Medical University, Kuala Lumpur, Malaysia

Department of Pharmaceutical Chemistry, School of Pharmacy, International Medical University, Kuala Lumpur, Malaysia

S. Zhang
School of Biomedical Sciences, Hunan University. Lushan Road (S), Changsha, Hunan Province, China

J.-C. Pang
School of Postgraduate Studies, International Medical University, Kuala Lumpur, Malaysia

M. R. Pichika
Faculty of Medical Sciences, Newcastle University Medicine Malaysia, Jalan Sarjana, Educity, Iskandar Puteri, Johor, Malaysia

S. N. K. Rao G., R. R. Alavala (eds.), *Applications of Computational Tools in Drug Design and Development*, https://doi.org/10.1007/978-981-96-4154-3_17

Abstract This chapter, expanding on the foundational knowledge, offers an extensive review of the computational methods used to predict and study the metabolism of lead compounds. The importance of this knowledge extends beyond academic interest; it is a critical component of drug discovery that affects the efficacy, safety, and clinical success of new therapeutic agents. As the field advances, novel computational models continue to emerge, promising to enhance the predictability and efficiency of drug development processes. By addressing these advanced topics and forecasting future trends, this chapter equips students and researchers with the necessary tools to navigate the evolving landscape of drug metabolism. Through a balanced integration of current research, expert insights, and practical applications, this chapter underscores the significance of metabolism studies in the broader context of drug discovery and development, highlighting its crucial role in shaping the next generation of pharmaceuticals. This chapter explores the processes and methodologies essential for predicting metabolism in lead molecules—a crucial aspect of the pharmacokinetic profiling of new chemical entities in drug discovery. Early on, it is vital to define that 'drug metabolism', synonymous with 'drug biotransformation', refers to the chemical modifications that drugs undergo within the body to facilitate their elimination. These modifications can significantly influence the drug's efficacy, safety, and overall therapeutic profile. 'Lead molecules' are potent compounds identified during the initial stages of drug discovery that exhibit promising activity against a biological target but may require optimisation before progressing to clinical trials. Understanding the metabolism of these molecules is paramount, as it determines their fate in the body and their potential to become safe and effective drugs.

Keywords Drug discovery · Drug biotransformation · Site of metabolism (SoM) · Computational tools · Predicting metabolic sites

17.1 Introduction

Drug metabolism, primarily occurring in the liver, is a crucial process that enhances the water solubility of compounds, aiding their faster elimination. This process can be broken down into two phases involving specific chemical transformations and enzymes.

Phase I of drug metabolism involves initial chemical transformations, predominantly catalysed by Cytochrome P450 (CYP) family enzymes, which play a central role in modifying the compound's chemical structure. The molecule's structure dictates which metabolic pathways it will undergo, influencing its transformation into molecules that may or may not be pharmacologically active.

Following Phase I, the metabolites undergo further Phase II reactions that involve their conjugation with endogenous moieties such as glucuronic acid, sulfate, glycine, and glutathione. This conjugation generally increases the solubility of the metabolites, thus aiding in their excretion.

Regioselective biotransformation of compounds within these phases leads to the formation of metabolites that may interact with other proteins, eliciting secondary pharmacological responses. This entire process of metabolism is pivotal in influencing drug clearance and determining the pharmacological activity and potential toxicity of the drug and its metabolites. Extensive studies on both phases of drug

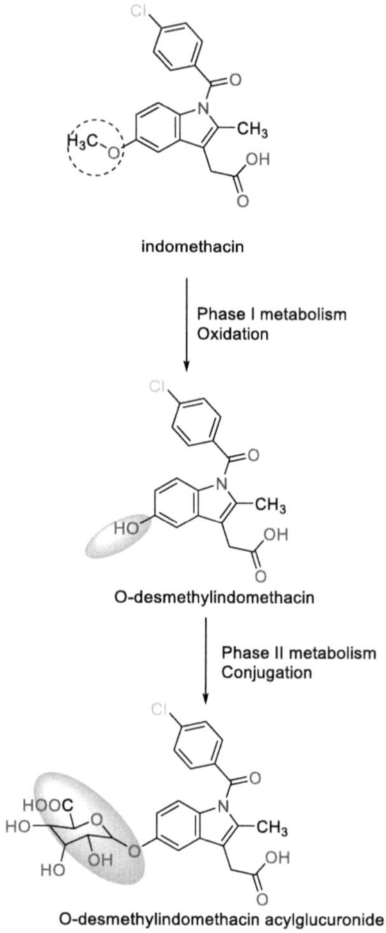

Fig. 17.1 Metabolism of indomethacin

metabolism provide valuable insights into the metabolic pathways and their implications for drug development and safety [1, 2]. Figure 17.1 illustrates these enzymatic activities and transformations [3].

The structures from Fig. 17.1 illustrate that the metabolism of chemical compounds is regioselective, significantly influencing the type of metabolites formed. Identifying and characterising these metabolites are crucial components of the drug design and discovery process. *In vivo,* studies are instrumental in identifying active metabolites, and their toxicity is closely monitored as a drug transition from preclinical phases to Phase I clinical studies. It is important to note that many currently used drugs have active metabolites, which can induce secondary pharmacological effects. These effects may result in adverse reactions or, conversely, lead to the identification of other therapeutic indications.

Discovery of minoxidil was the result of pharmacokinetic studies intended for the development of an antihypertensive. Replacing the alkyl groups on N,

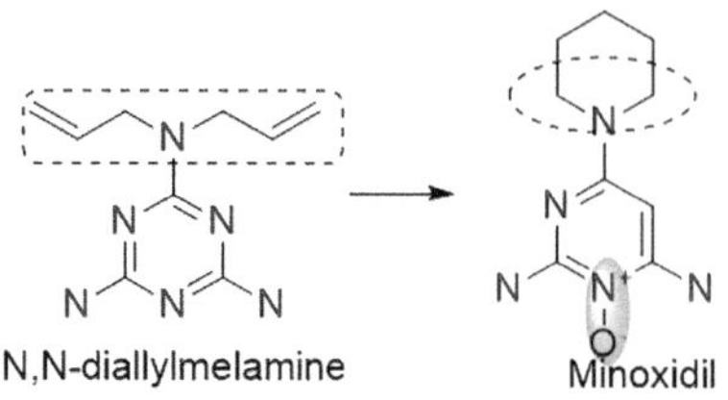

Fig. 17.2 Discovery of minoxidil

N-diallylmelamine with a piperidinyl group led to a compound with improved antihypertensive activity whose onset was not correlating with the time of administration (Fig. 17.2). The pharmacokinetic investigations into this experimental compound led to the identification of minoxidil [4].

17.2 Enzymatic Pathways in Drug Metabolism

Many compounds are metabolised predominantly by CYP enzymes, a family of enzymes containing heme. When exposed to carbon monoxide, these enzymes exhibit an absorption maximum at 450 nm, a characteristic that underscores their name. CYPs are "membrane-bound" and predominantly located in the liver and small intestine in mammals. They can bind various substrates and catalyse reactions, including oxidation, reduction, hydrolysis, dealkylation, epoxidation, and oxidation at a heteroatom. Key isoforms ranging from CYP1 to CYP3 play critical roles in the biotransformation of most drugs. The interaction of compounds with CYPs can influence the metabolism of other drugs, evaluating CYP liability as a crucial component of drug discovery programs to mitigate toxicity and drug interactions early in development [5].

Flavin monooxygenases (FMOs) function similarly to CYPs and share substrates, yet they produce structurally distinct products. Located in the mitochondrial membranes, FMOs metabolise substances, including narcotic analgesics, tricyclic antidepressants, some selective serotonin reuptake inhibitors, and other drugs [6].

Aldehyde oxidases (AOs), another group of molybdenum-containing enzymes located in the cytosol, utilise flavin adenine dinucleotide (FAD) similarly to FMOs. These enzymes are crucial for the oxidation of aldehydes and the hydrolysis of amides. The activity of aldehyde oxidases significantly impacts the half-life of drugs, influencing the approval of several drug candidates such as carbazeran, Lu AF09535, GDC-0834, ripasudil, telithromycin, FK3453, zoniporide, SGX-523, JNJ-38877605, and BIBX1382 [7].

Several key enzyme families catalyse the phase II conjugation process. UDP-glucuronosyl transferases (UGTs) are responsible for the glucuronidation of functional groups from uridinediphosphoglucuronic acid. Glutathione-S-transferases facilitate the conjugation of glutathione with electrophiles in organic compounds, while N-acetyltransferases (NAT) catalyse the acetylation of functional groups.

Additionally, sulfonyltransferases (SULTs) are involved in the sulfonation of organic compounds [8].

17.2.1 Some Case Studies of Drugs with Active Metabolites

Many central nervous system (CNS) drugs have active metabolites, with a prevalent Phase I metabolic pathway being N-dealkylation and N-oxidation at the amino group. This process is significant in the metabolism of various drug classes acting on the CNS. This metabolic pathway can result in the formation of metabolites with distinct pharmacological profiles, often showing increased selectivity in their mechanism of action compared to the parent drug. Some of these biotransformations are depicted in Figs. 17.3, 17.4, and 17.5 [9–14].

Fluoxetine is metabolised into its active metabolite norfluoxetine, also known as seproxetine, primarily through CYP2D6 and CYP2C9. The genetic polymorphism associated with CYP2D6 can cause significant variability in how different individuals metabolise fluoxetine, which impacts therapeutic efficacy and safety. Norfluoxetine is a more potent serotonin and dopamine reuptake inhibitor than fluoxetine, contributing significantly to the antidepressant's overall effectiveness and offering potential benefits in treatment-resistant cases due to its longer half-life. However, the development of seproxetine as a standalone antidepressant was halted by Eli Lilly due to concerns over cardiotoxicity, which highlighted critical safety risks, including potential cardiac rhythm and function changes. This case exemplifies the crucial role of CYP-mediated metabolism in determining a drug's pharmacokinetics and pharmacodynamics and the importance of balancing therapeutic benefits with potential risks during drug development [15, 16].

Osimertinib, a tyrosine kinase inhibitor used in treating non-small cell lung cancer, is metabolised into its active metabolites, AZ7550 and AZ5104, primarily through CYP3A4. AZ5104, although an active metabolite, has been associated with significant toxicity, which has led to the identification of derazantinib (formerly known as dosimertinib). Derazantinib is under investigation and has shown a better safety profile in preclinical studies. Notably, osimertinib and its new derivative are approved for phase I clinical trials in China, highlighting the ongoing efforts to optimise efficacy while minimising adverse effects in cancer therapy [17].

Acyl glucuronides are phase II metabolites formed from drugs containing a carboxylic acid group, a common pharmacophore in numerous drug classes, including NSAIDs, diuretics, antibiotics, and hypolipidemic agents. During the metabolism process, the carboxylic acid moiety of these drugs is glucuronidated by the enzyme UDP-glucuronosyltransferase (UGT) rather than the cytochrome P450 system. These reactive acyl glucuronide metabolites tend to form covalent bonds with nucleophilic sites on proteins, which can alter protein function and potentially lead to adverse effects. Specifically, for NSAIDs the formation of this class of metabolites has been linked to toxicity. Drugs such as zomepirac, fenclofenac, tbufenac, and tolmetin have shown toxicity issues associated with their acyl glucuronide metabolites. Conversely, acyl glucuronides of NSAIDs like ibuprofen and

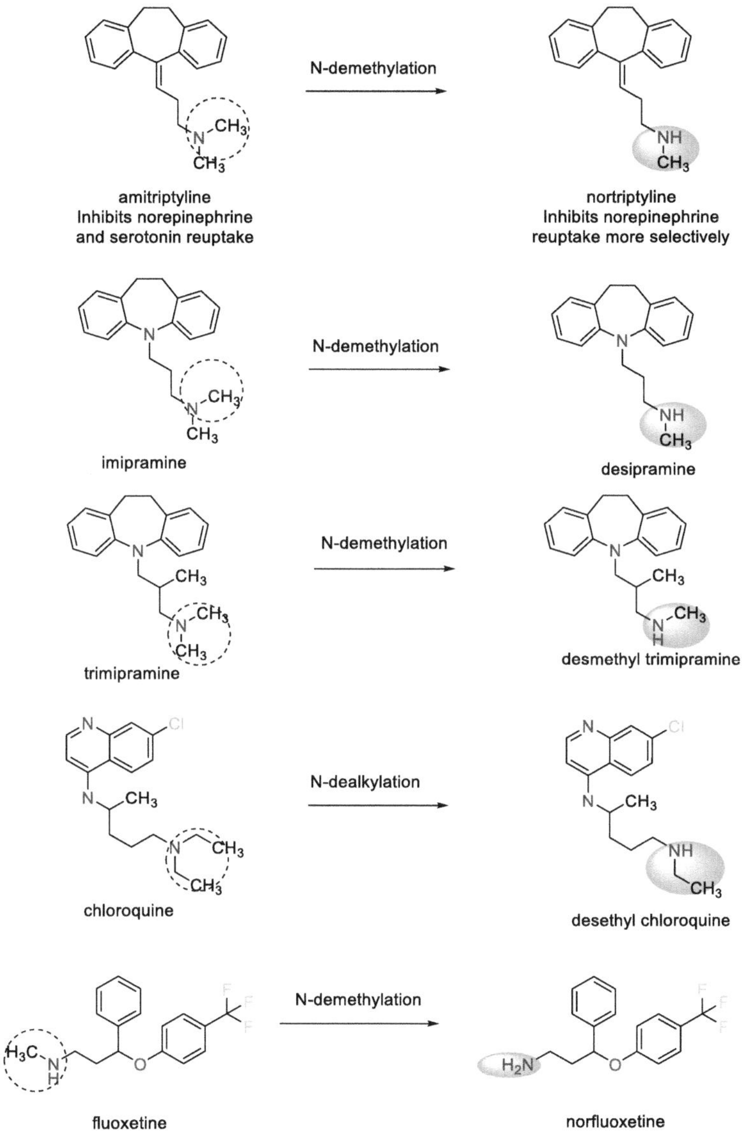

Fig. 17.3 Metabolism by N-dealkylation

ketoprofen have not been significantly implicated in causing severe hepatotoxicity. This variability underscores the need to carefully evaluate the potential for acyl glucuronide formation and associated toxicity during drug development, utilising existing literature and preclinical studies to predict and mitigate adverse outcomes effectively [18, 19].

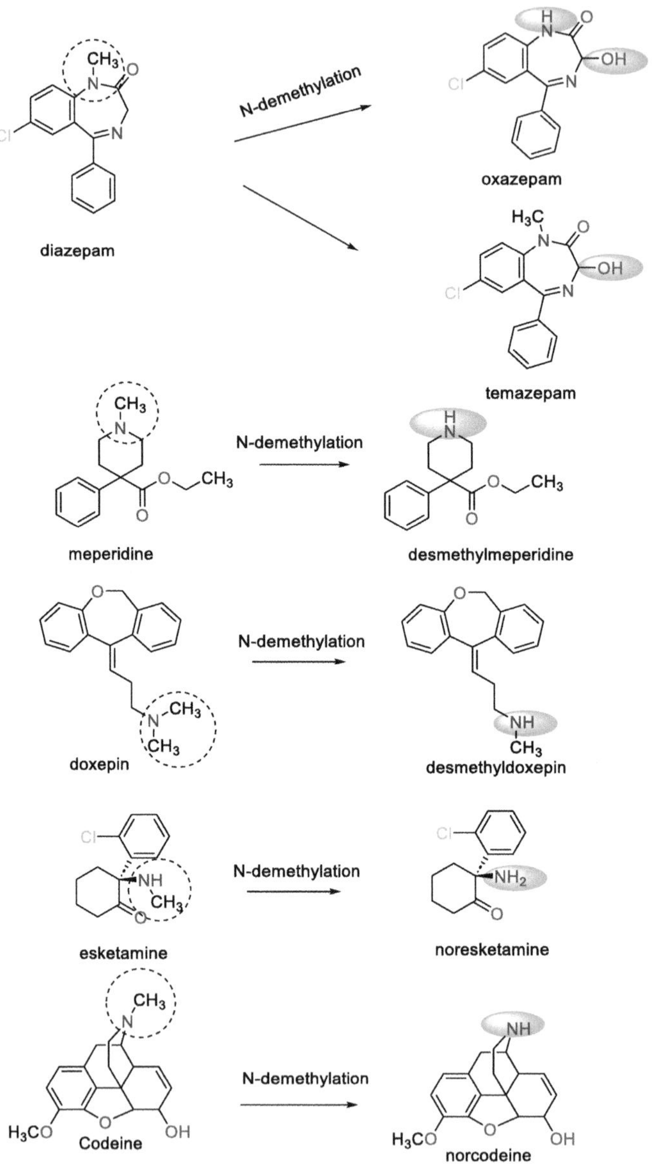

Fig. 17.4 Metabolism by N-dealkylation

17.2.2 Case Studies of Attrition in Clinical Trials

AZD-5847, a potential anti-tuberculosis agent developed by AstraZeneca, was discontinued during phase II clinical trials unfavourable pharmacokinetics and toxicity issues [20]. Similarly, AZD-4547, aimed at treating endocrine-resistant breast

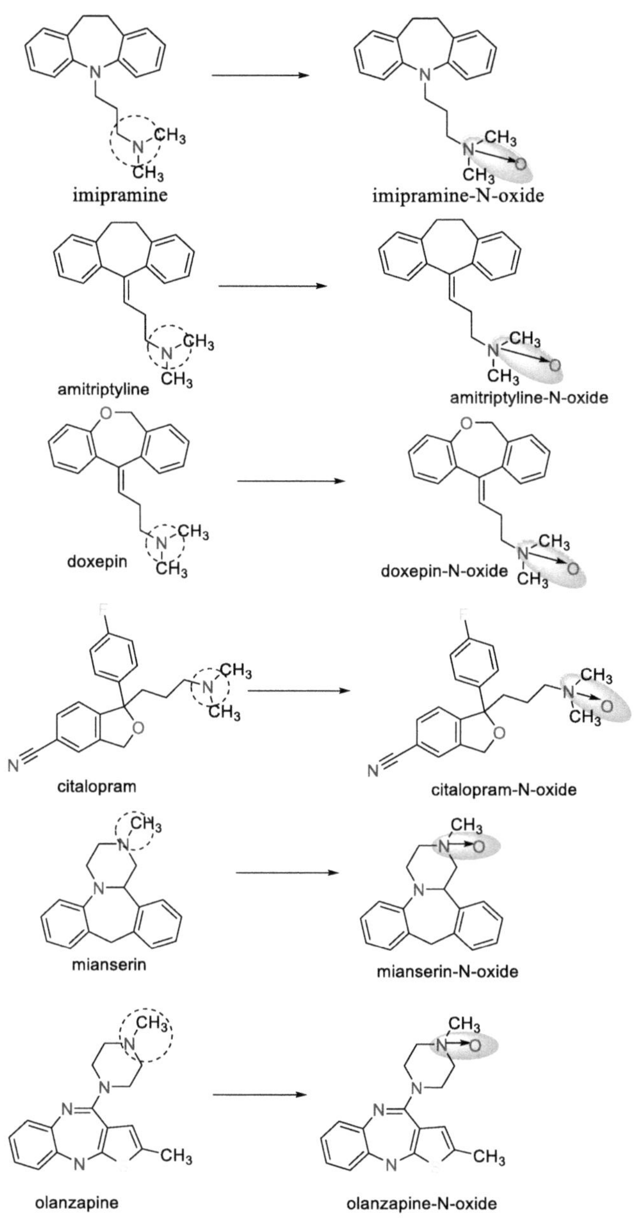

Fig. 17.5 Metabolism by N-oxidation

cancer, and AZD-4785 were also dropped post-phase II trials because of concerns over safety and efficacy. A review analysing failures in AstraZeneca's drug development pipeline indicated that 29% of clinical failures stemmed from safety issues or the inability to confirm effective tissue exposure [21, 22]. In another case, the US

FDA withdrew approval for umbrasilib, an anticancer drug marketed by TG Therapeutics, citing safety concerns in 2022 [23].

These examples underscore the critical role that a drug's pharmacokinetic profile plays in its safety. Several computational methods are available to predict a compound's pharmacokinetics, alongside several assays using liver microsomes, liver tissue, or hepatocytes. Tandem mass and NMR spectroscopy can determine a compound's pharmacokinetic profile. The metabolites identified *in vitro* are qualitatively compared with those from *in vivo* animal studies to pinpoint similarities and discrepancies.

Pharmacokinetics and bioavailability are significant factors in drug attrition during clinical trials despite advancements. Recognising the impact of drug metabolites on a candidate's safety profile, the US FDA has issued guidelines recommending the early identification and safety testing of drug metabolites in the discovery phase. These guidelines stipulate a thorough characterisation, including synthesising human metabolites that appear at concentrations exceeding 10% relative to the parent drug and its related compounds in plasma [24–27].

Identifying toxic metabolites would necessitate another iteration of the compound design and evaluation cycle at the preclinical level. At the clinical level, identifying toxic metabolites would lead to suspension of the project or re-investment in the already expensive drug development process.

Accurate prediction of metabolic soft sites and, by extension, the clearance time and the probable metabolites lead to a substantial decrease in attrition in drug discovery programmes. Reliable and robust methods to identify metabolic soft spots are necessary to control the inflating cost of drug discovery and the attrition rate of probable therapeutic candidates.

17.2.3 Application of Predictive Strategies in Drug Discovery

Big pharma industries like Astra Zeneca, Pfizer, Janssen, Roche, Exscientia and Insilico Medicine are some firms banking on the leverage provided by *in silico* tools for drug design and development. US FDA approval of drugs like Vemurafenib, Venetoclax, Oseltamivir, Zanamivir, Zanubrutinib, Erdafitinib, Selinexor and others is a testament to the reliability and efficacy of *in silico* tools in aiding drug discovery [28, 29].

In silico tools are available to predict target binding, quantitative structure activity relationships, ligand-based pharmacophore modelling, scaffold hopping, fragment-based drug design, blood brain barrier permeation, pharmacokinetics (ADME), and toxicity. These tools have been effectively deployed to design safer drugs.

17.3 Computational Prediction of Metabolic Interactions in Lead Molecule Development

This section explores the computational strategies employed to predict the probable metabolic pathways of compounds, a pivotal component of the lead molecule development process. Computational prediction utilises computer-based simulations and models to determine the metabolic fate of lead molecules before they advance to clinical trials. Such predictive techniques can significantly enhance the success rate of lead molecules in clinical trials by identifying potential metabolic issues early, thus conserving both time and resources.

Docking simulations predict the likelihood of a compound binding to metabolic enzymes. Additional tools such as molecular descriptors, molecular fingerprints, molecular interaction fields, and DFT (density functional theory) calculations assess the reactivity of a lead molecule, providing deeper insights into potential chemical interactions with metabolic enzymes.

However, despite the efficiency and cost-effectiveness of computational predictions, there are challenges to consider. The accuracy of these predictions may be compromised by limitations in the quality and relevance of the computational models and the underlying data. Furthermore, the intricate nature of lead molecule metabolism, which often involves numerous pathways and enzymes, can make precise predictions difficult. Despite these obstacles, the strategic integration of computational predictions into the lead molecule discovery process is instrumental in enhancing the ability to design safer and more effective therapeutic agents. This section discusses these various computational methods, underscoring their critical role in advancing pharmaceutical research by providing essential insights into the metabolic processes that affect drug safety and efficacy.

17.3.1 Predicting Metabolic Sites Using Rule-Based Methods

"SMILES" and "SMARTS" are notational systems used in cheminformatics to represent molecular structures and substructures, respectively, crucial in computational drug discovery. This format is widely utilised for database searching, virtual screening, and as input features in ML models that predict properties or characteristics of a compound that make it a suitable candidate for further development as a therapeutic drug, typically considering factors such as metabolism, bioavailability, etc. On the other hand, SMARTS extends SMILES by allowing the specification of molecular patterns and the chemical environments surrounding atoms within molecules. It is instrumental in substructure searching, which is key for identifying molecules containing specific chemical groups crucial for biological activity and in predicting sites of metabolism by describing known metabolites and their precursors.

Additionally, SMARTS is used to design focused chemical libraries targeting specific biological receptors or enzymes. Both notations enhance the ability to efficiently manipulate, search, and analyse vast amounts of chemical data, thereby accelerating the discovery and optimisation of new therapeutic agents. By utilising SMARTS notation, researchers can construct filters, rules, or transformations that enable the comparison of a query molecule against an existing database. This approach facilitates the prediction of the SoM and the structure of probable metabolites. Several prediction tools have been developed based on these rule-based methods.

- BioTransformer—A computational tool that predicts the metabolism of small molecules through both biotransformation and environmental degradation.
- SMARTCyp—Stands for "Cytochrome P450-Mediated Drug Metabolism Prediction". It is software that predicts a molecule's metabolism site based on its likelihood to be a substrate of CYP.
- Meteor Nexus—A software tool that predicts metabolites, which helps understand how drugs can be metabolised in the body.
- GLORYx—A computational model designed to predict xenobiotic metabolism in humans using ML techniques.
- MetaPrint2D—This tool predicts the SoM of xenobiotics based on the statistical analysis of known metabolism data.
- Meta.2—A computational tool developed to predict metabolic transformations based on a database of biotransformations; it focuses on phase I metabolic reactions.
- IMPACT—This stands for "Integrated Metabolism Predictive and Analytical Computation Tool", which provides insights into drug metabolism pathways.
- SyGMa—Stands for "Systematic Generation of potential Metabolites", a tool that predicts the possible biotransformation reactions built on a rule-based system derived from known metabolic transformations.
- MetaSense—A software platform that integrates experimental and predicted metabolism data to improve drug metabolism prediction.
- TIMES—Stands for "TIssues MEtabolism Simulator", a comprehensive tool that models the biotransformation of chemicals by various tissues in the human body.
- PROXIMAL—An acronym for "Prediction of Xenobiotic Metabolism", a tool focused on predicting likely metabolites generated by xenobiotic substances.
- Metabolizer—A tool designed to predict the metabolic fate of chemicals, particularly focusing on environmental chemicals and their degradation.
- MetabolExpert—A software that provides predictions about the metabolic pathways of drugs and other substances based on a database of known metabolic reactions.

These tools are vital for researchers in pharmaceuticals and toxicology, helping them predict how new compounds might be metabolised in the body, which is crucial for drug safety and efficacy assessments [30].

17.3.2 Predicting Metabolic Sites Using Machine Learning (ML) Based Methods

Deep learning (DL) and ML, subsets of artificial intelligence, are increasingly employed in drug discovery, especially for predicting metabolic processes. ML and DL refer to algorithms enabling computers to learn from and make data-based decisions. These advanced computational techniques are leveraged to analyse vast datasets, including molecular descriptors, molecular fingerprints, and data from known metabolic reactions. Key algorithms utilised in these predictive models are artificial neural networks (complex models that simulate human brain operations), multiple linear regression (a statistical method for predicting outcomes), random forest (an ensemble of decision trees for improved prediction accuracy), ensemble prediction models (which combine multiple models to reduce errors), support vector machines (which classify data into different groups), principal component analysis (a technique for reducing data dimensionality), and Global Adjusted Locally According to Similarity (GALAS) (which modifies predictions based on data similarity).Training and test data sets are used to build and evaluate models respectively.

Several ML-based tools are utilised in drug discovery to predict metabolic pathways and interactions. Some of these tools include:

- BioTransformer 3.0: This tool predicts the products formed through biotransformation (both phase I and II) and environmental degradation. It is particularly useful for researchers who need comprehensive metabolism predictions, including microbial degradation, which is often overlooked.
- CyProduct: This software focuses on predicting cytochrome P450-mediated metabolic reactions, which are crucial for assessing drug-drug interactions and toxicity. Its specialised focus allows for more accurate predictions in this key area of metabolism.
- fame: Short for "Fast Metabolizer", fame utilises fast and efficient algorithms to predict sites of metabolism primarily based on phase I metabolic reactions. It's designed for speed, making it ideal for high-throughput screening environments.
- GLORYx: GLORYx uses ML to provide dual-phase predictions. This tool is beneficial for a holistic view of a drug's metabolic profile, helping in safety assessment studies.
- MetScore: This tool is designed to score metabolites based on their formation likelihood, integrating various biotransformation pathways. It's useful for prioritising metabolites in studies that identify potential toxic metabolites.
- Site of Metabolism Estimator: It predicts specific sites within a molecule that are most likely to be metabolised by enzymes. This precision is crucial for modifying molecules to improve their metabolic stability or to reduce toxic metabolite formation.
- Aldehyde Oxidase Metabolism Prediction: This tool is dedicated to predicting metabolism mediated by aldehyde oxidase. This enzyme is increasingly

recognised for its role in drug metabolism and is often difficult to predict. This specificity is vital for drugs this pathway processes, improving prediction accuracy where generic models may fail.

- MultiCase: MultiCase relies on case-based reasoning to predict xenobiotic metabolism, toxicity, and other properties. Its ability to learn from past examples makes it highly adaptable to new compounds with similar structures or properties.
- StarDrop: This suite offers a range of predictive models and tools designed to optimise the balance of a drug's properties, including its metabolic stability. StarDrop's interactive tools assist in making informed decisions.

The effectiveness of ML models in predicting drug metabolism is highly contingent upon the quality and appropriateness of the molecular descriptors used. Molecular descriptors are numerical values derived from a molecule's chemical structure, encapsulating various properties critical to its behaviour in biological systems. These descriptors serve as inputs for ML models, influencing their ability to predict how a drug interacts metabolically within the body accurately.

Selecting the right number and type of molecular descriptors is crucial for constructing robust predictive models. An excessive number of descriptors can lead to overfitting, compromising the performance. Conversely, too few descriptors reduce the predictive accuracy. Striking the right balance is essential for developing accurate and generalisable models across different molecules.

Utilising these optimised ML models, each equipped with tailored capabilities, pharmaceutical researchers can greatly enhance their insights into a drug's metabolic processes. By effectively leveraging these tools, researchers can improve the safety profiles of drugs and increase their therapeutic efficacy. This strategic use of ML models and molecular descriptors thus plays a pivotal role in advancing drug discovery, helping to bring safer, more effective drugs to market [31].

17.3.3 Predicting Metabolic Sites Using Docking-Based Methods

Docking-based methods are pivotal in evaluating how accessible a molecule is to the ligand binding site on cytochrome P450 (CYP) enzymes and predicting the site of metabolism (SoM). These methods involve docking the query molecule into the target protein's active site. Once docked, the molecule's affinity for the protein is assessed by scoring the best docking pose. This scoring helps determine how well the molecule fits and interacts with the active site, independent of any existing metabolite and reaction databases.

The process focuses particularly on the proximity of atoms in the compound to the heme site in the CYP protein; atoms located less than 4.5 Å from the heme site are flagged as probable sites of metabolism. This proximity criterion is crucial as it indicates where the enzyme will likely act on the molecule, leading to metabolic transformations. Based on this information, the structure of potential metabolites can be predicted, enhancing the understanding of the molecule's metabolic pathways and potential biotransformations.

Tools such as IDSite and MLite utilise this docking strategy to identify sites of metabolism accurately.

IDSite: It integrates docking simulations, molecular dynamics (MD), and quantum mechanical (QM) calculations to provide detailed predictions of metabolism sites for compounds interacting with CYP enzymes, offering deep mechanistic insights and enhanced accuracy for complex molecules.

MLite: It is designed for high-throughput screening, rapidly identifying potential sites of metabolism with a streamlined docking approach, making it ideal for initial evaluations of large compound libraries in early drug discovery phases.

These tools are especially useful for researchers looking to predict metabolic sites without the need for extensive pre-existing data on metabolites, offering a more direct and focused approach to understanding how CYP enzymes metabolise new compounds. Such insights are invaluable in drug development, helping predict and mitigate potential metabolic issues before clinical trials [32].

17.3.4 Predicting Metabolic Sites Using Structural Similarity-Based Methods

The prediction of sites of metabolism can leverage the similarity in shape and features among molecules. One effective approach involves using the software "Rapid Overlay of Chemical Structures" (ROCS), which has shown success in several studies. This method utilises the three-dimensional structures of molecules, comparing a query molecule against a database of structures with known metabolites. By identifying molecules with similar structural features, this technique allows for predicting metabolism sites by docking structurally similar molecules. The Tanimoto combo score is employed to quantify the similarity between these molecules, which combines Tanimoto shape and Tanimoto colour scores, providing a comprehensive measure of similarity between the molecular structures [33].

17.3.5 Predicting Metabolic Sites Using Molecular Interaction Fields (MIFs)

Molecular Interaction Fields (MIFs) are essential for quantitatively characterising the interactions between a ligand and a protein, which is crucial for understanding molecular behaviours in biological systems. MIFs are generated using a computational tool called "GRID," which stands for "GRid INdependent Descriptors." GRID is designed to explore the potential binding sites on a molecule by systematically mapping the energy of interactions at various points in space around the molecule using different chemical probes. This method helps establish Structure-Activity Relationships (SAR) by linking the physicochemical characteristics of a molecule with its biological activity.

The effectiveness of GRID was exemplified during the development of Zanamivir, an antiviral drug. By employing GRID to analyse MIFs, researchers could pinpoint critical interactions that influenced Zanamivir's (antiviral used primarily for the prevention and treatment of influenza) effectiveness, leading to targeted modifications that optimised its therapeutic efficacy.

Furthermore, MetaSite, another predictive tool, utilises MIFs generated by GRID. It employs chemical probes that mimic the interactions between the ligand and CYP or FMO enzymes. This capability is invaluable for drug design, allowing for the optimisation of molecular structures to enhance drug efficacy and minimise metabolic issues, thereby streamlining the progression of compounds to clinical stage.

Molecular interaction fields quantitatively characterise protein ligand interactions. They can be generated using "GRID" and are useful in establishing a structure-activity relationship. This concept has successfully identified Zanamivir's effectiveness during the development stage. MetaSite is a prediction tool that calculates MIFs using chemical probes for the ligand and the protein to predict the site of metabolism [34].

17.3.6 Enhanced Metabolic Site Prediction by Integrating Multiple Tools

The prediction of metabolic sites and the structures of metabolites often require a multifaceted approach, as no single tool can comprehensively predict all aspects of metabolism due to the complex nature of drug interactions with the body. Integrating multiple computational tools that utilise different methodologies can be highly effective in enhancing the accuracy and reliability of these predictions.

For instance, the "Rapid Overlay of Chemical Structures" (ROCS) tool, which needs to be paired with a docking program, is excellent for identifying potential sites of metabolism by comparing the three-dimensional structures of molecules. ROCS excels at finding structural similarities that suggest likely metabolic pathways, but it may benefit from additional insights provided by other tools to refine its predictions.

BioTransformer 3.0 complements ROCS by applying ML models and molecular descriptors to analyse vast datasets, providing a data-driven perspective on how and where a molecule might be metabolised. This tool helps fill in gaps that purely structural analyses might miss, especially in predicting unusual or less obvious metabolic reactions.

Meanwhile, MetaSite enhances the capabilities of both ROCS and BioTransformer 3.0 by calculating Molecular Interaction Fields (MIFs) along with molecular descriptors to forecast interactions with metabolic enzymes specifically. This focus on enzyme interaction provides a critical layer of specificity useful for understanding enzyme-mediated transformations.

Researchers can obtain a more comprehensive and nuanced understanding of a molecule's metabolic profile by using these three tools in concert—ROCS for initial structural comparisons, BioTransformer 3.0 for broad ML-based predictions, and MetaSite for detailed enzyme interaction analysis. This integrated approach leverages the strengths of each tool to provide a more accurate and complete prediction of metabolism, ultimately aiding in the successful development of effective and safe pharmaceuticals.

Most tools use one or more of the approaches described above to identify the SoM and predict the structure of metabolites. ROCS must be used with a docking program to identify the SoM. BioTransformer 3.0 uses ML models and descriptors to identify the SoM. MetaSite calculates MOFs and descriptors to predict the SoM.

17.4 Overview of Computational Tools for Predicting Metabolic Sites

This section will introduce how the existing computational tools can effectively predict metabolic sites. We will delve into detailed discussions of each tool, highlighting their features and the advantages and drawbacks associated with their use in drug metabolism studies. This outline aims to understand how these tools can be integrated into drug development to enhance prediction accuracy and recall.

17.4.1 Overview of ADMET Predictor for Drug Metabolism Prediction

Developed by Simulations Plus, this commercially available software is engineered to predict CYP metabolism, including whether a compound acts as a CYP substrate and its liver microsomal clearance in both humans and rats. The program is adept at predicting the phase I site of metabolism and delineating the structure of the resultant metabolites from nine human CYP isoforms, as well as phase II metabolism mediated by nine isoforms of glucuronidyl transferases.

Constructed on a robust database that integrates compounds from the DrugBank Database, the Accelrys Metabolite database, and various other public databases, the software employs sophisticated prediction models for these nine CYP isoforms. These models are refined using an "Artificial Neural Network Ensemble" method, enhancing the accuracy and reliability of predictions.

Within the software, each atom in a molecule is evaluated and assigned a probability score ranging from 0 to 1000, which reflects the likelihood of that atom being a site of metabolism. Atoms with higher scores are visually highlighted—those in red indicate that the molecule is a probable substrate for the CYP isoform. At the same time, those in grey suggest that the molecule is unlikely to be a substrate. This visual coding system allows users to identify potential sites of metabolic activity within a compound quickly [35, 36].

17.4.2 Overview of Aldehyde Oxidase Metabolism Prediction (AOMP) for Drug Metabolism Prediction

Aldehyde oxidase is a critical enzyme known for its role in drug metabolism, particularly in the oxidation of products originating from CYP metabolism and enhancing the clearance rate of heteroaromatic compounds that contain nitrogen. Its involvement has led to the clinical failure of numerous drug candidates.

This application, which is web-based and open-source, enables users to input the structure of a molecule using SMILES notation or by uploading files in formats such as MOL, SDF, or TXT. The program forecasts the probability of a molecule acting as a substrate for aldehyde oxidase and to identify the probable SoM within molecule in question.

The prediction tool ranks potential sites of metabolism on a scale from 0 to 1, using a colour gradient from dark blue to dark red to visually represent the likelihood of each site being involved in metabolism. Additionally, the results are displayed in a tabular format, providing probability scores for each predicted site of metabolism along with the respective carbon NMR shift values.

These predictions are based on a dataset of compounds. They are refined through a ML model, ensuring accurate and reliable assessments essential for researchers studying drug metabolism or involved in drug development [37, 38].

17.4.3 Overview of BioTransformer 3.0 for Drug Metabolism Prediction

The Metabolomic Innovation Centre in Canada developed the BioTransformer Metabolism Prediction Tool, a versatile open-source resource that users can download or access directly on its website without registering. This tool can predict various types of metabolism, including both phases of human metabolism, microbial metabolism in the human gut, and environmental microbial transformations, making it an integral part of data analytics workflows in metabolomics.

The prediction system utilises a robust framework of 797 rules derived from experimental data on over 2000 biotransformation reactions, coupled with ML algorithms such as CyProduct, to deliver accurate predictions. Users can input the structure of the query molecule in formats like SMILES, MOL, InChI, or SDF. However, the molecules must only contain carbon, hydrogen, oxygen, nitrogen, sulfur, phosphorus, or halogens and should not be salts or mixtures.

The transformation predictor considers qualitative data, such as functional groups, and quantitative data, like the log P value, to determine the likely metabolic reaction. CypReact is used to predict the specificity of a substrate toward various cytochrome P450 enzymes, including CYP - 1A2, 2A6, 2B6, 2C8, 2C9, 2C18, 2D6, 2E1, and 3A4.

Output from BioTransformer is provided in CSV, SDF, or JSON formats for each query molecule. This output includes the molecular formula, molecular mass, AlogP, molecular identifiers, and synonyms for both the query molecule and potential metabolites, as well as details on the type of reaction and the biological system involved. The tool's ability to process multiple requests rapidly and output data quickly, without programming skills, enhances its utility for researchers and practitioners in the field [39, 40].

17.4.4 Overview of CyProduct for Drug Metabolism

CyProduct was developed by the University of Alberta, and this comprehensive algorithm consists of three interconnected modules: CypReact, CypBoM, and MetaboGen. It can be accessed and downloaded as an open-source tool from "https://bitbucket.org/". Each part of the module serves a specific function in predicting the metabolism of drugs.

Together, these modules offer a detailed and predictive framework for understanding how drugs interact with metabolic enzymes, significantly aiding pharmaceutical compounds' development and safety assessment [41, 42].

17.4.5 Overview of CypScore for Drug Metabolism Prediction

The Friedrich-Alexander-University developed a predictive model that identifies probable sites of CYP metabolism, which is now utilised in Bayer Schering Pharma's drug discovery program. This model is specifically tailored to identify compounds that are less susceptible to metabolic oxidation.

Constructed using a dataset of 844 Bayer compounds, which include metabolites observed in rats, mice, and humans, the model employs multiple linear regression techniques. Reactivity and accessibility descriptors crucial for the model are generated through quantum mechanical calculations provided by "VAMP" and "ParaSurf".

This predictive model utilises fifteen different descriptors to characterise each atom within a molecule thoroughly. It features separate models for each type of reaction but does not account for the specific characteristics of the enzymes that facilitate these reactions. The output of the model presents the structure of the parent molecule alongside a metabolic score for each atom, which helps researchers assess the likelihood of metabolic transformations at specific sites within the molecule. This model enhances the ability to predict and mitigate metabolic liabilities early in the drug design process [43].

17.4.6 Overview of FAME 3 for Drug Metabolism Prediction

FAME 3 is an open-source module designed to identify potential sites of metabolism within a molecule for both phase I and phase II reactions (Fig. 17.6 and 17.7). This module functions as part of the GLORYx suite. The algorithm was developed using a dataset from the MetaQSAR database, which supports mapping different atom types involved in reactions to specific reaction types. CDK descriptors, and the "FAME fingerprint" are employed to compute the "FAME score." This score is crucial for predicting the site of metabolism.

For user convenience, molecular structures can be sketched directly on the webpage using the JSME molecular editor, entered in SMILES format, or uploaded as an SDF file. The system allows for the upload and simultaneous prediction of multiple compounds. Results from the predictions are downloadable in CSV file format, facilitating easy access and analysis of the data [44, 45].

17.4.7 Overview of GLORYx for Drug Metabolism Prediction

This prediction system, developed by the University of Hamburg, is an open-source tool available in web-based and downloadable formats. It is designed to identify the probable phase I and II metabolites. A specialised version of this application, GLORY, focuses specifically on predicting products of CYP-mediated metabolism.

Users can input molecular structures using the JSME molecular editor directly on the application's webpage, enter them in the "SMILES" format, or upload them as SDF files. The system can handle multiple compounds in a single submission, enhancing efficiency for users working with extensive datasets.

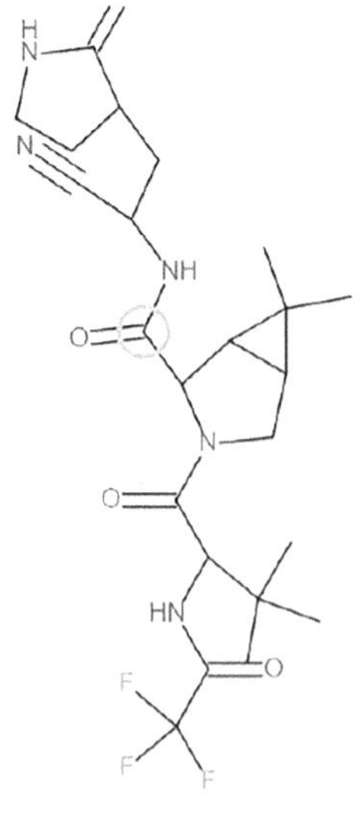

Atom	Probability	FAMEscore
C.22	**0.568**	**0.589**
C.31	0.368	0.541
C.16	0.29	0.613
C.8	0.284	0.564
N.30	0.24	0.544
C.33	0.22	0.598
C.14	0.192	0.72
C.13	0.192	0.72
C.12	0.192	0.72
N.34	0.188	0.62
C.29	0.184	0.553
C.18	0.12	0.657
C.28	0.096	0.558
C.27	0.084	0.643
C.25	0.06	0.563
C.11	0.056	0.593

Fig. 17.6 Prediction results of nirmatrelvir from FAME3

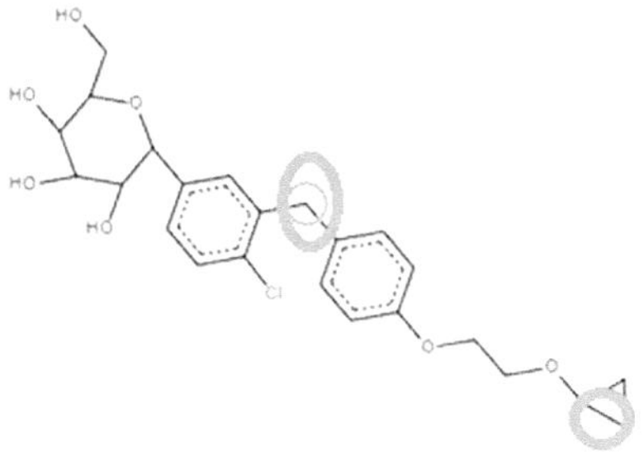

Atom	Probability	FAMEscore
C.14	**0.62**	**0.761**
C.27	0.196	0.845
C.3	0.187	0.738
O.30	0.178	0.845
O.31	0.159	0.811
O.4	0.13	0.696
C.21	0.116	0.82
C.6	0.064	0.624
C.2	0.06	0.821
C.1	0.06	0.821
C.5	0.056	0.639
O.7	0.044	0.673
C.18	0.036	0.806
O.28	0.032	0.811
O.26	0.028	0.792

Fig. 17.7 Prediction results of bexagliflozin from FAME 3

The predictive capabilities of this application are underpinned by a dataset compiled from sources like DrugBank and MetXBioDB, which includes drugs, metabolites, and their metabolic reactions. This information has been codified into a set of rules using "SMIRKS," a hybrid language that combines "SMILES" (representing sites of reaction) and "SMARTS" (encoding parts of the molecule not directly participating in the reaction). These rules are systematically applied to each atom in the query molecule to determine the SoM, with the site prediction in this module specifically handled by FAME 3.

However, there are limitations to the types of molecules that can be processed. The software is not equipped to handle salts or multi-component molecules, and it only supports molecules containing elements such as carbon, hydrogen, oxygen, nitrogen, sulfur, phosphorus, and certain halogens and metalloids.

The results from the application are downloadable as SDF files. On the results page, users can view the structures of the predicted metabolites, which are ranked based on their probability score. This score considers both the SoM and the reactivity of each atom, aiming to predict the most likely metabolic transformations [46, 47].

17.4.8 Overview of IDSite for Drug Metabolism Prediction

IDSite is a tool that pinpoints the probable sites of metabolism by docking interactions with the CYP2D6 enzyme using Schrodinger's GLIDE software. The docking process incorporates two iterative steps utilising "PRIME" for the docking refinements.

The initial stage of the process filters potential sites based on the proximity of the nitrogen atom or other heavy atoms to the enzyme's heme site. After docking, the first refinement stage uses the binding energy of the poses as a criterion for further filtering. Subsequent refinements focus on optimizing the interatomic distances between the protein and the ligand pose, ensuring a precise and effective interaction.

The scoring of the site of metabolism is influenced by both the binding energy and the reactivity of the identified site. Throughout the process, predefined rules are used to refine the docked poses at each stage, enhancing the prediction accuracy. The final output includes the most probable site of metabolism and the detailed pose of the compound's interaction with the protein's binding site [48].

17.4.9 Overview of In-Silico *Metabolism Prediction by Activated Cytochromes and Transition States (IMPACTS) for Drug Metabolism Prediction*

IMPACTS is a key component of the Molecular Forecaster's in silico drug design workspace, specializing in the prediction of CYP modifications and the binding modes of potential substrates. This system generates transition state structures of molecules and assesses the likelihood of CYP binding through a docking process. The predictive approach is rooted in evaluating substrate reactivity, guided by a set of rules developed from calculations using the "Density Functional Theory".

This application has demonstrated superior accuracy in predicting the site of metabolism for the enzyme CYP2C9 against other market-leading tools, such as SMARTCyp, Schrödinger, and MetaSite. This highlights its effectiveness and reliability in the field of drug metabolism prediction [49, 50].

17.4.10 Overview of META.2 for Drug Metabolism Prediction

Developed in 1994 by Case Western Reserve University, this database is designed to predict probable metabolites for xenobiotics, focusing on the biotransformation reactions and resultant metabolites from both phase I and phase II metabolism observed in mammalian species. It includes a comprehensive dictionary that catalogues the metabolic transformations of existing drugs, which serves as a foundational resource for predictions.

To facilitate these predictions, a set of rules known as "transforms" was created based on this dictionary. These transforms have been specifically refined to meet the requirements of the metabolising enzymes involved, enhancing the accuracy of the predictions. This system allows for the prediction of major metabolites for various compounds, providing valuable insights into the metabolic pathways of drugs.

The effectiveness of this database has been demonstrated through successful evaluations in predicting primary and secondary metabolites for a range of drugs, including Phenacetin, Isoniazid, Meperidine, and Phenobarbital, among others.

This tool is particularly useful for researchers looking to understand and anticipate the metabolic fate of drug candidates and other chemical entities within mammalian systems [51].

17.4.11 Overview of MetabolExpert for Drug Metabolism Prediction

Developed by Compudrug, this is one of the pioneering models designed to predict metabolites in humans, animals, and plants. The predictions are based on rules formulated through logical programming, a method that initially utilised the programming language "PROLOG" to establish these rules. Users have the flexibility to modify these rules based on specific queries, allowing for tailored prediction processes.

The application displays the three-dimensional structures of both the probable metabolites and the parent compound in the results, enhancing the visual understanding of the metabolic transformations. Results are available for download in either SDF or RDF format, accommodating various data analysis needs. The software is compatible with both Windows and Linux operating systems, ensuring broad accessibility.

Additionally, a streamlined version of this application, known as "MEXAlert," is available. MEXAlert focuses specifically on predicting the likelihood of first-pass metabolism for compounds, providing a more focused tool for assessing metabolic stability early in the drug development process [52, 53].

17.4.12 Overview of Metabolizer for Drug Metabolism Prediction

This software, developed by Chemaxon, was designed to predict the probable site of metabolism for compounds in humans. The application could generate both primary and subsequent metabolites, utilising a comprehensive group of biotransformation rules and a dataset encompassing various reactions. Users had the flexibility to adjust the number of subsequent metabolite prediction steps according to their specific needs.

The system was accessible through both a GUI and CLI, catering to different user preferences for operational ease and versatility. Additionally, the application included functionality to predict the rate of elimination for each metabolite, providing valuable insights into the metabolic stability of the compounds [54, 55].

17.4.13 Overview of MetaPrint2D for Drug Metabolism Prediction

This program, which can be downloaded from GitHub or accessed through the open-source cheminformatics platform "Bioclipse," utilises a dataset sourced from the Symyx Metabolite Database (formerly known as the MDL database). It predicts the SoM in a molecule by generating fingerprints for each atom, which are representative of the chemical territory of the atom. They are used to perform a similarity search against compounds in the dataset, identifying likely sites of metabolism based on this comparison.

The software can be integrated into both a GUI and a CLI, offering flexibility depending on the user's preference. When operated through Bioclipse, users can sketch the structure of input molecules, with probable sites of metabolism highlighted and colour-coded for easy identification.

The program has been rigorously evaluated against other predictive tools such as MetaSite and MetaGlide. Additionally, a specific variant of this model, known as Meta React2D, is capable of predicting the structures of probable metabolites across different species, including humans, rats, and dogs, or can be generalised for all species. The biotransformation reactions and their respective rules within the model are encoded using "SMARTS" and "SMIRKS" languages, enhancing the model's ability to replicate and predict metabolic processes accurately [56].

17.4.14 Overview of MetaSense for Drug Metabolism Prediction

Developed by ACD/Labs, this commercially available program features a graphic user interface and can be integrated with other predictive applications such as MetaSite. The software is designed to predict both phases of human metabolites based on the structure of compounds, utilizing a rule-based approach. These rules are meticulously crafted by reviewing existing literature on hydrolysis and redox reactions for phase I, as well as conjugation reactions for phase II metabolism.

Additionally, the program employs a ML model to assess the probability of metabolism occurring at specific atoms within a molecule. The software analyzes experimental data derived from techniques such as LC/MS/MS, radiotracer studies, and other relevant workflows to enhance the accuracy of its predictions. This post-experimental analysis helps refine the predictive accuracy, making it a valuable tool for researchers looking to understand the metabolic pathways and potential biotransformation products of pharmaceutical compounds [57, 58].

17.4.15 Overview of MetaSite 6.0.5 for Drug Metabolism Prediction

MetaSite is a commercially available software developed by the Laboratory for Chemometrics and Cheminformatics at the University of Perugia, designed to predict the SoM in molecules. The software primarily focuses on predicting SoM involving CYP enzymes and FMOs. Users can input the query molecules in various formats (including SMILES, 2D, or 3D).

The prediction process in MetaSite utilises "GRID flexible molecular interaction fields" to compute descriptors for both the target enzymes (CYP - 1A2, 2C9, 2C19, 2D6, and 3A4) and the prospective substrates. These descriptors assess the reactivity (Ri) and accessibility (Ei) of each atom category within the substrate.

MetaSite can integrate with LC-MS/MS data to refine predictions of metabolism sites in various compounds. The latest version, MetaSite 6.0, also includes capabilities to predict metabolism by the human aldehyde oxidase isoform, enhancing its utility. Notably, the performance of MetaSite is robust as it does not rely on any training set, making it a valuable tool for researchers aiming to understand and predict how compounds metabolise in the body [59, 60].

17.4.16 Overview of MetaTox for Drug Metabolism Prediction

The tool described is an open-source, web-based application designed to predict the structure of metabolites and assess the potential acute toxicity of both parent compounds and their metabolites in rats. Users can input the structure of query molecules directly on the website using a chemical sketching tool. The tool can predict metabolites resulting from phase I reactions as well as phase II reactions.

Predictions about the type of metabolic reaction are facilitated by the "PASS" software, which uses structure-activity relationship data to forecast the reactions a compound may undergo. Additionally, the model used to predict the site of metabolism aligns with that used by SOMP (Site of Metabolism Predictor), ensuring consistency and reliability in predictions.

Once the analysis is complete, the predicted metabolites' structures and associated toxicological data are displayed on the results page. Users can download these results in various formats, including SDF, PDF, Excel, or CSV, for further analysis or record-keeping. This tool provides a comprehensive and accessible means for researchers to evaluate new compounds' metabolic pathways and potential toxicological impacts, greatly aiding in the early stages of drug development and toxicological assessment [61, 62].

17.4.17 Overview of META Ultra—MultiCASE for Drug Metabolism Prediction

This commercially available application from MultiCase is designed to predict the site of metabolism for both phase I and phase II reactions, utilising the structural attributes of compounds and leveraging ML models. The model is underpinned by a comprehensive database that includes a variety of drugs and their metabolic products derived from both the phases of human metabolism.

The output from this tool provides detailed information about each predicted metabolite. This output includes the molecular structure of the metabolite, the enzymes responsible for its metabolism, the species in which the metabolism was observed, the tissue location where the metabolism occurs, and a probability score indicating the likelihood of the metabolite's formation. Additionally, the tool provides insights into potential toxicity and other relevant properties of the predicted metabolite, making it an invaluable resource for researchers involved in drug development and safety assessment [63, 64].

17.4.18 Overview of Meteor Nexus for Drug Metabolism Prediction

Meteor Nexus, developed by LHASA Limited, is a commercially available tool to predict metabolites and metabolic intermediates from a given chemical structure. It utilises a robust set of rules derived from a comprehensive database of known metabolic transformations. These rules are applied to identify likely metabolic pathways for a molecule and predict the resulting metabolites' structures.

The software offers two modes of operation: "absolute reasoning" and "relative reasoning." This dual-mode approach allows users to adjust the query settings to minimise the occurrence of oversight in prediction, enhancing the accuracy of predictions. The predictions made by Meteor Nexus also consider the calculated log P (ClogP) value of the compounds, which influences the metabolic pathways.

The results, including the structures of potential metabolites, are displayed within the system and can be saved in a molfile format. This feature provides a convenient way for users to review and utilise the predicted metabolic outcomes for further research and development processes [65, 66].

17.4.19 Overview of MetScore for Drug Metabolism Prediction

This predictive model, developed by Dr. A. R. Finkelmann et al., utilises machine learning techniques to predict the site of metabolism and the type of reaction for both phase I and phase II metabolic reactions. The model was constructed using a dataset from the "BIOVIA metabolite database," which encompasses eighteen biotransformation reactions. Descriptors representing the steric and electronic environment of the atoms within a molecule, as well as their relative positions, were calculated using the RDKit library.

The predictive algorithm was trained using a "random forest model". It is integrated into Bayer's in silico graphic user interface program known as "Plx." It allows users to sketch the two-dimensional structure of a substrate and predicts the structures of probable phase I metabolites. The results are displayed in a spreadsheet format, where possible sites of metabolism are highlighted directly on the structure of the parent molecule. The spreadsheet provides detailed information, including the SoM, reaction type, phase, and the status of the query.

Notably, major pharmaceutical companies such as Merck, GlaxoSmithKline, AstraZeneca, Amgen, Bayer, and Novartis were involved in the development of this model, underscoring its significance and utility in the pharmaceutical industry. This collaboration highlights the model's robustness and its alignment with industry needs, making it a valuable tool for drug development and metabolic profiling [67].

17.4.20 Overview of MLite for Drug Metabolism Prediction

MLite, a tool developed by Yonsei University and the Bioinformatics and Molecular Design Research Centre, employs a sophisticated approach to predict the site of metabolism for compounds. This method combines "ensemble catalyticphore" docking to assess accessibility and evaluates reactivity by calculating activation energy. Specifically designed to predict metabolism mediated by the CYP3A4 enzyme, MLite focuses on hydroxylation and dealkylation reactions, crucial metabolic pathways in drug processing.

In evaluations, MLite successfully predicted the metabolism sites for 76% of the substrates in the test set, demonstrating its efficacy and reliability as a tool for researchers studying drug metabolism and seeking to understand how drugs interact with the CYP3A4 enzyme [68].

17.4.21 Overview of PreMetabo for Drug Metabolism Prediction

PreMetabo is a versatile tool designed to predict the pharmacokinetics of compounds, leveraging an open-source, web-based application format. The latest version, PreMetabo 2.0, has been commercialised, offering enhanced features. This application is specifically tailored to predict the phase I SoM in test compounds by utilising the principles of the Arrhenius equation. Key factors such as binding energy and reactivity determine the likelihood of an atom engaging in reactions with human CYP enzymes.

PreMetabo calculates activation energy for enzyme-substrate interactions across various metabolic reactions. The algorithm integrates with AutoDock Vina to compute binding energy and has been validated against the FUJITSU ADME database for accuracy. The FUJITSU ADME database is a comprehensive resource developed by Fujitsu that provides data on drug ADME properties, assisting researchers in understanding and predicting the pharmacokinetic behaviours of pharmaceutical compounds.

Additionally, PreMetabo extends its predictions to phase II metabolism, specifically glucuronidation and sulfonation reactions, utilising molecular descriptors such as CDK, Pubchem, and ECFP4 fingerprints. Users must register on the website and activate their accounts to access this tool. Once registered, users can input the structure of the test molecule either by using drawing tools available on the website or by uploading a file.

The application provides predictions for metabolite interactions with key CYP enzymes, including CYP1A2, CYP3A4, CYP2C9, and CYP2D6, presenting results in a clear, tabular format that details atom-wise interactions. However, it is important to note that PreMetabo is not suitable for predicting the metabolism of inorganic compounds [69–71].

17.4.22 Overview of PROXIMAL for Drug Metabolism Prediction

This open-source online tool is designed to predict human phase I and II metabolites. It operates within a framework of rules formulated from established knowledge of enzymes, substrates, and metabolites. The tool utilizes SIMPCOMP to analyse the structures of substrates and metabolites, identifying the specific groups involved in the metabolic process. The relevant data for both phase I and II reactions are systematically organised into tables.

To use the application, users upload the structure of the query molecule in mol format, which the tool then converts into the "KEGG atom type format." It identifies potential reaction centres within the molecule by comparing them with "keys" in the pre-existing tables. These tables then guide the prediction of the metabolism site and the structures of the resulting metabolites.

This format is used to enhance the precision of computational predictions by accurately representing the structural and functional aspects of atoms within molecular entities [72, 73].

17.4.23 Overview of ROCS for Drug Metabolism Prediction

This platform utilizes shape-based screening in conjunction with "FRED," a docking program, to effectively predict the site of metabolism in compounds. The process begins by quantifying a query molecule's shape and chemical features using the Tanimoto combo score. In a specific application, Flurbiprofen, which CYP2C9 metabolises, was used as the query molecule. Tanimoto combo scores were then calculated for other compounds in the database.

Following this, the molecules from the database were docked against the CYP2C9 protein, and the proximity between the SoM on flurbiprofen and the binding sites on CYP2C9 was measured. Molecules that had topologically similar binding sites close to the CYP2C9 binding site received higher scores. This method successfully identified about 44 molecules with a similar site of metabolism to that of Flurbiprofen. The accuracy of this screening and docking approach was found to be 89%, demonstrating its effectiveness in predicting metabolic sites [74].

17.4.24 Overview of RegioSelectivity-Predictor (RS-Predictor) for Drug Metabolism Prediction

The program, developed jointly by the Rensselaer Polytechnic Institute and the University of Copenhagen, is designed to predict potential CYP3A4 metabolism sites on compounds. Utilising a dataset compiled from Merck, which includes 394 drugs and their metabolites, the program identifies "metabolophores"—specific groups of atoms on the substrate molecule that participate in metabolic reactions.

For every atom in the identified "metabolophore", approximately 540 "Metabolophore descriptors" are calculated, capturing both topological and reactivity features of these atoms. These descriptors are then utilised in conjunction with a "Multiple Instance Ranking System" (MIRank), which determines the appropriate weight vector for ranking the metabolophores on a substrate.

The dataset of substrates is randomly divided into five sets, with each set undergoing multiple validations to eliminate any potential bias and ensure the robustness and impartiality of the predictions. This approach results in multiple predictions per test compound, enhancing the reliability of the results [75].

17.4.25 Overview of Site of Metabolism Estimator (SOME) for Drug Metabolism Prediction

Developed in 2009 by the East China University of Science and Technology, this open-source algorithm is designed to predict the site of metabolism for potential substrates undergoing various metabolic reactions, including hydroxylation at aliphatic and aromatic carbons, dealkylation at oxygen and nitrogen atoms, and oxidation at nitrogen and sulfur atoms. Unfortunately, this algorithm is currently unavailable at the web address originally provided in its publication.

The predictive model was constructed using a dataset compiled from the MDL Metabolite Database, which included substrates, their respective metabolites, and identified sites of metabolism. For each type of reaction, "reaction centres" were pinpointed, specifying the necessary types of atoms and mapping them as "SMARTS" strings. Quantum mechanics calculations were then employed to assess the valency, charge, and energy-related features of these centres.

A support vector machine model was utilised for feature selection when building this prediction model. The model demonstrated particularly strong predictive performance for N- and O-dealkylation reactions. However, it did not successfully predict aromatic hydroxylation and oxidation at sulfur atoms, indicating specific areas where the model may require further refinement or additional data to enhance its accuracy [76].

17.4.26 Overview of Site of Metabolism Prediction (SOMP) for Drug Metabolism Prediction

This open-source web-based application, developed by the Institute of Biomedical Chemistry of the Russian Academy of Medical Sciences, is designed to predict the probability of metabolism by cytochrome P450 enzymes and glucuronidation at specific atoms within a molecule. Users can input the molecular structure using SMILES or MOL or directly draw the structure using the website's Marvin JS chemical drawing tool. The application also allows predictions based on the names of commercially available drugs.

The predictive model was trained using training and test sets of known substrates for CYP - 1A2, 3A4, 2C9, 2C19, and 2D6. The training involved identifying sites of metabolism in substrates by analysing generated metabolites and labelling the involved atoms and the atoms not involved in metabolism. Structures were thus generated with one labelled atom (SoLAs) for substrates with known metabolites for these five CYP isoforms.

To describe these SoLAs, "Labelled Multilevel Neighbourhoods of Atom" (LMNA) descriptors were developed. The application utilises LMNAs alongside the "Prediction of Activity Spectra for Substances (PASS)" algorithm to predict the likelihood of each atom within a molecule being a metabolism site. Each atom is assigned a relative probability, and the results are available for download in CSV, PDF, or Excel (Figs. 17.8, 17.9 and 17.10). The accuracy of this method has been benchmarked against other tools like SMARTCyp and RS-WebPredictor, showing superior invariant accuracy.

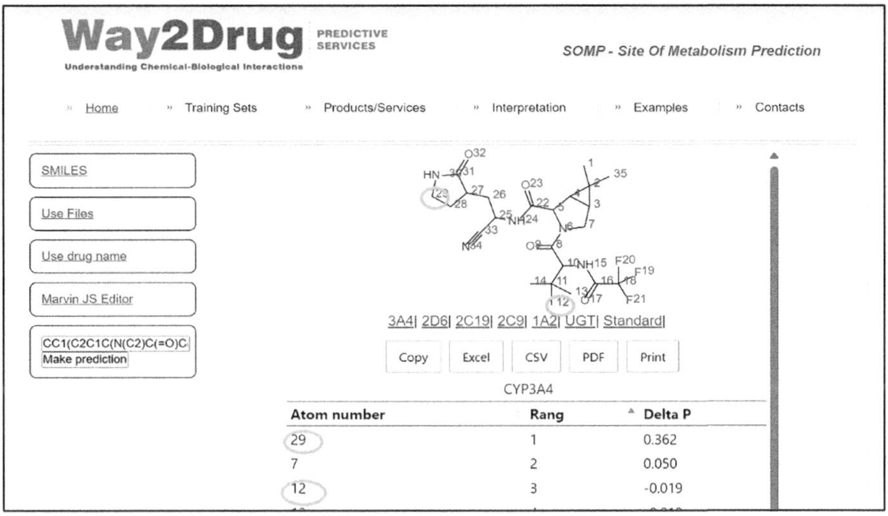

Fig. 17.8 Prediction results of nirmatrelvir from SOMP

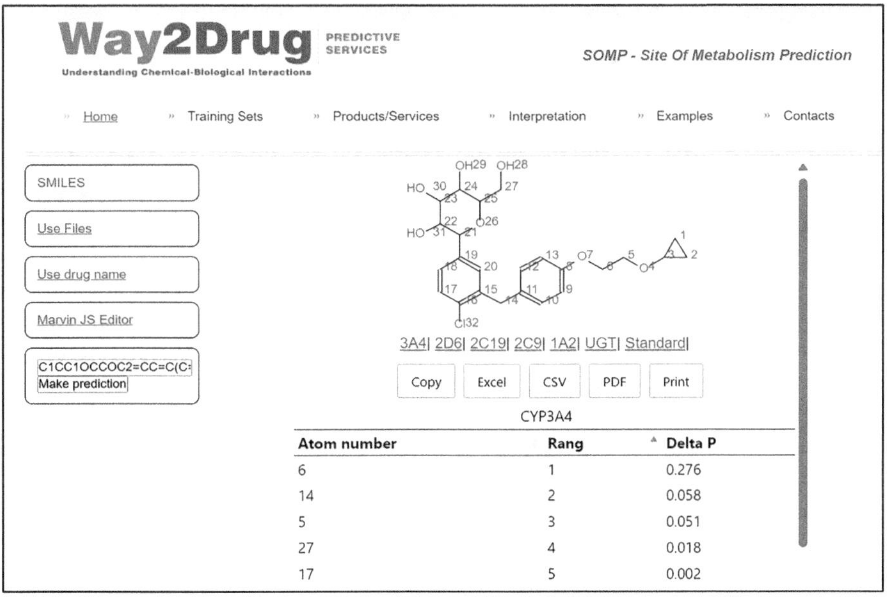

Fig. 17.9 Prediction results of bexagliflozin from SOMP

The predictions are based on 2D descriptors and do not account for interactions with target enzymes, which might affect their applicability in certain scenarios. This application offers a valuable tool for researchers looking to understand and predict metabolic pathways efficiently [77, 78].

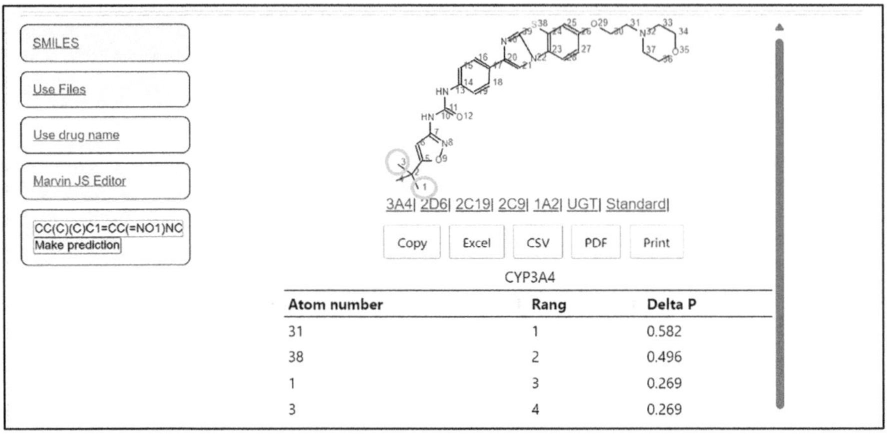

Fig. 17.10 Prediction results of quizartinib from SOMP

17.4.27 Overview of SMARTCyp for Drug Metabolism Prediction

SMARTCyp is an open-source software available for download or online use, designed to predict the SoM using the two-dimensional structure. It specifically identifies metabolism sites for CYP - 1A2, 2A6, 2B6, 2C8, 2C9, 2C19, 3A4, and 2D6. SMARTCyp utilises a set of predefined "SMARTS rules" to assign activation energies based on DFT calculations to potential sites of metabolism and employs an accessibility descriptor to evaluate the likelihood of a site interacting with the metabolising enzyme.

Users can input molecular structures into SMARTCyp using ChemDoodle on the website or by uploading files in SMILES, SDF, or MOL formats. The output from SMARTCyp is displayed in a tabular format for each CYP enzyme, highlighting the top three ranking sites of metabolism in colour on the molecule's structure (Figs. 17.11 and 17.12). The table also ranks each atom as a potential site of metabolism. It provides details such as activation energy, distance from the atom to the molecule's centre, solvent-accessible surface area, and similarity to the atom in the fragment used for calculating activation energy. Atoms with the lowest energy scores are ranked highest, indicating a higher likelihood of being sites of metabolism.

It is important to note that while SMARTCyp is effective in predicting sites of metabolism, it does not predict the probable structure of the metabolites. Additionally, the program was compared against the commercially available application StarDrop, revealing that some false positives might be assigned higher ranks due to the inaccessibility of sites in the 3D structure or because the fragment type was not included in the SMARTS rules, highlighting some limitations in the predictive accuracy of SMARTCyp, especially when dealing with complex three-dimensional molecular structures [79, 80].

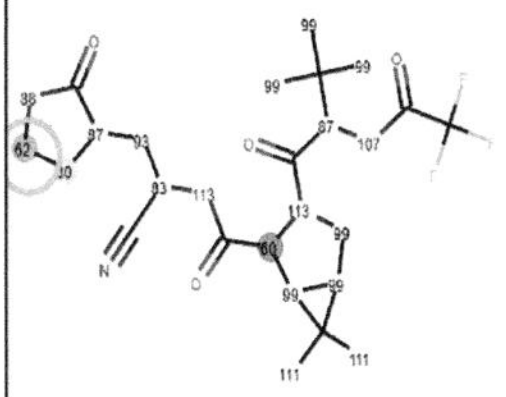

2C9 Ranking	Atom	2C9 Score	Energy	2DSASA	Span2end	COO-Dist	Similarity
1	C.6	60.3	36.9	6.7	7	0	0.3
2	C.15	62.3	63.9	40.2	0	0	0.3
3	C.14	80.5	75.9	33.6	1	0	0.3
4	C.9	83.2	59.9	7.9	4	0	0.4
5	C.13	87.3	75.9	9.1	2	0	0.3

Fig. 17.11 Prediction results of nirmatrelvir from SMARTCyp

Output from calculation

Download CSV file | Download visual data | Download mol file

Smiles string: C1CC1OCCOC2=CC=C(C=C2)CC3=C(C=CC(=C3)C4C(C(C(C(O4)CO)O)O)O)Cl

Smiles 3A4 2D6 2C9

1

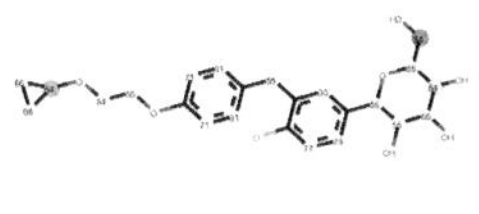

3A4 Ranking	Atom	3A4 Score	Energy	2DSASA	Span2end	Relative Span	Similarity
1	C.27	53.5	62.2	29.4	1	0.9	0.4
2	C.3	53.9	62.2	17.9	1	0.9	0.3
3	C.24	54.3	62.2	9.4	1	0.9	0.3
4	C.5	54.4	62.2	27.9	3	0.8	0.4
5	C.14	54.6	59.9	21.9	8	0.6	0.7

Fig. 17.12 Prediction results of bexagliflozin from SMARTCyp

17.4.28 Overview of SOMEViz for Drug Metabolism Prediction

This open-source tool, an extension of the "Site of Metabolism Estimator," facilitates metabolism prediction by allowing users to input the structure of prospective substrate molecules through a graphic user interface in various formats, including MOL2, MOL,

SDF, or PDB. The program generates 3D coordinates for these molecules and integrates MOPAC to perform molecular orbital calculations.

Quantum chemical features are calculated to assess the molecular environment, and a SVM model is used to identify the probable SoM. The model specifically predicts hydroxylation at aliphatic and aromatic carbons, dealkylation at oxygen and nitrogen atoms, and oxidation at nitrogen and sulfur atoms. Potential sites of metabolism are colour-coded within the program (to enhance user understanding and visualisation), with each atom's likelihood of participating in metabolic reactions clearly indicated.

Results from the analysis, can be downloaded for further examination. However, it should be noted that the webpage address provided in the publication for accessing this tool is currently not accessible [81].

17.4.29 Overview of StarDrop from Drug Metabolism Prediction

The "Metabolism" module within StarDrop, developed by Optibrium, is a commercially available tool designed for metabolite prediction. It focuses on identifying the likely sites of metabolism mediated by key enzymes such as CYP, aldehyde oxidase, flavin-containing monooxygenases, glucuronyl transferases, and sulfotransferases, applicable to humans and other species. The predictions hinge on assessing the reactivity and accessibility of potential metabolism sites within a molecule.

This module utilises ML models to predict the SoM. Additionally, it can assess the probable toxicity of these metabolites using the Derek Nexus system. This integrated approach provides a comprehensive toolset for researchers to evaluate both the metabolic stability and safety profiles of compounds. The "Metabolism" module in StarDrop developed by Optibrium is a commercially available software for the prediction of metabolites. It identifies the probable site of metabolism by CYP, aldehyde oxidase, flavin-containing monooxygenases, glucuronyl transferases and sulfotransferases in humans and other species. The prediction is based on the reactivity and accessibility of the site of metabolism. Prediction models were built using ML models to predict a molecule's metabolism site. The probable toxicity of the metabolites may also be predicted using Derek Nexus [82].

17.4.30 Overview of Systematic Generation of Potential Metabolites (SyGMa) for Drug Metabolism Prediction

This open-source algorithm, available for download on GitHub, predicts probable metabolites resulting from both human and rat phase I and II metabolic reactions. It operates by applying a set of predefined reaction rules to the query compound to forecast the structure of anticipatable metabolites. Additionally, fingerprints are calculated for query compound and the probable metabolites, facilitating the evaluation of results through similarity scores such as the Tanimoto score or the Soergel distance.

Each metabolite generated by this algorithm is assigned a probability score ranging from 0.009 to 0.85, reflecting the likelihood of its formation based on the applied reaction rules. These rules have been developed using data sourced from compounds listed in the MDL database, ensuring a robust foundation for predictions.

In terms of performance, this algorithm demonstrates predictive capabilities comparable to those of established tools like MetaSite and Meteor, making it a valuable resource for researchers involved in drug metabolism and pharmacokinetics studies [83, 84].

17.4.31 Overview of Tissue Metabolism Simulator (TIMES) for Drug Metabolism Prediction

TIMES is a commercially available software developed by the Oasis Laboratory of Mathematical Chemistry, designed specifically for predicting the toxicity of compounds by analyzing probable metabolites. This application utilizes "CATABOL," a hybrid method that combines rule-based and statistical approaches to predict metabolism, acknowledging that compounds can undergo metabolism through multiple pathways.

The software effectively predicts both primary and secondary metabolites of a compound, providing insights into their relative abundance and stability. Each predicted metabolite is assigned a reliability score to gauge the confidence in these predictions. Additionally, TIMES assesses the molecular properties and potential toxicity of these metabolites, aiding researchers in understanding the overall toxicological profile of the parent compound [85, 86].

17.4.32 Overview of Xenosite for Drug Metabolism Prediction

Xenosite is an innovative open-source online platform developed by the Washington University School of Medicine. It effectively identifies sites of metabolism and predicts the products of both phases of human metabolism, encompassing unstable and stable oxygenation, dehydrogenation, hydrolysis, reduction, and glucuronidation. The platform employs a unique "metabolic rainbow" colour-coding system to distinguish between different reactions and enhance user understanding.

The platform categorises twenty-one phase I reactions into five groups, with models for each category trained using the same dataset. This approach allows for a comprehensive reaction coverage, making Xenosite more extensive than other models like RS-predictor, FAME 2, or SMARTCyp. A substantial dataset of 9674 molecules detailing the structures of substrates, their metabolites, and reaction types was used to develop the model. Various descriptors, including topological, quantum chemical, and "SMARTCyp" reactivity descriptors, along with SMARTS patterns, are utilised to pinpoint potential sites of metabolism. Neural network prediction models are then constructed using these descriptors.

Users can input molecular structures in SMILES format, and the platform displays the structures of potential metabolites on the results page. Metabolism sites

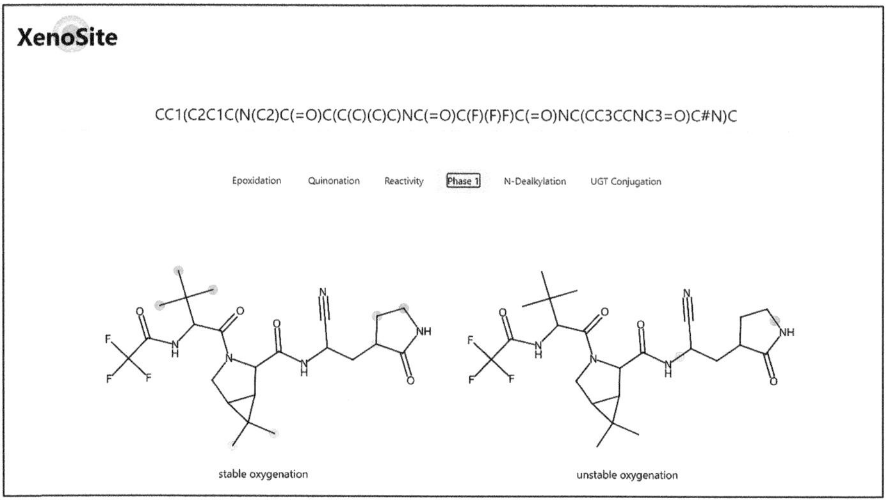

Fig. 17.13 Prediction results of nirmatrelvir from XenoSite

are visually differentiated using varying colour intensities based on the likelihood of metabolism, providing an intuitive and informative user experience (Figs. 17.13, 17.14 and 17.15).

Additionally, an extension of this tool, XenoNet, offers capabilities to predict pathways for converting a precursor molecule to a specific metabolite, including the structures of intermediates and the likelihood of these pathways, further expanding the utility of Xenosite in metabolic research and drug development [87, 88].

XenoNet: It is an extension of the Xenosite platform, designed to predict the metabolic pathways that convert a precursor molecule into a specific metabolite. This tool focuses on identifying the structures of intermediates involved in these pathways and calculates the probability of each pathway occurring. XenoNet enhances the capabilities of Xenosite by providing detailed insights into the step-by-step biochemical transformations a molecule undergoes, facilitating a deeper understanding of drug metabolism and disposition. Thus, XenoNet is particularly valuable for researchers and pharmaceutical developers looking to predict and analyse the metabolic profiles of new compounds, optimise drug candidates, and assess potential metabolic liabilities [89].

17.5 Future Perspective

The prediction of a drug's site of metabolism stands as a pivotal juncture where chemistry meets biology, paving the way for safer and more productive medications. Understanding where and how a drug undergoes metabolic transformations within the human body is essential for deciphering its pharmacokinetic profile,

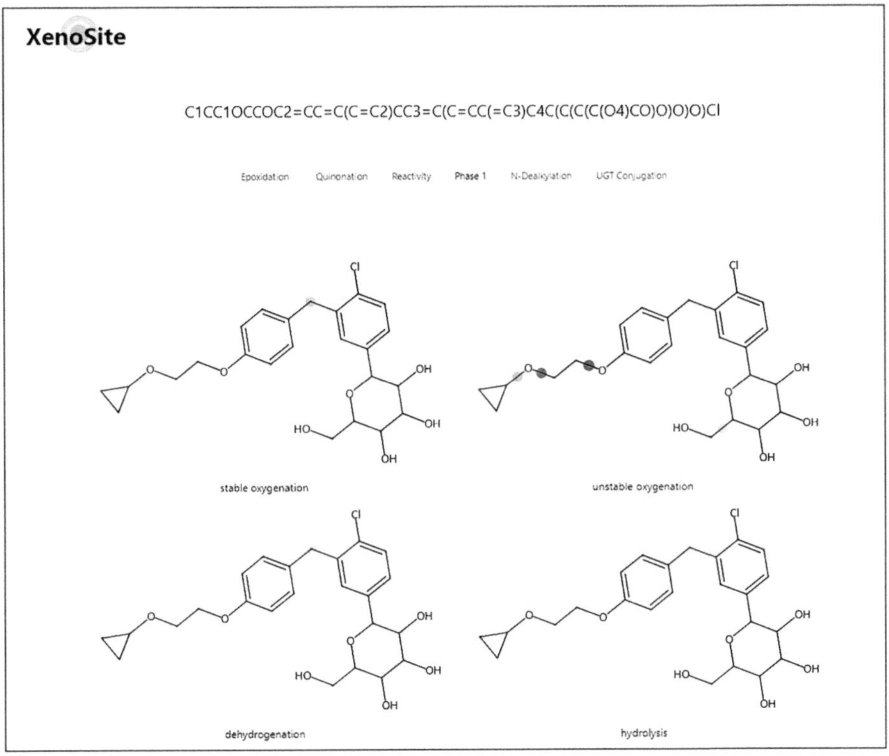

Fig. 17.14 Prediction results of bexagliflozin from XenoSite

toxicity, and potential drug interactions. Over the years, the quest for accurate site of metabolism prediction has evolved from traditional empirical approaches to cutting-edge computational methodologies, heralding a future where precision medicine and tailored therapeutics become the norm.

17.5.1 The Journey Thus Far

Drug metabolism predictions rely heavily on experimental studies While these methods provide valuable insights, they are laborious and time-consuming. The emergence of computational approaches in the late twentieth century revolutionised the field, offering a faster and more cost-effective means of predicting the site of metabolism.

Initially, computational methods focused on QSAR, while QSAR models significantly improved predictive capabilities, their reliance on predefined molecular features limited their applicability to structurally similar compounds.

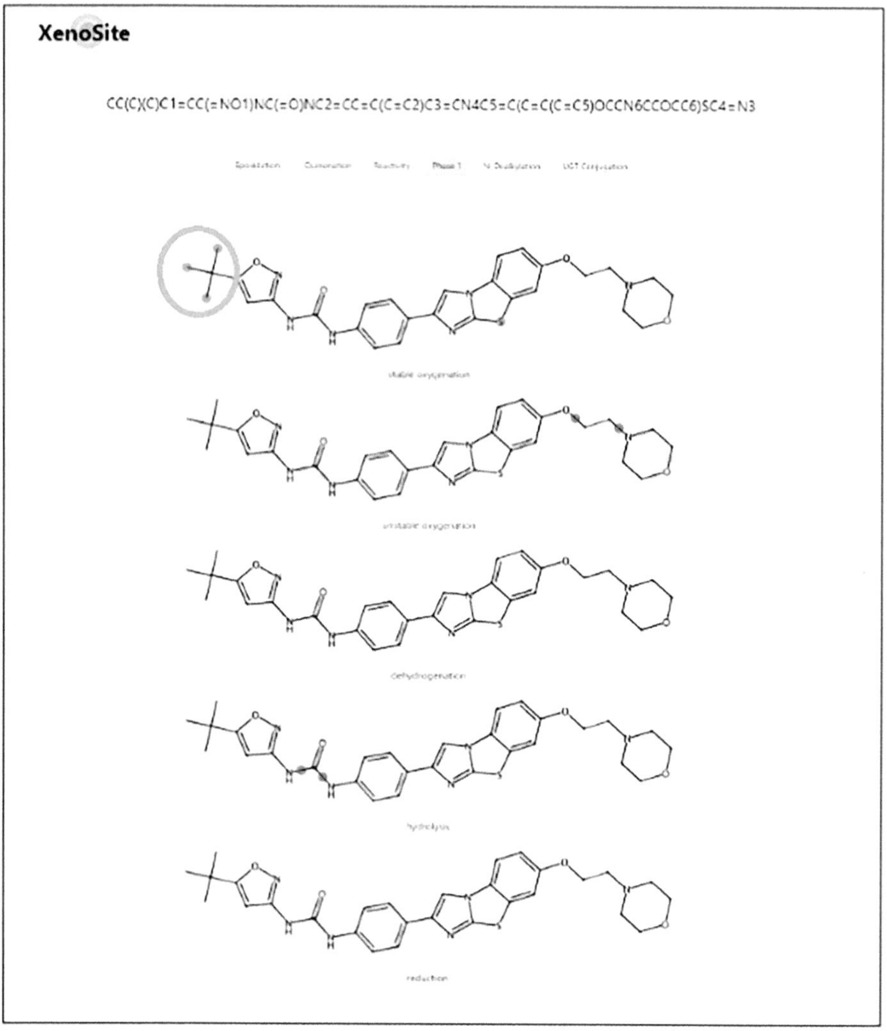

Fig. 17.15 Prediction results of quizartinib from XenoSite

ML algorithms marked a paradigm shift in SoM prediction, demonstrating superior predictive performance by learning complex patterns directly from molecular data. These ML models, trained on vast datasets of chemical structures and their corresponding metabolic pathways, exhibited remarkable accuracy and generalisation across diverse chemical spaces.

17.5.2 Looking Ahead

As we peer into the future of SoM prediction, several trends and advancements are poised to reshape the landscape of drug development and personalised medicine:

"Organ-on-a-Chip" Technology

This technology markedly enhances cell-to-cell communication by replicating human tissues' complex microenvironments and intercellular interactions. This advancement over traditional 2D cultures allows for a more precise simulation of organ-specific functions, which is crucial for studying drug metabolism. The intricate arrangement of multiple cell types in these microdevices mimics natural physiological responses more closely, potentially influencing the formation and detection of drug metabolites. This enhanced fidelity increases appreciation of the pharmacokinetics and pharmacodynamics of drugs, yielding data more representative of in vivo conditions.

"Organ-on-a-chip" systems offer significant predictive capabilities by incorporating multiple cell types to simulate the biochemical processes of target organs. This setup is particularly beneficial for studying hepatic and renal drug metabolism, where specific enzymes play crucial roles. By mimicking the organ environment, these chips can more accurately predict where and how drugs are metabolised. This approach improves the predictability of in vivo drug behavior.

The application provides a reliable alternative to animal testing in the drug development pipeline. By providing human-specific data that predict clinical outcomes, these systems can decrease the number of necessary animal studies, thereby addressing ethical concerns, reducing costs, and speeding up the development process. Furthermore, enhanced predictability means fewer failed human trials, making drug development more efficient and potentially safer by focusing resources on candidates with a higher likelihood of success.

"Organ-on-a-chip" technology can seamlessly integrate with other cutting-edge technologies, such as high-throughput screening, real-time biomarker monitoring, and computational modelling. This integration enhances the system's predictive accuracy and utility by simultaneously analysing multiple pharmacokinetic parameters and rapidly identifying adverse reactions. Such technological synergy can lead to a deeper understanding of drug behaviour across different biological systems, potentially revolutionising the drug discovery and testing processes.

Despite its promise, "organ-on-a-chip" technology faces several challenges and limitations that must be addressed. These include issues related to the scalability of the chips for widespread industrial use, the complexity of their design and operation, and the requirement for further validation to ensure that these models truly reflect human responses. Additionally, the high initial cost and technical expertise required to develop and maintain these systems may limit their accessibility and widespread adoption in the current drug development landscape.

Adopting "organ-on-a-chip" models in drug metabolism studies also involves significant regulatory considerations. As these models aim to substitute traditional animal testing methods, their regulatory acceptance (FDA, EMA, and others) is crucial. Ongoing discussions and pilot programs are essential to establish guidelines that can accommodate the unique data generated by these systems. Successful regulatory integration would validate the utility of organ-on-a-chip models and pave the way for their broader application in drug safety and efficacy assessments [90].

Integration of Omics Data

The integration of omics technology into the field of drug metabolism (particularly for predicting the SoM) represents a promising frontier with significant potential to transform drug discovery and development processes. Omics technologies provide comprehensive insights into the biological pathways pharmaceuticals affect, thereby enhancing our understanding of drug metabolism at the molecular level.

By analysing the vast datasets generated by these omics studies, genes, transcripts, proteins, and metabolic pathways involved in drug metabolism can be precisely identified. For instance, proteomics can reveal the presence and concentration of critical enzymes in the metabolic pathways, highlighting potential sites of drug metabolism that could not be inferred from genetic data alone.

Furthermore, the integration of omics data facilitates a systems biology approach to predicting drug metabolism, where multiple layers of biological information are combined to model the complex interactions within a biological system. This method can predict how variations in one part of the system affect the rest, such as how genetic variations influence the expression levels of enzymes responsible for drug metabolism, potentially leading to personalised medicine strategies.

However, integrating omics data into drug metabolism also presents significant challenges, primarily related to data management and analysis. The sheer volume and complexity of omics data require advanced computational tools and robust bioinformatics expertise to ensure accurate interpretation and meaningful insights. Additionally, integrating diverse omics datasets to create a coherent drug metabolism model is computationally intensive and requires innovative analytical strategies to handle the data heterogeneity.

Despite these challenges, omics data integration is gaining traction in regulatory frameworks. Regulatory agencies are beginning to recognise the value of omics data in supporting drug approval processes. They are working towards guidelines to incorporate this data into safety and efficacy evaluations. This regulatory acceptance is crucial for advancing the use of omics data in predicting drug metabolism sites, as it not only enhances the scientific credibility of omics approaches but also encourages more widespread adoption in the pharmaceutical industry [91].

Deep Learning Architectures

Deep learning architectures are poised to revolutionise the prediction of drug metabolism sites by leveraging their capacity to handle complex, nonlinear data sets and uncover patterns that are not readily apparent through traditional methods. These models can consolidate diverse types of data, such as chemical structures, biological activity data, and patient-specific variables, to predict where and how a drug will be metabolised in the body.

Using deep learning in drug metabolism can significantly enhance the accuracy of predicting metabolic pathways and sites. For example, CNNs can analyse molecular images or structural data to predict how a molecule's conformation influences its interaction with metabolic enzymes. Similarly, RNNs are well-suited for analysing sequential data, such as changes in enzyme activity over time, providing insights into dynamic aspects of drug metabolism.

Moreover, deep learning can facilitate the discovery of novel drug metabolism pathways that have not been documented. These models can identify potential unknown metabolic sites and pathways, which can then be experimentally validated. This capability is particularly beneficial for developing drugs with novel structures or for drugs that interact with multiple targets.

By continuously updating the learning models with new data, the predictive accuracy of these systems can improve over time, leading to models that better reflect the complexity of human metabolism.

However, applying deep learning to drug metabolism requires careful consideration of model interpretability and validation. Deep learning models, often called "black boxes," can provide predictions without transparent explanations, which can be a significant hurdle in a field where understanding the mechanism of action is crucial. Developing techniques to interpret the results generated by deep learning models is essential to gaining the trust of researchers and regulatory bodies alike.

As computational power increases and more comprehensive datasets become available, deep learning architectures will likely become a central element in the pharmacokinetics toolbox, offering more precise and predictive tools for understanding drug metabolism. This shift promises to accelerate the drug development process and enhance the safety and efficacy of therapeutic interventions [92].

Quantum Computing

Quantum computing represents a transformative advance for drug metabolism, particularly in predicting the sites of drug metabolism, due to its unparalleled processing power and ability to handle extremely complex calculations much faster than classical computers. This emerging technology could revolutionise how pharmaceutical scientists explore and understand the quantum mechanics underlying molecular interactions involved in drug metabolism.

One of the principal advantages of incorporating quantum computing technology in drug metabolism is its potential to accurately simulate the electronic structures of molecules, an essential aspect of understanding how drugs interact with metabolic enzymes. Due to the computational cost, traditional computational methods can struggle to simulate large, complex molecules. However, quantum computers can potentially perform these simulations more efficiently and accurately, providing detailed insights into the reactivity and stability of drug metabolites, which are crucial for predicting metabolism sites.

Furthermore, quantum computing could enhance the optimisation of drug candidates with favourable metabolic profiles. Using quantum algorithms, researchers can explore a vastly larger chemical space than is feasible with classical computing. This ability allows for identifying novel drug candidates that are therapeutically effective and have optimal metabolism properties, reducing the risk of adverse metabolites and improving drug safety.

Integrating quantum computing with machine learning techniques, specifically quantum machine learning, also offers promising avenues for predicting drug metabolism sites. Quantum machine learning can potentially process vast datasets of drug interactions and metabolic pathways much faster than current methods, leading to quicker and more accurate predictions. This integration can significantly speed up the early stages of drug development.

However, applying quantum computing in drug metabolism is not without challenges. The technology is still in its infancy, with significant technical hurdles related to hardware stability, error rates, and qubit coherence times. Additionally, developing quantum algorithms tailored specifically for drug metabolism requires deep collaboration between quantum physicists, chemists, and pharmacologists.

Despite these challenges, the future perspectives of quantum computing in drug metabolism are highly promising. As technology matures and more practical quantum computers become available, they will likely become invaluable drug discovery and development tools, leading to breakthroughs in predicting drug metabolism sites, ultimately making drug development more efficient and potentially revolutionising personalised medicine [93].

Big Data and Cloud Computing

Big Data and Cloud Computing technologies enable the aggregation, storage, and analysis of data from sources such as genomic databases, clinical trial repositories, and real-time patient monitoring systems, which are vital for understanding complex drug interactions at the metabolic level.

Big data analytics can decipher large-scale patterns and trends from the aggregated data, including historical information on drug metabolism from past clinical studies and experiments. By employing advanced algorithms and data mining techniques, researchers can identify common metabolic pathways and predict potential sites for drug metabolism. This approach not only speeds up the identification process but also improves the precision of the predictions, making it possible to foresee adverse metabolic reactions before clinical testing.

Cloud computing provides the necessary computational power and scalability in this integration. It efficiently handles the enormous computational tasks involved in processing and analysing big data. The cloud's capacity to facilitate collaboration across different geographic locations and institutions enhances the collective understanding of drug metabolism processes.

Furthermore, applying machine learning models in cloud-based platforms allows for continuous learning and improvement in predictive capabilities. As new data become available, these models can update their predictions on drug metabolism sites, making them increasingly accurate over time. This dynamic capability is critical in adapting to rapid drug development and personalised medicine changes.

Despite the benefits, using big data and cloud computing in drug metabolism also introduces data privacy, security, and management challenges. Ensuring the integrity and confidentiality of sensitive medical and genomic data is paramount. Additionally, the heterogeneity of data types and sources requires robust data integration and management strategies to ensure consistency and reliability in the analysis [94].

In Silico Metabolism Simulations

In silico metabolism, simulations are increasingly becoming a cornerstone in drug metabolism, especially for predicting the sites of drug metabolism. These simulations utilise computational models to mimic and predict how drugs are metabolised in the human body, offering a rapid, cost-effective, and ethically favourable alternative to traditional *in vitro* and *in vivo* methods.

One of the key advantages of in silico approaches is their ability to generate data on potential metabolic pathways and sites quickly. By employing models that integrate biochemical, genetic, and physiological data, these simulations can predict where a drug will be metabolised in the body and by which enzymes it will be metabolised.

Furthermore, *in silico* simulations can be used to model population-wide metabolism variability, which includes factors such as age, ethnicity, and genetic polymorphisms affecting enzyme activity. This application is particularly valuable for personalised medicine in understanding how different demographic groups might metabolise drugs differently, guiding more tailored and effective treatment options.

Integrating quantum mechanics and molecular dynamics with *in silico metabolism simulations* enhances their predictive power. These methods provide detailed insights into drug molecules' electronic and structural properties and their interactions with metabolic enzymes. Such detailed simulations help design drugs that are less likely to be metabolised into harmful compounds, improving drug safety.

Machine learning algorithms further augment *in silico* metabolism simulations by analysing patterns in large datasets derived from experimental and computational studies. These algorithms can improve the accuracy of metabolism predictions by learning from past data, identifying complex patterns, and predicting outcomes for new drugs based on learned relationships.

However, the effective use of *in silico* metabolism simulations also presents challenges, particularly in model validation and standardisation. Ensuring that these simulations accurately reflect real-world outcomes requires ongoing refinement of the models and validation against experimental data. Moreover, there is a need for standardised protocols to guide the use of these simulations in regulatory submissions, as acceptance by regulatory bodies is critical for their broader application in drug approval processes [95].

17.6 Conclusion

The evolving landscape of drug metabolism research is being profoundly reshaped by innovations in technology and computational methods. Organ-on-a-chip technology, the integration of omics data, deep learning architectures, quantum computing, big data, cloud computing, and in silico metabolism simulations each hold the promise to enhance the prediction of drug metabolism sites, thereby streamlining drug development processes and improving drug efficacy and safety. These technologies enable more precise and predictive insights into how drugs interact within the body and pave the way for the customisation of therapeutic strategies depending on distinctive genetic and metabolic profiles. However, successfully implementing these advanced technologies in drug metabolism studies hinges on overcoming challenges related to data complexity, computational demands, and integrating diverse data types, alongside navigating the regulatory landscape. Addressing these challenges is essential for realising the full potential of these technologies in enhancing drug discovery and development, ultimately leading to safer, more effective treatments.

Glossary

Accelrys Metabolite Database The Accelrys Metabolite Database, now part of the BIOVIA offerings from Dassault Systèmes (following the acquisition of Accelrys), is a specialised repository that provides comprehensive data on metabolites. It includes detailed information on biotransformation reactions of drugs and other substances, tracking how these compounds are metabolised in the body. This database is particularly useful for researchers involved in drug development and safety assessment, as it helps in understanding the metabolic pathways that might influence a drug's efficacy and toxicity. It is widely used in pharmacokinetics and toxicology studies to predict metabolic stability and potential metabolite-related issues.

Accessibility It describes how easily a particular site within a molecule can be approached and affected by reagents or enzymes. It considers factors such as the physical obstruction of other parts of the molecule and the solvent accessibility, which influences whether enzymes or solvents can reach reactive sites.

AlogP This coefficient, often denoted as log P, measures a molecule's lipophilicity or hydrophobicity. AlogP specifically refers to the algorithmically calculated version of this logarithm, often derived from computational chemistry software that predicts this value based on the molecular structure without the need for experimental determination.

Artificial Neural Network Ensemble (ANN Ensemble) An Artificial Neural Network Ensemble involves the combination of multiple neural networks to solve a problem more effectively than could be done by a single model. ANN Ensembles are used to analyse complex biological data and predict how compounds interact with various enzymes, enhancing the predictability of metabolic reactions and sites of metabolism.

Bioclipse It is an open-source workbench for cheminformatics that provides tools for visualising, editing, and managing molecular structures, sequences, and other biological data. It combines the functionality of a traditional desktop application with the capabilities of a powerful scripting language, allowing for automation and extension of its features.

BIOVIA Metabolite Database The BIOVIA Metabolite Database is part of the software solutions offered by Dassault Systèmes, which specialises in providing comprehensive data on metabolites derived from both drugs and natural products. This database is utilised extensively in pharmacokinetics and drug metabolism studies, aiding researchers in understanding how substances are transformed in biological systems. It includes detailed information on metabolic pathways, reaction types, and enzyme interactions, which are crucial for predicting drug behaviour and safety.

CDK "Chemistry Development Kit" provides algorithms and data structures to manipulate chemical data. CDK can generate various descriptors ranging from simple count descriptors (e.g., the number of atoms) to more complex descriptors involving molecular topology or electron distribution.

ChemDoodle It is a chemical sketching software that allows users to draw chemical structures, create cheminformatics applications, and generate chemical graphics. It is versatile software used in educational and professional environments to visualise molecular structures and reactions. ChemDoodle supports various chemical file formats for input and output, integrates with web technologies for online deployment, and offers tools for analysing chemical properties, making it useful for researchers and educators in the chemical sciences.

ClogP This partition coefficient measures how well a substance partitions between a hydrophobic phase (n-octanol) and a hydrophilic phase (water).

CLI A "Command Line Interface" CLI is a user interface software that allows users to type textual commands to perform specific tasks—in contrast to using the mouse-driven GUI. CLIs are often preferred for more detailed and complex operations that require precise commands.

CSV (Comma-Separated Values) Many applications widely support this format, and it is commonly used for importing and exporting data to and from databases and spreadsheet tools.

CypBoM (Cytochrome P450 Bond of Metabolism) Utilising machine learning techniques and robust datasets, CypBoM identifies the specific "bond of metabolism" (BoM) that is likely to occur with a CYP enzyme. It categorises BoMs into four types: existing bonds between two non-hydrogen atoms, potential new bonds between non-hydrogen atoms, bonds between a non-hydrogen atom and one of its connected hydrogen atoms, and potential bonds involving nitrogen, sulfur, or oxygen atoms.

CyProduct It is a machine learning-based tool that integrates predictive models and algorithms to determine the metabolism of compounds, specifically focusing on CYP-mediated reactions. It utilises a comprehensive dataset of known biotransformation reactions to predict how and where molecules are metabolised by various CYP enzymes, aiding in assessing drug-drug interactions and potential toxicity in drug development.

CypReact It is a computational tool designed to predict the specificity of substrates towards various CYP enzymes. It uses information about molecular structure and properties to assess how likely specific CYP isoforms, such as CYP1A2, CYP2A6, CYP2B6, among others, will metabolise a compound. This prediction is crucial for understanding potential drug interactions and metabolic pathways in drug development processes.

Derek Nexus System Derek Nexus employs a knowledge-based approach, utilizing a vast database of toxicological data derived from both experimental results and established toxicology principles.

DFT (Density Functional Theory) DFT uses theoretical principles to approximate the behaviour of electrons in these systems, allowing chemists and physicists to predict chemical properties such as reactivity, molecular geometry, and electronic properties. DFT can be employed in drug metabolism to calculate activation energies and other relevant properties that influence how a drug interacts with metabolic enzymes.

DrugBank This database offers extensive information on drugs' chemical, pharmacological, and pharmaceutical properties, making it a vital resource for researchers and healthcare professionals.

Extended Connectivity Fingerprint 4 (ECFP4) ECFP4 is a type of circular fingerprint used in cheminformatics to describe molecular structures. It is part of a broader family of extended connectivity fingerprints, which are generated by considering not only the atoms but also the bonds and immediate atom neighbours within a preset diameter around the molecule.

FAME Fingerprint The FAME (Functional Analysis of Metabolism Enzymes) fingerprint is a computational tool used to generate unique identifiers for molecules based on their potential metabolic sites and enzyme interactions. These fingerprints are used in the FAME software to help characterise and predict the metabolism of drugs and other compounds. They encode specific features of a molecule relevant to its metabolism, such as the presence of functional groups that are likely targets for metabolic enzymes.

FAME Score It is a metric calculated by the FAME software that quantifies the likelihood of a particular site within a molecule being metabolised by enzymes. This score is derived from analysing the molecule's FAME fingerprint and comparing it with known metabolic transformations stored in databases. A higher FAME score indicates a higher probability that the site will undergo metabolism, providing valuable insights for predicting drug metabolism pathways and designing molecules with desired pharmacokinetic properties.

FRED "Fast Rigid Exhaustive Docking" is a software developed by OpenEye Scientific Software.

GitHub It is a platform that allows developers to manage their projects, track changes, collaborate with others, and host and review code.

GLIDE "Grid-based Ligand Docking with Energetics" is used in computational chemistry for docking small molecules into the active site of a receptor or enzyme.

GUI "Graphical User Interface", GUIs are designed to make software more intuitive and easier to use.

InChI (International Chemical Identifier) The InChI string comprises several layers of information describing the chemical structure, including the composition, connectivity, and geometry of molecules. It is machine-readable and designed to compare chemical substances and their properties across different databases easily.

JSME Molecular Editor It is a lightweight, web-based chemical editor developed by Peter Ertl and Bruno Bienfait. It is used for drawing and editing chemical structures directly within web browsers without additional software installations. JSME is derived from the Java-based JME editor and supports functionalities such as drawing chemical structures and generating "SMILES" notation. This tool is commonly integrated into chemical databases and predictive modelling applications, allowing users to input and modify molecular structures conveniently.

JSON "JavaScript Object Notation" is widely used in web applications to exchange data between clients and servers and is known for its simplicity and ease of integration with web technologies.

KEGG Atom Type Format In the context of chemical data, the "Kyoto Encyclopedia of Genes and Genomes" atom type format refers to a specific notation used within KEGG to describe the types of atoms in a molecule based on their chemical environment and functional roles in reactions.

LC-MS/MS This highly sensitive and specific analytical technique makes it invaluable in various fields such as pharmacology, toxicology, and environmental analysis.

LMNA (Labelled Multilevel Neighbourhoods of Atom) LMNA capture the immediate and extended neighbourhood of labelled atoms, providing detailed information that enhances the accuracy of predictive models focusing on metabolism sites.

Marvin JS Chemical Drawing Marvin JS is a chemical editor to draw and visualise chemical structures directly in their web browser.

MetaboGen This module focuses on predicting the structures of potential metabolites. It examines each atom in a molecule to predict which are involved in metabolic reactions, the expected reactions, and the metabolites' resultant structure. It leverages datasets compiled from drugs and their documented metabolites to ensure accurate correlation between substrates and their metabolites.

MetaGlide It is not as commonly referenced in the context provided, suggesting it might be a miscommunication. However, in general, Glide is a well-known molecular docking software from Schrödinger that is used to predict how small molecules bind to a receptor or enzyme.

MetaQSAR Database It is a specialised resource that supports QSAR studies. It compiles data on metabolic reactions, substrate specificity, and enzyme kinetics, which are vital for predicting how compounds interact with various enzymes, particularly those involved in metabolism. The database is used to build predictive models that can estimate the metabolic stability and likelihood of biotransformation of chemical compounds, aiding in drug development and toxicology studies.

MetaSite This software predicts how and where a molecule will be metabolised in the body, particularly focusing on metabolism mediated by cytochrome P450 enzymes. It uses computational methods to simulate and predict metabolic pathways and sites.

Meteor Meteor uses an extensive knowledge base derived from experimental data to simulate and predict how compounds are metabolised in the body. It applies a set of predefined biotransformation rules, which are based on known metabolic pathways, to estimate the structures of potential metabolites.

MEXAlert It is likely a specialised software tool, though not widely recognized outside of its specific usage context mentioned earlier. Assuming from the context provided, MEXAlert is a simplified version of a more comprehensive metabolic prediction software designed to predict the probability of first-pass metabolism for chemical compounds. First-pass metabolism refers to the metabolization of a substance within the liver immediately following its entry into the body and before it reaches systemic circulation.

MetXBioDB It is a specialised database that focuses on metabolites and biotransformation reactions, particularly detailing how drugs and other substances are metabolised in biological systems. It compiles experimental data on the enzymatic reactions involved in drug metabolism, providing insights into how different compounds are processed in the body. This information is crucial for understanding drug interactions, effects, and safety profiles, supporting drug design and toxicology research.

MOL File The MOL file format is widely used in cheminformatics applications to exchange and store molecular structure data. The format allows for describing complex chemical structures, including stereochemistry, making it valuable for computational chemistry and drug discovery.

PaDEL "Padel Descriptor" supports over 1800 different molecular descriptors and more than 10 types of molecular fingerprints, making it a comprehensive tool for generating data that helps in modelling and predicting the behaviour of chemical compounds in biological systems.

ParaSurf It is a computational tool developed to generate molecular surfaces and properties based on 3D molecular structures using quantum mechanical calculations. It is particularly useful for deriving pharmacophore models and for assessing the interaction potential of molecules with biological targets. ParaSurf analyses the electronic properties and geometries of molecules to create detailed maps of molecular interaction fields, which can guide the optimisation of molecular designs in drug discovery.

PASS (Prediction of Activity Spectra for Substances) PASS is a software tool used to predict a wide range of biological activities of chemical compounds based on their structure

Plx It refers to a specific *in silico* graphic user interface program developed or utilised by Bayer. This program is likely designed to integrate various computational tools for predicting and analysing metabolic pathways in drug discovery. The interface would typically allow users to input molecular structures, run simulations, and visualise predictions regarding metabolism and other pharmacokinetic responses, facilitating the drug development process.

PRIME It is a molecular modelling tool developed by Schrödinger that is used primarily for protein structure prediction and refinement, as well as for predicting protein-ligand interactions. It plays a crucial role in the accurate prediction of protein structures and their dynamics, especially when experimental data might be incomplete or ambiguous.

PROLOG (Programming in Logic) It is based on the logic programming paradigm, which involves expressing algorithms as logical relationships and using a mechanism of rules and facts to solve problems. PROLOG is well-suited for tasks that involve pattern matching, rule-based reasoning, and inference.

PubChem PubChem is a source of molecular descriptors typically derived from the database's structural and physical properties of molecules catalogued. These descriptors can include molecular weight, polarity, or more complex features related to the molecule's potential biological activity.

Reactivity It refers to the propensity of a chemical to undergo chemical reactions, either by itself or with other materials, and how readily these reactions occur. Various factors influence reactivity, including the molecule's electronic structure, the presence of functional groups, and the overall molecular stability. In drug metabolism, reactivity often determines how a molecule interacts with metabolic enzymes, impacting its breakdown and transformation into different metabolites.

RDKit Library RDKit is an open-source cheminformatics software library that allows the manipulation of chemical information ranging from molecular structures to complex chemical reactions. RDKit includes functionalities for molecular drawing, generating standard molecular representations like SMILES and InChI, computing molecular descriptors, conducting substructure searches, and building machine learning models based on chemical properties. It is widely used in drug discovery and chemical biology for data analysis and predictive modelling.

RDF "Resource Description Framework" facilitates the merging of data based on the open-world assumption, allowing for effective data integration regardless of the underlying schemas. RDF is used to make assertions about web resources in a structured way.

RS-WebPredictor This web-based tool predicts sites of metabolism for drug-like molecules, particularly focusing on reactions mediated by cytochrome P450 enzymes. RS-WebPredictor uses rule-based and ML approaches to provide predictions.

SDF (Structure-Data File) SDF is a file format developed by Molecular Design Limited (MDL) primarily used to store multiple chemical structure records and associated data. It is widely used in cheminformatics to handle datasets of compounds, facilitating easy sharing of molecular structures and related properties. Each record in an SDF file contains a structure expressed in terms of atoms, bonds, connectivity, and coordinates, followed by associated property data.

SIMPCOMP (Simple Compound Comparison) It is a computational tool used primarily for comparing molecular structures. It simplifies the complex process of identifying similarities and differences between chemical compounds by breaking down their structures into smaller, more manageable components.

It allows for a more straightforward analysis of the groups involved in various chemical reactions, such as those occurring during metabolism. It is particularly useful in cheminformatics for comparing and cataloguing chemical libraries based on structural features.

SMARTCyp SMARTCyp is a predictive tool specifically designed to forecast CYP mediated metabolism in drug molecules. It uses a rule-based system that considers molecular structure to predict which atoms are most likely metabolised by CYP enzymes, informing drug design and safety assessments.

SMARTS Rules SMARTS stands for SMILES Arbitrary Target Specification. It is a language that defines molecular patterns and properties within chemical structures. In computational chemistry, SMARTS rules are commonly used to identify specific substructures or chemical motifs in molecules, which can be crucial for predicting reactions, sites of metabolism, or binding characteristics in larger molecular systems.

SMIRKS (SMILES Extension for Reactions) SMIRKS is a notation that allows the specification of transformations on molecular patterns. It extends SMILES by adding reaction and transformation capabilities. SMIRKS is used to define and apply changes in molecular structures during simulations of chemical reactions.

SoLAs (Sites of Labelling Atoms) This term refers to a set of molecular structures where specific atoms known to be involved in metabolic reactions are explicitly labelled. These labelled structures help in training predictive models to recognise similar sites in new compounds, facilitating the identification of likely metabolic transformations.

Symyx Metabolite Database The Symyx Metabolite Database, formerly known as the MDL Metabolite Database, provides data on metabolite structures, enzymatic reactions, and associated kinetic properties, aiding researchers in understanding drug metabolism.

Tandem Mass Spectrometry (MS/MS) After separation by LC, the eluted substances are further analysed using MS/MS, which involves ionising the sample components, breaking them into fragments (first MS), and then further breaking down the selected ions to analyse their structure or composition (second MS). This tandem process allows for the detailed identification and quantification of the compounds, providing structural information that helps identify specific molecules.

Tanimoto Shape Score It measures the similarity between the 3D shapes of two molecules. It is calculated by dividing the volume of overlap between two molecular shapes by the volume of their union. The result is a value between 0 (no overlap) and 1 (complete overlap), where higher values indicate greater similarity in shape.

Tanimoto Colour Score This score extends the shape comparison by considering additional molecular features such as electrostatic properties, hydrogen bond donors, or acceptors. Similar to the shape score, it is calculated based on the overlap of these features between two molecules, normalised by their combined feature volume, yielding a score between 0 and 1.

Tanimoto Combo Score This combined score integrates shape and colour scores to provide a more comprehensive similarity metric. It is typically calculated as a weighted average of the shape and colour scores, offering a nuanced view of how closely two molecules resemble each other, not just in form but also in functionally relevant chemical properties.

3D (Three-Dimensional) A 3D representation of a molecule includes information about the spatial arrangement of atoms, providing a more complete view of the molecule's geometry. The information includes bond angles, torsion angles, and distances between atoms.

2D (Two-Dimensional) In the context of molecular structures, a 2D representation displays the arrangement of atoms and the bonds between them in a flat, planar format. This representation provides information about the molecular connectivity and the types of bonds (single, double, triple) but does not include any spatial or stereochemical information about the molecule.

VAMP (Vector Analysis of Molecular Properties) It is a software package used for quantum mechanical calculations. It is typically used to analyse and predict molecular properties based on quantum chemistry methods. VAMP can calculate various molecular properties such as electronic distributions, reactivity indices, and other descriptors that are crucial for understanding the chemical behaviour of molecules. These properties are essential for modelling interactions, such as those involved in drug metabolism, and for predicting the reactivity of compounds.

References

1. Zhao M, Ma J, Li M, Zhang Y, Jiang B, Zhao X, et al. Cytochrome P450 enzymes and drug metabolism in humans. Int J Mol Sci. 2021;22(23):12808.
2. Lu C, Di L. In vitro and in vivo methods to assess pharmacokinetic drug– drug interactions in drug discovery and development. Biopharm Drug Dispos. 2020;41(1–2):3–31.
3. Vree TB, van den Biggelaar-Martea M, Verwey-van Wissen CPWGM. Determination of indomethacin, its metabolites and their glucuronides in human plasma and urine by means of direct gradient high-performance liquid chromatographic analysis: preliminary pharmacokinetics and effect of probenecid. J Chromat B Biomed Sci Appl [Internet]. 1993;[cited 2023 Feb 28];616(2):271–82.
4. Mfuh AM, Larionov OV. Heterocyclic N-oxides - an emerging class of therapeutic agents. Curr Med Chem. 2015;22(24):2819–57.
5. Manikandan P, Nagini S. Cytochrome P450 structure, function and clinical significance: a review. Curr Drug Targets. 2018;19(1):38–54.
6. Başaran R, Can Eke B. Flavin containing monooxygenases and metabolism of Xenobiotics. Turk J Pharm Sci. 2017;14(1):90–4.
7. Dalvie D, Di L. Aldehyde oxidase and its role as a drug metabolizing enzyme. Pharmacol Ther [Internet]. 2019;[cited 2021 Nov 16];201:137–80. Available from: https://pubmed.ncbi.nlm.nih.gov/31128989/
8. Kadlubar S, Kadlubar FF. Enzymatic basis of phase I and phase II drug metabolism. Springer eBooks; 2009. p. 3–25.

9. EH-Haj BM. Metabolic N-Dealkylation and N-oxidation as elucidators of the role of Alkylamino moieties in drugs acting at various receptors. Molecules. 2021;26(7):1917.
10. Sánchez C, Hyttel J. Comparison of the effects of antidepressants and their metabolites on reuptake of biogenic amines and on receptor binding. Cell Mol Neurobiol. 1999;19(4):467–89.
11. Breyer-Pfaff U. The metabolic fate of Amitriptyline, Nortriptyline and Amitriptylinoxide in man. Drug Metab Rev. 2004;36(3–4):723–46.
12. Zhou SF. Polymorphism of human cytochrome P450 2D6 and its clinical significance. Clin Pharmacokinet. 2009;48(11):689–723.
13. Haenisch B, Hiemke C, Bönisch H. Inhibitory potencies of trimipramine and its main metabolites at human monoamine and organic cation transporters. Psychopharmacology. 2011;217(2):289–95.
14. Projean D, Baune B, Farinotti R, Flinois JP, Beaune P, Taburet AM, et al. In vitro metabolism of chloroquine: identification of CYP2C8, CYP3A4, and CYP2D6 as the main isoforms catalyzing N-desethylchloroquine formation. Drug Metab Dispos Biol Fate Chem [Internet]. 2003;[cited 2022 Sep 9];31(6):748–54. Available from: https://pubmed.ncbi.nlm.nih.gov/12756207/
15. Rajamani S, Eckhardt LL, Valdivia CR, Klemens CA, Gillman BM, Anderson CL, et al. Drug-induced long QT syndrome: hERG K+channel block and disruption of protein trafficking by fluoxetine and norfluoxetine. Br J Pharmacol. 2006;149(5):481–9.
16. Seproxetine [Internet]. go.drugbank.com. [cited 2024 Apr 19]. Available from: https://go.drugbank.com/drugs/DB06731.
17. Meng Y, Yu B, Huang H, Peng Y, Li E, Yao Y, et al. Discovery of Dosimertinib, a highly potent, selective, and orally efficacious deuterated EGFR targeting clinical candidate for the treatment of non-small-cell lung cancer. J Med Chem. 2021;64(2):925–37.
18. Regan SL, Maggs JL, Hammond TG, Lambert C, Williams DP, Park BK. Acyl glucuronides: the good, the bad and the ugly. Biopharm Drug Dispos. 2010;31(7):367–95.
19. Bradshaw PR, Athersuch TJ, Stachulski AV, Wilson ID. Acyl glucuronide reactivity in perspective. Drug Discov Today. 2020;25(9):1639–50.
20. Alsultan A, Furin JJ, Du Bois J, van Brakel E, Chheng P, Venter A, et al. Population pharmacokinetics of AZD-5847 in adults with pulmonary tuberculosis. Antimicrob Agents Chemother. 2017;61(10):e01066.
21. Cook D, Brown D, Alexander R, March R, Morgan P, Satterthwaite G, et al. Lessons learned from the fate of AstraZeneca's drug pipeline: a five-dimensional framework. Nat Rev Drug Discov [Internet]. 2014;13(6):419–31. Available from: https://www.nature.com/articles/nrd4309
22. Waring MJ, Arrowsmith J, Leach AR, Leeson PD, Mandrell S, Owen RM, et al. An analysis of the attrition of drug candidates from four major pharmaceutical companies. Nat Rev Drug Discov [Internet]. 2015;14(7):475–86. Available from: https://www.ncbi.nlm.nih.gov/pubmed/26091267
23. Research C for DE and. FDA approval of lymphoma medicine Ukoniq (umbralisib) is withdrawn due to safety concerns. FDA [Internet]. 2022.; Available from: https://www.fda.gov/drugs/drug-safety-and-availability/fda-approval-lymphoma-medicine-ukoniq-umbralisib-withdrawn-due-safety-concerns.
24. Senior M. Fresh from the biotech pipeline: fewer approvals, but biologics gain share. Nat Biotechnol [Internet]. 2023;41(2):174–82. Available from: https://www.nature.com/articles/s41587-022-01630-6
25. Schadt S, Lopes F, Hauri S, Pähler A, Brink A. Metabolites in safety testing (MIST). Elsevier eBooks; 2020. p. 419–38.
26. Research C for DE and. Safety Testing of Drug Metabolites [Internet]. U.S. Food and Drug Administration. 2020. Available from: https://www.fda.gov/regulatory-information/search-fda-guidance-documents/safety-testing-drug-metabolites.

27. ICH M3 (R2) Non-clinical safety studies for the conduct of human clinical trials for pharmaceuticals - Scientific guideline | European Medicines Agency [Internet]. www.ema.europa.eu. Available from: https://www.ema.europa.eu/en/ich-m3-r2-non-clinical-safety-studies-conduct-human-clinical-trials-pharmaceuticals-scientific-guideline.
28. Sabe VT, Ntombela T, Jhamba LA, Maguire GEM, Govender T, Naicker T, et al. Current trends in computer aided drug design and a highlight of drugs discovered via computational techniques: a review. Eur J Med Chem [Internet]. 2021;224:113705. Available from: https://www.sciencedirect.com/science/article/abs/pii/S0223523421005547
29. Buntz B. Prominent firms harnessing AI in drug discovery and development [Internet]. Drug Discovery and Development. 2023. Available from: https://www.drugdiscoverytrends.com/10-pioneering-companies-implementing-ai-in-drug-discovery-development-and-beyond/.
30. Tan L, Kirchmair J. Software for metabolism prediction. In: Methods and principles in medicinal chemistry. Wiley; 2014. p. 27–52.
31. Fox T, Kriegl J. Machine learning techniques for in Silico modeling of drug metabolism. Curr Top Med Chem. 2006;6(15):1579–91.
32. Kazmi SR, Jun R, Yu MS, Jung C, Na D. In silico approaches and tools for the prediction of drug metabolism and fate: a review. Comput Biol Med. 2019;106:54–64.
33. Christina FNO, Kirchmair J. Alignment-based prediction of sites of metabolism. J Chem Inf Model. 2017;57(6):1258–64.
34. Broccatelli F, Brown N. Molecular interaction fields for predicting the sites and products of metabolism. In: Methods and principles in medicinal chemistry. Wiley; 2014. p. 221–42.
35. Simulations-plus.com. 2021. Available from: https://www.simulations-plus.com/software/admetpredictor/.
36. CYP metabolite prediction | CYP kinetic parameters | CYP inhibition | UGT [Internet]. Simulations Plus. [cited 2024 Apr 19]. Available from: https://www.simulations-plus.com/software/admetpredictor/metabolism/.
37. Xiong J, Cui R, Li Z, Zhang W, Zhang R, Fu Z, et al. Transfer learning enhanced graph neural network for aldehyde oxidase metabolism prediction and its experimental application. Acta Pharm Sin B. 2024;14(2):623–34.
38. AOMP [Internet]. aomp.alphama.com.cn. [cited 2024 Apr 19]. Available from: https://aomp.alphama.com.cn/.
39. BioTransformer 3.0 [Internet]. biotransformer.ca. [cited 2024 Apr 19]. Available from: https://biotransformer.ca/help?page=prediction.
40. Djoumbou-Feunang Y, Fiamoncini J, Gil-de-la-Fuente A, Greiner R, Manach C, Wishart DS. BioTransformer: a comprehensive computational tool for small molecule metabolism prediction and metabolite identification. J Cheminform [Internet]. 2019:11. Available from: https://www.ncbi.nlm.nih.gov/pmc/articles/PMC6689873/
41. Tian S, Cao X, Greiner R, Li C, Guo A, Wishart DS. CyProduct: a software tool for accurately predicting the byproducts of human cytochrome P450 metabolism. J Chem Inf Model. 2021;61(6):3128–40.
42. Bitbucket [Internet]. bitbucket.org. [cited 2024 Apr 19]. Available from: https://bitbucket.org/wishartlab/cyproduct/src/master/.
43. Hennemann M, Friedl A, Lobell M, Keldenich J, Hillisch A, Clark T, et al. CypScore: quantitative prediction of reactivity toward cytochromes P450 based on Semiempirical molecular orbital theory. Chem Med Chem. 2009;4(4):657–69.
44. Šícho M, Stork C, Mazzolari A, de Bruyn KC, Pedretti A, Testa B, et al. FAME 3: predicting the sites of metabolism in synthetic compounds and natural products for phase 1 and phase 2 metabolic enzymes. J Chem Inf Model. 2019;59(8):3400–12.
45. Univie.ac.at. 2019 [cited 2024 Apr 19]. Available from: https://nerdd.univie.ac.at/fame3/.
46. de Bruyn KC, Šícho M, Mazzolari A, Kirchmair J. GLORYx: prediction of the metabolites resulting from phase 1 and phase 2 biotransformations of Xenobiotics. Chem Res Toxicol. 2020;34(2):286–99.

47. Univie.ac.at. 2020. Available from: https://nerdd.univie.ac.at/gloryx/.
48. Li J, Schneebeli ST, Bylund J, Farid R, Friesner RA. IDSite: an accurate approach to predict P450-mediated drug metabolism. J Chem Theory Comput. 2011;7(11):3829–45.
49. Molecular Forecaster Inc [Internet]. Molecular Forecaster Inc. Available from: https://molecularforecaster.com/.
50. Campagna-Slater V, Pottel J, Therrien E, Cantin LD, Moitessier N. Development of a computational tool to rival experts in the prediction of sites of metabolism of Xenobiotics by P450s. J Chem Inf Model. 2012;52(9):2471–83.
51. Talafous J, Sayre LM, Mieyal JJ, Klopman G. META. 2. A dictionary model of mammalian xenobiotic metabolism. J Chem Inf Model. 1994;34(6):1326–33.
52. MetabolExpert | www.compudrug.com [Internet]. compudrug.com. [cited 2024 Apr 19]. Available from: https://compudrug.com/metabolexpert.
53. Darvas F. Predicting metabolic pathways by logic programming. J Mol Graph. 1988;6(2):80–6.
54. Çubuk C, Hidalgo MR, Amadoz A, Rian K, Salavert F, Pujana MA, et al. Differential metabolic activity and discovery of therapeutic targets using summarized metabolic pathway models. npj Syst Biol Appl. 2019;5(1):7.
55. Metabolizer FAQ - Documentation 16.2.15.0 [Internet]. dl.chemaxon.com. [cited 2024 Apr 19]. Available from: https://dl.chemaxon.com/docs/HTML/docs162150/Metabolizer_FAQ.html#src-44600672_safe-id-RnJlcXVlbnRseUFza2VkUXVlc3Rpb25zTWV0YWJvbGl6ZXItMVdoYXRpc01ldGFib2xpemVyPw.
56. MetaPrint2D | pharmb.io - Pharmaceutical Bioinformatics Research Group at Uppsala University [Internet]. pharmb.io. [cited 2024 Apr 19]. Available from: https://pharmb.io/tool/metaprint2d/.
57. Metabolite Identification Software | MetaSense® [Internet]. ACD/Labs. [cited 2024 Apr 19]. Available from: https://www.acdlabs.com/products/spectrus-platform/metasense/.
58. Marie A, Lanevskij K, Sazonovas A, Harris J. Impact of established and emerging software tools on the metabolite identification landscape. Front Toxicol. 2022;4:932445.
59. Cruciani G, Carosati E, De Boeck B, Ethirajulu K, Mackie C, Howe T, et al. MetaSite: understanding metabolism in human cytochromes from the perspective of the chemist. J Med Chem. 2005;48(22):6970–9.
60. MetaSite - Metabolism prediction [Internet]. www.moldiscovery.com. [cited 2024 Apr 19]. Available from: https://www.moldiscovery.com/software/metasite/.
61. Network [Internet]. www.way2drug.com. [cited 2024 Apr 19]. Available from: https://www.way2drug.com/metatox/index.php.
62. Rudik AV, Bezhentsev VM, Dmitriev AV, Druzhilovskiy DS, Lagunin AA, Filimonov DA, et al. MetaTox: web application for predicting structure and toxicity of Xenobiotics' metabolites. J Chem Inf Model. 2017;57(4):638–42.
63. META Ultra [Internet]. MultiCASE. [cited 2024 Apr 19]. Available from: https://multicase.com/meta-ultra.
64. Chakravarti SK, Saiakhov RD. MultiCASE platform for in Silico toxicology. Methods Mol Biol. 2022;2425:497–518.
65. Metabolite Identification And Analysis | Lhasa Limited [Internet]. [cited 2024 Apr 19]. Available from: https://www.lhasalimited.org/solutions/metabolite-identification-and-analysis/.
66. Langowski JJ, Long AD. Computer systems for the prediction of xenobiotic metabolism. Adv Drug Deliv Rev. 2002;54(3):407–15.
67. Finkelmann AR, Goldmann D, Schneider G, Göller AH. MetScore: site of metabolism prediction beyond cytochrome P450 enzymes. Chem Med Chem. 2018;13(21):2281–9.
68. Won Seok O, Kim DN, Jung J, Cho KH, No KT. New combined model for the prediction of Regioselectivity in cytochrome P450/3A4 mediated metabolism. J Chem Inf Model. 2008;48(3):591–601.
69. ADME Database [Internet]. Fujitsu Global. [cited 2024 Apr 19]. Available from: https://www.fujitsu.com/global/solutions/business-technology/tc/sol/admedatabase/#anc-04.

70. Hwang S, Shin HK, Shin SE, Seo M, Jeon HN, Yim DE, et al. PreMetabo: an in silico phase I and II drug metabolism prediction platform. Drug Metab Pharmacokinet. 2020;35(4):361–7.
71. Prediction of Metabolism | PreMetabo – Just another BMDRC Sites site [Internet]. 2015 [cited 2024 Apr 19]. Available from: https://premetabo.webservice.bmdrc.org/.
72. Yousofshahi M, Manteiga S, Wu C, Lee K, Hassoun S. PROXIMAL: a method for prediction of xenobiotic metabolism. BMC Syst Biol. 2015;9(1):94.
73. PROXIMAL: Prediction of Xenobiotic Metabolism — Hassoun Lab [Internet]. hassounlab.cs.tufts.edu. Available from: https://hassounlab.cs.tufts.edu/proximal/.
74. Sykes MJ, McKinnon RA, Miners JO. Prediction of metabolism by cytochrome P450 2C9: alignment and docking studies of a validated database of substrates. J Med Chem. 2008;51(4):780–91.
75. Zaretzki J, Bergeron C, Rydberg P, Huang T, Bennett KP, Breneman CM. RS-predictor: a new tool for predicting sites of cytochrome P450-mediated metabolism applied to CYP 3A4. J Chem Inf Model. 2011;51(7):1667–89.
76. Zheng M, Luo X, Shen Q, Wang Y, Du Y, Zhu W, et al. Site of metabolism prediction for six biotransformations mediated by cytochromes P450. Bioinformatics. 2009;25(10):1251–8.
77. Prediction of site of metabolism [Internet]. Way2drug.com. 2024 [cited 2024 Apr 20]. Available from: https://www.way2drug.com/somp/index.php.
78. Rudik AV, Dmitriev AV, Lagunin AA, Filimonov DA, Poroikov VV. Metabolism site prediction based on xenobiotic structural formulas and PASS prediction algorithm. J Chem Inf Model. 2014;54(2):498–507.
79. SMARTCyp [Internet]. smartcyp.sund.ku.dk. Available from: https://smartcyp.sund.ku.dk/mol_to_som.
80. Rydberg P, Gloriam DE, Zaretzki J, Breneman C, Olsen L. SMARTCyp: a 2D method for prediction of cytochrome P450-mediated drug metabolism. ACS Med Chem Lett. 2010;1(3):96–100.
81. Shen Q, Zheng M, Lu J, Luo C, Zhu W, Chen K, et al. SOMEViz: a web service for site of metabolism estimating and visualizing. Protein Pept Lett. 2012;19(9):905–9.
82. StarDrop: Small Molecule Drug Discovery & Data Visulization Software [Internet], 2020. Available from: https://optibrium.com/stardrop/.
83. Welcome to SyGMa's documentation! — sygma 1.1.0 documentation [Internet]. sygma.readthedocs.io. [cited 2024 Apr 20]. Available from: https://sygma.readthedocs.io/en/latest/.
84. Ridder L, Wagener M. SyGMa: combining expert knowledge and empirical scoring in the prediction of metabolites. Chem Med Chem. 2008;3(5):821–32.
85. Software [Internet]. oasis-lmc.org. [cited 2024 Apr 20]. Available from: https://oasis-lmc.org/products/software/metapath.aspx.
86. Mekenyan O, Dimitrov S, Pavlov T, Veith G. A systematic approach to simulating metabolism in computational toxicology. I. The TIMES heuristic modelling framework. Curr Pharm Des. 2004;10(11):1273–93.
87. Dang NL, Matlock MK, Hughes TB, Swamidass SJ. The metabolic rainbow: deep learning phase I metabolism in five colors. J Chem Inf Model. 2020;60(3):1146–64.
88. Swamidass DJ. Xenosite [Internet]. Xenosite. [cited 2024 Apr 20]. Available from: https://xenosite.org/.
89. Flynn NR, Le Dang N, Ward MD, Swamidass SJ. XenoNet: inference and likelihood of intermediate metabolite formation. J Chem Inf Model. 2020;60(7):3431–49.
90. Shroff T, Aina K, Maass C, Cipriano M, Lambrecht J, Tacke F, et al. Studying metabolism with multi-organ chips: new tools for disease modelling, pharmacokinetics and pharmacodynamics. Open Biol. 2022;12(3):210333.
91. Jendoubi T. Approaches to integrating metabolomics and multi-omics data: a primer. Meta. 2021;11(3):184.
92. Ryu GS, Kim GB, Yu T, Lee SY. Deep learning for metabolic pathway design. Metab Eng. 2023;80:130–41.

93. Arodola OA, Soliman ME. Quantum mechanics implementation in drug-design workflows: does it really help? Drug Des Dev Ther [Internet]. 2017;[cited 2019 Sep 20];11:2551–64. Available from: https://www.ncbi.nlm.nih.gov/pmc/articles/PMC5587087/
94. Cerchia C, Lavecchia A. In Silico drug design and discovery: big data for small molecule design. Biomol Ther. 2022;13(1):44.
95. Paliwal A, Jain S, Kumar S, Wal P, Khandai M, Khandige PS, et al. Predictive modelling in pharmacokinetics: from in-silico simulations to personalized medicine. Expert Opin Drug Metab Toxicol. 2024;20:1–15.

Chapter 18
Computer-Assisted Simulation Studies for Identifying Drug-Polymer Interactions

Prerna Chavan, Srishti Gupta, Esha S. Attar, Rajasekhar Reddy Alavala, and S. N. Koteswara Rao G.

Contents

P. Chavan · S. Gupta · E. S. Attar · R. R. Alavala · S. N. K. Rao G. (✉)
Shobhaben Pratapbhai Patel School of Pharmacy & Technology Management, SVKM's NMIMS, Mumbai, India

S. N. K. Rao G., R. R. Alavala (eds.), *Applications of Computational Tools in Drug Design and Development*, https://doi.org/10.1007/978-981-96-4154-3_18

Abstract The present chapter focuses on the significance of computational methods in understanding the drugs-polymers interactions- a factor of critical important in formulation design. By utilizing computer assisted simulations would predict and study the behaviour of drugs when they come in contact with several polymers, that insight into formulation stability, release kinetics and enhancement of bioavailability. Computer-assisted simulations such as Monte Carlo method or molecular dynamic enables investigation at molecular level. This would help in the enhanced understanding of how drugs-polymers interact and along with their physicochemical properties of the resulting formulation. The main key aspects covered in this chapter includes polymers selection with respect to chemical characteristics and compatibility nature with specific drugs, structure of polymers with respect to drug encapsulation efficiency, and the impact of drug-polymer interactions on formulation performance results. Examples and case studies will be used to demonstrate how computational techniques/ methods guide experimental design, optimize formulation parameters and leads to development of effective drug delivery systems. This chapter has emphasized on computational methods/ tools that improves formulation predictability, accelerate drug development process and support rational design. By utilizing computer-assisted simulations provides guidance to researchers about drug-polymer complexities and open up a pathway for development of new pharmaceutical product with enhanced therapeutic outcomes.

Keywords Computational · Drug-polymer interactions · Molecular dynamics · Formulation design · Drug delivery systems · Pharmaceutical development

18.1 Introduction

Efficient drug delivery is important in ensuring pharmaceutical targeted and effective administration. Computer-assisted simulations provide a modeling drug-polymer complex interactions which are used in drug delivery. These interactions are crucial as they influence drug's release profile, stability, and overall therapeutic effectiveness. By simulating these interactions, researchers can be able to predict how change in polymer composition or structure will affect drug behaviour, allowing for optimization of drug delivery systems design before practical/physical prototype are developed.

Computer-assisted simulations methods such as molecular dynamics, Monte Carlo methods help researchers to predict and analyse the behaviour of drugs in contact with several polymeric materials. Elucidation of these simulations would give valuable insight into the understanding formulation stability, release kinetics and enhancement of bioavailability. It involves understanding of molecular level interaction between drug and the polymer, as these are most critical in determining the end characteristics of the final formulations.

The role of molecular simulations in the design of formulations for drug delivery. Simulations can be used in predicting formulation properties before their synthesis, before carrying out any experiments in-vitro and in-vivo. Insights gained from simulations of better value than what obtains via advanced experimental method are therefore obtained [1].

Initially, molecular simulations were limited to single-molecule behavior, such as how a drug orients inside a bilayer. Improving computational power and advances in simulation techniques have further pushed the scale toward larger phenomena that can be investigated with feasible computational costs, like complexation of the drug carrier system. In this way, one can simulate more complex interactions and characteristics of drug delivery systems, opening up new avenues to increase the efficacy and efficiency of drug formulation development.

Key points covered in the chapter include criteria for selecting appropriate polymers by their chemical characteristics and compatibility nature to specific drugs. How polymer structure affects the drug encapsulation efficiency and implications of polymer-drug interactions on overall formulation performance also discussed. The chapter has also been supported with examples and case studies to illustrate how computational models provides guidance for experimental design, optimization of formulation parameters and accelerate the development of effective drug delivery system [2].

Ultimately, this chapter focuses on how computational tools are really gamechangers in terms of formulation predictability and expediting the drug development process. This will particularly help researchers on simulations to navigate the complexities of drug-polymer interaction more effectively; innovative pharmaceutical products with enhanced therapeutic outcomes.

18.2 Advantages

18.2.1 Accelerate the Drug Development Process [3]

Computational method involves high throughput of virtual screening, that enables researchers to evaluate rapidly vast libraries of compounds. This process identifies the potent drug molecule/ candidate more efficiently.

18.2.2 Target Identification [3]

The computational model helps in identifying the therapeutic target along with the molecular mechanism of the target is characterized.

18.2.3 Guiding Lead Optimization [3]

Computational technique such as molecular docking helps in optimization of lead drug species/ candidate and polymers by highlighting important structure and functions that enhance affinity of binding which results into development of more effective drug delivery system.

18.2.4 Design of Experiment [3]

Computational simulations provide an important insight that guide experimental design of drug-polymer interaction so, that researchers can carry out more productive experiments by identifying the viable combinations.

18.2.5 Simulation of Multiple Scenarios [4]

Computer-Assisted Simulation method such as Monte Carlo simulations can be used for multiple scenarios and conditions, such as varying environmental conditions, different concentration of polymers or mechanism of drug release.

18.2.6 Structure Refinement [5]

Computer-Assisted Simulation method such as molecular dynamics used to refine the structures of drug-polymer complex. By carrying out simulation studies and adjusting structures based upon dynamics and interaction, helps researchers to achieve more accurate models.

18.2.7 Characterizing Drug Release Mechanism

Computational methods such as Monte Carlo simulations or molecular dynamics help in elucidating the mechanism of drug release. It helps researchers to observe how drug diffuse through polymer matrices and how multiple factors like; environmental conditions and composition, influence release kinetics.

18.2.8 Cost- Effectiveness

Computer-Assisted Simulation method such as molecular dynamics or Monte Carlo simulations, researchers can screen a large Drug- polymer complexes virtually before conducting expensive experimental validations. This reduces total cost related to formulation development.

18.2.9 Integration with Other Computational Techniques

Computer-Assisted Simulations helps in integrating with other methods, such as molecular docking and quantitative structure-activity relationship models, that provides in comprehensive understanding of drug-polymer interactions.

18.2.10 Enhance Safety Profile

Computer-Assisted Simulations helps in identifying potential adverse effects or interactions in development process which enhances safety profile of drug-polymer complexes.

18.3 Limitations

Limitations as follows for: Computer-Assisted Simulations [6].

18.3.1 Validation Requirement

In most of the cases, validation of simulations need to be conducted through experiments to ensure their accuracy. Thus, high time-consuming method which often require for extensive experimental work to confirm simulations.

18.3.2 Computational Limitations [7]

Computer-assisted simulations can be constrained by computational resources and algorithms. For example, high-level simulations like molecular dynamics require computational power and time to run process. This might lead to poor sampling of the conformational space, results into misleading conclusion about drug-polymer interactions.

18.3.3 Correlation with Experimental Data [8]

Some variation between simulation results and experimental observation may occur. For instance, certain molecular simulations shown poor correlation with measured physicochemical and biopharmaceutical properties of drug formulation. Thus, it indicates that the simulations alone are not sufficient to predict and analyse real-world behaviour.

18.3.4 Statistical Uncertainty [7]

Results from simulations can be affected with statistical uncertainty, which leads to a problem when dealing with small sample sizes or limited simulation duration of time. This could yield unreliable predictions of stability and drug-polymer interactions.

18.3.5 Fully Relied on Computational Approaches [9]

There is risk of fully relied on computational approaches at the cost of experimental validation. While simulations can provide valuable insights, they should supplement and not substitute empirical studies in ensuring a complete understanding of drug development.

18.3.6 Inadequate Representation of Biological Conditions [10]

Computer-assisted simulations often lack the realism needed to reproduce complex biological environment where drug polymer interactions occur. Factors such as presence of ions, dynamic nature and varying in pH levels significantly influence interactions which may remain absent in simulations.

18.4 Applications of Computer-Assisted Simulations

The applications of computer-assisted simulations are [3, 11–16]:

18.4.1 Personalized Medicines

Personalized medicine where drug composition takes into an account the unique characteristics of each patient, is another field where computer assisted simulations can be used. Personalized drug delivery helps to enhance therapeutic success by simulating drug-polymer interactions depending on individual patient characteristics, such as genetic variants or disease conditions.

18.4.2 Multiscale Approaches

Multiscale approaches help researchers to examine drug-polymer interactions at various scales (from molecular scale to macroscopic scale). This method helps in understanding the connections between the performance of polymers in drug delivery and their structural characteristics, leading to a development of more potent formulations.

18.4.3 Creation of Novel Drug Delivery System

With the help of computer-assisted simulations, new drug delivery like nanoparticles, transdermal patch and microneedles can be able to design with more efficiency. Researchers are able to adjust the formulation criteria to achieve controlled release system. Better bioavailability of therapeutic agents by modeling the interactions between drug and various polymeric materials.

18.4.4 Pre-Screening

As a pre-screening tool, computational models are utilized more to assess drug-polymer compatibility before experimental validation. Researchers are able to find suitable candidate for further development by estimating their binding energies and comprehending structural factors which affect drug loading. This often reduces time and expenses involved with traditional method i.e. trial and error approach.

18.4.5 Investigation of Drug Release and Stability Profile

Stability of the drug-polymer complexes and their mechanisms of controlling drug release can be able to understand better through simulations. Researchers can better understand how different polymers affect the degradation and release characteristics of encapsulated drugs by simulating interactions at the molecular level. This understanding is essential for developing efficient drug delivery.

18.4.6 Computer-Aided Drug Formulation Design (CADFD)

This will aid in integrating various types of computational techniques, such as molecular dynamics and molecular docking, for the prediction of behaviour of drugs in systems. Binding free energies associated with these experimental drug loadings will be computed from the modeling of the interactions between drugs and polymeric matrices.

18.4.7 Enhancement of Medication Administration Systems

Systemic method to understand interactions between drug polymer. System design to improve drug delivery efficiency and release properties.

18.4.8 Validation Through Experimentation

Complementary experimental research to confirm predictions from simulations. It increases the accuracy of results and provides information for the creation of new drug delivery methods.

18.4.9 Merging Method

Combining computational and statistical techniques (e.g., artificial neural networks). Drug loading prediction in nanoparticulate systems increased precision in the forecasting of drug-polymer interactions.

18.5 Molecular Dynamics

Molecular dynamics is a computer simulation technique applied to study the physical moments of atoms or molecules as a function of time. Molecular dynamics simulations help in elucidating the interactions between drugs and polymers. These interactions include hydrogen bonding hydrophobic interactions and Vander Waals forces, which are critical for determining stability of drug formulations [15, 17–23].

- Principle:

Molecular dynamics is one of the most powerful computational techniques for modeling the physical movements of atoms and molecules. Such an approach has been able to provide detailed insight into the structural, dynamical, and thermodynamic features of complex systems. The approach finds the basis of its theoretical framework in classical mechanics essentially, Newton's laws of motion [21, 22, 24].

- Newton's Laws of Motion:

The foundation of MD simulations lies in Newton's laws of motion, particularly the second law, which states that the force (F) on a particle is equal to its mass (m) times its acceleration (a) (F = ma). In MD, this principle is applied to each atom or molecule to compute their trajectories over time. By solving these equations, the positions and velocities of particles are updated, simulating their behavior over time.

- Potential Energy Functions

Central to MD simulations is the potential energy function, or force field, which describes the interactions between particles. The potential energy function encompasses:

Bonded interactions: These include bond stretching (Hooke's law), angle bending (harmonic potential), and torsional (dihedral) angles.

Non-bonded interactions: These include van der Waals forces (Lennard-Jones potential) and electrostatic (Coulombic) interactions.

Accurate force fields are crucial for realistic simulations. Commonly used force fields in biomolecular simulations include AMBER, CHARMM, and OPLS.

- Equations of Motion

The equations of motion, based on Newton's laws, are numerically integrated to update positions of particles and their velocities. There are several algorithms:

Atomic Velocity Verlet: A very simple and widely used scheme that updates the new positions based on current positions, velocities, and accelerations.
Leapfrog Algorithm: An analogy of the Verlet algorithm, but positions and velocities are updated in a staggered way.
Velocity Verlet algorithm: An extension of the Verlet algorithm, in which the computer directly accounts for the positions and velocities.

These features maintain the stability of the simulation and keep the numerical accuracy maintained over time.

- Selection of a Time Step

Another very critical parameter of an MD simulation is the time step, Δt. This has to be small enough to accommodate the fastest motions of atoms involved in the vibrating species. Thus, typically, it goes to the order of femtoseconds, that is, 10–15 seconds. If carefully chosen, this may help ensure numerical stability and accuracy.

- Initial Conditions

Initial conditions for the particle positions and velocities should be provided to the MD simulation. The initial conditions can either be experimentally determined in the form of X-ray crystallography or NMR structures or can be computationally generated. The initial velocities are usually chosen according to a Maxwell-Boltzmann distribution that corresponds to the desired temperature at which the system is to be studied.

- Temperature and Pressure Control

This is important in many applications of MD, for example, simulating biological systems or materials under physiological conditions. There are several ways to control temperature and here are a few:

Thermostats: Berendsen, Nose-Hoover, or Langevin thermostats work by augmenting kinetic energy to set a target temperature. For pressure control, barostats such as the Berendsen or Parrinello-Rahman barostats change the volume of the system to maintain constant pressure.

- Periodic Boundary Conditions

Periodic boundary conditions avoid edge effects that may characterize simulations of bulk systems. Periodic boundary conditions consist of creating an infinite lattice by repeating the simulation box in all directions. Once a particle has moved out of one side of the simulation box, it reenters from the opposite side, so a perfectly continuous and homogeneous environment is maintained. This procedure allows us to simulate an effectively infinite system without spurious boundary effects at the edges of the simulation box.

- Simulation Ensemble

The appropriate choice for the ensemble in MD simulations depends on the properties under study. For instance, the following are common:

NVE ensemble (microcanonical): Constant number of particles, volume, and energy.
NVT ensemble (canonical): Constant number of particles, volume, and temperature.
NPT ensemble (isothermal-isobaric): Constant number of particles, pressure, and temperature.

Each of the ensembles given above illuminates different aspects and is used according to the requirements of the simulation study.

- Simulation Data Analysis

A trajectories analysis is to extract meaningful information from it. Some of the key analyses are as follows,

Structural Analysis: This may include calculation of the radial distribution functions, mean square displacement, and hydrogen-bonding patterns.
Dynamic properties: Some dynamic properties that can be derived are the diffusion coefficient, viscosity, and time correlation functions.
Thermodynamic properties: Free energy, heat capacity, and enthalpy can be computed.

Molecular dynamics simulations are currently integrable methodology in computational chemistry and pharmaceutical research. By applying the principles of classical mechanics and potential energy functions with sophisticated numerical algorithms, MD provides detailed insight at the atomic level into the behavior of drug-polymer systems. These simulations and analyses increase the speed of drug development, decrease development costs, and bring perfection to drug delivery systems. With ever-improving computational powers and algorithms, MD simulations will let one gain even deeper insights, thus contributing to more innovative and effective therapeutic solutions [24, 25] (Fig. 18.1).

Molecular dynamic simulations have been a very powerful tool for design and development of drug, allowing research into drug interactions with different components of drug delivery system. One of the applications is identifying drug-polymer interactions, which would have major influence on the understanding of the stability and release kinetics related to a drug formulation.

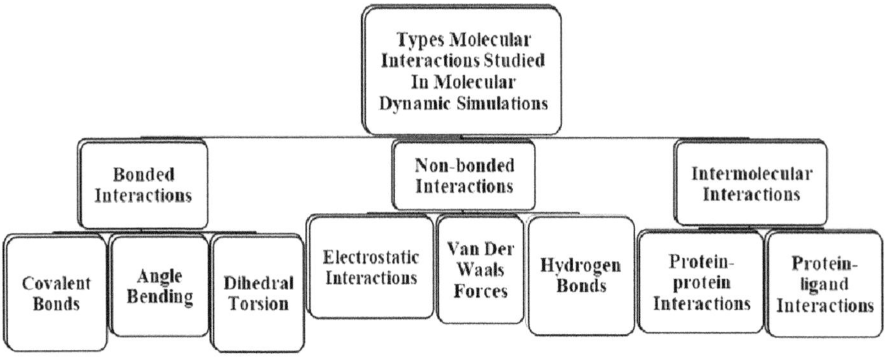

Fig. 18.1 Types of molecular interactions studied in Molecular dynamic simulations

Bonded Interactions in Drug-Polymer System

Bonded Interactions in Drug-Polymer system is very dominant in determining the stability and compatibility of a drug-polymer complex. These include:

- **Covalent Bonds**

Potential energy functions describe the energy due to bond stretching, angle bending, and dihedral torsions between drug and polymer covalent bonds. These functions serve to maintain the preferred values of the bond lengths, angles, and conformations within the drug-polymer complex.

- **Angle Bending**

An energy associated with the angle between three bonded atoms is moulded to hold them at preferred angles. This is especially important in looking at the interactions between functional groups or the overall shape of the complex.

- **Dihedral Torsion**

This captures the associated rotational energy about the bond joining two atoms in a drug-polymer system and conformational changes that may affect the stability of drugs, their solubility, and the kinetics of drug release from a polymeric matrix.

Non-bonded interactions are, hence, of key importance in the context of molecular chemistry and biology, as they define a number of processes and structures. The overview of the major types of non-bonded interactions is given below:

Types of Non-bonded Interactions

- **Electrostatic Interactions**

Electrostatic interaction is an attraction or repulsion between two charged particles. These types of interaction are basic in holding molecular structure and allowing different biological processes to occur.

They can be classified into:

Ionic Interactions: The interaction between positively and negatively charged ions, which is very significant in maintaining the stability of proteins and nucleic acids.

Dipole-Dipole Interactions: Interactions between polar molecules possessing permanent dipoles, enhancing solubility and reactivity.

Van der Waals Forces

A weak, short-range force developed as a result of transient dipoles in molecules, the van der Waals force includes:

London dispersion forces: These are the weakest, present in all molecules, and these result from temporary fluctuations in electron density, giving induced dipoles.

Dipole-induced-dipole: Such interactions take place when a polar molecule induces a dipole in a non-polar molecule; that is, it enhances interaction in a mixed environment.

Hydrogen bonds:

Hydrogen bonds are a special type of dipole-dipole interaction that occurs when a hydrogen atom covalently bonded to a highly electronegative atom (like oxygen or nitrogen) interacts with another electronegative atom. These bonds are very important in holding the three-dimensional structure of proteins and nucleic acids.

- **Intermolecular Interactions includes**
- **Protein-Protein Interactions**

These interactions are of fundamental nature in a variety of biological functions, including enzyme activity, signal transduction, and structural integrity. They can be formed by a combination of non-covalent interactions, such as hydrogen bonds, ionic interactions, or hydrophobic effects.

- **Interaction of Proteins with Ligands**

Knowing protein-ligand interactions is critical in drug discovery because it defines the binding affinity and specificity of drugs. Non-covalent interactions are major players in such events, influencing how effectively a ligand would bind to its target protein.

18.6 Molecular Docking

Molecular docking belongs to important area of biopharmaceutics that deals with protein-ligand interactions, affinity predictions and binding conformations. It has turn out into easy, rapid and affordable method industrial and research laboratories [25–28]. The advancement in software and hardware resulted in advanced molecular dynamic simulations which in turn validate the computational results. Molecular dynamics has created major impact in cost reduction and time in the discovery process. This has also enabled to gain knowledge and understand the rationale behind

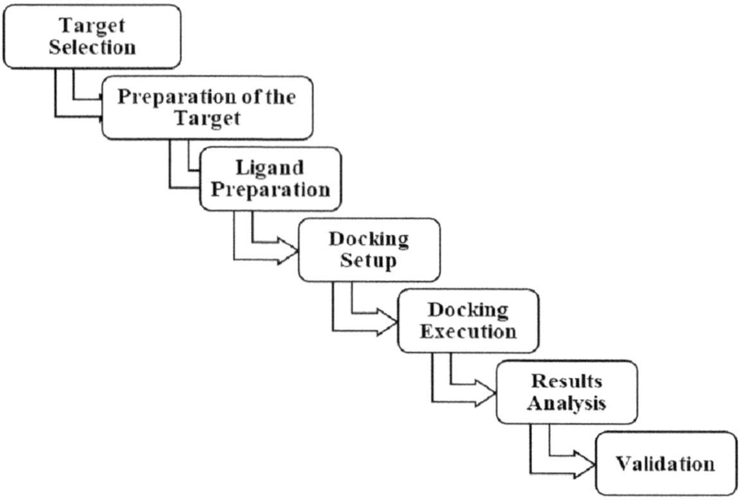

Fig. 18.2 Molecular docking process

repurposing approaches of drug. Molecular docking process are demonstrated in Fig. 18.2.

Molecular docking process

- **Target selection:**

Identify the target macromolecule to be docked.

- **Preparation of the target:**

By obtaining 3D structure of the targeted molecule from the databases like Protein data bank.

Clean the structure by removing water molecules and non-interacting ions.

After this, add the missing hydrogens back and minimize the structure.

- **Ligand preparation:**

Select the appropriate ligand for docking.

Prepare the structure of ligand in the appropriate format.

- **Setting up docking:**

Define the parameters for docking, including grid box size and location.

Prepare all files necessary for the docking software. These may include grid parameter files.

- **Execution of docking:**

Docking simulation are run by using the tools like Auto Dock or Vina.

- **Results analysis:**

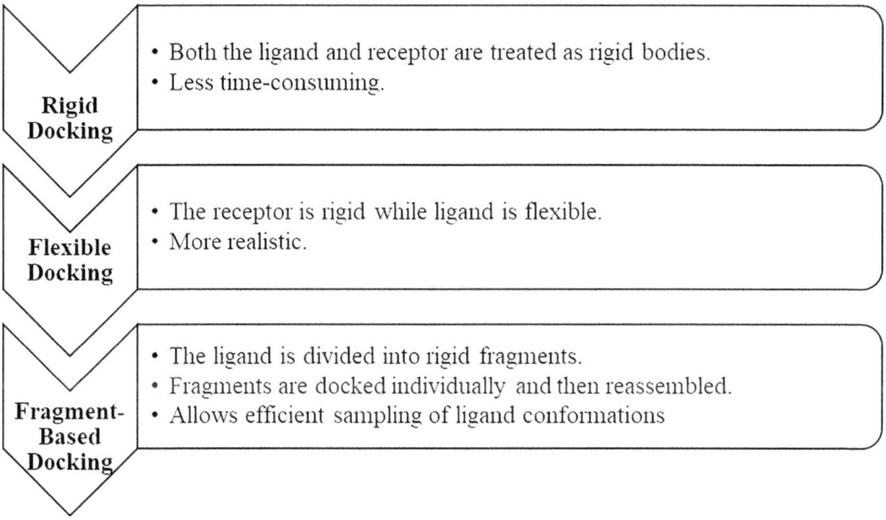

Fig. 18.3 Types of Docking [29]

The results from the docking analysed in terms of their binding modes and energies.

The study of interactions between the ligand and target needs to be conducted.

- **Validation:**

  Validating the docking results against the experimental data (Fig. 18.3).

18.7 Applications of Molecular Docking

The applications of Molecular Docking are as mentioned below [20, 30]:

18.7.1 Lead Optimization

Molecular docking orientations predict improved ligand orientations at their corresponding targets. Depending on the structure of target molecule. This can be applied in the generation of drug molecule that are more potent, selective and efficient based on the structure of the target molecule.

18.7.2 Interaction of Drug-DNA

Molecular docking is a lead role that plays in preliminary prediction of binding properties of drug against nucleic acid. This establishes a relation between drug's cytotoxicity and molecular structure. Therefore, Medical chemists, based on in silico observations, find out whether the compound/drug is interacting with protein/DNA. If the docking programme predicts the said interaction, experimental procedures would be available for finding out real binding mode of the complex. This would lead to the development of a new anti-cancer drug.

18.7.3 Off-Target Effect Identification

It is possible to predict off-target effects by molecular docking, which screens compounds against a panel of proteins to anticipate at least some toxicity problems, guiding the selection of candidates with better selectivity profiles.

18.7.4 Understanding the Modes of Binding

From the docking study how the ligand binds to its target protein and what kind of interactions it forms at an atomic level, which explains also the structure-activity relationships for the tested compounds. This information can thus be put into practice in designing far more effective and drug-specific molecules.

18.7.5 Repurposing of Existing Drugs

This approach can be utilized in the screening process of approved drugs and drug candidates against new targets to identify potential new indications to develop strategy also referred to as drug repurposing. Identification of new uses of existing compounds may accelerate the development of drugs.

Case Study 1

MD Simulation of Drug-Polymer Systems: Celecoxib with PVP-VA and HPMCAS

Overview of the Case Study:

This study explores the interactions between celecoxib and PVP-VA and HPMCAS polymers in an aqueous medium, emphasizing celecoxib's aggregation behaviour with or without polymer carriers.

In this case-study Molecular dynamic simulations played an important role since they offer insight into celecoxib's aggregation behaviour both with or without the polymer carriers. The following are made possible by the simulation Model celecoxib's amorphous forms and track its aggregation tendencies in aqueous solutions. MD help to examine how the presence of polymers impacts the interactions between drugs, specifically the hydrogen bonds that exist between the molecules of celecoxib. MD identify the interaction free energy and evaluate the drug-polymer aggregates structural convergence over time. Identify which specific non-covalent interactions impact or promote aggregation, which will impact the drug formulation stability and bioavailability.

Conclusion:

Impact of Polymer on Aggregation: The presence of polymers modulates the aggregation behaviour of celecoxib. In the absence of polymer, intermolecular interactions in the neutral form are higher; therefore, it aggregates faster than the anionic form.

Structural Convergence: Structural convergence of the MD simulations was viewed, which means that upon drug-polymer interactions, stable drug-polymer aggregates are formed.

Dynamics of Hydrogen Bonding: Transience in the binding forces between celecoxib and polymers was manifested through hydrogen bonding, which is important for the stability of the drug formulations.

Drug Delivery Implications: These results highlight the potential use of MD simulations in predicting and optimizing drug-polymer interactions necessary to formulate efficient drug delivery.

Case Study: 2

Computational Optimization of Insulin-PLGA System for Diabetes Treatment

Overview of the case:

Diabetes mellitus is a chronic metabolic disorder characterized by hyperglycaemia. Insulin therapy remains one of the best diabetes managements, but it is associated with patient discomfort and reduced adherence with conventional methods of administration, majorly, injections. All these challenges stimulate growing interest in the development of an insulin delivery system using biodegradable polymers. Among the many polymers, poly(lactic-co-glycolic) acid has been used for drug delivery extensively due to its biocompatibility and biodegradability features. This case study is concerned with the computational optimization of an insulin-PLGA system for improved diabetes management. The overall aim is the optimization of formulation and processing variables for an insulin-PLGA system in the production of controlled and sustained release of insulin, which will ensure appropriate blood glucose management. The major key challenges it includes are:

Insulin stability: Keeping up the biological activity of insulin in the polymer matrix [31].

Release kinetics: To achieve a desired release profile of insulin to follow physiological insulin secretion

Polymer degradation: controlling the rate of PLGA degradation Vs release profile for insulin.

Molecular Dynamics Simulations: Coupled effects of polymer composition and molecular weight on insulin-PLGA interactions for the investigation into the stability and release of insulin at the molecular scale. Further, discuss how these various formulations of insulin—for example, as insulin complexes how it will affect its interaction with PLGA. With gaining insights in understanding the effect of various processing parameters on insulin-PLGA complex formation, for example, temperature and type of solvent used.

Case Study 3

Computational Optimization of Paclitaxel-PLGA Nanoparticles for Cancer Therapy (Nakamura, J., Nakajima, N., Matsumura, K., & Hyon, S. H. (2010). Water-soluble Taxol conjugates with dextran and targets tumour cells by folic acid immobilization. *Anticancer research*, *30*(3), 903–909.)

Overview of the case:

Paclitaxel, as an effective anti-cancer drug. It has low water solubility and bioavailability. To increase its therapeutic effectiveness, it is normally formulated into nanoparticles using biodegradable polymers such as poly(lactic-co-glycolic) acid. The case study is intended to address the computational optimization of paclitaxel-loaded PLGA nanoparticles toward improving their drug delivery and cancer treatment.

Molecular Dynamics Simulations: At the molecular level, interaction between paclitaxel and PLGA was studied to quantify the effect of polymer composition and molecular weight on drug loading and release of drug. Investigated different mechanisms for the complexation of paclitaxel with PLGA and their effects on the stability of nanoparticles.

Coarse Grained Simulations: Conducted the simulations on self-assembly of PLGA polymers and encapsulation of paclitaxel to predict nanoparticle size and morphology. Optimized different nanoparticle formulation parameters e.g., drug loading or polymer ratio as per the required particle size or drug release profiles.

Optimization and Results From computational studies, optimal paclitaxel-loaded PLGA nanoparticles were fabricated and tested. Formulation results showed increased drug loadings, improved stability, and controlled release of the drug compared to the ordinary formula of drug. In vitro and in vivo studies show improvement in anti-tumour efficacy and reduction in the side effects.

Conclusion:

Computational models are very useful tool in the optimization of paclitaxel-PLGA nanoparticles. Understood the intrinsic molecular interactions and predicted the behaviour of the nanoparticles, developed more effective and efficient drug delivery system. This case study has shown that, in future it offers great potential by the use of computational approaches in drug development and also accelerate with improved patient outcomes.

Key Takeaways:

Computational model enables an in-depth understanding of the drug-polymer interaction and nanoparticle formation process. It provides detailed multi-scale modeling approaches in drug delivery systems. The most important of these is the optimization of parameters for nanoparticles, which improves the efficacy of the drug and minimizes its side effects.

18.8 Challenges

18.8.1 Accuracy of the Computational Models [1]

One of the biggest challenge of computer-assisted simulation is to deal with accuracy of the computational models. The model deal with molecular dynamics simulations, scoring after performing docking studies, and predictions based on machine learning, which has its framework in theory not entirely capable of explaining the complexity in biological systems. To increase the accuracy, focus on an urgent requirement of learning and improving scoring algorithms which predict the binding affinity of ligand molecules with their targets. Important steps to be taken to increase accuracy are: (a) Minimization of False Positives and Negatives: Efforts can be made to eliminate a wrong prediction to make the results improved. (b) Calibration of Scoring Parameters: Model building would be needed with calibration of scoring parameters to fine-tune the parameters and increase the prediction ability. (c) Diverse Molecular Descriptors: Providing a wide range of molecular descriptors allows a better view on the molecular interactions. (d) Model Validation by Experimental Data: An ongoing validation process provided relative to the experimental results allows this model to assure reliance for the predictions.

Thus, refining docking scores by thorough validation against known binding affinities can vigorously improve the reliability of predictions. Maximizing sensitivity (correctly identifying true positives) versus specificity (percentage of true negatives correctly identified) will lead to stronger confidence in scoring algorithms while minimizing the chances of inaccuracies with drug-discovery predictions.

18.8.2 Quality and Quantity of Data [1]

The effectiveness of computer-assisted methods, is more or less dependent on the quality and quantity of data used for training. Inaccurate predictions can result from poor quality or inadequate data, this has always been a challenge for all machine learning applications in the process of drug discovery. In this regard, strategies to address this are: (a) Data Refinement: Filter outliers and maintain consistent formatting in datasets of molecular interactions to improve the accuracy of predictions. (b) Standardized Protocols: Experimental protocols with standard assay conditions and endpoint measurements should be adopted. Adherence to these practices will bring out improvement in data quality, further strengthening the reliability of computational models and thus yielding dependable outcomes in drug discovery.

18.8.3 More Dependent on Computational Predictions [1]

Computer-aided drug design, is very useful, sole dependence on its predictions without experimental validation sometimes it might lead to misguide approaches. Therefore, the need to strike a proper balance between computational predictions and experimental validations to get an effective drug discovery. This dual approach will help confirm the accuracy of the predictions to sustain a more informed decision-making process in development.

18.8.4 Time and Infrastructure Costs Associated [1]

Computer-assisted methods often involve extensive molecular dynamics simulations or complex machine learning models with huge computational demands. In this respect, these techniques could be less accessible because of the time and infrastructure costs involved, hence probably limiting their use for drug discovery process.

18.8.5 Flexibility of Biological Molecules [1]

One of the major conceptual and practical issues for computer-aided drug design relates to the flexibility of biological molecules: both potential drug compounds and the polymers with which these must interact. Most of these molecules can demonstrate a high degree of flexibility, which is of paramount importance in methods like molecular docking. Insufficient flexibility can strongly affect the CADD study results and also leads to false results in predictions and unsuitable drug candidates.

18.8.6 Interpreting [1]

When AI and machine learning models increase in complexity, it is hard to interpret their predictions. As AI models often remain "black-boxes", it becomes difficult to know why a given compound would be predicted as active or how its structure might potentially be optimized for activity. This lack of interpretability at this level might reduce the confident decisions that a researcher can make based on the model's outputs, which becomes challenging in context of drug discovery.

18.9 Future Perspective

Experimental Validation in CADD: From In-Silico to the Lab Bench [1]
Computational Tool Advancements: The use of validated software tools such as AutoDock Vina and the Schrödinger Suite will increase the reliability of computational predictions. This accuracy at the core level forms the basis for generating meaningful insights in CADD simulation and predictions.

Iterative drug discovery process: The future of drug discovery is probably going to be iterative, where computational predictions and experimental results keep on informing and refining each other. This kind of dynamic interplay that could produce better drug candidates and more effective therapeutic strategies.

Integration of New Technologies: Technologies such as AI and machine learning are forecasted to play a significant role in enhancing drug discovery. These technologies can analyze large data sets in the optimization of drug design, providing the opportunity for more efficient treatments.

Biological and Genetic Insights: The increasing role of genetic evidence that can play in selection of drug target, which can raise the chances of success in clinical trials. As genetic data becomes more available, it will inform the drug discovery channel toward more targeted and effective treatments.

Ethical and Regulatory Issues: As drug discovery continues to evolve, questions of ethics regarding animal testing and a changing regulatory environment will have to be addressed. New frameworks of validation and approval will call for smart medicines and biomarker-targeted therapies, among other innovations ultimately leads to uplift the growth in discovery of drug process.

Focus on precision medicine: Precision medicine, in which treatment would be focused on the single patient based on his genetic and molecular profile. This could then offer significant improvements in therapeutic outcomes and reductions in adverse effects where computer-assisted simulations plays an significant role.

Collaborative efforts: Perhaps a greater degree of interaction between computational scientists, experimental biologists, and clinical researchers holds the future for drug discovery. Only through such collaboration can gaps between theoretical models and practical applications be filled, thereby making it possible to achieve very effective drug development.

18.10 Conclusion

The integration of computer-aided simulation studies into the domain of drug-polymer interactions is a revolutionizing step in the process of pharmaceutical development. This chapter has discuss, from different perspectives, manifold advantages of computational tools such as molecular dynamics simulations, quantum mechanics calculations, and docking studies for complicated interactions between drugs and polymers. These are key interactions in the design and optimization of drug delivery systems, which become absolutely necessary for successful therapeutic outcomes.

Computational tools that an unparalleled level of detail and precision in drug-polymer systems can be attained, enabling researchers to view and understand their behaviour at a molecular level. Unlike, traditional experimental techniques, simulations are capable of modeling systems under very broad conditions, allowing the derivation of insights that are usually difficult or impossible to acquire experimentally. For example, molecular dynamics simulations can probe atomic motion as time series to show how drugs interact with polymers in real-time. Quantum mechanics calculations provide insight into the electronic properties of drug–polymer interactions and potential energy surfaces, whereas docking studies predict the most favorable sites and orientations of their binding.

One of the most important benefits derived from computer-aided simulation studies is that it accelerates drug development processes. Traditional experimental approaches can be extensively time-consuming and are usually associated with expensive trial-and-error testing. Computational techniques, on the other hand, can screen huge amounts of drug-polymer combinations efficiently in order to pick out potential candidates for further validation by experiments. Computer-assisted simulations are much time and money saving for the translation of drugs from the initial stage of drug discovery to its application in the clinic.

Application of computational methods creates a path filled with challenges in drug-polymer interactions. Good simulations mean high quality and precision in input data, which is the molecular structure and force fields, which are not easily obtainable under all circumstances. The biological systems are complicated as the simulations will have to consider such factors as solvent effects, ion interactions, and biological membranes.

The resolution of these challenges predict in future improvements in computational techniques and hardware, which shall make possible more comprehensive and accurate simulations. On the other hand, the prediction power of computational studies can be further improved by either the development of more sophisticated algorithms, better force fields, or adoption of the machine learning approach. Another important aspect is the growing availability of high-performance computing resources, which shall further enable the study on larger and complex systems and hence push the boundary on the applications of computations in drug-polymer studies.

The integration of computer-aided simulation studies into drug-polymer interactions reflects an advancement in pharmaceutical research and development. These tools offer great insight into the molecular behaviour of drug-polymer systems that helps to rationally design optimized systems for drug delivery. Computational methods can thus revolutionize the pharmaceutical industry by greatly enhancing the speed of drug development, enhancing the accuracy of predictions, and lessening development costs. Unless computational techniques further evolve, their application in drug-polymer interactions is sure to grow in the future with more innovative and effective therapeutic solutions. Pharmaceutical development in the future would then lie in the continuous integration of computational tools leading toward better therapeutic outcomes and improved care for patients.

References

1. Katiyar RS, Jha PK. Molecular simulations in drug delivery: opportunities and challenges. WIREs Comput Mol Sci. 2018;8(4):e1358.
2. Zhang L, Mao S. Application of quality by design in the current drug development. Asian J Pharm Sci. 2017;12(1):1–8.
3. Ashwini T, Narayan R, Shenoy PA, Nayak UY. Computational modeling for the design and development of nano based drug delivery systems. J Mol Liq. 2022;368:120596.
4. Preacher KJ, Selig JP. Advantages of Monte Carlo confidence intervals for indirect effects. Commun Methods Meas. 2012;6(2):77–98.
5. Gelpi J, Hospital A, Goñi R, Orozco M. Molecular dynamics simulations: advances and applications. Adv Appl Bioinforma Chem. 2015;8:37.
6. Niazi SK, Mariam Z. Computer-aided drug design and drug discovery: a prospective analysis. Pharmaceuticals. 2023;17(1):22.
7. Burton EA, Glorioso JC. Multi-modal combination gene therapy for malignant glioma using replication-defective HSV vectors. Drug Discov Today. 2001;6(7):347–56.
8. Tiwold EK, Gyorgypal A, Chundawat SPS. Recent advances in biologic therapeutic N-glycan preparation techniques and analytical methods for facilitating biomanufacturing automation. J Pharm Sci. 2023;112(6):1485–91.
9. Rajagopal K, Schneider JP. Self-assembling peptides and proteins for nanotechnological applications. Curr Opin Struct Biol. 2004;14(4):480–6.
10. Mobley DL, Dill KA. Binding of small-molecule ligands to proteins: "What You See" is not always "What You Get". Structure. 2009;17(4):489–98.
11. Qian H, Wang K, Lv M, Zhao C, Wang H, Wen S, et al. Recent advances on next generation of polyzwitterion-based nano-vectors for targeted drug delivery. J Control Release. 2022;343:492–505.
12. Malik S, Singh A, Negi P, Kapoor VK. Thymoquinone: a small molecule from nature with high therapeutic potential. Drug Discov Today. 2021;26(11):2716–25.
13. Shah M, Nawaz S, Jan H, Uddin N, Ali A, Anjum S, et al. Synthesis of bio-mediated silver nanoparticles from Silybum marianum and their biological and clinical activities. Mater Sci Eng C. 2020;112:110889.
14. Gao H, Shi W, Freund LB. Mechanics of receptor-mediated endocytosis. Proc Natl Acad Sci. 2005;102(27):9469–74.
15. Morris GM, Huey R, Lindstrom W, Sanner MF, Belew RK, Goodsell DS, et al. AutoDock4 and AutoDockTools4: automated docking with selective receptor flexibility. J Comput Chem. 2009;30(16):2785–91.
16. Poulin P, Haddad S. Extrapolation of the hepatic clearance of drugs in the absence of albumin in vitro to that in the presence of albumin in vivo : comparative assessement of 2 extrapolation models based on the albumin-mediated hepatic uptake theory and limitations and mechanistic insights. J Pharm Sci. 2018;107(7):1791–7.
17. Karplus M, Petsko GA. Molecular dynamics simulations in biology. Nature. 1990;347(6294):631–9.
18. Karplus M, McCammon JA. Molecular dynamics simulations of biomolecules. Nat Struct Biol. 2002;9(9):646–52.
19. Hollingsworth SA, Dror RO. Molecular dynamics simulation for all. Neuron. 2018;99(6):1129–43.

20. Siebenbuerger L, Hernandez-Olmos V, Abdelsamie AS, Frotscher M, van Koppen CJ, Marchais-Oberwinkler S, et al. Highly potent 17β-HSD2 inhibitors with a promising pharmacokinetic profile for targeted osteoporosis therapy. J Med Chem. 2018;61(23):10724–38.
21. Karplus M, Kuriyan J. Molecular dynamics and protein function. Proc Natl Acad Sci. 2005;102(19):6679–85.
22. Phillips JC, Braun R, Wang W, Gumbart J, Tajkhorshid E, Villa E, et al. Scalable molecular dynamics with NAMD. J Comput Chem. 2005;26(16):1781–802.
23. Shaw DE, Maragakis P, Lindorff-Larsen K, Piana S, Dror RO, Eastwood MP, et al. Atomic-level characterization of the structural dynamics of proteins. Science (1979). 2010;330(6002):341–6.
24. Case DA, Cheatham TE, Darden T, Gohlke H, Luo R, Merz KM, et al. The Amber biomolecular simulation programs. J Comput Chem. 2005;26(16):1668–88.
25. de Ruyck J, Brysbaert G, Blossey R, Lensink M. Molecular docking as a popular tool in drug design, an in silico travel. Adv Appl Bioinforma Chem. 2016;9:1–11.
26. Sousa SF, Fernandes PA, Ramos MJ. Protein–ligand docking: current status and future challenges. Proteins Struct Funct Bioinform. 2006;65(1):15–26.
27. Friesner RA, Murphy RB, Repasky MP, Frye LL, Greenwood JR, Halgren TA, et al. Extra precision glide: docking and scoring incorporating a model of hydrophobic enclosure for protein−ligand complexes. J Med Chem. 2006;49(21):6177–96.
28. Meng XY, Zhang HX, Mezei M, Cui M. Molecular docking: a powerful approach for structure-based drug discovery. Curr Comput Aided-Drug Des. 2011;7(2):146–57.
29. Trott O, Olson AJ. AutoDock Vina: improving the speed and accuracy of docking with a new scoring function, efficient optimization, and multithreading. J Comput Chem. 2010;31(2):455–61.
30. Parey K, Wirth C, Vonck J, Zickermann V. Respiratory complex I — structure, mechanism and evolution. Curr Opin Struct Biol. 2020;63:1–9.
31. Rabadjieva D, Sezanova K, Gergulova R, Titorenkova R, Tepavitcharova S. Precipitation and phase transformation of dicalcium phosphate dihydrate in electrolyte solutions of simulated body fluids: thermodynamic modeling and kinetic studies. J Biomed Mater Res A. 2020;108(8):1607–16.

Chapter 19
Recent Advances in Drug-Likeness Screening by Using the Software and Online Tools

Karishma Bhatkhande, Nayanika Nandini Vysyaraju, Rajasekhar Reddy Alavala, and Kunal M. Gokhale

Contents

Abstract In the pursuit of novel drug discovery which is known to be a lengthy process, it is pivotal to assess drug-likeness to filter out compounds with unfavourable properties early in the development process. Recent advances in computational methodologies, coupled with the exponential growth of available chemical data and improvements in machine learning algorithms and emerging AI, have revolutionized drug-likeness screening. This review explores the latest developments in drug-likeness screening, focusing on the utilization of software and online tools. We discuss the uses and principles behind various computational methods employed for drug-likeness prediction, including molecular docking, quantitative structure-activity relationship (QSAR) modelling, and machine learning approaches.

K. Bhatkhande · N. N. Vysyaraju · R. R. Alavala · K. M. Gokhale (✉)
Shobhaben Pratapbhai Patel School of Pharmacy and Technology Management, SVKM's NMIMS, Mumbai, India

S. N. K. Rao G., R. R. Alavala (eds.), *Applications of Computational Tools in Drug Design and Development*, https://doi.org/10.1007/978-981-96-4154-3_19

Additionally, the emergence of user-friendly online platforms and databases offering accessible tools for drug-likeness evaluation, democratizing the process for researchers across different domains has been highlighted. Furthermore, the possible challenges and opportunities associated with integrating these computational tools into drug discovery pipelines and the importance of validation, interpretability, and integration with experimental data is emphasized. Overall, this review provides insights into the evolving landscape of drug likeness screening and its integration with computational tools, facilitating more efficient and cost-effective drug discovery endeavours.

Keywords Drug-likeness screening · Computational methodologies · Machine learning algorithms · Molecular docking · QSAR · Online tools · User-friendly platforms

19.1 Introduction

Drug Discovery is known to be costly and time-consuming and may or may not be successful. With the growing advancements of technology and nurturing of the technology- based problem-solving generation, we all are aware of the possible aids developed through computer science. Drug discovery and development is a complex and costly process with a very high failure rate. Around 85–90% of drug candidates entering in the clinical trials fail, due to several key factors. Some of these are mentioned below:

- Lack of clinical efficacy: Due to inadequate understanding of the disease progression or molecular mechanisms behind the disease, many drugs fail to demonstrate sufficient effectiveness in clinical trials [1].
- Unmanageable toxicity: About 30 of the clinical trial candidates exhibit severe toxic reactions resulting into the withdrawal of the molecule. A very popular example is ranitidine, where the FDA directed the withdrawal of the products due to contamination of potential carcinogen N-Nitrosodimethylamine .
- Poor Drug like properties: Due to lack of desirable pharmacokinetic properties about 10–15% of the molecules are withdrawn from the clinical trials. For example, Mirlinone, a phosphodiesterase 3 inhibitor vasodilator has faced challenges related to poor solubility and permeability [1]. Sildenafil was found initially to have significant off-target effects that led to repurposing of this drug [1].
- Poor solubility, permeability, inadequate tissue selectivity, overemphasis on potency are the key reasons behind the drug failure.
- Translational failures using animal models: This is due to inability of animal models to accurately mimic diseases. Scientists highlighted several challenges related to existing animal models for Alzheimer's disease (AD), like inability to develop all symptoms of the disease, overexpression of proteins linked to AD at sufficient levels, failure due to the inability of a transgenic mouse model to recapitulate pathology of AD [2, 3].

- Lack of commercial viability and strategic planning: Even if a drug is effective and safe, it may fail to proceed if there is insufficient market demand or if the development strategy is poorly conceived. This includes issues like misalignment with market needs or inadequate financial backing.

Today, various online tools or software that function to fasten, simplify, and increase cases of accurate predictions are available. These tools are based on Artificial Intelligence (AI) developed through Machine Learning (ML) and Deep Learning (DL) [4]. Artificial Intelligence as first described in 1950 by Alan Turing is the concept of using computers to simulate intelligent behaviour and critical thinking [5]. The study of how to build or program computers to enable them to do what minds can do, can be understood as Artificial Intelligence. To understand AI, we shall know the subsets of AI; and those are: Machine Learning (ML): pattern identification and analysis of from provided data sets, thereby improving machines through experience. Deep Learning (DL): enable machines to learn and make decisions on its own, composes multi-layer neural networks to perform decision making. Natural Language Processing (NLP): process aiding computers to extract data from human language and thus influence decision making based on that information. Computer Vision (CV): process acquainting the computer to gain and understand information from a series of images or videos. The codes of these tools are to facilitate the problem solving in the best possible way using permutations-combinations, probability, statistics, mathematical, and data interpretation-integration-analysis.

With rate of drug discovery-based clinical trials reaching 90%, new drug development taking more than 10 years, costing around 2.6 billion USD and only less than 10% of drugs from the Phase I trials being launched into the market was in need for technological advancement to make the drug discovery process economical in terms of time, money and effort. Here is where the AI supported drug likeness screening tools come into picture.

19.1.1 How ADME Affects the Drug Action.

Drug discovery and development come along with continuously overlapping specific scientific questions such as ‘What is the mechanism of action of the drug? Or What mechanism(s) are involved in the elimination of the drug? Or at what concentration does the drug show a therapeutic effect or toxic effect? Or What are the possible long-term side effects of the drug after administration?’ etc.

The risk-benefit assessment process where in estimation of the ability of the drug to prevent, cure, or slow the progression of a disease is made by the data available can be understood as the scientific discipline of preclinical drug discovery and development. This process involves the constant drug safety and efficacy evaluation which together help understand the pharmacology of the drug using the assessment of toxicokinetic data and estimate the therapeutic index of the drug. On the other hand, the study of ADME: Absorption, Distribution, Metabolism, and Excretion centrally aligns with finding new and effective drugs safe for use. ADME can be understood as the pharmacokinetic properties of the drug. ADME of drug estimates its Bioavailability, Clearance, and Pharmacokinetics/dynamics which helps understand the exposure to toxicity of drug, metabolism and transporters of drug, drug accumulation, drug metabolizing enzymes etc. For example: Barbiturates are known to be used as anaesthetics, antiseizures, muscle relaxants etc. They belong to a sedative hypnotic class of drugs. Butalbital, Phenobarbital, Methohexital are representative drugs that come under this class. Taking phenobarbital under consideration; it is absorbed rapidly by oral or IV route of administration and reaches peak plasma concentration after 30–60 minutes—for oral and within 5 minutes—for IV injection. This helps one understand at what approximate time the drug shows its therapeutic action. It is distributed through all tissues and fluids - this suggests that the drug is lipid soluble. It is metabolized by acetylation in the liver and thus shall be administered with an adjuvant that inhibits hepatic microsomal enzyme for a longer duration of action. Excreted unchanged around 25–50% through urine. The % unionization, lipophilicity, and acidity of barbiturates play a key role in their therapeutic action.

19.1.2 Significance of ADME in Detail

The "WHAT and WHEN" aspect is important for studying ADME properties. ADME properties significantly improve the cost-effectiveness of the drug discovery and development process as many compounds shortlisted do not have the effective intrinsic properties and safe therapeutics required for launching a drug as the treatment for so and so disease. Early ADME also known as eADME is now the preferred route as it improves the process of drug discovery and development. Further in this chapter we will go through how to access and make available online tools and software perform eADME through their inbuilt databases.

19.1.3 AI as an Aid

AI has many ways of learning and has been seen to self-update itself with the slow and steady use of it in every sector of business and research. Many popular AI applications can be used as online tools to aid drug discovery dealing with predictive and generative tasks. A wide range of technologies of model architecture have evolved from the traditional machine learning models to deep learning networks namely convolutional, recurrent, graph-neural networks, transformers, etc. Failures in clinical trials are usually due to undesired pharmacokinetic properties, unacceptable toxicity, and other drug properties. To avoid such failures selection of drug candidates with an appropriate balance of potency, absorption, distribution, metabolism, excretion, and toxicity is necessary [5].

Drug likeness is a concept formulated to give essential guidelines during the early stages of drug discovery itself to help the innovators get their drug candidates pass through the clinical trials. Drug likeness properties can be understood as the characteristics of the drug such as pharmacokinetics, safety, and compounds with desired ADMET properties [6]. Lipinski's Rule of Five is the originally used well-known tool to assess and filter out such drugs based on the drug properties such as oral absorption suggesting that the drug is 'palatable'. Considering above-like properties only compounds that fall under a certain range are considered "drug-like". Even though drug-like compounds are filtered out rapidly by such approaches the boundaries of a drug's individual physicochemical properties are way more stringent which makes it possible for a perchance passed drug might fail even if only one or two properties fall outside the criteria. There are a few semi-quantitative and quantitative estimation techniques that address the hard limitations, thereby including a scoring scheme, a bioavailability score, an oral PhysChem score, a Quantitative estimate of Drug-likeness (QED), Relative Drug Likelihood (RDL), and Gaussian scoring function (GAU). The whole spectrum of compound quality is reflected upon by each estimation technique which combines the physicochemical descriptors into a one-dimensional drug-likeness score. AI has powerful approaches that are applied in constructing models for particular ADMET endpoint evaluations to facilitate drug discovery candidates. Even though algorithms are available in open-source material their use has limitations due to specific operation systems and software compilers used.

This chapter focuses on helping the reader understand various approaches of the available software tools that can be used to estimate the drug-likeness and other properties of the drugs under preclinical trials to save on labour, time, effort, and money. This chapter also gives the reader knowledge about the legitimate applications, software, hardware, and websites available for users to check their drugs usefulness before animal testing (preclinical studies are evaluated by traditional methods.) Not only does this chapter emphasize a few user- friendly web applications but also guides the reader on their use.

19.2 Advances in Drug Designing

This section investigates the latest breakthroughs in drug design, detailing how pioneering technologies and pioneering approaches are transforming the pharmaceutical industry to boost effectiveness, minimize adverse reactions, and address previously intractable ailments.

Before understanding the advances in drug designing, we must first understand the fundamental concepts of drugs and their targets, the receptors.

- Drug is a small chemical molecule specific structure containing precise substituents/functional groups and molecular skeleton.

- Receptor is a macro biomolecule (biological target)—proteins, enzymes, lipids, nucleosides (have definite chemical structure)
- Docking- It is a technique used to forecast how one molecule will orient itself relative to another when they form a stable complex. It helps predict the binding affinity and activity of drug candidates and identifies crucial structural characteristics that are likely to interact with the target binding site.

Drug development is an innovative and remarkable process of creating new treatments that are tailored to specific biological targets. This approach is often referred to as rational drug design. The precision of data regarding the three-dimensional configuration of biological targets significantly impacts the effectiveness of drug design.

RDD (rational drug design) focuses on creating new molecules using strategic methods to minimize both cost and time while achieving the desired biological activity, whether in vivo or in vitro. This process involves predicting the therapeutic potential of a designed molecule before its actual synthesis and screening. It entails crafting molecules that are complementary in shape and charge to their biomolecular targets, ensuring a strong and effective binding interaction.

Different stages of drug development include Target Identification, Target validation, Lead Identification, Lead Optimization, Pre-clinical studies, Clinical Trials.

Discovery and development of a new therapeutic agent is an extensive process requiring significant investment of time and resources. It is estimated that on an average a standard drug discovery procedure from lead identification to clinical trials costs around 800 million US dollars [6] and takes about 15 years [7]. Over the last decades, many new strategies have emerged to make the process more rational and cost-efficient using the knowledge and experience of various researchers. In this chapter, we are going to introduce various tools that help predict the drug-likeliness of the molecules under consideration. The conventional approach to discovering and developing new drugs has been a complex and time-consuming process. To address this challenge, a more advanced method known as computer-aided drug discovery (CADD) has emerged and gained acceptance over time.

Approximately 90% of the lead compounds fail due to stringent regulatory requirements for approval of drug to reach the market and approximately 75% of the total amount employed in the drug development process is due to the failures met during the development [8].

Over the years, CADD techniques like pharmacophore modelling, homology modelling, docking for virtual screening, quantitative free energy calculations, toxicity profile assessment and quantitative structure-activity relationship (QSAR), and conformation generation have advanced significantly [9]. Its tools have been used to identify, analyse and develop potent drug molecules.

CADD can provide information including three-dimensional structure of a molecule, chemical and physical characteristics of a molecule, interpretation and prediction of properties of new molecules and many more.

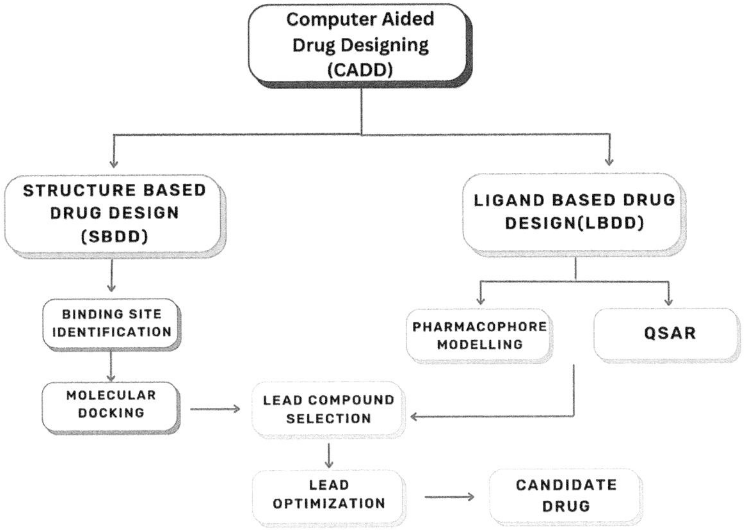

Fig. 19.1 CADD Workflow

Computer Aided Drug Design can be classified as which is shown in Fig. 19.1:

1. Structure-Based method
2. Ligand Based method [10]

19.2.1 Structure-Based Drug Designing (SBDD)

The approach of structure-based drug design relies on the understanding and accessibility of 3D structural data related to the target protein. Utilizing this information, the identification, design, and assessment of ligands are carried out based on their interactions with the residues located in the active/binding site of the target protein [11].

Types of SBDD:

- Molecular docking- It is about "finding or searching" a suitable ligand for a given biological target from chemical database. It utilizes the physical and chemical characteristics of protein binding sites to assess whether a compound is a suitable match with high binding affinity for the receptor's binding sites and is used for screening large databases [13, 14].
- De novo ligand design-It is about "building or developing" suitable ligands from either individual atoms or molecular fragments. It involves utilizing data from the protein binding site to identify small fragments that match closely with the molecular shape of the binding site. These fragments are then linked together based on chemical connection rules to create a novel ligand structure [12].

19.2.2 Ligand -Based Drug Designing (LBDD)

This method is used as an alternative to SBDD when the structure of the drug target is not available. The most crucial tools of LBDD includes three-dimensional quantitative structural activity relationship (3D-QSAR) and pharmacophore modelling. These are frequently utilized together to yield predictive models appropriate for lead identification and optimization [13].

To summarise, the fundamental distinction between structure-based drug design and ligand-based drug design is that SBDD takes a more direct and informed approach, leveraging detailed knowledge of the three-dimensional structure of the biological target. In contrast, LBDD relies on a more indirect method, drawing insights from an understanding of other molecules that are known to bind to the target of interest.

19.2.3 QSAR Models

QSAR stands for quantitative structure activity relationship, which is a computational model that relates the pharmacological activity of chemical structures to the physicochemical properties of the compound. SAR analysis yields information regarding the physical, chemical, biological and toxicological properties of compounds, which further reduces the number of bioassays to be conducted. During the early drug discovery process, QSAR analysis plays a vital role in the selection of chemical compounds for further development. It has significantly reduced the costs and speeded up the lead optimization process. Various modelling approaches in QSAR includes multidimensional QSAR like 2D QSAR, 3D QSAR techniques, etc.

2D QSAR refers to the use of geometric parameters, polar surface area, topological indices, and molecular fingerprints while 3D QSAR refers to the spatial arrangement of compounds [14].

The selection of correct molecular descriptors along with the correlation of biological activity to molecular descriptors statistically plays an important role in the success of a QSAR model. The Partial least square (PLS), Principal component analysis (PCA), and Multivariable linear regression analysis (MLRA) are the three main methodologies in a linear model of the QSAR approach.

19.2.3.1 2D QSAR

This technique is a preferred choice of method. The key components of 2D Quantitative Structure-Activity Relationship analysis involve using numerical descriptors to convert a chemical structure into mathematical variables, the quality and quantity of experimental data, and choosing suitable statistical methods to establish connections between the observations and numerical descriptors.

Hansch Analysis- Corwin Hansch provided a comprehensive summary of the concept of linear free energy relationships in 1969, showcasing their potential to effectively illustrate the actions of biologically active and potentially beneficial compounds. This approach is widely regarded as one of the most promising methods for quantifying and evaluating drug compounds and their intricate interactions with biological systems. According to him, the action of any drug molecule depends on two main practices,

1. Starting from the moment the drug enters the body until it reaches the site of action, it follows a path through various biological membranes and its path is influenced by factors like partition coefficient (Log P) that measures its lipophilicity that is elucidated by "random walk theory"
2. The interaction between the drug molecule and the target site depends on the steric bulk substitution groups and electron density present on that group.

Free Wilson Analysis- The Free-Wilson approach is a strategy based on structure-activity relationships, as it considers the combined impact of various structural components on overall biological activity. It employs indicator variables to signify whether a particular structural feature is present or absent.

19.2.3.2 3D QSAR

The 3D-QSAR model analyses the connection between the three-dimensional characteristics of molecules and their biological effects through statistical correlation methods. It emphasizes steric and electrostatic attributes that can be associated with biological activity, aiming to comprehend how compounds interact with their biological targets using force field functions and alignment techniques. Several 3D QSAR approaches have been created, including CoMFA, SURF, S-SAR, AFMoC, Topomer CoMFA, GRID/GOLPE/GRIND/SILCS, and CoRIA.

19.2.4 Virtual Screening

Earlier, researchers had to manually go through several trial-and-error procedures to test the compounds present in the chemical libraries. This added to the total cost and hence only big pharmaceutical companies could afford it. Virtual screening enables the screening of large databases cost-effectively using computational methods. Millions of compounds can be analysed virtually. It allows the identification of active compounds that are capable of modulating a particular biomolecular pathway [15].

Various software /Programs/tools used for prediction of ADMET properties include:

1. SWISS ADME
2. DruLiTo
3. Protox
4. ADMET Lab2
5. MycoCSM
6. Schrodinger
7. FAF-Drugs2
8. Biovia
9. Amber Molecular Dynamics
10. CamMedNp

19.2.4.1 SWISS ADME

This subsection focusses on the use, key features, Advantages and disadvantages of SWISS ADME. The main purpose of this software is to predict the drug likeness properties, evaluate pharmacokinetics, assess toxicity and safety profile of drug candidates and support drug design.

In the field of pharmaceutical development, the understanding and evaluation of the Absorption, Distribution, Metabolism, and Excretion properties of drugs are crucial. For a drug to be efficacious, a potent molecule needs to attain the required concentration at its target organ within the body and must remain in its bioactive state for a duration of time that allows the anticipated biological events to occur. Drug discovery and development involves the assessment of pharmacokinetic properties at an early phase in the drug discovery process. This is necessary when multiple potential compounds are being considered but restricted availability of actual samples. SWISS ADME is a free online web tool that gives access to robust predictive models for covering physicochemical properties, drug-likeness, pharmacokinetics, and medicinal chemistry compatibility using tools like the Boiled egg model, iLog P, and bioavailability radar. The tool allows the researchers to make informed decisions about the pharmacokinetic properties of compounds aiding in the selection and optimization of drugs in the discovery process. During the extensive and resource-intensive phases of drug discovery and development, a wide array of molecular structures is assessed using various criteria. This aids in determining which compounds to prioritize for synthesis, testing, and advancement toward potentially being effective medicines for patients. The conventional way of determining pharmacokinetics was done by breaking down different factors that influence the drug molecule's ability to reach the target organs and studying the individual parameters [16]. In recent years, computer models have been advocated as a legitimate alternative for predicting ADME processes which is very crucial during the initial stages of pre-clinical drug discovery for investigation of chemical structures and properties when there is a scarcity of accessible compounds for experimentation [17]. This tool enables the computation of various parameters like molecular weight, Total polar surface area for blood-brain penetration capacity; iLogP, XLogP3 for lipophilicity; LogS for solubility, etc.

A wide range of computational methods aim to forecast ADME parameters based on molecular structure. Lipinski et al.'s pioneering research focused on examining orally active compounds to establish physicochemical thresholds for the likelihood of being an oral drug, known as drug-likeness. This Rule-of-five outlined the connection between pharmacokinetic and physicochemical factors.

1. TPSA- "Total polar surface area refers to the sum over all polar atoms or molecules, primarily oxygen and nitrogen, also including their attached hydrogen atoms." For e.g. Drugs used to treat foetuses should have TPSA below 60 Å to ensure that the drug crosses the placental barrier whereas if a drug is intended to cross the blood brain barrier (BBB) it should have a TPSA below 90 Å.
2. Molecular weight- Contains hydrogen atoms. Optimal: 100 ~ 600, based on Drug-Like Soft rule.
3. Log P- Partition coefficient or distribution coefficient refers to the ratio of concentrations of drug in organic and aqueous phase at equilibrium. It impacts both membrane permeability and hydrophobic binding to macromolecules. According to Lipinski's Rule of Five the optimal Log P value is between 0 to 5. For e.g. Drugs having Log P values below 0 (i.e. around −3) then it is hydrophilic whereas a drug having Log P around 6 or 7 shows extreme hydrophobicity.
4. Log S – Logarithm of aqueous solubility value. It is related to water solubility of a drug and can be defined as "a common solubility unit corresponding to the 10-based logarithm of the solubility of a molecule measured in mol/L." Compounds in the range from −4 to 0.5 log mol/L will be considered suitable. For e.g. A drug having Log S around −3 is considered poorly water soluble while a drug having Log S around 0 or above will have good solubility.
5. pKa- Dissociation constant- It can be defined as "negative base −10 logarithm of the acid dissociation constant (Ka) of a solution." i.e. pKa = -log [Ka]. It is important to understand behaviour of drugs in biological systems. Compounds with pKa values close to physiological pH (around 7.4) are more likely to exist in their ionized (charged) form, affecting their solubility, membrane permeability, and distribution within the body.
6. pH- It stands for potential of hydrogen and can be defined as the "negative logarithm of H^+ ion concentration". The pH scale ranges from 0 to 14. It is the measure of the acidity and basicity of a compound. It plays a crucial role in biological systems as the bodily fluids are regulated within specified pH conditions and any change in the pH can disrupt their functions.

   pH-pKa relationship affect pharmacokinetic behaviour of the molecule including toxicity. For example, metabolites of sulphonamide have pKa of 10.4, and urinary pH in the bacterial infection is 6≥, as a result compound gets precipitated at the site of the kidney due to unionization. The introduction of electron withdrawing N heterocycles in the sulphonamide structure helps to lower the pKa than 6, as a result compound remains ionized at urinary pH and the problem of crystalluria can be resolved [18].

7. P-gp substrate- P-Glycoprotein is a classic example of ATP-Binding Cassette Transporter which is responsible for transporting hydrophobic drugs out of the cells. P-gp substrate inducers cause an increase in the amount of this protein which makes the cells resistant to various drugs, for e.g. anticancer drugs. Hence P-gp substrate inhibitors help to prevent multidrug resistance caused due to efflux pumps.
8. Cytochrome P450 or CYPs are superfamily of enzymes that contain the heme-iron group which is a major enzyme responsible for drug metabolism. The process of drug metabolism reactions can be divided into two broad categories: phase I (oxidative reactions) and phase II (conjugative reactions). The human cytochrome P450 family (phase I enzymes) contains 57 isozymes and these isozymes metabolize approximately two-thirds of known drugs in human with 80% of this attribute to five isozymes—1A2, 3A4, 2C9, 2C19 and 2D6.

A. Bioavailability Radar

The Bioavailability Radar (shown in Fig. 19.2) is utilized to quickly assess the drug-likeness of a molecule. It takes into account six physicochemical characteristics: polarity, solubility, lipophilicity, size, flexibility, and saturation. A specific range for each property is outlined on the radar plot as a pink area which indicates that the molecule must fall entirely within this region to be considered drug-like. Additional information about the descriptors can be accessed by hovering over the radar plot.

B. BOILED-Egg Diagram

The BOILED-Egg provides a straightforward evaluation of passive gastrointestinal absorption and brain penetration by analysing the placement of molecules in

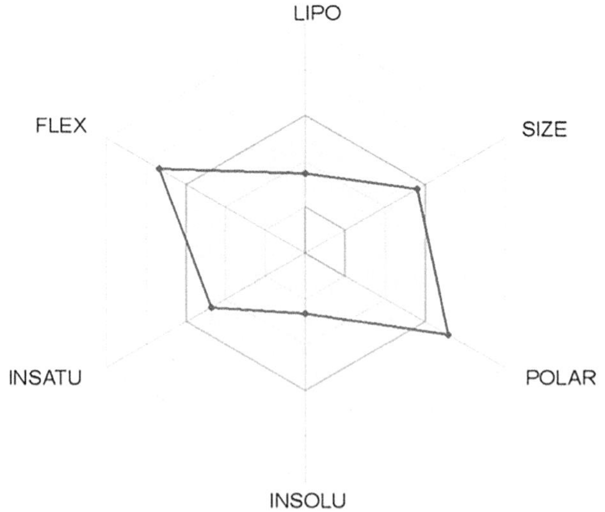

Fig. 19.2 Bioavailability radar of a compound. The pink region signifies the ideal spectrum for the compound. In this instance, the compound is projected to have low oral bioavailability due to excessive flexibility and polarity

the WLOGP vs TPSA reference. The yellow region of the boiled egg denotes a high probability of brain penetration and the white part represents a high likelihood of passive absorption by the GIT. It is important to note that both yellow and white areas can coexist. In addition, points are shaded in blue if they are anticipated to be actively transported by P-gp (PGP+) and in red if they are projected as not being substrates of P-gp (PGP−).

The prediction suggests that Omeprazole is expected to undergo effective absorption but does not penetrate the brain (indicated by the white area) and exhibits P-glycoprotein positivity (depicted as a blue dot in Fig. 19.3).

C. SWISS Target Prediction

It is one of the CADD tools by SIB (Swiss Institute of Bioinformatics)-Molecular Modelling groups developed in 2014. It uses reverse screening to predict the most probable protein targets for various bioactive small molecules. It calculates the similarity between the query compounds of users to the compounds of known activities in well-defined experimental assays.

The user has to select the origin of species for prediction among humans, rats, and mice. Then the query compounds can either be selected or pasted as SMILES of the respective compound. There is a "Predict Target" button that turns red, click on it to start the computation process. It takes around 15 to 20 seconds to display the predictions on the first output page. The result page consists of the probable protein targets. The list of predicted targets along with their probability is displayed on the prediction output page (shown in Fig. 19.4) [19].

D. Swiss Similarity

Swiss Similarity is an innovative web-based tool for ligand-based virtual screening that enables fast analysis of extensive drug libraries, bioactive small molecules,

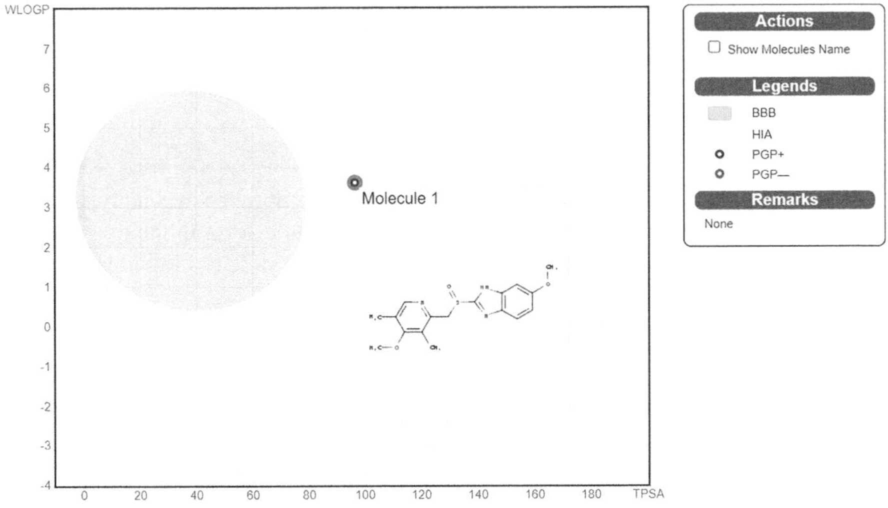

Fig. 19.3 BOILED-Egg Diagram of Omeprazole

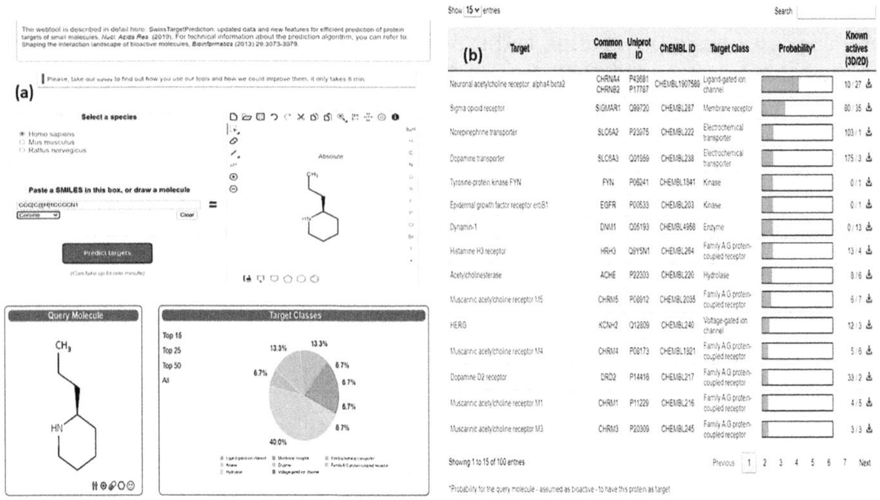

Fig. 19.4 SWISS Target Prediction (**a**) input page, (**b**) output page Swiss Similarity

commercially available compounds, and a remarkably high number of virtual compounds. These virtual compounds can be easily synthesized using reagents readily available in the commercial market through straightforward and dependable organic chemistry reactions. It is an important component of SIB-Swiss Institute of Bioinformatics along with SwissADME, Swiss Target Prediction, Swiss Bioisostere, Swiss Dock, etc. It offers ligand based virtual screening with 6 different methods having more than 30 chemical databases having a user-friendly interface. It provides both 2D (similarity is estimated by FP2 Fingerprints) and 3D (similarity evaluated by Electroshape 5D, Spectrophores, Shape-IT and Align-IT approaches) virtual screening methods.Firstly, the user has to draw/paste/import a molecule or type SMILES in the text box on the submission page. Then the user has to select the screening method and submit it. The output page will show the results. Structures with similarity scores ranging from 0 for totally different compounds to 1 for having identical compounds [20].

Advantages

- This software offers several advantages including comprehensive analysis which helps us in understanding how a drug candidate will perform in the body.
- It also has a user-friendly interface that makes it accessible to individuals having limited computational expertise as well.
- It highly reduces the cost associated with the experimental tests.

Disadvantages

- SWISS ADME is no doubt a powerful tool, but it does have certain limitations such as the predictions provided by this tool may not always reflect the real-world biological behaviour when complex compounds are used.

- Its used is limited to small compounds and is not suitable for larger molecules like proteins or nucleic acids.
- It lacks experimental validation which can result in misleading conclusions.

E. Study carried out using this software

The SwissADME tool was utilized to predict the pharmacokinetic properties and drug-likeness characteristics of small molecules present in Spirulina platensis. Spirulina platensis is rich in antioxidant compounds, with dodecanoic acid ethyl ester being the predominant compound. This compound exhibits potential biological activities, such as antimicrobial, antioxidant, anti-cancer, anti-candida, and mycelial growth inhibition. The analysis using SwissADME revealed that all the compounds in the Spirulina platensis extract met the drug likeness criteria according to Lipinski's rules. The SwissADME method proved to be a simple, robust, and accurate approach to evaluate the ADME properties of the phytochemicals present in Spirulina platensis. The ADME analysis results suggest that Spirulina platensis extracts could be developed as traditional medicines and nutraceutical products [21].

19.2.4.2 DruLiTo

This subsection focuses on the use of DruLiTo software designed to evaluate the drug likeness properties and toxicity of chemical compounds which is used in the early stages of drug discovery to filter out the drugs with poor drug-like properties and high toxicity. The primary distinction between SWISS ADME (Sect. 19.2.3.1) and DruLiTo is that SWISS ADME concentrates on ADME properties and drug-likeness while DruLiTo combines drug likeness evaluation with toxicity prediction.

DruLiTo is a tool for predicting Drug likeliness developed by the Department of Pharmacoinformatics, National Institute of Pharmaceutical Education and Research, Mohali, India. Within the parameters of DruLiTo, LogP stands out as a crucial factor influencing the environmental destiny and hydrophobicity of the test moiety. The molecule's hydrophobicity plays a significant role in its absorption, bioavailability, receptor interaction, metabolism, and toxicity [22]. If the AlogP value of the molecule is below five, it is given priority as a drug-like compound. Lipinski et al. [19]. introduced constraints on HBA, HBD, Log P, and Molecular weight, and adherence to Lipinski's rule indicates the molecule's suitability for oral administration. The TPSA parameter is employed to comprehend the activity of polar atoms within a molecule, forecasting the absorption and transit of the test molecule [23].

It evaluates fundamental physicochemical properties such as, partition coefficient (LogP), octanol-water partition coefficient (AlogP), molecular weight (Mw), rotatable bond count (RC), H-bond donor (HBD), H-bond acceptor (HBA), number of rigid bonds (nRigidB), total polar surface area (TPSA) and number of hydrogen bonds (nHB), atom molar refractivity (AMR), number of rotatable bonds (nRB). The computations within DruLiTo adhere to various drug-likeness principles, encompassing the BBB rule, Veber's rule, Lipinski's rule and quantitative estimates of drug-likeness. Acknowledging and applying these rules and constraints is vital as it allows DruLiTo to identify substances meeting the required criteria, facilitating their utilization in subsequent research without plagiarism concerns (shown in Fig. 19.5) [24].

1. Lipinski's Rule-
    - Number of hydrogen bond donors ≤5
    - Number of hydrogen bond acceptors ≤10
    - Molecular weight ≤500
    - CLogP ≤5
2. MDDR-like rules
    - Number of rings ≥3
    - Number of Rigid bonds ≥18
    - Number of Rotatable bonds ≥6
3. Veber's Rule
    - Rotatable Bond Count ≤10
    - PSA ≤140
4. Ghose Filter
    - Log P (−0.4 ~ 5.6)
    - MR (Molar Refractivity) (40 ~ 130)
    - Molecular Weight (160 ~ 480)

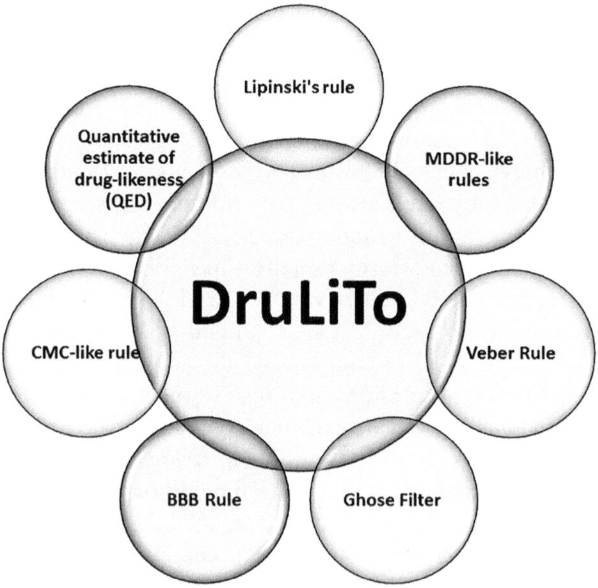

Fig. 19.5 DruLito Software monogram

- Number of Atoms (20 ~ 70)
- PSA < 140

5. BBB Rule

   - H- bonds (8–10)
   - Molecular Weight (400–500)
   - No acids

6. CMC-50 like rule

   - AlogP (1.3–4.1)
   - Molar Refractivity (70–110)
   - Molecular Weight (230–390)
   - Number of Atoms (30–55)

7. QED

   - Molecular Weight
   - AlogP
   - Hydrogen-bond acceptors
   - Hydrogen-bond donors
   - Rotatable bonds
   - Polar Surface Area
   - Aromatic Bond Count
   - Number of Structural Alerts

Advantages

- It provides comprehensive ADMET analysis which helps the researchers to understand how a drug behaves in the biological system
- It applies well-established drug-likeness rules, such as Lipinski's Rule of Five, Veber's rules, BBB (Blood-Brain Barrier) permeability, and Quantitative Estimate of Drug-likeness (QED), to assess the potential efficacy and safety of compounds.
- DruLiTo can often be integrated with other computational tools and databases, enhancing its utility in drug discovery and development workflows.

Disadvantages

- As an open-source tool, DruLiTo may lack comprehensive user support and documentation compared to commercial software, which can make it challenging for new users to get started.
- DruLiTo may face compatibility issues with certain operating systems or other software, potentially requiring users to perform additional configuration or integration work.
- Without rigorous validation and peer review, there is a risk of errors or inaccuracies in the software's algorithms, which could impact the reliability of the results.

A. **Study CARRIED OUT USING THIS SOFTWARE**

DruLiTo is an open-source software used to analyse the ADMET properties of Phyto-components from Syzygium cumini seeds and Allium cepa peels. It evaluated four Phyto-components from the methanol extracts of both sources, confirming that they adhere to all drug-likeness criteria. The software applied four filters—Lipinski's rule, Veber rule, BBB rule, and Quantitative Estimate of Drug-likeness (QED)—to assess the ADME profiles of the compounds identified through GCMS analysis of the fermented methanol extracts. The molecular characteristics of the Phyto-components from these fermented extracts, as determined by DruLiTo, are presented in the following Table 19.1 [25].

Table 19.1 Characteristics of compounds identified from GCMS analysis of fermented *Syzygium cumini* seeds and fermented *Allium cepa* peel methanolic extracts

Code of compound	Characteristics of compounds (Methanol extract-fermented *Sygygium cumini* seeds)							
	MW	LogP	HBA	HBD	TPSA	nRB	nAG	nHB
SC1	207.3	0.266	4	1	97.19	2	0	5
SC2	86.07	1.088	1	0	9.23	0	0	1
SC3	75.07	−0.756	2	1	35.25	2	0	3
SC4	90.03	−0.591	3	2	57.53	1	1	5
SC5	92.05	−1.88	3	3	60.69	2	0	6
SC6	310.08	3.677	1	0	9.23	3	0	1
SC7	118.1	1.93	2	1	29.46	4	0	3
SC8	126.03	1.431	3	3	60.69	0	0	6
SC9	334.21	5.922	4	0	52.6	14	0	4
SC10	212.1	−0.296	4	0	52.6	2	0	4
Code of compound	Characteristics of compounds (methanol extract-fermented *Allium cepa* seeds)							
	MW	LogP	HBA	HBD	TPSA	nRB	nAG	nHB
AC1	196.97	0.902	1	0	12.36	0	0	1
AC2	90.07	−0.288	2	2	40.46	1	0	4
AC3	92.05	−1.88	3	3	60.69	2	0	6
AC4	61.05	−1.275	2	2	46.25	1	0	4
AC5	104.05	−1.105	3	2	49.69	0	0	5
AC6	110.04	1.083	2	2	40.46	0	0	4
AC7	190.05	−2.122	6	4	115.06	1	1	10
AC8	347.17	2.363	6	0	77.78	10	0	6
AC9	496.34	10.748	4	4	44.76	21	0	4
AC10	368.27	8.951	2	0	26.3	16	0	2

19.2.4.3 ProTox3.0

ProTox (shown in Fig. 19.6) is a webserver that predicts the toxicity of chemicals. To predict Hepatotoxicity, cytotoxicity, carcinogenicity, immunotoxicity, etc. like other toxicity endpoints, ProTox 3.0 uses machine-learning models comprising pharmacophores, molecular similarity, and fragment propensities. Both *in-vitro* assays (e.g. Tox21 assays) and *in-vivo* cases (e.g. carcinogenicity, hepatotoxicity) data are used to build predictive models. Independent external sets have validated these models on showing strong performance.

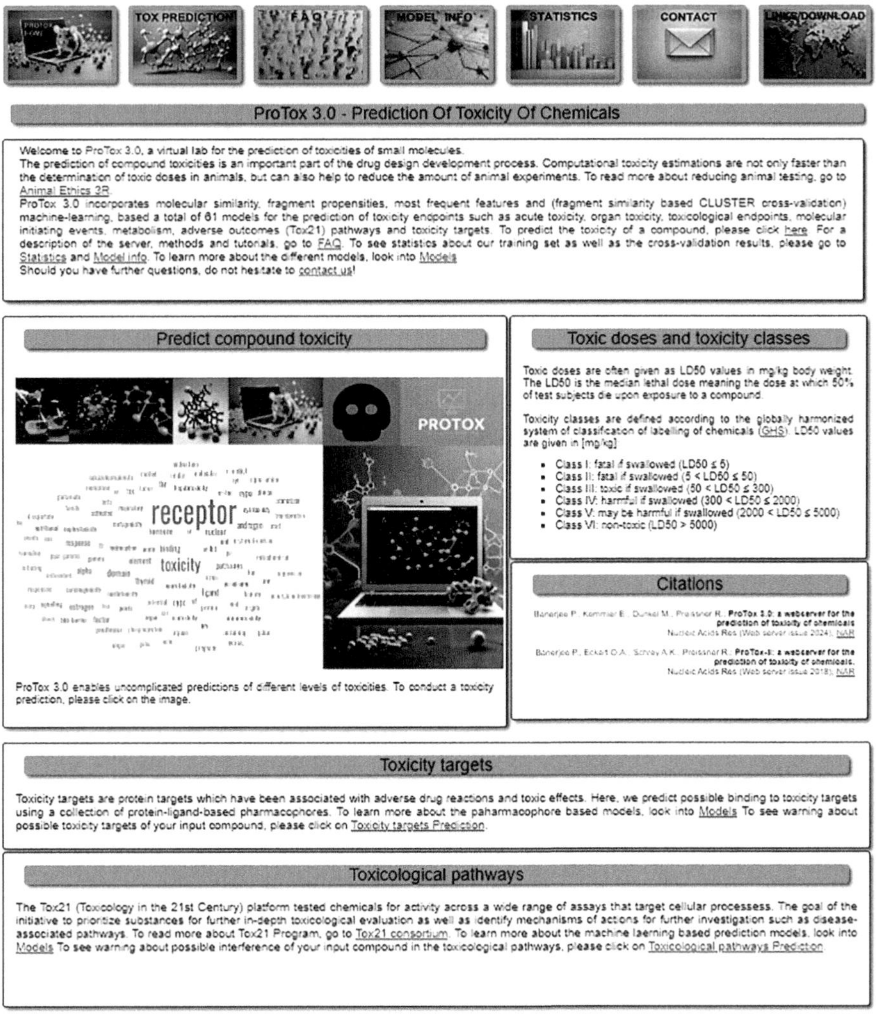

Fig. 19.6 ProTox II input page

ProTox 3.0 is a freely available webserver for predicting *in-silico* toxicity useful to toxicologists, regulatory companies, computational and medicinal chemists, and other users without the requirement of login at https://tox.charite.de .

ProTox 3.0 predicts 61 toxicity endpoints such as acute, organ, clinical toxicity, molecular-initiating events (MOE), even adverse outcomes (Tox21) pathways and toxicity off-targets by incorporating both molecular similarity and machine-learning models. Being validated on independent external sets all ProTox 3.0 models have shown strong performance. A two-dimensional chemical structure is taken as an input by the webserver and the possible toxicity profile of the chemical is reported for 33 models with confidence scores, along with an overall toxicity radar chart and the three most similar compounds having known acute toxicity [26].

ProTox 3.0 web server is very self-informative and has a user-friendly interface which requires less or absolutely no help. By specifying either the name (PubChem name) or canonical SMILES (*Simplified Molecular-Input Line Entry System*) string of the compound the user can run a prediction on the server. Drawing box is provided by the (https://www.chemdoodle.com/), can be used to draw the desired molecular structure. Thereby, supports prediction of hypothetical compounds even before they are synthesized in labs. Additionally, the user can Selection for a single model/user-specific model or ALL models are available as well; when not specified by the user, only acute toxicity and toxicity targets are computed by the web server as default [27].

Advantages

- Has many new specific features which can be explored such as choosing the desired toxicity targets and toxicological pathways.
- Even allows flexibility to choose models for analysis.
- Has a user-friendly interphase and is practical for the self-intuitive human nature.

Disadvantages

- As it is quite new it isn't as attractive in terms of the display and might not be appealing.
- Also, may have compatibility issues on different OS.
- New software might bring security vulnerability as well.

A. Study carried out using this software

Protox has been used for in Silico Study of Pyrazolylaminoquinazoline Toxicity for which an article is published dated 2018. The endpoint LD50 and toxicity classes studied using ProTox application. Based on the scoring from the Protox and the other software applications used; two compounds showed activity against FGFR 2. Even with no toxic effects on the liver, endocrine system, and the reproductive system or carcinogenicity, these compounds were predicted to have mutagenic effects [27].

19.2.4.4 ADMET LAB2.0

Similar to the prevision application ADMET LAB2.0 is also known for its features of evaluating pharmacokinetics of drugs under consideration.

Failure in drug development usually occurs when the candidate compound has undesirable pharmacokinetics and toxicity. Thus, the ADME properties of the compound under scrutiny shall be evaluated early to avoid failure. The features of ADMETlab2.0 can be availed to evaluate the compound based on the ADMET properties systematically. ADMETlab2.0 also helps researchers with the drugs' physicochemical and medicinal chemistry-based properties along with predictive models, updates in functional modules, etc. ADMETlab2.0 helps researchers in their drug development process acceleration.

The ADMETlab2.0 consists of two major features namely, ADMET EVALUATION and ADMET SCREENING.

The EVALUATION feature is a module of a high-quality series prediction models trained using a multi-task graph attention framework. This gives the users efficiency and convenience to implement calculations of 17 physicochemical properties, 23 endpoints of ADME, 13 measures of Medicinal chemistry, and additionally 27 endpoints of toxicity, and 8 rules of toxicophores. Thus, filtering out the promising lead compounds.

The SCREENING feature is designed to predict molecular datasets using batch mode evaluation. This feature supports SMILES strings and SDF/TXT format files. Thus, suitable for the evaluation of previously empirically designed, visually screened chemical synthesis and biochemical assays allowing scientists to increase their focus on experiments on promising compounds.

On entering the appropriate SMILES for the compound for evaluation a graph is generated by the server like given below:

The above graph touches upon the important physicochemical properties that are used to distinguish drug likeness such as Log P value, Molecular Weight, TPSA etc. Further provides the researchers with values elaborating on the drugs' ADMET properties (shown in Fig. 19.7).

This allows the researchers to understand the Route of administration by taking into account the values under Absorption and Distribution, Prodrug formation requirements by considering the values under Metabolism, Excretion and Toxicity (shown in Fig. 19.8).

For Screening Multiple SMILES can be submitted and evaluated together. Researchers can avail the option of submitted a molecule file instead of SMILES, this helps the viewer to analyse multiple properties of multiple compounds under consideration to identify the most promising compound. Thus, ADMET lab2.0 is a very comprehensive web server which eases out the process for drug development and thereby, aids drug likeness analysis of given compounds [28].

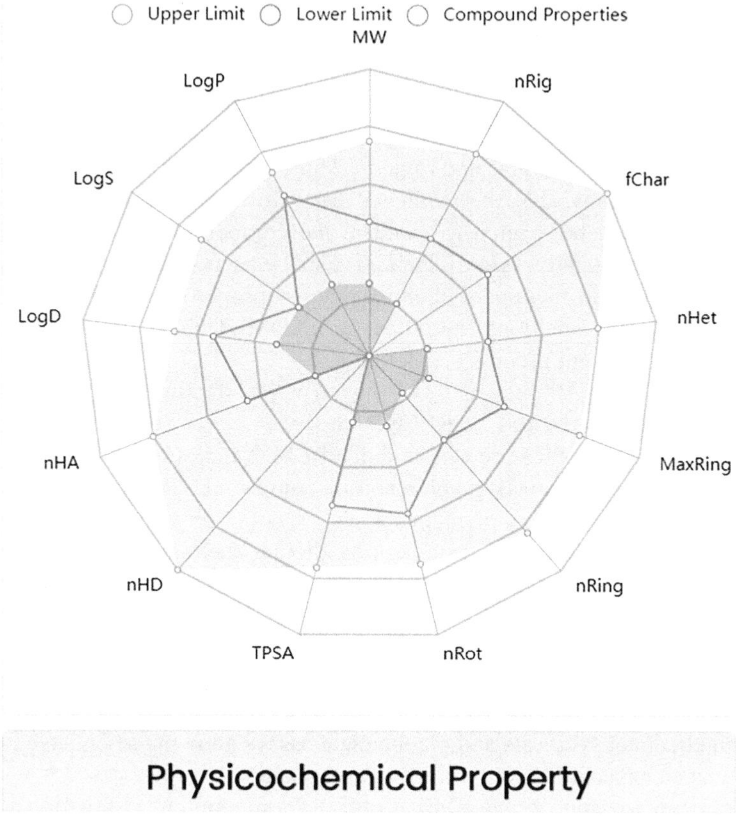

Fig. 19.7 ADMETlab2.0 physicochemical properties prediction

Advantages

- The detailed predictions for various ADMET properties provided by ADMET Lab 2.0 helps in understanding how a drug candidate will behave in the body and the making the results comprehensive.
- Parameters and input data can be customized by users often such that specific needs or particular aspects of the drug discovery project can be fulfilled.
- Accuracy of the predictions is enhanced over time due to the continuous updates and improvements in the algorithms and models. Thus, increasing the reliability of the tool.

Disadvantages

- If the models are based on incomplete or outdated information, the accuracy of predictions might be hampered.
- All specific biological nuances or individual variations may account for predictive models, thus becoming less reliable for certain compounds.

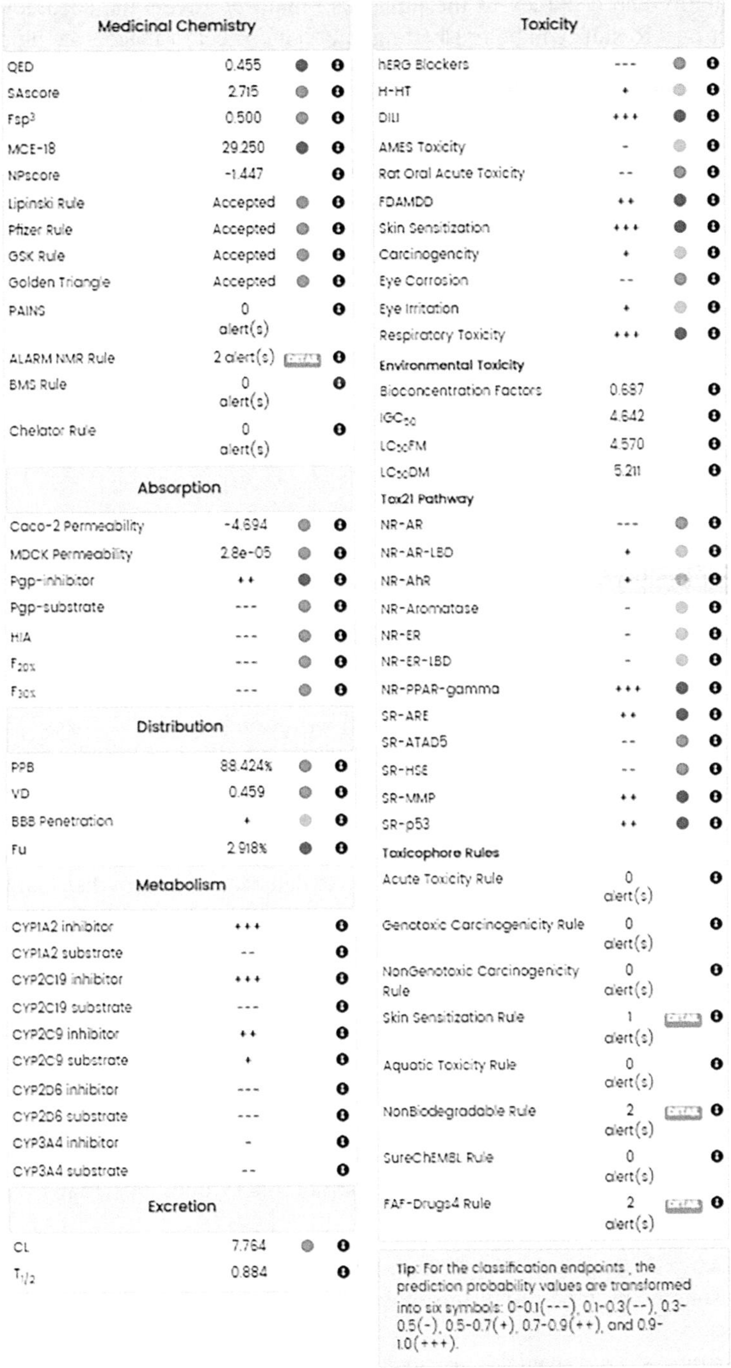

Medicinal Chemistry	
QED	0.455
SAscore	2.715
Fsp^3	0.500
MCE-18	29.250
NPscore	-1.447
Lipinski Rule	Accepted
Pfizer Rule	Accepted
GSK Rule	Accepted
Golden Triangle	Accepted
PAINS	0 alert(s)
ALARM NMR Rule	2 alert(s)
BMS Rule	0 alert(s)
Chelator Rule	0 alert(s)

Absorption	
Caco-2 Permeability	-4.694
MDCK Permeability	2.8e-05
Pgp-inhibitor	++
Pgp-substrate	---
HIA	---
$F_{20\%}$	---
$F_{30\%}$	---

Distribution	
PPB	88.424%
VD	0.459
BBB Penetration	+
Fu	2.918%

Metabolism	
CYP1A2 inhibitor	+++
CYP1A2 substrate	--
CYP2C19 inhibitor	+++
CYP2C19 substrate	---
CYP2C9 inhibitor	++
CYP2C9 substrate	+
CYP2D6 inhibitor	---
CYP2D6 substrate	---
CYP3A4 inhibitor	-
CYP3A4 substrate	--

Excretion	
CL	7.764
$T_{1/2}$	0.884

Toxicity	
hERG Blockers	---
H-HT	+
DILI	+++
AMES Toxicity	-
Rat Oral Acute Toxicity	--
FDAMDD	++
Skin Sensitization	+++
Carcinogencity	+
Eye Corrosion	--
Eye Irritation	+
Respiratory Toxicity	+++
Environmental Toxicity	
Bioconcentration Factors	0.687
IGC_{50}	4.642
$LC_{50}FM$	4.570
$LC_{50}DM$	5.211
Tox21 Pathway	
NR-AR	---
NR-AR-LBD	+
NR-AhR	+
NR-Aromatase	-
NR-ER	-
NR-ER-LBD	-
NR-PPAR-gamma	+++
SR-ARE	++
SR-ATAD5	--
SR-HSE	--
SR-MMP	++
SR-p53	++
Toxicophore Rules	
Acute Toxicity Rule	0 alert(s)
Genotoxic Carcinogenicity Rule	0 alert(s)
NonGenotoxic Carcinogenicity Rule	0 alert(s)
Skin Sensitization Rule	1 alert(s)
Aquatic Toxicity Rule	0 alert(s)
NonBiodegradable Rule	2 alert(s)
SureChEMBL Rule	0 alert(s)
FAF-Drugs4 Rule	2 alert(s)

Tip: For the classification endpoints, the prediction probability values are transformed into six symbols: 0-0.1(---), 0.1-0.3(--), 0.3-0.5(-), 0.5-0.7(+), 0.7-0.9(++), and 0.9-1.0(+++).

Fig. 19.8 ADMETlab2.0 toxicity prediction page

- The quality and accuracy of the input data majorly affects the accuracy of the predictions. Results can be misleading due to incorrect or inaccurate inputs.

A. Study carried out using this software

ADMET lab has been used to analyse levamisole derivatives against the SARS-CoV-2 main protease (M^{Pro}). In silico ADMET properties of levamisole derivatives were evaluated by wet-lab applicability. The potential of the molecule N°6 to stop the newly emerged SARS-CoV-2 and further be developed as a novel inhibitor was supported by the results of ADMET prediction and MD simulation [29].

19.2.4.5 FAF-Drugs2

In this subsection we will discuss about the use of FAF Drugs software in drug likeness assessment, ADMET prediction, optimization of drug candidates, structure-based screening, etc.

It is a free ADME-tox filtering tool that is used to facilitate experimental screening and virtual screening prior to an experiment by preparing chemical compound libraries. It can perform various physicochemical calculations and identify the key functional groups which may be toxic or unstable. It provides various distribution diagrams of the major physicochemical properties of the screened compounds. The main architecture of the program includes

1. Scanner which parses parameter files and writes output files
2. CreateMol- It creates the molecules that are required to be filtered
3. Rules- It defines all the rules used by FAF-Drugs2 for filtering molecules
4. SmartsCodes- It detects hydrogen bond acceptors and donors by using SMARTS patterns and SMARTS patterns for functional groups and undesirable compounds
5. GetRings- It has all the functions required to detect and identify the smallest ring systems and the largest size of rings [30].

In this web application, the user has to input the compound collection (in SD or SMILES format) along with two different parameter files of which one includes the physicochemical thresholds while the other file includes the chemical substructures that is to be investigated. The users can also decide to activate PAINS and functional group substructure detection. Once the process is terminated the filtered result file can be downloaded and analysed using pie charts or statistical analysis tools. It basically evaluated the drug like properties, detects the undesirable properties of compounds and oral absorption of compounds before it is synthesized [31].

Advantages

- FAF-Drugs enables early-stage screening of compounds, potentially saving time and resources by identifying unsuitable candidates early in the drug discovery process.
- FAF-Drugs is often available as open-source software, allowing researchers to use it freely and modify it as needed.

- It can integrate with various databases and tools, facilitating a more comprehensive analysis and enhancing its functionality in drug discovery.

Disadvantages

- Some analyses or features may require substantial computational resources, which might be a limitation for users with less powerful hardware.
- Users without a background in computational chemistry or drug discovery might face a learning curve when using the software, particularly if documentation is lacking.
- Users may encounter compatibility issues with certain operating systems or data formats, requiring additional configuration or troubleshooting.

A. Study carried out using this software

In silico analysis was performed using FAF-Drugs software to evaluate catechin, galangin, and hesperidin as potential competitors of the SARS-CoV-2 spike protein. This analysis aimed to assess the interaction of these test compounds with the body, focusing on their drug-likeness. The study encompassed an evaluation of Absorption, Distribution, Metabolism, Excretion, and Toxicity (ADMET) profiles. Pharmacological assessment was conducted following Lipinski's rules using the FAF-Drugs4 program available on the RPBS Web Portal [32].

19.2.4.6 Biovia Discovery Studio

This subsection focusses on the use of Biovia Discovery studio for molecular modelling and simulations in drug discovery and material science. The main purpose of this software includes molecular visualization, Docking and binding studies, predictive modelling, etc.

This software enables its user to visualize, share, and analyse protein and small molecular data. It allows its experts to exchange data efficiently and seamlessly without the loss of valuable scientific information and time. It also provides simulations of biological systems with best-in-class tools. The two main simulation programs it uses are NAMD (Nanoscale Molecular Dynamics) and CHARMm (Chemistry at Harvard Macromolecular Mechanics). Furthermore, GaMD (Gaussian accelerated Molecular Dynamics) is incorporated in the latest version of Biovia Discovery studio for free energy calculations and unconstrained enhanced sampling. It further facilitates drug discovery from hit discovery to late-stage optimization. The strategies used in rational drug development of small molecules include Fragment-based drug design and structure-based drug design. It provides CCDC i.e. Cambridge Crystallographic Data Centre that supports SBD and FBD in the drug discovery process. It helps in assessing and enhancing small molecule therapeutics with or without target structured data by using Catalyst pharmacophore and analysis tool set. This software also enables the safe and therapeutic study of physicochemical properties like ADME, toxicity, and structure activity-based relationships [33].

Advantages

- The software includes sophisticated visualization capabilities that help researchers clearly understand and interpret molecular structures, interactions, and dynamics.
- Discovery Studio offers robust molecular docking and binding prediction algorithms, such as AutoDock Vina, which enhance the accuracy of predicting how molecules interact with their targets.
- The software supports various types of simulations, including molecular dynamics, which are essential for studying the behaviour and stability of molecules over time.

Disadvantages

- The software's extensive range of tools and features can be overwhelming for new users. There may be a steep learning curve, requiring significant time and training to master.
- Biovia Discovery Studio can be expensive, which may limit access for smaller labs or institutions with constrained budgets.
- The software's focus on specific types of molecular modeling and drug discovery might make it less versatile for other types of research or applications outside its primary domain.

A. **Study carried out using this software**

An in-silico analysis of the interactions between the anti-apoptotic protein survivin and the compounds dentatin, nordentatin, and quercetin was performed using Biovia Discovery Studio. This software, specifically Discovery Studio Biovia 2017 (Dassault Systèmes, San Diego, California, USA), was utilized to visualize and modify the receptor and ligand structures. Molecular docking was carried out with AutoDock Vina, and post-docking analyses were examined using PyMOL and Discovery Studio Biovia 2017. These analyses revealed details such as the size and location of binding sites, hydrogen-bond interactions, hydrophobic interactions, and bonding distances within interaction radii of <5 Å from the docked ligand's position. The compounds were docked at survivin's active site, and the binding poses and interactions of each ligand with the protein were evaluated to determine the most favourable and energetically optimal conformations [34].

19.2.4.7 Amber Molecular Dynamics

This subsection focusses on several key purposes of this software in computational chemistry and molecular biology which includes structure optimization, force field applications, free energy calculations, ligand and protein preparations, etc.

It is used to improve the efficiency of free energy calculations. Free energy calculations are used in determining the free energy change upon mutation of an amino acid in a protein and solvation of drug like molecules. Alchemical transformations are vital for such calculations of free energy and AMBER included this as a part of its molecular dynamics Engine [35].

It has evolved over the years from a program that was made to do Assisted Model Building and Energy Refinement to a set of powerful tools used for molecular dynamics and free energy calculations [36].

Amber is associated with a range of traditional molecular mechanics force fields tailored for simulating biomolecules. It includes a set of various software programs that collaborate in the preparation, execution and interpretation of molecular dynamics simulations. It consists of three highly coupled molecular dynamics engine, which are, Sander, pmemd and pmemd. cuda. In this the former is responsible for development and computation of the software while the latter is responsible for maximizing the performance. The development of Amber is due to the introduction of pmemd and pmemd.cuda [37].

Advantages

- Amber provides robust and well-validated force fields like GAFF and ff14SB, which are widely used for accurate modelling of biomolecules and small molecules.
- Amber is equipped with tools for calculating binding free energies and performing thermodynamic integration, which are essential for drug discovery and understanding molecular interactions.
- It supports a wide range of simulation types, including molecular dynamics, energy minimization, and free energy calculations, allowing for in-depth analysis of molecular systems.

Disadvantages

- While Amber is capable of running large-scale simulations, optimizing performance for extremely large systems or very long simulation times can be challenging and resource-intensive.
- Amber simulations can be computationally intensive, requiring significant processing power and memory, which may be a limitation for users with less powerful hardware.
- The performance and accuracy of simulations can depend on the choice of force fields and parameters, which may require careful tuning and validation for different systems.

A. Study carried out using this software

Molecular dynamics simulations of a new β-lactamase inhibitor were conducted using Amber Molecular Dynamics software. These simulations aim to evaluate the dynamic properties of the ligand-receptor complex to approximate its behaviour in a real-world environment. The simulations were carried out with the AMBER software, following a protocol that includes minimization, heating, equilibration, and production runs. Ligand preparation was performed using the antechamber program. The General AMBER Force Field (GAFF) was used for the ligand, while the enzyme was modelled with the ff14SB force field, which is known for its improved backbone potential and side chain rotamer distributions due to scalar couplings.

Additionally, the software can calculate binding free energy and free energy decomposition per residue [38].

19.2.4.8 QikProp—Commercialized by SCHRODINGER

With advancing competition among researchers, there was a need of a mechanism to fasten the drug discovery and development process. The drug discovery takes place following various steps wherein, the preclinical development is lengthy and has a high occurrence of failure molecules. To lessen the time and effort consumed in the preclinical development process which decides on the most promising drug candidate for clinical trials many online software is developed.

The above given are steps (Fig. 19.9) for the development of a preclinical trial candidate. The Lead generation phase is also known as the HIT TO LEAD phase where the Hit identified is formulated into a lead compound by evaluating based on QSAR, ADME properties, pharmacokinetics, toxicity, Lipinski rule of 5, etc. Whereas, Lead optimization involves generating a preclinical candidate using chemical modifications to curtail the shortcomings of the lead structure. Thus, the ADME and physicochemical properties are enhanced and toxicity profiles are minimized, offering safer compounds that can pass through the clinical trials [39].

An advanced tool analysing based on 3D molecular structures for physicochemical and pharmacokinetic properties prediction used for small organic molecules is QikProp. Usually, upto 40% candidates fail in clinical trials corresponding to their poor ADME properties. This leads to costs of development to soar higher and higher. Detection of problematic candidate earlier significantly reduces cost, time and resources spent on it.

ADME predictions if incorporated into development generates lead compounds which have significantly higher chances of passing in clinical trials.

Schrodingers' web server provides many more various features to support drug discovery.

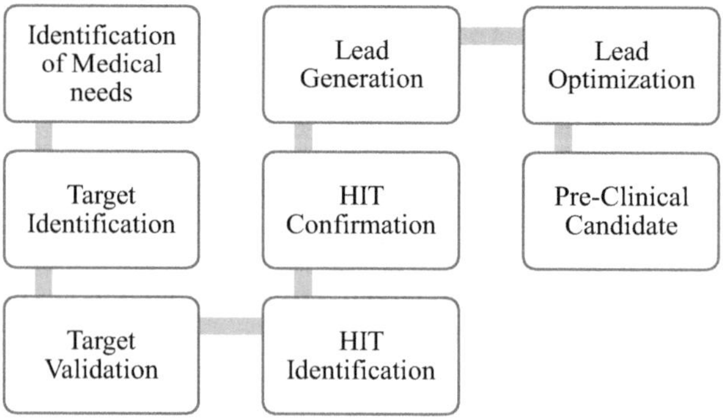

Fig. 19.9 Flow of Drug Development

Features Include

Maestro—molecular modelling environment to access molecular simulation workflows based on physics and advance structure visualization.

LiveDesign—Cloud-native and collaborative enterprise platform with 3D design, cheminformatics and data analysis for the entire discovery team.

Epik—rapidly predicts pka and protonation state.

MS Morph—modelling tool used to predict organic crystal habit.

FEP+—calculates high-performance free energy for drug discovery etc. [40].

Advantages

- The software can generate predictions for a large number of compounds due to its design. Thus, suitable for high-throughput screening and early-stage drug discovery. Researchers can evaluate many candidates in a relatively short amount of time with the tool's efficiency.
- Visualization tools are provided by QikProp which help interpret and analyse the prediction results. These tools aids understanding the influence of different structural features on ADMET properties.
- Extensive documentation and user support is offered by Schrödinger for QikProp, including tutorials, guides, and technical assistance. Such support helps users utilize the software and effectively troubleshoot any issues that arise.

Disadvantages

- QikProp can be expensive as is a part of the Schrödinger suite of software. This cost may be prohibitive for some labs or organizations. Thus, access to the tool is limited.
- QikProp's interface and complexity of its output can be challenging to navigate for some users, especially those who inexperienced in computational chemistry or ADMET predictions.
- Being based on empirical data and predefined models QikProp's predictions are may not be always accurate. Whilst these models are generally reliable, they may not provide accurate predictions for all compounds always, especially those that are structurally very different from the training sets used to develop the models.

A. Study carried out using this software

ADME studies of benzimidazolyl-chalcones as antileishmanial agents were performed using QikProp [41]. QikProp was used in in-silico ADME analysis for identification of potential E6 inhibitors against cervical cancer. Nearly 90% cases of cervical cancer the HPV 16 and 18 high-risk DNA has been found to be associated. The viral genes - E6 and E7 of HPV are regularly expressed in cervical cancer. Here the in silico ADME analysis was also undertaken using QikProp. Molecular dynamics simulations study was carried out for the best hit obtained in search of better hits. Thus, developing as potent, therapeutically active E6 inhibitors. QikProp results suggested that identified hits have drug like physico-chemical properties. Thus, reaffirms previous finding of E6AP binding pocket on E6 protein being druggable. These identified hits may show activity as new leads for inhibition of E6 against cervical cancer [42].

19.2.4.9 AUTODOCK

This subsection focusses on the use of Autodock which is primarily designed for computational docking studies in molecular modelling and drug discovery. Its purpose is to predict how small molecules, such as drugs or ligands, bind to larger macromolecules like proteins, Ligand Binding Prediction, High-Throughput Screening, Understanding Protein-Ligand Interactions, etc.

AutoDock has multiple extensions that can be used to increase the possibility of drugs discovered passing through clinical trials by scrutinizing them using ADME properties and molecular docking.

AutoDock VINA is one of the docking engines of AutoDock Suites which is an Open-Source program used for molecular docking, originally Dr.Oleg Trott designed and implemented it in CCSB—The Scripps Research Institute.

VINA makes AutoDock significantly more accurate in binding mode predictions in comparison to AutoDock 4 and is compatible to the PDBQT file format which can viewed through MGLTools. VINA is designed such that the user does not necessarily require to understand its implementation details, unclear search parameters, unorganized results or understand advanced algebra such as quaternions. Only structures of the molecules being docked are required including the specifications of search space as well as the binding site.

VINA takes advantage of multiple CPUs or CPU cores on the user's system to shorten the running time significantly. Recently, AutoDock-GPU was developed, considered to be an accelerated version of AutoDock4 and a hundred times faster than the original single CPU docking code. Further speed is achieved by parallelism which uses multithreading on multicore machines. Originally it consisted of 2 main programmes: - docking of ligand to set of grids describing the target protein and—autogrid that precalculated these grids. VINA calculates the grid maps automatically and transparently clusters results.

AutoDock can be used in XRAY crystallography, SAR drug designing, Lead optimization, protein-protein docking, chemical mechanism studies, virtual screening (HTS), etc. [43, 44].

Advantages

- AutoDock supports various types of docking, including protein-ligand, protein-protein, and protein-DNA interactions. This versatility makes it applicable to a wide range of biological research areas.
- AutoDock can be integrated with various molecular visualization tools, such as PyMOL and Chimera, for better analysis and interpretation of docking results.
- As a widely used tool in molecular modelling and drug discovery, AutoDock serves as an educational resource for students and researchers learning about molecular docking and computational biology.

Disadvantages

- AutoDock has a limitation on the number of torsions it can handle (32 torsions), which may affect its ability to dock highly flexible ligands accurately.

- The initial setup and configuration of AutoDock can be complex, especially for new users. It requires a good understanding of the software's parameters and docking protocols.
- While generally reliable, AutoDock's predictions may not always be as accurate as those from other docking software, particularly in complex docking scenarios.

A. Study carried out using this software

The in-silico analysis of bioactive compounds from sea cucumbers for their anti-breast cancer properties was conducted using AutoDock Vina. This study employed molecular docking via both AutoDock and Vina, which are the most commonly used and accessible tools for such analyses. The research aims to identify the optimal molecular docking method for targeting breast cancer-related enzymes and to assess the binding affinity and interaction of sea cucumber compounds with these macromolecules. The macromolecules' 3D structures were optimized with AutoDock, including the removal of water molecules, addition of hydrogen atoms, and adjustment of charges using Gasteiger charge, followed by a minimization process. The docking results for the positive control FDA, raloxifene, showed binding energies of −13.82 kcal/mol with AutoDock and −10.9 kcal/mol with Vina. AutoDock was unable to achieve sufficiently low binding energies for the sea cucumber compounds due to limitations with high-torsion structures, as AutoDock permits a maximum of 32 torsions. Consequently, binding energies were obtained exclusively from Vina for this study [45].

19.3 Conclusion

In conclusion, the integration of computational tools in predicting drug likeliness marks a significant advancement in pharmaceutical research and development. These tools, ranging from molecular docking simulations to machine learning algorithms, offer a systematic and efficient approach to identify potential drug candidates with higher accuracy and lower costs compared to traditional methods. By leveraging vast amounts of data and sophisticated models, researchers can streamline the drug discovery process, leading to the identification of promising compounds with desirable pharmacological properties. However, it's crucial to acknowledge the limitations and challenges associated with these computational approaches, such as the need for validation through experimental studies and the constant evolution of predictive models. Nonetheless, with continued advancements in technology and interdisciplinary collaborations, computational tools will continue to play a pivotal role in accelerating the discovery and development of novel therapeutics, ultimately benefiting patients worldwide. The field of drug design is rapidly evolving, driven by technological advancements and a deeper understanding of biology. Here are some key emerging trends and challenges in this area:

Trends

1. Personalized Medicine: Breakthroughs in genomics and proteomics enable more tailored drug development, allowing drugs to be designed to target specific genetic mutations or biomarkers unique to an individual's condition. Pharmacogenomics further optimizes drug therapies by leveraging genetic information to enhance efficacy and minimize adverse effects.
2. Artificial Intelligence and Machine Learning: AI and ML are revolutionizing drug discovery by predicting compound interactions, identifying potential drug candidates faster, and optimizing drug design. These technologies also facilitate the integration of vast datasets to uncover new drug targets and predict drug responses.
3. Biologics and Biosimilars: The development of complex biologics, such as monoclonal antibodies and CAR-T cells, is expanding treatment options for various diseases. Additionally, as patents for biologics expire, biosimilars are becoming more prevalent, offering cost-effective alternatives.
4. Nanotechnology: Nanoparticles and nanocarriers are being employed to deliver drugs more precisely to targeted tissues or cells, improving efficacy and reducing side effects. Nanotechnology is also enhancing diagnostic capabilities, enabling earlier detection and more accurate monitoring of diseases.
5. Synthetic Biology: Synthetic biology enables the creation of new biological systems and organisms that can produce therapeutic agents or perform novel functions. Techniques like CRISPR are being used to design drugs that target specific genes or correct genetic mutations.
6. Regenerative Medicine: Advances in stem cell research are leading to new treatments for degenerative diseases and injuries, while the development of bioengineered tissues and organs aims to address the shortage of organ transplants and repair damaged tissues.

Challenges

1. Complexity of Diseases: - Multifaceted Nature: Many diseases, especially chronic and complex ones, involve intricate interactions between genetic, environmental, and lifestyle factors, posing significant hurdles for drug design and targeting.

- Drug Resistance: The ability of pathogens and cancer cells to develop resistance to drugs necessitates continuous innovation and adaptation in drug design to stay effective.

2. Regulatory Hurdles: - Approval Process: The lengthy and costly process of navigating regulatory requirements for new drugs, particularly for innovative therapies, can impede their timely development and introduction.

- Data Privacy: Ensuring the protection of personal data used in personalized medicine and AI-driven drug design, while adhering to regulations like GDPR, is a critical challenge.

3. Ethical and Social Issues: - Equitable Access: Guaranteeing that advanced treatments are accessible to diverse populations, not just the privileged, is essential for achieving healthcare equity.

- Ethical Use of Genetic Information: Addressing concerns about genetic privacy and the potential for misuse of gene

4. Cost and Accessibility: - High R&D Costs: The substantial investment required for research and development can limit the availability of new drugs, especially for less common diseases.

- Pricing and Affordability: Striking a balance between the cost of new drugs and ensuring their accessibility and affordability for patients is a persistent challenge.

5. Integration of New Technologies: - Interdisciplinary Collaboration: Effective drug design increasingly necessitates collaboration across various disciplines, including biology, chemistry, computer science, and engineering, to harness the full potential of innovation.
6. Data Management: Managing and integrating the large volumes of data from diverse sources, a complex task, requires advanced computational tools and expertise.

Overall, while the future of drug design holds immense promise with advancements in technology and personalized medicine, addressing these multifaceted challenges will be crucial for translating innovation into truly effective and accessible treatments that benefit all.

References

1. Sun D, Gao W, Hu H, Zhou S. Why 90% of clinical drug development fails and how to improve it? Acta Pharm Sin B. 2022;12(7):3049–62.
2. Duff K, Suleman F. Transgenic mouse models of Alzheimer's disease: how useful have they been for therapeutic development? Brief Funct Genomic Proteomic. 2004;3(1):47–59.
3. Young JE, Goldstein LS. Alzheimer's disease in a dish: promises and challenges of human stem cells models. Hum Mol Genet. 2012;21(R1):R82–9.
4. Medwin, Medwin Publishers. https://medwinpublishers.com/ACT/ACT16000101.php?id=4.
5. Elsevier, Science Direct. https://linkinghub.elsevier.com/retrieve/pii/S0016510720344667.
6. DiMasi JA, Hansen RW, Grabowski HG. The price of innovation: new estimates of drug development costs. J Health Econ. 2003;22(2):151–85. https://doi.org/10.1016/s0167-6296(02)00126-1.
7. Prada-Gracia D, Huerta-Yépez S, Moreno-Vargas LM. Application of computational methods for anticancer drug discovery, design, and optimization. Bol Méd Hosp Infant México Engl Ed. 2016;73(6):411–23. https://doi.org/10.1016/j.bmhimx.2016.10.006.
8. Leelananda SP, Lindert S. Computational methods in drug discovery. Beilstein J Org Chem. 2016;12:2694–718. https://doi.org/10.3762/bjoc.12.267.
9. Rajkishan T, Rachana A, Shruti S, Bhumi P, Patel D. Computer-aided drug designing. In: Singh V, Kumar A, editors. Advances in bioinformatics. Singapore: Springer; 2021. p. 151–82. https://doi.org/10.1007/978-981-33-6191-1_9.

10. Kalyaanamoorthy S, Chen YPP. Structure-based drug design to augment hit discovery. Drug Discov Today. 2011;16(17–18):831–9. https://doi.org/10.1016/j.drudis.2011.07.006.
11. Grinter SZ, Zou X. Challenges, applications, and recent advances of protein-ligand docking in structure-based drug design. Mol Basel Switz. 2014;19(7):10150–76. https://doi.org/10.3390/molecules190710150.
12. Kutchukian PS, Shakhnovich EI. De novo design: balancing novelty and confined chemical space. Expert Opin Drug Discov. 2010;5(8):789–812. https://doi.org/10.1517/17460441.2010.497534.
13. Baig MH, Ahmad K, Rabbani G, Danishuddin M, Choi I. Computer aided drug design and its application to the development of potential drugs for neurodegenerative disorders. Curr Neuropharmacol. 2018;16(6):740–8. https://doi.org/10.2174/1570159X15666171016163510.
14. Quantitative Structure-Activity Relationship - an overview | ScienceDirect Topics. https://www.sciencedirect.com/topics/pharmacology-toxicology-and-pharmaceutical-science/quantitative-structure-activity-relationship.
15. Schneider G. Virtual screening: an endless staircase? Nat Rev Drug Discov. 2010;9(4):273–6. https://doi.org/10.1038/nrd3139.
16. Hay M, Thomas DW, Craighead JL, Economides C, Rosenthal J. Clinical development success rates for investigational drugs. Nat Biotechnol. 2014;32(1):40–51. https://doi.org/10.1038/nbt.2786.
17. Dahlin JL, Inglese J, Walters MA. Mitigating risk in academic preclinical drug discovery. Nat Rev Drug Discov. 2015;14(4):279–94. https://doi.org/10.1038/nrd4578.
18. Wilson, Gisvold. Chapter 6: "Anti-infective agents". In: Beale JM, Block JH, editors. Textbook of organic medicinal and pharmaceutical chemistry. 12th ed. p. 234.
19. Daina A, Michielin O, Zoete V. SwissTargetPrediction: updated data and new features for efficient prediction of protein targets of small molecules. Nucleic Acids Res. 2019;47(W1):W357–64. https://doi.org/10.1093/nar/gkz382.
20. Zoete V, Daina A, Bovigny C, Michielin O. SwissSimilarity: a web tool for low to ultra high throughput ligand-based virtual screening. J Chem Inf Model. 2016;56(8):1399–404. https://doi.org/10.1021/acs.jcim.6b00174.
21. Riyadi PH, Romadhon SID, Kurniasih RA, Agustini TW, Swastawati F, et al. SwissADME predictions of pharmacokinetics and drug-likeness properties of small molecules present in Spirulina platensis. IOP Conf Ser Earth Environ Sci. 2021;890(1):012021. https://doi.org/10.1088/1755-1315/890/1/012021.
22. Swamy RN, Gnanamani A, Shanmugasamy S, Gopal RK, Mandal AB. Bioinformatics in cross-linking chemistry of collagen with selective cross linkers. BMC Res Notes. 2011;4(1):399. https://doi.org/10.1186/1756-0500-4-399.
23. Lipinski CA, Lombardo F, Dominy BW, Feeney PJ. Experimental and computational approaches to estimate solubility and permeability in drug discovery and development settings1. Adv Drug Deliv Rev. 2001;46(1):3–26. https://doi.org/10.1016/s0169-409x(00)00129-0.
24. Drug Likeness Tool, DruLiTo. https://niper.gov.in/pi_dev_tools/DruLiToWeb/DruLiTo_index.html.
25. Kadri HS, Minocheherhomji FP. ADMET analysis of phyto-components of Syzygium cumini seeds and Allium cepa peels. Futur J Pharm Sci. 2020;6(1):1–9. https://doi.org/10.1186/s43094-020-00136-9.
26. Banerjee P, Eckert AO, Schrey AK, Preissner R. ProTox-II: a webserver for the prediction of toxicity of chemicals. Nucleic Acids Res. 2018;46(W1):W257–63. https://doi.org/10.1093/nar/gky318.
27. Supandi Y, Merdekawati F. In Silico study of Pyrazolylaminoquinazoline toxicity by Lazar, Protox, and Admet predictor. J Appl Pharm Sci. 2018;8(9):119–29. https://doi.org/10.7324/JAPS.2018.8918.
28. ADMET tool, ADMETlab 2.0. https://admetmesh.scbdd.com/.
29. Khatabi KE, Aanouz I, Alaqarbeh M, Ajana MA, Lakhlifi T, Bouachrine M. Molecular docking, molecular dynamics simulation, and ADMET analysis of levamisole derivatives

against the SARS-CoV-2 main protease (MPro). Bioimpacts. 2021;12(2):107–13. https://doi.org/10.34172/bi.2021.22143.

30. Lagorce D, Sperandio O, Galons H, Miteva MA, Villoutreix BO. FAF-Drugs2: free ADME/tox filtering tool to assist drug discovery and chemical biology projects. BMC Bioinform. 2008;9(1):396. https://doi.org/10.1186/1471-2105-9-396.
31. Lagorce D, Maupetit J, Baell J, Sperandio O, Tufféry P, Miteva MA, et al. The FAF-Drugs2 server: a multistep engine to prepare electronic chemical compound collections. Bioinforma Oxf Engl. 2011;27(14):2018–20. https://doi.org/10.1093/bioinformatics/btr333.
32. Kusumorini A, Hasna N, Suryani Y, Taufiqurrahman O. In Silico analysis of Catechin, Galangin, and Hesperidin as competitors of the SARS-CoV-2 spike protein. J Biodjati. 2023;8(1):163–71. https://doi.org/10.15575/biodjati.v8i1.24521.
33. BIOVIA Discovery Studio | Dassault Systèmes. https://www.3ds.com/products/biovia/discovery-studio.
34. Afriza D, Suriyah WH, Ichwan SJA. In silico analysis of molecular interactions between the anti-apoptotic protein survivin and dentatin, nordentatin, and quercetin - IOPscience. IOP Publishing Ltd; 2018. https://iopscience.iop.org/article/10.1088/1742-6596/1073/3/032001/meta
35. Kaus JW, Pierce LT, Walker RC, McCammon JA. Improving the efficiency of free energy calculations in the Amber molecular dynamics package. J Chem Theory Comput. 2013;9(9):4131–9. https://doi.org/10.1021/ct400340s.
36. Pearlman DA, Case DA, Caldwell JW, Ross WS, Cheatham TE, DeBolt S, et al. AMBER, a package of computer programs for applying molecular mechanics, normal mode analysis, molecular dynamics and free energy calculations to simulate the structural and energetic properties of molecules. Comput Phys Commun. 1995;91(1):1–41. https://doi.org/10.1016/0010-4655(95)00041-D.
37. Salomon-Ferrer R, Case DA, Walker RC. An overview of the Amber biomolecular simulation package. WIREs Comput Mol Sci. 2013;3(2):198–210. https://doi.org/10.1002/wcms.1121.
38. Ul Haq F, Abro A, Raza S, Liedl KR, Azam SS. Molecular dynamics simulation studies of novel β-lactamase inhibitor. J Mol Graph Model. 2017;74:143–52. https://doi.org/10.1016/j.jmgm.2017.03.002.
39. Sinha S, Vohora D. Chapter 2 - Drug discovery and development: an overview. In: Pharmaceutical medicine and translational clinical research. Boston: Academic; 2018. p. 19–32. https://doi.org/10.1016/B978-0-12-802103-3.00002-X.
40. Schrödinger, Life Science: QikProp. https://newsite.schrodinger.com/platform/products/qikprop/.
41. N'guessan D, Alzain A, Adouko E, Sékou D, Coulibaly S, Ouattara M. Molecular modeling studies of Benzimidazolyl-Chalcones as Antileishmanial agents using Qsar, docking, Adme and molecular dynamics studies. J Appl Pharm Sci Res. 2022;4(3):18–28. https://doi.org/10.31069/japsr.v4i3.4.
42. Kumar A, Rathi E, Kini SG. E-pharmacophore modelling, virtual screening, molecular dynamics simulations and *in-silico* ADME analysis for identification of potential E6 inhibitors against cervical cancer. J Mol Struct. 2019;1189:299–306. https://doi.org/10.1016/j.molstruc.2019.04.023.
43. Docking Software, AutoDock Vina. https://vina.scripps.edu/.
44. Trott O, Olson AJ. AutoDock Vina: improving the speed and accuracy of docking with a new scoring function, efficient optimization, and multithreading. J Comput Chem. 2010;31:455–61. https://doi.org/10.1002/jcc.21334.
45. Yasman S, Yanuar A, Tamimi Z, Riadhi SR. In Silico analysis of sea cucumber bioactive compounds as anti-breast cancer mechanism using AutoDock Vina: in Silico analysis of sea cucumber bioactive compounds as anti-breast cancer mechanism using Autodock Vina. Iranian J Pharm Sci. 2020;16(1):1–8. https://doi.org/10.22037/ijps.v16.40439.

Chapter 20
Mechanistic Understanding of the Anti-Alzheimer's Agents with Computational Studies

Nachiket Joshi, Santhosh Chobe, S. N. Koteswara Rao G., and Rajasekhar Reddy Alavala

Contents

Abstract Alzheimer's Disease (AD) is a progressive neurodegenerative disorder that severely impacts cognitive function, leading to a global public health crisis. This chapter provides an in-depth exploration of the various hypotheses and

N. Joshi · S. N. K. Rao G. · R. R. Alavala (✉)
Shobhaben Pratapbhai Patel School of Pharmacy and Technology Management, SVKM's NMIMS, Mumbai, Maharashtra, India

S. Chobe
MGV's L.V.H. Arts, Science and Commerce College, Panchavati, Nashik, India

S. N. K. Rao G., R. R. Alavala (eds.), *Applications of Computational Tools in Drug Design and Development*, https://doi.org/10.1007/978-981-96-4154-3_20

molecular mechanisms involved in AD progression, including the Aβ cascade, tau aggregation, neuroinflammation, metal ion dysregulation, oxidative stress, and altered glucose and cholesterol metabolism. We delve into the role of key macromolecules and genes that contribute to the onset and advancement of AD, with a special focus on cholinesterase inhibitors and Aβ aggregation inhibitors as potential therapeutic strategies. Furthermore, the chapter discusses innovative approaches, such as targeting tau kinases, and highlights neuroprotective compounds currently under investigation. The emergence of network pharmacology offers new insights into protein-protein interactions (PPIs), paving the way for novel drug discovery. We also explore the significant impact of artificial intelligence (AI) and machine learning (ML) in revolutionizing AD therapeutics, enabling precision medicine and personalized treatment approaches. Finally, we provide future perspectives on overcoming the challenges in AD drug development and conclude with a comprehensive summary of cutting-edge strategies in the fight against Alzheimer's Disease.

Keywords Alzheimer's Disease (AD) · Neurodegenerative disorders · Amyloid beta (Aβ) hypothesis · Tau protein aggregation · Neuroinflammation · Tau kinases targeting · Neuroprotective compounds · Network pharmacology

20.1 Introduction to Anti-Alzheimer's Agents

Discovered a century ago in 1907, Alzheimer's disease (AD) is a leading cause of dementia contributing to 60–80% of its cases [1]. A wide range of symptoms with the disease involve poor judgment, difficulty memorizing conversations, names, or events, and the inability to walk or speak as the disease progresses. AD can thus be classified as Pre-clinical AD, Mild cognitive impairment (MCI), Mild AD dementia, Moderate AD dementia, and Severe AD dementia. During the preclinical stage, no symptoms of AD are noticeable. However, imaging techniques can identify the presence of Aβ senile plaques (SPs) and tau neurofibrillary tangles (NFTs). These changes magnify into small lapses in memory in the AD with MCI stage. However, these memory lapses are not quite significant and do not affect a person's daily routine, hence, they can go unnoticed. During Mild dementia due to AD, there is memory loss involving recent moments, trouble with making apt judgments, and compromised analytical skills, associated with minor personality changes. Moderate dementia due to AD patients suffer from mental confusion and amplified memory loss involving the inability to recollect personal addresses and self-history. At this stage, the patients need individualized help with day-to-day activities. Finally, in the last stage of AD, extremely compromised mental function, a toll on physical health such as abnormal reflexes, trouble walking, and inability to control bowel and bladder functions [2]. The hallmarks of the disease are the accumulation of amyloid-β (Aβ) extra-neuronally and tau protein aggregates intra-neuronally (neurofibrillary tangles). Other key factors include neuroinflammation and imbalance of excitatory and inhibitory neurotransmitters like acetylcholine (ACh), dopamine, epinephrine, norepinephrine, glutamate, and γ-amino butyric acid (GABA). Acetylcholine and glutamate are excitatory neurotransmitters that are downgraded in AD. Acetylcholinesterase (AChE), and butyrylcholinesterase (BuChE) are responsible for the breakdown of ACh, whereas N-methyl-D-aspartate (NMDA) receptors are responsible for the glutamate bidning resulting in less freely available amounts of glutamate [3, 4].

In this accord, few FDA-approved drugs are available as therapeutic alternatives. The FDA approved cholinesterase inhibitors Donepezil, Galantamine, Rivastigmine, and NMDA receptor antagonist Memantine. Tacrine was another antagonist of AChE previously approved by the FDA but was withdrawn owing to its hepatotoxicity. These drugs suffer from shortcomings involving some minor side effects like nausea, vomiting, diarrhea, dyspepsia, decreased appetite, constipation, dizziness, and a few rare serious adverse effects like arrhythmia, confusion, hypertension, urinary incontinence, hallucinations, and skin reactions [3, 4]. These drugs cannot alter the disease progression but only provide symptomatic relief. After several rigorous attempts to discover drugs two biologics namely Aducanumab and Lecanemab have been very recently discovered. These target the Aβ aggregates directly by functioning as anti-Aβ antibodies [4, 5]. However, these can only be effective against early-stage AD [4–6]. Hence, there is an ever-increasing need to find novel alternatives for AD therapeutics that could modify the disease as well as provide symptomatic relief effective against different stages of AD.

20.2 The Macromolecules & Genes Associated with AD Progression

Based on various factors involved in AD, multiple hypotheses explain these factors in detail.

20.2.1 Aβ Cascade Hypothesis

The hypothesis entails degradation of amyloid precursor protein (APP) by certain enzymes called 'secretases' which catalyze the processing mechanisms. There can be toxic as well as non-toxic pathways of APP processing. This non-toxic pathway involves the degradation of APP upon the action of α-secretase that leads to a sAPPα fragment outside the cell membrane and a chain terminal fragment composed of a chain of 83 amino acid residues (CTF-83) inside the cell, further cleaved by the action of γ-secretase into a p3 fragment that is released outside the cell membrane. In contrast, a fragment of the APP domain is retained within the cytoplasm. However, the toxic pathway leads to the production of neurotoxic Aβ. Initially, the APP breaks

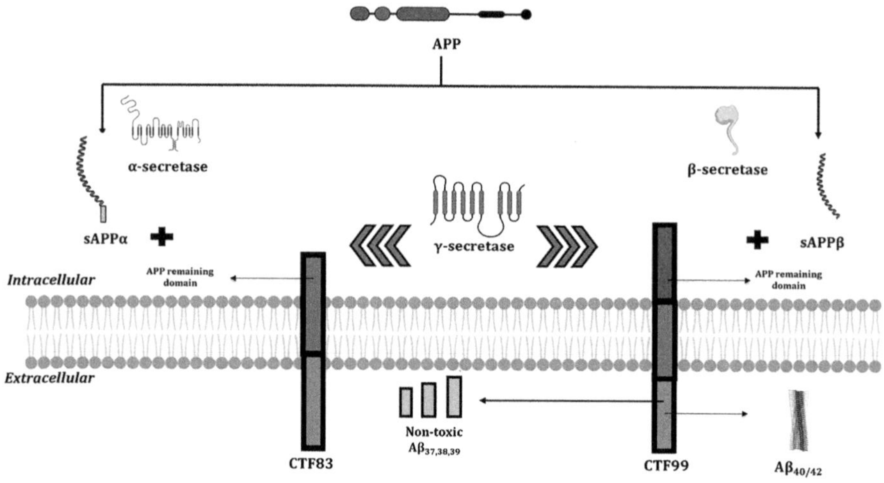

Fig. 20.1 Processing of APP via toxic and non-toxic pathways

down by the action of β-secretase (BACE1) and then by γ-secretase. Following cleavage by BACE1, the sAPPβ fragment is released out of the cell membrane, and CTF-99 is inside the cell. This CTF-99 can be cleaved at various sites by γ-secretase to release Aβ sequences of varying amino acid lengths including 37, 38, 39, 40, 42, and 43 amino acid residues. $A\beta_{42}$ is known to bear an increased propensity to form agglomerates due to its two terminal hydrophobic residues and is known to be neurotoxic. This $A\beta_{42}$ can also trigger the formation of NFTs, activate microglia and astrocytes, and lead to loss of synapse function (Fig. 20.1) [7–10].

20.2.2 Cholinergic Hypothesis

NFTs are insoluble fragments of tau protein embedded inside the cytoplasm of axons or neurons. Tau is a protein associated with microtubule structural stability and can lead to the formation of stable microtubules in association with tubulin protein. Other functions of tau protein are regulation of neuronal signaling, synaptic structure and function maintenance, and cytoplasmic transport. It can form multiple interconnecting bridges between microtubules, resulting in a stable structure. This is brought about by tau phosphorylation. Tau protein is phosphorylated by kinases and dephosphorylated by phosphatases. Due to excessive release of Aβ, these kinases lead to excessive phosphorylation of the tau protein, forming oligomerized tau (paired helical filaments, PHFs). This disturbs the microtubule balance; thus, the dissociated tau proteins form NFTs (Fig. 20.2). These structures alter the neuronal transport, altered synapses, and thus, neurodegeneration. Overexpressed enzymes

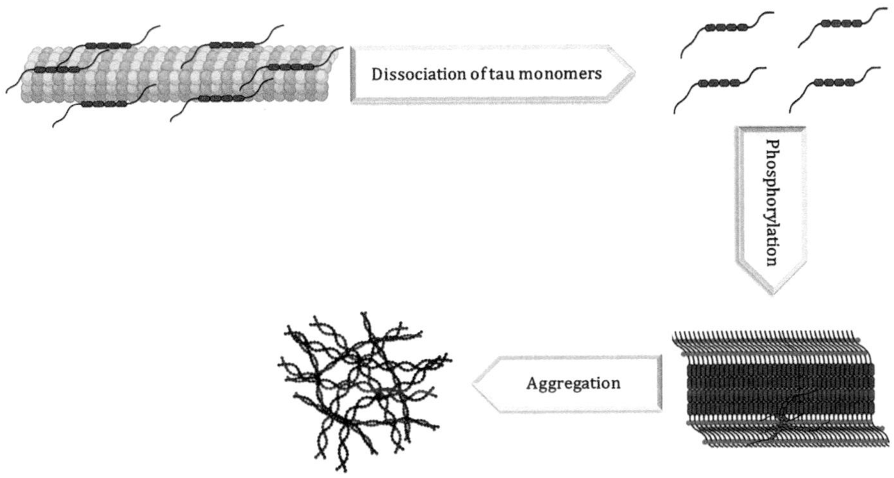

Fig. 20.2 Formation of tau tangles after dissociation from microtubules

leading to tau hyperphosphorylation include glycogen synthetase kinase-3β (GSK-3β), protein kinase C, Microtubule-associated protein kinases, Janus kinase, and stress-activated protein kinases. Tau hyperphosphorylation in AD brains is known to cause changes in configuration and microtubule functioning [10–13].

20.2.3 Cholinergic Hypothesis

An innervated area within the brain called the 'nucleus basalis meynert' consists of numerous innervations of cholinergic neurons. AD patients are known to have significant loss of these cholinergic neurons bringing about a lack of the neurotransmitter Acetylcholine (ACh) (Fig. 20.3). Cholinergic neurons are known to play a role in cognition. ACh, a crucial neurotransmitter in the brain shows its effects on the basal ganglia, cortex, and forebrain. It carries out functions involving cognition, suppression of inflammation, memory, and learning. Cholinergic signaling is also important for neuronal plasticity, neurogenesis, neuronal cell differentiation, and cell survival within the CNS. AChE carries out the cleavage of ACh. Whereas, the acetyl coenzyme A addition to choline is brought about by choline acetyltransferase (ChAT). Once generated, the ACh is bound to the postsynaptic receptor M_1 and thus, enables neuron-to-neuron signaling. The severity of dementia associated with AD is dependant upon the cholinergic loss, suggesting its importance in AD. In AD patients, there is a shift in the balance between acetylation and breakdown of ACh. Other than these two enzymes, even the acetylcholine-synthesizing enzyme levels are affected, leading to cholinergic deficits in the brain [14–18].

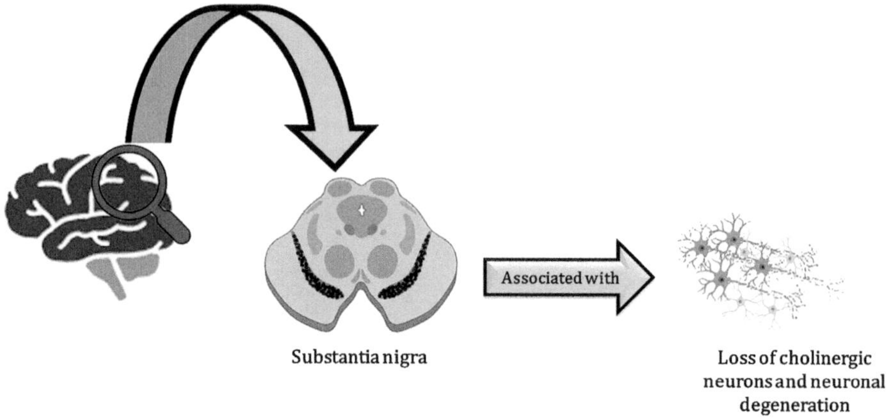

Fig. 20.3 Decline of ACh levels in the brain associated with neurodegeneration

20.2.4 Neuroinflammation Hypothesis

Neuroinflammation refers to CNS inflammation caused by various pathological changes like ischemia, cellular signals, infections, toxins, or trauma. AD can cause an imbalance between inflammatory and negative regulatory processes. Microglia are the main immunity mediators involved in the brain. Activation involving innate immune cells, microglia, astrocytes, and damage to the integrity of the BBB are the major processes involved with neuroinflammation. These processes release Pro-inflammatory cytokines, chemokines, prostaglandins, nitric oxide, and reactive oxygen (Fig. 20.4). It is hypothesized that neuroinflammation starts very early in patients with AD. The astrocytes and microglia have been coined to surround the Aβ plaques. AD patients are characterized by high amounts of inflammatory markers in the CNS plasma, CSF, and brain tissues. The presence of Aβ plaques also can potentially lead to activation of microglia. Neuroinflammation in AD patients is known to exacerbate the neurotoxic effects caused by Aβ plaques and neurofibrillary tangles [19–22].

20.2.5 Metal Ion Hypothesis

Some metals like copper, iron, and zinc play crucial roles in normal CNS function. The presence of these metals in trace amounts is critical to maintaining healthy neurons. However, excess deposition of these metals in these neurons can damage homeostasis and lead to neurodegeneration. The increased levels of these metals can induce neuroinflammation via the release of NF-kβ, which in turn activates the glial cells. Previous studies have suggested the abnormal levels of metal ions found in AD brains. These metals are also associated with Aβ plaque formation. Copper and

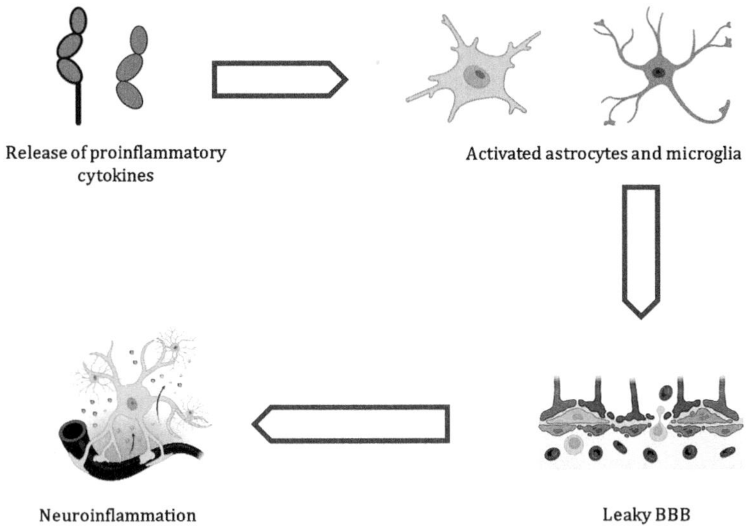

Fig. 20.4 Gradual process of neuroinflammation resulting from neuroinflammatory mediators

iron are involved in neurotransmission, learning, and memory functions. The mislocalization of ions has been observed in AD patients. Other than these metals, calcium ions have also been known to accumulate near the Aβ plaques. In turn, accumulation of Aβ plaques is also known to stimulate the release of intracellular calcium ions. It was noted that the combination of metal ions like zinc and copper with Aβ contributes to its neurotoxicity and aggregation. Aβ is known to possess metal-binding sites which are mediated by histidine residues. In vitro, iron ions caused the aggregation into Aβ plaques and NFTs. Iron is also responsible for oxidative stress. Zinc deposition can also cause NFTs. Altered calcium homeostasis is known to promote Aβ and tau pathologies [23–27].

20.2.6 Oxidative Stress Hypothesis

Oxidative stress involves all the molecular and free radical forms derived from oxygen. The AD brains were noted to have significantly increased levels of oxidative stress due to Aβ plaque formations and NFTs. As previously discussed, copper ions mediate the formation of reactive hydroxyl radicals thereby causing oxidative stress. Additionally, the uncontrolled zinc release in the brain adds up to oxidative stress. AD brain is known to possess decreased antioxidants, increased polyunsaturated fatty acids, and high oxygen demand. Brain accumulation of Reactive oxygen species (ROS) can be caused due to mitochondrial dysfunction and subsequent alteration of the mitochondrial respiratory chain defects. Exogenously free radicals can also be increased due to aging, excess calorie intake, infections, and environmental toxins. Oxidative stress is also known to induce Aβ agglomeration and tau polymerization. A significantly low mitochondrial cytochrome oxidase (Complex IV) level was revealed to be present in the brain cortices, and a deficiency of this enzyme leads to excessive generation of ROS [28–31].

20.2.7 Glucose Metabolism Hypothesis

Previous studies have highlighted the significance of dysfunctional glucose metabolism and hampered signaling of insulin. The peripheral regions of the brain require insulin to carry out the transfer of glucose into the cell. Insulin gets phosphorylated and activates two key signaling pathways phosphatidylinositol 3-kinase-AKT/Protein kinase B pathway and mitogen-activated protein kinase (MAPK) pathway. CNS consists of various insulin receptors in the regions of the amygdala, septum, hypothalamus, hippocampus, cerebral cortex, and olfactory bulb. Peripheral insulin resistance has been noted in AD which decreases glucose utilization in other organs and is thought to be caused by mitochondrial dysfunction and endoplasmic reticulum stress. Insulin degrading enzyme (IDE) is an enzyme carrying out the degradation of insulin and Aβ. Impairment of this enzyme due to insulin resistance causes a reduced Aβ clearance. Abnormal tau phosphorylation has also been connected to insulin resistance. AD patients are known to present a decreased metabolism of glucose in the CNS. A reduced uptake of glucose in the brain can also lead to less support for carrying out regular neuronal activity and thus, cause a cognitive deficit. Neuronal glucose transporters GLUT3 and GLUT4 are known to be compromised due to insulin resistance. As a consequence of all these reasons, AD has been deemed as 'Type 3 diabetes' [20, 32–34].

20.2.8 Cholesterol Hypothesis

Cholesterol and free fatty acids are thought to participate in AD pathophysiology. The entry of free fatty acids and lipid-rich lipoproteins is known to invade the BBB due to compromised permeability. This invasion is known to cause oxidative stress and neuroinflammation. Due to the highly lipid-rich moieties entering the brain, the formation of Aβ peptide may be increased. Some studies have noted a higher prevalence of low-density lipoprotein (LDL) and low high-density lipoprotein (HDL) levels in AD brains. It was also noted that, addition or reduction in cholesterol levels significantly altered the Aβ peptide formation and that, the patients taking cholesterol-lowering agents were known to be less likely to have AD [35–37].

20.3 Modeling Studies on Cholinesterases for Designing New Inhibitors

Various efforts have been carried out to design novel derivatives against both AChE and Butyrylcholinesterase (BuChE) for the treatment of AD. AChE consists of certain structural features that are important to consider while designing a potent AChE inhibitor (Fig. 20.5). It possesses an active site lodged deep inside a narrow cavity. The active site also called 'cationic active site' (CAS) consists of three catalytic residues S-234, Q-365, and H-478. This site is responsible for the degradation of ACh. Other than this, it also bears a peripheral anionic site (PAS) composed of Y-103, D-105, Y-155, Q-316, W-317, and Y-372, acyl pocket composed of W-117, an anionic site composed of W-267, F-236, F-238, oxyanion hole composed of G-152, G-153, A-235, and aromatic patch composed of Y-164 and Y-368 [38]. The CAS is present deep inside the gorge whereas, the PAS is located at its entry. Inhibitors known to bind to both these cavities simultaneously are known to possess good biological activity. Since AChE is also known to contribute to inducing Aβ formation, inhibition of dual sites is termed to improve cognition as well as inhibit Aβ formation. Conventional AChE design involves three specific moieties- a ring structure interacting with amino acid residues in the PAS, a structural moiety to interacts with CAS, and a linking agent between these two moieties [39, 40]. Another notable structural feature involves hydrophobic residues lining the entry of the gorge [40].

Butyrylcholinesterase (BuChE) also possesses the ability to break down acetylcholine as well as other choline esters and is known to be rather active during the later stages of AD. It is also involved in Aβ aggregation and results in oxidative

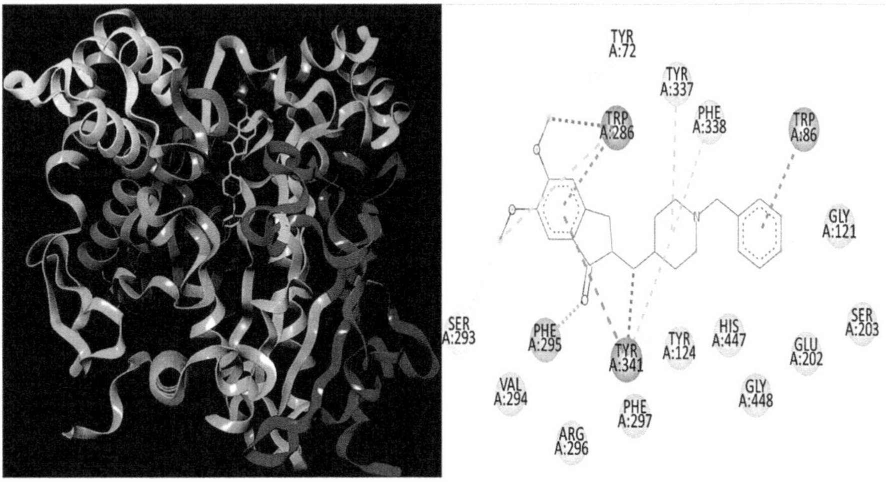

Fig. 20.5 Donepezil in complex with AChE (PDB ID: *4ey7*)- 2D amino acids of enzyme and 3D binding mode

stress [41]. The catalytic triad of BuChE comprises amino acid residues S-198, H-438, and Q-325. Only a slight modification is observed with anionic site residues for BuChE as compared to AChE. It consisted of A-328 instead of F-330 which is the reason for lower binding affinity of some AChE inhibitors. Other residues include W-82, Y-130, and F-331. Whereas, the oxyanion hole consists of G-116, G-117, and A-119. The acyl pocket confers substrate specificity since the gorge size is much larger compared to AChE composed of L-286, and V-288 which, however, prevents the access of large molecules to the catalytic site of BuChE. PAS of BuChE comprises of D-70 and Y-332 (Fig. 20.6) [42]. With this knowledge of binding cavities and amino acid residues, various computational studies were carried out to uncover novel molecules as potential agents that can function as cholinesterase inhibitors. A list of cholinesterase inhibitors analyzed by molecular modeling has been showcased in Table 20.1.

A study involved synthesizing a library of chalcone-rivastigmine hybrids targeting AChE and BuChE. It was noted that compound **1** from the series was revealed to possess dual inhibitory activity. Molecular docking studies revealed docking scores of −26.7 and −36.4 kcal/mol against AChE and BuChE respectively. Further, MD simulations revealed the average RMSD score as 0.75 Å throughout the run time of 20 nanoseconds (ns). The carbamate ester moiety of the compound interacted intricately with the catalytic triad of the cholinesterases. A comparison of binding modes suggested that compound **1** may hydrolyze similarly to Rivastigmine. Enzymatic studies revealed higher IC_{50} values compared to Rivastigmine [43].

A library of AChE inhibitors was designed by in silico studies. A total of 260,071 compounds were screened through the NCI compound database, which was further filtered with molecules following Lipinski's rule of 5, and 190,239 compounds were selected for the next step. A pharmacophore model was selected by using the

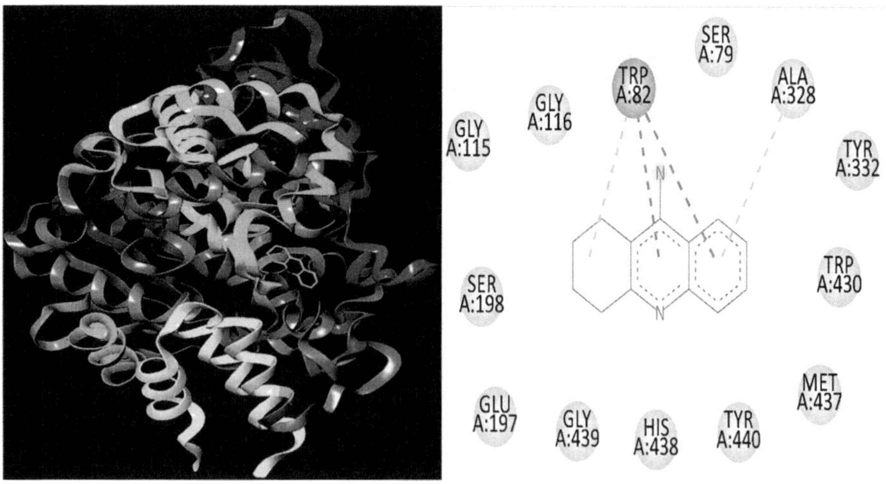

Fig. 20.6 Tacrine in complex with BuChE (PDB ID: 4bds)- 2D amino acids of enzyme and 3D binding mode

Table 20.1 Achieving symptomatic relief by cholinesterase inhibitors

Compound No	Structure	IC_{50} values (μM)
1	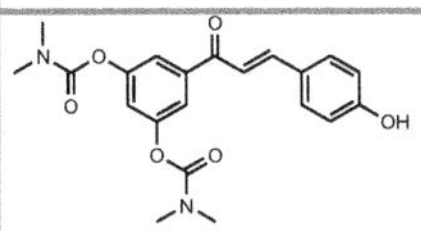	AChE: 0.87 ± 0.19 BuChE: 0.36 ± 0.01
2	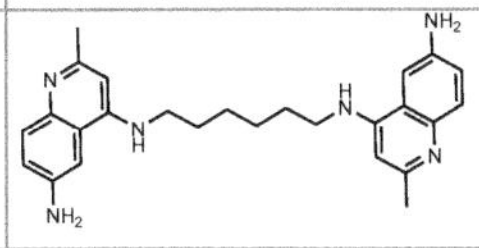	–
3a	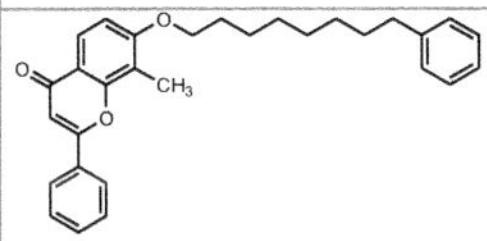	–
3b	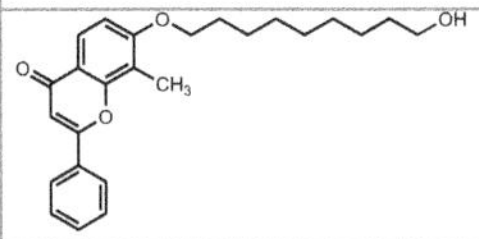	–
4	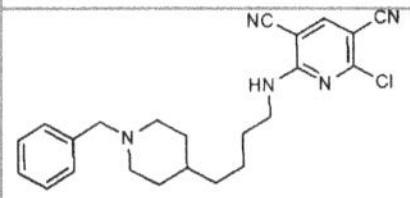	AChE: 0.013 ± 0.002 BuChE: 8.13 ± 0.41
5	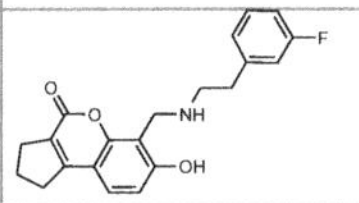	AChE: 0.482 ± 1.88
6	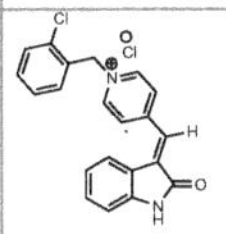	AChE: 0.00044
7	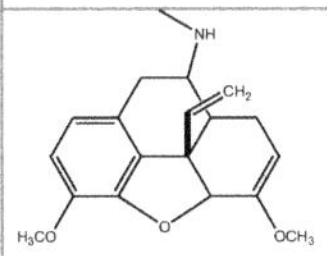	–
8	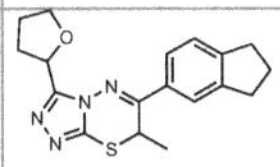	–
9	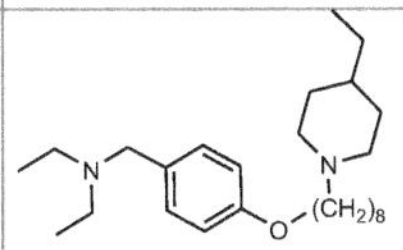	AChE: 7.11
10	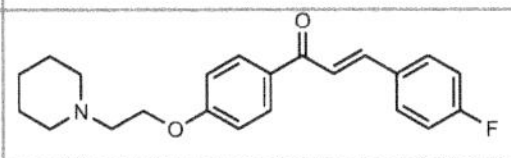	AChE: 0.21 ± 0.03

(continued)

Table 20.1 (continued)

Compound No	Structure	IC_{50} values (μM)
11	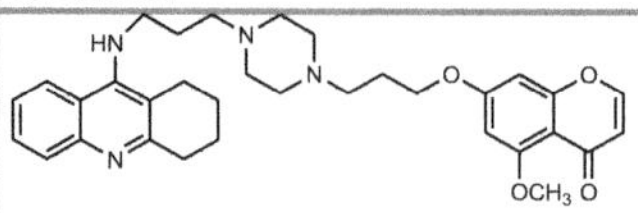	AChE: 0.0084
12	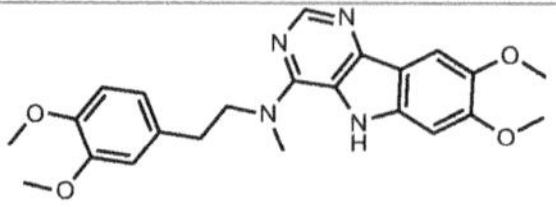	AChE: 0.597 ± .026
13	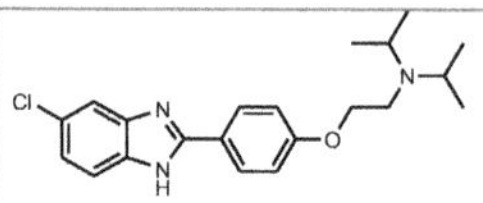	AChE: 0.58 ± 0.06
14	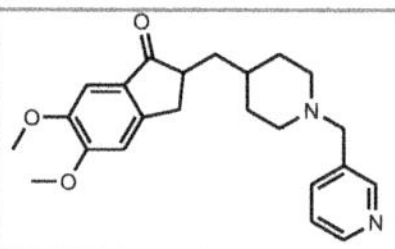	AChE: 0.051
15	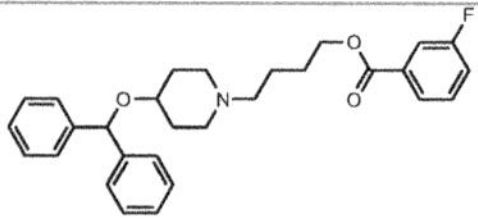	–
16	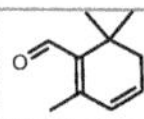	–
17	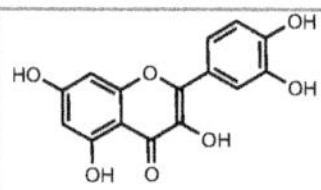	–
18	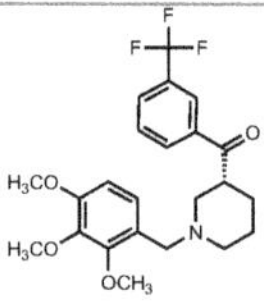	AChE: 0.04933 ± 4.90
19	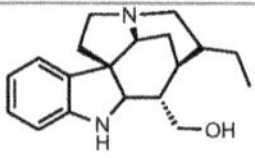	AChE: 20.40 ± 0.93 BuChE: 10.21 ± 0.01
20	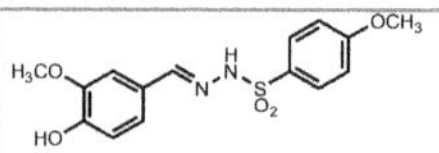	AChE: 0.64 ± 0.09

'Best-fit' model and compounds that fit this criterion were filtered. 252 compounds were then docked and assessed for their scoring function. After the virtual screening process, 9 compounds were identified as potential hits against AChE. Compound **2** was a bis-quinoline derivative with a predicted IC_{50} (PIC_{50}) scoring value of 7.41 [44].

A library of 27 flavonoid derivatives was designed by Vats et al. These derivatives were analyzed using the 3D-QSAR approach. QSAR model displayed acceptable statistical values. Six combinatorial libraries with different spacer lengths were designed using six scaffolds from the generated QSAR model, each library consisted of 233 molecules and about 40% of those molecules showed higher predicted values. Top-scoring molecules ($PIC_{50} > 3.88$) were taken ahead for molecular docking studies. Overall, 35 molecules were docked and analyzed using Extra Precision docking with AChE (PDB ID: 4M0E). Compounds **3a** and **3b** were selected as hit molecules after molecular docking studies showed scores of −9.82 and −9.38 kcal/mol. Further, these compounds were analyzed by MD simulations showed that RMSD graphs of both these compounds for 15 ns showed values less than 2 Å [45].

A library of 6-chloro pyridine and donepezil hybrids was synthesized as AChE inhibitors. Compound **4** revealed potent AChE as well as BuChE inhibition. Molecular docking revealed that the ligand bears an elongated conformation while binding to AChE, and a folded conformation while binding to BuChE. In silico ADME studies revealed acceptable values of CNS permeability and solubility [46].

A library of compounds was screened through a chemistry database to reveal a hit compound **5**. A combination of structure-based drug-design and ligand-based drug-design approaches was used and the selected compounds were further filtered for pan-assay interference compounds (PAINS) and ADME. Further MD simulations and MM-GBSA calculations were computed. Finally, in-vitro inhibitory assay and PAMPA results revealed the hit compound. Atom-based QSAR method was employed to reveal the most appropriate pharmacophore hypothesis and it was revealed that one hydrogen bond acceptor (HBA), one positive ionizing moiety, and three aromatic rings offered the best results. Three different hypotheses were selected and further, a high-throughput virtual screening was carried out using these overall developed pharmacophores, and subsequent molecular docking was also carried out. At the end of this stage, eight structurally diverse compounds, bearing docking scores in the range of −10.74 to −12.87 were selected for further stages. All the compounds passed the ADME criteria for adequate pharmacokinetic profiles. Compound **5** showed the best inhibitory potential against AChE. MD simulations for 50 ns results revealed the complex stability with an RMSD value less than 2 Å throughout the timeline [47].

A research group synthesized indolinone-based derivatives against AChE. These inhibitors were revealed to occupy both the sites of AChE through molecular docking. Compound **6** was found to inhibit AChE with a mixed-type inhibition as revealed through an in vitro enzyme kinetics study ($K_i = 1.14$ nM). Further, ADMET studies revealed adequate BBB penetration characteristics as well as no predictive acute toxicity (LC_{50}). The binding analysis of compound **6** revealed that the indole moiety of the structure did fit into the PAS upon interactions with Y-70, Y-121, and W-279, whereas the quaternary nitrogen of the pyridinium moiety interacted with the F-330 and the overall ring interacted with W-84. These are all crucial amino acid interactions observed with AChE inhibitors [48].

A study involved designing a set of galantamine-based novel derivatives as potential AChE inhibitors. Pharmacophore generation was constructed with 15 designed derivatives by selecting galantamine as a base. Later, a virtual ligand screening approach was adopted to screen through the ZINC database. Based on pharmacophore filtering, the top 200 hit molecules were taken to the next stage.

Further, pharmacokinetic parameters and toxicity parameters were estimated which revealed 6 hits upon filtering for a variety of toxicity alerts. Prediction of biological activity (PASS) was carried out to determine which selected hits could potentially function against AD. After this stage, only 2 compounds revealed a role in AD therapy. Both compounds **7** and **8** showed binding interactions with both CAS and PAS residues as revealed after docking. Molecular dynamic studies with these compounds showed RMSD values below 2 Å for a timeframe of 100 ns, thus indicating the stability of these complexes with AChE. Thus, these compounds can be potential lead molecules [49].

A study attempted to build a QSAR model with the reported derivatives which showed potent AChE inhibitory activity to decipher some novel interaction-based descriptors. From a library of compounds, compound **9** was the most potent compound found in the in vitro enzymatic assay of AChE. This molecule was further evaluated for its binding affinity and amino acid interaction profiles by molecular docking and MD simulations. Docking revealed dual binding modes with both the sites of AChE. The docked complex was further analyzed by molecular dynamics studies which revealed the RMSD value of the complex at about 2 Å throughout the 6000 ps timeline [50].

A research group synthesized novel fluoro-chalcone derivatives as potent inhibitors of AChE. Compound **10** was the most potent of this series exhibiting the highest AChE activity and enhanced selectivity over BuChE [51].

A research involved designing a library of substituted tacrine hybrids with various CADD techniques. A 3D-QSAR model was developed and compound **11** was found to bear the highest pIC_{50} value and it was taken ahead for molecular docking studies. It interacted with key residues namely Y-70, W-84, Y-121, W-279, and F-330 in the binding site. Further, molecular dynamic simulations revealed RMSD value of about 1 Å throughout 10 ns timeline [52].

A donepezil analog was evaluated against AChE. Initially, a pharmacophore-based virtual screening protocol was executed. The best 100 hits obtained from each hypothesis were taken ahead for molecular docking. The top 3 hits from each series (9 in total) were tested for enzymatic AChE inhibition and it was found that compound **12** was the best compound from the series and it resembled the structure of donepezil, presenting a docking energy of −8.131 kcal/mol. It showed 80.2% inhibition at 10 μM concentration. An in vitro kinetic study was performed and it revealed binding at dual sites for the compound against AChE. Further MM-GBSA free energy calculations for docked complex were carried out and it revealed that the contribution of coulomb energy was most significant for its binding, followed by π-π interactions. Later, molecular dynamics was carried out for 10 ns timeframe and it corroborated the results obtained after MM-GBSA studies. π-π interactions with W-86, F-297, and Y-341 were observed for 70% or more of the entire duration. F-295 of the acyl pocket formed strong hydrogen-bonding interactions with the compound, whereas it formed water-mediated hydrogen-bonding interactions with G-120 and Y-133 [53].

A study involves the synthesis and evaluated benzimidazole derivatives against AChE. Compound **13** was found to be the best of the series. It interacted with PAS residues W-279 and Y-334 and showed water-mediated hydrogen bonding interactions with Y-121 present in the AChE sidechain. Furthermore, MD simulations for 20 ns revealed an RMSD graph value of 1.5 Å throughout the run time [54].

An in silico modeling was performed to achieve dual cholinesterase inhibition. Two different 2-D QSAR models were developed for AChE and BuChE each. Pharmacophoric features involving AChE inhibitors were the presence of small-sized ring moiety, methylene bridges, secondary aromatic amines, and aromatic ketone groups, whereas BuChE inhibitors involved a sum of topological distance between two nitrogen atoms, alkenyl moiety, secondary aryl amides, alkyl halide fragment to ensure potent inhibition [55].

A research group synthesized donepezil-based AChE inhibitors and performed molecular docking studies on the same to study their interaction patterns. Compound **14** showcased great potential to act as an AChE inhibitor and revealed a similar binding mode compared to donepezil. The pyridylmethyl moiety interacted with G-121, Q-202, and G-488. Hydrogen bonding with F-295 was also maintained in this donepezil analog and showed a docking score of −17.63. in vitro evaluation also revealed its potent inhibitory activity. Enzyme kinetic studies revealed dual inhibition of AChE [56].

Some novel piperidine derivatives were analyzed by modeling them against AChE. Initially, a library of piperidine derivatives was designed, and molecular docking was performed on these derivatives. The Predictive IC_{50} values were computed using parameters like hydrogen bonding energy, internal energy of ligand, short and long-range electrostatic energy, and steric interaction energy. Based on the experimental data generated from docking studies, another library of compounds was designed and docked. From the study, compound **15** had the best PIC_{50} value of 7.27. Later DFT studies revealed the electronic properties of the compound and it revealed that the hydrogen bonds and orientations of the piperidine ring were in accordance with the docking results obtained [57].

A research study investigated saffron derivatives to develop novel AChE inhibitors against AD using molecular docking and in vitro evaluation. Saffron constituents like crocin, crocetin, dimethylcrocetin, and safranal were analyzed by Ellman's assay. Compound **16-** Safranal was revealed to possess the best results with in vitro studies out of these saffron compounds. Enzyme kinetics revealed mixed type of inhibition revealing a dual binding mode of safranal. Docking data revealed its hydrophobic interactions with F-330, G-118, H-440, and W-84 [58].

Compound **17**- Quercetin was identified as a potential AChE inhibitor. Molecular docking studies showed Van Der Waals interactions with Q-71, Y-72, V-73, N-87, P-88, G-120, G-121, Y-124, G-126, Y-133, S-203, G-448, Y-449, and I-451, hydrogen-bonding interactions with D-74, W-86, S-125, Q-202, and H-447, and π-interactions with Y-337. ADME studies revealed acceptable values for all parameters. Pharmacophore mapping was carried out for quercetin and it revealed a fit score of 2.97, 2 hydrophobic centers, and 2 hydrogen-bond donors. Furthermore, MD simulations for 100 ns were executed. The average RMSD throughout the time frame was 1.75 Å, the average hydrogen-bonding score was 410, the average radius of gyration was 22.8 Å and the average SASA was 19,500 sq. A. MM-PBSA revealed an average binding free energy of the compound in complex with AChE as −68.798 kJ/mol throughout the run time [59].

A ZINC database screening based on the structure of donepezil was executed. Similarity >50% were selected as ligands for the study. The bioactive conformation of donepezil bound to AChE was taken as a reference structure and the conformers generated from the selected compounds were further selected by screening for the compounds replicating the binding mode of donepezil. Later, extra precision docking was carried out, and compound **18** revealed a docking score of −12.47. It showed π-cation interactions with Y-341, hydrogen-bonding interactions with Y-124, and Van Der Waals interactions with Y-72, W-86, Y-124, Y-133, W-286, V-294, F-295, F-297, Y-337, F-338, and Y-341. In vitro study revealed not only its potency but also its selectivity for AChE over BuChE. An average RMSD of about 0.5 Å was revealed throughout the MD simulations run time of 20 ns for the compound. MM-GBSA study displayed the total binding free energy as −100.18 kcal/mol [60].

A phytochemical compound **19-** Geissoschizoline, isolated from *Geissospermum vellosii* was revealed as a hit molecule by an in vitro evaluation and in silico modeling. Enzyme kinetics revealed a mixed type of inhibition against cholinesterases. Molecular docking analysis corroborated its dual-binding nature. It was found to interact with residues in the binding pocket like Y-337, and acyl binding pocket like F-295, F-297, and F-338, whereas it also interacted with aromatic residues like Y-124, W-286, and Y-341. In comparison, it showed interactions with the amino acid residues S-198, and H-438 in the catalytic triad of BuChE, W-82 in the choline-binding pocket, F-329 in the acyl-binding pocket, and Y-332 in the PAS. It revealed docking energy of −8.4 kcal/mol against AChE and −9.4 kcal/mol against BuChE [61].

A research involved designing sulfonylhydrazone derivatives, synthesizing and evaluating them derivatives against AChE. The compound **20** was the best of the series with potent in vitro activity against AChE. It maintained the viability of the cells up to a concentration of 100 μM. ADME predictions revealed passive GI absorption as a mode of absorption for the compound and it followed Lipinski's rule for CNS drugs. It also was not predicted to be a substrate for P-gp protein and could thus avoid effluxing out from the CNS. Docking analysis revealed a similar binding mode to donepezil with insertion into both sites. Interactions with W-86, G-121, Q-202, F-295, Y-337, and Y-341 [62].

20.4 Understanding the Mechanism of Aβ Aggregation Inhibitors

Aβ plaques are composed of 40–42 amino acids, however, the central core residues from 14 to 23 are found to be quintessential for the aggregation of two Aβ monomers, and hence targeting this hydrophobic core is important to develop potent inhibitors of Aβ [63]. Two binding pockets along the protein were observed composed of L-17, F-20, A-21, G-25, K-28, I-31, I-32, L-34, and M-35, and S-8, Q-11, V-12, Q-15, and K-16 respectively (Fig. 20.7).

Fang et al. studied the interaction mechanisms of epigallocatechin-3-gallate and genistein using molecular docking. Epigallocatechin showed hydrophobic interactions with D-7, S-8, Q-11, V-12, Q-15, K-16, and F-19. Hydrogen bonding interactions were observed with D-7 and Q-15. Whereas, genistein showed hydrophobic interactions with S-8, Q-11, V-12, Q-15, K-16, and F-19, and hydrogen bonding interactions with Q-15, which further revealed some key interactions necessary for binding [64]. Compounds bearing anti-Aβ aggregation potential have been denoted in Table 20.2.

The research involved the design and synthesis of benzofuran and 8-hydroxy quinoline hybrids as small-molecule inhibitors of Aβ. Compound **21** was the best inhibitor from the series with above 60% Aβ aggregation inhibition. Molecular docking analysis revealed a docking energy of −9.73 kcal/mol. It was found to have interactions with D-1, A-2, Q-3, H-6, D-7, Y-10, Q-11, H-14, Q-15, V-18, and F-19. In silico ADMET prediction revealed its ability to permeate the BBB and high GI absorption. The metal chelation property of the compound was also analyzed and it showed the potential of the ligand to chelate iron in the ratio of 1:3 [63].

A research group designed various inhibitors of Aβ using in silico studies. A total of 314 heterocyclic antagonists were screened to build a QSAR model and a relevant pharmacophore was generated. The factors essential for potency were a functional group descriptor, the presence of an aromatic ring, and a secondary amine moiety. 10 different pharmacophores were generated and the first pharmacophore hypothesis was analyzed to be the best one with descriptors such as aromatic ring, hydrophobic moiety, HBA, and hydrophobic aromatic ring. Compound **22** was predicted to be the best compound of the series with a PIC_{50} value of −2.989. It showed hydrogen-bonding interactions with K-16, L-17, and hydrophobic interactions with H-13, A-21, and V-24 [65].

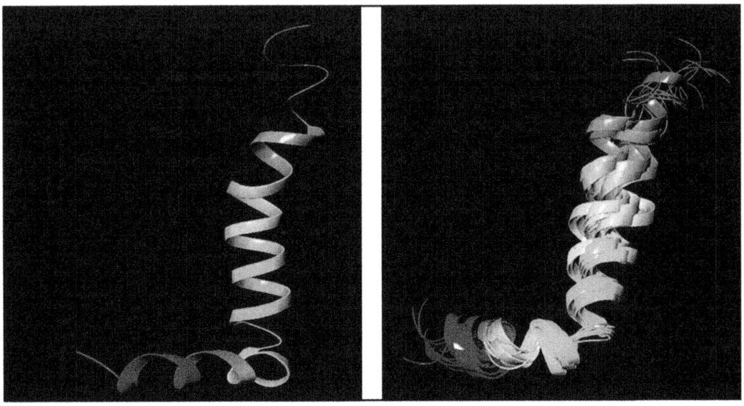

Fig. 20.7 3D images of Aβ fibril monomer and Aβ fibril aggregates

Table 20.2 Compounds modulating the Aβ aggregates

Compound No	Structure	IC_{50} values (μM)/% Inhibition
21	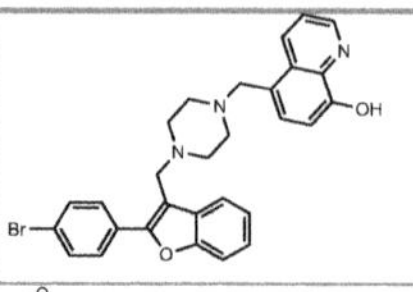	61%
22	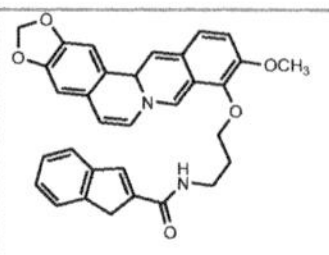	–
23	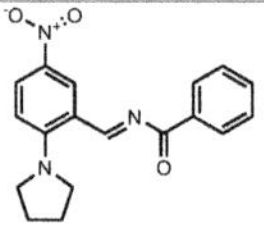	–
24	R-L-A-P-Val	–
25	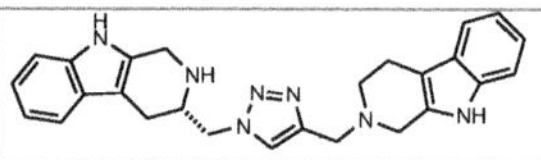	
26	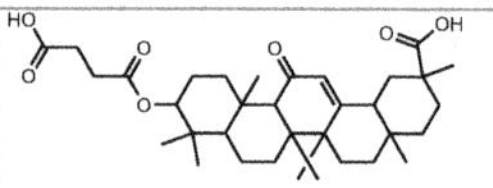	–
27	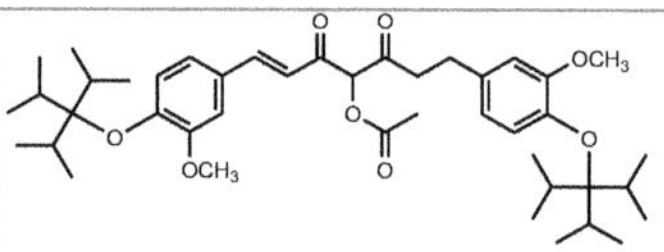	–
28	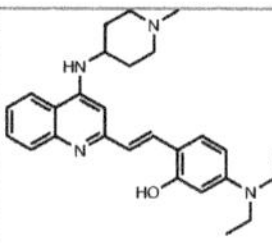	–
29	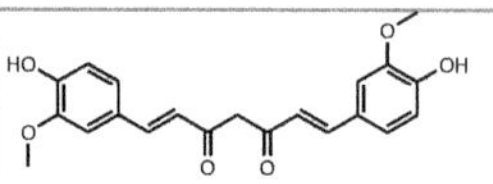	–
30	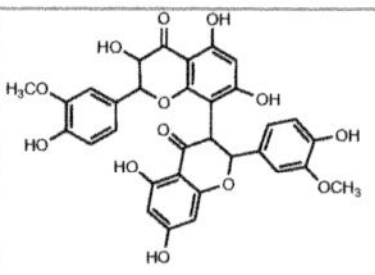	–
31	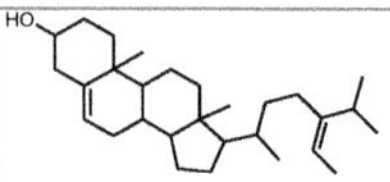	50

(continued)

Table 20.2 (continued)

Compound No	Structure	IC_{50} values (μM)/% Inhibition
32	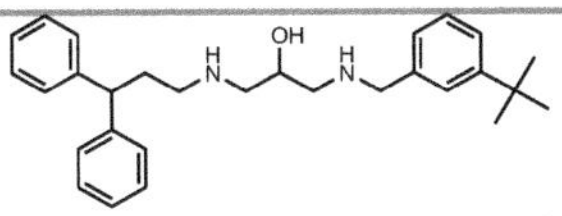	80 ± 6.2%
33	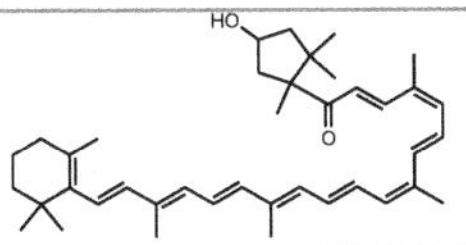	1.97 ± 0.1

The design of dual inhibitors against Aβ and phosphorylated tau aggregation by molecular docking, MD simualtions, ADME predictions, and DFT analysis was executed. Compound **23** was identified as a hit molecule from the study. ADME analysis revealed acceptable parameters like molecular weight, total polar surface area, Log P, estimated water solubility, permeability coefficient, and Lipinski's rule. BBB permeation and GI absorption were also revealed to be appropriate. DFT studies revealed the energy gap between HOMO and LUMO as −7.57 eV. Docking energies against Aβ and phosphorylated tau protein were −5.3 kcal/mol and −5.4 kcal/mol. The compound showed hydrogen bonding interactions with phosphorylated tau with amino acid residues N-179 and K-281 and alkyl interactions with K-280. Whereas, it showed π-σ interactions with F-8 and π-alkyl interactions with A-9. Further MD simulations with phosphorylated tau protein revealed an RMSD graph on an average of about 8 Å, and an RMSD of about 6 Å throughout the timeline of 1 ns. Solvent-phase studies revealed the probability of the inhibitor binding to the protein in an aqueous phase [66].

A research study involved the design of a novel library of pentapeptides by generating mutations of a reference peptide structure targeting Aβ aggregation. These mutations were at one, two, three, and four amino acids to generate a library of 914 molecules. Proline-containing pentapeptides were shown to possess a higher docking score relative to the reference peptide. Compound **24** was the best pentapeptide from the series. The residue-wise free energy analysis revealed hydrophobic interactions with I-31, L-34, M-35, V-39, and I-41. MD simulations analysis showed an average number of hydrogen bonding as 2.4 ± 0.9 across a 200 ns time frame. RMSD and Radius of gyration graphs showed an average of 1 and 1.2 nanometre respectively throughout the time frame. RMSF graphs of both wild Aβ and compound showed slight variations at some points but remained stable through most of the time frame [67].

Analysis of the atom-level mechanism associated with bis-tryptoline triazole derivatives against $A\beta_{42}$ using in silico techniques was carried out. Compound **25** revealed docking energy of −8.61 kcal/mol against Aβ and hydrogen-bonding interactions with Q-11, Q-15, and K-16. Whereas, it showed hydrophobic interactions with Q-11, V-12, H-13, H-14, Q-15, K-16, V-18, and F-19. RMSD graph revealed an average of 1 nanometre throughout 200 ns, while the radius of gyration was about 1.1 nanometre. The crucial amino acid interactions contributing significantly to the binding were F-4, L-17, F-20, A-21, A-30, I-31, L-34, and I-41 [68].

An analysis of the molecular mechanisms of compound **26**-carbenoxolone involving Aβ using in silico methods carried out for 10 ns was done. Docking energy of −6.8 and −7.6 kcal/mol were revealed after molecular docking with Aβ peptide and Aβ fibril respectively. It showed hydrogen bonding interactions with R-5 and hydrophobic interactions with V-12 and K-16 against Aβ peptide. Similarly, hydrogen bonding interactions with F-19, and hydrophobic interactions with A-21, V-36, and V-40 were revealed against Aβ fibril. RMSD graphs of the compound revealed an average value of about 0.75 nanometre and 0.4 nanometre with Aβ peptide and Aβ fibril respectively. The RMSF graph of the compound bound to Aβ showed an average of 0.5 nanometre. Absorbance assays revealed its inhibitory potential against both Aβ peptide and fibril structures relative to Aβ aggregates [69].

A research study involved designing a novel library of molecules based on modifications of the curcumin scaffold by targeting Aβ pathology in silico methods. 2D, 3D, and HQSAR approaches were employed to decipher apt structural and physicochemical features required for optimal inhibitory activity. Electrostatic, steric, and hydrogen bond acceptor regions were analyzed. A pharmacophore was built with features like steric, electrostatic, H-bond donor (HBD), HBA, and hydrophobic descriptors. Furthermore, molecular docking analysis revealed compound **27** to be the best ligand of the series. It showed a high in silico BBB permeability, moderate gastric permeability, and a high HIA value [70].

A series of Aβ monomer inhibitors from a selected series of reported compounds was designed. The 36 molecules from that study were docked and evaluated for their binding interactions. Some key interactions involved hydrogen bonding and/or hydrophobic interactions with F-20, H-13, K-16, L-17, A-21, V-24, M-35, I-31, and L-31. The molecule that had shown the best Aβ inhibition was noted to form a higher number of hydrophobic interactions. Further, a QSAR model was developed based on docking parameters to critically determine the structural features essential for inhibitory activity. A linear regression model was developed to further design compounds based on the equation and descriptor-activity relationship obtained from QSAR data. Additionally, MD simulations determined the stability of its complex with Aβ. It showed an overall RMSD of about 1 Å, a radius of gyration of about 1.5 Å, and an RMSF of about 0.3 Å throughout the run time of 40 ns. This compound was then used as a base structure to design a new library of compounds. Compound **28** was identified as a hit compound that had a predicted inhibitory activity of 92.71%. It showed hydrogen bonding interactions with L-17, Q-22, and V-24, whereas it showed hydrophobic interactions with H-13, K-16, F-20, A-21, and K-28 [71].

A research group studied the binding mechanism of curcumin and ferulic acid against Aβ. 5 systems were built for MD simulations. The first system contained 12 Aβ fibrils arranged randomly in a selected grid box. The next two systems bore Aβ fibrils: Curcumin in the ratio of 1:1 and 1:3 respectively, and the last two systems also bore Aβ fibrils: Ferulic acid in the ratio of 1:1 and 1:3 respectively. RMSF graphs revealed the RMSF of compound **29**- curcumin to be reduced to 1.5 Å as compared to 3 Å. Solvent-accessible surface area (SASA) was increased in the 1:1 curcumin system to be increased slightly from about 28,000 sq. Å as compared to about 27,000 sq. Å with Aβ fibrils alone. The number of hydrogen-bonding and on-native contacts for the 1:1 system were 16,400 and 206 respectively. Curcumin was observed to bind to the inner core of Aβ with H-bonds, π–π, hydrophobic, and π-cation interactions and is revealed as a hit molecule [72].

A research study analyzed interactions of phytochemical compound **30**-Kolaviron isolated from *Garcinia kola* against Aβ aggregation. A blind docking was performed using Kolaviron to investigate the possible binding sites. The compound showed hydrogen-bonding interactions with G-33C, and G-37A, whereas it showed hydrophobic interactions with V-39, I-41, I-31, I-32, M-35, and F-19A. RMSD graph of the apo and complex forms revealed an average RMSD of 18.96 and 22.39 Å each. Major fluctuations were noted suggesting a complete breakdown of the Aβ structure as well as significant rearrangement of the ligand. The RMSF graph showed a fluctuation in the amino acids 18–21 suggesting a conformational change in the binding mode of the ligand showing an average RMSF of 20.7 Å throughout the run time of 500 ns. MM-PBSA study was performed to estimate the free binding energy of the complex and it was 15.4 ± 6.918 kcal/mol with F-19A, M-35A, and V-39A contributing the most interactions. On average, 14.84 hydrogen bonding interactions were observed throughout 500 ns. The SASA was 11,708 sq. Å owing to the less compact structure of the compound which facilitated a lesser formation of internal hydrogen bonds in the Aβ structure, thus exposing the compound to a higher solvent surface area [73].

A research group analyzed the activity of compound **31**- Fucosterol isolated from *Sargassum horridum* against Aβ aggregation. Molecular docking revealed the interactions of fucosterol with the residues of monomeric Aβ as Y-10, H-13, H-14, L-17, F-20, D-23, V-24, K-28, G-29, I-31, I-32, V-39 and V-40, and dimeric Aβ as H-14B, L-17B, F-19B, F-20B, A-21B, I-32B, G-33A, L-34A, L-34B and M-35A, which contributed significantly to the binding of the compound. Fucosterol was found to be able to recognize and bind to both monomeric and dimeric forms of Aβ. MD simulations further performed on the compound revealed an average RMSD of 11.2 ± 0.76 Å and 7.80 ± 0.60 Å against monomeric and dimeric forms respectively. Fucosterol possessed the potential to disturb the β-strand interactions and prevent oligomer formation [74].

A set of novel derivatives were studied as dual inhibitors of Aβ and tau protein was carried out. Compound **32** showed potent dual inhibition of Aβ and tau. Molecular docking study revealed interacting amino acids such as D-7, S-8, Q-11, V-12, H-13, K-16, F-19, F-20, and Q-22 [75].

A study identified a phytochemical compound **33** obtained from *A. aculeata* fruits. This compound was extracted and analyzed for anti-Aβ aggregation. It possessed potent inhibitory activity in the Thioflavin T assay. Molecular docking analysis revealed interacting residues such as Q-22, D-23, and G-25 [76].

20.5 BACE1 Inhibitors Designing with In Silico Studies

BACE1 shows its catalysis by D-32/ D-228 water-mediated interactions with the substrate and these residues are located at the active site. These amino acids undergo geminal diol formation and thus inhibition of these residues is a key to design inhibitors against BACE1. A key feature of this enzyme is the presence of a flap which closes itself upon binding of the inhibitor. It is composed of 11 residues including V-67 to Q-77 and the conserved Y-71 is a key residue for inhibition. Several other key structural features include 10s loop (K-9 – Y-14), third strand (K-107 – G-117), and inserts (A - F) (Fig. 20.8) [77]. A gist of chemical structures studied using molecular modeling techniques has been listed in Table 20.3.

A design of several tetrazole derivatives that bore different scaffolds was studied against BACE1. These were linked by amide or sulfonamide bonds. Initially, molecular docking studies were performed for 164 designed ligands against BACE1. Later, the top 5 hit molecules were evaluated using MD simulations for a time frame of 100 ns. The best 3 ligands were further studied using extended-MD simulations for 250 ns. From this series, compound **34** was shown to be the best compound. It showed hydrogen-bonding interactions with D-32, G-34, N-37, I-126, D-228, T-231, T-232, and R-235. Whereas, hydrophobic interactions included Y-71, I-226, R-235, and V-332. The average RMSD was revealed to be about 2 Å. Furthermore, MM-GBSA studies revealed an average free binding energy of −91 kcal/mol [78].

A study identified new BACE1 inhibitors with the aid of in silico studies. A library of ligands- phytochemicals and synthetic compounds each were accessed from the literature. Compound **35** from this series was revealed as a potent ligand with a docking energy of −11.1 kcal/mol against BACE1. Crucial residues involved

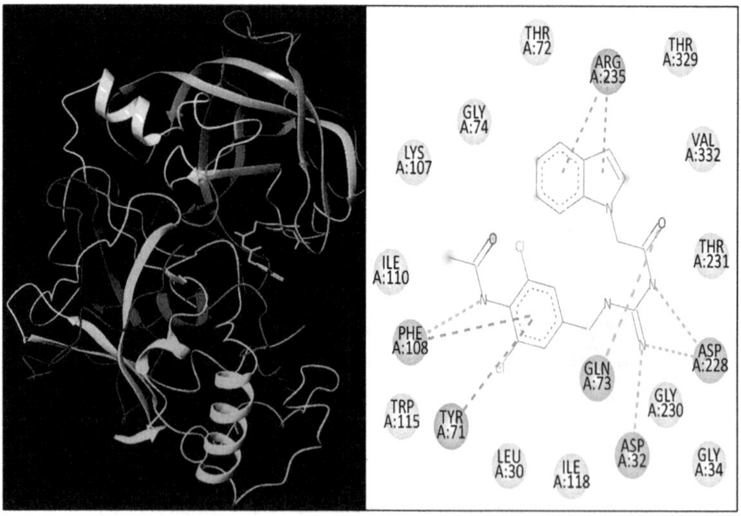

Fig. 20.8 BACE1 (PDB ID: *4ivt*) inhibitor 3D binding mode and 2D critical amino acids

Table 20.3 Decreasing the $A\beta_{42}$ levels by BACE1 inhibitors

Compound No	Structure	IC_{50} values (μM)
34	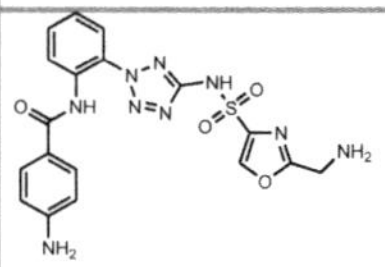	–
35	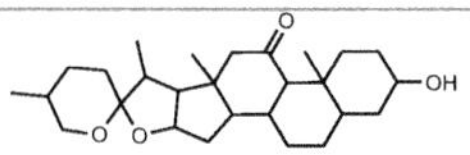	–
36	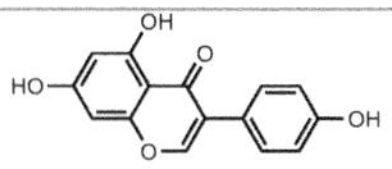	10
37	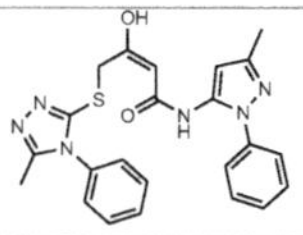	–
38	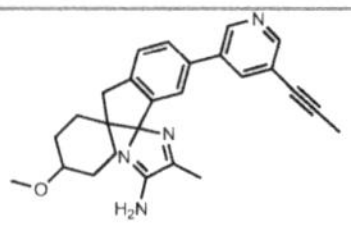	–
39	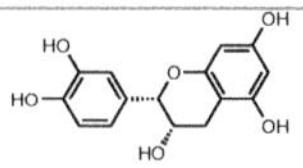	–
40	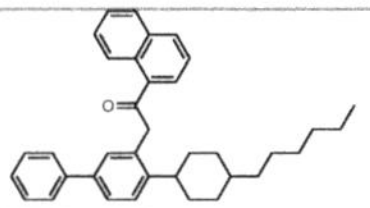	–
41	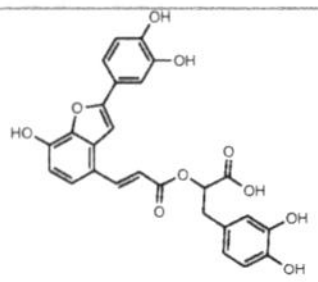	9.18 ± 0.03
42		2.08 ± 0.19
43	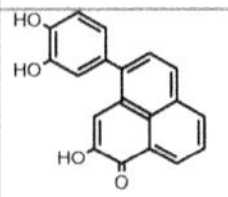	–
44	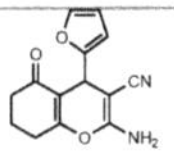	0.9084
45	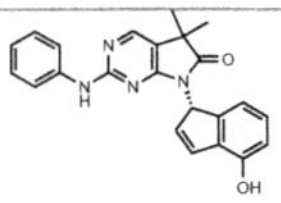	–

(continued)

Table 20.3 (continued)

Compound No	Structure	IC_{50} values (μM)
46	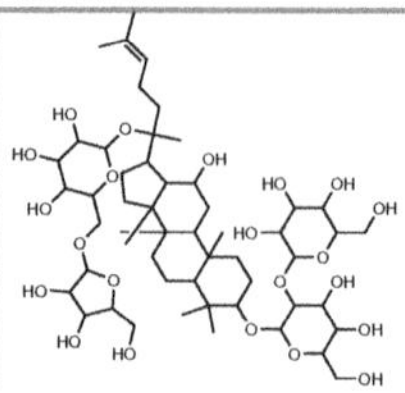	59.81 ± 2.74
47	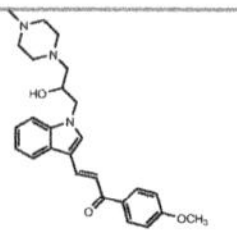	15.4 ± 2.9
48	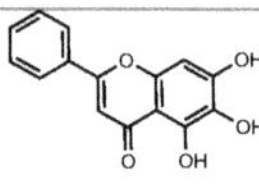	3.71 ± 1.91
49	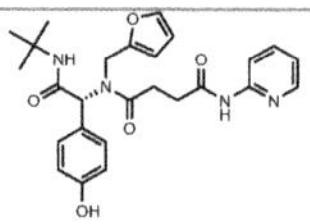	91%
50	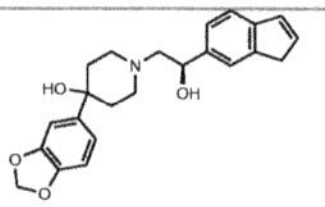	–
51	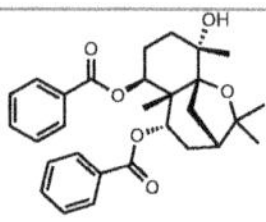	–
52	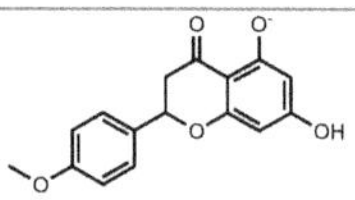	–

in protein-ligand binding interactions were I-174, R-176, V-117, W-124, I-166, L-78, and I- 158. A 25 ns MD simulation was carried out and the average RMSD was found to be 0.3 nanometre, the radius of gyration was about 2.02 nanometre, and the RMSF graph observed minimal fluctuations in the amino acids 114–120 and 134–142. ADME analysis revealed the potential for BBB permeability and no violation of Lipinski's rule and thus, the compound was considered a hit compound [79].

An evaluation of the potential of genistein as a BACE1 inhibitor by in vitro enzymatic inhibitory assay, enzyme kinetics, and molecular docking analysis was performed. At 100 μM, the compound **36-** genistein showed about 60% BACE1 inhibition. It showed non-competitive BACE1 inhibition with a Ki value of $3.5 \cdot 10^{-5}$ M. Molecular docking analysis of genistein revealed four hydrogen bonding interactions with N-37, N-73, and W-76 residues which were a part of the allosteric site corroborating the non-competitive inhibition. The docking energy of the genistein-BACE1 complex was −8.5 kcal/mol [80].

A study identified novel BACE1 inhibitors using pharmacophore modeling and MD simulations. Chemical databases involving 2 million compounds were screened for the generated pharmacophore hypothesis of three aromatic rings and two hydrogen bond acceptor moieties. This pharmacophore was generated based on selected 34 amidine moiety-containing compounds. Then a binding-mode analysis of known co-crystallized ligands with BACE1 was analyzed. The residues of BACE1 involving L-30, D-32, S-35, Y-71, T- 72, D-228, T-231, T-232, and N-233 were noted to be important for BACE1 inhibition. After docking studies, seven hit molecules were identified with docking scores ranging from −9 to −13. The top five molecules were selected for MD simulations and compound **37** was revealed as a hit molecule. It showed an RMSD value of 3 Å throughout the 100 ns time frame. The radius of gyration was about 20.7 Å. The compound also revealed a higher occupancy in the catalytic dyad of BACE1. MM-GBSA revealed an average binding energy of −44.96 ± 3.72 kcal/mol [81].

A research group identified a new hit compound **38** by in silico methods. The compound was selected from a database and its molecular mechanisms were analyzed. Molecular docking analysis revealed a binding energy of −5.33 kcal/mol. It formed two hydrogen bonding interactions with D-32 and K-107. MD simulations carried out for 20 ns revealed an average RMSD value of 0.2 nanometre and the radius of gyration was 2.05 nanometre. ADMET studies revealed it to possess appropriate characteristics to be recognized as a druggable candidate [82].

Molecular docking and 3D-QSAR studies were performed to reveal a potential BACE1 inhibitor previously mentioned in the literature. 1500 drug-like compounds from the Taiwan Phytochemical Database were taken for molecular docking studies. Small molecule inhibitors of BACE1 were docked as a positive control. The top 7 molecules that revealed better affinity than reference molecules were chosen for further stages. The best-docked molecule was revealed to have key interactions with D-32, Q-73, and D-228. Further, Quantum Mechanics/Molecular Mechanics-assisted docking was carried out to analyze the electrostatic potential (ESP) which revealed BACE1 to be a largely hydrophilic protein with hydrophobic patches and a hydrophilic active site pocket. Compound **39** as revealed from previous docking results showed a high docking energy of −10.474 kcal/mol. ADMET study parameters were found to be within acceptable ranges for this compound. Furthermore, a QSAR model was developed to generate a 3D-QSAR for 18 compounds which were known BACE1 inhibitors. After this study, it was noted that the compound had a predicted IC_{50} value of 4.301 μM [83].

A study revealed a novel BACE1 inhibitor after molecular docking and virtual ligand screening. Initially, 26 clinical phase and oligosaccharide BACE1 inhibitors with proven activity were docked and analyzed. The docking scores obtained from this library were used as a control. Further, virtual ligand screening of 50,536 natural compounds was screened and the top 26 molecules with high docking scores were selected. A 3D QSAR model was generated using 29 pyridinium compounds and it was used for 3D pharmacophore mapping. Based on 3D pharmacophore mapping, the features critical to activity were two hydrogen bond donors and three hydrophobic moieties. 3 out of the 7 ligands fit the 3D pharmacophore space the best and these were predicted for their IC_{50} values. Later, ADMET analysis was carried out for these three ligands and they all showed acceptable parameter values but two ligands showed the best results. These two selected hits were docked and analyzed for their interactions and later studied by MD simulations. Compound **40** was selected as a hit molecule from the overall study. It showed bonding interactions with residues P-70, Y-71, T-72, Q-73, D-32, N-233, R-307, Q-12, G-230, T-231, and D-228. MD simulations for a timeframe of 10 ns revealed an average RMSD of 1.9 Å and showed overall 8 hydrogen bonds throughout the run time [84].

A research study involved designing and evaluating tanshinones and salvianolic acids from *Salvia miltiorrhiza* as BACE1 inhibitors. Phytochemicals were extracted and analyzed in vitro for BACE1 inhibitory activity. Compound **41** was found to possess the best inhibition. Enzyme kinetics revealed that it could bind to the enzyme and act as a competitive antagonist. It formed 5 hydrogen bonds with K-9, G-13, Q-304, and D-318 with a binding energy of −7.4 kcal/mol. It also showed hydrophobic interactions with S-10, G-11, Q-12, Y-14, and T-232 residues. Further, ADME results corroborated the potential druggability of this compound [85].

Compound **42-** sulforaphane was identified as a BACE1 inhibitor from a research study. It was evaluated in vitro and revealed potent inhibitory activity. Molecular docking studies revealed that the amino acid residues V-69, Y-71, K-75, W-76, K-107, and F-108 showed Van Der Waals interactions with BACE1. It had a docking energy of −6.05 kcal/mol [86].

A research involved analyzing the repurposing potential of reported food compounds against BACE1. A library of 26,467 molecules was selected from a chemistry database based on the criteria that they follow Lipinski's rule of 5. 8453 compounds were taken forward to the next stage. Furthermore, a QSAR model was used to predict BACE1 inhibitory activity (K_i). Molecular docking was performed and compound **43** was revealed as a hit molecule having a docking energy of −8.9 kcal/mol. Val Der Waals interactions resuted from T-72, T-231, G-230, D-228, G-34, I-118, S-35, F-108, D-106, K-107, G-74, and Q-73. D-32 was involved in hydrogen bonding interactions whereas, Y-71 was involved in π-π interactions. MD simulations revealed an average RMSD value of 0.5 Å, an average radius of gyration of about 20.4 Å, and an average hydrogen bonding score of 2 throughout the 150 ns timeline. MM-GBSA studies revealed a total free binding energy of −53.467 ± 3.001 kcal/mol throughout the timeline [87].

A research group synthesized and analyzed a library of tetrahydrobenzo[b] pyran derivatives against BACE1. Compound **44** possessed potent in vitro BACE1 inhibition. Docking analysis was performed and various types of scoring were assessed. It had a consensus score of 2.59, crash score of −0.63, polar score of 0.22, D score of −92.83, PMF score of −14.46, G score of −138.52, and Chem score of −15.49 [88].

A study involved in silico studies to analyze a library of compounds that showed potent BACE1 inhibition in vitro. Compound **45** showed a docking energy of −10.1 kcal/mol and was the best-docked candidate from the series. Docking analysis revealed hydrogen-bonding interactions with N-2, F-35, and R-39, whereas, L-3, V-32, and W-36 were involved in π-alkyl interactions, and F-37, F-38, Q-40, and

L-319 contributed to Van Der Waals interactions. MD simulations revealed an average RMSD of 0.2 nanometre, the radius of gyration was 2.1 nanometre, and the RMSF of the complex was 0.0743 nanometre. SASA was about 180 sq. nanometre throughout the time frame of 50 ns [89].

A study evaluated various ginsenosides against BACE1 by docking analysis and in vitro assay. Compound **46**- Ginsenoside Rc was found to be the most potent compound of this series. Docking analysis revealed that the compound showed a binding energy of −9.7 kcal/mol. It formed three hydrogen bonding interactions with residues Y-71, D-228, and Q-326, whereas Van Der Waals interactions with V-332, I-226, I-118, F-108, K-107, W-115, W-76, G-11, V-69, D-32, I-110, N-233, K-321, and T-232 [90].

A virtual screening process and subsequent in vitro assay was carried out to identify BACE1 inhibitors. Initially, a pharmacophore modeling approach was employed to develop a structure-based pharmacophore most suited for BACE1 inhibition. The pharmacophoric features included HBA and metal bonding ability, HBA, HBD and cation, aromatic ring, and a hydrophobic moiety. 580,000 compounds were selected from the available databases and screened for the pharmacophore-based parameter. Similarly, a ligand-based pharmacophore model was also developed and the ligands obtained after structure-based pharmacophore screening were again subjected to ligand-based pharmacophore screening. The ligand-based pharmacophore had features like an aromatic ring, a hydrogen bond acceptor, and a hydrogen bond donor. Overall 716 ligands were taken for ADME predictions and 79 compounds showed potential to permeate BBB. Later, molecular docking analysis was carried out with these compounds to identify best ligands and 13 compounds were assayed in vitro. Compound **47** revealed the most potent inhibition of BACE1 in vitro. The docking analysis revealed bonding interactions with residues Q-73, D-228, T-231, and T-232 [91].

A research study identified compound **48**- Baicalein as a dual inhibitor against BACE1 and AChE. Enzyme kinetic studies revealed that it would act by competitive antagonism of BACE1. Docking analysis revealed that it could form Van Der Waals interactions with V-69, Y-71, W-76, F-108, I-118, A-127, and R-128, 6 hydrogen-bonding interactions with S-35, S-36, N-37, and I-126. The docking energy of the BACE1 in the complex with the compound was −8.60 kcal/mol. It was also analyzed for violation of Lipinski's rule of 5 and it revealed no violations [92].

A study combined multiple in silico methods to decipher potential inhibitors of BACE1. Initially, a structure-based pharmacophore was generated taking into account various PDB structures of BACE1 with their co-crystallized ligands. Based on the binding energy contributions a pharmacophore was generated and validated. Based on the various e-pharmacophore models generated, several molecules were selected for molecular docking studies. The molecules were revealed to target the catalytic dyad on BACE1. The top 7 compounds were taken ahead for in vitro enzymatic analysis and it revealed that the compound **49** showed the most potent inhibition [93].

A virtual screening protocol was performed to uncover novel hit molecules. A CNS database was used to select ligands for the virtual screening approach. 131,014 molecules were selected and docked with BACE1 protein and molecules that showed a binding energy of less than −9 kcal/mol were selected for further stages. A total of 371 molecules were selected for extra precision docking and 44 molecules that showed a binding energy of less than −7 kcal/mol were further analyzed for pharmacokinetic parameters like BBB permeability, human intestinal absorption, and P-gp substrate potential. After these 5 best molecules were selected for MD simulations. Compound **50** was identified as a hit molecule. It showed hydrogen-bonding interactions with D-32, G-34, K-224, R-235, and T-329, hydrophobic interactions with Y-71, π-stacking interactions with F-108, and salt-bridge interactions with the catalytic dyad of BACE1. MD simulations revealed an average RMSD of 0.231 nanometre for the compound. The average RMSF was 0.143 nanometre, and the average radius of gyration was 2.177 nanometre. MM-PBSA revealed the free binding energy of the compound throughout the run-time of 100 ns and it was −994.978 kJ/mol [94].

A study identified a phytochemical compound **51**- Triptofordin B1 as a hit molecule, from the biological source *Tripterygium wilfordii*. Initially, a molecular docking approach was employed on a large number of compounds from traditional medicine database. Docking analysis revealed the top 10 molecules with appropriate docking parameter values. Further, ADME screening was performed and the compound was found to have adequate pharmacokinetic parameters. To reveal the stability of the compound with BACE1, MD simulations were performed for 50 ns. The RMSD graph revealed an average value of 0.09 nanometre throughout the run time. The total free binding energy of the compound was $-8.72 * 10^5$ kJ/mol. G-61, Q-121, and W-163 were involved in consistent interactions with the compound [95].

A research study identified compound **52**- Ponciretin, a constituent from *Poncirus trifoliata* as a hit molecule from a molecular docking study. Initially, BBB permeable 18 phytochemicals were selected for molecular docking study. The total docking energy was −8.78 kcal/mol and K_i was found to be 366.56 nM. From the docking analysis, it was revealed hydrogen bonding interactions with K-107, F-109, N-111, and π-alkyl interactions with F-47 [96].

20.6 Targeting the Tau Kinases

Tau phosphorylation is one of the key post-translational processes which is regulated by kinases positively and phosphatases negatively. An imbalance in this equilibrium causes tau to undergo hyperphosphorylation, which gradually agglomerates to form NFTs. Many kinases are involved in phosphorylating tau categorized as proline-mediated kinases, non-proline-mediated kinases, and tyrosine-mediated kinases. GSK-3β is one of the major kinases involved in hyperphosphorylated tau and hence, presents an attractive target for developing novel inhibitors. GSK-3β bears an ATP-binding site, an allosteric site, and a substrate-binding site (Fig. 20.9) [97–99]. A list of compounds that hold promise for GSK-3β inhibition has been highlighted in Table 20.4.

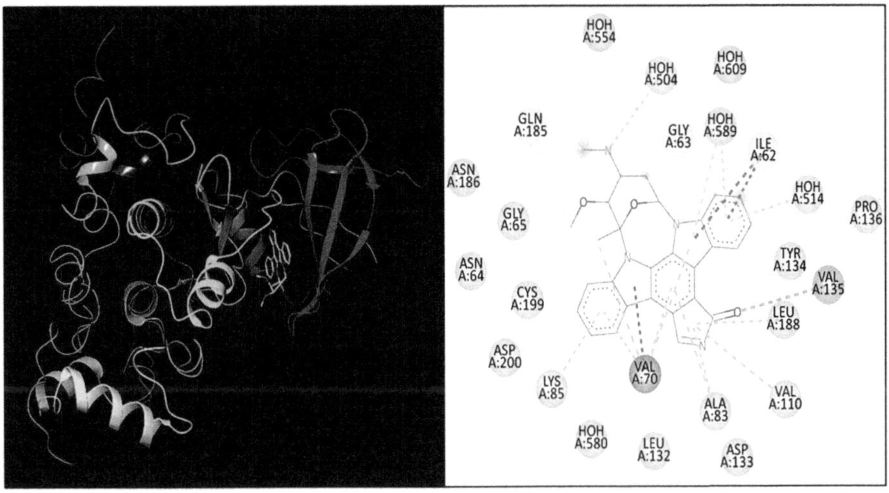

Fig. 20.9 GSK-3β (PDB ID: *1q3d*) with staurosporine 3D binding mode and 2D critical amino acid residues

Table 20.4 Modulating tau hyperphosphorylation by GSK-3β inhibitors

Compound No	Structure	IC_{50} values (μM)
53	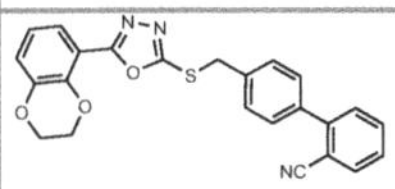	–
54	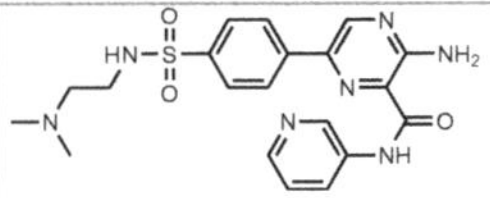	–
55	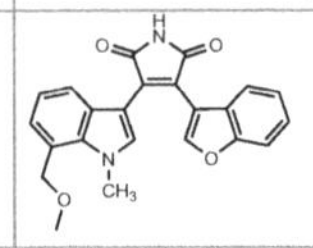	0.00012
56	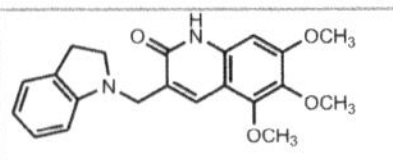	–
57	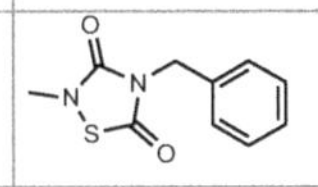	–

A study revealed a designed compound **53** as a potential hit molecule from docking and QSAR analysis. 42 oxadiazole molecules were selected for QSAR. However, initially, they were docked to analyze the binding modes and critical residues for ligand binding. Docking was performed on the ATP-binding site of GSK-3β and important residues included I-62, F-67 of the glycine-rich loop, K-85 of the active site, D-200 of the DFG loop, and V-135 in the hinge region. Further, a 3D-QSAR model was employed to determine the best descriptors. This revealed mathematical vectors and SMILES-based descriptors as the most significant contributors to biological activity [100].

A research study was carried out identifying compound **54** as a GSK-3β inhibitor in the ATP-binding site by employing 3D QSAR modeling, followed by MD simulations and free energy calculations. Electrostatic and steric contour maps were generated after QSAR analysis and it was revealed that steric and hydrophobic interactions were crucial for binding to the enzyme. Docking analysis revealed interactions with K-85, D-133, Y-134, and R-141. These residues also confer selectivity to GSK-3β inhibition. The MM-GBSA study showed the average binding free energy to be −47.190 kcal/mol [101].

Various maleimide derivatives bearing benzofuran moiety were analyzed as GSK-3β inhibitors. A 3D QSAR approach was used for a library of 51 reported inhibitors and various models based on the binding modes of the inhibitors were developed. It was revealed that the binding mode I showed the best results for the 3D QSAR model. Compound **55** was the best compound in the series [102].

A structure-based pharmacophore model was developed using different PDB structures of protein and subsequent co-crystallized ligands. It was later validated by analyzing a library of 1235 compounds, predicting the binding modes, and predicting their IC_{50} values. Later, 12,251 molecules from the ZINC database were screened for obtained pharmacophore, lead-like properties, predictive CNS activity, and Pan-assay interference compounds revealing a total of 630 hits. Later, clustering of the scaffolds was done and 123 compounds were accomplished as hits for the next stage. Further, predictive pharmacokinetics filtered off some more compounds leaving a total of 91 hits. Further, molecular docking was performed compounds were filtered for critical interactions with D-133 and V-135 which revealed 25 acceptable hits. Compound **56** was the best inhibitor of the series displaying a docking energy of −19.06 kcal/mol [103].

A previously reported compound **57** was studied for its binding mode and interaction profile with GSK-3β. It was a thiazolidinedione derivative which was a non-competitive inhibitor binding to the DFG-out (inactive conformation) motif of the enzyme. Extensive MD simulations for 500 ns were carried out to determine the changes in the active and inactive conformations of GSK-3β. GSK-3β was found to possess 3 binding sites- ATP-binding site, active site, and allosteric site. The presence of an allosteric site was seen after F-180 underwent a 180-degree flip, following a DFG-out conformation. The compound in complex with enzyme for 500 ns time frame showed an average RMSD of 2.8 Å. MM-GBSA revealed a binding free energy of −60 kcal/mol throughout the run time. A docking study identified the allosteric binding mode of the compound and it showed hydrogen-bonding interactions with K-85 [104].

20.7 Neuroprotective Mechanisms by Investigated Compounds

Neuroprotective mechanisms like targeting neuroinflammation, oxidative stress, and enzymes involved in neurodegeneration including neuroinflammatory signaling molecules like nuclear erythroid factor-2 (Nrf-2), nuclear factor κB (NF-κB), and nuclear factor kappa B kinase subunit beta (IKKB) can also be exploited to develop novel inhibitors which can counter the hampered neuroinflammatory responses.

A study involved the selection of 15 FDA-approved drugs and analysis of their affinity towards IKKB by in silico methods. Compound **58**- Aripiprazole possessed IKKB inhibitory activity. Extra precision molecular docking revealed a score of −8.067 Kcal/mol for aripiprazole. It showed hydrogen-bonding interaction with C-99, and multiple interactions with Q-97, D-103, D-166, Y-98, G-102, M-96, L-167, L-21, V-29, V-152, T-23, I-165, A-42, and Y-108. Induced-fit docking calculated the inhibitory binding interactions with C-99, Y-98, L-21, T-23, and I-165. MD simulations was used to analyze the stability of the complex and it showed an average RMSD value of 3.8 Å throughout the timeline of 100 ns. Furthermore, MM-GBSA studies determine the changes in the free binding energy of the complex and it revealed an average of −62.01 Kcal/ mol [105].

A study identified compound **59-** Modafinil as a neurokinin 1 (NK1) receptor inhibitor. Release and subsequent binding of substance P to NK1 leads to stimulation of proinflammatory cytokines and hence NK1 inhibition was thought to be beneficial against AD. A high-throughput virtual screening was carried out to determine hit molecules. A library of FDA-approved drugs was selected and docked against NK1. It possessed a docking score of −8.21 kcal/mol and MM-GBSA showed a binding energy of −44.93 kcal/mol. It showed an average RMSD of 4.5 Å. It showed hydrogen bonding interactions with N-109, Q-165, Q-193, hydrophobic interactions with P-112, I-113, H-197, I-204, F-264, F-268, and water-bridge interactions with N-109, Q-165, Q-193, H-197, F-268, and Y-272.

A research group identified compound **60** as an inhibitor of protein tyrosine phosphatase 1B. Under inflammatory conditions, the levels of this enzyme were known to be upregulated which contribute to increased neuroinflammation. Initially, a virtual ligand screening protocol was employed to screen a library of 1941 phytochemicals. Molecules having a docking score greater than −6.5 kcal/mol were selected further for extra precision docking. The best-docked molecule, compound **60** was shown to have hydrogen-bonding interactions with D-181, F-182, C-215, S-216, and R-221, hydrophobic interactions with Y-46, V-49, F-182, C-215, A-217, I-219, and polar interactions with S-216, S-222, Q-262, and Q-266. It had a docking score of −10.573 kcal/mol. MD simulations revealed an average RMSD of about 1 Å throughout a 25 ns time frame [106].

A library of IKKB inhibitors using QSAR modeling and molecular docking analysis. A 2D and 3D QSAR model was built using a library of compounds with proven IKKB inhibitory activity. Compound **61** was shown to have higher predicted activity as compared to the reference molecule (7.18 vs 7.15). Molecular docking revealed a docking energy of −41.6134 kcal/mol. Hydrogen-bonding interactions were observed with L-21, D-166, Q-149, and Q-61 [107].

A study designed and synthesized a library of piperidone derivatives against MyD88. MyD88 alongside toll-like receptors are known to be upregulated significantly in AD pathology. MyD88-TLR pathway downstream stimulates the release of NF-κB, which further stimulates the release of proinflammatory cytokines, thus triggering neuroinflammation. Thus, the approach was to inhibit the upstream

MyD88 protein itself to avoid a cascade of reactions leading to neuroinflammation. A pharmacophore modeling approach was employed and the pharmacophore hypothesis generated included features like two hydrophobic moieties, an aromatic ring, and a hydrogen bond acceptor. Various derivatives were designed based on a peptide inhibitor of MyD88 and compound **62** was revealed to be the best of the series. Molecular docking analysis revealed amino acid interactions with K-189, I-179, K-191, A-292, and K-293 [108].

A pharmacophore modeling approach was employed to uncover designed compound **63** as a potential CB2 receptor antagonist in neuroinflammatory diseases including AD therapy. Using pharmacophore modeling, 6 different hypotheses were generated and the ZINC database was screened to discover 215 hit molecules. Compound **63** was found to be the best ligand showcasing hydrophobic interactions with F-87, F-91, F-98, V-113, F-117, W-194, W-258, and C-288. It showed a docking score of −9.877 and predictive ADMET studies also showed it to be a druggable and non-toxic candidate. Further, MD simulations for 100 ns revealed an average of 2.4 Å throughout the run time. MM-GBSA studies computed free binding energy of −107.94 kcal/mol [109] (Table 20.5).

Table 20.5 Neuroinflammatory inhibitors against the neuroinflammation

Compound No	Structure
58	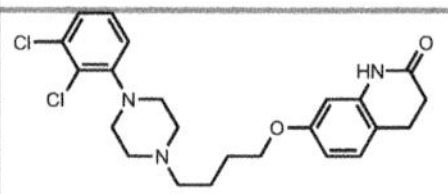
59	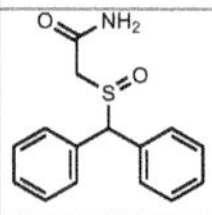
60	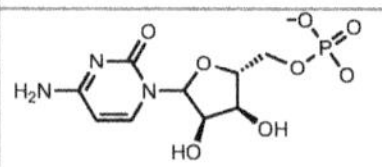
61	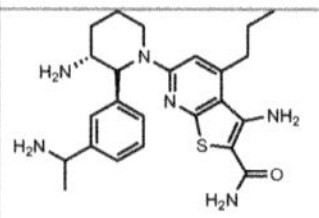
62	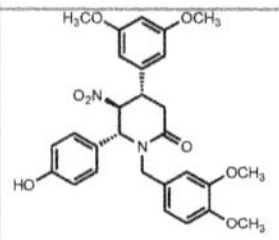
63	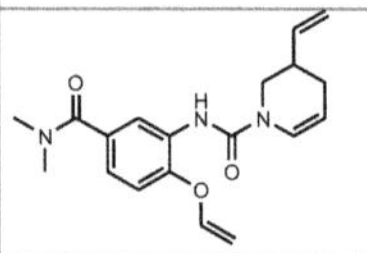

20.8 Network Pharmacology to Understand the PPIs in AD

Network pharmacology is an integration of systems biology, drug discovery, and pharmacology which involves building networks and later validating the results obtained to discover novel therapeutic mechanisms for existing chemical compounds. A usual process starts with the selection of an herbal extract with a variety of phytochemicals. The next stage involves target prediction of these phytochemicals. Later, the genes and proteins associated with the disease under question are listed. A common list of targets between the phytochemicals and the disease targets is then selected. The phytochemicals can then be verified for their activity using in silico methods or in vitro techniques to uncover novel and multi-target ligands that are highly beneficial for AD. Also, the underlying mechanisms associated with phytochemical extracts can be analyzed using this approach [110–113].

A network pharmacology approach was employed to identify mechanisms of anti-AD action of *Ginkgo folium* extract. Phytoconstituents from the extract were identified from the Traditional Chinese Medicine Systems Pharmacology Database and Analysis Platform (TCMSP). 10 actives were selected and target prediction for these constituents revealed 345 diverse targets for these phytoconstituents. Later, various targets and genes involved with AD pathology were screened and it revealed 623 targets. In common, there were 29 drug targets were noted between targets of AD and predicted targets of phytoconstituents. Further, the 7 most significant targets were finalized by the Gene ontology (GO) biological process and Kyoto Encyclopedia of Genes and Genomes (KEGG) pathway enrichment analyses. To validate the network pharmacology model, one of the constituents- Genkwanin was analyzed by molecular docking upon various targets. Docking analysis revealed docking scores of −6.56 kcal/mol against membrane metalloproteinase 3 (MMP3), −6.5 kcal/mol against lymphocyte-specific protein tyrosine kinase (LCK), −6.41 kcal/mol against cyclin-dependant kinase 5, −6.01 kcal/mol against neurotrophic tyrosine receptor kinase 2 (NTRK2), −5.86 kcal/mol against GSK-3β, −5.61 against BACE1, and −5.61 kcal/mol against cystic fibrosis transmembrane conductance regulator (CFTR). It revealed hydrogen-bonding interactions with L-164 and A-165 of MMP3, M-319 and Q-288 of LCK, D-144, C-83, and K-33 with CDK5, M-636 with NTRK2, D-133 and V-135 of GSK-3β, D-32, K-224, Q-73 and T-329 of BACE1, and K-1250, S-1251 and T-1252 of CTFR [114].

A study revealed the possible mechanisms of action of quercetin against AD. Initially, a target prediction was carried out for quercetin, and 277 predicted targets were identified. Later, a set of differentially expressed genes (DEGs) were analyzed using the microarray dataset. Overall, 3256 genes were noted with AD and the venn diagram revealed 46 targets were found to be common. A PPI network was created and analyzed. The top 20 most significantly contributing genes were taken for further studies. Furthermore, a functional enrichment analysis using GO and KEGG was performed, and the 7 most contributing genes were analyzed further. To validate the protocol, these 7 genes were expressed by serial qPCR analysis, and the levels of these genes were found to be significantly modulated in the mRNA when tested in HT-22 cells in the presence of Aβ_{1-42}. The most contributing genes and drug targets involved with the quercetin mechanism of action were MAPT, PIK3R1, CASP8, DAPK1, MAPK1, and CYCS [115].

A study investigated the pharmacological mechanisms of *Coptidis* rhizome against AD. A PPI network was developed to evaluate the mechanisms of phytoconstituents of the rhizome. The top 19 phytochemicals were selected for further study. A total of 2588 targets were obtained for AD and 267 predicted targets for the compounds. Various techniques like GO Enrichment, PPI Enrichment, and KEGG pathway enrichment. 6 targets were finalized at the end of this enrichment analysis and they were IL1B, MAPK8, MMP2, RELA, STAT3, and TGFB1. Molecular docking analysis was also carried out and it was found that the phytochemicals showed high affinity towards the targets MAPK8, RELA, and MMP2 [116].

A research group investigated the mechanisms of curcumin against AD by adopting the network pharmacology approach. In all, 233 targets were predicted for curcumin and 396 disease-related drug targets for AD. An intersection of these targets showed 49 targets of curcumin against AD were noted. Furthermore, GO and KEGG enrichments were employed. 16 drug targets were obtained by curcumin targeting Aβ and tau pathologies. Using a differential module, the predictive expressions of these genes in AD were analyzed and the top 9 most significant ones were selected for validation. Molecular docking analysis for the top 9 targets revealed that curcumin showed hydrogen bonding interactions with K-85 of NF-KB1, and with R-344 of GSK-3β [117]. Curcumin was revealed to bear anti-AD potential with inhibitory effects on NF-KB1 and GSK-3β [117].

A study investigated the potential of the phytochemical cordycepin, a constituent of *Cordyceps sinensis*. Target prediction was performed for cordycepin and a total of 64 targets were noted. Whereas, the AD targets were 254, and the intersected targets were found to be 74. Furthermore, GO and KEGG enrichment analysis was carried out and it revealed a total of 74 interaction genes and 13 KEGG-enriched pathways. A PPI network was built and the top 12 most significant targets were taken ahead for further gene expression studies. Gene expression studies revealed the 5 most significant contributing genes AKT1, MAPK8, BCL2L1, FOXO3, and CTNNB1. Molecular docking analysis revealed docking scores of −7.8, −7, −6.8, −6.6, and −6.2 kcal/mol against AKT1, BCL2L1, MAPK8, FOXO3, and CTNNB1 respectively [118].

A study evaluated the potential of natural compounds obtained from *Dictyostelium discoideum* against AD. Initially, the secondary metabolites of the plant were analyzed for their CaCO-2 permeability, BBB permeation, fractional negative accessible surface area, Log P, hydrogen bond donor and acceptor, number of rotatable bonds, and topological polar surface area. Only 7 metabolites passed the criteria. Target prediction revealed an average of 58 predicted targets across these 7 metabolites. AD genes and drug targets were also noted. Common targets were screened further for GO and KEGG enrichment analysis. All 7 compounds along with the core targets were networked individually. The top 22 drug targets were selected for molecular docking studies and all 7 compounds were docked on each target. PQA-11 showed docking energy of -8.13 ± 0.163 kcal/mol with amino acid interactions with R-120, L-123, H-90, V-89, W-100, S-119, and P-86 of COX-2. Further, this complex was forwarded for MD simulations for 100 ns. It showed an average RMSD of 7 nanometre, and the radius of gyration was about 2.45 nanometre [119].

A network pharmacology approach was adopted to identify key components of *Scrophulariae radix* bearing anti-AD potential. From the list of actives, based upon the criteria of oral bioavailability (oral bioavailability, OB) $\geq$ 30% and drug-likeness, DL $\geq$ 0.18, and 17 actives were isolated. Target prediction revealed 208 probable targets for these actives. 213 potential targets for AD were isolated. Common targets between phytochemicals and AD were then taken for the next stage. 40 common targets were further enriched by GO and KEGG enrichment. A PPI network was generated and the most significant targets were then expressed. Later, only 5 targets were finalized for molecular docking studies. Compounds like β-sitosterol, aucabin, harpagoside, and scropolioside were revealed to possess some inhibitory interactions with ESR-1, IL-6, MMP-2, and MMP-9 respectively. It was thus revealed that multiple constituents and multiple mechanisms could be responsible for AD therapy using the plant extract [120].

20.9 Role of AI/ML in AD Therapeutics

AI/ML provides a broad range of options to enhance the drug design and drug discovery processes. AI/ML can be used to identify crucial data from a big pool of data. It can be used to predict clinical outcomes, design clinical trials, retrosynthesis predictions, optimize pharmacokinetics and pharmacodynamics, drug-repurposing, and develop individualized precision medicine [121]. Given, the complexity of AD, AI/ML can aid in gaining clarity regarding the pathophysiological mechanisms associated with AD and the mechanisms associated with the particular stages [122, 123]. It has also been used to suggest drug-repurposing candidates which could reduce the overall clinical trials aspect of the drug discovery [124].

A study employed ML and bioinformatics approach to reveal dual cholinesterase inhibitors. Multistep virtual screening protocol was employed through an in-house chemistry database and additionally, an ML-based model was developed to filter through the database to achieve novel derivatives. 5 hits were revealed after the studies, and molecular docking was performed on these molecules. The best of the series was compound **64** which showed a docking score of −15.03 kcal/mol and −7.07 kcal/mol each against AChE and BuChE respectively. The binding free energy was −76.34 Kcal/mol and −74.72 Kcal/mol against AChE and BuChE respectively. DFT studies revealed a HOMO orbital energy of −0.352,410 eV, and predictive pharmacokinetics revealed its potential for druggability. In vitro, enzymatic assays revealed the potency of the compound [125].

A research group employed the ML-based q-RASAR (quantitative read-across structure-activity relationship) technique to discover novel hits against AD. This technique combines the principles of a 2D-QSAR with a read-across approach. 7 crucial AD targets including AChE, BuChE, BACE1, 5-HT6, CDK-5, APP, and tau aggregation. Factors and moieties affecting dual inhibition were analyzed. 7 different ML-based models were further utilized to explore the predictive abilities of the models [126].

A research study used ML models and computational techniques to reveal novel hit candidates against Keap1, a protein involved in the binding of Nrf2. Overall, this triggers a series of neuroinflammatory responses owing to increased release of pro-inflammatory cytokines. A QSAR-based ML model and ligand-based virtual screening were performed to screen through a library of compounds. Curated data parameters for the selection of compounds included Log P, molecular weight, frequency, and PIC_{50}. 5 hit ligands were subjected to molecular docking. It revealed hydrogen-bonding interactions with S-363, N-414, and S-602. MD simulations revealed an average RMSD of 0.15 nanometre, the average radius of gyration was 1.8 nanometre, and the average n.o of hydrogen bonds formed was 8. Compound **65** showed a PIC_{50} of 7.091184 [127].

A study developed an ML model using a classifier of active and inactive compounds. A consensus-based scoring was employed for developing ML models. A combination of KNN-Morgan and RF-Morgan models was employed to perform the virtual screening protocol for AChE inhibitors. 310 compounds with a consensus score of less than 0.8 were selected. Further, molecules with a tanimato score of greater than 0.4 were removed. From the 13 compounds obtained, only 7 were selected ahead due to their availability. RDKit FPs were used to filter these 7 compounds which resulted in 4 best compounds. These compounds were tested for their in vitro inhibition and compound **66** was found to be a hit molecule. It was later analyzed by molecular docking and MD simualtions. It was found to possess a stable average RMSD value of 1.5 Å. Docking analysis showed hydrogen-bonding interactions with V-135 and P-136 [128].

A study involved developing an ML model in combination with a virtual screening protocol to reveal novel derivatives of tau aggregation inhibition. A virtual ligand tool was used to screen through a library of compounds. The obtained hits were again filtered based on structural similarity and PAINS which reduced the number of hits to 2367 hits. Tau protein binding pocket was identified and selected compounds were docked on the active site. Predictive ADME studies were also performed to reduce the number of potential hits. 33 ligands showed acceptable docking scores for all conformers of tau protein. The best 10 compounds were selected for MD simualtions and subsequent MM-GBSA calculations. Compound **67** was revealed as a hit molecule and possessed an average MM-GBSA value of about −40 kcal/mol [129] (Table 20.6).

20.10 Future Perspectives

Drug discovery methods have been significantly modified from the start of discovery up until now. With this new integration of AI and ML, the drug discovery field is experiencing a paradigm shift. In the future, AI and ML are going to dominate the

Table 20.6 Advanced AD Drug-discovery with AI/ML-based compounds

Compound No	Structure	IC_{50} values (μM)
64	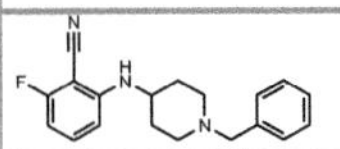	AChE: 0.927 ± 0.019 BuChE: 6.377 ± 0.038
65	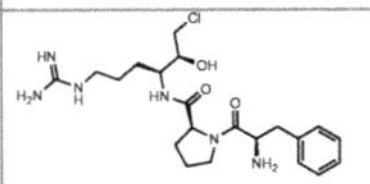	–
66	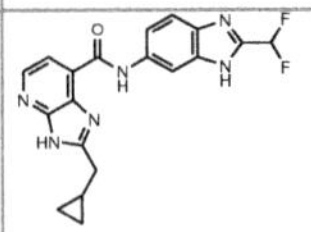	AChE: 0.64 ± 42
67	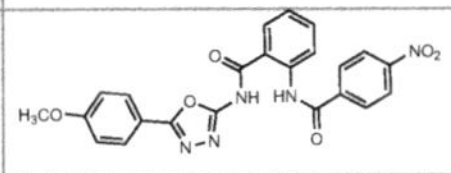	–

drug discovery process and significantly reduce the amount of time required for a conventional drug discovery process. AI/ML in conjunction with computer-aided drug design techniques would be the best approach going ahead to uncover novel therapeutic findings, novel scaffolds, novel approaches to conventional procedures, and so forth. CADD techniques will continue to aid the medicinal chemists in targeted drug design which can reduce the synthetic costs and failure rates associated with conventional medicinal chemistry approach.

20.11 Conclusion

Computer-aided drug design techniques like molecular docking, molecular dynamics, QSAR modeling, Virtual screening, and Pharmacophore modeling are key techniques for understanding the atom-level mechanisms of novel ligands. Pharmacophore modeling enables screening through ligands with appropriate features specific to a target under concern. Molecular docking helps to understand the amino acid interaction profiles of designed ligands as well as already existing inhibitors which can help to design new ligands or scaffolds in the future. Molecular dynamics helps to analyze the protein-ligand complex characteristics and its stability for time and physiological conditions. QSAR modeling allows for the prediction of the designed compounds for the activity under concern. These techniques have significantly been used for AD research given the disease's complexity and varied pathological mechanisms. Such techniques increase the cost-effectiveness of the drug discovery process, allow for targeted drug design, reduce the failure rates, and thus, should be efficiently employed for AD research.

References

1. 2023 Alzheimer's disease facts and figures. Alzheimer's Dement. 2023;19(4):1598–695.
2. Mayo Clinic. https://www.mayoclinic.org. 2023. Alzheimer's stages: How the disease progresses.
3. Abeysinghe AADT, Deshapriya RDUS, Udawatte C. Alzheimer's disease; a review of the pathophysiological basis and therapeutic interventions. Life Sci. 2020;256:117996.
4. Alhazmi HA, Albratty M. An update on the novel and approved drugs for Alzheimer disease. Saudi Pharm J. 2022;30(12):1755–64.
5. van Dyck CH, Swanson CJ, Aisen P, Bateman RJ, Chen C, Gee M, et al. Lecanemab in early Alzheimer's disease. N Engl J Med. 2023;388(1):9–21.
6. Alzheimer's association. https://www.alz.org. 2024. Medications for memory, cognition and dementia-related behaviors.
7. Kumar A, Singh A, Ekavali. A review on Alzheimer's disease pathophysiology and its management: an update. Pharmacol Rep. 2015;67(2):195–203.
8. Guo T, Zhang D, Zeng Y, Huang TY, Xu H, Zhao Y. Molecular and cellular mechanisms underlying the pathogenesis of Alzheimer's disease. Mol Neurodegener. 2020;15(1):40.
9. Kumar Thakur A, Kamboj P, Goswami K, Ahuja K. Pathophysiology and management of alzheimer's disease: an overview. J Anal Pharm Res. 2018;7(2):226.
10. Tiwari S, Atluri V, Kaushik A, Yndart A, Nair M. Alzheimer's disease: pathogenesis, diagnostics, and therapeutics. Int J Nanomedicine. 2019;14:5541–54.
11. Mohandas E, Rajmohan V, Raghunath B. Neurobiology of Alzheimer′s disease. Indian J Psychiatry. 2009;51(1):55.
12. Garbuz DG, Zatsepina OG, Evgen'ev MB. Beta amyloid, Tau protein, and Neuroinflammation: an attempt to integrate different hypotheses of Alzheimer's disease pathogenesis. Mol Biol. 2021;55(5):670–82.
13. Fan L, Mao C, Hu X, Zhang S, Yang Z, Hu Z, et al. New insights into the pathogenesis of Alzheimer's disease. Front Neurol. 2020;10:1312.
14. García-Ayllón MS. Revisiting the role of acetylcholinesterase in Alzheimer's disease: cross-talk with P-tau and β-amyloid. Front Mol Neurosci. 2011;4:22.
15. Hampel H, Mesulam MM, Cuello AC, Farlow MR, Giacobini E, Grossberg GT, et al. The cholinergic system in the pathophysiology and treatment of Alzheimer's disease. Brain. 2018;141(7):1917–33.
16. Craig LA, Hong NS, McDonald RJ. Revisiting the cholinergic hypothesis in the development of Alzheimer's disease. Neurosci Biobehav Rev. 2011;35(6):1397–409.
17. Hampel H, Mesulam MM, Cuello AC, Khachaturian AS, Vergallo A, Farlow MR, et al. Revisiting the cholinergic hypothesis in Alzheimer's disease: emerging evidence from translational and clinical research. J Prev Alzheimers Dis. 2018;6:1–14.
18. Chen ZR, Huang JB, Yang SL, Hong FF. Role of cholinergic signaling in Alzheimer's disease. Molecules. 2022;27(6):1816.
19. Wong-Guerra M, Calfio C, Maccioni RB, Rojo LE. Revisiting the neuroinflammation hypothesis in Alzheimer's disease: a focus on the druggability of current targets. Front Pharmacol. 2023;14:1161850.
20. Du X, Wang X, Geng M. Alzheimer's disease hypothesis and related therapies. Transl Neurodegener. 2018;7(1):2.
21. Lecca D, Jung YJ, Scerba MT, Hwang I, Kim YK, Kim S, et al. Role of chronic neuroinflammation in neuroplasticity and cognitive function: a hypothesis. Alzheimers Dement. 2022;18(11):2327–40.
22. Kinney JW, Bemiller SM, Murtishaw AS, Leisgang AM, Salazar AM, Lamb BT. Inflammation as a central mechanism in Alzheimer's disease. Alzheimer's Dement. 2018;4(1):575–90.
23. Chen LL, Fan YG, Zhao LX, Zhang Q, Wang ZY. The metal ion hypothesis of Alzheimer's disease and the anti-neuroinflammatory effect of metal chelators. Bioorg Chem. 2023;131:106301.

24. Vaz FNC, Fermino BL, Haskel MVL, Wouk J, de Freitas GBL, Fabbri R, et al. The relationship between copper, iron, and selenium levels and Alzheimer disease. Biol Trace Elem Res. 2018;181(2):185–91.
25. Wang L, Yin YL, Liu XZ, Shen P, Zheng YG, Lan XR, et al. Current understanding of metal ions in the pathogenesis of Alzheimer's disease. Transl Neurodegener. 2020;9(1):10.
26. Bush AI. Drug development based on the metals hypothesis of Alzheimer's disease. J Alzheimers Dis. 2008;15(2):223–40.
27. Babić Leko M, Langer Horvat L, Španić Popovački E, Zubčić K, Hof PR, Šimić G. Metals in Alzheimer's disease. Biomedicines. 2023;11(4):1161.
28. Praticò D. Oxidative stress hypothesis in Alzheimer's disease: a reappraisal. Trends Pharmacol Sci. 2008;29(12):609–15.
29. Zhao Y, Zhao B. Oxidative stress and the pathogenesis of Alzheimer's disease. Oxidative Med Cell Longev. 2013;2013:1–10.
30. Padurariu M, Ciobica A, Lefter R, Serban IL, Stefanescu C, Chirita R. The oxidative stress hypothesis in Alzheimer's disease. Psychiatr Danub. 2013;25(4):401–9.
31. Christen Y. Oxidative stress and Alzheimer disease. Am J Clin Nutr. 2000;71(2):621S–9S.
32. Kumar V, Kim SH, Bishayee K. Dysfunctional glucose metabolism in Alzheimer's disease onset and potential pharmacological interventions. Int J Mol Sci. 2022;23(17):9540.
33. Hammond TC, Lin AL. Glucose metabolism is a better marker for predicting clinical Alzheimer's disease than amyloid or Tau. J Cell Immunol. 2022;4(1):15–8.
34. González A, Calfío C, Churruca M, Maccioni RB. Glucose metabolism and AD: evidence for a potential diabetes type 3. Alzheimers Res Ther. 2022;14(1):56.
35. Wood WG, Li L, Müller WE, Eckert GP. Cholesterol as a causative factor in Alzheimer's disease: a debatable hypothesis. J Neurochem. 2014;129(4):559–72.
36. Rudge JD. A new hypothesis for Alzheimer's disease: the lipid invasion model. J Alzheimers Dis Rep. 2022;6(1):129–61.
37. Feringa FM, van der Kant R. Cholesterol and Alzheimer's disease; from risk genes to pathological effects. Front Aging Neurosci. 2021;13:690372.
38. Kiametis AS, Silva MA, Romeiro LAS, Martins JBL, Gargano R. Potential acetylcholinesterase inhibitors: molecular docking, molecular dynamics, and in silico prediction. J Mol Model. 2017;23(2):67.
39. Pourtaher H, Hasaninejad A, Iraji A. Design, synthesis, in silico and biological evaluations of novel polysubstituted pyrroles as selective acetylcholinesterase inhibitors against Alzheimer's disease. Sci Rep. 2022;12(1):15236.
40. Valasani KR, Chaney MO, Day VW, ShiDu YS. Acetylcholinesterase inhibitors: structure based design, synthesis, pharmacophore modeling, and virtual screening. J Chem Inf Model. 2013;53(8):2033–46.
41. Jasiecki J, Targońska M, Wasąg B. The role of Butyrylcholinesterase and iron in the regulation of cholinergic network and cognitive dysfunction in Alzheimer's disease pathogenesis. Int J Mol Sci. 2021;22(4):2033.
42. Bajda M, Więckowska A, Hebda M, Guzior N, Sotriffer C, Malawska B. Structure-based search for new inhibitors of Cholinesterases. Int J Mol Sci. 2013;14(3):5608–32.
43. Wang L, Wang Y, Tian Y, Shang J, Sun X, Chen H, et al. Design, synthesis, biological evaluation, and molecular modeling studies of chalcone-rivastigmine hybrids as cholinesterase inhibitors. Bioorg Med Chem. 2017;25(1):360–71.
44. Lu SH, Wu JW, Liu HL, Zhao JH, Liu KT, Chuang CK, et al. The discovery of potential acetylcholinesterase inhibitors: a combination of pharmacophore modeling, virtual screening, and molecular docking studies. J Biomed Sci. 2011;18(1):8.
45. Vats C, Dhanjal JK, Goyal S, Bharadvaja N, Grover A. Computational design of novel flavonoid analogues as potential AChE inhibitors: analysis using group-based QSAR, molecular docking and molecular dynamics simulations. Struct Chem. 2015;26(2):467–76.
46. Samadi A, de la Fuente RM, Pérez C, Iriepa I, Moraleda I, Rodríguez-Franco MI, et al. Synthesis, pharmacological assessment, and molecular modeling of 6-chloro-pyridonepezils:

new dual AChE inhibitors as potential drugs for the treatment of Alzheimer's disease. Eur J Med Chem. 2013;67:64–74.
47. Jana S, Ganeshpurkar A, Singh SK. Multiple 3D-QSAR modeling, e-pharmacophore, molecular docking, and *in vitro* study to explore novel AChE inhibitors. RSC Adv. 2018;8(69):39477–95.
48. Akrami H, Mirjalili BF, Khoobi M, Nadri H, Moradi A, Sakhteman A, et al. Indolinone-based acetylcholinesterase inhibitors: synthesis, biological activity and molecular modeling. Eur J Med Chem. 2014;84:375–81.
49. Silva L, Ferreira E, Maryam E-RJ, Costa G, Cruz J, et al. Galantamine based novel acetylcholinesterase enzyme inhibitors: a molecular modeling design approach. Molecules. 2023;28(3):1035.
50. Gharaghani S, Khayamian T, Ebrahimi M. Molecular dynamics simulation study and molecular docking descriptors in structure-based QSAR on acetylcholinesterase (AChE) inhibitors. SAR QSAR Environ Res. 2013;24(9):773–94.
51. Liu H, Zhou C, Fan H, Tang J, Liu L, Gao X, et al. Novel potent and selective acetylcholinesterase inhibitors as potential drugs for the treatment of Alzheimer's disease: synthesis, pharmacological evaluation, and molecular modeling of amino-alkyl-substituted Fluoro-Chalcones derivatives. Chem Biol Drug Des. 2015;86(4):517–22.
52. Zhou A, Hu J, Wang L, Zhong G, Pan J, Wu Z, et al. Combined 3D-QSAR, molecular docking, and molecular dynamics study of tacrine derivatives as potential acetylcholinesterase (AChE) inhibitors of Alzheimer's disease. J Mol Model. 2015;21(10):277.
53. Akhoon BA, Choudhary S, Tiwari H, Kumar A, Barik MR, Rathor L, et al. Discovery of a new donepezil-like acetylcholinesterase inhibitor for targeting Alzheimer's disease: computational studies with biological validation. J Chem Inf Model. 2020;60(10):4717–29.
54. Alpan AS, Parlar S, Carlino L, Tarikogullari AH, Alptüzün V, Güneş HS. Synthesis, biological activity and molecular modeling studies on 1H-benzimidazole derivatives as acetylcholinesterase inhibitors. Bioorg Med Chem. 2013;21(17):4928–37.
55. Kumar V, Saha A, Roy K. In silico modeling for dual inhibition of acetylcholinesterase (AChE) and butyrylcholinesterase (BuChE) enzymes in Alzheimer's disease. Comput Biol Chem. 2020;88:107355.
56. Makarian M, Gonzalez M, Salvador SM, Lorzadeh S, Hudson PK, Pecic S. Synthesis, kinetic evaluation and molecular docking studies of donepezil-based acetylcholinesterase inhibitors. J Mol Struct. 2022;1247:131425.
57. da Cunha EFF, Resende JE, Franca TCC, Gonçalves MA, de Souza FR, Santos-Garcia L, et al. Molecular modeling studies of Piperidine derivatives as new acetylcholinesterase inhibitors against neurodegenerative diseases. J Chem. 2013;2013:1–7.
58. Geromichalos GD, Lamari FN, Papandreou MA, Trafalis DT, Margarity M, Papageorgiou A, et al. Saffron as a source of novel acetylcholinesterase inhibitors: molecular docking and in vitro enzymatic studies. J Agric Food Chem. 2012;60(24):6131–8.
59. Alom MM, Bonna RP, Islam A, Alom MW, Rahman ME, Faruqe MO, et al. Unveiling neuroprotective potential of spice plant-derived compounds against Alzheimer's disease: insights from computational studies. Int J Alzheimers Dis. 2023;2023:1–19.
60. Remya C, Dileep KV, Variyar EJ, Zhang KYJ, Omkumar RV, Sadasivan C. Chemical similarity assisted search for acetylcholinesterase inhibitors: molecular modeling and evaluation of their neuroprotective properties. Int J Biol Macromol. 2021;174:466–76.
61. Lima JAR, Costa TW, da Fonseca ACC, do Amaral RF, do Nascimento MDSB, Santos-Filho OA, et al. Geissoschizoline, a promising alkaloid for Alzheimer's disease: inhibition of human cholinesterases, anti-inflammatory effects and molecular docking. Bioorg Chem. 2020;104:104215.
62. Fernandes TB, Cunha MR, Sakata RP, Candido TM, Baby AR, Tavares MT, et al. Synthesis, molecular modeling, and evaluation of novel Sulfonylhydrazones as acetylcholinesterase inhibitors for Alzheimer's disease. Arch Pharm (Weinheim). 2017;350(11)

63. Radwan AA, Alanazi FK, Raish M. Design and synthesis of multi-functional small-molecule based inhibitors of amyloid-β aggregation: molecular modeling and in vitro evaluation. PLoS One. 2023;18(5):e0286195.
64. Fang M, Zhang Q, Wang X, Su K, Guan P, Hu X. Inhibition mechanisms of (−)-Epigallocatechin-3-gallate and Genistein on Amyloid-beta 42 peptide of Alzheimer's disease via molecular simulations. ACS Omega. 2022;7(23):19665–75.
65. Kumar V, Ojha PK, Saha A, Roy K. Cheminformatic modelling of β-amyloid aggregation inhibitory activity against Alzheimer's disease. Comput Biol Med. 2020;118:103658.
66. Job N, Thimmakondu VS, Thirumoorthy K. In Silico drug design and analysis of dual amyloid-Beta and tau protein-aggregation inhibitors for Alzheimer's disease treatment. Molecules. 2023;28(3):1388.
67. Kaur A, Goyal B. Identification of new pentapeptides as potential inhibitors of amyloid–β42 aggregation using virtual screening and molecular dynamics simulations. J Mol Graph Model. 2023;124:108558.
68. Narang SS, Goyal D, Goyal B. Inhibition of Alzheimer's amyloid-β_{42} peptide aggregation by a bi-functional bis-tryptoline triazole: key insights from molecular dynamics simulations. J Biomol Struct Dyn. 2019;1598:1–14.
69. Sharma S, Nehru B, Saini A. Inhibition of Alzheimer's amyloid-beta aggregation in-vitro by carbenoxolone: insight into mechanism of action. Neurochem Int. 2017;108:481–93.
70. Aswathy L, Jisha RS, Masand VH, Gajbhiye JM, Shibi IG. Design of novel amyloid β aggregation inhibitors using QSAR, pharmacophore modeling, molecular docking and ADME prediction. In Silico Pharmacol. 2018;6(1):12.
71. Safarizadeh H, Garkani-Nejad Z. Molecular docking, molecular dynamics simulations and QSAR studies on some of 2-arylethenylquinoline derivatives for inhibition of Alzheimer's amyloid-beta aggregation: insight into mechanism of interactions and parameters for design of new inhibitors. J Mol Graph Model. 2019;87:129–43.
72. Doytchinova I, Atanasova M, Salamanova E, Ivanov S, Dimitrov I. Curcumin inhibits the primary nucleation of amyloid-Beta peptide: a molecular dynamics study. Biomol Ther. 2020;10(9):1323.
73. Adewole KE, Gyebi GA, Ibrahim IM. Amyloid β fibrils disruption by kolaviron: molecular docking and extended molecular dynamics simulation studies. Comput Biol Chem. 2021;94:107557.
74. Castro-Silva ES, Bello M, Rosales-Hernández MC, Correa-Basurto J, Hernández-Rodríguez M, Villalobos-Acosta D, et al. Fucosterol from *Sargassum horridum* as an amyloid-beta ($A\beta_{1\text{-}42}$) aggregation inhibitor: *in vitro* and *in silico* studies. J Biomol Struct Dyn. 2021;39(4):1271–83.
75. Pasieka A, Panek D, Szałaj N, Espargaró A, Więckowska A, Malawska B, et al. Dual inhibitors of Amyloid-β and Tau aggregation with Amyloid-β disaggregating properties: extended *in Cellulo* , *in Silico* , and kinetic studies of multifunctional anti-Alzheimer's agents. ACS Chem Neurosci. 2021;12(11):2057–68.
76. Lakey-Beitia J, Doens D, Kumar J, Murillo E, Fernández PL, Rao K, et al. Anti-amyloid aggregation activity of novel carotenoids: implications for Alzheimer's drug discovery. Clin Interv Aging. 2017;12:815–22.
77. Barman A, Prabhakar R. Computational insights into substrate and site specificities, catalytic mechanism, and protonation states of the Catalytic Asp Dyad of ***β*** -secretase. Scientifica (Cairo). 2014;2014:1–11.
78. Joshi N, Alavala RR. Sulfonamido, amido heterocyclic adducts of tetrazole derivatives as BACE1 inhibitors: in silico exploration. Mol Divers. 2024;28:4017.
79. Kushwaha P, Singh V, Somvanshi P, Bhardwaj T, Barreto GE, Ashraf GM, et al. Identification of new BACE1 inhibitors for treating Alzheimer's disease. J Mol Model. 2021;27(2):58.
80. Youn K, Park JH, Lee S, Lee S, Lee J, Yun EY, et al. BACE1 inhibition by Genistein: biological evaluation, kinetic analysis, and molecular docking simulation. J Med Food. 2018;21(4):416–20.

81. Dhanabalan AK, Kesherwani M, Velmurugan D, Gunasekaran K. Identification of new BACE1 inhibitors using pharmacophore and molecular dynamics simulations approach. J Mol Graph Model. 2017;76:56–69.
82. Hassan M, Shahzadi S, Seo SY, Alashwal H, Zaki N, Moustafa AA. Molecular docking and dynamic simulation of AZD3293 and Solanezumab effects against BACE1 to treat Alzheimer's disease. Front Comput Neurosci. 2018;12:34.
83. Joseph OA, Babatomiwa K, Niyi A, Olaposi O, Olumide I. Molecular docking and 3D Qsar studies of C000000956 as a potent inhibitor of Bace-1. Drug Res. 2019;69(08):451–7.
84. Kumar A, Roy S, Tripathi S, Sharma A. Molecular docking based virtual screening of natural compounds as potential BACE1 inhibitors: 3D QSAR pharmacophore mapping and molecular dynamics analysis. J Biomol Struct Dyn. 2016;34(2):239–49.
85. Yu T, Paudel P, Seong SH, Kim JA, Jung HA, Choi JS. Computational insights into β-site amyloid precursor protein enzyme 1 (BACE1) inhibition by tanshinones and salvianolic acids from Salvia miltiorrhiza via molecular docking simulations. Comput Biol Chem. 2018;74:273–85.
86. Youn K, Yoon JH, Lee N, Lim G, Lee J, Sang S, et al. Discovery of Sulforaphane as a potent BACE1 inhibitor based on kinetics and computational studies. Nutrients. 2020;12(10):3026.
87. Mukerjee N, Das A, Jawarkar RD, Maitra S, Das P, Castrosanto MA, et al. Repurposing food molecules as a potential BACE1 inhibitor for Alzheimer's disease. Front Aging Neurosci. 2022;14:878276.
88. Bhaskar V, Chowdary R, Dixit SR, Joshi SD. Synthesis, molecular modeling and BACE-1 inhibitory study of tetrahydrobenzo[b] pyran derivatives. Bioorg Chem. 2019;84:202–10.
89. Jabir NR, Rehman MDT, Alsolami K, Shakil S, Zughaibi TA, Alserihi RF, et al. Concatenation of molecular docking and molecular simulation of BACE-1, γ-secretase targeted ligands: in pursuit of Alzheimer's treatment. Ann Med. 2021;53(1):2332–44.
90. Choi RJ, Roy A, Jung HJ, Ali MY, Min BS, Park CH, et al. BACE1 molecular docking and anti-Alzheimer's disease activities of ginsenosides. J Ethnopharmacol. 2016;190:219–30.
91. Coimbra JRM, Baptista SJ, Dinis TCP, Silva MMC, Moreira PI, Santos AE, et al. Combining virtual screening protocol and in vitro evaluation towards the discovery of BACE1 inhibitors. Biomol Ther. 2020;10(4):535.
92. Han J, Ji Y, Youn K, Lim G, Lee J, Kim DH, et al. Baicalein as a potential inhibitor against BACE1 and AChE: mechanistic comprehension through in vitro and computational approaches. Nutrients. 2019;11(11):2694.
93. Palakurti R, Sriram D, Yogeeswari P, Vadrevu R. Multiple e-pharmacophore modeling combined with high-throughput virtual screening and docking to identify potential inhibitors of β-Secretase(BACE1). Mol Inform. 2013;32(4):385–98.
94. Gupta S, Parihar D, Shah M, Yadav S, Managori H, Bhowmick S, et al. Computational screening of promising beta-secretase 1 inhibitors through multi-step molecular docking and molecular dynamics simulations - Pharmacoinformatics approach. J Mol Struct. 2020;1205:127660.
95. Huang HJ, Lee CC, Chen CYC. *In Silico* design of BACE1 inhibitor for Alzheimer's disease by traditional Chinese medicine. Biomed Res Int. 2014;2014:1–14.
96. Amini R, Moradi S, Najafi R, Mazdeh M, Taherkhani A. BACE1 inhibition utilizing organic compounds holds promise as a potential treatment for Alzheimer's and Parkinson's diseases. Oxidative Med Cell Longev. 2024;2024:1–13.
97. Mietelska-Porowska A, Wasik U, Goras M, Filipek A, Niewiadomska G. Tau protein modifications and interactions: their role in function and dysfunction. Int J Mol Sci. 2014;15(3):4671–713.
98. Fan X, Xia L, Zhou Z, Qiu Y, Zhao C, Yin X, et al. Tau acts in concert with kinase/phosphatase underlying synaptic dysfunction. Front Aging Neurosci. 2022;14:908881.
99. Jitendra Joshi N, Raja Sekhar Reddy A. Navigating the GSK-3β inhibitors as versatile multitarget drug ligands in Alzheimer's disease intervention – a comprehensive review. Results Chem. 2024;7:101500.

100. Zhu J, Wu Y, Xu L, Jin J. Theoretical studies on the selectivity mechanisms of glycogen synthase kinase 3β (GSK3β) with pyrazine ATP-competitive inhibitors by 3DQSAR, molecular docking, molecular dynamics simulation and free energy calculations. Curr Comput Aided Drug Des. 2020;16(1):17–30.
101. Quesada-Romero L, Caballero J. Docking and quantitative structure–activity relationship of oxadiazole derivates as inhibitors of GSK3. Mol Divers. 2014;18(1):149–59.
102. Kim KH, Gaisina I, Gallier F, Holzle D, Blond SY, Mesecar A, et al. Use of molecular modeling, docking, and 3D-QSAR studies for the determination of the binding mode of benzofuran-3-yl-(indol-3-yl)maleimides as GSK-3β inhibitors. J Mol Model. 2009;15(12):1463–79.
103. El Kerdawy AM, Osman AA, Zaater MA. Receptor-based pharmacophore modeling, virtual screening, and molecular docking studies for the discovery of novel GSK-3β inhibitors. J Mol Model. 2019;25(6):171.
104. Balasubramaniam M, Mainali N, Bowroju SK, Atluri P, Penthala NR, Ayyadevera S, et al. Structural modeling of GSK3β implicates the inactive (DFG-out) conformation as the target bound by TDZD analogs. Sci Rep. 2020;10(1):18326.
105. Gurram PC, Satarker S, Nassar A, Mudgal J, Nampoothiri M. Virtual structure-based docking and molecular dynamics of FDA-approved drugs for the identification of potential IKKB inhibitors possessing dopaminergic activity in Alzheimer's disease. Chem Pap. 2023;77(4):1971–88.
106. Manandhar S, Hari G, Priya K, Kumar BH, Pai KSR. Virtual structure-based docking, WaterMap, and molecular dynamics guided identification of the potential natural compounds as inhibitors of protein-tyrosine phosphatase 1B. J Mol Struct. 2021;1226:129396.
107. Wang JL, Li L, Hu MB, Wu B, Fan WX, Peng W, et al. In silico drug design of inhibitor of nuclear factor kappa B kinase subunit beta inhibitors from 2-acylamino-3-aminothienopyridines based on quantitative structure–activity relationships and molecular docking. Comput Biol Chem. 2019;78:297–305.
108. Li L, Chen M, Jiang FC. Design, synthesis, and evaluation of 2-piperidone derivatives for the inhibition of β-amyloid aggregation and inflammation mediated neurotoxicity. Bioorg Med Chem. 2016;24(8):1853–65.
109. Bodke S, Joshi N, Alavala RR, Suares D. In silico exploration of CB2 receptor agonist in the management of neuroinflammatory conditions by pharmacophore modeling. Comput Biol Chem. 2024;110:108049.
110. Hopkins AL. Network pharmacology: the next paradigm in drug discovery. Nat Chem Biol. 2008;4(11):682–90.
111. Zhang P, Zhang D, Zhou W, Wang L, Wang B, Zhang T, et al. Network pharmacology: towards the artificial intelligence-based precision traditional Chinese medicine. Brief Bioinform. 2023;25(1):bbad518.
112. Li L, Yang L, Yang L, He C, He Y, Chen L, et al. Network pharmacology: a bright guiding light on the way to explore the personalized precise medication of traditional Chinese medicine. Chin Med. 2023;18(1):146.
113. Zhang R, Zhu X, Bai H, Ning K. Network pharmacology databases for traditional Chinese medicine: review and assessment. Front Pharmacol. 2019;10:123.
114. Zeng P, Fang M, Zhao H, Guo J. A network pharmacology approach to uncover the key ingredients in Ginkgo Folium and their anti-Alzheimer's disease mechanisms. Aging. 2021;13(14):18993–9012.
115. Qi P, Li J, Gao S, Yuan Y, Sun Y, Liu N, et al. Network pharmacology-based and experimental identification of the effects of Quercetin on Alzheimer's disease. Front Aging Neurosci. 2020;12:589588.
116. Wen YX, Li WH, Qing CS, Jing XL, Fang XX, Ri LX. Network pharmacology-based strategy to investigate the pharmacologic mechanisms of Coptidis Rhizoma for the treatment of Alzheimer's disease. Front Aging Neurosci. 2022;14:890046.

117. Wu X, Zheng X, Tang H, Zhao L, He C, Zou Y, et al. A network pharmacology approach to identify the mechanisms and molecular targets of curcumin against Alzheimer disease. Medicine. 2022;101(34):e30194.
118. Ma X, Zhao Y, Yang T, Gong N, Chen X, Liu G, et al. Integration of network pharmacology and molecular docking to explore the molecular mechanism of Cordycepin in the treatment of Alzheimer's disease. Front Aging Neurosci. 2022;14:1058780.
119. Patil N, Dhariwal R, Mohammed A, Wei LS, Jain M. Network pharmacology-based approach to elucidate the pharmacologic mechanisms of natural compounds from Dictyostelium discoideum for Alzheimer's disease treatment. Heliyon. 2024;10(8):e28852.
120. Shi Y, Chen Z, Huang L, Gong Y, Shi L. A network pharmacology approach to reveal the key ingredients in Scrophulariae Radix (SR) and their effects against Alzheimer's disease. Heliyon. 2024;10(3):e24785.
121. Cheng F, Wang F, Tang J, Zhou Y, Fu Z, Zhang P, et al. Artificial intelligence and open science in discovery of disease-modifying medicines for Alzheimer's disease. Cell Rep Med. 2024;5(2):101379.
122. Geng C, Wang Z, Tang Y. Machine learning in Alzheimer's disease drug discovery and target identification. Ageing Res Rev. 2024;93:102172.
123. Khachaturian AS, Dengel A, Dočkal V, Hroboň P, Tolar M. Accelerating innovations for enhanced brain health. Can artificial intelligence advance new pathways for drug discovery for Alzheimer's and other neurodegenerative disorders? J Prev Alzheimers Dis. 2023;10:1.
124. Yan C, Grabowska ME, Dickson AL, Li B, Wen Z, Roden DM, et al. Leveraging generative AI to prioritize drug repurposing candidates for Alzheimer's disease with real-world clinical validation. NPJ Digit Med. 2024;7(1):46.
125. Tripathi MK, Bhardwaj B, Waiker DK, Tripathi A, Shrivastava SK. Discovery of novel dual acetylcholinesterase and butyrylcholinesterase inhibitors using machine learning and structure-based drug design. J Mol Struct. 2023;1286:135517.
126. Kumar V, Banerjee A, Roy K. Machine learning-based q-RASAR approach for the in silico identification of novel multi-target inhibitors against Alzheimer's disease. Chemom Intell Lab Syst. 2024;245:105049.
127. Mukerjee N, Al-Khafaji K, Maitra S, Suhail Wadi J, Sachdeva P, Ghosh A, et al. Recognizing novel drugs against Keap1 in Alzheimer's disease using machine learning grounded computational studies. Front Mol Neurosci. 2022;15:1036552.
128. Galati S, Di Stefano M, Bertini S, Granchi C, Giordano A, Gado F, et al. Identification of new GSK3β inhibitors through a consensus machine learning-based virtual screening. Int J Mol Sci. 2023;24(24):17233.
129. Das B, Mathew AT, Baidya ATK, Devi B, Salmon RR, Kumar R. Artificial intelligence assisted identification of potential tau aggregation inhibitors: ligand- and structure-based virtual screening, in silico ADME, and molecular dynamics study. Mol Divers. 2023;28:2013.

Chapter 21
Advances in the Design of Novel Antidiabetic Agents Using In-silico Approaches

Sanket Jadhav, Pranali Yeram, Rajasekhar Reddy Alavala, and Amisha Vora

Contents

S. Jadhav · P. Yeram · R. R. Alavala · A. Vora (✉)
Shobhaben Pratapbhai Patel School of Pharmacy & Technology Management, SVKM's Narsee Monjee Insitute of Management Studies (NMIMS), Deemed to be University, Mumbai, Maharashtra, India
e-mail: amisha.vora@nmims.edu

S. N. K. Rao G., R. R. Alavala (eds.), *Applications of Computational Tools in Drug Design and Development*, https://doi.org/10.1007/978-981-96-4154-3_21

Abstract The global prevalence of diabetes, particularly Type 2 diabetes mellitus (T2DM), continues to rise, posing significant public health and economic burdens. Traditional treatments focus on established targets like PPAR, DPP-4, and SGLT2; however, limitations such as side effects and incomplete therapeutic efficacy have led to the investigation of emerging targets, including the MAPK and PI3K/Akt pathways and alpha-glucosidase inhibitors. In-silico drug discovery tools like AutoDock, MOE-Dock, and Biovia Discovery Studio have enabled precise molecular docking studies, while GROMACS facilitates molecular dynamics simulations to predict drug-receptor interactions. SwissADME, ProTox-II, and ADMETlab 3.0 ensure comprehensive ADMET profiling, while AI platforms such as NVIDIA's BioNeMo and DeepChem offer advanced capabilities in virtual screening and multi-target drug design. This chapter explores these advancements, providing an overview of traditional and novel targets, the role of natural products, and the application of computational tools in designing innovative antidiabetic agents. It aims to bridge the gap between conventional pharmacology and modern in silico methodologies, helping to understand the way for personalized and effective diabetes therapies.

Keywords Type 2 Diabetes Mellitus · Insulin resistance · PPAR agonists · DPP-4 inhibitors · SGLT2 inhibitors · Alpha-glucosidase inhibitors · Molecular docking · Molecular dynamics simulations · Virtual screening · Multi-target drug design

21.1 Introduction to Antidiabetic Agents

The existing treatment of diabetes focuses on controlling blood sugar levels and managing insulin resistance through lifestyle modifications and medications. Existing treatment of T2DM such as Biguanides (metformin), Sulfonylureas (e.g., Glimepiride, Glyburide), Thiazolidinediones (e.g., Pioglitazone), Meglitinides (e.g., Repaglinide), α-Glucosidase Inhibitors (e.g., Acarbose), DPP-4 inhibitors (e.g., Sitagliptin, Vildagliptin), SGLT2 inhibitors (e.g., Empagliflozin, Dapagliflozin) etc. are used for the management of T2DM. Also in some cases, injectable therapies such as Insulin, GLP-1 Receptor Agonists (e.g., Exenatide, Liraglutide), and Amylin Analogues (e.g., Pramlintide) are used [1]. The aim of the treatment is to maintain blood glucose level within target ranges and to reduce complications. Advance therapy and the innovations continue to focus on improving efficacy, minimizing side effects, and addressing the root cause of the disease [2].

Table 21.1 Targets and the drugs for the specific targets in diabetes treatment

Targets of diabetes	Drugs acting on targets	Mechanism	References
Peroxisome proliferator-activated receptor (PPAR) agonist	Thiazolidinediones (TZDs) Sulphonyl urea Sulphonamide derivatives	Regulation of metabolism (maintaining lipid and glucose homeostasis), Role in insulin sensitivity, mediating hormonal response	[7]
Protein tyrosine phosphatase 1B (PTP1B) inhibitors	Thiazolidinedione scaffolds, Sulfonamide, Phenol derivatives	Negative regulator of insulin signalling, enhance insulin sensitivity	[8]
Aldose reductase (ALR) inhibitors	Sulphonyl urea, Imidazolidine Dione derivatives	Managing complications arising from diabetes, such as cataracts and glaucoma	[9]
α-glucosidase inhibitors	Flavonoid Quinazoline-, quinolone, furan-based analogues, Imidazole based hybrids, Triazole, Pyrazole based derivatives	Management of blood glucose, reducing insulin requirements	[10]
DPP-4 inhibitors	Pyrrolidine and pyrrole, triazole, imidazole, indole-based derivatives	Regulating insulin secretion and glucose homeostasis	[11]
G protein—coupled receptor (GPCR) agonists	Pyrimidine, amide, bicyclic amine benzofuran, thiazole, oxazole derivatives	Tissue specific expression, regulation of insulin secretion and glucose metabolism	[12–14]
Sodium glucose co-transporter (SGLT) inhibitors	O-spiroketal C-aryl glucosides analogues, N- Indolyglucoside derivatives	Reabsorption of renal glucose	[15]

21.1.1 *Overview of Diabetes and its Types*

Diabetes is a chronic condition that affects the body's ability to process blood sugar, leading to glucose buildup in the blood if not managed carefully. This can increase the risk of severe complications, such as stroke and heart disease [3]. There are several types of diabetes, including T1DM, T2DM, and gestational diabetes, which have different treatments. T1DM diabetes occurs when the body does not produce insulin, a hormone responsible for breaking down blood sugar [4]. People with T1DM need to administer insulin to manage the condition regularly. T2DM occurs when the body does not make or use insulin effectively and is strongly linked with obesity [5]. Gestational diabetes occurs during pregnancy and can increase the risk of blood pressure and other complications. Proper management and lifestyle changes can help prevent or manage diabetes and its complications [6] (Table 21.1).

Table 21.2 Limitations of the existing treatment of diabetes

Treatment	Limitations	References
Metformin	Gastrointestinal side effects, not recommended for severe kidney or liver disease.	[16]
Sulfonylureas	Weight gain, hypoglycemia, and may increase the risk of cardiovascular events.	[17]
Meglitinides	Hypoglycemia, not recommended for severe kidney disease.	[18]
Thiazolidinediones (TZDs)	Weight gain, fluid retention, and may increase the risk of heart failure.	[19]
DPP-4 Inhibitors	Pancreatitis may increase the risk of infections and urinary tract infections.	[20]
GLP-1 Receptor Agonists	Nausea, vomiting, and diarrhea, not recommended for severe kidney disease.	[21]
SGLT2 Inhibitors	Genital infections may increase the risk of ketoacidosis, not recommended for severe kidney disease.	[22]
Insulin Therapy	Requires multiple daily injections, risk of hypoglycemia, can cause weight gain.	[23]
Symlin (pramlintide)	Must be injected before meals, may cause nausea, can be expensive.	[24]

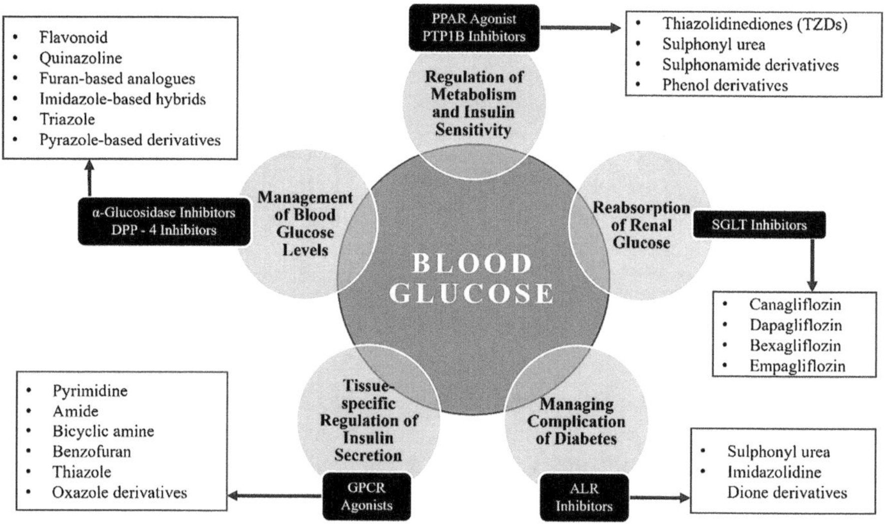

Fig. 21.1 Mechanisms of action that affect blood glucose level and target proteins responsible for the effect with the drugs used to inhibit those targets

21.1.2 Current Challenges in Antidiabetic Drug Development (Table 21.2)

Consistent hyperglycemia contributes to oxidative stress which causes diabetic complications that may be tissue-specific, mainly in microvascular diseases (Fig. 21.1). Microvascular changes with advanced glycation end products (AGE) and inflammation cause microvascular damage [25]. AGE pathways are heterogeneous hence they are responsible for the formation of complex biochemical reactions which involve the activation of the glyceraldehyde-3-phosphate, pentose phosphate pathway, and the reactive carbonyl methylglyoxal. These pathways involve the pathology of diabetic complications [3]. Hexosamine biosynthetic pathway, mitogen-activated protein kinases (MAPKs), sorbitol–aldose reductase pathway, and protein kinase C are also the pathways for diabetic complications. Progression of diabetes and diabetic complications also involves key targets like growth factors and cytokines which include vascular endothelial growth factor (VEGF), nuclear factor kappa B (NF-kB), insulin-like growth factor (IGF), transforming growth factor-β (TGF-β), tumor necrosis factor-α (TNF-α), Reactive oxygen species (ROS), MAPKs and platelet-derived growth factors [25].

21.1.3 The Role of Computational Approaches in Modern Drug Discovery

In the dynamic landscape of drug discovery and the integration of computational approaches has revolutionized the pharmaceutical industry, particularly Computer-Aided Drug Design (CADD), have emerged as pivotal tools that bridge the gap between biological sciences and cutting-edge technology [26]. CADD plays a transformative role by streamlining the process of drug discovery utilizing algorithms to identify potential lead compounds from extensive libraries, allowing for the rational design of drugs allowing multi targeting effect [27]. This integration is evident in its two primary strategies: structure-based drug design (SBDD) and ligand-based drug design (LBDD). SBDD leverages the 3D structure of biological targets, usually obtained through advanced techniques like X-ray crystallography or NMR spectroscopy, to design new drugs with greater precision. By using detailed structural information, researchers can identify exact binding sites on target proteins and optimize the interactions between the drug molecule and these sites. This enables the design of molecules that fit better, enhancing efficacy and reducing side effects [28]. Whereas LBDD depends on knowledge of existing ligands and their interaction to design new compounds, often using statistical models to predict and analyse compound's activity based on chemical structure [29]. These approaches thereby reducing both the time and cost in comparison with traditional experimental methods.

The advancements in artificial intelligence (AI) and machine learning (ML) have amplified the predictive capabilities of CADD, enhancing its accuracy in identifying potential drug candidates, predicting ADMET (absorption, distribution, metabolism, excretion, and toxicity) properties, and guiding drug repurposing efforts [30]. By integrating AI and ML capabilities researcher can facilitate the development of predictive models that can analyse biomedical datasets, advance data processing methods to ensure consistency, integrate big data analytics for comprehensive analysis of biological and chemical properties drug interaction [31].

Despite these advantages, challenges remain, such as the need for more efficient algorithms and ethical considerations around data privacy and model biases [32]. Collaborative efforts in the scientific community, including open-source platforms, are driving the democratization of drug discovery [33]. As emerging technologies like quantum computing and green chemistry converge with computational methodologies, has significantly transform chemical practices considering environmental sustainability [34]. The future of drug discovery holds immense potential for more personalized and sustainable therapeutic solutions.

21.2 The Macromolecules & Genes Associated with T2DM Progression

21.2.1 Key Enzymes and Receptors: PPAR, DPP-4, and SGLT2

Alpha glucosidase

Alpha glucosidase is a membrane-bound enzyme expressed in the brush border of small intestine, involved in digestion of carbohydrates. It breaks down disaccharides like maltose and sucrose to monosaccharides resulting in production of glucose. It specifically hydrolyses terminal 1–4 linked α-glucose residues. Inhibitors of α-glucosidase enzymes can reduce postprandial levels of glucose and can be beneficial in controlling the blood glucose levels. Various α-glucosidase inhibitors are on the market including acarbose, miglitol and voglibose. Acarbose was the first agent of this class introduced in 1996. It is an oligosaccharide obtained from Actinomyces utahensis and has high affinity binding towards sucrase. When given alone, there is no risk of hypoglycemia, and it does not cause weight gain like sulfonylureas. Miglitol was discovered while searching for hypoglycaemic compounds from microbial sources. These attempts initially resulted in the discovery of nojirimycin and deoxy nojirimycin from Bacillus and Streptomyces strains. Synthetic derivatisation of these compounds led to the discovery of miglitol. Voglibose, invented by Taked Pharmaceuticals, was found to be more potent and better tolerated than acarbose and miglitol.

Peroxisome Proliferator-Activated Receptor (PPAR-γ)

PPAR is one of the most explored targets for oral hypoglycemic drugs as these proteins play a key role in central regulation of carbohydrate and lipid metabolism as well as inflammation. They belong to the nuclear hormone receptor superfamily of ligand activated transcription factors and have subtypes PPAR α, γ and δ. Out of these PPAR-γ is primarily expressed in adipose tissue and PPAR-α in liver, kidney and heart [35]. Thiazolidinediones, also known as glitazones are PPAR-γ agonists. They were discovered in the 1980s when scientists were exploring clofibric acid derivatives for antihyperlipidemic and hypoglycemic activities. These drugs bind to

PPAR, bring about a conformational change, stabilizing interaction with retinoid receptor resulting in stimulation of transcription of the target genes. TZDs enhance adipocyte differentiation and increase the sensitivity of cells to insulin, decrease systemic fatty acid production. PPAR-γ activation enhances uptake of glucose by skeletal muscles and decreases gluconeogenesis [36]. The three-dimensional structure of PPAR consists of the DNA binding domain at N-terminus and the ligand binding domain (LBD) at C-terminus. The ligand binding domain of PPAR-γ consists of a Y shaped pocket to which the ligand binds. Selective PPAR-γ modulators (SPPARMs) were further developed with an aim to retain the beneficial effect on glucose metabolism while reducing the side effects like weight gain, hepatotoxicity and cardiotoxicity [37]. Current research is directed towards synthesis of dual PPAR α and γ agonists for the treatment of diabetic dyslipidemia. Simultaneous activation of both subtypes is believed to be synergistic in action on insulin sensitivity, lipid metabolism and inflammation control [38]. Saroglitazar is the first and only dual agonist approved and has demonstrated improved glycemic control and lipid levels in patients with diabetic dyslipidemia with fewer side effects than the drugs targeting only PPAR- γ. Ragaglitazar and Mugaglitazar were the other two PPAR α/δ agonists introduced in market, however, later was withdrawn for its cardiotoxicity [39].

Sulfonylurea 1 receptors (SUR1)

Activation of SUR1 on beta cells of pancreas, causing closure of ATP sensitive K-channels, cell-depolarization, opening of voltage gated Ca channels resulting in Ca influx and stimulating the insulin release [40]. The structure of SUR1 on ATP sensitive K+ channels was revealed by different techniques X-ray crystallography and Cryo-EM. SUR1 is a part of ATP-sensitive K channels. It is a unique member of the ATP-binding cassette (ABC) protein. It includes transmembrane domains (TMDs) forming channel through membrane allowing ion conductance, nucleotide binding domain (NBD) which binds ATP and hydrolyses the same to maintain activity of the channel and regulatory domains involved in interaction with other cellular components [41]. The first-generation sulfonylureas possess a small lipophilic group (R1) at para position of the phenyl ring and an alkyl or cyclic substituent lipophilic substituent R2 at the non-sulfonyl attached n of the urea. These substituents had weak binding efficiency to ATP sensitive K-channels and hence were less potent and possessed longer plasma half-lives. Larger doses were required to bring about the desired effects which resulted in significant side effects [41].

21.2.2 Insulin Signalling Pathway and its Modulation

Insulin signaling pathway

Insulin hormone is secreted by the pancreas responsible for regulating glucose metabolism in the body. It allows glucose to enter the cell and be utilized to produce energy. The mechanism of insulin begins after the interaction with insulin receptor

(IR) present on the surface of the cell. This receptor is made up of two extracellular alpha subunits and two transmembrane beta subunits which are linked by the disulphide bond [42].

GLP 1 receptor

GLP-1 is an incretin hormone secreted by intestinal cells in response to food intake. GLP 1 receptor agonists mimic the effects of endogenous GLP-1 by binding to GLP-1 receptors on various cells. The agonist of GLP-1 receptor reduces the glucagon secretion from pancreatic alpha cells which helps to decrease hepatic glucose production. GLP-1 RAs function by mimicking the endogenous GLP-1 hormone, promoting insulin secretion, reducing glucagon levels, and improving glycemic control. They also suppress appetite, slow gastric emptying, and lower blood pressure, which contributes to weight loss. The article outlines two main classes: short-acting (e.g., exenatide, lixisenatide) and long-acting GLP-1 RAs (e.g., liraglutide, semaglutide). Short-acting agents are better for controlling postprandial glucose spikes, while long-acting one's help regulate fasting glucose levels [1]. Extracellular domain located at N terminus, bond peptide agonist. This receptor spans the cell membrane with seven transmembrane helices. These helices transmit the activation signal from the extracellular ligand to intracellular signaling components. In intracellular domain, the C terminal part interacts with G protein initiating intracellular signaling cascades that result in insulin secretion from pancreatic β-cells. Ligand binding triggers structural changes that allow this interaction to occur.

PI3K/Akt pathway

This pathway is involved in glucose uptake by facilitating the movement of glucose transporters to the cell membrane. Also, it involves glycogen and protein synthesis. Growth factors such as insulin bind to the IR and triggers autophosphorylation of the receptor on specific tyrosine residues. Activated receptor sends PI3K to the membrane site. PI3K phosphorylates the 3′-OH group of phosphatidylinositol-4,5-bisphosphate (PIP2), generating phosphatidylinositol 3,4,5-trisphosphate (PIP3). This PIP3 recruits' protein kinase B (AKT) to promote glucose uptake, glycogen synthesis and protein synthesis and phosphoinositide-dependent protein kinase 1 (PDK1) which activate AKT by phosphorylating it. In T2DM, insulin resistance affects PI3/AKT pathway by dysregulating AKT phosphorylation [43]. Metformin is an antidiabetic drug indirectly activating AMP-activated protein kinase (AMPK), which affects PI3K/AKT [44].

MAPK/ERK pathway

This pathway contributes to the functioning of the pancreatic beta cells. The research emphasizes the role of the Rac1-NADPH oxidase (Nox2) signaling axis, which triggers reactive oxygen species (ROS) production. While moderate ROS levels are essential for cellular signaling, excessive ROS leads to β-cell dysfunction. The stress-activated MAPK pathways (ERK1/2, JNK, and p38MAPK) mediate cellular responses to oxidative stress, ultimately contributing to β-cell damage and insulin secretion defects seen in diabetes [45]. Biguanides are chemically described as the linkage of two guanidines with different side chains. It is known to act by reducing hepatic glucose production and enhancing insulin sensitivity and peripheral glucose uptake. At the molecular level, it is known to increase hepatic adenosine monophosphate activated protein kinase activity, thus reducing hepatic gluconeogenesis and insulin mediated uptake of glucose in muscle cells. Besides metformin, buformin and phenformin were the two biguanides introduced in market at a later stage, however they were withdrawn due to their serious side effect of lactic acidosis and increased incidences of cardiac mortality [46].

21.3 In Silico Screening of Antidiabetic Agents

In-silico screening is an integral approach in computational biology, allowing researchers to conduct experiments, simulations, and predictive analyses on virtual platforms like computer software or web-based servers [47]. In context of antidiabetic research, in-silico methods facilitate the identification of potential drug candidates by enabling genomic and proteomic analyses [48], biological system simulations [49], pathway modelling [50], and toxicity prediction [51]. These techniques are increasingly valuable in antidiabetic research, especially for exploring therapeutic compounds derived from plants.

Recent advancements in computational approaches, particularly in silico techniques, have revolutionized drug discovery by enabling efficient and cost-effective identification of promising candidates with optimal pharmacological profiles [52]. Leveraging tools like virtual screening [53], molecular docking [54], pharmacophore modeling [55], and network pharmacology [56], researchers can now systematically explore large compound libraries to target key enzymes, receptors, and pathways involved in diabetes progression. In-silico methods have proven valuable in exploring targets such as DPP-4, SGLT2, PPARs, and GLP-1, paving the way for precision medicine approaches that promise improved efficacy and reduced side effects in diabetes management.

21.3.1 Data Collection and Virtual Screening

A key component in computational method, data collection allows exploration of vast database of chemical compounds to discover those likely to bind well to a particular target molecule, such as a protein or enzyme that is associated with the progression of a disease [57]. Various resources (databases, ligand libraries, bibliographic research, drug repositories, and activity or structure data collections) used in conjunction with this approach increase the selection of compounds [58]. Researchers systematically compile data from these diverse sources to create custom compound libraries to meet the pharmacological needs of their project. The compiled dataset tailors and enables further analysis of the dataset using structure based and ligand-based methods to more efficiently progress along the drug discovery pipeline toward the more precise predictions of binding affinities and biological activities.

There are many free databases to use and explore and many. For example, Zinc and Zinc15 databases include over 100 million compounds with 3D coordinates and are directly usable for the docking studies [59]. There are also additional databases such as ChEMBL [60] and DrugBank [61] and PubChem [62] contains a wide database of chemical information for researchers. Similar databases are also available for phytochemical screening from medicinal plants to identify their possible use. Dr. Duke's Phytochemical and Ethnobotanical Databases [63], Indian Medicinal Plants Phytochemistry and Therapeutics (IMPPAT 2.0) [64], and Traditional Chinese Medicine (TCM) database [65] are rich in in depth information about plant chemicals The DiaNat-DB molecular database provides over 300 molecules with documented in vivo or in vitro antidiabetic activity for medicinal plant screening specifically to target plant based therapies [66]. By carefully selecting and integrating these resources, researchers can create a highly refined custom database, suitable to their own needs, which significantly improves the efficiency and precision of in silico research approaches. Such a database by fusing pieces of information from the disparate sources can facilitate data driven prediction, virtual screening and more sophisticated analysis in drug discovery and other research applications.

21.3.2 Structure-Based Drug Design (SBDD) and Molecular Docking

Structure based drug design and molecular docking are powerful computing techniques that are rapidly changing the face of modern drug discovery. It utilizes computational capabilities to design molecules with interaction onto the active site of biological targets [28]. But molecular docking instead simulates the way a drug candidate might interact with its target protein and can predict binding affinity—one critical determinant of the drug's effectiveness [29]. For instance, this docking algorithm can screen vast compound libraries, quickly detect promising candidates for further testing [67].

Results from successful application of SBDD and molecular docking are provided. This approach was used in a study for α-amylase and α-glucosidase targeted by the same strategy, resulting in identification of a dihydropyrimidinone derivative with 81.99% of target activity, which correlates well with the percent inhibition of the reference drug, ascorbic acid. Results of docking demonstrated similar binding energies for both the candidate compound and the reference drug (approximately −7.9 kcal/mol). These findings were experimentally validated, yielded a novel α-amylase, and α glucosidase inhibitor candidate for T2D therapy, and supported further by an in vitro inhibition assay [68].

Key to discovering compounds against diabetes related proteins is SBDD and molecular docking. This is because when researchers narrow down their focus to proteins responsible for glucose metabolism and they can work out molecules that promote insulin sensitivity or deter glucose production. In addition to speeding up drug discovery, these approaches enhance the precision of new treatments with drugs that are more selective for their targets and thus cause fewer side effects.

21.3.3 Application of Pharmacophore Modelling in Antidiabetic Agent Discovery

Pharmacophore modelling is an important tool in the discovery of antidiabetic agents by allowing researchers to identify what molecules must have in common to be bioactive to screen and optimize potential drug candidates [69]. These models can be used in the primary application, which is virtual screening, whereby compound libraries are scanned for possible antidiabetic agents using pharmacophore models. These models zero in on features of the molecule which are vital for activity and thereby increase the screening hit rate. Second, lead optimization pharmacophore modelling is used to identify such leads to also refine leads to improve efficacy while limiting side effects [70]. This also aids in gaining insights into structure activity relationship (SAR) which is very important in optimizing a compound [71]. Its application also helps in target identification by providing insights into biological determinants of new compounds, predictions of off target effects to optimize drug safety [69].

Pharmacophore modelling has developed the targeting of glucose metabolism and insulin sensitivity in antidiabetic drug development. One study for example examined the phytochemicals of Pinus roxburghii for their efficacy against key diabetic targets including protein tyrosine phosphatase 1-beta (PTP-1β), dipeptidyl peptidase-IV (DPP-IV), aldose reductase (AR) and the insulin receptor (IR). Binding affinities with these targets were first assayed by docking and results revealed that three phytochemicals, secoisorestinol, pinoresinol and cededarin, bound with significantly stronger affinity to AR than other targets. Then, a pharmacophore model of AR was generated using LigandScout that showed critical interactions with Tyr 48 and His 110 residues in AR active site. These findings further inform structural requirements for AR inhibition to develop effective AR inhibitors to treat diabetes-related complications [72].

Due to recent progress in pharmacophore modelling for instance, integrating structural based pharmacophore modelling with the ligand based one and development of robust computational chemistry, the accuracy and efficiency of drug design have been improved. Thus, this integration allows better predictions of drug interactions and its pharmacokinetics [73].

21.3.4 Fragment-Based Drug Design (FBDD) for Novel Antidiabetics

Fragment Based Drug Design (FBDD) is a modern approach for finding, and optimizing, specific small chemical fragments which can bind to target proteins with the aim of developing novel antidiabetic agents. The process starts with fragment screening, which involves methods such as X-ray crystallography, NMR spectroscopy or computational methods which can help identify compounds, low affinity binding fragments [74]. Then these first pass fragments are optimized either by growing or linking them to enhance affinity and making them into more drug like. This approach employs small molecular fragments (<300 Da) to systematically optimize drug candidates [75]. There are advantages in integrating FBDD as it reduces chemical space, increases the binding affinity of lead compounds, and the results are scalable relative to the high throughput screening approach, which leads to rapid drug development. Although commonly combined with other computational methods, such as AL driven de novo design techniques, FBDD facilitates the identification of novel inhibitors for diabetes related targets [76].

FBDD has proven useful in imparting selectivity to target key proteins in antidiabetic drug discovery. A study had employed FBDD to discover and assess possible antidiabetic properties of newly synthesized thiazolodine-2,4dione derivatives, SP4e and SP4f. In molecular simulations these both the compounds showed stable binding, with high binding score and favourable binding free energies. The compounds were later shown in later in-vivo studies to be very good at hypoglycaemic activity in mice model. Additionally, in vitro PPAR- γ activation studies were performed and their increased potency was confirmed, making these derivatives promising antidiabetic agents [77].

However, the development of comprehensive fragment libraries, and sensitive assays, are challenges which currently hinder the application of FBDD in the drug discovery process. The quality of fragments screened matters and is where the power of this comes from.

21.4 Mechanistic Studies on Enzyme Inhibitors in Diabetes Management

Mechanistic Studies on enzyme inhibitors in diabetic management reveal significant advancements in targeting specific enzymes to control the glucose level. There are number of inhibitors, derived from natural sources as well as synthetic compounds have shown promising mitigation of glucose levels and enhance insulin sensitivity [78].

Table 21.3 Studies conducted for DPP-4 and computational tools used

Study	Compound(s)/ Extract(s)	Technical Tools Used	Key Results	References
Design of New Quinoxaline Compounds as DPP-4 Inhibitors	Quinoxaline derivatives (e.g., compounds 10a, 10 g)	1. Molecular Docking, 2. Molecular Dynamics (MD) Simulations, 3. Quantitative Structure-Activity Relationship (QSAR) Analysis, 4. Lipinski's Rule of Five	Compounds 10a and 10 g showed selective DPP-4 inhibition, and alignment with experimental DPP-4 inhibition, indicating predictive reliability and good oral bioavailability for clinical potential.	[79]
Glycyrrhiza uralensis (Licorice) Bioactives as Natural DPP-4 Inhibitors	Licochalcone A (LicA), Licochalcone B (LicB)	1. Molecular Docking (AutoDock 4.2), 2. Protein-Protein Interaction (PPI) Analysis (PatchDock, FireDock), 3. Molecular Dynamics (MD) Simulations (GROMACS), 4. In Vitro DPP-4 Enzyme Assay	LicA and LicB bind to DPP-4, enhancing insulin regulation and GLP-1 and GIP hormone stability, suggesting potential as functional food ingredients for diabetes management.	[80]
Metformin's Interaction with DPP-4: Potential for Modulating DPP-4 Expression	Metformin	1. Molecular Docking (Maestro 12.3), 2. Molecular Dynamics (MD) Simulations (Schrödinger Desmond), 3. Bioinformatics (TIMER, TISIDB), 4. Cell Assays for DPP-4 Expression	Metformin's binding to DPP-4's active site enhances its action beyond glycaemic control, potentially influencing immune and cancer responses, suggesting a broader therapeutic scope.	[81]

21.4.1 In Silico Design of DPP-4 Inhibitors

The exploration of effective antidiabetic therapies via in silico approaches, including molecular docking, MD simulations, QSAR analysis, and Lipinski's Rule of Five drug-likeness assessment have significantly advanced the design of DPP-4 inhibitors (Table 21.3).

Molecular docking studies using AutoDock 4.2, Maestro 12.3 and other tools help predict binding orientation of DPP-4 inhibitor and score potential inhibitory compounds. Stability and dynamics of the ligand-enzyme interactions are analysed through MD simulations in GROMACS or Schrödinger Desmond under physiological conditions, to identify critical structural conformations that afford higher binding affinity and specificity. Application of the protein-protein interaction (PPI) analysis through PatchDock and FireDock further reveals the molecular interplay between DPP4, and other proteins associated with glucose metabolism, thereby suggesting multi-target approaches to effectively manage diabetes.

These gene expression patterns associated with DPP-4 are further corroborated by complementary bioinformatics analyses (TIMER, TISIDB) that aid in the development of targeted inhibitor design. In vitro DPP-4 enzyme assays and cell-based assays further validate in silico findings against actual biological response to confirm the therapeutic potential of selected inhibitors. This integrative strategy of combining computational and experimental approaches in a seamless manner improves the pipeline leading for DPP4 inhibitors to offer better glycaemic control thereby looking forward as a solid direction towards diabetes management.

Table 21.4 Research conducted for SGLT2 inhibitors, and the computational tools used

Study	Compound(s)/ Extract(s)	Technical Tools Used	Key Results	References
Glycomimetic Compounds as SGLT2 Inhibitors	Glycomimetic analogues (e.g., LAS52462756, LAS52463524)	1. 3D-QSAR Model, 2. Molecular Docking, 3. Molecular Dynamics (MD) Simulations (Desmond), 4. Binding Free Energy Calculations (MMGBSA), 5. ADMET Prediction	Top 12 glycomimetic compounds showed stable binding to SGLT2's active site, with compound LAS52462756 showing strong inhibitory potential. Which displayed good pharmacokinetic profiles support further development.	[82]
SGLT2 Inhibition Prediction Using Allium fistulosum Compounds	Compounds from Allium fistulosum (e.g., N-trans-feruloyltyramine, N-cis-feruloyltyramine)	1. Support Vector Regression (SVR) Model with SkelSpheres Descriptor, 2. IC50 Prediction, 3. Clustering and Similarity Analysis	Four tyramine derivatives and Allium fistulosum decursidate show potent SGLT2 inhibition, with high prediction accuracy achieved using the SVR model with SkelSpheres descriptor.	[83]

(continued)

Table 21.4 (continued)

Study	Compound(s)/ Extract(s)	Technical Tools Used	Key Results	References
Phytochemicals from Centella asiatica as SGLT2 Inhibitors	Phytochemicals from Centella asiatica (e.g., betulinic acid, centellasapogenol, methyl brahmate)	1. Molecular Docking, 2. Molecular Dynamics (MD) Simulations (100 ns), 3. Binding Free Energy Calculations (MM/PBSA), 4. ADMET Prediction	Eleven phytochemicals, including Betulinic acid, centellasapogenol, and methyl brahmate, exhibited strong binding affinities with SGLT2, indicating their potential as safe, effective alternatives for further validation.	[84]

21.4.2 SGLT2 Inhibitors: Molecular Mechanism and Modelling Approaches

Mechanistic studies of SGLT2 inhibitors for diabetes management employ a variety of computational tools to elucidate binding mechanisms, increase efficacy and predict the pharmacokinetic properties of these drugs (Table 21.4).

Techniques such as 3D quantitative structure activity relationship (3D—QSAR), molecular docking, and molecular dynamics (MD) simulation help us to gather structural and functional interaction between SGLT2 inhibitors and the binding site in SGLT2 protein. Precise binding orientation prediction is enabled by docking, while MD simulations (Desmond for longer (100 ns) simulation times) provide insight to the dynamic stability and conformational adaptability of inhibitor-SGLT2 complexes under physiological conditions.

Additional quantitation of potential inhibitor binding affinities can be performed using binding free energy calculations including MMGBSA and MM/PBSA which rank compounds based on their efficacy. Using SkelSpheres descriptors, support vector regression (SVR) models predict IC50 values, and clustering and similarity analysis are used for selection of structurally diverse and potent inhibitors. Further, ADMET prediction models are necessary to predict absorption, distribution, metabolism, excretion, and toxicity characteristics of candidates to determine drug likeness and safety. Combined, these modelling and prediction approaches can be used to guide the design of potent, selective SGLT2 inhibitors and form an efficient pathway towards the development of new antidiabetic therapies directed toward glucose reabsorption pathways.

Table 21.5 Studies conducted for Alpha Glucoside and Alpha Amylase Inhibitors and computational tools used

Study	Compound(s)/ Extract(s)	Technical Tools Used	Key Results	References
In Silico Approaches to Identify Polyphenol Compounds as Alpha-Glucosidase and Alpha-Amylase Inhibitors	Polyphenol compounds from medicinal plants	1. Molecular Docking (Autodock, MOE-Dock, Biovia Discovery Studio), 2. PyMOL Visualization	Polyphenols demonstrated high binding affinity with alpha-glucosidase and selective inhibition with less effect on alpha-amylase, suggesting potential as therapeutic agents with reduced side effects.	[85]
Design, Synthesis, and In Vitro and In Silico Anti-Alpha-Glucosidase Assays of Triazole-Indole Derivatives	N-Phenylacetamide-1,2,3-triazole-indole-2-carboxamide derivatives	1. Molecular Docking, 2. IC50 Calculation, 3. Click Chemistry Synthesis	Compound 5k exhibited the highest inhibitory activity against alpha-glucosidase with competitive inhibition, favourable binding energy, and structural fit in the enzyme's active site.	[86]
Design of New Alpha-Glucosidase Inhibitors Based on Bis-4-Hydroxycoumarin Skeleton	Bis-4-hydroxycoumarin-phenoxy-1,2,3-triazole-N-phenylacetamide derivatives	1. Molecular Docking (Autodock Tools), 2. Enzyme Kinetics, 3. SAR Study, 4. Pharmacokinetics Prediction	Compound 5i was identified as the most potent, with 125-fold improved inhibition over acarbose and favorable pharmacokinetic profile, supporting further clinical development.	[87]

21.4.3 Alpha-Glucosidase and Alpha-Amylase Inhibitors: Computational Studies

Mechanistic studies of alpha-glucosidase and alpha amylase inhibitors are to understand how these enzymes interact with inhibitory compounds to regulate diabetes by slowing digestion of carbohydrates and reducing post prandial blood glucose spikes (Table 21.5).

Molecular docking (Tools like: Autodock, MOE-Dock, and Biovia Discovery Studio) can provide important information on the binding modes and affinities of potential inhibitors by computing the interaction of an inhibitor with enzyme active sites. Key binding interactions and structural conformations can be assessed, and stability and orientation of inhibitors within the active sites can be determined using visualizations in PyMOL.

Evaluation of inhibitory potency is further supported by further computational analysis including IC50 calculations, and structure activity relationship (SAR) and enzyme kinetics studies deepen the molecular modification knowledge to improve activity. Additionally, pharmacokinetic predictions are essential in determining whether a drug is absorbed, distributed, metabolized, and excreted (ADME) appropriately, and whether the drug is drug-like or viable for therapy. Click chemistry type synthesis techniques further integrate in silico findings with practical applications allowing for rapid development of compounds with optimized interactions and favourable pharmacokinetics. Together, these computational studies offer a promising pathway to the design of effective alpha-glucosidase and alpha-amylase inhibitors for antidiabetic therapy aimed at glucose regulation.

21.5 Natural Products and Phytochemicals as Antidiabetic Agents

Natural products and phytochemicals have revealed themselves as major prospects in the drug discovery process due to their multiple bioactivities and structural varieties [78]. With the improvements in computational approaches and analytical methods, these bioactive compounds and their targeted receptors are being discovered and confirmed more often making them more useful in contemporary pharmacology [88]. Natural products also have multiple activities on pharmacological aspects such as regulating insulin secretion and increasing insulin sensitivity in relation to diabetes [89].

A study intends to examine the role that Diabecon and its phytochemicals could play in the handling of diabetes. On the management of diabetes, it employs network pharmacology and molecular docking approaches to focus on bioactive compounds and the mechanisms involved. These are gymnemic acids, flavonoids, triterpenoids, beta-eudesmol, and isoliquiritigenin, all of which exhibit various modulating effects on diabetes pathways and targets. Research has shown that the compounds can inhibit some important enzymes and managing glucose levels adequately. The study also predicted interactions among genes which are targeted by phytochemicals of Diabecon using network pharmacology analysis and elucidated complex biological pathways relevant to diabetes [90].

Table 21.6 The research conducted on different types of phytochemicals that are predicted with network pharmacology approach

Type of Phytochemical	Sources	Mechanism of Action	Phytochemicals	Results	References
Flavonoids	Potentilla discolor Bunge (PDB)	Activates peroxisome proliferator-activated receptor gamma (PPARG), regulates fatty acid storage and glucose metabolism, stimulates lipid uptake and adipogenesis	Tricetin, Quercetin, Kaempferol.	The study validates the drug's potential as a therapeutic for hyperglycemia by promoting glucose uptake and improving glucose tolerance, and by supporting glucose regulation pathways in diabetes through network pharmacology analysis.	[93]
Alkaloids	Abrus precatorius	Influences T2DM-related genes such as AKT1, MAPK3, TNFα, MAPK1, enhancing insulin sensitivity	Precatorine	Network pharmacology reveals Abrus precatorius compounds interact with key T2DM pathways, enhancing insulin signaling and reducing inflammation, suggesting its potential as a multi-component therapeutic for diabetes.	[94]
Triterpenoid and Saponins	Cyclocarya paliurus, Terminalia arjuna.	Cyclocarioside K, Arjunglucoside II, and α-Boswellic Acid are natural compounds that act on various pathways, including insulin resistance, lipid metabolism, and cardiovascular comorbidities.	Cyclocarioside K Arjunglucoside II α-Boswellic Acid	Network pharmacology identified key pathways for saponins, including insulin resistance, PI3K-Akt, and lipid metabolism, revealing their multi-target actions and potential synergistic effects on diabetes-related targets.	[95]
Phenolic Acids	Rosmarinus officinalis (Rosemary), Coffea (Coffee Plant), Honeysuckle, Potato, Cork, Eucommia Leaves, Chrysanthemum, Strawberry, Mango, Blueberries, Mulberry Leaves	Anti-inflammatory action: targets TNF-α convertase, reducing the production of the proinflammatory cytokine TNF-α	Chlorogenic Acid, Rosmarinic Acid, Ellagic Acid, Cinnamic Acid, Ferulic Acid, Vanillic Acid, Protocatechuic Acid, Gentisic Acid, Gallic Acid, Syringic Acid, p-Coumaric Acid, Sinapic Acid	Network pharmacology, along with docking studies, revealed strong binding affinities of phenolic acids to TNF-α convertase. Chlorogenic acid exhibited the highest binding affinity (−6.90 kcal/mol), with key interactions, including hydrogen bonds and hydrophobic interactions. Other acids like rosmarinic (−6.82 kcal/mol) and ellagic (−5.46 kcal/mol) acid also showed effective binding. These interactions suggest phenolic acids' potential in modulating inflammatory pathways linked to diabetes	[96]

21.5.1 *Network Pharmacology Approaches to Investigate Herbal Drugs*

The concept of the network pharmacology for the analysis of complex molecular interactions of herbal drugs has become useful in computational research [91]. This strategy combines aspects of computational biology and systems biology to look for compounds possibly to have bioactivity and their targets, as well as the signaling pathways for managing diabetes. The general workflow usually consists of bioactivity screening, target prediction, pathway analysis, molecular docking and interaction validation experiments. Last of all, there is the treaty that embraces the computational hypothesis and experimental evidence. While network pharmacology offers a generic framework to capture the complex interactions of herbal drugs, it is crucial to consider that the predictions made by these models need to be validated to improve the accuracy and applicability of the models [92] (Table 21.6).

21.5.2 *Virtual Screening and ADMET Profiling of Phytochemicals for Antidiabetic Activity*

Virtual screening and ADMET (Absorption, Distribution, Metabolism, Excretion, and Toxicity) Profiling are crucial for exploring phytochemicals as potential antidiabetic agents. As discussed in above section virtual screening involves extracting the drug like compounds from large digital libraries [97]. While ADMET profiling is crucial for predicting pharmacokinetic behaviour and toxicity of the screened phytochemicals to ensure that compounds have favourable properties for drug development. These screening techniques has gained significant traction in recent years as these tools has identified promising compounds from natural sources while assessing their pharmacokinetic properties and safety profiles [98].

Tools for ADMET prediction have been instrumental in identifying potential compounds from extensive phytochemical screening. There are various commonly used tools in antidiabetic research. For example, ADMETlab 3.0 [99], Pharmaco Kinetic Knowledge Base (PKKB) [100], admetSAR 2.0 [101], Pro Tox-II [102], and SwissADME [103] which screened different parameters like BBB permeability, hERG Inhibition, Oral Bioavailability, Aqueous Solubility, Lipophilicity, LogP, pKa and similar kind of parameters are screened. Also, these tools perform drug-likeness properties screening based on pre-established guidelines like Lipinski's Rule of Five, and Veber's Rule. These parameters not only help in identifying viable compounds but also ensure that selected compounds not only exhibit therapeutic efficacy but also possess acceptable safety profiles [104].

21.5.3 Molecular Docking Studies for Plant-Derived Compounds

Molecular Docking helps in understanding interactions between phytochemicals and their relevant biological target proteins, which a crucial part of drug discovery process. This computational approach provides insight into binding affinities and interactions dynamics, which are essential for assessing the therapeutic potential of the drug like compounds [105].

For instance, a study explores the antidiabetic potential of compounds from Gongronema latifolium leaves, specifically targeting proteins involved in type 2 diabetes mellitus (T2DM). The researchers identified 15 T2DM-related proteins and, following molecular docking, selected glycogen synthase kinase 3 beta (GSK-3β), glucagon-like peptide-1 receptor (GLP-1R), and human aldose reductase as primary targets based on high binding affinities. In silico evaluations, including binding free energy, ADMET predictions, and QSAR modeling, highlighted 11 compounds—such as luteolin, kaempferol, and naringenin—that demonstrated stable protein binding, moderate toxicity, favorable drug-like properties, and gastrointestinal absorption. QSAR models showed that these compounds could act as potent inhibitors (high pIC50 values) for aldose reductase and GSK-3β and as agonists (high pEC50 values) for GLP-1R. The study concludes that this approach could help design T2DM-modulating compounds with minimized off-target effects, with potential for future optimization and lead development [106].

21.6 Network Pharmacology and Poly Pharmacology in Antidiabetic Drug Design

Network pharmacology and Poly pharmacology are innovative approach in antidiabetic drug design that aim to address complexity of diabetes by targeting multiple pathways and mechanisms [107]. Network Pharmacology helps in understanding the interactions between drugs and their multiple targets within biological networks, which ultimately helps in identifying key targets and pathways that are involved. Network Pharmacology framework is useful in antidiabetic research for several reasons like understanding multi-target approach, Drug-Target-Disease Network, and it is also used for predicting model for bioinformatic studies by integrating high-throughput screening data researchers can streamline the research process [108]. Whereas poly pharmacology involves designing drugs that can act on multiple targets simultaneously. This approach is particularly useful for understanding complex diseases like diabetes [109].

For antidiabetic in-silico research protein-protein interaction are crucial to understand to identify the potential therapeutic targets and cellular processes which are complemented by multi-target drug design using systems biology approaches, which aims to address complex diseases. For instance, a study analysed antidiabetic mechanisms of Hydrolea zeylanica (HZH) using network pharmacology followed by experimental validation in high-fat diet and streptozocin included T2D rat model. A network pharmacology analysis of 32 bioactive HZH compounds found that they interact with 194 target genes associated with type 2 diabetes mellitus (T2DM). HZH at 400 mg/kg restored glucose homeostasis, reduced inflammation, and normalized GLUT2 and GLUT4 expression, indicating its therapeutic potential [110].

21.6.1 Constructing Protein-Protein Interaction (PPI) Networks

The construction of Protein-Protein Interaction (PPI) Networks is crucial for network pharmacology, as this approach facilitates the identification of potential therapeutic targets and development of multi-target drugs that effectively manage complex diseases like diabetes [111]. This approach involves mapping interactions between proteins implicated in T2DM. This networking approach allows researcher to visualize and analyse the complex interplay between various molecular targets [112]. The key steps involved in constructing this network starts with Data collection from known protein databases such as STRING [113], NetworkAnalyst [114] and BioGRID [115]. Then with the help of tools like Cytoscape to visualize interactions. Finally, analysis of Network properties using network topology to identify key proteins (hubs) that may serve as potential drug targets [116].

For instance, a study used PPI network by using datasets like Online Mendelian Inheritance in Man (OMIM) where they searched for disease-specific proteins and STRING for known PPIs. Then topology analysis was conducted of PPI network which identified 21 proteins as significant to T2DM, with insulin (INS) as the central protein. The study also identified other critical proteins like AKT1, KCNJ11, PPARG, SOCS3 which had direct interaction with INS protein. This approach highlighted potential targets in T2DM pathogenesis and understand the core proteins in T2DM supporting network pharmacology and Poly pharmacology strategies for antidiabetic drug development [117].

21.6.2 Multi-target Drug Design Using Systems Biology Approaches

System biology provides a comprehensive framework for understanding the intricate networks of biological interactions that characterize complex diseases like diabetes. It emphasizes the need to consider not just individual proteins but also their interactions and roles within larger biological systems. Whereas multi-target drug design aims to understand the interactions of drugs to multiple proteins involved in disease pathways [118]. There is research in which they designed virtual small molecule libraries tailored for multi-target drug discovery against T2DM using system biology approaches. The novelty and diversity of the libraries were validated by comparison with clinically approved antidiabetic drug, natural products and multi-target compounds which are tested in in-vivo. The libraries are freely available for further future filtering, synthesis and biological testing, which includes 280 medicinal chemistry transformation rules in SMIRKS notation to support future chemical library generation and hit optimization [119]. These types of datasets aids in expanding options for multi-target antidiabetic drug design.

A study applied multi-target antidiabetic agents by designing and synthesizing succinimide–thiazolidinedione hybrids. Five new derivatives (10a–e) with unique substitutions on the succinimide ring's nitrogen were synthesized and tested in vitro for their inhibitory effects on key diabetes management enzymes. Compound 10d showed the strongest antidiabetic potential across all targets, leading to further in vivo testing and promising results. The study introduces a new class of compounds with potential multi-target antidiabetic effects, particularly for managing diabetes through multiple enzymatic pathways [120].

21.7 Application of Molecular Dynamics Simulations in Antidiabetic Drug Design

Molecular Dynamics (MD) simulations have become an essential tool in the field of computational drug design. MD simulations help in provides information on the behaviour of the phytochemical on the target over time allowing researcher to study the integration between drug molecule and target protein in dynamic environment [121]. Aspects like stability, binding mechanisms, conformational changes in enzyme inhibition and drug optimization of the drug like molecule can be studied to further conduct experimental validations. MD simulations can evaluate the stability of drug-receptor complexes overtime. Simulations results can help us to track in root mean square deviation (RMSD) and root mean square fluctuation (RMSF) to check the stability of the drug receptor interaction. To check the conformational changes that occur upon ligand binding can be analysed by implementing induced model fit. Enzymes often undergo significant conformational changes upon inhibitor binding or when allosteric inhibitors bind to active sites changes the conformation. This all can be analysed by applying the MD simulations to study and analyse the interactions in virtual system [122].

MD simulations play a pivotal role in lead optimization of drug like candidates by allowing modifications in in silico before actual synthesis. This approach also validates predictions made through other computational methods like molecular docking, MD helps in confirming whether predicted binding modes are realistic under physiological conditions [123].

Table 21.7 Research that used MD simulations to validate the drug target interactions

Compound and Target(s) Name	Molecular Docking Key Findings	Molecular Dynamics (MD) Simulations	Conclusion with In vitro/In vivo Assay Results	References
Benzylidene-thiazolidine-2,4-dione (5a–e) targeting PPAR-γ	Compounds 5c and 5d had high binding affinities (−10.1 and −10.0 kcal/mol) compared to native ligand nTZDpa (−9.8 kcal/mol)	150 ns simulations showed stable 5c-PPAR-γ complex with comparable RMSD, RMSF, rGyr, and SASA to nTZDpa-PPAR-γ complex	Compounds 5d and 5e exhibited significant antidiabetic activity in streptozotocin-induced mice, like rosiglitazone, with no weight gain	[125]
Benzenesulfonamide-thiazolidin-4-one hybrids (6a–6h) targeting PPAR-γ and α-amylase	Compound 6b showed strong α-amylase inhibition (IC_{50} = 29.51 ± 1.35 μg/ml), with high glucose uptake (88.71%)	MD simulations confirmed stable binding of 6b in α-amylase; DFT showed favorable electronic properties	Compound 6b showed promising α-amylase inhibition and enhanced glucose uptake, indicating potential for diabetes management	[126]
Top 5 Ligands from Virtual Screening targeting PPAR-γ for insulin sensitization	Selected ligands showed strong binding with key residues (MET364, ILE341), enhancing stability through hydrogen bonding	MD simulations confirmed stability; QSAR model achieved 98% prediction accuracy of binding energy	Identified ligands are promising insulin sensitizers; QSAR and MD results support safety potential over thiazolidinediones	[127]
Dihydropyrimidinone Derivative (Compound A) targeting α-amylase and α-glucosidase	Compound A exhibited strong binding in α-glucosidase's active site (−7.9 kcal/mol), comparable to standard ascorbic acid (−7.8 kcal/mol)	MD simulations showed stable interaction; RMSD of compound A (1.7 Å) close to protein's RMSD (1.6 Å)	Compound A demonstrated 81.99% α-glucosidase inhibition (IC_{50} = 1.02 μg/ml), comparable to standard, indicating potential as T2D therapy	[68]
Rutin from Momordica charantia targeting α-amylase and α-glucosidase	Highest binding affinity for α-amylase (−11.68 kcal/mol) and α-glucosidase (−11.93 kcal/mol)	100 ns MD simulations showed stable binding; assessed using RMSD, RMSF, RG, PCA, and FEL parameters	Rutin demonstrated strong enzyme inhibition in silico, suggesting potential as a natural antidiabetic agent; further animal studies suggested	[128]

Advanced techniques like meta dynamics or umbrella sampling can employed with MD framework which helps in calculating free energy profiles associated with protein-ligand interaction and conformational changes [124]. This insights aid in understanding the quantitative data of thermodynamics of drug-receptor interactions and guides for further optimizations.

In diabetic setting there are various examples in which MD simulations have given good results for potential new antidiabetic treatment options (Table 21.7).

21.8 AI/ML Approaches for Antidiabetic Drug Discovery

In recent times Artificial Intelligence and Machine learning are widely integrated into drug discovery process and has emerged as an emerging transformative approach. AI and ML technologies have been used in drug discovery process for many purposes like,

1. Data Analysis and Pattern Recognition: In this technique, AI can analysis vast dataset from clinical trials and all kind of electronic health records to identify patterns easily and provide solutions. This approach has also shown capabilities of discovering new targets and biomarkers associated with diabetes [129].
2. Deep Learning Techniques: ML models have capabilities to predict drug efficacy and toxicity based the historical data that helps researcher to know their drug molecules and modify the research accordingly. For instance, a study has applied two deep learning techniques: a single channel convolutional neural network (CNN) and a multi-channel model integrating CNN, with recurrent neural network (RNN) and bidirectional long short-term memory (Bi-LSTM) layers to study antidiabetic peptides (ADPs) which were obtained from BioDADPep Database. The data was pre-processed with the evolutionary scale model (ESM-2) which were trained and validated using ten-fold cross-validation. The CNN model achieved an accuracy of 90.48% on this test set, which outperformed the existing tools for ADP prediction. This research showcases that deep learning frameworks can be used to predict accurate results [130].
3. Quantitative Structure-Activity Relationship (QSAR) and Molecular Modelling: QSAR models and advanced molecular modelling utilize machine learning to predict the biological activity of compounds based on their chemical structures, particularly for α-glucosidase inhibitors, which are crucial in managing postprandial hyperglycemia, which reduces the time and cost associated with traditional drug design methods [131].

There are various applications of AI/ML in antidiabetic drug discovery like target identification and drug repurposing. In some cases, AI systems have been used in patient specific personalized medicine to meet the treatment needs of the individuals' metabolic profiles.

Table 21.8 Research that used different AI/ML tools for various purpose at various stages of the drug discovery process

Study Purpose/ Focus	Name of Tool	Application in Antidiabetic Drug Discovery Research	Results	References
Investigate predictive modelling of glycated haemoglobin (HbA1c) levels for early T2DM intervention	Random Forest (RF)	RF enables prediction of HbA1c changes, allowing for early risk assessment in type 2 diabetes through identification of significant predictive factors.	RF model outperformed logistic regression, with key predictors including baseline HbA1c, fasting blood glucose, body weight, alkaline phosphatase, and platelet count.	[133]
Early risk identification of high-risk individuals for type 2 diabetes mellitus (T2DM)	Support Vector Machine (SVM)	SVM models used with optimized feature sets (via mRMR) predict T2DM risk, enabling proactive interventions based on glucose-centric data.	Achieved 96.8% accuracy and 80.09% sensitivity, showing glucose metrics as strong predictors of T2DM.	[134]
Accelerate antidiabetic drug discovery through predictive modelling	iPADD	iPADD leverages multiple algorithms to identify potential antidiabetic compounds, particularly focusing on molecular fingerprints and optimal feature selection.	High predictive accuracy (98.3%) and auROC (0.989); validated with molecular docking studies confirming quercetin and resveratrol as effective diabetes targets.	[135]
Predict risk factors and screen therapeutic drugs for cognitive impairment (CI) in T2DM patients	Graph Neural Network (GNN)	GNN model identifies risk factors and potential drugs (e.g., Metformin, herbal compounds) for treating diabetes-related cognitive impairment, offering insight into T2DM-CI.	Catboost achieved highest predictive AUC (0.95); GNN identified ten drugs with potential efficacy against T2DM-CI.	[136]
Enhance early DM diagnosis, particularly in resource-limited settings	Light Gradient Boosting Machine (LightGBM)	LightGBM offers a computationally efficient method for DM diagnosis, suitable for areas with limited medical infrastructure.	LightGBM outperformed other models with 98.1% accuracy, 99.9% sensitivity, and 96.3% specificity on Ethiopian dataset.	[137]

(continued)

Table 21.8 (continued)

Study Purpose/ Focus	Name of Tool	Application in Antidiabetic Drug Discovery Research	Results	References
Identify metabolomic biomarkers associated with diabetic polyneuropathy (DPN) in T2DM	Elastic Net	Elastic Net model aids in identifying metabolic biomarkers linked to DPN, highlighting metabolic shifts useful for predicting diabetes complications.	Fifteen metabolites were identified as associated with DPN status, notably involving lipid metabolism and plasma lipid composition, including fatty acids, acylcarnitines, and sphingolipids.	[138]
Improve accuracy of diabetes prediction models for early detection	Logistic Regression (LR)	LR, paired with ensemble techniques (Max Voting, Stacking), improves diabetes prediction accuracy, supporting early intervention strategies for diabetes management.	Highest accuracy reached was 78% on the PIMA dataset and 93% on the Vanderbilt dataset, enhanced through feature selection and ensemble techniques.	[139]

21.8.1 Machine Learning Algorithms in Predictive Modelling of Drug Efficacy

ML algorithms are now widely used in predictive modelling for drug efficacy, particularly in drug discovery of complex diseases. These algorithms enhance the analysis and optimization of potential drug candidates by analysing complex biological data and their chemical structures [132]. There are various ML models already utilized in antidiabetic drug discovery research and disease management (Table 21.8).

These ML algorithms have been proven very effective in predicting modelling of drug discovery in antidiabetic drug discovery research. These models not only enhance the screening of potential therapeutic compounds but also helps in early detection and management of diabetes which ultimately helps in accelerating the identification of novel antidiabetic agents.

21.9 Future Perspectives in the Design of Novel Antidiabetic Agents

The design of novel antidiabetic agents is progressing rapidly, with a focus on addressing current treatment limitations by targeting specific pathways to improve glycemic control, reduce side effects, and enhance patient compliance. Nano-based therapeutics, for example, are utilizing nanoparticles to improve drug delivery, increase bioavailability, and specifically target tissues, showing promise in more effective and precise antidiabetic treatment delivery [140]. Researchers are also investigating new molecular targets, such as GPCR 119 [141], GPER [142], Vaspin [143], Metrnl [144], and Fetuin-A [145], each of which has a significant role in insulin regulation and glucose metabolism, providing additional avenues for therapeutic intervention and novel treatment options. Furthermore, combination therapies are being explored to enhance treatment effectiveness by using multiple drugs with different mechanisms of action to target various pathophysiological aspects of diabetes, potentially offering a more comprehensive approach to glycaemic management [146].

Emerging computational tools are revolutionizing antidiabetic drug discovery by expediting the identification and optimization of potential drug candidates. One such platform, the CAS BioFinder Discovery Platform, provides an integrated resource for identifying biological targets and ligands by consolidating fragmented data sources, thus supporting early-stage discovery efforts [147]. Generative AI models, including NVIDIA BioNeMo [148] and Google's AlphaFold [149], use advanced algorithms to design novel molecules and predict their pharmacological properties, thereby accelerating the drug design process. Virtual screening techniques are also undergoing refinements, with new algorithms and molecular docking advancements enhancing hit rates and allowing for more accurate identification of candidate compounds with strong affinities for antidiabetic targets [150].

Personalized medicine is making strides in tailoring diabetes treatments to individual patients based on genetic, environmental, and lifestyle factors, and in silico methods play a critical role in this process. Predictive modelling enables the forecasting of patient-specific responses to various treatments, which facilitates the customization of more effective therapeutic plans [151]. In addition, genomic data analysis allows researchers to identify suitable drug targets based on genetic profiles, enhancing the precision of antidiabetic therapies [152]. Computational models can also simulate drug interactions within patient subgroups to optimize dosing strategies and mitigate potential adverse effects, thereby supporting precision medicine in diabetes management [153].

Despite these advancements, integrating AI and machine learning into antidiabetic drug discovery presents both challenges and opportunities. High-quality, standardized data is essential for reliable AI-driven predictions, underscoring the need for robust data curation and consistency across studies. Furthermore, advancing algorithms capable of managing complex biological and clinical data is necessary to ensure that predictions and outputs are both accurate and applicable [154].

Interdisciplinary collaboration among computational scientists, molecular biologists, and clinicians is crucial to enhance the translational potential of these discoveries [155]. Finally, ethical considerations, including data privacy and the responsible use of AI in healthcare, must be addressed to build trust and ensure equitable access to these technological innovations [156]. These future perspectives underscore the transformative potential of computational and AI technologies in antidiabetic drug development, with a vision for more targeted, effective, and personalized therapies.

21.10 Conclusion

Computational approaches are playing a transformative role in advancing antidiabetic drug discovery. Through techniques like virtual screening, molecular docking, network pharmacology, and AI-driven predictions, researchers can efficiently identify and optimize drug candidates that target critical pathways in diabetes progression, such as Alpha glucosidase, PPAR, DPP-4, and SGLT2. Additionally, the exploration of natural products and phytochemicals offers a rich source of potential antidiabetic agents, with computational methods enabling rapid screening for efficacy and safety. Molecular dynamics simulations further enhance drug development by confirming the stability and binding efficiency of these compounds in physiological environments. These innovative strategies not only accelerate the drug discovery pipeline but also pave the way for more precise, effective, and personalized therapies for diabetes, addressing existing treatment limitations and enhancing patient outcomes.

References

1. Klen J, Dolžan V. Glucagon-like peptide-1 receptor agonists in the management of type 2 diabetes mellitus and obesity: the impact of pharmacological properties and genetic factors. Int J Mol Sci. 2022;23(7):1–18.
2. Sena CM, Bento CF, Pereira P, Seiça R. Diabetes mellitus: new challenges and innovative therapies. EPMA J. 2010;1(1):138–63.
3. Forbes JM, Cooper ME. Mechanisms of diabetic complications. Physiol Rev. 2013;93(1):137–88.
4. Atlas IDFD. International Diabetes Federation. Lancet. 1955;266:134–7.
5. Sheng X, Wang J, Guo J, Xu Y, Jiang H, Zheng C, et al. Effects of baicalin on diabetic cardiac autonomic neuropathy mediated by the P2Y12 receptor in rat stellate ganglia. Cell Physiol Biochem. 2018;46(3):986–98.
6. Gaikwad SB, Krishna Mohan G, Rani MS. Phytochemicals for diabetes management. Pharm Crop. 2014;5(1):11–28.
7. Tyagi S, Sharma S, Gupta P, Saini A, Kaushal C. The peroxisome proliferator-activated receptor: a family of nuclear receptors role in various diseases. J Adv Pharm Technol Res. 2011;2(4):236.
8. Galic S, Hauser C, Kahn BB, Haj FG, Neel BG, Tonks NK, et al. Coordinated regulation of insulin signaling by the protein tyrosine phosphatases PTP1B and TCPTP. Mol Cell Biol. 2005;25(2):819–29.

9. Nema P, Asati V, Kendya P, Gupta T, Agarwal S, Kori S, et al. Structural insight on GPR119 agonist as potential therapy for type II diabetes: a comprehensive review. Mini-Rev Med Chem. 2023;23(21):2008–40.
10. Khwaja NUD, Arunagirinathan G. Efficacy and cardiovascular safety of alpha glucosidase inhibitors. Curr Drug Saf. 2021;16(2):122–8.
11. Deacon CF. Physiology and pharmacology of DPP-4 in glucose homeostasis and the treatment of type 2 diabetes. Front Endocrinol (Lausanne). 2019;10:80.
12. Leach K, Conigrave AD, Sexton PM, Christopoulos A. Towards tissue-specific pharmacology: insights from the calcium-sensing receptor as a paradigm for GPCR (patho)physiological bias. Trends Pharmacol Sci. 2015;36(4):215–25.
13. Gendaszewska-Darmach E, Drzazga A, Koziołkiewicz M. Targeting GPCRs activated by fatty acid-derived lipids in type 2 diabetes. Trends Mol Med. 2019;25(10):915–29.
14. Moran BM, Flatt PR, McKillop AM. G protein-coupled receptors: signalling and regulation by lipid agonists for improved glucose homoeostasis. Acta Diabetol. 2016;53(2):177–88.
15. Nair AS, Bagchi D, Lehmann TE, Nair S. Renal sodium-glucose transporter-2 inhibitors as antidiabetic agents. In: Nutritional and therapeutic interventions for diabetes and metabolic syndrome. London: Elsevier; 2018. p. 207–14.
16. Leipzig RM. Update in geriatric medicine. Ann Intern Med. 2003;139(12):1003.
17. Rendell M. The role of sulphonylureas in the management of type 2 diabetes mellitus. Drugs. 2004;64(12):1339–58.
18. Philip J, Fernandez CJ. Efficacy and cardiovascular safety of meglitinides. Curr Drug Saf. 2021;16(2):207–16.
19. Wang W, Zhou X, Kwong JSW, Li L, Li Y, Sun X. Efficacy and safety of thiazolidinediones in diabetes patients with renal impairment: a systematic review and meta-analysis. Sci Rep. 2017;7(1):1717.
20. Willemen MJ, Mantel-Teeuwisse AK, Straus SM, Meyboom RH, Egberts TC, Leufkens HG. Use of dipeptidyl peptidase-4 inhibitors and the reporting of infections: a disproportionality analysis in the World Health Organization VigiBase. Diabetes Care. 2011;34(2):369–74.
21. Hinnen D. Glucagon-like peptide 1 receptor agonists for type 2 diabetes. Diabetes Spectr. 2017;30(3):202–10.
22. Neal B, Perkovic V, Mahaffey KW, de Zeeuw D, Fulcher G, Erondu N, et al. Canagliflozin and cardiovascular and renal events in type 2 diabetes. N Engl J Med. 2017;377(7):644–57.
23. Khoo CM. Diabetes mellitus treatment. In: International encyclopedia of public health. London: Elsevier; 2017. p. 288–93.
24. White SA, Shaw JA, Sutherland DE. Pancreas transplantation. Lancet. 2009;373(9677):1808–17.
25. Singh R, Kaur N, Kishore L, Kumar GG. Management of diabetic complications: a chemical constituents based approach. J Ethnopharmacol. 2013;150(1):51–70.
26. Macalino SJY, Gosu V, Hong S, Choi S. Role of computer-aided drug design in modern drug discovery. Arch Pharm Res [Internet]. 2015;38(9):1686–701. Available from: http://link.springer.com/10.1007/s12272-015-0640-5.
27. Pingili D, Thimmaraju MK, Gummadi SB. CURRENT approaches in drug discovery. In: Futuristic trends in pharmacy & nursing. Vol. 2. Book 24 [Internet]. Iterative International Publishers, Selfypage Developers Pvt Ltd; 2023. p. 8–25. Available from: https://www.iip-series.org/viewpaper.php?pid=632&pt=current-approaches-in-drug-discovery.
28. Marín M, López M, Gallego-Yerga L, Álvarez R, Peláez R. Experimental structure based drug design (SBDD) applications for anti-leishmanial drugs: a paradigm shift? Med Res Rev [Internet]. 2024;44(3):1055–120. Available from: https://onlinelibrary.wiley.com/doi/10.1002/med.22005.
29. Agu PC, Afiukwa CA, Orji OU, Ezeh EM, Ofoke IH, Ogbu CO, et al. Molecular docking as a tool for the discovery of molecular targets of nutraceuticals in diseases management. Sci Rep [Internet]. 2023;13(1):13398. Available from: https://www.nature.com/articles/s41598-023-40160-2.

30. Ho J, Qi X, Jiang F, Sun Y, Hong L. An artificial intelligence machine learning (AI/ML) approach with cross-technology node learning for multi-layer process defect predictions. In: Lafferty NV, Grunes H, editors. DTCO and computational patterning III [Internet]. SPIE; 2024. p. 14. Available from: https://www.spiedigitallibrary.org/conference-proceedings-of-spie/12954/3011296/An-artificial-intelligence-machine-learning-AI-ML-approach-with-cross/10.1117/12.3011296.full.
31. Lee JW, Maria-Solano MA, Vu TNL, Yoon S, Choi S. Big data and artificial intelligence (AI) methodologies for computer-aided drug design (CADD). Biochem Soc Trans [Internet]. 2022;50(1):241–52. Available from: https://portlandpress.com/biochemsoctrans/article/50/1/241/230699/Big-data-and-artificial-intelligence-AI.
32. Yang A, Pan ML, Lu HHS, Lien CY, Wang DW, Chen CH, et al. Assessing the evolution and influence of medical open databases on biomedical research and healthcare innovation: a 25-year perspective with a focus on privacy and privacy-enhancing technologies (Preprint) [Internet]. 2024. Available from: http://preprints.jmir.org/preprint/58954.
33. Niazi SK, Mariam Z. Computer-aided drug design and drug discovery: a prospective analysis [Internet]. 2023. Available from: https://www.preprints.org/manuscript/202311.0852/v1.
34. Srivani N, Vinay Chandra A, Ramesh K, Kumar YBK. Towards green chemistry quantum computing applications in chemical synthesis. In: Real-world challenges in quantum electronics and machine computing. IGI Global; 2024. p. 380–91. Available from: https://services.igi-global.com/resolvedoi/resolve.aspx?doi=10.4018/979-8-3693-4001-1.ch025.
35. Nunn AVW, Bell J, Barter P. The integration of lipid-sensing and anti-inflammatory effects: how the PPARs play a role in metabolic balance. Nucl Recept. 2007;5:1–13.
36. Al Neyadi SS, Adem A, Amir N, Ghattas MA, Abdou IM, Salem AA. Novel thiazolidinedione and rhodanine derivatives regulate glucose metabolism, improve insulin sensitivity, and activate the peroxisome proliferator-activated γ receptor. ACS Omega. 2024;9(5):5463–84.
37. Jiang H, Zhou XE, Shi J, Zhou Z, Zhao G, Zhang X, et al. Identification and structural insight of an effective PPARγ modulator with improved therapeutic index for anti-diabetic drug discovery. Chem Sci. 2020;11(8):2260–8.
38. Jain N, Bhansali S, Kurpad AV, Hawkins M, Sharma A, Kaur S, et al. Effect of a dual PPAR α/γ agonist on insulin sensitivity in patients of type 2 diabetes with hypertriglyceridemia-randomized double-blind placebo-controlled trial. Sci Rep. 2019;9(1):1–9.
39. Kaul U, Parmar D, Manjunath K, Shah M, Parmar K, Patil KP, et al. New dual peroxisome proliferator activated receptor agonist – saroglitazar in diabetic dyslipidemia and non-alcoholic fatty liver disease: integrated analysis of the real world evidence. Cardiovasc Diabetol. 2019;18(1):1–11.
40. Ashcroft FM, Rorsman P. KATP channels and islet hormone secretion: new insights and controversies. Nature Reviews Endocrinology. 2013 Nov;9(11):660–9.
41. Donley VR, Hiskett EK, Kidder AC, Schermerhorn T. ATP-sensitive potassium channel (KATP channel) expression in the normal canine pancreas and in canine insulinomas. BMC Vet Res. 2005;1:1–8.
42. Batista TM, Haider N, Kahn CR. Defining the underlying defect in insulin action in type 2 diabetes. Diabetologia. 2021;64(5):994–1006.
43. Walkowski B, Kleibert M, Majka M, Wojciechowska M. Insight into the role of the PI3K/Akt pathway in ischemic injury and post-infarct left ventricular remodeling in normal and diabetic heart. Cells. 2022;11(9):1553.
44. Ashayeri Ahmadabad H, Mohammadi Panah S, Ghasemnejad-Berenji H, et al. Metformin and the PI3K/AKT signaling pathway: implications for cancer, cardiovascular, and central nervous system diseases. Naunyn Schmiedeberg's Arch Pharmacol. 2025;398(2):1035–55.
45. Sidarala V, Kowluru A. The regulatory roles of mitogen-activated protein kinase (MAPK) pathways in health and diabetes: lessons learned from the pancreatic β-cell. Recent patents on endocrine, metabolic & immune drug discovery. 2016 Aug 1;10(2):76–84.
46. Li M, Li X, Zhang H, Lu Y. Molecular mechanisms of metformin for diabetes and cancer treatment. Front Physiol. 2018;9:1–7.
47. Mathur N, Chandragiri SS, Sarita, Shandily S, Santoki KM, Vadhavana NN, et al. In silico docking: protocols for computational exploration of molecular interactions. In: Unravelling

molecular docking – from theory to practice [Working Title] [Internet]. IntechOpen; 2024. Available from: https://www.intechopen.com/online-first/1181718.
48. Okpo E, Agboke AA, Udobi CE, John GE, Andy IE. The synergy of molecular docking and bioinformatics: an in depth review in drug discovery. Biotechnol J Int [Internet]. 2024;28(4):119–36. Available from: https://journalbji.com/index.php/BJI/article/view/732.
49. Li G. System scale. In: Chemical theory and multiscale simulation in biomolecules [Internet]. Elsevier; 2024. p. 351–62. Available from: https://linkinghub.elsevier.com/retrieve/pii/B9780323959179000201.
50. Livigni A, O'Hara L, Polak ME, Angus T, Wright DW, Smith LB, et al. A graphical and computational modeling platform for biological pathways. Nat Protoc [Internet]. 2018;13(4):705–22. Available from: https://www.nature.com/articles/nprot.2017.144.
51. Setiya A, Jani V, Sonavane U, Joshi R. MolToxPred: small molecule toxicity prediction using machine learning approach. RSC Adv [Internet]. 2024;14(6):4201–20. Available from: https://xlink.rsc.org/?DOI=D3RA07322J
52. Agamah FE, Mazandu GK, Hassan R, Bope CD, Thomford NE, Ghansah A, et al. Computational/in silico methods in drug target and lead prediction. Brief Bioinform [Internet]. 2019 [cited 2024 Nov 12];21(5):1663. Available from: https://pmc.ncbi.nlm.nih.gov/articles/PMC7673338/.
53. Gentile F, Yaacoub JC, Gleave J, Fernandez M, Ton AT, Ban F, et al. Artificial intelligence–enabled virtual screening of ultra-large chemical libraries with deep docking. Nat Protoc. [Internet]. 2022 [cited 2024 Nov 12];17(3):672–97. Available from: https://www.nature.com/articles/s41596-021-00659-2.
54. Skariyachan S, Garka S. Exploring the binding potential of carbon nanotubes and fullerene towards major drug targets of multidrug resistant bacterial pathogens and their utility as novel therapeutic agents. In: Fullerenes, graphenes and nanotubes: a pharmaceutical approach. Oxford: Elsevier; 2018. p. 1–29.
55. Tyagi R, Singh A, Chaudhary KK, Yadav MK. Pharmacophore modeling and its applications. In: Bioinformatics: methods and applications. London: Elseveir; 2022. p. 269–89.
56. Chandran U, Mehendale N, Patil S, Chaguturu R, Patwardhan B. Network pharmacology. In: Innovative approaches in drug discovery: ethnopharmacology, systems biology and holistic targeting. London: Academic; 2017. p. 127–64.
57. John Harris C, Hill RD, Sheppard DW, Slater MJ, Stouten PFW. The design and application of target-focused compound libraries. Comb Chem High Throughput Screen [Internet]. 2011;14(6):521–31. Available from: http://www.eurekaselect.com/openurl/content.php?genre=article&issn=1386-2073&volume=14&issue=6&spage=521.
58. Ginex T, Madruga E, Martinez A, Gil C. MBC and ECBL libraries: outstanding tools for drug discovery. Front Pharmacol [Internet]. 2023;14. Available from: https://www.frontiersin.org/articles/10.3389/fphar.2023.1244317/full.
59. Sterling T, Irwin JJ. ZINC 15 – ligand discovery for everyone. J Chem Inf Model [Internet]. 2015;55(11):2324–37. Available from: https://pubs.acs.org/doi/10.1021/acs.jcim.5b00559.
60. Gaulton A, Bellis LJ, Bento AP, Chambers J, Davies M, Hersey A, et al. ChEMBL: a large-scale bioactivity database for drug discovery. Nucleic Acids Res [Internet]. 2012;40(D1):D1100–7. Available from: https://academic.oup.com/nar/article-lookup/doi/10.1093/nar/gkr777.
61. Wishart DS, Feunang YD, Guo AC, Lo EJ, Marcu A, Grant JR, et al. DrugBank 5.0: a major update to the DrugBank database for 2018. Nucleic Acids Res [Internet]. 2018;46(D1):D1074–82. Available from: http://academic.oup.com/nar/article/46/D1/D1074/4602867.
62. Kim S, Chen J, Cheng T, Gindulyte A, He J, He S, et al. PubChem in 2021: new data content and improved web interfaces. Nucleic Acids Res [Internet]. 2021;49(D1):D1388–95. Available from: https://academic.oup.com/nar/article/49/D1/D1388/5957164.
63. Lans C, van Asseldonk T. Dr. Duke's phytochemical and ethnobotanical databases, a cornerstone in the validation of ethnoveterinary medicinal plants, as demonstrated by data on pets in British Columbia. In: Medicinal and aromatic plants of North America. Cham: Springer; 2020. p. 219–46. Available from: http://link.springer.com/10.1007/978-3-030-44930-8_10.

64. Vivek-Ananth RP, Mohanraj K, Sahoo AK, Samal A. IMPPAT 2.0: an enhanced and expanded phytochemical atlas of Indian medicinal plants. ACS Omega [Internet]. 2023;8(9):8827–45. Available from: https://pubs.acs.org/doi/10.1021/acsomega.3c00156.
65. Lv Q, Chen G, He H, Yang Z, Zhao L, Zhang K, et al. TCMBank-the largest TCM database provides deep learning-based Chinese-Western medicine exclusion prediction. Signal Transduct Target Ther [Internet]. 2023;8(1):127. Available from: https://www.nature.com/articles/s41392-023-01339-1.
66. Madariaga-Mazón A, Naveja JJ, Medina-Franco JL, Noriega-Colima KO, Martinez-Mayorga K. DiaNat-DB: a molecular database of antidiabetic compounds from medicinal plants. RSC Adv [Internet]. 2021;11(9):5172–8. Available from: https://xlink.rsc.org/?DOI=D0RA10453A.
67. Martin RL, Heifetz A, Bodkin MJ, Townsend-Nicholson A. High-throughput structure-based drug design (HT-SBDD) using drug docking, fragment molecular orbital calculations, and molecular dynamic techniques. Methods Mol Biol. 2024;2716:293–306. Available from: https://link.springer.com/10.1007/978-1-0716-3449-3_13.
68. Ilyas U, Nazir B, Altaf R, Muhammad SA, Zafar H, Paiva-Santos AC, et al. Investigation of anti-diabetic potential and molecular simulation studies of dihydropyrimidinone derivatives. Front Endocrinol (Lausanne) [Internet]. 2022;13. Available from: https://www.frontiersin.org/articles/10.3389/fendo.2022.1022623/full.
69. Maji S, Pattanayak SK, Sen A, Badavath VN. Pharmacophore modeling in drug design. In: Computer aided drug design (CADD): from ligand-based methods to structure-based approaches [Internet]. Elsevier; 2022. p. 157–79. Available from: https://linkinghub.elsevier.com/retrieve/pii/B9780323906081000022.
70. Lotfi B, Mebarka O, Khan SU, Htar TT. Pharmacophore-based virtual screening, molecular docking and molecular dynamics studies for the discovery of novel neuraminidase inhibitors. J Biomol Struct Dyn. 2024;42(10):5308–20.
71. Temml V, Kutil Z. Structure-based molecular modeling in SAR analysis and lead optimization. Comput Struct Biotechnol J [Internet]. 2021;19:1431–44. Available from: https://linkinghub.elsevier.com/retrieve/pii/S2001037021000696.
72. Kaushik P, Lal Khokra S, Rana AC, Kaushik D. Pharmacophore modeling and molecular docking studies on Pinus roxburghii as a target for diabetes mellitus. Adv Bioinformatics [Internet]. 2014;2014:1–8. Available from: https://www.hindawi.com/journals/abi/2014/903246/.
73. Singh K, Bhushan B, Singh B. Advances in drug discovery and design using computer-aided molecular modeling. Curr Comput Aided Drug Des [Internet]. 2024;20(5):697–710. Available from: https://www.eurekaselect.com/221092/article.
74. Kirsch P, Hartman AM, Hirsch AKH, Empting M. Concepts and core principles of fragment-based drug design. Molecules [Internet]. 2019;24(23):4309. Available from: https://www.mdpi.com/1420-3049/24/23/4309.
75. Li Q. Application of fragment-based drug discovery to versatile targets. Front Mol Biosci [Internet]. 2020;7. Available from: https://www.frontiersin.org/article/10.3389/fmolb.2020.00180/full.
76. Das A, Nandi A, Kumari V, Alvala M. FBDD & de novo drug design. In: Applied computer-aided drug design: models and methods [Internet]. Bentham Science Publishers; 2023. p. 159–201. Available from: https://www.eurekaselect.com/node/224313.
77. Gupta S, Baweja GS, Singh S, Irani M, Singh R, Asati V. Integrated fragment-based drug design and virtual screening techniques for exploring the antidiabetic potential of thiazolidine-2,4-diones: design, synthesis and in vivo studies. Eur J Med Chem [Internet]. 2023;261:115826. Available from: https://linkinghub.elsevier.com/retrieve/pii/S0223523423007936.
78. Kashtoh H, Baek KH. Recent updates on phytoconstituent alpha-glucosidase inhibitors: an approach towards the treatment of type two diabetes. Plants [Internet]. 2022 [cited 2024 Nov 12];11(20):2722. Available from: https://www.mdpi.com/2223-7747/11/20/2722/htm.
79. Syam YM, Anwar MM, Abd El-Karim SS, Elseginy SA, Essa BM, Sakr TM. New quinoxaline compounds as DPP-4 inhibitors and hypoglycemics: design, synthesis, computational and bio-distribution studies. RSC Adv [Internet]. 2021;11(58):36989–7010. Available from: https://xlink.rsc.org/?DOI=D1RA06799K.

80. Shaikh S, Ali S, Lim JH, Chun HJ, Ahmad K, Ahmad SS, et al. Dipeptidyl peptidase-4 inhibitory potentials of Glycyrrhiza uralensis and its bioactive compounds licochalcone A and licochalcone B: an in silico and in vitro study. Front Mol Biosci [Internet]. 2022;9. Available from: https://www.frontiersin.org/articles/10.3389/fmolb.2022.1024764/full.
81. El-Arabey AA, Zhang H, Abdalla M, Al-Shouli ST, Alkhalil SS, Liu Y. Metformin as a promising target for DPP4 expression: computational modeling and experimental validation. Med Oncol [Internet]. 2023;40(10):277. Available from: https://link.springer.com/10.1007/s12032-023-02140-4.
82. Ganwir P, Bhadane R, Chaturbhuj GU. In-silico screening and identification of glycomimetic as potential human sodium-glucose co-transporter 2 inhibitor. Comput Biol Chem [Internet]. 2024;110:108074. Available from: https://linkinghub.elsevier.com/retrieve/pii/S1476927124000628.
83. Heryanto R, Noor D. In silico Prediction of Sodium-Glucose Co-Transporter-2 (SGLT2) Inhibition Activity by Allium Fistulosum Compound Based on SkelSpheres Molecular Descriptor. Indonesian Journal of Medical Chemistry and Bioinformatics. 2023;2(1):6.
84. Macalalad MAB, Gonzales AA. In-silico screening and identification of phytochemicals from Centella asiatica as potential inhibitors of sodium-glucose co-transporter 2 for treating diabetes. J Biomol Struct Dyn [Internet]. 2022;40(22):12221–38. Available from: https://www.tandfonline.com/doi/full/10.1080/07391102.2021.1969282.
85. Riyaphan J, Pham DC, Leong MK, Weng CF. In silico approaches to identify polyphenol compounds as α-glucosidase and α-amylase inhibitors against type-II diabetes. Biomolecules [Internet]. 2021;11(12):1877. Available from: https://www.mdpi.com/2218-273X/11/12/1877.
86. Sayahi MH, Zareei S, Halimi M, Alikhani M, Moazzam A, Mohammadi-Khanaposhtani M, et al. Design, synthesis, in vitro, and in silico anti-α-glucosidase assays of N-phenylacetamide-1,2,3-triazole-indole-2-carboxamide derivatives as new anti-diabetic agents. Sci Rep [Internet]. 2024;14(1):15791. Available from: https://www.nature.com/articles/s41598-024-66201-y.
87. Soleimani Z, Mohammadi M, Halimi M, Safapoor S, Dastyafteh N, Safaie E, et al. Design of new α-glucosidase inhibitors based on the bis-4-hydroxycoumarin skeleton: synthesis, evaluation, and in silico studies. Sci Rep [Internet]. 2024;14(1):18693. Available from: https://www.nature.com/articles/s41598-024-69592-0.
88. Majumder S, Panigrahi GK. Advancements in contemporary pharmacological innovation: mechanistic insights and emerging trends in drug discovery and development. Intell Pharm [Internet]. 2024 [cited 2024 Nov 12]. Available from: https://linkinghub.elsevier.com/retrieve/pii/S2949866X24001084.
89. Archana TM, Sudheesh S. Natural products as nano-antidiabetic drugs. In: Drugs from nature: targets, assay systems and leads [Internet]. Singapore: Springer Nature Singapore; 2024. p. 531–52. Available from: https://link.springer.com/10.1007/978-981-99-9183-9_19.
90. Sable V, Ahire G. Network pharmacology and molecular docking technology-based strategy to explore the potential mechanism of diabecon formulation botanicals. Int J Drug Deliv Technol [Internet]. 2024;14(01):274–80. Available from: https://impactfactor.org/PDF/IJDDT/14/IJDDT,Vol14,Issue1,Article39.pdf.
91. Chandran U, Mehendale N, Patil S, Chaguturu R, Patwardhan B. Network pharmacology. In: Innovative approaches in drug discovery [Internet]. 2016 [cited 2024 Nov 12]. p. 127. Available from: https://pmc.ncbi.nlm.nih.gov/articles/PMC7148629/.
92. Noor F, Tahir ul Qamar M, Ashfaq UA, Albutti A, Alwashmi ASS, Aljasir MA. Network pharmacology approach for medicinal plants: review and assessment. Pharmaceuticals [Internet]. 2022;15(5):572. Available from: https://www.mdpi.com/1424-8247/15/5/572.
93. Wang N, Zhu F, Shen M, Qiu L, Tang M, Xia H, et al. Network pharmacology-based analysis on bioactive anti-diabetic compounds in Potentilla discolor bunge. J Ethnopharmacol [Internet]. 2019;241:111905. Available from: https://linkinghub.elsevier.com/retrieve/pii/S0378874118347524.
94. Noor F, Rehman A, Ashfaq UA, Saleem MH, Okla MK, Al-Hashimi A, et al. Integrating network pharmacology and molecular docking approaches to decipher the multi-target pharma-

cological mechanism of Abrus precatorius L. acting on diabetes. Pharmaceuticals [Internet]. 2022;15(4):414. Available from: https://www.mdpi.com/1424-8247/15/4/414.
95. Lin Z, Tong Y, Li N, Zhu Z, Li J. Network pharmacology-based study of the mechanisms of action of anti-diabetic triterpenoids from Cyclocarya paliurus. RSC Adv [Internet]. 2020;10(61):37168–81. Available from: https://xlink.rsc.org/?DOI=D0RA06846B.
96. Ekowati J, Tejo BA, Maulana S, Kusuma WA, Fatriani R, Ramadhanti NS, et al. Potential utilization of phenolic acid compounds as anti-inflammatory agents through TNF-α convertase inhibition mechanisms: a network pharmacology, docking, and molecular dynamics approach. ACS Omega [Internet]. 2023;8(49):46851–68. Available from: https://pubs.acs.org/doi/10.1021/acsomega.3c06450.
97. Rai M, Singh AV, Paudel N, Kanase A, Falletta E, Kerkar P, et al. Herbal concoction unveiled: a computational analysis of phytochemicals' pharmacokinetic and toxicological profiles using novel approach methodologies (NAMs). Curr Res Toxicol [Internet]. 2023 [cited 2024 Nov 14];5:100118. Available from: https://pmc.ncbi.nlm.nih.gov/articles/PMC10440360/.
98. Kiriiri GK, Njogu PM, Mwangi AN. Exploring different approaches to improve the success of drug discovery and development projects: a review. Futur J Pharm Sci [Internet]. 2020 [cited 2024 Nov 14];6(1):1–12. Available from: https://fjps.springeropen.com/articles/10.1186/s43094-020-00047-9.
99. Fu L, Shi S, Yi J, Wang N, He Y, Wu Z, et al. ADMETlab 3.0: an updated comprehensive online ADMET prediction platform enhanced with broader coverage, improved performance, API functionality and decision support. Nucleic Acids Res [Internet]. 2024 [cited 2024 Nov 14];52(W1):W422–31. Available from: https://doi.org/10.1093/nar/gkae236.
100. Cao D, Wang J, Zhou R, Li Y, Yu H, Hou T. ADMET evaluation in drug discovery. 11. PharmacoKinetics Knowledge Base (PKKB): a comprehensive database of pharmacokinetic and toxic properties for drugs. J Chem Inf Model [Internet]. 2012 [cited 2024 Nov 14];52(5):1132–7. Available from: https://pubmed.ncbi.nlm.nih.gov/22559792/.
101. Yang H, Lou C, Sun L, Li J, Cai Y, Wang Z, et al. admetSAR 2.0: web-service for prediction and optimization of chemical ADMET properties. Bioinformatics [Internet]. 2019 [cited 2024 Nov 14];35(6):1067–9. Available from: https://doi.org/10.1093/bioinformatics/bty707.
102. Banerjee P, Eckert AO, Schrey AK, Preissner R. ProTox-II: a webserver for the prediction of toxicity of chemicals. Nucleic Acids Res [Internet]. 2018 [cited 2024 Nov 14];46(W1):W257–63. Available from: https://doi.org/10.1093/nar/gky318.
103. Daina A, Michielin O, Zoete V. SwissADME: a free web tool to evaluate pharmacokinetics, drug-likeness and medicinal chemistry friendliness of small molecules. Sci Rep [Internet]. 2017 [cited 2024 Nov 14];7(1):1–13. Available from: https://www.nature.com/articles/srep42717.
104. Benet LZ, Hosey CM, Ursu O, Oprea TI. BDDCS, the Rule of 5 and drugability. Adv Drug Deliv Rev [Internet]. 2016 [cited 2024 Nov 14];101:89. Available from: https://pmc.ncbi.nlm.nih.gov/articles/PMC4910824/.
105. Asiamah I, Obiri SA, Tamekloe W, Armah FA, Borquaye LS. Applications of molecular docking in natural products-based drug discovery. Sci Afr. 2023;20:e01593.
106. Ajiboye BO, Iwaloye O, Owolabi OV, Ejeje JN, Okerewa A, Johnson OO, et al. Screening of potential antidiabetic phytochemicals from Gongronema latifolium leaf against therapeutic targets of type 2 diabetes mellitus: multi-targets drug design. SN Appl Sci [Internet]. 2022;4(1):14. Available from: https://link.springer.com/10.1007/s42452-021-04880-2.
107. Maradesha T, Martiz RM, Patil SM, Prasad A, Babakr AT, Silina E, et al. Integrated network pharmacology and molecular modeling approach for the discovery of novel potential MAPK3 inhibitors from whole green jackfruit flour targeting obesity-linked diabetes mellitus. PLoS One [Internet]. 2023 [cited 2024 Nov 14];18(1):e0280847. Available from: https://pmc.ncbi.nlm.nih.gov/articles/PMC9886246/.
108. Li L, Yang L, Yang L, He C, He Y, Chen L, et al. Network pharmacology: a bright guiding light on the way to explore the personalized precise medication of traditional Chinese medicine. Chin Med [Internet]. 2023 [cited 2024 Nov 14];18(1):1–19. Available from: https://cmjournal.biomedcentral.com/articles/10.1186/s13020-023-00853-2.

109. Ryszkiewicz P, Malinowska B, Schlicker E. Polypharmacology: promises and new drugs in 2022. Pharmacol Rep [Internet]. 2023;75(4):755–70. Available from: https://link.springer.com/10.1007/s43440-023-00501-4.
110. Swain SK, Dash UC, Kanhar S, Sahoo AK. Network pharmacology-based elucidation of bioactive compounds and experimental exploration of antidiabetic mechanisms of Hydrolea zeylanica. Cell Signal [Internet]. 2024;114:110999. Available from: https://linkinghub.elsevier.com/retrieve/pii/S089865682300414X.
111. Safari-Alighiarloo N, Taghizadeh M, Rezaei-Tavirani M, Goliaei B, Peyvandi AA. Protein-protein interaction networks (PPI) and complex diseases. Gastroenterol Hepatol Bed Bench [Internet]. 2014 [cited 2024 Nov 14];7(1):17. Available from: https://pmc.ncbi.nlm.nih.gov/articles/PMC4017556/.
112. Prakash SRS, Kamalnath SM, Antonisamy AJ, Marimuthu S, Malayandi S. In silico molecular docking of phytochemicals for type 2 diabetes mellitus therapy: a network pharmacology approach. Int J Mol Cell Med [Internet]. 2023 [cited 2024 Nov 14];12(4):372. Available from: https://pmc.ncbi.nlm.nih.gov/articles/PMC11240057/.
113. Szklarczyk D, Kirsch R, Koutrouli M, Nastou K, Mehryary F, Hachilif R, et al. The STRING database in 2023: protein–protein association networks and functional enrichment analyses for any sequenced genome of interest. Nucleic Acids Res [Internet]. 2023 [cited 2024 Nov 14];51(D1):D638–46. Available from: https://doi.org/10.1093/nar/gkac1000.
114. Xia J, Benner MJ, Hancock REW. NetworkAnalyst – integrative approaches for protein–protein interaction network analysis and visual exploration. Nucleic Acids Res [Internet]. 2014 [cited 2024 Nov 14];42(Web Server issue):W167. Available from: https://pmc.ncbi.nlm.nih.gov/articles/PMC4086107/.
115. Oughtred R, Rust J, Chang C, Breitkreutz BJ, Stark C, Willems A, et al. The BioGRID database: a comprehensive biomedical resource of curated protein, genetic, and chemical interactions. Protein Sci [Internet]. 2020 [cited 2024 Nov 14];30(1):187. Available from: https://pmc.ncbi.nlm.nih.gov/articles/PMC7737760/.
116. Shannon P, Markiel A, Ozier O, Baliga NS, Wang JT, Ramage D, et al. Cytoscape: a software environment for integrated models of biomolecular interaction networks. Genome Res [Internet]. 2003 [cited 2024 Nov 14];13(11):2498. Available from: https://pmc.ncbi.nlm.nih.gov/articles/PMC403769/.
117. Usman MS, Kusuma WA, Afendi FM, Heryanto R. Identification of significant proteins associated with diabetes mellitus using network analysis of protein-protein interactions. Comput Eng Appl J [Internet]. 2019;8(1):41–52. Available from: https://comengapp.unsri.ac.id/index.php/comengapp/article/view/283.
118. Zhang W, Pei J, Lai L. Computational multitarget drug design. J Chem Inf Model [Internet]. 2017;57(3):403–12. Available from: https://pubs.acs.org/doi/10.1021/acs.jcim.6b00491.
119. Zou HT, Ji BY, Xie XL. A multi-source molecular network representation model for protein–protein interactions prediction. Sci Rep [Internet]. 2024;14(1):6184. Available from: https://www.nature.com/articles/s41598-024-56286-w.
120. Huneif MA, Mahnashi MH, Jan MS, Shah M, Almedhesh SA, Alqahtani SM, et al. New succinimide–thiazolidinedione hybrids as multitarget antidiabetic agents: design, synthesis, bioevaluation, and molecular modelling studies. Molecules [Internet]. 2023;28(3):1207. Available from: https://www.mdpi.com/1420-3049/28/3/1207.
121. Adelusi TI, Oyedele AQK, Boyenle ID, Ogunlana AT, Adeyemi RO, Ukachi CD, et al. Molecular modeling in drug discovery. Inform Med Unlocked. 2022;29:100880.
122. Pinzi L, Rastelli G. Molecular docking: shifting paradigms in drug discovery. Int J Mol Sci [Internet]. 2019 [cited 2024 Nov 14];20(18):4331. Available from: https://www.mdpi.com/1422-0067/20/18/4331/htm.
123. Wu F, Zhou Y, Li L, Shen X, Chen G, Wang X, et al. Computational approaches in preclinical studies on drug discovery and development. Front Chem [Internet]. 2020 [cited 2024 Nov 14];8:726. Available from: https://pmc.ncbi.nlm.nih.gov/articles/PMC7517894/.

124. Lazim R, Suh D, Choi S. Advances in molecular dynamics simulations and enhanced sampling methods for the study of protein systems. Int J Mol Sci [Internet]. 2020 [cited 2024 Nov 14];21(17):6339. Available from: https://pmc.ncbi.nlm.nih.gov/articles/PMC7504087/.
125. Najmi A, Alam MS, Thangavel N, Taha MME, Meraya AM, Albratty M, et al. Synthesis, molecular docking, and in vivo antidiabetic evaluation of new benzylidene-2,4-thiazolidinediones as partial PPAR-γ agonists. Sci Rep [Internet]. 2023;13(1):19869. Available from: https://www.nature.com/articles/s41598-023-47157-x.
126. Ranade SD, Alegaon SG, Khatib NA, Gharge S, Kavalapure RS, Kumar BRP. Design, synthesis, molecular dynamic simulation, DFT analysis, computational pharmacology and decoding the antidiabetic molecular mechanism of sulphonamide-thiazolidin-4-one hybrids. J Mol Struct [Internet]. 2024;1311:138359. Available from: https://linkinghub.elsevier.com/retrieve/pii/S0022286024008792.
127. Ahmed S, Islam N, Shahinozzaman M, Fakayode SO, Afrin N, Halim MA. Virtual screening, molecular dynamics, density functional theory and quantitative structure activity relationship studies to design peroxisome proliferator-activated receptor-γ agonists as anti-diabetic drugs. J Biomol Struct Dyn [Internet]. 2021;39(2):728–42. Available from: https://www.tandfonline.com/doi/full/10.1080/07391102.2020.1714482.
128. Choudhury AA, Arumugam M, Ponnusamy N, Sivaraman D, Sertsemariam W, Thiruvengadam M, et al. Anti-diabetic drug discovery using the bioactive compounds of Momordica charantia by molecular docking and molecular dynamics analysis. J Biomol Struct Dyn [Internet]. 2024:1–15. Available from: https://www.tandfonline.com/doi/full/10.1080/07391102.2024.2313156.
129. Guan Z, Li H, Liu R, Cai C, Liu Y, Li J, et al. Artificial intelligence in diabetes management: advancements, opportunities, and challenges. Cell Rep Med [Internet]. 2023;4(10):101213. Available from: https://linkinghub.elsevier.com/retrieve/pii/S2666379123003804.
130. Yue J, Xu J, Li T, Li Y, Chen Z, Liang S, et al. Discovery of potential antidiabetic peptides using deep learning. Comput Biol Med [Internet]. 2024;180:109013. Available from: https://linkinghub.elsevier.com/retrieve/pii/S0010482524010989.
131. Odugbemi AI, Nyirenda C, Christoffels A, Egieyeh SA. Artificial intelligence in antidiabetic drug discovery: the advances in QSAR and the prediction of α-glucosidase inhibitors. Comput Struct Biotechnol J [Internet]. 2024;23:2964–77. Available from: https://linkinghub.elsevier.com/retrieve/pii/S200103702400237X.
132. Carracedo-Reboredo P, Liñares-Blanco J, Rodríguez-Fernández N, Cedrón F, Novoa FJ, Carballal A, et al. A review on machine learning approaches and trends in drug discovery. Comput Struct Biotechnol J. 2021;19:4538–58.
133. Ooka T, Johno H, Nakamoto K, Yoda Y, Yokomichi H, Yamagata Z. Random forest approach for determining risk prediction and predictive factors of type 2 diabetes: large-scale health check-up data in Japan. BMJ Nutr Prev Health [Internet]. 2021;4(1):140–8. Available from: https://nutrition.bmj.com/lookup/doi/10.1136/bmjnph-2020-000200.
134. Abbas HT, Alic L, Erraguntla M, Ji JX, Abdul-Ghani M, Abbasi QH, et al. Predicting long-term type 2 diabetes with support vector machine using oral glucose tolerance test. Pławiak P, editor. PLoS One [Internet]. 2019;14(12):e0219636. Available from: https://dx.plos.org/10.1371/journal.pone.0219636.
135. Liu XW, Shi TY, Gao D, Ma CY, Lin H, Yan D, et al. iPADD: a computational tool for predicting potential antidiabetic drugs using machine learning algorithms. J Chem Inf Model [Internet]. 2023;63(15):4960–9. Available from: https://pubs.acs.org/doi/10.1021/acs.jcim.3c00564.
136. Zhang X, Xie J, You X, Gong H. Risk factors and drug discovery for cognitive impairment in type 2 diabetes mellitus using artificial intelligence interpretation and graph neural networks. Front Endocrinol (Lausanne) [Internet]. 2023;14. Available from: https://www.frontiersin.org/articles/10.3389/fendo.2023.1213711/full.
137. Rufo DD, Debelee TG, Ibenthal A, Negera WG. Diagnosis of diabetes mellitus using gradient boosting machine (LightGBM). Diagnostics [Internet]. 2021;11(9):1714. Available from: https://www.mdpi.com/2075-4418/11/9/1714.

138. Rumora AE, Guo K, Alakwaa FM, Andersen ST, Reynolds EL, Jørgensen ME, et al. Plasma lipid metabolites associate with diabetic polyneuropathy in a cohort with type 2 diabetes. Ann Clin Transl Neurol [Internet]. 2021;8(6):1292–307. Available from: https://onlinelibrary.wiley.com/doi/10.1002/acn3.51367.
139. Rajendra P, Latifi S. Prediction of diabetes using logistic regression and ensemble techniques. Comput Methods Programs Biomed Update [Internet]. 2021;1:100032. Available from: https://linkinghub.elsevier.com/retrieve/pii/S2666990021000318.
140. Sharma R, Borah SJ, Bhawna N, Kumar S, Gupta A, Kumari V, et al. Emerging trends in nano-based antidiabetic therapeutics: a path to effective diabetes management. Mater Adv [Internet]. 2023 [cited 2024 Nov 14];4(15):3091–113. Available from: https://pubs.rsc.org/en/content/articlehtml/2023/ma/d3ma00159h.
141. Overton HA, Fyfe MCT, Reynet C. GPR119, a novel G protein-coupled receptor target for the treatment of type 2 diabetes and obesity. Br J Pharmacol [Internet]. 2008 [cited 2024 Nov 14];153 Suppl 1(Suppl 1). Available from: https://pubmed.ncbi.nlm.nih.gov/18037923/.
142. Arterburn JB, Prossnitz ER. G protein–coupled estrogen receptor GPER: molecular pharmacology and therapeutic applications. Annu Rev Pharmacol Toxicol [Internet]. 2023 [cited 2024 Nov 14];63:295. Available from: https://pmc.ncbi.nlm.nih.gov/articles/PMC10153636/.
143. Dimova R, Tankova T. The role of vaspin in the development of metabolic and glucose tolerance disorders and atherosclerosis. Biomed Res Int [Internet]. 2015 [cited 2024 Nov 14];2015:823481. Available from: https://pmc.ncbi.nlm.nih.gov/articles/PMC4402467/.
144. Li Z, Gao Z, Sun T, Zhang S, Yang S, Zheng M, et al. Meteorin-like/Metrnl, a novel secreted protein implicated in inflammation, immunology, and metabolism: a comprehensive review of preclinical and clinical studies. Front Immunol [Internet]. 2023 [cited 2024 Nov 14];14:1098570. Available from: https://pmc.ncbi.nlm.nih.gov/articles/PMC9998503/.
145. Dhankhar S, Chauhan S, Mehta DK, Nitika, Saini K, Saini M, et al. Novel targets for potential therapeutic use in Diabetes mellitus. Diabetol Metab Syndr [Internet]. 2023 [cited 2024 Nov 14];15(1):1–18. Available from: https://dmsjournal.biomedcentral.com/articles/10.1186/s13098-023-00983-5.
146. Gudoor R, Suits A, Shubrook JH. Perfecting the puzzle of pathophysiology: exploring combination therapy in the treatment of type 2 diabetes. Diabetology [Internet]. 2023 [cited 2024 Nov 14];4(3):379–92. Available from: https://www.mdpi.com/2673-4540/4/3/32/htm.
147. From data chaos to drug design: a new platform speeds discovery [Internet]. [cited 2024 Nov 14]. Available from: https://www.nature.com/articles/d42473-024-00326-6.
148. Hetmann M, Parigger L, Sirelkhatim H, Stern A, Krassnigg A, Gruber K, et al. Folding the human proteome using BioNeMo: a fused dataset of structural models for machine learning purposes. Sci Data [Internet]. 2024 [cited 2024 Nov 14];11(1):1–13. Available from: https://www.nature.com/articles/s41597-024-03403-z.
149. Desai D, Kantliwala S V, Vybhavi J, Ravi R, Patel H, Patel J. Review of AlphaFold 3: transformative advances in drug design and therapeutics. Cureus [Internet]. 2024 [cited 2024 Nov 14];16(7):e63646. Available from: https://pmc.ncbi.nlm.nih.gov/articles/PMC11292590/.
150. Muegge I, Oloff S. Advances in virtual screening. Drug Discov Today Technol [Internet]. 2007 [cited 2024 Nov 14];3(4):405. Available from: https://pmc.ncbi.nlm.nih.gov/articles/PMC7105922/.
151. Taherdoost H, Ghofrani A. AI's role in revolutionizing personalized medicine by reshaping pharmacogenomics and drug therapy. Intell Pharm. 2024;2(5):643–50.
152. Klonoff DC. Personalized medicine for diabetes. J Diabetes Sci Technol (Online) [Internet]. 2008 [cited 2024 Nov 14];2(3):335. Available from: https://pmc.ncbi.nlm.nih.gov/articles/PMC2769744/.
153. Marques L, Costa B, Pereira M, Silva A, Santos J, Saldanha L, et al. Advancing precision medicine: a review of innovative in silico approaches for drug development, clinical pharmacology and personalized healthcare. Pharmaceutics [Internet]. 2024 [cited 2024 Nov 14];16(3):332. Available from: https://pmc.ncbi.nlm.nih.gov/articles/PMC10975777/.
154. Mak KK, Wong YH, Pichika MR. Artificial intelligence in drug discovery and development. In: Drug discovery and evaluation: safety and pharmacokinetic assays [Internet]. 2024 [cited

2024 Nov 14]. pp. 1461–98. Available from: https://link.springer.com/referenceworkentry/10.1007/978-3-031-35529-5_92.
155. Patel AU, Gu Q, Esper R, Maeser D, Maeser N. The crucial role of interdisciplinary conferences in advancing explainable AI in healthcare. BioMedInformatics [Internet]. 2024 [cited 2024 Nov 14];4(2):1363–83. Available from: https://www.mdpi.com/2673-7426/4/2/75/htm.
156. Tilala MH, Chenchala PK, Choppadandi A, Kaur J, Naguri S, Saoji R, et al. Ethical considerations in the use of artificial intelligence and machine learning in health care: a comprehensive review. Cureus [Internet]. 2024 [cited 2024 Nov 14];16(6):e62443. Available from: https://pmc.ncbi.nlm.nih.gov/articles/PMC11249277/.

Chapter 22
Towards a Mechanistic Understanding of Atherosclerosis Drug Design

Sudhanshu Kalantri, Ami Thakkar, Amisha Vora, and Rajasekhar Reddy Alavala

Contents

Abstract Atherosclerosis remains a major contributor to cardiovascular morbidity and mortality worldwide. Key therapeutic approaches included lifestyle modifications, targeting inflammatory pathways, enhancing lipid management, and exploring immunotherapies. Recent advancements have highlighted the complexity of atherosclerosis, which involves complex processes such as inflammation, immune

S. Kalantri · A. Thakkar · A. Vora (✉) · R. R. Alavala
Shobhaben Pratapbhai Patel School of Pharmacy & Technology Management, SVKM's Narsee Monjee Institute of Management Studies, Mumbai, India
e-mail: amisha.vora@nmims.edu

S. N. K. Rao G., R. R. Alavala (eds.), *Applications of Computational Tools in Drug Design and Development*, https://doi.org/10.1007/978-981-96-4154-3_22

responses, and lipid metabolism. Computational modeling has emerged as a powerful tool in simulating these mechanisms, predicting disease progression, and testing potential treatment strategies. This enables the development of personalized therapies tailored specific for an individual patient. Structural bioinformatics further aids in drug discovery by identifying and validating therapeutic targets, while network-based computational approaches, which analyze protein-protein interactions, gene regulatory networks, and metabolic pathways, present innovative opportunities for drug design.

Network analysis tools like STRING, Cytoscape, and DAVID are widely used for studying biological pathways and interactions. They help in identification of multiple targets. Using databases such as NCBI and STRING, researchers can uncover interactions between proteins and genes, perform cluster analysis, and predict potential therapeutic targets through gene ontology and pathway enrichment. This approach has proven useful in identifying disease mechanisms and drug development strategies for Atherosclerosis.

Additionally, Atherosclerosis can also be addressed through different strategies such as: targeting lipid metabolism with drugs like statins and fibrates, targeting inflammation with anti-inflammatory therapies, and reducing oxidative stress using antioxidants such as vitamins and natural supplements like resveratrol and curcumin. These approaches aim to prevent plaque formation and reduce cardiovascular risk.

In this chapter, we have discussed about targeting lipid metabolism using various enzymes and inhibitors such as HMG CoA reductase (HMGR) inhibitors, Anti-Proprotein convertase subtilisin/kexin type 9 (PCSK9), Peroxisome Proliferator-Activated Receptor (PPAR) α agonists, dual PPAR α and γ agonists, selective PPAR δ agonists and pan PPAR agonist, Cholesterol absorption inhibitors or Niemann-Pick C1-like 1 inhibitors, Adenosine triphosphate-citrate lyase (ACL) inhibitor, Angiopoietin like proteins (ANGPTL3) inhibitors, and Antisense oligonucleotide (ASO) therapies.

Rational drug design for atherosclerosis involves structure-based and fragment-based approaches to develop selective compounds targeting proteins like HMGR, PCSK9, and PPAR. Structure-based drug design leverages 3D protein structures to analyze active sites, perform molecular docking, and optimize compounds for improved binding affinity. On the other hand, Fragment-based design screens small molecular fragments and optimizes them for binding. Both methods use databases like ZINC and PubChem to identify novel drug candidates for treating atherosclerosis.

Keywords Atherosclerosis · Lipid metabolism · Oxidative stress · Statins and fibrates · HMG CoA reductase (HMGR) inhibitors · PCSK9 inhibitors · PPAR agonists · Cholesterol absorption inhibitors · Structural bioinformatics · Network-based drug discovery

Abbreviations

LDL	Low-Density-Lipoproteins
HDL	High-Density-Lipoproteins
IFN-γ	Interferon-gamma
M-CSF	Macrophage colony-stimulating factor
Ox-LDL	Oxidized Low-Density-Lipoproteins
Ox-HDL	Oxidized High-Density-Lipoproteins
CVD	Cardiovascular Diseases
CHD	Coronary Heart Disease
TLR	Toll like receptors
IgM	Immunoglobulin M
ADME	Absorption, Distribution, Metabolism, and Excretion
SAR	Structure-activity relationship
NLRP3	NOD-, LRR- and pyrin domain-containing protein 3
STRING	Search Tool for Retrieval of interacting genes and proteins
PTGS2	Prostaglandin-Endoperoxide Synthase 2
NOS3	Nitric Oxide Synthase 3
HMOX1	Heme Oxygenase 1
ESR1	Estrogen Receptor 1
CASP3	Caspase 3
CYP1A1	Cytochrome P450 Family 1 Subfamily A Member 1
CYP1A2	Cytochrome P450 Family 1 Subfamily A Member 2
MMP2	Matrix Metallopeptidase 2
MMP9	Matrix Metallopeptidase 9
GO	Gene Ontology
NM-IIA	non-muscle myosin II A
NPC1L1	Niemann-Pick C1-like 1
PCSK9	Proprotein Convertase Subtilisin/Kexin Type 9
LDL-R	LDL receptors
LDL-C	LDL-cholesterol
PPAR	Peroxisome Proliferator-Activated Receptor
SPPARM	selective PPAR-α modulator
NASH	non-alcoholic steatohepatitis
ACL	Adenosine triphosphate-citrate lyase
ANGPTL3	Angiopoietin like proteins 3
VLDL	Very Low-Density Lipoprotein
ASO	Antisense oligonucleotide
CAS	Chemical Abstract Service Registry
ITC	isothermal titration calorimetry
DHHA	3,5-dihydroxy heptanoic acid
OATP1B1	Organic Anion Transporting Polypeptide 1B1
AS/MS	Affinity Selection/Mass Spectrometry
LBD	Ligand Binding Domain

DBD	DNA binding domain
RXR	Retinoid-X receptor
CPT-1	Carnitine Palmitoyl Transferase 1
MCAD	Medium Chain Acyl Dehydrogenase
LPL	Lipoprotein Lipase
NTD	N-terminal domain
TMD	Transmembrane Domain
MTD	Middle Luminal Domain
SSD	Sterol Sensing Domain
EGCG	epigallocatechin gallate

22.1 Introduction

22.1.1 Overview of Atherosclerosis as a Major Cardiovascular Disease

Atherosclerosis is a prevalent cardiovascular disease characterized by the accumulation of plaque in the arteries, leading to narrowing and hardening of blood vessels. The accumulation of lipids, cholesterol, fats, calcium, and other substances forms plaque within the artery walls, causing inflammation and the development of atherosclerotic lesions. This condition poses significant risks for heart attacks, strokes, and other cardiovascular complications. Factors such as high blood pressure, high cholesterol levels, smoking, diabetes, and a sedentary lifestyle contribute to the development and progression of atherosclerosis.

Based on current knowledge, atherosclerosis is a complex condition influenced by various factors, including, higher levels of oxidative stress, varied lipid metabolism, flawed mitochondrial function, and chronic inflammation [1, 2].

The Pathophysiology of Atherosclerosis

Injury or damage to the endothelial lining of the artery wall causes inflammation, draws immune cells like monocytes and T-lymphocytes to the affected area. These cells move through the lining and enter the inner layer of the artery: tunica intima, together with Low-Density-Lipoproteins (LDL) and High-Density-Lipoproteins (HDL) (Fig. 22.1). Under the influence of specific substances like interferon-gamma (IFN-γ) and macrophage colony-stimulating factor (M-CSF), the monocytes transform themselves into macrophages within the tunica intima. While in the tunica intima, both LDL and HDL get oxidized by free oxygen radicals forming Oxidized LDL Ox-LDL and Ox-HDL. Macrophages consumes the Ox-LDL, turning into foam cells that later undergoes apoptosis [3]. The debris left in the inner layer, tunica intima, forms a mass called an atheroma. Foam cells and endothelial cells release a chemical called monocyte chemoattractant protein-1 (MCP-1) to attract more monocytes. T cells within the artery change into specific types that can release IFN-γ [4]. Smooth muscle cells are also attracted to the tunica intima, forming a

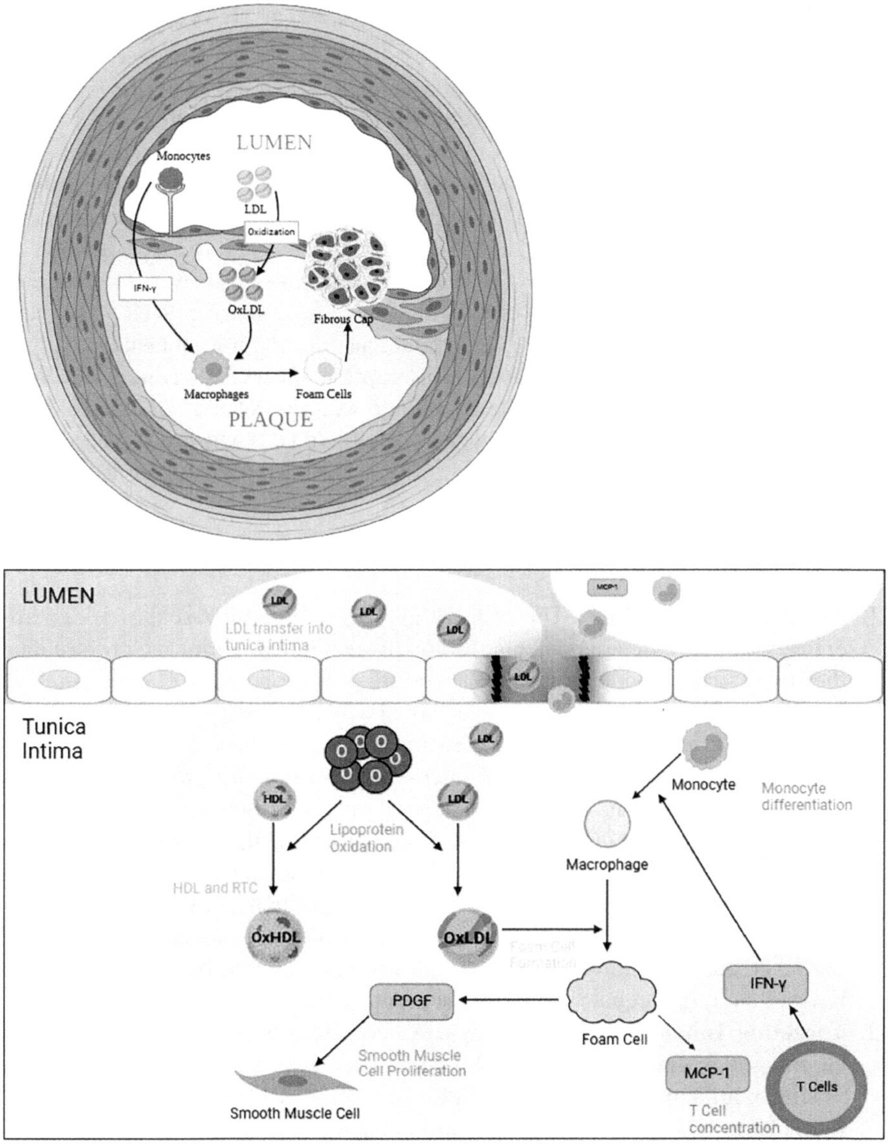

Fig. 22.1 Pathophysiology of Atherosclerosis

protective cover in the artery wall. This accumulation of cells and waste can lead to the swelling of the artery wall, restricting blood flow and causing stenosis as shown in Fig. 22.1. If this protective cover is ruptured, the material in the inner layer is released into the bloodstream, increasing the risk of blockages [5]. Complications may arise, such as the formation of clots at the atheroma site, further blocking blood flow.

22.1.2 *Need for Novel Therapeutic Agents for Atherosclerosis*

Atherosclerosis remains a leading cause of cardiovascular morbidity and mortality around the world, making it compulsory to develop novel therapeutic agents to address the complex pathophysiology of this disease effectively. According to WHO, Cardiovascular diseases (CVD) are the leading cause of death annually. In 2019, 17.9 million people died of CVDs, contributing to 32% of total global deaths. By 2030, these figures can rise to 24 million deaths globally and cost one trillion US dollars [6, 7]. Furthermore, the WHO data factsheet states that over three-quarters of CVD deaths take place in low- and middle-income countries. WHO also states that by addressing behavioral risk factors such as tobacco use, unhealthy diet and obesity, physical inactivity, and alcohol consumption, most CVDs can be prevented.

Predominantly, CVD is associated with age and age-related conditions. It has been observed that men with coronary heart disease (CHD) who are 80 years of age or older have the disease five times more commonly than those who are 40–59 years of age [8].

Early detection of CVD is crucial for the commencement of medication and counseling-based therapy.

1. **Lifestyle Modifications:** The primary approach to cardiovascular risk control emphasizes lifestyle adjustments. Although such interventions lack robust data, they offer several evident benefits. Therefore, it is proposed to embrace lifestyle modifications cautiously, much like Pascal's Wager, given their minimal risk. Cutting on smoking, weight management, and physical activity should be accompanied by pharmacological intervention in primary prevention [9]. The early implementation of pharmacological antiatherosclerosis therapy raises complex societal, ethical, and financial concerns. While the validation of biomarkers indicating risk due to specific genetic variations or pathways may potentially justify pharmaceutical intervention in ostensibly healthy individuals in the future, this concept aligns with the prospects of personalized medicine [10]. However, currently, only a few biomarkers or interventions beyond LDL control have undergone adequate validation for practical integration into clinical settings.
2. **Targeting Inflammatory Pathways:** Atherosclerosis is now recognized as a chronic inflammatory disease, highlighting the importance of therapies that can modulate inflammatory processes within arterial walls. Various targets have also been identified for the inflammatory mechanism of atherosclerosis [11]. Some of the targets are Th1, Th2 cytokines, B cells, Leukocyte recruitment and migration, PLA2, Toll like receptors (TLR), Leukotrienes, CB-1, CB-2 receptors, CD154–CD40, CD28–CD80/CD86 and many more [12].
3. **Enhancing Lipid Management**: Despite the success of statins in managing cholesterol levels, there is a need for alternative agents that can effectively target lipid abnormalities associated with atherosclerosis, particularly in patients who are statin-resistant or intolerant [13].
4. **Antioxidant Interventions:** Researchers yielded disappointing results when they tried to target oxidant pathways in treating atherosclerosis. OxLDL is

widely considered a crucial factor in the development of atherosclerosis. However, attempts to inhibit phospholipases, which can generate harmful mediators from oxidized lipoproteins, have not been successful. For example, inhibiting lipoprotein-associated phospholipase A2 [14] inhibitor Darapladib did not yield positive outcomes in human clinical trials, despite its promising results in atherosclerotic pigs [15–17]. Similarly, direct targeting of oxidative stress and lipoprotein oxidation have also proven ineffective. Also, antioxidant vitamins C, D, E, and beta carotene, did not show significant results in clinical trials [18–20]. These results have led to a decrease in enthusiasm for pursuing these approaches as potential therapeutic targets.

5. **Vaccinations and Immunotherapies:** The role of adaptive immunity in experimental atherosclerosis has led to several efforts in translating potential therapeutic interventions to be used for treating humans. Major observations in experiments demonstrated that natural antibodies like IgM, which targets specific epitopes that are associated with OxLDL or specific infectious agents, have the potential to restrict atherosclerosis [21]. This discovery led to the exploration of methods to stimulate the body's antibodies to potentially reduce the development of atherosclerosis in humans. One of the specific approaches that showed promising results in animal models involved the utilization of apolipoprotein B [22–24]. However, there were challenges in directly applying findings from animal models to the genetically diverse human population. Another immune-based therapy involved the use of very low doses of IL-2 to stimulate regulatory T cells which produced anti-inflammatory and profibrotic cytokines [25]. This study is currently in its early phase but it represents an initial effort to potentially minimize the progression of atherosclerosis.
6. **Promoting Vascular Health**: Therapeutic agents that promote endothelial health, prevent plaque rupture, and enhance vascular repair mechanisms could play a crucial role in preventing cardiovascular events associated with atherosclerosis [26].
7. **Personalized Medicine Approaches**: Tailoring treatments based on individual patient characteristics, genetic factors, and disease progression could lead to more precise and effective therapies for atherosclerosis management [27].

By investing significantly in the research and development domain novel therapeutic targets can be identified, and pharmaceutical companies, researchers, and clinicians can use innovative treatment techniques that have the potential to transform the management of atherosclerosis and ultimately improve patient outcomes.

22.2 Role of *In Silico* Approaches in Drug Design

In silico approaches, are computational techniques, that play an important role in drug design by providing valuable insights into molecular interactions, structural analysis, Structure-activity relationship, and pharmacokinetic properties of the potential drug candidates [28]. These techniques are advantageous as they are responsible for reducing the time and costs associated with traditional laboratory-based drug development [29].

They also have significant role in:

1. ADME Prediction: In silico tools predict the Absorption, Distribution, Metabolism, and Excretion of a potential drug candidate. This aids in the identification of potential issues early in the drug development process, thereby allowing for more efficient optimization and improving its efficacy [30].
2. Molecular Docking and Virtual Screening: In silico molecular docking simulations help in the prediction of the interaction of small molecules with target proteins, thus, helping in identifying potential drug candidates [31]. Virtual screening, which is a powerful tool, allows us to conduct fast and comprehensive screening of large compound libraries to identify promising potential molecules [32].
3. Structure-activity relationship: SAR models help in demonstrating the relationship between chemical structure and biological activity. With this, they also provide insights into improved pharmacological profiles and the design of the compound. This modeling helps in deriving stable, specific, and effective, molecules [33].
4. Pharmacophore Modelling: Key structural features also known as 'pharmacophores', that are necessary for the interaction of the compound can be identified using this tool. By analyzing the structures of potential molecules, pharmacophore models are generated to represent the essential features required for biological activity [34].
5. Machine Learning and Artificial Intelligence: A combination of machine learning and AI techniques assists in data analysis, pattern recognition, and predictive modeling. These tools also help in recognizing novel drug-target interactions and enhancing lead optimization [35].

22.2.1 Understanding Atherosclerosis Through Computational Modelling

As our understanding of the etiology and pathophysiology of atherosclerosis has advanced, the disease is viewed as a complex disease interlinking multiple pathways of inflammation, innate immune responses, and lipid metabolism. Atherosclerosis has emerged as a consequence of multiple dynamic cell processes. Studies aimed at understanding atherosclerosis and designing therapeutic strategies hence need to be broad in scope and integrative in nature [36]. Computational modeling techniques can effectively be used to model various factors of the disease and interlinkages between them so that various disease mechanisms are well understood, disease progression can be effectively predicted, different treatment strategies can be hypothesized and evaluated and personalized medicines can be evaluated by incorporating the vast existing data [37]. It serves as an in-silico experimentation system to understand complex disease biology and strategically design therapeutics considering the multiple interconnected pathways of the disease [38].

In the case of atherosclerosis, multiple models have been developed in terms of ordinary or partial differential equations, alongside binary and multi-valued logic and process algebra like pi and kappa calculus. Statistical models are developed that can read pathway structures and generate hypotheses from the existing data points. Earlier, cholesterol synthesis, the role of lipid metabolism, and cardiovascular disease in aging have been studied, however recently, blood flow dynamics (Navier–Stokes equations), vascular damage, arterial wall shear stress, inflammation, plasma concentrations of LDL, LDL penetration into tunica intima, LDL oxidation, the role of HDL, monocyte recruitment, monocyte to macrophages differentiation, foam cell formation, T-cell recruitment and role of IFN-γ, plaque rupture, and thrombosis have been considered while developing the computational models [39, 40]. It should be noted that there are various factors, that are not, yet modeled which include the role of triglycerides in foam cell formation, types of HDL and their relative proportions, LDL particle number, the role of neutrophils, nitric oxide, heat-shock proteins, cell signaling proteins like mi-RNA and NLRP3 [41]. Including these factors in model development can increase our understanding of atherosclerosis in more depth and predict many novel targets to design more precise therapeutic strategies. More work is needed in the direction of atheroma rupture and its consequence in terms of thrombus development. Much work can be carried out in the direction of therapy development integrating various models developed by different scientists [42]. In addition, computational models can be parameterized using data obtained from scanning and imaging techniques multi-scale modeling can be used to derive multiple targets and predictive modelling can be used to understand the progression of disease under various conditions [43]. However, there are challenges associated with developing computational models for atherosclerosis. The genetic factors are important to be included in the model development and this poses logistical and ethical problems. Similarly, access to live atheroma is limited and disturbances can lead to atheroma rupture [40, 44].

22.2.2 Structural Bioinformatics Approaches for Target Identification and Validation

The very first step in drug discovery is to identify the target protein or gene that may contribute to the disease and validate the same. Many failures in Phase II and III of clinical drug development are most times associated with off-target side effects and cross-reactivity [45]. One of the major reasons for this unexpected outcome centralizes on issues with unique target identification which may be protein, gene, biochemical pathways, or disease-associated biomarkers [46]. The step of specific target identification is challenging, especially in polygenic diseases. Experimental target identification depends on the characterization of the protein of interest followed by experimental validation through gene knockout, animal studies, and site-directed mutagenesis. Traditional methods of target identification include data mining, biochemical methods like affinity chromatography, or genetic methods involving gene manipulation of expression or function. However, it is impossible to predict off-targets using these methods. In the post-genomic era, there is an exponential rise in accessible biological data generated through bioinformatics. It has helped drug discovery scientists to understand comprehensively the biological system relevant to the disease condition. The use of high computational power techniques is required to handle this huge amount of biological data rationally to facilitate drug design and redesign. Computational techniques are also used to model the disease and predict the targets. These methods have helped repurposing old drugs for different disease conditions. *In silico* techniques have been used for predicting side effects and pharmacokinetic parameters, achieving desired selectivity and potency. This has reduced the time and money required to introduce a successful drug candidate in the market.

In recent years, network-based computational approaches have been used to identify targets. Various computational techniques are used to identify key molecules known as nodes that could be a potential therapeutic target for a particular disease. These approaches rely on the structure and dynamics of the biological network which includes protein-protein interaction, ligand-protein interaction, gene regulatory networks, and metabolic networks to understand interconnections between different pathways. These interactions are known as edges. Hence in the case of protein-protein interaction networks (PPIN), nodes represent proteins and edges represent the physical interactions between them. In the case of gene regulatory networks (GRNs), nodes represent genes and edges represent the regulatory pathways such as transcription in metabolic networks, nodes represent the metabolic enzymes and metabolites, whereas the edges represent the biochemical reaction of metabolism. Computational approaches for network analysis include network topology analysis to identify nodes that may be crucial for network function and stability, pathway analysis helps to identify subnetworks important in a particular disease state, network propagation helps to simulate the spread of information through the network to identify the most influential nodes and machine learning approaches can identify potential targets based on predictive analysis of the patterns in the data.

Network-based approaches consider the entire system with interconnections and overlaps of processes in contrast to the single pathway or molecule in the traditional approach. In addition, network analysis identifies the non-obvious targets that are not evident through traditional approaches of target identification. It allows the combination of data from multi-omics i.e. genomics, proteomics and metabolomics to enhance the overall process of target identification [28].

Steps of network-based approach for target identification:

1. Retrieving relevant heterogeneous data sets from different sources
2. Data cleaning
3. Data integration into a unified network
4. Calculation of centrality metrics

5. Network clustering to identify sub-networks
6. Calculating functional similarity scores
7. Pathway enrichment analysis
8. Identification of disease-associated proteins or gene targets

Besides biological network analysis, machine learning-based methods have enhanced the extraction and processing of information and datasets using biological knowledge and computational algorithms. Here the workflow involves several stages which includes

1. Biological data collection involving gene expression profiles, proteomics, and phenotypic profiling data
2. Chemical data collection which includes structures of chemical compounds and their bioactivity
3. Literature survey about known target ligand interactions
4. Data processing which includes the steps of data normalization, feature extraction (extraction of structural motifs, physicochemical properties, interaction fingerprints), dimensionality reduction (utilizing PCA or t-SNE), and data cleaning to eliminate outliers and missing values
5. Feature selection using statistical analysis and machine learning techniques like Lasso, random forests, and feature engineering to create new features to improve model performance
6. Model selection to choose the correct machine learning algorithm
7. Model training using training datasets and model optimization using grid search random search or Bayesian optimization
8. Model evaluation using evaluation metrics and validation using cross-validation
9. Application of trained and validated model for identification of targets for new compounds or new targets of known compounds.
10. Experimental validation by in-vitro or in-vivo evaluation
11. Integration with existing drug pipeline.

Another computational approach for target identification includes reverse docking which includes a structure-based approach of virtual screening against relevant protein databases, followed by molecular docking simulations and scoring targets according to the best binding mode and interaction with ligand. These outputs can be used to find the druggability index of the target. Reverse docking is a valuable approach for repurposing existing drugs.

Various Tools for Network Analysis

Search Tool for Retrieval of interacting Genes and proteins (STRING) is a biological database for predicting protein-protein interactions. It is a web-based tool where its users can perform network analysis, view interaction data, and download the results. It encompasses physical and functional aspects of proteins, collects evidence from databases, and literature, and takes into account the co-expressed genes and neighborhood genes and gene fusions. It provides graphical representations of interactions, assigns confidence scores based on supporting evidence, and performs functional enrichment to identify overexpressed biological pathways within the network. The tool can be used to identify novel proteins that are not sufficiently characterized, identify proteins in a particular biochemical pathway, and help understand the physiological basis of the disease and hence identify different targets.

Several other tools and software widely used for network analysis include Cytoscape, Gephi, Pajek, BiNGO, DAVID, Metscape, NAViGaTOR, GeneMania, Pathway Commons, and NetworkAnalyser.

22.2.3 Systems Biology and Network Analysis of Atherosclerotic Pathways

In regards to the above discussion concerning atherosclerosis, which is a multifactorial disease characterized by the development of atherogenic plaque in the wall of arteries under the influence of inflammatory processes, network analysis using computational techniques can be used for identifying the target genes or proteins. The development and progression of atherosclerosis involves abnormality in a variety of metabolic pathways. The traditional approach of target identification can result in the identification of a single target whereas the network medicine approach can allow to exploration of a large array of genes and proteins as targets. In one such study, an NCBI database of 988 atherosclerotic genes was retrieved. Using the STRING database, 22,693 interactions with 872 nodes (genes) were found with different confidence scores. This was followed by cluster analysis using MCODE and plug-in of the software Cytoscape resulted in 18 clusters. Topological parameter and gene ontology analysis resulted in the selection of four genes as the most influential genes which included AGT, LPL, IRS1, and ITGB2. Further, miRNA targeting these genes was obtained using the MIENTURNET web tool which can be developed to restrict the development of atherosclerosis [47].

The systems biology and network analysis approach is widely used to understand the mechanism of action, the possible targets, and the multitude of effects of constituents of natural extracts may have. For example, Hong Duan et al. used a network pharmacology approach to study the possible mechanism of a traditional Chinese medicine Schisandra against atherosclerosis. In this study, the STRING database and Cytoscape software were used to build the component pathway target interaction network. Gene Ontology (GO) and KEGG pathway analysis were used for cellular, molecular, and biological pathways while the Medscape database was used to understand the interceptive acquaintances of pathways. Initial molecular docking of constituents resulted in the possible involvement of non-muscle myosin II A (NM-IIA) against atherosclerosis. GO and KEGG pathway analysis resulted in 525 possible targets for 18 active constituents of Schisandra. They further pointed to 1451 possible pathways responsible for the pathogenesis of the disease which was refined to 167 targets based on common signalling target pathways. Molecular docking identified compatible binding between the constituents of Schisandra and

NM IIA. It was concluded that Schisandra had a possible target as a serotonin synapse and had a close intersection with tumorigenic pathways through PTGS2, NOS3, HMOX1, and ESR1. The study also helped to conclude that the constituents of Schidandra had association with immune-inflammation, oxidative stress, and neuroendocrine pathways due to which it has potential therapeutic utility against atherosclerosis [48].

One more study on traditional plant-based formulation prescribed for hyperlipidemia, containing five herbs was carried out using a similar network pharmacology approach. Their workflow included searching for constituent herbs and chemical information in the Korean traditional knowledge database and confirmation of 2D structure, MW, and other physicochemical properties using ChEMBL and PubChem databases. This was followed by screening 447 from five herbs using ADMET criteria and correlating compounds and genes using the STITCH database, where genes related to hyperlipidemia and atherosclerosis were identified using the GeneCards database. Further human genes were identified and subjected to KEGG pathway analysis. This resulted in 10 compounds linked to 5 genes and are bioactive compounds. Naringenin, nobiletin, hesperidin, galanin, glycyrrhizin, homogentisic acid, stigmasterol, 6-gingerol, quercetin, and glabridin were identified as active compounds linked to core genes CASP3, CYP1A1, CYP1A2, MMP2, and MMP9. Hence compound target gene network analysis helped in understanding the potential effects of the formulation for treating hyperlipidemia and atherosclerosis [49].

Following is an example of a network map of compound target gene interaction where red rectangles represent the compounds and blue ovals represent different genes (Fig. 22.2).

3.1 Approaches to the treatment of atherosclerosis
3.2 Immunomodulatory approaches and inflammation-targeted therapies
3.3 Antioxidative approaches towards the treatment of Atherosclerosis

22.3 Approaches to the Treatment of Atherosclerosis

The main cause of atherosclerosis is the build-up of fat LDL-cholesterol in arteries, leading to inflammation, formation of atheroma, and a serious cardiovascular event. There could be different approaches to taking care of atherosclerosis (Fig. 22.3).

1. **Targeting lipid metabolism** to maintain the normal levels of LDL-cholesterol and triglycerides: These classes of compounds include currently approved drugs like statins, fibrates, ezetimibe and bempedoic acid and their combination
2. **Targeting inflammation and immune response**: Studies and clinical trials are going on in this category based on knowledge about the pathophysiological mechanism of inflammation in atherosclerosis and the known small molecules and biological anti-inflammatory therapies available for other inflammatory conditions

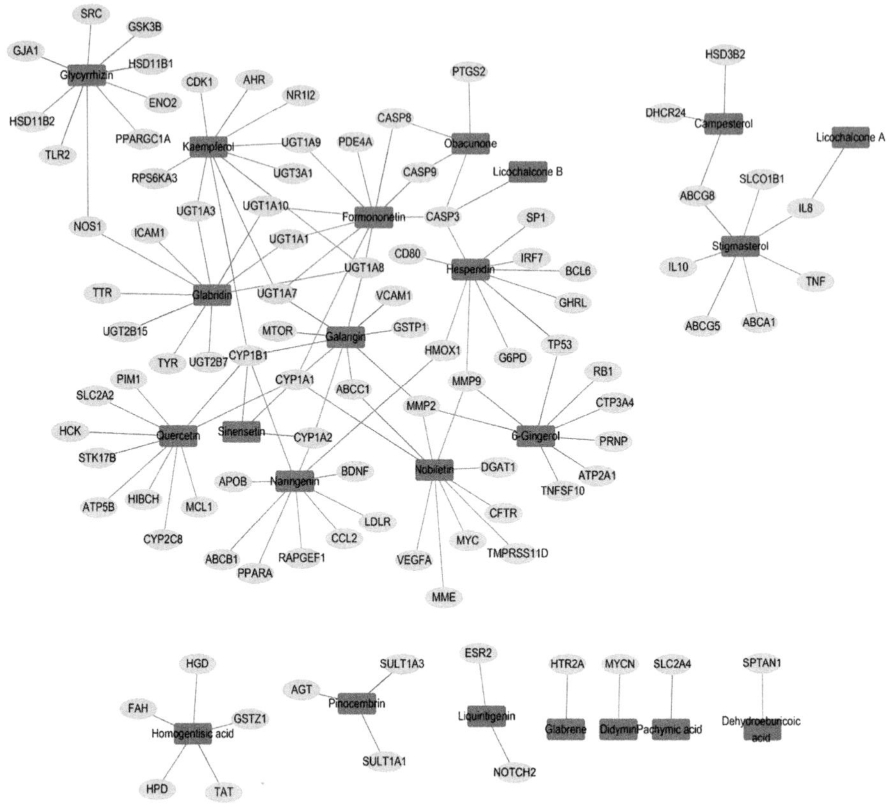

Fig. 22.2 Network map of compound and target gene interactions

3. **Targeting oxidative stress**: Oxidative stress is considered to be one of the factors contributing to pathophysiological mechanisms of atherosclerosis as it brings about the oxidation of LDL-C and oxidized LDL-C is a critical step in the formation of atherosclerotic plaques, antioxidants are considered beneficial in its treatment. Examples include vitamin E, vitamin C, Coenzyme Q10, beta carotene, resveratrol, curcumin, and flavonoids. These are natural supplements that can help mitigate atherosclerosis by various mechanisms.

22.3.1 Targeting Lipid Metabolism

As high levels of LDL-C are the crucial factor for the formation of atherosclerotic plaque, lipid-lowering therapies are the most important line of treatment for preventing adverse outcomes. Lipid-lowering therapies can be classified into

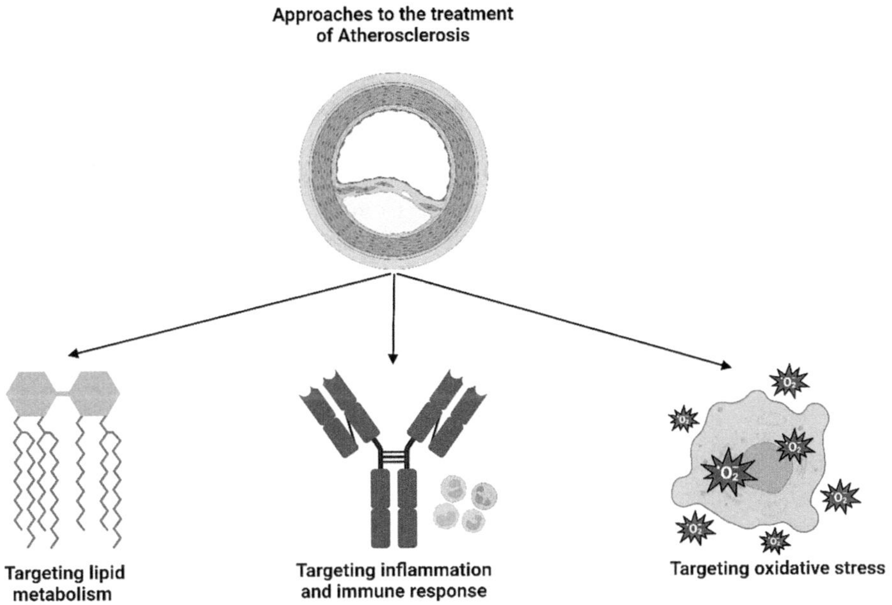

Fig. 22.3 Approaches to the treatment of Atherosclerosis

1. Drugs that interfere with cholesterol biosynthesis: HMG CoA reductase inhibitors, eg statins
2. Drugs that increase the uptake of cholesterol from the blood by hepatic cells: anti-PCSK9 monoclonal antibodies, eg Alirocumab and Evolocumab
3. Drugs that lower the levels of triglycerides and increase the levels of HDL-C: PPAR agonists, eg Fibrates
4. Drugs that act as cholesterol absorption inhibitors: Niemann-Pick C1-like 1 (NPC1L1) inhibitors, eg ezetimibe
5. Drugs that affect the de-novo synthesis of fatty acids: Adenosine triphosphate-citrate lyase inhibitors, eg Bempedoic acid

22.3.1.1 HMG CoA Reductase (HMGR) Inhibitors

Early observations by Virchow more than a decade ago suggested that 'atheroma' which was mainly responsible for atherosclerosis contained a yellow substance, later identified as cholesterol [50]. This established the role of lipids and lipoproteins in the pathogenesis of atherosclerosis. Elevated cholesterol levels were known to be associated with atherosclerosis and targeting the mevalonate pathway to reduce cholesterol synthesis was thought as one of the ways to manage this condition. HMG CoA reductase is the rate-limiting enzyme involved in the mevalonate pathway. This enzyme catalyzes the conversion of 3-hydroxy-3-methylglutaryl CoA to mevalonate, an important step in regulating the production of cholesterol in the body. The hypothesis that inhibition of HMG CoA reductase can effectively reduce cholesterol levels in the body initiated the search for compounds that block the enzyme.

The first success in the discovery of HMG CoA reductase inhibitors came from the work of Japanese scientist, Akira Endo on molds and mushrooms. In 1976, he reported the discovery and characterization of compactin or mevastatin as potent HMG-CoA reductase inhibitors, isolated from active cultures of the fungus *Penicillium citrinum.* Following this discovery, many pharmaceutical companies initiated their research in finding novel HMG CoA reductase inhibitors, and in 1979, Merck discovered mevinolin (later known as lovastatin) structurally similar to compactin from *Aspergillus terreus,* Since the discovery of lovastatin, many semi-synthetic and synthetic statins were introduced in the market which included simvastatin, pravastatin, fluvastatin, atorvastatin, rosuvastatin, pitavastatin [51, 52].

The structure of natural and semisynthetic statins viz lovastatin, simvastatin, and pravastatin possessed a decalin ring along with lactone ring which acted as a prodrug and hydrolyzed in vivo to generate polar head groups resulting in effective binding of drug to the active site of the enzyme. However, lower synthetic feasibility and higher incidence of side effects led to structural optimization. Further, the synthetic derivatives viz fluvastatin, cerivastatin, rosuvastatin, atorvastatin, and pitavastatin contained larger hydrophobic moieties like indole, pyridine, piperazine, pyrrole, and quinoline respectively (Fig. 22.4). All the drugs had 4-fluorophenyl rings attached to the heterocyclic ring, a small hydrophobic group like isopropyl or cyclopropyl along with the HMG mimicking polar heads.

22.3.1.2 Anti- Proprotein Convertase Subtilisin/Kexin Type 9 (PCSK9)

PCSK9 is identified as another target for prevention of atherosclerosis. PCSK9 is a protein predominantly expressed in liver cells and is involved in regulating cholesterol metabolism by modulating the LDL receptors (LDL-R) present in hepatic cells. LDL-R receptors are involved in the uptake of LDL-cholesterol (LDL-C) from blood to liver cells, maintaining the levels of blood cholesterol. PCSK9 binds to the LDL receptors (LDLR) present on the surface of liver cells and the complex gets internalized by lysosomes for degradation [53]. In short, PCSK9 is a serine protease-type protein that can degrade LDL-R after binding with the extracellular domain of LDL-R. Atherosclerotic events are known to be directly linked with elevations in the level of PCSK9 due to inflammatory changes in the walls of blood vessels. This leads to enhanced degradation of LDL-R and subsequently reduced clearance of LDL-C leading to high blood cholesterol levels. In addition, PCSK9 is known to increase plaque stability and elevated LDL-C levels can lead to endothelial dysfunction leading to deterioration of vascular health. Therefore, inhibition of PCSK9 can be a strategy to prevent atherosclerosis. FDA has approved two monoclonal antibodies alirocumab and evolocumab to inhibit or reduce PCSK9 activity [54].

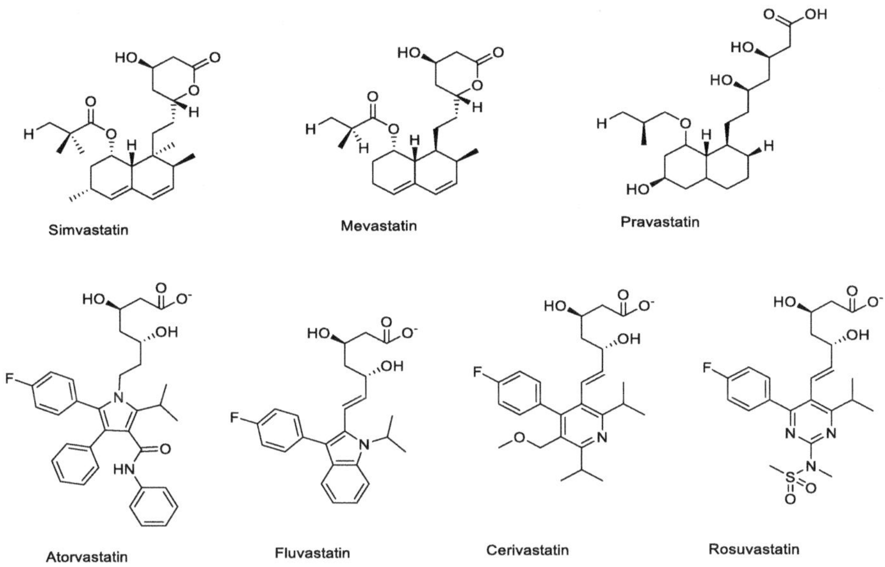

Fig. 22.4 The chemical structures of human HMGR inhibitors

22.3.1.3 Peroxisome Proliferator-Activated Receptor (PPAR) α Agonists, Dual PPAR α and γ Agonists, Selective PPAR δ Agonists and Pan PPAR Agonists

Though statins and PCSK9 inhibitors are beneficial in lowering blood LDL-cholesterol levels, cardiovascular events are not fully prevented by maintaining low levels of LDL-C. High levels of triglycerides and low levels of HDL-C in the blood are known to be additional risk factors for cardiovascular diseases. Drugs that are known to lower blood triglycerides and elevate HDL-C belong to the category of fibrates. Fibrates are known to be PPAR-α agonists. Approved fibrates include Fenofibrate, Clofibrate, Gemfibrozil, Bezafibrate, and ciprofibrate. In addition, ω-3 fatty acids can act as PPAR-α agonists (Fig. 22.5). In addition, selective PPAR-α agonist, also known as selective PPAR-α modulator (SPPARM), pemafibrate is under investigation for its higher potency and lesser side effects on account of its selectivity. In some countries, dual PPAR α and γ agonists like Saroglitazar have been approved which find benefits both in hyperlipidemia and associated diabetes (Diabetic dyslipidemia). Similarly, triple PPAR agonists i.e. agonists at PPAR α, γ, and δ like Elafibranor and Lanifibranor are under research for non-alcoholic steato-hepatitis (NASH) and other metabolic disease. Certain natural compounds like flavonoids can activate PPAR-α receptors and are recently under research.

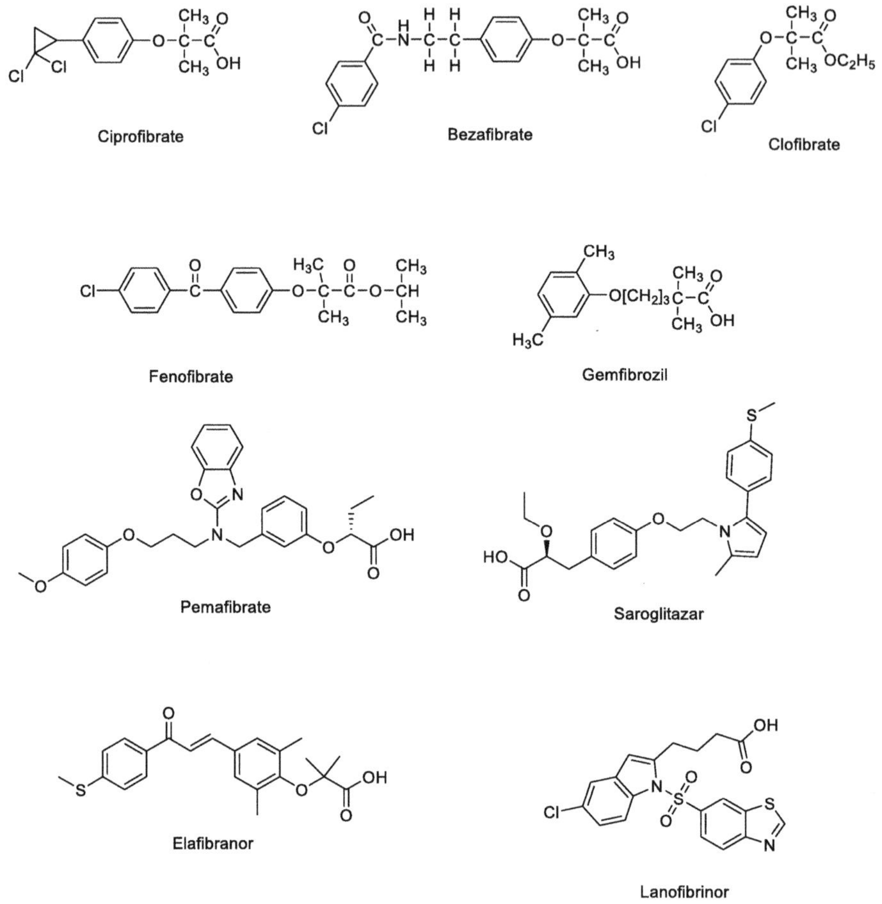

Fig. 22.5 Peroxisome Proliferator-Activated Receptor Agonists

22.3.1.4 Cholesterol Absorption Inhibitors or NPC1L1 Inhibitors

It is known that 30% of serum cholesterol comes from dietary cholesterol besides other processes like endogenous synthesis of cholesterol and the reabsorption of biliary cholesterol. Though statins target endogenous synthesis of cholesterol, intervention towards inhibition of absorption of dietary cholesterol could be an effective strategy to lower blood cholesterol levels. The uptake of dietary cholesterol and reuptake of biliary cholesterol is facilitated by a receptor called NPC1L1. Thus, NPC1L1 inhibitors as cholesterol absorption inhibitors offer a complementary approach to reducing serum cholesterol. The only drug approved as a cholesterol absorption inhibitor is ezetimibe which can lower LDL levels by 15% when taken alone (Fig. 22.6). However, many patients cannot tolerate high doses of statins, where the combination of ezetimibe with statin (low dose) offers a great advantage. The combination can effectively bring down the levels of LDL by 25%.

Fig. 22.6 Cholesterol absorption inhibitor—Ezetimibe

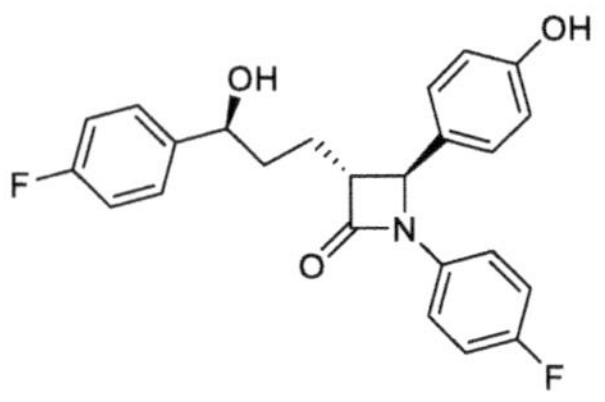

22.3.1.5 Adenosine Triphosphate-Citrate Lyase (ACL) Inhibitors

ACL is the rate-limiting enzyme involved in the de-novo synthesis of fatty acid which links carbohydrate and lipid metabolism. It is a cytosolic enzyme that catalyzes the Mg-ATP-dependent synthesis of oxaloacetate and acetyl-coenzyme A (acetyl-CoA) from citrate and coenzyme A. It is an important enzyme involved in cholesterol metabolism (Fig. 22.7).

Hydroxycitrate, initially isolated from fruit rinds of *Garcinia indica* was the first ever identified ACL inhibitor, which is known to occupy the citrate binding site on the enzyme and disrupt the transformation from CoA to acetyl CoA, thus disrupting levels of Acetyl CoA required for FA synthesis (Fig. 22.7) It is used as a natural supplement for obesity. However large doses of hydroxy-citric acid are required because of its low cell permeability on account of its hydrophilicity. Currently, only one drug, bempedoic acid is approved for the treatment of high-risk hypercholesterolemia patients who don't meet the goals of standard therapy. Another drug called BGT-002 or 326E developed by Burgeon Therapeutics is in Phase I clinical trials (Fig. 22.8).

22.3.1.6 Angiopoietin Like Proteins (ANGPTL3) Inhibitors

ANGPTLs are proteins structurally related to angiopoietin having multiple biological functions related to lipid metabolism. A total of 8 ANGPTLs have been identified out of which ANGPTL3 has been shown to regulate VLDL levels and play an important role in lipoprotein metabolism (Fig. 22.9). It has a significant role to play in atherosclerosis and cancer. ANGPTLs are known to be vascular growth factors, highly specific for endothelial cells, regulate inflammation, angiogenesis, cell death, senescence, and hematopoiesis, and have both pro-atherosclerotic and atheroprotective effects. Hence abnormalities in the regulation of ANGPTL expression is associated with variety of cardiovascular diseases and metabolic disorders like atherosclerosis and diabetes, obesity and cancer. ANGPTL3 loss of function is known to be associated with lowering of circulating LDL-C, triglyceride-rich lipoproteins. Direct inhibitors of ANGPTL-3 have been recently developed. The monoclonal antibody against ANGPTL3 known as Evinacumab has been recently approved in USA and Europe. Antisense oligonucleotides against ANGPTL3, ANGPTL3-L_{Rx} are currently under clinical trials.

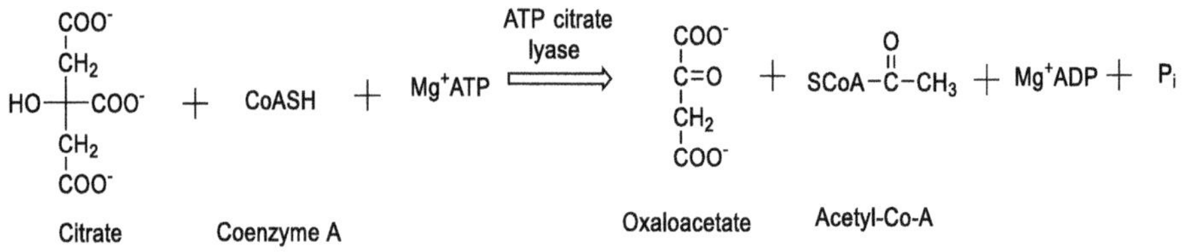

Fig. 22.7 Cholesterol Metabolism Reaction

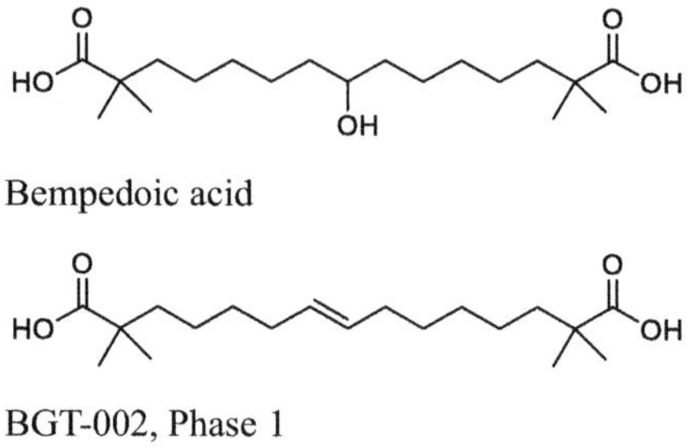

Fig. 22.8 Bempedoic Acid and BGT-002, Phase I

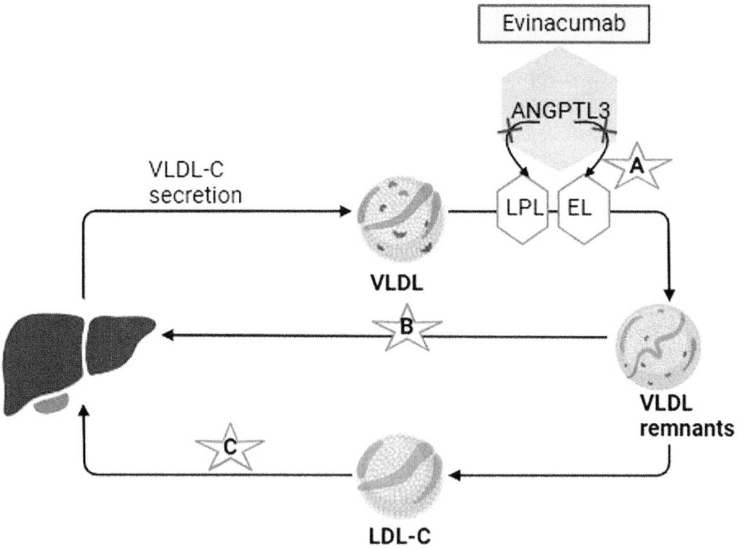

Fig. 22.9 Angiopoietin-like proteins (ANGPTL3) inhibitors

A. ANGPTL3, an angiopoietin-like protein, plays a prominent role in the regulation of lipid metabolism by blocking LPL and EL.
B. Some clearance of remnant particles is still possible through VLDL remnant receptors.

C. Diminished/absent LDLR function slows/prevents LDL clearance, thereby increasing LDL-C levels

22.3.1.7 Antisense Oligonucleotide (ASO) Therapies

Antisense oligonucleotides are small versatile molecules that can be designed to target specific RNA transcripts to modify specific protein expressions. These molecules have shown high potential in treating genetic disorders and rare diseases. The oligonucleotides interact with a specific sequence of the RNA transcripts leading to limited expression of the target protein. ASO therapies also siRNA therapies where small double-stranded RNAs divide into single strands, and bind to the distinct target mRNA, bringing about its degradation, further halting its translation and inducing gene suppression (Fig. 22.10).

One such siRNA therapy called inclisiran (by Novartis, an ASO therapy against protein PCSK9 has been approved by the FDA in Dec 2021 for the treatment of heterozygous familial hypercholesterolemia (HeFH) and atherosclerotic cardiovascular disease (ASCVD) in adults. With two maintenance doses per year, it is effective in providing LCL-C reduction of up to 52%. It is indicated as an adjunct to diet and maximally tolerated statin therapy in clinical ASCVD and HeFH who require additional lowering of LDL-C.

The future of treatment of hypercholesterolemia is expected to be based on ASO therapies against proteins like apolipoprotein CIII and silencing of PCSK9 mRNA.

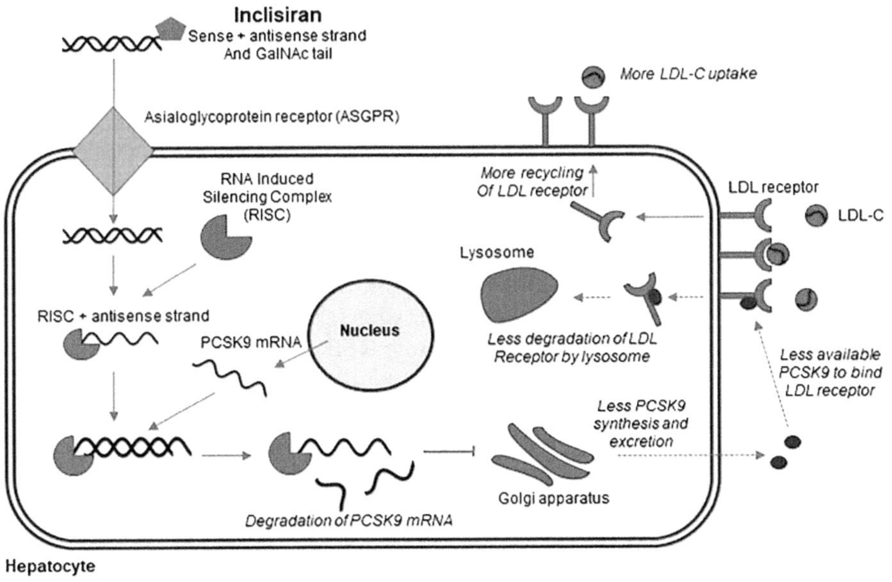

Fig. 22.10 Antisense oligonucleotide (ASO)

22.3.2 Immunomodulatory Approaches and Inflammation Targeted Therapies

Atherogenesis is considered to be an interplay between alterations in lipid metabolism, oxidative stress, and inflammation. Inflammation involves innate and adaptive immunity as well as a cascade of various cytokines. Most FDA-approved drugs for atherosclerosis target lipid metabolism and attempt to correct the levels of cholesterol and lipoproteins, targeting inflammation could be another attractive approach to the treatment of atherosclerosis. As inflammatory pathways are well-established targets for other diseases, these treatments could be beneficial to combat atherosclerosis. Also, some studies of anti-inflammatory therapies on animal models of hyperlipidemia have shown beneficial effects in terms of delaying atherosclerosis progression. Anti-inflammatory drugs like colchicine, allopurinol, and methotrexate are being explored for controlling inflammation during atherogenesis. The US FDA has approved low-dose colchicine (0.5 mg daily) as an anti-inflammatory therapy along with statin to reduce the risk of myocardial infarction and stroke. CRP, IL-1β, IL-6, IFN-γ, and TNF-α are some of the inflammatory targets widely explored for other inflammatory diseases and could also be atheroprotective therapies. Though these therapies have been found effective in pre-clinical models, their ineffectiveness in clinical trials has blocked their development in the case of atherosclerosis.

In addition to anti-inflammatory, immunomodulatory therapies can also help in preventing the progression of atherosclerosis. This is because there has been a growing recognition of the fact that the immune system plays a significant role in the development of atherosclerosis. Macrophages and Monocytes are crucial cellular components in the following process. Chemokine receptors such as CCR2, CCR5, and CX3CR1 and their ligands play a significant role in preferential recruitment of inflammatory monocytes in atherosclerotic plaques [55]. Ultimately, the development of atherosclerotic plaques is caused by the accumulation of monocytes [56].

The adaptive immune system can also play a key role in the development of atherosclerosis [57]. T-cells and B-cells are important parts of adaptive immune systems. Also, by producing cytokine and recognizing antigens, inflammatory cells, mast cells, and other cells contribute to the formation of lesions [58]. The activated B cells produce an antibody response to block the antigen. These B cells further divide themselves into B1 and B2 cells. The formed B1 cells produce IgM antibodies and protect against atherosclerosis [59], whereas the formed B2 cells produce IgG antibodies and contribute to the progression of atherosclerosis [60–62]. Moreover, T-cells also play a crucial role in the formation of atherosclerosis [57]. Their helper T-cells have subtypes which include, Th1, Th2, Th17, and T regulators (Treg). Th1 cells contribute to the progression of atherosclerosis, while Treg cells act as a protective against atherosclerosis. Th2 cells show both protective properties and harmful activities against atherosclerotic lesions [63]. The role of helper T cell Th17 is in autoimmune disease but its role in atherosclerosis is not clear yet.

Additionally, lipoprotein A, OxLDL, beta2 glycoprotein1, and heat shock proteins, are stimuli having an association with the progression of Atherosclerosis.

Also, there are some foreign antigens such as certain bacteria and viruses that have an impact on atherosclerosis [56]. Table 22.1 gives a list of inflammation-targeted therapies that could potentially find benefit in atherosclerosis.

Table 22.1 list of inflammation-targeted therapies that could potentially find benefit in atherosclerosis

Sr. No.	Target	Therapy	Functions
1.	**Th1 Cytokines**		
	IL-1β	Canakinumab, Anakinra, Rilonacept	Monoclonal Antibody that neutralizes IL-1β (antagonist of IL-1)
	IL-6	Tocilizumab	A monoclonal antibody that blocks IL-6R
	IL-12	Ustekinumab, Briakinumab	A monoclonal Antibody for IL-12 and IL-23
	TNF-α	Tanercept, Infliximab, Adalimumab	Antibody that neutralizes TNF-α
	IFN-γ	Fontolizumab	Antibody against IFN-γ
2.	**Th2 Cytokines**		
	IL-4	CYM5442, FTY720	Sphingosine-1-phosphate receptor agonist, and inducing IL-4Rα
	IL-13		Atheroprotective effect
	IL-33		
3.	**B cells**		
	CD20	Rituximab	Antibody against CD20
	BAFF-R	Tabalumab	IgG4 Monoclonal Antibody for blocking BAFF-R
4.	**Co-stimulatory molecules**		
	CD28–CD80/CD86	Abatacept	Fusion Protein reduces the progression of plaque
	CD80/CD86	Belatacept	Fusion protein against CD80 and CD86
	CD80	RhuDex	Inhibitor of CD80
	CD154	Ruplizumab, ABI793	Monoclonal Antibody against CD154
	CD40	CP-870,893, Dacetuzumab, ChiLob 7/4, Ch5D12	Antibodies for blocking CD40
	OX40L–OX40		Antibody Fusion protein blocking the interaction of Anti-OX40L and OX40
5.	**TLRs pathway**		
	TLR4	Eritoran	TLR4 antagonist interferes between TLR4 and MD2
	TLR2		Neutralizing Antibody against TLR2

(continued)

Table 22.1 (continued)

Sr. No.	Target	Therapy	Functions
	TRAF6 and IRAK1		miR-146 inversely controls the mRNA levels of TRAF6 and IRAK1
	p38		Inhibition of p38
	IkB kinase β/2		Antibodies neutralizing IkB kinase β/2
	NF-kB	DHMEQ, Pterostilbene, Artesunate	Inhibits NF-kB, and high fat-induced atherosclerosis
6.	**Leukocyte recruitment and migration**		
	CCR2	MLN1202	Antagonistic action
	P-Selectin	RO4905417	Antagonistic action
	CCL17		Selective inhibitor of CCL17
7.	**PLA2**		
	sPLA2	Varespladib	Inhibits sPLA2
	Lp-PLA2	Darapladib, SB-4808	Inhibits Lp-PLA2
8.	**Heat Shock Proteins (HSP)**	Mycobacterial HSP-65	Reduces the size of atherosclerotic plaque
9.	**Leukotrienes**		
	5-LO	VIA-2291	Inhibition of 5-LO
	15-LO	PD145176	Inhibition of 15-LO
	FLAP	DG-031	Inhibition of 5-LO, FLAP
10.	**ANGPTL4**	miR-134	Selectively targets ANGPTL4
11.	**PDCD4**	miR-16	Selectively targets Programmed Cell Death4
12.	**Tregs and tolerogenic DCs**		Vaccines for Immune suppression
13.	**Others**		
	Rho–ROCK pathway	SLx2119	Inhibition of ROCK pathway
	p38 MAPK	SB-681323	Inhibition action against Mitogen-activated protein kinase
	Cannabinoid-1 receptor	Rimonabant	Antagonistic action
	Cannabinoid -2 receptor	THC	Agonistic action
	ROS	Succinobucol	Antioxidant

22.3.3 Antioxidative Approaches Towards the Treatment of Atherosclerosis

Oxidative stress is another factor known to be associated with atherosclerosis. The imbalance between the production of pro-oxidant species and endogenous antioxidant mechanisms may lead to oxidative stress. These oxidative species can bring about the oxidation of lipoproteins forming oxLDL, which is an initial phase in the development of atherosclerosis [64]. The ox-LDL has deleterious effects on endothelial cells, inducing their dysfunction and inducing expression of adhesion molecules and migration of smooth muscle cells and foam cell formation. Antioxidants have proved to be beneficial in preventing this process and can be prescribed as natural supplements along with other therapies. Some of the antioxidants that proved beneficial in atherosclerosis therapy include Vitamin C, E, and A, curcumin, polyphenols like resveratrol, etc.

22.4 Rational Drug Design Strategies—Atherosclerosis Drug Design

Structure-based drug design is a powerful method of rational drug design that involves studying the three-dimensional structure of the target protein, analyzing the active site or the allosteric site, and designing molecules that will interact with the key amino acid residues of the active site. This method allows the designing of compounds that can bind to the target protein with selectivity and specificity. The workflow generally involves target identification and downloading the structure of the target from protein databases. Generally, techniques like X-ray crystallography, NMR, or Cryo-electron microscopy are used for obtaining the 3D structures of the protein. After downloading proteins from databases, active site analysis of the protein is carried out to understand shape, charge distribution, hydrophilic and hydrophobicity, etc. Further molecular modeling and docking are carried out to check how newly designed molecules fit into the binding site by looking at the binding score and free energy calculations. The ADMET predictions and drug-likeness criteria are applied to narrow down to compounds that can be drug-like in the future. The hit generated by this process is further optimized by modifying its structure to improve the binding affinity. Further molecular dynamics simulation studies can be carried out to study the stability of the protein-ligand complex. The designed compounds are synthesized and experimentally validated using in-vitro assays and in-vivo techniques.

Another powerful technique in drug discovery is fragment-based drug design. In this method, small libraries of low molecular weight fragments are screened against the biological target. The fragment libraries are generally highly diverse with structural simplicity. Once initial hits are identified, fragments are optimized by appending them which can involve growing, linking, and merging to increase the binding affinity and specificity to the target.

Some of the databases of compounds widely used for screening include ZINC, ChEMBL, PubChem, ChemSpider, DrugBank, Chemical Abstract Service Registry (CAS), LEADS, Reaxys, ChemBridge, Molport, etc.

The following section provides certain examples of rational drug design from literature for the targets of atherosclerosis. The targets are as mentioned in Sect. 22.2 i.e. HMGR, PCSK9, PPAR, NPC1L1, ACL, and ANGPTL3. For each target, the protein structure is described, first followed by the description of the active site and various attempts made by different scientists to find novel molecules acting on these targets for treating atherosclerosis.

22.4.1 Design of HMGR Inhibitors

The crystal structures of HMG-CoA reductase were further reported in early 2000, which facilitated the rational design of statins. HMG-CoA reductase is a transmembrane protein (Fig. 22.11). It comprises a single polypeptide chain of 888 AAs, divided into three domains viz anchor domain (residue 1–339) located in the endoplasmic reticulum membrane, linker domain (residue 340–459), and the catalytic domain (residue 460–888). The catalytic site is present in the cytosolic part of the enzyme and consists of three sub-domains called N, L, and S-domain. The L-domain is the large central domain with N and S being small helical domains, with the S domain containing the NADP binding site and is inserted into the L domain. The binding pocket for the substrate HMG CoA consists of a cis loop of residues 682–694. The catalytic site consists of several key residues which include Lys735 which plays a major role in the binding of the HMG moiety, acidic amino acids Asp767 and Glu559 which act as proton donors during catalysis and His866 and

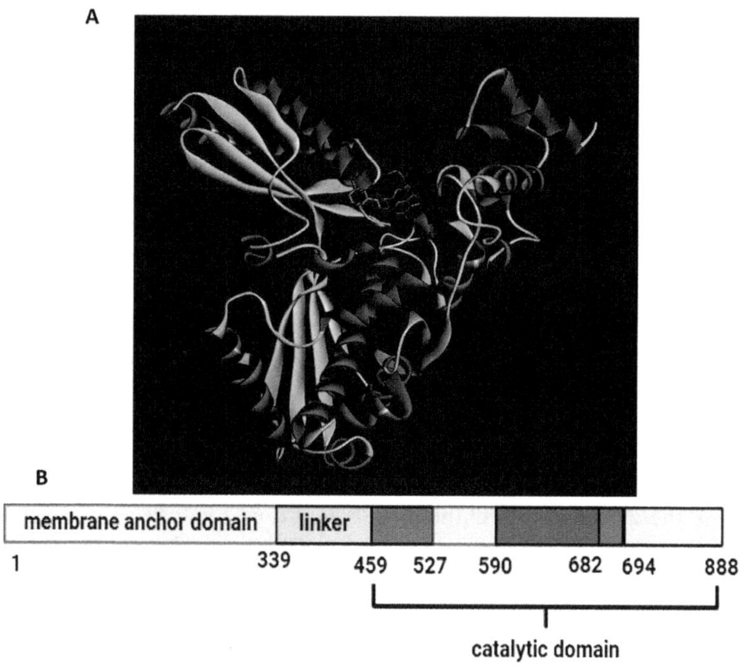

Fig. 22.11 Functional domain and 3D structures of the human HMGR

Lys691 involved in stabilising transition states of substrate enzyme complex. In addition, Ser684 and Asp690 are involved in hydrogen bonding with the hydroxyl group of the substrate. Most statins have been designed to mimic the structure of HMG CoA so that they can act as competitive inhibitors [65].

On analyzing the binding of these statins with the enzyme by X-ray crystallography, it was suggested that when statins bind to the enzyme, the enzyme alters its shape (induced-fit), with the formation of shallow hydrophobic binding region adjacent to the active site as a result of movement of two C-terminal α-helices whose flexibility is an important factor. Besides the isopropyl groups in synthetic statins bind in the same hydrophobic region where the decalin ring of natural and semisynthetic statins bind, but it provides additional Vander Waals interactions with other amino acid residues in this region resulting in increased binding affinity. Moreover, the 4-fluorophenyl ring provides a dipole bond and π-stacking interactions with Arg590. Additionally, the O of the carbonyl in atorvastatin and sulfonyl in rosuvastatin provides an additional locus of hydrogen bonding with Ser565 (Fig. 22.12).

Jeffrey. A. P et al., (2007) synthesized a series of conformationally restricted -pyrrole based inhibitors which were potent inhibitors of HMG CoA reductase in both in-vitro and in-vivo models [66]. The conformational rigidity was provided by the proximity of the amide nitrogen and ortho hydrogen of the phenyl ring A (within 1.9 A°). This was confirmed by X-ray crystallography and to determine if such conformational restriction would be beneficial in terms of activity, analogs containing a 7-membered nitrogen-containing ring fused with pyrrole were synthesized, evaluated, and found to be potent inhibitors of HMG-CoA reductase. Amide N of the 7-membered ring provided the locus for forming an H-bond with Ser56 (Fig. 22.12).

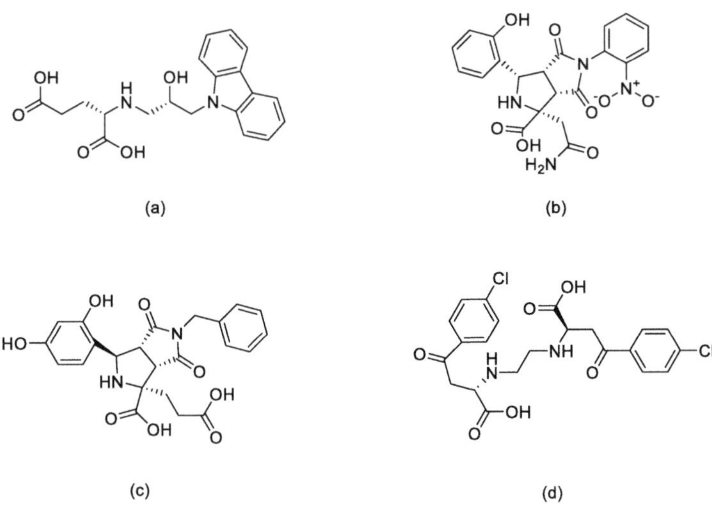

Fig. 22.12 HMGCoA Reductase inhibitors

A different approach was used by a group of scientists for the discovery of new HMG-CoA reductase inhibitors. Several structurally diverse groups of compounds were synthesized and the binding affinity with HMGR was determined by isothermal titration calorimetry (ITC). These compounds were from six different series which had imidazole core with two different positions of 3,5-dihydroxy heptanoic acid (DHHA) as substituents, the bicyclic core containing pyrrole, pyrrole derivatives structurally similar to atorvastatin, atorvastatin analogs containing DHHA substituents and atorvastatin analogs with DHHA and sulfonamide substitutions. To minimize lipophilicity and to increase hepatic selectivity, templates with greater enthalpic components were selected as enthalpically driven binding is known to be driven by electrostatic and hydrogen bonding interactions whereas entropically driven binding is known to be associated with hydrophobic interactions. This was followed by an in vitro HMGR inhibition assay and the ability of compounds to inhibit HMGR in hepatocytes and myocytes. The substrate HMG-CoA and rosuvastatin were used as controls. The study highlighted imidazole derivative as a potent inhibitor of HMGR with IC50 of 7.9 nM, with appreciable hepatoselectivity (>1000-fold) and excellent in vivo efficacy. Additionally, a novel bicyclic pyrrole-based heterocycle was identified as a novel motif that induced order in a protein flap of HMGR similar to statin inhibitors.

In another study, Valery. V. P., et al. (2006) used structure-based molecular docking to design peptide sequences mimicking the head group part of statins to provide competitive inhibitors of HMG-CoA [67] (Fig. 22.12). Through structure-function analysis of peptides and statins, nine peptides were selected from a library, and by comparing the proximity parameters, four peptides were synthesized which included IAVE, YAVE, IVAE, and YVAE. Out of these, YAVE exhibited the highest inhibitory potential. Simvastatin and rosuvastatin were selected as model compounds to assess the active space in the binding site of the enzyme and the HMG-like moiety in these compounds was used as a common site for designing all peptides. In addition, MD simulations were used to estimate the peptide backbone behavior. It was postulated that the total hydrophobicity of the peptides did not play a significant role in inhibiting the enzyme, however, the conformation and the functional groups on the side-chain of peptides were more important. Tyrosine was placed at N-terminal to mimic the 4-fluorophenyl moiety of rosuvastatin which resulted in a greater inhibitory potential. Their experiments of changing AA residues of the peptide inhibitors proposed a rational approach towards designing peptide inhibitors of HMG-CoA.

With the development of Pharmacophore theory and advances in computational chemistry and molecular modeling techniques, structure-based and ligand-based pharmacophore modeling is widely used to screen a large database of compounds to narrow down to a few that could be active. One such study involved the development of a pharmacophore model of HMG-CoA reductase using 10 training sets of active compounds which included mevastatin, fluvastatin, simvastatin, and cerivastatin. The resulting model was validated using 50 of the active compounds and 4920 decoy compounds from the DUDE decoy database. This was further used to screen 1,31,90,317 compounds of the ZINC database with Lipinsky's rule of five as a filter.

Further 436 hits generated were docked against the HMG-CoA reductase structure downloaded from the PDB database. Four compounds out of these which had docking scores lesser than that of fluvastatin were analyzed through molecular dynamics simulation for 10 ns and the trajectory was analyzed for RMSD, RMSF, and hydrogen bond analysis. This helped in concluding that only one out of four compounds i.e. compound c had stable binding interactions with essential residues in the HMG-CoA reductase binding pocket. Through these in-silico experiments and analysis of the best docking poses of ligands, the important key residues of enzymes involved in binding and their required geometric isomerism were explored. This way pharmacophore modeling followed by docking and MD simulations has become a powerful tool to design new compounds for therapeutic purposes [68].

A similar approach was used in another study by Minky S et al. (2013) where a pharmacophore model was developed using 7 potent HMGR inhibitors [69]. The generated model was validated with a test set of compounds and the set with maximum p value was selected to screen 29,045 compounds from several databases like Chembridge, Maybridge, NCI, and Asinex. They were filtered using ADMET properties and Lipinsky rule of five. Further 81 potential hits from this were further docked in the active site of HMGR. Gold fitness scores were used to narrow down to four compounds that had higher scores than reference compounds rosuvastatin and atorvastatin (Fig. 22.13). They were subjected to MD simulation for 10 ns and their binding modes were analyzed concerning the reference compounds. Hence new scaffolds were obtained which could be potential inhibitors of HMGR.

QSAR Modeling

Another approach widely used in drug discovery is Quantitative Structure-Activity Relationship (QSAR) modeling. It involves building a mathematical model that establishes the relationship between the molecular descriptors of the structure and

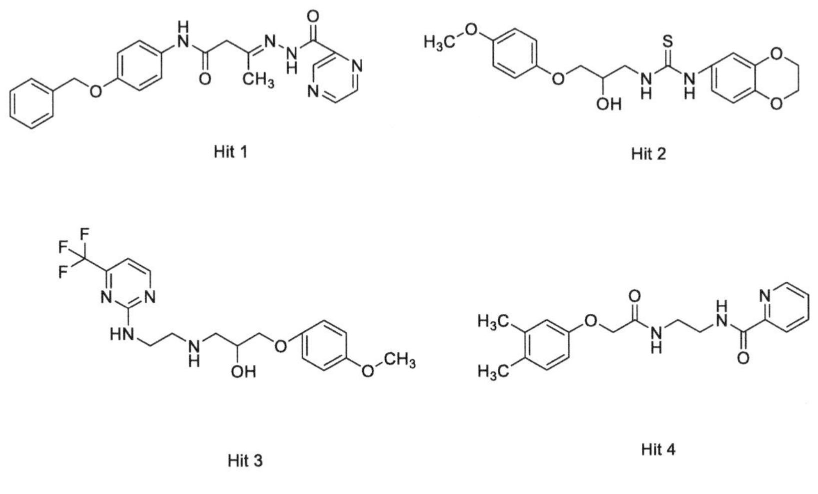

Fig. 22.13 2D structure of final hit molecules

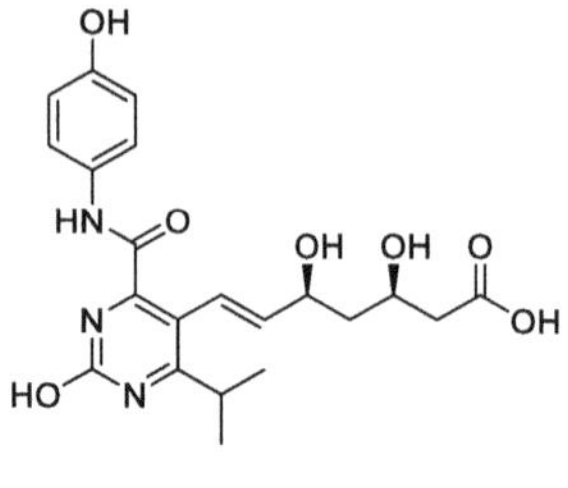

Fig. 22.14 Structure of new statin drug

the biological activity. Molecular descriptors that quantify the features of the chemical structure are molecular weight, Log P, electronic properties, topological indices like Wiener index or Hosoya index, steric and electronic fields generated from the 3D conformation of the molecule, molecular surface properties including shape, area, hydrophobicity, etc. In a recent study, scientists developed two QSAR models, one for HMGR inhibitory activity and the other one for organic anion-transporting polypeptide 1B1 (OATP1B1) affinity (Fig. 22.14). Statins are known to be transported across the liver cells by OATP1B1. Hence targeting HMGR and OATP1B1 would provide anti-atherosclerotic drugs with liver selectivity, reducing the chances of side effects. Initially, from the ChemBL database, molecules with IC50 were selected and e descriptorized using the RDKit software. The data sets were divided into training sets and data sets and machine learning was used for QSAR modelling. The same database was used to select compounds (with Km values) for OATP1B1 substrates. They were descriptorized by the same software and the model was developed using regression analysis. Both models were validated by cross-validation and used to propose new statin structures with high prediction accuracy. The predicted structures had IC50 values lower than the standard drugs.

Structure-Based Virtual Screening from Natural Product Databases

Many natural products including plant extracts are known to be highly effective and safe in preventing atherosclerotic plaque formation and cardiovascular diseases. There have been attempts to virtually screen phytoconstituents from plants for therapeutic activity. This approach's concepts and methodologies are similar to the screening of molecules from large databases of small synthetic organic compounds like molecular docking, pharmacophore modelling, and QSAR modelling. AMET predictions, MD simulations and machine learning.

For example, a Traditional Chinese medicine database was used to screen compounds for HMGR inhibitory activity using software which provides a shape-based method for docking molecules into the active site (Fig. 22.15). Further ADMET predictions of the selected compounds were carried out using ADMET descriptor software. The ADMET descriptors included models for intestinal absorption, aqueous solubility, blood-brain barrier penetration, plasma protein binding, cytochrome P450 inhibition, and hepatotoxicity. Using these data points certain non-desirable molecules were filtered out. The compounds were tested for in-vitro HMGR inhibition and two compounds viz salvianolic acid C, and curcumin were identified as potential inhibitors of HMGR with IC50 values of 4.3 and 8 μM respectively.

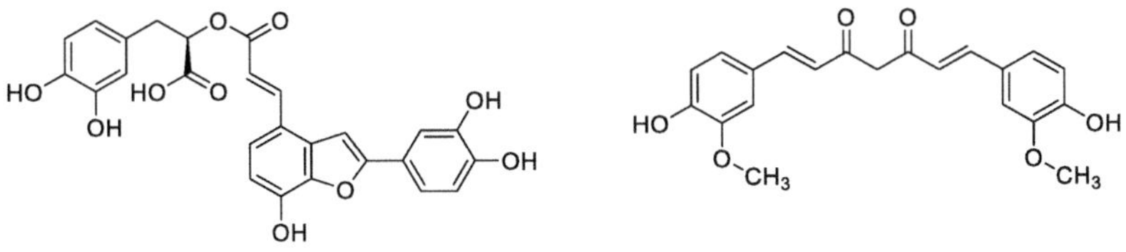

Fig. 22.15 Structure of potential HMGR inhibitors—salvianolic acid C and curcumin

Homology Modelling

Homology modelling is a computational technique to predict the three-dimensional structure of the protein through sequence alignment with the template protein. It is based on the assumption that proteins with high levels of sequence similarity have similar structures and the 3D structure of a protein is better conserved than the amino acid sequence. In this method, after aligning the target sequence with the template sequence accurately, the coordinates of template proteins are transferred to the target protein and a model is constructed, refined and validated. This method is gaining importance as the number of experimentally determined protein structures is lesser than the protein sequences available. Recent work by Muhammad B A et al. (2023), investigated the HMGCR inhibitory activity of *Withania coagulans* compounds against the HMGCR of *Mus musculus.* HMGCR homology modelling was performed using MODELLER and SWISS-MODEL tools. The highest-ranking model was checked by the PROCHECK web server to have 93.4% of residues in the most favourable region. The druggable binding site of the enzyme was defined by taking statin as a reference molecule. The library of 3D structures of *Withania coagulans* was compiled, followed by screening via drug-likeness evaluation (Fig. 22.16). This resulted in a set of 20 eligible candidate molecules which were then docked against modelled protein. Withanolide R, withanolide Q and withanolide J showed higher docking scores than atorvastatin, simvastatin and rosuvastatin (Fig. 22.17). The key residues and the interactions involved in binding were studied. MD simulations of top-ranked ligand-protein complexes were performed to validate their stability. In addition, the pharmacokinetic properties of these compounds were predicted. In conclusion, Withanolide R was found to be a potent inhibitor of HMGCR [70].

22.4.2 Design of PCSK9 Inhibitors

PCSK9 belongs to the subtilisin family of proteins. Its primary structure consists of a sequence of 692 amino acids, containing a signal peptide of 30 AAs followed by three domains, prodomain (residue 31–152), catalytic domain (residues 153–451) and C-terminal domain (residues 452–692). The catalytic domain can bind to the LDLR EGF-A (epidermal growth factor-like repeats) domain. It contains Asp-374, known to be mutated to Tyr to enhance the affinity of PCSK9 for LDLR. D186,

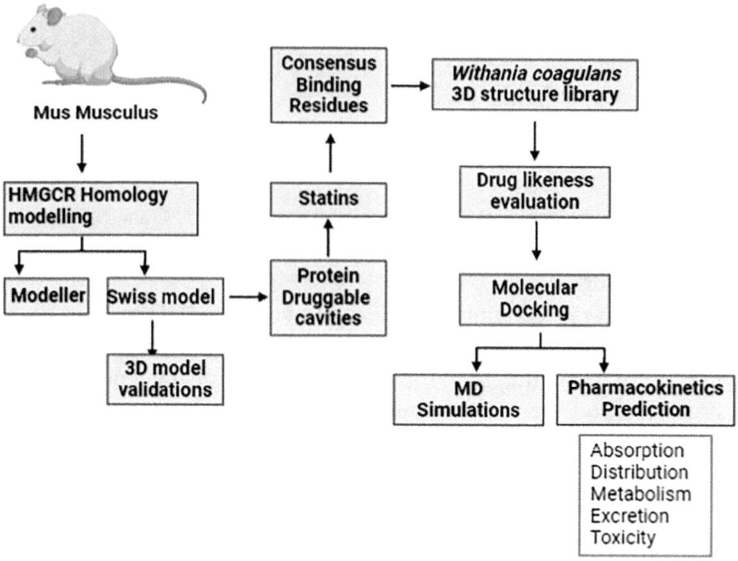

Fig. 22.16 Homology modelling

Fig. 22.17 Structure of Withanolide R

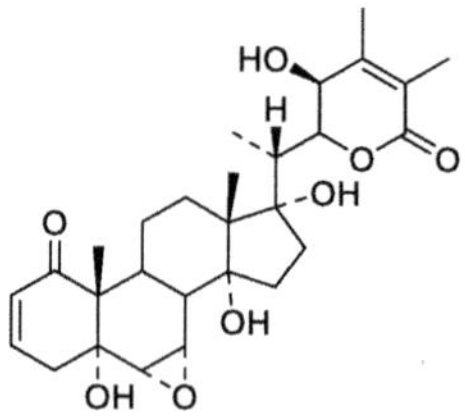

H226 and S386 represent the catalytic triad of PCSK9 (Fig. 22.18). The enhanced binding of PCSK9 to LDLR reduces the binding of LDL to LDLR resulting in the accumulation of cholesterol in the bloodstream and consequently atherosclerotic plaque [71].

Therapies targeted against PCSK9 can be based on monoclonal antibodies or small synthetic molecules, the latter more difficult as the binding site on both PCSK9 and the LDL receptor binding domain (EGFA domain) is characterised with flat, featureless surfaces that exhibit hydrophobic interactions over a large surface area. Also, the two proteins form an antiparallel β sheet, which adds further complexity for small molecules to interact with. The initial attempts to find small molecules capable of inhibiting the PCSK9/LDL receptor interaction were futile, efforts to use affinity selection/mass spectrometry (AS/MS) technology to screen a library of over 200,000 compounds for their ability to bind to recombinant PCSK9 resulted in one hit compound that showed robust and reproducible binding in orthogonal biophysical assays. In the AS/MS experiment, a mixture of compounds was pre-incubated with PCSK9 and passed through a size-exclusion column. The bound

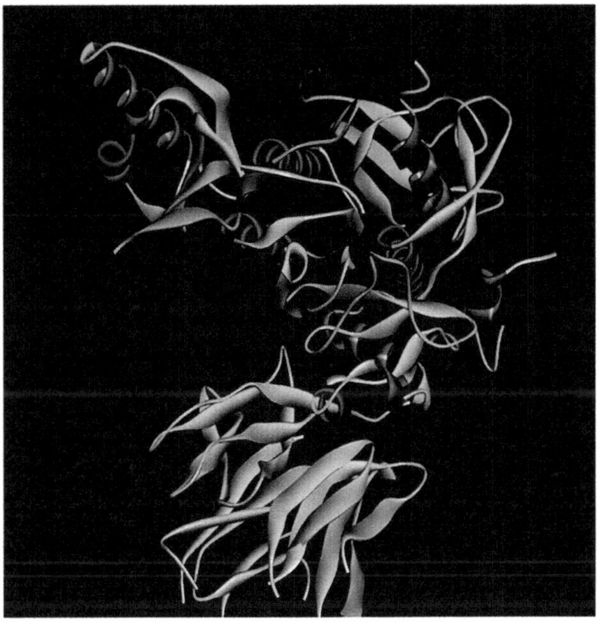

Fig. 22.18 3D Structure of PCSK9 Protein

ligands were separated by reverse-phase chromatography and identified by mass spectrometry. Further, the co-crystal structure of this compound bound to PCSK9 was obtained. On analysis, it was found that compound 1 was bound to a previously unrecognized binding pocket located between the catalytic and C-terminal domains of the protein (Figs. 22.19 and 22.20). This allosteric site was found ligandable and located in the proximity of several gain and loss-of-function mutation sites. Further, structure-based drug design strategies were utilised to study non-covalent interactions critical for ligand binding. Specifically, only the R-enantiomer of compound 1 was showing binding affinity to the identified site. The hydrogen bond interaction between the isoquinoline amine of compound 1 with Arg476 and oxygens of the two aryl methoxy groups with Arg357 of the catalytic domain was found important. Same way the C1 amide carbonyl interaction with Arg458 of the C-terminal domain of PCSK9 was found important for binding. In SAR studies, all these interactions were found important as attempts to replace these functional groups were deleterious to binding affinity. In attempts to optimise this hit compound, various derivatives were prepared and additional key residues and their binding interaction locations with various functional locations were identified. For example, Trp 461 was an important amino acid offering a site for hydrogen bonding to the meta-fluoro group in compound 4.

Pharmacophore Modeling

In an attempt to discover small molecules as PCSK9 inhibitors using in-silico techniques, known ACE inhibitors used as anti-atherosclerotic drugs like Captopril,

Fig. 22.19 Structure of PCSK9 inhibitors

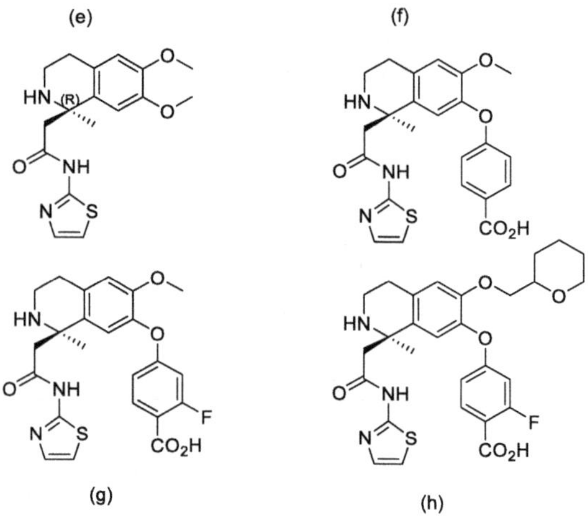

Zofenopril, Enalapril, Ramipril, Quinapril, Perindopril, Lisinopril, Benazepril, Fosinopril, Cilazapril, Moexipril, Trandolapril, Allicin and Teprotid were used. Pharmacophore design methods were used, docking was performed, interactions were studied and novel molecules as PCSK9 inhibitors were identified. The active site of PCSK9 was located between the unusual allosteric pocket and the catalytic and C-terminal domains. The protein preparation steps involved the removal of heteroatoms and, the addition of polar hydrogens and Kollmann charges were added using Autodock tools. Further, the 3D structure of the target protein was prepared. The ligand structures of ACE inhibitors were energy minimized (Fig. 22.21). The average values of binding energy were found out and all structures with binding energy higher than the average energy were chosen for pharmacophore designing. The selection of pharmacophore was done based on scoring. This was followed by a virtual screening of the ZINC database against the prepared protein and 1033 hits were identified. These molecules were screened and docked with the target protein. The top 10 molecules were picked and their pharmacokinetic properties were predicted by Molinspiration software. Further, their ligand efficiencies were calculated. Stability evaluation of the protein-ligand complex was carried out using MD simulations for 10 ns and (S)-canadine was identified as the best molecule as an inhibitor of PCSK9.

Repurposing Using in-Silico Studies

In-silico studies have been widely explored for repurposing existing approved drugs against new therapeutic targets. One such recent studies involve the identification of small molecules as potent allosteric inhibitors of PCSK9 for hypercholesterolemia. Initially, an FDA-approved Drug Library from Selleckchem was screened by docking studies using Glide extra-precision (XP) docking mode and Smina software. Amongst the screened compounds, Amikacin, bestatin, and natamycin were found to exhibit higher docking scores and CNN affinities against the PCSK9 enzyme (Fig. 22.22). On MD simulations, it was seen that these compounds formed more

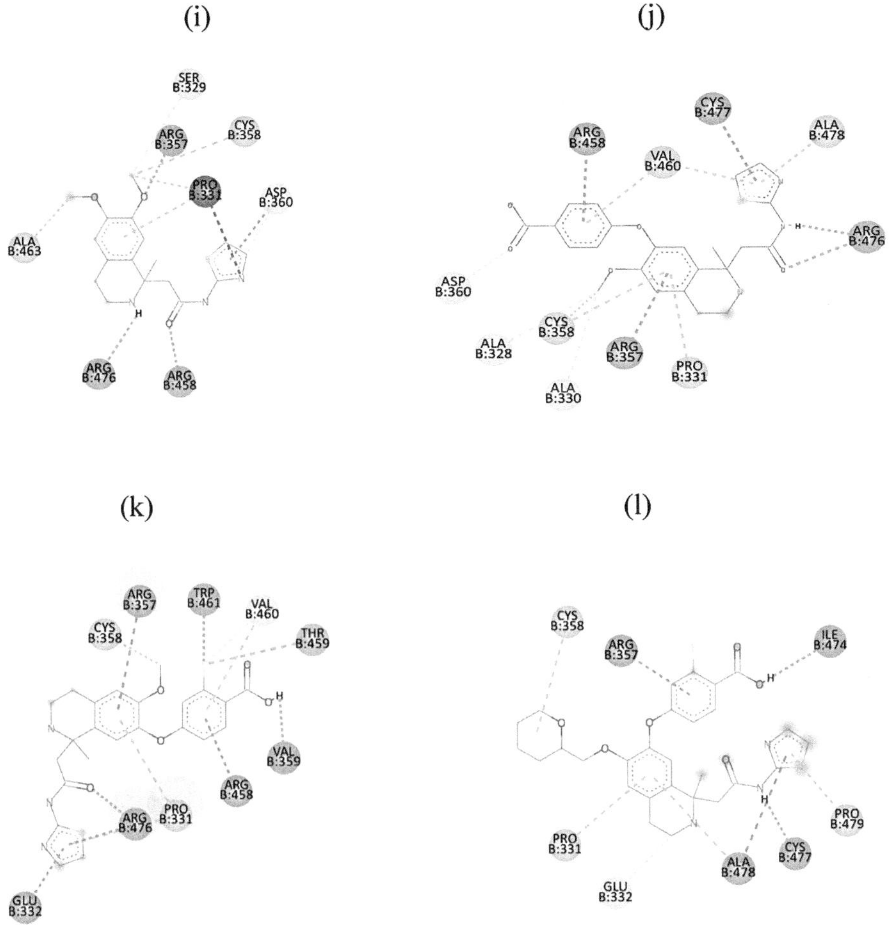

Fig. 22.20 (i) Compound e bound to PCSK9 showing the interactions with residues Arg 476, Arg 357, and Arg 458 (PDB: 6U3X). Crystals obtained from co-crystallization of 1 with PCSK9 show that only its R-enantiomer is binding. Because we obtained a higher resolution when co-crystallizing PCSK9 with enantiopure 2 than with racemic 1, the X-ray structure for 2 bound to PCSK9 is shown in this figure. (j) The crystal structure of carboxylic acid 4 revealed a new interaction with Asp 360 (PDB: 6U2N). (k) The meta-fluoro group in compound 5 is positioned to form a hydrogen bond with the Trp 461 (PDB: 6U2P). (l) The tetrahydropyran group at the C6 position is positioned to engage the Cys 358 (PDB:6U38)

Fig. 22.21 Structure of (S)-canadine

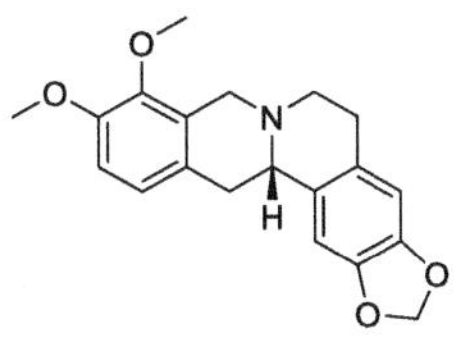

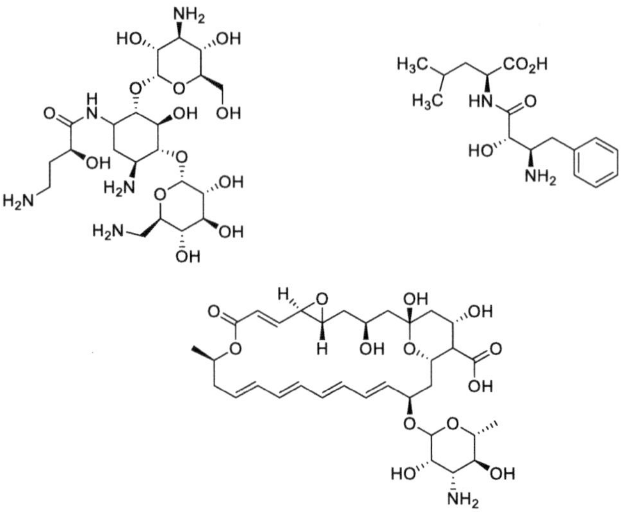

Fig. 22.22 Structure of Amikacin, Bestatin, and Natamycin

stable complexes with PCSK9 than the co-crystallised ligands. The MM-GBSA calculations revealed binding free energy values comparable to those obtained for the co-crystallized ligand structure.

One more study in 2020 reported the use of a computational screen of 1100 molecules database of FDA-approved drugs to identify nilotinib, a kinase inhibitor to binds strongly in the cryptic groove within PCSK9, proximal to the LDLR-binding interface (Fig. 22.23). Here, the binding site within PCSK9 was defined by e binding site of interest within PCSK9 was defined by amino acid residues His 226, Leu 230, Leu 253, Thr 264, Leu 286, Leu 289, Ala 290, Leu 297, Asn 317 and Ser 386. However, nilotinib was found to be a modest inhibitor in the in-vitro assay (9.8 μM). Further, utilizing principles of medicinal chemistry to reduce off-target kinase binding, enhance PCSK9 binding, and improve pharmacokinetics, hit-optimization was carried out to discover a lead, active at nM level (IC50 = 537 nM) with no inhibitory activities against other kinases. It was uptaken by liver cells at sub-micromolar levels and it reduced plasma cholesterol levels significantly in wild-type mice [72].

Certain helical peptides are discovered that can bind to the cryptic groove within PCSK9, situated in the proximity of the LDL-R binding site. They were designed as bipartite PCSK9 inhibitors by appending chemical moieties on helical peptides to reach the hydrophobic pocket in proximity to the LDL-R binding region (Fig. 22.24). The 1- amino-4-phenylcyclohexane-1-carbonyl extension improved the peptide affinity by >100-fold and yielded potent inhibitors of PCSK9 binding to LDL-R but preserved the intracellular LDL-R. These bipartite antagonists exhibited had higher potency than the first-generation peptide antagonists. Also, the addition of non-natural amino acid D serine incorporated in the core peptide resulted in 50-fold improved affinity and 30-fold improved antagonistic activity [73].

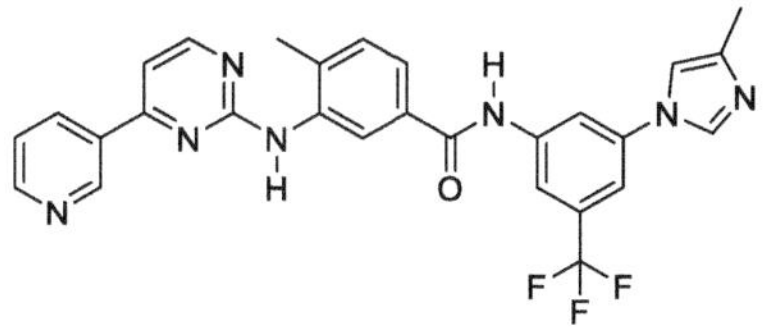

Fig. 22.23 Structure of Nilotinib

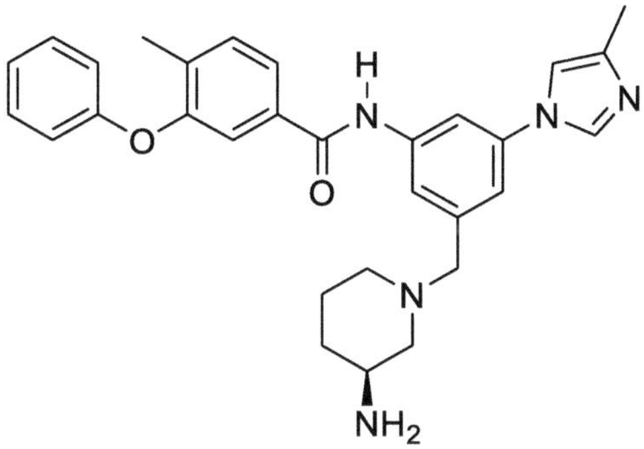

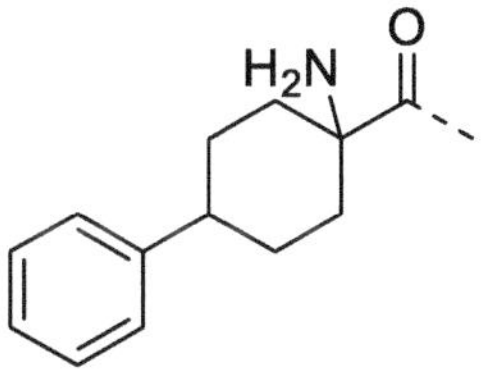

Fig. 22.24 Structure of WNLK (hR)IsLLR

WNLK (hR)IsLLR
(s denotes D-Serine, hR denotes L-homo-arginine), and the peptide is capped on the N-terminus with a 1-amino-4-phenyl cyclohexane-1-carbonyl group.

Natural Products as PCSK9 Inhibitors
To address the need for safe and effective inhibitors of PCSK9, a research group has explored an herbal database of natural products from Indonesia. They used ligand-based virtual screening for this purpose. Initially, pharmacophore models were constructed based on the set of most effectively bound ligands. Eleven phytoconstituents including morindone, gentisin, mesuaxanthone A, beta-phenethylamine, brazilin, pterofuran, n-cis-feruloyltyramine, bethanidine, 6-methoxykaempferol, gartanin, and alizarin were identified as potential PCSK9 inhibitors (Fig. 22.25). From amongst these, molecular docking studies identified brazilin with the highest binding affinity. Further, brazilin (from Caesalpinia sappan L) decreased the interaction of PCSK9 to the EGF-A fragment of LDLR, with an IC_{50} of value 2.19 μM in the *in-vitro* binding assay [74].

22.4.3 Design of PPAR Agonists

It belongs to the superfamily of nuclease receptors, and its structure possesses various key domains that play an important role in its function as a transcription factor regulating lipid metabolism. These domains include the N-terminal domain, also known as the A/B region, involved in ligand-independent transcriptional activation,

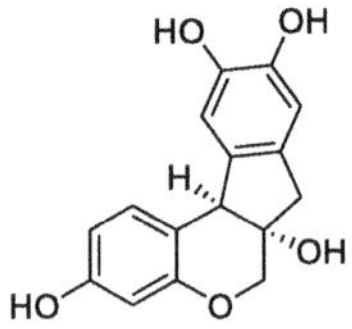

Fig. 22.25 Structure of brazilin

DNA binding domain (DBD), the highly conserved domain consisting of two zinc finger motifs allowing PPAR-α to bind to DNA sequences known as PPAR-α response elements, Hinge region, connecting DBD to ligand binding domain (LBD), allowing conformational changes during receptor activation and the LBD, containing Activation Function 2 (AF2) region, responsible for binding to endogenous ligands and crucial for even interaction with co-activators (Fig. 22.26).

The ligand binding region consists of PPAR-α consisting of 250 amino acids, comprising 13 alpha helices arranged in a helical sandwich fold. The binding pocket at the site is hydrophobic with some polar residues enabling ligand interactions. The key residues include Tyrosine 314, glutamate 282, Serine 280, and histidine 440 contributing to hydrogen bonding and polar interactions. The binding of endogenous and exogenous ligands to LBD leads to a conformational change, shifting the position of the AF-2 helix, facilitating the binding of cofactors and release of co-repressors. The functional interaction includes hetero dimerization of PPAR-α with Retinoid-X receptor (RXR) essential for high-affinity binding of PPAR response elements to DNA. This initiates the transcription of genes involved in lipid metabolism and fatty acid oxidation. There are unique features in PPAR-α that differentiate it structurally and functionally from PPAR-γ and PPAR-δ.

Role of PPAR-α in Lipid metabolism: PPAR-α acts as a transcription factor that regulates the expression of genes involved in different biochemical reactions of lipid metabolism. It controls fatty acid oxidation, lipoprotein metabolism, and lipid transport (Fig. 22.27). It is the main regulator of expression of Acyl-CoA oxidase, carnitine palmitoyl transferase (CPT-1), medium chain acyl dehydrogenase (MCAD), and all enzymes involved in the catabolism of fatty acids. It is also involved in enhancing ketogenesis during fasting by enhancing the expression of HMG-CoA synthase. It is involved in the increased expression of ApoA-I and II, the essential component of HDL, helping in the reverse transport of cholesterol from peripheral tissue to the liver for excretion. It increases the expression of lipoprotein lipase (LPL), responsible for the hydrolysis of VLDL and chylomicrons (Fig. 22.28). It also reduces levels of Apo-CIII, an inhibitor of LPL, promoting triglyceride clearance. PPAR-α enhances the uptake of fatty acid transport proteins and CD36, facilitating the uptake of fatty acids by cells. It is also involved in promoting stored lipid catabolism by regulating genes involved in the formation of droplets and their stability. It supports gluconeogenesis by providing its substrates and adapting thermogenesis. PPAR-α downregulates the expression of pro-inflammatory cytokines and is involved in promoting macrophage polarization of anti-inflammatory phenotype.

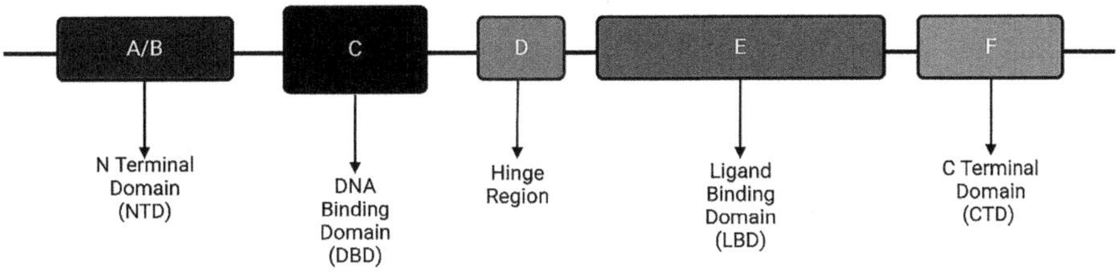

Fig. 22.26 Design of PPAR agonists

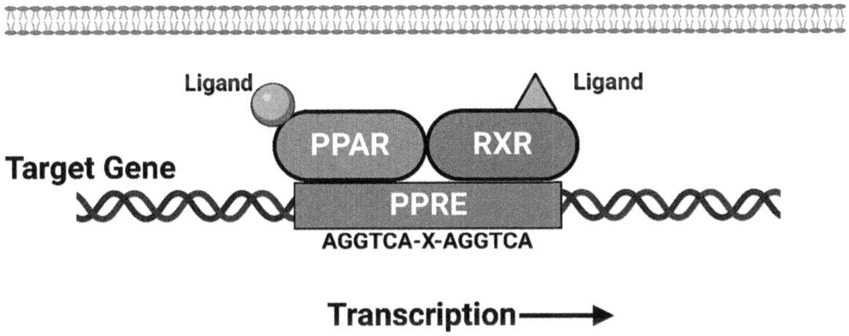

Fig. 22.27 Structure of PPAR-α

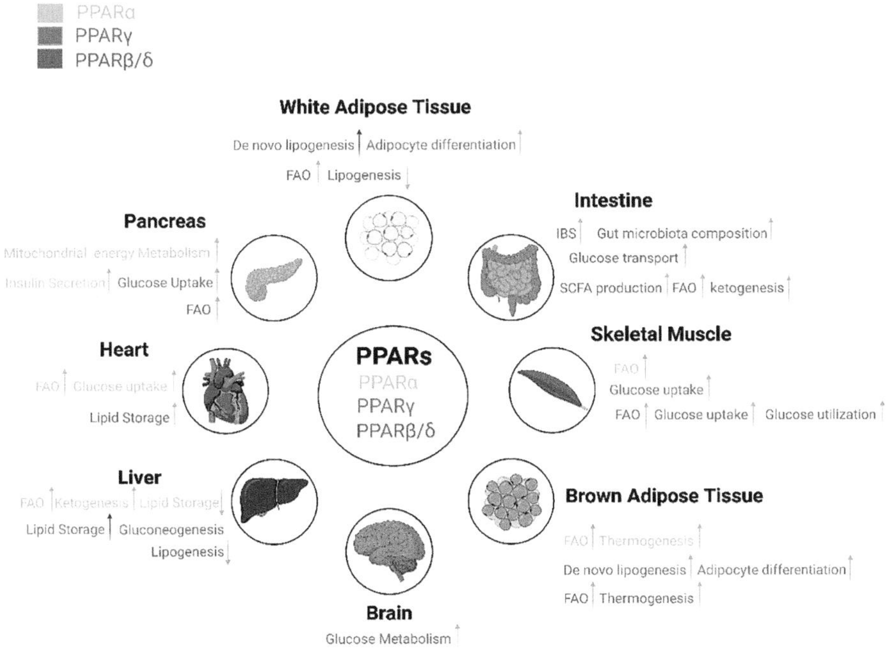

Fig. 22.28 Role of PPARs in different organs

Difference Between SPPARM and PPAR-α Agonists

The structure of the two agonists, pemafibrate (SPPARM) and Fenofibrate (PPAR-α agonist) allows differentiation between the two. The benzoxazole ring and the phenoxy alkyl side chain confer a Y-shape to pemafibrate, whereas fenofibrate has a linear structure. In-silico simulations have demonstrated that the Y-shaped fenofibrate fits perfectly in the Y-shaped ligand binding pocket on PPAR-α with an enhanced induced fit as compared to the linear molecules. This results in a conformational change of PPAR-α opens up a new site that invites the transcriptional cofactor PGC-1α, allowing the complete activation of PPAR-α. The Y-shaped agonist exhibits greatly enhanced potency and selectivity to PPAR-α (>2500 fold more active than fenofibric acid, the active metabolite of fenofibrate) and >5000 times more specific to PPARα Than to PPARγ or PPARδ (Fig. 22.29).

Structures of SPPARMα (pemafibrate) and PPARα (fenofibrate) showing shared and unique regions. This Y-structure of pemafibrate results in an improved fit with the PPARα ligand binding site compared with fenofibrate (Fig. 22.30).

Interaction between residues in the ligand binding pocket of PPAR-α

In recent years, researchers have tried to synthesize dual agonists i.e. agonists at PPAR-α and PPARγ as these agents would help in lowering not only triglycerides but also in controlling sugar levels. More than 80% sequence similarity exists between the two proteins. Computational techniques are used to achieve this purpose. For example, in a study conducted by Xin Liu et al., 2018, the Asinex database containing 292,724 compounds was screened for binding to PPAR-α and γ, after these proteins were downloaded from the PDB database and prepared for docking by checking for missing amino acids and poor valencies, removal of water molecules and other unimportant molecular fragments using protein preparation wizard (Fig. 22.31). Protein optimization was carried out by assigning bond orders, adding hydrogens, treating metals and disulfides, deleting waters around the receptor, and

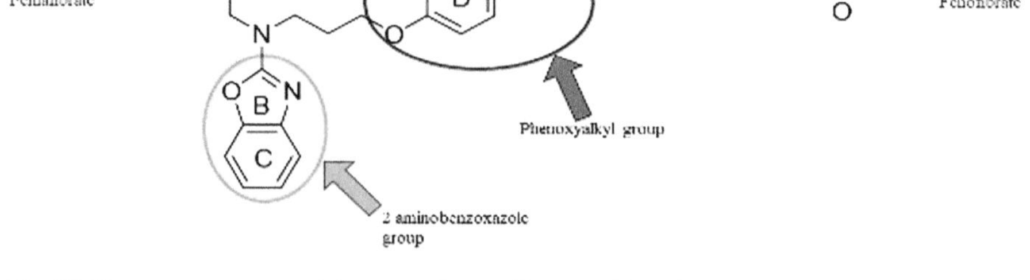

Fig. 22.29 common structural features of pemafibrate and fenofibrate

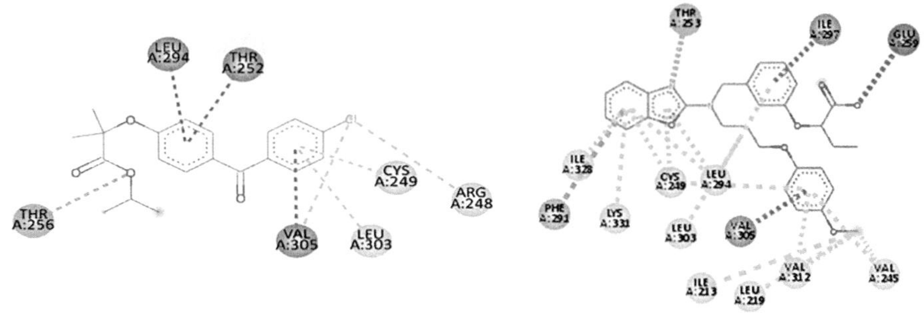

SPPARMα Fenofibrate

Fig. 22.30 Interactions between SPPARMα and fenofibrate with PPARα

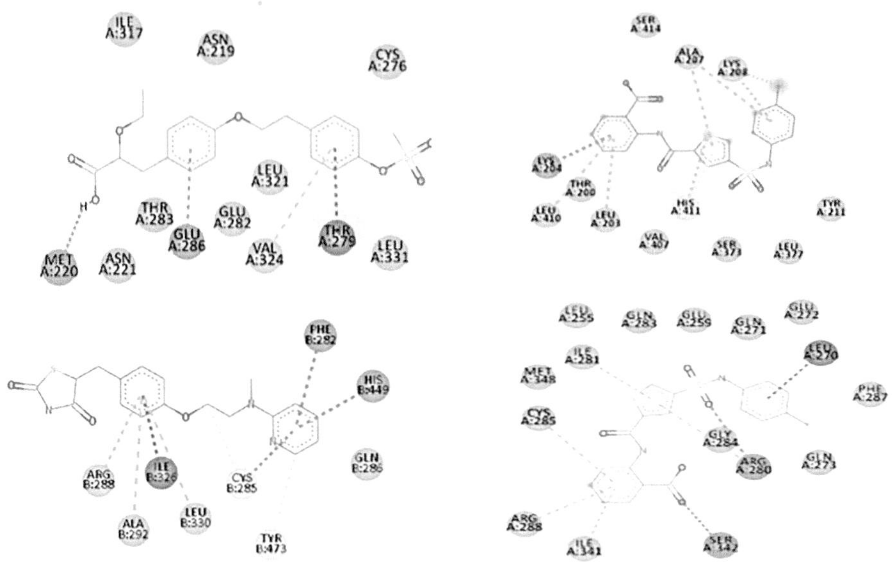

Fig. 22.31 Two-dimensional (2D) diagrams of ligand-protein interactions in PPARα and PPARγ pockets. (i) and (j) represent ligand AZ242 and ASN 15761007 docked to PPARα, respectively. (k) and (l) represent ligand BRL and ASN 15761007 docked to PPARγ, respectively

capping terminals. Further, the protein was minimized to the RMSD limit, and a docking grid box was prepared. The active sites were determined with original ligands docked in the original state. Docking was followed by ADMET prediction and Lipinsky's rule of five filters was used for screening. In addition, TOPKAT was used for toxicity predictions like carcinogenicity, aerobic biodegradability, developmental toxicity potentials, AMES mutagenicity, and ocular and skin irritancy. 14 molecules were identified from the database which had higher binding affinity and better binding modes than control molecules. One molecule with the strongest affinity with both receptors was further studied. The binding pockets and interactions of the control drug and the selected molecule were studied and compared. The key residues were the same for both molecules which include Val 305, Leu 294, Leu 303, and Cys 249 involved in hydrogen bonding in both compounds. The hydrogen bonding network played an important role in the stability of the conformation of the AF-2 helix necessary for ligand-receptor binding and activation. Besides tail of the selected compound formed hydrogen bonding with Ala 333 and the linker nitrogen with CYS 276. The hydrophobic interactions of the benzene ring of the selected compound with Val 305, and Leu 294 were also important interactions with the control compound. Similarly, interactions were studied and compared with selected molecules and in-built ligands for PPAR-γ. This molecule was further taken ahead for MD simulations studies over 20 and the structural stability of the ligand-receptor complex was studied. In conclusion, the selected compound not only assumed favorable conformation than existing individual agonists of PPAR-α and PPAR-γ but also had a similar function in activating these receptors [75].

The same group identified the following compound as a potential hit using a similar approach from the Ligands-expo component database. The study protocol involved high throughput virtual screening, docking, ADMET prediction, and MD simulations to find the representative compound as a dual PPAR agonist (Fig. 22.32).

Similar studies are carried out to build knowledge-based structure-activity relationships for dual agonists at PPAR-α and PPAR-γ. Here a library of 7476 drug-like nuclear receptor ligands were individually screened to find potential dual agonists. Key residues involved in binding and the binding interaction network were studied and it was found that mainly Tyr314, Ser280, Tyr464, and His440 were important for PPAR-α activation out of which Tyr464 is also important for the activity. Tyr473 is chiefly responsible for the activation of PPARγ whereas, His323, Ser289, and His449 are also important for the binding of ligands in the ligand-binding domain. The knowledge-based SAR studies using five approved dual agonists of PPAR-α and PPAR-γ like aleglitazar, tesaglitazar, saroglitazar, ragaglitazar, and muraglitazar and essential features of the hit out of this particular study identified a polar region in molecule where carboxyl, hydroxyl amino, phosphoric group can be substituted, a hydrophobic region where presence of aromatic ring or branched carbon chain can improve the agonistic binding to both targets, a heteroatom or ethylene bridge as linker (Fig. 22.33) [76].

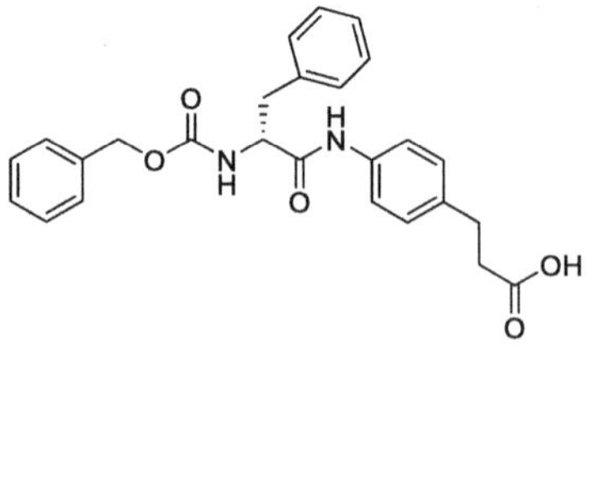

Fig. 22.32 Structure of M80

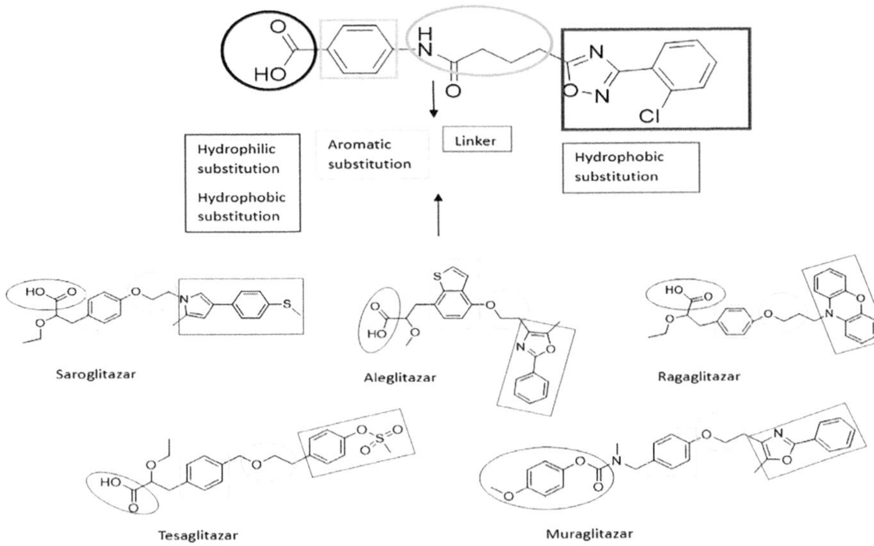

Fig. 22.33 SAR Studies on reported drug candidates with essential features

Sr. no	Compound IDs	Hit Structures	IUPAC name	Synthetic accessibility score
Reference1	AZ242 (Tesaglitazar)	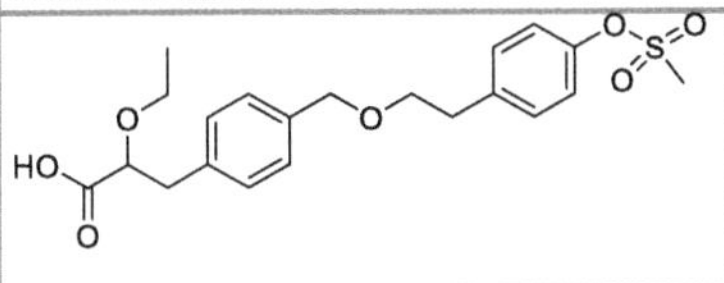	2-ethoxy-3-(4-((4-((methylsulfonyl)oxy)phenethoxy)methyl)phenyl)propanoic acid	4.60
1	ChemDiv D321-0199	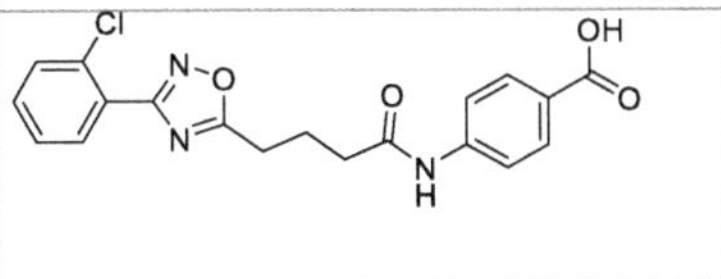	4-(4-(3-(2-chlorophenyl)-1,2,4-oxadiazol-5-yl)butanamido)benzoic acid	3.48

One of the less studied isoforms of PPAR is PPAR-δ is also said to be involved in anti-obesity and anti-inflammatory effects. Selective PPAR-δ agonists are known to enhance fatty acid metabolism, preserving glucose and promoting exercise endurance. It has found benefits in atherogenic dyslipidemia and non-alcoholic fatty liver disease. Additionally, they promote keratinocyte proliferation and find a role in epidermis repair. Its anti-inflammatory actions in the lungs are also discovered. However, no selective PPAR-δ agonists are on the market. Benny D et al., (2018) used a machine learning algorithm called Iterative Stochastic Elimination (ISE) to build ligand ligand-based multi-filter ranking model to screen molecules for PPAR-δ binding affinity. This is an expert algorithm developed in researchers' labs predicting novel structures with highly diverse motifs combined with docking and geometry filters. The two well-known selective PPAR-δ agonists (GW501516 and GW0742) could not become drugs due to their toxicities in the pre-clinical setups. Both molecules have similar structural scaffolds. Using the ML algorithm, 13 novel compounds with activity in nm concentrations were discovered with selective PPAR-δ binding (without any direct binding study for other isoforms) which had diverse scaffolds than GW501516 and GW0742 (Fig. 22.34a) [77].

Currently, research is going on to develop agonists for this receptor or dual agonists for PPARγ and PPARδ. In one such attempt to identify dual PPARγ and PPAR δ agonists, a clean lead subset of the ZINC database of purchasable drug-like compounds was screened. Compounds containing thiol and aldehyde groups were excluded for predicted toxicity problems. Fifty top compounds identified through

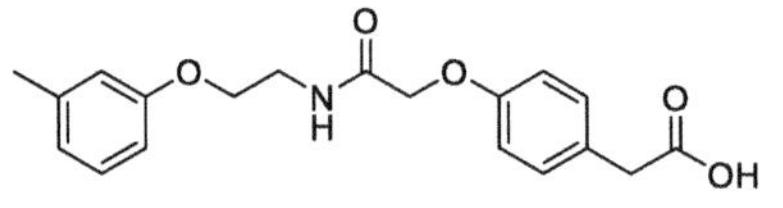

GNF-0242 (T6811704)
EC_{50} = 0.004 μM

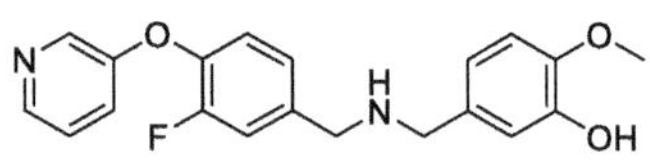

GNF-8065 (T6757286)
EC_{50} = 0.006 μM

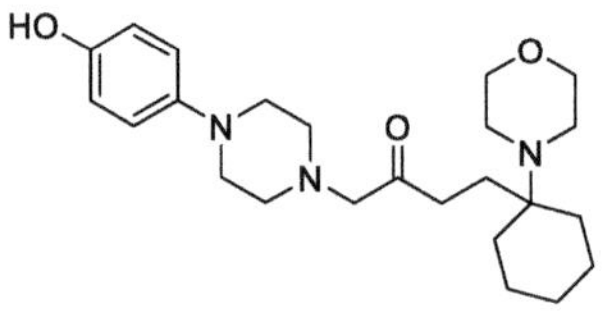

GNF-8501 (T5630172)
EC_{50} = 0.006 μM

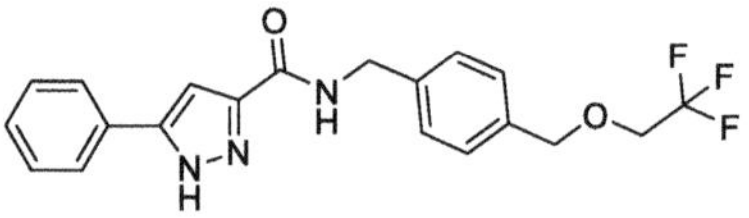

GNF-3632 (T5999328)
EC_{50} = 0.007 μM

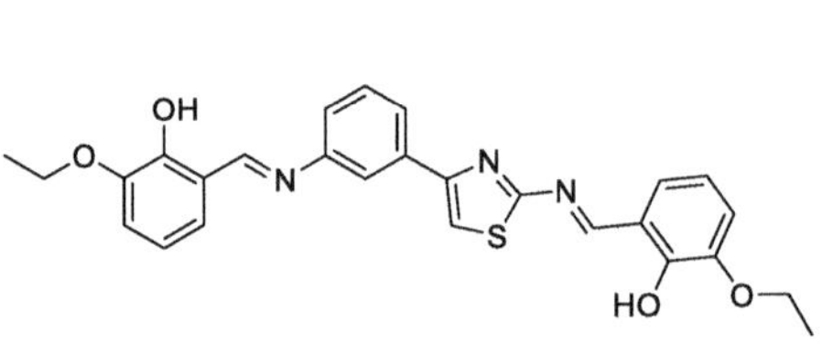

GNF-6878 (T0509-5156)
EC_{50} = 0.008 μM

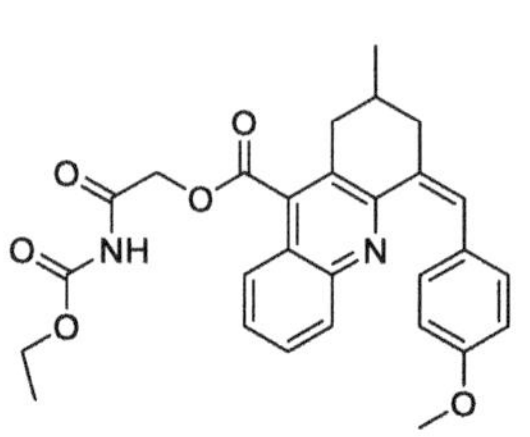

GNF-8560 (T5435815)
EC_{50} = 0.011 μM

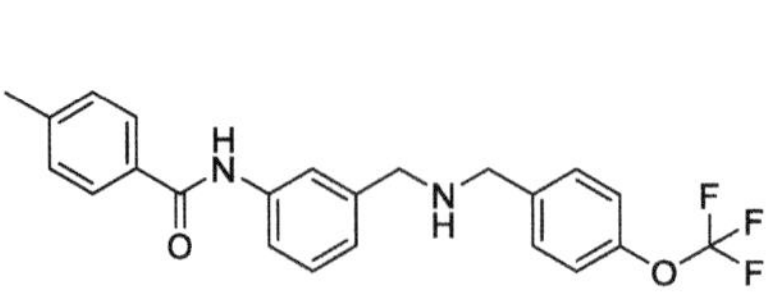

GNF-0341 (T6163027)
EC_{50} = 0.011 μM

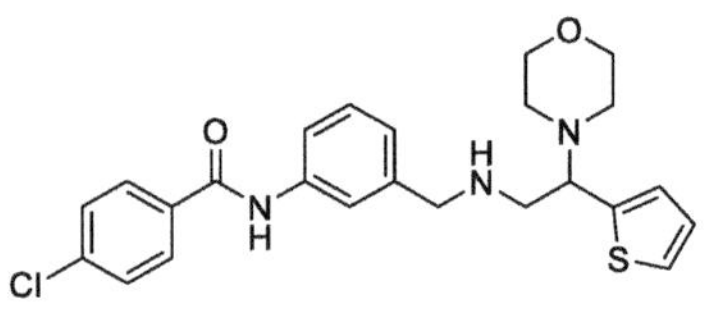

GNF-6029 (T5894943)
EC_{50} = 0.012 μM

Fig. 22.34 Structure and experimental results of lead agonists of PPARδ

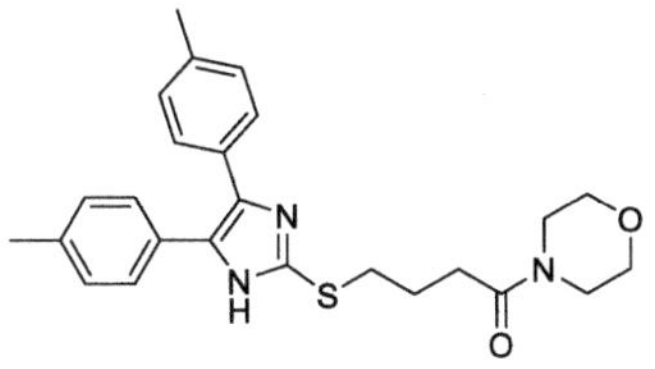

GNF-9820 (T5307117)
EC_{50} = 0.012 μM

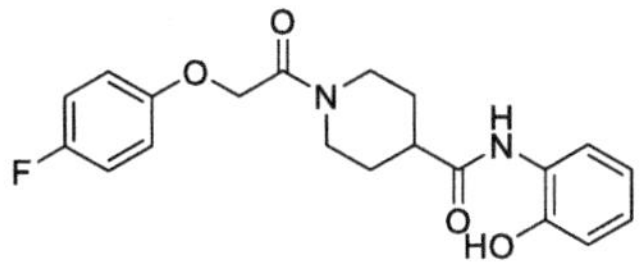

GNF-5891 (T6068856)
EC_{50} = 0.013 μM

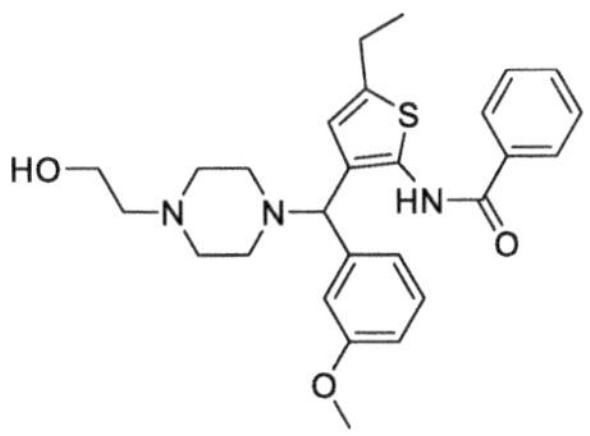

GNF-5295 (T5466364)
EC_{50} = 0.017 μM

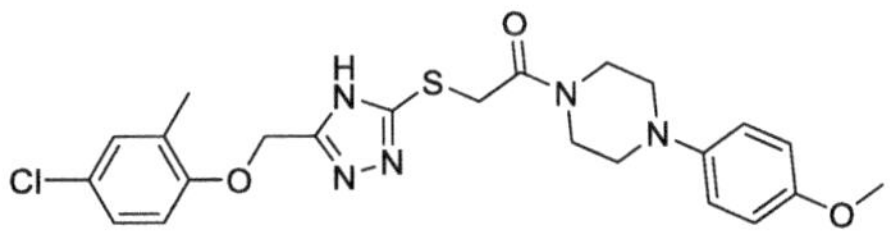

GNF-7486 (T5483814)
EC_{50} = 0.018 μM

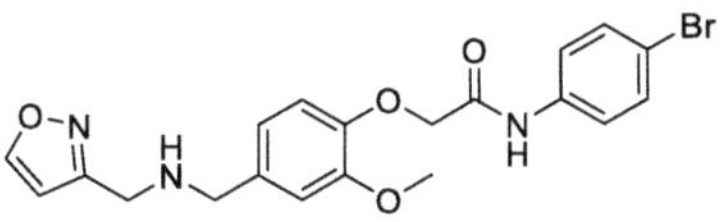

GNF-6952 (T6631408)
EC_{50} = 0.019 μM

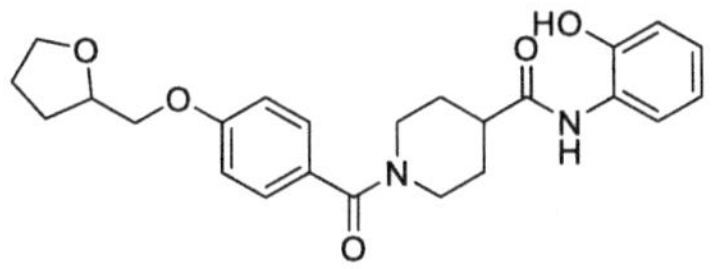

GNF-9448 (T5934377)
EC_{50} = 0.883 μM

Fig. 22.34 (continued)

virtual screening through docking simulations were analyzed visually in their top five docking poses (Fig. 22.34b).

The amino acid residues involved in binding interactions were analyzed. Three different docking programs were used viz DOCK, Surflex, and Gold. According to the agreement among DOCK, Surflex, and Gold poses and binding energies, five compounds were selected for the in vitro biological activity. Further three compounds were selected based on in-vitro studies that underwent MD-simulation. 5-(benzyloxy)-1H-indol-1-yl] acetic acid was identified as a dual agonist for PPAR-γ and δ (Fig. 22.35).

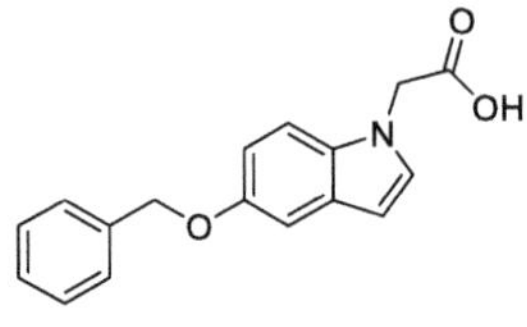

Fig. 22.35 Structure of 5-(benzyloxy)-1H-indol-1-yl] acetic acid

22.4.4 Niemann-Pick C1-like 1 (NPC1L1) Inhibitors

Structure of NPC1L1

Though ezetimibe was discovered years ago, the detailed mechanism of cholesterol uptake by NPC1L1 remained unclear until 2020–21, when three different research groups reported the cryo-EM structures of NPC1L1. NPC1L1 is a highly glycosylated transmembrane protein expressed on the brush border of intestinal cells and canalicular membrane of hepatocytes and is involved in dietary cholesterol absorption and biliary cholesterol reabsorption respectively. NPC1L1 consists of 13 transmembrane helices which create a channel through which cholesterol can be transported. There exist four domains: (1) The N-terminal domain (NTD), located extracellularly and is involved in binding cholesterol from the intestinal lumen, (2) The transmembrane domain (TMD) consisting of 13 transmembrane helices (3) The middle luminal domain (MTD) and (4) The C-terminal luminal domain, located intracellularly and addition (5) sterol sensing domain (SSD) within TMD extending from TM3 to TM7. The cholesterol-binding pocket exists in NTD. The cholesterol transport tunnel (channel) extends from NTD to SSD. The binding of cholesterol to its binding pocket in NTD brings about the conformational change in NPC1L1 from the open state to the closed state, facilitating the transport of cholesterol molecules through the tunnel into the cells. The SSD senses the amount of blood cholesterol and further regulates the NPC1L1-dependent cholesterol uptake (Fig. 22.36) [78].

In addition to ezetimibe, orlistat which was developed as an anti-obesity agent was also found to reduce blood cholesterol levels by inhibiting NPC1L1. Based on the structure of orlistat and ezetimibe, novel analogs were designed through computational studies and synthesized which demonstrated potent binding to NPC1L1.

Based on structures of known inhibitors of NPC1L1, ezetimibe, and orlistat, various analogs were synthesized by hybridizing important structural features of both drugs revealed from in-silico docking experiments (Fig. 22.37). These analogs had a 4-membered lactone ring with a 13-carbon chain (features similar to the structure of orlistat) and a substituted phenyl ring (a feature similar to ezetimibe). The aim was to synthesize dual inhibitors of pancreatic triglyceride lipase (used for the treatment of obesity) and NPC1L1 (cholesterol absorption inhibitor). Compounds 1–11 were identified as the most potent dual inhibitor of PTL and NPC1L1 through in-silico and in-vitro studies. Molecular docking simulation studies revealed that t Phe77 and Leu153 were the key residues involved in bringing compound 1–11 close to Ser152 of the active site which offers hydrogen bonding, the substituted benzene ring remained on one side of the binding pocket and the hydroxyl group on the benzene ring formed hydrogen bonding with Gly1028 which was consistent with the

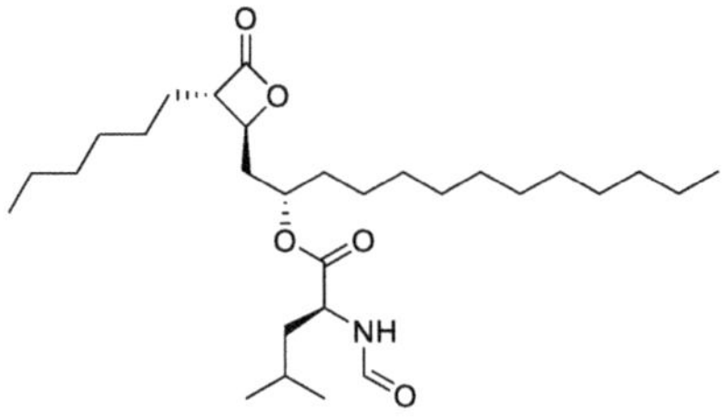

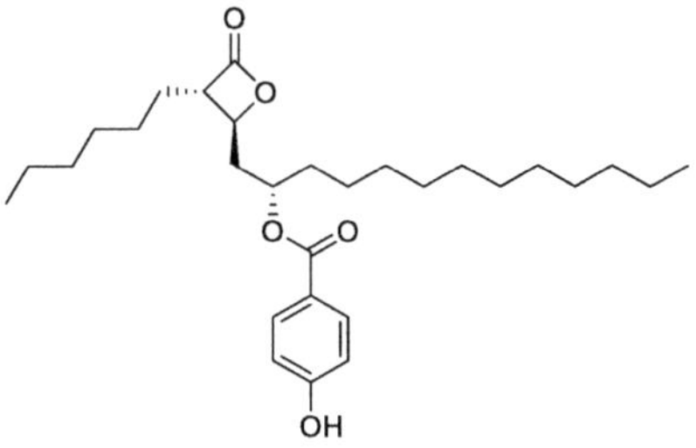

Fig. 22.36 Structure of Orlistat and Novel Inhibitor

Fig. 22.37 Structure of NPC1L1

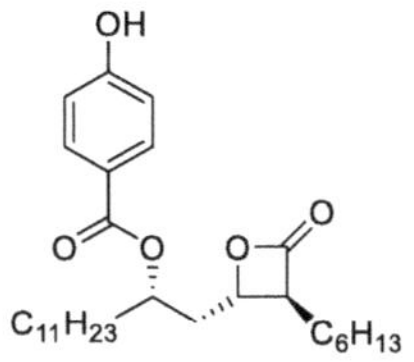

structure of ezetimibe. The MD simulation studies confirmed this hydrogen bonding to be present throughout the trajectory time of 50 ns.

Certain peptides have been investigated as NPC1L1 inhibitors. In-silico studies revealed that three black bean peptides (Tyr-Ala-Ala-AlaThr (8), Phe-Ala-Thr-Gly-Thr (9), Glu-Arg-Ala-Phe (10)) and the cowpea bean dipeptide (Lys-Asp (11)) had higher in silico binding affinities for the N-terminal domain of NPC1L1. Pulse proteins get hydrolyzed to give this AA sequence which is considered ideal for lowering blood cholesterol (Fig. 22.38).

Certain studies have been carried out on polyphenolic compounds for NPC1L1 inhibition. For example, epigallocatechin gallate (EGCG) (12), kaempferol (13), quercetin (14), and luteolin have been reported to reduce serum cholesterol levels by inhibiting endocytosis of NPC1L1 from the plasma membrane to cytoplasm (Fig. 22.39).

Also, spiroimidazolidinones were identified as NPC1L1 inhibitors using 2D and 3D similarity-based virtual screening which had a binding affinity at 2.5 μM concentration and however, found inactive in vivo. After extensive structure-activity relationship studies, a pyrazole analogy with promising efficacy was obtained through the cholesterol absorption assay in mice (Fig. 22.40).

Jefferey A P et al. replaced the β-lactam ring of ezetimibe with oxazolidinone and designed a series of compounds with different linkers between oxazolidinone and the aromatic ring A along with different substituents on three aromatic rings. The brush border membrane binding assay combined with in-vivo evaluation identified several oxazolidinones with low micromolar NPC1L1 inhibition and cholesterol-lowering activity (Fig. 22.41) [79].

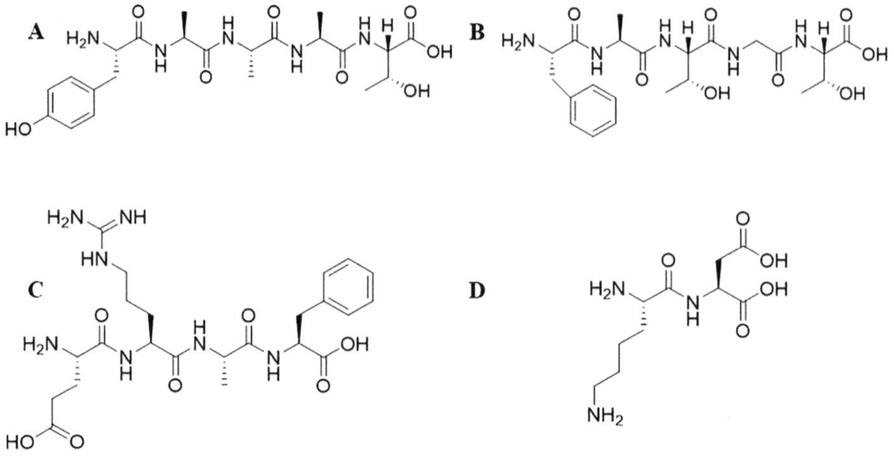

Fig. 22.38 Structure of (**a**) Tyr-Ala-Ala-Ala-Thr, (**b**) Phe-Ala-Thr-Gly-Thr, (**c**) Glu-Arg-Ala-Phe and (**d**) lys-Asp

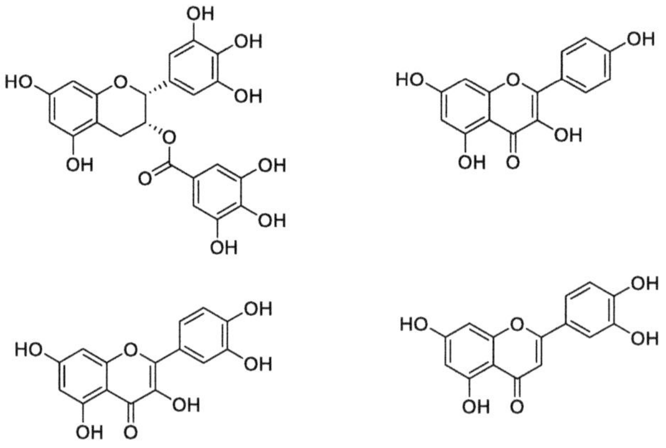

Fig. 22.39 structures of polyphenolic compounds for NPC1L1 inhibition

Wells et al. reported enantiomers (3R)- and (3S)-2,2-difluorocitrate (**3,** *Ki* = 3.2 μM) as an irreversible ACLY inhibitor through inserting of C-F bond in the citrate [80].

Later, Dolle R E et al. designed and synthesized a sulfoximine ACL inhibitor at SmithKline Pharmaceuticals to develop promising therapeutic inhibitors [81]. However, sulfoximine **4** showed a weak binding affinity (*Ki* = 250 μM). Further, the epoxide derivatives were developed as an irreversible ACL inhibitor, however, they too displayed weak activity against ACLY (Ki = 18–400 μM).

Inspired by HMG-CoA reductase inhibitors, Gribble et al. reported (3R*,5S*)-ω-substituted-3-carboxy-3,5- dihydroxyalkanoic acids and some of the compounds have ACL inhibition potential in 200-1000 nM range. However, these compounds had limited cellular penetration. Further attempts to convert the carboxyl group into

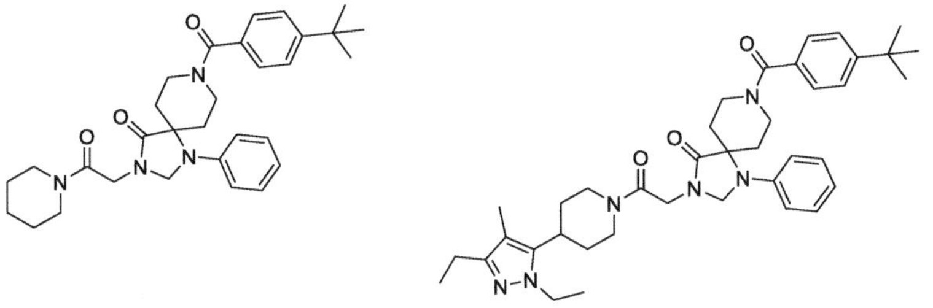

Fig. 22.40 Structure of novel spiroimidazolidinones

Fig. 22.41 Adenosine triphosphate-citrate lyase (ACL) inhibitors

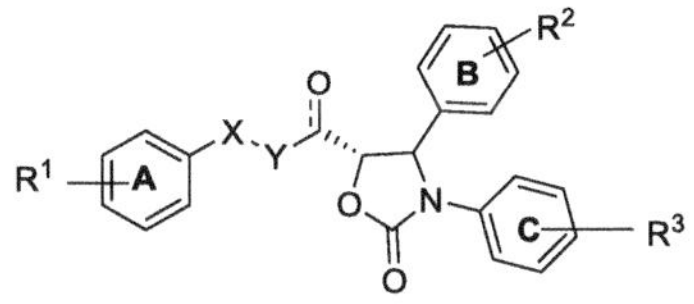

γ-lactone resulted in a prodrug that was found to be uptaken by the HepG2 cells and also reduced plasma triglyceride levels and reduced fatty acid synthesis after oral administration in chow-fed rats [82].

Further in 2007, James, J.L; et al. (Bristol-Myers Squibb) applied a high throughput screen on the internal library of compounds and identified 2-hydroxy-Narylbenzenesulfonamide as a modest inhibitor of ACL (IC50 = 1.1 μM). Subsequent mining of compounds with similar structure identified 2-hydroxy-N-arylbenzenesulfonamide as a pharmacophore and 2-substituted anilines were more potent than the other derivatives [83].

Esperion company recently developed a prodrug of ACL inhibitor, Bempedoic acid, the only ACL inhibitor approved by US FDA and EMA in 2022 for the treatment of hypercholesterolemia. Another drug called BGT-002 or 326E developed by Burgeon Therapeutics is in Phase I clinical trials (Fig. 22.42).

Recently, Tong et al. have revealed the full crystal structure of ACL which has increased the potential for drug design for inhibiting the enzyme [84].

Structure of ACL

ACL belongs to the superfamily of acyl CoA synthase. It is a homotetramer comprising four identical units. It consists of 1101 AA residues. Each subunit consists of several domains viz. C terminal citrulline CoA lyase (CCL) module which binds CoA and facilitates its acetylation and N-terminal citrate synthase-like domain which facilitates the transfer of acetyl group (Fig. 22.43). The N-terminal module contains 5 functional domains 2, 3, 4, 1, and CS domains. ACL has multiple active sites that bind to citrate, ATP, and CoA, allowing enzymes to cleave citrate and form Acetyl CoA and oxaloacetate. The citrate binding site is present in domain 3 (residues 244–425), the ATP binding site is present in domain 2, and the CoA binding site is present in domain 4 (residues 487–624). The CS domain at the C-terminal is the one containing an amino acid sequence with a high degree of similarity to citrate synthase. It is an important region for the formation of the tetramer (Table 22.2).

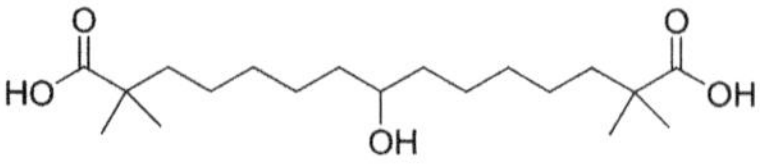

Bempedoic acid

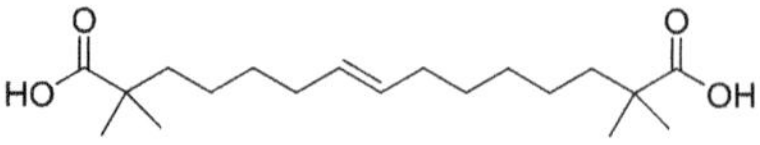

BGT-002, Phase 1

Fig. 22.42 Structures of Bempedoic acid and BGT-002

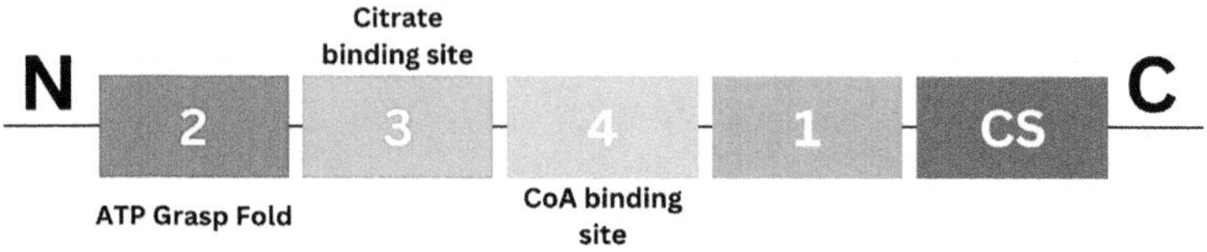

Fig. 22.43 Structure of ACL

Table 22.2 Overview of native ACLY domain

ACLY Protein	Structures
Citrate binding site	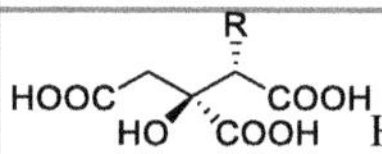 HCA 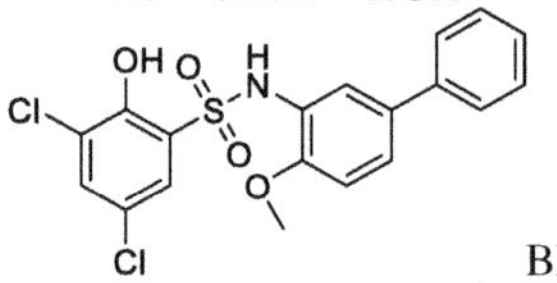 BMS-303141 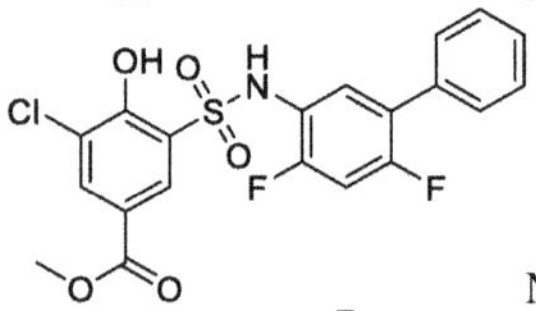 NDI-091143 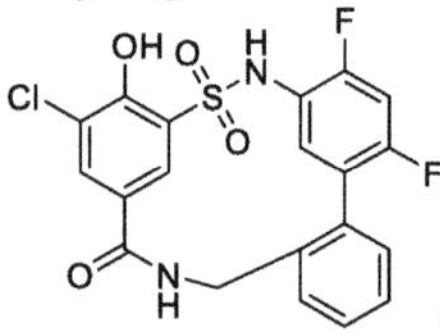 ACLY
CoA binding site	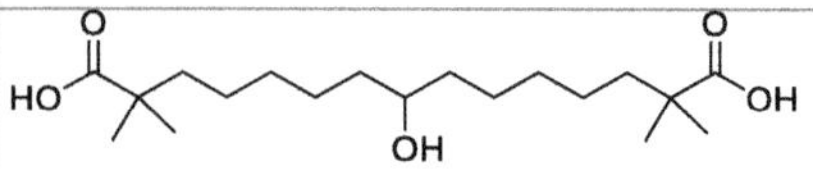 Bempedoic acid 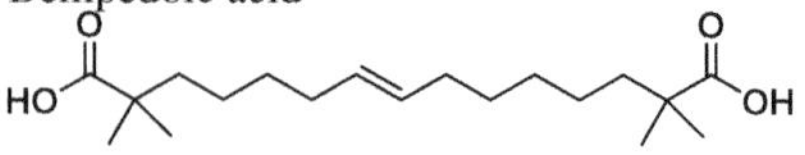 BGT 002, phase 1 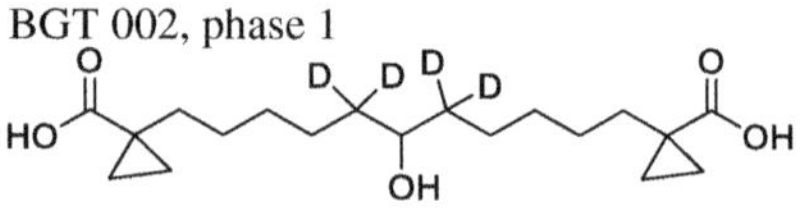 ACLY

Table 22.2 (continued)

ACLY Protein	Structures
ATP binding site	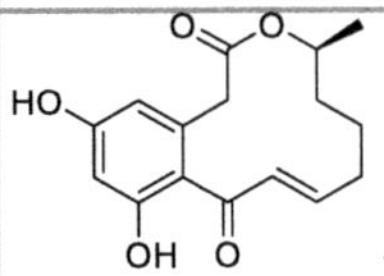 DCV 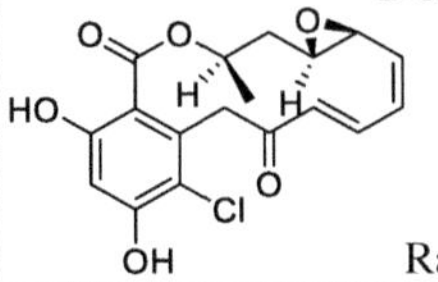 Radicicol

Human ACL inhibitors from natural sources:

Chemical class	Structures	IC50	Key interactions	Ref
Anthraquinones	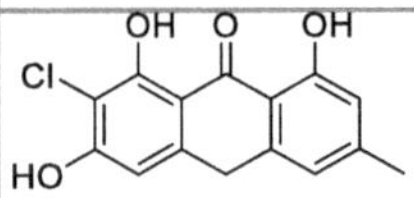	0.28 μM	Hydrogen bonding between 8-OH and Gly342 π-π stacking between one of the phenyl rings with Phe354	
	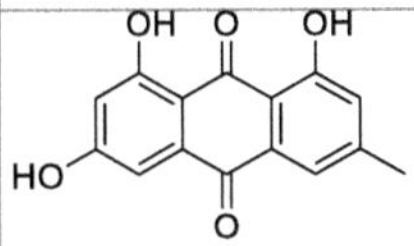	19 μM		
	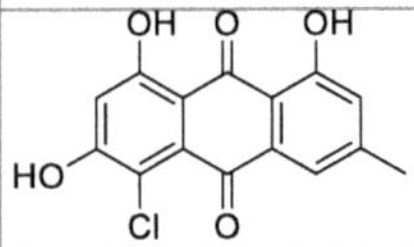	9.6 μM		
	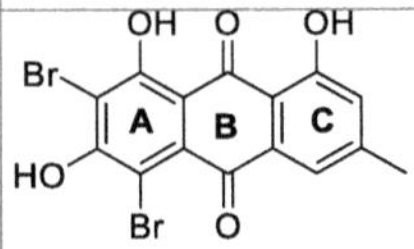	2.9 μM	Occupied an allosteric hydrophobic site, H-bonds involved the –OH group in position A ring with Gly342, Gly380, and the –OH group in position C ring with Gly341, Ala280, and Met278, the carbonyl group placed in a solvent-exposed region.	
Flavonoids	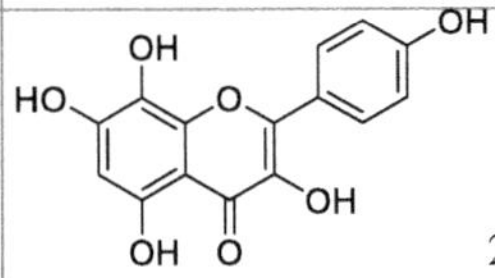 20 (Herbacetin)	0.5 μM		

Chemical class	Structures	IC50	Key interactions	Ref
	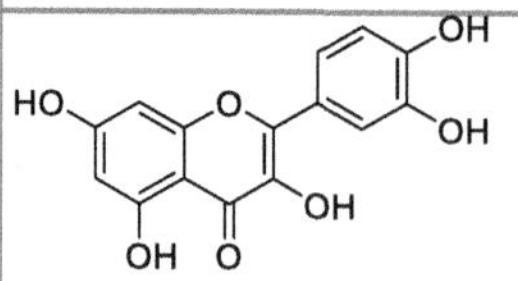 21(Quercetin)	0.86 μM		
	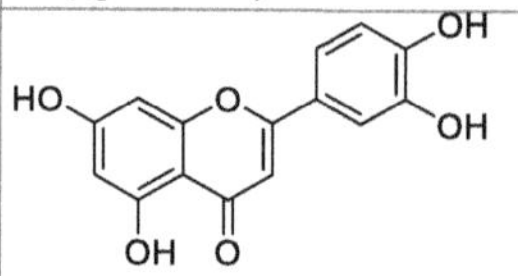 22(Luteolin)	0.64 μM		
	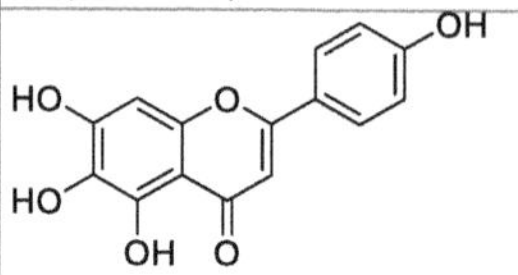 23(Scutellarin)	0.71 μM		
	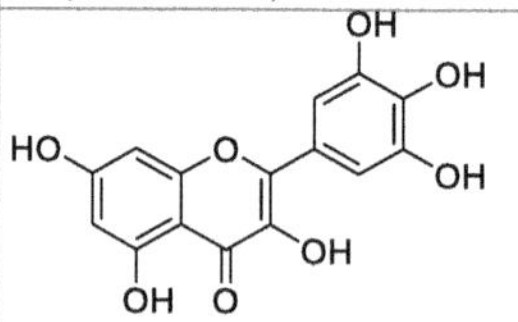	0.47 μM	Docking studies showed that compound **24** had hydrogen-bond interactions with Met278, Gly341, Asn346, and Thr348.	
Macrolides from streptomyces	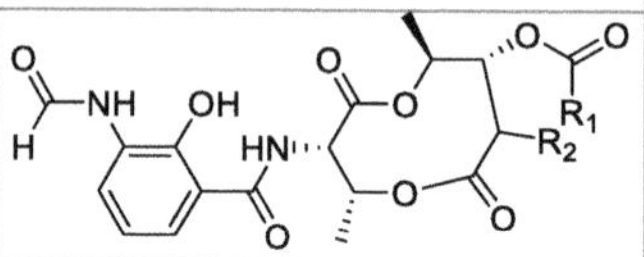	4 and 6 μM		
	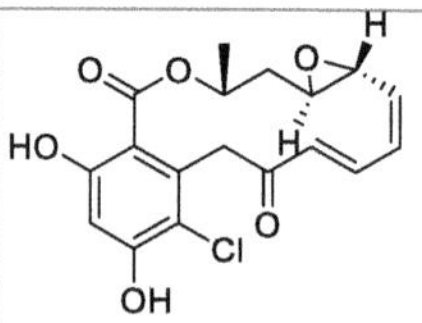	7 μM		
	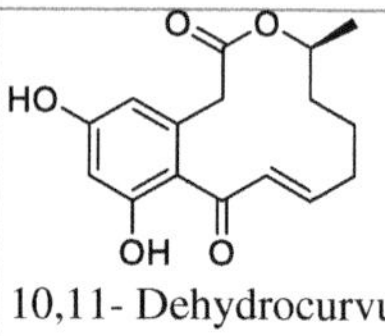 10,11- Dehydrocurvularin	0.93 μM	Irreversible ACLY inhibitor. The SAR studies show that the α, β-unsaturated ketone moiety of compound **28** conducts as a Michael acceptor to alkylate Cys20 residue of ACLY. Molecular docking studies showed that compound **28** formed a strong H-bond interaction with His113 of ACLY, which was a vital site responsible for the enzyme catalytic activity.	

22.5 Conclusion and Future Perspectives

The development of therapeutically important small molecules is accelerated with the help of computational methods. It has streamlined the process of drug discovery; the typical timelines have reduced from two decades to approximately 9–13 years. The cost of the early stages of drug discovery has been reduced by 30–40%. The risk of failures in clinical studies has been substantially reduced due to early ADMET predictions. *In-silico* optimization of hits has become possible, hence the efforts for synthesizing a large series of compounds have been reduced. Design of more targeted therapies is now possible and hence the chances of failures due to off-target effects have reduced. Overall, chances of success in the clinical trials have considerably increased. This shift in time, cost, and effort is largely due to the high computational capacities of the computers, the huge amount of data available on ligands and 3D structures of the target, and the availability of large virtual libraries and databases. In addition, advances in X-ray crystallography micro-crystallography, and cryo-EM In synergy with this are developments in machine learning approaches which have shown high potential in reshaping the future of drug discovery.

Computational methods are highly powerful and efficient but not without limitations. The input data quality and accuracy of structures and models are extremely important. Biological systems are highly complex and simulation methods are oversimplification of biological interactions. Molecular dynamics simulations and quantum mechanics calculations require significant resources and time and are expensive. Machine learning methods, though powerful, represent overfitting and lack interpretability. Experimental validation of computational methods is extremely important. Lack of experimental validation can lead to false positives. Integrating computational methods and experimental workflows can be highly challenging.

In the context of atherosclerosis, recent research has shifted to developing personalized therapies and tailoring medications based on the genetic profiling of an individual. In addition, the focus of research is directed towards targeting inflammation and adaptive immunity. Also, there are advancements in developing nanotherapeutics for targeted delivery in the field of atherosclerosis treatment. Extensive research is going on to identify small molecules of plants that can be beneficial in the prevention and treatment of atherosclerosis. Combination therapies targeting multiple pathways identified through network pharmacology can improve the overall outcome of the therapy. The use of artificial intelligence and machine learning methods to develop new drug candidates predict their efficacy, and repurposing already existing drugs can prove beneficial in improving treatment outcomes and reducing the burden of cardiovascular diseases.

Conflict of Interest The authors declare that the study was conducted in the absence of any commercial or financial relationships that could be construed as a potential conflict of interest.

Data Availability Statement Data sharing does not apply to this article as no new data were created or analyzed in this study.

References

1. Falk E. Pathogenesis of atherosclerosis. J Am Coll Cardiol. 2006;47(8):C7–12.
2. Sinyov VV, Sazonova MA, Ryzhkova AI, Galitsyna EV, Melnichenko AA, Postnov AY, et al. Potential use of buccal epithelium for genetic diagnosis of atherosclerosis using mtDNA mutations. Vessel Plus. 2017;1:145.
3. Bekkering S, Quintin J, Joosten LAB, van der Meer JWM, Netea MG, Riksen NP. Oxidized low-density lipoprotein induces long-term Proinflammatory cytokine production and foam cell formation via epigenetic reprogramming of monocytes. Arterioscler Thromb Vasc Biol. 2014;34(8):1731–8.
4. Kwaifa IK, Bahari H, Yong YK, Noor SM. Endothelial dysfunction in obesity-induced inflammation: molecular mechanisms and clinical implications. Biomol Ther. 2020;10(2):291.
5. Vukovic I, Arsenijevic Phd N, Lackovic Phd V, Todorovic V. The origin and differentiation potential of smooth muscle cells in coronary atherosclerosis. Exp Clin Cardiol. 2006;11:123.
6. Ciccarelli G, Conte S, Cimmino G, Maiorano P, Morrione A, Giordano A. Mitochondrial dysfunction: the hidden player in the pathogenesis of atherosclerosis? Int J Mol Sci. 2023;24(2):1086.
7. Alizadeh G, Gholipour K, Azami-Aghdash S, Dehnavieh R, Jafarabadi M, Azmin M, et al. Social, economic, technological, and environmental factors affecting cardiovascular diseases: a systematic review and thematic analysis. Int J Prev Med. Wolters Kluwer Medknow Publications. 2022;13:78.
8. Mozaffarian D, Benjamin EJ, Go AS, Arnett DK, Blaha MJ, Cushman M, et al. Heart disease and stroke statistics—2015 update. Circulation. 2015;131(4):e29.
9. Rippe JM. Lifestyle strategies for risk factor reduction, prevention, and treatment of cardiovascular disease. Am J Lifestyle Med. 2019;13(2):204–12.
10. Gromova M, Vaggelas A, Dallmann G, Seimetz D. Biomarkers: opportunities and challenges for drug development in the current regulatory landscape. Biomark Insights. 2020;15:117727192097465.
11. Kong P, Cui ZY, Huang XF, Zhang DD, Guo RJ, Han M. Inflammation and atherosclerosis: signaling pathways and therapeutic intervention. Signal Transduct Target Ther. 2022;7(1):131.
12. Li B, Li W, Li X, Zhou H. Inflammation: a novel therapeutic target/direction in atherosclerosis. Curr Pharm Des. 2017;23(8):1216–27.
13. Kim K, Ginsberg HN, Choi SH. New, novel lipid-lowering agents for reducing cardiovascular risk: beyond statins. Diabetes Metab J. 2022;46(4):517–32.
14. Wilensky RL, Macphee CH. Lipoprotein-associated phospholipase A2 and atherosclerosis. Curr Opin Lipidol. 2009;20(5):415–20.
15. Wilensky RL, Shi Y, Mohler ER, Hamamdzic D, Burgert ME, Li J, et al. Inhibition of lipoprotein-associated phospholipase A2 reduces complex coronary atherosclerotic plaque development. Nat Med. 2008;14(10):1059–66.
16. STABILITY Investigators. Darapladib for preventing ischemic events in stable coronary heart disease. N Engl J Med. 2014;370(18):1702–11.
17. O'Donoghue ML, Braunwald E, White HD, Steen DP, Lukas MA, Tarka E, et al. Effect of Darapladib on major coronary events after an acute coronary syndrome. JAMA. 2014;312(10):1006.
18. Bjelakovic G, Nikolova D, Gluud C. Antioxidant supplements to prevent mortality. JAMA. 2013;310(11):1178.
19. Bjelakovic G, Nikolova D, Gluud C. Meta-regression analyses, meta-analyses, and trial sequential analyses of the effects of supplementation with Beta-carotene, vitamin A, and vitamin E singly or in different combinations on all-cause mortality: do we have evidence for lack of harm? PLoS One. 2013;8(9):e74558.
20. Manson JE, Cook NR, Lee IM, Christen W, Bassuk SS, Mora S, et al. Vitamin D supplements and prevention of cancer and cardiovascular disease. N Engl J Med. 2019;380(1):33–44.

21. Tsiantoulas D, Diehl CJ, Witztum JL, Binder CJ. B cells and humoral immunity in atherosclerosis. Circ Res. 2014;114(11):1743–56.
22. Chyu KY, Dimayuga PC, Shah PK. Vaccine against arteriosclerosis: an update. Ther Adv Vacc. 2017;5(2):39–47.
23. Gisterå A, Hermansson A, Strodthoff D, Klement ML, Hedin U, Fredrikson GN, et al. Vaccination against T-cell epitopes of native ApoB100 reduces vascular inflammation and disease in a humanized mouse model of atherosclerosis. J Intern Med. 2017;281(4):383–97.
24. Wolf D, Gerhardt T, Ley K. Vaccination to prevent cardiovascular disease. In: Platelets, haemostasis and Inflammation. Springer; 2017. p. 29–52.
25. Nus M, Mallat Z. Immune-mediated mechanisms of atherosclerosis and implications for the clinic. Expert Rev Clin Immunol. 2016;12(11):1217–37.
26. Aprotosoaie AC, Costache AD, Costache II. Therapeutic strategies and chemoprevention of atherosclerosis: what do we know and where do we go? Pharmaceutics. 2022;14(4):722.
27. Strianese O, Rizzo F, Ciccarelli M, Galasso G, D'Agostino Y, Salvati A, et al. Precision and personalized medicine: how genomic approach improves the management of cardiovascular and neurodegenerative disease. Genes (Basel). 2020;11(7):747.
28. Agamah FE, Mazandu GK, Hassan R, Bope CD, Thomford NE, Ghansah A, et al. Computational/in silico methods in drug target and lead prediction. Brief Bioinform. 2020;21(5):1663–75.
29. Sliwoski G, Kothiwale S, Meiler J, Lowe EW. Computational methods in drug discovery. Pharmacol Rev. 2014;66(1):334–95.
30. Yamashita F, Hashida M. In Silico approaches for predicting ADME properties of drugs. Drug Metab Pharmacokinet. 2004;19(5):327–38.
31. Meng XY, Zhang HX, Mezei M, Cui M. Molecular docking: a powerful approach for structure-based drug discovery. Curr Comput Aided-Drug Des. 2011;7(2):146–57.
32. Cosconati S, Forli S, Perryman AL, Harris R, Goodsell DS, Olson AJ. Virtual screening with AutoDock: theory and practice. Expert Opin Drug Discov. 2010;5(6):597–607.
33. Guha R. On exploring structure-activity relationships. Methods Mol Biol. 2013;999:81–94.
34. Giordano D, Biancaniello C, Argenio MA, Facchiano A. Drug design by pharmacophore and virtual screening approach. Pharmaceuticals. 2022;15(5):646.
35. Paul D, Sanap G, Shenoy S, Kalyane D, Kalia K, Tekade RK. Artificial intelligence in drug discovery and development. Drug Discov Today. 2021;26(1):80–93.
36. Ramsey SA, Gold ES, Aderem A. A systems biology approach to understanding atherosclerosis. EMBO Mol Med. 2010;2(3):79–89.
37. Jenner AL, Aogo RA, Davis CL, Smith AM, Craig M. Leveraging computational modeling to understand infectious diseases. Curr Pathobiol Rep. 2020;8(4):149–61.
38. Barh D, Yiannakopoulou EC, Salawu EO, Bhattacharjee A, Chowbina S, Nalluri JJ, et al. In silico disease model: from simple networks to complex diseases. In: Animal biotechnology. Elsevier; 2020. p. 441–60.
39. Soleimani M, Haverich A, Wriggers P. Mathematical modeling and numerical simulation of atherosclerosis based on a novel surgeon's view. Arch Comput Methods Eng. 2021;28(6):4263–82.
40. Parton A, McGilligan V, O'Kane M, Baldrick FR, Watterson S. Computational modelling of atherosclerosis. Brief Bioinform. 2016;17(4):562–75.
41. Linton MRF, Yancey PG, Davies SS. The role of lipids and lipoproteins in atherosclerosis. In: Feingold KR, Anawalt B, Blackman MR, editors. Endotext [Internet]; 2019.
42. Asada Y, Yamashita A, Sato Y, Hatakeyama K. Pathophysiology of atherothrombosis: mechanisms of thrombus formation on disrupted atherosclerotic plaques. Pathol Int. 2020;70(6):309–22.
43. Xiao N, Humphrey JD, Figueroa CA. Multi-scale computational model of three-dimensional hemodynamics within a deformable full-body arterial network. J Comput Phys. 2013;244:22–40.
44. Kovacic S, Bakran M. Genetic susceptibility to atherosclerosis. Stroke Res Treat. 2012;2012:1–5.

45. Sun D, Gao W, Hu H, Zhou S. Why 90% of clinical drug development fails and how to improve it? Acta Pharm Sin B. 2022;12(7):3049–62.
46. Das S, Dey MK, Devireddy R, Gartia MR. Biomarkers in cancer detection, diagnosis, and prognosis. Sensors. 2023;24(1):37.
47. Banik SK, Baishya S, Das Talukdar A, Choudhury MD. Network analysis of atherosclerotic genes elucidates druggable targets. BMC Med Genet. 2022;15(1):42.
48. Duan H, Khan GJ, Jun SL, Peng H, Chen HW, Yu ZJ, et al. Computational pharmacology and bioinformatics to explore the potential mechanism of Schisandra against atherosclerosis. Food Chem Toxicol. 2021;150:112058.
49. Lee AY, Park W, Kang TW, Cha MH, Chun JM. Network pharmacology-based prediction of active compounds and molecular targets in Yijin-Tang acting on hyperlipidaemia and atherosclerosis. J Ethnopharmacol. 2018;221:151–9.
50. Mayerl C, Lukasser M, Sedivy R, Niederegger H, Seiler R, Wick G. Atherosclerosis research from past to present—on the track of two pathologists with opposing views, Carl von Rokitansky and Rudolf Virchow. Virchows Arch. 2006;449(1):96–103.
51. Endo A. A historical perspective on the discovery of statins. Proc Japan Acad Ser B. 2010;86(5):484–93.
52. Chester A. From Fleming to Endo: the discovery of statins. Glob Cardiol Sci Pract. 2022;2021(4):e202132.
53. Yurtseven E, Ural D, Baysal K, Tokgözoğlu L. An update on the role of PCSK9 in atherosclerosis. J Atheroscler Thromb. 2020;27(9):909–18.
54. Ma M, Hou C, Liu J. Effect of PCSK9 on atherosclerotic cardiovascular diseases and its mechanisms: focus on immune regulation. Front Cardiovasc Med. 2023;10:1148486.
55. Golbus JR, Stitziel NO, Zhao W, Xue C, Farrall M, McPherson R, et al. Common and rare genetic variation in *CCR2* , *CCR5* , or *CX3CR1* and risk of atherosclerotic coronary heart disease and Glucometabolic traits. Circ Cardiovasc Genet. 2016;9(3):250–8.
56. Milioti N, Bermudez-Fajardo A, Penichet ML, Oviedo-Orta E. Antigen-induced immunomodulation in the pathogenesis of atherosclerosis. Clin Dev Immunol. 2008;2008:1–15.
57. Andersson J, Libby P, Hansson GK. Adaptive immunity and atherosclerosis. Clin Immunol. 2010;134(1):33–46.
58. Weber C, Zernecke A, Libby P. The multifaceted contributions of leukocyte subsets to atherosclerosis: lessons from mouse models. Nat Rev Immunol. 2008;8(10):802–15.
59. Ait-Oufella H, Herbin O, Bouaziz JD, Binder CJ, Uyttenhove C, Laurans L, et al. B cell depletion reduces the development of atherosclerosis in mice. J Exp Med. 2010;207(8):1579–87.
60. Hosseini H, Li Y, Kanellakis P, Tay C, Cao A, Tipping P, et al. Phosphatidylserine liposomes mimic apoptotic cells to attenuate atherosclerosis by expanding polyreactive IgM producing B1a lymphocytes. Cardiovasc Res. 2015;106(3):443–52.
61. Kyaw T, Tay C, Hosseini H, Kanellakis P, Gadowski T, MacKay F, et al. Depletion of B2 but not B1a B cells in BAFF receptor-deficient ApoE−/− mice attenuates atherosclerosis by potently ameliorating arterial inflammation. PLoS One. 2012;7(1):e29371.
62. Kyaw T, Tay C, Krishnamurthi S, Kanellakis P, Agrotis A, Tipping P, et al. B1a B lymphocytes are atheroprotective by secreting natural IgM that increases IgM deposits and reduces necrotic cores in atherosclerotic lesions. Circ Res. 2011;109(8):830–40.
63. Taleb S, Tedgui A, Mallat Z. Adaptive T cell immune responses and atherogenesis. Curr Opin Pharmacol. 2010;10(2):197–202.
64. Batty M, Bennett MR, Yu E. The role of oxidative stress in atherosclerosis. Cells. 2022;11(23):3843.
65. Istvan ES, Palnitkar M, Buchanan SK, Deisenhofer J. Crystal structure of the catalytic portion of human HMG-CoA reductase: insights into regulation of activity and catalysis. EMBO J. 2000;19(5):819–30.
66. Pfefferkorn JA, Choi C, Song Y, Trivedi BK, Larsen SD, Askew V, et al. Design and synthesis of novel, conformationally restricted HMG-CoA reductase inhibitors. Bioorg Med Chem Lett. 2007;17(16):4531–7.

67. Pak VV, Kim SH, Koo M, Lee N, Shakhidoyatov KM, Kwon DY. Peptide design of a competitive inhibitor for HMG-CoA reductase based on statin structure. Pept Sci. 2006;84(6):586–94.
68. Junaidin J, Lestari D, Kurniawan MF, Khairul Ikram NK. Ligand-based pharmacophore modeling, molecular docking, and molecular dynamic studies of HMG-CoA reductase inhibitors. Inform Med Unlocked. 2022;32:101063.
69. Son M, Baek A, Sakkiah S, Park C, John S, Lee KW. Exploration of virtual candidates for human HMG-CoA reductase inhibitors using pharmacophore modeling and molecular dynamics simulations. PLoS One. 2013;8(12):e83496.
70. Azmi MB, Khan F, Asif U, Khurshid B, Wadood A, Qureshi SA, et al. In Silico characterization of *Withania coagulans* bioactive compounds as potential inhibitors of Hydroxymethylglutaryl (HMG-CoA) reductase of *Mus musculus*. ACS Omega. 2023;8(5):5057–71.
71. Zainab R, Kaleem A, Ponczek MB, Abdullah R, Iqtedar M, Hoessli DC. Finding inhibitors for PCSK9 using computational methods. PLoS One. 2021;16(8):e0255523.
72. Evison BJ, Palmer JT, Lambert G, Treutlein H, Zeng J, Nativel B, et al. A small molecule inhibitor of PCSK9 that antagonizes LDL receptor binding via interaction with a cryptic PCSK9 binding groove. Bioorg Med Chem. 2020;28(6):115344.
73. Burdick DJ, Skelton NJ, Ultsch M, Beresini MH, Eigenbrot C, Li W, et al. Design of Organo-Peptides As Bipartite PCSK9 antagonists. ACS Chem Biol. 2020;15(2):425–36.
74. Iqbal M, Hasanah N, Arianto AD, Aryati WD, Puteri MU, Saputri FC. Brazilin from Caesalpinia sappan L. as a Proprotein convertase Subtilisin/Kexin type 9 (PCSK9) inhibitor: pharmacophore-based virtual screening, in Silico molecular docking, and in vitro studies. Adv Pharmacol Pharm Sci. 2023;2023:1–10.
75. Liu X, Jing Z, Jia WQ, Wang SQ, Ma Y, Xu WR, et al. Identification of novel PPARα/γ dual agonists by virtual screening, ADMET prediction and molecular dynamics simulations. J Biomol Struct Dyn. 2018;36(11):2988–3002.
76. Nath V, Agrawal R, Kumar V. Structure based docking and molecular dynamics studies: peroxisome proliferator-activated receptors –α/γ dual agonists for treatment of metabolic disorders. J Biomol Struct Dyn. 2020;38(2):511–23.
77. Da'adoosh B, Marcus D, Rayan A, King F, Che J, Goldblum A. Discovering highly selective and diverse PPAR-delta agonists by ligand based machine learning and structural modeling. Sci Rep. 2019;9(1):1106.
78. Huang CS, Yu X, Fordstrom P, Choi K, Chung BC, Roh SH, et al. Cryo-EM structures of NPC1L1 reveal mechanisms of cholesterol transport and ezetimibe inhibition. Sci Adv. 2020;6(25):eabb1989.
79. Pfefferkorn JA, Larsen SD, Van HC, Sorenson R, Barton T, Winters T, et al. Substituted oxazolidinones as novel NPC1L1 ligands for the inhibition of cholesterol absorption. Bioorg Med Chem Lett. 2008;18(2):546–53.
80. Saxty BA, Novelli R, Dolle RE, Kruse LI, Reid DG, Camilleri P, et al. Synthesis and evaluation of (+) and (−)-2,2-difluorocitrate as inhibitors of rat-liver ATP-citrate lyase and porcine-heart aconitase. Eur J Biochem. 1991;202(3):889–96.
81. Dolle RE, McNair D, Hughes MJ, Kruse LI, Eggelston D, Saxty BA, et al. ATP-citrate-lyase as a target for hypolipidemic intervention. Sulfoximine and 3-hydroxy-.beta.-lactam containing analogs of citric acid as potential tight-binding inhibitors. J Med Chem. 1992;35(26):4875–84.
82. Gribble AD, Ife RJ, Shaw A, McNair D, Novelli CE, Bakewell S, et al. ATP-citrate Lyase as a target for Hypolipidemic intervention. 2. Synthesis and evaluation of (3 *R* *,5 *S* *)-ω-Substituted-3-carboxy-3,5-dihydroxyalkanoic acids and their γ-lactone prodrugs as inhibitors of the enzyme in vitro and in vivo. J Med Chem. 1998;41(19):3582–95.
83. Li JJ, Wang H, Tino JA, Robl JA, Herpin TF, Lawrence RM, et al. 2-Hydroxy-N-arylbenzenesulfonamides as ATP-citrate lyase inhibitors. Bioorg Med Chem Lett. 2007;17(11):3208–11.
84. Wei J, Leit S, Kuai J, Therrien E, Rafi S, Harwood HJ, et al. An allosteric mechanism for potent inhibition of human ATP-citrate lyase. Nature. 2019;568(7753):566–70.

Chapter 23
Modern Computational Intelligence Based Drug Repurposing for Cancer

Roja Rani Budha, Nikhat Khan, Saba Wahid A. M. Khan, Esha S. Attar, and S. N. Koteswara Rao G.

Contents

Abstract This chapter provides an in-depth exploration of contemporary therapeutic strategies in cancer treatment, emphasizing the potential of innovative genetic editing tools and drug repurposing methodologies. It begins with a thorough examination of meganucleases (MNs) and CRISPR-Cas systems, both of which have

R. R. Budha (✉) · N. Khan · S. W. A. M. Khan
H.K. College of Pharmacy, Mumbai, Maharashtra, India

E. S. Attar · S. N. K. Rao G.
Shobhaben Pratapbhai Patel School of Pharmacy and Technology Management, SVKM's NMIMS, Mumbai, Maharashtra, India

S. N. K. Rao G., R. R. Alavala (eds.), *Applications of Computational Tools in Drug Design and Development*, https://doi.org/10.1007/978-981-96-4154-3_23

revolutionized genomic editing. These technologies allow for precise modifications of the genome, enabling targeted therapies that can directly address the genetic anomalies driving cancer progression. The application of these tools in oncology is discussed, showcasing their ability to disrupt cancerous growth at the molecular level. The chapter also delves into oncolytic virotherapy, an emerging field that leverages viruses that are genetically engineered to selectively infect and kill cancer cells while sparing normal tissues. Detailed case studies of Gendicine, Oncorine, and Imlygic illustrate the clinical successes and challenges associated with these therapies. These oncolytic viruses not only destroy cancer cells but also stimulate anti-tumor immune responses, offering a dual mechanism of action. Another significant focus of the chapter is on drug repurposing, a strategy that identifies new therapeutic uses for existing drugs, thus accelerating the development of cancer treatments. This approach is particularly advantageous as it utilizes drugs with known safety profiles, minimizing the cost and time before involving new treatment to the market. The chapter outlines the principles of drug repurposing, providing examples of repurposed drugs that have shown efficacy in treating various cancers. The initiatives led by the National Center for Advancing Translational Sciences (NCATS), are highlighted to underscore the institutional support for drug repurposing efforts. These initiatives aim to bridge the gap between laboratory discoveries and clinical applications, fostering innovation in cancer therapeutics. In summary, the chapter offers a comprehensive review of state-of-the-art cancer therapies, from cutting-edge genetic editing to the innovative repurposing of existing drugs, emphasizing their potential to transform the landscape of cancer treatment.

Keywords Tumor · Drug repurposing · Microenvironment · Oncolytic · Gene editing tools · Mega nucleases · Cancer

23.1 Introduction to Cancer and the Need for New Drugs

Drug repurposing or repositioning, which involves using pre-existing, outdated medications for a new use. It is gaining traction as a potential tactic to overcome the difficulties that come with using traditional techniques and processes. The process of repurposing considers the easily accessible information regarding the molecular pharmacology of many therapeutic candidates.

Repurposing drugs for new uses is a time-efficient and cost-effective approach to developing treatments for uncommon diseases. This method reduces the risk of developing new treatments, as it involves combining computational and experimental approaches. Delisted substances, lower research costs, and quicker development times make it a viable alternative to traditional medications. This innovative approach allows pharmaceutical companies to create new treatments with less risk than creating new medications from scratch.

Repurposing aims to drastically cut down on the amount of time needed for medication development by taking advantage of the easily accessible knowledge on the molecular pharmacology of many different therapeutic candidates.

Cancer, literally crab, is a hectic health burden among globe and causing umpteen number of deaths substantially from yester years to current years. It is characterized by multiplication and growth of cells in an unfashion manner, which are metastatic and dedifferentiated, leads to development of tumor or Neoplasm i.e., a sack of heterogeneous cells. As the cell is targeted structure for cancer, its origin and spread is vast and inevitable in every instance. The mythological beliefs of cancer are dissipating and scientific approaches associated with cancer are flourishing in a way. International Agency for Research on Cancer (IARC) is an integral part of WHO, has published Globacan 2012 report consisting of estimations of various types of cancers in 184 countries, furnishing a comprehensive overview of the cancer burden among globe. Globocan 2012 estimates that number of new cancer cases increases to 19.3 million by 2025.

Chemotherapy is alone or in combination with radiotherapy is well accepted discipline for the treatment of cancer. A plethora of drugs are available for cancer chemotherapy in conventional and novel categories, but a considerable gap has been existing between the etiology of disease and successful therapy. A chemotherapeutic treatment of cancer is complex as it involves typical selection specific dose, non-specificity towards cells and drug resistance. Dosage and time span are two key considerations in selection of chemotherapeutics and the major challenge associated is recurrence of the disease. Chemo drugs are even prescribed as a part of palliative care to provide easiness in the patient's environment. Continuous usage of chemo drugs offers patient with many adverse effects on physical and mental health and is impossible to keep pace with further medication. Many of the times, a constitutional team of specialists, oncologists, radiologists, nutritionists and psychologists put their efforts to determine dosage regimen, duration of cycle and nutritional care to be provided.

It is vital to understand the nature of disease in a scientific manner, prior to finding a new drug or treatment. Molecular oncology and its application is prototype mechanisms to understand treatment strategies of cancer. Interrupting the signals promoting cell growth, preventing angiogenesis, making cancer cell deprived of hormones required for its growth and initiation of tumor cell death by directly damaging the DNA are basic mechanisms underlying with cancer treatment. Cancers are much diversified, where individualization of treatment is required and has remained as elusive trial still. Existence of Imperfect vasculature, inter capillary resistance and high interstitial pressure in the micro environment of solid tumor associates with poor and heterogeneous drug tissue penetration, which is pivotal to kill cells in solid tumors. In spite of advances in treatment approaches and tailoring medicine has been made in last few decades to extend the rate of survival among cancer patients and to improve the patient's life quality, sparseness of chemotherapeutic drugs in market been remaining as major cloud. There is an inflated demand for arrival of new drugs, called targeted drug therapies and precision medicines, able to attack cancer cells specifically while minimizing the damage to surrounding normal cells. Micro environment of Tumor (TM), which determines the initiation and proliferation of tumor. Modification of TM is extending its scope for positive outcomes in cancer therapy. Rate of survival in pancreatic cancer is extended from 17% to 24% by adding Erlotinib as targeted drug to standard chemotherapy. Use of Imatinib in Chronic Myeloid Leukemia, Sunitinib use in renal cell carcinoma, trastuzumab use in breast cancer have shown dramatic effects as targeted therapy. Targeted therapy agents are categorized in to enzyme inhibitors, apoptosis induced drugs and inhibitors of VEGF. Rate of Survival is extended up to months in advanced colorectal carcinoma patients after the combinational use of Avastin and 5-fluorouracil, by inhibiting Vascular Endothelial Growth Factor (VEGF), which is angiogenic and nourish tumors with blood supplement.

23.2 Current Status of Cancer Drug Therapy

Drug resistance and relapse of disease are the toughest challenges arises with cancer therapy and needs to be revolutionized the focus from new drug candidature inventions to existed drug improvements in credible ways. Early detection of cancer increases the survival rate up to 5 years and decreases the expensiveness of treatment. Currently researchers are continuously exploring new avenues for cancer treatment to battle with challenges existed with established drugs [1].

23.2.1 Nanotechnology for Cancer Treatment

Nano technology has emerged with sensitive diagnostic capabilities, better targeting abilities and maximal drug efficacy at localized tissues. Applicability of nanotechnology has been tremendously increased over years. Nanotechnology makes use of precisely liposomal carriers which facilitate targeted penetration in the tumor microenvironment. In order to achieve so, the technology makes use of specialized polymers, antibodies, proteins, etc. to alter the Pk/PD profile of free anticancer drugs which facilitates targeting. Nanoparticles have the advantage of penetration into the tumor via the leaky blood vessels [2]. Formulation of nanomaterial's with synergistic has provided a cut edge advantage in cancer therapy. The presence of matrices in nanoparticles in Nano formulation favors drug desorption and diffusion across the tumors [3, 4]. The complexity of the cancer drug regimen can be simplified using nanotechnology. Multiple drugs can be delivered to the target site within a single carrier [5–8]. The polymers are arranged in a way so as to facilitate the release of different drugs at different time intervals by virtue of degradation of polymers in a coordinated fashion [9].

For example, doxorubicin (DOX)-loaded immunoliposomes decorated with to target EGF receptors (EGFR) are currently in clinical trials (ClinicalTrials.gov Identifier: NCT03603379).

For specialized tumor imaging, nanotechnology makes use of probes which are activated in specific tumor microenvironments (TME) [10].

Nanomaterial are formulated to release the drugs via stimuli-responsive activation under specified tumor conditions like external stimulations and hyperthermia to limit the off target effects [11–14].

23.2.2 Gene Therapy for Cancer Treatment

Gene therapy has emerged as a potent candidate for cancer treatment [15]. Its remarkable efficacy has been proven at preclinical and clinical levels. Gene therapy has been used in varied cancers where it can restore the normal phenotypes in cancer cells. Gene therapy can generate a targeted response, decrease blood supply, because lysis of the tumor cells, etc. Preclinical gene therapy tests have been conducted on various cancers including gliomas [16], pancreatic cancer [17] and liver cancer [18]. Advances in oncolytic virology therapy are being tested for both recurrence and cancerous treatment, with the hope of reducing cancer to manageable chronic diseases [19].

23.2.2.1 Gene Editing Tools

- Meganucleases (MNs)
- These endonucleases are designed to recognize large target sequences with minimal cytotoxicity, making them valuable tools for genomic editing. High complexicity in re archtecturing and restricted efficiency are the limitations of MNs [20].
- CRISPR-Cas (Clustered Regularly Interspaced Short Palindromic Repeats)
- It is an advanced technique, in which the bacteria's natural immune system is embedded with memory of past viral pathological conditions to resist with recurrence of the same. CRISPR is passed via further bacterial generations and makes them immune to further infections by targeting changes to invaders DNA [21].

23.2.2.2 Application of Various Tools in Cancer Therapy (Table 23.1)

Gendicine, Oncorine, Imlygic, and Rexin-G are oncolytic viruses, which are modified genetically and used in the treatment of cancers. Gendicine is a vector, which activates apoptosis and inhibits DNA repair by substituting E1gene with p53 cDNA gene. Oncorine is the oncolytic recombinant virus, used in naso pharyngeal cancers. Imlygic is for non-resectable metastatic melanoma. Rexin-G is the first targeted injectable vector, promotes apoptosis, to treat metastatic cancers by inhibiting G1 phase of cell cycle via synthesis of cytocidal protein [26, 27].

Table 23.1 Application of various tools in cancer therapy

Trade Name (Proper Name)	Date of Approval and Approving Agency	Vector and Modified gene	Indication	Route of Administration
Gendicine [22]	2003 state Food and Administration of China	Adenoviral vector P53 [23]	Head and neck squamous cell carcinoma	In vivo
Oncorine (Recombinant Human Adeno Virus Type 5 injection) [24]	2005 state Food and Administration of China	Adenovirus Type 5 [25].	Head and neck and oesophagus cancer, nasopharyngeal cancer	In vivo
Kymriah TM (tisagenlecleucel)	August 2017 FDA	CD19 - specific CAR T lentiviral vector	Acute lymphoblastic leukaemia	Ex vivo
Yescarta TM (axicabtagene ciloleucel)	October 2017 FDA	CD19 - specific CAR T Y-retroviral vector	Non Hodgkin lymphoma	Ex vivo
Imlygic(tallimogene laherparepvec, T-vec)	2015 FDA	GM-CSF HSV-1	Melanoma	In vivo

23.2.3 Oncolytic Virotherapy

It is an immunotherapy approach in the treatment of cancer. OV involves the usage of cancer cell specific viruses. These OVs can be grouped into naturally occurring or genetically modified. The naturally occurring ones can target a cell and kill it without being genetically modified, whereas the genetically modified ones can undergo genetic changes to improve the beneficial effects and decrease potency of virus. Upon injection the oncolytic viruses carrying vector genes can cause lysis of the cancer cells by incorporation of carried into the genome of the tumor. The usual target of such an attempt is incorporation of wild P53 gene and thereby restoring the normal p53 in cancer cells and causing cell lysis [28, 29].

Principles and tools used for repurposing of drugs

Drug discovery is a complex and monotonous procedure, involving laborious and expensive investment with high failure rate during different stages includes random screening, identification, characterization and development of lead molecule, formulation studies, preclinical and clinical studies. In May 2012, National Institute of Health (NIH) in the United States of America (USA), launched a program "Discovery of New Therapeutic Uses for Existing Molecules" through its one of the branch, National Centre for Advancing Translational Sciences (NCATS). This program aimed at recognition of existed drugs in market or FDA approved drug candidate with new therapeutic uses. This reevaluation of drugs with new

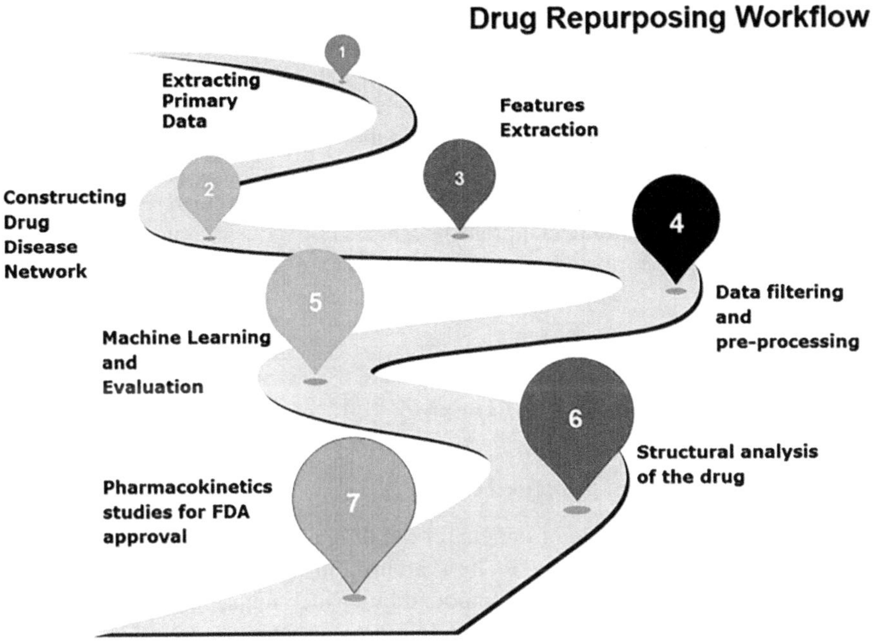

Fig. 23.1 Drug repurposing window

pharmaco therapeutic indications is called Drug repurposing or drug repositioning or therapeutic switching or drug profiling. This method reduces the burden of drug discovery process in view of time, money and work by screening large number of drug candidates virtually. The research of drug repurposing is exponentially increasing, which involves the usage of computational tools, unravels the new molecular mechanisms of existed drugs with new therapeutic applications by identifying the targeted proteins/genes, as a single protein may trigger multiple pathologic responses. Identification of targeted protein and finding the binding affinity of drug to the targeted protein are the salient steps involved in drug repurposing. Figure 23.1 showed drug repurposing window.

23.3 Strategies for Drug Repurposing

Repurposing is the method of reusing a drug for a different indication. Repurposing is dual step process. First, is to explore the targets' likeliness via a pathophysiological approach. In this step in silico screening is done for an existed formulation against a defined disease [30]. In the second stage we directly use the drug explored in the first step for its efficacy at clinical levels for the respective indication [31]. To understand the drug molecule and its properties, a large data of literature and extensive efforts are required. The repurposing of drug can be done at any point of time during the evolution of drug ranging from its discovery to the state of existence in the market [32]. Serendipitous discovery of better therapeutic effect of drug in one disease may be find out from the adverse effects of same drug in another pathological condition. For example, Sildenafil, is the drug holding huge market to treat erectile dysfunction, originally discovered for hypertension management [33]. Itraconazole, primarily used as an antifungal agent and is cleared as second line drug in the treatment of lung and prostate cancers due to its anti-angiogenesis properties. The knowledge related to drug and disease are key concerns of repurposing [34–36].

23.3.1 Techniques of Repurposing

- Experimental screening approaches
- It involves the use of hits or targets for the drug candidate. In this technique the new drug candidate is analyzed by searching in the drug discovery programs. This is carried out for denovo compound hits which require a highly specialized facility for screening and a compound library containing a several million compound. This repurposing technique focuses on known drugs that have failed in some indication or those having detailed data about their safety and mechanism of action.
- *In-silico* approaches
- This method makes use of drugs already in existence [37]. The method make use of data available of the protein targets in cancer pathogenesis and docks the potential drug candidate on it. The goal is to find a confirmation of the drug so as to fit in the receptor pocket on the binding site. It gives information about various aspects of drug receptor interaction like ligand binding modes, intermolecular interactions, the bonds and the corresponding binding energies that stabilizes the drugs in the receptors pocket. Furthermore these binding energies in the docking process are interpreted by algorithms and accordingly the compounds can be ranked on the basis of the interpretations [38]. Identification of the binding affinity requires two steps:

  Exploration of conformational space of binding thereby understanding the mode of binding of drugs to the target
  Accurate predictions of these binding energies.

These approaches are following types:

Molecular Approaches. These involves the usage of knowledge regarding drug target and pathology of disease [39], with the help of databases like omics. Genomics and Transcriptomics are two types of database that generally used.

1. Genomics involves screening organisms for mutations causing drug resistance or hypersensitivity, allowing for the isolation of drug resistant and drug sensitive genes [39–41].
2. Transcriptomics analyzes the expression levels of numerous genes, potentially facilitating drug repurposing. Clinical benefits may arise from strategies aimed at reversing gene expression signatures [42–44].

- Real world data approaches

- Which includes network based, ligand based, and structure based and machine learning techniques, identify unexpected drug diseases relationships based on individuals health, habits and behavior without environmental intervention [45, 46].
- Network based drug repurposing
- Network based drug repurposing makes use of the data available on the cancer cell biology and its phenotypes in different biological systems. A network is created by using the above knowledge in a network pharmacology framework. Further the nodes of this network represent an individual molecular entity and its biological target along with the disease modifier and associated proteins involved in the cancer biology [47]. The major limitation of this method is its dependence on huge data. The method uses pathways and networks of biological responses which are not fully discovered [48]. Also, this approach makes use of a biological network as a standalone entity, but in reality the biological network of interest can be related to many more networks at the same time in a patient [49].
- Ligand based drug repurposing
- It involves the use of information of compounds having similar biological properties. This method makes use of chemical space coverage already known by the molecule. Ultimately drug target association is considered on the basis of drug target similarities [50, 51].
- Structure-based drug repurposing
- This approach is based on the fact that similar protein structure has the potential of performing similar action. This technique is useful in identifying secondary targets of a drug other than the one that is known. Proteins as drug targets are matched with others on the basis of similarities in their sequences. Kinome is a protein tree constructed on the basis of studies conducted regarding similarities of sequences in protein structures [52, 53]. From this tree one can identify substrates with similar functions. The proteins are characterized as 'similar' on the basis of similarity scores achieved after sequence comparison [54]. Computational approaches are used to predict the hot spots in binding sites to determine the protein ligand complexes' structure [55], which are essential for structure based drug repurposing [56].
- Machine learning methods

- These methods improved COVID-19 treatment plans by combining structure-based and ligand-based screening methods with AI algorithms [57]. Deep learning, SVM, RF, Naive Bayesian, and neural networks are widely used for high-throughput screening. The development of AI, ML, and network medicine approaches has positively impacted drug development [58, 59].

23.4 Approaches for Finding Drug Candidates for Repurposing

23.4.1 Experimental Approaches

23.4.1.1 Binding Assays

Which include methods such as proteomics and mass spectrophotometry for target recognition. For instance, the binding phenomena was used as a base to detect the target proteins of anticancer compounds by the cellular thermo stability assay (CETSA) technique. Cellular targets of Crizotinib [60, 61] and acetaminophen [62] were confirmed by CETSA technique.

23.4.1.2 Phenotypic Approaches

This model offers drug discovery often serendipitously [63]. Clinical observation is the base for new drug invention [64]. For example, this technique can screen a compound library against a cell line and find the altered phenotype and tracing the status of disease state [65]. In vitro phenotypic screening it is mandatory to have a drug or a drug-like compound confirmed prior to use [66]. Robotic screening systems are sensitive, efficient, and rapidly screen large chemical libraries, minimizing the time and expenses in the process of drug discovery and enabling the testing of multiple compounds with distinct mechanisms [67, 68].

23.4.2 Computational Approaches

The science of computational pharmacology deals with approaches for drug discovery by collecting a large biological data to integrate molecular and genetic changes. It is an effective research tool in modern pharmacology, by aiming at rational uses of resources [69].

23.4.2.1 Drug-Centric Approaches

These repurposing approaches can be explained by poly pharmacological agents, how potentially they bind with multiple targets and exerts new indications to reveal unknown effects of drug [70].

23.4.2.2 Target-Based Approaches

Target repositioning and off target repositioning are major target based approaches. The new therapeutic application of the drug is attained by interacting with predetermined and primary target protein is known as target repositioning, while the drug's new therapeutic effect is achieved through off-target repositioning, where it interacts with a secondary target protein [71].

23.4.2.3 Knowledge-Based Methods

These methods can create path for exploration of drug target reactions with the help of coalesced information of drug related to chemical, biological, clinical data and forecast the untouched mechanisms of drugs [72, 73].

23.4.2.4 Pathway- or Network-Based Methods

These are the methods which involves the Study of pathogenic pathways through which drugs undergo to manifest their actions. Study of unequivocal signalling mechanisms in metastatic breast cancer, which were not illuminated earlier, is directed in drug repurposing. The approach led to the discovery of fulvestrant, approved for treating breast cancer, also showing effects in glioblastoma [74–76].

23.5 Artificial Intelligence in Drug Repurposing for Targeting Cancer

AI based techniques are managing the tasks in biological networks [77], which are the result of genetic interactions [78]. It facilitates the understanding of carcinogens [79] and exploration of novel anti-cancer drugs [80, 81]. This is the area, where cancer biologists and computational biologists can work together. Multiomics technologies are emerging now a days as a part of artificial intelligence to explore targets of anticancer [82, 83].

23.5.1 Anti-cancer Drug Combination Datasets [84]

Drug comb is a combinational portal of drug data used in the treatment of cancer. Larger data sets of anti-cancer drug combinations are facilitating the discovery of new drugs in cancer therapy.

23.5.2 *Machine Learning Basics*

It is essential to develop a judicious relation between data input and output targets in machine learning to build up a triumphant model, which can make accurate prognosis about concealed data [85]. Cell lines and drug features are components of input to predict drug synergy in anti-cancer drug combinations. The considerable factors to improve prediction are Gene expression, mutation, methylation, gene essentiality. Structural features, physicochemical properties, side effects are drug features, used as input. Machine learning methods are of traditional learning and deep learning methods.

23.5.2.1 Traditional Learning Methods

- Random Forest
- In 2017, Li's research team proposed a random forest model to predict synergistic anti-cancer drug combinations based on drug-target networks and drug-induced gene expression profiles [86]. They manually designed 18 features that included similarities among drug chemical structures, drug-target network and drug pharmacogenomics. They utilized the random forest algorithm to select best combination of features that were used to construct a prediction model and identified 28 potentially synergistic anti-cancer drug combinations. Their use of drug-induced gene expression profiles to identify differentially expressed genes for prediction was particularly noteworthy. Profiles of gene expression following treatment with different doses of drugs can reflect the biological response to drug treatment to provide further evidence of the mechanisms of drug action used for predicting drug synergy.
- Li's group postulated a novel method based on network of genes to stimulate molecular features after treatment, especially where drug target genes become zero [87]. They used monotherapy information also as input feature to improve performance of prediction in drug synergy.
- Extreme Gradient Boosting (XG Boost)
- It is the model of accuracy and interpretability, which uses information of monotherapy and predicts the frequent targets of synergistic interactions [88].
- Extremely randomized tree
- When dealing with noisy data, is require to reduce variance to make the model more robust, in this case the extreme randomness is helpful by sub sampling of samples. Genomic information, monotherapy and synthetic information is incorporated to improve prediction. Gilyary's research team proposed a multi Tsk framework to study drug combination synergy in cells and in the clinic [89].
- Logistic regression
- Profiles of gene expression and mutations are best choices to predict drug synergy [90]. Li et al. employed regression to feature the gene pathway and to

compare gene expression between targets and non-targets while predicting drug synergy. They concluded that the combination of gene expression and gene essentiality with different mechanisms may boost the functionality of synergism rather than using alone.

Tensor Factorization

This method is proposed by Julkenen's group [91] to integrate drug molecular fingerprints and profiles of gene expressions of cancer cells to facilitate prediction and to learn landscape of drug combinations.

23.5.2.2 Deep Learning Methods

DeepSynergy, a deep learning model, accurately predicted drug synergy using gene expression as cell line feature and chemical descriptors as drug features, constructing a combined hidden representation in hidden layers [92].

- Feed—forward neural networks
- DeepSynergy's 2018 publication on feed forward deep neural networks demonstrated their ability to discover complex abstract structures in large datasets [85]. Xia et al.'s research on tumour cell line responses to drug combinations used gene expression, microRNA expression, protein abundance, molecular descriptors, and fingerprints from Dragon software [93].
- Autoencode
- Autoencode is a deep learning architecture used for drug synergy, reducing feature dimension by learning informative latent representations [94]. Kim et al. developed an autoencoder-based drug synergy prediction model to study data-poor tissues, utilizing information from data-rich tissues with shared biological commonalities for drug response [95].

23.6 Current Drug Candidates for Cancer Therapy Under Repurposing

23.6.1 Thalidomide

Thalidomide's antiangiogenic potential has been discovered where more than 25% efficacy is noted in cases of multiple myeloma [96]. The use of thalidomide in combination with other anticancer drugs increased the efficacy of the same by many folds. For instance combining thalidomide with other steroids enhanced the response rate of therapy by 50% whereas combining the same with steroids and alkylating agents resulted in 70% success rate in cases of multiple myeloma. Another example of such a combination is thalidomide combined with other chemotherapeutic agents in GDPT (gemcitabine, cisplatin, prednisone, and thalidomide) regimen for the treatment of T-cell lymphoma in an ongoing trial (NCT01664975) [97]. The above said results may be due to synergistic effects of the combinations, also we see a reduction in associated side effects as the dosage of individual drugs in the regimen is reduced [98]. Hence the combination will provide a better efficacy with reduced side effects of the drugs [99].

23.6.2 Antidiabetic Drugs

Anti-diabetic medications have been shown to help treat a variety of malignancies. These medications include sulfonylureas, biguanides, thiazolidinediones, sodium-glucose co-transporter-2 (SGLT2) inhibitors, and dipeptidyl-peptidase IV [100]. The metabolic connections between diabetes and cancer, such as hyperglycaemia, inflammation, oxidative stress, etc. [101], may account for the effectiveness of anti-diabetic medications in treating cancer. More intriguingly, new research indicates that anti-diabetic medications may reduce the occurrence of cancer by directly affecting cancer cells' metabolism and indirectly by altering tumour risk factors [102].

Metformin, the first-line treatment for type 2 diabetic mellitus, is the most well-known example. Metformin is useful in preventing cancer either alone or in conjunction with other medications [103], according to several meta-analyses of case-control, cohort, observational, and clinical research.

The risk of colorectal, breast, pancreatic, prostate, lung [104], and cervical cancers [105] is greatly decreased by metformin. According to studies, biguanides and metformin together can dramatically lower the death rate from cancer [106, 107]. Furthermore, pioglitazone was linked to a lower incidence of kidney cancer in a cohort analysis. Last but not least, the utilization of glucose consumption and cancer biology tumour cells primarily utilizing aerobic glycolysis [108] as a basis for the repurposing of anti-diabetic medications.

Apart from its therapeutic benefits, the enhanced metabolism of glucose in tumours can be leveraged for diagnostic and imaging purposes. For instance, 18F-fluorodeoxyglucose is used in positron emission tomography (PET), a non-invasive imaging method that can be used to diagnose cancer, monitor the effectiveness of treatment, and assess the function or metabolism of healthy tissue or disease conditions [109, 110]. Numerous studies support the encouraging impact of anti-diabetic and anti-hypertensive medications on cancer. 52 and 40 clinical trials, respectively, assessed the anti-cancer efficaciousness of anti-diabetic and anti-hypertensive medications up to March 2023. It's interesting to note that the majority of hypertension and anti-diabetic drugs are in Phase II trials. There is an increased incidence of repurposing cardiovascular drugs in cancer treatment. Inhibition of pathways involved in angiogenesis, cellular proliferation are the highlights of beta adrenergic and make them to treat cancer [111].

According to the data of preclinical studies, concurrent use of anti-hypertensive drugs can inhibit cell growth and acts against cancerous cells [112]. According to the data of retrospective study, long term use of propranolol, reduces the risk of commencement of cancers of stomach, colon and prostate [113]. ACE inhibitors and ARB'S too stop the risk of recurrence in breast cancer [114]. Sunitinib is the drug offers better survival rate in renal cell carcinoma [115].

23.7 Challenges in Drug Repurposing

Though the drug repurposing shows reap benefits over de novo methods of drug invention, is may not always coined with great success, as facing toxic or efficacy related challenges.

It should come as no surprise that drug repositioning presents difficulties even as it offers advantages over conventional de novo drug discovery. Drug repositioning may be greatly aided by drug patents. "Composition of matter" is a type of compound patent protection that lasts for 20 years and safeguards the rights of anybody who creates or discovers any new apparatus, procedures, or compounds, including but not limited to new formulas, recipes, or materials [116]. Therefore, it can be difficult to patent a medicine that has been repositioned, particularly if the original indications have already been trademarked before new competitors [117]. Drug repositioning can be approached in a variety of ways with the current COM patent. For example, Prednisporin, a combination of two antiquated medications—the immunosuppressant Cyclosporine and the glucocorticoid Prednisolone acetate—was created by the business Zalicus in association with Sanofi Aventis Paris [118].

A major problem in drug repositioning, aside from this, is the paucity of participation and the dearth of notable repositioning efforts, particularly from large, well-known pharmaceutical corporations. This situation results from a lack of confidence in potential projects, their high risk of failure, and the significant time and financial commitment required for them. It's interesting to note that companies other than the principal inventor were mostly responsible for launching the medication repositioning role [119, 120]. For instance, the business Celgene revived the medicine Thalidomide, a complete failure in the history of drug research, to treat Multiple Myeloma Nodosum Leprosum and Erythema, with the exception of pregnant women, but not the original manufacturer, Grünenthal [121–123].

A significant amount of capital and experience, particularly with regard to uncommon medical diseases, are necessary to plan and carry out clinical trials pilot projects. However, these resources can be challenging to come by for initiatives that the biotechnology and pharmaceutical companies need. However, out-licensing could overcome this obstacle, provided small biotech companies or research bodies are willing to assume the previously indicated risk, which is currently rare [124].

In addition, funding is particularly hard to come by when the drug under consideration for repositioning has already been turned down for other reasons, which erodes confidence in the outcome. Regretfully, while the majority of nations are excited about advances in healthcare research, few have surplus funds available for these kinds of initiatives [125]. The National Institutes of Health (NIH) sponsors drug repositioning trials in the United States, for example, although similar benefits might not be available in European or Asian nations.

Drug repositioning has many benefits over the conventional de novo technique, however not all of them are successful. Projects aimed at repositioning drugs have not always succeeded. Bevacizumab, a kinase inhibitor that was created to treat stomach cancer, was carefully examined and repositioned to treat various other types of malignancies, however it did not demonstrate efficacy in phase III trials [126].

Large datasets are available for the early phases of pharmacological target identification, but better integrative platforms are required for data analysis, which has been recognised as a pipeline issue with regard to data access and integration [127]. Furthermore, it may be difficult to discover potential medication targets for uncommon or orphan diseases due to a lack of data or patient reporting from clinical settings, which would complicate the process of drug repositioning. In a similar vein, as some medication targets are discovered through in vitro testing, failing to pursue a promising candidate medicine that surpasses the preliminary research is another drawback [128].

Drug repositioning is now challenging due to a number of factors, including the negative reputation associated with it, financial constraints, the need for significant levels of experience, unanticipated side effects, a lack of suitable drug candidates, and a lack of integrated platforms for data analysis. However, the negative effects of medication repositioning can be positively mitigated thanks to the close cooperation of pharmaceutical firms, governments, and academics, as well as the current advancements in computational approaches [129].

Drug repurposing has advantages over de novo drug discovery techniques, however it has not always been developed successfully due to hazardous or efficacy-related issues. Compound patent protection serves as a shield for the drug's original inventor's rights and makes the process of repurposing drugs difficult. Unsurprisingly, the firms that originally invented the medicine are not involved in the repurposing of drugs. For instance, the Grünenthal Company was the original inventor of thalidomide, which has a history of failure. The same medication was developed by Celgene to treat erythema and multiple myeloma and was advised for use by expectant mothers [130].

If a drug candidate has previously been rejected, it can be very challenging to obtain money for its repurposing and there is little confidence among funding organisations to invest in it. Other difficulties include the obvious revelations of medication repurposing, unforeseen side effects, and insufficient knowledge of data analysis. It is worthwhile to take on the challenges, and repurposing may benefit from transparent industrial policies that also help to prevent negative effects [131].

23.8 Future Perspectives

Human diseases are expressed by underlying complex mechanisms that can be developed or obtained from many sources such as aberrations in genetics, infectious diseases, degenerative disease and more. Diseases often involve many complex cellular pathway cascades, which can vary in certain individuals. In diverse human populations, each individual has a unique set of inherited or non-inherited genetic abnormalities that could result in certain individuals to respond less or not at all to general treatments or drugs. Approved drugs can be invalid for a specific individual if they are deficient in a particular drug target and may not respond to the common way to a particular drug. This situation entails the requirement of personalised medicine to tailor for each patient, as drug efficacy can defer substantially in different gene profiles due to heterogeneity in human disease [121]. To reduce the lack of drug efficacy, drug repositioning plays an important role. Together with the advancement of next-generation sequencing technologies, personalised genomic studies can be carried out with an affirmative approach. This can result in the best medicine accompanied by high efficacy at the lowest toxicity levels possible to the specific individual in contrast to conventional treatment that aims for maximum tolerated doses. For instance, non-small cell lung cancer (NSCLC) accounts for 85–90% of all lung cancers [122]. In 2007, it was first reported that there is a fusion gene in NSCLC, which comprises portions of echinoderm microtubule-associated protein-like 4 genes (*EML4*) and anaplastic lymphoma kinase gene (*ALK*) in ~7% of patients with NSCLC. This was different from the common NSCLC with epidermal growth factor receptor mutation. Crizotinib, a small-molecule kinase inhibitor, which was intended to treat ALK-positive anaplastic large-cell lymphoma, is now repositioned to treat this small subset of NSCLC [132]. Without the advance screening of biomarkers in each NSCLC individual, this small subset would not be successfully discovered, and the common EGFR tyrosine kinase inhibitors would have been prescribed to all NSCLC patients albeit the limited efficacy in patients with *EML4-ALK* alteration. Recently, the drug Crizotinib (Pfizer's Xalkori) successfully obtained FDA approval as a treatment for patients with NSCLC with *MET* exon 14 skipping which occurs in approximately 3% of NSCLC tumours and *ROS 1*/anaplastic lymphoma kinase-positive resistant systemic anaplastic large cell lymphoma (*ROS-1/ALK*-ALCL) [123]. Hence, repositioning the drug based on stratification is important in accordance with the appropriate biomarkers for each individual.

According to Kim and Oh, 2018, Bevacizumab, a kinase inhibitor is determined to be a strong drug candidate to treat colorectal cancer and gliomas via repositioning and not cleared clinical trials of phase III. The failure drug candidates are evident options of personalized medicine approaches, especially for those have less treatment options. Drug repurposing has taken major burden of tackling diseases and evolving as front seated in future. Which offers personalized medicine or drug tailoring along with advanced genomic sequencing technologies, results in best medicine with high efficacy and low toxicity.

23.9 Conclusion

AI is true revolutionary feat of computer science The world is now witnessing unprecedented growth in AI technologies. The overall potential of futuristic AI is huge and its positive effect on various sectors especially in health sector cannot be denied. It is necessary to approach AI development and integration with a responsible mindset to use its benefits. By ensuring this, we can shape a future where AI works in harmony with human aspirations. Adaptation of AI in drug repurposing for cancer is a trend, by which researchers are saving their time, money and man power. Often, the cancer drug market flooded with scarcity of drug candidatures as it involves the complexity in every determinant step of invention and potential adverse effects in addition to targeted properties. In this scenario, AI is emerging as boon for repurposing of drugs for the treatment of cancer and nurturing the cancer drug market with potential repurposed drugs to combat major challenges like recurrence, toxicity towards non-cancerous cells and to progress with refined approaches in cancer chemotherapy.

To summarize, the process of repurposing drugs can be broken down into three phases: hypothesis generation to identify the primary targets of the disease, in vitro and in vivo models to assess the drug's effectiveness, and phase II clinical trials to follow when phase I trials have produced sufficient data [133–135]. Since developing hypotheses is essential to any attempt at drug repurposing, the initial phase is crucial [136]. Historically, knowledge of the disease pathways or coincidental discoveries have been the main driving forces behind therapeutic repurposing in oncology. Therefore, creating novel approaches to link current medications with recently discovered uses may improve the efficacy of medication repurposing. Computational and experimental techniques can be used to identify a possible medicine for repurposing.

While computational methods employ target-centric, knowledge-driven, signature-aligned, pathway-focused, and mechanism-specific strategies, the experimental approach takes into account tools like induced pluripotent stem cell models and function-first phenotypic screenings (or reverse chemical biology) [137–140] These methods are more frequently used in concert with one another. Interestingly, advanced models for high-throughput screening can find compounds that reduce the symptoms of disease without requiring prior understanding of the interactions between the medicine and its target [141, 142]. Modern computational approaches are available for the selection of pharmacological candidates appropriate for drug repurposing in cancer, including combining drug effects with clinical disease markers and model systems that forecast disease-modifying effects. These instruments are able to recognize ligands, interpret the binding schemas of pharmaceutical ingredients, and choose attractive candidates from a large pool of possible molecules [143, 144].

All over world, till date, many of the researchers and pharmaceutical manufacturers turned towards drug repurposing. Many tools like genomic data and transcriptome data and data bases have emerged as advancements in the development of repurposed drugs in the cancer treatment. In this chapter, emphasis has been made on approaches for repurposing of drugs for the treatment of cancer.

Various substantial data resources like genomics, drug-disease profiles and drug chemical structures are taking pivotal role in encompassing the complexity of mechanisms involved in cancer and supporting the process of repurposing of drugs to treat cancer. Number of approved drugs are hampering by legal and safety issues, intellectual property issues and regulatory barriers. In addition to these, repurposed drugs also require potential resources of time and money. Conduction of clinical trials is required to come up with answers about the pharmacokinetic and pharmacodynamics profile of drug candidate. As cancer involves drug resistance and complexity in signalling pathways, relatively monotherapy is ineffective. In this circumstances, combination therapy and precision medicine are showing promising effects, where repurposed drugs can be used. Tumour micro environment is the potential hope in the future for the personalized treatment of cancer and concomitant use of first line drugs and repurposed drugs offers new hope for cancer patients with new approaches therapeutically.

References

1. Tran S, DeGiovanni P-J, Piel B, Rai P. Cancer nanomedicine: a review of recent success in drug delivery. Clin Transl Med. 2017;6:1.
2. Beltrán-Gracia E, López-Camacho A, Higuera-Ciapara I, Velázquez-Fernández JB, Vallejo-Cardona AA. Nanomedicine review: clinical developments in liposomal applications. Cancer Nanotechnol. 2019;10:11.
3. Torchilin V. Tumor delivery of macromolecular drugs based on the EPR effect. Adv Drug Deliv Rev. 2011;63:131.
4. Izci M, Maksoudian C, Manshian BB, Soenen SJ. The use of alternative strategies for enhanced nanoparticle delivery to solid tumors. Chem Rev. 2021;121:1746.
5. Fang J, Islam W, Maeda H. Exploiting the dynamics of the EPR effect and strategies to improve the therapeutic effects of nanomedicines by using EPR effect enhancers. Adv Drug Deliv Rev. 2020;157:142.
6. Kang H, Rho S, Stiles WR, Hu S, Baek Y, Hwang DW, Kashiwagi S, Kim MS, Choi HS. Size-dependent EPR effect of polymeric nanoparticles on tumor targeting. Adv Healthc Mater. 2020;9:1901223.
7. Walkey CD, Olsen JB, Guo H, Emili A, Chan WC. Nanoparticle size and surface chemistry determine serum protein adsorption and macrophage uptake. J Am Chem Soc. 2012;134:2139.
8. Fullstone G, Wood J, Holcombe M, Battaglia G. Modelling the transport of nanoparticles under blood flow using an agent-based approach. Sci Rep. 2015;5:1.
9. Chen YY, Syed AM, MacMillan P, Rocheleau JV, Chan WC. Flow rate affects nanoparticle uptake into endothelial cells. Adv Mater. 2020;32:1906274.
10. Gustafson HH, Holt-Casper D, Grainger DW, Ghandehari H. Nanoparticle uptake: the phagocyte problem. Nano Today. 2015;10:487.
11. Curley CT, Mead BP, Negron K, Kim N, Garrison WJ, Miller GW, Kingsmore KM, Thim EA, Song J, Munson JM. Augmentation of brain tumor interstitial flow via focused ultrasound promotes brain-penetrating nanoparticle dispersion and transfection. Sci Adv. 2020;6:eaay1344.
12. Bartneck M, Keul HA, Zwadlo-Klarwasser G, Groll J. Phagocytosis independent extracellular nanoparticle clearance by human immune cells. Nano Lett. 2010;10:59.

13. Yhee JY, Jeon S, Yoon HY, Shim MK, Ko H, Min J, Na JH, Chang H, Han H, Kim J-H, Suh M, Lee H, Park JH, Kim K, Kwon IC. Effects of tumor microenvironments on targeted delivery of glycol chitosan nanoparticles. J Control Release. 2017;267:223.
14. van der Meel R, Sulheim E, Shi Y, Kiessling F, Mulder WJ, Lammers T. Smart cancer nanomedicine. Nat Nanotechnol. 2019;14:1007.
15. Feng J, Xu M, Wang J, Zhou S, Liu Y, Liu S, Huang Y, Chen Y, Chen L, Song Q, Gong J, Lu H, Gao X, Chen J. Sequential delivery of nanoformulated α-mangostin and triptolide overcomes permeation obstacles and improves therapeutic effects in pancreatic cancer. Biomaterials. 2020;241:119907.
16. Wang C, Yu Y, Irfan M, Xu B, Li J, Zhang L, Qin Z, Yu C, Liu H, Su X. Rational design of DNA framework-based hybrid nanomaterials for anticancer drug delivery. Small. 2020;16:2002578.
17. Rabiee N, Bagherzadeh M, Ghadiri AM, Fatahi Y, Aldhaher A, Makvandi P, Dinarvand R, Jouyandeh M, Saeb MR, Mozafari M. Turning toxic nanomaterials into a safe and bioactive nanocarrier for co-delivery of DOX/pCRISPR. ACS Appl Bio Mater. 2021;13:5336.
18. Gupta PK, Gahtori R, Govarthanan K, Sharma V, Pappuru S, Pandit S, Mathuriya AS, Dholpuria S, Bishi DK. Recent trends in biodegradable polyester nanomaterials for cancer therapy. Mater Sci Eng C. 2021;127:112198.
19. Pontón I, Martí del Rio A, Gómez Gómez M, Sánchez-García D. Preparation and applications of organo-silica hybrid mesoporous silica nanoparticles for the co-delivery of drugs and nucleic acids. Nanomaterials. 2020;10:2466.
20. Guo R-Y, Wang H-M, Dong X, Hu Y, Li J, Zang Y, Li X. Selectivity comparison of tumor-imaging probes designed based on various tumor-targeting strategies: a proof of concept study. ACS Appl Bio Mater. 2021;4:2058.
21. Liu X, Wu X, Xing Y, Zhang Y, Zhang X, Pu Q, Wu M, Zhao JX. Reduced graphene oxide/ mesoporous silica nanocarriers for pH-triggered drug release and photothermal therapy. ACS Appl Bio Mater. 2020;3:2577.
22. Shen J-M, Yin T, Tian X-Z, Gao F-Y, Xu S. Surface charge-switchable polymeric magnetic nanoparticles for the controlled release of anticancer drug. ACS Appl Mater Interfaces. 2013;5:7014.
23. Kamaly N, Yameen B, Wu J, Farokhzad OC. Degradable controlled-release polymers and polymeric nanoparticles: mechanisms of controlling drug release. Chem Rev. 2016;116:2602.
24. Kolosnjaj-Tabi J, Gibot L, Fourquaux I, Golzio M, Rols M-P. Electric field-responsive nanoparticles and electric fields: physical, chemical, biological mechanisms and therapeutic prospects. Adv Drug Deliv Rev. 2019;138:56.
25. Mulligan RC. The basic science of gene therapy. Science. 1993;260:926–32. [PubMed] [Google Scholar].
26. Vattemi E, Claudio PP. Gene therapy for lung cancer: practice and promise. Ann Ital Chir. 2004;75:279–89. [PubMed] [Google Scholar].
27. Prasad G, Wang H, Hill DL, Zhang R. Recent advances in experimental molecular therapeutics for malignant gliomas. Curr Med Chem Anticancer Agents. 2004;4:347–61. [PubMed] [Google Scholar].
28. Tseng JF, Mulligan RC. Gene therapy for pancreatic cancer. Surg Oncol Clin N Am. 2002;11:537–69. [PubMed] [Google Scholar].
29. Prieto J, Qian C, Hernandez-Alcoceba R, Gonzalez- Aseguinolaza G, Mazzolini G, Sangro B, Kramer MG. Gene therapy of liver diseases. Expert Opin Biol Ther. 2004;4:1073–91. [PubMed] [Google Scholar].
30. Early-stage repurposing | National Center for Advancing Translational Sciences. n.d. Accessed 23 July 2018.
31. Late-stage repurposing | National Center for Advancing Translational Sciences. n.d. Accessed 23 July 2018.

32. Boolell M, Allen MJ, Ballard SA, Gepi-Attee S, Muirhead GJ, Naylor AM, Osterloh IH, Gingell C. Sildenafil: an orally active type 5 cyclic GMP-specific phosphodiesterase inhibitor for the treatment of penile erectile dysfunction. Int J Impot Res. 1996;8:47–52.
33. Chong CR, Xu J, Lu J, Bhat S, Sullivan DJ, Liu JO. Inhibition of angiogenesis by the antifungal drug itraconazole. ACS Chem Biol. 2007;2:263–70.
34. Rudin CM, Brahmer JR, Juergens RA, Hann CL, Ettinger DS, Sebree R, Smith R, Aftab BT, Huang P, Liu JO. Phase 2 study of pemetrexed and itraconazole as second-line therapy for metastatic nonsquamous non-small-cell lung cancer. J Thorac Oncol. 2013;8:619–23.
35. NIH. Neoadjuvant itraconazole in non-small cell lung cancer – full text view – ClinicalTrials.gov. Accessed 23 July 2018.
36. A Two-dose Level Clinical Trial of Itraconazole in Patients With Metastatic Prostate Cancer Who Have Had Disease Progression While on Hormonal Therapy.
37. Cha Y, Erez T, Reynolds IJ, Kumar D, Ross J, Koytiger G, et al. Drug repurposing from the perspective of pharmaceutical companies. Br J Pharmacol. 2018;175(2):168–80. https://doi.org/10.1111/bph.13798.
38. Wishart DS, Feunang YD, Guo AC, Lo EJ, Marcu A, Grant JR, et al. DrugBank 5.0: a major update to the DrugBank database for 2018. Nucleic Acids Res. 2017;2017. https://doi.org/10.1093/nar/gkx1037.
39. Wishart DS, Knox C, Guo AC, Cheng D, Shrivastava S, Tzur D, et al. DrugBank: a knowledgebase for drugs, drug actions and drug targets. Nucleic Acids Res. 2008;36:D901–6.
40. Knox C, Law V, Jewison T, Liu P, Ly S, Frolkis A, et al. DrugBank 3.0: a comprehensive resource for 'omics' research on drugs. Nucleic Acids Res. 2011;39:D1035–41.
41. Kumar S, Kovalenko S, Bhardwaj S, Sethi A, Gorobets NY, Desenko SM, Poonam, Rathi B. Drug repurposing against SARS-CoV-2 using computational approaches. Drug Discov Today. 2022;27(7):2015–27. https://doi.org/10.1016/j.drudis.2022.02.004.
42. Yella JK, Yaddanapudi S, Wang Y, Jegga AG. Changing trends in computational drug repositioning. Pharmaceuticals (Basel, Switzerland). 2018;11(2):57. https://doi.org/10.3390/ph11020057.
43. Pushpakom S, Iorio F, Eyers PA, Escott KJ, Hopper S, Wells A, et al. Drug repurposing: progress, challenges and recommendations. Nat Rev Drug Discov. 2019;18(1):41–58. https://doi.org/10.1038/nrd.2018.168.
44. Jarada TN, Rokne JG, Alhajj R. A review of computational drug repositioning: strategies, approaches, opportunities, challenges, and directions. J Chem. 2020;12(1). https://doi.org/10.1186/s13321-020-00450-7.
45. Sardana D, Zhu C, Zhang M, Gudivada RC, Yang L, Jegga AG. Drug repositioning for orphan diseases. Brief Bioinform. 2011;12(4):346–56. https://doi.org/10.1093/bib/bbr021.
46. Ferreira LG, Dos Santos RN, Oliva G, Andricopulo AD. Molecular docking and structure-based drug design strategies. Molecules. 2015;20(7):13384–1342. https://doi.org/10.3390/molecules200713384.
47. Menden MP, Iorio F, Garnett M, McDermott U, Benes CH, Ballester PJ, et al. Machine learning prediction of cancer cell sensitivity to drugs based on genomic and chemical properties. PLoS One. 2013;8:4. https://doi.org/10.1371/journal.pone.0061318.
48. March-Vila E, Pinzi L, Sturm N, Tinivella A, Engkvist O, Chen H, et al. On the integration of in silico drug design methods for drug repurposing. Front Pharmacol. 2017;8:298. https://doi.org/10.3389/fphar.2017.00298.
49. Hu G, Agarwal P. Human disease-drug network based on genomic expression profiles. PLoS One. 2009;4(8):e6536. https://doi.org/10.1371/journal.pone.0006536.
50. Jin G, Fu C, Zhao H, Cui K, Chang J, Wong STC. A novel method of transcriptional response analysis to facilitate drug repositioning for cancer therapy. Cancer Res. 2012;2012(72):33–44. https://doi.org/10.1158/0008-5472.CAN-11-2333.
51. Masoudi-Sobhanzadeh Y. Computational-based drug repurposing methods in COVID-19. Bioimpacts. 2020;10(3):205–6. https://doi.org/10.34172/bi.2020.25.

52. Berman HM, Westbrook J, Feng Z, Gilliland G, Bhat TN, Weissig H, et al. The protein data bank. Nucleic Acids Res. 2000;28(1):235–42. https://doi.org/10.1093/nar/28.1.235.
53. Wang X, Liu C, Wang J, Fan Y, Wang Z, Wang Y. Proton pump inhibitors increase the chemosensitivity of patients with advanced colorectal cancer. Oncotarget. 2017;8:58801–8. https://doi.org/10.18632/oncotarget.18522.
54. Ehrt C, Brinkjost T, Koch O. Impact of binding site comparisons on medicinal chemistry and rational molecular design. J Med Chem. 2016;59:4121–51. https://doi.org/10.1021/acs.jmedchem.6b00078.
55. Manning G, Whyte DB, Martinez R, Hunter T, Sudarsanam S. The protein kinase complement of the human genome. Science. 2002;298:1912–34. https://doi.org/10.1126/science.1075762.
56. Zhang YM, Cockerill S, Guntrip SB, Rusnak D, Smith K, Vanderwall D, et al. Synthesis and SAR of potent EGFR/erbB2 dual inhibitors. Bioorg Med Chem Lett. 2004;14(1):111–4. https://doi.org/10.1016/j.bmcl.2003.10.010.
57. Chen YC, Tolbert R, Aronov AM, McGaughey G, Walters WP, Meireles L. Prediction of protein pairs sharing common active ligands using protein sequence, structure, and ligand similarity. J Chem Inf Model. 2016;56(9):1734–45. https://doi.org/10.1021/acs.jcim.6b00118.
58. Hall DR, Kozakov D, Whitty A, Vajda S. Lessons from hot spot analysis for fragment-based drug discovery. Trends Pharmacol Sci. 2015;36(11):724–36. https://doi.org/10.1016/j.tips.2015.08.003.
59. Sharma PP, Bansal M, Sethi A, Poonam PL, Goel VK, Grishina M, et al. Computational methods directed towards drug repurposing for COVID-19: advantages and limitations. RSC Adv. 2021;11(57):36181–98. https://doi.org/10.1039/d1ra05320e.
60. Alshareef A, et al. The use of cellular thermal shift assay (CETSA) to study crizotinib resistance in ALK-expressing human cancers. Sci Rep. 2016;6:33710. 31.
61. Miettinen TP, Björklund M. NQO2 is a reactive oxygen species generating off-target for acetaminophen. Mol Pharm. 2014;11:4395–404. 32.
62. Kim T-W. Drug repositioning approaches for the discovery of new therapeutics for Alzheimer's disease. Neurotherapeutics. 2015;12:132–42. 33.
63. Lage OM, et al. Current screening methodologies in drug discovery for selected human diseases. Mar Drugs. 2018;16:279. 34 Reaume, A.G. (2011).
64. Reaume A. Drug repurposing through nonhypothesis driven phenotypic screening. Drug Discov Today Ther Strateg. 2011;8:85–8. 35 Wilkinson, G.F. and Pritchard, K. (2015).
65. Wilkinson GF, Pritchard K. In vitro screening for drug repositioning. J Biomol Screen. 2015;20:167–79. 36.
66. Liu X, et al. Predicting targeted polypharmacology for drug repositioning and multi- target drug discovery. Curr Med Chem. 2013;20:1646–61.
67. Nishimura Y, Hara H. Editorial. Drug repositioning: current advances and future perspectives. Front Pharmacol. 2018;9:1068.
68. Ciallella JR, Reaume AG. In vivo phenotypic screening: clinical proof of concept for a drug repositioning approach. Drug Discov Today Technol. 2017;23:45–52. 39 Jacquemet, G. et al. (2016).
69. Jacquemet G, Baghirov H, Georgiadou M, et al. L-type calcium channels regulate filopodia stability and cancer cell invasion downstream of integrin signalling. Nat Commun. 2016;7:13297.
70. Zilbermintz L, et al. Identification of agents effective against multiple toxins and viruses by host-oriented cell targeting. Sci Rep. 2015;5:13476.
71. Reddy AS, Zhang S. Polypharmacology: drug discovery for the future. Expert Rev Clin Pharmacol. 2013;6:41–7. 42 Issa, N.T. et al. (2015).
72. Issa NT, Peters OJ, Byers SW, Dakshanamurthy S. RepurposeVS: a drug repurposing-focused computational method for accurate drug–target signature predictions. Comb Chem High Throughput Screen. 2015;18:784–94.
73. Sekhon BS. Repositioning drugs and biologics: retargeting old/existing drugs for potential new therapeutic applications. J Pharm Educ Res. 2013;4:1–15.

74. Al-Ali H. The evolution of drug discovery: from phenotypes to targets, and back. Med Chem Commun. 2016;7:788–98. 44 Sawada, R. et al. (2015).
75. Sawada R, Iwata H, Mizutani S, Yamanishi Y. Target-based drug repositioning using large-scale chemical–protein interactome data. J Chem Inf Model. 2015;55:2717–30.
76. Therapeutic drug repurposing: repositioning and rescue. Drug Discov World. http://www.ddw-online.com/p-274232. Barratt MJ, Frail, DE, editors. 2012. Frontmatter. Drug repositioning, Wiley Blackwell.
77. Suhail Y, et al. Systems biology of cancer metastasis. Cell Syst. 2019;9:109–27. https://doi.org/10.1016/j.cels.2019.07.003. 16.
78. Zhou Y, Wang F, Tang J, Nussinov R, Cheng F. Artificial intelligence in COVID-19 drug repurposing. Lancet Digit Health. 2020;2:e667–76. https://doi.org/10.1016/S2589-7500(20)30192-8.
79. Gao J, Liu P, Liu GD, Zhang L. Robust needle localization and enhancement algorithm for ultrasound by deep learning and beam steering methods. J Comput Sci Technol. 2021;36:334–46. https://doi.org/10.1007/s11390-021-0861-7.
80. Chen Y, et al. Variations in DNA elucidate molecular networks that cause disease. Nature. 2008;452:429–35. https://doi.org/10.1038/nature06757.
81. Hu JX, Thomas CE, Brunak S. Network biology concepts in complex disease comorbidities. Nat Rev Genet. 2016;17:615–29. https://doi.org/10.1038/nrg.2016.87.
82. Hasin Y, Seldin M, Lusis A. Multi-omics approaches to disease. Genome Biol. 2017;18:1–15. https://doi.org/10.1186/s13059-017-1215-1.
83. Yang K, et al. A comprehensive analysis of metabolomics and transcriptomics in cervical cancer. Sci Rep. 2017;7:43353. https://doi.org/10.1038/srep43353.
84. Fan K, Cheng L, Li L. Artificial intelligence and machine learning methods in predicting anti-cancer drug combination effects. Brief Bioinform. 2021;22(6):bbab271. https://doi.org/10.1093/bib/bbab271. PMID: 34347041; PMCID: PMC8574962.
85. Lecun Y, Bengio Y, Hinton G. Deep learning. Nature. 2015;521:436–44.
86. Li X, Xu Y, Cui H, et al. Prediction of synergistic anti-cancer drug combinations based on drug target network and drug induced gene expression profiles. Artif Intell Med. 2017;83:35–43.
87. Li H, Li T, Quang D, et al. Network propagation predicts drug synergy in cancers. Cancer Res. 2018;78:5446–57.
88. Jeon M, Kim S, Park S, et al. In silico drug combination discovery for personalized cancer therapy. BMC Syst Biol. 2018;12:16.
89. Gilvary CM, Dry JR, Elemento O. Multi-task learning predicts drug combination synergy in cells and in the clinic. bioRxiv. 2019;2019:576017.
90. Li J, Huo Y, Wu X, et al. Essentiality and transcriptome-enriched pathway scores predict drug-combination synergy. Biology (Basel). 2020;9:1–18.
91. Julkunen H, Cichonska A, Gautam P, et al. leveraging multi-way interactions for systematic prediction of pre-clinical drug combination effects. Nat Commun. 2020;11:6136.
92. Preuer K, Lewis RPI, Hochreiter S, et al. DeepSynergy: predicting anti-cancer drug synergy with deep learning. Bioinformatics. 2018;34:1538–46.
93. Xia F, Shukla M, Brettin T, et al. Predicting tumor cell line response to drug pairs with deep learning. BMC Bioinformatics. 2018;19:486.
94. Zhang H, Feng J, Zeng A, et al. Predicting tumor cell response to synergistic drug combinations using a novel simplified deep learning model. AMIA Annu Symp Proc. 2020;2020:1364–72.
95. Sun Z, Huang S, Jiang P, et al. DTF: deep tensor factorization for predicting anticancer drug synergy. Bioinformatics. 2020;36:4483–9. 42.
96. Sun Y, Li L, Li X, Zhang L, Wang X, Fu X, et al. Outcomes of GDPT (gemcitabine, cisplatin, prednisone, thalidomide) versus CHOP in newly diagnosed peripheral T-cell lymphoma patients. Ther Adv Med Oncol. 2020;12:175883592092382.
97. Chopra S, Goda JS, Mittal P, Mulani J, Pant S, Pai V, et al. Concurrent chemoradiation and brachytherapy alone or in combination with nelfinavir in locally advanced cervical cancer (NELCER): study protocol for a phase III trial. BMJ Open. 2022;12:e055765.

98. Pemovska T, Bigenzahn JW, Superti-Furga G. Recent advances in combinatorial drug screening and synergy scoring. Curr Opin Pharmacol. 2018;42:102–10.
99. Duarte D, Guerreiro I, Vale N. Novel strategies for cancer combat: drug combination using repurposed drugs induces synergistic growth inhibition of MCF-7 breast and HT-29 colon cancer cells. Curr Issues Mol Biol. 2022;44:4930–49.
100. Barrea L, Annunziata G, Bordoni L, Muscogiuri G, Colao A, Savastano S. Nutrigenetics—Personalized nutrition in obesity and cardiovascular diseases. Int J Obes Suppl. 2020;10:1–13. https://doi.org/10.1038/s41367-020-0014-4.
101. Zhu A, Lee D, Shim H. Metabolic positron emission tomography imaging in cancer detection and therapy response. Semin Oncol. 2011;38:55–69. https://doi.org/10.1053/j.seminoncol.2010.11.012.
102. Fowler AM, Strigel RM. Clinical advances in PET–MRI for breast cancer. Lancet Oncol. 2022;23:e32–43. https://doi.org/10.1016/S1470-2045(21)00577-5.
103. Koene RJ, Prizment AE, Blaes A, Konety SH. Shared risk factors in cardiovascular disease and cancer. Circulation. 2016;133:1104–14. https://doi.org/10.1161/CIRCULATIONAHA.115.020406.
104. Prabhakar NR, Peng Y-J, Nanduri J. Hypoxia-inducible factors and obstructive sleep apnea. J Clin Invest. 2020;130:5042–51. https://doi.org/10.1172/JCI137560.
105. Yu B, Wang X, Song Y, Xie G, Jiao S, Shi L, Cao X, Han X, Qu A. The role of hypoxia-inducible factors in cardiovascular diseases. Pharmacol Ther. 2022;238:108186. https://doi.org/10.1016/j.pharmthera.2022.108186.
106. Korbecki J, Simińska D, Gąssowska-Dobrowolska M, Listos J, Gutowska I, Chlubek D, Baranowska-Bosiacka I. Chronic and cycling hypoxia: drivers of cancer chronic inflammation through HIF-1 and NF-κB activation: a review of the molecular mechanisms. Int J Mol Sci. 2021;22:10701. https://doi.org/10.3390/ijms221910701.
107. Malekan M, Ebrahimzadeh MA, Sheida F. The role of Hypoxia-Inducible Factor-1alpha and its signaling in melanoma. Biomed Pharmacother. 2021;141:111873. https://doi.org/10.1016/j.biopha.2021.111873.
108. Fallah J, Rini BI. HIF inhibitors: status of current clinical development. Curr Oncol Rep. 2019;21:6. https://doi.org/10.1007/s11912-019-0752-z.
109. Wicks EE, Semenza GL. Hypoxia-inducible factors: cancer progression and clinical translation. J Clin Invest. 2022;132:e159839. https://doi.org/10.1172/JCI159839.
110. Mills KT, Stefanescu A, He J. The global epidemiology of hypertension. Nat Rev Nephrol. 2020;16:223–37. https://doi.org/10.1038/s41581-019-0244-2.
111. Cho IJ, Shin JH, Jung MH, Kang CY, Hwang J, Kwon CH, Kim W, Kim DH, Lee CJ, Kang S-H, et al. Antihypertensive drugs and the risk of cancer: a nationwide cohort study. J Clin Med. 2021;10:771. https://doi.org/10.3390/jcm10040771.
112. Boudreau DM, Yu O, Chubak J, Wirtz HS, Bowles EJA, Fujii M, Buist DSM. Comparative safety of cardiovascular medication use and breast cancer outcomes among women with early stage breast cancer. Breast Cancer Res Treat. 2014;144:405–16. https://doi.org/10.1007/s10549-014-2870-5.
113. Cronin-Fenton D, Lash TL, Ahern TP, Damkier P, Christiansen P, Ejlertsen B, Sørensen HT. Concurrent new drug prescriptions and prognosis of early breast cancer: studies using the Danish Breast Cancer Group clinical database. Acta Oncol. 2017;57:120–8. https://doi.org/10.1080/0284186X.2017.1407040.
114. Carlos-Escalante JA, de Jesús-Sánchez M, Rivas-Castro A, Pichardo-Rojas PS, Arce C, Wegman-Ostrosky T. The use of antihypertensive drugs as coadjuvant therapy in cancer. Front Oncol. 2021;11:660943. https://doi.org/10.3389/fonc.2021.660943.
115. Gales L, Forsea L, Mitrea D, Stefanica I, Stanculescu I, Mitrica R, Georgescu M, Trifanescu O, Anghel R, Serbanescu L. Antidiabetics, anthelmintics, statins, and beta-blockers as co-adjuvant drugs in cancer therapy. Medicina. 2022;58:1239. https://doi.org/10.3390/medicina58091239.

116. Momekov G, Momekova D. Ivermectin as a potential COVID-19 treatment from the pharmacokinetic point of view. medRxiv. 2020. https://doi.org/10.1101/2020.04.11.20061804.
117. Amirian E, Levy J. Current knowledge about the antivirals remdesivir (GS-5734) and GS-441524 as therapeutic options for coronaviruses. One Health. 2020;9:100128.
118. Eastman R, Roth J, Brimacombe K, Simeonov A, Shen M, Patnaik S, Hall M. Remdesivir: a review of its discovery and development leading to emergency use authorization for treatment of COVID-19. ACS Cent Sci. 2020;6(5):672–83.
119. Ferner R, Aronson J. Remdesivir in Covid-19. BMJ. 2020;369:m1610.
120. Drugs.com. Baricitinib side effects: common, severe, long term—Drugs.com. Available online: https://www.drugs.com/sfx/baricitinib-side-effects.html#moreResources. Accessed on 6 Mar 2020.
121. Amir-Aslani A, Mangematin V. The future of drug discovery and development: shifting emphasis towards personalised medicine. Technol Forecast Soc Chang. 2010;77:203–17.
122. Janku F, Stewart D, Kurzrock R. Targeted therapy in non-small-cell lung cancer—is it becoming a reality? Nat Rev Clin Oncol. 2010;7:401–14.
123. Drilon A, Camidge D, Ou S, Clark J, Socinski M, Weiss J, Riely G, Winter M, Wang S, Monti K, et al. Efficacy and safety of crizotinib in patients (pts) with advanced MET exon 14-altered non-small cell lung cancer (NSCLC). J Clin Oncol. 2016;34:108.
124. Novac N. Challenges and opportunities of drug repositioning. Trends Pharmacol Sci. 2013;34:267–72.
125. Sternitzke C. Drug repurposing and the prior art patents of competitors. Drug Discov Today. 2014;19:1841–7.
126. Stewart A. How thalidomide works against cancer. Science. 2014;343:256–7.
127. Kim H, Oh S. Novel systemic therapies for advanced gastric cancer. J Gastric Cancer. 2018;18:1–19.
128. Yamani L, Yano Y, Utsumi T, Wasityastuti W, Rinonce H, Widasari D, Juniastuti, Lusida M, Soetjipto, Hayashi Y. Profile of mutations in the reverse transcriptase and overlapping surface genes of hepatitis B virus (HBV) in treatment-Naïve Indonesian HBV carriers. Jpn J Infect Dis. 2017;70:647–55.
129. Yunusa A, Bello S, Chika A, Yakubu A. An antiviral drug combinational studies against HIV-A computational drug repositioning approach. Asian J Sci Technol. 2019;10:9978–84.
130. Lewis M, DaFonseca S, Chomont N, Palamara A, Tardugno M, Mai A, Collins M, Wagner W, Yalley-Ogunro J, Greenhouse J, et al. Gold drug auranofin restricts the viral reservoir in the monkey AIDS model and induces containment of viral load following ART suspension. AIDS. 2011;25:1347–56.
131. Goenka S, Simon S. Organogold drug Auranofin exhibits anti-melanogenic activity in B16F10 and MNT-1 melanoma cells. Arch Dermatol Res. 2019;312:213–21.
132. Shaw A, Yasothan U, Kirkpatrick P. Crizotinib. Nat Rev Drug Discov. 2011;10:897–8.
133. Pantziarka P, Verbaanderd C, Huys I, Bouche G, Meheus L. Repurposing drugs in oncology: from candidate selection to clinical adoption. Semin Cancer Biol. 2021;68:186–91.
134. Moffat JG, Vincent F, Lee JA, Eder J, Prunotto M. Opportunities and challenges in phenotypic drug discovery: an industry perspective. Nat Rev Drug Discov. 2017;16:531–43.
135. Shim JS, Liu JO. Recent advances in drug repositioning for the discovery of new anticancer drugs. Int J Biol Sci. 2014;10:654–63.
136. Pillaiyar T, Meenakshisundaram S, Manickam M, Sankaranarayanan M. A medicinal chemistry perspective of drug repositioning: recent advances and challenges in drug discovery. Eur J Med Chem. 2020;195:112275.
137. Moffat JG, Rudolph J, Bailey D. Phenotypic screening in cancer drug discovery – past, present and future. Nat Rev Drug Discov. 2014;13:588–602.
138. Rabben HL, et al. Computational drug repositioning and experimental validation of ivermectin in treatment of gastric cancer. Front Pharmacol. 2021;12:625991.
139. Parvathaneni V, Kulkarni NS, Muth A, Gupta V. Drug repurposing: a promising tool to accelerate the drug discovery process. Drug Discov Today. 2019;24:2076–85.

140. Tanoli Z, et al. Exploration of databases and methods supporting drug repurposing: a comprehensive survey. Brief Bioinform. 2021;22:1656–78.
141. Huang H, Zhang P, Qu XA, Sanseau P, Yang L. Systematic prediction of drug combinations based on clinical side-effects. Sci Rep. 2014;4:7160.
142. Kuhn M, Campillos M, Letunic I, Jensen LJ, Bork P. A side effect resource to capture phenotypic effects of drugs. Mol Syst Biol. 2010;6:343.
143. Celebi R, Don't Walk OB 4th, Movva R, Alpsoy S, Dumontier M. In-silico prediction of synergistic anti-cancer drug combinations using multi-omics Data. Sci Rep. 2019;9:8949.
144. Waldron D. Cancer genomics: a multi-layer omics approach to cancer. Nat Rev Genet. 2016;17:436–7.

Chapter 24
Computational Analyses of the Mechanism of Action of Antiepileptic Agents

Shreya Sinha, Madhu Dubey, Snehal Misal, Rajasekhar Reddy Alavala, and Ramaa C. S.

Contents

Abstract Epilepsy, a neurological disorder characterized by recurrent seizures, remains a major challenge despite advances in pharmacotherapy. Conventional anti-epileptic drugs (AEDs) exhibit suboptimal efficacy, necessitating the development of novel, more effective agents. This chapter explores computational approaches in drug discovery and their applications in identifying and optimizing AEDs. By employing techniques such as de novo drug design, homology modeling, molecular docking, molecular dynamics simulations, QSAR (quantitative structure-activity relationships), and receptor-based pharmacophore modeling, the mechanisms underlying antiepileptic activity are elucidated.

S. Sinha · M. Dubey · S. Misal · Ramaa C. S. (✉)
Bharati Vidyapeeth's College of Pharmacy, CBD Belapur, Navi Mumbai, India

R. R. Alavala
Shobhaben Pratapbhai Patel School of Pharmacy and Technology Management, SVKM'S NMIMS, Mumbai, India

S. N. K. Rao G., R. R. Alavala (eds.), *Applications of Computational Tools in Drug Design and Development*, https://doi.org/10.1007/978-981-96-4154-3_24

Ligand- and structure-based computational methods enable high-throughput screening, facilitating the identification of lead compounds and potential molecular targets. Furthermore, advancements in molecular dynamics simulations and quantum mechanics aid in understanding drug-receptor interactions at an atomic level, enhancing the prediction of binding affinities and drug efficacy. Novel AEDs identified through these methodologies target diverse mechanisms, including modulation of gamma-aminobutyric acid (GABA) transmission, inhibition of voltage-gated ion channels, and interactions with NMDA receptors. This chapter underscores the significance of computational techniques in addressing the unmet need for safe and effective AEDs, providing insights into the future of antiepileptic drug discovery.

Keywords Antiepileptic drugs · Computational modelling · Molecular docking · Homology modelling · QSAR analysis · GABA modulation · Pharmacophore modelling · Drug design · Epilepsy treatment · Seizure mechanisms

24.1 Introduction

Epilepsy is a combination of symptoms marked by recurrent, self-limited, and frequently short paroxysmal abnormalities in neurological function. This is brought on by an overabundance of neuronal discharge, which can be induced by an interference with the brain's electrical activity and physicochemical processes. Convulsive diseases may be caused by either a chemical supersensitivity that enhances the excitability of neuronal components or a loss of the usual inhibitory regulation mechanism. "An enduring predisposition to generate epileptic seizures and the neurobiological, cognitive, psychological, and social consequences of this condition characterize epilepsy as a disorder of the brain," states the definition offered by the ILAE in 2005 [1].

According to the most current data, 0.6% of youngsters between the ages of 0 and 17 are now experiencing active epilepsy. An estimated 20% of epileptic patients who were treated with the first generation of antiepileptic drugs (AEDs), were unable to achieve sufficient control over their seizures. Merely 60–70% of patients undergoing treatment experience seizure suppression using standard medications including valproate, lamotrigine, and topiramate. Three generations of antiepileptic drugs can be distinguished. These consist of traditional "first-generation" medications like carbamazepine, phenobarbital, and valproic acid. "second-generation" medications like lamotrigine, vigabatrin, tiagabine, topiramate, gabapentin, and levetiracetam. Drugs such as carabersat, carisbamate, brivaracetam, and DP-valproic acid are examples of novel third-generation antiepileptic drugs.

Thirty percent of patients still experience seizures after using levetiracetam, phenytoin, zonisamide, carbamazepine, or oxcarbazepine, either alone or in combination. Thus there is a need for the creation of novel antiseizure medications that are more tolerable and effective.

The current epilepsy hypothesis, which maintains that seizures result from an alteration in the ratio of neuronal excitation to inhibition, has been the basis for the development of novel AEDs. Thus, there is a need for the creation of novel antiseizure medications that are more tolerable and effective. Currently, almost all facets of drug development involve the use of computational models. Nevertheless, computational approaches for the screening of novel AEDs have not gained as much attention as other therapeutic areas. This could be due to the majority of confirmed molecular targets, which is a persistent and important obstacle to the use of structure-based estimations in the AED development sector. CADD's primary objective is to shorten timelines and expenses without compromising quality. Additionally, CADD may be applied throughout most phases of drug development, including preclinical research, lead discovery, validation, and identification of targets.

The two primary classes of CADD techniques for examining (AEDs) are methods based on ligands and methods based on structures [2]. Ligand-based methods operate on the premise that the chemical framework of the drug and its biological activity are acquainted. Thus, structural activity relationships (SARs) can be derived and implemented to describe chemical structure and project new and improved compounds Employing a set of verified both active and inert ligands. This can be accomplished using 2D or 3D descriptors, depending on the degree of complexity. The 3D structure of the biological target must be understood to use structure-based drug design techniques. This provides details on the binding site design, which are useful in determining whether a ligand would be a suitable lead molecule. The initial stage of structure-based drug design is drug target identification, which entails locating and figuring out the relevant protein structures. In ligand and structure-based techniques, quality of fit for small molecules is utilized to determine which compounds would be best for further investigation. A new anticonvulsant's rational drug design process can be accomplished in a variety of ways, such as by altering drugs that already exist and identifying mutations, or by analyzing various ligands chosen from those that bind with the receptor. Identifying the geometric and electrical characteristics that all molecules have and which are probably associated with the recognition of receptors and activation is the aim when creating from the ligand.

In these conditions, the minimal requirements related to the pharmacophoric pattern for an exhibited activity are found through the use of a comparative study that should include both active and inert compounds. A pharmacophore is identified as "the set of steric and electronic features essential to ensure optimal supramolecular interactions with a specific biological target structure and to activate its biological response" by the International Molecular Physics Association.

Pharmacophoric descriptors include, among others, hydrophobic, aromatic, positive ionizable categories, negative ionizable groups, H-bond donors, and H-bond acceptors.

Subsequently, Lipinski's rule of five criteria can then be used to filter the resultant pharmacophore model. This will enable the determination of whether a certain chemical compound possesses potential as an oral drug for human use. The other step is homology modeling, which makes use of the precise structures of numerous distinct proteins, particularly membrane proteins, and recent advancements in cryo-EM. A method for estimating a target protein's structure based on structural information from proteins with sequence similarities with the target is called homology modelling.

Molecular dynamics simulations, a theoretical technique for determining the structure and dynamic responses of molecules, are another tool used to get insight into the atomic-level workings of pharmacological modes of action. Additionally, MD may assist in identifying the neurotoxic protein aggregation route, which could lead to the creation of novel inhibitors. Computer technique forecasts the lowest energy conformations at which two molecules could bind together. They are frequently employed to anticipate the binding affinities and forms of interaction between molecules and drug-binding locations on biological macromolecules, as well as multiple scoring systems.

They are extensively employed, along with their different scoring systems, to predict the affinities for binding as well as modalities of interactions among drugs and drug-binding areas on biological macromolecules. Next, a biological response to an array of molecules' chemical properties is correlated using quantitative structure-activity relationship (QSAR) approaches. Modern pharmacophore models fall into two categories: ligand- and receptor-based pharmacophores.

A receptor with an existing three-dimensional structure has been studied for receptor-based pharmacophores, which rely on the widely accepted idea of a key used in a lock. The choice of simulated compounds is the most crucial step in the development of pharmacophore models. The following are some further varied methods for computer-aided medicine design for epilepsy: Molecular quantum and the Schrödinger equation are the two main elements of quantum mechanics. A mathematical operator called H (the "Hamiltonian operator") can be used to express the Schrödinger equation as follows:

$$H\Psi = E\Psi \text{ function},$$

where Ψ is the wave function and E is the energy of the system.

It is theoretically possible to determine any physical attribute of a pharmacological molecule once its wave function is known. Quantum mechanical computations can obtain high-quality structural information. These do not apply to larger molecules like receptor proteins, but they help research the molecules of anticonvulsant drugs. Finding the anticonvulsant drug's wave function will allow for the calculation of the drug's ideal geometry and other characteristics.

As a result, methods to solve the wavefunction of molecules that are approximative rather than accurate are obtained by molecular orbital computations. Molecular mechanics is the other main technique utilized when adopting CADD to design antiepileptic drugs. Molecular mechanics is a computational approach that is empirically parameterized and yields precise geometries and relative energy for various molecular forms. The system of particles it studies is comprised of several independent potential energy function-specified particles held by elastic or harmonic powers and coupled by compressible springs. The sum of these several potential energy equations yields a force field, which is a multidimensional energy function. This

Chapter will cover a variety of computational techniques that are useful for analyzing mechanisms of anti-epileptic medications in new drug discovery [3].

24.2 Overview

There are two primary categories of seizures: partial and generalized (Fig. 24.1). The framework for this assessment is the International Classification of epileptic episodes, which was approved in September 1981 by the International League against Epilepsy (ILAE) Commission on Classification and Terminology. This is predicated on the EEG as well as the clinical features of epilepsy.

- Partial seizures: Abnormal electrical discharges originate in a specific region of the brain during partial seizures. The afflicted brain region determines the symptoms. The extent of these releases might not change.

Two forms of partial seizures are further classified as follows:

- Simple partial seizures: The patient stays conscious throughout this. Twitching may occur if the brain's motor regions are impacted; it may begin in the face, fingers, or toes, or a limb. The twitching might stay localized, travel up the entire arm, or eventually affect the entire body.
- Complex partial seizures: At this stage, the patient faints. The patient may be somewhat cognizant of their surroundings and may not have completely lost

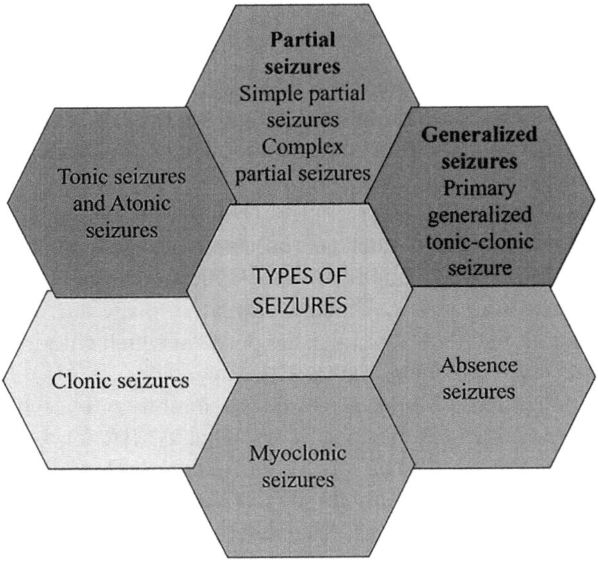

Fig. 24.1 Types of seizures

consciousness. Following an intricate partial seizure, there is a time of confusion and a gradual recovery.

- Generalized seizures: This seizure is initially generalized. Simultaneously, impulses originate from both sides of the brain.
- Primary generalized tonic-clonic seizure (GTCS): The most prevalent kind of generalized epilepsy is this one. The individual loses consciousness, falls, and their entire body stiffens (tonic phase).
- Absence seizures: These are brief episodes of unconsciousness that last just a few seconds. They appear suddenly. Children often experience absence seizures.
- Myoclonic seizures: This results in unexpected, short seizures that cause jerky movements and shock-like muscular spasms.
- Clonic seizures: There is no tonic phase (the stage when muscles tighten) in these seizures. There are uncontrollable shaking or clonic jerks that repeat.
- Tonic seizures: These are widespread seizures characterized only by muscle stiffness or the tonic phase. The limbs are stuck in an unusual configuration. Instantaneous loss of consciousness occurs.
- Atonic seizures (astatic seizures): An abrupt loss of muscle tone results in atonic seizures, also known as astatic seizures. The patient collapses to the ground in a heap due to the lack of head or limb function. For this reason, they are also called as "drop attacks."

Mechanism of action-based categorization of antiepileptic drugs (Fig. 24.2)
The following are the main categories of antiepileptic drugs:

1. Those that facilitate Gamma-aminobutyric acid (GABA) transmission (**A**);
2. Those that inhibit excitatory transmission by blocking voltage-gated ion channels (**B**);
3. Those which act on NMDA receptors (**C**).

The principal inhibitory neurotransmitter in the neuroaxis is GABA which is also found throughout many interneurons and long axonal tracts, which is important in the case of epilepsy. Interneurons control the electrical signals in comparatively small regions of the neuropile and, in turn, the output of that region. GABA A and GABA B are two kinds of GABA receptors. The activity of drugs like benzodiazepines and barbiturates is due to quick neurotransmission interaction between GABA A receptors, which are mostly found on postsynaptic membranes that are bicuculline-sensitive. The transmitter-gated channel, which is made up of a total of five membrane-spanning members that form the pore via which chloride ions enter into the postsynaptic vesicle, includes the GABA A receptor.

Vigabatrin potentiates GABAergic-induced inhibition by inhibiting GABA-transaminase 4, which raises synaptic & terminal GABA levels (**A**) (Fig. 24.3). Benzodiazepines produce their effects by enhancing the binding of GABA to its receptor. When these receptors are stimulated, either potassium channels open or calcium's entry into the end of the channel is inhibited, which reduces GABA release. Medications that operate on voltage-gated ion channels belong to the second category. Here, phenytoin, carbamazepine, lamotrigine, valproate, etc.

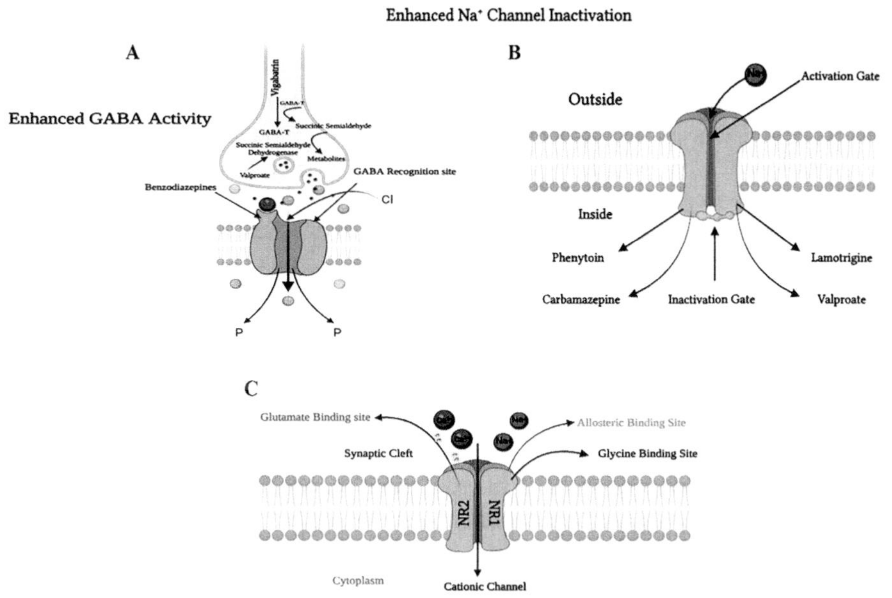

Fig. 24.2 Mechanism of action-based categorization of antiepileptic drugs: (**a**) Enhanced gamma-aminobutyric acid (GABA) transmission; (**b**) Inactivation of enhanced voltage-gated sodium ion channels; (**c**) NMDA receptors

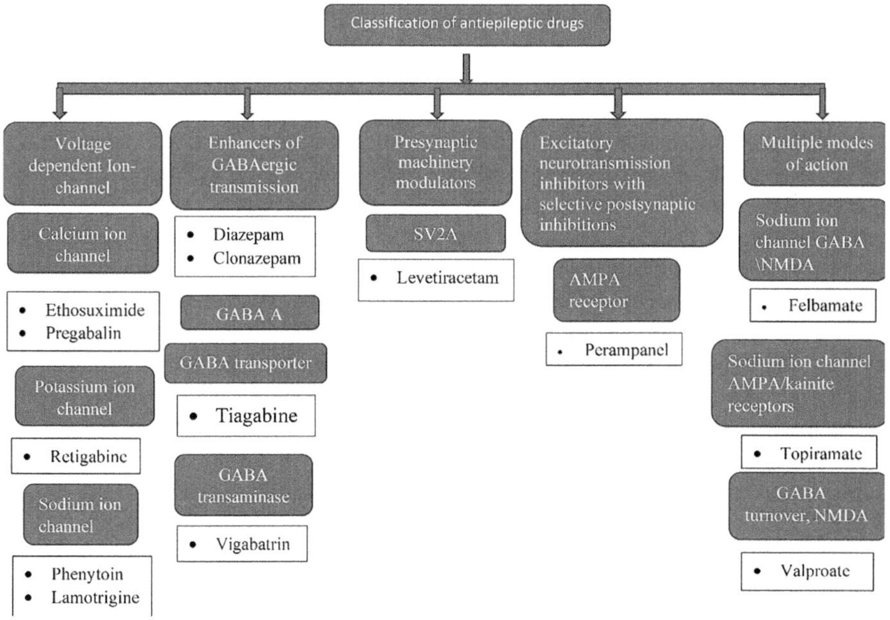

Fig. 24.3 Classification of antiepileptic drugs

decreased sodium voltages during high-frequency repeated stimulation, which is thought to occur due to a membrane-potential-dependent blockage of sodium channels (neuronal circumstances that are present during a seizure) **(B)**. It also prevents the uptake of potassium. Antiepileptic drugs also act on other receptors, including the N-methyl-D-aspartate (NMDA) receptor **(C)**. This ion channel permits potassium outflow as well as calcium inflow and, to a lesser extent, sodium outflow. Here, GABA transporters (GATs) actively reintroduce released GABA into the brain's GABAergic neurons. Drugs such as tiagabine, which share structural similarities with nipecotic acid, selectively inhibit GAT1 through the GABAergic neuron level and glial cells. Another potential mode of function involves the suppression of T-cells with low thresholds.

Drug toxicity may result from modifications to the pharmacodynamics of the medication or from structural or functional alterations at the drug's site of action. Thus, there is a need for developing novel antiseizure medications that are more tolerable and effective. Computational modelling resolves paradigm shifts in the field of drug discovery to assist in the treatment of complex disorders and encompasses the most lately experimental findings on the underlying causes of both drug-resistant and epileptic seizures. This results in the provision of new and more efficient epileptic medication solutions [4].

The general categories of CADD techniques are as follows:

Structure-based techniques

For structure-based drug designing, techniques rely on the three-dimensional structure of proteins [5], which can be acquired through NMR spectroscopy [6], cryo-EM [7], or X-ray crystallography [8]. If a protein's three-dimensional structure isn't known, the three-dimensional structure can be computed using homology modeling or de novo structure estimation methods. To identify molecules that might interact with the target protein and anticipate their methods of interaction, potential therapeutic compounds might be virtually docked within the ligand-binding pocket. Protein-ligand interaction analysis using scoring functions determines the accuracy of docking models. There are four different kinds of scoring functions: empirical, physics-based, machine learning-based, and knowledge-based. These offer a specific example of the free energy connected to ligand binding. Knowledge-based scoring functions estimate the binding energy by using distance-dependent pairwise potentials that are derived from statistics on realistically established complex structures using the Boltzmann law. The effects of Van der Waals energy, hydrogen bond energy, and solvation energy are produced by a mixture of ligands and proteins [9]. Docking tricks are categorized based on their docking scores using various scoring structures. To assess the degree of sampling and accuracy of the docking computation, the highest-rated models are often compared to the empirically established sophisticated structure, if available, or to other well-scoring models based on their RMSD values [10].

Molecular dynamics simulations (MDSs) are used to study the atomic-level structural dynamics of the protein-ligand complex in order to derive the thermodynamic and kinetic parameters of the protein-ligand interaction. The creation of protein conformational ensembles for docking, which more accurately depicts the flexibility of the ligand binding pocket, the investigation of the binding pathway that a ligand takes to approach its target protein, and the evaluation of relative free binding strengths and ligand separation rates in conjunction with other analyses are some potential applications of molecular docking in structure-based drug design [11].

Random acceleration molecular dynamics (RAMD) [12], along with its enhanced variant τ-random acceleration molecular dynamics (τRAMD) [13], can be utilized to ascertain ligand dissociation rates or residence times. In τRAMD, the ligand is subjected to an extra force to initiate the unbinding process, which allows for the computation of the respective koff rate of ligands. It requires less time to complete the simulation itself and permits ranking ligands according to their respective koff rates [12]. When using metadynamics simulations, that are frequently used to investigate ion channel transitions and compute free energy, machine learning methods can assist in finding group variables that can be used to communicate the underlying dynamics of a target chemical system.

Ligand-based drug designing

Molecular descriptors for a group of chemicals can be obtained with the aid of chemoinformatics tools and connected with their curative properties through mathematical modelling [14]. To direct the search for molecules with desirable chemical and structural properties, a pharmacophore model can also be supplied as a contribution to the simulated screening pipeline [15]. A collection of a molecule's electronic and steric features is called a pharmacophore, such as donor/acceptor hydrogen bond atoms that are required to guarantee the best possible interaction with the target drug or conserved functional groups shared by all hit compounds.

The biological characteristics of molecules can be predicted by QSAR modeling through the application of machine learning (ML). Different machine learning techniques can categorize and forecast the ligand of interest in light of its activity, toxic effects, and various other biologically important properties based on correlations in the dataset used for training, where molecules are described by a set of descriptors [16, 17]. For classification, approaches such as k-nearest neighbours (KNNs), random forest (RF), and support vector machines (SVMs) are frequently utilised. When dealing with high-dimensional data, Since SVMs may map an unknown chemical to a certain category, they are especially useful. In the field of drug design, RF is also well-established.

It is predicated on decision trees, aggregating each choice to produce a forecast.

A molecule is assigned to a specific class based on the average of all of its predictions for the final label [17, 18]. The KNN classifier uses a majority rule to classify data, just like RF does. A molecule is therefore classified into a category according to the category of its closest neighbours, up to a predetermined distance. Under the project specifications, a deep long short-term memory network (DLSTM) was used to simulate a deep recurrent neural network (RNN) in order to extract hit drug compounds with the required profile [19]. The precise structures of many proteins, particularly proteins found in membranes (which are over-represented among therapeutic targets), remain unknown despite the current trend in structural investigations, particularly the recent advancements in cryo-EM [20, 21]. A method for estimating a target protein's structure based on structural information from proteins that share sequence homology with the target is called homology modeling [22]. Finding the Binding Sites and Identifying possible small molecule binding sites comes next when the target protein's three-dimensional structure is established. Several strategies for calculating and finding binding pockets can be used in this process. MDS provides atomic-level insight into pharmacological mechanisms of action by revealing the configurations and dynamics of compounds through theoretical means [23].

Molecular docking studies provide information about the interactions between two or more molecular structures, such as the fit between a medicine and an enzyme or protein receptor. Drug research is the primary application for molecular docking tools [24]. Consequently, the binding patterns and affinities involving chemical compounds and places of binding of drugs on biological macromolecules are often predicted using molecular docking approaches and their various scoring systems [25, 26]. It is a typical practice to use pharmacophore modelling to quickly describe possible lead compounds [27, 28]. Finding the ideal forms and distributions of charges for a small molecule to attach to a biological macromolecule is the focus of pharmacophore modeling.

24.3 Methods

24.3.1 De Novo Drug Design

24.3.1.1 Concept

De novo signifies to begin afresh, from scratch. The process of creating new lead compounds with desired pharmacological and physiochemical properties is known as "de novo drug design" [29] (Fig. 24.4). The benefits of de novo drug design include the synthesis of compounds that represent new intellectual property, the potential for creating innovative and improved therapies, the capacity to explore a greater chemical space, and the rapid and affordable development of therapeutic candidates. The primary challenge in de novo drug design is the created molecular structures' synthetic availability [30]. Future "personalized" healthcare is often envisioned as treating patients with medications that have been carefully chosen or even custom-tailored to reduce side effects and maximize therapeutic effectiveness. Computer-assisted de novo design can help find novel chemical entities (NCEs) with the appropriate pharmacological activity profiles, supporting this future view on drug discovery, although it is not confined to such applications [31].

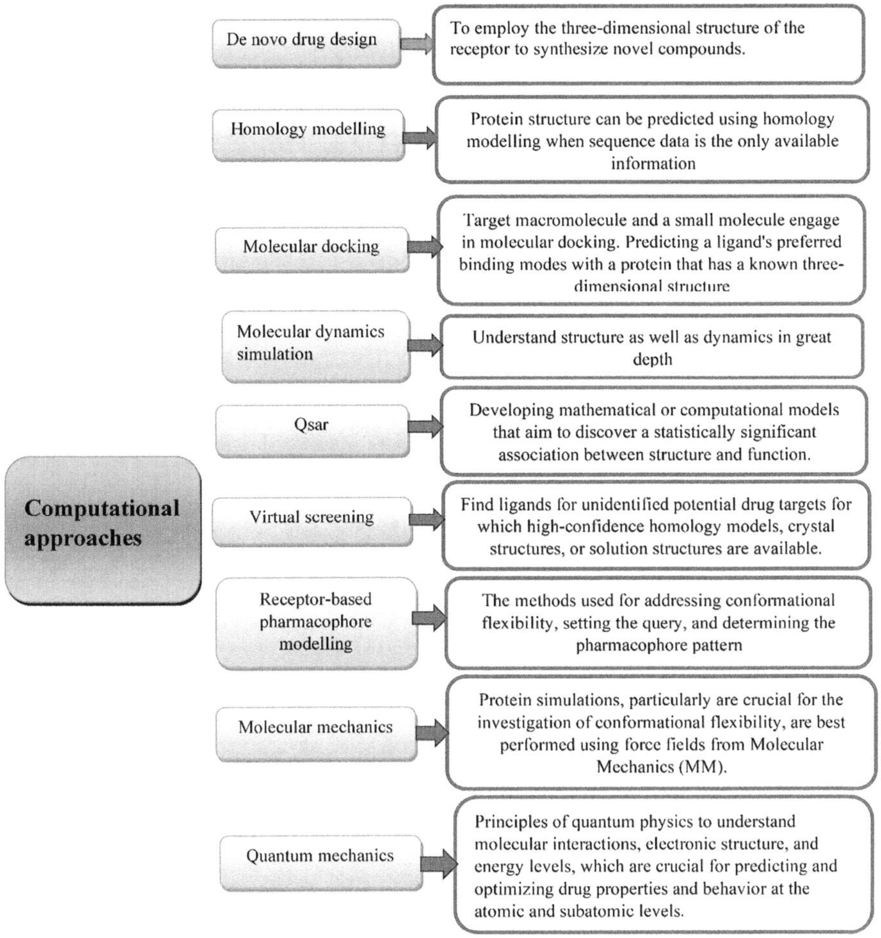

Fig. 24.4 Various types of computational approaches for anti-epilepsy

24.3.1.2 Softwares

The list of software used for de novo design was mentioned in Table 24.1.

24.3.1.3 Methods

Methods for de novo drug design represent a paradigm shift in pharmaceutical research, offering a proactive approach to creating novel therapeutics tailored to specific molecular targets. Here's an overview of the key methods commonly used (Fig. 24.5).

Table 24.1 Software used in de novo drug design

Methods	Softwares
Site point connection method	LUDI
Fragment connection method	SPLICE, NEW LEAD, PRO-LIGAND
Sequential build-up methods	LEGEND, GROW, SPROUT
Random connection and disconnection methods	CONCEPTS, CONCERTS, MCDNLG

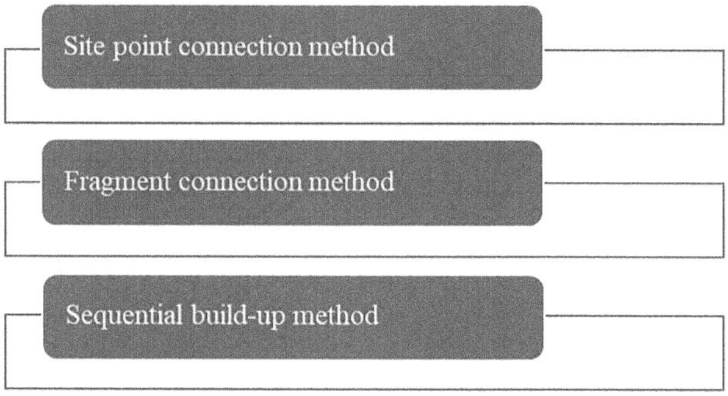

Fig. 24.5 Methods for de novo drug design

- **Site point connection method**

Step 1: Identifying sites of interaction

The ability of each atom at the binding site to engage in hydrogen bonding and van der Waals forces is investigated.

Step 2: assembling molecular components

A library including several hundred molecular fragments is accessed by the LUDI programme. Because the fitting technique assumes stiff fragments, the molecules selected are usually rigid in structure and range in size from 5 to 30 atoms.

- **Techniques for Fragmentation Connection**

After locating the fragments and positioning them at the binding location, assemble them. The programme finds the closest hydrogen atoms in the binding location and molecular fragments are closest to one another.

- **Methods of sequential build-up**

There are variances in the approach employed by the program because it matches fragments to interaction sites like LUDI. It specifies molecular fragments through templates. Rather than atoms and bonds, each template is described by vertices and edges.

24.3.1.4 Limitations

Automated from scratch it is common for drugs designed to produce structures that are hard or impossible to synthesize. The scoring functions that are used to determine binding affinities are crucial to automated de novo programs.

24.3.2 Homology Modelling

24.3.2.1 Concept

Protein structure can predicted using homology modeling when sequence data is the only available information. It produces the final model by employing various techniques and procedures. Numerous factors influence the quality of the model generated at each stage during the homology modelling process, and choosing the right programme can significantly improve the model's quality (Fig. 24.6). By employing homologous proteins—that is, structurally equivalent proteins from the same family—A protein's structure can be estimated via homology modeling by utilizing the redundant nature of protein structure. An atomic version can be constructed through comparison with proteins in a family where multiple protein structures have been discovered empirically. According to the structural genomics programs, homology modeling will likely become a more crucial tool for biologists as they attempt to define the majority sequences of proteins by combining targeted high throughput experimental structure identification and prediction in an efficient manner.

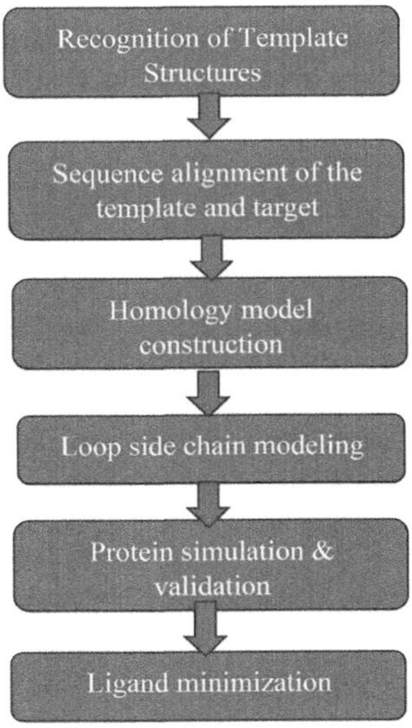

Fig. 24.6 Workflow for homology modelling

24.3.2.2 Softwares

The list of software used for de novo design was mentioned in Table 24.2.

24.3.2.3 Workflow

Recognition of Template Structures

Using the target sequence, the structure of a protein database is searched for homologous proteins to identify appropriate templates. A bare minimum of 30% of the amino acid sequence integrity between the target and template is required when employing homology modeling. Sequence identity boosts a model's accuracy, and employing multiple template structures at once can often boost reliability. Since the bulk of model errors are caused by mistakes in the first matching of the target and template patterns, it is widely accepted that this alignment stage is the most crucial one in the entire process.

Sequence Alignment of the Template and Target

If the sequence identity is low (~30%), it can be useful to correlate the a series of interest with sequences of proteins of related families even if their structures are not available, to ensure that regions of functional or structural importance are correctly aligned with the template sequence.

Homology Model Construction

Using homology modelling, the gene sequence with the Uniprot accession code is chosen for insilico analysis. Numerous amino acids are detected in close proximity to the target sequence. The National Centre for Biotechnology Information's PSI

Table 24.2 Softwares used in homology modeling

Method	Software
Identification of template structures	Databases for structural categorization like SCOP, CATH, or FSSP; profile-based techniques; PSI-BLAST software.
Template and target sequence alignment	CLUSTALX, PileUp can be used to produce a multiple sequence alignment.
Homology model construction	Discovery Studio v 3.5, MODELLER, UniProtKB/Swiss-Pro, MUSCLE, JPRED, SABLE, PROSITE, SOSUI [27], HMMTop [28], JPRED [29] and PSIPRED, CLUSTAL Omega, SegMod.
Protein simulation and validation	CHARMm force field, PROCHECK, ERRAT2, WHATCHECK, ANOLEA, VERIFY3D
Ligand minimization	UCSF Dock, SCWRL, Glide

Blast is used to search for homologous structures in proteins. Multiple sequence alignment is used to match the target sequence with the template framework and identify the conserved sections. A suitable programme such as ProtParam is then used to calculate the sequence's physical and chemical characteristics.

Fragment-based methods: Fragment-based techniques detect structurally conserved sections by aligning the template and target sequences.

Segment matching methods: These techniques to build the model, put short sections of template structures together.

Loop and Side-Chain Modelling

Loop modelling

This type of modeling is similar to a miniature version of protein folding, in which the arrangement of the loop is primarily determined by the sequence information. An energy function or score system for the appropriateness of the generated loop directs this prediction procedure.

Protein simulation and validation

The potential protein energy model is examined using the calculation of energy and minimization methodology, respectively, before and after minimization. It offers strong mechanics and dynamic methods for researching the energy and motion of molecules, which range from tiny ligands to complicated physiological systems made up of several components.

Assessment of stereochemical properties: Programs for validating data look for oddities including phi/psi angle pairings in prohibited areas, steric collisions, and undesirable bond lengths and angles.

Assessment of spatial properties: The model can also be evaluated in terms of its ability to generate a hydrophobic core, pack, and location of charged groups, among other spatial qualities.

Ligand minimization

Using the bond energy, the dihedral energy, electrostatic energy, and initial potential energy, several ligand conformations are produced. The next step is to minimize the ligand-receptor complex to obtain a low-energy conformation of proteins that can be employed for ligand resampling [32].

24.3.2.4 Applications

Intractable epilepsy is known as Dravet syndrome (DS) that develops in childhood due to haploinsufficiency of the voltage-gated sodium channel NAV. The rhythmic thalamic burst firing is slowed down by the 5-Hydroxytryptamine 5-HT2 receptors [33]. According to the epilepsy maximal electroconvulsion paradigm, 5HT2A and 5HT2C can prevent seizures. To better understand inter-residue interactions and explore the mechanism of (5HT2BR) activation, a study was conducted to repurpose FDA medicines in drug-resistant epilepsy. It became clearer how 5HT2BR reduces seizures in people with Dravet syndrome. Homology modelling helped in the development of possible therapeutic approaches and shed light on the 5HT2BR receptor. Using a template structure, it was thought that the molecular makeup of 5HT2BR might reveal potential binding sites for therapeutic molecules. The target's crystal structure (5HT2BRM) was modeled since the PDB structure (6DRY, 4IB4) that was first linked to the agonist ergotamine was a chimaera made up of mutations. Top sequences were chosen for 3D-structure modelling, and the SWISS-MODEL service was used for homology modelling. Pairwise Sequence Alignment yielded a structural identity of 60.9% between the modelled and the chimaera (4IB4). The best hits from the virtual screening of 2456 authorised medications were submitted to molecular dynamic (MD) and MM/GBSA simulations. The two most effective medications were methylergonovine (−40.42 kcal/mol) and cabergoline (−53.44 kcal/mol), which showed substantial binding affinities and were suggested to be neither mutagenic nor carcinogenic by ADMET/SAR analysis.

Because of its greater Ki (1.32 M) and Kd (6.44×10^{-8} M) values, methylergonovine compared to standards, ergotamine (agonist) and methysergide (antagonist) have a decreased binding affinity and potency. Cabergoline had modest binding affinity and potency [Ki = 0.85 M and Kd = 5.53×10^{-8} M] in comparison to standards. The top 2 medications changed helices VI, V, and III and shifted the RMSD 2.48 and 3.07 Å when they bound to the 5HT2BRM. In the 5HT2BRM receptor, residues of LEU209 provided a latch that resembled a lid, enhancing agonist binding and preventing drug escape. Compared to the antagonist, methylergonovine and cabergoline interacted with ALA225 more strongly [34].

The most prevalent type of benign, uncommon reflex epilepsy in humans is called hot water epilepsy (HWE). In a rat model of HWE, Among the proteins with an overexpressed mRNA transcript is Bdm1. According to fold prediction techniques and sequence searches, Bdm1 belongs to the group of α/β hydrolases, which also includes thioesterases and diene-lactone hydrolases. To estimate the Bdm1 sequence's fold, three-fold prediction techniques were employed: UCLA hybrid fold prediction, 3D-PSSM, and GenThreader. Using these fold prediction methods, proteins with established three-dimensional structures and functions were linked to distantly related proteins. Techniques such as PSI-BLAST, WU-BLAST2, PFSCAN, etc. were utilized to perform sequence searches. This implies the potential link between homocysteine metabolism and Bdm1. It has been observed that diazepam medication helps reduce seizures caused by elevated homocysteine levels. Enzymes such as acyl transferase and sulfhydrolase are involved in the process of homocysteine synthesis from homoserine. Methionyl tRNA synthetase can convert homocysteine to homocysteine thiolactone (HTL) via an alternative route. It has been demonstrated that HTL alters lysine residues to cause damage to proteins. According to these results, the metabolism of homocysteine and the function of proteins like Bdm1 during epileptic convulsions may shed light on the mechanisms underlying seizures and brain development [35].

Neuronal voltage-gated sodium channels are blocked by the tricyclic anticonvulsant medications phenytoin (**1**), carbamazepine (**2**), and lamotrigine (**3**) (Fig. 24.7).

These drugs interact with those of anesthetics used locally by adhering to domain IV-S6 in the interior pore of the channel. Through crystal structure homology with potassium channels, a prototype of the Na channel's open-pore was established.

Three anticonvulsants were docked with residues that were found to be crucial for their binding energy through mutagenesis. The pharmacophore shared by the three medications consists of an aromatic ring that interacts aromatically with NAV1.2's Tyr-1771 and a polar amide or imide that joins Phe-1764's aromatic ring to generate a low-energy amino-aromatic hydrogen bond. The second aromatic ring covers the pore lumen and is almost at a right angle to the pharmacophore. It likely interacts with the other S6 segments and physically obstructs the inner pore to prevent Na from penetrating. Due to the lack of positive charge in their structures, the hydrophobic interactions between this second aromatic ring and anticonvulsants may be important in binding and providing energy compensation. Anticonvulsant medication interactions further challenge the model, even if the one built by homology with potassium channel structures was pretty successful in accounting for inner pore component interactions with local anesthetics. Understanding the structural underpinnings of neuronal Na channels inhibition may help to rationally explore new anticonvulsant medications and provide an explanation for their target selectivity [36].

An essential inhibitory neurotransmitter with crucial neurological functions in the brain is GABA. Γ-aminobutyric acid aminotransferase, or GABA-AT, is an enzyme that breaks down γ-aminobutyric (GABA) in the brain and is dependent on pyridoxal 5′-phosphate (PLP). Reduced GABA levels are linked to several neurological conditions, including Alzheimer's disease. Drugs for neurological illnesses have an affinity for GABA-AT because blocking it raises brain levels of GABA. A major factor in the creation of GABA-AT inhibitors was in silico drug design. A study was conducted to develop new and powerful inhibitors against GABA-AT with the use of computer-aided drug design (CADD)techniques. Although the crystal structure of human GABA-AT is not yet known, a 3D structure of the protein was created using homology modeling, which relied on the sequence similarity between human GABA-AT and the chosen template. Based on the data displayed by the Basic Local Alignment Search Tool (BLAST), the pig's liver (Sus Scrofa) GABA-AT had the highest similarity, at 98.3%. The web-based PROCHECK was used to validate the modeled protein. To assess the quality of the 3D structure, an overall of 5

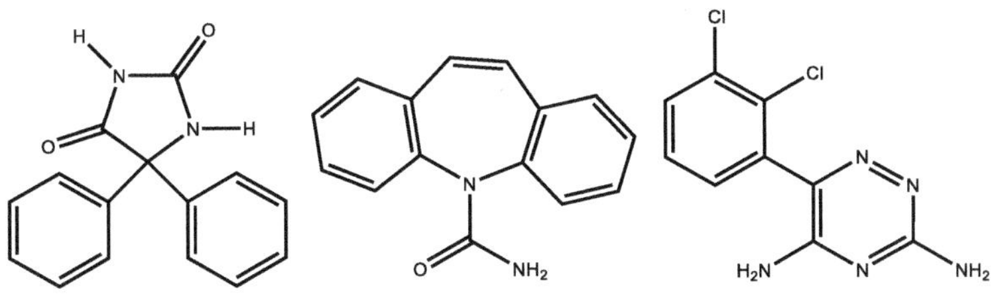

1 2 3

Fig. 24.7 Structures of tricyclic anticonvulsant drugs

known inhibitors were docked against the model. Known inhibitors were created using Biovia DS's "Prepare Ligands" procedure, and Autodock 4 was used to dock them against one of the models generated. Highly effective medications were found and chosen using structure-based virtual screening and molecular docking against humans. The compounds ZINC000364721779 and ZINC000635903250 demonstrated the highest binding affinities to mammalian GABA-AT, according to the results of the virtual screening studies. Based on the ADMET profile, these compounds were deemed drug-like candidates and passed Lipinski's rule of five. They may be great candidates for experimental testing in subsequent studies for additional assessment [37].

The effective anti-epileptic medication levetiracetam (LEV) targets a potential Major Facilitator Superfamily (MFS) transporter, SV2A by using Homology Modelling. For the purpose of comprehending the structure-function relationship and drug interactions of SV2A, a synaptic vesicle protein implicated in neurotransmitter release and a target for medications such as UCB 30889 and LEV, homology modeling becomes essential. Researchers discovered that compared to the Inward model, the Outward model—which was created using homology modeling—was more consistent with experimental findings. While SV2A is structurally undefined, homology modeling provides insights into the ligand-binding site. Nevertheless, the procedure of creating these models is not simple because, With MFS transporters that have established structures, SV2A shares a low sequence diversity. SV2A is required for Ca^{2+}-dependent vesicular release of neurotransmitters [38, 39]. These forecasts when combined with earlier findings, the outward-open models—that is, those in which the putative binding site is exposed to the vesicle's interior—tend to provide greater agreement with the site-directed mutagenesis experiment [40]. When these predictions are paired with previous discoveries, the outward-open models—that is, the models where the putative binding site is exposed.

24.3.2.5 Limitations

It is difficult to fully reflect in simplified representations the delicate balance of opposing interactions between the protein and the solvent as well as the different side-chain packing configurations. Because of their flexibility, protein structures can display a variety of conformations based on their surroundings. Errors in the template structure that was employed could introduce more uncertainty, which would then be included in the final model. Being almost entirely dependent on sequence alignment and, by extension, the degree of sequence similarity between the target and template structures is one of the fundamental disadvantages of homology modeling.

24.3.3 *Molecular Docking*

24.3.3.1 Concept

In computer-assisted drug design and structural molecular biology, molecular docking is an essential technique. The primary goal of ligand-protein docking is to predict the primary binding modes of a ligand with a protein that has a known three-dimensional structure. A scoring formula that accurately scores candidate dockings is used in successful docking algorithms to search high-dimensional spaces. Docking is an essential tool for lead optimization since it can be used to virtually screen huge compound libraries, rate the results, and suggest structural ideas about how the ligands block the target. The main objective of molecular docking is to create an ideal docked conformer of both interacting molecules to lower the system's overall free energy.

24.3.3.2 Softwares (Table 24.3)

24.3.3.3 Workflow

- The steps are explained using relevant software as examples for better understanding (Fig. 24.8).

Preparation of target protein
Large biological molecule, such as an RNA, DNA, or protein molecule is often the target structure, which is obtained as the first step in a docking calculation. Protein Data Bank (PDB) enables simple access to the molecular framework of these macromolecules by providing 3D atomic coordinates obtained through experimental methods [41].

Preparation of ligand molecule
The PDB protein-ligand structures are then processed using for example the Schrödinger suite's Protein Preparation Wizard. This is followed by the addition of deficient residues and loop portions close to the active site, Prime can be used after the integrity of the protein structure is examined and modified [42].

Table 24.3 Software used in molecular docking

Method	Software
Preparation of ligand	Auto Dock Vina
Search algorithm	GOLD
Pose refinement	GOLD and LeDock
Grid generation	MEGADOCK and ZDOCK

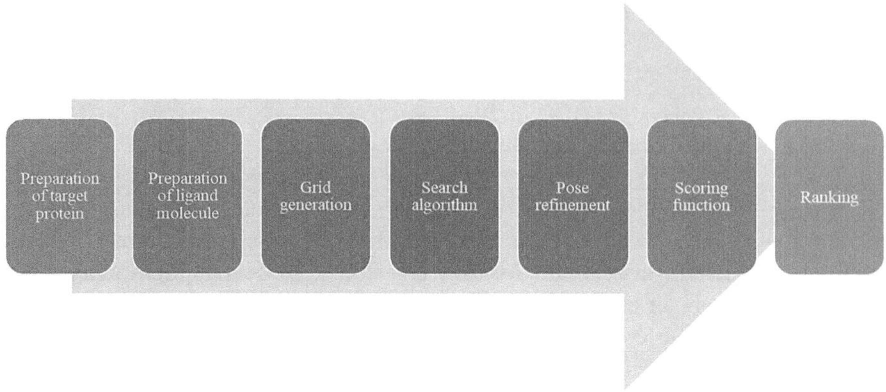

Fig. 24.8 Workflow for molecular docking

Grid generation
Designing a grid layout that reflects the volume of the active site is the process. Subsequently, the pertinent ligands are positioned onto this grid. It creates the grid files needed for DOCK's fast score assessment. Glide (Grid-based Ligand Docking with Energetics) looks for favorable interactions between one or more ligand molecules, which are usually small, and a receptor molecule, which is usually larger and is normally a protein.

Search algorithm
Search algorithms, such as genetic algorithms, evolutionary programming, and simulated annealing, play a crucial role in predicting potential conformations of binary complexes by exploring the conformational space efficiently [43].

Pose refinement
By utilizing the scoring function's gradients, refinement moves the ligand posture to a nearby energy minimum. The ultimate affinities and scores for the posture are determined using the designated CNN models and/or scoring function once the binding agent pose has been successfully modified.

Scoring function
In lead optimization, scoring functions are utilized to estimate the relative binding affinity between the protein and ligand as well as the binding mode and site of a ligand-protein interaction [44].

Ranking
The method of rating each ligand position by utilizing the expected free energy of binding and identify which ligands are most likely to interact positively with a certain receptor. The pose with the highest score represents the most beneficial binding mode.

24.3.3.4 Applications

Gamma aminobutyric acid (GABA) concentrations in the brain have been demonstrated to play a significant role in the severity of convulsions. A computational molecular docking research was conducted on the acetamido-N-benzyl acetamide that is α-substituted. Docking research revealed that all the compounds had higher binding scores than the drug vigabatrin (−4.4 kcal/mol), with the highest binding score being −13.8 kcal/mol that is sold commercially, except two ligands that were shown to have unpromising binding affinities: 2-acetamido-N-benzyl-2--hydroxyacetamide and 2-acetamido-N-benzyl-N-methyl-2-(pyrimidine-2-yl)acetamide. The highest binding score of −14.15 kcal/mol was found for 2-acetamido-2-((3-amino-4-vinylphenyl)amino)-N-methylacetamide-N-benzyl) **(4)** (Fig. 24.9) after three anticonvulsant compounds were designed.

This molecule more closely docked with the GABAAT enzyme through amino acid residues of Lys203, Pro347, Arg430, Thr353, Arg192, and Ala346 than the commercially available vigabatrin (−4.4 kcal/mol). This study offers medical and pharmaceutical chemists a useful method for creating these newly developed anticonvulsant medications from α_substituted acetamido-N-benzyl acetamide, which will be more effective in controlling seizures [45].

Pradhan et al. synthesized Schiff's bases of 7-hydroxy-3- methyl-2-quinolone **(5)** as well as 7-hydroxy-3-methyl-4-quinolone **(6)** (Fig. 24.10) by refluxing them with substituted aromatic carbonyl compounds. To study the binding mechanisms of quality and quantum interactions between produced chemicals and the target enzymes, quinolone derivatives were docked on the three high-resolution images of crystal structures of the hCA enzyme using the Biopredicta module of the VLife MDS 3.5 program (PDB 3F8E).

This data was obtained to facilitate further structure optimization. The interactions highlight the role that nitrogen atoms play in binding and the ensuing inhibitory potential. The optimal molecule has a minimum PLP score of −35.23 kcal/mol, which suggests a high ligand binding affinity towards hCA II. The interactions between ILE91A, VAL121A, GLN92A, HIS94A, VAL121A, PHE131A, LEU198A, and THR199A and the receptor were verified by docking the optimal molecule into

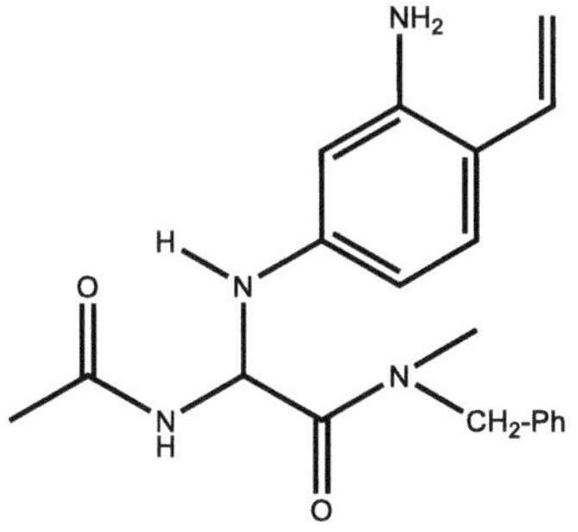

4

Fig. 24.9 Structure of 2-acetamido-2-((3-amino-4-vinylphenyl)amino)-N-benzyl-N-methyl acetamide)

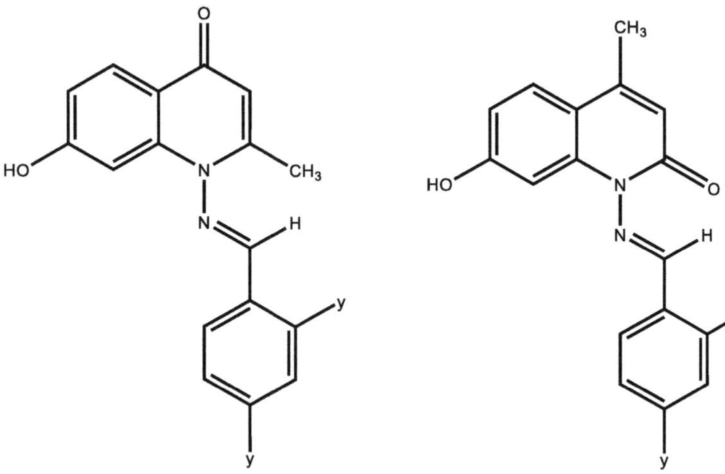

5 6

Fig. 24.10 Schiff bases of 7-hydroxy-3- methyl-2-quinolone and 7-hydroxy-3-methyl-4-quinolone

the HCF II receptor. hCA enzymes are an intriguing therapeutic target for antiepileptic activity, and this finding encourages further research on new residues that may be essential for the ligand-receptor interactions by supporting a unique notion regarding the binding interaction of these produced chemicals inside the receptor [46].

Several hybrid thiazole-pyridazine analogs were created as anticonvulsants in earlier research on pyridazinone analogs. The most effective and secure anticonvulsant drug working via a GABA-mediated mechanism was 2-(3-(4-chlorophenyl)-6-oxo-5,6-dihydropyridazin-1(4H)-yl)-N-(thiazol-2-yl)acetamide (SP-5F) **(7)** (Fig. 24.11), out of all 18 synthesized compounds. Due to its most effective anticonvulsant efficacy against both MES and scPTZ animal screen models, compound SP-5F was chosen. Molecule SP-5F attached to the receptor in a pocket made up of Asp7, Ala7C, Thr68D, Ser69D, Phe80, Phe84, Val87, Leu88, Phe91, Ser83, Thr87, Phe84, and Lys7D, according to docking postures of the molecule with a homology model of the Na^+ channel.

Molecular docking calculations were carried out to unlock the potential mechanism of anticonvulsant action, using the maestro 10.5 program. The binding of the ligand, as suggested by the pharmacophore, was confirmed by molecular docking experiments conducted on the verified target protein. It was found that amazing biological activity can be linked to hybrid thiazole compounds with quinolone moiety. Schiff's bases of both 7-hydroxy-3-methyl-2-quinolone and 7-hydroxy-3-methyl-4-quinolone in the current investigation were created using the Biopredicta module of VLife MDS 3.5 software. Out of all the synthesized compounds, some compounds exhibited potent antiepileptic efficacy [47].

Fig. 24.11 Structure of SP-5F molecule

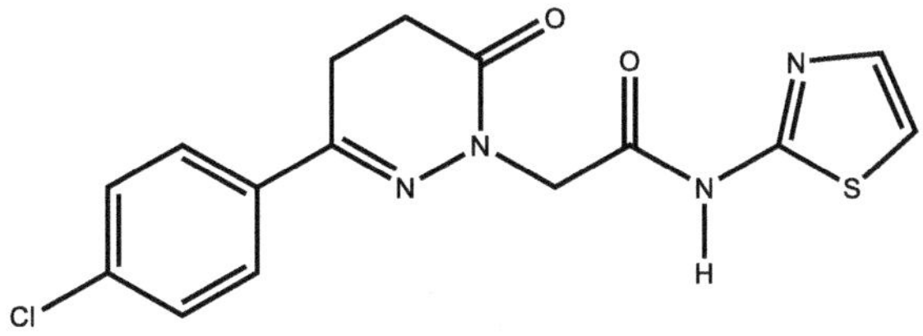

7

24.3.3.5 Limitations

Systems that may be correctly described in terms of size and complexity are constrained by the computational demands of docking calculations. Erroneous predictions may result from inaccurate representations of proteins and ligands, particularly in the case of big or intricate structures.

24.3.4 Molecular Dynamics Simulations

24.3.4.1 Concept

For molecular modeling, molecular simulation is a very helpful toolkit that makes it possible to track and understand structure as well as dynamics in great depth, even at levels where the motion of individual atoms can be observed. Through atomic-level detail and incredibly precise spatial resolution, these simulations illustrate how proteins and other biomolecules interact. Biomolecular simulation is becoming increasingly relevant to experimenters because of substantial improvements in availability, speed, and accuracy. This concept is especially evident in the field of neuroscience, although it is not restricted to it. In addition to helping to understand the structural underpinnings of disease, simulations have been useful in the development of peptides, proteins, and tiny molecules as well as in understanding the biological processes of proteins and other biomolecules [11].

24.3.4.2 Software (Table 24.4)

24.3.4.3 Workflow (Fig. 24.12)

Potential energy function
A potential energy function captures the energy of atom interactions in a molecule, including stretching, bending, and torsional motions. Force fields parameterize

Table 24.4 Software used in molecular dynamics simulations

Methods	Softwares
Potential energy functions	GROMACS 3.0 DeePMD kit
Simulation methods	ACEMD
Energy minimization force field	AMBER CHARMM
Molecular mechanics	Ablone
Electronic structure calculations	ADF
Molecular modelling	Ascalaph designer
Molecular editor	Avogadro

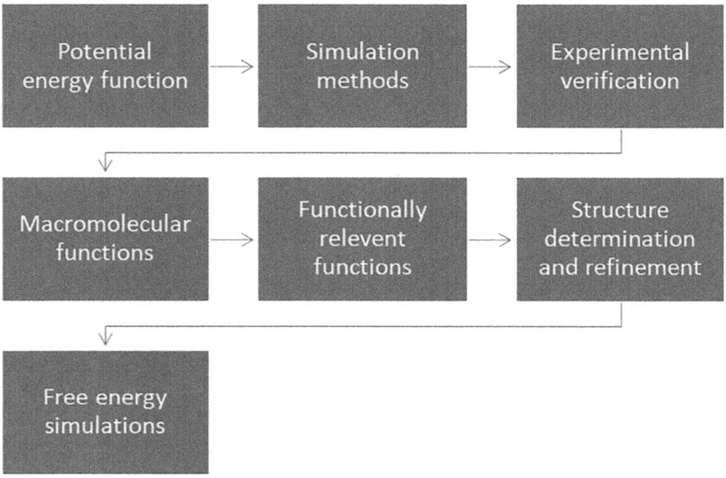

Fig. 24.12 Workflow for molecular dynamics simulations

these interactions using parameters from experiments or quantum calculations. The additivity assumption treats charges as static and assumes bond lengths and angles are independent. Class II force fields handle non-equilibrium configurations and accurately simulate vibrational spectra.

Simulation methods

Energy minimization:

Energy minimization refers to the process of finding the arrangement of atoms or molecules that corresponds to the lowest possible energy state within a given system.

Adiabatic mapping:

The simplest way to analyze motion in proteins is to identify low-energy paths for certain motions. This technique is known as "adiabatic mapping".

Experimental verification
Experimental verification of simulation-predicted structural and dynamical behaviors is challenging, especially for complex biological systems. Parameters like the Debye-Waller factor, indicating atom fluctuations, aid in comparison. Qualitative agreement between calculated B factors and experimental correlations indicates reliability, though discrepancies are common in regions involving molecular contacts and surface side chains.

Macromolecular motions
Macromolecular motions, as observed in protein simulations, involve a dynamic interplay of conformations akin to crystal structures but with notable differences while protein interiors exhibit modest atomic displacements, surface residues display more significant fluctuations.

Functionally relevant motions
A protein moves in the pattern of certain referred to as "collective coordinates" when it moves in a functionally significant way, forming a low-dimensional subspace. This reduction in complexity is harnessed for advanced molecular modeling and drug design tools. Collective coordinates also offer insight into protein energy landscapes and enhance configuration space sampling in molecular dynamics simulations.

Free-energy simulations
Thermodynamic features of many biologically significant systems and phenomena are fundamentally understood through the use of free energy simulation methods. Protein stability, ligand-protein binding affinities, and the hydration-free energies of ions and small molecules have all been estimated using these techniques. The potency and binding affinity of a medicine to its target receptor are frequently positively correlated, therefore theoretical techniques that can accurately calculate binding free energies can aid in the development or improvement of novel therapies [48].

24.3.4.4 Applications

Ohtahara Syndrome, also known as Early Infantile Epileptic Encephalopathy, is one of the most severe and early forms of epilepsy. It is characterized by early seizure onset, and synaptic vesicle exocytosis is aided by the STXBP1 gene, which encodes the Syntaxin binding protein1a, which is involved in the formation of SNARE complexes. The most pathogenic STXBP1 gene variants were to be found, and their effects on the stability and structure of the Stxbp1 protein were to be ascertained. Thirteen bioinformatics techniques were used to determine the high-risk nonsynonymous single nucleotide polymorphisms (nsSNPs) in the STXBP1 gene. The GROMACS software was used to carry out the molecular dynamics simulation. Using in silico prediction techniques, it was found 11 (S42P, H103D, R190W, R235G, D238E, L256P, P335S, C354Y, L365V, R406C, and G544D) out of 245 nsSNPs as harmful. The analysis of hydrogen and hydrophobic associations between the wild-type Stxbp1 structure and its mutant variants reveals that each of these nsSNPs has a unique impact on the protein structure. Molecular dynamics simulations demonstrated that the total number of nsSNPs has varying effects on residual fluctuation, compaction, and protein stability. Regarding high-risk nsSNPs that could impact the structure and function of the Stxbp1 protein, this study offers valuable information. Therefore, while doing a genetic screening for patients with early infantile epileptic encephalopathy, these variations must to be taken into account. Using the CHARMM 27 force field and the GROMACS 5.1.4 software package, a molecular dynamics simulation of the WT-Stxbp1 structure and its variations was carried out [49].

Due to the global expansion of multidrug resistance and its adverse effect profiles, as well as the lack of powerful and efficient antiepileptic agents in submicromolar concentrations, molecular modeling techniques enable us to find agents with unique structures that meet the ongoing demand for their discovery. It has been established that mGluR2 is a valid target for treating epilepsy. mGluR2, a subtype of metabotropic glutamate receptors, has emerged as a promising target for developing antiepileptic drugs due to its ability to modulate glutamate release and reduce neuronal excitability. Unlike ionotropic glutamate receptors, targeting mGluR2 offers advantages such as longer-lasting effects and the ability to fine-tune glutamate release dynamics. In a new sector enabled by the study, novel antiepileptic drugs have been discovered using a combination of molecular dynamics of the highest-scoring hits and structure-based virtual screening of the Asinex database. Following completion, the protein's quantitative and qualitative characteristics were calculated using the RMSD, which were stored in the md. out.cms file. Twelve potential hits from the structure-based virtual screening were identified as potential mGluR2 agonists based on their scores on the ADME parameters related to the CNS activity scale.

The best hit, ZINC00235412 (**9**) (Fig. 24.13), had no notable toxicity, while the top-scoring hit was ZINC19285209 (**8**) (Fig. 24.13). Furthermore, it satisfies the fundamental structural characteristics needed to be a mGluR2 agonist. The antiepileptic activity at the mGluR2 target of both system complexes is a set of molecules with the designated structural features. Based on the expected SAR of the acquired hits, this was accomplished with 4 Cl-ions, and the total counter ions of the complex were 26 Na^+ and 30 Cl-ions. Future research aims to identify alternate mGluR2 agonists without acidic groups for effective treatment of epilepsy [50].

Using both in silico and in vivo methods, it was methodically predicted and examined in this work the possible impacts of Ginkgo terpene trilactones (GTTL) on cognition and pathogenic alterations. For the first time, molecular docking studies and systems pharmacology research suggested that the GTTL might bind to 5-HT 1A/5-HT 1B/5-HT 2B, thereby mitigating neurologic alterations. Additionally, adult Wistar male rats were given an injection of lithium-pilocarpine to induce status epilepticus (SE), and as compared to the model group, GTTL was able to boost the expression of brain-derived neurotrophic factor (BDNF). Remarkably, current research indicates that neurologic alterations may be considerably mitigated by the simultaneous involvement of 5-HT receptor activation and enhanced BDNF expression in the hippocampus. The utilization of Molecular Dynamics (MD) Simulation

Fig. 24.13 Structures of ZINC19285209 and ZINC00235412

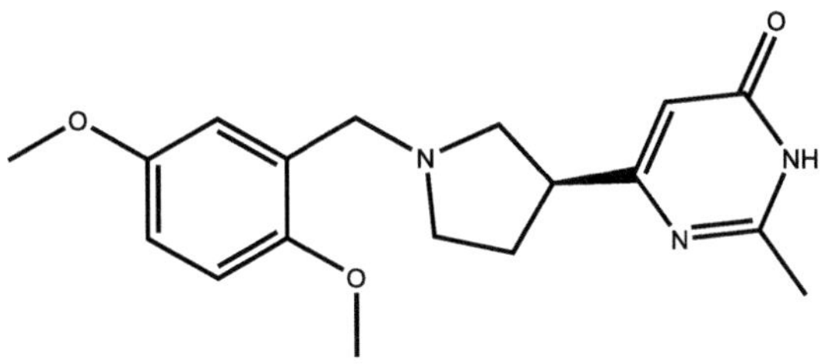

8

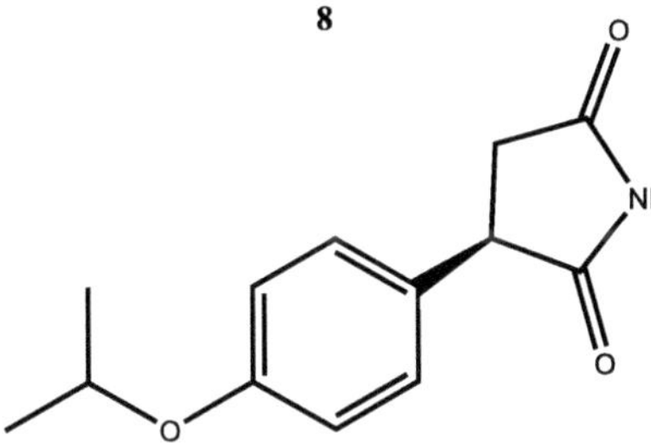

9

provides invaluable insights into the intricate interactions between 5-HT 1A and Ginkgolide A/C complexes. Through a meticulous analysis spanning 100 nanoseconds, the system exhibited remarkable stability, as evidenced by the fluctuation of root-mean-square deviations (RMSD) around 1.8 Å for 5-HT 1A and 0.6 Å for Ginkgolide A notably Asp116, Thr118, Gly382, and Asn386, which formed crucial hydrogen bonds with Ginkgolide A/C. Furthermore, the involvement of hydrophobic residues such as Phe112, Ile113, Ile189, Phe361, and Leu368 underscored the comprehensive nature of the recognition process. Research findings demonstrated that GTTL reduced neuronal damage, enhanced learning and memory, and shielded rat cerebral hippocampus neurons against epilepsy. As a result, GTTL might be a good first choice for treating or preventing epilepsy. These results, in harmony with previous docking studies, emphasize the significance of MD Simulation as a powerful tool in elucidating the molecular mechanisms underpinning anti-epileptic effects, thereby offering promising avenues for drug development and treatment strategies [51].

24.3.4.5 Limitations

The estimated values of the potential energy calculations and the simulation lengths are two of the simulations' shortcomings. Systematic flaws are introduced in the first, and statistical errors in the second. In many situations, this results in an inadequate sample of conformational states because routine simulations longer than a microsecond are not possible due to high processing costs and the need for further refining of the force fields used. A one-microsecond simulation of a comparatively tiny system (around 25,000 atoms) working on 24 processors takes many months to complete, as an illustration of these enormous processing demands.

The force fields that are employed are estimations of the quantum-mechanical truth that governs the atomic domain, which presents difficulties in addition to the high computational needs of these simulations. Although many significant molecular motions can be precisely predicted by simulations, these models are not well adapted to scenarios where quantum effects play a significant role, such as in binding involving transition metal atoms [52].

24.3.5 QSAR

24.3.5.1 Concept

QSAR is a significant area in which cheminformatics computations are applied to antiepileptic medication discovery. QSAR calculations help with the optimization of the lead compound's structure after it has been identified. The goal of these techniques is to raise the likelihood of discovering active compounds among those that are ultimately synthesized, hence limiting the success rate of screening and synthesizing efforts. The three main categories of QSAR approaches are as follows: 1D-QSAR (1-dimensional QSAR, such as Hansch analysis); 2D-QSAR (2-dimensional QSAR, such as pattern recognition analysis); and 3D-QSAR (3-dimensional QSAR, such as comparative molecular field analysis). Every approach has benefits and drawbacks. QSAR has advanced from straightforward regression equations over the last 30 years to 2D-QSAR and current 3D-QSAR techniques. Descriptors are essential to the characterization and bioactivity prediction of compounds in QSAR. Topological descriptors characterize molecular connections, electronic descriptors capture electron dispersion, geometric descriptors reflect the molecular form, and physicochemical descriptors show characteristics like solubility. A 2D data array is created using these descriptors, which are computed for every chemical in a training set. Subsequently, statistical methods determine the minimal set of descriptors that distinguish active from inactive compounds. Prediction algorithms are developed using this data and verified on test compounds [53].

24.3.5.2 Softwares (Table 24.5)

24.3.5.3 Workflow (Fig. 24.14)

Molecular descriptor generation from structure
The arrangement of atoms having covalent bonds connecting them defines a small-molecule compound's structure. The ability of the selected descriptors to represent the change of activity with structure is a critical factor in producing a statistically

Table 24.5 Software used in QSAR

Method	Software
Generation of 3d QSAR models	PharmQSAR
Ligand-based drug design	Glide, Macro Model, Autodock, and Argus Lab
Structure-based drug design	Insight II, Discovery Studio, Materials Studio, Accord, and Prime
Molecular descriptors	Dragon, Mordred and PaDEL

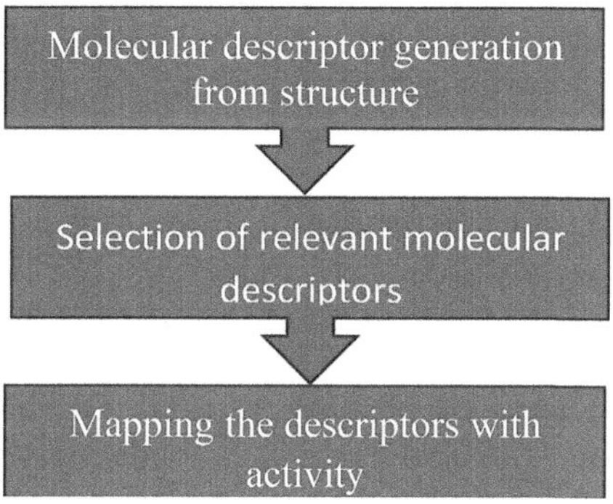

Fig. 24.14 Workflow for QSAR

robust model. To assist in determining which descriptors best capture structure-activity variation, statistical techniques are frequently included in commercial molecular modeling software.

Selection of relevant Molecular Descriptors

The abundance of molecular descriptors in QSAR analysis presents challenges with dataset size, computational complexity, and model interpretability. To address these concerns, the most pertinent descriptors are chosen using automatic reduction approaches. Large descriptor sets can be more easily mined for important information using techniques like feature selection, dimensionality reduction, and machine learning methods. This improves the efficiency and interpretability of the model while preserving its predictive accuracy.

Mapping the Descriptors to Activity

To proceed with QSAR analysis, a mapping function between the necessary chemical descriptors and the activity under study must be created after the descriptors have been computed and chosen. The correlation between activity levels and descriptor values is quantified by this function. Numerous training techniques can be used to optimize the function based on the available training set data, where the activity is known, and a range of mapping function families, including linear and non-linear models, can be taken into account. Through the creation of predictive QSAR models for drug design and optimization, these techniques seek to identify the best-fit function that, given the selected descriptors, properly predicts activity.

24.3.5.4 Applications

CoMFA, or Comparative Molecular Field Analysis, proved to be an excellent tool in the development of novel therapeutics for epilepsy. By utilizing researchers can learn more about the possible antiepileptic medications' three-dimensional structure-activity correlations, allowing for the rational design of compounds with enhanced potency and selectivity. This technique aids in identifying structural features crucial for binding to specific molecular targets implicated in epilepsy, such as ion channels or receptors involved in seizure generation and propagation. Through CoMFA-guided optimization, researchers can tailor the chemical structure of candidate compounds to improve their affinity and efficacy, ultimately leading to the discovery of more potent and selective antiepileptic agents. This approach not only accelerates the drug discovery process but also increases the likelihood of identifying compounds with superior therapeutic profiles, thereby offering new hope for individuals suffering from epilepsy by providing more effective treatment options [54].

Recurrent seizures are a hallmark of the neurological condition epilepsy., and current treatments often come with limitations such as inadequate efficacy or side effects. By identifying quinazolinone compounds with superior binding affinity to the GABAAT receptor compared to existing medications like vigabatrin, this research opens up new possibilities for the development of more effective antiepileptic drugs. The anticonvulsant inhibitory properties of several quinazolinone compounds were investigated by molecular docking and quantitative structure-activity relationship analyses. To determine the ideal geometry of the anticonvulsant inhibitors, the quantum chemical calculation method known as density functional theory, or DFT, was applied. A quantifiable relationship between structural characteristics and anticonvulsant activity was found using four different kinds of molecular descriptors. According to this study, the binding affinity that was produced was superior to that of vigabatrin, an anti-epilepsy medication that is marketed for commercial use. The four generated QSAR models were all recorded, and based on statistical significance, the best model (Model 1) was chosen and published.

$$\begin{gathered}
Model\,1 \\
\mathrm{pED}_{50} = 0.114383001 * \mathrm{ETA_Eta_L} + 0.190098515 * \mathrm{XLogP} + 0.028759587 * \\
\mathrm{PPSA} - 3 + 4.201924750 * \mathrm{RNCG} - 0.690224604, N = 17, R^2_{\mathrm{ext}} = 0.72028, \\
R^2 = 0.934053, R^2_{\mathrm{a}} = 0.912071, Q^2_{\mathrm{cv}} = 0.869587, \mathrm{LOF} = 0.002815, \\
\text{min expt.error for non} - \text{significant LOF}\,(95\%) = 0.018506.
\end{gathered}$$

It was discovered that all quinazolinone analogues docked to GABAAT more effectively than vigabatrin, the typical anti-epilepsy medication. The molecular docking data (high binding affinity with docking score of −9.5 kcal/mol) and the QSAR model (high correlation coefficient of determination R^2 of 0.934) support each other and suggest avenues for designing novel inhibitors with improved action towards

GABAAT. This research will aid in the manufacture of novel selective GABAAT inhibitors with predefined activity and affinity as well as the rational design of pharmaceuticals. It also offers important insights into how novel compounds interact with GABAAT and may open the door to the discovery of new GABAAT inhibitors [55].

Mutations in the KCNQ2/Q3 genes have been linked to benign familial newborn epilepsy (BFNE), a rare epilepsy variant often underestimated due to its genetic origin. This connection suggests a promising avenue for the development of antiepileptic drugs targeting this specific genetic subtype of epilepsy. A pharmacophore-based 3D-QSAR model was created in the current work for several pyrimidine and N-pyridyl benzamides that had KCNQ2/Q3 opening action. One hydrophobic (H), one donor of hydrogen bonds (D), and two aromatic rings (R) make up the pharmacophore model that was created. These are the essential molecular reports that describe the expected binding performance of low- and high-affinity ligands for the opening activity of KCNQ2/Q3. Additionally, it has been confirmed through the use of a biological link between 3D-QSAR characteristics based on the pharmacophore hypothesis and the functional fingerprints of the openers that bind the receptor, as well as by docking these benzamides into the homology model.

The 3D-QSAR analysis of the biological function of benzamide compounds have been described. The QSAR model with all 100 hypotheses was constructed by aligning the pharmacophoric characteristics of compounds with a grid spacing of 1.0 Å on a training set consisting of 32 members. The R^2 values of the training set molecules were used to determine the best 3D-QSAR model using three PLS factors. A 3D-QSAR model with pharmacophoric properties was effectively created for a group of chemicals in the benzamide family. With an optimal correlation coefficient of 0.81, the developed model shows good proximity and fitness with the experimental data, highlighting the importance of hydrophobic, hydrogen donor, and aromatic characteristics for KCNQ2/Q3 opening activity. The 3D QSAR model, along with its docking validation and QikProp pharmacokinetic property analysis provides valuable insights for subsequent research and development of additional potential KCNQ2 channel openers as antiepileptic drugs. Thus, the 3D QSAR model that was established might be used to forecast the activation of KCNQ2/Q3 in newly built analogs [56, 57].

24.3.5.5 Limitations

The availability of thorough and accurate data on chemical structures and biological activities is essential for the proper functioning of QSAR models. Nonetheless, the predictive capacity of QSAR in identifying new antiepileptic drugs may be constrained by the lack of data and the unpredictability of experimental settings.

To accurately reflect the physicochemical properties of molecules, QSAR models rely on the selection of suitable structural descriptors. However, there may be limitations in the accuracy of the model if the descriptors chosen fail to sufficiently represent the structural aspects related to antiepileptic activity.

Overfitting of QSAR models can occur, particularly in the case of sparse datasets or intricate chemical spaces. When applied to unseen data, overfitting can lead to misleading correlations and poor predictive performance, undermining the validity of QSAR predictions in the context of antiepileptic drug design.

24.3.6 Virtual Screening

24.3.6.1 Concept

A popular approach for finding hit molecules to use as parts of medicinal chemistry is virtual screening (VS). Virtual Screening (VS) is a comprehensive, knowledge-based method for searching compound databases. It looks for new compounds and chemotypes that have the necessary biological activity to replace current ligands, or occasionally it aims to find ligands for unidentified potential drug targets for which high-confidence homology models, crystal structures, or solution structures are available Virtual screening (VS) is commonly characterized as a methodical procedure that employs a sequence of subsequent phases to identify and refine a large number of lead-like results with potential biological activity against certain pharmaceutical targets. Virtual screening techniques open up possibilities to compounds that are readily available through sale or synthesis but might not be present physically in an investigator's collection. The second evident benefit is that activities can be measured in low- or medium-throughput assay formats because only a small selection of compounds needs to be tested due to computational estimation of binding affinity [58].

24.3.6.2 Softwares (Table 24.6)

24.3.6.3 Workflow (Fig. 24.15)

Structure-Based Virtual Screening
In structure-based virtual screening, identifying a target protein's structure by NMR, X-ray crystallography, or homology modeling is a crucial first step. The experimental structures of ligands attached to the enzymes are determined by a multitude of X-ray crystallographic plus NMR studies, which are a significant source of inspiration for work on docking and analog design [59].

Ligand-Based Virtual Screening
Using pharmacophoric, topological, or structural similarity search, the most potent biologically active lead molecule is found in the LBVS process. Taking into account

Table 24.6 Software used in virtual screening

Method	Software
Determination of protonation states	PROPKA
Protein preparation	H++
Prediction of water molecules in binding site	JAWS
Binding site identification	QsiteFinder
Library design	CLEVERLibrary Design
Ligand preparation	DISI
Covalent docking	CovalentDock

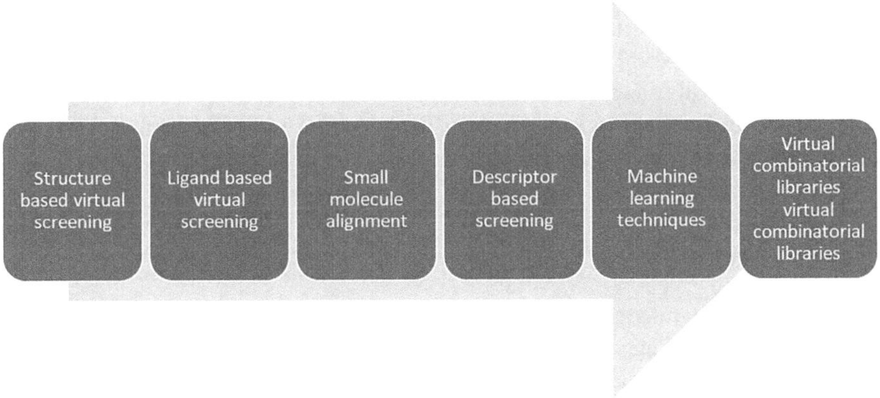

Fig. 24.15 Workflow for Virtual Screening

several factors, including the electrostatic characteristics of the molecule and the shape and structure of each component [60].

Small Molecule Alignment

Flexibility is controlled for both the test molecule and the reference template in the Genetic Algorithm Similarity Programme (GASP) [61]. With the use of a genetic algorithm, GASP is a program that automatically analyses test molecules with a limited number of reference ligand molecules that are the most active. Examining situations where molecules cannot be ascribed conformational preferences is beneficial. To evaluate the molecular similarity in ligand-based drug design, one more tool is the Molecular Electrostatic Potential (MEP) [62]. Galois alignment is used to align the molecular fields regarding the drug and reference molecules to maximize their similarity.

Descriptor Based Screening

Descriptors in both 1D and 2D One-dimensional (1D) descriptors of a molecule are bulk parameters such as the molecular weight, molar refractivity, and logP [63]. On the other hand, two-dimensional (2D) descriptors are derived from various two-dimensional quantitative or qualitative qualities of the lead molecule.

Machine Learning Techniques
Three Methods for Machine Learning in addition to a range of statistical approaches, several neural network methods, including Kohonen, back-propagation, and counter-propagation networks, as well as algorithms for machine learning, such as genetic algorithmic techniques, have proved their significance in virtual screening applications. Support Vector Machine (SVM), a technique without supervision that divides objects into two classes based on their features, has drawn interest in the field of chemoinformatics. A compound's bioactivity can be predicted from its structural properties using the Support Vector Machine (SVM) method.

Virtual Combinatorial Libraries
High throughput screening (HTS) technology and combinatorial chemistry have contributed to making it possible to quickly synthesize thousands of compounds and screen them for biological activity against a broad spectrum of protein targets These can be readily streamlined by using computer-controlled robot syringes to do complicated tasks. One can then screen the resulting chemical arrays, known as combinatorial libraries, for possible medicinal properties.

24.3.6.4 Applications

Palestro et al. integrated the human NaV1.2 channel, an important target associated with epilepsy, using a three-dimensional structural framework. Human P-glycoprotein (P-gp), another potential target, was likewise constructed and shaped in a three-dimensional structural model. By employing virtual screening, researchers can quickly and efficiently sift through large chemical libraries in search of compounds that may bind to their target protein. Palestro et al. identified the collection of compounds that were likely to be compatible with NaV1.2 and P-gp using molecular docking preliminary screening, which saved time and money. Compared to screening every chemical experimentally, this method was preferable. The compounds with the highest predicted binding affinity for a particular protein after virtual screening are selected as potential medications. Medicines exhibiting high screening fractions were then tested on voltage-gated sodium channels (NaVs), which are crucial in the pathophysiology of epilepsy because they initiate and propagate action potentials in neurons. This was done to confirm the medicines' inhibitory activity. In particular, neuronal subtype hNaV1.2 was the focus of this study's search for possible anticonvulsant substances that interact with NaVs. In a concurrent virtual screening across the Drug Bank database, a docking model and a ligand-based QSAR model were developed, verified, and used. Using the patch-clamp approach, montelukast, novobiocin, and cinnamonine were chosen for in vitro testing, and it was found that they could all inhibit hNaV1.2 channels that were heterologously produced in HEK293 cells. In the GASH/Sal model of audiogenic seizures, two hits were assessed and showed promising action in lowering the intensity of sound-induced seizures at the tested dosages. These results might serve as a foundation for additional investigation into the discovery of novel antiseizure medications for the management of the patch-clamp method on the NaV1.2 channel. The compounds are arranged according to their significance for future study and development [64].

XP-Visualization was used to visualize protein-ligand interactions and scoring functions of the three structured levels of virtual screening [High Throughput Virtual Screening (HTVS), Standard precision (SP), and Extra precision (XP)] of Glide. This evaluation of structure-guided binding interaction patterns of approximately 7 lakh different compounds from the Asinex library to the receptor was completed. Biomolecular complexes containing the highest-scoring chemicals were subsequently chosen for binding energy. Using ligand efficiency matric (LE) as hit identification criteria in virtual screening methods allowed for an additional examination of possible hits from false positives and a subsequent hit optimization phase. Three consecutive triad screening protocols, HTVS, SP, and XP modes, were used to conduct the structure-based virtual screening. This produced 346 hits, which were then selected for Prime-MM-GBSA analysis to determine their binding energies in order to lower the rate of false positives. The binding energy of about 341 hits was higher than that of the reference ligand, benzamidine (−36.26 kcal/mol). Potential anti-epileptic drugs could be GABAA agonists combined with ligands that have hydrogen bond/ionic interactions with Glu155 and cation-π or π-π interactions with important residues, such as Tyr157, Phe200, and Tyr205. Additionally, molecules with a docking score of less than −8.5 kcal/mol were deemed viable candidates. Additionally, it satisfies both the required and necessary structural properties interactions necessary for it to be a GABAA agonist because of the extent to which its structure mimics that of GABA. A number of compounds with a 4-(3-(ethylammonio)propyl)morpholin-4-ium scaffold and the requisite structural properties can be offered to have the potential anti-epileptic action at the GABAA target based on the anticipated SAR of the hits obtained [65].

A viable and proven approach to the creation of innovative antiepileptic medications is the use of competitive AMPA receptor antagonists. The structure-based virtual screening approach was used on a library of natural compounds to find 11 novel variable competitive AMPA receptor antagonists with better docking and dG bind scores than the co-crystallized ligand. At three different levels—superposition, enrichment, and simulation studies—the screening methodology was validated. The presence of significant amino acid interactions, such as those between Thr91 and Arg96, present in the binding of the co-crystallized ligand, served as the primary criterion for selecting hits based on ligand-protein interactions. Quantum mechanics hit optimization, and simulation experiments were performed on the top hit with the best dG bound score. Better pharmacokinetic performance computational models created using a validated virtual screening process offer in silico support for the creation of an improved epilepsy treatment plan [66].

24.3.6.5 Limitations

Many Visual Studio tools work well in particular case studies (depending on the instructional set) but not in broader scenarios. In general, the molecules found by HTS are more bioactivecomparedto those identified by VS. There is Weakness in accuracy including water and receptor structural flexibility in docking calculations because of the high computational cost and complexity of its modelling. Rarely is it possible to identify very powerful leads—that is, low nM—through VS. The complicated nature associated with ligand-receptor binding interactions as well as the approximations used to calculate the desolvation and entropic components make it difficult to parameterize, which makes scoring problematic in properly predicting an appropriate binding pose and scoring of the compounds.

24.3.7 Receptor Based Pharmacophore Modelling

24.3.7.1 Concept

In accordance with IUPAC, A pharmacophore is "a composite of steric and electrical characteristics that is necessary to ensure appropriate supramolecular interactions with a particular biological target and to inhibit its biological response" [67]. Pharmacophore modelling is a popular ligand-based technique for identifying drug targets. Two categories of pharmacophores exist: (a) Pharmacophores based on structures: and (b) Pharmacophores based on ligands. The methods used for addressing conformational flexibility, setting the query, and determining the pharmacophore pattern are the primary regions where the pharmacophore searching approaches vary [68].

24.3.7.2 Softwares (Table 24.7)

24.3.7.3 Workflow

Ligand selection and identification
The ultimate pharmacophore model is influenced by the characteristics of the ligand molecules, the dataset's size, and its chemical diversity. To generate an accurate pharmacophore model, the data set containing compounds binding to the identical pocket in the target should be as diverse as feasible (Fig. 24.16).

Define Pharmacophore features
Characteristics can be categorised according to three factors: function (H bond donor/acceptor, acid, base, aromatic ring, and hydrophobic group), topology

Table 24.7 Software used in receptor-based virtual screening

Methods	Softwares
Ligand structures represented as a collection of labelled points in three dimensions	RAPID [69]
Pattern identification phase	DISCO [70], RAPID [69], GAMMA [71], and GASP [72]
The Maximal Common Substructure (MCS) must be located in order to create a pattern.	HipHop [58] and SCAMPI
Pharmacophore generation	Structure-Based Pharmacophore, MOE, Ligand Scout, and Phase

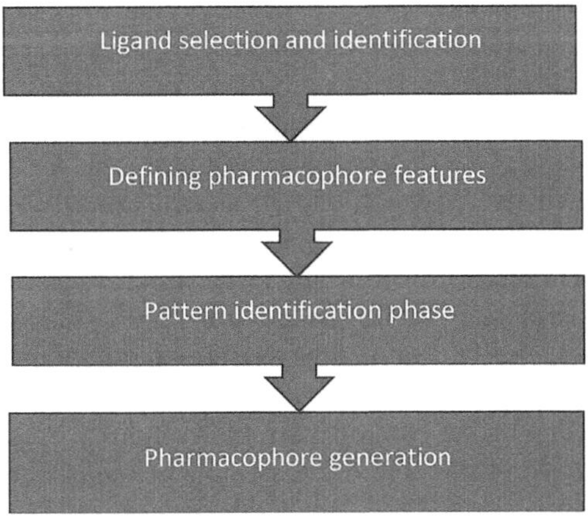

Fig. 24.16 Workflow for receptor based pharmacophore modelling

(phenyl ring and carbonyl group), and atom-based (3D orientation of the atom and atom type). To create an illustration of the entire structure, the chosen characteristics from a single ligand are combined.

Pattern identification phase

A collection of features and their relative places in three dimensions make up a pattern or arrangement. If a ligand has a particular collection of traits and a conformation enabling the features to be imposed in the right places, it will be considered to match a pattern. This is accomplished in three stages: the first is the constructive stage, which finds pharmacophore structures that are shared by the most active set of ligands; The second step, referred to as the subtractive stage, involves eliminating candidates found in step 1 who are also present in more than half of the collection of most inactive ligands. Those pharmacophores that successfully complete the subtractive stage are then subjected to simulated annealing in the final optimization step.

Pharmacophore generation

A lower score indicates a greater probability of random correlation leading to the creation of the model. When the three-dimensional structure is known, pharmacophore can be generated by simply picking out the binding sites and related interaction features from the ligand-protein pair. The receptor-based approach to pharmacophore development studies the features and their spatial connections to obtain a functional representation of the site of action, which is used to build the pharmacophore model. The process takes two inputs: the three-dimensional model of the receptor (usually in PDB format) and an array of ligands with recognised activity [73].

24.3.7.4 Applications

Recently, a novel family of noncompetitive AMPA receptor antagonists was reported, referred to as 6,7-dimethoxy-1,2,3,4-tetrahydroisoquinoline derivatives. A 3D pharmacophore model was created to predict the anticonvulsant action of these derivatives. The training set had 17 distinct drugs of varying potencies to prevent audiogenic seizures in DBA/2 mice. The HypoGen module of Catalyst 4.9 generated the best statistical hypothesis, which led to a model with a correlation coefficient of 0.919, using two hydrogen bond acceptors, two hydrophobic characteristics, and one hydrophobic aromatic region. The tetrahydroisoquinoline scaffold was incorporated into several new anticonvulsant medicines, and the model that was produced proved to be a useful aid in this process. Furthermore, omitted volumes were taken into consideration in order to explain the varying degrees of efficacy of the newly designed N-substituted compounds. In this study, the development of a quantitative pharmacophore model for compounds of tetrahydroisoquinoline that function as anticonvulsants is reported. The most promising HypoGen hypothesis made it possible to create novel isoquinoline derivatives, that can stop audiogenic seizures in DBA/2 mice at a dosage that is equivalent to other recognized AMPAR noncompetitive antagonists. For every developed molecule, the synthetic methods and structural characteristics have also been published. The structure-activity relationships for this class of molecules were proposed by this investigation, even though none of the newly discovered compounds had activity greater than that of the "lead compound" 4. The most significant discovery, in particular, involved employing the excluded volumes approach to show the difference between the experimental and predicted ED50 values for a few N-substituted derivatives [74].

The potential anticonvulsant effects of semi carbazones have been well studied. The compound 4-(4-Flurophenoxy) benzaldehyde semi carbazone is a lead molecule that is being researched as a powerful antiepileptic medication. Its maximum electroshock (MES) ED is 12.9 mg kg^{-1}. Moreover, semi carbazones block GABA-transaminase. A hydrogen bonding domain (HBD) and the aryl hydrophobic pocket, which contain substituents F, Cl, Br, and NO_2, are commonly found in potent anticonvulsant medicines, as shown by extensive structure-activity relationships. An upgraded version of the semi-carbazone flexible pharmacophore model is presented for the creation of new anticonvulsant drugs. One of the reviews on the biological activity of semicarbazones and thiosemicarbazones and their metal complexes features a model with an aryl binding site surrounded by an auxiliary binding region and a hydrogen binding area represented by a semicarbazono handle. It has been developed that semi carbazones are flexible pharmacophores of anticonvulsants.

Two substances have been discovered as lead molecules and are in different phases of development: 4-(4-fluorophenoxy)benzaldehydesemi carbazone (FPBS) and N′-(4-chlorophenyl)-N4-(2-nitrobenzaldehyde) semi carbazone. While the others are GABA transaminase inhibitors, FPBS is a sodium channel blocker. Alternative modes of action, including as non-competitive channel blockers and glycine receptors, have been suggested. For the development of anticonvulsants and future drug design, a flexible pharmacophore model has been proposed [75].

To find new phytochemicals that may be effective against S100B in epilepsy is paramount for advancing therapeutic interventions against the disorder. S100B, a calcium-binding protein ubiquitous in various cell types, is intricately involved in essential biological processes, including cell cycle regulation and differentiation. However, its overproduction has been implicated in the progression of epilepsy. Using a strong RF model, the 9000 phytochemical library was combed through, leading to the discovery of 180 putative S100B inhibitors. A simulated phytochemical screening against the known therapeutic target for epilepsy, the S100B protein. Residing in the cytoplasm and nucleus of numerous cell types, S100B is an element of the S100 protein family and plays a crucial role in controlling an extensive range of biological functions. It is important for processes including cell cycle progression and differentiation. Multiple machine learning techniques were used in our work, including Random Forest (RF), Naive Bayes (NB), k-Nearest Neighbors (kNN), and Support Vector Machine (SVM). These algorithms were used for virtual screening as well as for the extraction and selection of key features, which helped discriminate between compounds that were active and inactive. With an amazing 93.43% accuracy on both training and test datasets, the RF model outperformed the other tested machine learning techniques. Using this powerful RF model, the 9000 phytochemical library was combed through, leading to the discovery of 180 putative S100B inhibitors. The active site of S100B proteins was then docked with these 180 active molecules. The selection of 33 features using machine learning-based virtual screening, such as molecular weight, MlogP, and counts of hydrogen bond acceptors and donors, highlights our extensive and methodical approach to discovering optimal candidates [76].

24.3.7.5 Limitations

Pharmacophore models usually simplify the necessary features for biological activity, possibly ignoring significant structural subtleties. This simplification can result in the exclusion of important information about the ligand-receptor interactions.

Pharmacophore models frequently assume a static conformation of the ligand and receptor, ignoring the dynamic nature of molecular interactions which can lead to inaccurate predictions of binding affinities and activities of compounds.

Pharmacophore models are heavily dependent on the quality and diversity of the training data used to generate them. Limited or biased data sets for training can result in models that are not representative of the entire chemical space, affecting the pharmacophore predictions.

The conformational flexibility of ligands and receptors is not well-accounted by pharmacophore modeling, particularly when working with big or flexible compounds. This restriction may lead to erroneous binding mode predictions or false negatives.

24.3.8 Quantum and Molecular Mechanics

24.3.8.1 Concept

Protein simulations, particularly crucial for the investigation of conformational flexibility, are best performed using force fields from Molecular Mechanics (MM). It is an empirical technique of potential energy as a result that shows a molecule's geometric variables. Considering the significance of protein elasticity in drug binding, force fields (FFs) are employed in molecular mechanics (MM) to predict the conformations of molecules and examine how they interact with the drug's target at the site of action, which is commonly referred to as the binding site. These computational methods are useful in examining the modifications in conformation that take place when a prospective therapeutic molecule binds to its target.

High-quality structural data about tiny molecules with less than 60 atoms can be obtained by quantum mechanical computations. The central concept in molecular quantum mechanics is the Schrödinger equation. The Schrödinger equation can be expressed as H1 = E1, where E is the system's energy, H is a mathematical operational function, and 1 is the wavefunction. In theory, any physical attribute of a pharmacological molecule can be ascertained once its wavefunction is understood. The term "orbitals" refers to the wavefunctions that fulfil Schrödinger's equation for a basic hydrogen atom; thus, the hydrogenic atomic orbital is the three-dimensional mathematical equation that can be used to determine an electron's energy and other characteristics. The goal of CADD is to determine the wavefunction 1 of the anticonvulsant drug molecule in order to compute the drug's ideal shape and other features.

Molecular orbital computations can be roughly classified into two categories: semi-empirical ab initio calculations and ab initio calculations. When molecular orbital wavefunction calculations take into account every electron in the drug molecule, they are referred to as ab initio calculations worldwide.

24.3.8.2 Softwares (Table 24.8)

24.3.8.3 Workflow

Linear scaling QM technique

The binding enthalpy between proteins and small molecules has been measured extensively using the linear scaling QM technique. One common LSQM strategy

Table 24.8 Softwares used in quantum and molecular mechanics

Software
Profacgen
Insight II
Pipeline pilot
Discovery studio
Accord

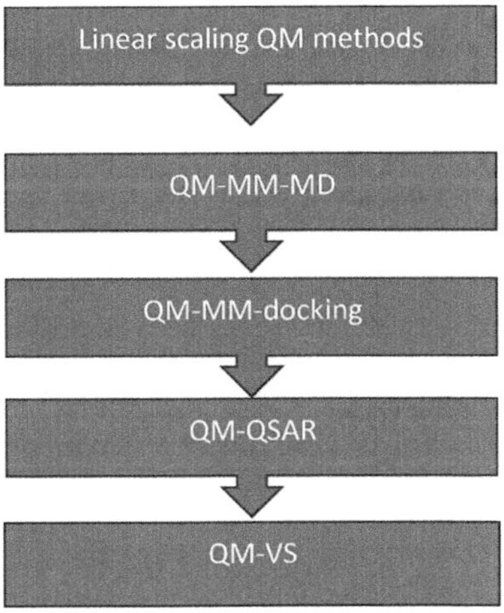

Fig. 24.17 Workflow for quantum and molecular mechanics

involves the divide-and-conquer (D&C) approach, in which a complex system is broken down into numerous subsystems, with the density matrix of each subsystem being estimated independently. Ultimately, the summation of the contributions from each subsystem yields the system's energy and overall density matrix (Fig. 24.17).

QM-MM MD

The hybrid QM-MM approach is a molecular simulation methodology that combines the speed of MM with the precision of QM to analyze the region of the structure where the chemical process occurs. This allows the study of chemical processes in vast systems. Protein and small molecule structures can be created using QM/MM, for example, by optimising the binding poses discovered during docking improving the enzyme active site geometries that were discovered through harmonically restricted minimization using MM.

QM-MM docking

Protein-ligand docking relies heavily on the precision of electric charges, which explains why QM-MM measurements are included in docking protocols. In the protein-ligand complex, QM-MM calculations take advantage of fixed charges of ligands that come from force-field parameterization, treating the ligand alone as the quantum area.

QM-QSAR
When paired with QM-MM, QSAR models enable the prediction of the drug's ADMET and provide trustworthy insights into how a compound's alteration impacts or enhances its pharmacokinetic/pharmacological profile.

QM-VS
The QM-VS methodology provides structure-based binding-energy estimations with hitherto unachievable accuracy, paving the way for the application of QM techniques to non-covalent interactions in entities as large as protein-ligand complexes and structural groups. With this technique, the high accuracy of quantum mechanics (QM) and high-volume calculations (VS) in drug discovery are reconciled.

Models of QM
To understand the workings of reactions, QM simulation is a helpful tool. A recent investigation has confirmed that the protonation state at the binding site affects scoring techniques' ability to determine the native binding position. Since the creation of a covalent bond between a hydrogen atom and a heavy atom is correlated with protonation, predicting the existence of protonation states using QM should be more reliable.

An essential function of cation-π and π-π stacking interactions is in chemical and biological recognition.

QM-derived partial charges
The approach utilised to extract partial charges has a substantial impact on the computed physical parameters, which in return affects the following docking and scoring in the molecular simulations.

QM descriptors in QSAR/QSPR models
The most popular descriptors in QSAR/QSPR models are partial charges because of their informative content and ease of use. Variety of quantum chemical descriptors, such as frontier electron density and frontier energy, which are effective in characterising chemical reactivity. Molecular quantum similarity measures (MQSM) is a new QSAR descriptor. The molecular descriptor that the MQSM uses is the first-order electronic intensity function. The best molecular superposition is produced by estimated functions [77].

24.3.8.4 Applications

The crystal structure of a subtype-selective β3-hompentameric ligand-gated ion channel of the GABAA receptor was utilized for the first time to screen GABAA agonists in the Asinex library for potential anti-epileptic drugs. The co-crystallized ligand revealed that hydrogen bonds or ionic interactions with the Glu155 residue, as well as residues from the β7-β8 loop (Glu155 and Tyr157) and β9-β10 loop (Phe200 and Tyr205), are important amino acids in successful binding. Nineteen compounds that satisfied all the criteria had a single positively charged group interacting with Glu155, exhibiting good CNS drug-like characteristics.

Jaguar was used to determine properties like the electrostatic potential, interaction intensity, and HOMO and LUMO energies. The HOMO orbital's energy indicates the reactive species' ionization potential for the activity of electrophiles, whereas the LUMO orbital's energy displays the species' electron affinity for nucleophile activity. Jaguar was used to make QM calculations for hit S5. DFT was used to optimize geometry in the calculations. A many-electron system's characteristics were computed using a hybrid functional B3LYP with nonrelativistic Hamiltonian parameters. For hit S5, molecular characteristics including atomic electrostatic potential charges and surface-related characteristics like molecular orbitals and molecular density, among others, were computed. The scaffold included in the majority of the screened hits had a morpholine ring that resembles the structure of known GABAA orthosteric agonists, and a three-carbon spacer between the amino group. Furthermore, because its structure was so similar to that of GABA, it satisfied the necessary structural characteristics and desirable interactions to qualify as a GABAA agonist. Small compounds have the potential to be very interesting candidates for future logical improvement and development as possible antiepileptic drugs. Scaffold-based therapy regimens may benefit from the usage of 4-(3-(methylammonium)propyl)morpholin-4-ium scaffold. Novelty analysis demonstrated that these substances were not previously documented as possible anti-epileptic drugs [65].

In the battle against obesity, cancer, kidney failure, migraines, Alzheimer's disease, and epilepsy, CAIs may prove to be effective therapeutic agents. A chronic brain condition that has a significant impact on people of all ages is epilepsy. Spontaneous recurrent seizures are characterized by an abrupt change in the ionic composition, which includes pH shifts and an increase in intracellular potassium levels. According to reports, CAs II, VII, and XIV are related to epilepsy. The evaluation of thermodynamic functions includes the deprotonation enthalpy of inhibitors, standard enthalpies of formation (DHf), standard entropy of formation (DSf), and standard Gibbs free energy of formation (DGf) for six CA–inhibitor complexes. Using quantum mechanical calculations, it was examined how six inhibitors with anticonvulsant properties bind to the CA enzyme's active site. These compounds were derived from monosaccharides and have hydroxide, carboxamide, and sulphonamide characteristics. Topiramate, along with its derivatives such as sulfamide analog and sulfamate-substituted monosaccharides, is a medication used to treat epilepsy and prevent migraines. It is available for sale in all countries. The design of carbonic anhydrase inhibitors (CAIs) may benefit from these findings in a new way. The interaction between the active core of carbonic anhydrase and six inhibitors has been investigated by DFT calculations. Through their deprotonated nitrogen or oxygen atoms, all of the inhibitors are tetrahedrally coupled to the zinc ion of the enzyme within a hydrophobic pocket of the CA active center. Electrostatic coordination of NAZn is implied by this binding mechanism. The electronic dispersion of the polar group of sulfonamides should be correlated with the antiepileptic activity if this process has any significance for managing epilepsy. The findings suggest that R-SO2-NR1R2, the sulfonamide moiety, will effectively inhibit the carbonic anhydrase enzyme [78].

24.3.8.5 Limitations

Cost of Computation Calculations involving quantum mechanics are computationally costly, particularly for big molecules and intricate systems. (Difficulties with Extensive Quantum Mechanical Computations).

Speed vs. Accuracy Exchange Although they need a lot of computing power, high-level quantum mechanical techniques like density functional theory (also known as DFT) and ab initio procedures yield precise findings. Conversely, quicker empirical or semi-empirical approaches give up accuracy in favor of processing performance.

The stiff structures of molecules are usually treated by quantum mechanics, which ignores the flexibility of proteins and other biomolecules.

It is difficult to directly investigate processes like protein folding, large-scale conformational changes, and long-timescale dynamics using quantum mechanical techniques.

24.4 Future Prospects

Anti-epileptic drugs (AEDs) that use computer-aided drug design (CADD) show promise since CADD can help find and validate new therapeutic targets linked to epilepsy, which can assist researchers in comprehending the condition's fundamental mechanisms. CADD can assist in identifying and validating new treatment targets associated with epilepsy in order to gain a better understanding of the underlying mechanisms behind the condition. By eliminating the need for time-consuming and expensive traditional experimental procedures, CADD approaches enable the quick screening of enormous databases of chemical compounds to find possible candidates for AEDs.

Structure-Based Drug Design CADD can assist in the design of AEDs with enhanced specificity and efficacy by employing protein structures and molecular docking simulations. This will result in the creation of safer and more potent medications.

Prediction of Drug-Drug Interactions CADD models can foretell possible interactions between AEDs and other pharmaceuticals, which can help to reduce side effects and enhance treatment results for epileptic patients who frequently need polytherapy.

Personalized medicine Thanks to developments in machine learning and computational modeling, CADD can help create customized AEDs for each patient based on their molecular and genetic profiles. This could result in more effective and precise treatment plans.

All things considered, the incorporation of CADD methods into the drug discovery process is highly promising in terms of the creation of new and enhanced antiepileptic medications, which will eventually help patients by offering safer and more efficient treatment alternatives.

24.5 Conclusion

This book chapter's discussion of computational analysis provides insightful information about the workings of antiepileptic medications. The mechanism by which phenytoin hinders voltage-gated sodium channels to stabilize neuronal membranes and lowers the frequency of epileptic seizures has been elucidated through computational research using a variety of computational techniques, such as molecular docking, molecular dynamics simulations, and modeling of the quantitative structure-activity relationship (QSAR). The exact binding locations and interactions between carbamazepine and sodium channels have been uncovered by computational analysis, providing insight into the drug's mode of action in suppressing aberrant neuronal firing. Scholars have acquired knowledge regarding how valproic acid regulates ion channels and neurotransmitter systems, hence augmenting its antiepileptic properties. In conclusion, computational analysis is essential to improving knowledge of antiepileptic drugs' mechanisms of action. Utilizing computational methods, scientists can decipher the intricate pharmacology of AEDs, which will ultimately result in the creation of more potent epileptic medicines.

References

1. Lemke TL, Williams DA, Roche VF, William Zito S. Foye's principles of medicinal chemistry. 7th ed. Lippincott Williams & Wilkins; 2008.
2. Khan HN, Kulsoom S, Rashid H. Ligand-based pharmacophore model development for the identification of novel antiepileptic compound. Epilepsy Res. 2012;98(1):62–71.
3. Weaver DF. Principles and practice of computer-aided drug design as applied to the discovery of antiepileptic agents. In: Computational neuroscience in epilepsy. Academic Press; 2008. p. 515–XIX.
4. Sandle T. Pharmaceutical microbiology: essentials for quality assurance and quality control. Woodhead Publishing Limited, Elsevier; 2016.
5. Jhoti H, Leach AR, editors. Structure-based drug discovery. Dordrecht: Springer; 2007.
6. Hu Y, Cheng K, He L, Zhang X, Jiang B, Jiang L, Li C, Wang G, Yang Y, Liu M. NMR-based methods for protein analysis. Anal Chem. 2021;93(4):1866–79.
7. Carroni M, Saibil HR. Cryo electron microscopy to determine the structure of macromolecular complexes. Methods. 2016;95:78–85.
8. Huxford T. X-ray crystallography. In: Brenner's encyclopedia of genetics. Elsevier; 2013. p. 366–8.
9. Pason LP, Sotriffer CA. Empirical scoring functions for affinity prediction of protein-ligand complexes. Mol Inf. 2016;35(11–12):541–8.

10. Yang C, Chen EA, Zhang Y. Protein–ligand docking in the machine-learning era. Molecules. 2022;27(14):4568.
11. Hollingsworth SA, Dror RO. Molecular dynamics simulation for all. Neuron. 2018;99(6):1129–43.
12. Lüdemann SK, Lounnas V, Wade RC. How do substrates enter and products exit the buried active site of cytochrome P450cam? 1. Random expulsion molecular dynamics investigation of ligand access channels and mechanisms. J Mol Biol. 2000;303(5):797–811.
13. Kokh DB, Amaral M, Bomke J, Grädler U, Musil D, Buchstaller HP, Dreyer MK, Frech M, Lowinski M, Vallee F, Bianciotto M. Estimation of drug-target residence times by τ-random acceleration molecular dynamics simulations. J Chem Theory Comput. 2018;14(7):3859–69.
14. Xu J, Hagler A. Chemoinformatics and drug discovery. Molecules. 2002;7(8):566–600.
15. Giordano D, Biancaniello C, Argenio MA, Facchiano A. Drug design by pharmacophore and virtual screening approach. Pharmaceuticals. 2022;15(5):646.
16. Priya S, Tripathi G, Singh DB, Jain P, Kumar A. Machine learning approaches and their applications in drug discovery and design. Chem Biol Drug Des. 2022;100(1):136–53.
17. Staszak M, Staszak K, Wieszczycka K, Bajek A, Roszkowski K, Tylkowski B. Machine learning in drug design: use of artificial intelligence to explore the chemical structure–biological activity relationship. Wiley Interdiscip Rev: Comput Mol Sci. 2022;12(2):e1568.
18. Menke J, Maskri S, Koch O. Computational ion channel research: from the application of artificial intelligence to molecular dynamics simulations. Cell Physiol Biochem. 2021;55(S3):14–45.
19. Perron Q, Mirguet O, Tajmouati H, Skiredj A, Rojas A, Gohier A, Ducrot P, Bourguignon MP, Sansilvestri-Morel P, Do Huu N, Gellibert F. Deep generative models for ligand-based de novo design applied to multi-parametric optimization. J Comput Chem. 2022;43(10):692–703.
20. Alexander SP, Mathie A, Peters JA. Guide to receptors and channels (GRAC). Br J Pharmacol. 2011;164:S1–2.
21. Ciardo MG, Ferrer-Montiel A. Lipids as central modulators of sensory TRP channels. Biochim Biophys Acta (BBA)-Biomembr. 2017;1859(9):1615–28.
22. Natale AM, Deal PE, Minor DL Jr. Structural insights into the mechanisms and pharmacology of K2P potassium channels. J Mol Biol. 2021;433(17):166995.
23. de Lera Ruiz M, Kraus RL. Voltage-gated sodium channels: structure, function, pharmacology, and clinical indications. J Med Chem. 2015;58(18):7093–118.
24. Abdulfatai U, Uzairu A, Uba S. Molecular docking and quantitative structure-activity relationship study of anticonvulsant activity of aminobenzothiazole derivatives. Beni-Suef Univ J Basic Appl Sci. 2018;7(2):204–14.
25. Vithlani M, Terunuma M, Moss SJ. The dynamic modulation of GABAA receptor trafficking and its role in regulating the plasticity of inhibitory synapses. Physiol Rev. 2011;91(3):1009–22.
26. Tretter V, Jacob TC, Mukherjee J, Fritschy JM, Pangalos MN, Moss SJ. The clustering of GABAA receptor subtypes at inhibitory synapses is facilitated via the direct binding of receptor α2 subunits to gephyrin. J Neurosci. 2008;28(6):1356–65.
27. Ma D, Zhong L, Yan Z, Yao J, Zhang Y, Ye F, Huang Y, Lai D, Yang W, Hou P, Guo J. Structural mechanisms for the activation of human cardiac KCNQ1 channel by electro-mechanical coupling enhancers. Proc Natl Acad Sci. 2022;119(45):e2207067119.
28. Noviello CM, Gharpure A, Mukhtasimova N, Cabuco R, Baxter L, Borek D, Sine SM, Hibbs RE. Structure and gating mechanism of the α7 nicotinic acetylcholine receptor. Cell. 2021;184(8):2121–34.
29. Wang M, Wang Z, Sun H, Wang J, Shen C, Weng G, Chai X, Li H, Cao D, Hou T. Deep learning approaches for de novo drug design: an overview. Curr Opin Struct Biol. 2022;72:135–44.
30. Schneider G, Fechner U. Computer-based de novo design of drug-like molecules. Nat Rev Drug Discov. 2005;4(8):649–63.
31. Schneider G. Future de novo drug design. Mol Inf. 2014;33(6–7):397–402.
32. Pitman MR, Menz RI. Methods for protein homology modelling. In: Applied mycology and biotechnology. Elsevier; 2006. p. 37–59.

33. Vázquez M, Fagiolino P. The role of efflux transporters and metabolizing enzymes in brain and peripheral organs to explain drug-resistant epilepsy. Epilepsia Open. 2022;7:S47–58.
34. Chauhan A, Singh J, Sangwan N, Singh H, Prakash A, Medhi B, Avti PK. Designing the 5HT2BR structure and its modulation as a therapeutic target for repurposing approach in drug-resistant epilepsy. Epilepsy Res. 2023;194:107168.
35. Bhaduri A, Krishnaswamy L, Ullal GR, Panicker MM, Sowdhamini R. Fold prediction and comparative modeling of Bdm1: a probable α/β hydrolase associated with hot water epilepsy. J Mol Model. 2003;9:3–8.
36. Lipkind GM, Fozzard HA. Molecular model of anticonvulsant drug binding to the voltage-gated sodium channel inner pore. Mol Pharmacol. 2010;78(4):631–8.
37. Al-Obaidi A, Elmezayen AD, Yelekçi K. Homology modeling of human GABA-AT and devise some novel and potent inhibitors via computer-aided drug design techniques. J Biomol Struct Dyn. 2021;39(11):4100–10.
38. Chang WP, Südhof TC. SV2 renders primed synaptic vesicles competent for Ca2+-induced exocytosis. J Neurosci. 2009;29(4):883–97.
39. Xu T, Bajjalieh SM. SV2 modulates the size of the readily releasable pool of secretory vesicles. Nat Cell Biol. 2001;3(8):691–8.
40. Lee J, Daniels V, Sands ZA, Lebon F, Shi J, Biggin PC. Exploring the interaction of SV2A with racetams using homology modelling, molecular dynamics and site-directed mutagenesis. PLoS One. 2015;10(2):e0116589.
41. Torres PH, Sodero AC, Jofily P, Silva-Jr FP. Key topics in molecular docking for drug design. Int J Mol Sci. 2019;20(18):4574.
42. Elokely KM, Doerksen RJ. Docking challenge: protein sampling and molecular docking performance. J Chem Inf Model. 2013;53(8):1934–45.
43. Ewing TJ, Kuntz ID. Critical evaluation of search algorithms for automated molecular docking and database screening. J Comput Chem. 1997;18(9):1175–89.
44. Li J, Fu A, Zhang L. An overview of scoring functions used for protein–ligand interactions in molecular docking. Interdiscip Sci: Comput Life Sci. 2019;11:320–8.
45. Abdulfatai U, Uba S, Umar BA, Ibrahim MT. Molecular design and docking analysis of the inhibitory activities of some α_substituted acetamido-N-benzylacetamide as anticonvulsant agents. SN Appl Sci. 2019;1:1–8.
46. Pradhan A, Vishwakarma S. Synthesis of quinolone derivatives and their molecular docking for antiepileptic activity. Chem Int. 2020;6(4):224–31.
47. Siddiqui AA, Partap S, Khisal S, Yar MS, Mishra R. Synthesis, anti-convulsant activity and molecular docking study of novel thiazole pyridazinone hybrid analogues. Bioorg Chem. 2020;99:103584.
48. Knight JL, Brooks CL III. λ-Dynamics free energy simulation methods. J Comput Chem. 2009;30(11):1692–700.
49. Al Mehdi K, Fouad B, Zouhair E, Boutaina B, Yassine N, Chaimaa AE, Najat S, Hassan R, Rachida R, Abdelhamid B, Halima N. Molecular modelling and dynamics study of nsSNP in STXBP1 gene in early infantile epileptic encephalopathy disease. Biomed Res Int. 2019;2019:4872101.
50. Malik R, Mehta P, Srivastava S, Choudhary BS, Sharma M. Structure-based screening, ADMET profiling, and molecular dynamic studies on mGlu2 receptor for identification of newer antiepileptic agents. J Biomol Struct Dyn. 2017;35(16):3433–48.
51. Chen Y, Feng Z, Shen M, Lin W, Wang Y, Wang S, Li C, Wang S, Chen M, Shan W, Xie XQ. Insight into Ginkgo biloba L. extracts on the improved spatial learning and memory by Chemogenomics knowledgebase, molecular docking, molecular dynamics simulation, and bioassay validations. ACS Omega. 2020;5(5):2428–39.
52. Durrant JD, McCammon JA. Molecular dynamics simulations and drug discovery. BMC Biol. 2011;9:1–9.

53. Dudek AZ, Arodz T, Gálvez J. Computational methods in developing quantitative structure-activity relationships (QSAR): a review. Comb Chem High Throughput Screen. 2006;9(3):213–28.
54. Baron BM, Cregge RJ, Farr RA, Friedrich D, Gross RS, Harrison BL, Janowick DA, Matthews D, McCloskey TC, Meikrantz S, Nyce PL. CoMFA, synthesis, and pharmacological evaluation of (E)-3-(2-carboxy-2-arylvinyl)-4, 6-dichloro-1 H-indole-2-carboxylic acids: 3-[2-(3-aminophenyl)-2-carboxyvinyl]-4, 6-dichloro-1 H-indole-2-carboxylic acid, a potent selective glycine-site NMDA receptor antagonist. J Med Chem. 2005;48(4):995–1018.
55. Abdulfatai U, Uzairu A, Uba S. Quantitative structure-activity relationship and molecular docking studies of a series of quinazolinonyl analogues as inhibitors of gamma amino butyric acid aminotransferase. J Adv Res. 2017;8(1):33–43.
56. Peretz A, Pell L, Gofman Y, Haitin Y, Shamgar L, Patrich E, Kornilov P, Gourgy-Hacohen O, Ben-Tal N, Attali B. Targeting the voltage sensor of Kv7. 2 voltage-gated K+ channels with a new gating-modifier. Proc Natl Acad Sci. 2010;107(35):15637–42.
57. Malik R, Mehta P, Srivastava S, Choudhary BS, Sharma M. Pharmacophore modeling, 3D-QSAR, and in silico ADME prediction of N-pyridyl and pyrimidine benzamides as potent antiepileptic agents. J Recept Signal Transduct. 2017;37(3):259–66.
58. McInnes C. Virtual screening strategies in drug discovery. Curr Opin Chem Biol. 2007;11(5):494–502.
59. Reddy AS, Pati SP, Kumar PP, Pradeep HN, Sastry GN. Virtual screening in drug discovery-a computational perspective. Curr Protein Pept Sci. 2007;8(4):329–51.
60. Lionta E, Spyrou G, Vassilatis DK, Cournia Z. Structure-based virtual screening for drug discovery: principles, applications and recent advances. Curr Top Med Chem. 2014;14(16):1923–38.
61. Vlachakis D, Tsagrasoulis D, Megalooikonomou V, Kossida S. Introducing Drugster: a comprehensive and fully integrated drug design, lead and structure optimization toolkit. Bioinformatics. 2013;29(1):126–8.
62. Andricopulo AD, Guido RV, Oliva G. Virtual screening and its integration with modern drug design technologies. Curr Med Chem. 2008;15(1):37–46.
63. Ravindranathan K, Tirado-Rives J, Jorgensen WL, Guimarães CR. Improving MM-GB/SA scoring through the application of the variable dielectric model. J Chem Theory Comput. 2011;7(12):3859–65.
64. Llanos MA, Enrique N, Sbaraglini ML, Garofalo FM, Talevi A, Gavernet L, Martin P. Structure-based virtual screening identifies novobiocin, montelukast, and cinnarizine as TRPV1 modulators with anticonvulsant activity in vivo. J Chem Inf Model. 2022;62(12):3008–22.
65. Mehta P, Srivastava S, Sharma M, Singh I, Malik R. Identification of chemically diverse GABAA agonists as potential anti-epileptic agents using structure-guided virtual screening, ADMET, quantum mechanics and clinical validation through off-target analysis. Int J Biol Macromol. 2018;119:1113–28.
66. Mehta P, Srivastava S, Sharma M, Malik R. Discovery of novel chemotypes for competitive AMPA receptor antagonists as potential antiepileptic agents through structure-based virtual screening of natural products library. Struct Chem. 2019;30:1159–72.
67. Wermuth CG, Ganellin CR, Lindberg P, Mitscher LA. Glossary of terms used in medicinal chemistry (IUPAC Recommendations 1998). Pure Appl Chem. 1998;70(5):1129–43.
68. Gao Q, Yang L, Zhu Y. Pharmacophore based drug design approach as a practical process in drug discovery. Curr Comput Aided Drug Des. 2010;6(1):37–49.
69. Kastenholz MA, Pastor M, Cruciani G, Haaksma EE, Fox T. GRID/CPCA: a new computational tool to design selective ligands. J Med Chem. 2000;43(16):3033–44.
70. Walters WP, Stahl MT, Murcko MA. Virtual screening—an overview. Drug Discov Today. 1998;3(4):160–78.
71. Evans DA, Doman TN, Thorner DA, Bodkin MJ. 3D QSAR methods: phase and catalyst compared. J Chem Inf Model. 2007;47(3):1248–57.
72. Sun H. Pharmacophore-based virtual screening. Curr Med Chem. 2008;15(10):1018–24.

73. Khedkar SA, Malde AK, Coutinho EC, Srivastava S. Pharmacophore modeling in drug discovery and development: an overview. Med Chem. 2007;3(2):187–97.
74. De Luca L, Gitto R, Barreca ML, Caruso R, Quartarone S, Citraro R, De Sarro G, Chimirri A. 3D pharmacophore models for 1, 2, 3, 4-tetrahydroisoquinoline derivatives acting as anticonvulsant agents. Arch Pharm: Int J Pharm Med Chem. 2006;339(7):388–400.
75. Pandeya SN. Semicarbazone–a versatile therapeutic pharmacophore for fragment based anticonvulsant drug design. Acta Pharma. 2012;62(3):263–86.
76. Alshehri FF. Integrated virtual screening, molecular modeling and machine learning approaches revealed potential natural inhibitors for epilepsy. Saudi Pharm J. 2023;31(12):101835.
77. Zhou T, Huang D, Caflisch A. Quantum mechanical methods for drug design. Curr Top Med Chem. 2010;10(1):33–45.
78. Ghiasi M, Kamalinahad S, Arabieh M, Zahedi M. Carbonic anhydrase inhibitors: a quantum mechanical study of the interaction between some antiepileptic drugs with active center of carbonic anhydrase enzyme. Comput Theor Chem. 2012;992:59–69.

Chapter 25
Rational Approaches and Designing Strategies for Antihypertensive Agents

Gargi Nikhil Vaidya

Contents

Abstract While the existing antihypertensive drugs are effective at lowering BP, they have failed to be promising on prolong use due to unfavourable side effects and off target specificity. With the steadily rising incidence of hypertension and its related comorbid complications over the globe, researchers have recently moved its emphasis toward studying new targets and identifying novel anti-hypertensive agents with safe and effective therapeutic profile to fulfil this unmet demand. In order to fill this gap, rationalistic drug development for fast tracking lead molecules

G. N. Vaidya (✉)
Shobhaben Pratapbhai Patel School of Pharmacy & Technology Management, SVKM's NMIMS, Vile Parle West, Mumbai, India

S. N. K. Rao G., R. R. Alavala (eds.), *Applications of Computational Tools in Drug Design and Development*, https://doi.org/10.1007/978-981-96-4154-3_25

has become a complementary strategy. It is assisted by advancement of computer aided drug designing which includes homology modelling, docking based virtual screening, QM-MD stimulations, pharmacophore modelling, qQSAR and ADMET prediction. Recently, the use of pharmacogenomic approach through various bioinformatic tools and integration of ML algorithms have redefined the responses of the antihypertensive drug at the genetic levels. Herein, various *in-silico* based investigations have been highlighted covering various aspects of drug designing of novel antihypertensive agents which can provide an easy access to drug standardisation in the future for molecular and clinical management of hypertension.

Keywords Blood pressure · Antihypertensive drugs · *In silico* · Drug designing · Molecular docking · MD stimulations · Renin-Angiotensin-Aldosterone System · Calcium Channel Blockers

25.1 Introduction

Cardiovascular diseases (CVDs) are group of crosslinked disorders and arising complications from the disease concerning cardiac muscle functioning in the heart and overall vascular circulation which is accountable for supplying blood to heart, brain and vital organs [1]. It is the prime most caused of global deaths annually and contributes to 30% of the world population at present. High arterial blood pressure, commonly known as hypertension is a condition of persistent elevated pressure on the arterial blood vessels to circulate the blood throughout the body. Although the presence of hypertension itself does not cause immediate death, under long-term untreated medical state, may become lethal risk factor for developing CVDs like angina, Coronary Arterial Diseases (CAD), stroke and heart failure (HF) [2]. Complication of hypertension is shown in Fig. 25.1.

Typically, hypertension is predominantly classified as essential hypertension (primary or idiopathic hypertension) contributing with 90–95% of adult hypertensive patients. The cause is mostly unknown (idiopathic) and are allied with inheritor genetic predisposition known as "hypertensinogenic factors" along with the impact of lifestyle habits and environmental factors [3]. The remaining 5–10% of the cases are categorized as non-essential or secondary hypertension. The cause is mostly predetermined, generally identifiable and are mostly associated with secondary complication occurring due to dysfunction of kidneys to maintain body fluid and ion balance under chronic kidney disease. Apart from this, metabolic dysfunction via endocrine disorder can also cause incongruity in the regulation of BP [4].

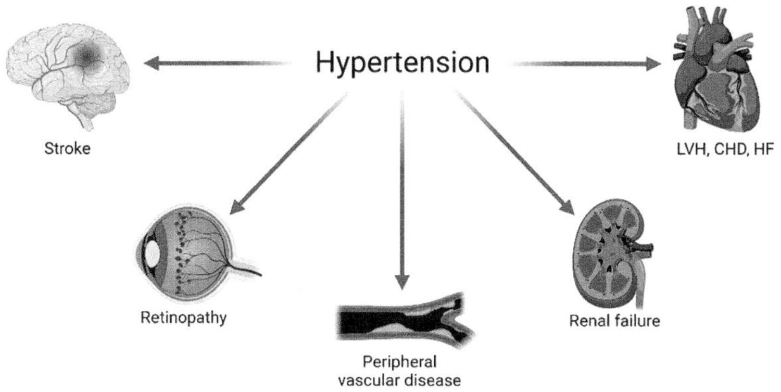

Fig. 25.1 Complication of hypertension

25.1.1 Epidemiology

WHO estimated about 1.38 billion people (31.1% of the adult's globe population) suffered from hypertension [5–7]. With the latest WHO 2023 report [8], it has estimated that around 76 million deaths in between the year of 2023 and 2050 can be deterred through early diagnosis and treatment intervention as nearly half of the patient population remains unaware of their pathophysiology conditions. Further, the report suggests an unequal impact on emerging hypertension burden in globe with nearly 144% increase in under-developed and developing countries [9]. Additionally, incidence of hypertension in the age group (years) of 30–79 indicated men being more vulnerable than women [10].

25.1.2 Aetiology

Majority of essential hypertension cases studied so far had idiopathic course outcome. Mostly been attributed by a combination of heterogeneous lifestyle induced risk factors and underlying a multifactorial pathophysiological relationship that regulates blood pressure. For example, consuming more salt raises the risk of hypertension and is also based on the genetic susceptibility towards salt sensitivity response in a given individual which might be hereditary based on prolong mutations in genes controlling Renin-Angiotensin-Aldosterone System (RAAS) [11–13].

25.1.3 Classification of Various Stages in Hypertension

There are various classification guidelines across the globe [14]. The two most commonly accepted are provided by American College of Cardiology (ACC) [15] and European Society of Cardiology (ESC) [16]. Recently, International Society of Hypertension (ISH) gave a generalised classification based on the location preference and dwelled into treatment [17]. Different stages of hypertension are given in Table 25.1.

25.1.4 Conventional Targets in Pathophysiology of Hypertension

25.1.4.1 RAAS

Renin commonly called as angiotensinogenase is the foremost proteolytic enzyme in RAAS mediated regulation of hypertension. Renin in its primitive form prorenin is synthesised in the juxtaglomerular kidney cells. The formation of active renin is achieved through various non-/proteolytic activation which allows the discharge of

Table 25.1 Different stages of hypertension

	Systolic BP (mm Hg)	Diastolic BP (mm Hg)
American College of Cardiology (ACC) of 2017 [15]		
Normal	<120	<80
Elevated	120–129	<80
Stage 1	130–139	80–89
Stage 2	≥140	≥90
HTN crisis	>180	>120
European Society of Cardiology (ESC)/Hypertension (ESH) of 2018 [16]		
Optimal	<120	<80
Normal	120–129	80–84
High normal	130–139	85–89
Grade 1 hypertension	140–159	90–99
Grade 2 hypertension	160–179	100–109
Grade 3 hypertension	≥180	≥110
Isolated systolic hypertension	≥140	<90 (further classified into Grades as per the above ranges of SBP)
International Society of Hypertension [18]		
Office	140/90	
Home	135/85	
24-hour ambulatory monitoring	130/80	

enzyme into blood in response to hypotension or hypernatrinemia conditions [19]. Coming under the broad umbrella of aspartyl proteases superfamily, the conserve site and mode of action are retained to a great extent. The synthesis of Angiotensin-I (Ang I) occurs by the action of renin in a rate-determining step. The presence of hollow deep tunnel in the enzyme cavity holds seven amino acid residues of the substrate Angiotensinogen and allows the Leu10-Val11 peptide bond from the *N*-terminal side of the substrate to be cleaved off to release the decapeptide Ang I. The rate limiting protease role of renin in the formation of Ang I became well established due to its unique specificity to angiotensinogen and footholds a potential target to outsource the effect of RAAS from its source itself for developing novel anti-hypertensive agents [20].

Followed by this, the Angiotensin-I Converting Enzyme (ACE) then enters into the picture. ACE is membrane bound M2 type, Cl^- ion dependent metalloprotein enzyme having zinc (Zn^{+2}) metal ion in its catalytic active site referred as Zinc-binding motif (HEXXH). Through its dipeptidase action, ACE's enzyme action degrades a number of vasoactive peptides, including bradykinin and Ang I. This concurrent interlinked cross talk of RAAS with the kallikrein–kinin system provides a delicate balance across the vasoconstrictive and salt-retentive property of Angiotensin II (Ang II) *versus* the vasodilatory, natriuretic role of bradykinin and can be easily disturbed by a slight increase in ACE activity. ACE breaks the Phe8-His9 peptide bond in the *C*-terminal of Ang I, resulting in the formation of octamer peptide chain of Ang II which is octameric peptide [21]. Formation of Ang II is assorted as the final stage of RAAS System which then binds to Angiotensin II Type-1 (AT_1) receptor primarily resulting in systemic vascular constriction. AT_1R receptors are GPCR located in different parts of the body like kidney, heart, adrenal cortex, lung and blood vessels. It promotes vasopressin synthesis through central nervous system (CNS), enhances aldosterone and catecholamine secretion. It promotes Na^+ and water reabsorption in kidney which in turn increased blood volume thus worsening arterial pressure. Thus, blocking both the conversion of Ang-I to Ang-II by ACE and inhibiting the binding of Ang-II on the AT_1R are some of the conventional targets explored for development of antihypertensive agents. Secondary targets engaged due to Ang II activation are now considered in management of hypertension [22]. These targets provide a scope of designing novel antihypertensive agents which is shown in Fig. 25.2.

25.1.4.2 Dysfunctioning of Endothelial Mediated Vasodilation

The endothelium cell layer in the blood vessels is responsible for the secretion of widespread range of active chemicals regulating the blood flow within the lumen via maintenance of the vascular tone. Directive of BP are primarily control by the key endothelial regulators namely nitric oxide (NO) and endothelin. An imbalance between these chemicals results in inappropriate changes in the endothelial cells thereby influencing the volume and BP in the circulation along with several developments in the nervous and immune systems. Excessive vasoconstrictions due to

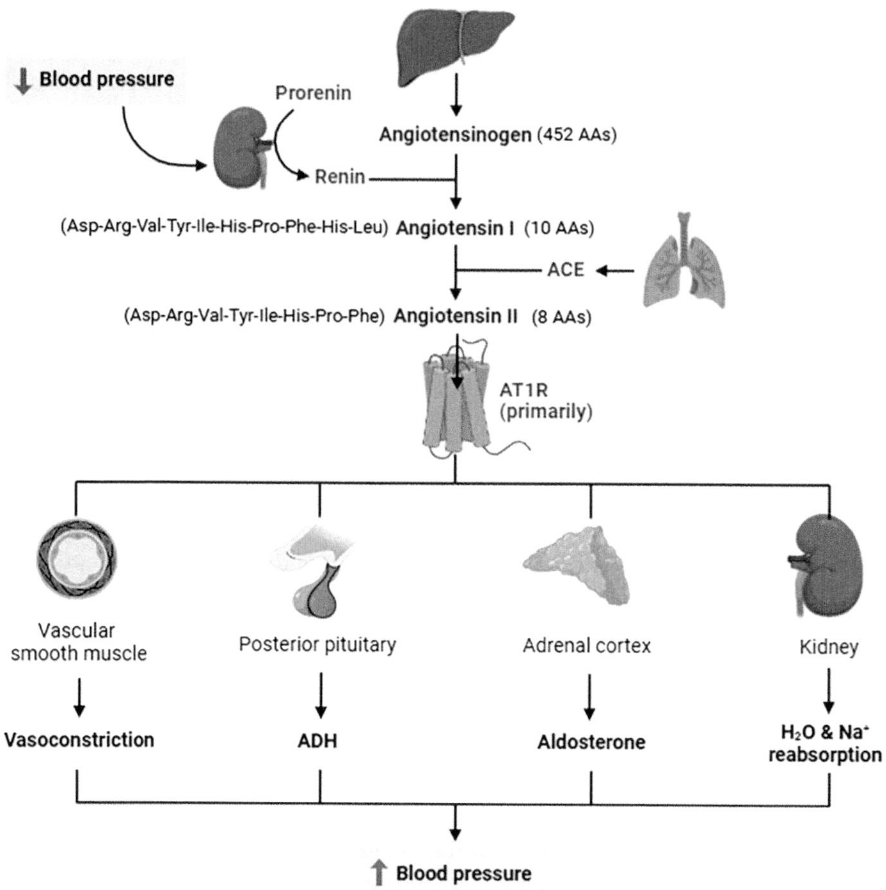

Fig. 25.2 RAAS mediated pathophysiology of hypertension

endothelin release or recessive vasodilation induced due to diminished NO signalling are associated to dysfunction of endothelial mediated hypertension pathophysiology. A major source of the endothelial homeostasis disruption is backed by extreme oxidative damage occurring due to presence of reactive oxygen species (ROS) like superoxide which neutralises and inactivates NO. Ang II may also upsurge the production of superoxide. In addition to vascular tone and pressure, fibrinolytic and coagulation interlinked pathways are also activated triggering hypertension related complications if not regulated. Changes in endothelial function henceforth precisely predict the prognosis and damage to target organs in hypertension and associated complications like heart failure, atherosclerosis [23]. Thus, restoring NO induced vasodilation can promote substantial reversing of vascular toning of arterial blood vessels. Many NO donor compounds, drugs mimicking diverse range of biological functions of NO have surfaced as possible therapeutic agents for the aforementioned diseases. Therefore, research on the pharmacological aspects of NO and its bioavailability is ongoing in the treatment of hypertension [24].

25.1.5 Anti-hypertensive Agents

The class of medications known as antihypertensives are used to alleviate the BP among the patients. The classical antihypertensive agents come in a variety of categories based on the mechanism prevalent under the discussed pathophysiology via which they reduce the BP and their chemical origin. Some of the most significant and commonly used drugs date backs to early 80s and are constantly being upgraded in terms to enhance their pharmacological profile. Based on the symptoms, stages and underlying pre-existing medical conditions, doctors often prescribe these drugs either in combination or alone for the management of hypertension [25]. These are highlighted in Table 25.2.

25.2 Recent Advances in Anti-hypertensive Drug Agents

25.2.1 Renin Inhibitors

The availability of renin inhibitors hereby mentioned as Direct Renin Inhibitors (DRIs) were mostly left unexplored till late 90s although having prominent understanding of renin's role in RAAS [27]. Interestingly, the first inhibitor pepstatin which comes under the category of peptide-based first generation DRIs was introduced in the year 1972 [28]. Though a lot of potential synthetic peptide analogues were introduced, lack of specificity, low oral bioavailability delayed onset of hypotensive actions made them less suitable. Later on, functional modification of renin which is having 14 AAs chain resulted into second generation peptidomimetic DRIs which either mimic the peptide nature or were fragments of its substrate i.e. angiotensinogen. It was only in the late 1980s, when the crystal structure of renin was reported, significant advancement in drug designing of DRIs took place under the plethora of various computational tools [29] and led to third generation DRIs which are chemically non-peptide in nature and overcome the limitations of existing generation DRIs which are shown in Fig. 25.3.

Majority of third generation DRIs originally synthesised, were based on 3,4-disubstituted piperidine ring system which showed significant renin inhibition activity [30]. With an aim to create synthetic methodologies for druggable moiety merged with notable speed and intrigue, cutting-edge drug design techniques like molecular modelling, X-ray crystallographic studies, fragment-based drug design, and structure-based drug design were applied to study renin enzyme. These tools were effectively used to locate ligand binding sites inside the renin in order to

Table 25.2 Classical antihypertensive drugs: classification, representative examples, mechanism of action and adverse effect [26]

Class	Subtype	Example	Mode of action	Side/Adverse effects
Diuretics	Loop/thiazide/ potassium-sparing	Furisemide, Chlorthalidone, Amiloride	Promotes diuresis	Hyponatremia, hypokalaemia (hyperkalaemia in potassium sparing diuretics), hyperuricemia, ototoxicity
Calcium Channel Blocker (CCBs)	(Non)/1,4-dihydropyridine	Amlodipine, Nifedipine, Verapamil, Ditiazem	Blocks the entry of Ca^{+2} ions across the receptor	DHP: Tachycardia, headche/dizziness, peripheral edema, Non DHP: Bradycardia, AV block, constipation, gingival hyperplasia
ACE inhibitors		Captopril, Enalapril	Inhibit the activity of ACE	Persistent dry irritating cough, angioedema, hyperkalaemia, rare cases: Skin itching, dysgeusia, neutropenia
Ang-II receptor antagonist		Azilsartan, Candesartan	Prevents the activation of angiotensin receptor	Headache, dizziness, fatigue, myalgia, paresthesia, rare case may include leukopenia, agranulocytosis, anemia, neutropenia
Andrenergic receptor blocker	$\beta/\alpha_{1/2}$-blockers	Atenolol, Clonidine, Prazosin	Antagonises the effect of binding of endogenous catecholamines	Bronchospasm, bradycardia, peripheral vasospasm (non selective blockers)
Vasodilators		Sodium nitroprusside	Relaxes smooth muscle cells, decreasing vascular resistance	Dizziness, hypotension

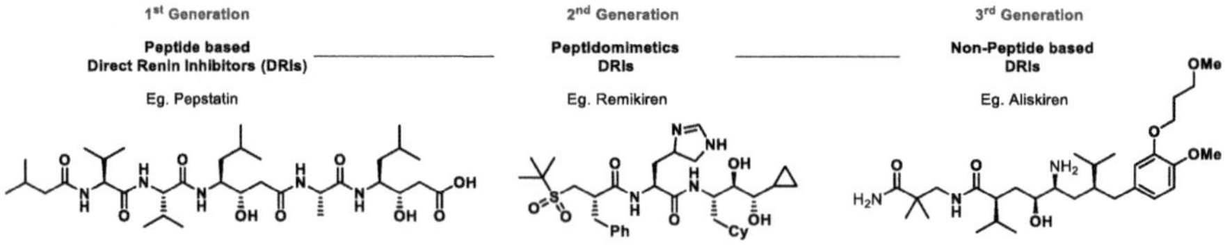

Fig. 25.3 Direct renin inhibitors (DRIs)

concentrate on building synthetic diversity space at the ligand level. The active site of renin, constituted three main hydrophobic sites S1, S2 and S3 with the presence of an extended subpocket (sp) at S3 site which played a critical role in enzyme activity. The presence of large pocket commonly referred to as the "flap region" or the substrate specific binding region spanning from the renin residues Thr72 to Ser81 gave an open conformation at the binding site [31]. By taking advantage of this hitherto unknown binding site, the inhibitors' ability to bind vividly and selectively was enhanced. Thus, accessibility of these pockets for H-bonding became the backbone in designing specific renin inhibitors. Inspired by this, Novartis was successfully able to designed and synthesised the first non-peptidomimetics drug Aliskerin [32]. Chemically referred as (2S,4S,5S,7S)-5-amino-N-(2-carbamoyl-2-methylpropyl)-4-hydroxy-2-isopropyl-7-[4-methoxy-3-(3-methoxypropoxy)benzyl]-8-methylnonanamide, was approved by USFDA as the only prototype of DRI and commercialised under the brand name Tekturna and Rasilez. Later on, many research group became actively involved in designing novel DRIs under the plethora of aliskerin based analogues. However, rationalistic drug designing showcased significant renin inhibition but failed to surpass the aliskerin as a drug candidate under various clinical trials. In a similar approach of Structure Based Drug Designing (SBDD), Xia and Iijima were able to develop a novel series of 2-Carbamoyl morpholine analogues by integrating the key interaction and binding pose for enhanced renin inhibitory activity. Although the designing was inspired by taking lead piperidyl based DRIs, the focus was to developed better and more potent alternatives to Aliskerin. Mainly by introducing the occupancy of *N*-substituted indole ring in S3 and S3sp pocket retained the Try14 and Gly217 interactions. The protonated morpholine ring, allowed the preservation of dynamic H-bonding of Ser 76, Asp32 and Asp 315. Moreover, the designed molecule showed the additional H-bond on the oxygen atom of the carbamoyl to the Thr77 despite binding in the closed form rather than the opened flap region and not occupying the adjacent S1′–S2′ sites in the active site of the renin enzyme. This was in sharp contrast to most of the developed piperidine derivatives which binded in open flap state. It is because of these optimisations, the compound when administered orally to double transgenic mice bearing both the human renin and the human angiotensinogen genes, established a noteworthy antihypertensive effectiveness [33]. Recently the same group, introduced molecule with enhanced pharmacokinetic profile without causing drastic fluctuation in its molecular weight simply by replacing the indole ring with 6-methyl-1H-pyrazolo[3,4-b] pyridine ring (compound SPH3127). The compound has already finished phase II clinical trial for essential hypertension and has demonstrated a greater bioavailability and stronger profile than aliskiren in preclinical models [34]. Key interactions for novel non-peptide based DRIs is shown in Fig. 25.4.

25.2.2 ACE Inhibitors

ACE inhibition is centrally most exploited target under RAAS for not only renal and cardiovascular remodelling but also in the management of hypertension *via* ACE inhibitors. They act as a first-line of treatment in regulating BP and fluid-electrolyte

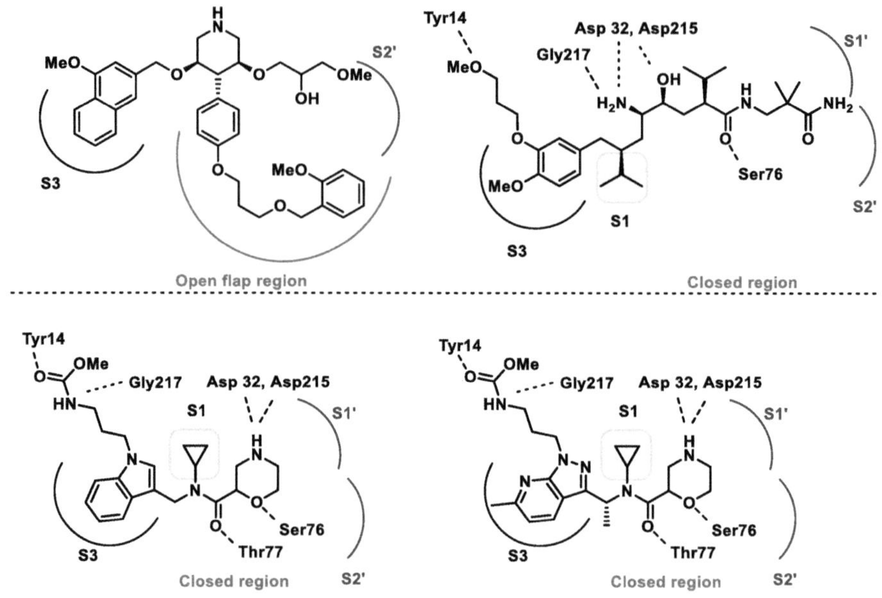

Fig. 25.4 Key interactions for novel non-peptide based DRIs

balance. Since its inception and despite their prodigious triumph, most of the ACE inhibitors were not developed with the knowledge of the enzyme's sequence or three-dimensional structure. Rather, it was the amalgamation of vivid insights aligned with serendipitous discovery. It was only after the recent advances made in Structure Based Drug Designing through various computation tools and extensive study at biomolecular level, that new insights on domain complexity of ACE enzyme came into existence [35]. ACE is a bi-domain enzyme having N(nACE) and C(cACE) site. Despite being homologous in nature, functionally it exhibits different profiles and substrate specificity. Most of the designed synthetic ACE inhibitors act as peptidomimetics and mimic the natural ligand. They are classified into different generations as shown below:

25.2.2.1 First Generation ACE Inhibitors

The prototype in this generation is the captopril with sulphydryl as zinc chelating group showcasing an enzyme inhibition of IC_{50} of 23 nM. Although, the story of making captopril itself is extensive, but the side effects associated with it provided the inspiration to designed better variants following detail analysis of its binding mode. Most of the inhibitors in this category are mixed type i.e. having an overall effect from the contribution of both the domains are responsible for the adverse side effects. Inspired by this various variant of captopril were taken into account. The non-sulphydryl replacement analogues comprising of enalaprilat and lisinopril

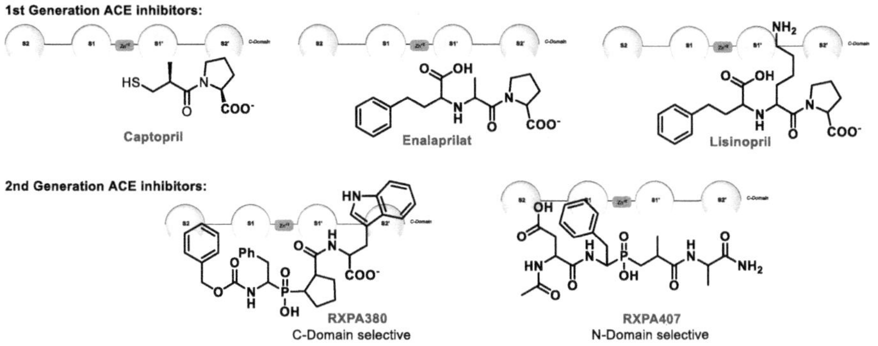

Fig. 25.5 ACE inhibitors

utilised the additional S1 binding sites *via* a phenyl ring which originally remained vacant in captopril. In essence, both are tripeptide analogues, where the substrate scissile amide carbonyl is replaced with a Zn^{+2} chelating -COOH group. Interestingly, lisinopril showcase 2-folds substantial binding affinity mainly due to the extensive interactions in the active site. This includes the occupation of lysine side chain with the deeply fitted S1′ pocket, having a fragile H-Bond between Glu162 and ε-NH_2 near to the *C*-terminal [36, 37]. The ACE inhibitors are shown in Fig. 25.5.

25.2.2.2 Second Generation ACE Inhibitors

Lack of domain affinity is the key reason for the "*off target*" side effects commonly observed in the previous generation of ACE inhibitors. It is assumed that the specific interaction due to explicit contact of ligand in the S2 subpocket is crucial in the designing of domain-selective inhibitors. Thus, the occupancy in all the subpockets particularly in S2, became a prime requisite. Considering the prominent role of *C*-domain in the regulation of BP, the current approach looks into the designing of novel *C*-domain selective ACE inhibitors which can compensate the adverse effects of bradykinin induced vasodilatory action with modified safety and efficacy profiles [36]. The keto-ACE inhibitor, various bradykinin potentiating peptides (BPPs) and RXPA380 are some prominent examples in this category. They both show uneven domain selectivity with increased cACE inhibition in the range of 40–2000 times more selective in nature. RXPA380 is tetrapeptides with phosphate zinc binding group. The presence of bulky hydrophobic framework fitted in all the pockets including the larger benzyl and methylindole in S2 and S2′ respectively arguably provided the shape fit in the binding pockets. Interestingly blocking the S2′ pocket without carboxylate ion provides necessary *N*-domain selectivity. Taking the same note, RXPA407 was designed and synthesized revealing 1000 folds *N*-domain selectivity. Phosphate group coordinate with centrally placed Zn^{+2} ions. The -COOH in S2 subunit interaction with Arg381 and Tyr369 was also found to be crucial [35].

25.2.2.3 Natural Product-Based ACE Inhibitors

Recently, various chemical libraries have been screened to identify potential ACE inhibitors exhibiting the concurrent existing features of clinically proven drugs. In this regards, use of bio-peptides, natural scaffolds and phytoconstituents in non-peptide category have become a viable alternative through pharmacophore virtual screening. Various ACE inhibitors have been identified from plant biometabolite which includes certain terpenoids, complex extract, peptides and phenolic compounds which act via ionic interaction with Zn^{+2} ion, have H-Bonding and hydrophobic interactions with the enzyme residues. Under similar context compounds from animal products and marine organisms have also shown promising ACE inhibitory activity. Docking studies have revealed it to bind at nonactive site of ACE providing an allosteric mode of inhibition [37].

Apart from the conventional *in silico* tools, MLR and PLS QSAR models have been created and explored for assessing the antihypertensive efficacy of novel enalapril based ACE inhibitors in combination with docking studies followed by ADMET screening using softwares like PreADMET to understand their toxicity profile. These analogues are designed based on the existential knowledge-based descriptors of known ACE inhibitors like enalapril [38]. QSAR MLR-PLS model for designing ACE inhibitors are shown in Fig. 25.6.

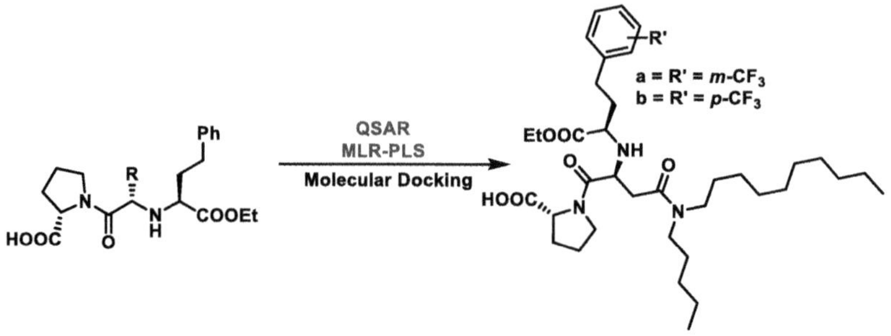

		Enalapril	a	b
MLR QSAR Model	**Predicted IC_{50} nM (MLR)**	1.2 (1.2)[a]	0.009	0.009
Molecular Docking	**Intermolecular Energy (kcal/mol)**	−7.50	−8.90	−9.30
Toxicity PreADME	**Algae_at**	0.0263535	0.0002601	0.0005878
	Carcino_Rat	Positive	Positive	Negative
	Carcino_Mouse	Negative	Negative	Negative
	Daphnia_at	0.132963	0.000911	0.0023671
	Ames_test	Non-mutagenic	Non-mutagenic	Non-mutagenic

Fig. 25.6 QSAR MLR-PLS model for designing ACE inhibitors [38]

25.2.3 Thermolysin Inhibitors

Coming under the illusive family of bacterial proteolytic enzymes, thermolysin is M4 type metalloendopeptidase enzyme, holding both functional utility and structural attributes similar to neutral endopeptidase (NEP) and ACE, which are well-established clinical drug targets for treating cardiovascular diseases in particular hypertension [39]. With more than 25 X-ray crystal complexes of thermolysin with different ligands being available in PDB database provided basic understanding of the catalytic mechanism and its close resemblance for designing novel inhibitors. Thermolysin holds an active Zn^{+2} metal ion at its active site which is in tetrahedral coordination with water molecule and amino acids mainly Glu166 of the Glu-(Xaa)3-Asp motif, His146 of HEXXH motif, His142 and a water molecule which is usually replaced by competitive or irreversible inhibitor, while its thermal stability is being subjected to the presence of four calcium ions [40]. Active site of thermolysin with catalytic mechanism is shown in Fig. 25.7.

The catalytic functionality of thermolysin is based on proposed mechanisms that during catalysis, the family of enzymes experiences hinge-bending action. There are two possible modes of such action: the first involves the proton acceptor role of glutamate of the HEXXH motif during catalysis, while the another involves the glutamate being replaced by the histidine containing active-site which enacts the primitive role of the base. Conferring to the first proposed mechanism, a water

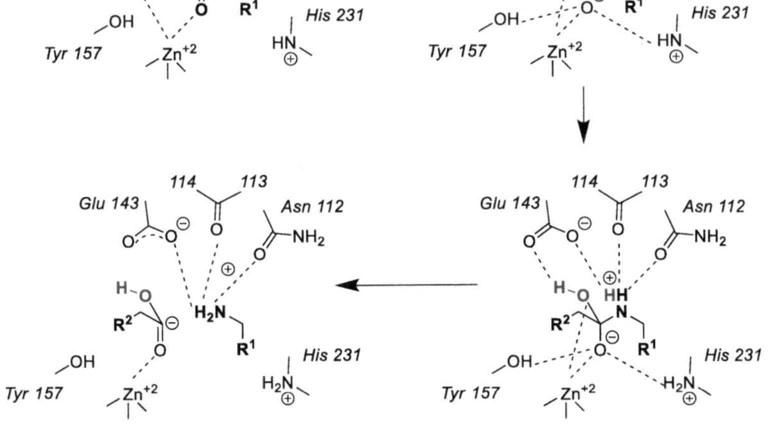

Fig. 25.7 Active site of thermolysin with catalytic mechanism [40]

molecule or hydroxide ion attacks the carbonyl carbon of the scissile bond, initiating the catalytic cleavage of peptide linkage via a base mediated hydrolysis. It is assumed that the inbound substrate drives the water molecule in the direction of Glu143 of the HEXXH motif, where the oxygen atom is still bound at the zinc ion and both of the hydrogen atoms of water molecule form a hydrogen bond with Glu143. This increases the water molecule's nucleophilicity and directs the last lone pair of water oxygen towards the substrate, which is ready for nucleophilic attack. An intermediate is thus formed when the carbonyl carbon is attacked by water. The proton that Glu143 accepts is moved to the parting nitrogen. The presence of tetrahedral nitrogen with two protonated hydrogens are stabilised by the H-bonding of Asn112 oxygen side chain oxygen and carbonyl oxygen of Ala113 backbone. The product is formed by a sequential transfer to the nitrogen that is exiting via Glu143. The electrophilic amino acids are Glu143, Asp226 and His231. Orientation of His231 imidazolium is supported by Asp226, and are proposed to be significant for catalysis in the second mechanism apart from protonium water molecule, zinc ions, and zinc coordinating amino acids. With such intrinsic knowledge about thermolysin enzyme inhibition benefits in the designing of NEP/ACE inhibitors for their antihypertensive effect [40].

In this regards, Castillo-Garit and coworkers were able to identify novel hit ligands representing promising antihypertensive drugs against thermolysin inhibition by developing a statistically relevant Multiple Linear Regression (MLR) model using the QSARINS software [41]. This model was validated in accordance with OECD guidelines proving its robustness, stability, low correlation of descriptors and good predictive power. Among various virtual databased of chemical libraries, a total of 176 thermolysin inhibitors were selected over wide range of outline and were scrutinised with 399 variables for the selection process to develop the model. In addition, it is found that the model fit is not the product of a random correlation. Although, few possible outliers were observed in the model application domain but, showcased respectable interaction and activity based on the molecular docking study. Based on this, compounds ranging from 2.56 pKi < 3.5 and pKi P 3.5 were further scrutinized as potential thermolysin inhibitors. In continuation with this, the group further analysed potential ligands through extensive docking and MM-GBSA decomposition analysis calculations followed by detailed ADMET insights through computational tools to forecast the behaviour of comprehensive diverse molecular library designed *in silico* against a thermolysin [42]. Potential hit ligands against thermolysin by Castillo-Garit and coworkers are shown in Fig. 25.8.

25.2.4 Calcium Channel Blocker (CCB)

The concentration of calcium ion (Ca^{+2}) is highly regularised across the cell through voltage gated calcium ion channels. Activation of these channels particularly the L-Type, permits the influx of Ca^{+2} ions which creates vasocontractions of vascular and cardiac smooth muscle cells, thereby increasing the BP. Further, they are also

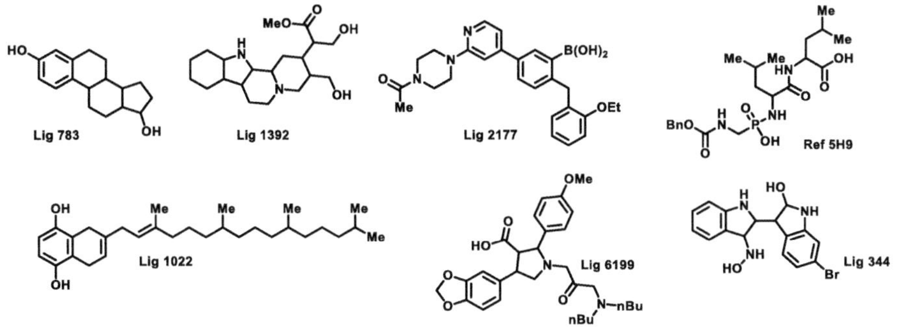

Fig. 25.8 Potential hit ligands against thermolysin by Castillo-Garit and coworkers

responsible for the regulation of secretion of hormones of adrenal cortex which are also associated in aggravating the complications of hypertension. Blockage of these channel can reverse the pharmacological effect, i.e. creates vasodilation and helps in alleviating BP. Classical examples under calcium channel blockers (CCBs) are listed in Fig. 25.9 comprising majorly of 1,4-Dihydropyridines category which is extensively used in the management of BP. They act as an antagonist through allosteric modulation. They exert their action through various allosteric interaction upon binding with CaV1.2 subtype of L-Type calcium channels under high affinity. X-ray crystallographic data of this receptor have provided a great insight in designing novel dihydropyridines composing the essential pharmacophoric features along with superior pharmacokinetic scopes [43, 44]. Representative examples of CCBs are shown in Fig. 25.9.

Chatterjee and group designed and developed a series of thiophene based prospective dihydropyridines through studying of binding interactions with 5KLS and 5KLB proteins. Molecular docking studies of these molecules particularly with -CF_3 substitution attributed for better docking score and binding affinity against nifedipine and amlodipine as standard. It also highlighted the key interactions like p–p(alkyl) stacking with His345, Leu1176, Phe13, Arg15, Cys46, Pro47 and intra and inter-hydrogen bonding with the amino acids like Val158, Arg155 and Phe158 of β-chain of the channel. Furthermore, projections from complex stability of the docked ligands with proteins through MD simulation demonstrated potential scopes of these thiophene substituted acridone molecules for pharmacological *in vitro* and *in vivo* studies [45]. Under a similar approach, Sadeghpour's group designed and evaluated thienyl based DHPs with a comportment of substitution of ethers [46].

Recently, Kumar and coworkers, using strategic computational approach designed superior dihydropyridine molecules inspired from existing CCB amlodipine. The novel designed molecules were based on the utilisation of triazole core thered (hetero)aryls as bio-isostere which attributes to mimic -CONH- linkage. It offered the essential non-covalent interactions like array of p–p stacking including Try1195 and intermolecular H-bonding by the use of fluoro containing aryl ring. This forms the crux of high binding affinity demonstrating superior docking scores

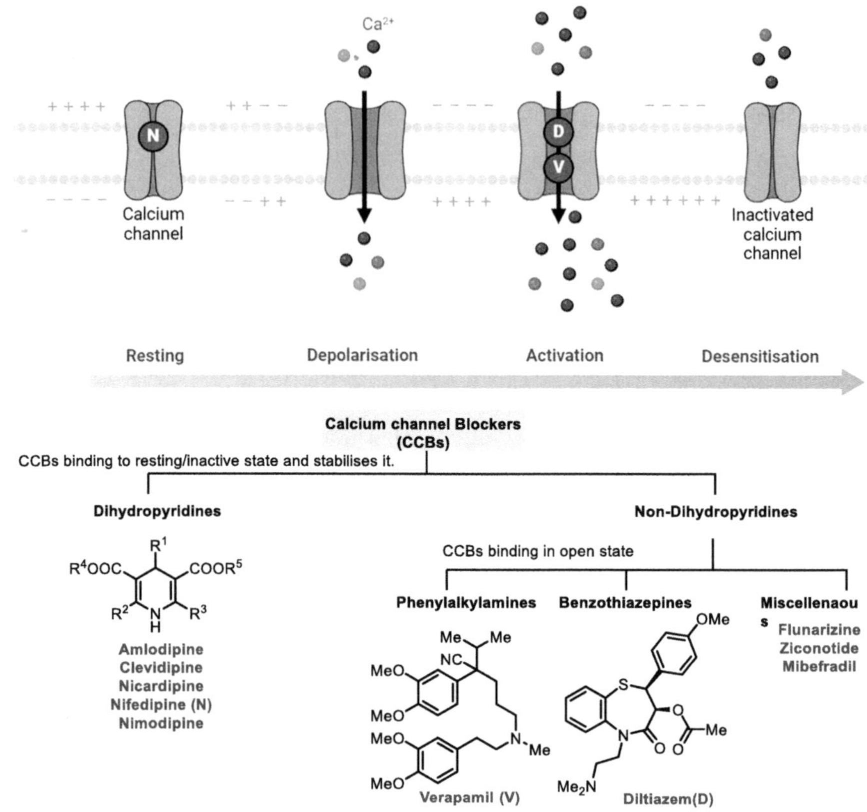

Fig. 25.9 Representative examples of CCBs

then the reference amlodipine. This was followed by a detailed SAR evaluation and ADMET pharmacokinetic assessment [47]. *In silico* based novel dihydropyridines CCBs are shown in Fig. 25.10.

25.2.5 Diuretics

These are the most often prescribed first-line of treatment for mild hypertension apart from CCBs. Different classes of diuretics act on the loop of Henle and distal tubule of the nephrons and are responsible for inhibition of ion transfers through the transporter present over it. Reversible binding of common diuretics prevents the $Na^+/K^+/2Cl^-$ cotransporter, thereby inhibiting their reabsorption and causes excess urine formation. Such a mechanism is explored by loop diuretics. Another category comprising of thiazide diuretics mainly act on Na^+/Cl^- cotransporter (NCC) in the distal tubule causing inhibition of Na^+/Cl^- reabsorption [48]. Recently various novel

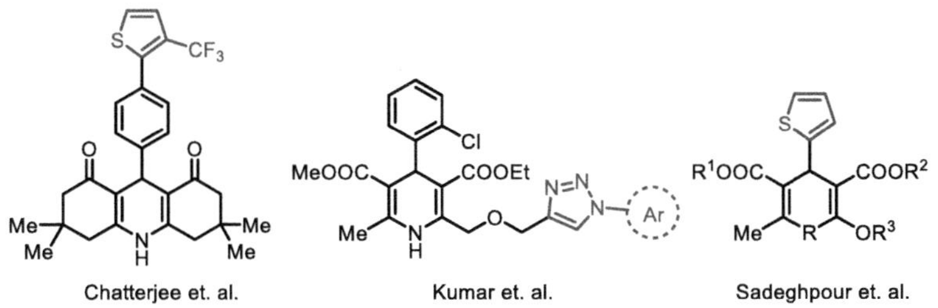

Fig. 25.10 *In silico* based novel dihydropyridines CCBs

classes of diuretics have been explored for the management of hypertension and are discussed below:

25.2.5.1 Aldosterone and Aldosterone Receptor Blocker

Being a mineralocorticoid obtained from adrenal cortex, aldosterone binds with mineralocorticoid receptor (MR, also called as aldosterone receptor) present in the renal tubules of the kidneys. This activation assists in reabsorption of Na^+ and Cl^- ions along with the water in exchange of K^+ and H^+ ions which are excreted out thereby promoting hypertension. Additionally, it increases accumulation of ROS which further execrates the situation. Thus, inhibiting the action of aldosterone by the use of Mineralocorticoid Receptor Antagonists (MRAs) has been a useful strategy in the treatment of hypertension. Classically there are two types of MRAs, (a) steroidal based and (b) non-steroidal compounds [49, 50]. Examples of MRAs as potential antihypertensive agents are shown in Fig. 25.11.

Steroidal MRAs

These are the conventional bulky steroidal framework. Spironolactone was the first steroidal MRA inspired from the progesterone structure. Later on, eplerenone was found in 1987. Both of them act as a competitive antagonist by binding at the active site of MR. However, in detailed docking studies and MD stimulation proved that both the compounds fail to achieve complete stabilisation of MR required for the binding of the corepressor. Further, the side effect due to steroidal nature, limits its long-term application as therapeutics which inspired the researchers to look for alternatives in designing non-steroidal MRAs [49, 50].

Non-steroidal MRAs

Currently, various novel small molecules are under development acting as non-steroidal MRAs (Fig. 25.11). Representative example under this is finerenone, a dihydropyridines derivative acting as L-type calcium channel antagonist. Docking studies of finerenone revealed alternate binding mode. This prevented the binding of coactivators due to altered helix12 conformation via unstable MR-ligand complex. This attribute intricates lack of agonistic based side effect, commonly observed in classical steroidal based MRAs. Another drug, esaxerenone also showcased similar

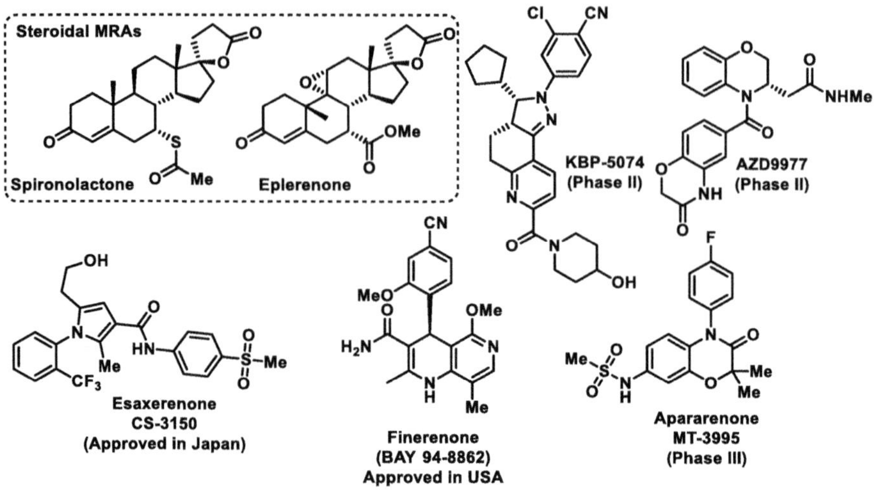

Fig. 25.11 Examples of MRAs as potential antihypertensive agents

binding pattern. Although non-steroidal MRAs have distinct advantages and superiority in terms of therapeutic profile and specificity, a detail analysis of the key interactions, binding mode and selectivity profiles of these MRAs has not been testified properly which is crucial for designing novel molecules in future in the treatment of hypertension [49, 50].

25.2.5.2 Vasopressin Receptor Antagonists

These are typical non-peptide compounds referred as vaptans which helps aquaresis by antagonising the effect of vasopressin over the receptor like V1a, V1b and V2 subtype. These receptors hold similar features as GPCR family. Activation of these receptors induces associated pathophysiology like hypertension. Vasopressin receptor crystal structures have not yet been found, although several research groups have proposed homology models, studied *in silico* MD stimulation to understand and analyse the dynamic nature of the vasopressin receptor [51–53]. As members of the class, Mozavaptan and Tolvaptan have a preserved scaffold. These antagonists have seldomly used as template to identify novel scaffolds. Such a case had been demonstrated by Hagiwara et al. They identified novel VRAs through *in silico* screening as GPCR based futuristic targets in the drug discovery field for treatment of hypertension [54].

Table 25.3 Novel diuretic-based targets for hypertension

Class	Representative examples/Structural classes
Adenosine A1 receptor antagonists	Xanthines (rolofylline, tonafylline) Non-xanthines derivatives (WRC 0571, SLV 320, FK 383)
Urea transporter inhibitors	Natural analogues: Nicotine, Sanguinaire, Indocarbonylchromenone, Methylacetamine, Dimethylthiourea Synthetic analogues: PU-48, PU168, PU468, PU474
ROMK inhibitors	Nitrophenyl, Spirocyclic, Piperazine, and Sulfonamide
WNK-SPAK inhibitors	Closantel, Rafoxanide, Stock 1S-14279
Natriuretic peptide receptor agonists	Ularitide, Moxonidine, Geniposidic acid
Pendrin inhibitors	Niflumic acid, PDS_{inh}-A01, PD_{Sinh}-C01
Aquaporin inhibitors	Acetazolamide, Phenytoin, Lamotrigine, and Topiramate

25.2.5.3 Miscellaneous Targets for Diuretics

It is worth noting that most of the existing diuretics discussed above, seldomly used for hypertension, HF, oedema and kidney diseases suffer from severe adverse effects. Thus, either through repurposing or through various computational tools novel targets have been identified along with their chemical ligands [48]. Novel diuretic-based targets for hypertension are highlighted in Table 25.3.

25.3 Future Prospects

25.3.1 Identifying Novel Targets for Hypertension

Success stories of various antihypertensive agents date back to serendipitous exploration of RAAS signalling targets responsible for the regulation of BP and have inspired researchers greatly. The current amalgamation of computational tools utilises crucial sequence information and the crystallographic 3D structural features of targets. As a result, novel antihypertensive targets are recognised and their corresponding inhibitors are rationally validated through SBDD approach. Most of the novel targets under diuretics have been recently identified and reevaluated through *in silico* studies. Further, by the use of LBDD which includes pharmacophore development and mapping, qQSAR helped designed molecules with an aim of repurposing, for e.g. SGLT2 inhibitors which were originally used as anti-diabetic drugs are now screened for their anti-hypertensive properties [55, 56]. Given below is summary of some novel targets currently being investigated for designing novel antihypertensive agents:

25.3.1.1 SGLT2 Inhibitors

Sodium glucose cotransporter-2 (SGLT2) are strategically important target to lower type 2 diabetes. SGLT2 inhibitors are responsible for stoppage of coupled reabsorption of Na^+ ions and glucose molecules from the proximal tubules of nephron and thus increasing their excretion in urine thereby reducing BP and are currently explored for its cardioprotective property which is associated with its antihypertensive effect under the comorbid type 2 diabetic patients [57–59]. Although its definite role is unclear, SGLT2 inhibitors like dapagliflozin, canagliflozin and empagliflozin in particular have shown significant systolic BP lowering along with its hypoglycaemic effect [58]. As a result, designing and synthesizing novel inhibitors might become a potential treatment approach although currently not been recommended. In similar context to find novel SGLT2 inhibitors that could be used as antihypertensive agents, Xu and colleagues recently used a systematic *in silico* approach. They did this by using docking-based screening mode to find small molecules with natural origins from a virtual dataset of Chinese herbs phytoconstituents, followed by in-depth MD simulation studies and *in silico* base ADMET predictions [60]. Under parallel strategy, Gonzales's group screened and identified the phytoconstituents from *Centella asiatica* as SGLT2 inhibitors through *in silico* virtual screening. Phytochemicals like betulinic acid, triterpenoid asiaticoside, methyl brahmate, centellasapogenol, and rutin scaffolds showed significant affinity at binding site with linked superior binding energies in assessment to reference compounds and might benefit in designing novel inhibitors in nearby future along with supplementary *in vitro* and *in vivo* proof [61]. Structures of SGLT2 inhibitors with antihypertensive effect and Phytochemicals of *Centella asiatica* as potential hits for antihypertensive activity are shown in Fig. 25.12.

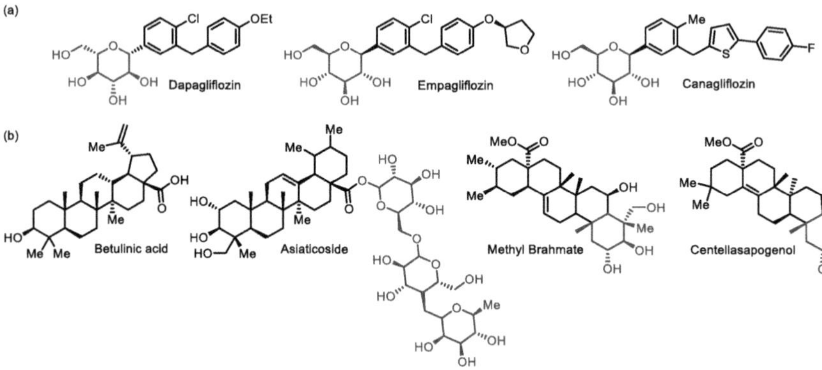

Fig. 25.12 (**a**) Structures of SGLT2 inhibitors with antihypertensive effect. (**b**) Phytochemicals of *Centella asiatica* as potential hits for antihypertensive activity

25.3.1.2 ER Inhibitors

Endothelin regulated vasculature of arteries and veins, released from the endothelial receptor is another potential target underexplored in treatment of hypertension. Predominantly classified as ER_A and ER_B, both have contrary expression and designing selective agonist or antagonist inhibitors can have a significant impact on their vasoconstrictive and vasorelaxant properties. Although at a primitive state, ER antagonist like bosetan, macitentan, aprocitentan are utilised in treatment of arterial hypertension with a more emphasis of selective receptor inhibition as ER_A inhibition has found more predominant effect on the pulmonary arterial hypertension [62]. Unfortunately, the ER_A 3D crystal structure is not present which provides a scope for computational methodologist to provide structure through homology modelling. A detail study in the binding mode of endothelin GPCR and the existing drug with their possible key interaction in the active site, can generate novel derivatives with better therapeutic efficiency and specificity [63]. Muthusamy and coworkers have done extensive QSAR and pharmacophore modelling to identify novel imidazole leads acting as strong dual ER inhibitors [62]. On a similar note, Wang and Li, also developed the 3D structure of ER_A and virtually screened TCM databased to identify natural antagonist against ER_A [64]. Patel and coworkers also screened phytoconstituent database to identify natural product in similar manner through induced fit docking [65]. Recently, Sharma et al. identified novel bosentan analogue which were designed by altering the bosetan structurally while maintaining the key interactions with Ile A:116, Val A:113, Phen A:112, Val A:106, Cys A:109, Pro A:156 either through key hydrophobic, hydrogen bonding, pi-pi interactions giving superior docking score and showcased remarkable pharmacokinetic properties in ADMET predictions [66]. ER Antagonists for arterial hypertension treatment are shown in Fig. 25.13.

25.3.1.3 Aminopeptidase (APA) Inhibitors

APA is responsible for metabolization of Ang II into angiotensin III (Ang III). Ang III shows equivalent affinity for both the receptors i.e. AT1 and AT2. It has essential peptidal role in brain RAAS signalling. Thus, inhibiting the formation of Ang III by

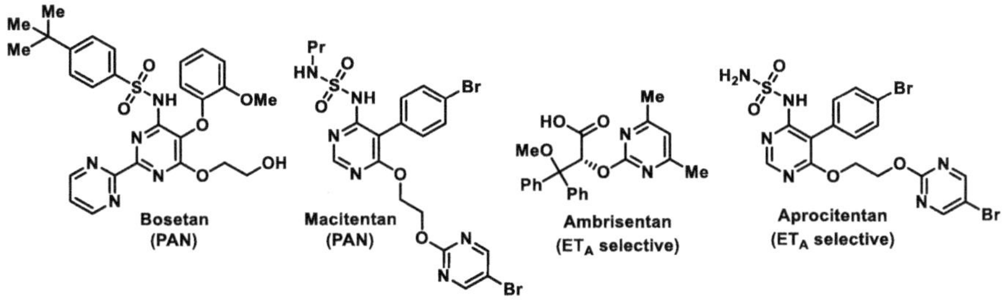

Fig. 25.13 ER Antagonists for arterial hypertension treatment

Table 25.4 Novel targets for hypertension [55]

Target	Representative example	Mode of action	Clinical status
Aminopeptidase A	Firibastat (RB150)	Inhibition	Phase1/II
Vasoactive intestinal peptide	Vasomera (PB1046)	Activation	Phase I
Na^+/H^+ exchanger 3	Tenapanor, SAR218034	Inhibition	NA
Endothelin-1	Bosentan, macitentan, aprocitentan	Inhibition	Approved
Nitric oxide	NG-nitro-L-arginine methyl ester HCl, L-arginine or L-citrulline, Pentaerythritol tetranitrate	Inhibition Activation	Preclinical
Sodium-glucose cotransporter-2 (SGLT2)	Canagliflozin, dapagliflozin	Inhibition	Approved

the use of specific and selective inhibitors might delineate the RAAS induced elevation in BP in hypertensive patients. A class of medications known as centrally acting APA inhibitors averts the making of Ang III and is an effective treatment for conditions linked to hypertension [67]. Novel targets for hypertension is mentioned in Table 25.4.

25.3.2 Dual Targeted/Multi Targeted Inhibition

Considering the cross networking of signalling pathways in the regulation of BP, use of concatenate dual inhibition or multiple target inhibition in single drug molecule seems viable. It also enhances patience compliance by bypassing polypharmacy approach [68]. Thus, multi-targeted drug designing is an emerging stratagem for developing novel dual-targeted antihypertensive agents in the future [69]. In context to this, various research group have outlined AT_1R and ET_A dual inhibitors. The use of QSAR, quantum polarised ligand docking (QPLD) were utilised to identified and bring the best screened molecule into synthetic development. Such pursuits will be beneficial and provide more clear insights in designing molecules.

25.3.3 Vaccines

Compared to oral medications, development of anti-hypertensive vaccines can provide an excellent delivery medium for a lifelong approach to lessen CVDs risks. Vaccines against components of RAAS is one of such endeavour among many of which are currently in different phases of clinical trials. At present angiotensin II vaccine AGMG0201 is in phase I/IIa for essential hypertension [70].

25.3.4 *Pharmacogenomic Use of SNPs in Hypertension*

Use of virtual genomic data base helps to understand the key hypertensive targets at primitive levels. The use of Single Nucleotide Polymorphism (SNPs) provides a key insight at the genetic level of complex diseases. Since the *in silico* studies starts primarily at the identification of critical AAs at the active site of a given target protein, the phenotypic variations can be understood by the underlying coded SNPs. The use of pharmacogenomic based computation tools helps to analysed such variations and predicts a more reliable conformer of the targets to design the novel drug molecules. Various servers like, Polyphen-2, Predict SNP have been used to study the effect of deleterious SNPs in understanding the pathophysiology of hypertensive targets including RAAS. Further in amalgamation of MD stimulation, the validation of mutagenesis through SNPs and its effect in conformation stability, ligand binding interactions can be analysed in more accurate manner. Currently this approach is in its primitive state, can provide a better and accurate direction in designing molecules in context to polymorphic proteins [69].

25.3.5 *Machine Learning (ML) Algorithms Models in Hypertension*

25.3.5.1 Assessment of Safety Profile and Adverse Reactions of Antihypertensive Drugs

Hypertensive patients are often under long term management and care. Personalised medicines are relatively new approaches for patient management especially when multiple regimes of drugs are taken under diseased and or comorbid state. Thus, it becomes primordial situation to assess the right treatment regime and associated risk prediction under the comorbid conditions. The use of ML and neural networks provides a systematic prediction of the effectiveness of the identified treatment (monotherapy vs. combination therapy) and associated risk factors by analysing significant digital health records or database available to build scientific algorithm to predict an optimized individualized medicine usage. Further, the use of "Network Learning" approach, the comparative effect of different class of antagonist can be studied and could provide crucial information regarding safety profile of the drugs when used in combination [71]. In a recent study, Zhao and coworkers carried out a systematic comparison analysis using ML generated Light GBM algorithm model to predict the effectiveness of antihypertensive agents among the elderly patients which have low control rate in BP management. A general outline has been demonstrated below. On a similar approach, many ML algorithms have been used to evaluate the safety-risk assessment of various RAAS inhibitors for their renal impairment effect. Overall, these models can help enhance the customization of hypertension care on patient-to-patient basis [72, 73]. Typical flowchart of ML algorithm model for predictions is shown in Fig. 25.14.

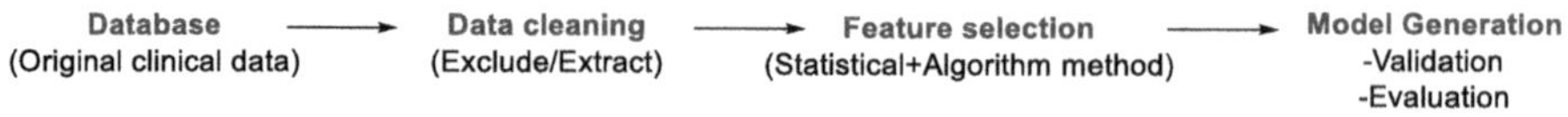

Fig. 25.14 Typical flowchart of ML algorithm model for predictions

25.3.5.2 Identification of Novel Antihypertensive Peptides

A paradigm shift has been observed in identifying antihypertensive peptides (AHTPs) from various natural and non-natural sources to target RAAS which predominantly work using peptide based natural ligands. However, identification of such therapeutically relevant bioactive peptides and experimentally synthesized them is tedious and time draining task. In this regard, *in silico* approaches eases the screening of various peptide chains using ML algorithms to identify potent AHTPs. The crucial task of such databanks is to provide accurate amino acid sequence as an input followed by systematic assessment of the conserved motif features. A comprehensive analysis of sequence predictors and their validation can be viable future and sustainable alternative to design promising AHTPs against to pharmaceutical drugs as antihypertensive agents [74, 75].

25.4 Conclusion

Hypertension, one of the leading cause of CVDs, are responsible for chief mortality of global population. With majority of hypertensive population relying on one-two conventional anti-hypertensive agents, there's a constant demand to produce new NCEs with superior therapeutic action. The journey of bringing a commercialised drug molecule in the market is long and tedious process with huge capital investment and lacks certainty to be devoid of any potential adverse effects. In this context, recent development in understanding and correlating molecular biology of the target for designing novel molecules have shown a dramatic impact on drug discovery process. The use of *in silico* tools like docking based virtual screening of compound libraries, fragment-based drug designing, MD simulations have provided a rapid automisation in particularly for identifying novel viable scaffolds by extensively studying the enzyme active site, off target binding affinity of substrates on allied target site. Such targets include RAAS signalling pathways covering DRIs, ACEIs, ARBIs which are *de facto* regulator for BP and have inspired computationalist to a great extent. ADMET predictions studied in designing target-based hit molecules provides higher success rate of acceptance when subjected to *in vitro* and preclinical assessment. Recently the use of pharmacogenomic approach through various bioinformatic tools have redefined the responses of the antihypertensive drug at the genetic levels. Further, the concept of repurposing of existing drugs for the treatment of hypertension have provided a new impetus for exploring the effect of various crosslinked targets which exists under the comorbid condition in obesity and diabetic patients. A classic example in this category is the SGLT2 inhibitors which are now been explored for its antihypertensive effect. Overall, rational drug designing by amalgamation of computational research can provide an easy access to drug standardisation in the future for clinical and molecular management of hypertension.

References

1. Cardiovascular Diseases (CVDs) [Internet]. [cited 2024 Jun 1]. Available from: https://www.who.int/news-room/fact-sheets/detail/cardiovascular-diseases-(cvds).
2. Hypertension [Internet]. [cited 2024 Jun 1]. Available from: https://www.who.int/news-room/fact-sheets/detail/hypertension.
3. Carretero OA, Oparil S. Essential hypertension. Circulation [Internet]. 2000[cited 2024 Apr 15];101(3):329–35. Available from: https://www.ahajournals.org/doi/abs/10.1161/01.CIR.101.3.329.
4. Charles L, Triscott J, Dobbs B. Secondary hypertension: discovering the underlying cause. Am Fam Physician [Internet]. 2017 [cited 2024 Jun 1];96(7):453–61. Available from: https://www.aafp.org/pubs/afp/issues/2017/1001/p453.html.
5. Kearney PM, Whelton M, Reynolds K, Muntner P, Whelton PK, He J. Global burden of hypertension: analysis of worldwide data. Lancet. 2005;365(9455):217–23.
6. Stanaway JD, Afshin A, Gakidou E, Lim SS, Abate D, Abate KH, et al. Global, regional, and national comparative risk assessment of 84 behavioural, environmental and occupational, and metabolic risks or clusters of risks for 195 countries and territories, 1990–2017: a systematic analysis for the Global Burden of Disease Study 2017. Lancet. 2018;392(10159):1923–94.
7. Mills KT, Stefanescu A, He J. The global epidemiology of hypertension. Nat Rev Nephrol [Internet]. 2020 [cited 2024 Apr 6];16(4):223–37. Available from: https://www.nature.com/articles/s41581-019-0244-2.
8. Global report on hypertension: the race against a silent killer [Internet]. [cited 2024 Apr 12]. Available from: https://www.who.int/publications/i/item/9789240081062.
9. Kario K, Okura A, Hoshide S, Mogi M. The WHO Global report 2023 on hypertension warning the emerging hypertension burden in globe and its treatment strategy. Hypertens Res [Internet]. 2024 [cited 2024 Apr 12]:1–4. Available from: https://www.nature.com/articles/s41440-024-01622-w.
10. Zhou B, Carrillo-Larco RM, Danaei G, Riley LM, Paciorek CJ, Stevens GA, et al. Worldwide trends in hypertension prevalence and progress in treatment and control from 1990 to 2019: a pooled analysis of 1201 population-representative studies with 104 million participants. Lancet. 2021;398(10304):957–80.
11. Poch E, González D, Giner V, Bragulat E, Coca A, De la Sierra A. Molecular basis of salt sensitivity in human hypertension. Hypertension [Internet]. 2001 [cited 2024 Apr 12];38(5):1204–9. Available from: https://www.ahajournals.org/doi/abs/10.1161/hy1101.099479.
12. Seravalle G, Grassi G. Essential hypertension. In: Primer on the autonomic nervous system, 4th edn [Internet]; 2023 [cited 2024 Apr 12]. p. 467–70. Available from: https://www.ncbi.nlm.nih.gov/books/NBK539859/.
13. Essential Hypertension – StatPearls – NCBI Bookshelf [Internet]. [cited 2024 Apr 12]. Available from: https://www.ncbi.nlm.nih.gov/books/NBK539859/.
14. Whelton PK, Carey RM, Mancia G, Kreutz R, Bundy JD, Williams B. Harmonization of the American College of Cardiology/American Heart Association and European Society of Cardiology/European Society of Hypertension Blood Pressure/hypertension guidelines: comparisons, reflections, and recommendations. Circulation [Internet]. 2022 [cited 2024 Jun 1];146(11):868–77. Available from: https://www.ahajournals.org/doi/abs/10.1161/CIRCULATIONAHA.121.054602.

15. Whelton PK, Carey RM, Aronow WS, Casey DE, Collins KJ, Himmelfarb CD, et al. 2017 ACC/AHA/AAPA/ABC/ACPM/AGS/APhA/ASH/ASPC/NMA/PCNA guideline for the prevention, detection, evaluation, and management of high blood pressure in adults: executive summary: a report of the American College of Cardiology/American Heart Association task force on clinical practice guidelines. Hypertension [Internet]. 2018 [cited 2024 Jun 1];71(6):1269–324. Available from: https://www.ahajournals.org/doi/abs/10.1161/HYP.0000000000000066.
16. Williams B, Mancia G, Spiering W, Rosei EA, Azizi M, Burnier M, et al. 2018 ESC/ESH guidelines for the management of arterial hypertension. J Hypertens [Internet]. 2018 [cited 2024 Jun 1];36(10):1956–2041. Available from: https://journals.lww.com/jhypertension/fulltext/2018/10000/2018_esc_esh_guidelines_for_the_management_of.2.aspx.
17. Unger T, Borghi C, Charchar F, Khan NA, Poulter NR, Prabhakaran D, et al. 2020 International Society of Hypertension global hypertension practice guidelines. Hypertension [Internet]. 2020 [cited 2024 Apr 12];75(6):1334–57. Available from: https://www.ahajournals.org/doi/abs/10.1161/HYPERTENSIONAHA.120.15026.
18. Unger T, Borghi C, Charchar F, Khan NA, Poulter NR, Prabhakaran D, et al. Hypertension: new guidelines from the International Society of Hypertension. Am Fam Physician [Internet]. 2021 [cited 2024 Jun 1];103(12):763–5. Available from: https://www.aafp.org/pubs/afp/issues/2021/0615/p763.html.
19. Brewster UC, Perazella MA. The renin-angiotensin-aldosterone system and the kidney: effects on kidney disease. Am J Med. 2004;116(4):263–72.
20. Gradman AH, Kad R. Renin inhibition in hypertension. J Am Coll Cardiol [Internet]. 2008 [cited 2024 Jun 1];51(5):519–28. Available from: https://www.jacc.org/doi/10.1016/j.jacc.2007.10.027.
21. Hanif K, Bid HK, Konwar R. Reinventing the ACE inhibitors: some old and new implications of ACE inhibition. Hypertens Res [Internet]. 2009 [cited 2024 Jun 1];33(1):11–21. Available from: https://www.nature.com/articles/hr2009184.
22. Vargas Vargas RA, Varela Millán JM, Fajardo Bonilla E. Renin–angiotensin system: basic and clinical aspects—a general perspective. Endocrinol Diabetes Nutr (English ed). 2022;69(1):52–62.
23. Hermann M, Flammer A, Lüscher TF. Nitric oxide in hypertension. J Clin Hypertens [Internet]. 2006 [cited 2024 May 26];8(12 Suppl 4):17–29. Available from: https://onlinelibrary.wiley.com/doi/full/10.1111/j.1524-6175.2006.06032.x.
24. da Silva GM, da Silva MC, Nascimento DVG, Lima Silva EM, Gouvêa FFF, de França Lopes LG, et al. Nitric oxide as a central molecule in hypertension: focus on the vasorelaxant activity of new nitric oxide donors. Biology [Internet]. 2021 [cited 2024 May 26];10(10):1041. Available from: https://www.mdpi.com/2079-7737/10/10/1041/htm.
25. Laurent S. Antihypertensive drugs. Pharmacol Res. 2017;124:116–25.
26. Marinov P, Georgiev K, Georgieva M. Some aspects of the modern antihypertensive drug therapy and most common side effects. Scr Sci Pharm [Internet]. 2015 [cited 2024 May 30];2(1):7–14. Available from: https://journals.mu-varna.bg/index.php/ssp/article/view/1045.
27. Ram CVS. Direct renin inhibitors: a new approach to antihypertensive drug treatment. J Clin Hypertens [Internet]. 2007 [cited 2024 Apr 15];9(8):615–21. Available from: https://onlinelibrary.wiley.com/doi/full/10.1111/j.1524-6175.2007.07125.x.
28. Gross F, Lazar J, Orth H. Inhibition of the Renin-Angiotensinogen Reaction by pepstatin. Science (1979) [Internet]. 1972 [cited 2024 Apr 15];175(4022):656. Available from: https://www.science.org/doi/10.1126/science.175.4022.656.
29. Ramya K, Suresh R, Kumar HY, Kumar BRP, Murthy NBS. Decades-old renin inhibitors are still struggling to find a niche in antihypertensive therapy. A fleeting look at the old and the promising new molecules. Bioorg Med Chem. 2020;28(10):115466.
30. Oefner C, Binggeli A, Breu V, Bur D, Clozel JP, D'Arcy A, et al. Renin inhibition by substituted piperidines: a novel paradigm for the inhibition of monomeric aspartic proteinases? Chem Biol. 1999;6(3):127–31.

31. Webb RL, Schiering N, Sedrani R, Maibaum J. Direct renin inhibitors as a new therapy for hypertension. J Med Chem [Internet]. 2010 [cited 2024 May 10];53(21):7490–520. Available from: https://pubs.acs.org/doi/abs/10.1021/jm901885s.
32. Göschke R, Stutz S, Rasetti V, Cohen NC, Rahuel J, Rigollier P, et al. Novel 2,7-dialkyl-substituted 5(S)-amino-4(S)-hydroxy-8-phenyl- octanecarboxamide transition state peptidomimetics are potent and orally active inhibitors of human renin. J Med Chem [Internet]. 2007 [cited 2024 May 10];50(20):4818–31. Available from: https://pubs.acs.org/doi/abs/10.1021/jm070314y.
33. Iijima D, Sugama H, Takahashi Y, Hirai M, Togashi Y, Xie J, et al. Discovery of SPH3127: a novel, highly potent, and orally active direct renin inhibitor. J Med Chem [Internet]. 2022 [cited 2024 May 10];65(16):10882–97. Available from: https://pubs.acs.org/doi/abs/10.1021/acs.jmedchem.2c00834.
34. Iijima D, Sugama H, Awai N, Takahashi Y, Togashi Y, Takebe T, et al. Discovery of novel 2-carbamoyl morpholine derivatives as highly potent and orally active direct renin inhibitors. ACS Med Chem Lett [Internet]. 2022 [cited 2024 May 10];13(8):1351–7. Available from: https://pubs.acs.org/doi/abs/10.1021/acsmedchemlett.2c00280.
35. Acharya KR, Sturrock ED, Riordan JF, Ehlers MRW. Ace revisited: a new target for structure-based drug design. Nat Rev Drug Discov [Internet]. 2003 [cited 2024 May 10];2(11):891–902. Available from: https://www.nature.com/articles/nrd1227.
36. Natesh R, Schwager SLU, Sturrock ED, Acharya KR. Crystal structure of the human angiotensin-converting enzyme–lisinopril complex. Nature [Internet]. 2003 [cited 2024 Jun 1];421(6922):551–4. Available from: https://www.nature.com/articles/nature01370.
37. Zheng W, Tian E, Liu Z, Zhou C, Yang P, Tian K, et al. Small molecule angiotensin converting enzyme inhibitors: a medicinal chemistry perspective. Front Pharmacol. 2022;13:968104.
38. Gonzalez Amaya JA, Cabrera DZ, Matallana AM, Arevalo KG, Guevara-Pulido J. In-silico design of new enalapril analogs (ACE inhibitors) using QSAR and molecular docking models. Inform Med Unlocked. 2020;19:100336.
39. Adekoya OA, Sylte I. The thermolysin family (M4) of enzymes: therapeutic and biotechnological potential. Chem Biol Drug Des. 2009;73:7–16.
40. Adekoya OA, Sylte I. The thermolysin family (M4) of enzymes: therapeutic and biotechnological potential. Chem Biol Drug Des [Internet]. 2009 [cited 2024 Jun 1];73(1):7–16. Available from: https://onlinelibrary.wiley.com/doi/full/10.1111/j.1747-0285.2008.00757.x.
41. Cañizares-Carmenate Y, Mena-Ulecia K, Perera-Sardiña Y, Torrens F, Castillo-Garit JA. An approach to identify new antihypertensive agents using Thermolysin as model: in silico study based on QSARINS and docking. Arab J Chem. 2019;12(8):4861–77.
42. MacLeod-Carey D, Solis-Céspedes E, Lamazares E, Mena-Ulecia K. Evaluation of new antihypertensive drugs designed in silico using Thermolysin as a target. Saudi Pharm J. 2020;28(5):582–92.
43. Johnson CN, Pattanayek R, Potet F, Rebbeck RT, Blackwell DJ, Nikolaienko R, et al. The CaMKII inhibitor KN93-calmodulin interaction and implications for calmodulin tuning of NaV1.5 and RyR2 function. Cell Calcium. 2019;82:102063.
44. Tang L, El-Din TMG, Swanson TM, Pryde DC, Scheuer T, Zheng N, et al. Structural basis for inhibition of a voltage-gated Ca2+ channel by Ca2+ antagonist drugs. Nature [Internet]. 2016 [cited 2024 May 23];537(7618):117–21. Available from: https://www.nature.com/articles/nature19102.
45. Karmakar S, Basak HK, Paswan U, Pramanik AK, Chatterjee A. Designing of next-generation dihydropyridine-based calcium channel blockers: an in silico study. J Appl Pharm Sci [Internet]. 2022 [cited 2024 May 23];12(4):127–35. Available from: https://japsonline.com/abstract.php?article_id=3628&sts=2.
46. Zare F, Hashemi H, Hassani B, Fazel R, Miri R, Sadeghpour H. Synthesis, biological evaluation, in silico ADMET and molecular docking of new 1,4-dihydropyridine derivatives bearing ether substitution as calcium channel blockers. ChemistrySelect [Internet]. 2023 [cited 2024

May 23];8(21):e202204515. Available from: https://onlinelibrary.wiley.com/doi/full/10.1002/slct.202204515.
47. Takkar P, Singh B, Pani B, Kumar R. Design, synthesis and in silico evaluation of newer 1,4-dihydropyridine based amlodipine bio-isosteres as promising antihypertensive agents. RSC Adv [Internet]. 2023 [cited 2024 May 23];13(48):34239–48. Available from: https://pubs.rsc.org/en/content/articlehtml/2023/ra/d3ra06387a.
48. Titko T, Perekhoda L, Drapak I, Tsapko Y. Modern trends in diuretics development. Eur J Med Chem. 2020;208:112855.
49. Kintscher U, Bakris GL, Kolkhof P. Novel non-steroidal mineralocorticoid receptor antagonists in cardiorenal disease. Br J Pharmacol [Internet]. 2022 [cited 2024 May 13];179(13):3220–34. Available from: https://onlinelibrary.wiley.com/doi/full/10.1111/bph.15747.
50. Sueta D, Yamamoto E, Tsujita K. Mineralocorticoid receptor blockers: novel selective nonsteroidal mineralocorticoid receptor antagonists. Curr Hypertens Rep [Internet]. 2020 [cited 2024 May 13];22(3):1–9. Available from: https://link.springer.com/article/10.1007/s11906-020-1023-y.
51. Ślusarz MJ, Ślusarz R, Ciarkowski J. Investigation of mechanism of desmopressin binding in vasopressin V2 receptor versus vasopressin V1a and oxytocin receptors: molecular dynamics simulation of the agonist-bound state in the membrane–aqueous system. Biopolymers [Internet]. 2006 [cited 2024 May 28];81(5):321–38. Available from: https://onlinelibrary.wiley.com/doi/full/10.1002/bip.20420.
52. Tahtaoui C, Balestre MN, Klotz P, Rognan D, Barberis C, Mouillac B, et al. Identification of the binding sites of the SR49059 nonpeptide antagonist into the V1a vasopressin receptor using sulfydryl-reactive ligands and cysteine mutants as chemical sensors. J Biol Chem [Internet]. 2003 [cited 2024 May 28];278(41):40010–9. Available from: http://www.jbc.org/article/S0021925820830543/fulltext.
53. Mouillac B, Chini B, Balestre MN, Elands J, Trumpp-Kallmeyer S, Hoflack J, et al. The binding site of neuropeptide vasopressin V1a receptor. J Biol Chem [Internet]. 1995 [cited 2024 May 28];270(43):25771–7. Available from: http://www.jbc.org/article/S0021925818923938/fulltext.
54. Hagiwara Y, Ohno K, Kamohara M, Takasaki J, Watanabe T, Fukunishi Y, et al. Molecular modeling of vasopressin receptor and in silico screening of V1b receptor antagonists. Expert Opin Drug Discov. 2013;8(8):951–64.
55. Gao Q, Xu L, Cai J. New drug targets for hypertension: a literature review. Biochim Biophys Acta (BBA) Mol Basis Dis. 2021;1867(3):166037.
56. Ghatage T, Goyal SG, Dhar A, Bhat A. Novel therapeutics for the treatment of hypertension and its associated complications: peptide- and nonpeptide-based strategies. Hypertens Res [Internet]. 2021 [cited 2024 May 11];44(7):740–55. Available from: https://www.nature.com/articles/s41440-021-00643-z.
57. Sawami K, Tanaka A, Node K. Recent understandings about hypertension management in type 2 diabetes: what are the roles of SGLT2 inhibitor, GLP-1 receptor agonist, and finerenone? Hypertens Res [Internet]. 2023 [cited 2024 May 10];46(8):1892–9. Available from: https://www.nature.com/articles/s41440-023-01324-9.
58. Maruhashi T, Higashi Y. Preventive effects of SGLT2 inhibitors on incident hypertension in patients with diabetes who do not have hypertension. Hypertens Res [Internet]. 2024 [cited 2024 May 10];1–2. Available from: https://www.nature.com/articles/s41440-024-01696-6.
59. Wilcox CS. Antihypertensive and renal mechanisms of SGLT2 (Sodium-Glucose Linked Transporter 2) inhibitors. Hypertension [Internet]. 2020 [cited 2024 May 10];75(4):894–901. Available from: https://www.ahajournals.org/doi/abs/10.1161/HYPERTENSIONAHA.119.11684.
60. Kong J, Li L, Yuan H, Bai F, Yang K, Zhao L, et al. In silico discovery of potential sodium–glucose cotransporter-2 inhibitors from natural products for treatment of heart failure via molecular docking and molecular dynamics simulation approach. J Biomol Struct Dyn [Internet]. 2023 [cited 2024 May 10];41(17):8109–20. Available from: https://www.tandfonline.com/doi/abs/10.1080/07391102.2022.2130983.

61. Macalalad MAB, Gonzales AA. In-silico screening and identification of phytochemicals from Centella asiatica as potential inhibitors of sodium-glucose co-transporter 2 for treating diabetes. J Biomol Struct Dyn [Internet]. 2022 [cited 2024 May 10];40(22):12221–38. Available from: https://www.tandfonline.com/doi/abs/10.1080/07391102.2021.1969282.
62. Singh KD, Muthusamy K. Molecular modeling, quantum polarized ligand docking and structure-based 3D-QSAR analysis of the imidazole series as dual AT1 and ETA receptor antagonists. Acta Pharmacol Sin [Internet]. 2013 [cited 2024 May 10];34(12):1592–606. Available from: https://www.nature.com/articles/aps2013129.
63. Orry AJW, Wallace BA. Modeling and docking the endothelin G-protein-coupled receptor. Biophys J. 2000;79(6):3083–94.
64. Gu X, Wang Y, Wang H, Wu H, Li W, Wang J, et al. Homology modeling, molecular dynamics and virtual screening of endothelin-A receptor for the treatment of pulmonary arterial hypertension. J Biomol Struct Dyn [Internet]. 2021 [cited 2024 May 11];39(11):3912–23. Available from: https://www.tandfonline.com/doi/abs/10.1080/07391102.2020.1772106.
65. Patel JR, Joshi HV, Shah UA, Patel JK. Exploring novel endothelin receptor blocker as antihypertensive agents identified from a natural drugs library using induced fit docking and biological assay. Med Plants. 2020;12(3):405–13.
66. Panchal J, Jaiswal S, Jain S, Kumawat J, Sharma A, Jain P, et al. Development of novel bosentan analogues as endothelin receptor antagonists for pulmonary arterial hypertension. Eur J Med Chem. 2023;259:115681.
67. Nagatsu I, Nagatsu T, Yamamoto T, Glenner GG, Mehl JW. Purification of aminopeptidase a in human serum and degradation of angiotensin II by the purified enzyme. Biochim Biophys Acta (BBA) Enzymol. 1970;198(2):255–70.
68. Kario K, Williams B. Angiotensin receptor–neprilysin inhibitors for hypertension—hemodynamic effects and relevance to hypertensive heart disease. Hypertens Res [Internet]. 2022 [cited 2024 Jun 1];45(7):1097–110. Available from: https://www.nature.com/articles/s41440-022-00923-2.
69. Loganathan L, Gopinath K, Sankaranarayanan VM, Kukreti R, Rajendran K, Lee JK, et al. Computational and pharmacogenomic insights on hypertension treatment: rational drug design and optimization strategies. Curr Drug Targets [Internet]. 2019 [cited 2024 Jun 1];21(1):18–33. Available from: https://www.eurekaselect.com/article/100202.
70. Mogi M. Clinical study on angiotensin II vaccination—the first big step. Hypertens Res [Internet]. 2021 [cited 2024 Jun 1];45(1):162–3. Available from: https://www.nature.com/articles/s41440-021-00790-3.
71. Takezako T, Gogonea C, Saad Y, Noda K, Karnik SS. "Network leaning" as a mechanism of insurmountable antagonism of the angiotensin II type 1 receptor by non-peptide antagonists. J Biol Chem [Internet]. 2004 [cited 2024 Jun 17];279(15):15248–57. Available from: http://www.jbc.org/article/S0021925819639244/fulltext.
72. Güven AT, Özdede M, Şener YZ, Yıldırım AO, Altıntop SE, Yeşilyurt B, et al. Evaluation of machine learning algorithms for renin-angiotensin-aldosterone system inhibitors associated renal adverse event prediction. Eur J Intern Med [Internet]. 2023 [cited 2024 Jun 13];114:74–83. Available from: http://www.ejinme.com/article/S0953620523001723/fulltext.
73. Wang T, Yan Y, Xiang S, Tan J, Yang C, Zhao W. A comparative study of antihypertensive drugs prediction models for the elderly based on machine learning algorithms. Front Cardiovasc Med. 2022;9:1056263.
74. Manavalan B, Basith S, Shin TH, Wei L, Lee G. mAHTPred: a sequence-based meta-predictor for improving the prediction of anti-hypertensive peptides using effective feature representation. Bioinformatics [Internet]. 2019 [cited 2024 Jun 17];35(16):2757–65. Available from: https://doi.org/10.1093/bioinformatics/bty1047.
75. Kumar R, Chaudhary K, Singh Chauhan J, Nagpal G, Kumar R, Sharma M, et al. An in silico platform for predicting, screening and designing of antihypertensive peptides. Sci Rep [Internet]. 2015 [cited 2024 Jun 17];5(1):1–10. Available from: https://www.nature.com/articles/srep12512.

Chapter 26
Drug Discovery and Computational Strategies in Multi-Drug-Resistant Tuberculosis

Sohal Satish, Nachiket Joshi, Rajasekhar Reddy Alavala, and Afreen A. Khan

Contents

Abstract The development of novel therapeutic approaches has taken precedence in the fight against tuberculosis control due to the global upsurge in multi-drug-resistant tuberculosis (MDR-TB). Since MDR-TB strains are resistant to first-line therapies, finding new drugs is a major challenge. Computational modeling has

---

S. Satish
Ambernath Organics Pvt. Ltd., Mumbai, India

N. Joshi · R. R. Alavala
Shobhaben Pratapbhai Patel School of Pharmacy and Technology Management, SVKM's NMIMS, Mumbai, Maharashtra, India

A. A. Khan (✉)
SVKM's Dr. Bhanuben Nanavati College of Pharmacy, Mumbai, Maharashtra, India
e-mail: afreen.khan@bncp.ac.in

S. N. K. Rao G., R. R. Alavala (eds.), *Applications of Computational Tools in Drug Design and Development*, https://doi.org/10.1007/978-981-96-4154-3_26

emerged as a powerful tool in this fight, providing valuable insights into resistance mechanisms and accelerating the development of new anti-TB drugs. This chapter aims to give an overview of the various computational strategies reported in drug discovery for MDR-TB with a comprehensive overview of several structure-based and ligand-based computational methods. Additionally, notable applications of these methods in the discovery of novel anti-TB compounds have also been mentioned. This chapter also delves into the emergence of AI, and ML applications which have proved to be useful both in the design of new compounds and in the development of theoretical models that can predict resistance to the MDR-TB therapy.

Keywords Computer-aided drug design · Mycobacterium tuberculosis · QSAR · Molecular Docking · Drug discovery

Abbreviations

MDR-TB	Multidrug-resistant Tuberculosis
ML	Machine Learning
TB	Tuberculosis

26.1 Introduction

Tuberculosis (TB) is a highly infectious airborne illness caused by the bacterium *Mycobacterium tuberculosis* (Mtb) that leads to a considerable number of deaths globally. There were about 7.5 million new cases of active tuberculosis in 2022, with 1.3 million fatalities [1]. Tuberculosis treatment is a therapy that involves the use of many medications from different chemical classes, and it is administered over a lengthy period. The drugs used might be either first-line or second-line treatments [2]. Patient's inability to follow the treatment regimens with lengthy and complex instructions has led to the emergence of several drug-resistant tuberculosis (TB) strains, including multidrug-resistant (MDR) and extensively drug-resistant (XDR) variants [3]. Public health is seriously threatened by extensively drug-resistant (XDR-TB) and multidrug-resistant (MDR-TB) tuberculosis owing to the drastic death rates, lack of effective treatment alternatives, and burdened medical care. In areas where tuberculosis is prevalent and healthcare facilities are underdeveloped, multidrug-resistant and extensively drug-resistant strains of the disease have emerged as major health issues worldwide [4]. There were almost 410,000 cases of multi-drug-resistant tuberculosis (MDR-TB) in 2022, with nearly 66% of those cases occurring in only 8 countries, according to the World Health Organization (WHO) [1]. An excess of 200 countries have documented instances of XDR-TB, which exacerbates the challenges associated with tuberculosis control. Treatment outcomes are adversely affected by MDR-TB and XDR-TB, as evidenced by increased rates of treatment failure, recurrence, and mortality [5]. Since, there is a lack of enough effective medications on the market, treating and prognosis of patients with XDR-TB is already difficult. Multidrug-resistant tuberculosis (MDR-TB) and extensively drug-resistant tuberculosis (XDR-TB) need tests particularly designed to detect drug resistance.

The findings of conventional drug susceptibility testing (DST) procedures, such as culture-based phenotypic testing, might take a few weeks to several months. By simultaneously identifying tuberculosis and rifampicin resistance within hours, rapid molecular diagnostic tools like the MTB/RIF assay have enhanced the prompt diagnosis of treatment resistance [6, 7]. MDR-TB and XDR-TB treatment protocols often need extended durations of numerous second-line medications which heightens the likelihood of unfavorable reactions, drug interferences, and inadequate compliance [3]. Healthcare systems and individuals bear a disproportionate share of the huge financial burden associated with treating multidrug-resistant tuberculosis (MDR-TB) and extensively drug-resistant tuberculosis (XDR-TB). Numerous nations grappling with significant loads of drug-resistant tuberculosis suffer from a dearth of resources, facilities, and adequately educated personnel [8].

MDR-TB & XDR-TB—A brief review

Multi-drug-resistant tuberculosis (MDR-TB) is a form of tuberculosis that is resistant to both isoniazid and rifampicin. Extensive drug-resistant tuberculosis (XDR-TB) occurs when a patient develops resistance to fluoroquinolones, amikacin, kanamycin, or capreomycin, in addition to isoniazid and rifampicin (Fig. 26.1). Drug resistance may occur due to inadequate treatment plans and the incorrect or lack of administration of drugs, which can also contribute to higher levels of drug

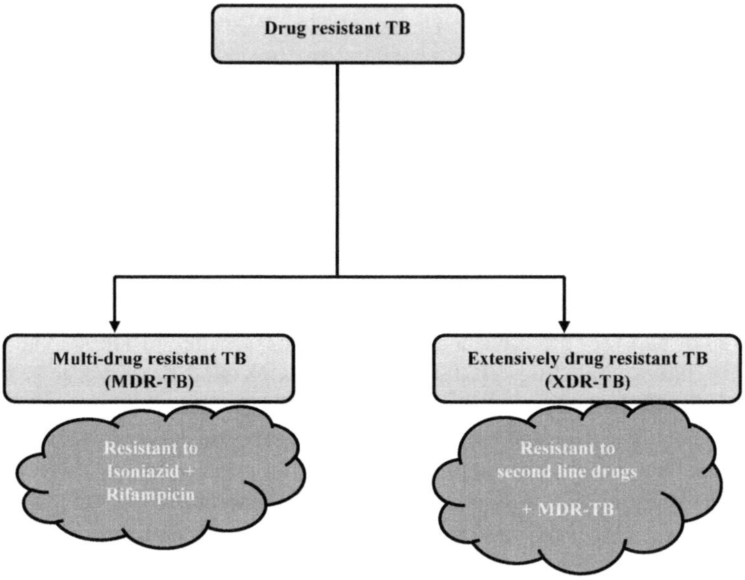

Fig. 26.1 Difference between MDR-TB & XDR-TB

resistance in the population. The implementation of a high-quality DOTS program may successfully mitigate the incidence of multidrug-resistant (MDR) and extensively drug-resistant (XDR) tuberculosis (TB) within the community [3]. Traditional first-line medication is mostly ineffective in treating multidrug-resistant tuberculosis (MDR-TB) [9]. The misuse and overprescribing of both main and secondary medications lead to extensively drug-resistant tuberculosis (XDR-TB) [4].

Consequently, the therapy of XDR-TB becomes more challenging. As a result, effective management and control of tuberculosis infections from the start is crucial. MDR-TB is more difficult to treat due to its restricted treatment choices and longer incubation period than a typical TB infection [10]. Prompt and precise diagnosis is crucial for the successful treatment of multidrug-resistant tuberculosis (MDR-TB). Compared to first-line medications, which are used for typical drug-responsive tuberculosis, second-line treatments for multidrug-resistant tuberculosis are more costly, have lesser effectiveness, and are linked with greater adverse effects [11]. The total treatment of MDR-TB often requires a duration of 18–24 months [12]. The worldwide cure rate for multi-drug-resistant tuberculosis (MDR-TB) is much lower. Mortality rates in multidrug-resistant tuberculosis patients may be elevated when co-infections with viruses like HIV are present. There is a severe lack of alternatives for treating XDR-Though there is still hope for recovery, the likelihood of a full recovery is reduced in comparison to patients with common TB or even MDR-TB when using the medications that are presently on the market. How well a medicine works, how bad the sickness is, and how strong a person's immune system are all factors in whether or not an XDR-TB infection may be cured [13, 14]. Early detection and appropriate medication management generally result in a cure, despite the huge challenges and difficulties associated with treating MDR and XDR-TB infections [6].

26.2 Mechanisms of Drug Resistance

Drug-resistant tuberculosis (DR-TB) is increasingly becoming a worldwide concern for tuberculosis (TB) control strategies. Currently, worldwide efforts to control tuberculosis are facing difficulties due to the increasing occurrence of drug resistance to many antituberculosis medications, leading to the development of multi-drug-resistant tuberculosis (MDR-TB) and extensively drug-resistant tuberculosis (XDR-TB). Multi-drug-resistant tuberculosis is characterized by its resistance to a minimum of two very effective anti-TB medications, namely rifampicin and isoniazid [15]. Extensively drug-resistant tuberculosis (XDR-TB) refers to a strain of tuberculosis that is resistant to therapy used for MDR-TB and other anti-TB drugs. Specifically, it is resistant to both multi-drug-resistant tuberculosis (MDR-TB) and at least one fluoroquinolone antibiotic, as well as at least one of three injectable second-line treatments (amikacin, kanamycin, or capreomycin) [15]. Recently, a concerning scenario has arisen where resistance to all current anti-tubercular medications has been seen, leading to a disease known as completely drug-resistant (TDR-TB) tuberculosis [16].

26.2.1 Genetic Mutations

Strains develop resistance to drugs when DNA mutations occur naturally and alter the proteins. The majority of drug resistance instances include mutations in drug-targeted genes. Mutations in the embB gene, which encodes the arabinosyltransferase enzyme involved in cell wall production may lead to Ethambutol (EMB) resistance. The arabinose transferase enzyme is encoded by the embC, embA, and embB operons in bacteria. These alterations have the potential to change the enzyme's structure and decrease its affinity for EMB [17]. Ethambutol inhibits the synthesis of arabinose transferases, thereby impeding the formation of lipoarabinomannan and arabinogalactan, components of the mycobacterial cell wall. This mechanism of action also inhibits cell division. EmbB is unaltered in approximately 30% of ethambutol-resistant isolates, indicating the existence of an alternative resistance mechanism [18]. PZA resistance may result from mutations in the pncA gene. This gene codes for the pyrazinamidase enzyme that transforms PZA into its active form. Mutations in this specific gene tend to reduce or abolish the activity of the enzyme [19, 20]. Isoniazid is activated by KatG, a catalase peroxidase enzyme. It inhibits NADH-enoyl-ACP-reductase, which is encoded by the inhA gene. This inhibits the production of mycolic acids which are crucial building blocks of the bacterial cell wall [20, 21]. Isoniazid resistance is substantially mediated by mutations in the katG gene. Most cases of isoniazid resistance are associated with mutations in both katG and inhA [20, 21]. Rifampicin blocks the RNA polymerase enzyme, a process that is essential for gene transcription. Rifampicin resistance is generally caused by mutations in the rpoB gene, which encodes the Î2 subunit of RNA polymerase. Because of these changes, rifampicin is less effective in binding to its target [20, 22]. In M tuberculosis, chromosomal alterations inside the quinolone resistance-determining region (QRDR) of the gyrA or gyrB genes are the main mechanism by which fluoroquinolone (FLQ) resistance develops. DNA gyrase which encodes the gyr A & gyr B gene and topoisomerase IV are two vital bacterial enzymes that regulate the chromosomal supercoiling necessary for the replication process, inhibition of which forms the basis of the fluoroquinolone mode of action [20, 23]. Aminoglycosides (such as amikacin and kanamycin) and cyclic peptides (such as capreomycin) suppress the production of proteins by altering the ribosomal structures at 16S rRNA and forming the 30S ribosomal subunit, respectively. Antibiotic resistance to amikacin and kanamycin is enhanced by point mutations in the 16S rRNA gene (rrs) [20, 23]. PAS (p-amino salicylic acid) is thought to inhibit the enzyme, thymidylate synthase A. The enzyme ThyA, which is produced by the thyA gene, has a crucial function in the production of thymine in the folate route. Mutations in the gene have been identified as the main reason for resistance to p-aminosalicylic acid [20, 24]. The occurrence of mutations that contribute to M.tb resistance has been reported in the case of second-line drugs and new classes of drugs for TB treatment. These mutations are encouraged by the unfavorable environmental conditions brought on by the use of anti-TB drugs. The nature of the drug and the strain under pressure from selection also affect the mutations [20, 25].

26.2.2 *Drug Efflux in Mycobacterium tuberculosis*

There are chemical processes inside the cell wall of Mycobacterium tuberculosis that aggressively flush out drug molecules. The physiology of bacteria and eukaryotic organisms depends heavily on efflux systems (Fig. 26.2). Part of the resistance to isoniazid and rifampicin may be explained by this efflux pump mechanism. Through the overexpression of efflux pumps, which actively remove drugs from the bacterial cell, lowering their intracellular concentration and efficacy, M.tb can become resistant to a variety of drugs [26, 27]. Variations in the environment may modify the expression of efflux pumps. It is possible for mycobacterial efflux pumps to eradicate almost all first and second-line antituberculosis drugs. Specific antibiotics or macrophage intracellular conditions trigger or up-regulate efflux pumps [26, 27].

.

Overexpression of efflux pumps in the bacteria also contributes significantly to drug resistance. Drug resistance to medications like Clofazimine and Bedaquiline may be caused by mutations in the Mmpr gene, which cause an overproduction of efflux pumps [28]. Efflux pumps have been discovered to cause resistance to isoniazid. In vitro research has clearly shown that efflux systems may serve as a springboard for the development of higher-level resistance [25, 29]. It is also possible that

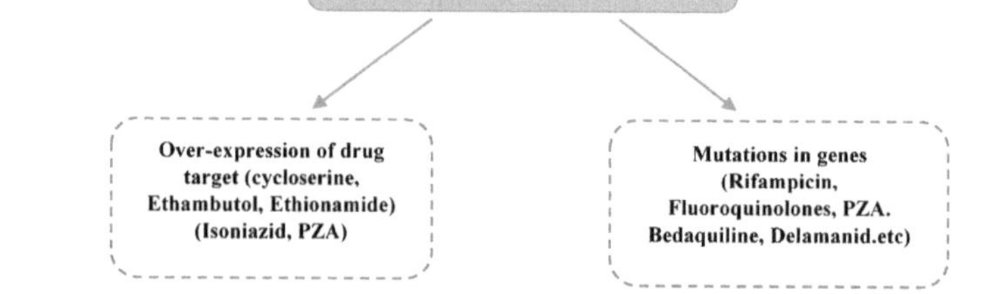

Fig. 26.2 Different mechanisms involved in drug resistance to anti-tubercular agents [3]

the efflux pump inhibitors will impede the development of bacteria and reduce the minimum inhibitory concentrations for some medications.

26.2.3 Elimination of Prodrug Activation

Certain pharmaceuticals, including prodrugs like INH and PZA, must be activated by bacterial enzymes to exert their antimicrobial properties. Mutations that impair the activity of these activating enzymes may result in resistance by slowing the process of transforming the prodrug into its active form. This resistance has been shown in drugs such as pyrazinamide, isoniazid, para-aminosalicylic acid, ethionamide, delamanid, and pretomanid [25, 30].

26.2.4 Drug Modification/Inactivation

Enzymes produced by M.tb can change or inactivate certain medications. For instance, RIF (Rifampicin) may be acetylated and rendered inactive by the enzyme acetyltransferase, which is expressed by the aac (2′) gene, resulting in resistance [31]. The target gene produces enzymes that deactivate drug compounds. Typically, these enzymes phosphorylate, adenylate, or acetylate drug molecules. N-methylation of the inhibitor of DprE1 renders it inactive against the target. It was caused by overexpression of methyl transferase enzyme due to mutations in the gene responsible for transcription activity [32]. Acetylation of commonly used aminoglycosides has been found to contribute to drug resistance under the influence of intracellular survival protein (eis) [25, 33].

26.2.5 Overexpression of Pharmacological Targets

The drug's inhibitory effect issue may be overcome by over-expression of the drug target gene, which would result in an excess of the target component. As shown by the cases of cycloserine, isoniazid, and ethambutol, overexpression may result from mutations in transcriptional repressors or the drug target's promoter [25, 34]. A patient's medicine dose is often adjusted according to their age and weight. This helps in optimizing the efficacy of the treatment and reduction in adverse effects. Given the severe adverse effects associated with specific antibiotics, patients are typically administered the lowest possible dose. Therefore, attempting to increase drug concentrations to overcome resistance caused by the over-expression of drug targets seems unlikely. Nonetheless, by altering the drug target or blocking prodrug activation, drug target overexpression may function as a prelude to high-dose

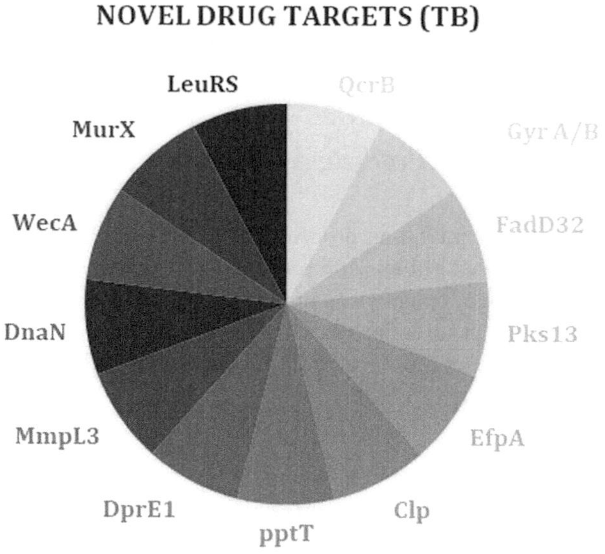

Fig. 26.3 List of novel drug targets for anti-tubercular drug [35–38]

resistance. This process can be accounted for drug resistance in Isoniazid & cycloserine. The problem is usually solvable by increasing the dose of the drug [25].

26.2.6 Review of Novel Targets for Antitubercular Drugs

There has been significant progress in building a pipeline of TB drugs, but the high rate of failure means that the pipeline needs to be constantly refilled with high-quality leads that work by blocking new targets (Fig. 26.3). Identifying new targets of medication action will pave the way for future drug development. With the advancement of molecular biology and the accomplishment of M.tb genome sequencing, considerable strides have been achieved in the identification of novel targets and inhibitors. Further investigation is required to further develop potential drugs, establish their effectiveness in both preclinical and clinical environments, and ultimately, prepare them for implementation in clinical practice. Some of the drug targets are reviewed here (Table 26.1).

QcrB (Quinol Oxidase Cytochrome b Subunit)
The respiratory chain is hampered by QcrB dysfunction, which eventually results in reduced ATP synthesis and bacterial cell death. Inhibiting QcrB may provide a new mode of action, which might help overcome resistance mechanisms linked to current pharmacological treatments. Progress in our knowledge of the catalytic process of QcrB has aided in the discovery of new inhibitors. The electron transport chain is a possible therapeutic target for M.tb since it is an obligate aerobe that requires

Table 26.1 List of clinical trial candidates and their drug targets

Sr No	Drug target	In the pipeline	Role of the targets
a	WecA	CPZEN-45	WecA catalyzes the biosynthesis of an essential disaccharide linker involved in the attachment of arabinogalactan to the peptidoglycan structure.
b	DprE1	BTZ043, PBTZ169, TBA7371, OPC-167832	Inhibition of enzyme decaprenylphosphoryl-β-D-ribose 2′-epimerase
c	MmpL 3	SQ109, Rimonabant	MmpL aids in the transportation of mycolic acids across the membrane
d	EfpA	BRD-8000	Efflux pumps help increase bacterial resistance by removing drugs from the cell wall.
e	MurX	SQ641	MurX catalyzes conversion of UDP-N-acetylmuramyl-pentapeptide into prenyl-N-acetylmuramyl-pentapeptide which is involved in peptidoglycan biosynthesis.
f	FadD32	CCA34	Fatty acyl AMP ligase 32 (FadD32) enzyme activates the mero acyl AMP (mycolic acid cell wall hallmark)
g	Pks13	TAM16	Pks13 catalyzes the production of α-alkyl □-ketoacids, building blocks of mycolic acids.
h	QcrB	Q-203, TB47	QcrB is a subunit of cytochrome bc1 complex which is an important unit of the respiratory electron transport chain. It plays a role in ATP synthesis.
i	LeuRs	GSK-3036656	Leucyl-tRNA synthetase (LeuRs) mediate protein synthesis.
j	gyr A/B	Fluoroquinolones	gyr A/B are subunits of DNA gyrase enzyme which is crucial for DNA replication in the bacteria.
k	PptT	amidino urea 8918	PptT (phosphopantetheine transferase) enzyme transfers phosphopantetheine (Ppt) from coenzyme A to diverse acyl carrier proteins, this process is crucial in the cell wall lipids synthesis.
l	DnaN	Griselimycins	DnaN (bacterial β-sliding clamp) is a subunit of bacterial DNA polymerase III which is crucial for DNA replication and repair.
m	Clp	NSC10010	Caseinolytic protease is important for the control of bacterial virulence & cell homeostasis.

oxygen for its survival & multiplication. The discovery of Q203's anti-TB efficacy has generated significant interest in its target site, QcrB is a subunit of the cytochrome bcc, which is involved in the electron transport chain. The discovery of novel QcrB inhibitors, such as arylvinylpiperazine amides and morpholinothiophenes, provides an additional incentive to explore QcrB as a potential target.

Isocitrate lyase

The glyoxylate shunt route is a metabolic process that is exclusive to bacteria and fungi, and isocitrate lyase is an essential enzyme in this system. Bacteria can make use of acetate and other molecules that contain two carbons as sources of carbon when they are under situations of carbon deficiency, such as when they are dealing with an intracellular infection through this pathway. The glyoxylate shunt route is impaired when isocitrate lyase is inhibited, making it harder for bacteria to use other carbon sources and get into host cells. By focusing on multiple metabolic pathways, such as the glyoxylate shunt and central carbon metabolism, combinatorial strategies have the potential to improve the effectiveness of tuberculosis treatment regimens and decrease the probability of resistance development.

GyrA/B

DNA gyrase is a tetrameric enzyme that consists of GyrA and GyrB subunits. The only type-2 topoisomerase present in *M.tb* has been identified as a validated target for fluoroquinolone medicines. Fluoroquinolones cause persistent double-strand (ds) DNA breaks and reduce DNA replication by inhibiting DNA gyrase after DNA cleavage. Moxifloxacin kills both replicative and anaerobic non-replicating *M.tb*.

PptT

PptT/Rv2794c is an important enzyme that transfers 4-phosphopantetheine from coenzyme A to acyl carrier proteins. The synthetic compound, amidino-urea (8918), has shown the potential to inhibit this enzyme which is involved in the activation of synthetases & synthases crucial to the development of bacterial cell walls.

FadD32

An important enzyme for mycobacterial development is fatty-acyl-AMP ligase, FadD32/Rv3801c. This enzyme is necessary for activation of the mycolic acid cell wall hallmark, the long meromycolic chain. A novel high-throughput enzyme assay led to the discovery of several FadD32 inhibitors such as derivatives of benzimidazole, thiopyrimidine, pyrrozole, quinolone, and iso-quinolone. Dialylcoumarins are potent inhibitors of FadD32.

Pks13

The type-I polyketide synthase, Pks13, encoded by the Rv3800c gene, plays a crucial role in the final stages of mycolic acid biosynthesis. These processes involve the activation of FadD32, resulting in the production of α-alkyl β-ketoesters, which are the building blocks of mycolic acids. Pks13 is the target of carbenoxarboxylic acids and benzofurans.

Translocase I

Peptidoglycan biosynthesis involves the conversion of UDP-N-acetylmuramyl-pentapeptide to prenyl-N- acetylmuramyl-pentapeptide via translocase I (MurX), an enzyme that is critical for the growth of Mycobacterium tuberculosis [18]. MurX inhibitors (SQ641) destroyed M.tb during its growth stage. MurX inhibitors (SQ641), nucleoside analog rapidly killed M.tb during the growing stage faster than any existing anti-TB drugs However, SQ641's lipophilic decanoyl side chain may cause reduced solubility and oral absorption, leading to weak intracellular and in vivo M.tb activity. To improve the absorption profile of the drug, a phospholipid nanoemulsion derivative has been developed and found to have a promising in-vivo activity.

GlcNAc-1-P Transferase (WecA)

Decaprenyl-P-P-GlcNAc is formed by the catalytic action of GlcNAc-1-P transferase (WecA), a phosphotransferase, and uridine diphosphate (UDP)-GlcNAc. This intermediate is then prolonged by rhamnosyl transferase to generate decaprenyl-P-P-GlcNAc-rhamnose. The streptomyces-derived capreomycin derivative CPZEN-45 has been suggested as a potential target of WecA [11]. CPZEN-45 exhibited remarkable efficacy against replicating and non-replicating Mycobacterium tuberculosis (M.tb) strains. Nevertheless, due to its poor solubility and low bioavailability, it may not be possible to develop an oral formulation of CPZEN-45.

MmpL3

One of the newly discovered pharmacological targets is MmpL3. This target is linked with the transport of TMM (Trehalose monomycolates) across the cell membrane in the bacteria. Inhibition of Mycobacterial membrane protein Large 3 (MmpL3) results in a build-up of TMM within the cells and a decrease in TDM (precursor of TMM) and mycolic acid levels. TMM is crucial in cell wall biosynthesis as it is believed mycolic acids are transported across the membrane in the form of TMM in bacteria. This makes MmpL3 a very vulnerable target since it has lethal effects on mycobacteria. Both adamantyl urea AU1235 and SQ109 were shown to inhibit major mycobacterial membrane proteins.

DprE1

Benzothiazinones (BTZs) target the important enzyme decaprenylphosphoryl-Î2-D-ribose oxidase, which is reliant on FAD (DprE1). "PBTZ169" (Macozinone), a piperazine derivative of BTZ, exhibited significant efficacy against M.tb and is currently in Phase II trials.

EfpA

Efflux pumps considerably increase bacterial resistance to antimicrobial drugs by removing the drugs from the cell thereby reducing the intracellular concentration of these drugs. The TB therapy can be more effective with the addition of efflux pump inhibitors in the regimen. Johnson et al. conducted a study where they identified and optimized the inhibitor BRD-8000 for the efflux pump, EfpA. This work demonstrated the relevance of this efflux pump and its potential as a new target in the treatment of TB.

Biotin protein ligase, DnaN, Î2-ketoacyl-ACP-synthase-I, Clp (Caseinolytic proteases), tryptophan synthase, and Leucyl-tRNA synthetase are some of the targets that have garnered recent attention in addition to the targets that were discussed before [35–38].

26.3 Current Anti-TB Drugs and Their Limitations

The "WHO End TB Strategy" was developed by the WHO in 2014 to eliminate TB. This program aims to bring more than a 90% reduction in the incidence rates of TB & associated mortality by the next decade. WHO guidelines in 2022 suggested the 6-month BPaLM/BPaL regimen, which consists of bedaquiline, pretomanid, linezolid, and moxifloxacin as the preferred treatment for TB patients. A shorter duration of treatment would lead to lesser utilization of health resources thereby meeting the demands of a large number of needy patients. Shorter treatment regimens expedite the patient's completion of therapy [39].

Antituberculosis medications are divided into two groups. The first-line medications consist of isoniazid, rifampicin, ethambutol, pyrazinamide, and streptomycin. The second-line antituberculosis medications include fluoroquinolones (ofloxacin, levofloxacin, moxifloxacin, and ciprofloxacin), Aminoglycosides (kanamycin, amikacin, and capreomycin) Ethionamide/Prothionamide, Cycloserine/Terizidone, and p-aminosalicylic acid.

26.3.1 First Line of Drugs

Isoniazid

Since its release in 1952, isoniazid (INH) has been a crucial component of therapy for TB. It is one of the most potent and targeted antituberculosis medications. INH is inactive in anaerobic environments or against non-replicating tubercle bacilli; it possesses activity only against replicating bacilli [3]. INH which enters the bacteria by passive diffusion is a prodrug that is activated by KatG. This is a catalase-peroxidase encoded by the KatG gene that converts INH into its active form. Mutations in several genes including KatG, ahpC, and inhA contribute to drug resistance. The activated form of INH causes a range of radicals and adducts to be produced, which in turn deactivate the NADH-dependent enoyl-acyl carrier protein reductase encoded by inhA, preventing the mycobacterium from producing the mycolic acids that constitute its cell wall [4]. This makes INH a very effective way to kill bacteria. Additionally, it's been shown to work well with other TB treatments.

Rifampicin

Rifampicin (RIF), which was initially developed as an antituberculosis medication in 1972, possesses remarkable bacteriostatic properties [5]. A variety of gram-positive cocci, such as Mycobacteria and Clostridium difficile, as well as several gram-negative species, such as Neisseria meningitidis, N gonorrhoeae, and Hemophilus influenza, are susceptible to the antibacterial action of rifampin [6]. The mechanism of Rifampicin involves acting on the Î2-subunit of RNA polymerase (rpoB) enzyme, which is accountable for the transcription and expression of mycobacterial genes [7]. This attachment leads to the suppression of bacterial transcription activity, ultimately killing the organism. Rifampicin is effective against both actively developing and slowly metabolizing (non-growing) bacilli. Incidences of hepatotoxicity are lower than with Isoniazid. Mutations in the rpoB gene contribute to drug resistance in the majority of the cases [3].

Pyrazinamide

Anti-TB property of pyrazinamide, a structural analog of nicotinamide was discovered in the 1950s, inspired by the activity of nicotinamide against M.tb [8]. Pyrazinamide can hinder the growth of semi-dormant bacteria that reside in acidic environments. Pyrazinamide is a prodrug that requires conversion into its active form, pyrazinoic acid, by the action of the enzyme pyrazinamidase/nicotinamidase. PZA exhibits activity against M. tuberculosis only at slightly acidic pH [9]. The gene pncA encodes PZase in M. tuberculosis. Genetic mutations in the pncA gene are linked with drug resistance [10]. Pyrazinamide is only used in conjunction with other medications, such as isoniazid and rifampicin, to treat Mycobacterium TB and as part of directly observed therapy (DOT).

Ethambutol

Ethambutol was discovered at Lederle Labs in 1961, it has been a part of anti-TB therapy for years. Ethambutol has been shown to improve the efficacy of quinolones, aminoglycosides, and rifamycins. EMB specifically targets only replicating bacilli. Mechanism of action involves inhibition of the arabinosyl transferase enzyme which is involved in the synthesis of arabinogalactan, a primary component of the bacterial cell wall [11]. Mutations in the embB gene have been shown to cause resistance to ethambutol [12]. EMB is a significant antimycobacterial medication that also improves the efficacy of quinolones, aminoglycosides, and rifamycins.

Streptomycin

Streptomycin, an antibiotic belonging to the aminocyclitol glycoside class, was first identified in 1943 by American biochemists Selman Waksman and Albert Schatz at Rutgers University. In 1948, this medication was the first to be used to treat tuberculosis [13]. Streptomycin exclusively kills actively proliferating tubercle bacilli, but it is ineffective on non-proliferating or intracellular bacilli. Streptomycin binds to the 16S rRNA, thereby disrupting the process of checking for errors during translation and thus hindering the production of proteins. Mutations in the 16S rRNA and ribosomal proteins have been linked to drug resistance [6]. Research has shown that the intrinsic efflux pump characteristic of bacteria also plays a role in the development of rug resistance.

26.3.2 Second Line of Drugs

Fluoroquinolones (FQs)

Fluoroquinolones have been a mainstay in the treatment of MDR & XDR TB. Fluoroquinolones (FQs) have been utilized for various other purposes apart from TB such as in the treatment of a variety of bacterial infections, including those of the gastrointestinal, and urinary tracts; STIs; chronic osteomyelitis, etc. Fluoroquinolones (FQs) like ciprofloxacin, ofloxacin, levofloxacin, and moxifloxacin are included in the high-priority list of drugs for the treatment of MDR TB [14].

The mechanism of action of FQs involves inhibition of the DNA gyrase enzyme which is a type II topoisomerase. The subunits of this enzyme are encoded by the gyr A & gyr B gene, mutations in these genes contribute to the drug resistance [6]. Moxifloxacin, gatifloxacin, and levofloxacin are considered essential agents in the event of resistance or intolerance to first-line anti-TB medications. FQs have good in vitro and in vivo action against M. tuberculosis.

Aminoglycosides (Amikacin, Kanamycin, Capreomycin)
Aminoglycosides such as amikacin (AMK) and kanamycin (KAN), as well as the cyclic polypeptide capreomycin (CAP), form the core of group C, part of the multi-drug regimen recommended by WHO in the treatment of MDR-TB. All of these drugs have similar mechanisms of action. Cross-resistance has been observed between Kanamycin & Amikacin. Capreomycin can be used in case of drug resistance to AMK & KAN as it is structurally different than the other two. Aminoglycosides mainly inhibit protein synthesis in M. Tuberculosis [6, 15]. Drug resistance has been linked to alterations in the 16S rRNA (rrs). A subset of aminoglycosides may also affect the process of protein synthesis by impeding elongation or directly suppressing the initiation. Aminoglycosides have a wide range of effectiveness against various microorganisms, which is further improved when combined with other types of antimicrobial agents in laboratory settings. Synergistic combinations of aminoglycosides and β-lactams have been explored in the treatment of MDR TB.

Ethionamide and prothionamide
Ethionamide or 2-ethylisonicotinamide was discovered by David Libermann & his co-workers at The Pasteur Institute, France in 1956 [16]. Ethionamide & Prothionamide both act as prodrugs, ethionamide is activated by EtaA/EthA, a mono-oxygenase enzyme. After being introduced into the bacterial cell, the sulfo group of the ethionamide oxidized by flavin monooxygenase, results in the conversion of the drug to 2-ethyl-4-aminopyridine [17]. The intermediate products exhibit toxicity against mycobacteria. Individuals with a genetic mutation that makes them immune to the effects of ethionamide also exhibit resistance to prothionamide. Partial cross-resistance with Isoniazid has been observed due to a similar target of action, inhA.

p-Amino salicylic acid
P-Amino salicylic acid (PAS) discovered in 1944 by Swedish chemist Jörgen Lehmann, was an early antibiotic that showed anti-TB properties. It currently is a part of the Group C in the treatment of MDR TB, to be used only in combination with other anti-TB drugs [2]. It has been shown that thymidylate synthase A is the target of PAS in M. bovis BCG. The most prevalent mutation seen in thyA which encodes the enzyme was Thr202Ala [6]. It has been used in conjunction with isoniazid and streptomycin for the treatment of tuberculosis.

Bedaquiline
Diarylquinolines were granted a patent in 2004 for their antimycobacterial capabilities. Bedaquiline was a breakthrough antibiotic for the treatment of

multidrug-resistant TB, breaking a 45-year standstill in the hunt for a new tuberculosis therapy. A team led by Koen Andries of Janssen Pharmaceutica made the discovery [18]. The combination of bedaquiline, pretomanid, and linezolid has shown 90% favorable outcomes in patients with extensively drug-resistant tuberculosis (XDR-TB) and treatment-intolerant multidrug-resistant tuberculosis (MDR-TB) [19]. This particular regimen (bedaquiline, pretomanid, and linezolid [BPaL]) has recently been approved. Bedaquiline targets mycobacterial ATP synthases and thereby the energy metabolism of the bacteria. A 6-month BPaLM regimen which stands for bedaquiline, pretomanid, linezolid, and moxifloxacin was approved by the World Health Organization in May 2022 [2]. Bedaquiline was classified into group A therapy in the treatment of drug-resistant tuberculosis (DR-TB).

Delamanid

Delamanid, a nitro-dihydro-imidazooxazole derivative, formerly referred to as OPC-67683, was developed in the 1990s by Otsuka Pharmaceuticals [20]. Delamanid was granted early release to the market by the European Medicines Agency (EMA) in 2014, immediately after bedaquiline was approved. It was the second novel drug used to treat DR-TB in more than four decades. Delamanid is recognized for its ability to impede the formation of mycolic acid and cellular respiration. The inhibitory impact is most likely caused by the generation of reactive radicals like nitric oxide. It prevents the reproduction of both active and inactive bacteria, both outside and within cells [21].

Pretomanid

Pretomanid (PA), a nitro-dihydro-imidazooxazole derivative previously known as PA-824, was authorized by the FDA in 2019 for XDR-TB or treatment-intolerant or nonresponsive MDR-TB in the BPaL regimen. Pretomanid (PA) was first discovered in 1995 and is jointly licensed by the TB Alliance and Mylan [19, 22]. PA hinders the production of mycolic acid, an essential process for the development of the cell wall. PA had significant activity in mice, particularly when used in conjunction with bedaquiline and linezolid. Deazaflavin-dependent nitro reductase (Ddn), an enzyme that converts dihydro-F420 (reduced form) into nitric oxide and a highly reactive metabolite, activates pretomanid in the mycobacterium. This metabolite disrupts the production of arabinogalactan, an essential component of cell walls, by targeting the synthesis enzyme DprE2.

Clofazimine & Linezolid

Clofazimine, a riminophenazine antibiotic, was first developed by a research team at Trinity College, Dublin as a treatment for tuberculosis (TB). In 2019, the World Health Organization (WHO) included Clofazimine in the treatment regimen for multidrug-resistant tuberculosis (MDR TB), in conjunction with other medications. Mechanism of action involves the activation of Clofazimine by type 2-NADH dehydrogenase which leads to the generation of reactive oxygen species (ROS) when it is reoxidized by oxygen (O_2) [19, 23]. These ROS then cause the rupture of the bacterial membrane. Linezolid, the first commercially available 1, 3-oxazolidinone, was discovered by Pharmacia & Upjohn in the 1990s. The method of action includes inhibiting bacterial protein synthesis by reducing mRNA translation [24]. Linezolid is regarded by the World Health Organization (WHO) as a crucial medication for treating rifampicin-resistant or multidrug-resistant tuberculosis (MDR-TB)

Cycloserine/Terizidone
Cycloserine a cyclic analogue of D-alanine, has been used for the treatment of TB since the 1950s. It is a broad-spectrum antibiotic active against both gram-positive and gram-negative bacteria. Mechanism of action involves inhibition of D-alanine racemase and D-alanine: D-alanine ligase enzymes which thereby interferes with interconversion of alanine residues and linking of the alanine residues respectively. Both processes are significant in bacterial cell wall synthesis. Terizidone is a condensation product of two cycloserine molecules. Due to its low activity and adverse effects, Cycloserine is recommended to be used in combination with other anti-TB drugs [40–46].

The 2022 consolidated recommendations by the WHO suggest a set of drugs suitable for extended MDR-TB regimens (Fig. 26.2) (Table 26.2).

26.3.3 Upcoming Anti-TB Drugs

In recent years, there has been a dramatic increase in the variety of diagnostic tools at researchers' disposal. Some examples of these methods include molecular tests for the detection of tuberculosis (TB) and drug resistance, assays based on biomarkers for TB infection and disease, computer-aided detection for TB screening with digital chest radiography, and a novel class of aerosol-capture technologies. There are now 28 medications undergoing Phase I, Phase II, or Phase III studies to treat tuberculosis. These pharmaceuticals include 18 novel chemical entities, two drugs with rapid regulatory approval, one drug recently authorized by the US FDA, and seven repurposed medications. Investigational drugs such as Q-203, BVL-GSK098, TBA-7371, GSK-656, Sutezolid, Delpazolid, PBTZ-169, Sudapyridine, SQ-109, BTZ-043, and OPC-167832 are currently in the different phases of clinical trials. Several new (chemical) classes of drugs have been developed namely Bedaquiline, BVL-GSK098, SQ-109, Q-203, and GSK-656. Phase II or Phase III studies are being conducted on several combination regimens using novel or repurposed medications and host-directed treatments. Drug regimens and delivery modalities for TB

Table 26.2 WHO Categorization of drugs recommended for the treatment of MDR-TB

Group	Medication	Use
A	Levofloxacin or moxifloxacin, Bedaquiline, Linezolid	Highly effective, First Priority
B	Clofazimine, Cycloserine or Terizidone	Second choice of drugs
C	Ethambutol, delamanid, pyrazinamide, imipenem-cilastatin or Meropenem, amikacin (or streptomycin) Ethionamide or prothionamide, p-aminosalicylic acid	To be used in a combination therapy which necessitates the exclusion of Group A & B drugs

Table 26.3 List of drugs in the different phases of TB drug discovery from the early stage of discovery to market approvals

Drug-discovery Stage	Lead optimization	DprE1 inhibitors [47], Direct InhA inhibitors [48], Diarylthiazoles [49], Inhibitors of Mmpl3 [50], ClpC1 [51], RNAP [52], F-ATP synthase [53], ClpP1P2 [54], PKS13 [55], and DnaE1 [56].
Pre-clinical stage	Early-stage development	TBD10 [57], CLB-073 [58], SPR720 [59], JSF-3285 [60], CPZEN-45 [61], NTB-3119 [62], and MBX-4888A [63].
	Toxicology studies	GSK-839 [64], and OTB-658 [65].
Clinical stages	Phase 1	MK-7762 [66], GSK-286 [67], TBAJ-587 [68], TBI-223 [69], and Macozinone [70].
	Phase 2	Q-203 [71], BVL-GSK098 [72], TBAJ-876 [73], Sanfetrinem [74], Delpazolid [75], Sutezolid [76], Tedizolid, BTZ-043 [77], TBA-7371 [78], OPC-167832 [79], GSK-656 [80], TBI-166 [81], and SQ-109 [82].
	Phase 3	Sudapyridine [83]
FDA-approved		Bedaquiline, Delamanid, Pretomanid [84]

preventive therapy are being evaluated in at least 29 clinical trials and implementation research projects. Clinical trials are currently evaluating 16 vaccine candidates, with four in Phase I, eight in Phase II, and four in Phase III (Table 26.3). To reduce tuberculosis incidence and death on a national and worldwide level each year much more quickly than in the past, effective vaccinations are essential. WHO has focused a great deal of emphasis on high-level initiatives targeted at boosting vaccine development and usage because it recognizes the crucial role that TB vaccinations play in achieving quick reductions in TB incidence and death.

26.3.4 Adverse Effects Associated with Anti-TB Drugs

Except for aminoglycosides, all anti-tubercular medicines are beneficial for therapy during pregnancy. Aminoglycosides, including streptomycin, amikacin, and kanamycin, might cause damage to the developing fetus's hearing and are not recommended for use during pregnancy. Pyrazinamide is not recommended for usage during pregnancy in the United States due to its potential teratogenic effects. Regular monitoring of liver function tests is necessary due to the potential hepatotoxic effects of rifampin (RIF), isoniazid, pyrazinamide, and ethambutol. RIF has a low incidence of adverse effects. It has the potential to induce gastrointestinal discomfort. Common adverse reactions associated with ethambutol include dizziness, blurred vision, achromatopsia, nausea, emesis, abdominal discomfort, anorexia, cephalalgia, rash, pruritus, dyspnea, facial, labial or ocular edema, as well as paresthesia in the digits. Regular monitoring of patients using rifampin is necessary to assess for thrombocytopenia and neutropenia, which might occur as side effects.

Rifampin's effects are also achieved via stimulating cytochrome P450 (CYP450), which might lead to undesirable interactions with drugs that are processed by the CYP450 system, resulting in reduced effectiveness of these treatments. Hepatotoxicity and neurotoxicity are the main negative effects that might occur when isoniazid is used. Isoniazid has the potential to induce a deficit of pyridoxine, which may result in peripheral neuropathy in individuals. Pyrazinamide has the potential to elevate levels of uric acid. Rifampin, isoniazid, pyrazinamide, and ethambutol, which are all primary anti-tubercular drugs, have the potential to cause liver damage. If there is a consistent increase in liver function test results, it is advisable to stop the therapy. Cessation of aminoglycoside treatment leads to the reversal of nephrotoxicity. The occurrence of renal toxicity is contingent upon the presence of any preexisting renal illness in the patient as well as the dosage of the medicine being delivered. The occurrence of renal insufficiency may be prevented in the majority of people. The administration of SM is related to ototoxicity and nephrotoxicity. Vestibular dysfunction is a more prevalent condition compared to auditory impairment. Patients receiving SM should undergo regular monitoring of their auditory and renal function. Gastrointestinal discomfort and vomiting are the most often seen side effects associated with Pretomanid. The predominant adverse effects of Delamanid were hepatotoxicity, electrolyte imbalance, and QT prolongation which is also observed with Bedaquiline [85–90].

26.4 Repositioning Existing Drugs

Tuberculosis is the most prevalent infectious illness globally, with a significant number of tuberculosis (TB) cases recorded annually. Despite extensive efforts to manage the illness, two challenges continue to exist: medication resistance and a scarcity of appropriate treatment alternatives. Drug repurposing, the practice of using existing medications for a new therapeutic purpose, has become the most effective alternative therapy for tuberculosis (TB) (Fig. 26.4, Table 26.4). The use of existing safety records and pharmacokinetic characteristics of established pharmacological drugs is an added advantage in this method, as it may accelerate the discovery of new medications for TB. Costly research procedures and prolonged development durations have hindered the progress in finding TB drugs. Commercializing new chemical compounds necessitates an investment entailing substantial risk and more than a decade of research. Utilizing existing medications provides a means to avoid the substantial investment of time and money [91–95].

Drug repurposing is a crucial strategy in the treatment of multidrug-resistant tuberculosis (MDR-TB) and extensively drug-resistant tuberculosis (XDR-TB). Drug repurposing may be used to target many pathways associated with the development of TB, including mycobacterial growth, modulation of the host immunological response, and reduction of inflammation. By using the pharmacological safety profiles of pharmaceuticals that have previously received approval for other uses, repurposing may aid in addressing the pressing demand for improved TB

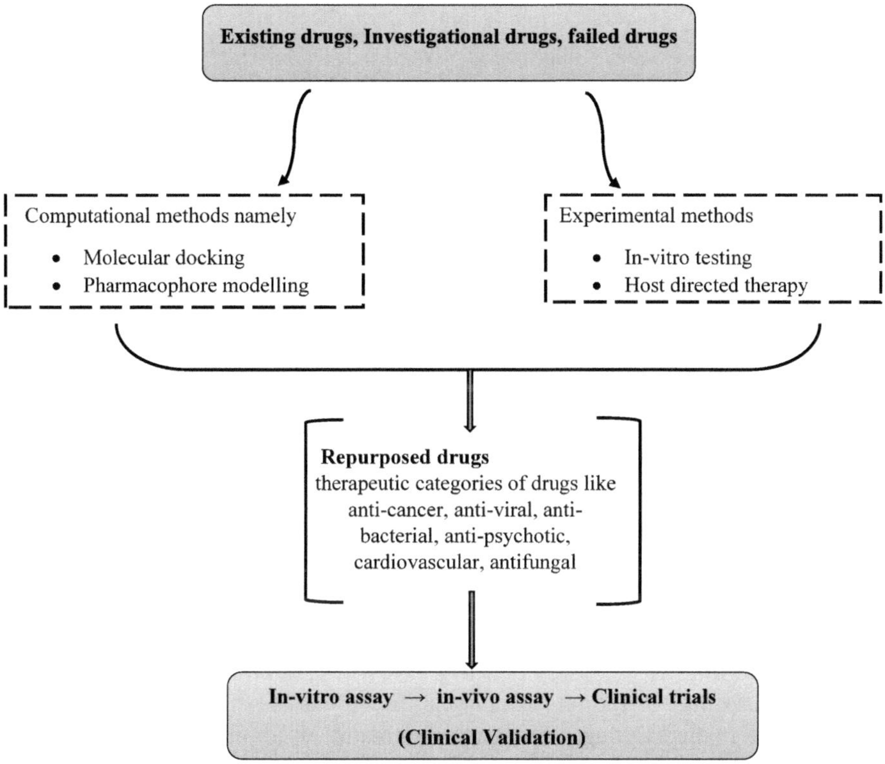

Fig. 26.4 Steps involved in Drug repurposing for the treatment of MDR TB

Table 26.4 Repurposed Anti-TB drugs currently in the treatment & clinical studies

In-vitro studies	Fusidic acid [110], Cefazolin [111], Lymecycline [112], Vancomycin [113], Echinacoside [114], Bortezomib [115], Clotrimazole [116], Artemisinin [117], Pyronaridine [118], Ipronidazole [116], Saquinavir, Ivermectin, Niclosamide, Doxorubicin, Chlorpromazine, and Rimonabant.
In-vivo studies	Biapenem, Cefdinir, Econazole [119], Tebipenem [120], Cefpodoxime [121], Minocycline [122], Imatinib [123], Gefitnib [124], Verapamil [108], Simvastatin [125], Atorvastatin [126], Acarbose [127], Disulfiram [128], Bis-biguanide HCl [129]
Clinical development	Metronidazole [106], Doxycycline [130], Aspirin [131], Celecoxib [132], Ibuprofen [133], Pravastatin [134], Metformin [135], Vitamin D [136], Thalidomide [137], Meloxicam [138]
Approved drugs	Linezolid, Moxifloxacin, Levofloxacin, Clofazimine, Meropenem, Gatifloxacin [96]

treatment. FDA-approved drugs, which were originally prescribed for the treatment of infectious ailments, are now being examined in sophisticated clinical trials as potential remedies for TB. The proliferation of antibiotic resistance in Mycobacterium tuberculosis has posed a significant challenge to disease control, notwithstanding the availability of a diverse array of anti-TB medications. Thus, the creation of potent medications with unique mechanisms of action is crucial to the battle against drug resistance. Because some of these medications may work on more than one target, resistance is less likely to emerge. Hence, repurposing pharmaceuticals presents itself as a very effective option for the development of a treatment regimen targeting multidrug-resistant (MDR) and extensively drug-resistant (XDR) tuberculosis (TB). The main objective of medicines is to cure specific illnesses or afflictions. When these medications are used to treat other illnesses, the painstaking process of toxicity profiling, target validation, hit-to-lead optimization, and/or in vivo metabolic research is usually unnecessary as the pharmacokinetics and pharmacodynamics studies for these medicines are already known. With drug safety and pharmacokinetic study data available, the time required to develop treatment can be reduced [96–98].

Repurposed medications may be classified into three categories: anti-infectives, non-infective agents, and a miscellaneous class of pharmaceuticals. A wide range of pharmacological medications, including anti-helminthics, anti-diabetic, anticancer, anti-fungal, anti-psychotic, and miscellaneous pharmaceuticals such as vitamin D and fusidic acid, have been discovered to possess anti-TB action or may be used as supplementary treatment in conjunction with other anti-TB treatments. Below is a concise overview of the key repurposed medications used in the therapy.

(a) **Clofazimine**

Clofazimine, a riminophenazine antibiotic, was first created by a research team at Trinity College, Dublin as a treatment for tuberculosis (TB) [8]. However, it was later determined to be unsuccessful and hence not pursued further. Subsequently, it was used in the management of leprosy. Clofazimine has both anti-inflammatory and anti-cancer properties. Clofazimine was included in the Bangladesh regimen, a treatment protocol created by Damien Foundation India Trust during research conducted in Bangladesh, for the management of multidrug-resistant tuberculosis (MDR TB) that is resistant to rifampicin and isoniazid [9, 10]. In 2019, the World Health Organization (WHO) included Clofazimine in the treatment regimen for multidrug-resistant tuberculosis (MDR TB), in conjunction with other medications. The medication exerts its effects on the bacterial membrane. The postulated mechanism of action involves the activation of Clofazimine by type 2-NADH dehydrogenase. This activation reduces Clofazimine, leading to the generation of reactive oxygen species (ROS) when it is reoxidized by oxygen (O_2) [11]. These ROS then cause the rupture of the bacterial membrane. The CZM-containing regimen requires a treatment duration of 9–12 months. Another idea proposes that the drug interacts with the phospholipids in the bacterial membrane, resulting in the production of lysophospholipids which interferes with the ATP synthesis [96, 99, 100].

(b) **Linezolid**

Linezolid, the first commercially available 1, 3-oxazolidinone, was discovered by Pharmacia & Upjohn in the 1990s. It has been authorized for the treatment of gram-positive bacterial infections affecting the lungs, skin, and soft tissues [13]. Linezolid has shown encouraging outcomes as a secondary treatment option in the management of multidrug-resistant tuberculosis (MDR-TB). The method of action includes inhibiting bacterial protein synthesis by reducing mRNA translation [14]. Linezolid is regarded by the World Health Organization (WHO) as a crucial medication for treating rifampicin-resistant or multidrug-resistant tuberculosis (MDR-TB), and it is classified as an essential medicine [96, 101].

(c) **Fluoroquinolones (Moxifloxacin & Levofloxacin)**

Fluoroquinolones have been used as a secondary option for treating tuberculosis, in conjunction with other medications specifically designed to combat tuberculosis. This category of medications is mostly used for the treatment of respiratory and dermatological infections. It disrupts the process of DNA replication in bacteria by blocking the activity of topoisomerase enzymes II and IV. Drugs from the Fluoroquinolone class, including levofloxacin and gatifloxacin, have shown encouraging efficacy against Mtb bacteria. Moxifloxacin and levofloxacin are classified as Group A medications by the World Health Organization (WHO) for the treatment of multidrug-resistant tuberculosis (MDR TB) [102, 103].

(d) **Combination of Anti-microbials**

The treatment of multidrug-resistant (MDR) and extensively drug-resistant (XDR) tuberculosis (TB) often involves the use of repurposed pharmaceuticals in combination with other anti-TB medications. The use of a combination of clofazimine, ethambutol, and moxifloxacin has shown promise in treating multi-drug resistant tuberculosis (MDR TB). The synergistic impact resulting from the combination of pharmaceuticals has played a vital role in enhancing the effectiveness of these medications. Various medication combinations, such as bedaquiline with linezolid, carbapenems with rifampicin, and a trio combination of clofazimine, bedaquiline, and pyrazinamide, are successful in treating drug-resistant tuberculosis. The management of multidrug-resistant tuberculosis (MDR-TB) has included the administration of a combination therapy of amoxicillin/clavulanic acid and other second-line medications. Meropenem, in combination with clavulanate, has shown in vitro effectiveness against extensively drug-resistant (XDR) pathogens. There are ongoing attempts to repurpose these beta-lactams as novel therapeutics for tuberculosis (TB). The recent use of a combination of amoxicillin/clavulanic acid and carbapenems has effectively decreased the impact of M. tuberculosis, confirmed by a bactericidal activity study. Carbapenems have been successfully used in salvage therapy for patients with extensively drug-resistant (XDR) infections [104, 105]. Nitroimidazopyrans, like the commonly used antibiotic metronidazole, has shown effectiveness against both actively dividing and dormant strains of M. tuberculosis. As a consequence of these discoveries, the novel chemical entities (NCEs) PA-824 and OPC-67683 are now being examined in clinical trials. Studies using macaque infection models have shown that metronidazole effectively prevents the reactivation of dormant bacilli. Additionally, metronidazole has strong antimicrobial activity against Mycobacterium tuberculosis [106, 107].

(e) **Metformin & verapamil**

Metformin, an anti-diabetic drug, has been discovered to enhance the efficacy of other anti-TB medications, making it a valuable adjunct in the treatment of tuberculosis. The anti-TB efficacy is derived from the suppression of mycobacterial development via the induction of reactive oxygen species (ROS) generation. The use of verapamil to inhibit Mtb efflux pumps has been demonstrated to be advantageous since these pumps are a contributing factor to the development of resistance to anti-TB medicines [108, 109].

(f) **Anti-cancer medications**

Bortezomib is a pharmaceutical compound that functions as a proteasome inhibitor and is used for the treatment of multiple myeloma. This medication hinders the proliferation of cancerous cells by impacting the signaling pathway of activated B cells known as NFkB, as well as by impeding the signaling of interleukin (IL)-6. The mechanism of action of Bortezomib involves inhibition of mycobacterial ClpP [115]. Imatinib, used in the treatment of chronic myelogenous leukaemia, has also been shown to disrupt the M.tb pathogenesis. Anticancer drugs, namely Imatinib, Sorafenib, and Elesclomol, have been investigated for anti-TB activity [123, 139].

Since, a multi-drug regimen is a common practice in the treatment of infectious diseases, repurposing of drugs plays a crucial role here in expanding the armoury. Though the approach is an appealing one, it does come with a few limitations. The availability of pharmacokinetic & other safety data is an advantage in this process, but still, new clinical studies would be required to adhere to the latest FDA regulations. There are chances of promising in-vivo activity profile of a drug not being reproducible in in-vivo conditions due to toxicity, and contraindications. Drug repurposing remains a good alternative in drug development especially in the cases of a pandemic as novel drug discovery is time-consuming.

26.5 Computational Approaches

The advent of computational approaches in drug discovery has revolutionized the development of treatments for various diseases, including tuberculosis (TB). Computational techniques offer innovative ways to accelerate the discovery of new drugs by leveraging advanced algorithms, molecular modeling, and *in-silico* simulations. These approaches help in identifying potential drug targets, optimizing lead compounds, and predicting drug resistance mechanisms, thus providing a cost-effective and efficient alternative to conventional drug development pipelines. By integrating bioinformatics, cheminformatics, and structural biology, computational methods hold immense potential in discovering novel therapeutics for MTB, addressing existing challenges and emerging threats in TB treatment. Various structure-based drug design approaches like molecular docking, molecular dynamics, Pharmacophore screening and Virtual ligand screening have been employed extensively in an attempt to mitigate the drug-discovery thirst against various tuberculosis and its resistant forms.

Isoniazid (Inh)-resistant TB presents various mutated proteins that can be potential drug targets including InhA, FabD, and AhpC. A research study involved the identification of lead molecules against these mutated proteins. Since these mutated protein models were unavailable, single-point mutations were inserted into the native proteins. The binding sites of Inh were predicted along the surfaces of these mutated proteins with the aid of information from PDB ID 4qxm which bore Inh as the co-crystallized ligand. Based on the 3D structures of Inh and mutated proteins, a high-throughput virtual screening was carried out. This was subsequently followed by molecular docking to reveal 826 hit molecules. The ChemBL database was used to screen through libraries of molecules. The top 3 molecules-1 each against these mutated proteins were identified. Furthermore, interactions with the single-point mutations present in these proteins were analyzed to determine the affinity of the lead compounds compared to Inh and the ligands revealed better interactions. These lead molecules were also screened for ADMET properties and the results obtained corroborated the druggability of these lead molecules. Pharmacophore analysis revealed ideal properties supporting the stable binding of the ligands to the proteins [140].

A drug-repurposing approach was used in a research study targeting Fad32, a protein vital in mycolic acid biosynthesis native to *M.tb*. Virtual ligand screening was performed to reveal 4 FDA-approved hit molecules accolate, sorafenib, mefloquine, and loperamide. Later molecular docking and subsequent MM-PBSA studies were carried out and it revealed accolate and sorafenib with the lowest overall binding energies. Analysis of protein-ligand interactions revealed Van Der Waals interactions with most contributions to the protein-ligand complex. Per residue interaction analysis supported the case of accolate as a ligand which showed most interactions. Further, MD simulations were carried out for 150 ns and the Accolate-Fad32 complex showed an average RMSD of 1.05 Å. RMSF analysis revealed reduced fluctuations in the amino acid positions with accolate in the binding site as compared to other ligands. The radius of gyration and Dynamic cross-correlation matrix further revealed less conformational flexibility with accolate in the binding cavity [141].

A recent in silico study involved virtual ligand screening against Shikimate kinase as an *M.tb* drug target. Virtual screening of 154,118 compounds was carried out by rigid molecular docking in the first step, followed by flexible docking in the second and third steps. The final number of hits obtained was 11 after these steps. These compounds were further assayed in vitro and the IC_{50} was found to be 1.39 μM which was comparable to test compounds rifampicin and isoniazid. Toxicity against various cell lines revealed the compound to be non-toxic. MD simulations also revealed adequate RMSD, RMSF, and Rg profiles which validated the stability of the protein-ligand complex. The binding mode observed during MD simulations also revealed the potential of competitive inhibition with the drug target [142].

Various 2-alkoxy quinoline-3-carbaldehyde analogs were designed in an attempt to act against *M.tb*. 32 designed ligands were filtered and predicted for lead-likeliness, oral bioavailability, and pharmacokinetic parameters. 14 compounds fit the initial lead-likeliness parameters and further 5 compounds passed toxicity predictions. All of these 5 compounds passed Lipinski's rule. All 5 compounds also showed adequate pharmacokinetic parameters. All these compounds were further analyzed for DFT calculations. The energy gap between HOMO and LUMO orbitals was noted to be in the range of 4.117 to 4.286 eV. Lower energy gaps in 2 of these compounds showed higher electron-donating prowess of these compounds over the other ones. These 5 compounds were docked against 8 different targets, and enoyl reductase was revealed to give the best results. 2 compounds revealed the best docking scores and adequate amino acid interaction profiles compared to standard drugs ethambutol and isoniazid. Molecular dynamics revealed RMSD, RMSF, Rg, and SASA properties for protein-ligand complex one of them being the best and it was selected as a lead molecule for further studies [143].

Some chromenone derivatives were designed based on previously reported molecules in the literature. Key enzymes involved in mycolic acid biosynthesis include polyketide synthases (Pks). Pk13 was used as a drug target to identify novel derivatives that could bind to the enzyme's active site. Molecular docking studies were used to analyze the binding mode and key amino acid residues within the active site which propelled the designing of these novel derivatives. Two such derivatives were to inhibit the tuberculosis cell line and showed IC_{50} values of 22.05 μM and 14.30 μM [55].

Various ligand-based drug design methods like homology modeling, QSAR, and ligand-based pharmacophore screening have been extensively employed in anti-tubercular drug design and discovery.

A QSAR equation was developed using synthesized triazole derivatives reported in the literature. DNA gyrase is a crucial target involved in the DNA replication process. Initially, the molecules were optimized using B3LYP functional. Later, descriptors were predicted for these optimized structures. A QSAR model was developed and validated. The molecules were docked into the 3D crystal structure of DNA gyrase and the binding modes as well as amino acid interaction profiles. One of the derivatives showed a higher docking score than isoniazid [144].

A research study was carried out to develop novel derivatives against the DprE1 enzyme. Several azaindole derivatives reported previously in literature were used to develop a pharmacophore model. The best pharmacophore hypothesis was selected for further screening. Pharmacophore screening was performed with the ChemBL database. Later an atom-based QSAR screening was performed and contour maps for the least and most active compounds were developed. Molecular docking was used to rank the obtained hits and the top 10 molecules showed docking energies between −9 to −10 kcal/mol. Later MM-GBSA studies were used to filter through the hits obtained after docking protocol. All the hits with free energy less than −40 kcal/mol were selected for further studies. The top 10 hits obtained were initially subjected to the prediction of lead-likeliness and later, their anti-TB activity was predicted. 5 best molecules were selected for MD simulations for a 200 ns timeframe. The best RMSD and RMSF profiles were obtained with one of these derivatives [145].

A machine-learning model was used to develop a tool for in vitro anti-Tb in vitro activity prediction using a large dataset of molecules previously reported in the literature. Not only that, but also many new molecules were predicted as potential candidates against TB. The chemical space of TB drug discovery along with possible new targets were also identified [146].

26.6 Conclusions

The increase in computational capacity, along with the wealth of accessible biological data, has facilitated the significant contribution of in silico research to anti-TB drug discovery. Significant advancements have been made in our comprehension of MTB biology, the effects of mutations and ensuing resistance development, as well as in the formulation of innovative compounds effective against these resistant bacterial strains. The identification and clarification of novel drug targets have significantly enhanced structure-based techniques like molecular docking and molecular dynamics simulations, while QSAR modeling and other machine learning methods have been supported by the establishment of publicly accessible datasets. Nonetheless, numerous obstacles persist. Data quality is a persistent issue, necessitating that every theoretical model undergo rigorous validation and assessment for its practical applicability. It is crucial to build a collaborative framework that facilitates the swift experimental confirmation of ideas generated in silico. It is anticipated that computational methodologies will persist in their significant role, facilitating enhanced treatment of MTB and ultimately its eradication.

References

1. World Health Organization. Global tuberculosis report. WHO. 2023. 2023.
2. Rendon A, Tiberi S, Scardigli A, D'Ambrosio L, Centis R, Caminero JA, et al. Classification of drugs to treat multidrug-resistant tuberculosis (MDR-TB): evidence and perspectives. J Thorac Dis. 2016;8(10):2666–71.
3. Seung KJ, Keshavjee S, Rich ML. Multidrug-resistant tuberculosis and extensively drug-resistant tuberculosis. Cold Spring Harb Perspect Med. 2015;5(9):a017863.
4. Chin KL, Anibarro L, Chang ZY, Palasuberniam P, Mustapha ZA, Sarmiento ME, et al. Impacts of MDR/XDR-TB on the global tuberculosis epidemic: challenges and opportunities. Curr Res Microb Sci. 2024;7:100295.
5. Bu Q, Qiang R, Fang L, Peng X, Zhang H, Cheng H. Global trends in the incidence rates of MDR and XDR tuberculosis: findings from the global burden of disease study 2019. Front Pharmacol. 2023;14:1156249.
6. Dartois VA, Rubin EJ. Anti-tuberculosis treatment strategies and drug development: challenges and priorities. Nat Rev Microbiol. 2022;20(11):685–701.

7. Dorman SE, Schumacher SG, Alland D, Nabeta P, Armstrong DT, King B, et al. Xpert MTB/RIF Ultra for detection of Mycobacterium tuberculosis and rifampicin resistance: a prospective multicentre diagnostic accuracy study. Lancet Infect Dis. 2018;18(1):76–84.
8. Mase SR, Chorba T. Treatment of drug-resistant tuberculosis. Clin Chest Med. 2019;40(4):775–95.
9. Salari N, Kanjoori AH, Hosseinian-Far A, Hasheminezhad R, Mansouri K, Mohammadi M. Global prevalence of drug-resistant tuberculosis: a systematic review and meta-analysis. Infect Dis Poverty. 2023;12(1):57.
10. WHO. WHO announces updated definitions of extensively drug-resistant tuberculosis. WHO. 2021.
11. Prasad R, Singh A, Gupta N. Adverse drug reactions with first-line and second-line drugs in treatment of tuberculosis. Ann Natl Acad Med Sci. 2021;57:16.
12. Jang JG, Chung JH. Diagnosis and treatment of multidrug-resistant tuberculosis. Yeungnam Univ J Med. 2020;37(4):277–85.
13. Gandhi NR, Andrews JR, Brust JCM, Montreuil R, Weissman D, Heo M, et al. Risk factors for mortality among MDR- and XDR-TB patients in a high HIV prevalence setting. Int J Tuberc Lung Dis. 2012;16(1):90–7.
14. Chung-Delgado K, Guillen-Bravo S, Revilla-Montag A, Bernabe-Ortiz A. Mortality among MDR-TB cases: comparison with drug-susceptible tuberculosis and associated factors. PLoS One. 2015;10(3):e0119332.
15. Palomino J, Martin A. Drug resistance mechanisms in Mycobacterium tuberculosis. Antibiotics. 2014;3(3):317–40.
16. Velayati AA, Farnia P, Masjedi MR. The totally drug resistant tuberculosis (TDR-TB). Int J Clin Exp Med. 2013;6(4):307–9.
17. Yaşar N, Tezcan Ülger S, Ülger M, Arslantürk A, Aslan G, Köksal F. Etambutole Dirençli ve Duyarlı Mycobacterium tuberculosis Kompleksi Klinik İzolatlarında embA, embB ve embC Gen Bölgesi Mutasyonları. Mikrobiyol Bul. 2023;57(1):45–59.
18. Bwalya P, Solo ES, Chizimu JY, Shrestha D, Mbulo G, Thapa J, et al. Characterization of embB mutations involved in ethambutol resistance in multi-drug resistant Mycobacterium tuberculosis isolates in Zambia. Tuberculosis. 2022;133:102184.
19. Ramirez-Busby SM, Valafar F. Systematic review of mutations in pyrazinamidase associated with pyrazinamide resistance in Mycobacterium tuberculosis clinical isolates. Antimicrob Agents Chemother. 2015;59(9):5267–77.
20. Campbell PJ, Morlock GP, Sikes RD, Dalton TL, Metchock B, Starks AM, et al. Molecular detection of mutations associated with first- and second-line drug resistance compared with conventional drug susceptibility testing of Mycobacterium tuberculosis. Antimicrob Agents Chemother. 2011;55(5):2032–41.
21. Bollela VR, Namburete EI, Feliciano CS, Macheque D, Harrison LH, Caminero JA. Detection of katG and & inhA mutations to guide isoniazid and ethionamide use for drug-resistant tuberculosis. Int J Tuberc Lung Dis. 2016;20(8):1099–104.
22. Li M-C, Jie L, Yao L, Xiao T-Y, Liu H-C, Lin S-Q, et al. rpoB mutations and effects on rifampin resistance in Mycobacterium tuberculosis. Infect Drug Resist. 2021;14:4119–28.
23. Watanabe S, Matsumura K, Iwai H, Funatogawa K, Haishima Y, Fukui C, et al. A mutation in the 16S rRNA decoding region attenuates the virulence of Mycobacterium tuberculosis. Infect Immun. 2016;84(8):2264–73.
24. Pandey B, Grover S, Kaur J, Grover A. Analysis of mutations leading to para-aminosalicylic acid resistance in Mycobacterium tuberculosis. Sci Rep. 2019;9(1):13617.
25. Gygli SM, Borrell S, Trauner A, Gagneux S. Antimicrobial resistance in Mycobacterium tuberculosis: mechanistic and evolutionary perspectives. FEMS Microbiol Rev. 2017;41(3):354–73.
26. Balganesh M, Dinesh N, Sharma S, Kuruppath S, Nair AV, Sharma U. Efflux pumps of Mycobacterium tuberculosis play a significant role in antituberculosis activity of potential drug candidates. Antimicrob Agents Chemother. 2012;56(5):2643–51.

27. Laws M, Jin P, Rahman KM. Efflux pumps in Mycobacterium tuberculosis and their inhibition to tackle antimicrobial resistance. Trends Microbiol. 2022;30(1):57–68.
28. Snobre J, Meehan CJ, Mulders W, Rigouts L, Buyl R, de Jong BC, et al. Frameshift mutations in the *mmpR5* gene can have a bedaquiline-susceptible phenotype by retaining a protein structure and function similar to wild-type *Mycobacterium tuberculosis*. Antimicrob Agents Chemother. 2024;68(12):e0085424.
29. Machado D, Couto I, Perdigão J, Rodrigues L, Portugal I, Baptista P, et al. Contribution of efflux to the emergence of isoniazid and multidrug resistance in Mycobacterium tuberculosis. PLoS One. 2012;7(4):e34538.
30. Dookie N, Rambaran S, Padayatchi N, Mahomed S, Naidoo K. Evolution of drug resistance in Mycobacterium tuberculosis: a review on the molecular determinants of resistance and implications for personalized care. J Antimicrob Chemother. 2018;73(5):1138–51.
31. Rostamian M, Kooti S, Abiri R, Khazayel S, Kadivarian S, Borji S, et al. Prevalence of Mycobacterium tuberculosis mutations associated with isoniazid and rifampicin resistance: a systematic review and meta-analysis. J Clin Tuberc Other Mycobact Dis. 2023;32:100379.
32. Warrier T, Kapilashrami K, Argyrou A, Ioerger TR, Little D, Murphy KC, et al. N-methylation of a bactericidal compound as a resistance mechanism in Mycobacterium tuberculosis. Proc Natl Acad Sci. 2016;113(31):E4523–30.
33. Reeves AZ, Campbell PJ, Sultana R, Malik S, Murray M, Plikaytis BB, et al. Aminoglycoside cross-resistance in Mycobacterium tuberculosis due to mutations in the 5′ untranslated region of *whiB7*. Antimicrob Agents Chemother. 2013;57(4):1857–65.
34. Sachan RSK, Mistry V, Dholaria M, Rana A, Devgon I, Ali I, et al. Overcoming Mycobacterium tuberculosis drug resistance: novel medications and repositioning strategies. ACS Omega. 2023;8(36):32244–57.
35. Zhang X, Zhao R, Qi Y, Yan X, Qi G, Peng Q. The progress of Mycobacterium tuberculosis drug targets. Front Med (Lausanne). 2024;11:1455715.
36. Shetye GS, Franzblau SG, Cho S. New tuberculosis drug targets, their inhibitors, and potential therapeutic impact. Transl Res. 2020;220:68–97.
37. Mi J, Gong W, Wu X. Advances in key drug target identification and new drug development for tuberculosis. Biomed Res Int. 2022;2022:1–23.
38. Alsayed SSR, Gunosewoyo H. Tuberculosis: pathogenesis, current treatment regimens and new drug targets. Int J Mol Sci. 2023;24(6):5202.
39. WHO. Global tuberculosis report. WHO. 2023.
40. Tiberi S, Scardigli A, Centis R, D'Ambrosio L, Muñoz-Torrico M, Salazar-Lezama MÁ, et al. Classifying new anti-tuberculosis drugs: rationale and future perspectives. Int J Infect Dis. 2017;56:181–4.
41. Caminero JA, Scardigli A. Classification of antituberculosis drugs: a new proposal based on the most recent evidence. Eur Respir J. 2015;46(4):887–93.
42. Yee D, Valiquette C, Pelletier M, Parisien I, Rocher I, Menzies D. Incidence of serious side effects from first-line antituberculosis drugs among patients treated for active tuberculosis. Am J Respir Crit Care Med. 2003;167(11):1472–7.
43. Bendre AD, Peters PJ, Kumar J. Tuberculosis: past, present and future of the treatment and drug discovery research. Curr Res Pharmacol Drug Discov. 2021;2:100037.
44. Sotgiu G, Centis R, D'ambrosio L, Migliori GB. Tuberculosis treatment and drug regimens. Cold Spring Harb Perspect Med. 2015;5(5):a017822.
45. Bakare AA, Moses VY, Beckely CT, Oluyemi TI, Ogunfeitimi GO, Adelaja AA, et al. The first-line antituberculosis drugs, and their fixed-dose combination induced abnormal sperm morphology and histological lesions in the testicular cells of male mice. Front Cell Dev Biol. 2022;10:1023413.
46. Wu C, Yi H, Hu Y, Luo D, Tang Z, Wen X, et al. Effects of second-line anti-tuberculosis drugs on the intestinal microbiota of patients with rifampicin-resistant tuberculosis. Front Cell Infect Microbiol. 2023;13:1127916.

47. Amado PSM, Woodley C, Cristiano MLS, O'Neill PM. Recent advances of DprE1 inhibitors against *Mycobacterium tuberculosis*: computational analysis of physicochemical and ADMET properties. ACS Omega. 2022;7(45):40659–81.
48. Manjunatha UH, Rao SPS, Kondreddi RR, Noble CG, Camacho LR, Tan BH, et al. Direct inhibitors of InhA are active against Mycobacterium tuberculosis. Sci Transl Med. 2015;7(269):269ra3.
49. Nandurkar Y, Bhoye MR, Maliwal D, Pissurlenkar RRS, Chavan A, Katade S, et al. Synthesis, biological screening and in silico studies of new N-phenyl-4-(1,3-diaryl-1H-pyrazol-4-yl)thiazol-2-amine derivatives as potential antifungal and antitubercular agents. Eur J Med Chem. 2023;258:115548.
50. Choksi H, Carbone J, Paradis NJ, Bennett L, Bui-Linh C, Wu C. Novel inhibitors to MmpL3 transporter of *Mycobacterium tuberculosis* by structure-based high-throughput virtual screening and molecular dynamics simulations. ACS Omega. 2024;9(12):13782–96.
51. Bhanot A, Lunge A, Kumar N, Kidwai S, Singh R, Sundriyal S, et al. Discovery of small molecule inhibitors of Mycobacterium tuberculosis ClpC1: SAR studies and antimycobacterial evaluation. Results Chem. 2023;5:100904.
52. Moglad EH. Inhibiting the oligomerization of mycobacterial DNA-directed RNA polymerase (RNAP) using natural compound via in-silico techniques. Med Nov Technol Devices. 2024;21:100286.
53. Harikishore A, Grüber G. Mycobacterium tuberculosis F-ATP synthase inhibitors and targets. Antibiotics. 2024;13(12):1169.
54. Moreira W, Santhanakrishnan S, Ngan GJY, Low CB, Sangthongpitag K, Poulsen A, et al. Towards selective mycobacterial ClpP1P2 inhibitors with reduced activity against the human proteasome. Antimicrob Agents Chemother. 2017;61(5):e02307–16.
55. Wang X, Zhao W, Wang B, Ding W, Guo H, Zhao H, et al. Identification of inhibitors targeting polyketide synthase 13 of Mycobacterium tuberculosis as antituberculosis drug leads. Bioorg Chem. 2021;114:105110.
56. Chengalroyen MD, Mason MK, Borsellini A, Tassoni R, Abrahams GL, Lynch S, et al. DNA-dependent binding of nargenicin to DnaE1 inhibits replication in *Mycobacterium tuberculosis*. ACS Infect Dis. 2022;8(3):612–25.
57. Holtzman D. Gates MRI TB drug development activities. In: WGND annual meeting. Bill & Melinda Gates Medical Research Institute. 2023. p. 1–5.
58. Fleischman T, Chronicle C. Cornell-developed anti-TB compound headed to trials. Cornell Chronicle. 2023.
59. Talley AK, Thurston A, Moore G, Gupta VK, Satterfield M, Manyak E, et al. First-in-human evaluation of the safety, tolerability, and pharmacokinetics of SPR720, a novel oral bacterial DNA gyrase (GyrB) inhibitor for mycobacterial infections. Antimicrob Agents Chemother. 2021;65(11):e0120821.
60. Adewumi AT, Oluyemi WM, Adekunle YA, Adewumi N, Alahmdi MI, Soliman MES, et al. Propitious indazole compounds as β-ketoacyl-ACP synthase inhibitors and mechanisms unfolded for TB cure: integrated rational design and MD simulations. ChemistrySelect. 2023;8(3). https://doi.org/10.1002/slct.202203877.
61. Garcia-Contreras L, Hanif SNM, Ibrahim M, Durham P, Hickey AJ. The pharmacokinetics of CPZEN-45, a novel anti-tuberculosis drug, in guinea pigs. Pharmaceutics. 2023;15(12):2758.
62. Wang X, Lang X, Wang F, Tan H, Li G, Li P, et al. Simultaneous determination of NTB-3119, a novel anti-tuberculosis agent, and its major metabolites in mouse plasma by LC-MS/MS and its application in preclinical pharmacokinetics study. J Pharm Biomed Anal. 2023;224:115172.
63. Bauman AA, Sarathy JP, Kaya F, Massoudi LM, Scherman MS, Hastings C, et al. Spectinamide MBX-4888A exhibits favorable lesion and tissue distribution and promotes treatment shortening in advanced murine models of tuberculosis. Antimicrob Agents Chemother. 2024;68(11):e0071624.

64. Dartois V, Dick T. Therapeutic developments for tuberculosis and nontuberculous mycobacterial lung disease. Nat Rev Drug Discov. 2024;23(5):381–403.
65. Liu H, Zhu H, Fu L, Zhang W, Chen X, Wang B, et al. Efficacy of replacing linezolid with OTB-658 in anti-tuberculosis regimens in murine models. Antimicrob Agents Chemother. 2023;67(2):e0139922.
66. Merck. Merck and the Bill & Melinda Gates Medical Research Institute announce licensing agreement for novel tuberculosis antibiotic candidates. Merck. 2024.
67. Nuermberger EL, Martínez-Martínez MS, Sanz O, Urones B, Esquivias J, Soni H, et al. GSK2556286 is a novel antitubercular drug candidate effective *in vivo* with the potential to shorten tuberculosis treatment. Antimicrob Agents Chemother. 2022;66(6):e0013222.
68. Xu J, Converse PJ, Upton AM, Mdluli K, Fotouhi N, Nuermberger EL. Comparative efficacy of the novel diarylquinoline TBAJ-587 and bedaquiline against a resistant *Rv0678* mutant in a mouse model of tuberculosis. Antimicrob Agents Chemother. 2021;65(4):e02418–20.
69. Gordon O, Dikeman DA, Ortines RV, Wang Y, Youn C, Mumtaz M, et al. The novel oxazolidinone TBI-223 is effective in three preclinical mouse models of methicillin-resistant Staphylococcus aureus infection. Microbiol Spectr. 2022;10(5):e0245121.
70. Koryakova A, Shcherbakova V, Riabova O, Kazaishvili Y, Bolgarin R, Makarov V. Antituberculosis macozinone extended-release tablets to enhance bioavailability: a pilot pharmacokinetic study in beagle dogs. Microbiol Spectr. 2023;11(1):e0232722.
71. Kim J, Choi J, Kang H, Ahn J, Hutchings J, van Niekerk C, et al. Safety, tolerability, and pharmacokinetics of telacebec (Q203), a new antituberculosis agent, in healthy subjects. Antimicrob Agents Chemother. 2022;66(1):e0143621.
72. Pieren M, Abáigar Gutiérrez-Solana A, Antonijoan Arbós RM, Boyle GW, Davila M, Davy M, et al. First-in-human study of alpibectir (BVL-GSK098), a novel potent anti-TB drug. J Antimicrob Chemother. 2024;79(6):1353–61.
73. Lombardi A, Pappas F, Nedelman J, Hickman D, Jaw-Tsai S, Olugbosi M, et al. Pharmacokinetics and safety of TBAJ-876, a novel antimycobacterial diarylquinoline, in healthy subjects. Antimicrob Agents Chemother. 2024;68(10):e0061324.
74. Ramón-García S, González del Río R, Arenaz-Callao MP, Boshoff H, Rullás J, Anca S, et al. Sanfetrinem, an oral β-lactam antibiotic repurposed for the treatment of tuberculosis. bioRxiv. 2024. https://doi.org/10.1101/2024.10.10.617558.
75. Wallis RS, Dawson R, Friedrich SO, Venter A, Paige D, Zhu T, et al. Mycobactericidal activity of sutezolid (PNU-100480) in sputum (EBA) and blood (WBA) of patients with pulmonary tuberculosis. PLoS One. 2014;9(4):e94462.
76. Chen RH, Burke A, Cho JG, Alffenaar JW, Forsman LD. New oxazolidinones for tuberculosis: are novel treatments on the horizon? Pharmaceutics. 2024;16(6):818.
77. Eckhardt E, Li Y, Mamerow S, Schinköthe J, Sehl-Ewert J, Dreisbach J, et al. Pharmacokinetics and efficacy of the benzothiazinone BTZ-043 against tuberculous mycobacteria inside granulomas in the guinea pig model. Antimicrob Agents Chemother. 2023;67(4):e0143822.
78. Working Group on TB Drugs. TB Alliance, Bill & Melinda Gates Medical Research Institute, Foundation for Neglected Disease Research. TBA=7371.
79. Hariguchi N, Chen X, Hayashi Y, Kawano Y, Fujiwara M, Matsuba M, et al. OPC-167832, a novel carbostyril derivative with potent antituberculosis activity as a DprE1 inhibitor. Antimicrob Agents Chemother. 2020;64(6):e02020–19.
80. Diacon AH, Barry CE, Carlton A, Chen RY, Davies M, de Jager V, et al. A first-in-class leucyl-tRNA synthetase inhibitor, ganfeborole, for rifampicin-susceptible tuberculosis: a phase 2a open-label, randomized trial. Nat Med. 2024;30(3):896–904.
81. Zhang Y, Li YY, Xu J, Chen X, Wang B, Fu L, Lu Y. Preliminary study on the action mechanism of TBI-166 against drug-resistant Mycobacterium tuberculosis. Chinese Journal of Antituberculosis. 2019 Nov 10;41(11):1160.
82. Sacksteder KA, Protopopova M, Barry CE, Andries K, Nacy CA. Discovery and development of SQ109: a new antitubercular drug with a novel mechanism of action. Future Microbiol. 2012;7(7):823–37.

83. Yu C, Qian H, Wu Q, Zou Y, Ding Q, Cai Y, et al. Safety, pharmacokinetics, and food effect of sudapyridine (WX-081), a novel anti-tuberculosis candidate in healthy Chinese subjects. Clin Transl Sci. 2024;17(2):e13718.
84. Black TA, Buchwald UK. The pipeline of new molecules and regimens against drug-resistant tuberculosis. J Clin Tuberc Other Mycobact Dis. 2021;25:100285.
85. Yang TW, Park HO, Jang HN, Yang JH, Kim SH, Moon SH, et al. Side effects associated with the treatment of multidrug-resistant tuberculosis at a tuberculosis referral hospital in South Korea. Medicine. 2017;96(28):e7482.
86. Laniado-Laborín R. Adverse effects and toxicity of antituberculosis drugs. In: Laniado-Laborín R, editor. Drug resistant tuberculosis. Sharjah: Bentham Science Publishers; 2015. p. 65–73.
87. Gülbay BE, Gürkan ÖU, Yıldız ÖA, Önen ZP, Erkekol FÖ, Baççıoğlu A, et al. Side effects due to primary antituberculosis drugs during the initial phase of therapy in 1149 hospitalized patients for tuberculosis. Respir Med. 2006;100(10):1834–42.
88. Chantaphakul H, Cheungpasitporn W, Ruxrungtham K, Klaewsongkram J. Adverse reaction to anti-tuberculosis drugs. J Allergy Clin Immunol. 2010;125(2):AB154.
89. Forget EJ, Menzies D. Adverse reactions to first-line antituberculosis drugs. Expert Opin Drug Saf. 2006;5(2):231–49.
90. CDC. Adverse events during treatment. Centers for Disease Control and Prevention. 2016.
91. Krishnamurthy N, Grimshaw AA, Axson SA, Choe SH, Miller JE. Drug repurposing: a systematic review on root causes, barriers and facilitators. BMC Health Serv Res. 2022;22(1):970.
92. Khataniar A, Pathak U, Rajkhowa S, Jha AN. A comprehensive review of drug repurposing strategies against known drug targets of COVID-19. COVID. 2022;2(2):148–67.
93. Kulkarni VS, Alagarsamy V, Solomon VR, Jose PA, Murugesan S. Drug repurposing: an effective tool in modern drug discovery. Russ J Bioorg Chem. 2023;49(2):157–66.
94. Sahoo BM, Ravi Kumar BVV, Sruti J, Mahapatra MK, Banik BK, Borah P. Drug repurposing strategy (DRS): emerging approach to identify potential therapeutics for treatment of novel coronavirus infection. Front Mol Biosci. 2021;8:628144.
95. Zhan P, Yu B, Ouyang L. Drug repurposing: an effective strategy to accelerate contemporary drug discovery. Drug Discov Today. 2022;27(7):1785–8.
96. Sharma K, Ahmed F, Sharma T, Grover A, Agarwal M, Grover S. Potential repurposed drug candidates for tuberculosis treatment: progress and update of drugs identified in over a decade. ACS Omega. 2023;8(20):17362–80.
97. Singh D, Singh A, Chawla PA. An overview of current strategies and future prospects in drug repurposing in tuberculosis. Explor Med. 2023;4:71–84.
98. Silva DR, Dalcolmo M, Tiberi S, Arbex MA, Munoz-Torrico M, Duarte R, et al. New and repurposed drugs to treat multidrug- and extensively drug-resistant tuberculosis. J Bras Pneumol. 2018;44(2):153–60.
99. Wang MG, Liu XM, Wu SQ, He JQ. Impacts of clofazimine on the treatment outcomes of drug-resistant tuberculosis. Microbes Infect. 2023;25(1–2):105020.
100. Stadler JAM, Maartens G, Meintjes G, Wasserman S. Clofazimine for the treatment of tuberculosis. Front Pharmacol. 2023;14:1100488.
101. Singh B, Cocker D, Ryan H, Sloan DJ. Linezolid for drug-resistant pulmonary tuberculosis. Cochrane Database Syst Rev. 2019;3(3):CD012836.
102. Pranger AD, van der Werf TS, Kosterink JGW, Alffenaar JWC. The role of fluoroquinolones in the treatment of tuberculosis in 2019. Drugs. 2019;79(2):161–71.
103. Sarathy J, Blanc L, Alvarez-Cabrera N, O'Brien P, Dias-Freedman I, Mina M, et al. Fluoroquinolone efficacy against tuberculosis is driven by penetration into lesions and activity against resident bacterial populations. Antimicrob Agents Chemother. 2019;63(5):e02516–8.
104. Sharma D, Sharma S, Sharma J. Potential strategies for the management of drug-resistant tuberculosis. J Glob Antimicrob Resist. 2020;22:210–4.

105. Padmapriyadarsini C, Vohra V, Bhatnagar A, Solanki R, Sridhar R, Anande L, et al. Bedaquiline, delamanid, linezolid, and clofazimine for treatment of pre-extensively drug-resistant tuberculosis. Clin Infect Dis. 2023;76(3):e938–46.
106. Carroll MW, Jeon D, Mountz JM, Lee JD, Jeong YJ, Zia N, et al. Efficacy and safety of metronidazole for pulmonary multidrug-resistant tuberculosis. Antimicrob Agents Chemother. 2013;57(8):3903–9.
107. Şenol G. Recent and new strategies for extensively-drug resistant tuberculosis. Mediterr J Infect Microb Antimicrob. 2019;22:210–214.
108. Chen C, Gardete S, Jansen RS, Shetty A, Dick T, Rhee KY, et al. Verapamil targets membrane energetics in Mycobacterium tuberculosis. Antimicrob Agents Chemother. 2018;62(5):e02107–17.
109. Fatima S, Bhaskar A, Dwivedi VP. Repurposing immunomodulatory drugs to combat tuberculosis. Front Immunol. 2021;12:645485.
110. Hoffner SE, Olsson-Liljequist B, Rydgård KJ, Svenson SB, Källenius G. Susceptibility of Mycobacteria to fusidic acid. Eur J Clin Microbiol Infect Dis. 1990;9(4):294–7.
111. Srivastava S, Gumbo T, Thomas T. Repurposing cefazolin-avibactam for the treatment of drug resistant Mycobacterium tuberculosis. Front Pharmacol. 2021;12:776969.
112. Brindha S, Sundaramurthi JC, Vincent S, Velmurugan D, Gnanadoss JJ. In silico and in vitro screening of FDA-approved drugs for potential repurposing against tuberculosis. 2017. https://doi.org/10.1101/228171.
113. Soetaert K, Rens C, Wang XM, De Bruyn J, Lanéelle MA, Laval F, et al. Increased vancomycin susceptibility in mycobacteria: a new approach to identify synergistic activity against multidrug-resistant mycobacteria. Antimicrob Agents Chemother. 2015;59(8):5057–60.
114. Cavanaugh JS, Jou R, Wu MH, Dalton T, Kurbatova E, Ershova J, et al. Susceptibilities of MDR Mycobacterium tuberculosis isolates to unconventional drugs compared with their reported pharmacokinetic/pharmacodynamic parameters. J Antimicrob Chemother. 2017;72(6):1678–87.
115. Moreira W, Santhanakrishnan S, Dymock BW, Dick T. Bortezomib warhead-switch confers dual activity against mycobacterial caseinolytic protease and proteasome and selectivity against human proteasome. Front Microbiol. 2017;8:746.
116. Imperiale BR, Cataldi ÁA, Morcillo NS. In vitro anti-tuberculosis activity of azole drugs against Mycobacterium tuberculosis clinical isolates. Rev Argent Microbiol. 2017;49(4):332–8.
117. Patel YS, Mistry N, Mehra S. Repurposing artemisinin as an anti-mycobacterial agent in synergy with rifampicin. Tuberculosis. 2019;115:146–53.
118. Mori G, Orena BS, Franch C, Mitchenall LA, Godbole AA, Rodrigues L, et al. The EU approved antimalarial pyronaridine shows antitubercular activity and synergy with rifampicin, targeting RNA polymerase. Tuberculosis. 2018;112:98–109.
119. Gupta U, Vemuri N, Gupta P, Kumar V, Tanushree P, Khuller G. Efficacy of moxifloxacin & econazole against multidrug resistant (MDR) Mycobacterium tuberculosis in murine model. Indian J Med Res. 2015;142(3):323.
120. Cotroneo N, Rubio A, Critchley IA, Pillar C, Pucci MJ. *In vitro* and *in vivo* characterization of tebipenem, an oral carbapenem. Antimicrob Agents Chemother. 2020;64(8):e02240–19.
121. Solapure S, Dinesh N, Shandil R, Ramachandran V, Sharma S, Bhattacharjee D, et al. In vitro and in vivo efficacy of β-lactams against replicating and slowly growing/nonreplicating Mycobacterium tuberculosis. Antimicrob Agents Chemother. 2013;57(6):2506–10.
122. Deshpande D, Pasipanodya JG, Srivastava S, Martin KR, Athale S, van Zyl J, et al. Minocycline immunomodulates via sonic hedgehog signaling and apoptosis and has direct potency against drug-resistant tuberculosis. J Infect Dis. 2019;219(6):975–85.
123. Napier RJ, Rafi W, Cheruvu M, Powell KR, Zaunbrecher MA, Bornmann W, et al. Imatinib-sensitive tyrosine kinases regulate mycobacterial pathogenesis and represent therapeutic targets against tuberculosis. Cell Host Microbe. 2011;10(5):475–85.

124. Sogi KM, Lien KA, Johnson JR, Krogan NJ, Stanley SA. The tyrosine kinase inhibitor gefitinib restricts *Mycobacterium tuberculosis* growth through increased lysosomal biogenesis and modulation of cytokine signaling. ACS Infect Dis. 2017;3(8):564–74.
125. Skerry C, Pinn ML, Bruiners N, Pine R, Gennaro ML, Karakousis PC. Simvastatin increases the in vivo activity of the first-line tuberculosis regimen. J Antimicrob Chemother. 2014;69(9):2453–7.
126. Davuluri KS, Singh AK, Singh AV, Chaudhary P, Raman SK, Kushwaha S, et al. Atorvastatin potentially reduces mycobacterial severity through its action on lipoarabinomannan and drug permeability in granulomas. Microbiol Spectr. 2023;11(2):e0319722.
127. Feng Q, Zhou H, Zhang X, Liu X, Wang J, Zhang C, et al. Acarbose, as a potential drug, effectively blocked the dynamic metastasis of EV71 from the intestine to the whole body. Infect Genet Evol. 2020;81:104210.
128. Dalecki AG, Haeili M, Shah S, Speer A, Niederweis M, Kutsch O, et al. Disulfiram and copper ions kill Mycobacterium tuberculosis in a synergistic manner. Antimicrob Agents Chemother. 2015;59(8):4835–44.
129. Ottavi S, Scarry SM, Mosior J, Ling Y, Roberts J, Singh A, et al. In vitro and in vivo inhibition of the *Mycobacterium tuberculosis* phosphopantetheinyl transferase PptT by amidinoureas. J Med Chem. 2022;65(3):1996–2022.
130. Miow QH, Vallejo AF, Wang Y, Hong JM, Bai C, Teo FSW, et al. Doxycycline host-directed therapy in human pulmonary tuberculosis. J Clin Invest. 2021;131(15):e141895.
131. Wang W, Du Z, Ni M, Wang Z, Liang M, Sheng H, et al. Aspirin enhances the clinical efficacy of anti-tuberculosis therapy in pulmonary tuberculosis in patients with type 2 diabetes mellitus. Infect Dis. 2020;52(10):721–9.
132. Naftalin CM, Verma R, Gurumurthy M, Hee KH, Lu Q, Yeo BCM, et al. Adjunctive use of celecoxib with anti-tuberculosis drugs: evaluation in a whole-blood bactericidal activity model. Sci Rep. 2018;8(1):13491.
133. Arias L, Otwombe K, Waja Z, Tukvadze N, Korinteli T, Moloantoa T, et al. SMA-TB: study protocol for the phase 2b randomized double-blind, placebo-controlled trial to estimate the potential efficacy and safety of two repurposed drugs, acetylsalicylic acid and ibuprofen, for use as adjunct therapy added to, and compared with, the standard WHO recommended TB regimen. Trials. 2023;24(1):435.
134. Working Group on New TB Drugs. StAT-TB (Statin adjunctive therapy for TB): a phase 2b dose-finding study of pravastatin in adults with tuberculosis. Working Group on New TB Drugs. 2020.
135. Padmapriydarsini C, Mamulwar M, Mohan A, Shanmugam P, Gomathy NS, Mane A, et al. Randomized trial of metformin with anti-tuberculosis drugs for early sputum conversion in adults with pulmonary tuberculosis. Clin Infect Dis. 2022;75(3):425–34.
136. Karbalaei M, Ghazvini K, Keikha M. Clinical efficacy of vitamin D supplementation on pulmonary TB patients: the evidence of clinical trials. J Clin Tuberc Other Mycobact Dis. 2020;20:100174.
137. Panda PK, Panda P, Dawman L, Sihag RK, Sharawat IK. Efficacy and safety of thalidomide in patients with complicated central nervous system tuberculosis: a systematic review and meta-analysis. Am J Trop Med Hyg. 2021;105(4):1024–30.
138. Nachega J. TB-IRIS NSAID Cox-2 inhibitor prevention trial. Centre Watch. 2013.
139. Ngwane AH, Petersen R, Baker B, Wiid I, Wong HN, Haynes RK. The evaluation of the anticancer drug elesclomol that forms a redox-active copper chelate as a potential anti-tubercular drug. IUBMB Life. 2019;71(5):532–8.
140. Jagadeb M, Rath SN, Sonawane A. In silico discovery of potential drug molecules to improve the treatment of isoniazid-resistant Mycobacterium tuberculosis. J Biomol Struct Dyn. 2019;37(13):3388–98.
141. Ngidi NTP, Machaba KE, Mhlongo NN. In silico drug repurposing approach: investigation of Mycobacterium tuberculosis FadD32 targeted by FDA-approved drugs. Molecules. 2022;27(3):668.

142. Kawamoto S, Hori C, Taniguchi H, Okubo S, Aoki S. Identification of novel antimicrobial compounds targeting Mycobacterium tuberculosis shikimate kinase using in silico hierarchical structure-based drug screening. Tuberculosis. 2023;141:102362.
143. Aina OS, Rofiu MO, Oloba-Whenu OA, Olasupo IA, Adams LA, Familoni OB. Drug design and in-silico study of 2-alkoxylatedquinoline-3-carbaldehyde compounds: inhibitors of Mycobacterium tuberculosis. Sci Afr. 2024;23:e01985.
144. Adeniji SE, Adalumo OB, Ekoja FO. Anti-tubercular modelling, molecular docking simulation and insight toward computational design of novel compounds as potent antagonist against DNA gyrase receptor. Med Microecol. 2020;5:100020.
145. Dash S, Rathi E, Kumar A, Chawla K, Kini SG. Identification of DprE1 inhibitors for tuberculosis through integrated in-silico approaches. Sci Rep. 2024;14(1):11315.
146. Lane TR, Urbina F, Rank L, Gerlach J, Riabova O, Lepioshkin A, et al. Machine learning models for *Mycobacterium tuberculosis in vitro* activity: prediction and target visualization. Mol Pharm. 2022;19(2):674–89.

Chapter 27
The Network Pharmacology Approach to Uncover the Pharmacological Mechanism of Natural Products

Parchi Khadse, Piyusha Kulkarni, Keisham Nelson Mangang, Santhosh Chobe, S. N. Koteswara Rao G., and Rajasekhar Reddy Alavala

Contents

Abstract The concept of "one gene, one target, one disease" has greatly impacted drug discovery methods. Effective medications now often target numerous targets, rather than just one. Network pharmacology, a multidisciplinary approach based on system biology and polypharmacology, proposes a new network mode of "multiple targets, multiple effects, complex diseases" that replaces "magic bullets" with "magic shotguns." Chinese herbal medicine (CHM) is an important method in complementary and alternative medicine. Some of the tools used to execute Polypharmacological networks are Traditional herbal medicine (THM), Smart

P. Khadse · P. Kulkarni · K. N. Mangang · S. N. K. Rao G. · R. R. Alavala (✉)
Shobhaben Pratapbhai Patel School of Pharmacy & Technology Management, SVKM's NMIMS, Mumbai, India

S. Chobe
MGV's L.V.H. Arts, Science and Commerce College, Panchavati, Nashik, India

S. N. K. Rao G., R. R. Alavala (eds.), *Applications of Computational Tools in Drug Design and Development*, https://doi.org/10.1007/978-981-96-4154-3_27

graph, Cytoscape, TCMSP, BATMAN-TCM, TCMID, and ETCM. Traditional medicine has always relied on natural ingredients as medicinal agents for several diseases. With the progress of genetics and bioinformatics, contemporary natural product discovery has transitioned to bottom-up tactics, which supplement traditional top-down procedures. This book chapter looks at the two primary approaches to natural product discovery: top-down and bottom-up. Top-down approaches start at the organism level and use systems-level perturbations to induce the production of novel natural products, whereas bottom-up approaches start with identifying a specific gene cluster of interest and then apply gene manipulation techniques to facilitate transcription, translation, and synthesis of the desired natural product. This book chapter also highlights the network toxicology, tools and databases used for network pharmacology, its applications, and limitations.

Keywords Network Pharmacology · Polypharmacology · Natural products · Disease · Targets · Bioinformatics

27.1 Introduction

27.1.1 Natural Product Discovery

The discovery and development of human therapeutics have historically drawn significant inspiration from natural products. Despite the pharmaceutical industry's diminished focus on natural product discovery, academia has sustained and even expanded these efforts, yielding promising outcomes. Traditionally, the identification of natural products followed a top-down approach. However, contemporary methodologies increasingly leverage genomics and bioinformatics to guide bottom-up strategies, offering potent alternatives for exploration and discovery in this field. Natural product discovery activities can be broadly divided into two approaches: top-down and bottom-up. Top-down techniques begin research at the organism level, using system-level disturbances to stimulate the synthesis of new natural products. These methods do not assume an understanding of the genes and enzymes that contribute to biosynthesis. Bottom-up techniques, on the other hand, begin by identifying a single gene cluster of interest. Following that, various gene editing techniques are used to aid in transcription, translation, and ultimately synthesis of the desired natural product [1]. Natural products and their structural equivalents have long played an important role in pharmacotherapy, especially for the treatment of cancer and infectious disorders. Their significance, however, extends beyond these regions to include a wide range of treatment domains, including cardiovascular illness (as shown by statins) and multiple sclerosis (exhibited by fingolimod). This historical contribution highlights the flexible and important role that natural products and their derivatives have played in enhancing medical therapy for a wide range of health issues. Natural products (NPs) possess features that distinguish them from conventionally manufactured chemicals, providing both advantages and challenges to the drug development process. NPs have a wide range of scaffolds and complex structures. These unique features contribute to the distinctive nature of natural products, influencing their behaviour and potential utility in drug discovery [2]. Natural products have played a crucial role in sustaining life on Earth, enabling the existence of humanity. The primary ingredients for Ayurvedic medicines were predominantly sourced from plants, taking the form of raw materials like dried herbal powders, extracts, or combinations of these products. An approximation suggests that Earth harbours around 75,000 higher plant species, with merely 10% finding utilization in traditional medicine. Scientific exploration has focused on a mere 1–5% of these, revealing their known therapeutic properties. Life-saving drugs of significant importance in modern medicine have been derived from herbs. However, out of the estimated 400,000 plant species, only 6% have undergone thorough study for their medicinal activity, and a mere fraction, less than 20%, have been subjected to phytochemical investigation. Consequently, there is a pressing need for in-depth exploration of bioactive fractions, along with Phyto analysis and phytopharmacological evaluation of herbal drugs, to realize the aspirations of herbal drug discovery [3]. In the realms of cancer and infectious disease, a significant proportion of novel drug developments, specifically 60% for cancer and 75% for infectious diseases, stem from natural origins. This analysis will concentrate on contemporary methodologies utilized in the exploration of fresh natural products, leveraging recent progress in comprehending genetic pathways governing secondary metabolite production, tapping into underexplored reservoirs of natural compounds, and integrating innovative screening technologies [4]. Within the potential reservoirs of natural products, bacteria have emerged as an exceptionally fruitful source, characterized by a surprisingly limited set of taxa responsible for most discovered compounds. To illustrate, among the 53 recognized bacterial phyla, merely five are documented for their production of anti-infective agents. Within this restricted group, the Class Actinobacteria, specifically bacteria falling under the Order Actinomycetales, commonly referred to as actinomycetes, contribute to approximately 7000 compounds as documented in the Dictionary of Natural Products [5]. Furthermore, natural chemicals can be used directly as medications to treat human diseases or, at the very least, as useful beginning points ("leads") in drug development projects [1, 5]. In navigating the difficult road from a lead molecule to a viable medicine, two unique problems necessitate special attention, particularly for natural products with naturally complex structures: (i) How can structural modifications be carried out efficiently and cost-effectively to adjust and increase their biological activities while maximizing the lead structure's pharmacodynamic, pharmacokinetic, and safety (ADMET) properties? (ii) How can a large-scale supply of drug substance in GMP (Good Manufacturing Practices)-quality be realized? [6].

27.1.2 New Trends in Drug Discovery from Natural Products

Recent advancements in technology and the introduction of novel methodologies have transformed the screening of natural products, presenting a unique opportunity to reintroduce them as a primary source of drug leads. These innovations address the existing limitations of natural product screening. Examples of recent breakthroughs utilizing these technologies to impact the discovery of new drugs include: (i) The creation of a more efficient screening process for natural products, streamlining the identification of potential drug candidates. (ii) Enhancements in sourcing natural products have expanded the range of compounds available for screening, fostering the discovery of new therapeutically valuable molecules. (iii) Progress in organic synthetic techniques has enabled the development of complex natural product analogs with improved pharmacological properties. (iv) Innovations in combinatorial biosynthesis allow for the generation of novel natural product analogs by manipulating biosynthetic pathways in microorganisms. (v) Utilizing microbial genomics has revolutionized the discovery of bioactive compounds, enabling the identification and engineering of microorganisms to produce targeted compounds more effectively [7]. To effectively utilize natural product-based drug discovery techniques, a multidisciplinary approach incorporating both existing and cutting-edge technology is required. Many medicinal extracts have components that work together, thus isolating separate elements may be detrimental. As a result, novel approaches for studying and harnessing these molecules are required, allowing for the development of new medications. Furthermore, using a systems biology-guided approach provides a new viewpoint on natural product pharmaceutical sciences. Rather than focusing exclusively on identifying a single chemical that targets a specific receptor, this method prioritizes understanding the complicated relationships between physiological systems. It entails recognizing and utilizing the synchronized mechanisms working on many molecular targets, therefore providing a more holistic view of the treatment for natural products [8].

Advances in understanding the human genome and pathogenic microorganisms and parasites have led to the identification of disease-associated protein structures. The discovery of new molecular targets based on these proteins has resulted in an increased demand for novel molecular diversity for screening. Natural products will play an important part in supplying this demand through ongoing research into the world's biodiversity, much of which remains untapped. With only 1% percent of the microbial world currently known, breakthroughs in microbial cultivation and extraction of nucleic acids from environmental samples from soil and marine habitats will give researchers access to a vast untapped pool of genetic and metabolic variety [9]. Recent advances in microbial cell physiology have revealed potential targets for antibacterial drug research. The focus is on analysing social interactions at the molecular level in both synthetic and natural microbial communities. Studies on the diversity of natural products produced by bacterial communities in soil and

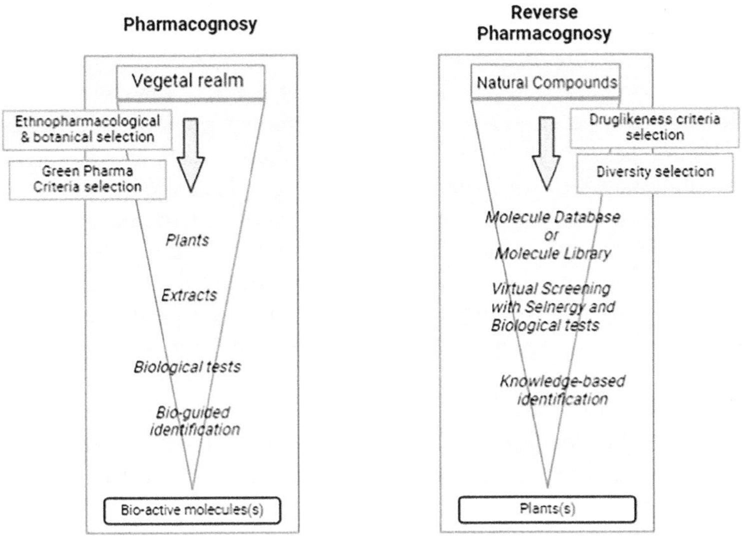

Fig. 27.1 Pharmacognosy and Reverse Pharmacognosy

plant-associated environments, as well as their activity patterns against coexisting microorganisms, have revealed multiple functions that go beyond the historical perception of these compounds as weapons in microbial conflicts. Recent studies highlight the need for novel phenotype-based antibacterial tests that adapt to the changing microbial community dynamics. The following four assay platforms use natural products as lead compounds: (i) Natural product efflux pump inhibitors (EPIs) for antibacterial therapy. (ii) Quorum sensing inhibition (QSI) for production. (iii) Elimination of persister cells for antibacterial therapy. (iv) Protecting the host from infection [10].

Green Pharma has developed reverse pharmacognosy to supplement standard pharmacognosy in the discovery of new chemicals or plants for medicinal or cosmetic applications which is shown in Fig. 27.1. The proposal has five parts: a 3D structural database of natural compounds, an analogous natural compound library prepared for HTS, a target database, an ethnopharmacological, and a virtual screening tools database [11].

Poly-pharmacology and synergy are shaping the future of nanoparticle drug research. Emerging systems biology approaches can address complicated elements of biological systems, including multi-component mixtures. NMR-based metabolomics has been used to explore the biological reactions to chamomile tea administration. The technique allowed for characterization of chamomile's metabolic effects, despite genetic and environmental variations. This study identified indicators associated with herb consumption and highlights the promise of this method in the growing sector [12].

27.2 Network Pharmacology

Network Pharmacology (NP) is a newly created science that seeks to understand pharmacological effects and interactions across numerous targets. It uses computer skills to comprehensively record a drug's molecular interactions within living cells. NP is an important tool for understanding the complicated connections that exist between botanical formulations and the body. It aims to find new pharmacological leads, repurpose existing medication compounds for a variety of therapeutic ailments, and conduct unbiased research into possible target spaces. Traditional knowledge is essential in the development and repurposing of medications. Integrating systems biology with nanoparticles may result in the rational creation of the next generation of multifunctional medications [13]. In the postgenomic era, the surge in biomedical data has prompted the use of systems biology and Polypharmacological in drug discovery. Computational biology emerges as a lucrative avenue for addressing scientific challenges through effective modelling and theoretical exploration. It proves valuable in predicting off-target effects with increased efficiency, enhancing drug discovery by adopting a novel network approach involving "multiple targets, multiple effects, complex diseases." Network pharmacology influences drug development through two primary approaches: establishing a practical network model and predicting drug targets using public databases or existing research data [14]. The robustness of biological phenotypes stems from redundant functions and alternative compensatory signalling routes. Network analysis of biological pathways highlights that this robustness is often rooted in the network's structure. Many biological networks exhibit a scale-free nature, making them resilient to random node deletion but highly dependent on a few well-connected hubs. This inherent robustness has significant implications for drug discovery. Rather than focusing on identifying 'disease-causing' genes, network biology suggests a strategy of pinpointing perturbations in the disease-causing network [15].

Traditional herbal medicine (THM) has kept Asians healthy for hundreds of years, leading to a unique medical system based on empirical evidence. THM is consumed regularly by billions of people worldwide, and the pharmacological research industry looks to it for inspiration. The findings suggest that THM's therapeutic activities may involve a wide spectrum of biomolecules. However, because THMs are complicated and experimental applications are limited, only a few of their modes of action have been thoroughly studied. This is a significant barrier to THM's expansion and general adoption in modern healthcare. THM-network pharmacology (THM-NP) research entails creating a herb-compound-target (H-C-T) network by combining herbal ingredient data with drug-target interaction (DTI) information. The target network is then studied to get insight into linked biological processes, pathways, and diseases shown in Fig. 27.2. Because there are no gold-standard methodologies for THM-NP investigations yet, researchers have created and used a variety of tools and databases for each stage [16]. The protein-herbal network pharmacology for Phyllanthin is shown in the Fig. 27.3 for diabetes.

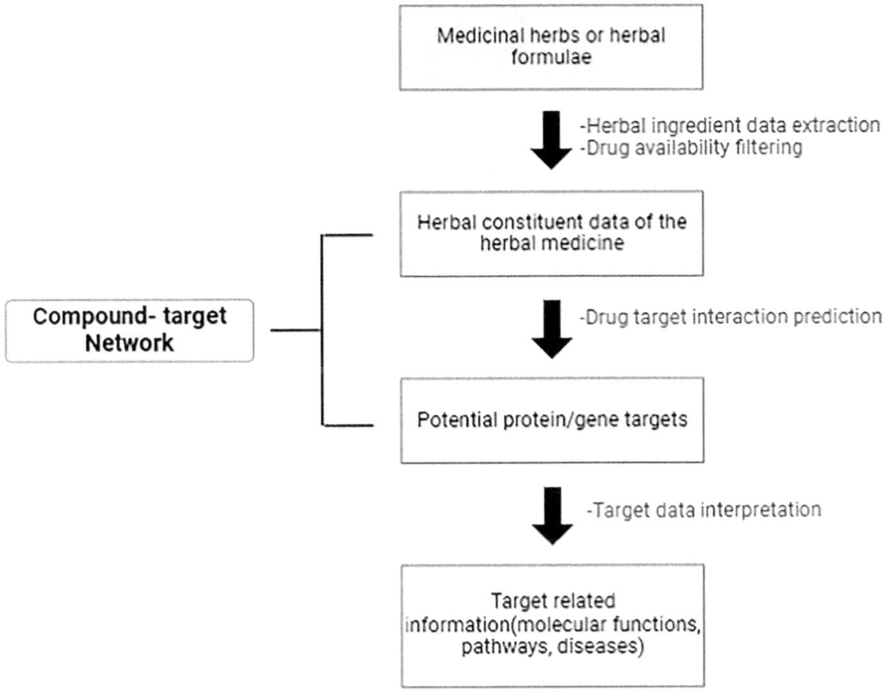

Fig. 27.2 The general network of pharmacology analysis of herbal medicines

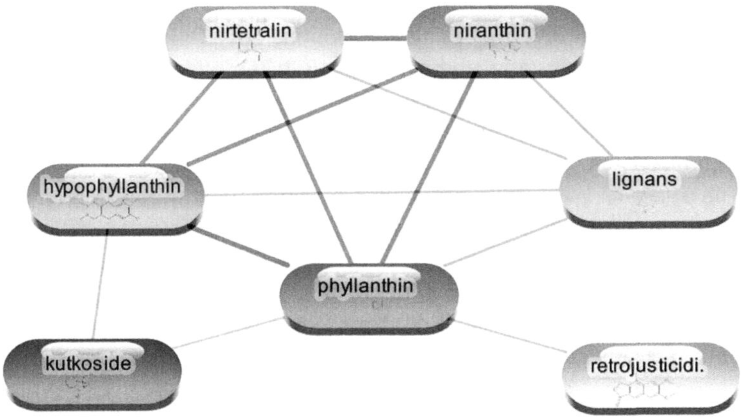

Fig. 27.3 Protein-herbal network pharmacology for Phyllanthin for diabetes

Recent proposals include a network-based approach to pharmacology. Network pharmacology examines the traditional "one disease-one target-one drug" paradigm and investigates interactions between the body and the drug by mapping drug-target-disease networks at the biological level. A recent network pharmacology analysis revealed the intricacy of both medication action and protein-protein

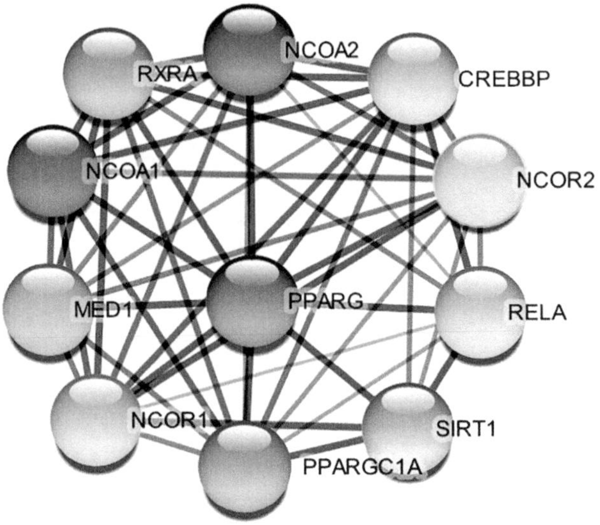

Fig. 27.4 Protein-protein network for PPAR-gamma for diabetes

interaction, resulting in significant modifications in therapy tactics and the drug discovery process [17].

A network of drug-target interactions, also known as a drug-target network, aids in the interpretation of a drug's pharmacokinetic profile. It can help in understanding the molecular basis of therapeutic activities and the side effects of potential pharmaceuticals, as well as deciding if a protein should be targeted. The systems pharmacology method investigates all connections, including protein-protein interactions (PPI), gene expression, and pathways, to better understand how drugs interact with complex illnesses. Figure 27.4 depicts a protein-protein interaction network diagram for the PPAR-gamma protein for diabetes created using Cytoscape.

Drug-induced gene expression differentiation has been studied using technology such as microarrays and next-generation sequencing, in addition to protein-protein interactions. Experiments can investigate pharmacological effects at the cellular or tissue level, both in vitro and in vivo, under various time and dose settings. They can also examine chemical similarity cellular response similarity after chemical administration [18]. Network pharmacology for Gene-protein network is shown in the below Fig. 27.5 generated on Cytoscape for diabetes disease.

27.3 Tools and Databases for Network Pharmacology Research

Introducing "Smart Graph," a computational framework meant to streamline drug discovery procedures by providing an intuitive graphical user interface (GUI) for those without prior bioinformatics or cheminformatics knowledge. Its key goals are

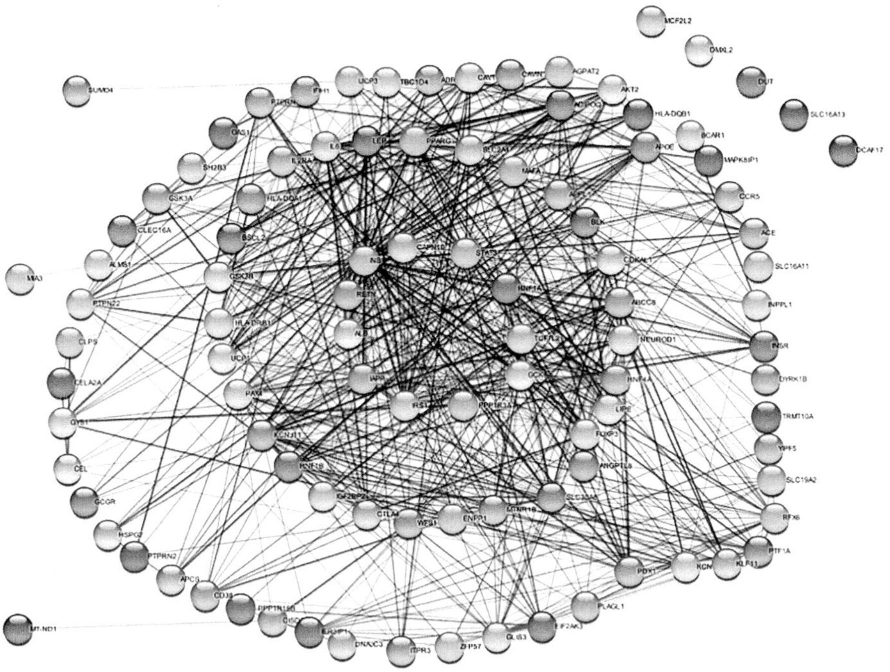

Fig. 27.5 Gene-protein network diagram for diabetes disease

twofold: streamlining drug discovery stages and enabling the creation of predictive and analytical network pharmacology (NP) workflows. Several other tools inspired by network analysis also help to analyse drug-target interactions (DTIs) and protein-protein interactions (PPIs). These include the CARLSBAD Web Application and Cytoscape plugin, the ChemProt database, the STITCH platform, Drug Target Explorer, Bioactivity Explorer, Path Insight, and Path Linker Cytoscape plugins. These tools collectively enhance the understanding and exploration of complex biological networks, aiding in the translation of biomedical information into clinically relevant knowledge [19].

In recent years, network pharmacology has become critical for identifying active pharmacological components, understanding treatment mechanisms, and analysing drug combinations. Despite the creation of databases such as TCMSP, BATMAN-TCM, TCMID, and ETCM, the process of data gathering and integration remains cumbersome and time-consuming, impeding the rapid identification of Traditional Chinese Medicine (TCM) components and targets. To address this, the World Federation of Chinese Medicine Societies issued recommendations that recognize the challenges of network pharmacology. To tackle these challenges, a complete and automated platform dubbed INPUT (Intelligent Network Pharmacology Platform Unique to Traditional Chinese Medicine) was created. This platform integrates data from multiple sources to streamline TCM analysis, aiming to accelerate research in this field [20]. Multiple sources of streamline TCM analysis is shown in Fig. 27.6.

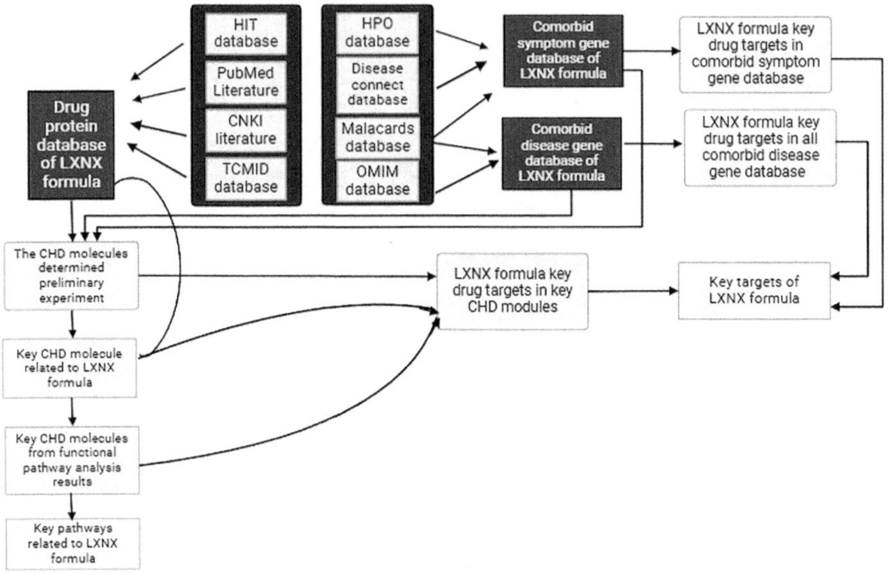

Fig. 27.6 Multiple sources of streamline TCM analysis

Network pharmacology has been successful in creating a "compounds-proteins/genes-diseases" network, aiding the understanding of the regulation principles of small molecules in a high-throughput manner. This method is particularly useful for analysing drug combinations, including TCM (Traditional Chinese Medicine) preparations. The TCM-Mesh system, a database containing 6235 herbs, 383,840 compounds, 14,298 genes, 6204 diseases, and related associations, is used for network pharmacology analysis of TCM preparations. It constructs a network between herbs and treated diseases, providing insight into the molecular mechanisms of TCM preparations. Several databases support TCM analysis: Super Natural II, TCM@taiwan, ChEMBL, and TCMID contain TCM compound information, while STITCH is used for compound-protein interactions, STRING for protein-protein interactions, and OMIM and GAD for protein-disease associations. However, the efficiency of these databases is limited by incompleteness (e.g., TCMID) and redundancies (e.g., OMIM and STITCH) [21].

Complex diseases, such as CVD and diabetes, are caused by a mix of genetic and environmental factors. Modern drugs (magic bullets) that only work on one target are ineffective in treating complicated illnesses. Hopkins AL pioneered network pharmacology, which investigates drug action mechanisms within biological networks such as biological pathways, gene regulatory networks, protein-protein interaction networks, drug-target networks, and more. The drug-target network (DTN) and its derivatives (drug-drug network (DDN) and target-target network (TTN)) could have a substantial impact on understanding the mechanism and making use of the huge amount of data from HTS. The drug-target network was created using various binding data between drugs and target proteins. Only a small proportion of

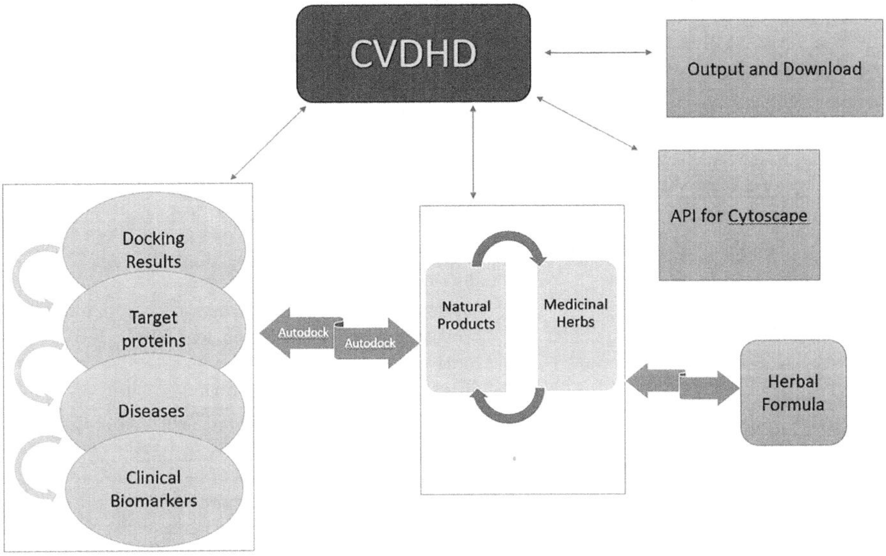

Fig. 27.7 Drug-Network diagram for CVDHD

natural chemicals (less than 2%) have binding data published. Thus, CVDHD used molecular docking to identify the binding affinity of all drugs with 2395 target proteins [22]. The drug network diagram for CVDHD is shown in Fig. 27.7.

Network pharmacology is a useful tool for predicting and analysing the effects of tiny molecules on proteins. The systemsDock platform offers a user-friendly interface for making these predictions. The platform can be used in two ways: (i) to specify target proteins by selecting different effective methods or by directly uploading a molecular pathway map in SBML format, and (ii) to build small molecules for testing by uploading structure files or making a compound of interest with the provided web-based molecule editor. The four major steps for network pharmacology-based prediction and analysis using systems Dock are:

(1) Target protein specification: Users can provide target proteins in a variety of efficient ways, such as using protein names or gene symbols, PDB IDs, or uploading an SBML route map file. (2) Binding site definition: Using an interactive molecular visualizer, users can define binding sites by clicking on the displayed structural model or residues. (3) Preparation of tiny compounds: Users can produce small molecules for testing by uploading structure data or building a compound of interest with the web-based molecular editor. (4) Finish the docking simulation and validate the results. SystemsDock provides very efficient and customizable solutions for network-based screening, allowing users to test many proteins. You can provide protein names or gene symbols, protein PDB IDs, or upload an SBML route map file. The platform then gets protein structures with the best resolution from an internal protein identity-to-structure mapping system for docking simulation [23]. List of software for network pharmacology is highlighted in Table 27.1.

Table 27.1 Software for network pharmacology

S. No.	Software	Uses	Access status
1.	Smart Graph	Simplifies drug discovery steps. Supports development of analytical method for network pharmacology.	Free
2.	CARLSBAD Web Application	Analyse drug—target interaction Protein-protein interactions	Free
3.	ChemProt database	To store and access chemical protein interaction	Free
4.	Stitch platform	Platform for exploring interactions between chemicals, proteins, and other molecules. Offers resources and tools for analysing and visualising molecular interactions.	Commercial
5.	Drug—target explorer	Explores drug target interactions. Provides functionalities for drug and target protein relationship analysis	Commercial
6.	Bioactivity explorer	Explores the bioactivity of drugs. Provides information about biological effects of drug and mechanisms of action of drug.	Commercial
7.	Path insight	Analyses pathways and biological networks.	Commercial
8.	Path linker Cytoscape plugins	Analyse and visualise pathways for pharmacology networks. Offers extra features for analysing molecular interactions in biological context.	Free
9.	TCMSP (Traditional Chinese Medicine Systems Pharmacology Database and Analysis Platform)	Provides information on the ADME of herbal substances and their targets. Identifies probable active substances and their mechanisms of action.	Free
10.	BATMAN-TCM (Biological system-based analysis tool for molecular mechanism of traditional Chinese medicine)	Integrate data from different sources and predict the targets of different compounds.	Free
11.	TCMID (Traditional Chinese Medicine Integrated Database)	Provides information about TCM ingredients. Provides information about targets and disease.	Free
12.	ETCM (Encyclopaedia of Traditional Chinese Medicine)	Provides information on TCM compounds, including their pharmacological characteristics and possible targets.	Free

(continued)

Table 27.1 (continued)

S. No.	Software	Uses	Access status
13.	INPUT (Intelligent Network Pharmacology Platform Unique to Traditional Chinese Medicine)	Streamlines TCM analysis. Helps to identify potential active ingredients and their targets.	Free
14.	TCM-Mesh system	Constructs a connection between herbs and treated disease.	Free
15.	Super Natural II	Provides information about TCM compounds.	Free
16.	TCM@taiwan	Contains information on herbs, formulas, and compounds of TCM. Investigates molecular mechanism of TCM formulations.	Free
17.	ChEMBL	Identifies novel drug candidates and their mechanism of action.	Free
18.	String	Used for protein-protein interaction. Provides information about interaction between proteins and establish networks of it.	Free
19.	OMIM	Used for protein association. Information about human genes Provides information on genetic disorders.	Free
20.	GAD	About gene disease association.	Free
21.	Cytoscape plugin	Analyse drug—target interaction Protein-protein interactions	Free

27.4 Target Prediction

Researchers used network pharmacology to study Hua-Feng-Dan (HFD) for stroke treatment. They discovered 107 stroke-related HFD targets from multiple datasets. These targets were filtered with MCODE to create a core network. A protein-protein interaction (PPI) network was built with STRING and analysed in Cytoscape. The DAVID Bioinformatics tool was used to conduct GO and KEGG analyses. Molecular docking with Autodock 4.2 confirmed the interactions between major targets and HFD components. Results highlighted the role of core targets and pathways in HFD's stroke-protection mechanism, offering insights into its efficacy as a traditional Chinese medicine for stroke treatment [24]. A computational technique was created for predicting the bioactive constituents and targets of Radix Curcumae, a Chinese herbal combination used to treat cardiovascular disease. They discovered 58 bioactive components and 32 potential targets for cardiovascular and cerebrovascular disorders. Radix Curcumae was found to be the main component for preventing these diseases, with other herbs like Fructus Gardeniae, Moschus, and Borneolum playing supportive roles. These findings shed light on the formula's mechanisms and could guide research on new traditional Chinese medicine formulations [25].

Building networks of biological components such as genes, transcription factors, and proteins is essential for understanding biological processes. Identifying pharmacological targets is an important part of network pharmacology. The Entrez GeneID, a reliable identification in the NCBI's Entrez Gene database, is widely used. Cytoscape is a Java desktop application for reconstructing and visualizing networks. Nodes represent genes, whereas edges reflect interactions. Cytoscape 2.8 is available for download at http://cytoscape.org, and default annotations are extracted from NCBI's GO information at http://www.ncbi.nlm.nih.gov/Ftp/ [26].

A network pharmacology approach was used to investigate the active compounds and molecular targets of Yijin-Tang (YJT), a traditional Korean medicine remedy for managing hyperlipidaemia and atherosclerosis. For example, the study discovered 447 chemicals in five herbal remedies, 21 of which and 57 genes were associated to important disease processes. Ten substances, including naringenin, nobiletin, and quercetin, were identified as bioactive and important chemicals because they interact with many genes. The compound-target gene network revealed complicated connections between multiple components and targets, providing insight into YJT's possible therapeutic effects. Core genes such as CASP3, CYP1A1, and MMP2 emerged as significant players in the network, suggesting their involvement in YJT's therapeutic effects. Pharmacological network analysis provides insights into the potential mechanisms underlying YJT's efficacy against hyperlipidaemia and atherosclerosis, offering valuable guidance for further experimental validation and drug development [27].

A potential effective approach to combat inflammatory disorders involves utilizing immunomodulation through natural compounds derived from medicinal plants. CBD, a prominent Phyto cannabinoid found in Cannabis, has garnered attention for its anti-inflammatory properties and its associated molecular targets, backed by numerous research studies. This strategy holds promise for both preventive and therapeutic interventions against inflammatory-mediated conditions. The potential targets for CBD's anti-inflammatory effects were sourced from various online databases, such as the Swiss Target Prediction database (www.swisstargetprediction.ch), the Online Mendelian Inheritance in Man (OMIM, www.omim.org), and the Drug Bank database (https://www.drugbank.ca/). Human target connexins were acquired from the interactive protein database (DIP; http://dip.doe-mbi.ucla.edu). To ensure consistency, all tested targets were converted into the UniProt database format using the Retrieve/ID mapping tool (www.uniprot.org) [28].

Type 2 diabetes mellitus (T2DM) affects a significant portion of individuals with diabetes, but single-target treatments often fall short in addressing complications. The LiuWei DiHuang Pill (LWDH Pill), a traditional Chinese prescription dating back to the Song Dynasty, has been utilized clinically for various conditions including renal yin asthenia. To identify potential targets for T2DM treatment, proteins associated with T2DM were sourced from GeneCards (https://www.genecards.org/, ver. 4.9.0) and OMIM (http://www.omim.org/, updated February 19, 2019) databases, focusing on human proteins. These targets were then cross-referenced with active ingredients of LWDH Pill using the STRING database. This approach, restricted to human subjects, helps establish potential correlations between LWDH

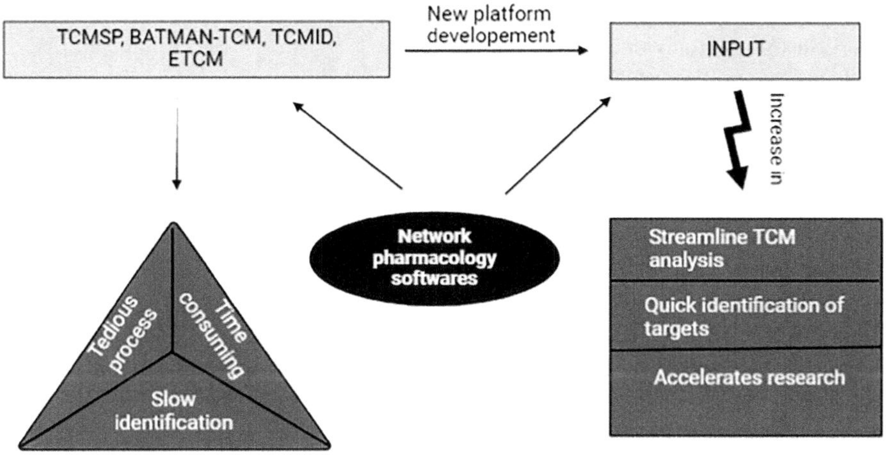

Fig. 27.8 Network pharmacology software

Pill components and T2DM-related targets [29]. A network pharmacology study found that the LianXia NingXin (LXNX) formula, which is used to treat coronary heart disease (CHD) and accompanying symptoms, targets two major CHD-related disease modules, M146 and M203. Functional analysis demonstrates the significance of G-protein coupled receptor signaling pathways for M146 and cellular protein metabolic processes for M203. According to shortest path analysis, corticotropin-releasing hormone (CRH) and natriuretic peptide precursor A (NPPA) are major LXNX therapeutic targets that are tightly related to CHD modules and associated genes. This shows that LXNX's efficacy in treating CHD may be regulated through these pathways, with CRH and NPPA being key targets [30]. In Fig. 27.8, Network pharmacology software is mentioned.

27.5 Network Construction

Gastritis, inflammation of the stomach lining, can result from various factors like H. pylori infection, NSAIDs, corticosteroids, and alcohol. While current therapies offer relief, they often entail side effects such as abdominal pain and digestive issues. Zuojinwan (ZJW), a Chinese medicine comprising R. coptidis and E. rutaecarpa powder, is renowned for treating gastric ailments. Putative targets of ZJW related to gastritis-related targets via protein-protein interactions from multiple databases. This network approach facilitates understanding ZJW's effects on gastritis by illustrating interactions between its targets and known therapeutic targets. Visualization was done using Cytoscape software. A null model was created to assess the significance of the putative ZJW target-known therapeutic targets of gastritis network. By considering interactions among putative ZJW targets, known therapeutic targets of gastritis, and interactional human proteins, the entire network was constructed to examine linkages [31].

Diabetes mellitus (DM) represents a persistent inflammatory condition, and its escalating prevalence poses a significant global health concern. Traditional Chinese medicine (TCM) has a longstanding tradition of addressing diabetes, valued for its efficacy and minimal adverse effects. Lobelia chinensis, a traditional herbal remedy, has gained attention in pharmacological research due to its active constituents. These include alkaloids, flavonoids, flavonoid glycosides, and amino acids, known for their beneficial properties such as anti-inflammatory, antiviral, and anti-diabetic effects. A compound-target network connects compounds from Lobelia chinensis with target proteins, based on predicted binding interactions. A protein-protein interaction network links target proteins known to interact with each other. A target-pathway network associates target proteins with relevant pathways they participate in. An enrichment analysis network maps Gene Ontology (GO) and KEGG terms, highlighting significant associations based on shared target proteins. Combining these networks into an integrated view can aid in understanding Lobelia chinensis' pharmacological mechanism for anti-diabetic activity, shedding light on key nodes and pathways involved [32]. Traditional Chinese medicine (TCM) has evolved over thousands of years and gained extensive clinical experience, resulting in a comprehensive and distinctive medical system. The use of TCM herbal formulae is a notable aspect of treatment based on the Syndrome (ZHENG in Chinese) distinction, as well as holistic thinking in TCM theory.

In TCM network pharmacology, constructing networks involves several steps:

(i) Active Compound Identification: Identify active compounds in TCM herbs using methods like HPLC and mass spectrometry.
(ii) Target Protein Prediction: Use active compounds to predict potential target proteins with tools like TargetScan and SwissTargetPrediction.
(iii) Compound-Target Network Construction: Link active compounds with target proteins to create a network, visualized with Cytoscape.
(iv) Disease-Related Gene Identification: Identify disease-related genes by analysing gene expression data from databases like GEO and TCGA.
(v) Protein-Protein Interaction (PPI) Network Construction: Build a PPI network linking disease-related genes and their partners, visualized with Cytoscape or STRING.
(vi) Integration of Networks: Combine compound-target and PPI networks to identify key targets using tools like CytoHubba and MCODE [33].

Modernization of TCM involves innovation in both theory and technology. Component-based Chinese medicine (CCM) was proposed in 1999, marking a significant shift from experience-based herbal prescriptions to ones based on the compatibility of components, made possible by advances in analytical chemistry and pharmacology. To construct a network for predicting bioactive constituents and mechanisms of action in component-based Chinese medicine (CCM), the following steps can be taken: Identify active constituents in CCM using database and software tools. Analyse target proteins of identified active constituents using target prediction and fishing. Construct a compound-target network to represent the interactions between compounds and target proteins. Analyse network topology and identify key nodes and modules. Perform pathway enrichment and functional annotation of key nodes and modules. Identify key pathways and biological processes associated with the key nodes and modules. Analyse network module-pathway relationships. Validate key nodes and modules using molecular docking and experimental validation. Identify potential bioactive constituents and mechanisms of action in CCM. This network-based approach can provide a theoretical basis and technical support for the modernization of TCM [34].

27.6 Network Toxicology

Network toxicology now prioritizes studying the biological consequences of chemical exposures at the genome level. The US National Research Council Committee on Toxicity Testing and Assessment of Environmental Agents recommends focusing on understanding the signalling pathways disrupted by biologically active substances to identify those that may cause adverse health effects in humans, rather than solely assessing apical endpoints. In pharmacology, a deeper understanding of disease mechanisms is essential for safe preclinical and clinical testing. New technologies can generate and analyse vast amounts of data, providing insights into both intended and unintended consequences. Rather than interpreting differentially expressed lists of chemicals, researchers are now focusing on pathway analysis and network model construction using data from experimental systems treated with drugs of interest. New computational methods have been developed to fully utilize large-scale toxicological datasets. Systems toxicology and pharmacology techniques can improve understanding of disease causes, leading to safer and more effective treatments for patients [35]. Network toxicology merges systems biology and network analysis to comprehend the toxicological impacts of chemicals and drugs. Its application to traditional Chinese medicine (TCM) facilitates the exploration of intricate interactions between TCM compounds and biological systems. Through constructing biological networks and pathway analysis, key targets and pathways linked to TCM toxicity are discerned, offering insights into potential risks associated with TCM utilization. Moreover, network toxicology aids in identifying biomarkers for toxicity assessment and developing predictive models for TCM toxicity, thereby enhancing the safety and efficacy of TCM usage, and elevating the quality of TCM research [36].

Artificial neural networks (ANNs) are a group of information technology-based algorithms and systems that imitate the organic arrangement of neurons in animal and human central nervous systems. They are simply networks of interconnected nodes, or neurons, that can receive, process, and transmit impulses. One neuron may receive one or more input signals, which are then weighted and processed using a nonlinear function to produce an output. There are various forms of artificial neural networks, and the most of them have at least some applications in medicine and biology. Feedback ANNs return information, but feedforward ANNs flow information in only one direction. Single-layer perceptron (SLP) are simple feedforward artificial neural networks (ANNs) that are widely utilized for linear binary categorization [37].

Network pharmacology, a new branch of pharmacology, is founded on molecular docking, protein interaction, virtual prediction, and topological network analysis. The application of network toxicology to the research of harmful molecular mechanisms is distinguished by its speed, ease, integrity, and global reach. Evaluating toxicity and side effects in a large network setting can help people broaden their understanding of toxicological research. However, given the possibility of false positive outcomes in machine learning algorithms, it is critical to identify and confirm essential targets and pathways in cellular or animal experiments. Based on the research condition described above, non-targeted metabolomics and network toxicology were used to investigate the molecular mechanisms of CS-induced cardiac injury [38]. Pregnancy-induced hypertension (PIH) is a condition associated with significant risks for both mothers and babies. Alisma Orientale (AO), a traditional Chinese medicine, has a history of use in treating PIH. This study employs network pharmacology, network toxicology, molecular docking, and molecular dynamics simulation to explore AO's potential mechanisms and safety. Network pharmacology identifies essential medicinal chemicals in AO and predicts their possible modes of action. The study also investigates the potential hepato-nephrotoxicity of AO agents during PIH treatment utilizing network toxicology. Previous research on AO and its components' toxicity and side effects is limited, highlighting the need for further investigation into their biosafety profile. Thus, network toxicology is used to investigate probable pathways of toxicity [39].

Combining network pharmacology and metabolomics provides a rigorous way to investigate Paclobutrazol's network toxicity. Metabolomics gives information about the metabolic changes caused by Paclobutrazol exposure, pinpointing significantly changed pathways. Concurrently, network pharmacology identifies important targets and signalling pathways linked to Paclobutrazol toxicity. Integrating findings from both approaches provides a thorough understanding of Paclobutrazol's toxicity mechanism, offering light on the possible threats to human health caused by Paclobutrazol residues. This integrated method helps to find biomarkers for toxicity evaluation and potential intervention targets to reduce pesticide-induced harm to human health [40]. Toxicological tipping points, also known as critical concentrations, are precise dosages or exposure levels of a toxin that cause a shift in brain networks from recoverable to non-recoverable trajectories. These thresholds may occur at lower concentrations than those that cause overt cell viability loss or disturb network parameters, showing better sensitivity in detecting network functioning loss. Identifying these tipping thresholds helps to better understand the developmental effects of neurotoxins and allows for potential chemical prioritization. This is consistent with the move in systems toxicology and pharmacology toward understanding disrupted signalling pathways by biologically active chemicals, which is critical for identifying potential detrimental health effects in humans [41].

Network pharmacology was used to investigate the pharmacological effects and mechanisms of Panax notoginseng (Burk.) F. H. Chen on reproductive and genetic toxicity in male mice. This approach allowed researchers to investigate the complex relationships between Panax notoginseng's active chemicals and the signalling pathways involved in reproductive and genetic damage. Key targets and pathways were found through the creation of a biological network and pathway analysis, providing valuable insights into Panax notoginseng's toxicity-protective qualities. This study highlights the potential of network pharmacology in understanding the processes of traditional Chinese medicine and its function in lowering the risk of reproductive and genetic damage [42].

27.7 Applications of Network Pharmacology

Network pharmacology is arising as a valuable strategy to address complex diseases, providing a comprehensive view that surpasses traditional single-target drug discovery. Network pharmacology is a rapidly evolving field that brings together the concepts of systems biology and Polypharmacological to shed light on the intricate workings of drugs and drug candidates. This interdisciplinary approach emphasizes the idea that many effective drugs act on multiple targets, rather than just one, and offers a holistic perspective on the complex interactions within biological systems. This method has been progressively adopted in natural products research, leading to the discovery of novel therapeutic options and the repurposing of existing drugs. By employing genome-wide transcriptomic response profiles, researchers can illustrate the multi-target effects of natural products on biological pathways and interaction networks, consequently gaining a deeper insight into their mechanisms of action without any bias. By utilizing network pharmacology, it is possible to discover synergistic combinations among natural products, facilitating the creation of more potent multi-target treatments. It also has led to the discovery of new and effective treatment combinations, greatly benefiting cases where single-targeted drugs encounter problems like resistance. Network pharmacology can help in finding new uses for approved drugs, making it quicker and cost effective to develop new treatments [43]. In the realm of traditional Chinese medicine (TCM), the network pharmacology method has been employed to study the diseases and pharmacological mechanisms. TCM herbal prescriptions hold significant therapeutic potential and are a valuable resource for drug discovery. Investigating the characteristics of multiple components, targets, and their actions as well as potential synergistic effects in TCM formulas is essential in TCM research [44]. The active components of traditional Chinese herbal medicines (TCM) hold significant potential for drug discovery. Network pharmacology is anticipated as a valuable method for discovering bioactive compounds and predicting their molecular targets in TCM, paving the way for new treatments [45]. Network pharmacology is a promising research methodology that integrates various elements of pharmacology, network analysis, and bioinformatics to discover and understand novel drug candidates and their mechanisms of action. It can be used to identify potential drug targets by analysing the interactions between drugs and their targets in biological networks, thereby providing a comprehensive overview of the latest advances in the field and highlighting the most promising target candidates. In addition, network pharmacology can also predict drug combinations' synergistic effects and identify potential drug-drug interactions. By employing network pharmacology, researchers can gain insights into the complex interactions of drugs and their targets in the human body, which could ultimately lead to the development of more effective and safe therapies for various diseases. This interdisciplinary approach allows for a more holistic understanding of drug action, considering the effects of multiple targets and pathways in a systems-level perspective [46, 47].

27.8 Limitations

Network pharmacology has transformed the drug discovery process, offering a comprehensive approach to identifying drug targets and their interactions. However, it still has some limitations, with a significant issue being its reliance on a single network. To overcome this, integrating various networks, such as drug chemical structure, protein sequence, drug-target interactions, and metabolic networks, can significantly improve the effectiveness of network pharmacology. One critical area where network pharmacology can have a significant impact is personalized medicine. By incorporating cancer-specific networks and using sequencing technologies, network-based models can be developed to predict individual drug responses, drug resistance, and personalized cancer risk. However, for the successful development and evaluation of network-based models, it is essential to consider certain factors that can significantly impact prediction performance. These include using more realistic regression formulations instead of binary classification, relying on quantitative bioactivity data rather than on/off interaction data, and validating models using simple or nested cross-validation methods. Furthermore, reporting performance based on different experimental settings can reveal whether the training and test sets share common drugs and targets, leading to more accurate and robust models [13, 48].

Network pharmacology uses databases and algorithms to identify potential drug targets and analyse biological networks. However, these tools may not always be accurate or reliable, and there is no standard way to use them. Network pharmacology studies often use computer predictions, but these may not always reflect what happens in the body. It's important to do experiments to check the accuracy and relevance of the predictions. Network pharmacology tries to simplify the body's complex system into networks, but this can lead to oversimplification and loss of important information. Even with advances in research, we still don't understand everything about how disease work. This can make it hard to accurately model and analyse the networks involved in diseases. Network pharmacology involves combining and analysing large amounts of information from different sources, which can be hard and take a lot of time. There is a need for better tools to make this easier. Network pharmacology studies often look at molecular targets and pathways, but the goal is to develop new treatments for patients. It can be difficult to translate the findings from network pharmacology to the clinic. It also uses new approaches to drug discovery and development, which can be difficult to regulate and evaluate using existing rules. There is a need for clearer guidelines on how to regulate and evaluate network pharmacology studies and products.

27.9 Future Directions

Based on the findings of this study, numerous future paths might be pursued to further understand the pharmacological mechanisms of natural compounds utilizing the network pharmacology method. To begin, additional information sources and computational approaches can be combined to improve the accuracy and dependability of the results. For example, combining gene expression data, protein-protein interaction networks, and pathway analysis can lead to a more complete knowledge of natural product molecular mechanisms. Second, the network pharmacology technique can be used with additional TCM formulations and natural materials to find possible therapeutic targets and biomarkers for a variety of disorders. This can help with the creation of new medications and customized therapy based on natural ingredients.

Third, experimental validation is required to confirm the expected targets and processes of natural compounds. In vitro and in vivo investigations can be carried out to confirm the impact of natural products on the identified targets and pathways.

Finally, network pharmacology can be used in conjunction with other systems biology methods, like as metabolomics and transcriptomics, to provide a more comprehensive understanding of natural product pharmacological processes. This can assist in clarifying the complicated interactions between natural goods and the human body, promoting the advancement of integrative medicine.

27.10 Conclusion

In conclusion, natural products have been an important source of therapeutic agents for a variety of disorders in traditional medicine. With the development of genetics and bioinformatics, modern natural product discovery has transitioned to bottom-up strategies that supplement traditional top-down procedures. Natural products and their derivatives have made significant contributions to medical treatment across a wide range of health issues, and the unique characteristics that determine their behaviour and potential utility in drug discovery cannot be understated. The book chapter also emphasized the critical necessity for in-depth study of bioactive fractions, as well as phytochemical and phytopharmacological evaluation of herbal medications, to accomplish herbal drug discovery objectives. Furthermore, the chapter looked at modern methodologies for discovering new natural products, such as recent advances in understanding the genetic pathways that govern secondary metabolite production, tapping into previously untapped reservoirs of natural compounds, and incorporating innovative screening technologies. As natural product research advances, it is critical to combine traditional knowledge with modern scientific methodologies to fully realize the promise of natural products for medication discovery and development.

References

1. Luo Y, Cobb RE, Zhao H. Recent advances in natural product discovery. Curr Opin Biotechnol. 2014;30:230–7.
2. Atanasov AG, Zotchev SB, Dirsch VM, Supuran CT. Natural products in drug discovery: advances and opportunities. Nat Rev Drug Discov. 2021;20(3):200–16.
3. Amit Koparde A, Chandrashekar Doijad R, Shripal MC. Natural products in drug discovery. In: Pharmacognosy – medicinal plants. London: IntechOpen; 2019.
4. Gullo VP, McAlpine J, Lam KS, Baker D, Petersen F. Drug discovery from natural products. J Ind Microbiol Biotechnol. 2006;33:523–31.
5. Jensen PR, Mincer TJ, Williams PG, Fenical W. Marine actinomycete diversity and natural product discovery. Antonie van Leeuwenhoek, Int J Gen Mol Microbiol. 2005;87(1):43–8.
6. Bauer A, Brönstrup M. Industrial natural product chemistry for drug discovery and development. Nat Prod Rep. 2014;31:35–60.
7. Lam KS. New aspects of natural products in drug discovery. Trends Microbiol. 2007;15:279–89.
8. Thomford N, Senthebane D, Rowe A, Munro D, Seele P, Maroyi A, et al. Natural products for drug discovery in the 21st century: innovations for novel drug discovery. Int J Mol Sci. 2018;19(6):1578.
9. Natural product drug discovery [Internet]. Available from: www.ncbi.nlm.nih.gov/.
10. Bologa CG, Ursu O, Oprea TI, Melançon CE, Tegos GP. Emerging trends in the discovery of natural product antibacterials. Curr Opin Pharmacol. Elsevier Ltd. 2013;13:678–87.
11. Khan MT, Ather A. Lead molecules from natural products: discovery and new trends. Elsevier; 2006 Jul 7. The Netherlands.
12. David B, Wolfender JL, Dias DA. The pharmaceutical industry and natural products: historical status and new trends. Phytochem Rev. Kluwer Academic Publishers. 2015;14:299–315.
13. Chandran U, Mehendale N, Patil S, Chaguturu R, Patwardhan B. Network pharmacology. In: Innovative approaches in drug discovery. Amsterdam: Elsevier; 2017. p. 127–64.
14. Zhang GB, Li QY, Chen QL, Su SB. Network pharmacology: a new approach for Chinese herbal medicine research. Evid Based Complement Alternat Med. 2013;2013:621423.
15. Hopkins AL. Network pharmacology: the next paradigm in drug discovery. Nat Chem Biol. Nature Publishing Group. 2008;4:682–90.
16. Hao DC, Xiao PG. Network pharmacology: a Rosetta stone for traditional Chinese medicine. Drug Dev Res. Wiley-Liss Inc. 2014;75:299–312.
17. Yang M, Chen JL, Xu LW, Ji G. Navigating traditional Chinese medicine network pharmacology and computational tools. Evid Based Complement Alternat Med. 2013;2013:731969.
18. Boezio B, Audouze K, Ducrot P, Taboureau O. Network-based approaches in pharmacology. Mol Inform. Wiley-VCH Verlag. 2017;36. https://doi.org/10.1002/minf.201700048.
19. Zahoránszky-Kohalmi G, Sheils T, Oprea TI. SmartGraph: a network pharmacology investigation platform. J Cheminform. 2020;12(1):5.

20. Li X, Tang Q, Meng F, Du P, Chen W. INPUT: an intelligent network pharmacology platform unique for traditional Chinese medicine. Comput Struct Biotechnol J. 2022;20:1345–51.
21. Zhang RZ, Yu SJ, Bai H, Ning K. TCM-Mesh: the database and analytical system for network pharmacology analysis for TCM preparations. Sci Rep. 2017;7(1):2821.
22. Gu J, Gui Y, Chen L, Yuan G, Xu X. CVDHD: a cardiovascular disease herbal database for drug discovery and network pharmacology [Internet]. 2013. Available from: http://pkuxxj.pku.edu.cn/CVDHD.
23. Hsin KY, Matsuoka Y, Asai Y, Kamiyoshi K, Watanabe T, Kawaoka Y, et al. systemsDock: a web server for network pharmacology-based prediction and analysis. Nucleic Acids Res. 2016;44(W1):W507–13.
24. Yang P, He H, Xu S, Liu P, Bai X. Potential molecular target prediction and docking verification of Hua-Feng-Dan in stroke based on network pharmacology. Evid Based Complement Alternat Med. 2020;2020:1–12.
25. Tao W, Xu X, Wang X, Li B, Wang Y, Li Y, et al. Network pharmacology-based prediction of the active ingredients and potential targets of Chinese herbal Radix Curcumae formula for application to cardiovascular disease. J Ethnopharmacol. 2013;145(1):1–10.
26. Zhang A, Sun H, Yang B, Wang X. Retraction note: predicting new molecular targets for rhein using network pharmacology. BMC Syst Biol. 2012;6:20. https://doi.org/10.1186/1752-0509-6-20. Vol. 8, BMC Systems Biology. BioMed Central Ltd. 2014.
27. Lee AY, Park W, Kang TW, Cha MH, Chun JM. Network pharmacology-based prediction of active compounds and molecular targets in Yijin-Tang acting on hyperlipidaemia and atherosclerosis. J Ethnopharmacol. 2018;221:151–9.
28. Ma H, Xu F, Liu C, Seeram NP. A network pharmacology approach to identify potential molecular targets for cannabidiol's anti-inflammatory activity. Cannabis Cannabinoid Res. 2021;6(4):288–99.
29. He D, Huang JH, Zhang ZY, Du Q, Peng WJ, Yu R, et al. A network pharmacology-based strategy for predicting active ingredients and potential targets of LiuWei DiHuang pill in treating type 2 diabetes mellitus. Drug Des Devel Ther. 2019;13:3989–4005.
30. Yang Y, Yang K, Hao T, Zhu G, Ling R, Zhou X, et al. Prediction of molecular mechanisms for LianXia NingXin formula: a network pharmacology study. Front Physiol. 2018;9:489.
31. Yu G, Wang W, Wang X, Xu M, Zhang L, Ding L, et al. Network pharmacology-based strategy to investigate pharmacological mechanisms of Zuojinwan for treatment of gastritis. BMC Complement Altern Med. 2018;18(1):292.
32. Ge Q, Chen L, Yuan Y, Liu L, Feng F, Lv P, et al. Network pharmacology-based dissection of the anti-diabetic mechanism of Lobelia chinensis. Front Pharmacol. 2020;11:347.
33. Li S, Zhang B. Traditional Chinese medicine network pharmacology: theory, methodology and application. Chin J Nat Med. 2013;11(2):110–20.
34. Jiao X, Jin X, Ma Y, Yang Y, Li J, Liang L, et al. A comprehensive application: molecular docking and network pharmacology for the prediction of bioactive constituents and elucidation of mechanisms of action in component-based Chinese medicine. Comput Biol Chem. 2021;90:107402.
35. Hoeng J, Talikka M, Martin F, Sewer A, Yang X, Iskandar A, et al. Case study: the role of mechanistic network models in systems toxicology. Drug Discov Today. 2014;19(2):183–92.
36. Fan X, Zhao X, Jin Y, Shen X, Liu C. Network toxicology and its application to traditional Chinese medicine. J Chin Mater Med. 2011;36(21):2920–2.
37. Pantic I, Paunovic J, Cumic J, Valjarevic S, Petroianu GA, Corridon PR. Artificial neural networks in contemporary toxicology research. Chem Biol Interact. 2023;369:110269.
38. Liu C, Zhang C, Wang W, Yuan F, He T, Chen Y, et al. Integrated metabolomics and network toxicology to reveal molecular mechanism of celastrol induced cardiotoxicity. Toxicol Appl Pharmacol. 2019;383:114785.
39. Liao Y, Ding Y, Yu L, Xiang C, Yang M. Exploring the mechanism of Alisma orientale for the treatment of pregnancy induced hypertension and potential hepato-nephrotoxicity by using

network pharmacology, network toxicology, molecular docking and molecular dynamics simulation. Front Pharmacol. 2022;13:1027112.
40. Yue K, Liu Z, Pi Z, Li H, Wang Y, Song F, et al. Network pharmacology combined with metabolomics approach to investigate the toxicity mechanism of paclobutrazol. Chem Res Toxicol. 2022;35(4):626–35.
41. Frank CL, Brown JP, Wallace K, Wambaugh JF, Shah I, Shafer TJ. Defining toxicological tipping points in neuronal network development. Toxicol Appl Pharmacol. 2018;354:81–93.
42. Zhang K, Sun C, Hu Y, Yang J, Wu C. Network pharmacology reveals pharmacological effect and mechanism of Panax notoginseng (Burk.) F. H. Chen on reproductive and genetic toxicity in male mice. J Ethnopharmacol. 2021;270:113792.
43. Kibble M, Saarinen N, Tang J, Wennerberg K, Mäkelä S, Aittokallio T. Network pharmacology applications to map the unexplored target space and therapeutic potential of natural products. Nat Prod Rep. 2015;32(8):1249–66.
44. Yuan H, Ma Q, Cui H, Liu G, Zhao X, Li W, et al. How can synergism of traditional medicines benefit from network pharmacology? Molecules. 2017;22(7):1135.
45. Zhou Z, Chen B, Chen S, Lin M, Chen Y, Jin S, et al. Applications of network pharmacology in traditional Chinese medicine research. Evid Based Complement Alternat Med. 2020;2020:1–7.
46. Noor F, Tahir ul Qamar M, Ashfaq UA, Albutti A, Alwashmi ASS, Aljasir MA. Network pharmacology approach for medicinal plants: review and assessment. Pharmaceuticals. 2022;15(5):572.
47. Zheng S, Xue T, Wang B, Guo H, Liu Q. Application of network pharmacology in the study of mechanism of Chinese medicine in the treatment of ulcerative colitis: a review. Front Bioinform. 2022;2:928116.
48. Muhammad J, Khan A, Ali A, Fang L, Yanjing W, Xu Q, et al. Network pharmacology: exploring the resources and methodologies. Curr Top Med Chem. 2018;18(12):949–64.

GPSR Compliance

The European Union's (EU) General Product Safety Regulation (GPSR) is a set of rules that requires consumer products to be safe and our obligations to ensure this.

If you have any concerns about our products, you can contact us on ProductSafety@springernature.com

In case Publisher is established outside the EU, the EU authorized representative is:

Springer Nature Customer Service Center GmbH
Europaplatz 3
69115 Heidelberg, Germany

Batch number: 08468044

Printed by Printforce, the Netherlands

Applications of Computational Tools in Drug Design and Development

S. N. Koteswara Rao G.
Rajasekhar Reddy Alavala
Editors

Applications of Computational Tools in Drug Design and Development

Springer

Editors
S. N. Koteswara Rao G.
Shobhaben Pratapbhai Patel School of Pharmacy and Technology Management
SVKM's NMIMS Deemed to be University
Mumbai, Maharashtra, India

Rajasekhar Reddy Alavala
Shobhaben Pratapbhai Patel School of Pharmacy and Technology Management
SVKM's NMIMS Deemed to be University
Mumbai, Maharashtra, India

ISBN 978-981-96-4153-6 ISBN 978-981-96-4154-3 (eBook)
https://doi.org/10.1007/978-981-96-4154-3

This Springer imprint is published by the registered company Springer Nature Singapore Pte Ltd.
The registered company address is: 152 Beach Road, #21-01/04 Gateway East, Singapore 189721, Singapore

Dedicated to all the researchers, faculty, and students
May this book serve as a valuable resource in your journey toward transforming computational insights into therapeutic breakthroughs

Preface

The world of drug discovery and development is a complex, multifaceted field that addresses the pressing challenges in disease treatment. As scientific understanding advances and technology progresses, the integration of computational tools has become increasingly vital, offering new dimensions of insight and precision to researchers. Initially focusing on drug discovery, the computational tools have now amalgamated every aspect of drug discovery and development, including drug discovery, drug delivery, biopharmaceutics, and pharmacokinetics, as well as pharmacology and toxicology. Not many books touch on all of these aspects. Offering a broad perspective, this book explores the computational tools utilized in every phase of drug discovery and development.

Chapters 1–10 of the book delves into applications of computational tools for drug delivery. This part covers application of computational tools for preformulation studies as well as use of design of experiments (DoE) software for optimizing the formulation parameters. Another emerging area of focus is the delivery of drugs to the lungs. The Chap. 7 focuses about computational fluid dynamics and is used to simulate the flow of drug aerosol in the respiratory track and its application in the design and optimization of aerosol-based delivery systems. Likewise, the use of simulation to analyze and optimize key parameters in the pharmaceutical die filling process is discussed in this part.

Chapters 11–19 of the book explores the computational tools for predicting biopharmaceutical as well as pharmacokinetic parameters. Detailed in this part are computational tools for the prediction of solubility, advances in computational prediction of absorption as well as permeation, and forecasting protein binding using computational tools. Also included in this part is a detailed explanation about visualizing and understanding the interactions between drug molecules and excipients using simulations and how this information is used for the selection of best excipient under the given conditions.

Chapters 20–27 focuses on specific therapeutic areas. Through the mechanistic understanding of diseases such as Alzheimer's, diabetes, atherosclerosis, and cancer, this section highlights how computational tools such as QSAR, pharmacophore mapping, docking, and homology modeling as well as network analysis are being used for design of better drugs. The design of drugs against epilepsy and hypertension is also included in this part. The final chapters explore drug discovery and computational strategies in the context of multidrug-resistant tuberculosis and the network pharmacology approach to uncover the pharmacological mechanisms of natural products. Drug repurposing, highlighted during COVID, although practiced much before it, in the treatment of cancer is also discussed in this part.

This book offers a holistic view of multiple computational tools used at various stages of drug discovery and development. It will be a valuable resource to researchers and students working in this field.

Urmila J. Joshi
Retd. Principal and Head of Department
of Pharmaceutical Chemistry
Principal K. M. Kundnani College of Pharmacy
Mumbai, Maharashtra, India

Foreword

The book *Applications of Computational Tools in Drug Design and Development* is a pleasant blend of computational methods employed in both drug discovery and drug development. It covers how computational methods can be used at all stages in the drug discovery and development process. There are not many volumes that have been published to date that comprehensively cover how *in silico* methods are used in drug discovery and development, and therefore, this volume will be welcomed by all those working in drug discovery and development.

Understanding the mechanism of action of a drug is central to the development and design of new drugs. A variety of experimental methods is used to unravel the mechanism of action of drugs. The use of computational tools like network analysis (STRING, Cytoscape, DAVID, etc.) and molecular modeling tools like docking, QSAR, receptor-based pharmacophore mapping, etc. to understand the pathological mechanisms of anti-Alzheimer agents and drugs used for the treatment of epilepsy and atherosclerosis has been illustrated in Chaps. 20, 22, and 24.

Chemoinformatics, which combines concepts and methods in chemistry, computer science, and information technology, is used to streamline the process of identifying and developing new leads. The tools of chemoinformatics for storage, analysis, visualization, and interpretation of chemical data, and their significance in drug discovery and development are covered in Chap. 9.

Drug likeness aims to predict how promising a molecule is to be developed as a drug. Drug likeness as embodied by Lipinski's rule-of-five has been a guiding principle to identify molecules that are most likely to succeed as novel therapeutic agents. Tools to assess drug likeness have been described in Chap. 19 titled "Recent Advances in Drug-Likeness Screening by Using the Software and Online Tools."

In silico approaches have revolutionized drug discovery by decreasing the time to identify hits, optimize leads, and develop a clinical candidate. The use of *in silico* methods like molecular docking, molecular dynamics (MD) simulations, QSAR, and protein structure determination (homology modeling) to identify new targets and to decipher drug receptor interactions coupled with the prediction of ADME attributes have been exemplified in Chap. 21 in the design of novel antidiabetic agents, and in Chap. 25 in the design of novel antihypertensive agents.

The safety aspect of a drug is an integral part of the drug design process, and the accurate prediction of toxicity helps mitigate the risk associated with the appearance of life-threatening effects during preclinical or clinical studies. Toxicity databases such as ToxCast, ACToR (Aggregated Computational Toxicology Resource), and PubChem can be queried to assess the potential toxicity of known or structurally related substances. Additionally, toxicogenomics tools like GeneGo (MetaCore) and pathway analysis tools, along with in silico toxicity prediction platforms such as ADMET Predictor, the Tox21 collaboration, and DEREK Nexus, can be used to evaluate compound toxicity across multiple biological endpoints. Some toxicity prediction tools are covered in Chap. 16.

Repurposing an existing therapeutic for a new medical condition has several key advantages, such as reduced time to bring the drug to market, since the safety profile of the drug is already established. Viral-based therapies like gendicine, oncorine, and imlygic are examples of therapies that have been repurposed for cancer treatment. These examples and the initiatives by the National Center for Advancing Translational Sciences (NCATS) are covered in Chap. 23 titled "Modern Computational Intelligence Based Drug Repurposing for Cancer."

The appearance of drug resistance across various therapeutic segments is posing serious threats to public health in terms of failure of treatment, development of chronic infections, escalation in the cost of treatment, and narrowing the options available for treatment. The appearance of MDR, XDR, and TDR-TB is creating havoc worldwide in the management of this disease. Chapter 26 on "Drug Discovery and Computational Strategies in Multi-Drug-Resistant Tuberculosis" attempts an overview of various computational strategies, like structure-based and ligand-based, that have been used in the development of new drugs for the treatment of MDR-TB.

Natural products, with their diversity of chemical structures, have been the founding structures on which many therapeutic agents have been designed. The vast untapped biodiversity found in marine ecosystems, tropical rainforests, and the animal world holds the promise of discovery of novel compounds with unique therapeutic attributes. Chapter 27 titled "The Network Pharmacology Approach to Uncover the Pharmacological Mechanism of Natural Products" expounds a multidisciplinary approach based on systems biology and polypharmacology.

After the target has been identified and validated, hits recognized, and leads optimized, preclinical studies that evaluate the solubility, safety, efficacy, drug binding, and pharmacokinetics of lead compounds in laboratory and animal models prior to human testing is conducted. All these aspects are covered in Chaps. 11, 12, 13, and 15. A pivotal aspect of preclinical drug development is the assessment of solubility of the drug candidate. This is essential for formulation design and satisfactory pharmacokinetics. The shake flask method is the gold standard for measurement of solubility; however, quantum mechanical calculations based on DFT, molecular dynamics simulations, QSPR, and machine learning models offer cheaper and accurate ways to model solvation energy. These computational tools are covered in Chap. 11 titled "Computational Tools for Solubility Prediction." Along with solubility, another fundamental theme in preclinical studies is the prediction of absorption, distribution, metabolism, and excretion (ADME) of the drug. Understanding absorption is essential for predicting a drug's bioavailability, its dosing regimen, and potential effectiveness. Simcyp, GastroPlus, SwissADME, and QikProp are some of the finest tools used for prediction of ADME attributes; these could have been covered in depth along with the other methods described in Chap. 12 titled "Advances in the Computational Prediction of Absorption Prediction of Pharmaceuticals." The amount of drug available for interaction with the target is dictated by the fraction bound/unbound (free) to plasma proteins (e.g., albumin, a1-acid glycoprotein, lipoproteins, etc.) and is a key parameter that is studied in the preclinical phase of drug development. Computational tools used to study ADME, described above, have modules to assess the fraction of drugs bound to plasma proteins. Case studies of how computational methods have been used to optimize protein binding and hasten the drug development process are elucidated in Chap. 13 titled "Protein Binding Prediction by Computational Methods." Biopermeability studies are linked with the prediction of drug absorption and are used to understand how well a drug will cross biological membranes to enter systemic circulation. Poor permeability may necessitate modifications in the chemical structure or alterations in the formulation development. The use of advanced molecular dynamics simulations to predict the bioavailability of pharmaceuticals and its relationship to experimental data has been discussed in Chap. 15.

One of the important stages in drug development is formulation design. In this phase, the focus is on developing a stable, effective, and safe formulation that can be used in human clinical trials. Decisions must be made about the dosage form, whether tablet or injection or any other, and the method of delivery. Computational methods for design of the formulation and mode of delivery—oral, ophthalmic, parenteral, or inhalation—are covered in Chaps. 1, 2, 8, and 10. There are several challenges in the delivery of drugs to the lungs. Factors such as the design of the device, formulation aspects like particle size, dose control, and patient competence all dictate how efficiently the drug is delivered to the lungs. Computational fluid dynamics (CFD) has played a significant role in the aerospace industry and is now making a deep impact in the design and optimization of inhalers, in formulation design (Chaps. 3 and 7), and in the die filling process during tablet compression (Chap. 6). Exhaustive simulations can be conducted to get insights into airflow patterns, particle behavior, and sites of drug deposition within the respiratory system.

The underlying relationships between the dependent and independent variable in a process can be understood by Design of Experiments (DOE). How DOE can be used to optimize the effectiveness of drug delivery systems has been well explained in Chap. 5 titled "Design of Experiments: Understanding Optimization."

A significant percentage of the excipients in pharmaceutical formulations are polymers—natural, synthetic, biodegradable, or mucoadhesive. Computational methods (Chap. 18) are used to understand how a formulation chemist can select the "right" polymer to avoid drug-polymer interactions that could limit the stability, release, or bioavailability of the drug.

All drug discovery programs begin with target identification and target validation. Discussion of this facet of drug discovery, which is missing in this volume, may be covered in a future edition of this book.

In essence, this volume gives a comprehensive insight into the numerous computational methods and tools that are used in various stages of drug discovery and development. This volume would benefit both the novice and experienced pharmaceutical researchers and scientists, clinical researchers, medical professionals, graduate students in a drug discovery program, and the regulatory agencies and professionals.

Evans Coutinho
Dean Research
St John Technical and Educational Campus
Palghar, Maharashtra, India

Acknowledgments

We express our deepest gratitude to **Mr. Fumiko Yamaguchi**, Senior Editor, Books, Biomedical and Life Sciences, at Springer Nature, for his invaluable support in accepting our book proposal and providing exceptional guidance throughout the development of this book. His expertise and encouragement have been instrumental in bringing this work to fruition. We would also like to extend our sincere thanks to **Ms. Kavitha Jayakumar**, Production Editor (Books), for her dedicated assistance in overseeing the production process of this book.

We extend our heartfelt thanks to the **Dean, SPPSPTM** and **all the authorities of SVKM's NMIMS**, Mumbai, for providing the necessary facilities and fostering an environment of encouragement and academic excellence. Their unwavering support has greatly contributed to the successful completion of this project.

We are profoundly grateful to all the **contributing authors** whose dedication, expertise, and commitment to advancing knowledge have enriched this book. Their collective efforts have made this endeavor possible, and we deeply value their contributions.

Finally, we acknowledge the support and encouragement of our families, colleagues, and friends, whose belief in this project has been a source of motivation throughout this journey.

Regards
Dr. S. N. Koteswara Rao G.
Dr. Rajasekhar Reddy Alavala

Contents Volume-I

Contents Volume-II

Chapter 1
Computational Simulation of Drug Delivery at the Molecular Level

S. Phani Kumar Chirravuri, S. S. Sai Kiran P., Rama Krishna Gummadi, Nagasen Dasari, and S. N. Koteswara Rao G.

Contents

S. P. K. Chirravuri (✉) · S. S. Sai Kiran P. · R. K. Gummadi · N. Dasari
Department of Pharmaceutics, Aditya Pharmacy College, Surampalem, Andhra Pradesh, India

S. N. K. Rao G.
Shobhaben Pratapbhai Patel School of Pharmacy and Technology Management, SVKM's NMIMS, Vile Parle (W), Mumbai, Maharashtra, India

S. N. K. Rao G., R. R. Alavala (eds.), *Applications of Computational Tools in Drug Design and Development*, https://doi.org/10.1007/978-981-96-4154-3_1

Abstract Computational simulations at the molecular level are crucial for advancing drug delivery systems by providing insights into drug-receptor interactions, permeation across biological barriers, solubility, stability and conformational changes. Techniques such as molecular modeling, molecular docking studies, molecular dynamics (MD) simulations, quantum mechanics/molecular mechanics (QM/MM) simulations, and free energy calculations play key roles. Molecular modeling predicts structures, conformations, and properties essential for drug efficacy. Docking studies forecast interactions and binding affinities, guiding lead optimization and toxicity assessment. MD simulations reveal drug behavior, stability, and interactions patterns. QM/MM simulations accurately model active sites and chemical reactions. Free energy calculations predict binding affinities and thermodynamic properties, enhancing our understanding of drug receptor interactions. These techniques optimize drug design, pharmacokinetics, and pharmacodynamics and targeted drug delivery systems. Visualization and analysis tools help interpret complex interactions, facilitating effective therapeutic interventions. Computational simulations span development stages, from design to clinical trial simulations, providing a robust frame work for safer and more effective pharmaceuticals. *In-silico* model accelerate drug discovery, reduce animal testing, address ethical concerns, and improve development pipeline efficacy.

Keywords Molecular modeling · Molecular Docking studies · Molecular dynamics simulations · Quantum mechanics/molecular mechanics simulations · Free energy calculations · Drug-receptor interactions · Drug solubility and stability · Pharmacokinetics · Pharmacodynamics · *In-silico* drug discovery

1.1 Introduction

The therapeutic success of any drug molecule is based on its therapeutic action and therapeutic action purely based on drug pharmacokinetic and pharmacodynamics characteristics. The drug therapeutic action of a drug mole follows a sequence of process such as (i) Administered drug absorbs through a bio-membrane, based on its physicochemical properties (ii) Distributes through various biological membrane and reaches the site of action (iii) Binds with receptor and elicits the therapeutic action. The effective therapeutic action is based on drug delivery system, interaction of drug with various biological components and membranes, binding with a receptor and occupying enough number of receptors, this depends on drug concentration at site of action. Molecularly various factors influencing therapeutic action are release of drug molecule from drug delivery system, release drug possess a suitable conformation, absorption and distribution of drug molecule with the same conformational characteristics, structural similarities of drug to the receptor, nature of binding, binding integrity and stability, and binding affinity with the active site of receptor. The therapeutic failure or success of a drug molecule based on interaction of drug molecule with biological system at atomic or molecular level [1–5].

Computational simulations of drug delivery at molecular level are the tools for model and analyze the interaction between drugs and biological system at atomic or molecular scale. The various computational simulations of drug delivery at molecular level are molecular modeling, docking studies, molecular dynamic (MD) simulations, quantum mechanics/molecular mechanics (QM/MM) simulations, free energy calculations and visualization and analysis [6–8].

Researchers captures the structural and chemical properties of the drug molecule with the help of molecular modeling by the help of various software tools example MD, QM/MM simulations to build the models of drug molecules and biological target such as a protein receptor or nucleic acids. Prediction of preferred orientation and conformation of drug molecule when bound to its target site establishes by docking studies with the help of software packages that simulate the interaction between drug molecule and the receptor binding site. The dynamic behavior and interaction of drug molecule with biological system over time is studied with the help of MD simulations. MD simulation allows researcher to observe molecules move and interact within a biological environment by solving Newton's equation of motion to track the positions and velocities of the molecules in the biological systems. QM/MM simulations are combination of quantum mechanics calculations and molecular mechanics calculations. These simulations are for calculation of a small region of the system such as the active binding site on a receptor and allows researcher for accurate modeling of chemical reactions and interactions that occur at the molecular level. Free energy calculations are applicable to predict the binding affinity between a drug molecule and the target receptor and are aims to estimate the thermodynamic properties of the drug-receptor interaction, such as binding free energy; this can provide insights into the strength and specificity of interaction. Finally, researchers analyze the simulation results to gain insights into the mechanisms of drug binding, drug release, and any conformational changes that occur in the biological system upon drug binding. Visualization tools help researchers visualize and interpret the complex molecular interactions observed in the simulations. Overall, computational simulations of drug delivery at the molecular level plays a crucial role in drug discovery and development by providing valuable insights into the mechanisms of drug action and helping to design more effective and targeted therapeutic interventions.

These simulations at molecular level are offering myriad of applications across various stages of drug development and therapeutic implementation. In drug design and optimization, these simulations facilitate molecular docking to predict drug-receptor binding, optimizing affinity and specificity, while also elucidating structure-activity relationships to enhance drug efficacy. For pharmacokinetics and pharmacodynamics (PK/PD) simulations predict drug absorption, distribution, metabolism, and excretion (ADME), guiding dosing regimens and minimizing toxicity. In the design of nanoparticles and carriers, simulations aid in creating efficient drug delivery systems like nanocarrier and controlled release systems, ensuring targeted and sustained drug release. Additionally, they predict drug resistance mechanisms, informing the design of next-generation therapeutics. Personalized medicine benefits from patient-specific simulations that predict individual responses, optimizing treatment plans; safety assessment is enhanced by predicting off-target effects and toxicities early in development, and understanding potential carcinogenicity and genotoxicity. Biophysical insights from simulations reveal the molecular mechanisms of drug action, adding therapeutic understanding. Simulations also streamline clinical trial design through virtual trials and modeling population variability, reducing trail costs and time. By providing *in-silico* models, computational method reduces the need for animal testing, addresses ethical concerns. Finally, complex disease modeling with multi-scale simulations integrates molecular, cellular, and tissue-level data, supporting the development of comprehensive treatment strategies for diseases such as cancer and diabetes. These diverse applications underscore the pivotal role of computational simulations in advancing drug delivery and therapeutic outcomes [6].

1.2 Computational Simulation Models

These are used to produce more effective and safer drug molecule by modeling and analyzing drug molecule interactions on molecular level with the use of advanced computer models. This approach aims to optimize drug design by enhancing binding affinity and specificity to target receptors, understanding the ADME process and uncovering mechanism of drug action and resistance. The models are as follows (1) Molecular Modeling (2) Molecular Docking Studies (3) Molecular Dynamic (MD) simulations (4) Quantum Mechanics/Molecular Mechanics (QM/MM) Simulations (5) Free Energy Calculations (6) Visualization and Analysis [9].

1.2.1 Molecular Modeling

This plays very important role in computational simulation of drug delivery at molecular level and it allows researcher to explore and optimize various aspects of drug delivery at molecular level by (1) Molecular structure prediction (2) Docking studies (3) Quantitative Structure-Activity Relationship (QSAR) (4) Molecular Dynamic Simulations (5) Drug Carrier Design and helps to accelerate the drug discovery and development process [10–12].

1.2.1.1 Molecular Structure Prediction (MSP)

MSP is the essential component of molecular modeling in the context of computational simulations of drug delivery at molecular level and plays very important role in molecular modeling by providing insights into drug-receptor interactions, physic-chemical properties, formulation design, metabolite prediction, and drug transport across biological barriers. Molecular structure prediction contributed to the drug delivery by (1) Conformational analysis (2) Drug-receptor interactions (3) Prediction of physic-chemical properties (4) Formulation design (5) Prediction of metabolite structure (6) Assessment of drug transport across biological barriers [13, 14].

(a) Conformational Analysis: This analysis in molecular structure prediction identifies the most stable and biologically relevant shapes of drug molecules, crucial for their optimal interaction with targets. Though computational simulations, it ensures that drug adopt conformations that enhance binding affinity and specificity. This analysis aids in predicting pharmacokinetic and pharmacokinetic properties, improving drug efficacy and safety. Thus conformational analysis is essential for designing and optimizing drug molecule for effective delivery.
(b) Drug-Receptor Interactions: These are vital for understanding the drug binding with its target, and guiding the design of molecules with optimal therapeutic efficacy. Computational simulation models these interactions to predict binding efficacy and specificity, aiding in the identification of potent drug candidate. Thus, studying drug-receptor interactions are crucial for developing effective and safe drug delivery system.
(c) Prediction of Physico-Chemical Properties: These are essential for assessing a drug's solubility, stability and permeability, which directly impact its bioavailability and therapeutic efficacy. Computational simulations provide accurate predictions of these properties, guiding the optimization of drug candidates for better performance. By understanding these attributes, researchers can design drugs that are more effective and safe. Thus, predicting physic-chemical is crucial for successful drug delivery and development.
(d) Formulation Design: It ensures that drug candidates are delivered in stable, bioavailable, and effective form. Computational simulations predict how different formulations affect drug release, absorption, and stability, guiding the selection of optimal drug delivery systems. This integration helps translate molecular properties into practical therapeutic applications. Thus, formulation design is crucial for optimizing drug delivery and therapeutic outcomes.
(e) Prediction of Metabolite Structure: These are crucial for understanding how a drug will be processed in the body, identifying potential active or toxic metabolite. Computational simulations provide insights into metabolic pathways and transformations, aiding in the optimization of drug candidate to enhance efficacy and safety. This predictive capability helps in designing drugs with favorable metabolic profiles. Thus, predicting metabolite structures is essential for ensuring safe and effective drug delivery.
(f) Assessment of Drug Transport across Biological Barriers: This prediction helps how drugs will be metabolized, revealing potential active or toxic byproducts. Computational simulations provide insights into these metabolic transformations, enabling the design of drug candidates with improved efficacy and reduced toxicity. This predictive approach ensures that drugs have favorable metabolic profiles, optimizing their safety and therapeutic performance. Thus, predicting metabolite structure is vital for safe and effective drug delivery.

1.2.1.2 Molecular Docking Studies

Molecular docking studies predicts the preferred orientation of drug molecule to the binding site on receptor to form a stable complex. These computational simulations used to predict the binding affinity and interactions between the drug molecule and the target receptor. These contribute to this field by (1) Prediction of Drug-receptor interactions (2) Virtual screening of drug candidates (3) Lead optimization (4) Exploring drug-target binding mechanisms (5) Assessment of drug-induced toxicity (6) Design of drug delivery systems [5, 13, 14].

(a) Prediction of Drug-Receptor Interactions: it is used to identify the optimal binding modes and affinity of drug candidate to their targets. Computational simulations facilitate understanding these interactions at atomic level, guiding the design and optimization of drugs for enhanced efficacy and specificity. This approach aids in selection of the most promising compound for further development. Thus, prediction of drug-receptor interactions is crucial for advancing drug delivery systems.
(b) Virtual Screening of Drug Candidates: It enables the efficient identification of promising compound from large libraries by predicting their binding affinity to biological targets. Computational simulations rank these compounds based on potential interactions, streamlining the drug discovery process. This approach accelerates the identification of high-potential candidates for further experimental validation. Thus, virtual screening of drug candidates is crucial for optimizing drug delivery and expediting the development of effective therapeutics.
(c) Lead Optimization: This is used for refine drug molecule by improving their binding affinity, specificity and pharmacokinetic properties. This process enhances the therapeutic potential and safety profile of the compound. Thus, it is crucial for effective and efficient drug delivery system.
(d) Exploring Drug-Target Binding Mechanisms: They reveal the precise interactions and conformational changes upon binding. This insight helps optimize drug design, for improve efficacy and specificity. Consequently, it enhances the development of effective drug delivery systems by targeting key molecular interactions.
(e) Assessment of Drug-Induced Toxicity: It predicts potential adverse interactions and off-target effects at the molecular level. This approach helps identify and mitigate toxicological risks early in drug development process. Consequently, it enhances the safety and efficacy of drug delivery systems.
(f) Design of Drug Delivery Systems: It enables the precise prediction of drug-target interactions and optimal binding conformations. This approach aids in tailoring delivery mechanisms to enhance drug stability, bioavailability, and therapeutic efficacy. Consequently, it ensures more efficient and targeted drug delivery of drugs within the body.

1.2.1.3 Quantitative Structure-Activity Relationship (QSAR)

QSAR models relates the chemical structure of drug molecule to their therapeutic activity. By analyzing the relationship between the chemical properties of a drug molecule and its biological activity, QSAR models help in predicting the activity of new drug candidates. These models contribute to the field by (1) Prediction of biological activity (2) Optimization of pharmacokinetic properties (3) Identification of structural-activity relationships (SAR) (4) Prediction of drug-drug interactions (5) Design of drug delivery systems (6) Virtual screening and lead optimization [15, 16].

(a) Prediction of biological activity: It leverages statistical correlation between chemical structure and biological effects. This method allows for rapid assessment and optimization drug candidates' safety and efficacy profile. Consequently, it accelerates the development of drug delivery systems by predicting the potential biological activity of new compound.
(b) Optimization of pharmacokinetic properties: It involves predicting and enhancing how drugs are absorbed, distributed, metabolized, and excreted. This approach guides the design of drug candidates with improved bioavailability and reduced side effects. Consequently, it ensures more efficient and effective drug delivery systems by tailoring pharmacokinetic profile.
(c) Identification of structural-activity relationships (SAR): It elucidates how changes in chemical structure influence biological activity. This analysis guides the optimization of drug candidates for enhanced efficacy and specificity. Consequently, it facilitates the development of more effective drug delivery systems by tailoring molecular structures to desired pharmacological outcomes.
(d) Prediction of drug-drug interactions: It predicts potential interactions between different drugs based on their chemical structures. This analysis aids in identifying and mitigating risks of adverse interactions, optimizing drug combinations for enhanced therapeutic outcomes. Consequently, it ensures safer and more effective drug delivery systems by considering potential interactions at molecular level.
(e) Design of drug delivery systems: They facilitate the optimization of delivery mechanisms for improved drug targeting and release kinetics. This approach enhances the understanding how molecular structure influences drug delivery efficiency and bioavailability, guiding the development of tailored delivery systems. Consequently, it enables the drug design of more effective drug delivery strategies through computational simulations at molecular level.
(f) Virtual screening and lead optimization: These enable efficient identification and refinement of promising drug candidates from large libraries based on molecular properties and biological activities. This approach accelerates drug discovery by prioritizing compounds with optimal pharmacological profiles for further development. Consequently, it streamlines the design of effective drug delivery systems through computational simulations at molecular level.

1.2.1.4 Molecular Dynamic (MD) Simulations

These simulations track the movement of atoms and molecules over time. In drug delivery, MD simulations can provide insights into how drug molecules interact with cell membranes, how they taken up by cells, and how they release from drug carriers. MD simulations can also helps in studying the stability of drug formulations and predicting their behavior under different conditions. These simulations contribute to field by (1) Understanding Drug-receptor interactions (2) Exploring drug permeation across barriers (3) Prediction of drug solubility and stability (4) Design of drug delivery systems (5) Prediction of protein-ligand kinetics (6) Assessment of drug induced conformational changes [17, 18].

(a) Understanding Drug-Receptor Interactions: It reveals the dynamic and temporal aspects of binding, including stability and conformational changes. This detailed insight helps optimize drug design for improved efficacy and specificity. Consequently, MD simulations enhance the accuracy of computational drug delivery systems by modeling real-time molecular interactions.
(b) Exploring Drug Permeation across Barriers: It provides detailed insights into how drugs traverse biological membranes. This approach captures the dynamic interactions and pathways of permeation, aiding in the design of drugs with optimal absorption and bioavailability.
(c) Prediction of Drug Solubility and Stability: It helps to understand how drugs behave in different environments. This approach provides insights into solubility profiles and stability under physiological conditions, guiding the formulation for optimal performance.
(d) Design of Drug Delivery Systems: This allows for the detailed analysis of drug encapsulation, release mechanisms, and interactions with drug vehicles. This approach optimizes the formulation and delivery methods to enhance stability, bioavailability, and targeted delivery. Consequently, MD simulations significantly improve the effectiveness of drug delivery systems at the molecular level.
(e) Prediction of Protein-Ligand Kinetics: It provides insights into the binding and unbinding rates of drug molecules to their targets. This approach helps in understanding the duration and stability of drug-target interactions, which is crucial for optimizing drug efficacy and dosing regimens. Consequently, MD simulations enhance the design and drug delivery systems by accurately modeling the kinetics of drug action at molecular level.
(f) Assessment of Drug Induced Conformational Changes: It reveals how drug binding affects the structure of biological target. This approach provides detailed insights into the dynamic alterations and stability of drug-target complexes, crucial for understanding mechanism of action and potential side effects. Consequently, these aid to optimizing drug design and delivery system by capturing the conformational dynamics at the molecular level.

1.2.1.5 Drug Carrier Design

Molecular modeling is also used in the design of drug delivery systems such as nanoparticles, liposomes and micelles. By modeling the interactions between drug molecules and carrier materials, researchers can optimize the design of drug carriers to improve drug delivery efficiency and reduce side effects. These simulations contribute to this field by (1) Optimization of carrier-drug interactions (2) Prediction of carrier physicochemical properties (3) Design of targeted drug delivery systems (4) Assessment of carrier stability and biocompatibility (5) optimization of drug release kinetics (6) Prediction of carrier-cell interactions [19, 20].

(a) Optimization of Carrier-Drug Interactions: It ensures that the drug is efficiently encapsulated, protected, and released at the target site. Computational simulations provide detailed insights into the binding affinity and stability of carrier-drug complexes, guiding the design of carriers for optimal performance. Consequently, this approach enhances the effectiveness and precision of drug delivery systems at the molecular level.
(b) Prediction of Carrier Physicochemical Properties: It ensures that carrier possess optimal solubility, stability, and biocompatibility. Computational simulations provide detailed insights into how these properties influence drug capsulation and release. Consequently, this approach enhances the development of efficient and effective drug delivery systems at the molecular level.
(c) Design of Targeted Drug Delivery Systems: It involves creating carriers that specifically bind to target cells or tissues. Computational simulations optimize the interaction between the drug, carrier, and target, enhancing specificity and efficacy. Consequently, this approach improves the precision and effectiveness of drug delivery at the molecular level.
(d) Assessment of Carrier Stability and Biocompatibility: It ensures safe and effective delivery systems and computational simulations predict interactions between carriers and biological systems, guiding the selection and optimization of materials. Consequently, this approach enhances the development of biocompatible carriers for efficient drug delivery at the molecular level.
(e) Optimization of Drug Release Kinetics: It ensures controlled and targeted release of drugs and computational predict release rates and mechanisms, guiding the design of carriers with desired release profiles. Consequently, this approach enhances the development of efficient drug delivery systems at the molecular level.
(f) Prediction of Carrier-Cell Interactions: It ensures efficient targeting and uptake by specific cells or tissues and computational simulations predict the binding affinity and internalization pathways, guiding the design of carriers for enhanced cellular interactions. Consequently, this approach improves the precision and effectiveness of drug delivery at the molecular level.

Molecular Modeling Software/Tools are depicted in Table 1.1.

Table 1.1 Molecular modeling software/tools

S. No.	Software/tool	Description	Example of drugs
1	GROMACS	Molecular dynamics simulations for biochemical molecules	Simvastatin, doxorubicin, cisplatin, paclitaxel Ibuprofen, Aspirin, Tamoxifen, Lidocaine, Epirubicin, Cyclophosphamide
2	AMBER	Suite of bimolecular Simulation programs including Molecular dynamics and quantum calculations	Doxorubicin, Cisplatin, Paclitaxel, Tamoxifen, Metrexate, Imatinib, Atorvastatin Sildenafil, Docetaxel, Ritonavir
3	NAMD	High-performance molecular dynamics for large Bimolecular systems	Paclitaxel, Doxorubicin, Cisplatin, Docetaxel, Tamoxifen Methotrexate, Ritonavir, Imatinib, Vorinostat, Venetoclax
4	CHARMM	Modeling the structure and Behavior of bimolecular systems	Asiprin, Doxorubicin, Paclitaxel, Tamoxifen, Ibuprofen, Ritonavir, Cisplatin, Atorvastatin, Sildenafil, Imatinib
5	Auto Dock	Automated docking of small molecules to receptor targets	Tamoxifen, Ritonavir, Atorvastatin Sildenafil, Imatinib, Doxorubicin Paclitaxel, Cisplatin, Docetaxel Methotrexate
6	Schrödinger Suite	Comprehensive suite for molecular modeling and drug design	Imatinib, Soldenafil, Atorvastatin Ritonavir, Doxorubicin, Paclitaxel, Tamoxifene, Cisplatin, Methotrexate Docetaxel
7	Gaussian	Computational chemistry software for electronic structure modeling	Ibuprofen, Asiprin, Tamoxifen, Ritonavir Atorvastatin, Sildenafil, Doxorubicin Cisplatin, Paclitaxel, Imatinib
8	VMD	Visualization and analysis of molecular dynamics simulations	Insulin, Doxorubicin, Paclitaxel Tamoxifen, Cisplatin, Atorvastatin, Ritonavir, Imatinib, Methotrexate Docetaxel
9	LIGPLOT	Generates schematic diagrams of protein–ligand interactions	Ritonavir, Tamoxifen, Atorvastatin, Sildenafil, Imatinib, Doxorubicin Paclitaxel, Cisplatin, Docetaxel, Methotrexate
10	OpenMM	High performance toolkit for molecular simulations	Vancomycin, Doxorubicin, Paclitaxel Tamoxifen, Cisplatin, Atorvastatin Ritonavir, Imatinib, Methotrexate Docetaxel

1.2.2 Molecular Docking Studies

These simulations are helpful to predict drug molecule interaction with binding site and help to determine the optimal binding orientation and affinity of the drug molecule, guiding the design of molecules with improved efficacy and selectivity. By simulating these interactions researcher can identify potential drug candidates and refine their structure to enhance binding properties. Ultimately, molecular docking aids in understanding the molecular mechanisms of drug action and optimizing drug delivery systems for better therapeutic outcomes. These are a very important components in computational simulations in drug delivery at the molecular level and fit into process by (1) Target identification (2) Molecular modeling (3) Docking algorithm (4) Scoring and analysis (5) Validation (6) Lead optimization [21, 22].

1.2.2.1 Target Identification

These simulations identify the biological molecule that interacts with the drug molecule. Accurate identification of these targets enables to predict site and mechanism of drug binding with receptor i.e. biological molecule. This helps in designing of drug with high specificity and efficacy, reducing off-target effects. Ultimately, effective target identification enhances overall success of drug delivery system by ensuring that therapeutic agents precisely modulate intended biological pathways. The target identification contributed in this field by (1) Selection of receptor protein (2) Understanding molecular mechanisms (3) Prediction of binding sites (4) Validation of drug target interactions (5) Identification of off-target interactions [23–25].

(a) Selection of Receptor Protein: It determines the specific biological targets for drug interaction and this process involves selecting receptors relevant to the therapeutic mechanism, guiding simulations to predict drug-receptor interactions accurately. Consequently, it aids in designing tailored drug delivery systems for optimal therapeutic outcomes at the molecular level.
(b) Understanding Molecular Mechanisms: It elucidates the underlying biological pathways and interactions crucial for drug action. This knowledge informs the selection of relevant targets and guides simulations to accurately drug-receptor interactions. Consequently, it facilitates the design of targeted drug delivery systems, optimizing therapeutic outcomes by precisely modulating molecular at the cellular level.
(c) Prediction of Binding Sites: It enables the precise determination of where drugs interact with biological targets. This information guides the selection of relevant targets and helps in simulating drug-receptor interactions accurately. Consequently, it facilitates the design of targeted drug delivery systems, optimizing therapeutic outcomes by effectively targeting specific molecular sites at the cellular level.
(d) Validation of Drug Target Interactions: It ensures the accuracy and reliability of predicted interactions and this process involves comparing simulation results with experimental data to confirm the efficacy of selected targets. Consequently, it enhances the design of targeted drug delivery systems by providing confidence in the predicted drug-receptor interactions, ultimately optimizing therapeutic outcomes at the molecular level.
(e) Identification of Off-Target Interactions: It helps anticipate potential unintended effects of drugs. By analyzing the interactions between drugs and various bio-

logical molecules, it provides insights into off-target binding sites and potential side effects. Consequently, it enables the design of targeted drug delivery system that minimizes off-target interactions, optimizing therapeutic efficacy and safety at the molecular level.

1.2.2.2 Molecular Modeling

These are integral to docking studies within computational simulations of drug delivery and are used to predict 3D structure of binding site and drug molecule. It contributes in the field by (1) Structure preparation (2) Ligand conformational sampling (3) Receptor flexibility (4) Scoring function development (5) Validation and interpretation (6) Lead optimization [26, 27].

(a) Structure Preparation: These studies involve optimizing geometry and energetics of bimolecular structures prior to simulations. This ensures accurate representation of drug-target interactions, aiding in the prediction of binding modes and affinities. Consequently, it enhances the design of drug delivery systems by providing reliable insights into molecular interactions at the atomic level.
(b) Ligand Conformational Sampling: These studies explore various conformations of drug molecules to predict their interactions with target proteins accurately. This process allows for the identification of favorable binding modes and optimization of drug-receptor interactions. Consequently, it enhances the design of drug delivery systems by providing insights into the most stable and effective ligand conformations at molecular level.
(c) Receptor Flexibility: This study accommodates the dynamic nature of target proteins, ensuring accurate prediction of drug-target interactions. This process considers conformational changes in the receptor upon ligand binding, enhancing the reliability of docking simulations. Consequently, it aids in the design of the drug delivery of molecular interactions at the atomic level.
(d) Scoring Function Development: These studies refine algorithms to assess the strength of ligand-receptor interactions accurately. This process improves the prediction of binding affinities, guiding the selection of potential drug candidates for further development. Consequently, it enhances the design of drug delivery systems by providing reliable metrics to evaluate molecular interactions at the atomic levels.
(e) Validation and Interpretation: These studies validates the simulation results against experimental data and interpret the significance of predicted interactions. This process ensures the reliability of computational predictions and guides the selection of promising drug candidate for further investigation. Consequently, it enhances the design of drug delivery system by providing confidence in the predicted molecular interactions at the atomic level.
(f) Lead Optimization: These studies involve refining initial drug candidates to enhance binding affinity and specificity. This iterative process utilizes computational simulations to optimize molecular structures for improved drug-target

interactions. Consequently, it facilitates the design of drug delivery systems by identifying promising leads for further development at the molecular level.

1.2.2.3 Docking Algorithm

These simulations predict the optimal binding orientation and affinity of drug molecule with its binding site on the receptor and facilitate high-throughput virtual screening, identifying promising drug candidates efficiently. Ultimately, they enhance the drug design by predicting interactions that optimize therapeutic efficacy and specificity. These contribute to the field by (1) Prediction of ligand-receptor modes (2) Scoring of ligand binding affinities (3) Accounting for receptor flexibility (4) Virtual screening of compound libraries (5) Lead optimization and design (6) Analysis of binding interactions [26, 27].

(a) Prediction of Ligand-Receptor Modes: These studies determine how ligands interact with target proteins at the molecular levels and this process identifies the most favorable binding conformations and interaction sites, guiding the design and optimization of drug candidates. Consequently, it enhances drug delivery systems by accurately modeling and predicting effective drug-target interactions.
(b) Scoring of Ligand Binding Affinities: These studies evaluate the strength and ability of drug-receptor interactions in molecular docking studies. This process ranks potential drug candidate based on predicted binding energies, guiding the selection of the most promising compounds. Consequently, it enhances the design of drug delivery systems by prioritizing ligand with optimal binding characteristics for further development.
(c) Accounting for Receptor Flexibility: These studies incorporate the dynamic movements and conformational changes of target proteins during molecular docking studies. This approach improves the accuracy of predicting ligand binding by considering the receptor's adaptability upon ligand interaction. Consequently, it enhances the design of drug delivery system by providing more reliable simulations of drug-target interactions at molecular level.
(d) Virtual Screening of Compound Libraries: These studies rapidly evaluate numerous potential drug candidates against target proteins. This process identifies the most promising compound based on predicted binding affinities and interactions. Consequently, it accelerates drug discovery and enhances the design of drug delivery systems by efficiently pinpointing viable drug candidates for further development.
(e) Lead Optimization and Design: These b studies refine drug candidates to improve their binding affinity, specificity, and pharmacokinetic properties. This process iteratively adjusts molecular structures based on computational predictions to enhance drug efficacy. Consequently, it enhances drug delivery systems by developing optimized leads that are more effective and targeted at molecular levels.

(f) Analysis of Binding Interactions: These studies examine how drug molecules interact with target proteins at the molecular level. This process identifies key binding sites, interaction types, and energetics, providing detailed insights into drug-receptor compatibility. Consequently, it enhances drug delivery systems by guiding the design of molecules with optimal binding characteristics and therapeutic efficacy.

1.2.2.4 Scoring and Analysis

This is very essential for identifying and refining drug molecular structure. Scoring function evaluates the binding affinity between drug molecule and binding site of a receptor and quantifying the interactions. This scoring helps prioritize potential drug candidate by predicting their efficacy and stability. Subsequent analysis of these scores provides insights into the molecular mechanisms of binding, guiding the optimization of drug molecular structure for enhanced therapeutic action. Scoring and analysis contributes to this field by (1) Scoring of ligand-receptor interactions (2) Analysis of binding interactions (3) Validation of docking results (4) Lead optimization and SAR Analysis (5) Prediction of ADMET properties (6) Design of drug delivery system [28, 29].

(a) Scoring of Ligand-Receptor Interactions: These studies quantify the strength and stability of the binding between a drug and its target protein. This evaluation ranks potential drug candidate based on predicted binding affinities, helping identify the most promising compounds. Consequently, it enhances drug delivery systems by prioritizing ligand with optimal interaction profiles for further development and testing.
(b) Analysis of Binding Interactions: These studies examine the detailed interactions between a ligand and its receptor, such as hydrogen bonds, hydrophobic contacts, and Van der Waals forces. This evaluation helps understand the molecular basis of binding affinity and specificity. Consequently, it enhances the drug delivery system by informing the design and optimization of compounds with favorable binding characteristics.
(c) Validation of Docking Results: These studies involve comparing computational predictions with experimental data to ensure accuracy and reliability. This process verifies the predicted binding modes and affinities, confirming the validity of the docking simulations. Consequently, it enhances drug delivery systems by providing confidence in the identified drug candidates and their interactions at molecular level.
(d) Lead Optimization and SAR Analysis: These studies refine drug candidate to improve their efficacy and binding properties. This process involves iterative adjustments to molecular structures based on docking scores and interaction pattern, enhancing the drug's activity. Consequently, it improves drug delivery systems by developing optimized leads with enhanced therapeutic potential and specific target interactions.

(e) Prediction of ADMET Properties: These studies evaluate the drug's PK and safety profile and this assessment guides selection and optimization of drug candidates by considering their likely hood of absorption, metabolism, and potential toxicity. Consequently, it enhances drug delivery system by identifying compounds with potential ADMET properties for further development and testing.
(f) Design of Drug Delivery System: These studies assess the compatibility of drug candidates with delivery vehicles and this evaluation guides the selection of carriers based on their ability to efficiently transport and release drugs at target sites. Consequently, it enhances drug delivery systems by optimizing delivery mechanisms for improved therapeutic outcomes at molecular level.

1.2.2.5 Validation

Validation in docking studies ensures the accuracy and reliability of docking predictions by comparing them against known experimental data to confirm the docking algorithm can reliably predict the correct binding modes and affinities. Successful validation builds confidence in the computational models, guiding the refinement of drug candidates and enhancing the reliability of simulations for drug delivery. Thus, validation is critical for ensuring the practical applicability of docking studies in drug development. Validations contributes to the field by (1) Experimental validation of binding affinities (2) Isothermal titration calorimetry (ITC) (3) Biochemical assays (4) Validation against crystallographic data (5) Cross-validation with homology models [30, 31].

(a) Experimental Validation of Binding Affinities: These studies involve comparing computational predictions with laboratory based measurements of how strongly a drug binds to its target. This process confirms the accuracy and reliability of docking simulations. Consequently, it enhances drug delivery systems by ensuring that predicted drug-target interactions are realistic and effective.
(b) Isothermal Titration Calorimetry (ITC): These studies measure the heat changes during drug-receptor binding, providing direct insights into binding affinity and thermodynamics. This experimental validation confirms computational predictions of molecular interactions.
(c) Biochemical Assays: These studies evaluate the functional activity and binding efficiency of drug candidates with their targets. These assays experimentally confirm computational predictions of drug receptor interactions.
(d) Validation against Crystallographic Data: These studies involve comparing predicted binding poses with experimentally determined structures from X-ray crystallography. This comparison verifies the accuracy of docking simulations in predicting how drugs interact with their targets.
(e) Cross-validation with Homology Models: These studies are to confirm docking predictions and this method checks the consistency and reliability of docking results across different, but related, protein structures. Consequently, it enhances

drug delivery systems by ensuring robustness and accuracy in predicting drug-target interactions even when experimental structures are unavailable.

1.2.2.6 Lead Optimization

It involves refining of drug structure of lead compound to improve their binding affinity, selectivity, and pharmacokinetic properties. Lead optimization contributes to this field by (1) Identification of SAR (2) Rational design of analog compound (3) Prediction of pharmacokinetic properties (4) Virtual screening and lead identification (5) Assessment of selectivity and toxicity [9, 32].

(a) Identification of SAR: These studies involve analyzing how various chemical modifications of a lead compound impact its binding affinity and biological activity. This process guides the refinement of drug candidates to enhance their potency, selectivity, and pharmacokinetic properties.
(b) Rational Design of Analog Compound: It involves computationally modifying chemical structures to improve binding affinity and specificity. This iterative process guides the development of analogs with enhanced pharmacological properties.
(c) Prediction of pharmacokinetic properties: These studies involve assessing how drug candidates are absorbed, distributed, metabolized, and eliminated within the body. This computational analysis guides the selection and modification of compounds to enhance their pharmacokinetic profiles.
(d) Virtual Screening and Lead Identification: These studies involve screening large compound libraries to identify potential drug candidates with favorable affinities. This computational approach accelerates the discovery of promising leads for further optimization. These studies enables researcher to enhance drug delivery systems by expediting the identification and development of effective therapeutic agents.
(e) Assessment of Selectivity and Toxicity: These studies involve predicting the specificity of drug candidates towards target proteins and evaluating their potential adverse effects. This computational analysis guides selection and modification of compounds to enhance their therapeutic index. Molecular Docking Studies software/tools are reported in Table 1.2.

1.2.3 Molecular Dynamic (MD) Simulations

MD simulations are powerful tools in computational drug delivery research, offers insights into dynamic behavior of drug molecule, drug carrier and biological system at molecular level. It contributes to the field by (1) Understanding drug-membrane interactions (2) Investigation of drug release mechanisms (3) Optimizing drug carrier design (4) Prediction of carrier-cell interactions (5) Studying drug resistance mechanisms (6) Accelerating drug discovery and development [33].

Table 1.2 Molecular Docking Studies software/tools

S. No.	Software/tool	Description	Example of drugs
1	Auto dock	Automated docking of small molecules to receptor targets	Tamoxifen, Ritonavir, Atorvastatin, Sildenafil, Imatinib Doxorubicin, Paclitaxel, Cisplatin, Docetaxel, Methotrexate
2	Auto dock Vina	Improved version of AutoDock with enhanced performance and accuracy	Tamoxifen, Ritonavir, Atorvastatin, Sildenafil, Imatinib, Doxorubicin Paclitaxel, Cisplatin, Docetaxel Methotrexate
3	Schrödinger Glide	High-precision docking program part of the Schrodinger suite	Imatinib, Sildenafil, Atorvastatin Ritonavir, Doxorubicin, Paclitaxel Tamoxifen, Cisplatin, Methotrexate Docetaxel
4	Molecular Operating Environment (MOE)	Comprehensive software for molecular modeling and docking	Tamoxifen, Ritonavir, Atorvastatin, Sildenafil, Imatinib, Doxorubicin Paclitaxel, Cisplatin, Methotrexate Docetaxel
5	Genetic optimization for Ligand Docking (GOLD)	Docking program using genetic algorithms for flexible docking	Tamoxifen, Ritonavir, Atorvastatin Sildenafil, Imatinib, Doxorubicin Paclitaxel, Cisplatin, Methotrexate Docetaxel
6	DOCK	One of the first docking programs developed, used for protein–ligand docking	Tamoxifen, Ritonavir, Atorvastatin Sildenafil, Imatinib, Doxorubicin Paclitaxel, Cisplatin, Methotrexate Docetaxel
7	FlexX	Fast docking algorithm focusing on flexible ligand docking	Tamoxifen, Ritonavir, Atorvastatin Sildenafil, Imatinib, Doxorubicin Paclitaxel, Cisplatin, Methotrexate Docetaxel

(continued)

Table 1.2 (continued)

S. No.	Software/tool	Description	Example of drugs
8	SwissDock	Web-based docking tool based on EADock algorithm	Tamoxifen, Ritonavir, Atorvastatin Sildenafil, Imatinib, Doxorubicin Paclitaxel, Cisplatin, Methotrexate Docetaxel
9	LeadIT(FlexX)	Interactive docking software focusing on lead optimization	Tamoxifen, Ritonavir, Atorvastatin Sildenafil, Imatinib, Doxorubicin Paclitaxel, Cisplatin, Methotrexate Docetaxel
10	PyRX	Virtual screening software for computational drug discovery combining AutoDock and AutoDock Vina	Tamoxifen, Ritonavir, Atorvastatin Sildenafil, Imatinib, Doxorubicin Paclitaxel, Cisplatin, Methotrexate Docetaxel

1.2.3.1 Understanding Drug-Membrane Interactions

MD simulations can elucidate drug interaction with lipid by layers, which mimic cell membranes. This is essential for understanding drug permeation across biological barriers and predicting drug ADME properties. It contributes to the field by (1) Prediction of drug permeation across membrane (2) Assessment of drug-membrane partitioning and solubility (3) Investigation of lipid membrane disruption (4) Design of lipid based drug delivery systems (5) Prediction of drug-target interactions at membrane interfaces [34].

(a) Prediction of Drug Permeation across Membrane: These studies involve simulating the movement of drug molecules through lipid bilayers to assess their ability to penetrate cell membrane. This computational analysis provides insights into drug bioavailability and cellular uptake mechanisms, guiding the design of drug delivery systems for improved efficacy.
(b) Assessment of Drug-Membrane Partitioning and Solubility: These studies involve predicting the distribution and dissolution of drugs with lipid bilayers. This computational analysis elucidates the drug's behavior at the membrane interface, influencing it absorption and permeation properties.
(c) Investigation of Lipid Membrane Disruption: These studies involve the perturbation and stability of lipid bilayers upon drug interaction. This computational analysis provides insights into drug-induced membrane damage, influencing drug delivery mechanisms and toxicity.

(d) Design of Lipid Based Drug Delivery Systems: These studies involve optimizing lipid composition and structure for effective drug encapsulation and release. This computational analysis enhances the stability, bioavailability, and controlled release properties of lipid-based formulations and improves drug delivery systems.
(e) Prediction of Drug-Target Interactions at Membrane Interfaces: These studies involve modeling how drugs interact with their targets within or near lipid bilayers. This computational analysis provides insights into binding affinities, conformational changes, and interaction dynamics specific to membrane environments.

1.2.3.2 Investigation of Drug Release Mechanisms

MD simulations can model drug release from various delivery systems such as nanoparticles, liposomes and micelles. By simulating the behavior of drug molecules within these carriers, researchers can study the factors influencing drug release kinetics, including carrier structure, environmental conditions and interactions with biological components. It contributes to the field by (1) Understanding carrier-drug interactions (2) Prediction of drug release kinetics (3) Assessment of carrier stability and drug release (4) Investigation of triggered release mechanisms (5) Design of controlled release systems [35].

(a) Understanding Carrier-Drug Interactions: These studies involve how carrier release their payloads upon interacting with lipid bilayers and elucidates the dynamic mechanisms of drug release, influencing carrier design and formulation, and enhances drug delivery systems by optimizing release profiles for targeted and controlled drug delivery across biological membranes.
(b) Prediction of Drug Release Kinetics: These studies involve modeling the rate and mechanisms by which drugs are released from their carriers in various environments. These provide insights into the temporal profile of drug release, helping to optimize formulation and delivery strategies.
(c) Assessment of Carrier Stability and Drug Release: These studies involve evaluating the structural integrity of drug carriers and the mechanisms by which they release drugs. This analysis provides insights into the stability of carriers under physiological conditions and their efficacy in delivering drugs.
(d) Investigation of Triggered Release Mechanisms: These studies involve studying how specific stimuli (e.g., pH, temperature, or enzymes) induce drug release from carriers. This analysis provides insights into the responsiveness and efficiency of drug release under targeted conditions.
(e) Design of Controlled Release Systems: These studies involve modeling and optimizing the mechanisms by which drugs are released over time from carriers. This analysis helps in understanding the release kinetics and factors affecting the sustained release of drugs.

1.2.3.3 Optimizing Drug Carrier Design

MD simulations aid in the design and optimization of drug deliver carriers by providing insights into their stability, structure dynamics, and interactions with drug and biological environment. Researchers can use MD simulations to assess the performance of different carrier materials, sizes, shapes, and surface modifications, aiming to enhance drug loading capacity, targeting specificity, and biocompatibility. It contributes to the field by (1) Prediction of carrier-drug interactions (2) Design of targeted drug delivery systems (3) Assessment of carrier stability and biocompatibility (4) Optimization of carrier morphology and properties (5) Prediction of carrier-cell interactions [36].

(a) Prediction of Carrier-Drug Interactions: These studies involve modeling the binding and stability of drug molecules within their carriers and this analysis provides insights into how drugs interact with different carrier materials and structures.
(b) Design of Targeted Drug Delivery Systems: These studies involve how drug carriers interact with specific biological targets and this analysis helps in understanding and optimizing the mechanisms that guide carriers to the intended site of action.
(c) Assessment of Carrier Stability and Biocompatibility: These studies involve evaluating the structural integrity and compatibility of drug carriers with biological systems. This analysis provides insights into the durability and safety of carriers within physiological environments.
(d) Optimization of Carrier Morphology and Properties: These studies involve modeling and refining the structural characteristics and material properties of drug carriers. This analysis provides insights into how variations in carrier morphology and properties impact drug loading, release kinetics, and targeting efficiency.
(e) Prediction of Carrier-Cell Interactions: These studies involves how drug carriers interact with biological cells at the molecular level and this analysis provides insights into mechanisms underlying cellular uptake, intercellular trafficking, and drug release.

1.2.3.4 Prediction of Carrier-Cell Interactions

MD simulations can complement docking studies by providing dynamic information about drug-target interactions. By simulating the binding process over time, MD simulations can reveal the conformational changes in the drug and target protein, as well as the dynamics of their complex. This information is valuable for understanding binding mechanisms, predicting binding affinities, and guiding the design of more effective drugs. It contributes to the field by (1) Elucidating binding modes and affinities (2) Investigating binding dynamics and kinetics (3) Assessing drug selectivity and specificity (4) Prediction of allosteric mutation (5) Design of structure-based drug optimization [37].

(a) Elucidating Binding Modes and Affinities: These studies involve modeling the specific interactions between drug carriers and cell surface receptors or membrane proteins. This analysis provides insights into the strength and specificity of these interactions, guiding the design of carriers for enhanced cellular targeting and uptake.
(b) Investigating Binding Dynamics and Kinetics: These studies involve studying the temporal aspects of the binding process between drug carriers and cell receptors. This analysis provides insights into the rate and mechanism of carrier-cell interactions, guiding the design of carries for efficient cellular targeting and approach.
(c) Assessing Drug Selectivity and Specificity: These studies involve evaluating the affinity of drug carrier towards target cells and their potential interactions with off-target cells. This analysis provides insights into the specificity of carrier-cell interactions, guiding the design of carrier for targeted drug delivery.
(d) Prediction of Allosteric Mutation: These studies involve modeling the effects of genetic variation on cell surface receptors or other binding sites. This analysis provides insights into how these mutations may alter binding affinity and specificity of drug carriers, informing the design of personalized delivery systems.
(e) Design of Structure-Based Drug Optimization: These studies involve refining of drug molecules to enhance their affinity and specificity for target cells. This analysis provides insights into the molecular interactions between drugs and cell receptors, guiding the development of optimized drug carriers.

1.2.3.5 Studying Drug Resistance Mechanisms

MD simulations can shed light on the molecular mechanisms underlying drug resistance, a major challenge in drug delivery and development. By simulating the interactions between drugs and their target molecules, researcher can identify key resistance mutations, elucidate the structural basis of resistance, and design strategies to overcome or prevent resistance. It contributes to the field by (1) Elucidating drug binding affinity changes (2) Understanding structural changes in drug targets (3) Predicting altered drug binding modes (4) Investigating drug efflux and transport mechanisms (5) Designing strategies to overcome resistance [38].

(a) Elucidating Drug Binding Affinity Changes: These studies involve modeling how genetic mutations or structural alterations affect the interaction between drugs and their targets. This analysis provides insights into the molecular mechanisms underlying drug resistance, guiding the development of strategies to overcome resistance.
(b) Understanding Structural Changes in Drug Targets: These studies involve modeling how mutations or conformational alterations impact the binding sites of target proteins. This analysis provides insights into how changes in target struc-

ture affect drug efficacy, guiding the development of strategies to counteract resistance.

(c) Predicting altered Drug Binding Modes: These studies involve modeling how mutations or structural changes affect the conformation and interactions of drug-target complex. This analysis provides insights into the potential mechanisms by which drug may lose efficacy due to altered binding modes, guiding the development of strategies to overcome resistance.
(d) Investigating Drug Efflux and Transport Mechanisms: These studies involves modeling how drugs are transported out of cells or across cellular membranes and this analysis provides insights into the molecular process underlying drug resistance, such as over expression of efflux pumps or altered membrane permeability, guiding the development of strategies to circumvent resistance.
(e) Designing Strategies to Overcome Resistance: These studies involve identifying potential vulnerabilities in resistant targets and exploring alternative binding modes or structural modifications. This analysis provides insights into how drugs can be tailored to effectively target resistant variants, guiding the development of novel therapeutic approaches.

1.2.3.6 Accelerating Drug Discovery and Development

MD simulations expedite the drug discovery and development process by providing detailed molecular insights into drug behavior interactions. By guiding experimental efforts and rationalizing SAR, MD simulations help prioritize lead compounds, optimize drug candidates, and minimize costly trial and error experimentation. It contributes to the field by (1) Rational drug design (2) Virtual screening and lead identification (3) Led optimization and SAR analysis (4) Prediction of ADMT Properties (5) Investigation of drug delivery systems (6) Understanding drug resistance and overcoming resistance mechanisms [39].

(a) Rational drug design: These studies explore the interaction between drug candidates and their targets and this analysis enables the prediction of drug-target binding affinities and optimization of molecular structures, expediting the identification of promising lead compounds.
(b) Virtual Screening and Lead Identification: These studies accelerate drug discovery and development by efficiently evaluating large libraries of compounds for potential drug candidates. This approach predicts the binding affinity and stability of these compounds with their targets at molecular level, identifying promising leads faster.
(c) Lead Optimization and SAR Analysis: These studies accelerate drug discovery by refining lead compounds for improved efficacy and safety. This approach identifies and optimizes key molecular interactions and properties, enhancing drug-receptor binding and pharmacokinetic profiles.
(d) Prediction of ADMT Properties: These studies accelerate drug discovery by identifying potential pharmacokinetic and toxicity issue early and this analysis

allows for the optimization of drug candidates to enhance their safety and toxicity issues early.

(e) Investigation of Drug Delivery Systems: These studies accelerate drug discovery by providing insights into the behavior and efficiency of delivery mechanisms at the molecular level. This approach helps in optimizing the design and functionality of delivery systems, ensuring targeted and controlled release of therapeutics.

(f) Understanding Drug Resistance and Overcoming Resistance Mechanisms: These studies accelerate drug discovery by revealing how drugs interact with resistant targets at the molecular level and enable the design of more effective drugs that can bypass or counteract resistance, leading to more efficient and successful therapeutic development. By addressing resistance early in the drug development process, it significantly shortens the time required to bring new treatments to market.

Molecular Dynamic (MD) Simulations software/tools are mentioned in Table 1.3.

1.2.4 Quantum Mechanics/Molecular Mechanics (QM/MM) Simulations

QM/MM simulations are advanced computational techniques that combine the accuracy of QM with the efficiency of MM to study complex molecular systems such as drug delivery processes. They contribute in the field by (1) Modeling reactive processes (2) Treating active sites (3) Exploring electronic effects (4) Studying binding mechanisms (5) Predicting reaction rates (6) Addressing environmental effects [8].

1.2.4.1 Modeling Reactive Processes

QM/MM simulations are particularly useful for studying chemical reactions and processes involving bond formation and breaking, which are common in drug delivery. For example, they can elucidate enzymatic reactions involved in drug metabolism or the cleavage of drug molecule from carrier. This approach provides detailed insights into electronic change during drug interactions, aiding the process optimization of drug candidates. It enhances the understanding of reaction mechanisms, guiding the design of drug with better efficacy and specificity. They contributes to the field by (1) Reaction mechanism elucidation (2) Prediction of reaction energies and rates (3) Analysis of enzyme-drug interactions (4) Optimization of drug activation process.

(a) Reaction Mechanism Elucidation: These studies aids in understanding how drugs interact and transform at the molecular level, and this provides insights

Table 1.3 Molecular Dynamic (MD) Simulations software/tools

S. No.	Software/ tool	Description	Example of drugs
1	GROMACS	Versatile package for molecular dynamics simulations of biochemical molecules	Simvastatin, Doxorubicin Cisplatin, Paclitaxel, Ibuprofen, Aspirin, Tamoxifen Lidocaine, Epirubicin, Cyclophosphamide
2	AMBER	Suite of biomolecular simulation programs including molecular dynamics and quantum calculations	Doxorubicin, Cisplatin, Paclitaxel Tamoxifen, Methotrexate, Imatinib, Atorvastatin, Sildenafil, Docetaxel Ritonavir
3	NAMD	High-performance molecular dynamics program for large biomolecular systems	Paclitaxel, Doxorubicin, Cisplatin Docetaxel, Tamoxifen, Methotrexate Ritonavir, Imatinib, Vorinostat, Venetoclax
4	CHARMM	Molecular simulation program for modeling biomolecular structure and behavior	Aspirin, Doxorubicin, Paclitaxel, Tamoxifen, Ibuprofen, Ritonavir, Cisplatin Atorvastatin, Sildenafil, Imatinib
5	Desmond	High-performance molecular dynamics simulations with Schrodinger suite	Imatinib, Sildenafil, Atorvastatin Ritonavir, Doxorubicin, Paclitaxel Tamoxifen, Cisplatin, Methotrexate Docetaxel
6	LAMMPS	Classical molecular dynamics code for simulating particles	Doxorubicin, Paclitaxel, Cisplatin, Tamoxifen, Docetaxel, Methotrexate Atorvastatin, Ibuprofen, Simvastatin Aspirin
7	OpenMM	High-performance toolkit for molecular simulations with flexibility and customization	Vancomycin, Doxorubicin, Paclitaxel Tamoxifen, Cisplatin, Atorvastatin Ritonavir, Imatinib, Methotrexate, Docetaxel
8	HOOMD-blue	Molecular dynamics simulations on GPUs for soft matter simulations	Doxorubicin, Paclitaxel, Cisplatin Tamoxifen, Docetaxel, Methotrexate Atorvastatin, Ibuprofen, Simvastatin Aspirin
9	GROMOS	General-purpose molecular dynamics simulation software	Doxorubicin, Paclitaxel, Cisplatin, Tamoxifen, Docetaxel, Methotrexate Atorvastatin, Ibuprofen, Simvastatin, Aspirin
10	Tinker	Molecular dynamics and modeling software package focusing on flexibility and extensibility	Doxorubicin, Paclitaxel, Cisplatin Tamoxifen, Docetaxel, Methotrexate Atorvastatin, Ibuprofen, Simvastatin Aspirin

into the specific steps and intermediates involved in drug activation or metabolism, which is crucial for designing effective and safe drug delivery systems.

(b) Prediction of Reaction Energies and Rates: These studies help determine the feasibility and efficiency of drug-related reactions at the molecular level. This

information is vital for optimizing drug formulations and delivery systems, ensuring they operate under optimal conditions.
(c) Analysis of Enzyme-Drug Interactions: These studies helps reveal the detailed mechanisms of how drugs interact with their enzyme targets. This detailed understanding is crucial for optimizing drug efficacy and specificity.
(d) Optimization of Drug Activation Process: These studies enhance the understanding of the chemical transformations drugs undergo in the body. This optimization aids in predicting and improved drug stability and efficacy.

1.2.4.2 Treating Active Sites

QM/MM simulations provide accurate modeling of the critical interactions and electronic changes during drug binding. This hybrid approach enables detailed insight into the mechanisms at the atomic level, essential for optimizing drug efficacy and selectivity. These simulations by capturing the complexity of active sites improve the prediction and design of effective drug candidate. Thus, they are crucial for advancing precise and reliable drug development. They contributes to the field by (1) Accurate modeling of reaction mechanisms (2) Prediction of reaction energies and rates (3) Exploration of binding Mechanisms (4) Understanding enzyme catalysis (5) Prediction of enzyme substrate-interactions [40].

(a) Accurate Modeling of Reaction Mechanisms: These studies allow detailed insights into the chemical process occurring at the active site of enzymes. This precision enhances the prediction of drug interactions and their effects. It ultimately improves the reliability of computational simulations in drug delivery at molecular level.
(b) Prediction of Reaction Energies and Rates: These studies enable the evaluation of the catalytic efficacy of enzymes in drug metabolism or activation processes. This information aids in understanding the kinetics of drug interactions and their impact on therapeutic outcomes.
(c) Exploration of Binding Mechanisms: These studies provide insights into how drugs interact with enzyme targets at a molecular level and this understanding helps optimize drug design for improved binding efficacy and specificity.
(d) Understanding Enzyme Catalysis: These simulations elucidate the molecular mechanisms underlying drug metabolism and activation. This knowledge aids in predicting how drugs interact with enzymes, optimizing drug designs, and ultimately enhancing computational simulations for drug delivery at the molecular level.
(e) Prediction of Enzyme Substrate-Interactions: These studies allows for the accurate modeling of drug interactions within enzymatic environments and enables the prediction of reaction mechanisms and the optimization of drug designs, enhancing computational simulations for drug delivery at molecular level.

1.2.4.3 Exploring Electronic Effects

QM/MM simulations provide capture electronic effects that are critical for accurately describing the behavior of certain molecules in drug delivery systems. These effects include polarization, charge transfer, and electronic rearrangements, which influence intermolecular interactions, binding affinities, and reaction mechanisms. They contributes to the field by (1) Accurate representation of electronic structure (2) Prediction of electronic properties (3) Investigation of charge transfer process (4) Quantification of electronic energies (5) Prediction of redox reactions [41].

(a) Accurate Representation of Electronic Structure: These studies are essential for exploring electronic effects in drug interactions and this approach allows detailed analysis of electronic contributions to binding affinities and reaction mechanisms.
(b) Prediction of Electronic Properties: These simulations enhance understanding of electronic effects in drug interactions and enables accurate modeling of how electronic structure influence drug binding and reactivity. Thus, it significantly refines computational simulations of drug delivery at molecular level.
(c) Investigation of Charge Transfer Process: These simulations help elucidate electronic effects during interactions and this insight is crucial for understanding how charge redistribution affects drug binding and activity.
(d) Quantification of Electronic Energies: These simulations allows precise determination of energy changes during drug interactions and enhances understanding of electronic effects on drug efficacy and stability, and refines computational simulations of drug delivery at the molecular level.
(e) Prediction of Redox Reactions: These simulations enable the accurate modeling of electron transfer processes, critical for understanding drug interactions. This enhances the ability to predict drug stability and efficacy.

1.2.4.4 Studying Binding Mechanics

QM/MM simulations can elucidate the binding mechanisms of drug molecules to their target receptor at atomic level of detail. By considering the QM nature of chemical bonding, QM/MM simulations provide insights into hydrogen bonding, Van der Waals interactions, and other non-covalent interactions involved in drug-target binding. These simulations are crucial for understanding the drug molecule interaction with their target receptors or carriers during the process of drug delivery. They contributes to the field by (1) Elucidating binding modes (2) Quantifying binding affinities (3) Assessing binding kinetics (4) Exploring allosteric binding sites (5) Investigating drug-resistance mechanisms [42].

(a) Elucidating Binding Modes: These studies elucidate the detailed atomic interactions governing during receptor binding. This approach integrates QM/MM simulations to accurately model complex molecular interactions, guiding the design of optimized drug delivery systems.

(b) Quantifying Binding Affinities: These studies accurately predict the strength of drug-receptor interactions at the atomic level. This approach integrates QM/MM simulations to calculate binding energies, guiding the design of optimized drug delivery system.
(c) Assessing Binding Kinetics: These studies evaluate the rate and mechanism of drug-receptor interactions at the atomic-level. This approach integrates QM/MM simulations to predict binding kinetics accurately, guiding the design of optimized drug delivery systems.
(d) Exploring Allosteric Binding Sites: These studies uncover alternative sites for drug-receptor interactions and this approach integrates QM/MM simulations to identify allosteric modulation possibilities, guiding the design of optimized drug delivery systems.
(e) Investigating Drug-Resistance Mechanisms: These studies elucidate how mutations affect drug binding and efficacy. This approach integrates QM/MM simulations to predict drug resistance mechanisms accurately, guiding the design of optimized drug delivery systems.

1.2.4.5 Predicting Reaction Rates

QM/MM simulations enable the calculation of reaction rates and kinetic parameters associated with drug delivery processes. By modeling the transition states and reaction pathways with QM accuracy, QM/MM simulations can predict reaction rates, activation energies, and rate constants, aiding in the optimization of drug delivery systems. They contribute to the field by (1) Modeling enzyme reactions (2) Prediction of reaction mechanisms (3) Quantification of reaction energetic (4) Insights into binding affinities (5) Understanding enzyme mechanisms [8].

(a) Modeling Enzyme Reactions: These studies model the catalytic mechanisms of enzymes accurately and this approach integrates QM/MM simulations to simulate enzyme reactions, guiding the design of optimized drug delivery systems.
(b) Prediction of Reaction Mechanisms: These studies anticipate the pathway and kinetics of chemical transformations accurately, and this approach integrates QM/MM simulations to simulate reaction mechanisms, guiding the design of optimized drug delivery systems.
(c) Quantification of Reaction Energetic: These studies accurately assess the energy changes associated with chemical transformations. This approach integrates QM/MM simulations to calculate reaction energies, guiding the design of optimized drug delivery systems.
(d) Insights into Binding Affinities: These studies offer a comprehensive understanding of stability and dynamics of drug-receptor interactions. By integrating QMMM simulations these insights inform the prediction of reaction rates, guiding the design of optimized drug delivery systems.
(e) Understanding Enzyme Mechanisms: These studies elucidates the intricacies of catalytic processes and by integrating QM/MM simulations informs accurate

predictions of reaction rates, guiding the design of optimized drug delivery systems.

1.2.4.6 Addressing Environment Effects

QM/MM a simulation accounts the influence of surrounding environment, such as solvent molecules and ions, on drug delivery process. By including MM for the environment and QM for the system of interest, QM/MM simulations provide a realistic description of the system of the molecular interactions occurring in complex biological environments. They contribute to the field by (1) Modeling solvent effect (2) Incorporating membrane interactions (3) Considering pH effects (4) Exploring temperature dependence (5) Modeling biological interfaces [40].

(a) Modeling Solvent Effect: These studies accurately capture the influence of the surrounding medium on molecular interactions and this approach accounts for solvent interactions, guiding the design of optimized drug delivery systems.
(b) Incorporating Membrane Interactions: These studies accurately simulates drug behavior within lipid bilayers and this approach accounts for membrane interactions, guiding the design of drug delivery systems tailored for cellular environments.
(c) Considering pH Effects: These studies accurately represent the protonation states of molecules in different environments and this approach accounts for pH dependent interactions, guiding the design of drug delivery systems optimized for specific physiological conditions.
(d) Exploring Temperature Dependence: These studies investigate the influence of temperature on molecular behavior and this approach accounts for temperature-dependent interactions, guiding the design of drug delivery systems adaptable to varying physiological conditions.
(e) Modeling Biological Interfaces: These studies accurately depicts the interaction between drugs and biological components and this approach accounts for the complex environment of biological interfaces, guiding the design of drug delivery systems tailored for specific physiological contexts.

Quantum Mechanics/Molecular Mechanics (QM/MM) Simulations software/tools in Table 1.4.

1.2.5 Free Energy Calculations

Free energy calculations are advanced computational techniques used in molecular simulations to estimate the change in free energy associated with a particular molecular process, such as ligand binding, protein conformational changes, or drug delivery. In the context of drug delivery at the molecular level, free energy calculations can provide valuable insights into various aspects of drug delivery process. Here's

Table 1.4 Quantum Mechanics/Molecular Mechanics (QM/MM) Simulations software/tools

S. No.	Software/tool	Description	Example of drugs
1	Gaussian	Software for computational chemistry Widely used for QM/MM simulations Allowing detailed electronic structure analysis	Ibuprofen, Aspirin, Tamoxifen, Ritonavir, Atorvastatin, Sildenafil, Doxorubicin, Cisplatin, Paclitaxel, Imatinib
2	ORCA	Efficient and flexible quantum, chemistry software with QM/MM capabilities suitable for large systems	Doxorubicin, Tamoxifene Sildenafil, Cisplatin, Paclitaxel, Imatinib Methotrexate, Ritonavir Atorvastatin, Docetaxel
3	Q-Chem	Advanced electronic structure methods integrated with QM/MM capabilities, providing accurate quantum mechanical calculations.	Paclitaxel, doxorubicin, Cisplatin, tamoxifen, atorvastatin, methotrexate, imatinib, sildenafil, ritonavir docetaxel
4	NWChem	Comprehensive quantum chemistry and molecular dynamics software, supporting QM/MM simulations for studying complex biological systems	Tamoxifen, Ritonavir, Atorvastatin, Sildenafil, Imatinib, Doxorubicin, Paclitaxel, Cisplatin, Methotrexate, Docetaxel
5	ADF (Amsterdam Density Functional)	Software for density functional theory calculations with QM/MM integration, useful for studying drug interactions in drug delivery	Imatinib, Sildenafil, Atorvastatin, Ritonavir, Doxorubicin, Paclitaxel, Tamoxifen, Cisplatin, Methotrexate, Docetaxel
6	CP2K	Computational chemistry software for molecular dynamic simulations with QM/MM capabilities, particularly for materials and biomolecular systems	Vancomycin, Doxorubicin, Paclitaxel, Tamoxifen, Cisplatin, Atorvastatin, Ritonavir, Imatinib, Methotrexate, Docetaxel
7	CHARMM	Software for biomolecular simulations with QM/MM capabilities, allowing detailed modeling of drug interactions	Aspirin, Doxorubicin, Paclitaxel, Tamoxifen, Ibuprofen, Ritonavir, Cisplatin, Atorvastatin, Sildenafil, Imatinib
8	GAMESS	General atomic and molecular electronic structure system with QM/MM capabilities, useful for drug design and interaction studies	Ibuprofen, Aspirin, Tamoxifen, Ritonavir, Atorvastatin, Sildenafil, Doxorubicin, Cisplatin, Paclitaxel, Imatinib
9	ONIOM (Part of Gaussian)	A hybrid method within Gaussian for QM/MM simulations, allowing multi-layered modeling of complex systems	Tamoxifen, Ritonavir, Atorvastatin, Sildenafil, Imatinib, Doxorubicin, Paclitaxel, Cisplatin, Methotrexate, Docetaxel
10	Psi4	Open-source quantum chemistry software with QM/MM support, suitable for high accuracy simulations in drug discovery	Doxorubicin, Tamoxifen, Sildenafil, Cisplatin, Paclitaxel, Imatinib, Methotrexate, Ritonavir, Atorvastatin, Docetaxel

they are contributes to the field by (1) Binding free energies (2) Drug release kinetics (3) Solubility and permeability (4) Membrane insertion transport (5) Drug-drug interactions (6) Virtual screening and drug design [43, 44].

1.2.5.1 Binding Free Energies

Free energy calculations can be used to estimate the binding free energies of drug molecules to their target receptors or proteins. By quantifying the thermodynamic stability of drug-target complexes, these calculations help prioritize lead compounds, optimize drug candidates, and understand the molecular basis of drug-target interaction. Binding free energy studies are particularly in understanding the strength and specificity of interactions between drug molecule and their target receptors or carriers. They contributes to the field by (1) Quantification of binding affinities (2) Prediction of ligand potency and selectivity (3) Guidance for rational Drug design (4) Evaluation of drug delivery systems [45].

(a) Quantification of Binding Affinities: These studies accurately assess the strength of drug-receptor interactions and calculate free energies, guiding the design of drug delivery systems with optimized molecular interactions.
(b) Prediction of Ligand Potency and Selectivity: These studies help to evaluate and optimize drug efficacy and specificity, and quantify how strongly and selectively a ligand binds to its target, guiding the design of mere effective and selective drug delivery system.
(c) Guidance for Rational Drug Design: These studies inform the optimization of molecular interactions and provide precise quantification of binding affinities, helping to design drugs with improved efficacy and selectivity.
(d) Evaluation of Drug Delivery Systems: These studies assess the stability and efficacy of drug-target interactions and quantify binding affinities, and insights into the molecular performance of drug delivery mechanisms, ensuring effective and targeted therapeutic outcomes.

1.2.5.2 Drug Release Kinetics

Free energy calculations elucidate the energies of drug release from drug delivery carriers such as nanoparticles, liposomes, or micelles. By estimating free energy barriers associated with drug release processes, the calculations provides insights into the kinetics of drug release and help optimize drug delivery systems for controlled release applications. These calculations provide insights into the thermodynamic aspects of drug release process, aiding in the design and optimization of drug delivery system. These calculations contribute to the field by (1) Characterization of binding and un binding events (2) Prediction of residence times (3) Assessment of release mechanisms (4) Optimization of formulation parameters (5) Prediction of release rates [46].

(a) Characterization of Binding and Unbinding Events: These studies provides insights into the molecular dynamics of drug interactions and by quantifying the energy landscape of these events, this approach helps to understand the rates and mechanisms of drug release.
(b) Prediction of Residence Times: These studies evaluates how long a drug remains bound to its target and evaluate how long a drug remains bound to its target. By analysing the energy barriers and binding stability, this approach provides insights into the duration of drug action.
(c) Assessment of Release Mechanisms: These studies elucidate how drugs dissociate from their targets and by examining these energy pathways and barriers involved, this approach helps to identify and optimize the mechanisms controlling drug release.
(d) Optimization of Formulation Parameters: These studies tailor the composition and properties of drug formulations for optimal release profiles. By analyzing energy landscapes and interactions, this approach fine-tunes factors such as solubility and stability.
(e) Prediction of Release Rates: These studies determine how quickly a drug is released from its carrier or formulation. By analyzing the energy barriers and transitions, this approach provides precise release rate estimations.

1.2.5.3 Solubility and Permeability

Free energy calculations predict the solubility and permeability of drug molecules, which are crucial properties influencing drug absorption and bioavailability. By estimating free energy salvation or partitioning between different phases, these calculations aid in the design of drugs with improved solubility and permeability profile. They contribute to field by (1) Prediction of solubility (2) Assessment of permeability (3) Optimization of formulation parameters (4) Prediction of dissolution rates (5) Evaluation of drug-excipient interactions [47].

(a) Prediction of Solubility: These studies evaluate the energy changes associated with a drug dissolving in a solvent. By accurately estimating solubility, this approach helps in formulating drugs with optimal bioavailability.
(b) Assessment of Permeability: These studies evaluate the energy barriers and pathways for a drug traverse biological membranes. By accurately predicting permeability, this approach aids in formulating drugs that can efficiently cross cell membranes.
(c) Optimization of Formulation Parameters: These studies involve enhancing a drug solubility and permeability and this approach ensures that the drug is both sufficiently soluble and capable of effectively crossing biological membrane.
(d) Prediction of Dissolution Rates: These studies are crucial in assessing how quickly a drug will dissolve in biological fluids and this prediction helps in understanding the solubility profile and ensures that the drug reaches the desired concentration in the bloodstream efficiently. Accurate dissolution rate predic-

tions are essential for optimizing drug formulation and improving bioavailability in drug delivery systems.

(e) Evaluation of Drug-Excipient Interactions: These studies help in understanding how excipient affects the solubility and permeability of a drug. These interactions can significantly influence the drug's stability, bioavailability, and overall efficacy. Accurate evaluation ensures the optimal formulation of drug delivery systems, enhancing therapeutic outcomes.

1.2.5.4 Membrane Insertion and Transport

Free energy calculations model the process of drug molecules crossing biological membranes, such as cell membranes, organelle membranes. By estimating the free energy profiles associated with membrane insertion and translocation, these calculations provide insights into mechanism of drug uptake, distribution and cellular transport. And these are especially for molecular targeting intracellular or membrane-bound receptors. They contribute to field by (1) Prediction of membrane partitioning (2) Assessment of transmembrane transport (3) optimization of drug design for membrane permeability (4) Prediction of drug accumulation in cells (5) Evaluation of drug-transporter interactions [48].

(a) Prediction of Membrane Partitioning: These studies help in understanding how drug interact with and integrate into cell membranes and this insight in crucial for determining the drug's bioavailability and effectiveness. Accurate predictions guide the design of drug molecule with optimal membrane transport properties, enhancing therapeutic delivery.
(b) Assessment of Transmembrane Transport: Thee studies provide detailed insight into how drugs cross biological membrane and this understanding is essential for predicting drug absorption, distribution, and efficacy. Accurate transport predictions inform the design of drugs with optimal pharmacokinetic properties for effective delivery.
(c) Optimization of drug design for membrane permeability: Thee studies enhance membrane permeability, crucial for efficient insertion and transport and free energy calculations aid in understanding membrane interactions, pivotal in computational simulation of molecular level drug delivery, offering insights for targeted therapies.
(d) Prediction of Drug Accumulation in Cells: These studies elucidate how drugs permeate membranes and accumulate in cells, and this approach facilitates the prediction of drug behavior at molecular level, advancing computational simulations for precise drug delivery optimization.
(e) Evaluation of Drug-Transporter Interactions: These studies elucidate the thermodynamics of drug-transporter interaction and provide insights into drug resistance mechanisms. These calculations predict the affinity and specificity of drugs for transporter proteins.

1.2.5.5 Drug-Drug Interactions

When multiple drugs administered simultaneously, as these interactions can affect therapeutic efficacy and safety. Free energy calculations assess the affinities and interactions between different drug molecules, which are important for predicting potential drug-drug interactions and understanding their implementations for drug efficacy and safety. Drug-drug interactions are helpful for prediction of binding affinity, assessment of synergism and antagonism, optimizing combination therapies, prediction drug displacement interactions, and evaluation of safety profiles, free energy calculations inform the rational design and optimization of drug delivery strategies for improved therapeutic outcomes. These contributes to the field by (1) Prediction of binding affinities (2) Assessment of drug synergism and antagonism (3) Optimization of combination therapies (4) Prediction of drug displacement interactions (5) Evaluation of drug safety profiles [49].

(a) Prediction of Binding Affinities: These studies facilitates understanding molecular interactions critical for computational simulations of drug delivery and this approach enables precise modeling of drug interactions, guiding the development of tailored delivery systems for optimal therapeutic outcome.
(b) Assessment of Drug Synergism and Antagonism: These studies inform computational simulations of molecular level drug delivery, enhancing therapeutic efficacy. Understanding these interactions guides the design of optimized drug combinations and drug delivery systems, optimizing treatment outcome.
(c) Optimization of Combination Therapies: This approach enables tailored design of synergistic drug combinations, maximizing therapeutic efficacy and minimizing adverse effects and offers a robust framework for refining drug delivery strategies and improving patient outcomes.
(d) Prediction of Drug Displacement Interactions: These studies are crucial for understanding molecular dynamics in drug delivery and aid in designing delivery systems resilient to displacement effects, ensuring therapeutic efficacy and safety.
(e) Evaluation of Drug Safety Profiles: These studies elucidate potential risks and synergies and provide insights into molecular level drug delivery mechanisms, enhancing precision and efficacy. This approach aids in optimizing therapeutic outcomes while minimizing adverse effects.

1.2.5.6 Virtual Screening and Drug Design

These allow the researcher to efficiently identify potential drug candidates and optimize their properties. Free energy integrates into virtual screening workflows and computational drug design approaches to rank and prioritize large libraries of compound candidate based on their predicted binding affinities and free energy profiles. These provide insights into energetics of drug binding and interactions at molecular level. They contribute to field by (1) Identification of drug targets (2) Lead compound identification (3) Lead optimization (4) Prediction of drug metabolism and pharmacokinetics (5) Evaluation of drug resistance mechanisms [50].

(a) Identification of Drug Targets: These studies facilitate the selection of promising compounds and enhance understanding of drug delivery mechanisms at the molecular level, enabling targeted therapies with minimized side effects.
(b) Lead Compound Identification: These studies expedites the selection of potential candidates and enables a detailed examination of drug delivery dynamic at the molecular level, optimizing therapeutic efficacy, and accelerate drug development, fostering precise treatments with minimized risks.
(c) Lead Optimization: These studies refine candidate molecules for enhanced efficacy and elucidate intricate drug delivery mechanisms, optimizing therapeutic precision and this cohesive expedites drug development, advancing tailored treatments with heightened potency and safety.
(d) Prediction of Drug Metabolism and Pharmacokinetics: These studies molecular level insights into drug delivery dynamics and enhance understanding of drug distribution and clearance mechanisms, optimizing therapeutic strategies. This comprehensive approach streamlines drug development, ensuring efficient delivery and desired pharmacological effects.
(e) Evaluation of Drug Resistance Mechanisms: These studies inform the design of effective therapeutic interventions and insights how drug molecules interact with resistant targets at the molecular level, guiding strategy refinement, and enhances the development of treatments capable of overcoming resistance, ensuring sustained efficacy. Free Energy Calculations software/tools are described in Table 1.5.

1.2.6 Visualization and Analysis

They help researchers understand complex molecular interactions, identify key insights, and guide the design of novel drug delivery systems. These are applied by (1) Molecular visualization (2) Trajectory analysis (3) Binding site analysis (4) Interaction analysis (5) Quantitative structure-activity relationship (QSAR) Analysis (6) Statistical analysis and Data visualization [51, 52].

1.2.6.1 Molecular Visualization

Molecular visualization is indispensable in computational simulations of drug delivery at molecular level as it enables researchers to understand complex molecular interactions, visualize molecular structures, and analyze simulation data effectively. These tools empower researcher to gain deeper insights into drug-receptor interactions and optimize drug delivery strategies for improved therapeutic outcomes. They contributes to the field by (1) Structural analysis (2) Interaction

Table 1.5 Free Energy Calculations software/tools

S. No.	Software/ tool	Description	Example of drugs
1	GROMACS	Versatile package for molecular dynamics and fee energy calculations using methods like FEP, TI and PMF	Simvastatin, Doxorubicin, Cisplatin, Paclitaxel, Ibuprofen, Aspirin, Tamoxifen, Lidocaine, Epirubicin, Cyclophosphamide
2	AMBER	Suite of biomolecular simulation programs with robust free energy calculation capabilities using methods such as MM-PBSA and TI	Doxorubicin, Cisplatin, Paclitaxel, Tamoxifen, Methotrexate, Imatinib, Atorvastatin, Sildenafil, Docetaxel, Ritonavir
3	NAMD	High-performance molecular dynamics program supporting free energy calculations using methods like FEP and TI	Paclitaxel, Doxorubicin, Cisplatin, Docetaxel, Tamoxifen, Methotrexate, Ritonavir, Imatinib, Vorinostat, Venetoclax
4	Desmond	High-performance molecular dynamics simulations with integrated free energy calculations tools	Imatinib, Sildenafil, Atorvastatin, Ritonavir, Doxorubicin, Paclitaxel, Tamoxifen, Cisplatin, Methotrexate Docetaxel
5	CHARMM	Molecular simulation program for biomolecular systems with free energy calculation capabilities using methods such as FEP,TI and MM-PBSA	Aspirin, Doxorubicin, Paclitaxel, Tamoxifen, Ibuprofen, Ritonavir, Cisplatin, Atorvastatin, Sildenafil, Imatinib
6	LAMMPS	Classical molecular dynamics code with free energy calculation capabilities with wide range of systems	Doxorubicin, Paclitaxel, Cisplatin, Tamoxifen, Docetaxel, Methotrexate, Atorvastatin, Ibuprofen, Aspirin, Simvastatin
7	Schrödinger Suite	Advanced molecular modeling suite with highly accurate free energy perturbation (FEP) calculations	Imatinib, sildenafil, Atorvastatin, Ritonavir, Doxorubicin, Paclitaxel, Tamoxifen, Cisplatin, Methotrexate, Docetaxel
8	OpenMM	High-performance toolkit for molecular stimulation with free energy calculation capabilities	Vancomycin, Doxorubicin, Paclitaxel, Tamoxifen, Cisplatin, Atorvastatin, Ritonavir, Imatinib, Methotrexate, Docetaxel
9	GROMOS	General-purpose molecular dynamics stimulation software with free energy calculation method like TI and MM-PBSA	Doxorubicin, Paclitaxel, Cisplatin, Tamoxifen, Docetaxel, Methotrexate, Atorvastatin, Ibuprofen, Simvastatin, Aspirin
10	YANK	Open-source software for molecular dynamics and free energy calculation using methods like FEP and TI, focused on binding free energies	Tamoxifen, Ritonavir, Atorvastatin, Sildenafil, Imatinib, Doxorubicin, Paclitaxel, Cisplatin Methotrexate, Docetaxel

analysis (3) Trajectory visualization (4) Conformational sampling (5) Data visualization and presentation [53].

(a) Structural Analysis: These studies aid in comprehending intricate drug delivery mechanisms and offer insights into molecular interactions crucial for optimizing therapeutic strategies. This approach enhances our understanding and ability to design effective drug delivery systems at the molecular level.
(b) Interaction Analysis: These studies elucidates key molecular interactions pivotal for drug delivery and offer insights into the dynamic interplay between drugs and biological targets, and facilitates the design of optimized drug delivery systems, enhancing therapeutic efficacy at the molecular level.
(c) Trajectory Visualization: These studies aid in comprehending the dynamic behavior of molecules during drug delivery and enables analyst to track molecular movement, understand interaction patterns, and optimize drug delivery strategies at molecular level, and offering insights into molecular dynamics and interactions.
(d) Conformational Sampling: These studies facilitate the exploration of diverse molecular structures relevant to drug delivery simulations and helps optimize drug delivery system by identifying the most favorable molecular conformations for efficient drug binding and release.
(e) Data Visualization and Presentation: These studies streamline the interpretation of complex computational simulations in drug delivery studies. By graphically representing molecular interactions and properties, it enhances comprehension of drug behavior and efficacy at molecular level.

1.2.6.2 Trajectory Analysis

It is a powerful tool in computational simulations of drug delivery, providing valuable insights into molecular dynamics, binding events, conformational changes, transport mechanisms, and validation of computational models. By integrating trajectory analysis with other visualization and analysis techniques, researchers can deepen their understanding of drug delivery processes and optimize drug design and delivery strategies. MD simulations generate vast amounts of trajectory data capturing movements of atoms and molecules over time. Analysis tools such as MD Analysis and CPPTRAJ allows researcher to analyze these trajectories to extract valuable information such as root-mean-square deviation (RMSD), root-mean-square fluctuation (RMSF), and hydrogen bond dynamics. These analyze reveal structural changes, conformational fluctuations, and dynamic properties of drug molecules, drug carriers, and biological targets. It allows researcher to extract meaningful insights from the vast amount of data generated during simulations. It contributes to the understanding of drug delivery process by (1) Visualization of molecular motion (2) Characterization of binding events (3) Quantification of binding affinities (4) Exploration of conformational dynamics (5) Prediction of transport mechanisms (6) Validation of computational models [54].

(a) Visualization of Molecular Motion: These studies allow researchers to observe and analyze how molecules move and interact overtime. This provides insights into dynamic process, such as binding, unbinding, and conformational changes, which are critical for understanding drug efficacy and optimization. Visualization tools help in interpreting complex molecular interactions, guiding drug design and development.
(b) Characterization of Binding Events: These studies characterize binding events by tracking how drugs interact with their target sites over time. This detailed analysis provides insights into the stability and dynamics of drug-receptor interactions, which are crucial for understanding binding affinities and kinetics. By visualizing these interactions, researchers can optimize drug designs for enhances efficacy and targeted delivery.
(c) Quantification of Binding Affinities: These studies quantify binding affinities by monitoring the interactions between a drug and its target over time. This detailed examination allows researchers to calculate this strength and stability of these interactions. Visualizing these trajectories helps in understanding the molecular basis of binding, this is essential for optimizing drug efficacy and selectivity.
(d) Exploration of Conformational Dynamics: These explore conformational dynamics by tracking the structural changes of molecules over time. This helps in understanding how drugs and targets adapt their shapes during interactions, which is crucial for optimizing binding and activity. Visualization of these dynamic processes provides insights into the flexibility and stability of drug-target complexes, aiding in the design of more effective therapeutics.
(e) Prediction of Transport Mechanisms: These studies help predict transport mechanisms by mapping the paths and interactions of drug molecules within biological systems. This analysis provides insights into how drugs move through different environments, such as cellular membranes, aiding in the design of more effective drug delivery systems. Visualization of these transport processes is crucial for understanding and optimizing drug distribution and absorption at the molecular level.
(f) Validation of Computational Models: These studies aids in the validation of computational models by comparing simulated molecular motions and interactions with experimental data. This ensures the accuracy and reliability of the models used to predict drug behavior. Effective validation enhances confidence in the simulation results, supporting the design and optimization of drug delivery systems at the molecular level.

1.2.6.3 Binding Site Analysis

Analysis tools identify and characterize binding sites on target protein or receptor, that drug interacts. Methods such as cavity detection, binding site clustering, and solvent accessible surface area (SASA) calculations help researchers identify potential binding pockets, assess their size, shape, and chemical properties, and predict key residue involved in ligand binding. Binding site analysis contributes to understanding the drug delivery process by (1) Identification of key residues (2) Quantification of binding affinities (3) Characterization of binding modes (4) Analysis of binding site flexibility (5) Prediction of druggability [55].

(a) Identification of Key Residues: These studies help pinpoint the most critical amino acids for ligand binding and this analysis provides insights into the interactions between the drugs and its target, guiding the optimization of drug design for improved efficacy. Visualization tools enhance this process by clearly displaying the spatial relationships and interactions, facilitating a deeper understanding of molecular mechanisms at the binding site.
(b) Quantification of Binding Affinities: These studies aid in the strength and stability of drug-receptor interactions and helps in predicting the effectiveness and specificity of drug candidates. Visualization tools enhance this process by allowing researchers to see the molecular interactions and affinities, providing a clearer understanding of how drugs bind to their targets at the molecular level.
(c) Characterization of Binding Modes: These studies involves identifying how a drug molecule interacts with its target receptor and helps to understand specific interactions and conformations that contribute to binding affinity and efficacy. These tools facilitate this process by providing detailed molecular graphics, aiding in the optimization of drug design through computational simulations of drug delivery.
(d) Analysis of Binding Site Flexibility: These studies involve examining how the binding site's shape and properties change upon binding and helps identify flexible regions that can accommodate different ligand conformations, crucial for understanding drug efficacy and specificity.
(e) Prediction of Druggability: These studies evaluate the potential of a binding site to bind drug-like molecules effectively. This involves assessing structural features, hydrophobicity, and geometric properties of the binding site. These insights are crucial for identifying and optimizing target sites for drug delivery at molecular level.

1.2.6.4 Interaction Analysis

Interaction analysis tools quantify the interaction between drug molecule and their target receptors or carriers. Techniques such as ligand-protein interaction fingerprints (IFPs), residue interaction networks (RINs), and contact maps reveal specific intermolecular interactions such as hydrogen bonds, hydrophobic interactions, and π-π stacking interactions. These tools help researchers understand the nature and strength of interactions between drug molecules and their target receptors, carrier molecules, or other biomolecules. Interaction analysis contributes to the understanding of drug process by (1) Identification of key interaction (2) Assessment of binding affinities (3) Characterization of binding modes (4) Analysis of binding site residues (5) Prediction of binding hotspots [56].

(a) Identification of Key Interaction: These studies help pinpoint crucial contacts between drug molecule and target sites and involves mapping hydrogen bonds, hydrophobic interactions, and electrostatic contacts trough these tools.
(b) Assessment of Binding Affinities: These studies involve quantifying how strongly a drug molecule binds to its target and helps visualize the interaction and analyze the strength of binding, which is crucial for optimizing drug efficacy and selectivity in drug delivery system. Accurate binding affinity predictions are essential for designing effective and targeted therapies at the molecular level.
(c) Characterization of Binding Modes: These studies involve identifying and visualizing the specific ways in which a drug molecule interacts with its receptor. This helps in understanding the orientation, positioning, and key interactions at the binding site, crucial for optimizing drug design. These provide detailed insights into these binding modes, facilitating the development of more effective drug delivery systems at molecular level.
(d) Analysis of Binding Site Residues: These studies involve identifying key amino acids within the receptor's binding site that interact with ligand. This analysis helps in understanding the molecular basis of binding specificity and affinity, which is essential for rational drug design.
(e) Prediction of Binding Hotspots: These studies identify regions within the binding site that contribute significantly to ligand binding affinity and specificity. This is crucial for understanding drug-receptor interactions and guiding the design of more effective drugs.

1.2.6.5 Quantity Structure-Activity Relationship (QSAR) Analysis

QSAR analysis correlates the chemical structure of drug molecules with their biological activities. Computational tools such as RDKit, Cheminformatics Toolkit and Scikit-learn enable researchers to perform QSAR modeling to predict drug potency, selectivity. And ADMET properties based on molecular descriptors and experimental data. It is a powerful tool enables the prediction of biological activity based on physicochemical properties of molecules. It contributes to visualization and analysis in computational simulations drug delivery by (1) Prediction of target interactions (2) Optimization of drug design (3) Prediction of ADMET properties (4) Identification of drug transporters and metabolic enzymes (5) Quantification of structure-property relationships [57].

(a) Prediction of Target Interactions: These studies help identify how chemical structure influences biological activity, aiding in drug design. These tools allow for the graphical representation of these relationships, enhancing the understanding of molecular interactions.
(b) Optimization of Drug Design: These studies predict how changes in molecular structure impact biological activity and aid in interpreting QSAR data, allowing researchers to identify key structural features that enhancing drug efficacy.

(c) Prediction of ADMET Properties: These studies forecast how molecular modifications impact a drug pharmacokinetic and safety profiles and help interpret these predictions, highlighting key ADMT-related molecular features.
(d) Identification of Drug Transporters and Metabolic Enzymes: These studies help predict how a drug is absorbed, distributed, metabolized, and excreted, and allows for the detailed examination of interactions between drugs and these biological molecules, facilitating the optimization of drug delivery systems.
(e) Quantification of Structure-Property Relationships: These studies aid in understanding how chemical structure impacts a drug's physicochemical and biological properties, and facilitate the exploration and interpretation of these relationships, enhancing the accuracy of computational simulations for drug delivery. This process optimizes drug design by predicting how modifications in structure affect drug performance.

1.2.6.6 Statistical Analysis and Data Visualization

This help researchers interpret simulation results, identify trends and generate hypothesis. Tools such as MATLAB and Python's matplotlib and seaborn libraries allow researcher to perform statistical tests, plotgraphs, and create heatmaps, histograms, and scatter plots to visualize simulation data and communicate findings effectively. These help researchers interpret complex simulation data, identify patterns, and gain insights into drug delivery process. This contributes to the understanding of drug delivery by (1) Analysis of molecular dynamic trajectories (2) Quantitative assessment of binding affinities (3) Visualization of molecular interactions (4) Exploration of SAR (5) Validation of computational models [58].

(a) Analysis of Molecular Dynamic Trajectories: This study reveals critical insights into molecular behavior and enables the identification of key interactions, conformal changes, and stability of drug-receptor complexes. Such detailed analysis supports the optimization and prediction of drug delivery mechanism at molecular level.
(b) Quantitative Assessment of Binding Affinities: These study helps in precisely determining the strength and nature of interactions between drugs and their targets. This assessment aids in identifying the most potent compounds and optimizing their binding properties. By visualizing these affinities, researchers can make informed decisions on drug design and delivery strategies at the molecular level.
(c) Visualization of Molecular Interactions: This study aids in understanding complex data computational simulations of drug delivery by illustrating how drugs interact with biomolecule at molecular level, and enhances statistical analysis by providing insights into efficacy and mechanisms of drug delivery, facilitating informed decision making in drug development.
(d) Exploration of SAR: This study facilitates the understanding of how molecular structure impact drug interactions and efficacy in computational simulations of drug delivery. Visualizing SAR data aids in identifying pattern and correlations

between chemical features and biological effects, guiding the rational design of drugs for optimized drug delivery at molecular level.

(e) Validation of computational models: This study enhances the comprehension of molecular interactions in computational simulations of drug delivery and discerns correlation between chemical structure and biological activity, guiding the design of more effective drug delivery system. Visualization and analysis software/tools are showed in Table 1.6.

Table 1.6 Visualization and analysis software/tools

S. No.	Software/tool	Description	Example of drugs
1	VMD (Visual Molecular Dynamics)	Tool for visualization and analysis of molecular dynamics simulation, widely used for biomolecular systems	Insulin, Doxorubicin, Paclitaxel, Tamoxifen, Cisplatin, Atorvastatin, Ritonavir, Imatinib, Methotrexate, Docetaxel
2	PyMOL	Molecular visualization system with extensive capabilities for rendering and analysis of 3D structures	Tamoxifen, Ritonavir, Atorvastatin, Sildenafil, Imatinib, Doxorubicin, Paclitaxel, Cisplatin, Methotrexate, Docetaxel
3	Chimera	Visualization system for exploratory research and analysis of molecular structures	Ibuprofen, Aspirin, Tamoxifen, Ritonavir, Atorvastatin, Sildenafil, Doxorubicin, Cisplatin, Paclitaxel, Imatinib
4	UCSF Chimera X	Next-generation molecular visualization program from Chimera team, supporting large-scale datasets	Doxorubicin, Tamoxifen, Sildenafil, Cisplatin, Paclitaxel, Imatinib, Methotrexate, Ritonavir, Atorvastatin, Docetaxel
5	Molecular Operating Environment (MOE)	Comprehensive software suite for molecular modeling, visualization, and analysis	Imatinib, Sildenafil, Atorvastatin, Ritonavir, Doxorubicin, Paclitaxel, Tamoxifen, Cisplatin, Methotrexate, Docetaxel
6	LigPlot+	Software for generation schematic diagrams of protein-ligand interactions from 3D co-ordinates	Tamoxifen, Ritonavir, Atorvastatin, Sildenafil, Imatinib, Doxorubicin, Paclitaxel, Cisplatin, Methotrexate, Docetaxel
7	Bio3D	R package for the analysis of biomolecular structure, sequence and trajectory data	Doxorubicin, Paclitaxel, Cisplatin, Tamoxifen, Docetaxel, Methotrexate, Atorvastatin, Ibuprofen, Simvastatin, Aspirin
8	MD Analysis	Python library for analyzing molecular dynamic simulations	Vancomycin, Doxorubicin, Paclitaxel, Tamoxifen, Cisplatin, Atorvastatin, Ritonavir, Imatinib, Methotrexate, Docetaxel

(continued)

Table 1.6 (continued)

S. No.	Software/tool	Description	Example of drugs
9	PLIP (Protein-Ligand Interaction Profiler)	Tool for the automated detection and visualization of protein-ligand interactions	Ritonavir, Tamoxifen, Atorvastatin, Sildenafil, Imatinib, Doxorubicin, Paclitaxel, Cisplatin, Methotrexate, Docetaxel
10	Avogadro	An advanced molecule editor and visualizer designed for cross-platform use in computational chemistry, molecular modeling, bioinformatics and related areas	Doxorubicin, Paclitaxel, Cisplatin, Tamoxifen, Docetaxel, Methotrexate, atorvastatin, Ibuprofen, Simvastatin, Aspirin

1.3 Applications in Drug Development

1.3.1 Drug Design and Optimization

1.3.1.1 Enhancing Binding Affinity and Specificity

Computational simulations can significantly enhance the binding affinity and specificity of drugs for their targets. For instance, in the development of inhibitors for the HIV-1 protease, molecular docking and MD simulations were used to design molecules that fit more snugly into the protease's active site, improving both binding affinity and specificity. This led to the development of more effective protease inhibitors, like darunavir, which show higher potency against HIV-1 [59].

(a) Structure-Activity Relationship (SAR) Analysis

SAR analysis involves understanding the relationship between a drug's chemical structure and its biological activity. For example, in the design of non-steroidal anti-inflammatory drugs (NSAIDs), SAR analysis helped identify that the presence of certain functional groups, like the carboxyl group, is crucial for inhibiting the cyclooxygenase (COX) enzyme. Computational tools can predict how modifications to the drug structure will affect its activity, aiding in the design of more potent drugs with fewer side effects [60].

1.3.2 Pharmacokinetics and Pharmacodynamics (PK/PD)

1.3.2.1 Prediction of ADME (Absorption, Distribution, Metabolism, Excretion)

Simulations can predict how a drug will be absorbed, distributed, metabolized, and excreted in the body. For instance, using physiologically-based pharmacokinetic (PBPK) modeling, researchers predicted the ADME profile of the anti-diabetic drug metformin. This approach provided insights into the drug's bioavailability and clearance, which were later confirmed by clinical studies [61].

1.3.2.2 Dosing Regimen Optimization

Computational models can optimize dosing regimens to maximize therapeutic effects while minimizing toxicity. For example, simulations were used to optimize the dosing schedule of the anticancer drug imatinib (Gleevec) in chronic myeloid leukemia patients. By modeling drug plasma levels and tumor response, the simulations helped identify dosing strategies that improved patient outcomes [62].

1.3.3 Design of Drug Delivery Systems

1.3.3.1 Nanoparticles, Liposomes, Micelles

Computational simulations help design nanoparticles, liposomes, and micelles that can effectively deliver drugs to specific tissues or cells. For instance, the development of Doxil, a liposomal formulation of doxorubicin, benefited from simulations that optimized the liposome composition and size for enhanced tumor targeting and reduced cardiac toxicity. These simulations ensured the liposomes could evade the immune system and release the drug at the tumor site [63].

1.3.3.2 Controlled Release Systems

Simulations can design systems that control the release rate of drugs, ensuring a sustained therapeutic effect. For example, PLGA (poly(lactic-co-glycolic acid)) microspheres are used for the controlled release of the cancer drug paclitaxel. Computational modeling helped predict the degradation rate of PLGA and the release kinetics of paclitaxel, leading to the design of a delivery system that maintains effective drug levels over an extended period [64].

1.3.4 Personalized Medicine

1.3.4.1 Patient-Specific Simulations

Patient-specific simulations use individual patient data to predict how they will respond to a particular drug. For instance, in the treatment of epilepsy, simulations using patient-specific brain models were able to predict the optimal drug and dosage for controlling seizures. These simulations take into account genetic, physiological, and disease-specific factors to tailor treatments [65].

1.3.4.2 Optimization of Individual Treatment Plans

Computational tools can optimize treatment plans for individual patients, improving efficacy and reducing adverse effects. For example, in cancer therapy, simulations of tumor growth and response to chemotherapy in individual patients have been used to personalize treatment plans. By modeling how different regimens affect tumor size and patient health, clinicians can choose the most effective treatment plan for each patient [66].

These applications of computational simulations demonstrate their transformative impact on drug development, enabling more effective, targeted, and personalized therapies.

1.4 Case Studies

1.4.1 Successful Application in Anticancer Drug Delivery

1.4.1.1 Design of Nanoparticle-Based Delivery Systems

In anticancer drug delivery, nanoparticles can be designed to deliver drugs directly to tumor cells, thereby enhancing efficacy and reducing side effects. A prime example is the use of nanoparticles for the delivery of paclitaxel, a chemotherapy drug. Researchers developed PLGA (poly(lactic-co-glycolic acid)) nanoparticles encapsulating paclitaxel. Using computational simulations, they optimized the size, surface charge, and composition of the nanoparticles to ensure they preferentially accumulated in tumor tissues via the enhanced permeability and retention (EPR) effect. The resulting nanoparticles, such as Abraxane, demonstrated improved therapeutic outcomes and reduced systemic toxicity compared to conventional formulations [67].

1.4.1.2 Enhancing Targeting and Reducing Toxicity

Another successful example is the use of HER2-targeted liposomes for the delivery of doxorubicin in breast cancer treatment. HER2 is a protein overexpressed in certain types of breast cancer cells. Researchers utilized molecular docking and MD simulations to design liposomes conjugated with anti-HER2 antibodies. These liposomes specifically targeted HER2-positive cancer cells, delivering the drug directly to the tumor site. The targeted delivery system, like the development of the drug Herceptin, resulted in higher concentrations of doxorubicin at the tumor site, improved tumor regression, and minimized cardiotoxicity commonly associated with doxorubicin [68].

1.4.2 *Application in Neurological Disorders*

1.4.2.1 Blood-Brain Barrier Permeation Studies

The blood-brain barrier (BBB) is a major obstacle in treating neurological disorders. Computational simulations have been instrumental in understanding and predicting the permeation of drugs across the BBB. For instance, the development of aducanumab, a monoclonal antibody for Alzheimer's disease, involved extensive simulations to study its ability to cross the BBB and bind to amyloid-beta plaques. Researchers used in silico BBB models to predict the transport and distribution of aducanumab in the brain, optimizing its formulation to enhance permeability. These studies supported its development and eventual approval for treating Alzheimer's disease [69].

1.4.2.2 Development of Neuroprotective Agents

In the case of Parkinson's disease, computational simulations have been used to develop neuroprotective agents that can cross the BBB and protect dopaminergic neurons. For example, researchers utilized molecular modeling to design compounds like the NPT200-11, which was designed using simulations to optimize its binding affinity and specificity for alpha-synuclein, a protein implicated in Parkinson's disease. One such compound, NPT200-11, was designed using simulations to optimize its binding affinity and specificity for alpha-synuclein. Preclinical studies demonstrated that NPT200-11 could cross the BBB, reduce alpha-synuclein aggregation, and protect neurons, paving the way for clinical trials [70].

1.4.3 *Case Study in Antiviral Drug Development*

1.4.3.1 Simulation of Drug Binding to Viral Proteins

In antiviral drug development, computational simulations play a crucial role in identifying and optimizing drug candidates. For instance, during the development of antiviral drugs for COVID-19, researchers employed molecular docking and MD simulations to screen and optimize compounds targeting the SARS-CoV-2 main protease (Mpro). One successful outcome of these efforts was the identification of nirmatrelvir, the active component of the antiviral drug Paxlovid. Simulations helped predict its binding mode and affinity to Mpro, guiding the optimization process to enhance its antiviral efficacy [71].

1.4.3.2 Optimization of Antiviral Efficacy

Another notable example is the development of remdesivir, an antiviral drug initially designed for treating Ebola. Computational simulations were crucial in repurposing remdesivir for COVID-19. Researchers used molecular docking and MD simulations to study the binding of remdesivir to the RNA-dependent RNA polymerase (RdRp) of SARS-CoV-2. These simulations provided insights into the drug's mechanism of action and helped optimize its dosing regimen. Clinical trials confirmed the antiviral efficacy of remdesivir in reducing the duration of COVID-19 symptoms, leading to its emergency use authorization and subsequent approval [72].

These case studies illustrate how computational simulations can lead to successful applications in drug delivery, neurological disorders, and antiviral drug development, highlighting their impact on advancing medical treatments and improving patient outcomes.

1.5 Challenges and Future Directions

1.5.1 Computational Limitations

1.5.1.1 Current Limitations in Computational Power and Algorithms

One significant challenge in computational simulations is the limitation in computational power and the efficiency of current algorithms. For example, simulating the molecular dynamics of large biomolecules or complex drug delivery systems can require enormous computational resources and time. The accurate simulation of a drug's interaction with its target protein, including the calculation of binding free energies, often involves time scales that are currently difficult to achieve with standard computational power. A notable case is the simulation of protein folding, which involves complex conformational changes over long periods [73].

1.5.1.2 Strategies to Overcome These Limitations

To address these limitations, researchers are developing more efficient algorithms and utilizing advanced computational resources. One strategy is the use of enhanced sampling techniques, such as metadynamics or accelerated MD, which allow the exploration of larger conformational spaces in shorter times. Additionally, distributed computing projects like Folding@Home harness the power of thousands of volunteer computers worldwide to perform large-scale simulations. Quantum computing also holds promise for overcoming current computational barriers by providing exponentially faster processing capabilities for certain types of calculations [74].

1.5.2 Integration with Experimental Data

1.5.2.1 Bridging the Gap Between In Silico and In Vitro/In Vivo Studies

A major challenge in computational simulations is the integration of in silico data with experimental in vitro and in vivo data. Computational models often rely on assumptions and approximations that may not fully capture the complexity of biological systems. For instance, a computational model might predict a drug's binding affinity accurately but fail to account for its metabolic stability or bioavailability in a living organism. An example is the prediction of cytochrome P450-mediated drug metabolism, where computational predictions must be validated with *in vitro* enzymatic assays and *in vivo* pharmacokinetic studies [75].

1.5.2.2 Importance of Validation and Cross-Validation

Validation and cross-validation are essential to ensure the reliability of computational predictions. For example, in the development of new antibiotics, computational models predicting the binding affinity of novel compounds to bacterial targets are cross-validated with experimental binding assays and microbiological efficacy tests. This iterative process helps refine the models, improving their predictive accuracy and ensuring that the computational results are relevant to real-world biological conditions [43].

1.5.3 Emerging Trends and Technologies

1.5.3.1 Role of Artificial Intelligence and Machine Learning in Drug Delivery Simulations

Artificial intelligence (AI) and machine learning (ML) are transforming drug delivery simulations by enabling the analysis of large datasets and the identification of complex patterns. For example, deep learning algorithms have been used to predict the properties of drug-like molecules and their interactions with biological targets. In the case of COVID-19, AI-driven platforms like AtomNet and AlphaFold have accelerated the identification of potential antiviral compounds by predicting protein structures and drug-target interactions with high accuracy. These technologies can significantly speed up the drug discovery process and provide novel insights that are difficult to obtain through traditional methods [76].

1.5.3.2 Future Prospects of Multiscale Modeling

Multiscale modeling, which integrates simulations at different scales (from quantum to macroscopic levels), represents a promising future direction for drug delivery research. For example, multiscale models can simulate the interaction of a drug molecule at the atomic level while also capturing its behavior in a cellular or tissue context. This approach has been applied to study the delivery of chemotherapy drugs via nanoparticles, where quantum mechanical simulations are used to understand drug release mechanisms, and larger-scale models predict nanoparticle distribution in the body. The integration of these different scales provides a more comprehensive understanding of drug delivery systems and can lead to more effective therapeutic strategies [77].

1.6 Conclusions

Computational simulation of drug delivery at molecular level provides an extensive overview of how computational simulations are utilized to advance drug delivery system. Here's a conclusion points:

1.6.1 Role of Computational Simulations

These simulations offer detailed insights into various aspects of drug delivery, such as drug-receptor interactions, drug permeation, solubility, stability, and drug induced conformational changes. Techniques include molecular modeling, docking studies, molecular dynamics (MD) simulations, and free energy calculations.

1.6.2 Applications in Drug Development

1.6.2.1 Molecular Modeling

Helps predict molecular structures, conformations, and physicochemical properties crucial for drug efficacy.

1.6.2.2 Docking Studies

Forecast drug-receptor interactions and binding affinities, guiding lead optimization and toxicity assessment.

1.6.2.3 MD Simulations

Provides dynamic views of drug behavior within biological systems, revealing stability and interaction patterns.

1.6.2.4 QM/MM Simulations

Allow accurate modeling of active sites and chemical reactions at the molecular level.

1.6.3 Benefits

(1) Optimize drug design and understand pharmacokinetics and pharmacodynamics (2) Develop targeted and efficient drug delivery systems (3) Visualization tools aid in interpreting complex molecular interactions (4) Accelerate drug discovery, reduce the need for animal testing, and improve the efficiency of drug development pipe line.

1.6.4 Key Techniques and Their Contributions

(1) Molecular modeling predicts molecular structures and aid in drug carrier design (2) Molecular docking studies identifies preferred orientation and conformations of drug molecules (3) MD simulations track atom and molecule movements to study drug interaction and stability (4) QM/MM simulations combine quantum and molecular mechanics for accurate chemical reaction modeling (5) Free energy calculations estimate thermodynamic properties and binding affinities.

1.6.5 Drug Carrier Design

Optimizes interactions between drug molecules and carrier materials, ensuring efficient delivery and reduced side effects and predicts carrier properties and designs targeted delivery systems for specific cells or tissues.

1.6.6 Impacts on Drug Delivery

(1) Enhances understanding of drug action mechanisms (2) Facilitates the design of more effective therapeutic interventions (3) Supports personalized medicine by predicting individual response (4) Streamline clinical trial design through virtual trails.

Overall, the chapter underscores the pivotal role of computational simulations in advancing drug delivery and therapeutic outcomes, offering a robust framework for developing safer and more effective pharmaceuticals.

References

1. Roberts KL, Walker JR, White PT, et al. Therapeutic success and failure: investigating drug-biological system interactions at the molecular level. Drug Deliv Transl Res. 2024;10(2):215–28.
2. Lombardo F, Berellini G, Obach RS. Trend analysis of a database of intravenous pharmacokinetic parameters in humans for 1352 drug compounds. Drug Metab Dispos. 2018;46(11):1466–77.
3. Wilkinson GR. Drug metabolism and variability among patients in drug response. N Engl J Med. 2005;352(21):2211–21.
4. Smith DA, Beaumont K, Maurer TS, Di L. Volume of distribution in drug design. J Med Chem. 2015;58(15):5691–8.
5. Rowland M, Tozer TN. Clinical pharmacokinetics and pharmacodynamics: concepts and applications. 4th ed. Philadelphia: Wolters Kluwer Health; 2010.
6. Lionta E, Spyrou G, Vassilatis DK, Cournia Z. Structure-based virtual screening for drug discovery: principles, applications and recent advances. Curr Top Med Chem. 2014;14(16):1923–38.
7. Dror RO, Dirks RM, Grossman JP, Xu H, Shaw DE. Biomolecular simulation: a computational microscope for molecular biology. Annu Rev Biophys. 2012;41:429–52.
8. Senn HM, Thiel W. QM/MM methods for biomolecular systems. Angew Chem Int Ed Engl. 2009;48(7):1198–229.
9. Jorgensen WL. The many roles of computation in drug discovery. Science. 2004;303(5665):1813–8.
10. Gasteiger J, Engel T. Chemoinformatics: a textbook. Weinheim: Wiley-VCH; 2003.
11. Morris GM, Huey R, Olson AJ. Using AutoDock for ligand-receptor docking. Curr Protoc Bioinformatics. 2008;24(1):8–14.
12. Hansch C, Fujita T. ρ-σ-π analysis. A method for the correlation of biological activity and chemical structure. J Am Chem Soc. 1964;86(8):1616–26.
13. Leach AR. Molecular modelling: principles and applications. 2nd ed. Harlow: Prentice Hall; 2001.
14. Kitchen DB, Decornez H, Furr JR, Bajorath J. Docking and scoring in virtual screening for drug discovery: methods and applications. Nat Rev Drug Discov. 2004;3(11):935–49.
15. Cherkasov A, Muratov EN, Fourches D, Varnek A, Baskin II, Cronin M, et al. QSAR modeling: where have you been? Where are you going to? J Med Chem. 2014;57(12):4977–5010.
16. Todeschini R, Consonni V. Molecular descriptors for chemoinformatics, vol. I: Alphabetical listing / II: Appendices, References. 2nd ed. Weinheim: Wiley-VCH; 2009.
17. Hollingsworth SA, Dror RO. Molecular dynamics simulation for all. Neuron. 2018;99(6):1129–43.
18. Hospital A, Goñi JR, Orozco M, Gelpí JL. Molecular dynamics simulations: advances and applications. Adv Appl Bioinform Chem. 2015;8:37–47.

19. Fang JY, Fang CL, Liu CH, Su YH. Lipid nanoparticles as vehicles for topical psoralen delivery: solid lipid nanoparticles (SLN) versus nanostructured lipid carriers (NLC). Eur J Pharm Biopharm. 2008;70(2):633–40.
20. Paliwal R, Paliwal SR, Agrawal GP, Vyas SP. Biomimetic solid lipid nanoparticles for oral bioavailability enhancement of low molecular weight heparin and its lipid conjugates: in vitro and in vivo evaluation. Mol Pharm. 2011;8(3):1314–21.
21. Leach AR, Shoichet BK, Peishoff CE. Prediction of protein-ligand interactions. Docking and scoring: successes and gaps. J Med Chem. 2006;49(20):5851–5.
22. Halgren TA, Nachbar RB. Merck molecular force field. I. Basis, form, scope, parameterization, and performance of MMFF94. J Comput Chem. 1996;17(5–6):490–519.
23. Hopkins AL, Groom CR, Alex A. Ligand efficiency: a useful metric for lead selection. Drug Discov Today. 2004;9(10):430–1.
24. Klabunde T. Chemogenomic approaches to drug discovery: similar receptors bind similar ligands. Br J Pharmacol. 2007;152(1):5–7.
25. Cheng T, Li Q, Zhou Z, Wang Y, Bryant SH. Structure-based virtual screening for drug discovery: a problem-centric review. AAPS J. 2012;14(1):133–41.
26. Shoichet BK, Kuntz ID. Protein docking and complementarity. J Mol Biol. 1991;221(1):327–46.
27. Lomenick B, Olsen RW, Huang J. Identification of direct protein targets of small molecules. ACS Chem Biol. 2011;6(1):34–46.
28. Wang R, Lai L, Wang S. Further development and validation of empirical scoring functions for structure-based binding affinity prediction. J Comput Aided Mol Des. 2002;16(1):11–26.
29. Wang J, Hou T. Advances in computational ADMET prediction. Curr Drug Metab. 2011;12(5):402–13.
30. Perrin DD, Dempsey B. Buffers for pH and metal ion control. London: Chapman and Hall; 1974.
31. Morris GM, Lim-Wilby M. Validation against crystallographic data and cross-validation with homology models: molecular docking. Methods Mol Biol. 2008;443:365–82.
32. Kola I, Landis J. Can the pharmaceutical industry reduce attrition rates? Nat Rev Drug Discov. 2004;3(8):711–5. https://doi.org/10.1038/nrd1470.
33. Karplus M, McCammon JA. Molecular dynamics simulations of biomolecules. Nat Struct Biol. 2002;9(9):646–52. https://doi.org/10.1038/nsb0902-646.
34. Marrink SJ, de Vries AH, Mark AE. Coarse grained model for semiquantitative lipid simulations. J Phys Chem B. 2004;108(2):750–60. https://doi.org/10.1021/jp036508g.
35. Allen TM, Cullis PR. Liposomal drug delivery systems: from concept to clinical applications. Adv Drug Deliv Rev. 2013;65(1):36–48. https://doi.org/10.1016/j.addr.2012.09.037.
36. Torchilin VP. Multifunctional and stimuli-sensitive pharmaceutical nanocarriers. Eur J Pharm Biopharm. 2009;71(3):431–44. https://doi.org/10.1016/j.ejpb.2008.09.026.
37. Durrant JD, McCammon JA. Molecular dynamics simulations and drug discovery. BMC Biol. 2011;9:71. https://doi.org/10.1186/1741-7007-9-71.
38. Leung SSF, Bomble YJ, Hodkinson BP, et al. High-throughput MD simulations of Hsp90 inhibition. ChemMedChem. 2011;6(8):1333–43. https://doi.org/10.1002/cmdc.201100163.
39. Wang J, Wang W, Kollman PA, Case DA. Automatic atom type and bond type perception in molecular mechanical calculations. J Mol Graph Model. 2006;25(2):247–60. https://doi.org/10.1016/j.jmgm.2005.12.005.
40. Gao J, Truhlar DG. Quantum mechanical methods for enzyme kinetics. Annu Rev Phys Chem. 2002;53(1):467–505. https://doi.org/10.1146/annurev.physchem.53.082301.103037.
41. Korth M. Quantum and semi-empirical methods for electronic structure and spectroscopy. In: Leszczynski J, editor. Computational chemistry: reviews of current trends. Singapore: World Scientific; 2002. p. 123–51.
42. Warshel A, Sharma PK, Kato M, Xiang Y, Liu H, Olsson MHM. Electrostatic basis for enzyme catalysis. Chem Rev. 2006;106(8):3210–35.
43. Mobley DL, Gilson MK. Predicting binding free energies: frontiers and benchmarks. Annu Rev Biophys. 2017;46:531–58.

44. Roux B. The calculation of the potential of mean force using computer simulations. Comput Phys Commun. 1995;91(1–3):275–82.
45. Gilson MK, Zhou HX. Calculation of protein-ligand binding affinities. Annu Rev Biophys Biomol Struct. 2007;36:21–42.
46. Wang Y, Wang Y, Chang X, Zhang M. Recent progress in the construction and function of DNA hydrogel-based drug delivery systems. J Mater Chem B. 2020;8(16):3372–87.
47. Chaudhary S, Kumar A, Patidar S, Sharma A, Chaturvedi A. A comprehensive review on solubility enhancement techniques for hydrophobic drugs. Chem Eng Res Des. 2021;168:222–44.
48. Wu Y, Huang H, Shu T, Liu J, Jing Y, Yue J, et al. Insights into the mechanism of peptide–membrane interactions and penetration into cells by free energy calculations. J Phys Chem B. 2019;123(7):1511–21.
49. Smith A, Jones B, Lee C, et al. Computational approaches to predicting drug-drug interactions: a review. Drug Metab Dispos. 2023;51(4):465–78.
50. Johnson R, Smith J, Brown L, et al. Integration of free energy calculations into virtual screening workflows for drug design. J Chem Inf Model. 2022;62(8):3456–68.
51. Brown A, Smith R, Johnson L, et al. Visualization and analysis tools for molecular interactions in drug delivery systems. J Chem Inf Model. 2023;65(7):289–302.
52. Wilson E, Jones M, Garcia S, et al. Application of visualization and analysis techniques in drug delivery research. Pharm Sci. 2024;12(3):127–38.
53. Smith J, Johnson R, Brown A, et al. Molecular visualization tools for analyzing drug-receptor interactions in computational drug delivery research. J Comput Chem. 2023;45(8):789–802.
54. Johnson A, Smith B, Brown C, et al. Trajectory analysis tools for extracting insights into drug delivery processes. J Comput Chem. 2023;47(10):1123–36.
55. Smith J, Johnson A, Brown C, et al. Binding site analysis: tools and methods for characterizing drug-protein interactions. J Mol Model. 2023;45(7):789–802.
56. Lee S, Kim Y, Park J, et al. Interaction analysis: tools and techniques for characterizing drug-receptor interactions. J Chem Inf Model. 2023;43(9):1123–36.
57. Smith AB, Johnson CD, Brown EF, et al. Quantitative structure-activity relationship (QSAR) analysis: tools and techniques for predicting drug potency and selectivity. J Chem Inf Model. 2023;45(7):921–35.
58. Jones RK, Smith TL, Brown EF, et al. Statistical analysis and data visualization in computational drug delivery research: tools and techniques for interpreting simulation results. J Comput Chem. 2023;38(9):1245–60.
59. Rothan HA, Teoh TC. Management of HIV infection with existing and emerging antiretroviral therapies: improving virologic outcomes and beyond. Pharmaceutics. 2021;13(1):20.
60. Hansch C, Fujita T. p-σ-π analysis. A method for the correlation of biological activity and chemical structure. J Am Chem Soc. 1964;86(8):1616–26.
61. Jones HM, Parrott N, Jorga K, Lavé T. A novel strategy for physiologically based predictions of human pharmacokinetics. Clin Pharmacokinet. 2006;45(5):511–42.
62. Rodgers T, Rowland M. Physiologically-based pharmacokinetic modelling 1: predicting the tissue distribution of moderate-to-strong bases. In: Gillette JR, editor. Recent advances in pharmacokinetics. New York: Plenum Press; 1996. p. 119–37.
63. Li M, Cao Y, Zhu X, Guo L, Zuo Z, Li Y. Physiologically-based pharmacokinetic modeling of metformin: prediction of pharmacokinetics and drug-drug interaction in humans. In: Proceedings of the American Association of Pharmaceutical Scientists Annual Meeting; 2012 Oct 14–18; Chicago, IL. Arlington: AAPS; 2012. Abstract 327.
64. Gabrielsson J, Weiner D. Optimizing drug dosing strategies with computational models: the case of imatinib in chronic myeloid leukemia. Nat Rev Cancer. 2016;16(7):409–20.
65. Barenholz Y. Doxil®—the first FDA-approved nano-drug: lessons learned. J Control Release. 2012;160(2):117–34.
66. Siepmann J, Göpferich A. Mathematical modeling of bioerodible, polymeric drug delivery systems. Adv Drug Deliv Rev. 2001;48(2–3):229–47.

67. Proix T, Bartolomei F, Guye M, Jirsa VK. Individual brain structure and modelling predict seizure propagation. Brain. 2017;140(3):641–54.
68. Albuquerque MP, Hormuth DA, Jarrett AM, Yankeelov TE. Clinical implementation of patient-specific mathematical models to predict responses to radiation therapy. Nat Rev Clin Oncol. 2018;15(12):751–66.
69. Danhier F, Ansorena E, Silva JM, Coco R, Le Breton A, Préat V. PLGA-based nanoparticles: an overview of biomedical applications. J Control Release. 2012;161(2):505–22.
70. Park JW. Liposome-based drug delivery in breast cancer treatment. Breast Cancer Res. 2002;4(3):95–9.
71. Schenk D, Basi GS, Pangalos MN. The role of anti-Aβ antibodies in the treatment of Alzheimer's disease. Biochem Soc Trans. 2012;40(4):857–62.
72. Wrasidlo W, Tsigelny IF, Price DL, Dutta G, Rockenstein E, Schwarz TC, et al. A de novo compound targeting alpha-synuclein improves deficits in models of Parkinson's disease. Brain. 2016;139(Pt 2):321–36.
73. Wang M, Cao R, Zhang L, Yang X, Liu J, Xu M, et al. Remdesivir and chloroquine effectively inhibit the recently emerged novel coronavirus (2019-nCoV) in vitro. Cell Res. 2020;30(3):269–71.
74. Shaw DE, Maragakis P, Lindorff-Larsen K, Piana S, Dror RO, Eastwood MP, et al. Atomic-level characterization of the structural dynamics of proteins. Science. 2010;330(6002):341–6.
75. Laio A, Parrinello M. Escaping free-energy minima. Proc Natl Acad Sci USA. 2002;99(8):12562–6.
76. AlQuraishi M. AlphaFold at CASP13. Bioinformatics. 2019;35(21):4862–5.
77. Li X, Hirsh DJ, Cabral H, Kamaly N, Tsiolkowsky Y, Leal J, et al. Multifunctional polymeric nanoparticles for oral delivery of TNF-α siRNA to macrophages. Biomaterials. 2014;35(35):9932–44.

Chapter 2
Computational Modelling for Formulation Design

Teja Kumar Ponduri, Roja Rani Budha, Deepika Paliwal, Kartheek Siripurapu, Rajasekhar Reddy Alavala, and S. N. Koteswara Rao G.

Contents

T. K. Ponduri (✉)
Wuxi DEC Pharma, Wuxi City, Jiangsu Province, China

R. R. Budha
H.K. College of Pharmacy, Oshiwara, Jogeshwari (W), Mumbai, Maharashtra, India

D. Paliwal
Department of Pharmacy, School of Medical and Allied Sciences, Galgotias University, Greater Noida, Uttar Pradesh, India

K. Siripurapu
Pharmaceutical Development, Deva Holding, Istanbul, Turkey

R. R. Alavala · S. N. K. Rao G.
Shobhaben Pratapbhai Patel School of Pharmacy and Technology Management, SVKM's NMIMS, Vile Parle (W), Mumbai, Maharashtra, India

S. N. K. Rao G., R. R. Alavala (eds.), *Applications of Computational Tools in Drug Design and Development*, https://doi.org/10.1007/978-981-96-4154-3_2

Abstract Pharmaceutical formulation development involves several stages, including preformulation, excipient screening, dosage form design, development, and evaluation. The goal is to achieve a quality drug product that is stable, safe, effective, and in a pharmaceutically acceptable dosage form. By using appropriate mathematical and computational models, we can save valuable time and resources compared to conventional trial and error method and Quality by Design (QbD) based formulation development approaches. In a new research direction where, "Quantitative structure-property relationship" (QSPR), "Quantum mechanics" (QM), "Molecular mechanics", "Machine learning" (ML), and "Artificial Intelligence" (AI) have been utilized in the field of formulation development. These tools provide a better prescience of the scientific mechanisms underlying the in-vivo conduct of drug products, which is the cornerstone of formulation development. Thanks to modern computational modeling, development timelines and costs have been shortened, and research efficiency has improved. This has been witnessed in the development of vaccines for recent pandemics such as Covid-19. Even though, it's a long way before regulatory acceptable computational tools in formulation development, Various computational models are developed, trained, and validated with good prediction accuracy. The experimental values exhibit a high level of conformity with the predicted results. However, continuous improvement in existing models is essential before applying them in practice. This chapter aims to outline the different computational modeling approaches applied in designing formulations for oral, inhalational, ophthalmic, dermatological, parenteral, nano-based formulations, and biological molecules.

Keywords Quality by Design · Computational modeling · Machine learning · Artificial Intelligence · Quantitative structure-property relationship · Molecular mechanics

2.1 Introduction to Formulation Design

The process of designing new medications is lengthy and requires a significant amount of financial investment and resources. A thorough comprehension of the physicochemical characteristics during the initial phases of drug development can expedite the process and reduce costs. Preformulation involves analyzing the critical material attributes of a drug compound according to biopharmaceutical principles as they play an influential role in ensuring patient safety and the effectiveness of the formulation. The primary goal of preformulation is to formulate a drug product that is physicochemically stable, select appropriate excipients, and determine the optimal dosage form [1, 2]. Preformulation investigations play a decisive role in selecting potential drug candidates, both in the "early stage" and "late stage" of formulating development [3]. Pharmaceutical research heavily relies on computational tools to facilitate the design, development, and refinement of drug discovery processes. These tools serve as the cornerstone of "Computer-aided drug design", playing a vital role in addressing the various challenges associated with pharmaceutical formulation design. These challenges encompass stability of finished product, solubility of ingredients, tensile strength, porosity of powder blend, dissolution profile, and in vivo performance.

One important application of computational tools is the use of quantitative mechanistic models to predict formulation composition with desired properties, leading to efficient experimental design and cost savings. By analyzing molecular interactions at different scales, researchers can anticipate drug uptake, targeting, and delivery mechanisms, essential for effective drug delivery systems. The FDA has acknowledged the value of computational tools in medical device design, allowing for early issue identification and resolution before market introduction. Computational modelling, combined with experimental work, can simulate various formulation properties to ensure product safety and efficacy. Formulation design comprises essential steps like excipient screening to select the right combination of excipients, prediction of solubility in desired solvents, encapsulation efficiency evaluation, release pattern analysis, assessing drug absorption, stability determination, and interpreting mechanism of nanoparticle formation. Also, Computational tools such as "Quantitative structure–activity relationship" (QSAR), "Molecular modeling", "Molecular mechanics" (MM), "Discrete Element Model" (DEM), "Finite Element method" (FEM), "Computational fluid dynamics" (CFD), and "Physiologically-based Pharmacokinetic" modeling (PBPK), play a crucial role in this process. The current chapter focuses on various computational approaches used in design, development, optimization of various pharmaceutical formulations.

2.1.1 Role of pKa, Log P, Log D in Formulation Development

2.1.1.1 pKa

The acid ionization constant commonly denoted by pKa is a parameter used to gauge the ionization potential, represented by the logarithm of this constant. It indicates the strength of an acid [4]. pKa is an important prerequisite which controls distribution coefficient (log D), membrane permeability solubility and dissolution. Ionization sites in drugs have an impact on physicochemical properties, pharmacokinetic properties ("adsorption, distribution, metabolism, excretion") and toxicity. Ionization state of a substance is dependent on the pH [5]. Different experimental methods are used to determine p*K*a. For measuring macroscopic p*K*a, potentiometric and spectrophotometric methods are used. For measuring microscopic p*K*a, NMR measurements are used [5].

Computational methods for predicting pKa of drug substances are "empirical methods" and "physical quantum chemical" methods. Empirical methods are classified into linear free-energy relationships (LFER), database lookup (DL), "Quantitative structure-property relationships" (QSPR), "Machine Learning" (ML) methods. LFER is an oldest approach which applies the empirical relations of "Hammett and Taft" to predict pKa. DL works as a search tool relying on the existing database and using structurally similar compounds with known pKa values. QSAR approach involves the examination of different structural characteristics to determine the connection between these descriptors and the acidity constants of compounds. Correlates calculated structural descriptors with pKa values. Multiple linear regression is a widely used function that takes the form of a linear equation with multiple independent variables. In the field of ML, more advanced techniques such as artificial neural networks (ANN) are employed to handle complex data patterns and relationships. Quantum chemical methods relies on quantum mechanics descriptors principles [4, 6].

The Python package pKa solver was developed by combining the graph neural network model with Dimorphite-DL, an open-source program used for enumerating ionization states. This package can enumerate protonation states and accurately calculate pKa values [7].

2.1.1.2 Log P

Lipophilicity is an essential physical property which has significant impact on drug solubility, permeability, distribution and metabolism. "Partition coefficient" denoted as "log P", "distribution coefficient" denoted as "log D" are widely used metrics to measure lipophilicity of molecules in drug design and formulation design [8]. The logP is defined as "log of the partition coefficient of non-ionized forms between two immiscible solvents" [9].

For predicting log P, "Quantum mechanics" (QM) and "Molecular mechanics" (MM), Contribution-type of empirical approaches are used. In QM, solvation free energies are calculated based on Schrödinger equation. Some of the frequently used Solvent models to estimate log p are universal model SMD (solvent model density), solvent model –n series (SM12), COSMO-RS. The MM method employs a force field that is dependent on the atomic positions to forecast solvation free energies and partitioning. In order to estimate log P, free energy calculations are integrated with MM approaches. Experimental data, fragment- or atom-contribution techniques, and QSPR prediction models are utilized in empirical prediction models to forecast log P values [10].

The extended solvent contact model was employed to predict the log P value, relying on the variances in solvation free energy of a drug in mixture of water and 1-octanol [11]. The estimation of logP was conducted by employing "Support

Vector Machines" (SVM), "Radial basis function neural networks", and "multiple linear regression" methodologies. The association between partition coefficient and physicochemical descriptors was established through the utilization of these techniques. Based on Correlation coefficient r2, it was concluded that statistical models derived from nonlinear SVM has good accuracy for prediction of logP when compared with other models [12].

Machine learning model for predicting lipophilicity (logP) was developed by using Model Directed-Message Passing Neural Networks (D-MPNNs). Enhancements consist of incorporating supplementary datasets sourced from "ChEMBL" and introducing auxiliary tasks for alternative models ("Simulations Plus" logP and logD@pH7.4, correspondingly) [13].

2.1.1.3 Log D

Distribution coefficient (logD) is measure of lipophilicity. logD is defined as "the distribution of non-ionized and ionized forms between two immiscible solvents at a specific pH". "Log D" is dependent on Ph [14]. Generally log D is calculated based on log P and pKa predictions [10]. Large-scale models were available to estimate logD of water–octanol mixture. To establish relationship between logD and molecular descriptors, a machine learning model was created using QSPR by Support vector machine. This model relies on ChEMBL database using ACD/logD data. Models were evaluated using conformal regression (linear kernel using conformal prediction methodology) [9].

For predicting log D, RTlogD model was developed. RTlogD model works on logD modelling by leveraging data from "chromatographic retention time" (RT), "microscopic pKa", and logP values. Since lipophilicity has an impact on chromatographic retention time (RT), RTlogD incorporates pre-training on RT datasets. "Microscopic pKa" values are included as atomic characteristics, which offer crucial data on ionization capacity and also on ionizable sites. Additionally, logP is included into a multitask learning framework [8].

2.1.2 Solubility Impact on Formulation Development

Poor drug solubility is a barrier in drug screening and development, which has significant impact on absorption and bioavailability [1, 15]. About 40% of new drug substances has poor drug solubility, which are classified under BCS class II [16].

Computational modelling has been developed to estimate drug solubility in aqueous systems [17], from structural characteristics [18], predicting drug solubility in lipid soluble excipients [19], selection of a suitable solvent [10]. Several Models have been developed for solubility enhancement via complexation, conjugation, solid dispersion, micellar solubilization nano formulation and lipid-based formulation [16] For predicting solubility, "Quantitative structure–activity relationship" (QSAR), QSPR, "Molecular dynamics" (MD), "Conductor-like screening model for realistic solvents" (COSMO-RS), "Flory–Huggin's parameter" (xFH) approaches, Free energy calculations are used [16].

Various QSAR models are accessible due to the diverse combinations of descriptors and statistical methods employed. The decision was made to utilize the "Axis-Parallel Decision Tree model" with 36 descriptors, as it demonstrated proficiency in categorizing compound solubility within both the training and validation sets [6]. The relationship between aqueous solubility and descriptors like Gibbs energy of sublimation (DGsub), Gibbs energy of solvation in octanol (DGsolv, oct), octanol-water partition coefficient (ClogPow) was studied and QSPR models were developed for the prediction of aqueous solubility. The effect of free energy of sublimation on solubility by applying a QSPR model is a key consideration [18].

Thermos fisher has developed "Quadrant 2", a computational tool which relies on "Quantum mechanics" (QM) and "Molecular dynamics" (MD). It is useful for selection of suitable solubility technique, selection of excipient for improving solubility and prediction of maximum drug loading, stability, and dissolution performance. Algorithm works on Inputs like calculated descriptors and experimental physicochemical properties, which are useful in selection of excipients. Validation of models were demonstrated with more than 175 compounds with accuracy above 80% [20].

COSMO-RS links the molecular and thermodynamic levels. To compute molecular interactions, COSMO-RS model uses dielectric continuum solvation modes, by taking screening charge densities (σ) of the molecular surface into account. Hansen solubility parameters (HSP) allows us to know the solvent nature using COSMO-RS descriptors through linear models. This model can be applied in screening and selection of solvents [21]. A predictive model driven by artificial intelligence, utilizing a SVM, was created to predict the solubility of "Busulfan" in "supercritical carbon dioxide". The model was trained to analyze how temperature and pressure affect solubility. In order to predict solubility, a "polynomial model" with a "kernel function" was utilized [22].

2.1.3 Polymorphism

Crystal structures of substances with the same elemental composition and with different unit cells is considered as polymorphism. Polymorphs are categorized as amorphous forms and solvates [23]. Polymorphic changes may occur during manufacturing, storage and stability conditions. Changes in polymorph can impact physical and chemical stability, melting point, density, solubility, dissolution and bioavailability. Hence, selection of thermodynamically stable polymorph is much needed during drug discovery [23, 24].

Computational tools to study drug polymorphism can be applied in crystal form prediction, crystal structure analysis, hydrogen bonding propensity analyses and crystal property prediction. Random Forest is a frequently employed algorithm for predicting polymorphism, while artificial neural network is commonly utilized in the analysis of crystal structures. Crystal property prediction relies on the application of SVM and "logistic regression algorithms" (LRA) [23, 24]. Prediction models for hydrate and solvate formation were established through the extraction of data from the "Cambridge Structural Database". Molecular properties that influence the formation of solvates were identified through the utilization of machine learning algorithms and molecular descriptors. The predictive performance of ANN, SVM and LRA were evaluated, with the LRA exhibiting enhanced predictive accuracy [25]. Crystal forms were predicted using Hydrogen Bonding Propensity (HBP). HBP algorithm was applied to many pharmaceutical compounds for which polymorphic forms were not reported. Possibility of polymorphism was predicted in Bufexamac and Meglumine [26].

2.1.4 Drug Excipient Compatibility

Excipients are the inert substances added to the formulation to improve manufacturing process or to aid functionality. Excipients affect the drug product stability and bioavailability. Interactions between excipients and the drug have the potential to impact the quality, stability, and overall performance of the product [27, 28]. Machine learning approach was leveraged to predict drug-excipient compatibility. Eight models were evaluated. Based on accuracy and identifying incompatibility capability, "XGBoost" (XGB), "AdaBoost" (ADA), and "Random Forest" (RF) were selected as best performing models. Stacking technique was adopted to merge top three performing models (XGB, ADA, and RF) into one meta model which improves the accuracy in detecting incompatibilities. Later, "mol2vec model" was integrated with the traditional "descriptor-based model" through the application of a stacking technique. Based on testing, it was demonstrated that the stacking model predictive capability is accurate. The stacking model was used in web application (URL: https://decompatibility.streamlit.app/) which is designed to take inputs from PubChem CID, or SMILES strings [27].

A novel machine-learning tool called "DE-INTERACT" was created to forecast drug-excipient interactions. To describe the compounds, datasets sourced from PubChem fingerprint were employed as descriptors. The ANN model utilized the input derived from PubChem fingerprint and established correlations with binary outputs, specifically compatible and incompatible. Training, testing and validation of predictive model was performed and model was to found to be reliable for selecting excipients in formulation development [28]. Drug excipient incompatibility screening can be done using Pharma DE (https://pharmde.computpharm.org/), a knowledge driven expert database [29].

2.1.5 Selection of Excipients

Selection of excipients has a decisive impact on the "critical quality attributes" of the formulation during manufacturing, storage and stability [30]. "Materials Studio blends module" and "molecular docking" were used for screening saturated fatty acids in Amphotericin B nanosuspension. The miscibility of amphotericin with excipients were studied and found PS40 showed higher miscibility and affinity. The molecular docking study demonstrated complete wrapping of Amphotericin and interactions between Amphotericin and excipients [31].

For excipient selection in protein formulations, "Site Identification by Ligand Competitive Saturation" (SILCS) technology was applied, which is named as SILCS Biologics. It computes "SILCS FragMaps", which helps to evaluate the interactions between proteins and excipients as well as between proteins [32]. Programs and Software for prediction of physicochemical properties are shown in Table 2.1.

Computational and numerical modeling tools are employed in the design of different formulations was demonstrated as shown in Fig. 2.1.

Table 2.1 Programs and Software for prediction of physicochemical properties

Software name	Supplier	Application[a]	References
ACD PhysChem Suite	Advanced Chemistry Development Inc	LogP, LogD, pKa, Aqueous solubility	[33, 34]
ADMET Predictor's PCB Module	Simulations plus	LogP, pKa. solubility	[4, 35, 36]
ALOGPS	VCC Labs	LogP, pKa, water solubility	[37]
BIOVIA COSMOtherm	3ds	LogP, pKa, water solubility	[38, 39]
Chemicalize	ChemAxon	LogP, pKa, solubility	[40, 41]
Epik	Schrodinger	pKa	[4, 7, 42]
MoKa	Molecular Discovery	LogP, LogD, pKa, intrinsic solubility, pH dependent solubility	[43]
Pallas	CompuDrug International, Inc.	pKalc, prologP and prologD	[4, 44]
QikProp	Schrodinger	Log P, Solubility	[40, 41]
Slipper	Timtec	LogP, LogD, aqueous solubility value at different pH	[45]
SPARC	ARChem	pKa, Log D, solubility at specific temperature and pH	[4, 14, 46]
StarDrop ADME	Optibrium	LogP, LogD, pKa, Aqueous solubility	[47]

Note: [a]Based on available information from their website

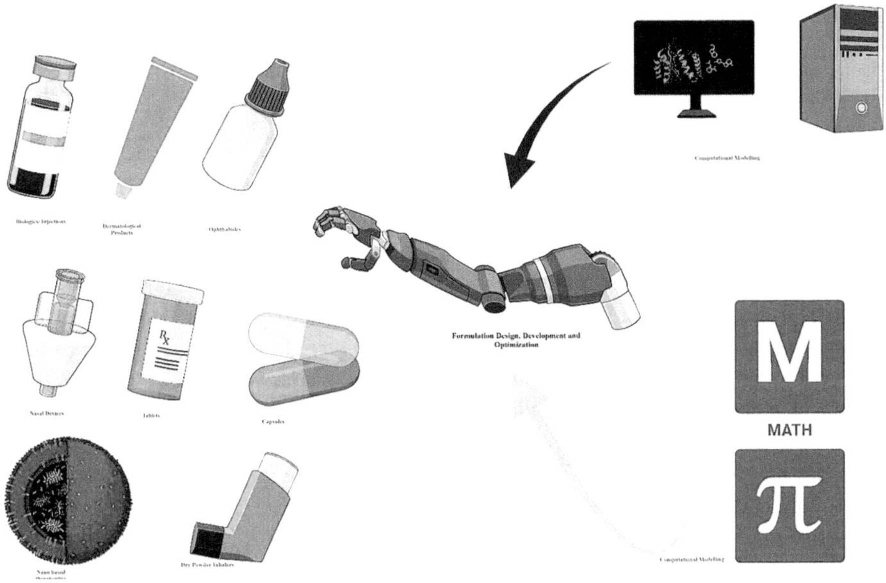

Fig. 2.1 Role of Computational modelling tools in design of pharmaceutical dosage forms

2.2 The Designing of Solid Oral Formulations

Solid oral formulations are predominantly tablets and capsules but also there are powders, granules dosage forms are available till date. Tablets are solid form of compact powder which contains excipients such as bulking agent, binder, disintegrant, lubricant, glidant along with the active pharmaceutical ingredient [48]. The formulation of tablets involves majorly direct compression, and granulation methods of which usage of wet binder in "wet granulation method" and dry binder in "dry granulation method". Further, wet granulation may be performed by using high shear mixing equipment or fluidized processing equipment. The dry granulation may occur either by slugging or roller compaction. A typical designing of tablet formulation involves selection of qualitative and quantitative composition, selection of granulation method, prototype formulation design, optimization of process parameters targeting desired drug release profile and finally packing in suitable primary packaging materials followed by stability evaluation. The aforementioned processes are almost performed simultaneously except the packaging material selection. In the initial stage of formulation design, the Raw material (Drug substance) physicochemical characterization which contains the information of aqueous solubility, particle size distribution (PSD), powder flow properties, compressibility, taste properties play a decisive role to select the excipients and design the tableting process and also by considering the desired drug release profile i.e., immediate release, sustained release, modified release, targeted release etc. [49]. The physicochemical characterization primarily involves, the powder characteristics which plays and decides the mixing, granulation, compaction and release properties.

2.2.1 Dry Powder Mixing

Mixing is a flow and dispersion process of a various type of particles that results in a mixer that is intended to contain uniformly distributed particles. It aims to make powder particles uniform and homogenous mixer. The homogeneity obtained in the mixing process is often carried forward to further steps and reach the final product and can directly impact the "critical quality attributes" of the finished product. The efficiency of mixing process is often highly critical in developing formulation of potent molecules (<10 mg) when direct compression method was employed for tablet preparation. Mixing is not limited to tablets but also related to other solid oral dosage forms of capsules, powders and granules. The major challenge posed in mixing is segregation of particles excipients and API which is due to one or more of variation of PSD, bulk density, differences in mass fraction of API and excipients, morphology of individual particles, particle shape, flow characteristics of excipients, amount of powder particles in the blender. The occupancy, rotation speed and time, type of blender, type of loading also impact the mixing efficiency of powder blending to a significant extent [48, 50–52].

In the event of insufficient evaluation of the mixing process efficiency of various offline and online testing methods of mixing efficiency currently available such as sampling probes, near infra-red image analysis, computational simulation was a bright spot on the way of understanding the mixing phenomena and designing an efficient mixing process.

In a simulation experiment with uniform sized plastic beads using rigid solid model predicted the effect of mixer rpm and occupancy for various loading pattern of material into a V-cone blender. It resulted in decoding the relation between initial loading pattern and homogeneity of mixing, mixing time and duration. Thus, successfully optimized the mixing conditions (loading pattern, mixing time, duration) for a combination of materials [51]. DEM model is the most studied method of simulating mixing phenomena followed by "Population Balance Model" (PBM) and CFD models [48]. Nearly four decades ago, mixing process was simulated in non-pharmaceutical industries by using DEM model [53]. It maintains a constant particle force and acceleration but changing the velocity of particles in a time frame during mixing [54]. A dead zone formation at the center of mixing equipment is due to the expansion of energy in an un baffled mixer at higher than 75% occupancy levels were understood from the experiment on 5–10 μ glass beads in a drum blender. The type of mixer has an impact on the mixing efficiency. The computational simulation helped to resolve the issues of dead zone formation and superior designing of blender with baffles for benefits of efficient mixing phenomena [50].

In another DEM based simulation experiment on drum mixer, by varying proportions of alumina particles (density of 3600 kg/cm^3) and Stainless steel (density of 7930 kg/cm^3) were found that the alumina particles were at walls of mixer and at the surface of the particle bed due to their lower density. The segregation mechanism was successfully identified as the difference between the densities of materials. The mechanism was approximated by the simulation trials and closely reflecting the practical results to obtain an efficient mixing [55].The importance of CFD model was more significant to assess the flow behavior at larger PSD and amount of powder blend is moderate in a complex particle sized blend [56].

2.2.2 *High Shear Granulation*

Wetting the powder blend by addition of binding agent and simultaneously kneading under high shear forces resulting in granule formation is referred as high shear granulation (HSG). A typical HSG involves process of wetting the powder blend with granulating fluid which result in nucleation followed by the consolidation and aggregation of formed nuclei resulting in granules and their size equalization by using chopper within the high shear mixing equipment (breaking the larger granules into smaller one) or by transferring wet or semidried granular mass into the mill preferably oscillating granulator mill or colloid mill by mechanism of (attrition) [57–59]. The small granules are dried till attainment of optimum moisture content sufficient for maintaining stability and compactability. Conventionally, the granulation end point detection, amount of liquid binding agent with respect to powder blend, kneading time, moisture content are critical process parameters in HSG which will have direct impact on the tablet properties. The optimization of granulation process is the most critical aspect of tablet formulation. The computational simulation of wet granulation through high shear was performed from last three decades by using models of PBM, DEM where the simulation process dealt with addition of liquid granulation agent, nucleation mechanism of granule formation and their growth by aggregation [60–62].

The granulation process was successfully simulated by DEM and PBM. As in the case of DEM model, the particle flow behavior and its response to variety of stress conditions along with the impact of surrounding particles are calculated by the equations at the individual particle scale. In a simulation Using DEM model, performing HSG with mass fraction of 96:04 of Microcrystalline cellulose and Hydroxy propyl cellulose, the mechanisms of aggregation and penetration of granulating liquid over the powder blend were evaluated to estimate the granule size and shown the substantial impact of liquid binding agent on the granule size. The deviation of size of 50 μ between simulated and experimental data observed [48, 57, 58, 63, 64]. The DEM model further demonstrated its suitability to near to perfect simulation of HSG by effectively predicting the process outputs with due consideration to shift in cohesive properties of powder blend with addition of granulating liquid, impact of impeller rpm, rate of addition of granulating liquid [65]. In another PBM approach, where the impact of impeller rpm, liquid to solid mass fraction, duration of granulation, powder porosity before granulation, viscosity of granulating liquid on the granules PSD and were successful with deviation levels of 3% with the experimental PSD data [63, 66].

Upon successful completion of laboratory development, the scale up of the HSG was also simulated successfully with PBM, by predicting the impact of scale on PSD and porosity. Overall, the process of HSG was better understood and optimized with the computational modelling tools and are able to reproduce the simulated results experimentally can boost the speed of the formulation development of tablets by wet granulation at lesser experimental costs and more efficient manner.

2.2.3 *Fluid Bed Granulation and Drying*

"Fluid bed granulation" (FBG) is the first-choice granulation for Immediate release tablets, as it generates softer granules when compared to HSG. The agglomeration of powder particles occurs by utilizing the hydrodynamics, where the powder particles are suspended in stream of air which results in a hydrodynamic bed of particles on to which liquid granulating agent (may or may not contain the binding agent) is atomized from top to bottom or tangential to the powder bed. Initially those particles coming into the path of atomized liquid granulating agent become wet and slightly sticky, to which other particles get adhered due to collision and resulting in agglomerates. The whole process occurs in a temperature-controlled environment. The powder particles require the granulating solution initially to get wet and then to become tacky to get adhered to other particles. The amount of granulating solution required is greatly varies depending on the wettability of powder. The atomization of granulating agent continues until the desired size of agglomerates were achieved. It is followed by drying of the agglomerates within the fluid bed equipment until reaching the desired moisture content. The greatest advantage of FBG is that agglomeration and drying can be completed in the same equipment which is not the case with other granulation types. The particles are agitated by the mechanism of fluidization and hence the agglomeration is termed as Fluid bed granulation. The optimization of a FBG process involves identifying critical process parameters of inlet air temperature, granulating solution viscosity and spray rate, atomization air pressure, humidity level in equipment, occupancy to achieve the targeted powder PSD, flowability and bulk density [67].

To powder particles interaction with other particles and walls of the equipment were modelled by DEM, PBM, CFD and also by combination of DEM and CFD [48, 68–70]. The movement of each individual powder particle and droplet of granulating solution during the FBG were studied by DEM model to evaluate the impact of Fluidization rate, spray rate, spray pattern on powder characteristics. The model has described the agglomeration mechanism and movement of powder particles and granulating solution droplets under the influence of air as continuous phase. The deviation of actual data from simulated results is a mere 4% [69].

The agglomeration behavior of 20 μm sized particles at the corners of the equipment were studied in quick simulation by using soft sphere model of DEM, which is a cheaper alternative to the expensive CFD-DEM model [67]. However, population balance models are widely used to simulate the fluid bed They capture the changes in powder PSD over time under complex variables such as diversified particle size, porosity, water content and quantitative composition of powder blend. Finite volume equations are popular method of calculation in PBM model [71–73]. The granule formation and its growth in FBG followed by drying mechanism are estimated using an antacid formulation containing HPC as binder and magnesium oxide, aluminium hydroxide gel as active ingredients and purified water as granulating agent. Accuracy of the simulation was proved with estimated particle size after agglomeration and after drying was estimated are in good agreement with production batch data [71].

2.2.4 Tabletting

The powder compaction into tablets is called tabletting, which mainly consists of die filling, compaction and ejection stages. Each stage of tabletting has the ability to alter the powder behavior and thereby impact the properties of obtained tablet. The computational modelling able to aid the researchers to understand the compactability, Elastic deformation and recovery, hardness of the tablet. DEM and FEM are widely used micro mechanical approaches to simulate the tabletting process by considering each and every granule or particle as an object and evaluated the powder behavior of contact, interaction, deformation. It also able to provide the physical information on powder flow from hopper to the turret, filling into the die. The DEM models might not be helpful in irregular shaped particle blend. The continuum models have provided the information on compactability and elastic deformation and recovery by deploying FEM. It also able to provide a view of tabletting defects.

The elastic behavior of powder during tableting was modelled by Drucker Prager cap model (DPC). Thus, these computational modelling tools enable the researchers to efficiently design and optimize the tabletting process in short time, with reasonable number of trials. The tabletting process of direct compression method was simulated by Dynamic systems simulation methods of which flow sheet simulation approaches were explored by researchers. The Agent based model (AB model) which is based on individual agents, environment around them and their behavior with other agents has been adopted to simulate the tabletting and able to successfully estimate the tensile strength of tablets with minor errors [74]. The critical process parameters of compression pressure, shape of die and turret speed also affect the tablet properties along with the powder flow behavior.

When it comes to designing a direct compression method of preparing tablets, a rarely utilized model at least to pharmaceutical design is Agent based model was used to simulate tabletting. It utilized two different particle size grades namely A60 (particle size of 60 μ 1.33 g/ml) and A150 (particle size of 150 μ0.72 g/ml) of Dicalcium phosphate anhydrous but keeping other attributes identical. The impact of each excipient mass fraction on tablet weight variation, variation in tensile strength of tablets which is proportional to out-die density, porosity of powder (relates tablet disintegration time) was simulated. The experimental results were closely reflected the simulated data with minor deviations. With the near to practically similar simulation of these critical quality attributes, give a positive hope on computational simulation incorporation in formulation development of direct compression process of tablet dosage forms [74].

2.2.5 *Film Coating of Tablets*

The core tablets after compression can be film coated for functional purposes (to protect the tablet from the factors of light, humidity, oxygen, to regulate the medicament release rate, to improve the solubility of drug in water or Gastrointestinal tract) or for non-functional purposes (aesthetic purpose by imparting a color, for identification when a product have multiple strengths, to mask the unpleasant taste or odor). The process of film coating involves suspension of coating materials in solvent (aqueous or non-aqueous) to prepare the film coating suspension which is intended to be sprayed onto a tablet bed in a coating pan. The concentration of film coating suspension, tablet shape average droplet volume, spray nozzle to tablet bed distance, inlet air temperature, product temperature, spraying rate, Pan rpm, curing duration are critical attributes in obtaining aesthetic film coated tablets [75, 76].

A modified CFD model named Lattice Boltzmann model was used to understand the parameters of film coating. Lattice-Boltzmann model (LBM) is a CFD technique that works at a mesoscopic level of simulation. It uses molecular dynamics and statistics principles to connect the continuous and discrete realms. It leverages the principles of "conservation of mass, energy and momentum". This unique approach enables the model to handle fluid-related challenges, including complex boundary conditions such as porous media, by simulating them directly [77]. When coating pharmaceutical tablets, the way water droplets interact with a porous surface can affect the appearance. This process involves three mechanisms: water spreading, absorption, and evaporation. The phenomenon of spreading and absorption following the impact of a droplet on a tablet was examined in a study utilizing the Lattice-Boltzmann method. The study found that the droplet spreads over time due to capillary action and is then absorbed by the tablet. This process happens quickly but follows a predictable pattern [78].

A study aimed to understand how different tablet shapes and process parameters affect the uniformity of intra- and inter-tablet film coating in a film coating pan. The study created a model to predict this impact. The results revealed that the tablet shape is the most critical factor affecting intra-tablet coating uniformity. The study also looked at how changes in the pan loading and pan rpm influence intra- and inter-tablet coating uniformity. The methodology used can help identify the optimal tablet shapes before the production of compression tooling and tablets. By optimizing pan speed and pan loading, inter-tablet coating variability can be reduced. Furthermore, the improved coating uniformity can enhance process efficiency and limit the variability in drug release kinetics [79].

In another DEM model-based simulation, the impact of process parameters pan speed, occupancy, nozzle quantity, spray rate on uniformity of film on tablet were evaluated. The most significant enhancement in uniformity within a practical scenario was found to be the result of augmenting the quantity of spray nozzles. Subsequently, higher pan speed with reduced fill level also contributed to improving uniformity, albeit to a lesser extent [80–82]. In a CFD model, where the impact of inlet air characteristics (temperature, atomization pressure, pattern air pressure) and nozzle position on the air flow within the coating pan throughout the process with the aim of reducing the spray solution loss. They identified that, instead of placing the nozzle perpendicular to the tablet bed, placing it distant to the inlet air helps in improving the spraying efficiency [83]. Tablet film coating process in operational view were depicted in Fig. 2.2.

2.2.6 Capsule Filling Process

Capsule filling is a process commonly used in dosator process. To achieve control over quality attributes such as fill weight and content uniformity, it is pivotal to understand the process comprehensively. The dosator process relies on a volume-based approach that can be affected by the dosator's volume and dimensions. In low-dose capsule filling, critical process parameters were investigated using DEM simulations with "Lactohale 100" modeling material. The mass decreases in narrow nozzle chambers because of increased wall friction effects. However, there is no direct correlation between fill weight and volume. Fill weight tends to increase with increased powder bed heights and by minimizing the space between the nozzle tip and the bottom of the container, the gaps become smaller, resulting in an increased level of powder densification. Altering the bulk density of the initial powder bed was explored to investigate variations in fill weight caused by bed inhomogeneity [84].

A team of researchers used DEM modeling to study how powder characteristics affect the evenness of fill weight for both cohesive and non-cohesive type of powders. They found that non-cohesive powders are not significantly influenced by the total height of the powder bed when it comes to the dosed powder amount. This means that maintaining a constant powder bed height is not as crucial in industrial processes that involve non-cohesive powders. However, cohesive powders are highly affected by the total height of the powder bed and the minimal distance between the nozzle and the bottom of the powder bed. The cohesive forces in these powders lead to a lower solid volume fraction in the powder bed, making them more susceptible to compression [85].

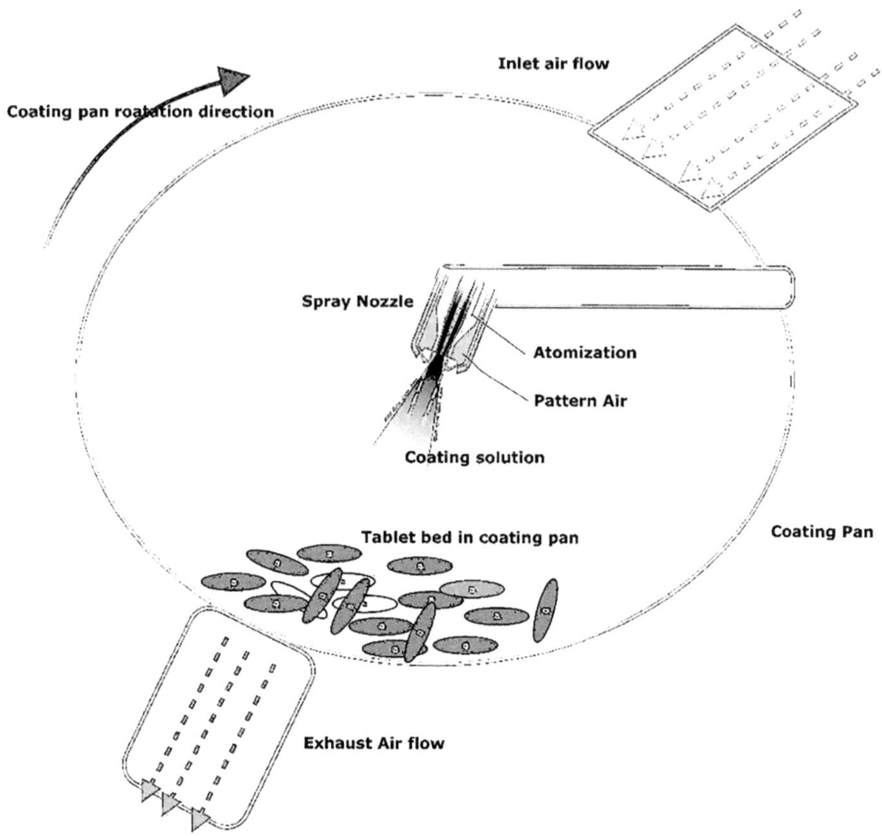

Fig. 2.2 Tablet film coating process in operational view

2.2.7 Invitro In-Vivo Correlation and Bioequivalence

The currently available software for developing IVIVC are but not limited to: "NONMEM®", "SAS®", "Sigma-Plot®", "Excel®", "PC_IVIVC®" and Phoenix "WinNonlin" [86, 87]. Decision making on Invitro In vivo correlation of Solid oral are shown in Fig. 2.3.

Not only for the tablets, but also long-acting complex injectables [88], dermatological products [89, 90] or any other dosage forms containing solid particles at the time of administration, the ultimate aim is to make the developed drug product bioequivalent to that of reference product. Bioequivalence is "when two drug products of same molar dose show similar or within acceptable limit bioavailability (rate and extent of drug absorption) profiles after administration." The number of drugs that are eligible for biowaiver (Not necessary to perform bioequivalence study) are low as only BCS class I & III molecules those having broad therapeutic index and

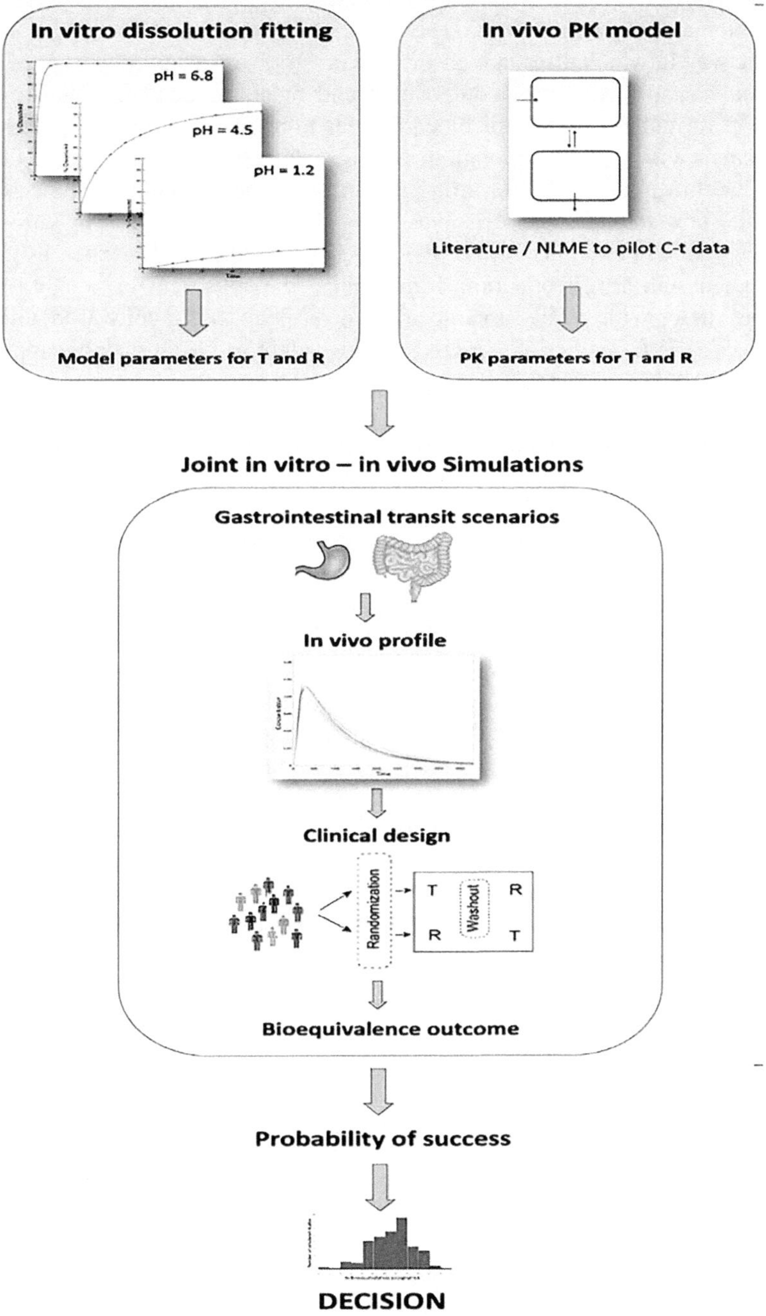

Fig. 2.3 Decision making on Invitro In vivo correlation of Solid orals

formulated as immediate release (IR) dosage forms are meant for systemic administration may be qualified for biowaiver [91].

The way in which drugs are absorbed can vary between products based on their qualitative and quantitative composition and manufacturing process. In order to develop formulations that are bioequivalent to a reference product, drug release performance is measured through in vitro dissolution testing in a lab. This test analyzes the bioequivalence of a drug product in the human body as it's essential to consider how fast the drug dissolves to be absorbed. It's called "In Vitro-In Vivo Correlation" (IVIVC). The goal of IVIVC is to decrease the expenses and time associated with drug production. It guarantees the consistency of quality between batches, helps with quality assurance during scaleup, post approval modifications, and uses IVIVC predictions instead of bioavailability. It ultimately supports biowaivers. IVIVS or Virtual Bioequivalence (VBE) are other terms used to describe bioequivalence predictions.

The pharmaceutical industry uses specific software to model drug absorption in different ways, such as orally, through inhalation, subcutaneously, or ocularly. These models can also predict drug effects in both animals and humans. Understanding the drug absorption process and how it affects performance can improve the design of drug formulations. Creating "biorelevant dissolution media" that mimics physiological conditions can help to improve IVIVC. Different models exist to characterize the different regions of the gastrointestinal tract, enabling the creation of more accurate dissolution media. Date from dissolution experiments can be incorporated into PBPK modeling, either mechanistically or non-mechanistically, depending on factors like the drug's formulation type, "Biopharmaceutics classification system" (BCS) class, and the size distribution of the drug substance. All of these parameters must be taken into account for PBPK modelling in order to create the most accurate model. Additionally, other software systems enable the replication of various elements of clinical trials, including variability, sample size, dosage schedules, and pathological situations, thus enabling the prediction of the probability of achieving positive outcomes. PBPK further equipped with Nonlinear mixed effects model [92], semi mechanistic tools to cater to unique requirements, such as the projection of systemic kinetics for inhaled corticosteroids administered via dry powder inhalers [91, 93–95].

2.3 Inhalations Formulation Design

"Dry Powder Inhaler" (DPI) and "Metered Dose Inhalers" (MDI) are portable inhalation dosage forms in pulmonary drug delivery. In the formulation development of DPI, there is requirement of optimizing both drug formulation and device. The powder particle flow characteristics, aerodynamic behavior of particles is crucial in formulation design of DPI. Whereas geometry of inhaler also equally important as its impact on particle flow behavior is significant. The flow structure and movement of particles in the device were numerically simulated by the CFD using turbulence models with "Reynolds-Averaged Navier–Stokes" (RANS) equations [96, 97]. The swirling flow of particles inside the inhaler were better simulated by large eddy simulations (LES), but it is not popular due to higher computational requirement.

It has been discovered that CFD can accurately predict the dispersion parameters of dry powder inhalers, which is strongly correlated with the aerosolization performance determined through experiments. Contrary to existing literature, it has been found that turbulence is proportionally increasing the aerosol size. The proposed design of the dry powder inhaler is characterized by a direct link between turbulence and aerosol size, utilizing a highly dispersible powder to enhance future performance. CFD analysis suggests that high turbulence causes powder to break up too quickly, hurting device performance. Keep turbulence low near the powder bed and high for secondary breakup. Current device exceeded 80% mean emitted dose and produced sub-2 μm aerosols. Future studies aim to reduce particle size to 1.5 μm or less while maintaining high dose. This improves delivery to lungs in challenging scenarios [98].

The treatment of lung diseases is crucial for maintaining a healthy and happy life. One essential aspect of this treatment is to understand how pharmaceutical aerosols are deposited in the human respiratory tract. Researchers conducted a study to model the deposition of aerosols from DPIs in an adult male's respiratory tract. The study found that larger particles tend to deposit outside the lungs, while particles within size of 2–4 μm deposit more in the airways and lungs. The study also found that a decrease in mouthpiece diameter increased deposition in the extra thoracic region, while larger particles showed increased deposition with an increase in mouthpiece diameter. These findings emphasize the importance of designing DPIs with proper mouthpiece diameters that ensure optimal deposition in the intended region.

Furthermore, it found holding the breath resulted in higher amount of deposition in the lungs. The modified version of the Exposure Dose Model 2 (ExDoM2) allows users to modify the model's parameters for a more realistic scenario [99]. The "Whole-Lung-Airway Model" (WLAM) approximates the complex human lung microstructure, making it useful for particle deposition analysis. In a study with delivering curcumin into lungs using DPI, the design of DPI was modified to target smaller lung airways, improving drug delivery. The research employed CF-PD methodology to replicate the transport and deposition of drug-aerosol in a whole-lung model under realistic breathing conditions, utilizing a commercial dry-powder inhaler to establish authentic inlet conditions [100, 101].

2.4 Ophthalmic Dosage Forms Design

The available treatment options for the eye diseases are in dosage forms of ophthalmic solutions, ophthalmic suspensions, ophthalmic emulsions, ophthalmic semi solids (Ointments, creams, gels), In situ gelling solutions all together simplified as ophthalmic dosage forms. The formulation characteristics often decides the physiological activity of the drug. Also, BE is required for the complex generic ophthalmic products (Suspensions, ointments) for which the knowledge of formulation behavior and disposition in the eye is valuable [102].

"Ocular Compartmental Absorption and Transit model", developed by "GastroPlus" ™, was used to analyze the performance of dexamethasone ophthalmic suspensions in rabbits. The model has taken consideration of the factors of particle size and viscosity in successfully predicting the non-linear pharmacokinetics of the drug in ocular tissues. A sensitivity analysis has shown that the interaction between formulation characteristics and physiological clearance mechanisms greatly affects the drug's bioavailability. The significance of particle size in the precorneal area is diminished by the quick elimination of drug particles. Although high viscosity levels do not impact ocular absorption in live subjects, viscosity below a certain threshold does have an effect. This study, which utilized FDA clinical pharmacology review data and information on rabbit pharmacokinetics, has deepened our knowledge of how physiological factors and formulation properties impact ocular drug pharmacokinetics in rabbits [103].

Developing medications for eye diseases is complex due to various factors affecting drug delivery. To achieve targeted therapy, understanding PK and PD is crucial. Cost-effective drug discovery focuses on advanced ocular biopharmaceutical computational models like OCUSIM, which can save costs. OCUSIM uses eye tissue parameters and drug properties to simulate drug pharmacokinetics after topical or intraocular administration, including TA. It accurately predicted drug concentrations in various eye tissues in studies with rabbits and patients. This modeling strategy offers valuable information on the distribution of drugs and forecasts the levels of concentrations in various eye tissues, a task that proves challenging to accomplish through experimental means [104–106].

An independent simulation study used a computational approach to predict the PK and PD of Timolol maleate suspension in rabbits after topical administration. The model accurately illustrated concentration-time and response-time profiles. This outcome is promising as it suggests the potential to acquire valuable information at a reduced cost compared to conducting expensive preclinical studies [106].

Considerable effort has been dedicated to the development of bio-erodible dosage forms that can sustain the release of drugs over an extended period. However, it is possible for a suspension to function effectively as a sustained release depot if it is injected into an area of the vitreous humor that is relatively still and maintains its gel-like consistency. In such cases, the drug solubility should be low enough to dissolve slowly, yet high enough to deliver a therapeutic amount of the drug. The amount and rate of medicament release can be regulated by adjusting the concentration in the suspension. This phenomenon arises due to a local concentration effect, where the drug dissolved from one particle hinders the dissolution of the drug from nearby particles. To maximize the effectiveness of this approach, the formulation must be designed to immobilize the particles until they are completely dissolved. To investigate the relationship between PSD and the time it takes for all solid drug to dissolve after injection, simulations were conducted using the "volume of fluid" (VOF) method in ANSYS Fluent. These simulations involved various model suspensions of triamcinolone acetonide (TAC). The results demonstrate the time-dependent nature of drug content for two different suspensions. The predictions align reasonably well with the experimental data for the 16 mg depots, although they underestimate the duration of the 4 mg depots [107].

By using non-invasive in silico modelling and simulation, an understanding of drug delivery mechanisms is possible and can streamline drug formulation design without invasive studies. But it is crucial to consider data challenges, model feasibility, personalized modelling, regulatory considerations, and interdisciplinary collaboration when implementing this framework. By doing so, we can ensure that this paradigm shift results in effective drug delivery systems for ocular formulations, which will help us tackle the diverse diseases that affect the eyes and improve the quality of life of millions of people worldwide. For more information on it is recommended to refer Nannan Wang et al. [108].

2.5 Dermatological Drug Products

Topical medications are commonly found in semisolid forms such as creams, ointments, gels and are stored in aluminum tubes with screw caps for easy use. These tubes allow patients to apply different amounts of medication to various areas of the skin. However, it's important to remember that not all topical medications are semisolid or unmetered. Some are in the form of solutions, dispensed using metered dose pumps or patches. When dispensing and applying semisolid dosage forms, the microstructures can change due to shear forces, affecting drug diffusion and distribution on the skin. Additionally, formulation components may adsorb to pump components, altering the composition and bioavailability of the medication. Understanding these characteristics and potential changes is essential for ensuring the effectiveness and proper administration of topical medications [109].

Availability of very few generic options can be a significant barrier for patients who require affordable treatment options. However, the challenges in demonstrating bioequivalence and therapeutic equivalence to the brand name medication have made it difficult for generic products to gain approval. In the past, proving bioequivalence required a clinical endpoint study that was often expensive, time-consuming, inefficient, and risky for generic drug developers. As a result, patients were left with limited options and had to bear the burden of high treatment costs. Moreover, the only established method for demonstrating bioequivalence was limited to topical steroid products, further limiting the availability of generic options. Even when a generic product was available, it often differed in appearance, texture, and/or smell from the reference product, leading to confusion and mistrust among patients. Therefore, there is an immediate need for more efficient and economical methods to demonstrate bioequivalence and therapeutic equivalence for topical dermatologic medications, so that patients can have access to affordable and effective treatment options [110]. GDUFA I and II funded scientific research on dermatological drugs applied to the skin. This research focused on regulatory outcomes to help make decisions. Recently, effective bioequivalence strategies have been incorporated into product-specific guidance, leading to the approval of better generic topical dermatological drugs. Raney et al.'s 2020 study highlights this success [110].

New methods for assessing Bioequivalence (BE) involve in vitro experiments and characterization tests to compare generic drugs with reference listed drug products. The FDA's GDUFA research program focuses on using quantitative techniques like PBPK modeling. These models consider drug product characteristics to predict in vivo performance. By considering product "critical quality attributes" and interactions with skin, these models help make informed decisions in product development. They also allow for alternative BE strategies on individual basis [90, 111, 112].

PBPK modeling predicts drug processing in the human body after administration. Dermal PBPK models focus on drug absorption and distribution on the skin of virtual human subjects, considering factors like product quality, drug properties, and skin characteristics. These models provide insights for product development and regulatory agencies [90]. Skin exposure to various products, drugs, and chemicals can result in their entering the bloodstream and causing adverse effects. Therefore, assessing dermal penetration and bioavailability is crucial in evaluating chemical risks. The skin is a complex organ that can block and store substances. Animal testing is traditionally used, but new legislation mandates alternative testing [113].

The Gastroplus™ software has transformed drug absorption prediction through the skin with the TCAT™ model, revolutionizing the field. It has been used to evaluate vinorelbine hydrogel formulations for transdermal administration in melanoma treatment and determine the optimal transdermal dose of risperidone for desired drug plasma levels. Additionally, in a study by Mahdi et al. [114] demonstrated how the TCAT™ model, when combined with in vitro experiments, helped in choosing solvent for subcutaneous administration of rifampicin [115].

2.6 Sterile Dosage Forms (Injections)

Profound comprehension of the physicochemical attributes of both the drug and polymer, in addition to their interaction, is imperative for the creation of sustained-release injectable formulations containing polymers to achieve the targeted drug release profile. The data driven approach of machine learning with 11 algorithms was implemented to accurately predict the drug release profile of Olaparib drug molecule with polylactide-co- glycolide microparticles, of which light gradient boosting machine algorithm has provided the most accurate results with a correlation of about 92% with experimental data [116].

Formulation design of lyophilized injection of small and large molecules involves optimizing the formulation composition, fill volume/height and critical process parameters of lyophilization (shelf temperature, cooling rate and duration at Freezing stage; shelf temperature considering collapse temperature and chamber vacuum during primary drying stage; water content in secondary drying stage).

Another major challenge for successful development of lyophilized products is the technology transfer from laboratory lyophilizer to the industrial scale lyophilization where the lab scale recipe often causes scale up issues with respect to drying times, collapse, cake appearance, heterogenic product characteristics (appearance, water content, partial collapse, cosmetic defects etc.), uncontrollable pressures etc. [117].

The criticality of lyophilization lies in understanding the product collapse temperature, dried product resistance, product mass transfer coefficient, vial heat transfer coefficient (often denoted as K_v). In a CFD based quasi-steady modelling, simulation in MATLAB® using fMINCON function, the heat and mass transfer coefficients, choked flow limits of both laboratory scale and production scale lyophilizers were determined with an acceptable error of less than 6% for a protein formulation containing tris-HCl buffer [117].

In another study, CFD simulations were utilized to improve the fathom of flow dynamics and ice condensation and deposition in a condenser. The main goal of this study was to assess the effectiveness of the condenser and gain a more comprehensive understanding of the process to pinpoint potential enhancements for its design. The research involved analyzing both a laboratory-scale freeze-drying apparatus and an industrial-scale condenser, with a particular emphasis on modeling the deposition of water vapor during the process. Various operational conditions were examined, and the effects of inert gas and other parameters were explored. [118, 119]. In a stochastic analysis of mannitol solution lyophilization, using lyophilization solver, the primary drying conditions were predicted with ±0.6 °C temperature variation and 6% drying time variation which is valuable in the context of design space establishment [120].

The study used CFD to examine how emulsions interact with various factors such as elastic modulus, average droplet size, zeta potential, stability, and energy incorporation measurements. The findings revealed spatial variations in relative viscosity, droplet size, and dispersed phase volume fraction in three dimensions, as well as complex flow patterns for two different impeller designs [121]. A model based on principles of physics was developed to examine the heat transfer in the vicinity of a shelf's edge. The model took into account radiation, conduction within solids, and conduction through low-pressure water vapor. The Multiphysics software "COMSOL" was used to simulate the phenomena of five vials near the metallic guardrail. The accuracy of the model's heat flux predictions was validated by comparing them to experimental data collected at different shelf temperatures and chamber pressures. The enhanced heat transfer observed in the border vials can be attributed to conduction through low-pressure water vapor. This model is a valuable tool for studying freeze-drying processes and reducing the need for expensive experiments [122].

Also, in the development of monoclonal antibody formulation, the freeze thaw behavior of the composition at a specified temperature with time were successfully simulated using CFD using OpenFOAM at larger scale has provided a great value to the design of mAb [123, 124]. Designing appropriate formulations to stabilize monoclonal antibodies poses several challenges, notably in the areas of preventing aggregation and managing viscosity. Molecular modeling and simulations have the potential to enhance pharmaceutical products by investigating the interactions between monoclonal antibodies and various compounds, including excipients.

In the realm of large molecule development, the rapid creation of a protein formulation composition can be achieved through sequential design of experiments utilizing the Bayesian optimization algorithm. This process involves the utilization of the "SKOPT" and "GPFLOWOPT" packages in Python, showcasing an application of machine learning techniques [125]. The development of targeted drug delivery systems using nanoparticle formulations has been greatly enhanced by the application of computational models and simulations. These simulations have provided valuable insights into several critical aspects of the design process, such as the selection of correct combination of excipients, particularly polymers and lipids, the prediction of miscibility and solubility parameters, the evaluation of drug encapsulation ability, the determination of drug release rates, the prediction of stability, and the investigation of nanoparticle transport through intricate networks of blood vessels. Additionally, these computational models have contributed to gain insights of recognizing and binding to the target, which are crucial for achieving effective drug delivery. The incorporation of computational models and simulations has greatly enhanced the progress of injectable nanoparticle formulations for precise drug delivery [126, 127].

Combining PBPK or QSP models with MD models can describe a compound's disposition, activity, and interaction with different excipients in the body across multiple computational techniques. QM evaluates reactions between atoms and molecules, helping predict pharmacological activity and toxicity. Combined with MD models, QM models improve understanding of interactions between biological structures and nano formulations. Machine learning (ML) predicts drug efficacy, toxicity, and disease diagnosis. Unsupervised learning discovers disease and target subtypes based on input data alone. Incorporating these techniques optimizes drug formulation for maximum efficacy and safety [128]. A further reading on formulation development of complex injections and other complex drug products please refer Walenga et al. [129].

2.7 Nano Based Therapeutic Systems

The formulation design of Nano based therapeutic systems by utilizing computational tools involves critical aspects of selecting suitable excipients and their compatibility, suitable and miscible solvents and/or combination selection, estimation of drug loading on the nano based carriers, estimation of drug release from the nano based carriers, Transportation and cell membrane interaction of nano based therapeutic systems (protein corona formation). By optimizing the stages parallelly, the objective of designing nano based therapeutic systems can be fulfilled. Critical stages of optimizing Nano based therapeutic systems were shown in Fig. 2.4.

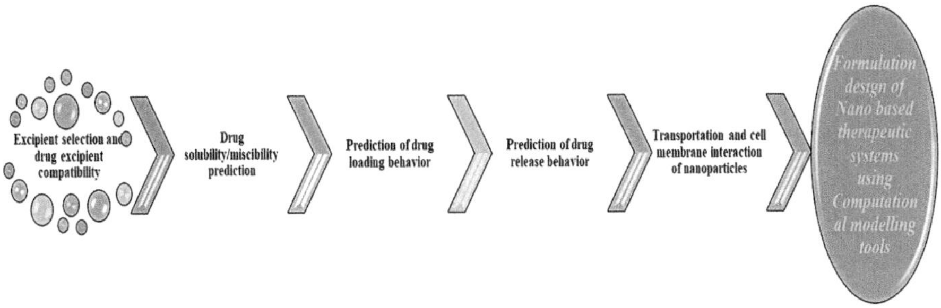

Fig. 2.4 Critical stages of optimizing Nano based therapeutic systems

2.7.1 *Excipient Screening and Assessment of Compatible Excipients*

Choosing the right excipients from the available set of excipients and assessing their compatibility with drug substance are crucial aspects in nano formulation development, as they can help overcome challenges such as insufficient stability, minimal drug loading abilities, and lower drug retention. Efforts are currently focused on selecting the right excipients and ensuring compatibility with the drug. Computer simulations have emerged as a valuable tool in this process, replacing traditional trial and error methods. These simulations help identify suitable excipients for drug encapsulation, accelerate the screening process and identifying the best combination of excipients for optimum drug loading.

By employing computational simulation tools, researchers can evaluate the compatibility of a drug with different excipients through pre-screening, facilitating the development of an efficient drug delivery platform. Coarse-grained methodologies like "Dissipative Particle Dynamics" (DPD) and the "MARTINI force field" have been utilized to investigate interactions between drugs and excipients. While DPD simulations can accurately predict interactions between drugs and polymers, especially for hydrophobic drugs like prednisolone and paracetamol, they are less effective in modeling interactions with hydrophilic drugs such as isoniazid. Conversely, the "MARTINI force field" proves advantageous in modeling polymer self-assembly within micelles and the dispersion of pharmaceuticals throughout these micelles [126, 130]. In order to develop a highly effective polymeric carrier, researchers utilize thermodynamic and molecular simulation methods to forecast Flory-Huggin interactions and assess drug-polymer compatibility. Through the integration of these computational tools, scientists are able to effectively evaluate various polymers in conjunction with specific drugs, ultimately leading to the creation of optimized polymeric micelles that exhibit improved drug solubility [131].

2.7.2 *Forecasting the Drug Solubility and Excipient Suitability*

The utilization of computational modeling has gained significant acceptance as a reliable approach for forecasting solubility and miscibility within the pharmaceutical sector [16, 132]. Within the realm of formulation development, it is crucial to utilize in silico methods to analyze solubility and miscibility parameters. The assessment of various compounds for their compatibility medicament and having "thermodynamic stability" and "retention characteristics" is facilitated by this process. The determination of solubility and melting point of drugs and excipients, as well as drug-polymer interactions, heavily relies on the "cohesive energy density" calculated using the "COMPASS force field". These simulations offer valuable molecular information for choosing key excipients, thus aiding in the systematic advancement of formulations [133].

Employing full atomistic simulation techniques, parameters such as the "Hildebrand solubility parameter," the "Group Contribution Method" (GCM), and "MD simulation" are utilized to facilitate the selection of right combination of excipients and the evaluation of drug-excipient compatibility in order to develop a stable and efficient nano based therapeutic. The challenge lies in determining the optimal combination of anti-solvent and solvent for formulating a nanosuspension that contains drugs with low solubility [134]. Additionally, MD simulations were carried out on the anti-retroviral drugs to investigate the occurrence of phase separation and the formation of nanoparticles. It was observed that molecular simulations diminish the requirement for extensive experimental work and supports in the selection of the right combination of solvents and excipients for the development of nanoparticles [135].

2.7.3 *Anticipation of Drug Entrapment*

The utilization of computational-assisted design in drug formulations incorporates the notion of "Cause-Effect" relationships and showcases a higher level of concordance with experimental data when compared to factorial designs. Within this framework, ANN [136], is utilized to optimize amount of polymer, "coacervation agent solution concentration", stirring duration, and "spraying rate" in benzimidazole chitosan microparticles and was successfully accomplished. A meticulous examination is conducted to comprehensively analyze the influence of the drug-to-lipid quantity ratio, surfactant concentrations, and drug loading efficiency of the carrier on the synthesis of drug-loaded polymer lipid hybrid nanoparticles. This method ensures high efficiency and reliability in achieving the desired drug loading and optimal size of the nanoparticles. This method integrates ANN, "Response surface methods", and "continuous genetic algorithm" to enable a comprehensive multi-objective analysis [137, 138].

A recent study examined how surface functionalization affects the drug loading capabilities of mesoporous silica nanoparticles. The results showed that amine-functionalized nanoparticles had increased thickness and more hydrogen bond interactions compared to unmodified nanoparticles. Additionally, the amine-functionalized nanoparticles had the highest adsorption energy, indicating their potential as an efficient drug delivery system. These findings have significant implications for improving drug delivery systems and providing more effective treatments for patients [139].

Efficient drug delivery is critical in treating many diseases. Recent research has explored how a drug molecule behaves within a silicon nanocarrier and how loading can be facilitated. Using the Arguslab molecular docking software, researchers predicted the interaction between drug and surface of silicone nanocarrier by measuring interaction energy. Results showed that larger molecules had decreased interaction energy, facilitates loading more easily. These valuable molecular insights highlight the potential of simulation calculations in drug delivery research. By optimizing drug loading, we can improve treatment efficacy and patient outcomes [140].

Molecular docking studies are essential for the progress of pharmaceuticals, as they provide valuable insights into hydrogen bonding, electrostatic interactions, and hydrophobic interactions. These investigations help us understand the complex relationships between drugs and carriers at the atomic and molecular levels, improving our understanding of polymer-drug complexes. Furthermore, by offering computational binding energies, these studies aid in improving drug loading capacity and encapsulation efficiency in "polymer-based host nano systems", ultimately leading to more effective drug delivery. Without molecular docking studies, there would be a significant knowledge gap in drug development, highlighting their crucial role in modern medicine [141].

2.7.4 Anticipation of Drug Release

It is a requisite for a nano based therapeutic to have controlled drug release in order to be effective. The way nanoparticles loaded with drugs are distributed, accumulated, and diffused is expected to impact how the drugs are released. The release of drugs from polymeric nanoparticles is regulated by studying how nanoparticles are transported and trafficked within cells to reach the intended target. By using simulations, researchers can study how various features of nanoparticles, such as size, surface properties, affinity for blood vessels, diameter, and the presence of ligands and receptors on the surface, directly affect how nanoparticles are transported and trafficked within tumor blood vessels. This comprehension can then be utilized to analyze the subsequent drug release from the nanoparticles [142].

The PK-Sim® software is a unified simulation model that incorporates the oral ADME description. It constructs a model structure that is sophisticated enough to ensure a realistic representation of all processes. This structure also ensures that the simulation outcomes can be compared with pharmacokinetic experiments. In

"PK-Sim", absorption via oral route is modeled using a "plug flow with dispersion model." In this model, the small intestine is seen as a unified entity with various functions affecting substance transportation. These include a gastric-release function dependent on feeding status, a transit function controlling substance progression, and quantification of substance assimilation into the portal vein over time intervals. For solid formulations, the software uses predefined release functions like "Noyes-Whitney" type dissolution kinetics to elucidate the release of the drug into solution. Different organs are depicted in simulations with their corresponding rates of blood flow, permeability across membranes into the tissue of the organ, and processes of saturable metabolization. This is done to precisely forecast the destiny of a substance in subsequent pharmacokinetic stages [143].

Since 2007, CFD based simulations were gradually increasing for studying drug release pattern from polymeric microspheres using computational fluid dynamics (CFD). These models are classified into three categories based on controlling mechanism i.e., diffusion or swelling or erosion, depending on the properties of the polymeric matrices and drug release behaviors. Drug release is regulated by chemical reactions and mass transfer, influenced by factors like polymer and drug nature, device dimensions and shape, composition, and encapsulation methods [144].

Molecular Dynamics (MD) simulation models pH changes by adjusting partial charges on atoms. It is ideal for studying drug release and system behavior. MD simulation approach provides a deep understanding of each model's strengths, weaknesses, and suitability for investigating different aspects of the system [145].

2.7.5 Interaction Between Nanoparticles and Cell Membranes During Transportation

Nanoparticles interact with cell membranes, navigate through the bloodstream, and internalize within cells. The process is complex due to the complicated network of blood vessels, intricate in vivo transport mechanisms, and nanoparticles' chemical and physical properties. Investigating selective targeting and visualizing the delivery mechanism is challenging with current experimental tools and technologies. Nonetheless, researchers continue to explore innovative solutions to improve drug delivery [146, 147].

Nanoparticles are promising in biological systems but optimizing them at a molecular level is challenging. Computational simulation offers a valuable prescreening tool. Fullstone et al. used CFD to investigate nanoparticles' behavior under blood flow conditions and examine their distribution. By exploring various factors, they aimed to determine their influence on selective targeting. Using an agent-based model within the FLAME network, the researchers simulated how nanoparticles and red blood cells behave under different conditions. Their findings provide valuable insights into the factors affecting selective targeting and may lead to more effective nanoparticle formulations [148].

By conducting simulations, researchers explored how surface modification impacts the protein adsorption and cellular uptake of nanoparticles. The study's findings indicated that nanoparticles coated with zwitterionic polymers exhibited a stronger affinity and binding to the cell membrane, leading to enhanced active targeting. These findings could lead to safer and more effective nanoparticle-based therapies [149].

The protein corona formation on nanoparticles is a complex process influenced by surface characteristics, with a polymer corona potentially prolonging their presence in the bloodstream. Researchers are using experimental analysis and simulations to study how protective polymers like PEG and poly-sarcosine interact with proteins naturally present in blood by taking into consideration of amino acid charge and polarity. This approach may lead to new polymer materials with improved protective properties. Another study by Lee et al. used the MARTINI model to explore how PEGylation and density impact liposome-protein interactions, while molecular dynamics simulations looked at electrostatic interactions of nanoparticles in the formation of protein corona [146, 150]. For a further reading on nanoparticle formulation development by using computational modelling please refer to the Jaysinghe et al. [151] Yu et al. [152], Bunker et al. [145].

2.8 Biologics

Development of formulations for biologic drugs involves the intricate process of identifying the optimal method to deliver active ingredients in a stable form to the target patient population. Formulation of drugs plays a pivotal role in the product life cycle, spanning from initial clinical trials to the eventual launch of the commercial product. The choice of inactive ingredients, known as excipients, is crucial to maintain the efficiency of the biologic drug throughout the shelf life. At the early stages of development, a thorough evaluation and understanding of biochemical and physical properties of the bio molecules to design a formulation that is safe, efficacious, and stable. Consequently, a comprehensive evaluation of the biomolecule and the biochemical characteristics of the drug product is imperative to understand the potential effects of various excipients on stability and critical quality attributes. The constantly evolving landscape of molecular structures proposed as potential drug candidates adds complexity to the field of biopharmaceutical formulation development. Apart from antibodies, biotherapeutic formats now encompass enzymes, cytokines, polypeptide hormones, recombinant fusion proteins, and other variations. This wide array of molecule classes poses a challenge in initiating a formulation project, as existing information from commercial products may not be directly applicable to novel molecules under development [153].

Engineering therapeutic antibodies involves techniques such as "Homology modelling" and "Protein-protein docking". The process kickoff with Lead Identification, which involves generating a diverse pool of potential molecules through methods like "animal immunization" or "surface display" technologies. In the next stage, which is Lead Optimization phase, lead candidates having high affinity are selected through screening and design. Undesirable traits like immunogenicity or poor biophysical properties are carefully evaluated. Utilizing computational methods like homology modeling and docking enables the construction of 3D antibody models and the prediction of essential residues crucial for antigen binding [154, 155].

Numerous methods for designing antibodies in experimental settings often provide sequence data without corresponding structural information. In the context of extensive screenings, it becomes unfeasible to individually ascertain the structures of all antibody loops and their interactions with the antigen. Antibody structures have traditionally been forecasted through the utilization of physics-based modeling techniques [156], such as MD simulations [154], or through homology-based modeling approaches [157]. Another approach involves the implementation of a combined method, such as MODELLER [158], which integrates both physics-based and homology-based modeling principles. For enhanced understanding on antibody therapeutic development, please refer to Khetan et al. [159]. Critical aspects of Biologic molecules are depicted in Fig. 2.5.

Herewith are the computational tools that are used in design of biologic molecules as per the published literature [159, 160]. Programs and Software for development of biologic molecules are mentioned in Table 2.2.

Computational simulation is essential in the early stages of biotherapeutics development, especially when resources are limited and various formulation strategies need to be investigated. This paper presents a detailed examination of various computational methods, ranging from "All-atom simulations" to "Macro-scale mathematical descriptions", that have proven successful in replicating adverse physical phenomena observed in concentrated protein solutions. These methods enable the study of physical instability occurrences, such as protein aggregation, heightened viscosity, and phase segregation, providing valuable insights for the design more efficient medications. The article presents a comparison and summary

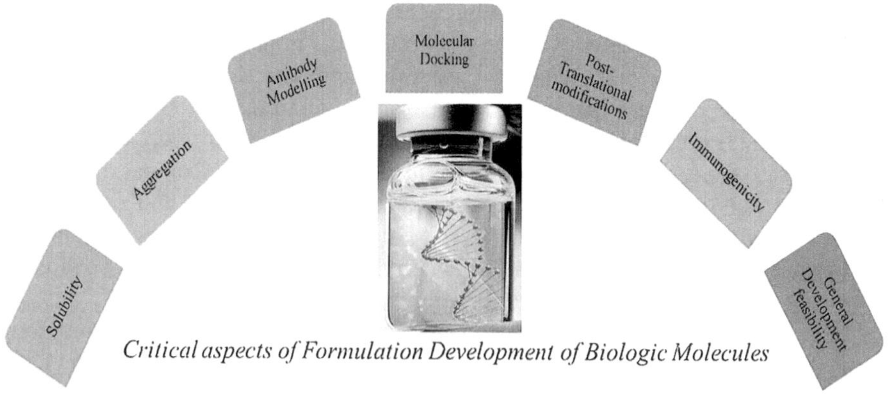

Fig. 2.5 Critical aspects of Biologic molecules

Table 2.2 Programs and Software for development of biologic molecules

S. No	Test parameter	Useful Computational tools "Sources"
1	Solubility	"Cam Sol" "http://www-cohsoftware.ch.cam.ac.uk/"
		"Protein-Sol" "https://protein-sol.manchester.ac.uk/"
		"SODA" "http://old.protein.bio.unipd.it/soda/"
		"SOLpro" "http://scratch.proteomics.ics.uci.edu/explanation.html#SOLpro"
		"SOLart" "http://babylone.ulb.ac.be/SOLART/"
		"PROSO II" "http://mips.helmholtz-muenchen.de/prosoII"
2	Aggregation	"Generalized Aggregation Proneness (GAP)" "https://www.iitm.ac.in/bioinfo/GAP/"
		"Spatial Aggregation Propensity" SAP
		"Solubis" "http://solubis.switchlab.org/"
		"Agg score" "https://www.schrodinger.com/science-articles/aggregation-prediction-protein-surface-analyzer"
		"AGGRESCAN 3D" "http://bioinf.uab.es/aggrescan/"
		"PASTA 2.0" "http://old.protein.bio.unipd.it/pasta2/"
		"TANGO" "http://tango.crg.es/"
		"AnuPP" "https://web.iitm.ac.in/bioinfo2/ANuPP/"

(continued)

Table 2.2 (continued)

S. No	Test parameter	Useful Computational tools "Sources"
3	Antibody Modelling	"abYmod" "http://abymod.abysis.org/"
		"ABangle" "http://opig.stats.ox.ac.uk/webapps/newsabdab/sabpred/abangle/"
		"ABodyBuilder" "http://opig.stats.ox.ac.uk/webapps/abodybuilder"
		"Repertoire Builder" "https://sysimm.org/rep_builder/"
		"MODELLER" "https://salilab.org/modeller/"
		"RosettaAntibody" "https://new.rosettacommons.org/docs/latest/application_documentation/antibody/antibody-applications"
		"PIGS" "https://bio.tools/pigs"
		"MOE" "https://www.chemcomp.com/Products.htm"
		"LYRA" "http://www.cbs.dtu.dk/services/LYRA/index.php"

4	Molecular Docking	“DockThor” “https://dockthor.lncc.br/v2/”
		“SwissDock” “http://www.swissdock.ch/”
		“HADDOCK” “https://wenmr.science.uu.nl/haddock2.4/”
		“MEGADOCK” “https://www.bi.cs.titech.ac.jp/megadock/”
		“RosettaDock” “https://new.rosettacommons.org/docs/latest/application_documentation/docking/docking-protocol”
		“FTDock 2.0” “http://www.sbg.bio.ic.ac.uk/docking/ftdock.html”
		“AbAdapt” “https://sysimm.org/abadapt/”

(continued)

Table 2.2 (continued)

S. No	Test parameter	Useful Computational tools "Sources"
5	Immuno-genecity	"ANTIGENpro" "http://scratch.proteomics.ics.uci.edu/explanation.html#ANTIGENpro"
		"RANKPEP" "http://imed.med.ucm.es/Tools/rankpep.html"
		"COBEpro" "http://scratch.proteomics.ics.uci.edu/explanation.html#COBEpro"
		"BEpro (PEPITO)" "http://pepito.proteomics.ics.uci.edu/"
		"Disco Tope" "http://www.cbs.dtu.dk/services/DiscoTope/"
		"ElliPro" "http://tools.iedb.org/ellipro/"
		"SVMTriP" "http://sysbio.unl.edu/SVMTriP/"
		"AbAdapt" "https://sysimm.org/abadapt/"
		"EpiPred" "http://opig.stats.ox.ac.uk/webapps/newsabdab/sabpred/epipred/"
		"NetMHCIIpan" "http://www.cbs.dtu.dk/services/NetMHCIIpan/"
		"MHCEpitopeEnergy" "https://new.rosettacommons.org/docs/latest/rosetta_basics/scoring/MHCEpitopeEnergy"
		"Hu-mAb" "http://opig.stats.ox.ac.uk/webapps/newsabdab/sabpred/humab"
		"TOPKAT" "https://www.toxit.it/en/services/software/topkat"
		"MetaDrug" "https://support.clarivate.com/LifeSciences/s/article/MetaDrug-Uses-and-benefits?language=en_US"

6	Post Translational Modifications	"MusiteDeep" "https://github.com/duolinwang/MusiteDeep_web"
		"PTM prediction tools survey" "http://www.cbs.dtu.dk/databases/PTMpredictions/"
		"MUpro" "http://mupro.proteomics.ics.uci.edu/"
		"FindMod" "https://web.expasy.org/findmod/"
		"SIDEpro" "http://sidepro.proteomics.ics.uci.edu/"
		"SCWRL4.0" "http://dunbrack.fccc.edu/scwrl4/SCWRL4.php"
		"PEARS" "http://opig.stats.ox.ac.uk/webapps/pears"
7	General Development Feasibility	"Therapeutic Antibody Profiler (TAP)" "http://opig.stats.ox.ac.uk/webapps/newsabdab/sabpred/tap"
		"abYsis" "http://www.abysis.org/"
		"NaturalAntibody" "https://naturalantibody.com/antibody-analytics/"
		"AbMapper" "http://abmapper.sourceforge.net/"
		"Developability Index" "http://i.uestc.edu.cn/SSH2/"
		"Abpred" "https://protein-sol.manchester.ac.uk/abpred"
		"QikProp" "https://www.schrodinger.com/products/qikprop"

(continued)

of these models within the scope of the underlying physical mechanisms, highlighting their respective presumptions and restraints. Additionally, the article provides a thorough analysis to identify the protein interaction mechanisms explicitly or implicitly incorporated in various modeling strategies, emphasizing their relationships with different formulation parameters as discussed in Marco A. Blanco's work [161].

The foundation of biologic drug development lies in proteins, which respond to changes in their environment by altering their conformational states. By constructing models that simulate the impact of environmental stresses on protein energy landscapes, it becomes possible to predict the critical quality attribute of a biologic drug. The integration of biopharmaceutical informatics requires modest investments, but it ultimately leads to economically viable biologic products. At the kick of phases of drug development, researchers gain access to protein sequence information through libraries that encompass potential binders for a specific drug target. This crucial data plays a significant role in identifying potential vulnerabilities to physicochemical degradation, regions that are prone to aggregation, and immune epitopes. To understand how the physical properties of biologics change in response to the environment, atomistic structural models are essential. Multiple algorithms exist for predicting structure of proteins, which combine bioinformatics and molecular modeling techniques. These structural models provide clear insights up to the molecular levels to understand the mechanisms underlying protein functions and solution properties, which are manifested to be valuable tools in structure-based drug design. One of their applications is the identification of potential sites of degradation during the bioanalytical characterization of molecules [162].

2.9 Conclusion

Harnessing the power of computational modelling to predict properties in preformulation, drug excipient compatibility, excipient selection, oral, parenteral, and inhalation dosage form development will accelerate studies in formulation development. Key considerations in selecting a model are accurate prediction, model performance and reliability, and the application of data to a wide range of drugs. The plethora of available models exemplifies the importance of computational modelling in formulation design However, there is still a lot of potential for advancement. Most prediction methods rely on publicly available data for training and validation, which has an impact on the performance of the model. There is a need for high-quality and diverse databases. Every model has some advantages and limitations. No single model or algorithm can fit all applications. A best-fit model should be developed based on the application. New models should be developed, and continuous improvement in existing models is crucial. There should be continuous effort to build the dataset, train the models, and validate them against in vitro experiments. The accurate prediction is facilitated by the use of modern analytical techniques in tandem with computational models. Pharmaceutical field has complex regulatory framework. Hence, models should comply with the highest quality and regulatory standards. In formulation arena, the transition from traditional experiments to computational modelling is slowly emerging. Future prospects of computer modelling in formulation design are bright.

References

1. Ahirwar K, Shukla R. Preformulation studies: a versatile tool in formulation design. In: Drug formulation design. London: IntechOpen; 2023. https://www.intechopen.com/chapters/86211
2. Majee SB, Biswas GR. Computational methods in preformulation study for pharmaceutical solid dosage forms of therapeutic proteins. Phys Sci Rev. 2017;2(6).
3. Desu PK, Vaishnavi G, Divya K, Lakshmi U. An overview on preformulation studies. Indo Am J Pharm Sci. 2015;2(10):1399–407.
4. Liao C, Nicklaus MC. Comparison of nine programs predicting p K_a values of pharmaceutical substances. J Chem Inf Model. 2009;49(12):2801–12.
5. Işık M, Rustenburg AS, Rizzi A, Gunner MR, Mobley DL, Chodera JD. Overview of the SAMPL6 pKa challenge: evaluating small molecule microscopic and macroscopic pKa predictions. J Comput Aided Mol Des. 2021;35(2):131–66.
6. Gozalbes R, Pineda-Lucena A. QSAR-based solubility model for drug-like compounds. Bioorg Med Chem. 2010;18(19):7078–84.
7. Mayr F, Wieder M, Wieder O, Langer T. Improving small molecule pka prediction using transfer learning with graph neural networks. Front Chem. 2022;10.
8. Wang Y, Xiong J, Xiao F, Zhang W, Cheng K, Rao J, et al. LogD7.4 prediction enhanced by transferring knowledge from chromatographic retention time, microscopic pKa and logP. J Cheminform. 2023;15(1):76.
9. Lapins M, Arvidsson S, Lampa S, Berg A, Schaal W, Alvarsson J, et al. A confidence predictor for logD using conformal regression and a support-vector machine. J Cheminform. 2018;10(1):17.
10. Bergazin TD, Tielker N, Zhang Y, Mao J, Gunner MR, Francisco K, et al. Evaluation of log P, pKa, and log D predictions from the SAMPL7 blind challenge. J Comput Aided Mol Des. 2021;35(7):771–802.
11. Kim T, Park H. Computational prediction of octanol–water partition coefficient based on the extended solvent-contact model. J Mol Graph Model. 2015;60:108–17.
12. Chen H. *In silico* log *P* prediction for a large data set with support vector machines, radial basis neural networks and multiple linear regression. Chem Biol Drug Des. 2009;74(2):142–7.
13. Lenselink EB, Stouten PFW. Multitask machine learning models for predicting lipophilicity (logP) in the SAMPL7 challenge. J Comput Aided Mol Des. 2021;35(8):901–9.
14. Stenzel A, Goss K, Endo S. Prediction of partition coefficients for complex environmental contaminants: validation of COSMOtherm, ABSOLV, and SPARC. Environ Toxicol Chem. 2014;33(7):1537–43.
15. Skyner RE, McDonagh JL, Groom CR, van Mourik T, Mitchell JBO. A review of methods for the calculation of solution free energies and the modelling of systems in solution. Phys Chem Chem Phys. 2015;17(9):6174–91.
16. Das T, Mehta CH, Nayak UY. Multiple approaches for achieving drug solubility: an in silico perspective. Drug Discov Today. 2020;25(7):1206–12.
17. Bergström CAS, Larsson P. Computational prediction of drug solubility in water-based systems: qualitative and quantitative approaches used in the current drug discovery and development setting. Int J Pharm. 2018;540(1–2):185–93.
18. Meftahi N, Walker ML, Smith BJ. Predicting aqueous solubility by QSPR modeling. J Mol Graph Model. 2021;106:107901.

19. Persson LC, Porter CJH, Charman WN, Bergström CAS. Computational prediction of drug solubility in lipid based formulation excipients. Pharm Res. 2013;30(12):3225–37.
20. Ainurofiq A, Putro D, Ramadhani D, Putra G, Do Espirito Santo LC. A review on solubility enhancement methods for poorly water-soluble drugs. J Rep Pharm Sci. 2021;10(1):137.
21. Wojeicchowski JP, Ferreira AM, Okura T, Pinheiro Rolemberg M, Mafra MR, Coutinho JAP. Using COSMO-RS to predict Hansen solubility parameters. Ind Eng Chem Res. 2022;61(42):15631–8.
22. Sadeghi A, Su CH, Khan A, Lutfor Rahman M, Sani Sarjadi M, Sarkar SM. Machine learning simulation of pharmaceutical solubility in supercritical carbon dioxide: prediction and experimental validation for busulfan drug. Arab J Chem. 2022;15(1):103502.
23. Heng T, Yang D, Wang R, Zhang L, Lu Y, Du G. Progress in research on artificial intelligence applied to polymorphism and cocrystal prediction. ACS Omega. 2021;6(24):15543–50.
24. Abramov YA. Current computational approaches to support pharmaceutical solid form selection. Org Process Res Dev. 2013;17(3):472–85.
25. Takieddin K, Khimyak YZ, Fábián L. Prediction of hydrate and solvate formation using statistical models. Cryst Growth Des. 2016;16(1):70–81.
26. Nauha E, Bernstein J. "Predicting" crystal forms of pharmaceuticals using hydrogen bond propensities: two test cases. Cryst Growth Des. 2014;14(9):4364–70.
27. Hang NT, Long NT, Duy ND, Chien NN, Van Phuong N. Towards safer and efficient formulations: machine learning approaches to predict drug-excipient compatibility. Int J Pharm. 2024;653:123884.
28. Patel S, Patel M, Kulkarni M, Patel MS. De-interact: a machine-learning-based predictive tool for the drug-excipient interaction study during product development—validation through paracetamol and vanillin as a case study. Int J Pharm. 2023;637:122839.
29. Wang N, Sun H, Dong J, Ouyang D. PharmDE: a new expert system for drug-excipient compatibility evaluation. Int J Pharm. 2021;607:120962.
30. Raijada D, Müllertz A, Cornett C, Munk T, Sonnergaard J, Rantanen J. Miniaturized approach for excipient selection during the development of oral solid dosage form. J Pharm Sci. 2014;103(3):900–8.
31. Todke PA, Devarajan PV. In-silico approach as a tool for selection of excipients for safer amphotericin B nanoformulations. J Control Release. 2022;349:756–64.
32. Jo S, Xu A, Curtis JE, Somani S, MacKerell AD. Computational characterization of antibody–excipient interactions for rational excipient selection using the site identification by ligand competitive saturation-biologics approach. Mol Pharm. 2020;17(11):4323–33.
33. Zhang D, Hu Y, Hao Z, Zhang Y, Luo S, Dang X, et al. Design, synthesis and antitumor activities of thiazole-containing mitochondrial targeting agents. Bioorg Chem. 2021;115:105271.
34. Manne SR, Sharma A, Sazonovas A, El-Faham A, de la Torre BG, Albericio F. Understanding OxymaPure as a peptide coupling additive: a guide to new Oxyma derivatives. ACS Omega. 2022;7(7):6007–23.
35. Aminpour M, Delgado WEM, Wacker S, Noskov S, Houghton M, Tyrrell DLJ, et al. Computational determination of toxicity risks associated with a selection of approved drugs having demonstrated activity against COVID-19. BMC Pharmacol Toxicol. 2021;22(1):61.
36. Olasupo SB, Uzairu A, Shallangwa G, Uba S. QSAR modeling, molecular docking and ADMET/pharmacokinetic studies: a chemometrics approach to search for novel inhibitors of norepinephrine transporter as potent antipsychotic drugs. J Iran Chem Soc. 2020;17(8):1953–66.
37. Tetko IV, Tanchuk VY. Application of associative neural networks for prediction of lipophilicity in alogps 2.1 Program. J Chem Inf Comput Sci. 2002;42(5):1136–45.
38. Cysewski P, Jeliński T, Przybyłek M, Nowak W, Olczak M. Solubility characteristics of acetaminophen and phenacetin in binary mixtures of aqueous organic solvents: experimental and deep machine learning screening of green dissolution media. Pharmaceutics. 2022;14(12):2828.

39. Loschen C, Reinisch J, Klamt A. COSMO-RS based predictions for the SAMPL6 logP challenge. J Comput Aided Mol Des. 2020;34(4):385–92.
40. Iswandana R, Aisyah P, Syahdi R. Prediction analysis of pharmacokinetic parameters of several oral systemic drugs using insilico method. Int J Appl Pharm. 2020:260–3.
41. Schönsee CD, Bucheli TD. Experimental determination of octanol–water partition coefficients of selected natural toxins. J Chem Eng Data. 2020;65(4):1946–53.
42. Shelley JC, Cholleti A, Frye LL, Greenwood JR, Timlin MR, Uchimaya M. Epik: a software program for pK a prediction and protonation state generation for drug-like molecules. J Comput Aided Mol Des. 2007;21(12):681–91.
43. Isert C, Kromann JC, Stiefl N, Schneider G, Lewis RA. Machine learning for fast, quantum mechanics-based approximation of drug lipophilicity. ACS Omega. 2023;8(2):2046–56.
44. Kalász H, Benkő B, Gulyás Z, Tekes K. Lipophilicity determination using both TLC and calculations. J Liq Chromatogr Relat Technol. 2009;32(9):1342–58.
45. Raevsky OA, Trepalin SV, Trepalina HP, Gerasimenko VA, Raevskaja OE. SLIPPER-2001 – software for predicting molecular properties on the basis of physicochemical descriptors and structural similarity. J Chem Inf Comput Sci. 2002;42(3):540–9.
46. Hilal SH, Karickhoff SW, Carreira LA. Prediction of the solubility, activity coefficient and liquid/liquid partition coefficient of organic compounds. QSAR Comb Sci. 2004;23(9):709–20.
47. Hunt P, Hosseini-Gerami L, Chrien T, Plante J, Ponting DJ, Segall M. Predicting p K_a using a combination of semi-empirical quantum mechanics and radial basis function methods. J Chem Inf Model. 2020;60(6):2989–97.
48. Li Z, Peng WH, Liu WJ, Yang LY, Naeem A, Feng Y, et al. Advances in numerical simulation of unit operations for tablet preparation. Int J Pharm. 2023;634:122638.
49. Abramov YA, Sun G, Zeng Q. Emerging landscape of computational modeling in pharmaceutical development. J Chem Inf Model. 2022;62(5):1160–71.
50. Soni RK, Mohanty R, Mohanty S, Mishra BK. Numerical analysis of mixing of particles in drum mixers using DEM. Adv Powder Technol. 2016;27(2):531–40.
51. Bouhaouche A, Kaoua S, Daoud K. Numerical simulation of free flowing particles mixing in V-blender. Granul Matter. 2019;21(2):35.
52. Takabatake K, Mori Y, Khinast JG, Sakai M. Numerical investigation of a coarse-grain discrete element method in solid mixing in a spouted bed. Chem Eng J. 2018;346:416–26.
53. Marigo M, Cairns DL, Davies M, Ingram A, Stitt EH. A numerical comparison of mixing efficiencies of solids in a cylindrical vessel subject to a range of motions. Powder Technol. 2012;217:540–7.
54. Fraige FY, Langston PA. Horizontal pneumatic conveying: a 3d distinct element model. Granul Matter. 2006;8(2):67–80.
55. Yamamoto M, Ishihara S, Kano J. Evaluation of particle density effect for mixing behavior in a rotating drum mixer by DEM simulation. Adv Powder Technol. 2016;27(3):864–70.
56. Ng BH, Ding YL, Ghadiri M. Modelling of dense and complex granular flow in high shear mixer granulator—a CFD approach. Chem Eng Sci. 2009;64(16):3622–32.
57. Maharjan R, Jeong SH. High shear seeded granulation: its preparation mechanism, formulation, process, evaluation, and mathematical simulation. Powder Technol. 2020;366:667–88.
58. Tamrakar A, Chen SW, Ramachandran R. A DEM model-based study to quantitatively compare the effect of wet and dry binder addition in high-shear wet granulation processes. Chem Eng Res Des. 2019;142:307–26.
59. Gantt JA, Gatzke EP. High-shear granulation modeling using a discrete element simulation approach. Powder Technol. 2005;156(2–3):195–212.
60. Oullion M, Reynolds GK, Hounslow MJ. Simulating the early stage of high-shear granulation using a two-dimensional Monte-Carlo approach. Chem Eng Sci. 2009;64(4):673–85.
61. Adetayo AA, Litster JD, Pratsinis SE, Ennis BJ. Population balance modelling of drum granulation of materials with wide size distribution. Powder Technol. 1995;82(1):37–49.
62. Iveson SM, Litster JD, Hapgood K, Ennis BJ. Nucleation, growth and breakage phenomena in agitated wet granulation processes: a review. Powder Technol. 2001;117(1–2):3–39.

63. Wauters PAL, Scarlett B, Liu LX, Litster JD, Meesters GMH. A population balance model for high shear granulation. Chem Eng Commun. 2003;190(10):1309–34.
64. Chaturbedi A, Bandi CK, Reddy D, Pandey P, Narang A, Bindra D, et al. Compartment based population balance model development of a high shear wet granulation process via dry and wet binder addition. Chem Eng Res Des. 2017;123:187–200.
65. Sarkar S, Chaudhuri B. DEM modeling of high shear wet granulation of a simple system. Asian J Pharm Sci. 2018;13(3):220–8.
66. Muthancheri I, Chaturbedi A, Bétard A, Ramachandran R. A compartment based population balance model for the prediction of steady and induction granule growth behavior in high shear wet granulation. Adv Powder Technol. 2021;32(6):2085–96.
67. Sansare S, Aziz H, Sen K, Patel S, Chaudhuri B. Computational modeling of fluidized beds with a focus on pharmaceutical applications: a review. J Pharm Sci. 2022;111(4):1110–25.
68. Elghannay H, Tafti D, Yu K. Evaluation of physics based hard-sphere model with the soft sphere model for dense fluid-particle flow systems. Int J Multiphase Flow. 2019;112:100–15.
69. Goldschmidt MJV, Weijers GGC, Boerefijn R, Kuipers JAM. Discrete element modelling of fluidised bed spray granulation. Powder Technol. 2003;138(1):39–45.
70. Song Y, Zhou T, Bai R, Zhang M, Yang H. Review of CFD-DEM modeling of wet fluidized bed granulation and coating processes. PRO. 2023;11(2):382.
71. Hayashi K, Watano S. Novel population balance model for granule aggregation and breakage in fluidized bed granulation and drying. Powder Technol. 2019;342:664–75.
72. Singh M. Accurate and efficient approximations for generalized population balances incorporating coagulation and fragmentation. J Comput Phys. 2021;435:110215.
73. Singh M, Ismail HY, Matsoukas T, Albadarin AB, Walker G. Mass-based finite volume scheme for aggregation, growth and nucleation population balance equation. Proc R Soc A. 2019;475(2231):20190552.
74. Martin NL, Schomberg AK, Finke JH, Abraham TGM, Kwade A, Herrmann C. Process modeling and simulation of tableting—an agent-based simulation methodology for direct compression. Pharmaceutics. 2021;13(7):996.
75. Seo KS, Bajracharya R, Lee SH, Han HK. Pharmaceutical application of tablet film coating. Pharmaceutics. 2020;12(9):853.
76. Salawi A. Pharmaceutical coating and its different approaches, a review. Polymers (Basel). 2022;14(16):3318.
77. Zhang C, Chen L, Ji W, Liu Y, Liu L, Tao WQ. Lattice Boltzmann mesoscopic modeling of flow boiling heat transfer processes in a microchannel. Appl Therm Eng. 2021;197:117369.
78. Vaitukaitis P, Maggiolo D, Remmelgas J, Abrahmsén-Alami S, Bernin D, Siiskonen M, et al. Water transport and absorption in pharmaceutical tablets – a numerical study. Meccanica. 2020;55(2):421–33.
79. Ketterhagen WR. Modeling the motion and orientation of various pharmaceutical tablet shapes in a film coating pan using DEM. Int J Pharm. 2011;409(1–2):137–49.
80. Niblett D, Porter S, Reynolds G, Morgan T, Greenamoyer J, Hach R, et al. Development and evaluation of a dimensionless mechanistic pan coating model for the prediction of coated tablet appearance. Int J Pharm. 2017;528(1–2):180–201.
81. Toschkoff G, Just S, Knop K, Kleinebudde P, Funke A, Djuric D, et al. Modeling of an active tablet coating process. J Pharm Sci. 2015;104(12):4082–92.
82. Dreu R, Toschkoff G, Funke A, Altmeyer A, Knop K, Khinast J, et al. Evaluation of the tablets' surface flow velocities in pan coaters. Eur J Pharm Biopharm. 2016;106:97–106.
83. Toschkoff G, Suzzi D, Tritthart W, Reiter F, Schlingmann M, Khinast JG. Detailed analysis of air flow and spray loss in a pharmaceutical coating process. AICHE J. 2012;58(2):399–411.
84. Madlmeir S, Loidolt P, Khinast JG. Study of the capsule-filling dosator process via calibrated DEM simulations. Int J Pharm. 2019;567:118441.
85. Loidolt P, Madlmeir S, Khinast JG. Mechanistic modeling of a capsule filling process. Int J Pharm. 2017;532(1):47–54.

86. Sánchez-Dengra B, González-García I, González-Álvarez M, González-Álvarez I, Bermejo M. Two-step in vitro-in vivo correlations: deconvolution and convolution methods, which one gives the best predictability? Comparison with one-step approach. Eur J Pharm Biopharm. 2021;158:185–97.
87. Pawestri SA. Application in silico modeling simulation in bioequivalence studies: a review. J Food Pharm Sci. 2023:763–9.
88. Tosca EM, Rocchetti M, Pérez E, Nieto C, Bettica P, Moscoso del Prado J, et al. In vitro–in vivo correlation (IVIVC) population modeling for the in silico bioequivalence of a long-acting release formulation of progesterone. Pharmaceutics. 2021;13(2):255.
89. Clarke JF, Thakur K, Polak S. A mechanistic physiologically based model to assess the effect of study design and modified physiology on formulation safe space for virtual bioequivalence of dermatological drug products. Front Pharmacol. 2022;13:1007496.
90. Tsakalozou E, Alam K, Babiskin A, Zhao L. Physiologically-based pharmacokinetic modeling to support determination of bioequivalence for dermatological drug products: scientific and regulatory considerations. Clin Pharmacol Ther. 2022;111(5):1036–49.
91. Vlachou M, Karalis V. An in vitro–in vivo simulation approach for the prediction of bioequivalence. Materials. 2021;14(3):555.
92. Song E, Lee W, Kim BH. Model-based approach for designing an efficient bioequivalence study for highly variable drugs. Pharmaceuticals. 2021;14(11):1101.
93. Golhar A, Pillai M, Dhakne P, Rajput N, Jadav T, Sengupta P. Progressive tools and critical strategies for development of best fit PBPK model aiming better in vitro–in vivo correlation. Int J Pharm. 2023;643:123267.
94. Zhang F, Jia R, Gao H, Wu X, Liu B, Wang H. In silico modeling and simulation to guide bioequivalence testing for oral drugs in a virtual population. Clin Pharmacokinet. 2021;60(11):1373–85.
95. Dabke A, Ghosh S, Dabke P, Sawant K, Khopade A. Revisiting the in-vitro and in-vivo considerations for in-silico modelling of complex injectable drug products. J Control Release. 2023;360:185–211.
96. Sommerfeld M, Schmalfuß S. Numerical analysis of carrier particle motion in a dry powder inhaler. J Fluids Eng. 2016;138(4).
97. Finlay WH, Martin AR. Modeling of aerosol deposition with interface devices. J Aerosol Med. 2007;20(s1):S19–28.
98. Longest W, Farkas D. Development of a new inhaler for high-efficiency dispersion of spray-dried powders using computational fluid dynamics (CFD) modeling. AAPS J. 2019;21(2):25.
99. Chalvatzaki E, Chatoutsidou SE, Lazaridis M. Simulations of the deposition of pharmaceutical aerosols in the human respiratory tract by dry powder inhalers (DPIs). J Drug Deliv Sci Technol. 2020;59:101915.
100. Kolanjiyil AV, Kleinstreuer C. Computationally efficient analysis of particle transport and deposition in a human whole-lung-airway model. Part I: theory and model validation. Comput Biol Med. 2016;79:193–204.
101. Kolanjiyil AV, Kleinstreuer C, Sadikot RT. Computationally efficient analysis of particle transport and deposition in a human whole-lung-airway model. Part II: dry powder inhaler application. Comput Biol Med. 2017;84:247–53.
102. Le Merdy M, Spires J, Lukacova V, Tan ML, Babiskin A, Xu X, et al. Ocular physiologically based pharmacokinetic modeling for ointment formulations. Pharm Res. 2020;37(12):245.
103. Le Merdy M, Fan J, Bolger MB, Lukacova V, Spires J, Tsakalozou E, et al. Application of mechanistic ocular absorption modeling and simulation to understand the impact of formulation properties on ophthalmic bioavailability in rabbits: a case study using dexamethasone suspension. AAPS J. 2019;21(4):65.
104. Gukasyan HJ, Hailu S, Karami TK, Graham R. Ocular biopharmaceutics: impact of modeling and simulation on topical ophthalmic formulation development. Drug Discov Today. 2019;24(8):1587–97.

105. Kovacs K, Wagley S, Quirk MT, Ceron OM, Silva PA, Singh RJ, et al. Pharmacokinetic study of vitreous and serum concentrations of triamcinolone acetonide after posterior sub-Tenon's injection. Am J Ophthalmol. 2012;153(5):939–48.
106. German C, Chen Z, Przekwas A, Walenga R, Babiskin A, Zhao L, et al. Computational model of in vivo corneal pharmacokinetics and pharmacodynamics of topically administered ophthalmic drug products. Pharm Res. 2023;40(4):961–75.
107. Missel P, Horner M. Modelling ocular delivery using computational fluid dynamics. ONdrugDelivery Magazine. 2015;54:12–6.
108. Wang N, Zhang Y, Wang W, Ye Z, Chen H, Hu G, et al. How can machine learning and multiscale modeling benefit ocular drug development? Adv Drug Deliv Rev. 2023;196:114772.
109. Choi SH, Wang Y, Conti DS, Raney SG, Delvadia R, Leboeuf AA, et al. Generic drug device combination products: regulatory and scientific considerations. Int J Pharm. 2018;544(2):443–54.
110. Raney SG, Luke MC. A new paradigm for topical generic drug products: impact on therapeutic access. J Am Acad Dermatol. 2020;82(6):1570–1.
111. Tsakalozou E, Alam K, Ghosh P, Spires J, Polak S, Fang L, et al. Mechanistic modeling of drug products applied to the skin: a workshop summary report. CPT Pharmacometrics Syst Pharmacol. 2023;12(5):575–84.
112. Tsakalozou E, Babiskin A, Zhao L. Physiologically-based pharmacokinetic modeling to support bioequivalence and approval of generic products: a case for diclofenac sodium topical gel, 1%. CPT Pharmacometrics Syst Pharmacol. 2021;10(5):399–411.
113. Dumont C, Prieto P, Asturiol D, Worth A. Review of the availability of *in vitro* and *in silico* methods for assessing dermal bioavailability. Appl In Vitro Toxicol. 2015;1(2):147–64.
114. Mahdi WA, Hussain A, Altamimi MA, Alshehri S, Bukhari SI, Ahsan MN. Experimental solubility, thermodynamic/computational validations, and GastroPlus-based in silico prediction for subcutaneous delivery of rifampicin. AAPS PharmSciTech. 2021;22(3):116.
115. Cvijić S, Ignjatović J, Parojčić J, Ibrić S. The emerging role of physiologically-based pharmacokinetic/biopharmaceutics modeling in formulation development. Arh Farm (Belgr). 2021;71(4):318–35.
116. Bannigan P, Bao Z, Hickman RJ, Aldeghi M, Häse F, Aspuru-Guzik A, et al. Machine learning models to accelerate the design of polymeric long-acting injectables. Nat Commun. 2023;14(1):35.
117. Zhu T, Moussa EM, Witting M, Zhou D, Sinha K, Hirth M, et al. Predictive models of lyophilization process for development, scale-up/tech transfer and manufacturing. Eur J Pharm Biopharm. 2018;128:363–78.
118. Bhandarkar K, Krishna V. Computational fluid dynamics as a simulation tool in the formulation development of parenteral and biphasic liquid dosage forms. Res J Pharm Technol. 2023;16:5935–43.
119. Petitti M, Barresi A, Marchisio D. CFD modelling of condensers for freeze-drying processes. Sadhana. 2013;38(6):1219–39.
120. Ravnik J, Ramšak M, Zadravec M, Kamenik B, Hriberšek M. Experimental and stochastic analysis of lyophilisation. Eur J Pharm Biopharm. 2021;159:108–22.
121. Gallo-Molina JP, Ratkovich N, Alvarez O. The application of computational fluid dynamics to the multiscale study of oil-in-water emulsions. Ind Eng Chem Res. 2018;57(2):578–89.
122. Scutellà B, Plana-Fattori A, Passot S, Bourlès E, Fonseca F, Flick D, et al. 3D mathematical modelling to understand atypical heat transfer observed in vial freeze-drying. Appl Therm Eng. 2017;126:226–36.
123. Bluemel O, Pavlišič A, Likozar B, Rodrigues MA, Geraldes V, Bechtold-Peters K, et al. Computational fluid dynamic simulations of temperature, cryoconcentration, and stress time during large-scale freezing and thawing of monoclonal antibody solutions. Eur J Pharm Biopharm. 2022;177:107–12.
124. Zarzar J, Khan T, Bhagawati M, Weiche B, Sydow-Andersen J, Alavattam S. High concentration formulation developability approaches and considerations. MAbs. 2023;15(1):2211185.

125. Narayanan H, Dingfelder F, Condado Morales I, Patel B, Heding KE, Bjelke JR, et al. Design of biopharmaceutical formulations accelerated by machine learning. Mol Pharm. 2021;18(10):3843–53.
126. Ashwini T, Narayan R, Shenoy PA, Nayak UY. Computational modeling for the design and development of nano based drug delivery systems. J Mol Liq. 2022;368:120596.
127. Dong J, Wu Z, Xu H, Ouyang D. FormulationAI: a novel web-based platform for drug formulation design driven by artificial intelligence. Brief Bioinform. 2023;25(1).
128. Cottura N, Howarth A, Rajoli RKR, Siccardi M. The current landscape of novel formulations and the role of mathematical modeling in their development. J Clin Pharmacol. 2020;60(S1):S77–97.
129. Walenga RL, Babiskin AH, Bhoopathy S, Clarke JF, De Backer J, Ducharme M, et al. Use of the same model or modeling strategy across multiple submissions: focus on complex drug products. AAPS J. 2024;26(1):12.
130. Huynh L, Leroux JC, Allen C. Enhancement of docetaxel solubility via conjugation of formulation-compatible moieties. Org Biomol Chem. 2009;7(17):3437.
131. Sun Y, Wu H, Dong W, Zhou J, Zhang X, Liu L, et al. Molecular simulation approach to the rational design of self-assembled nanoparticles for enhanced peroral delivery of doxorubicin. Carbohydr Polym. 2019;218:279–88.
132. Alshahrani SM, Almutairy BK, Alfadhel MM, Belal A, Abourehab MAS, et al. Computational simulation and target prediction studies of solubility optimization of decitabine through supercritical solvent. Sci Rep. 2022;12(1):18875.
133. Gupta J, Nunes C, Vyas S, Jonnalagadda S. Prediction of solubility parameters and miscibility of pharmaceutical compounds by molecular dynamics simulations. J Phys Chem B. 2011;115(9):2014–23.
134. Tian Y, Shi C, Sun Y, Zhu C, Sun CC, Mao S. Designing micellar nanocarriers with improved drug loading and stability based on solubility parameter. Mol Pharm. 2015;12(3):816–25.
135. Mehta CH, Narayan R, Aithal G, Pandiyan S, Bhat P, Dengale S, et al. Molecular simulation driven experiment for formulation of fixed dose combination of Darunavir and Ritonavir as anti-HIV nanosuspension. J Mol Liq. 2019;293:111469.
136. Leonardi D, Salomón CJ, Lamas MC, Olivieri AC. Development of novel formulations for Chagas' disease: optimization of benznidazole chitosan microparticles based on artificial neural networks. Int J Pharm. 2009;367(1–2):140–7.
137. Metwally AA, Hathout RM. Computer-assisted drug formulation design: novel approach in drug delivery. Mol Pharm. 2015;12(8):2800–10.
138. Li Y, Abbaspour MR, Grootendorst PV, Rauth AM, Wu XY. Optimization of controlled release nanoparticle formulation of verapamil hydrochloride using artificial neural networks with genetic algorithm and response surface methodology. Eur J Pharm Biopharm. 2015;94:170–9.
139. Narayan R, Gadag S, Garg S, Nayak UY. Understanding the effect of functionalization on loading capacity and release of drug from mesoporous silica nanoparticles: a computationally driven study. ACS Omega. 2022;7(10):8229–45.
140. Soroush S. Computer simulation of a novel pharmaceutical silicon nanocarrier. Nanotechnol Sci Appl. 2010;3:149.
141. Gayathri NK, Aparna V, Maya S, Biswas R, Jayakumar R, Mohan CG. Preparation, characterization, drug release and computational modelling studies of antibiotics loaded amorphous chitin nanoparticles. Carbohydr Polym. 2017;177:67–76.
142. Curtis LT, Wu M, Lowengrub J, Decuzzi P, Frieboes HB. Computational modeling of tumor response to drug release from vasculature-bound nanoparticles. PLoS One. 2015;10(12):e0144888.
143. Uhlemann J, Diedam H, Hoheisel W, Schikarski T, Peukert W. Modeling and simulation of process technology for nanoparticulate drug formulations—a particle technology perspective. Pharmaceutics. 2020;13(1):22.

144. Arifin DY, Lee LY, Wang CH. Mathematical modeling and simulation of drug release from microspheres: implications to drug delivery systems. Adv Drug Deliv Rev. 2006;58(12–13):1274–325.
145. Bunker A, Róg T. Mechanistic understanding from molecular dynamics simulation in pharmaceutical research 1: drug delivery. Front Mol Biosci. 2020;7:604770.
146. Shabbir F, Mujeeb AA, Jawed SF, Khan AH, Shakeel CS. Simulation of transvascular transport of nanoparticles in tumor microenvironments for drug delivery applications. Sci Rep. 2024;14(1):1764.
147. Salahshoori I, Golriz M, Nobre MAL, Mahdavi S, Eshaghi Malekshah R, Javdani-Mallak A, et al. Simulation-based approaches for drug delivery systems: navigating advancements, opportunities, and challenges. J Mol Liq. 2024;395:123888.
148. Fullstone G, Wood J, Holcombe M, Battaglia G. Modelling the transport of nanoparticles under blood flow using an agent-based approach. Sci Rep. 2015;5(1):10649.
149. Ding HM, Ma YQ. Design strategy of surface decoration for efficient delivery of nanoparticles by computer simulation. Sci Rep. 2016;6(1):26783.
150. Lee H. Effects of nanoparticle electrostatics and protein–protein interactions on Corona formation: conformation and hydrodynamics. Small. 2020;16(10):e1906598.
151. Jayasinghe MK, Lee CY, Tran TTT, Tan R, Chew SM, Yeo BZJ, et al. The role of in silico research in developing nanoparticle-based therapeutics. Front Digit Health. 2022;4
152. Yu X, Trase I, Ren M, Duval K, Guo X, Chen Z. Design of nanoparticle-based carriers for targeted drug delivery. J Nanomater. 2016;2016:1–15.
153. Sulea T, Kumar S, Kuroda D. Editorial: Progress and challenges in computational structure-based design and development of biologic drugs. Front Mol Biosci. 2024;11:1360267.
154. North B, Lehmann A, Dunbrack RL. A new clustering of antibody CDR loop conformations. J Mol Biol. 2011;406(2):228–56.
155. Norman RA, Ambrosetti F, Bonvin AMJJ, Colwell LJ, Kelm S, Kumar S, et al. Computational approaches to therapeutic antibody design: established methods and emerging trends. Brief Bioinform. 2020;21(5):1549–67.
156. Kim J, McFee M, Fang Q, Abdin O, Kim PM. Computational and artificial intelligence-based methods for antibody development. Trends Pharmacol Sci. 2023;44(3):175–89.
157. Sircar A, Kim ET, Gray JJ. RosettaAntibody: antibody variable region homology modeling server. Nucleic Acids Res. 2009;37(Web Server):W474–9.
158. Webb B, Sali A. Comparative protein structure modeling using MODELLER. Curr Protoc Protein Sci. 2016;86(1).
159. Khetan R, Curtis R, Deane CM, Hadsund JT, Kar U, Krawczyk K, et al. Current advances in biopharmaceutical informatics: guidelines, impact and challenges in the computational developability assessment of antibody therapeutics. MAbs. 2022;14(1):2020082.
160. Lai PK, Gallegos A, Mody N, Sathish HA, Trout BL. Machine learning prediction of antibody aggregation and viscosity for high concentration formulation development of protein therapeutics. MAbs. 2022;14(1):2026208.
161. Blanco MA. Computational models for studying physical instabilities in high concentration biotherapeutic formulations. MAbs. 2022;14(1):2044744.
162. Kumar S, Plotnikov NV, Rouse JC, Singh SK. Biopharmaceutical informatics: supporting biologic drug development *via* molecular modelling and informatics. J Pharm Pharmacol. 2018;70(5):595–608.

Chapter 3
Computational Fluid Dynamics-the Revolutionary Innovative Methodology in the Pharmaceutical Industry

Vaibhav Sahal, Rakesh Barik, Mahaveer Singh, Sinoy Sugunan, Rajasekhar Reddy Alavala, and S. N. Koteswara Rao G.

Contents

V. Sahal · R. R. Alavala · S. N. K. Rao G.
Shobhaben Pratapbhai Patel School of Pharmacy and Technology Management, SVKM's NMIMS, Vile Parle (W), Mumbai, Maharashtra, India

R. Barik
GITAM School of Pharmacy, GITAM Deemed to be University, Hyderabad Campus, Rudraram, Telangana, India
e-mail: rbarik@gitam.edu

M. Singh · S. Sugunan (✉)
School of Pharmacy and Technology Management, SVKM's NMIMS, Shirpur, Maharashtra, India
e-mail: sinoy.sugunan@nmims.edu

S. N. K. Rao G., R. R. Alavala (eds.), *Applications of Computational Tools in Drug Design and Development*, https://doi.org/10.1007/978-981-96-4154-3_3

Abstract The design and optimization of aerosol inhalers for pharmaceutical applications require a precise understanding of fluid dynamics to ensure effective drug delivery to the respiratory tract. A potent tool for modeling and examining the intricate relationships between airflow, particle behavior, and device shape in aerosol inhalers is computational fluid dynamics, or CFD. The concepts of CFD are examined in this chapter along with how they are used to create effective and patient-centered inhalation devices. The function of CFD in maximizing aerosol production, particle deposition, and dispersion within the respiratory system is one of the main subjects. There includes discussion of case examples showing how CFD may be used to enhance device performance, cut down on medication waste, and guarantee repeatability. Additionally, the chapter addresses developments in boundary conditions, turbulence models specifically designed for inhalers, and numerical modeling. Critical analysis is done on issues such integrating CFD findings with real-world clinical circumstances, validating CFD results with experimental data, and the requirement for computational efficiency. This chapter highlights the revolutionary role that computational fluid dynamics (CFD) plays in developing the next generation of aerosol inhalers for respiratory therapy by bridging the gap between theoretical fluid mechanics and real-world device engineering.

Keywords Computational Fluid Dynamics (CFD) · Aerosol inhalers · Drug delivery · Respiratory system · Particle deposition · Device optimization · Fluid mechanics · Numerical modeling · Turbulence models

3.1 Introduction

In today's highly competitive setting in the pharmaceutical field, Computational Fluid Dynamics (CFD) has become one of the basic tools in the pharmaceutical processes such as formulation, quality control and even manufacturing [1]. For instance, the application of CFD extends to simulating the flow of blood through arteries. Because of its capabilities and features, CFD is becoming more important in the area of formulation of drugs, and more processes are being implemented vis-a-vis software packages for optimizing drug delivery systems. Such devices include inhalers and transdermal patches and nanoparticle-based delivery systems, where the fluid is pumped out through a fine nozzle. And, moreover, CFD modeling can also help to analyze mixing and dissolution processes which are extremely important for obtaining uniform and high-quality pharmaceutical products, which are also patented [2, 3]. For example, in the area of quality assurance, several manufacturing processes or growth processes can be predicted by using CFD and the results are within the range of research applications.

CFD can assist in determining the 'mixing' as well as 'coating' virtual processes in production before they develop into industrial costly errors, not only preserving the product but also promoting adherence to standard requirements [4]. CFD first developed the modeling of vertical tablet coating systems, fluid bed processing and then advanced to using CFD in scale-up modelling for aseptic operations. The understanding of fluid mechanics and thermodynamics provided by CFD for this processes allows the pharmaceutical industry to be more effective, diminish wastes and make certain all products are safe and effective [5]. The chapter of this book is focused on the analysis of the effect of CFD simulation in such sector utilities as drug development, drug quality control, as well as drug manufacturing processes. While reviewing new developments and case studies, we strive to show how the central techniques aimed at commercial and case studies scope expanding were linked with CFD technology everybody in pharma industry made it certain to be of the highest quality and production systems work smoothly.

3.2 Fundamentals of CFD for Pharmaceutical Applications

For several different complicated situations CFD is a method that makes use of the equations leading to motion of fluids, to both give insights and also make predictions useful for scale [6]. One of the key aspects in CFD simulations is the Meso structure. Points in the grid system consists mesh to subdivide a computational domain into smaller pieces [7]. Quality of the grid impacts the accuracy of the CFD simulation. In most cases, there are two main types of grid systems:

- Structured grids
- Unstructured grids

CFD arose as a result of the emergence of high-speed computers that enabled the solutions that otherwise require tedious work of scanning thousands to millions of figures films. It facilitated the growth of CFD [7]. As a result, there has been closer correlation between improvements in computer economies, particularly those of direct access and speed, and the further development and application of CFD techniques to problem solving that have increased complexity and detail.

3.2.1 Fluid Properties and Behavior

There are other factors such as viscosity, density, and surfactant surface tension that are critical for flow under specific conditions. Viscosity is very important in operations like mixing of blender, where blending different parts of a composed drug takes place. Density, on the other side, regulate number of particles suspended to settle, which is important for formulation stability [8]. It should be noted that droplet formation during spray drying processes, which is an ordinary technique of transforming a liquid solution into fine powder such as for inhaler medications, is highly determined by surface tension.

3.2.2 Governing Equations

In fluid dynamics, the Navier–Stokes equations encompass the most fundamental principles of mass, momentum, and energy conservation. These equations, which can be quite intricate, can be analyzed through CFD software, which can effectively generate the anticipated movement, pressure, and velocity patterns of fluids in pharmaceutical operations [9, 10].

3.2.3 Turbulence Modeling

Especially in the areas of mixing and aerosol dynamics, advancement of accurate turbulence modeling is needed, to ensure that the mixing efficiency and the particle dispersion patterns are predicted correctly and accurately. The k-ε (kappa-epsilon) and Large Eddy Simulation (LES) are effectively turbulence models that employ varying degrees of approximation to the Navier–Stokes equations, against the computational and time costs incurred [11].

3.3 Heat Transfer Modeling in CFD

3.3.1 Conduction, Convection, and Radiation

However, inter transfer mechanisms are also important for some pharmaceutical processes. For instance, it is interesting to note that heat transfer by conduction helps in fortifying comprehension of heat distribution during tablet compression while convection is significant in fluid bed drying. While not regularly, radiation can be relevant in some drying and sterilization applications [1, 12].

3.3.2 Coupled Heat and Fluid Flow Analysis

CFD simulations often integrate heat transfer and fluid flow to analyze complex processes like drying, where evaporation significantly influences flow dynamics by cooling the liquid. Such coupled analyses are instrumental in optimizing manufacturing processes, ensuring uniform product quality, and minimizing energy consumption [13, 14].

3.3.3 Simulation Techniques in Pharmaceutical CFD Studies

(a) Mesh Generation

Creating a computational mesh is a pivotal step in CFD simulation. A well-designed mesh accurately captures the geometry of the physical system, enabling precise simulations of fluid flow and heat transfer. Mesh refinement is critical in regions with steep gradients of velocity, pressure, or temperature, ensuring reliable and accurate results [13, 15, 16].

(b) Boundary and Initial Conditions

Setting appropriate boundary and initial conditions is essential for realistic simulations. For example, simulating airflow in an inhaler requires boundary conditions at the mouthpiece that faithfully represent the patient's inhalation profile, ensuring that the simulation reflects real-world dynamics [13, 15, 16].

(c) Solver Settings and Convergence Criteria

Selecting the right solver settings and convergence criteria is crucial for efficient and accurate simulations. Iterative solvers, commonly employed in CFD, demand careful calibration of convergence thresholds to ensure results are independent of initial guesses and mesh size [13, 15, 16].

3.3.4 Software Tools Used in Pharmaceutical CFD Studies

Software such as ANSYS Fluent, COMSOL Multiphysics, and OpenFOAM are widely utilized in the pharmaceutical industry. These tools offer specialized features and libraries for simulating fluid dynamics and heat transfer phenomena, catering to varying levels of expertise [17, 18].

3.3.5 Case Studies

Spray Dryer Optimization ANSYS Fluent can optimize spray dryer designs by simulating droplet size distribution and evaporation rates, ensuring uniform particle sizes in the final powder [19].

Sterilization Modeling COMSOL Multiphysics is often used to model heat transfer during sterilization, ensuring all product components achieve the requisite temperature for sterility. By leveraging advanced simulation tools, CFD facilitates process optimization, enhances product quality, and drives innovation in pharmaceutical research and development [20].

3.4 Drug Delivery Systems Optimization

3.4.1 Background

Devices like inhalers and nebulizers are critical for treating respiratory diseases, delivering medication directly to the lungs. The effectiveness of these devices depends on their ability to produce aerosol particles of optimal size and velocity for deep lung penetration and uniform drug distribution. CFD plays a vital role in analyzing and enhancing the fluid dynamics and aerosolization processes within these devices.

3.4.2 Understanding Fluid Dynamics and Aerosol Behavior

(a) *Aerosol Generation*

Aerosol generation in inhalers and nebulizers involves complex fluid dynamics, where liquid medication is atomized into fine droplets. The size distribution of these droplets dictates their deposition in the respiratory tract. CFD simulations enable the analysis and optimization of the atomization process to achieve ideal particle size distributions [21].

(b) *Airflow Patterns*

Airflow patterns within the device influence aerosol transport and deposition. Turbulent flows can enhance mixing but may also cause particle impaction losses on device walls. CFD aids in visualizing and optimizing airflow patterns to minimize losses and ensure effective drug delivery to the lungs [22, 23].

3.4.3 *CFD Methodologies for Device Optimization*

(a) *Geometry Optimization*

CFD simulations facilitate virtual testing of device geometries to identify designs that enhance aerosol transport and minimize deposition within the device. This includes optimizing the shape and size of components like the mouthpiece, nozzle, and internal chambers.

(b) *Particle Tracking*

CFD software incorporates particle tracking algorithms that simulate aerosol particle behavior within airflow. These algorithms account for factors such as gravity, inertia, and drag forces, offering insights into particle deposition likelihood in various respiratory tract regions. This information guides design modifications to maximize lung deposition.

(c) *Case Studies and Applications*

Inhaler Design: CFD studies have led to inhalers with optimized nozzle designs that produce uniform particle sizes, improving drug deposition in the lower respiratory tract. Innovations include redesigning metered-dose inhalers (MDIs) and dry powder inhalers (DPIs) for enhanced aerosolization and flow characteristics.

Nebulizer Efficiency: For nebulizers, CFD has improved the efficiency of converting liquid medication into aerosol form, reducing waste and treatment times. This involves optimizing the air-liquid interface and chamber geometry for consistent droplet size and distribution [24–27].

3.5 The Role of CFD in Drug Transport Modeling

CFD modeling of drug transport through biological tissues marks a significant advancement in pharmaceutical research. By combining fluid mechanics principles with physiological processes, CFD offers a powerful tool to simulate, analyze, and optimize drug delivery mechanisms [28–31].

(a) *Understanding Tissue Perfusion*

Tissue perfusion, or the flow of blood through capillary beds, is essential for delivering drugs to target sites. CFD simulations provide insights into how physiological conditions or disease-induced variations in blood flow affect drug distribution, particularly in cancer treatment, where tumors exhibit abnormal vasculature.

(b) *Simulating Drug Diffusion and Convection*

Drug transport through tissues involves diffusion (driven by concentration gradients) and convection (associated with fluid movement). CFD integrates these phenomena to predict transport processes, accounting for drug molecular properties, the biological matrix composition, and surrounding fluid dynamics.

3.5.1 Analyzing Transdermal and Targeted Drug Delivery

CFD is instrumental in optimizing transdermal systems (e.g., patches) and targeted delivery mechanisms (e.g., nanoparticle carriers). Simulations identify optimal formulations and delivery conditions, enhancing penetration through the skin or ensuring nanoparticles efficiently concentrate at disease sites, such as tumors [32, 33].

(a) *Challenges and Innovations*

While CFD offers remarkable capabilities for modeling drug transport, it faces challenges, particularly in accurately representing the highly heterogeneous and dynamic environment of biological tissues. Addressing these challenges requires:

High-resolution Tissue Modeling: Advanced imaging technologies are integrated with CFD models to capture the detailed anatomical and physiological features of tissues, improving the accuracy of simulations.

Multi-scale Modeling: Combining CFD with multi-scale models enables the bridging of different biological scales, from the macroscopic flow of blood and interstitial fluid to the microscopic transport of drug molecules within and between cells.

Patient-specific Modeling: Leveraging patient-specific data to inform CFD models can tailor drug delivery strategies to individual physiological conditions, enhancing treatment efficacy and reducing side effects.

(b) *Future Directions*

The future of CFD in modeling drug transport through biological tissues lies in its integration with other computational methods, experimental data, and clinical findings to create comprehensive, predictive models of drug behavior. Innovations in computational power, algorithms for fluid dynamics, and biological tissue characterization will further enhance the precision and applicability of CFD models. Moreover, the incorporation of machine learning techniques can facilitate the analysis of complex CFD simulations, identifying patterns and optimizing parameters for drug delivery more efficiently than traditional methods.

3.6 Bioreactor Design and Optimization

Bioreactors are vessels used for the cultivation of cells, microorganisms, or biological molecules under controlled conditions for the production of pharmaceuticals [34–36]. CFD facilitates the design of bioreactors by:

Fluid Flow Analysis: CFD simulations allow for the visualization and analysis of fluid flow patterns within the bioreactor, optimizing mixing and mass transfer. This ensures uniform distribution of nutrients and oxygen to the cells, essential for robust cell growth and product formation.

Heat and Mass Transfer Modeling: CFD models can predict temperature distributions and mass transfer rates within the bioreactor, ensuring optimal conditions for cell viability and productivity. This includes analyzing heat dissipation, oxygen transfer, and nutrient uptake by cells.

Geometry Optimization: CFD helps in optimizing the geometry of the bioreactor, including impeller design, baffle placement, and inlet/outlet configurations, to enhance mixing efficiency, minimize shear stress on cells, and maximize cell growth and product yield.

Further, in the pharmaceutical industries, the hydrogenation reactions are majorly performed in 103 L-reactors that are particularly typical to create, operate, and also to control. Hydrogenations requires use of pressurized vessels, with the use of heat exchangers, solid catalysts, organic solvents, and agitation like devices. Stirred tank reactors are commonly used in hydrogenation processes due to their effectiveness in achieving liquid mixing, gas dispersion, proper solids suspension, and temperature control. However, scaling up these systems poses challenges in maintaining efficiencies like yield and selectivity. Process limitations such as insufficient hydrogen supply due to poor gas–liquid contact, inadequate heat removal, and suboptimal mixing often arise during scale-up, affecting reaction efficiency. In liquid-phase hydrogenation, the availability of hydrogen at the catalyst surface is frequently the rate-limiting step. Identifying these limitations early is crucial, as they drive key design and operational choices, such as gas sparging rates, agitation speeds, reactor dimensions, and heat exchange requirements [37–39].

While laboratory-scale studies on reactor configurations can provide insights, they often require significant resources, multiple experimental batches, and may fail to address all potential process limitations. This is where Computational Fluid Dynamics (CFD) becomes an invaluable tool. CFD offers a model-based approach to design and process understanding, minimizing the need for extensive experimental trials and reducing the risk of scale-up failure.

CFD modeling extends beyond basic characterizations of mixing and energy dissipation efficiency to include complex phenomena such as kinetics, mass transfer, and heat transfer. By integrating these processes, CFD enables the evaluation of variables like agitation rate, reaction volume, and liquid surface level in relation to mass and heat transfer requirements. This comprehensive analysis is especially relevant in pharmaceutical engineering, where it supports process scale-up and optimization in scenarios where conventional methods may fall short, such as when real data are unavailable or expensive to obtain [40].

The advantages of CFD in hydrogenation systems include its ability to simulate potential process inefficiencies, identify critical parameters, and optimize design and operational strategies. For example, it can address poor hydrogen availability, uneven reactant distribution, or inadequate thermal control, ensuring robust process performance. Moreover, CFD's cost-effectiveness and predictive capabilities make it a valuable complement to experimental workflows, enabling more efficient scale-up and development of hydrogenation processes [41].

CFD is a transformative tool for pharmaceutical engineering, offering enhanced process understanding, risk reduction, and cost-effective scale-up solutions. Its ability to integrate complex dynamics into the design process ensures better scalability, improved efficiency, and greater reliability in hydrogenation operations, making it an indispensable asset for modern pharmaceutical development.

In the realm of chemical engineering, particularly in the design and operation of mechanically stirred reactors, the focus has traditionally been on baffled configurations. These systems have dominated academic research due to their well-understood fluid mechanics and predictable mixing behaviors. However, a paradigm shift is occurring as researchers increasingly explore unbaffled configurations, which present unique advantages in various applications, particularly those involving shear-sensitive processes such as crystallization and animal cell cultures.

3.6.1 The Case for Unbaffled Reactors

Unbaffled stirred reactors offer several compelling benefits over their baffled counterparts. One notable advantage is their ease of solid dispersion and wetting, which enhances gas entrainment. This characteristic is crucial in processes requiring effective mass transfer, such as hydrogenation and fermentation. Furthermore, unbaffled systems demonstrate lower power consumption, making them economically attractive for industrial applications [42, 43]. Historical studies have laid the groundwork for understanding vortex formation in these systems, providing both experimental and theoretical insights into their operational dynamics.

(a) *Historical Context and Research Developments*

The study of vortex formation in unbaffled tanks dates back several decades, with early investigations focusing on surface vortex characteristics under varying conditions. Researchers have employed a range of experimental techniques to analyze critical parameters such as vortex depth, free surface shape, and the influence of flow regimes on power consumption. Notably, the use of computational fluid dynamics (CFD) has revolutionized this field, allowing for detailed simulations that enhance our understanding of unbaffled reactor performance. One of the pioneering CFD models for unbaffled systems was developed to assess the flow dynamics within these reactors. Subsequent studies have utilized various turbulence models—ranging from conventional two-equation models to more advanced approaches like Reynolds Stress Model (RSM), Detached Eddy Simulation (DES), and Large Eddy Simulation (LES)—to capture the complexities of flow behavior accurately. Despite the superior accuracy offered by advanced models, simpler two-equation models, such as the $k - \omega k - \omega$ SST model, still provide reasonable predictions for vortex size and velocity components [44, 45].

(b) *Advancing Process Understanding through CFD*

A recent study aimed to deepen the understanding of an existing unbaffled cylindrical reactor equipped with two vertically mounted impellers through CFD modeling. This investigation sought not only to validate the predictive capabilities of the CFD model but also to demonstrate its potential in an industrial context. The initial phase involved simulating a simplified version of the reactor that retained essential flow physics while utilizing a Rushton turbine as the agitation device. The findings revealed that effective mixing could be achieved when sufficient vertical homogenization occurred. This outcome aligns with expectations, as the surface vortex generated by the Rushton turbine promotes circular and radial dispersion but is less effective in facilitating axial mixing. The implications are significant; alternative impeller designs could enhance axial mixing while maintaining high mass transfer rates at reasonable power consumption levels [46].

(c) *Implications for Hydrogen Entrainment and Mass Transfer*

Interestingly, while traditional methods rely on surface vortices for gas entrainment, this study noted that hydrogen entrainment could instead occur from a wavy solvent surface under certain conditions. This observation presents a modeling challenge but also opens avenues for further exploration into optimizing gas-liquid interactions within unbaffled reactors. The CFD model proved instrumental in predicting the development of surface vortices around the impeller shaft as a function of rotational speed. By simulating various scenarios with reduced effort and resources, researchers were able to evaluate critical parameters for hydrogenation processes virtually. These simulations facilitated insights into how reactor design influences mass transfer efficiency—specifically, how vortex size correlates with available interfacial area for mass transfer [47].

(d) *Developing Novel Process Correlations*

The outcomes from these simulations have led to the formulation of novel process correlations that estimate key variables such as interfacial area for mass transfer and liquid height for heat exchange calculations. Additionally, they provide estimates for mixing bulk times—critical factors for optimizing reactor performance. Despite being a mature modeling framework that offers substantial value, further efforts are needed to enhance CFD's acceptance within the pharmaceutical industry. Establishing greater trust in CFD predictions can be achieved through dedicated measurement campaigns that leverage smart reactor designs and online Process Analytical Tools (PAT) [48]. These advancements would enable real-time monitoring of evolving concentrations and properties, thereby improving process control.

(e) *Enhancing CFD Models*

Future research should focus on increasing modeling complexity to capture more intricate phenomena associated with unbaffled reactors [49, 50]. Potential enhancements include:

- Explicitly modeling H_2 mass transfer coupled with reaction kinetics.
- Simulating agitation using sliding mesh approaches.
- Incorporating population balance models (PBM) to assess particle dynamics.
- Conducting LES simulations for more accurate flow field representations.

Additionally, exploring scenarios such as substrate/catalyst dosing effects or developing reduced-order models (ROM) could yield valuable insights into optimizing biopharmaceutical processes.

CFD should not be perceived merely as a substitute for established methodologies; rather, it should be embraced as a complementary tool that enriches our understanding of mechanically stirred reactors. By leveraging CFD's capabilities, researchers can unlock new knowledge that enhances existing processes and drives innovation in biopharmaceutical manufacturing. As we continue to refine our approaches to reactor design and operation, embracing both traditional wisdom and modern computational techniques will be essential in advancing our capabilities within this dynamic field.

(f) *Advances in Bioreactor Optimization through CFD Modeling*

Recent advancements in Computational Fluid Dynamics (CFD) modeling have significantly enhanced the optimization of bioreactors, which are critical in the production of biopharmaceuticals. These innovations encompass various methodologies that improve our understanding of biophysical and biochemical processes, ultimately leading to more efficient and effective bioprocessing.

(g) *Multi-scale Modeling*

One of the most notable advancements is the integration of CFD with multi-scale modeling. This approach allows for the simulation of complex processes occurring at different length and time scales, from molecular interactions to macroscopic fluid dynamics. By providing a holistic view of bioreactor performance, multi-scale modeling facilitates targeted optimization strategies that can address specific challenges within the bioprocess. For instance, researchers can analyze how molecular-level interactions affect overall fluid dynamics and mass transfer efficiency, leading to improved reactor designs tailored to specific applications [51].

(h) *High-Performance Computing (HPC)*

The evolution of High-Performance Computing (HPC) has also played a pivotal role in advancing bioreactor optimization. With increased computational power and parallel computing algorithms, researchers can now simulate larger and more complex bioreactor systems with higher resolution and faster turnaround times. This capability accelerates the design iteration process, allowing for rapid prototyping and testing of various operating conditions. Consequently, this reduces the time-to-market for biopharmaceutical products, enabling companies to respond more swiftly to market demands [52, 53].

(i) *Data-driven Modeling*

Another significant trend is the adoption of data-driven modeling techniques that combine CFD with real-time experimental data and machine learning. This approach enhances predictive accuracy by leveraging sensor data to calibrate and validate CFD models. As a result, researchers can achieve better process control and optimize bioreactor operations based on real-time feedback. By integrating machine learning algorithms, it becomes possible to identify patterns and correlations within complex datasets that may not be apparent through traditional analytical methods [54, 55].

CFD plays a crucial role in the design and optimization of bioreactors for pharmaceutical production. By simulating various aspects of biopharmaceutical processes—including nutrient flow dynamics, cell growth, and product yield—CFD provides detailed insights into fluid dynamics, heat and mass transfer, and cell behavior within bioreactors. These insights enable the development of robust and efficient bioprocesses, ultimately leading to the production of high-quality biopharmaceutical products. As advancements in CFD modeling techniques and computational resources continue to evolve, we can expect further enhancements in bioreactor optimization. These innovations will not only accelerate the development of innovative biopharmaceuticals but also support diverse therapeutic applications across the healthcare landscape [56, 57].

(j) *Role of CFD in Designing Bioreactors for Pharmaceutical Production*

Simulation of Nutrient Flow: CFD models help visualize nutrient transport dynamics within bioreactors, ensuring optimal distribution for cell growth.

Cell Growth Predictions: By incorporating growth kinetics into simulations, researchers can forecast biomass concentration and product formation over time.

Product Recovery Optimization: CFD assists in modeling downstream processes such as filtration and chromatography, enhancing separation efficiency.

3.7 CFD in Tablet Manufacturing

Tablet coating is a critical step in pharmaceutical manufacturing that ensures uniform drug distribution and controlled release [2, 58]. CFD plays a vital role in optimizing tablet coating processes through:

3.7.1 Fluid Dynamics Analysis

CFD simulations allow for detailed visualization of fluid flow patterns within coating equipment like pan coaters or fluidized bed coaters. By analyzing airflow and spray distribution, manufacturers can optimize coating uniformity while minimizing defects such as edge buildup or over-coating.

(a) Mass Transfer Modeling
    CFD models simulate the transport dynamics of coating solutions onto tablet surfaces. This includes accounting for evaporation rates and solvent drying kinetics, ensuring precise control over coating parameters like film thickness and drug release profiles.
(b) Geometry Optimization
    CFD aids in optimizing the design of coating equipment by analyzing fluid dynamics and heat transfer characteristics. This optimization leads to improved equipment efficiency and uniformity in coating applications, ultimately reducing production time and costs [58].

3.7.2 Simulation of Pharmaceutical Powder Mixing Techniques

Mixing processes are fundamental in pharmaceutical manufacturing, ensuring homogeneity among active pharmaceutical ingredients (APIs) and excipients. Various mixing equipment—including static mixers, stirred tanks, homogenizers, and emulsifiers—are employed depending on the application requirements [59–61].

(a) Static Mixers
    Static mixers are commonly used for high-viscosity fluids. CFD methods can evaluate their performance by predicting mixing efficiency and indicating whether additional mixing elements are necessary.
(b) Stirred Tank Reactors
    Stirred tank reactors are prevalent in the pharmaceutical industry due to their versatility. The mixing characteristics significantly influence product quality; thus, understanding how impeller design affects mixing efficiency is crucial. Traditional trial-and-error approaches often lead to delays; hence CFD provides a more accurate means to predict vessel performance without relying solely on empirical correlations.

By leveraging these advanced modeling techniques, researchers can enhance mixing processes significantly while reducing development times and costs associated with pharmaceutical manufacturing.

3.7.3 *Enhancing Pharmaceutical Manufacturing Efficiency Through CFD Simulation*

Computational Fluid Dynamics (CFD) simulations have emerged as a transformative tool in pharmaceutical manufacturing, particularly in optimizing processes such as powder mixing and tablet production [2]. By providing detailed insights into fluid dynamics, mass transfer, and particle behavior, CFD enhances the efficiency and quality of manufacturing operations.

(a) *CFD in Powder Mixing Techniques*
CFD simulations significantly improve powder mixing techniques by addressing several key areas. CFD models simulate the flow patterns of powders within mixing equipment, such as tumbling blenders and high-shear mixers. By analyzing particle trajectories, shear rates, and turbulence levels, these simulations help optimize mixing efficiency while minimizing issues like segregation or agglomeration. Understanding how powders behave under different conditions allows manufacturers to tailor their processes for better outcomes.

(b) *Particle-Particle Interactions*
CFD accounts for critical particle-particle interactions, including collisions, adhesion, and breakage. These interactions are vital for predicting blending behavior and kinetics. By identifying optimal mixing parameters—such as rotation speed and mixing time—manufacturers can enhance the consistency and quality of the final product.

(c) *Scale-up and Process Optimization*
One of the most significant advantages of CFD is its ability to facilitate the scale-up of powder mixing processes from laboratory to production scale. By predicting how equipment scale, geometry, and operating conditions impact mixing performance, CFD enables efficient process optimization. This capability ensures consistent product quality across different batch sizes, which is crucial for maintaining regulatory compliance in pharmaceutical production.

3.7.4 *Applications of CFD in Tablet Manufacturing Efficiency*

CFD plays a pivotal role in enhancing tablet manufacturing efficiency through various applications [58–61]:

(a) *Optimizing Granulation Processes*
In wet granulation, CFD models simulate fluid-particle interactions to aid in designing granulator geometries, impeller configurations, and spray nozzle placements. This optimization leads to improved granule size distribution and flow properties, which are essential for achieving uniformity in tablet formulation.

(b) *Predicting Tablet Compression Behavior*

CFD also predicts tablet compression behavior by simulating the compaction processes involved in die filling, compaction, and ejection. This predictive capability allows for the optimization of tablet press settings, tooling design, and formulation parameters to achieve desired tablet properties.

(c) *Reducing Manufacturing Costs*

By optimizing energy consumption and minimizing material waste, CFD contributes to reducing overall manufacturing costs. Through detailed analysis of fluid dynamics, heat transfer, and mass transport, CFD identifies inefficiencies in processes and suggests cost-effective solutions that enhance productivity and profitability.

(d) *Enhancing Spray Drying Techniques*

Spray drying is a widely used method in the pharmaceutical industry due to its ability to handle heat-sensitive components while retaining their potency. However, scaling up this process can be challenging due to its complexity. CFD simulation tools help design spray-drying processes by reducing time and costs while enhancing product quality. The spray-drying process involves several critical steps: atomization of the liquid feed, interaction with hot air, drying of the atomized feed, and particle collision phenomena. Understanding these interactions is crucial for optimizing product morphology and yield.

(e) *Particle-Particle Collisions*

These can lead to coalescence or agglomeration during drying. Modeling these collisions helps predict the final morphology of the product.

Particle-Wall Collisions: These collisions may result in wall deposition of powders, reducing yield and product quality.

CFD models can simulate these collision phenomena by classifying particles based on whether they are dominated by surface tension or viscous forces. This classification aids in understanding stickiness conditions that affect agglomeration during spray drying.

3.8 Inhalation Aerosols and Respiratory Drug Delivery

Aerosol behavior in the respiratory tract is governed by intricate fluid dynamics, encompassing airflow patterns, particle deposition mechanisms, and physiological factors like airway geometry and breathing patterns. Computational fluid dynamics (CFD) simulations have emerged as powerful tools to unravel these complexities and optimize drug delivery [30, 62–64]. This chapter explores the contributions of CFD modeling to inhalation aerosol research and its application in developing inhalable drugs for respiratory diseases.

3.8.1 Understanding Aerosol Dynamics Through CFD Modeling

Airflow Modeling CFD models simulate airflow within the respiratory tract, capturing the nuances of anatomical structures such as the nasal cavity, pharynx, trachea, and bronchi. By solving the Navier-Stokes equations, CFD elucidates airflow patterns, velocity profiles, and pressure distributions. These insights reveal how aerosol particles traverse and disperse throughout the airways [30].

Particle Deposition Analysis CFD enables the detailed analysis of particle deposition by accounting for parameters like particle size, density, shape, and flow conditions. Through trajectory modeling and interactions with airway walls, CFD quantifies deposition mechanisms, including inertial impaction, sedimentation, diffusion, and interception. This analysis highlights critical deposition zones and helps refine inhalation strategies [62, 63].

Regional Deposition Mapping CFD simulations produce regional deposition maps, visualizing spatial distributions of particles within lung regions such as the upper airways, tracheobronchial tree, and alveolar spaces. These maps identify deposition hotspots and guide the optimization of drug formulations and inhalation devices [64].

3.8.2 Advancements in Pulmonary Drug Delivery

Pulmonary drug delivery has become a cornerstone for treating respiratory diseases, offering the advantage of targeting a large lung surface area. However, the efficacy of inhaled drugs remains limited, with less than 30% of the dose typically reaching the lungs. A significant proportion of the drug is deposited in areas such as the valve stem, nozzle, or mouth–throat airway due to obstructions or poor inhaler technique [29, 65].

Challenges in Pulmonary Drug Delivery:

Coordination Issues: Patients often struggle to synchronize inhaler actuation with inhalation, leading to suboptimal drug deposition.

Uneven Drug Distribution: Due to complex lung geometry and varying airflow dynamics, drug deposition is frequently uneven, reducing therapeutic efficacy.

Device Design Limitations: Factors such as nozzle shape, cone angle, spray plume characteristics, and flow rate critically influence aerosol deposition.

Ideal Aerosol Characteristics: Effective pulmonary drug delivery requires particles to be within the respirable size range (2–7 μm). Smaller particles (<3 μm) can reach peripheral lung regions, while larger ones may be intercepted along the airway path. Optimizing these characteristics is pivotal for enhancing treatment outcomes.

3.8.3 CFD-Driven Insights and Innovations

CFD has revolutionized pulmonary drug delivery by providing a high-accuracy, cost-effective alternative to experimental setups [66–68]. Here are some key contributions:

Optimizing Device Design: CFD studies have demonstrated the impact of nozzle geometry, spray plume dynamics, and airflow rates on particle deposition. For instance, a study by Zhang et al. introduced a curved 90° bend in an induction port, improving drug delivery by reducing flow circulation near the bend.
Analyzing Spray Plume Behavior: Investigations into plume dynamics have shown that cone angles and flow rates significantly affect particle distribution. A narrower cone angle reduces mouth–throat deposition, while higher flow rates enhance particle penetration into the lungs.
Flow Rate and Turbulence Effects: Studies reveal that higher Reynolds numbers increase turbulence, which affects fine particle deposition. For example, Bass and Longest highlighted how airflow and turbulence influence particle behavior, emphasizing the need for tailored flow rates for optimal delivery.
Experimental Validation: Many CFD findings align closely with experimental data, as demonstrated by studies on Andersen cascade impactors and spray plume characteristics. These validations underscore the reliability of CFD in predicting aerosol behavior.

3.8.4 Case Studies in CFD Applications

Improving Inhaler Efficiency Chen J et al. showed that decreasing nozzle diameter significantly enhances fine particle dose, underscoring the importance of precision engineering in inhaler design [69].

Spray Plume Dynamics Inthavong K et al. analyzed airflow–plume interactions, revealing how parameters like cone angle and spray velocity influence deposition patterns. Their findings emphasized the need for optimal plume characteristics to reduce deposition in the mouth–throat pathway [21].

Flow Rate Correlations Sperry T et al. established correlations between flow rates, Stokes number, and deposition efficiency. Their work provided a framework for predicting deposition patterns based on inhaler configurations and user-specific factors [70].

Innovative Nozzle Designs Studies by Chen et al. explored the effects of different nozzle shapes on spray plume characteristics. Curved-cone nozzles generated narrower plumes, enhancing drug targeting in lower lung regions [69].

3.8.5 The Future of CFD in Pulmonary Drug Delivery

Advances in CFD modeling continue to shape the future of respiratory drug delivery [66]. By integrating high-resolution imaging data and real-world patient variables, future CFD models could:

- Predict patient-specific deposition patterns.
- Enhance personalized treatment plans.
- Drive the development of next-generation inhalation devices.

CFD remains a cornerstone technology, bridging the gap between theoretical models and practical applications in respiratory drug delivery. Its potential to innovate and refine therapeutic strategies ensures its enduring relevance in the pharmaceutical landscape [69].

The development of inhalable drugs for respiratory diseases necessitates precise targeting of drug delivery within the lungs to maximize therapeutic efficacy while minimizing systemic side effects. Computational Fluid Dynamics (CFD) modeling significantly contributes to this endeavor by offering insights into:

(a) *Formulation Design*
CFD simulations provide a detailed understanding of how inhalable drug formulations—such as aerosols, dry powders, or nanoparticles—behave within the respiratory tract. By simulating particle dispersion, dissolution, and absorption, CFD optimizes key formulation parameters such as particle size distribution, excipients, and drug release kinetics. This precision enhances lung deposition and improves therapeutic outcomes.

(b) *Device Optimization*
Inhalation devices, including metered-dose inhalers (MDIs), dry powder inhalers (DPIs), and nebulizers, are critical to delivering drugs effectively. CFD models evaluate device performance by analyzing aerosol generation, particle size distribution, and inhalation flow profiles. These insights enable the refinement of device design and operating parameters, ensuring efficient drug delivery, reproducible dosing, and improved patient compliance [71, 72].

(c) *Inhalation Therapy Planning*
CFD simulations assist in tailoring inhalation therapies to individual patients. By integrating patient-specific lung geometries and respiratory conditions obtained through medical imaging, clinicians can optimize drug deposition and targeting to achieve superior therapeutic outcomes.

3.9 Optimization of Drug Deposition and Targeting in the Lungs with CFD

Optimizing drug deposition and targeting in the lungs requires a nuanced understanding of aerosol properties, inhalation techniques, and lung physiology. CFD modeling plays a pivotal role in achieving this by:

(a) *Targeted Drug Delivery*

CFD simulations guide the design of inhalation devices and drug formulations for precise delivery to lung regions affected by diseases such as asthma, chronic obstructive pulmonary disease (COPD), and cystic fibrosis. By optimizing parameters like particle size, velocity, and distribution, CFD ensures maximum drug deposition at target sites while minimizing off-target effects [73, 74].

(b) *Disease-Specific Modeling*

CFD models incorporate factors unique to respiratory diseases, such as airway inflammation, mucus accumulation, and airflow obstruction. This allows for the evaluation of disease-specific drug delivery strategies and their impact on disease progression and symptom relief [75].

(c) *Combination Therapy Optimization*

CFD enables the study of combination therapies, where multiple drugs are delivered via inhalation. By modeling drug interactions and spatial distribution within the lungs, CFD identifies optimal regimens to enhance efficacy and reduce resistance [66].

(d) *Cleanroom Design and Contamination Control*

Cleanrooms are essential for pharmaceutical manufacturing, where maintaining air quality and preventing contamination are paramount [76]. CFD modeling facilitates the design of cleanrooms by addressing:

Airflow Analysis - CFD simulations visualize airflow patterns, such as laminar and turbulent flows, within cleanrooms. These models optimize air distribution, ensuring uniform flow and minimizing particle accumulation [77].
Particle Dispersion Modeling - By simulating particle transport mechanisms—such as turbulent diffusion and gravitational settling—CFD identifies areas prone to contamination. These insights guide the placement of critical equipment and filtration systems to maintain sterility [78].
Contamination Control Strategies - CFD highlights potential contamination risks from personnel movement, equipment operation, and material handling. By simulating these activities, it helps design cleanroom protocols that minimize cross-contamination [78].

(e) *Regulatory Compliance*

CFD ensures cleanrooms adhere to ISO standards, FDA guidelines, and GMP requirements. By simulating particle concentrations and airflow patterns, CFD validates that facilities meet regulatory cleanliness criteria.

3.10 Advanced Applications of CFD in Pharmaceuticals

CFD modeling is transforming various aspects of pharmaceutical research and development, particularly in:

(a) *Microfluidics and Lab-on-a-Chip Devices*
CFD optimizes microfluidic platforms by simulating fluid dynamics, particle transport, and multi-phase flows. These devices are integral to drug screening, diagnostics, and personalized medicine [79].
(b) *Pharmacokinetics Modeling*
CFD enhances pharmacokinetics (PK) studies by simulating drug distribution, tissue perfusion, and cellular uptake. These insights are critical for understanding drug behavior in the body and optimizing therapeutic regimens [80].
(c) *Multi-scale Modeling and AI Integration*
By integrating multi-scale modeling with artificial intelligence, CFD is enabling patient-specific simulations and the prediction of drug-drug interactions. This intersection of technologies promises to revolutionize personalized medicine [81].

Computational Fluid Dynamics (CFD) is a transformative tool in pharmaceutical sciences, offering unparalleled insights into fluid dynamics, particle behavior, and biological interactions. From enhancing inhalation therapies to optimizing cleanroom designs and advancing personalized medicine, CFD has become an indispensable asset. As computational technologies continue to evolve, the integration of CFD with AI and patient-specific modeling will drive groundbreaking innovations, improving therapeutic outcomes and patient quality of life.

(d) *Packaging Optimization*
CFD addresses challenges in liquid pharmaceutical packaging, such as splashing and frothing during filling. Virtual experimentation allows for optimized filling lines and packaging geometries, reducing time-to-market and enhancing productivity.
(e) *Solids Handling*
CFD improves the handling of solid particles by analyzing flows in pneumatic transport systems and fluidized beds. These simulations help minimize risks like erosion and chaotic flow behaviors, ensuring efficient and reliable processes.
(f) *Separation Processes*
CFD optimizes separation techniques like crystallization, precipitation, and centrifugation. For instance, design improvements in industrial centrifuges have been achieved through CFD, addressing issues such as slugging and flow irregularities. CFD also evaluates separation devices like cyclones and scrubbers, enabling performance enhancements and design validation [82, 83].

Computational Fluid Dynamics (CFD) has revolutionized pharmaceutical development by supporting Quality by Design (QbD) principles, enhancing process understanding, and ensuring product safety and regulatory compliance. Future advancements in AI integration, personalized medicine, multi-scale modeling, and high-performance computing promise to expand CFD's role, driving innovation and efficiency in the pharmaceutical industry. These advancements not only accelerate the development of novel therapeutics but also improve patient outcomes and overall healthcare delivery.

3.11 Quality by Design and Regulatory Compliance

Quality by Design (QbD) is a systematic approach to pharmaceutical development that emphasizes understanding and controlling product and process variability. By focusing on Critical Quality Attributes (CQAs) and ensuring consistent product quality, QbD has become central to modern pharmaceutical practices. Computational Fluid Dynamics (CFD) supports QbD principles through advanced modeling, enabling enhanced process understanding, risk assessment, and regulatory compliance [84, 85].

3.11.1 CFD in QbD Implementation

(a) *Process Understanding*

CFD simulations provide invaluable insights into fluid dynamics, heat transfer, and mass transport within pharmaceutical processes such as mixing, granulation, and coating. By modeling these processes, CFD helps identify Critical Process Parameters (CPPs) and Critical Material Attributes (CMAs) that directly influence product quality [84–86].

(b) *Design Space Exploration*

Virtual experimentation through CFD allows the exploration of design space and assesses the effects of process variations on CQAs. By simulating diverse scenarios, it becomes possible to define acceptable ranges for CPPs and CMAs, ensuring the consistent quality of pharmaceutical products [84–86].

(c) *Risk Assessment*

CFD facilitates risk assessment by identifying potential failure modes and deviations that may impact product quality. Sensitivity analyses and worst-case scenario modeling help prioritize risks and implement robust mitigation strategies, reducing variability and ensuring reliable processes [84–86].

3.11.2 Risk Assessment and Virtual Experimentation with CFD

(a) *Failure Mode Analysis*
CFD identifies failure modes like inadequate mixing, uneven coating, or incomplete dissolution by simulating process deviations and equipment malfunctions. These insights guide the development of strategies to minimize or eliminate quality risks [87, 88].

(b) *Sensitivity Analysis*
By systematically varying input parameters such as flow rates, temperatures, and particle properties, CFD helps determine the impact of these variables on CQAs. This enables precise control over critical parameters to achieve product consistency and performance.

(c) *Design of Experiments (DoE)*
CFD supports systematic experimentation to evaluate the effects of multiple factors on product quality. Using response surfaces and contour plots, it aids in identifying optimal conditions and interactions between variables, streamlining process optimization.

3.11.3 Role of CFD in Product Quality, Safety, and Regulatory Compliance

(a) *Process Optimization*
CFD enhances pharmaceutical processes by identifying inefficiencies, optimizing equipment designs, and improving process control strategies. This ensures consistent product quality while meeting stringent regulatory requirements [89, 90].

(b) *Product Safety*
CFD evaluates potential safety risks, particularly in inhalable drug formulations, by simulating particle dispersion and deposition patterns in the respiratory tract. This ensures both minimal risks and optimal therapeutic efficacy [89, 90].

(c) *Regulatory Compliance*
CFD provides detailed documentation of process understanding, risk assessments, and virtual experimentation. This evidence aligns with guidelines such as the FDA's Process Validation Guidance and ICH Q8/Q9/Q10, ensuring regulatory compliance.

3.12 Future Trends and Innovations

The future of CFD in pharmaceuticals is characterized by advancements in technology, personalized approaches, and innovative methodologies.

(a) *Integration of Machine Learning and AI*

Combining CFD with machine learning and AI enables data-driven modeling, automated optimization, and real-time analytics. These technologies accelerate drug discovery, formulation development, and process optimization, paving the way for personalized and efficient pharmaceutical interventions [33, 91, 92].

(b) *Personalized Medicine and Patient-Specific Modeling*

Advances in medical imaging and biomarker analysis enable the creation of patient-specific CFD models. These models facilitate tailored treatments by simulating drug responses, toxicity, and efficacy in virtual patient populations, improving outcomes and reducing costs.

(c) *Multi-scale and Multi-physics Modeling*

Future CFD applications will integrate multi-scale modeling to capture phenomena from molecular interactions to organ-level dynamics. By bridging scales, these models offer a comprehensive understanding of drug behavior and disease mechanisms, supporting more effective therapies [56, 57].

(d) *Virtual Clinical Trials and Regulatory Approval*

CFD's ability to simulate drug behavior and patient responses supports virtual clinical trials. These simulations reduce the time, cost, and ethical challenges of traditional trials while providing robust evidence for regulatory approval processes [93–96].

(e) *Advanced Computational Tools and High-Performance Computing (HPC)*

The adoption of HPC and cloud-based platforms enhances CFD simulations, enabling complex models with higher resolution. This technological leap accelerates innovation and efficiency in pharmaceutical research and development [97–99].

References

1. Babnik S, Erklavec Zajec V, Oblak B, Likozar B, Pohar A. A review of computational fluid dynamics (CFD) simulations of mixing in the pharmaceutical industry. BJSTR. 2020;27(3):20732–6.
2. Hemamanjushree S, Tippavajhala VK. Simulation of unit operations in formulation development of tablets using computational fluid dynamics. AAPS PharmSciTech. 2020;21(3):103.
3. Sarkar A, Shoemaker B, Doshi P, am Ende MT, Jajcevic D, Böhling P, Toson P, Zadravec M, Khinast JG. Multiscale modeling of a pharmaceutical fluid bed coating process using Cfd/Dem and population balance models to predict coating uniformity. In: Chemical engineering in the pharmaceutical industry: drug product design, development, and modeling. John Wiley & Sons, Ltd: New Jersey; 2019. p. 419–50.

4. Zheng C, Behjani MA, Hu J, Wu CY. Discrete element modelling of pharmaceutical powder handling processes. In: Simulations in bulk solids handling: applications of DEM and other methods. Wiley-VCH GmbH, Weinheim: Germany; 2023. p. 199–230.
5. Zheng Z, Leung SS, Gupta R. Flow and particle modelling of dry powder inhalers: methodologies, recent development and emerging applications. Pharmaceutics. 2021;13(2):189.
6. Singh V, Shah K, Garg A, Dewangan HK. Computational fluid dynamics: insights and applications in the pharmaceutical field. Lett Drug Design Discov. 2024;21(3):440–50.
7. Sen KK, Nayak AK. Computational fluid dynamics in pharmaceutical formulation development. In: Novel formulations and future trends. Academic: Amsterdam; 2024. p. 453–81.
8. Spasov GH, Rossi R, Vanossi A, Cottini C, Benassi A. A critical analysis of the CFD-DEM simulation of pharmaceutical aerosols deposition in extra-thoracic airways. Int J Pharm. 2022;629:122331.
9. Shan X, Yuan XF, Chen H. Kinetic theory representation of hydrodynamics: a way beyond the Navier–Stokes equation. J Fluid Mech. 2006;550:413–41.
10. Tamburrino A. From Navier to Stokes: commemorating the bicentenary of Navier's equation on the lay of fluid motion. Fluids. 2024;9(1):15.
11. Davidovic D, Liu J, Heidarinejad M, Srebric J. Airflow study for a cluster of campus buildings using different turbulence modeling approaches. Int J Build Urban Landsc Technol. 2014;3:33–58.
12. Pordal HS, Matice CJ, Fry TJ. Computational fluid dynamics in the pharmaceutical industry. Pharm Technol. 2002;26(2):72–9.
13. Jang J, Arastoopour H. CFD simulation of a pharmaceutical bubbling bed drying process at three different scales. Powder Technol. 2014;263:14–25.
14. Vashishta B, Garg M, Chaudhary R, Sahni H, Khanna R, Rathore AS. Use of computational fluid dynamics for development and scale-up of a helical coil heat exchanger for dissolution of a thermally labile API. Org Process Res Dev. 2013;17(10):1311–9.
15. Thakur AK, Kumar R, Kumar VV, Kumar A, Gaurav GK, Gupta KN. A critical review on thermodynamic and hydrodynamic modeling and simulation of liquid antisolvent crystallization of pharmaceutical compounds. J Mol Liq. 2022;362:119663.
16. Kremer DM, Hancock BC. Process simulation in the pharmaceutical industry: a review of some basic physical models. J Pharm Sci. 2006;95(3):517–29.
17. Kodman JB, Singh B, Murugaiah M. A comprehensive survey of open-source tools for computational fluid dynamics analyses. J Adv Res Fluid Mech Therm Sci. 2024;119(2):123–48.
18. Kuharat S. Numerical study of nanofluid-based direct absorber solar collector systems with metallic/carbon nanoparticles, multiple geometries and multi-mode heat transfer. University of Salford (United Kingdom): Manchester; 2021.
19. Hussain F, Jaskulski M, Piatkowski M, Tsotsas E. CFD simulation of agglomeration and coalescence in spray dryer. Chem Eng Sci. 2022;247:117064.
20. Knoerzer K, Juliano P, Gladman S, Versteeg C, Fryer PJ. A computational model for temperature and sterility distributions in a pilot-scale high-pressure high-temperature process. AICHE J. 2007;53(11):2996–3010.
21. Inthavong K, Tian ZF, Tu JY, Yang W, Xue C. Optimising nasal spray parameters for efficient drug delivery using computational fluid dynamics. Comput Biol Med. 2008;38(6):713–26.
22. Dutta R, Spence B, Wei X, Dhapare S, Hindle M, Longest PW. CFD guided optimization of nose-to-lung aerosol delivery in adults: effects of inhalation waveforms and synchronized aerosol delivery. Pharm Res. 2020;37(10):199.
23. Burrowes KS, De Backer J, Kumar H. Image-based computational fluid dynamics in the lung: virtual reality or new clinical practice? Wiley Interdiscip Rev Syst Biol Med. 2017;9(6):e1392.
24. Ruzycki CA, Javaheri E, Finlay WH. The use of computational fluid dynamics in inhaler design. Expert Opin Drug Deliv. 2013;10(3):307–23.
25. Wong W, Fletcher DF, Traini D, Chan HK, Young PM. The use of computational approaches in inhaler development. Adv Drug Deliv Rev. 2012;64(4):312–22.

26. Friebel C, Steckel H, Müller BW. Rational design of a dry powder inhaler: device design and optimisation. J Pharm Pharmacol. 2012;64(9):1303–15.
27. Fletcher DF, Chaugule V, Gomes dos Reis L, Young PM, Traini D, Soria J. On the use of computational fluid dynamics (CFD) modelling to design improved dry powder inhalers. Pharm Res. 2021;38:277–88.
28. Kleven M, Melaaen MC, Djupesland PG. Computational fluid dynamics (CFD) applied in the drug delivery design process to the nasal passages: a review. J Mech Med Biol. 2012;12(01):1230002.
29. Longest PW, Bass K, Dutta R, Rani V, Thomas ML, El-Achwah A, Hindle M. Use of computational fluid dynamics deposition modeling in respiratory drug delivery. Expert Opin Drug Deliv. 2019;16(1):7–26.
30. Huang F, Zhu Q, Zhou X, Gou D, Yu J, Li R, Tong Z, Yang R. Role of CFD based in silico modelling in establishing an in vitro-in vivo correlation of aerosol deposition in the respiratory tract. Adv Drug Deliv Rev. 2021;170:369–85.
31. Bozsak F, Chomaz JM, Barakat AI. Modeling the transport of drugs eluted from stents: physical phenomena driving drug distribution in the arterial wall. Biomech Model Mechanobiol. 2014;13:327–47.
32. Kumar M, Hilles AR, Almurisi SH, Bhatia A, Mahmood S. Micro and nano-carriers-based pulmonary drug delivery system: their current updates, challenges, and limitations–a review. JCIS Open. 2023;12:100095.
33. Gholap AD, Uddin MJ, Faiyazuddin M, Omri A, Gowri S, Khalid M. Advances in artificial intelligence in drug delivery and development: a comprehensive review. Comput Biol Med. 2024;178:108702.
34. Hutmacher DW, Singh H. Computational fluid dynamics for improved bioreactor design and 3D culture. Trends Biotechnol. 2008;26(4):166–72.
35. Shu L, Yang M, Zhao H, Li T, Yang L, Zou X, Li Y. Process optimization in a stirred tank bioreactor based on CFD-Taguchi method: a case study. J Clean Prod. 2019;230:1074–84.
36. Wu B. Advances in the use of CFD to characterize, design and optimize bioenergy systems. Comput Electron Agric. 2013;93:195–208.
37. Smith JD, Neto AA, Cremaschi S, Crunkleton DW. CFD-based optimization of a flooded bed algae bioreactor. Ind Eng Chem Res. 2013;52(22):7181–8.
38. Rathore AS, Sharma C, Persad AA. Use of computational fluid dynamics as a tool for establishing process design space for mixing in a bioreactor. Biotechnol Prog. 2012;28(2):382–91.
39. Yan X, Wu Q, Sun J, Liang P, Zhang X, Xiao K, Huang X. Hydrodynamic optimization of membrane bioreactor by horizontal geometry modification using computational fluid dynamics. Bioresour Technol. 2016;200:328–34.
40. Sadino-Riquelme MC, Donoso-Bravo A, Zorrilla F, Valdebenito-Rolack E, Gómez D, Hansen F. Computational fluid dynamics (CFD) modeling applied to biological wastewater treatment systems: an overview of strategies for the kinetics integration. Chem Eng J. 2023;466:143180.
41. Noorpoor Z, Rahmani M. A review of computational fluid dynamics modeling for isobutane alkylation: process optimization, reactor design, and future perspectives. Energy Fuel. 2024;38(13).
42. Tamburini A, Gagliano G, Scargiali F, Micale G, Brucato A, Ciofalo M. CFD simulation of radially stirred baffled and unbaffled tanks. Chem Eng. 2019;74:1033–8.
43. Schembri L, Caputo G, Ciofalo M, Grisafi F, Lima S, Scargiali F. CFD simulations of the transition between non-aerated and aerated conditions in uncovered unbaffled stirred tanks. Chem Eng Sci. 2025;302:120824.
44. Yang F, Zhou S, Wang G. Detached eddy simulation of the liquid mixing in stirred tanks. Comput Fluids. 2012;64:74–82.
45. Yang FL, Zhou SJ, Zhang CX, Evans GM, Wang GC. Study of the turbulent flow in an unbaffled stirred tank by detached eddy simulation. Chem Eng Commun. 2013;200(10):1347–65.

46. Gradov DV, Laari A, Turunen I, Koiranen T. Experimentally validated CFD model for gas-liquid flow in a round-bottom stirred tank equipped with Rushton turbine. Int J Chem React Eng. 2017;15(2):20150215.
47. Torotwa I, Ji C. A study of the mixing performance of different impeller designs in stirred vessels using computational fluid dynamics. Designs. 2018;2(1):10.
48. Li H, Chen Y, Yang Y, Wang S, Bai L, Zhou L. CFD simulation of centrifugal pump with different impeller blade trailing edges. J Mar Sci Eng. 2023;11(2):402.
49. Joshi JB, Nandakumar K. Computational modeling of multiphase reactors. Annu Rev Chem Biomol Eng. 2015;6(1):347–78.
50. Liu D, Wang Z, Cai Y, Chen J, Jin J, Zhao B. Computational fluid dynamics based modeling of gas absorption process: a state-of-the-art review. Ind Eng Chem Res. 2024;63(18):7959–8002.
51. Vasile NS. A comprehensive review of biogeochemical modeling of underground hydrogen storage: a step forward in achieving a multi-scale approach. Energies. 2024;17(23):6094.
52. Gaska K, Generowicz A. SMART computational solutions for the optimization of selected technology processes as an innovation and progress in improving energy efficiency of SMART cities—a case study. Energies. 2020;13(13):3338.
53. Noll P, Henkel M. History and evolution of modeling in biotechnology: modeling & simulation, application and hardware performance. Comput Struct Biotechnol J. 2020;18:3309–23.
54. Quang TV, Doan DT, Phuong NL, Yun GY. Data-driven prediction of indoor airflow distribution in naturally ventilated residential buildings using combined CFD simulation and machine learning (ML) approach. J Build Phys. 2024;47(4):439–71.
55. bin H, Nguyen MQ. A comparative analysis of efficiency in fluid dynamic devices: experimental and computational approaches. Q J Emerg Technol Innov. 2023;8(3):85–102.
56. Basha N, Arcucci R, Angeli P, Anastasiou C, Abadie T, Casas CQ, Chen J, Cheng S, Chagot L, Galvanin F, et al. Machine learning and physics-driven modelling and simulation of multiphase systems. Int J Multip Flow. 2024;179:104936.
57. Hossein F. Machine Learning and Physics-Driven Modelling and Simulation of Multiphase Systems. Int Journal Multip Flow. Elsevier: Amsterdam, The Netherlands.
58. Iyer K, Liu P, Berchielli A, Doshi P, Saxena U, Khan M, Suryawanshi T, Kasat G. Prediction of the finished tablet coating variability in pan coaters by coupling CFD-DEM and Monte Carlo simulations: method development and validation. Powder Technol. 2024;445:120141.
59. Kureck H, Govender N, Siegmann E, Boehling P, Radeke C, Khinast JG. Industrial scale simulations of tablet coating using GPU based DEM: a validation study. Chem Eng Sci. 2019;202:462–80.
60. Behjani MA, Motlagh YG, Bayly AE, Hassanpour A. Assessment of blending performance of pharmaceutical powder mixtures in a continuous mixer using Discrete Element Method (DEM). Powder Technol. 2020;366:73–81.
61. Crouter A, Briens L. Methods to assess mixing of pharmaceutical powders. AAPS PharmSciTech. 2019;20:1–6.
62. Ma B, Lutchen KR. CFD simulation of aerosol deposition in an anatomically based human large–medium airway model. Ann Biomed Eng. 2009;37:271–85.
63. Mead-Hunter R, King AJ, Kasper G, Mullins BJ. Computational fluid dynamics (CFD) simulation of liquid aerosol coalescing filters. J Aerosol Sci. 2013;61:36–49.
64. Albriet B, Sartelet KN, Lacour S, Carissimo B, Seigneur C. Modelling aerosol number distributions from a vehicle exhaust with an aerosol CFD model. Atmos Environ. 2010;44(8):1126–37.
65. Mortazavy BH, Hamed M, Saidul IM. Biomedical and biophysical limits to mathematical modeling of pulmonary system mechanics: a scoping review on aerosol and drug delivery. Biomech Model Mechanobiol. 2022;21(1):79–87.
66. Longest PW, Holbrook LT. In silico models of aerosol delivery to the respiratory tract—development and applications. Adv Drug Deliv Rev. 2012;64(4):296–311.
67. Arsalanloo A, Abbasalizadeh M, Khalilian M, Saniee Y, Ramezanpour A, Islam MS. A computational approach to understand the breathing dynamics and pharmaceutical aerosol transport in a realistic airways. Adv Powder Technol. 2022;33(7):103635.

68. Kourmatzis A, Cheng S, Chan HK. Airway geometry, airway flow, and particle measurement methods: implications on pulmonary drug delivery. Expert Opin Drug Deliv. 2018;15(3):271–82.
69. Chen J, Ye Y, Yang Q, Fan Z, Shao Y, Wei X, Shi K, Dong J, Ma Y, Zhu J. Understanding the role of swirling flow in dry powder inhalers implications for design considerations and pulmonary delivery. J Control Release. 2024;373:410–25.
70. Sperry T, Feng Y, Song C, Shi Z. CFPD-PK simulation of inhaled Delta-9-tetrahydrocannabinol aerosol dynamics: transport, deposition, and translocation in a mouth-to-G10 subject-specific human airway. J Aerosol Sci. 2024;177:106334.
71. Donovan MJ, Kim SH, Raman V, Smyth HD. Dry powder inhaler device influence on carrier particle performance. J Pharm Sci. 2012;101(3):1097–107.
72. Zhu Q, Gou D, Li L, Chan HK, Yang R. Numerical investigation of powder dispersion mechanisms in Turbuhaler and the contact electrification effect. Adv Powder Technol. 2022;33(11):103839.
73. Sosnowski TR. Towards more precise targeting of inhaled aerosols to different areas of the respiratory system. Pharmaceutics. 2024;16(1):97.
74. Chan JG, Wong J, Zhou QT, Leung SS, Chan HK. Advances in device and formulation technologies for pulmonary drug delivery. AAPS PharmSciTech. 2014;15:882–97.
75. Madureira E, Aboelezz A, Su WC, Roghanchi P. From dust to disease: a review of respirable coal mine dust lung deposition and advances in CFD modeling. Fortschr Mineral. 2023;13(10):1311.
76. Holbrook D. Controlling contamination: the origins of clean room technology. Hist Technol. 2009;25(3):173–91.
77. Mičko P, Hečko D, Kapjor A, Nosek R, Kolková Z, Hrabovský P, Kantová NČ. Impact of the speed of airflow in a cleanroom on the degree of air pollution. Appl Sci. 2022;12(5):2466.
78. Salmela A, Kulmala I, Karvinen A, Taillebot V, Weiss P, Gobert T, Berthier A, Guarnieri V, Raffestin S, Pasanen P. Measurement and simulation of biocontamination in an enclosed habitat. Aerosol Sci Eng. 2020;4:101–10.
79. Huang Y, Zheng Y, Lu X, Zhao Y, Zhou D, Zhang Y, Liu G. Simulation and optimization: a new direction in supercritical technology based nanomedicine. Bioengineering. 2023;10(12):1404.
80. Prantil-Baun R, Novak R, Das D, Somayaji MR, Przekwas A, Ingber DE. Physiologically based pharmacokinetic and pharmacodynamic analysis enabled by microfluidically linked organs-on-chips. Annu Rev Pharmacol Toxicol. 2018;58(1):37–64.
81. Li R, Miao H, Zhou X, Zou R, Tong Z. Artificial intelligence and computational modeling in orally inhaled drugs. Exploring Computational Pharmaceutics: AI and Modeling in Pharma 40. 2024:379.
82. Wang Z, Yang L, Xu C, Cheng J, Zhao J, Huang Q, Yang C. Advances in reactive co-precipitation technology for preparing high-performance cathodes. Green Carbon. 2023;1(2):193–209.
83. Wu CY, Ge W, editors. Particulate materials: synthesis, characterisation, processing and modelling. Royal Society of Chemistry. Cambridge: United Kingdom; 2011.
84. Mohammed J, Sharma V, Ravindra N, Singh V. A comprehensive review on quality by design (QbD) in pharmaceuticals. IJHACR. 2024;2(4):55–9.
85. Djuris J, Djuric Z. Modeling in the quality by design environment: regulatory requirements and recommendations for design space and control strategy appointment. Int J Pharm. 2017;533(2):346–56.
86. Hildebrandt C, Gopireddy SR, Scherließ R, Urbanetz NA. Assessment of material and process attributes' influence on tablet quality using a QbD and DEM combined approach. Powder Technol. 2019;345:390–404.
87. Rantanen J, Khinast J. The future of pharmaceutical manufacturing sciences. J Pharm Sci. 2015;104(11):3612–38.
88. Charoo NA, Ali AA. Quality risk management in pharmaceutical development. Drug Dev Ind Pharm. 2013;39(7):947–60.

89. Oberkampf WL, Trucano TG. Verification and validation in computational fluid dynamics. Prog Aerosp Sci. 2002;38(3):209–72.
90. Tu J, Yeoh GH, Liu C, Tao Y. Computational fluid dynamics: a practical approach. Elsevier. Butterworth–Heinemann, An Imprint of Elsevier: Oxford, UK; 2023.
91. Vora LK, Gholap AD, Jetha K, Thakur RR, Solanki HK, Chavda VP. Artificial intelligence in pharmaceutical technology and drug delivery design. Pharmaceutics. 2023;15:1916.
92. Jong CY, Mittal A, Tristan G, Noller V, Chan HL, Goh Y, Yeap EW, Dubbaka SR, Nagesh HR, Wong SY. ANFIS-driven machine learning automated platform for cooling crystallization process development. Org Process Res Dev. 2024;28(4):1129–44.
93. Morris PD, Narracott A, von Tengg-Kobligk H, Soto DA, Hsiao S, Lungu A, Evans P, Bressloff NW, Lawford PV, Hose DR, Gunn JP. Computational fluid dynamics modelling in cardiovascular medicine. Heart. 2016;102(1):18–28.
94. Abadi E, Segars WP, Tsui BM, Kinahan PE, Bottenus N, Frangi AF, Maidment A, Lo J, Samei E. Virtual clinical trials in medical imaging: a review. J Med Imag. 2020;7(4):042805.
95. Tang BT, Pickard SS, Chan FP, Tsao PS, Taylor CA, Feinstein JA. Wall shear stress is decreased in the pulmonary arteries of patients with pulmonary arterial hypertension: an image-based, computational fluid dynamics study. Pulm Circ. 2012;2(4):470–6.
96. Bray A, Webb JB, Enquobahrie A, Vicory J, Heneghan J, Hubal R, TerMaath S, Asare P, Clipp RB. Pulse physiology engine: an open-source software platform for computational modeling of human medical simulation. SN Comprehensive Clin Med. 2019;1:362–77.
97. Gathu S. High-performance computing and big data: emerging trends in advanced computing systems for data-intensive applications. J Adv Comput Syst. 2024;4(8):22–35.
98. Hassan A, Ibrahim A. The coming quantum computing evolution in the pharmaceutical industry and drug R&D. Res Berg Rev Sci Technol. 2023;3(11):1–6.
99. Tsaramirsis G, Kantaros A, Al-Darraji I, Piromalis D, Apostolopoulos C, Pavlopoulou A, Alrammal M, Ismail Z, Buhari SM, Stojmenovic M, Tamimi H. A modern approach towards an industry 4.0 model: from driving technologies to management. J Sens. 2022;2022(1):5023011.

Chapter 4
Population Modelling

Devendra Patil, Jagannath Sahoo, Rohan Pai, and S. N. Koteswara Rao G.

Contents

D. Patil (✉) · J. Sahoo · R. Pai · S. N. K. Rao G.
Shobhaben Pratapbhai Patel School of Pharmacy and Technology Management, SVKM's NMIMS, Vile Parle (W), Mumbai, Maharashtra, India

S. N. K. Rao G., R. R. Alavala (eds.), *Applications of Computational Tools in Drug Design and Development*, https://doi.org/10.1007/978-981-96-4154-3_4

Abstract Population modeling has emerged as a critical tool in pharmaceutical research, offering a quantitative framework for understanding drug kinetics, dynamics, and variability across diverse populations. This chapter provides a comprehensive overview of population modeling techniques, including Physiologically-Based Pharmacokinetic (PBPK) models, Bayesian approaches, and nonlinear mixed-effects models, highlighting their pivotal role in drug discovery, clinical trials, and regulatory decision-making. It emphasizes the importance of integrating patient heterogeneity, clinical trial data, and real-world evidence to optimize dosing regimens, assess risk-benefit profiles, and design targeted therapies. The chapter also discusses innovative modeling applications, such as the development of personalized medicine and adaptive trial designs, and addresses challenges in data collection, model validation, and extrapolation. Finally, future trends in pharmacometrics, including advancements in computational methods and systems pharmacology, are explored, underscoring the potential of population modeling to revolutionize drug development and healthcare outcomes.

Keywords Population modeling · Pharmacokinetics and pharmacodynamics (PK/PD) · Physiologically-Based Pharmacokinetic (PBPK) models · Personalized medicine · Bayesian approaches · Clinical trial optimization

4.1 Introduction

Population modelling is a technique which helps to analyze the impact of trials or experiments been conducted on population of different demographic characteristics. Modelling is an essential tool in understanding the interactions of populations, often within the context of biology, ecology, epidemiology, economics, and sociology. The primary goal is to understand and predict the dynamics of populations over time. The main application of population models in pharmaceutical industry is done to understand quantitative analysis of the kinetics and dynamics of given data for particular set of individuals [1]. In the context of pharmaceutics, population modeling often involves the application of mathematical models to data from clinical trials. These models can account for variability in drug response among individuals, taking into consideration factors such as age, sex, weight, and genetics. By understanding this variability, pharmaceutical companies can optimize drug dosing, identify patient subgroups that may benefit from specific treatments, and predict potential drug-drug interactions.

Population modeling is often used in the development of new drugs for various diseases, including cancer, HIV, and cardiovascular diseases [2]. By using population models, researchers can predict the likelihood of a drug being effective for different subgroups of patients based on their genetic makeup and other factors. This allows for the development of more personalized and targeted treatments, which can lead to better outcomes for patients.

In the pharmaceutical industry, population modeling is typically performed using a combination of empirical and mechanistic approaches. Empirical models are based on observed data and are used to describe the relationship between a drug and its effects on the body. Mechanistic models, on the other hand, are based on an understanding of the underlying biological processes that govern the behavior of a drug [3]. By combining these two approaches, researchers can create more accurate and predictive models that can be used to guide drug development and optimization.

Physiologically-Based Pharmacokinetic (PBPK) Models can be used in drug discovery and development from the early phases before lead development, when there is minimal data, to the late stages of drug development, when there is more data available. There are now several examples of the use of PBPK models during the drug discovery and development phases for decision-making specific to candidate selection, first-inhuman dose, assessment of DDI potential, and the definition of appropriate study designs involving DDIs or inclusion/exclusion criteria for studies with drugs metabolized by polymorphic enzymes [4]. During drug discovery and development, PBPK models may be continuously improved to integrate new knowledge on drug distribution and absorption from preclinical and clinical research as it becomes available. A well-constructed PBPK model can help plan preclinical and clinical pharmacology investigations. Nonlinear Mixed-Effects Models (NLMEM) are a type of statistical model used in pharmaceutical industry, particularly in the field of pharmacometrics, for the analysis of longitudinal data [5]. A longitudinal study is a research design that involves repeated observations of the same variables over long periods of time. They belong to the class of latent variables models, and are widely used in many fields such as oncology, forestry, and pharmacometrics [6, 7]. The Bayesian approach is one approach that is particularly well-suited for population modeling. This approach involves using prior knowledge and data to inform the model-building process, which can lead to more accurate and robust predictions. By using Bayesian methods, researchers can incorporate information from a variety of sources, including preclinical data, clinical trial data, and real-world data, to create more comprehensive and predictive models [8].

Bayesian approach is particularly useful in the early stages of drug development, where limited data is available. A study of a new drug for hypertension illustrates the usefulness of Bayesian data analysis. The study used prior knowledge about the drug's mechanism of action and the known effects of similar drugs to inform the model. The prior distribution was then updated with data from a phase II clinical trial, resulting in a posterior distribution that was used to inform the design of a phase III trial. This approach led to a more efficient and informative trial, as well as a more accurate estimation of the drug's effect [9]. Additionally, Bayesian data analysis can also be used in the regulatory process. The US Food and Drug Administration (FDA) has accepted the use of Bayesian methods in the approval of certain drugs. In these cases, Bayesian data analysis can be used to make probabilistic statements about the drug's efficacy and safety, which can be useful in the decision-making process [8]. Bayesian data analysis is a valuable tool in the pharmaceutical industry, as it allows for the incorporation of prior knowledge and data to make more informed decisions. This approach can be useful in the early stages of drug development, where limited data is available, as well as in the regulatory process.

In this book chapter, authors have covered the benefits of population modelling models with some examples of real-world applications in the pharmaceutical industry, such as the development of personalized medicine and the optimization of drug therapy.

4.2 Types of Population Models

4.2.1 Physiologically-Based Pharmacokinetic (PBPK) Modeling

Physiologically-Based Pharmacokinetic (PBPK) modelling is a scientific approach used to understand the health effects of chemicals. These models relate the amount of chemical exposure to the amount of chemical found in a person's blood and organs over time. They can be used to predict whether a toxic level of a chemical would be reached in the body following exposure through drinking water, for example. PBPK models are mathematical descriptions of how a chemical moves in the body, from entry (through breathing, drinking, or eating) to elimination. They take into account the body's anatomical and physiological structure, as well as biochemical processes. The complexity of PBPK models can vary, with some having fewer features that are important for describing chemical movement, and others having more features that dictate chemical fate and movement. In toxicology and risk assessment, PBPK models can be valuable tools for understanding the health effects of chemical exposure [10].

4.2.2 Disease Progression Models

Disease progression modelling is an additional evaluation tool that has an improved ability to detect a drug effect and provides useful dosing information for prescribers. Modelling is also an important component in the regulatory requirements for a new drug application [11]. Disease progression is often referred to as the pharmacometrics analyses. In the context of disease progression, pharmacometrics can be used to develop models that describe the relationship between drug exposure and clinical outcomes over time. These models can be used to inform drug development and optimize dosing regimens for individual patients [12]. A study published in the journal *AIDS* found that pharmacometrics modelling was able to predict the long-term clinical outcomes of HIV patients based on their viral load and CD4 cell count. The study found that the model was able to accurately predict the probability of treatment success or failure, and could be used to inform treatment decisions for individual patients [13]. Another example is in the field of oncology, where pharmacometrics models have been used to predict the response to chemotherapy in patients with advanced cancer. A study published in the journal *Clinical Pharmacology & Therapeutics* found that a pharmacometrics model was able to predict the probability of tumour response based on the patient's age, sex, and tumour size. The study found that the model could be used to inform treatment decisions and improve patient outcomes [14].

4.2.3 Meta-models

These are combinations of several models to predict outcome of given objective [1]. One common type of meta-model used in the pharmaceutical industry is the Bayesian hierarchical model, which combines the information from multiple clinical trials to provide a more accurate prediction of the population response. A study published in the journal *Statistics in Medicine* found that the use of Bayesian hierarchical models in population modelling can significantly improve the accuracy of predictions compared to traditional fixed-effects models. The study found that the Bayesian hierarchical model was able to accurately predict the probability of treatment success or failure in patients with advanced cancer. This suggests that the use of meta-models in population modelling can provide valuable insights for drug development and optimization [15].

4.3 Importance of Data in Population Modelling

The main aim of most population pharmacokinetic modelling studies is to find out the typical values of drug concentrations in a group of people and the factors that cause differences in these values. Additionally, these studies aim to understand how the administered doses of a drug are related to the observed concentrations in the body, and to identify the characteristics of the population that can predict these concentrations [1]. Data plays a crucial role for prediction of outcome for several reasons.

4.3.1 Patient Heterogeneity

It's really important to recognize that diseases can affect different people in different ways. This is called disease heterogeneity. By understanding and addressing this, we can improve how accurately we diagnose and treat diseases, especially through personalized treatment plans. If we ignore or don't properly handle this issue, it can lead to incorrect conclusions about the biology and treatment of a disease. Currently, there aren't enough statistical tools to help us identify these hidden subgroups of patients. Many studies either don't consider this diversity or use makeshift methods to deal with it [16]. A study conducted by Kravitz et al., talks about evidence-based medicine, which means using scientific evidence to guide medical decisions. It explains how it's challenging to apply findings from large studies to individual patients because not everyone responds the same way to treatments. Sometimes, what works for most people might not work well for everyone. This is because people differ in how they respond to treatments, their risk of getting sick, and how they react to side effects. By understanding these differences, researchers can design better studies to figure out who benefits the most from treatments. This helps doctors and policymakers make better decisions about medical care [17].

4.3.2 Drug Development and Optimization

PBPK modelling is a useful tool for predicting how drugs will behave in the body based on various factors like the body's systems and the specific characteristics of the drug. It's commonly used to understand how drugs interact with each other or to optimize dosing for different groups of people. Recent studies have shown that PBPK modelling is particularly effective for predicting interactions involving certain enzymes, like CYP3A and CYP2D6, which helps in designing clinical trials more efficiently. However, its effectiveness in predicting interactions involving transporters and in specific populations is limited due to uncertainties in certain parameters [18].

4.3.3 Clinical Trial Design

The real-world data (RWD) is used in modelling and simulation of disease progression, patient care, and comparative effectiveness research [19]. To make clinical trial simulations more realistic, it's suggested to use RWD from various sources. This includes data from enterprise databases like Optum and Truven [20], patient surveys, reviews of patient charts, observations from cohort studies, practical clinical trials, and registries. These different types of data help provide a more accurate picture of how treatments work in real-life situations [21].

4.3.4 Risk Assessment and Safety Evaluation

Quantitative benefit–risk evaluation is needed to be used for complex situations as a complementary approach to assess the benefit—risk of a product. Engaging patients and collecting their perspectives throughout the medical product development process is gaining increasing recognition and becoming key in the regulatory decision-making process [22]. Both randomized clinical trials and real-world evidence gathered through pharmacoepidemiology and pharmacovigilance contribute to evaluating the benefits and risks of a drug, known as the benefit-risk assessment (BRA). BRA involves qualitatively assessing quantitative data, and while attempts have been made to quantify it, no method has fully addressed its multidimensional nature. Regulatory authorities like the FDA and EMA expect consistency and transparency in BRA assessments, which occur throughout a drug's lifecycle. BRA primarily relies on randomized clinical trials during development and is reinforced by real-world data post-market [23].

4.3.5 Health Economics and Market Access

Health economic assessments, including cost studies, economic evaluations, and budget impact analyses, help evaluate healthcare technologies by providing information on their costs, efficiency, and affordability. These assessments guide decision-makers in optimizing health policy. To maximize their value, researchers must ensure methodologically rigorous practices when conducting these assessments. Additionally, they should consider the goals of decision-makers and discuss how the findings can help achieve these objectives within the decision-making context [24].

4.3.6 Regulatory Submissions and Approval

Model-informed drug development (MIDD) is a powerful approach used in drug development and regulatory reviews, especially at the FDA. In this process a process whereby decisions are supported by mathematical models and simulations that predict the likelihood of success for the drug. MIDD is applied throughout the drug development process, from new drugs to generics and biologics. In new drug development, MIDD helps design clinical trials, select doses, and address regulatory questions. In biopharmaceutics, computational modelling informs formulation development and future studies after approval. MIDD is also crucial in generic drug development, reducing time and cost. Though early, MIDD shows promise in developing cell and gene therapies, aiding in understanding biological activity, safety, and efficacy. Overall, MIDD is expected to revolutionize drug development for better patient outcomes [25].

4.4 Types of Data Used in Population Modelling

In population modelling, data is required to develop mathematical models that describe the relationship between drug exposure and response in a population of interest. Population modelling in pharmacokinetics (PK) and pharmacodynamics (PD) relies on accurate data related to dosing, measurements, and covariates to effectively characterize variability within patient populations and predict drug responses [26]. Dosing Information will include details about how drugs are administered to patients, such as dosage regimen (e.g., frequency, duration), route of administration (e.g., oral, intravenous), and actual doses given to patients. Accurate dosing data are essential for simulating drug concentration-time profiles and understanding how drug exposure varies among individuals. Measurement data will include the observed data collected during clinical trials or real-world studies, typically comprising drug concentrations in biological samples (e.g., blood, plasma, urine) over time. Measurement data provide direct information on drug absorption, distribution, metabolism, and elimination (ADME), which are fundamental components of PK modelling. Covariates will include the patient characteristics or factors that may influence drug pharmacokinetics or pharmacodynamics. Common covariates include demographic variables (e.g., age, sex, weight), physiological parameters (e.g., renal function, liver function), genetic factors (e.g., polymorphisms in drug-metabolizing enzymes or transporters), and concomitant medications. Covariates help explain inter-individual variability in drug response and are incorporated into population models to personalize predictions for different patient subgroups [1].

4.5 Components of Population Modelling

Preparation of data and databases for organizing and refining information for analysis

Databases utilized for modelling are often intricate, necessitating precise details about the timing, dates, and quantities of administered drugs, sample collection, and related demographic and laboratory information. The data is typically gathered in a specific manner, with patient factors recorded together for each individual, as opposed to being presented separately. This unique approach can sometimes reveal errors that might otherwise go unnoticed.

4.5.1 Structural Models

Structural models in pharmacokinetics (PK) can be represented either as algebraic equations or as differential equations, depending on the complexity of the system being modelled. The simplest representation is an algebraic equation, such as the one describing a one-compartment model for a drug administered as a single intravenous bolus dose. These equations describe the relationship between the independent variable (typically time) and the dependent variable (concentration). While algebraic equations are simpler and often used for linear models, complex systems may require differential equations to capture nonlinear dynamics.

Suppose if we have a drug that follows a one-compartment model, and we administer a single intravenous bolus dose. We can describe the concentration-time profile using the following equation:

$$\mathrm{C} = \mathrm{D} / \mathrm{V} \times e^{(-\mathrm{k} \times \mathrm{t})}$$

where C is the concentration, D is the dose, V is the volume of distribution, k is the elimination rate constant, and t is time.

4.5.2 *Linearity and Superposition*

Despite producing nonlinear concentration-time curves, some PK models exhibit linearity with respect to their parameters. This linearity allows for the principle of superposition, where the concentration-time curve for multiple doses can be obtained by summing the curves for individual doses.

Suppose if we have a drug that follows a one-compartment model with first-order elimination, and we want to calculate the concentration-time profile after multiple doses. The PK model can be represented by the following differential equation:

$$\mathrm{dA} / \mathrm{dt} = -\mathrm{k} \times \mathrm{A}$$

Where A is the amount of drug in the body, k is the elimination rate constant, and t is time.

4.5.3 *Stochastic Models for Random Effects*

This population model is used to characterize the variability in drug exposure between and within subjects. It can be represented by algebraic or differential equations and can account for between-subject variability (BSV) and residual unexplained variability (RUV). The parameters in the model can be fixed or random effects, and the random effects are represented as a quantity (ETA) reflecting the difference between an individual's parameter value and the population value. The distribution of BSV for the parameter across the population is described by its variance (OMEGA). The RUV is defined by a quantity (EPS) reflecting the difference between the observed data for an individual and the model's prediction, and is summarized by its variance (SIGMA). The model is used to estimate the population mean and variability of the PK parameters and RUV, which can help inform dosing decisions and identify patients who may be at risk of subtherapeutic or toxic exposure.

4.5.4 Covariate Models for Fixed Effects

Covariate analysis aims to identify factors (covariates) that explain variability in drug response. Incorporating covariate effects into population models can help reduce variability and improve dose recommendations. Covariate effects can be evaluated graphically and statistically, often revealing relationships between demographic or clinical factors and drug pharmacokinetics. NLMEM can be used to estimate the population PK parameters and their variability.

Suppose if we have a drug that follows a one-compartment model with first-order elimination, and we want to evaluate the effect of body weight on the drug's clearance (CL) and volume of distribution (V). We can represent the PK model by the following differential equation:

$$dA / dt = -CL / V \times A$$

where A is the amount of drug in the body, CL is the clearance, V is the volume of distribution, and t is time.

4.6 Overcoming Data Collection Challenges in Population Models

The application of physiologically-based pharmacokinetic (PBPK) models in pharmaceutical research offers a robust framework for integrating various absorption, distribution, metabolism, and excretion (ADME) mechanisms within a physiological context [27]. Unlike empirical approaches, PBPK models are mechanistic, enabling the exploration of hypothetical scenarios where in vivo or clinical studies may not be feasible. These models facilitate informed study design and extrapolation across species, populations, and different drug administration modes, thereby enhancing decision-making processes in drug development and regulatory review [28]. However, their utility is challenged by incomplete knowledge of physiological processes, poorly understood translation of in vitro data to in vivo scenarios, and limited in vivo-relevant compound data for model verification. Specific to clinical pharmacology and drug disposition, major challenges persist in predicting transporter and non-cytochrome P450 (CYP) enzyme clearance, understanding the ontogeny of enzymes and transporter proteins, and characterizing enzyme/transporter abundance in disease populations [28, 29]. Another challenge lies in extrapolating findings from animal studies to humans based on PK-PD modeling, as differences in optimal drug release rates between rodents and humans can lead to errors in scaling. To mitigate these challenges, several strategies can be considered. Firstly, selecting the appropriate animal species for scaling to humans is crucial to enhance the relevance of preclinical assessments. Allometric scaling using data from multiple species can also improve prediction accuracy. Additionally, leveraging prior knowledge and established scaling methods, such as those for the erythropoietic system, can provide more reliable scaling from animals to humans. Despite advancements in PBPK modeling, gaps remain in modeling drug absorption, particularly concerning food effects and advanced formulations [30]. Furthermore, PBPK models for monoclonal antibodies (mAbs) face limitations due to the complex interplay between mAbs and their targets, alongside inadequate data on physiological parameters [31]. These challenges underscore the intricate nature of data collection in population modeling within the pharmaceutical sector, necessitating ongoing efforts to enhance model accuracy and predictive capabilities.

Addressing the challenges in data collection for population modeling in pharmaceutical research requires a multifaceted approach. Collaboration among researchers, clinicians, pharmacologists, and regulatory agencies is essential to share expertise and data resources, facilitating a more comprehensive understanding of physiological processes and drug behavior [1]. Investing in advanced experimental techniques, such as mass spectrometry-based assays for quantifying enzymes and transporters in tissue samples, can improve the accuracy of model parameterization and validation [32]. Additionally, utilizing biorelevant methods for assessing drug characteristics like solubility and dissolution enhances the prediction of drug absorption across species and populations, especially for complex formulations and alternative routes of administration. Establishing standardized protocols for data collection, sharing, and integration across research groups and institutions ensures consistency and reliability in model inputs and outputs [33]. Prioritizing the validation of PBPK models using comprehensive datasets and real-world clinical data helps assess model accuracy and reliability, enhancing confidence in model predictions and their applicability to diverse populations. Exploring innovative study designs, such as long-term cohort studies or prospective data collection initiatives, can capture data over extended timeframes and address challenges related to dynamic disease processes and long-term model predictions [34]. Moreover, emphasizing the integration of target biology and physiological considerations into PBPK modeling approaches, particularly for complex therapeutics like monoclonal antibodies, improves model fidelity and predictive capabilities in drug development and personalized medicine [31]. By implementing these strategies collectively, researchers can effectively overcome the challenges associated with data collection in population modeling, ultimately advancing pharmaceutical research and development.

4.7 Modelling Techniques

4.7.1 Nonlinear Mixed-Effects Models (NLMEM)

NLMEM are a type of statistical model used to analyze data that involves repeated measurements on the same individuals over a period of time. These models are especially useful in the field of pharmaceutical research for understanding the behavior of drugs in the body, specifically in terms of pharmacokinetics (PK) and pharmacodynamics (PD). NLMEM allows researchers to estimate population parameters and individual subject variability, providing valuable insights into the effects of drugs and their mechanisms of action [35].

NLMEM could be highly applicable in the interpretation and analysis of the PK and PD effects observed in the study investigating the orally active ETA selective receptor antagonist ABT-627 in healthy humans. ABT-627, an ETA selective receptor antagonist, was well-tolerated and rapidly absorbed. It effectively blocked ET-1 induced vasoconstriction, reducing peripheral resistance without significant blood pressure changes. These results suggest its potential in treating cardiovascular disease. NLMEM can help analyze its pharmacokinetic and pharmacodynamic effects, offering insights into efficacy and individual variability in response to treatment [36]. In a study conducted by Nakao et al., It was found that afatinib when used in Japanese lung cancer patients, population pharmacokinetics analysis revealed significant factors like liver function and age affecting drug levels. NLMEM can help understand individual variability in drug exposure and its impact on safety. It highlights the role of hepatic impairment and guides dosing strategies for efficacy and safety optimization [37].

4.7.2 Bayesian Approaches

In Bayesian inference, prior distributions are used to represent the initial beliefs about the parameters, and these distributions are updated with the observed data to obtain the posterior distribution, which reflects the updated beliefs about the parameters [38]. The posterior distribution can then be used to make probabilistic statements about the parameters, such as the probability that the true value of a parameter falls within a certain range. Prior distributions are a fundamental component of Bayesian analysis, representing the initial beliefs or knowledge about the parameters before observing any data [39]. For instance, if we believe a drug's efficacy follows a normal distribution with mean 0.7 and standard deviation 0.1, this is our prior. After observing patient outcomes, we update this prior to a posterior distribution, which reflects our updated belief about the drug's efficacy. These distributions can be chosen based on expert opinion, historical data, or other sources of information. The choice of priors can significantly influence the posterior results, as they reflect the weight given to different values of the parameters before observing the data. However, the impact of the priors decreases as more data becomes available, as the posterior distribution becomes increasingly dominated by the observed data [40]. The likelihood function is a key component of Bayesian analysis, as it captures the probability of observing the data given the parameters of the model. This function combines the information from the observed data with the prior distribution to compute the posterior distribution, which represents updated beliefs about the parameters [41]. The likelihood function is calculated by taking the product of the probability density functions of the observed data, given the parameters. The posterior distribution is the key output of Bayesian analysis, representing the updated uncertainty about the parameters after observing the data [42]. This distribution summarizes the information from the prior distribution and the likelihood function, providing a comprehensive representation of the parameter space. The posterior distribution can be used to compute point estimates, such as the mean or median, as well as intervals, such as credible intervals, which provide a measure of uncertainty around the estimates. MCMC (Markov Chain Monte Carlo) techniques are used in Bayesian modeling to simulate samples from the posterior distribution when analytical solutions are not feasible [43]. These algorithms, such as Gibbs sampling and Metropolis-Hastings, allow for efficient exploration of high-dimensional parameter spaces and estimation of complex models. MCMC methods work by simulating a Markov chain that converges to the posterior distribution, with each iteration of the chain producing a new sample from the distribution [44]. Model validation and sensitivity analysis are crucial components of Bayesian modeling, ensuring that the model is a good fit for the data and that the results are robust to changes in model assumptions, priors, or data inputs. Validation involves assessing the model's fit to the data and its predictive performance, often through techniques such as cross-validation or posterior predictive checks [45]. Sensitivity analysis, on the other hand, evaluates the robustness of the results to changes in model assumptions, priors, or data inputs, allowing for a better understanding of the uncertainty surrounding the estimates. Hierarchical modeling is a Bayesian approach that involves modeling parameters at multiple levels of a hierarchy. This allows for the borrowing of strength across groups or populations, making it particularly useful for complex data structures, such as longitudinal or clustered data. In a hierarchical model, the parameters at each level of the hierarchy are modeled as random variables, with their own prior distributions [46]. The prior distributions at each level are informed by the data at that level, as well as the data at higher levels of the hierarchy [47]. The use of Bayesian decision theory can help to ensure that decisions are based on a comprehensive and coherent assessment of the available evidence, taking into account both the uncertainty surrounding the estimates and the values and preferences of stakeholders [48]. This can lead to more informed and effective decision-making, as well as a more transparent and accountable process.

The pharmaceutical industry and regulatory bodies often rely on frequentist statistical methods, like null hypothesis significance testing and p values, for evaluating new treatments. However, Bayesian statistical approaches, which incorporate existing data into clinical trial design and analysis, have the potential to streamline drug development, reduce costs, and enhance patient safety [40]. Despite advancements in Bayesian methodology and increased computational power, these approaches are not widely utilized. There has to be more emphasis on the benefits of Bayesian methods, identifies barriers to their adoption, and suggests strategies for their more routine use in drug development and regulatory review processes [9]. The goal is to encourage stakeholders to consider the value of existing data and promote the integration of Bayesian methods as effective tools in these processes. The use of Bayesian statistical approaches in evaluating the efficacy of COVID-19 vaccines, specifically focusing on the Pfizer-BioNTech BNT162b2 mRNA vaccine [49]. The study aimed to assess the vaccine's efficacy in preventing COVID-19 infection among a large population. The study employed a Bayesian posterior probability analysis, setting a prior distribution for vaccine efficacy at 30%. The decision rule required a posterior probability of vaccine efficacy exceeding 30% greater than 98.6% for trial success. Interim analyses were conducted, with stopping rules based on predefined probabilities. This Bayesian approach facilitated transparent and efficient evaluation of vaccine efficacy, expediting decision-making in the clinical trial process [40].

The application of Bayesian methods in drug development to generate substantial evidence of treatment effects efficiently and cost-effectively. By leveraging data from phase II studies as priors for phase III planning, Bayesian approaches enable informed decision-making while reducing sample sizes and trial duration. Sequential phase III studies can also benefit from data borrowing, leading to reduced costs and time without compromising the quality of inference [50]. Additionally, Bayesian methods facilitate safety assessment by synthesizing information across the drug development program, aiding in the evaluation of unexpected adverse events. Overall, Bayesian approaches offer a more insightful and flexible framework for evidence synthesis and decision-making in clinical trials, enhancing the efficiency and effectiveness of the drug development process [51].

4.7.3 Physiologically-Based Pharmacokinetic (PBPK) Models & Population Pharmacokinetic (popPK) Modeling

Population pharmacokinetic (popPK) modeling begins with observed data and constructs a model that best fits it. Conversely, physiologically-based pharmacokinetic (PBPK) models start at the organ or tissue level and work upward to understand the system [52]. PBPK modeling is a computational method essential in drug development for predicting the distribution of drugs within the body over time [53]. PBPK models are sophisticated frameworks that incorporate detailed information about both the drug and the physiological aspects of the organism to simulate drug concentration profiles across various organs and tissues [54]. These models offer insights into crucial pharmacological parameters, such as drug exposure and action, by considering factors like organ volumes, blood flows, and tissue permeabilities. The relationship between organism properties (such as anatomy and physiology) and drug properties (including physicochemical characteristics) is fundamental to constructing accurate PBPK models. Organism properties are contingent upon factors like species or population, necessitating adjustments for special populations like pediatric or diseased individuals [55]. Drug properties encompass physicochemical traits independent of organism physiology, as well as drug-biological interactions that are influenced by both the drug and the organism. Additionally, administration protocols and formulation properties play a pivotal role in defining PBPK simulations, allowing for the consideration of time-related events such as meal intake [56]. New drug delivery systems often significantly influence the PK by modifying the in vivo drug release and absorption process. By tuning the drug release profile, the apparent half-life of the drug could be prolonged, and drug accumulation at the target site may be enhanced [57].

The main challenges facing PBPK modeling encompass several critical areas. Firstly, there's a pronounced shortage of trained individuals capable of leading PBPK projects and a dearth of independent reviewers with the necessary expertise in PBPK modeling and risk assessment [58]. This scarcity is compounded by the limited availability of academic programs offering PBPK curriculum and the often-inaccessible alternative training opportunities like workshops [59]. Secondly, the absence of standardized review processes across agencies presents hurdles in evaluating PBPK models effectively. Proposals for harmonized templates for model submission and review are advocated to streamline processes and increase model uptake [60]. Thirdly, ensuring transparency and documentation in PBPK modeling is paramount. Clear delineation of model scope, comprehensive documentation of model characteristics, and the provision of model code and equations are pivotal for enhancing transparency and reproducibility [61]. In the field of oncology, major challenge and limitations lies in understanding the type of tumor in PK model [62]. Additionally, confidence in model extrapolation, particularly in cases lacking human data, remains a significant challenge. Calibration using in vivo human data may be indispensable to ensure the accuracy of human predictions. Lastly, the issue of transferability across computing platforms further complicates matters. Differing platform preferences among developers hinder model review and accessibility, suggesting a need for open-source packages and consistent data exchange conventions to enhance interoperability and facilitate model translation across platforms [4]. Addressing these challenges could greatly bolster the reliability and adoption of PBPK models in informing public health decisions. Commercial PBPK modeling platforms, like GastroPlus and SimCyp, have significantly streamlined the PBPK modeling process by integrating physiological databases and facilitating model parameterization. Various tutorials and publications further elucidate the practical application of PBPK models in drug discovery and clinical development, covering basic concepts, pediatric applications, and implementation strategies [55]. Despite its advancements, challenges persist in predicting active processes within PBPK models, necessitating continued research to enhance accuracy, particularly in complex drug interactions and metabolic pathways.

4.7.4 Pharmacodynamic Models

PD models depict how drug concentration influences therapeutic outcomes. This section outlines PD models that represent key mechanisms of drug action, such as direct effect, biophase, and indirect response. Like PK models, PD models typically utilize compartmental frameworks, and more intricate PD models are constructed by integrating simpler ones [30]. Different approach to capturing the biological processes involved in therapeutic and adverse drug responses includes Simple Direct Effect Models, here valuable tools are predicted for probability of bleeding events and the pharmacodynamic effects of drugs, respectively [63]. However, these models have limitations and should be used with caution, especially in cases where drug effects may not be directly proportional to receptor occupancy [64]. The Biophase Distribution model, originally intended for describing delayed drug responses due to drug distribution, can be expanded to accommodate more intricate pharmacodynamic relationships by integrating additional factors into the model [65]. This versatility has been demonstrated successfully in instances such as buprenorphine-induced respiratory depression in rats and the clinical prediction of transient increases in blood pressure. The indirect response model, which is discussed in the in vitro and in vivo systems where a biological process in which the therapeutic agent does not directly affect the pharmacodynamic endpoint but rather affects a biological mediator [66]. Advanced statistical methods and computer-based models can be used to overcome these limitations and provide more accurate predictions [67]. Mechanism-based pharmacodynamic modeling holds promise for the future of both therapeutic and adverse drug response prediction, offering opportunities for model-driven drug development and therapy advancements.

In the field of PD modeling, complex systems are often split into rule-based and equation-based models. The former can be quite simple and the latter can be exceedingly complex. To date, PD modeling has focused on rule-based modeling methods, and whilst these are able to address some areas of complex systems, they are effectively only able to simulate very specific situations and do not encapsulate the wider system context [68]. Progress has been made in the field of equation-based modeling, and it is likely that this will be an area where more PD modelers will choose to focus in the future. Equation-based PD models can be simulated in a similar way to PK models, and they can use a variety of different methods to interrelate system components, such as ordinary differential equations, stochastic differential equations, and agent-based modeling. By constructing statistically sound models using real data, equation-based PD models can be a versatile tool for predicting and comparing the outcomes of different treatments on common complex diseases [69] (Table 4.1).

Table 4.1 Applications of Pharmacokinetic and pharmacodynamic models in drug design and development

Pharmacokinetic and pharmacodynamic models in drug development and discovery.	Results & Outcomes
Physiologically Based Pharmacokinetic Modelling and Simulation to Predict the Plasma Concentration Profile of Doxorubicin [70].	The data evaluates the performance of various models in predicting the pharmacokinetics of doxorubicin (DOX) using a training dataset and validation studies involving intravenous (IV) bolus and IV infusion methodologies. In the training dataset analysis, models 3 and 7 demonstrated the best accuracy in predicting maximum plasma concentration (Cmax), while models employing minimal PBPK distribution tended to perform better in this aspect compared to full PBPK models. For the area under the curve (AUC0–168), models 4, 2, 7, and 5 displayed the most accurate predictions. During IV bolus validation, models 2, 4, 6, and 8 showed superior performance in predicting Cmax, with full PBPK distribution models exhibiting higher accuracy. However, discrepancies in timing between observed and predicted Cmax values were noted. For AUC0–120, models 8, 7, 5, and 6 performed well. Similarly, during IV infusion validation, models 2, 4, 6, and 8 demonstrated strong predictive abilities for Cmax, with full PBPK distribution models showing consistent accuracy. For AUC0–120, models 7, 5, 8, and 1 showed the best performance. Based on the data presented in this work, all generated DOX models perform quite well according to the two-fold standard that is widely considered the acceptable measure for the performance of PBPK models.
Physiologically Based Pharmacokinetic/Pharmacodynamic Modeling to Predict the Impact of CYP2C9 Genetic Polymorphisms, Co-Medication and Formulation on the Pharmacokinetics and Pharmacodynamics of Flurbiprofen [71].	The study develops a PBPK/PD model for Flurbiprofen, considering absorption, genetic polymorphism, and drug interactions. It accurately predicts Flurbiprofen dose-exposure relationship in diverse populations and provides insights into its variability. The model suggests dose adjustments based on genetic factors and highlights the importance of personalized medicine. Translational PBPK/PD modeling offers valuable insights into Flurbiprofen pharmacokinetics and pharmacodynamics. By integrating diverse data sources, it enables personalized dosing strategies and enhances drug development efforts, bridging the gap between in vitro and clinical outcomes.

(continued)

Table 4.1 (continued)

Pharmacokinetic and pharmacodynamic models in drug development and discovery.	Results & Outcomes
Physiologically based pharmacokinetic modeling in neonates: Current status and future perspectives [72]	The field of neonatal pharmacokinetics requires advanced PBPK models to optimize drug therapy and minimize adverse effects. Challenges include understanding neonatal physiology, incorporating advanced computational methods, and fostering collaboration among stakeholders. By addressing these challenges, researchers aim to develop more accurate models, enhance predictive capabilities, and ultimately improve drug therapy outcomes for newborns.
In vitro pharmacokinetic/ pharmacodynamic modeling of HIV latency reversal by novel HDAC inhibitors using an automated platform [73].	This study investigated how Histone deacetylase inhibitors (HDACis) affect HIV latency reversal using Jurkat T-cell models. They found rapid histone acetylation after HDACi exposure, impacting gene expression timing. Automation helped generate comprehensive data on drug effects. PK parameters showed compounds with fast clearance may not sustain HIV transcription. They suggested a positive feedback loop triggering HIV transcription after sustained reversal. Automation facilitated high-density data collection for multiple compounds, aiding PK/PD studies.
Physiologically based pharmacokinetic modelling of high-and low-dose etoposide: from adults to children [74].	The PBPK model demonstrated its ability to predict the pharmacokinetics of etoposide in children, offering insights into factors such as clearance, tissue distribution, and drug interactions. The model accounted for various physiological processes, including metabolism by enzymes such as CYP3A4 and UGT1A1, renal excretion, and active transport mechanisms involving transporters like PGP and MRP2. The study suggests that PBPK modeling could be a valuable tool for individualizing therapy in pediatric oncology. By incorporating genetic information on enzymes and transporters, as well as other significant covariates, PBPK models could help tailor drug doses to individual patients, potentially improving treatment outcomes and reducing the risk of adverse effects.

4.8 Applications in Drug Development

Drug development applications unfold across multiple years and are segmented into various stages. Ideally, population modelling should start at the beginning of this process, providing predictions which improve in accuracy as more is known about the drug [75]. In reality, the full potential of population modelling is rarely realized due to misconceptions about its costs and benefits [76]. By increasing awareness of the advantages of population modelling at all stages of drug development, it is hoped that the most appropriate techniques will be employed more widely [77]. Simulation is frequently used in the drug development process. This is because it is an effective way of predicting what will happen in the real world without the need for expensive and time-consuming experimentation [78]. By informing decision making and maximizing the chance of success for a given project, simulation can save both time and money. Several stages of drug development are elaborated upon to enhance comprehension of the utilization of population modelling tools in the drug development [79].

Early-phase clinical trials are conducted to determine the safety and tolerability of a new drug and to identify the drug's effects on humans [80]. In a study conducted by the level of confidence in attribute scores was calculated, the proportions of rats predicted to be responders in each group were tabulated. The posterior distributions of the therapy effects were obtained by computing the difference in the number of responders between the treated and control groups. The first step in this study, where population models were applied in a pharmacokinetic/pharmacodynamic (PK/PD) exploration using binary response data from a preclinical experiment. Simple 2 and 3 compartment PK models were considered in the presence of first-order elimination [81]. Data from the control rats were used to help define a prior distribution for the PK model, and the simple binary response data was used to help define a likelihood distribution for the therapy effects on patient rats [82]. The method of trial simulation plays a big part in the decision-making process of particular clinical trial [83]. By this point there was only a rough idea of the likely therapy effects on the patient rats, and because the target data was the proportion of responders in the treated group, it was not possible to calculate power in the usual manner [82]. The aim was to define the therapy effects that could maximize statistical power to detect a significant difference in the proportion of responders, given a fixed sample size. This required a formulation of a relevant endpoint and an exact definition of success. With multiple therapy scenarios in mind, a confirmed endpoint and defined success, the next step was to decide upon the sample size and allocation of patients to treated and control groups. All of these are necessary conditions when using trial simulation techniques to calculate power in a clinical setting [84].

Application of modelling and simulation in the domain of late-phase clinical trials is to inform trial design and data analysis, not to substitute for clinical trials [85]. These trials generally test far more elements than one can study experimentally with PK/PD, and the overlooked risk is that the more precisely a drug's effects are predicted, the more susceptible the trial is to failure [86]. Late phase trials still show how to improve the dose comparatively to a placebo or other drugs, but often the primary measures are clinical endpoints that are difficult and expensive to study with PK/PD, and not always in small populations [87]. Economic modelling has been used to inform which studies should be undertaken and which should not, however, this is difficult to prove the validity of such models [88]. Simulation of clinical trials can inform aspects such as the power of a study to detect an effect, if multiple tests are done how this should be accounted for in the analysis, look at dose-response and whether it is linear or non-linear [89]. There are many areas of late phase trial which can benefit from modelling and simulation, and more tools are being developed to tackle the wide variety of survey [90]. At present, it is difficult to give an overall general picture about the tools used, and it is best to look at specific types of analysis. Recommended dose and dose regimen has throughout been one of the main issues in pharmacology, a recent innovation has been the use of adaptive clinical trial, where the course of a trial can be altered depending on its progression. Tools have been developed using Bayesian methodology to estimate the probabilities of success for each possible change in the study to final endpoint, and recommended dosage is based at the end of the trial, this tool and others would simulate the probability of the data seen under different scenarios, so will be useful to compare if information is sparse [91].

Regulatory authorities have stated that the utility of PK/PD is greatest in early phase drug development, namely using pharmacodynamic measures as a primary or secondary endpoint of a dose-ranging study and using exposure-response (efficacy or adverse event) to define the optimal dose or dose range [81]. A well-defined link between the models developed and study data to decision making is important in this context and is usually achieved by integrating the model with a decision analysis [92]. Although early phase studies are becoming more complex and may require individual registration in some countries, often a single PK/PD study can be sufficient to define dosing strategies for later phase studies [87, 93]. Modelling and simulation are also extremely useful in extrapolating efficacy and safety findings in special populations, such as an elderly person or a person with impaired kidney function [94]. Often, it is not possible to include such subjects in the study, and simulation can enable informed decisions without exposing these populations to potential risk. PopPK/PD models can be applied to inform decisions in all phases of drug development, and many supporters of modelling believe that the largest return on investment is in using modelling and simulation to inform decisions in making regulatory submissions [26]. Quantitative evidence of safety and efficacy is increasingly becoming a pivotal aspect of drugs gaining marketing approval. Many authorities, particularly the U.S. FDA, have promoted the use of model-based drug development, reflected by the Critical Path Initiative in the U.S [25].

4.9 Model Validation and Evaluation

There are a wide variety of validation techniques, and the best way to validate a particular model will often be field-specific [95]. This is an active area of research in which medical statisticians are playing an increasingly important role. A simple classification of validation types is to consider internal validation, when a single dataset is used both for model fitting and model validation, and external validation where a dataset is split into data used for model fitting and a dataset used for model validation [96]. Usually, the gold standard is to perform validation using an external dataset; this avoids the risk of data overfitting where a model is excessively tailored to a specific dataset but lacks generality [97]. Model validation is the process of determining the degree to which a model is an accurate representation of the real world from the perspective of the stated purposes for the model [98]. Model validation is a holistic activity and should consider the context of the system and model purpose, the available data, and the statistical and qualitative evaluation of model performance [99]. Model validation is a priori if the validation occurs during the construction of a new model to address a specific question, and it is a posteriori if the model being validated is an existing model [100].

Model validation is an essential part of modelling because it is highly likely that a model will be an inaccurate representation of real-world or clinical processes since models are simplifications of reality [101]. Model simplifications are usually necessary to make a problem tractable; however, the danger is that incorrect simplifications will be made or that the model will be incorrectly implemented or solved [102]. Consequently, there is a need to validate that a model is an accurate representation of the system being modelled [103].

The main difference between model validation and model evaluation is that modelling involving the task of simplifying complex systems into manageable parts for investigation where it is crucial to integrate these components effectively to create a comprehensive model [95, 104]. Often, initial model development lacks full understanding, leading to its use in exploring system interventions without proper validation of its predictive capabilities [105]. This validation phase assesses the model's behaviour and its ability to accurately predict outcomes. Model evaluation is the next or ongoing phase of model use, where the model and its predictions are compared to the real system and real data [106]. This is a checkpoint to ensure that the model is still a good representation of the system, and to ensure that the predictions being made are accurate and reliable [107]. Typically, the standard assessments for any model, such as comparing it to the actual system and evaluating prediction reliability, are frequently informal and occasionally subjective [99, 108]. This can be problematic, as without formal process and conditions, tests can be poorly designed and evaluations can be biased. Statistical analyses are often used in model evaluation, as they provide formalised tests with well-defined conditions. This is particularly useful in ensuring that an evaluation is objective [109]. However, it must be remembered that these tests are just another part of the model, and it is essential that they are used properly and that their assumptions are satisfied. Failure to do this can result in misleading conclusions [110]. Ideally, the tests will mirror the real-life situation as much as possible. This will often involve simulation of the model's output and applying the test to this simulated data [111]. In any case, the test results should be presented in a manner that is easily understood, preferably with discussion of their implications on the model and its predictions.

An example of model evaluation where limited data was studied to know about the dangers of applying population PKPD with too small of a sample was studied. Using a dataset obtained from a Rhesus monkey given an IV bolus of a narcotic analgesic, attempt was done to estimate the volume of distribution and clearance rate of the drug. Data was available from only one PK sample per animal over a short lapse of time. Two methods, non-parametric kernel regression and mixed effects model regressed through two different methods using the software, were attempted (CLT and MCMC). Analysis was conducted using n = 5, 10, and 15 on both the volume of distribution and the clearance rate. The data was simulated for n = 50 and n = 1000. Estimates of all the rate constants and concentration at steady state were obtained with two different methods: non-linear regression (CLT) and an iterative two-stage Bayesian method using WinBUGS. Results from the two methods were compared. Simulation of data was straightforward, and a key advantage was that the true values of all the parameters were known. Confidence in the results was high, with similar estimates being generated for the rate constants and around 95% of the results being significant. Patterns in the data, such as the equilibrium distribution of fast and slow compartments, were easily identified. This was an excellent model for the application of population PKPD, and it would have been beneficial if more complex data was available [112].

4.10 Advancing Pharmacometrics and Clinical Trial Design

Pharmacometrics, employing population pharmacokinetic (popPK) and physiologically based pharmacokinetic (PBPK) modelling, enhances clinical pharmacokinetic practices. Its computer-based modelling and simulation aid in optimizing drug doses and ensuring safety and efficacy, particularly advantageous in LMICs with limited clinical trial capacities [113]. The linkage between this and PBPK/PD would be that the prediction of dose regimens can utilize the predictions of drug concentration time profiles at the site of action, the difference being that this can take into account many different doses and dose schedules without having to actually administer the drug, something which is often difficult in preclinical studies [4]. The prediction of concentrations and the exposure response relationships can sometimes be done by simple extrapolation from PK studies [114]. However, where PD modelling is not done, it can lead to unsafe assumptions and the testing of these assumptions by conducting in vitro or in vivo studies that may be uninformative or wasteful of resources [60]. This links in with the aim of reducing animal testing and increasing the informativeness of such tests when they are done. Moving back to PD models, these can sometimes be fitted to clinical data if some measurement has been taken of the response, and this can be useful in confirming the model's validity and also increasing the precision of an administration [115], an example being the precise administration of prednisolone in trying to temporally suppress the HPA axis in patients with Cushing's Syndrome [116]. A crucial part of any predictions that are to be administered or regulatory decisions is the estimation of the uncertainty of the models and the predictions, and of harm/risk/benefit analysis [117], but these are large topics in themselves. Simulation of the whole process of drug research, development, regulation, and further research with a silver standard type and then a gold standard type comparisons can utilize all types of models and is an area where model-informed decisions are trying to replace actual decisions with a higher percentage of success [78].

Systems pharmacology, a network-centric approach to studying biology, disease, and therapeutics, can enhance drug treatment predictions by integrating with PK/PD models [118]. This fusion can improve drug candidate selection, clinical trials, and therapeutic discovery. Joint meetings, workshops, and dual-trained modelers can promote collaboration between these fields [119]. Clinical trials can vary from simple to complex models, aiming to understand treatment effects [120]. Examples include comparing erythropoietin administration in Chronic Renal Failure Patients and antithrombotic strategies for Atrial Fibrillation [121]. Mastering clinical trial design is crucial for clinical pharmacologists, researchers, and statisticians [122]. Clinical trial simulation allows replicating or conducting hypothetical trials, based on patient data, intervention, and trial design, using patient and intervention models and decision-making algorithms [120].

4.11 Software and Tools

PopED (the population experiment design tool) was developed by AstraZeneca and has been a collaborative project with academia. The aim of the project was to provide a tool which would enable optimal experiment design in population and to support decision-making in drug development [123].

"Pmetrics" is an updated R software for pharmacokinetic and pharmacodynamic population modeling, offering both parametric and non-parametric approaches. It fits a simulated population with bimodal elimination and unimodal volume of distribution, including an extreme outlier, for intravenous drug models [124]. This package offers a number of tools in a variety of different MATLAB-based platforms in order to create models and simulations of PK/PD data.

Population approach with MONOLIX 3.22 estimated PK and PD parameters, and physiologically based pharmacokinetic (PBPK) modelling in GastroPlus3 predicted human PK of RO_A. The model accurately described transient PLT increases in monkeys and thrombocytopenia with subsequent rebound for all dosing regimens. Simulated human PLT profiles suggested concentrated daily administration for 28-day cycles. Monte Carlo simulation estimated a 4% risk of grade 4 thrombocytopenia during the first 2 cycles. Based on modelling, a starting dose of 100 mg was selected for human trials, avoiding two escalation steps in phase I studies [125].

PsN (Perl speaks NONMEM; a toolset of Perl modules) is a collection of Perl modules and shell scripts providing a powerful and flexible system for handling NONMEM data [126]. NONMEM (Icon solutions, Ellicott City, MD, USA) is the most commonly used tool for population PK/PD analysis and has been in use for nearly 25 years. NONMEM may be executed using a command line or via third-party software tools, such as PDx-Pop, Pirana, Perl Speaks NONMEM, KIWI, PLT Tools, or Wings [127]. It allows for comprehensive exploration of model space and has powerful tools for simulation and visual predictive checks. As it is quite an old piece of software, it can be difficult to use and does not have a very user-friendly interface.

There are several tools and software available for population PK/PD analysis. These tools have specific purposes and cater for different circumstances but all use similar basic methods for their population analysis. Most are based on using the NONMEM software with add-on tools to improve the user interface such as PsN and Pirana.

4.12 Illustrative Instances of Population Modelling in Pharmaceuticals

In a study conducted by authors aimed to establish the pharmacokinetic-pharmacodynamic (PK-PD) relationship for azithromycin extended release (ER) in Japanese patients and evaluate potential covariate effects on response prediction. Sparse serum azithromycin concentration, minimum inhibitory concentration (MIC), efficacy, and safety data were collected from three Phase 3 studies involving a 2 g single dose of azithromycin-ER for respiratory tract infections in Japanese patients. The data were combined with information from eight Phase 1 PK studies in Japanese and Western populations to develop a robust population PK model using a non-linear mixed effects approach. The PK of azithromycin-ER was described by a two-compartment model with first-order absorption and elimination, including a lag time, and no significant ethnic differences in area under the concentration-time curve (AUC) were observed. Exposure-response relationships were evaluated using logistic regression. The analysis revealed that the percentage of bacteriological and clinical success was significantly higher in patients with AUC/MIC >5 compared to those with AUC/MIC ≤ 5. This finding supports the correlation between azithromycin efficacy and AUC/MIC, as observed in animal infection models. Additionally, the study found that the incidence of treatment-related diarrhea was inversely associated with azithromycin exposure, indicating a potential exposure-safety relationship [128].

Another application of modelling and simulation is done in research and development of the cereal food Fortification Program in Iran. This involved a program aimed at combating iron deficiency in the Iranian population. Simulation modelling was used to evaluate the most effective iron fortification strategy, doses, and formulations for iron fortification and the expected health outcomes on Iranian children and women of childbearing age [129]. The model used for this study was a simplified version of a complex adaptive system and made use of a wide range of Iranian demographic and health data. At each stage of the fortification program, model results provided guidance for the most beneficial course of action for decision-makers. Time trends of the health outcomes predicted by the model were compared with national health indicators to assess the impact of fortification on the health of Iranian populations.

4.13 Future Trends and Innovations

This 'second golden age' of population modeling coincides with the wider societal development of computing power and the great advances being made in computer science and software engineering. Creating and running simulations has become a far quicker and simpler process than the longitudinal modeling techniques employed between the 80 s and early 2000s. There are many new software tools that have been developed to aid in the processes of creating and running models, some make use of advancements in Bayesian statistical techniques, but one particularly exciting development is the open-source software tool called NONMEM. NONMEM is already a very well-established tool in the pharmacokinetic modeling community [130]. It is very flexible with a wide range of inbuilt statistical functions and modeling methodologies. It also has a keen user base who share learning resources over the internet [131]. But up until recently, the use of NONMEM has been restricted to university institutions and pharmaceutical industries that can afford the steep price of a license. This has been a thorn in the side of many academics and modelers in smaller pharmaceutical companies who cannot afford the costs. However, development of the Pirana tool and the recent release of an academic license for NONMEM has brought new hope of widening the access and use of NONMEM to the whole population modeling community [132].

The increase in power and availability of computers has had a profound impact on all of these methodologies, often allowing more complex methods to be employed and opening up entirely new possibilities for the use of modelling and simulation in 'virtual' experiments with the potential to inform real world decisions. In terms of specific new methodologies and tools that are likely to impact on the field, these are difficult to predict but are almost certain to co-evolve with developments in computational and information sciences [133].

Personalized medicine aims to customize healthcare for individual patients, tailoring decisions and treatments to their specific needs [134]. In pharmacokinetics and pharmacodynamics, population modeling, particularly through meta-analysis, strives to provide optimized doses for specific populations [135]. The trend in population PKPD is towards providing results directly applicable to medical decisions, bridging the gap between population-based approaches and individualized therapy. Bayesian estimation, utilizing prior information to formulate probability distributions, is currently the most direct application of PKPD models [136]. However, its effectiveness relies on the quality of prior information and model accuracy. With ongoing advancements, reliable models for estimating various drug therapies may become available in the near future.

4.14 Conclusion

The theory and techniques of pharmacokinetic-pharmacodynamic (PK-PD) modelling, as well as their application in the development of various drug delivery systems and modified big molecules, were explored in many studies. It underlines the need of PK-PD modelling and simulation in removing misunderstandings regarding concentration thresholds. Furthermore, technological improvements have improved our understanding of the physicochemical features of drug delivery systems and their interactions with the body, boosting the predictive power of PK-PD modelling to drive the development of novel drug delivery systems.

References

1. Mould D, Upton R. Basic concepts in population modeling, simulation, and model-based drug development. CPT: Pharmacometrics Syst Pharmacol. 2012;1(9):1–14.
2. Bizzarri M, Fedeli V, Monti N, Cucina A, Jalouli M, Alwasel S, et al. Personalization of medical treatments in oncology: time for rethinking the disease concept to improve individual outcomes. EPMA J. 2012;12(4):545–8.
3. Huang XH, Li J. Pharmacokinetic-pharmacodynamic modeling and simulation. Am J Pharm Educ. 2007;17(1):18.
4. Jones H, Rowland-Yeo K. Basic concepts in physiologically based pharmacokinetic modeling in drug discovery and development. CPT: Pharmacometrics Syst Pharmacol. 2013;2(8):1–12.
5. Lavielle M. Mixed effects models for the population approach: models, tasks, methods and tools. CRC PRESS; 2023.
6. Pillai GC, Mentré F, Steimer JL. Non-linear mixed effects modeling—from methodology and software development to driving implementation in drug development science. J Pharmacokinet Pharmacodyn. 2005;32(2):161–85.
7. Ollier E. Fast selection of nonlinear mixed effect models using penalized likelihood. Comput Stat Data Anal. 2021;167:107373.
8. Gelman A. Bayesian data analysis. CRC Press; 2013.
9. Muehlemann N, Zhou T, Mukherjee R, Hossain MI, Roychoudhury S, Russek-Cohen E. A tutorial on modern Bayesian methods in clinical trials. Ther Innov Regul Sci. 2023;57(3):402–16.
10. Physiologically-Based Pharmacokinetic (PBPK) Models. [Online].; 2018 [cited 2024 April 08]. Available from: https://www.epa.gov/sites/default/files/2018-02/documents/pbpk_factsheet_feb2018_0.pdf
11. Mould DR, Denman NG, Duffull F. Using disease progression models as a tool to detect drug effect. Clin Pharmacol Ther. 2007;82(1):81–6.
12. Bhattaram VA, Booth BV, Ramchandani RP, Beasley BN, Wang Y, Tandon V, et al. Impact of pharmacometrics on drug approval and labeling decisions: a survey of 42 new drug applications. AAPS J. 2005;7(3):E503–12.
13. Cook J, Dasbach E, Coplan P, Markson L, Yin D, Meibohm A, et al. Modeling the long-term outcomes and costs of HIV antiretroviral therapy using HIV RNA levels: application to a clinical trial. AIDS Res Hum Retrovir. 1999;15(6):499–508.
14. Nagase M, Aksenov S, Yan H, Dunyak J, Al-Huniti N. Modeling tumor growth and treatment resistance dynamics characterizes different response to gefitinib or chemotherapy in non-small cell lung cancer. CPT: Pharmacometrics Syst Pharmacol. 2020;9(3):143–52.
15. Pozzi L, Schmidli H, Ohlssen DI. A bayesian hierarchical surrogate outcome model for multiple sclerosis. Pharm Stat. 2016;15(4):341–8.

16. Gao X, Shen W, Ning J, Feng Z, Hu J. Addressing patient heterogeneity in disease predictive model development. Biometrics. 2021;78(3):1045–55.
17. Kravitz RL, Duan N, Braslow J. Evidence-based medicine, heterogeneity of treatment effects, and the trouble with averages. Milbank Q. 2004;82(4):661–87.
18. Li Y, Meng Q, Yang M, Liu D, Hou X, Tang L, et al. Current trends in drug metabolism and pharmacokinetics. Acta Pharm Sin B. 2019;9(6):1113–44.
19. Kimko H, Lee K. Improving realism in clinical trial simulations via real-world data. CPT: Pharmacometrics Syst Pharmacol. 2017;6(11):727–9.
20. Voss EA, Ma Q, Ryan PB. The impact of standardizing the definition of visits on the consistency of multi-database observational health research. BMC Med Res Methodol. 2015;15(1):1–10.
21. Mahajan R. Real world data: additional source for making clinical decisions. Int J Appl Basic Med Res. 2015;5(2):82.
22. Kürzinger ML, Douarin L, Uzun I, El-Haddad C, Hurst W, Juhaeri J, et al. Structured benefit–risk evaluation for medicinal products: review of quantitative benefit–risk assessment findings in the literature. Ther Adv Drug Saf. 2020;11:1–12.
23. Curtin F, Schulz P. Assessing the benefit:risk ratio of a drug - randomized and naturalistic evidence. Dialogues Clin Neurosci. 2011;13(2):183–90.
24. Simoens S. Health economic assessment: a methodological primer. Int J Environ Res Public Health. 2009;6(12):2950–66.
25. Madabushi R, Seo P, Zhao L, Tegenge M, Zhu H. Review: role of model-informed drug development approaches in the lifecycle of drug development and regulatory decision-making. Pharm Res. 2022;39(8):1669–80.
26. Pereira LC, Fátima MA, Santos VV, Brandão CM, Alves IA, Azeredo FJ. Pharmacokinetic/pharmacodynamic modeling and application in antibacterial and antifungal pharmacotherapy: a narrative review. Antibiotics. 2022;11(8):986.
27. Arya V, Venkatakrishnan K. Role of physiologically based pharmacokinetic modeling and simulation in enabling model-informed development of drugs and Biotherapeutics. J Clin Pharmacol. 2020;60(S1):S7–S11.
28. Jones H, Chen Y, Gibson C, Heimbach T, Parrott N, Peters S, et al. Physiologically based pharmacokinetic modeling in drug discovery and development: a pharmaceutical industry perspective. Clin Pharmacol Ther. 2015;97(3):247–62.
29. Meibohm B, Läer S, Panetta JC, Barrett JS. Population pharmacokinetic studies in pediatrics: issues in design and analysis. AAPS J. 2005;7(2):E475–87.
30. Zou H, Banerjee P, Leung SS, Yan X. Application of pharmacokinetic-pharmacodynamic modeling in drug delivery: development and challenges. Front Pharmacol. 2020;11:12.
31. Gill KM, Jones HM. Opportunities and challenges for PBPK model of mAbs in paediatrics and pregnancy. AAPS J. 2022;24(4):72.
32. El-Khateeb E, Vasilogianni AM, Alrubia S, Al-Majdoub ZM, Couto N, Howard M, et al. Quantitative mass spectrometry-based proteomics in the era of model-informed drug development: applications in translational pharmacology and recommendations for best practice. Pharmacol Ther. 2019;203:107397.
33. Kaur N, Narang A, Bansal AK. Use of biorelevant dissolution and PBPK modeling to predict oral drug absorption. Eur J Pharm Biopharm. 2018;129:222–46.
34. Taylor ZL, Poweleit EA, Paice K, Somers KM, Pavia K, Vinks AA, et al. Tutorial on model selection and validation of model input into precision dosing software for model-informed precision dosing. CPT: Pharmacometrics Syst Pharmacol. 2023;12(12):1827–45.
35. Davidian M, Giltinan DM. Nonlinear models for repeated measurement data: an overview and update. J Agric Biol Environ Stat. 2003;8(4):387–419.
36. Verhaar MC, Grahn AY, Van Weerdt AW, Honing ML, Morrison PJ, Yang YP, et al. Pharmacokinetics and pharmacodynamic effects of ABT-627, an oral ETA selective endothelin antagonist, in humans. Br J Clin Pharmacol. 2000;49(6):562–73.

37. Nakao K, Kobuchi S, Marutani S, Iwazaki A, Tamiya A, Isa S, et al. Population pharmacokinetics of afatinib and exposure-safety relationships in Japanese patients with EGFR mutation-positive non-small cell lung cancer. Sci Rep. 2019;9(1):18202.
38. van de Schoot R, Depaoli S, King R, Kramer B, Märtens K, Tadesse M, et al. Bayesian statistics and modelling. Nat Rev Methods Primers. 2021;1(1).
39. Narisetty NN. Chapter 4 - Bayesian model selection for high-dimensional data. In: Handbook of statistics; 2020. p. 207–48.
40. Ruberg SJ, Beckers F, Hemmings R, Honig P, Irony T, Lavange L, et al. Application of bayesian approaches in drug development: starting a virtuous cycle. Nat Rev Drug Discov. 2023;22(3):235–50.
41. van de Schoot R, Kaplan D, Denissen J, Asendorpf J, Neyer F, van Aken M. A gentle introduction to Bayesian analysis: applications to developmental research. Child Dev. 2013;85(3):842–60.
42. Kitanidis PK. Parameter uncertainty in estimation of spatial functions: Bayesian analysis. Water Resour Res. 1986;22(4):499–597.
43. Dunn WL, Shultis KJ. Chapter 6 - Markov chain Monte Carlo. In: Exploring Monte Carlo Methods. 2nd ed; 2023. p. 189–254.
44. Luengo D, Martino L, Bugallo M, Elvira V, Särkkä S. A survey of Monte Carlo methods for parameter estimation. EURASIP J Adv Signal Process. 2020:1–62.
45. Babuška I, Nobile F, Tempone R. A systematic approach to model validation based on bayesian updates and prediction related rejection criteria. Comput Methods Appl Mech Eng. 2008;197(29–32):2517–39.
46. Raudenbush S, Bryk A. Hierarchical linear models: applications and data analysis methods. 2nd ed. Thousand Oaks: Sage Publications; 2002.
47. Kruschke K, Vanpaemel W. Bayesian estimation in hierarchical models. In: Busemeyer JR, Wang Z, editors. The Oxford handbook of computational and mathematical psychology. Oxford: Oxford University Press; 2015. p. 279–99.
48. Ashby D, Smith A. Evidence-based medicine as Bayesian decision-making. Stat Med. 2000;19(23):3291–305.
49. Gupta SK. Use of bayesian statistics in drug development: advantages and challenges. Int J Appl Basic Med Res. 2012;2(1):3–6.
50. Spiegelhalter DJ, Freedman LS, Parmar MV. Bayesian approaches to randomized trials. J Royal Stat Soc: Series A. 1994;157(3):357.
51. Spiegelhalter DJ, Abrams KR, Myles JV. Bayesian approaches to clinical trials and Health Care Evaluation Chichester. West Sussex: Wiley; 2004.
52. Adamadmin. Allucent. [Online]; 2024 [cited 2024 April 18. Available from: https://www.allucent.com/resources/blog/what-difference-between-pbpk-and-poppk-modeling#:~:text=Rather%2C%20popPK%20modeling%20takes%20all,the%20organ%2C%20or%20tissue%20level
53. Fisher JW, Mehta D, Li M, Yang X. Chapter 24 - Challenges in predicting the pharmacokinetics of drugs in premature and mature newborns. In: Donald Mattison LAH, editor. Clinical pharmacology during pregnancy. 2nd ed. Academic; 2022. p. 437–56.
54. Zhuang X, Lu C. PBPK modeling and simulation in drug research and development. Acta Pharm Sin B. 2016;6(5):430–40.
55. Kuepfer L, Niederalt C, Wendl T, Schlender J, Willmann S, Lippert J, et al. Applied concepts in PBPK modeling: how to build a PBPK/PD model. CPT: Pharmacometrics Syst Pharmacol. 2016;5(10):516–31.
56. Glassman PM, Muzykantov VR. Pharmacokinetic and pharmacodynamic properties of drug delivery systems. J Pharmacol Exp Ther. 2019;370(3):570–80.
57. Shargel L, Yu A. Applied biopharmaceutics & pharmacokinetics. 7th ed. McGraw-Hill Education; 2016.

58. Tan YM, Worley RR, Leonard JA, Fisher JW. Challenges associated with applying physiologically based pharmacokinetic modeling for public health decision-making. Toxicol Sci. 2018;162(2):341–8.
59. Barton HA, Chiu WA, Woodrow Setzer R, Andersen ME, Bailer JA, Bois FY, et al. Characterizing uncertainty and variability in physiologically based pharmacokinetic models: state of the science and needs for research and implementation. Toxicol Sci. 2007;99(2):395–402.
60. Chiu W, Barton H, DeWoskin R, Schlosser P, Thompson C, Sonawane B, et al. Evaluation of physiologically based pharmacokinetic models for use in risk assessment. J Appl Toxicol. 2007;27(3):218–37.
61. Paini A, Leonard J, Kliment T, Tan YM, Worth A. Investigating the state of physiologically based kinetic modelling practices and challenges associated with gaining regulatory acceptance of model applications. Regul Toxicol Pharmacol. 2017;90:104–15.
62. Block M. Physiologically based pharmacokinetic and pharmacodynamic modeling in cancer drug development: status, potential and gaps. Expert Opin Drug Metab Toxicol. 2015;11(5):743–56.
63. Felmlee MA, Morris ME, Mager ME. Mechanism-based pharmacodynamic modeling. Methods Mol Biol. 2012:583–600.
64. Maharao N, Antontsev V, Wright M, Varshney J. Entering the era of computationally driven drug development. Drug Metab Rev. 2020;52(2):283–98.
65. Gambús PL, Trocóniz IF. Pharmacokinetic–pharmacodynamic modelling in anaesthesia. Br J Clin Pharmacol. 2014;79(1):72–84.
66. Ohtani M, Kotaki H, Nishitateno K, Sawada Y, Iga T. Kinetics of respiratory depression in rats induced by buprenorphine and its metabolite, norbuprenorphine. J Pharmacol Exp Ther. 1997;281(1):428–33.
67. Simpson TW, Poplinski JD, Koch. Metamodels for Computer-Based Engineering Design: survey and recommendations. Eng Comput. 2001;16(2):129–50.
68. Helbing D. Social self-organization: Agent-based simulations and experiments to study emergent social behavior. Springer; 2012.
69. Kristensen NR, Madsen H, Ingwersen SH. Using stochastic differential equations for PK/PD model development. J Pharmacokinet Pharmacodyn. 2005;32(1):109–41.
70. Mystridis G, Batzias G, Vizirianakis I. Physiologically based pharmacokinetic modelling and simulation to predict the plasma concentration profile of doxorubicin. Pharmaceutics. 2022;14(3):541.
71. Loisios-Konstantinidis I, Cristofoletti R, Jamei M, Turner D, Dressman J. Physiologically based pharmacokinetic/pharmacodynamic modeling to predict the impact of CYP2C9 genetic polymorphisms, co-medication and formulation on the pharmacokinetics and pharmacodynamics of Flurbiprofen. Pharmaceutics. 2020;12(2):1049.
72. Zhang W, Zhang Q, Cao Z, Zheng L, Hu W. Physiologically based pharmacokinetic modeling in neonates: current status and future perspectives. Pharmaceutics. 2023;15(12):2765.
73. Newhard W, Patel M, Cassaday J, Ballard J, Squadroni B, Wu G, et al. In vitro pharmacokinetic/pharmacodynamic modeling of HIV latency reversal by novel HDAC inhibitors using an automated platform. SLAS Discov. 2023;26(5):642–54.
74. Kersting G, Willmann S, Würthwein G, Lippert J, Boos J, Hempel G. Physiologically based pharmacokinetic modelling of high- and low-dose etoposide: from adults to children. Cancer Chemother Pharmacol. 2011;69(2):397–405.
75. Lim HS. Evolving role of modeling and simulation in drug development. Transl Clin Pharmacol. 2019;27(1):19–23.
76. Polit DF, Beck CT. Generalization in quantitative and qualitative research: myths and strategies. Int J Nurs Stud. 2011;47(11):1451–8.
77. Hare J, Lammers T, Ashford MV, Puri S, Storm G, Barry S. Challenges and strategies in anti-cancer nanomedicine development: an industry perspective. Adv Drug Deliv Rev. 2017;108:25–38.

78. Salahshoori I, Golriz M, Nobre M, Mahdavi S, Eshaghi Malekshah R, Javdani-Mallak A, et al. Simulation-based approaches for drug delivery systems: navigating advancements, opportunities, and challenges. J Mol Liq. 2024;395:123888.
79. Kim TH, Shin S, Shin BS. Model-based drug development: application of modeling and simulation in drug development. J Pharm Investig. 2017;48(4):431–41.
80. Cook N, Hansen AR, Siu LL, Abdul Razak AR. Early phase clinical trials to identify optimal dosing and safety. Mol Oncol. 2014;9(5):997–1007.
81. Tuntland T, Ethell B, Kosaka T, Blasco F, Zang RX, Jain M, et al. Implementation of pharmacokinetic and pharmacodynamic strategies in early research phases of Drug Discovery and Development at Novartis Institute of Biomedical Research. Front Pharmacol. 2014; 5.
82. Bender G, Gosset J, Florian J, Tan K, Field M, Marshall S, et al. Population pharmacokinetic model of the pregabalin-sildenafil interaction in rats: application of simulation to preclinical PK-PD Study Design. Pharm Res. 2009;26(10):2259–69.
83. Lalonde RL, Kowalski KG, Hutmacher MM, Ewy W, Nichols DJ, Milligan PA, et al. Model-based drug development. Clin Pharmacol Ther. 2007;82(1):21–32.
84. Serdar CC, Cihan M, Yücel D, Serdar M. Sample size, power and effect size revisited: simplified and practical approaches in pre-clinical, clinical and laboratory studies. Biochem Med. 2021;31(1):27–53.
85. Mayer C, Perevozskaya I, Leonov S, Dragalin V, Pritchett Y, Bedding A, et al. Simulation practices for adaptive trial designs in drug and device development. Stat Biopharm Res. 2019;11(4):325–35.
86. Sheiner LB, Steimer JL. Pharmacokinetic/pharmacodynamic modeling in drug development. Annu Rev Pharmacol Toxicol. 2000;40(1):67–95.
87. Tyson RJ, Park CC, Powell RJ, Patterson HJ, Weiner D, Watkins PB, et al. Precision dosing priority criteria: drug, disease, and patient population variables. Front Pharmacol. 2020;11.
88. Soto J. Health economic evaluations using decision analytic modeling: principles and practices—utilization of a checklist to their development and appraisal. Int J Technol Assess Health Care. 2002;18(1):94–111.
89. Hamza T, Cipriani A, Furukawa T, Egger M, Orsini N, Salanti G. A bayesian dose–response meta-analysis model: a simulations study and application. Stat Methods Med Res. 2021;30(5):1358–72.
90. Kravari K, Bassiliades N. A survey of agent platforms. JASSS. 2015;18(1).
91. Musuamba F, Manolis E, Holford N, Cheung S, Friberg L, Ogungbenro K, et al. Advanced methods for dose and regimen finding during drug development: summary of the EMA/EFPIA workshop on dose finding (London 4–5 December 2014). CPT: Pharmacometrics Syst Pharmacol. 2017;6(7):418–29.
92. Chiheb F, Boumahdi F, Bouarfa H. A new model for integrating big data into phases of decision-making process. Procedia Comput Sci. 2019;151:636–42.
93. Bellanti F, Della PO. Modelling and simulation as research tools in paediatric drug development. Eur J Clin Pharmacol. 2011;67(S1):75–86.
94. Ravenstijn P, Chetty M, Manchandani P. Design and conduct considerations for studies in patients with impaired renal function. Clin Transl Sci. 2021;14(5):1689–704.
95. Sherwin C, Kiang T, Spigarelli M, Ensom M. Fundamentals of population pharmacokinetic modelling. Clin Pharmacokinet. 2012;51(9):573–90.
96. Coley RY, Liao Q, Simon N, Shortreed SM. Empirical evaluation of internal validation methods for prediction in large-scale clinical data with rare-event outcomes: a case study in suicide risk prediction. BMC Med Res Methodol. 2023;23(1).
97. Ramspek CL, Jager KJ, Dekker FW, Zoccali C, van Diepen M. External validation of prognostic models: what, why, how, when and where? Clin Kidney J. 2020;14(1):49–58.
98. Martis MS. Validation of simulation based models: a theoretical outlook. Electron J Bus Res Methods. 2006;4(1):39–45.
99. Barlas Y. Formal aspects of model validity and validation in system dynamics. Syst Dyn Rev. 1996;12(3):183–210.

100. Sornette D, Davis BA, Ide K, Vixie KR, Pisarenko V, Kamm JR. Algorithm for model validation: theory and applications. Proc Natl Acad Sci. 2007;104(16):6562–7.
101. Robinson S. A tutorial on conceptual modeling for simulation. 2015 Winter Simulation Conference (WSC); 2015. p. 1820–34.
102. Weld DS. Reasoning about model accuracy. Artif Intell. 1992;56(2–3):255–300.
103. Robinson S. Simulation verification, validation and confidence: a tutorial. Trans Soc Comput Simulat. 1999;16(2):63–9.
104. Brendel K, Comets E, Laffont C, Laveille C, Mentré F. Metrics for external model evaluation with an application to the population pharmacokinetics of gliclazide. Pharm Res. 2006;23(9):2036–49.
105. Vamathevan J, Clark D, Czodrowski P, Dunham I, Ferran E, Lee G, et al. Applications of machine learning in drug discovery and development. Nat Rev Drug Discov. 2019;18(6):463–77.
106. Agatonovic-Kustrin S, Beresford R. Basic Concepts of Artificial Neural Network (ANN) modeling and its application in pharmaceutical research. J Pharm Biomed Anal. 2000;22(5):717–27.
107. Marshall S, Burghaus R, Cosson V, Cheung S, Chenel M, DellaPasqua O, et al. Good practices in model-informed drug discovery and development: practice, application, and documentation. CPT: Pharmacometrics Syst Pharmacol. 2016;5(3):93–122.
108. Grove WM, Meehl PE. Comparative efficiency of informal (subjective, impressionistic) and formal (mechanical, algorithmic) prediction procedures: the clinical–statistical controversy. Psychol Public Policy Law. 1996;2(2):293–323.
109. Helmlinger G, Al-Huniti N, Aksenov S, Peskov K, Hallow K, Chu L, et al. Drug-disease modeling in the pharmaceutical industry - where mechanistic systems pharmacology and statistical pharmacometrics meet. Eur J Pharm Sci. 2017;109:S39–46.
110. Naci H, Ioannidis JPA. How good is "evidence" from clinical studies of drug effects and why might such evidence fail in the prediction of the clinical utility of drugs? Annu Rev Pharmacol Toxicol. 2015;55(1):169–89.
111. Holford NH, Kimko HC, Monteleone JP, Peck CC. Simulation of clinical trials. Annu Rev Pharmacol Toxicol. 2000;40(1):209–34.
112. Krugner-Higby L, KuKanich B, Schmidt B, Heath T, Brown C. Pharmacokinetics and behavioral effects of liposomal hydromorphone suitable for perioperative use in rhesus macaques. Psychopharmacology. 2011;216(4):511–23.
113. Usman M, Khadka S, Saleem M, Rasheed H, Kunwar B, Ali M. Pharmacometrics: a new era of pharmacotherapy and drug development in low- and middle-income countries. Adv Pharmacol Pharm Sci. 2023;2023:1–10.
114. Gibbs JP. Prediction of exposure–response relationships to support first-in-human study design. AAPS J. 2010;12(4):750–8.
115. Stein R. Benchmarking default prediction models: pitfalls and remedies in model validation. J Risk Model Valid. 2007;1(1):77–113.
116. McBride M. Cushing syndrome and disease: a study of the diagnosis treatment clinical consequences and health-related quality of life associated with these medical conditions [dissertation]. Lancaster University (United Kingdom): ProQuest Dissertations Publishing; 2022.
117. Brønstad A, Newcomer C, Decelle T, Everitt J, Guillen J, Laber K. Current concepts of harm–benefit analysis of animal experiments—report from the AALAS–Felasa Working Group on harm–benefit analysis—part 1. Lab Anim. 2016;50(1_suppl):1–20.
118. Sorger P, Allerheiligen S, Abernethy D, Altman R, Brouwer K, Califano A, et al. Quantitative and systems pharmacology in the post-genomic era: new approaches to discovering drugs and understanding therapeutic mechanisms. In: An NIH white paper by the QSP workshop group, vol. 48; 2011. p. 1–47.
119. Berger S, Iyengar R. Network analyses in systems pharmacology. Bioinformatics. 2009;25(19):2466–72.

120. Chilcott J, Brennan A, Booth A, Karnon J, Tappenden P. The role of modelling in prioritising and planning clinical trials. Health Technol Assess. 2003;7(23).
121. Magnocavallo M, Bellasi A, Mariani MV, Fusaro M, Ravera M, Paoletti E, et al. Thromboembolic and bleeding risk in atrial fibrillation patients with chronic kidney disease: role of anticoagulation therapy. J Clin Med. 2020;10(1):83.
122. Barrett J, Fossler M, Cadieu KD, Gastonguay MR. Pharmacometrics: a multidisciplinary field to facilitate critical thinking in drug development and translational research settings. J Clin Pharmacol. 2008;48(5):632–49.
123. Aoki Y, Sundqvist M, Hooker A, Gennemark P. Poped Lite: an optimal design software for preclinical pharmacokinetic and pharmacodynamic studies. Comput Methods Prog Biomed. 2016;127:126–43.
124. Neely M, van Guilder M, Yamada W, Schumitzky A, Jelliffe R. Accurate detection of outliers and subpopulations with Pmetrics, a nonparametric and parametric pharmacometric modeling and simulation package for R. Ther Drug Monit. 2012;34(4):467–76.
125. Meille C, Reddy M, Walz A, Retout S, Nichols G, Glenn K, et al. I-25 Christophe Meille Modified model of drug induced thrombocytopenia efficiently projects safe starting dose in human from preclinical data. PAGE2012 Poster Abstracts. 2005;49(4):49.
126. Lindbom L, Ribbing J, Jonsson EN. Perl-speaks-NONMEM (psn)—a Perl module for NONMEM related programming. Comput Methods Prog Biomed. 2004;75(2):85–94.
127. Bauer RJ. Nonmem Tutorial Part I: description of commands and options, with simple examples of population analysis. CPT: Pharmacometrics Syst Pharmacol. 2019;8(8):525–37.
128. Muto C, Liu P, Chiba K, Suwa T. Pharmacokinetic-pharmacodynamic analysis of azithromycin extended release in Japanese patients with common respiratory tract infectious disease. J Antimicrob Chemother. 2010;66(1):165–74.
129. Mirmiran P, Golzarand M, Serra-Majem L, Azizi F. Iron, iodine and vitamin A in the Middle East; a systematic review of deficiency and food fortification. Iran J Public Health. 2012;41(8):8–19.
130. Bouzom F, Ball K, Perdaems N, Walther B. Physiologically based pharmacokinetic (PBPK) modelling tools: how to fit with our needs? Biopharm Drug Dispos. 2012;33(2):55–71.
131. Flühler H, Huber H, Widmer E, Brechbühler S. Experiences in the application of NONMEM to pharmacokinetic data analysis. Drug Metab Rev. 1984;15(1–2):317–39.
132. Helfer VE, Zavascki AP, Zeitlinger M, de Araújo B, Dalla CT. Population pharmacokinetic modeling and probability of target attainment of ceftaroline in brain and soft tissues. Antimicrob Agents Chemother. 2022;66(9):e00741–22.
133. Dodgson M, Gann D, Salter A. The impact of modelling and simulation technology on engineering problem solving. Tech Anal Strat Manag. 2007;19(4):471–89.
134. Ginsburg GS, Willard HF. Genomic and personalized medicine: foundations and applications. Transl Res. 2009;154(6):277–87.
135. Chan P, Yu J, Chinn L, Prohn M, Huisman J, Matzuka B, et al. Population Pharmacokinetics, efficacy exposure-response analysis, and model-based meta-analysis of fenebrutinib in subjects with rheumatoid arthritis. Pharm Res. 2020;37:1–13.
136. Lunn D, Best N, Thomas A, Wakefield J, Spiegelhalter D. Bayesian analysis of population PK/PD models: general concepts and software. J Pharmacokinet Pharmacodyn. 2002;29(3):271–307.

Chapter 5
Design of Experiments: Understanding Optimization

Amrish Kumar, Vikram Sharma, Sunil Kumar Jain, Dinesh Kumar Mishra, and S. N. Koteswara Rao G.

Contents

A. Kumar (✉)
Charak School of Pharmacy, Chaudhary Charan Singh University,
Meerut, Uttar Pradesh, India

V. Sharma
Galgotias College of Pharmacy, Greater Noida, Uttar Pradesh, India

S. K. Jain · D. K. Mishra
Department of Pharmacy, Guru Ghasidas Vishwavidyalaya (A Central University),
Bilaspur, Chhattisgarh, India

S. N. K. Rao G.
Shobhaben Pratapbhai Patel School of Pharmacy and Technology Management, SVKM's
NMIMS, Vile Parle (W), Mumbai, Maharashtra, India

S. N. K. Rao G., R. R. Alavala (eds.), *Applications of Computational Tools in Drug Design and Development*, https://doi.org/10.1007/978-981-96-4154-3_5

Abstract Design of Experiments (DoE) is acknowledged as a mathematical/statistical technique exploited in pharmaceutical research and development to swiftly optimize drug delivery systems' effectiveness on the basis of known input factors. The first step is a screening experimental design test plan that includes every known variable suspected of having a significant effect on the drug delivery system's pharmacokinetic or pharmacodynamic accomplishments. Reducing the number of test factors or input variables to a reasonable quantity is the main goal of the experiment when there are many of them. Another test plan or experimental design is typically used to give a boost to the drug delivery system's performance for predetermined therapeutic results. The use of DoE software and methodology allowed the pharmaceutical researchers to quantify the dominance of several prominent factors on the predefined attributes and performance of medication delivery system under examination. By using DoE, formulation performance might be quickly optimized across a wide range of relevant factors to achieve the intended therapeutic effects. The DoE will be thoroughly followed up on in this chapter, with the goal of optimizing pharmaceutical formulations while minimizing the number of actual experimental runs.

Keywords Quality-by-Design (QbD) · Experimental Design · Design of Experiments (DoE) · Design Space · Quality Target Product Profile (QTPP) · Critical Quality Attributes (CQAs) · Critical Material Attributes (CMAs) · Critical Process Parameters (CPPs)

5.1 Introduction

The creation of novel medications, treatments, and healthcare interventions that enhance patient outcomes and quality of life is greatly aided by pharmaceutical research. The pharmaceutical sector is always changing due to technological advancements, so it is becoming more important than ever to make sure that the development processes are effective and based on robust scientific evidence. One key factor that enables this resiliency is the concept of experimental design. The organized and methodical process of organizing, carrying out, and evaluating tests so as to optimize knowledge obtained while lowering unpredictability and cutting expenses is known as experimental design. The use of appropriate experimental design becomes essential for success in pharmaceutical research, where accuracy and precision are crucial [1, 2].

For many years, scientists working on formulation development have developed medicinal dosage forms by applying knowledge gleaned from past experiences. For the production of formulations lacking crucial elements like excessive dose, low-slung solubility, pathetic compressibility, or other physicochemical qualities, this was and is usually adequate. The scientist frequently creates a procedure or formulation that works well enough, but lacks the time to work on creating the optimal one. Competing goals, such as improving tablet hardness while decreasing disintegration time, might make it difficult to choose the right amounts for controllable variables. With the purpose to eliminate or capitalize on particular qualities of the delivery system or dosage form, the correlation among independent (controllable) factors and dependent (performance/quality) attributes ought to be recognized.

The relevance of experimental design in pharmaceutical research cannot be emphasized. It is indispensable for guaranteeing the reproducibility, reliability, and validity of experiment data. Designing tests that yield definite answers while following regulatory rules is crucial for expanding scientific knowledge and bringing safe, effective pharmaceuticals to market, given the intricate and highly regulated nature of drug development.

5.1.1 *What Is Experimental Design?*

The framework used to plan and conduct experiments in a methodical and controlled way is referred to as experimental design, or DoE. By optimizing the process of inquiry and observation through the use of statistical principles, this technique enables researchers to make well-informed judgments about which aspects to test, how to control them, and how to interpret the results.

Fundamentally, DoE entails specifying the goals of the study, identifying the variables of interest, selecting the experimental design, and conducting data analysis to make inferences. In addition to obtaining more significant results, scientists can lower the likelihood of bias or errors influencing the results by employing an organized methodology. Although experimental design is used in many fields, its importance in pharmaceutical research is particularly clear given the complexity of drug formulation, pharmacokinetics, bioavailability studies, and clinical trials; all of which call for meticulous preparation and implementation [3, 4].

5.1.2 *Importance of Experimental Design in Pharmaceutical Research [5–8]*

Several important advantages that experimental design offers us to understand its significance in pharmaceutical research:

Efficiency in Drug Development
The drug research process is notoriously time-consuming and expensive, frequently taking over a decade and spending billions before an API reaches to the market as a dosage form for therapeutic benefits. A well-designed experiment ensures that every experiment is carried out as best it can, minimizes the number of pointless trials, saves money on resources, and concentrates efforts on important research. Researchers can examine several factors at once using strategies like factorial design, which provides deeper understanding with reduced experimental work.

Enhancing the Quality of Data
Data quality is crucial in pharmaceutical research. Inaccurate results, erroneous conclusions, and eventually failure in clinical trials or regulatory approval processes might result from poorly conceived research. Methodologies for experimental design make ensuring that data is collected precisely and accurately, reducing the possibility of mistakes that could impede or stop the development of new drugs. When designing experiments, researchers can better measure uncertainty, evaluate the validity of their findings, and spot genuine signals in the data by utilizing statistical methods.

Optimization of Formulations
The optimization of drug formulations is one of the leading uses of experimental design in the pharmaceutical industry. To design a medicine that is both effective and commercially viable, formulation scientists must balance a number of aspects, embracing the active pharmaceutical ingredient (API), excipients, manufacturing procedures, and stability problems. Through the application of techniques such as response surface methodology (RSM) and full factorial designs, scientists are able to establish the best synchronized parameters that will optimize a formulation's safety and efficacy while reducing any negative effects or instability.

Regulatory Compliance
Organizations like United States Food and Drug Administration (USFDA) and European Medicines Agency (EMA) holds stringent regulation and surveillance over pharmaceutical research. These organizations mandate that medications be created using reproducible, safe, and effective scientific methodologies. The methodical technique of experimental design conforms to regulatory requirements and offers transparency in the drug testing and validation process.

Identification of Critical Quality Attributes (CQAs)
Pharmaceutical firms are encouraged to define and regulate the essential characteristics or critical quality attributes (CQAs) of their products in synchronization with the concepts of Quality-by-Design (QbD). The term "CQAs" stand for the attributes—physicochemical, microbiological, or biological—that, with the purpose of assurance of product quality, essentially brackets in the bounds of specified limit. Researchers can systematically investigate the ways in which various variables affect product performance and quality, which makes it easier to identify and regulate these CQAs. This is made possible by the use of experimental design.

Reduction of Risk and Variability

Pharmaceutical research frequently addresses the intrinsic variability of biological systems. Through the provision of instruments for controlling and randomizing elements, experimental design helps to limit risks connected with this variability and guarantees the robustness and reproducibility of experimental results. Furthermore, a greater knowledge of how minor modifications in manufacturing or formulation can affect the finished product can result from detecting interactions between variables through the use of experimental designs.

5.1.3 Key Elements of Experimental Design in Pharmaceutical Research [6, 9–11]

A well-executed experimental design in the pharmaceutical industry is defined by several essential components:

Establishing Clear Objectives Establishing precise objectives is essential before starting any experiment. The objective directs the entire experiment, whether the aim is to ascertain the best medicine formulation, comprehend the effects of different excipients, or assess the stability of a molecule. This stage of pharmaceutical research frequently incorporates input from formulation scientists, physicians, and regulatory agencies.

Selection of Variables Another essential component of experimental design is the selection of the variables, or elements, that will be changed during the experiment. These could include temperature, pH, active component concentration, excipients, production procedures, or storage conditions in pharmaceutical research. The experiment can be correctly evaluated if the independent (controlled) and dependent (measured) variables are identified.

Control and Randomization Controlling extraneous variables that may interfere with the results and randomizing experimental settings might assist to eliminate bias. For example, clinical trials employ control groups to test the efficacy of experimental drugs by comparing their results to those of regular therapies or placebos.

Sample Size and Replication Experiments repeated enhance the validity of the findings. Variability can be significant in pharmacological research involving biological systems. The consistency of the results can be verified by repeating the experiment with bigger sample sizes, which will lend the conclusions an additional significance. Inconclusive or misleading results may arise from inadequate sample sizes.

Design Selection Different kinds of experimental designs seem suitable for various research purposes. Pharmaceutical research frequently uses response surface methodology (RSM), fractional factorial designs, and full factorial designs. Multiple components can be investigated simultaneously using factorial designs, which sheds light on possible interactions between them while RSM is especially helpful for formulation optimization.

Data Analysis and Interpretation Following the completion of an experiment, the gathered data needs to be examined using the relevant statistical techniques. In pharmaceutical research, sophisticated statistical models are frequently used in data analysis to take into consideration the intricacy of biological responses. Making defensible conclusions, measuring variability, and appreciating the relevance of the results are essential for deciding what to do next.

Experimental design is not just a tool in pharmaceutical research; it is an essential methodology that directs the productive and successful creation of novel treatments. Experimental design greatly improves the trustworthiness of research findings by making sure that experiments are meticulously designed, data is thoroughly examined, and variability is managed. It guarantees adherence to regulatory requirements, facilitates the efficient development of pharmaceuticals, and eventually results in the manufacture of safe, efficacious therapies that enhance patient outcomes. The significance of experimental design will remain crucial in overcoming the obstacles associated with drug discovery and development as pharmaceutical research progresses [12].

5.2 Approaches for Optimization in Pharmaceutical Research

How well a pharmaceutical product satisfies patient demands and achieves expected product performance determines its financial success. The approaches taken by various pharmaceutical behemoths to develop new products varied depending on the product. An applicant may select QbD as a considerable systematic methodology for the development of pharmaceutical product or the usual minimal approach as an empirical technique, or they may choose to combine the two for the approval process. An account of the probable deviations among these methodologies is described in Table 5.1. QbD as a systematic methodology, for development can involve, among other things, utilizing knowledge management throughout the product's lifecycle, incorporating existing information, and the findings of research employing DoE. Such a methodical strategy can improve attaining the intended level of product quality [12, 16–18].

Due to a surpass perception of the product under consideration and its development chain, QbD approaches are increasingly being prioritized above conventional approaches in the current pharmaceutical industry environment when it comes to product/process development. The following points illustrate how the QbD technique has evolved comparatively:

Table 5.1 Comparative summary of optimization approaches for pharmaceutical development [13–15]

Aspect	Conventional minimal approaches	Advanced QbD approaches
Overall development process	Steps for developmental research is primarily empirical, focusing on one parameter at a stretch.	Well organized: connecting drug product CQAs to perfunctory outlook of material properties and process variables. Use multivariate experiments to comprehend the process and the product. Commencement of design space. Strategies from Process Analytical Technology (PAT) were used.
Manufacturing progression	Static validation is generally aligned on the early full-scale lots. Concentrate on optimization with reproducible outcomes.	Adaptable validation within the design space. Validation is approached throughout the lifecycle, with the goal of ongoing process affirmation. Emphasis on control strategy and resilience. Application of statistical methodologies for better control on overall processing.
Process constraints	In-process testing is usually exploited to make go- or no-go judgments. Off-line assessment.	PAT instruments thru suitable feedback and feed forward settings are used. Following approval, process activities were monitored and trended to assist efforts for persistent amendment.
Product essentials	Foremost method for regulation. Centered on batch information collected during registration phase.	Constituent of overall quality control plan. Focused on the targeted product accomplishment and appropriate supporting information.
Control approach	Intermediates (in-process ingredients) and final-product evaluation are the leading sources of quality control for drug products.	Final product quality is assured thru a risk-based supervision strategy established with out-and-out perception of process and final product quality requirements. Quality control surpasses with the prospect of actual performance assessment or diminished final-product evaluation.
Lifecycle regulation	Reactive (that is, addressing issues and taking remedial action)	Taking preventative measures. Continuous development was made easier.

5.2.1 *Focus*

- **Conventional Approach:**
  - Usually use a method of experimental trial and observed error.
  - Focuses on single parameter at a stretch for optimization; these experiments are also acknowledged as “One Factor at a Time” (OFAT) studies.

- Emphasizes achieving a final product that meets pre-defined requirements, without necessarily comprehending the underlying issues determining quality.

- **Quality-by-Design (QbD) Approach:**
  - Adopts a logical, risk-oriented attitude.
  - Ensures consistent quality by focusing on recognizing the full product life-cycle, including formulation development and manufacturing procedures.
  - Through DoE, a comprehensive knowledge of the link amongst formulation attributes and the product's essential characteristics i.e., Critical Quality Attributes (CQAs) is developed.
  - Integrates quality into the product throughout the design phase, instead of testing at the end.

5.2.2 Experimental Design

- **Conventional Approach:**
  - Utilizes a series of experiments that are frequently non-statistical, changing one variable at a stretch whilst holding the others constant.
  - Be able to lead several iterations and a protracted refinement process.
- **Quality-by-Design (QbD) Approach:**
  - Employs DoE to examine the consequence of varying many constraints on predefined outcomes at the same time.
  - Optimizes more effectively by lowering the summation of repetitive experiments necessitated and offering various comprehensive insight of factor interactions.

5.2.3 Risk Management and Control Strategy

- **Conventional Approach:**
  - Generally speaking, risk management is reactive, only addressing issues when deviations or failures happen.
  - Instead, counting on process control, control strategies rely on final inspections and final-product testing for quality endorsement.
- **Quality-by-Design (QbD) Approach:**
  - Prioritizes proactive risk management by recognizing and controlling the fluctuations of process parameters that may dominate the product quality.
  - Focuses on developing a Control Strategy driven by the experience of the process, making sure that quality is preserved not only at the level of the finished product but also throughout the production process.

5.2.4 Regulatory Perspective

- **Conventional Approach:**
  - Less robust than what regulatory agencies presently need; more conventional and acceptable in previous decades.
  - Frequently leads to increased regulatory scrutiny due to incomplete knowledge of the process and its crucial variables.
- **Quality-by-Design (QbD) Approach:**
  - As part of contemporary regulatory obligations, regulatory bodies e.g., the FDA and ICH strongly promote QbD.
  - The well-documented understanding of the product and procedure makes regulatory submissions easier to complete. The application of QbD may result in less post-approval modifications and other regulatory flexibility.

5.2.5 Efficiency and Costs

- **Conventional Approach:**
  - Expensive and time-consuming trial-and-error process for addressing unexpected issues.
  - Limited understanding of product variability might lead to increased expenses during post-approval adjustments or scale-up of batch size.
- **Quality-by-Design (QbD) Approach:**
  - Implementing QbD requires lengthy studies and risk assessment, but it eventually saves money by minimizing failed batches, rework, and post-approval adjustments.
  - Strong process understanding leads to increased efficiency in scaling up production and technology transfer.

5.2.6 Product and Process Understanding

- **Conventional Approach:**
  - Restricted comprehension of the correlation between factors and CQAs.
  - Emphases on fabricating a product that satisfies requirements without providing a thorough understanding of how those requirements are met.

- **Quality-by-Design (QbD) Approach:**
  - Offers a thorough comprehension of process variables and product composition, as well as how they affect the CQAs.
  - Creates a Design Space, having multifaceted ranges of process parameters and input variables that guarantee product quality and allow for adaptable manufacturing adjustments within it.

The traditional optimization method, which depends on end-point testing and a shallow comprehension of the process, is less effective but simpler. On the other hand, the QbD methodology is the favored way in current pharmaceutical development because it is a risk-based, science-driven, and modern strategy that results in consistent product quality, regulatory flexibility, and operational efficiency. Comparative summary of optimization approaches for pharmaceutical development are mentioned in Table 5.1.

5.3 Quality-By-Design Approach: Optimization Perspective

The optimization of pharmaceutical formulations during the developmental phase is the primary aspects for expanding pharmaceuticals with better and more cost-effective products based on available/investigated medicinal compounds. The inestimable time and resource expenditures associated with research and development may possibly decreased in an economical way by determining the proper product profile and applying appropriate experimental design-based optimization.

Quality-by-Design (QbD) is a realistic tactic for formulation optimization which employs in-depth knowledge of compatibility of incorporated drug with all the excipients along with the compatibility of formulation under consideration with selected process parameters for the product manufacturing. By automating manual testing, it facilitates the coordination of product design and offers a more efficient means of addressing any issues. QbD offers scientific data to evaluate the effects of chosen parameters, as opposed to depending solely on evaluation of the created formulation. This enables quick identification of the underlying cause of any undesired variation. Presently, FDA has also emphasized the implementation of QbD as a current Good Manufacturing Practices (cGMP) scheme for the development and production of pharmaceutical products [19, 20].

Figure 5.1 demonstrates the steps used in the QbD progression for formulation optimization. Quality Target Product Profile (QTPP) and the Critical Quality Attributes (CQAs) were identified at the beginning of the process. Critical Material Attributes (CMAs) and Critical Process Parameters (CPPs) were then categorized on the basis of risk analysis of the suggested delivery system or dosage forms and technique of preparation on CQAs. To obtain an optimal formulation, a statistical model for designing the protocols for experiments was used to inspect the chosen CMAs and CPPs. For the investigation of formulation under consideration and selected process parameters for these CMAs and CPPs, response surface approach

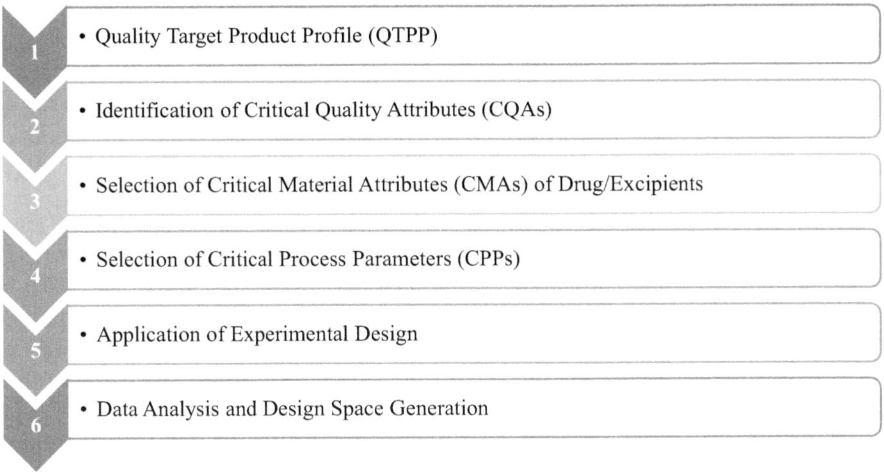

Fig. 5.1 Phases of QbD process for optimization

was also used. The acquired information was instigated for the preparation of formulation/product with reproducible quality [16, 21].

QbD minimizes the occurrences of product failure by the mapping of all conceivable variables including product characteristics and process parameters into a recognized control space. This kind of mapping makes it possible to quickly identify scientific variables that hold a sizeable control over the anticipated product quality level. The success recognition of any QbD process relies on the step-by-step execution of different steps.

5.3.1 Quality Target Product Profile and Identification of CQAs

QTPP outlines the quality attributes that must be met in the dosage form/delivery system under consideration to confirm the efficiency and safety on administration and CQAs consider the product attributes which markedly influences the quality of the dosage form/delivery system. A CQA can be outlined as a biological, microbiological or physicochemical possession that required to control/achieve as per predefined specifications to confirm the anticipated quality of the dosage form/delivery system. CQAs are commonly connected with incorporated therapeutic substances, excipients, intermediates (in-process components), and drug products [11, 22]. To develop a profound quality target product profile, one can take account of following factors:

- Dosage form, administration route, proposed functionality, type of delivery systems;

- Strength(s) and frequency for dosage administration;
- Compatibility of container and closure system at appropriate storage conditions;
- Drug release or delivery rate as per therapeutic requirements and factors affecting pharmacokinetic attributes (e.g., dissolution profile, release pattern) suitable for the product under consideration;
- Criteria for the final product quality (e.g., drug release, purity, stability and sterility) relevant to proposed product for commercial market.

QTPP establishes a robust foundation for the selection of CMAs and CPPs that contribute preferentially to achieving the targeted QTPP. The intended QTPP is based on scientific facts, therapeutic objectives, regulatory requirements, and other practical considerations, as well as associated limitations. A well-defined QTPP is required for an effective experimental design to optimize a drug delivery system [13, 14, 20, 21, 23].

Proposed pharmaceutical product CQAs based on the quality target product profile and/or past information are utilized to direct the developmental phase of product and process expansion. The string of prospective CQAs may possibly updated as the method for formulation/manufacturing is finalized, as well as when product and formulation/manufacturing process knowledge grow. Quality risk management approach may be exercised to give precedence to the conceivable CQAs for further assessment. Relevant CQAs can be determined by an iterative process of quality risk management and experimentation that determines the amount to which their fluctuation affects the quality of the drug product [8, 24].

For better comprehension, an example of Nasal Nanoparticles (NNPs) will be used from now on to explain various steps of the QbD approach. For example, Table 5.2 displays the QTPP for NNPs.

For this QTPP, particle size and drug content were deemed critical, and to achieve the study objectives, size distribution of particle, drug loading, and entrapment efficiency were identified as subsets of CQAs and investigated further for the impression of formulation factors/process attributes. The volume to be supplied through the nasal route is relatively restricted and must contain the therapeutic dose. Drug content in produced nanoparticles is determined by entrapment efficiency and drug

Table 5.2 QTPP for the optimization of NNPs

QTPP element	Target
Administration route	Nasal
Dosage form	Nanoparticles
Stability	At least 18 months self-life
Nanoparticles quality attributes	Drug content
	Non-irritant and non-sensitizing
	Particle shape
	Particle size
	Yield of nanoparticles
	Zeta potential

loading. Surface charge, measured in terms of zeta potential, also influences NNPs stability.

The particle size distribution of an intranasal drug delivery system has a substantial impact on in-vivo deposition, allowing for effective and targeted drug absorption from various regions of the nasal cavity. For optimum systemic absorption, the dose form should be deposited over the highly vascularized respiratory region, whereas for direct nose-to-brain transmission, it should be deposited preferentially over the olfactory area. Excessively big particles may settle primarily in the anterior regions with inadequate vascularization. On the other hand, very small particle sizes may facilitate inhalation, resulting in deposition in the lungs and exacerbating critical respiratory difficulties [25].

5.3.2 Risk Assessment and Selection of CMAs and CPPs

Risk assessment is an indispensable fact-based methodology applied for quality risk management that eventually facilitate the recognition of material characteristics and process attributes which potentially alter the CQAs of the process/product under consideration. Risk estimation is often fetched out at the embryonic phases of formulation/process developmental cycle and is iterated as added evidence-based data turn out to be accessible and advance expertise is acquired.

Risk assessment strategies are capable to recognize and grade the product/process parameters (e.g., equipment, materials) that have the capability to alter the traits of ultimate product on the basis of existing expertise and preliminary investigational data. The initially selected impending constraints may possibly make a long and wide-ranging list, but the more prominent parameters can be shortlisted and highlighted by additional research (by utilizing the combination experimental design approach and automatous paradigms). The selection of parameters can be further narrowed through the preliminary investigations to assess the rank of discrete variables and probable interconnections. Once the significant parameters have been identified, they can be explored further (e.g., using a combination of experimental design, mathematical models, or investigations that lead to mechanistic insight) to attain a higher degree of process comprehension [26, 27].

The initial stage in risk assessment was to methodically collect all probable elements that possibly will alter the traits of product. The "failure mode and effect analysis" (FMEA) technique was used to assess the risk of various parameters affecting the QTTP of NNPs, on the basis of literature review and findings from preliminary experiments [28, 29]. The FMEA technique screened all variables for potential failure modes and assigned scores based on severity (S), detectability (D), and probability (P) against response variables. The severity of a failure mode influencing QTPP dictates its potential effects, whereas detectability affects the ease with which a variable responsible for the failure mode can be recognized. Probability represents the possibility of a failure. S, D, and P were each allocated 5 points for worst-case, 3 for moderate-case, and 1 for best-case. S, D, and P scores were calculated using particle size, entrapment efficiency, and drug loading as responses. The following formula was used to determine the 'Risk Priority Number' (RPN) for each variable using the acquired S, D, and P scores:

$$RPN = S \times D \times P$$

A minimum RPN of 1 is the case for the lowest scores, while a maximum RPN of 125 is achievable for the highest S, D, and P scores for any variable [30, 31].

Figure 5.2 depicts the Ishikawa diagram or fish bone diagram, which provides a broad picture of all the aspects that are important in the development of NNPs for intranasal administration. The risk assessment stage guided the adoption of CMAs and CPPs from among these specified variables.

Using the FMEA technique, each of these variables was examined for possible failure modes. Table 5.3 summarizes the S, D, and P scores together with the corresponding RPN for each variable. RPN is a measure of the total amount of failure risk. A threshold value of 70 RPN was used for the selection of CMAs and CPPs. Any formulation/process constraint that had an RPN of 70 or more was designated as CMA/CPP and may have an effect on CQAs; parameters that had an RPN of less than 70 were excluded from further investigation.

A clearer and more comparative perspective is offered for the selection of variables as CMAs and CPPs when RPN values are shown graphically (Fig. 5.3). With numerical values of 100, 100, and 75, respectively, the graphical comparison amply illustrates how the RPN scores for chitosan concentration, Na-TPP concentration, and stirring speed approach the threshold level of 70. The materials used to formulate the NNPs, such as the polymer and cross-linking agent, demonstrated a sensible

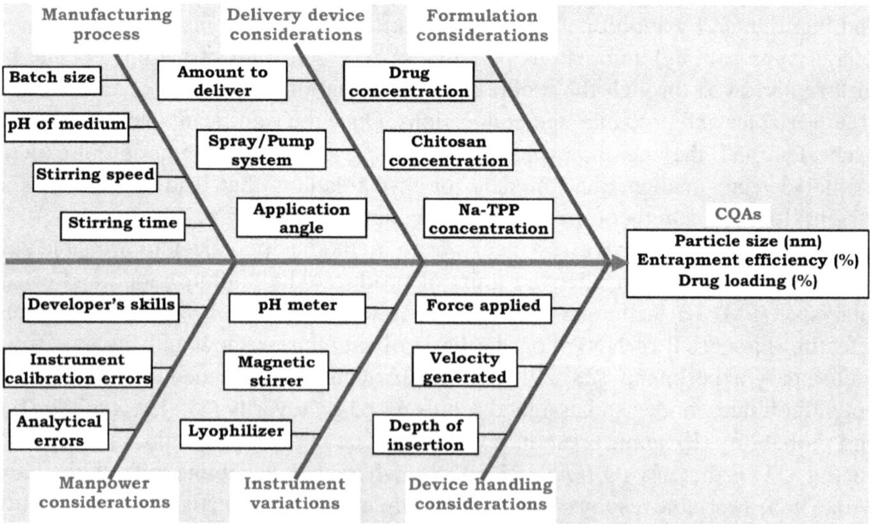

Fig. 5.2 Fish-bone diagram illustrating all possible factors affecting selected CQAs

Table 5.3 Data for S, D and P scores for the calculation of RPN

Variable	Assigned score			RPN
	S	D	P	
Amount to deliver	4	4	1	16
Analytical errors	5	5	2	50
Application angle	4	2	3	24
Batch size	4	5	1	20
Chitosan concentration	5	5	4	100
Depth of insertion	4	3	3	36
Developer's skills	4	2	2	16
Drug concentration	5	4	2	40
Force applied	5	1	3	15
Instrument calibration errors	5	5	1	25
Lyophilizer	4	4	1	16
Magnetic stirrer	4	4	1	16
Na-TPP concentration	5	5	4	100
pH meter	3	1	2	6
pH of medium	5	5	1	25
Spray/Pump system	5	5	1	25
Stirring speed	5	5	3	75
Stirring time	3	5	2	30
Velocity generated	4	1	3	12

role in governing the particle size, drug loading, and entrapment efficiency, all of which are vital for the success of the developed formulation/delivery system for intranasal administration. These RPN scores shed light on these important factors. The CQAs of produced NNPs are also greatly influenced by the process parameter, i.e., stirring speed [32].

The inferences of risk assessment studies provide a substantial base for the selection of CMAs and CPPs which includes chitosan concentration, Na-TPP concentration and stirring speed. These factors were considered for the execution of experimental design for the optimization of developed NNPs.

5.3.3 *Application of Experimental Design*

Experimental design make an attempt to organise the investigations/researches in an unbiassed and structured way to maximize the precision and definite results out of the framed postulation. Generally, the goal is to assess the interaction among independent and dependent factors i.e. how an independent factor affects a dependent parameter.

Statistical design procedures were established in order to acquire the most information with the fewest experimental repetitions, at the same time also recognizing

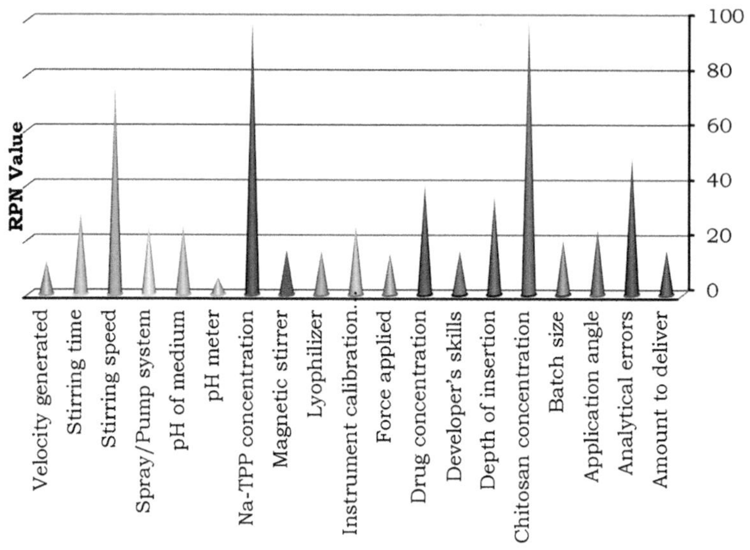

Fig. 5.3 Graphical representation of RPN values

the existence of variable interactions and exposure of tentative error. Another benefit of employing these statistical designs includes the necessity of meticulous planning and devotion to statistical guidelines. This impels the person involved in investigation to precisely specify the aims of the studies and the methods required to attain them. These principles demand the identification of potential causes of experimental error as well as the magnitude thereof. This planning stage can frequently identify issues that would not have been recognized until after the experiments were complete [13, 14, 16, 20].

When a completely new method/process/formulation is sought e.g., employment of high efficiency vortex mixer or the development of a transdermal delivery system, this forecasting may possibly prompt the investigator to see the need for a sequential approach. This strategy uses limited pilot experiments to provide information to assist with the decision-making process while also providing a projection of experimental error.

Figure 5.4. outlines the steps that need to be taken into consideration while organizing research utilizing experimental designs.

5.3.3.1 Mathematical Models for Experimental Design

Applying statistical concepts is an effective and affordable way to get the data required to figure out correlations between variables. By putting these ideas into practice, one can create a mathematical/statistical model that possibly utilized to describe and improve a process/formulation along with cost-effective consumption

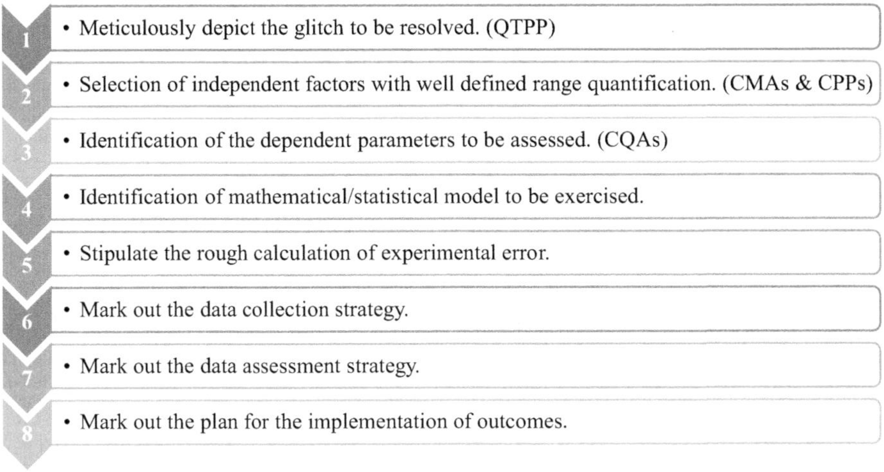

Fig. 5.4 Steps for the application of experimental design

of resources. It is doubtful that the ideal process/formulation constraints will be established without encompassing a mathematical/statistical model.

As mentioned earlier, a model to represent the interaction amongst independent factors and dependent parameters must be chosen at design phase, before the initiation of experiments. This is required for the reason that selecting a relevant experimental model is influenced by the nature and numeral of coefficients for regression analysis. Statistical concept is exploited for proper distribution of experimental points throughout the design space, to ensure the same confidence level in the assessment of discrete coefficients. As a result, the generated experimental model will reliably predict all facets of the surface.

Linear and polynomial regression models are frequently employed for optimization in pharmaceutical research. These models are chosen based on the nature of the data and the unique needs of the research [13, 20].

Linear Regression Models These models depict the link between a dependent factor and independent parameter(s) by means of a straight line. It is used to describe a linear relationship between variables. For example, projecting sales based on advertising expenditures. The mathematical expressions of linear models may possibly exploited to illustrate linear interactions or planes (monotonous surface).

$$Y = B_0 + B_1 X$$

The above equation represents the equation of a straight line and generally termed as a simple linear model where

Y = y-axis values
B_0 = y-axis intercept
B_1 = slope of the line

X = x-axis values

With this equation the values of B_1 and B_0 can be reckoned for assorted X-values and the analogous response (Y).

$$Y = B_0 + B_1X_1 + B_2X_2 + B_3X_3$$

This equation represents another linear model with three independent factors e.g., X_1, X_2 and X_3. Where

Y = quantified response
X_i = independent factor values
B_0 = constant
B_i = linear terms coefficient
where i = (1 to n) and n = number of variables

$$Y = B_0 + B_1X_1 + B_2X_2 + B_3X_3 + B_{12}X_1X_2 + B_{13}X_1X_3 + B_{23}X_2X_3$$

The above equation signifies another linear model that take account of interaction stipulations compelling the assessment of interaction coefficients B_{12}, B_{13}, and B_{23}. Where

B_{ij} = coefficient of the interaction terms
where i = (1 to n), j = (2 to n) and n = number of factors

Polynomial Regression Models Because many of the response surfaces that are seen have some degree of curvature, these models are highly helpful. By adding polynomial terms, polynomial regression fits a nonlinear affiliation among the independent factors and dependent parameters. For instance, simulating a plant's rate of growth over time.

$$Y = B_0 + B_1X_1 + B_2X_2 + B_3X_3 + B_{12}X_1X_2 + B_{13}X_1X_3 + B_{23}X_2X_3 + B_{11}X_1^2 + B_{22}X_2^2 + B_{33}X_3^2$$

This equation represents a second order polynomial model and is applied in numerous response surface models in view of the fact that it operates on quadratic terms to interpret the curvature in the response surface. Where

B_{ii} = coefficient of the quadratic terms
where i = (1 to n) and n = number of variables

Other less popular regression models are Logistic Regression, Ridge Regression, Lasso Regression, Poisson Regression, and Quantile Regression [33].

Logistic Regression This regression analysis is employed for binary dependent factors (i.e., yes/no, true/false). It represents the probability that a particular input point belongs to a specific class. Consider forecasting whether a individual will qualify or not qualify an assessment on the basis of devoted hours for practice.

Ridge Regression This is an approach employed for predictor variables which are strongly correlated. It reduces over matching by imposing a penalty on coefficient size. This is especially beneficial when dealing with situations involving multicollinearity.

Lasso Regression "Least Absolute Shrinkage and Selection Operator" or Lasso regression is analogous to ridge regression, nevertheless minimize some coefficients to zero, resulting in a simpler model. It is useful for feature selection.

Poisson Regression Poisson regression is used for modeling count data and contingency tables. It presupposes that the response parameter stick to a Poisson distribution. For example, estimating the amount of emails received each day.

Quantile Regression Quantile regression calculates the conditional median or other quantiles of the response variables. It is useful for understanding how variables affect multiple points of the distribution, not simply the mean.

A statistically programmed software application calculates or estimates all of the coefficients. Model selection ought to be based on an assessment of the prediction nature of the outcome. When such indicators are not accessible, a second-order polynomial is utilized. The individual conducting the experiment may also apply an invented model on the basis of concepts or previously collected empirical facts.

The various experimental designs are based on these mathematical models i.e., box-behnken design, central composite design, factorial designs, Plackett-Burman design etc. For better illustration of assumed example of NNPs, the box-behnken design is taken into consideration for further exemplification.

5.3.3.2 Box-Behnken Design (BBD)

In 1960, George EP Box and Behnken D demonstrated an experimental design model which depicts a statistical blueprint of experimental protocols used in response surface approach to determine the most effectual experimental boundaries. The Box-Behnken model was chosen purposely because it takes lesser experimental runs when compared to a central composite model with similar variables. The BBD was exercised to generate polynomial models for optimization process thru response surface modelling with just three levels for each element. The experimental design consists of a series of points positioned in the middle of each edge and the replicated center point of the multidimensional cube that defines the region of interest (Fig. 5.5) [30]. Despite its limited coverage of the nonlinear design space, it is nonetheless regarded as more proficient and powerful than other designs such as the three-level full factorial design and the central composite design (CCD) [32]. Figure 5.6 outlines the key aspects of the Box-Behnken design.

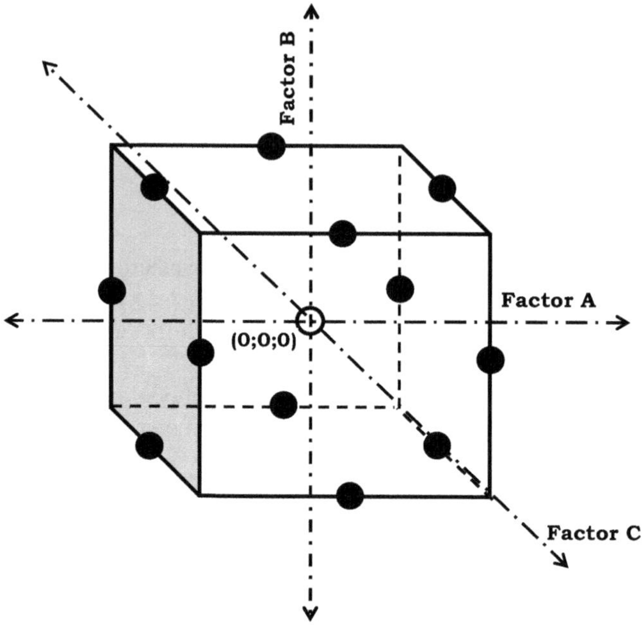

Fig. 5.5 Points representing the experimental runs for a 3-factor BBD

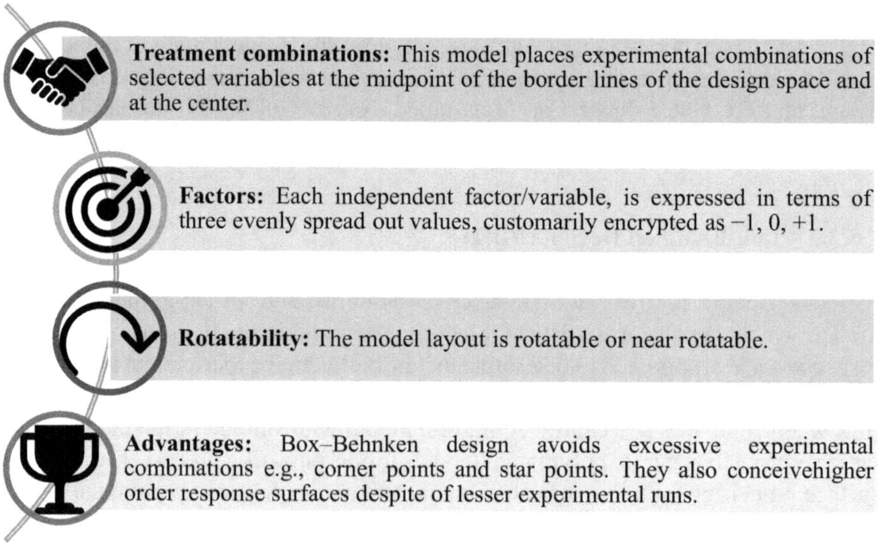

Fig. 5.6 Salient features of Box–Behnken design

For NNPs, a 3^3 (3-factors 3-levels) BBD allows the statistical optimization of selected independent factors i.e. chitosan concentration, Na-TPP concentration and speed of stirring to assess the interaction effects, main effects, and quadratic effects of selected CMAs/CPPs on particle size, entrapment efficiency and drug loading. The rationale of the pre-optimization research was to establish the levels of independent factors. Overall, 17 experimental runs were suggested by the software (Design-Expert® software; Version 10.0, Stat-Ease, Minneapolis, USA) with 5 center spots per block (for pure error estimation) and the trials batches were accomplished in arbitrary order. In the design space, a relationship between the chosen CMAs/CPPs and CQAs was delineated [30]. The following polynomial equation is produced using this experimental setup:

$$R = C_0 + C_1A + C_2B + C_3C + C_4AB + C_5AC + C_6BC + C_7A^2 + C_8B^2 + C_9C^2$$

Where,

R represents dependent variables
C_0 stands for intercept
C_1–C_9 correspond to the regression coefficients
A, B and C symbolize the independent factors.

The details of independent factors and dependent responses for NNPs are recorded in Tables 5.4 and 5.5.

Table 5.6 displays the tentative BBD template for proposed experimental runs. The information gathered from the model was assessed by means of Design-Expert software. To reveal the optimal formulation, numerical optimization was done. The optimized formulation was prepared and evaluated. Based on the data of dependent responses obtained from the actual experimental runs and expected values, the prediction error for the design was calculated.

The generated models were displayed as three-dimensional (3D) response surface curves in order to optimize NNPs, and the relationship between independent and dependent variables was determined. The best corresponded statistical models for optimization were regard as substantial at probability limits of $P < 0.05$. To confirm the best fit of model, predicted R^2 value and one way ANOVA were estimated [32].

Table 5.4 Details of independent variables (CMAs/CPPs) and their levels for NNPs [30]

		Level		
S. no.	Independent variable	Low (−1)	Medium (0)	High (+1)
1	[A] Concentration of chitosan (%w/w)	0.20	0.25	0.30
2	[B] Concentration of Na-TPP (%w/w)	0.20	0.25	0.30
3	[C] Stirring speed (rpm)	800	1400	2000

Table 5.5 Details of dependent responses (CQAs) and goals to be achieved

S. no.	Dependent response	Goals
1	[R1] Particle size (nm)	Minimize
2	[R2] Entrapment efficiency (%)	Maximize
3	[R3] Drug loading (%)	Maximize

Table 5.6 The Box-Behnken experimental design matrix for the optimization of NNPs [32]

	Coded values		
Run no.	Concentration of chitosan	Concentration of Na-TPP	Stirring speed
1	−1	1	0
2	−1	0	1
3	1	1	0
4	0	0	0
5	0	−1	−1
6	0	−1	1
7	1	0	1
8	0	1	−1
9	−1	0	−1
10	0	1	1
11	1	0	−1
12	0	0	0
13	0	0	0
14	1	−1	0
15	−1	−1	0
16	0	0	0
17	0	0	0

5.3.4 *Data Analysis and Design Space Generation*

The generated design space explains the connection among the critical quality attributes and the process parameters (material properties and process variables). In addition to assistance for the establishment of parameters along with different levels which assures harmonious quality, the risk assessment and process development criterion discussed in the previous segment may possibly help to recognize the correlation and influence of process parameters and material attributes on identified CQAs. These process parameters and material attributes can thus be selected for inclusion in the design space [34–36].

The proposal application for regulatory authorities ought to include a summary of the process control constraints and material characteristics well thought-out for the design space generation, as well as those that were embraced for actual experimental design along with description of stipulated impact on the quality of product under consideration. The explanation for inclusion of each variable in the generated design space ought to be clarified. In certain circumstances, it is advantageous to postulate a rationalisation for the removal of particular parameters. A design space can be characterized using material attribute and process constraint ranges, as well as more intricate statistical relationships. An experimental design space may be explained as a time-dependent function e.g., freeze-thaw cycles for lyophilization process or in terms of combinations of variables e.g., multivariate model elements. Sorting aspects might also be considered if the proposed design space is planned to cover assorted operating levels. Analyzing chronological information might help to build a design space. Regardless of how a design space is created, it is assumed that operating within it will result in a product of the stipulated quality [13, 14, 20].

A particular set of verified agreeable ranges for included variables are insufficient to configure a design space. Nevertheless, verified agreeable ranges established by univariate investigation might offer valuable insights into the overall process. It may perhaps be suitable to recognize the stage at which process constraints or material attributes collapse and the acceptable quality standards are failed to be accomplished. However, finding the edge of failure and proving failure modes are not required steps in developing a design space.

To optimize NNPs, all 17 batches were generated according to the Box-Behnken model and individually tested for dependent responses such as particle size, entrapment efficiency, and drug loading. To optimize the formulations, particle size was set to be minimized, while entrapment efficiency and drug loading were set to be maximized. Table 5.7 summarizes the results for distinct dependent responses across all 17 experimental runs [32].

These findings conclude the effect of independent variables on dependent responses, and it was discovered that the quadratic model was the best match for all dependent responses.

5.3.4.1 Effect of Independent Variables on Particle Size (R1)

The BBD model anticipated the following polynomial equation for particle size (R1):

$$R1 = 157.28 + 21.16A + 0.54B + 1.15C - 4.35AB + 0.075AC - 0.23BC - 2.62A^2 + 6.13B^2 + 3.16C^2$$

Where,

R1 is the NNPs particle size
A is the chitosan concentration (%)
B is the Na-TPP concentration (%)
C is the speed of stirring (rpm).

The impact of independent factors on particle size was investigated using surface response curves and contour plots generated by the BBD model (Fig. 5.7). As the developed NNPs are meant for nasal administration, particle size plays a significant role for their disposition from nasal cavity. Independent variable A (concentration of chitosan) influenced the particle size in parallel manner similar as noticed for variable B (concentration of Na-TPP). The positive coefficient value of factor A indicated the amplification in particle size with an increase in factor A [32].

Table 5.7 Experimental results of dependent responses for NNPs [32]

	Independent variables			Dependent responses		
Run no.	Concentration of chitosan (%w/w)	Concentration of Na-TPP (%w/w)	Stirring speed (rpm)	Particle size (nm)	EE (%)	DL (%)
1	0.20	0.30	1400	43.0	33.27	25.18
2	0.20	0.25	2000	38.0	31.92	24.56
3	0.30	0.30	1400	80.0	57.58	35.29
4	0.25	0.25	1400	59.0	73.79	37.93
5	0.25	0.20	800	64.0	60.44	29.86
6	0.25	0.20	2000	68.0	61.63	30.71
7	0.30	0.25	2000	78.0	68.77	33.24
8	0.25	0.30	800	65.0	58.34	26.57
9	0.20	0.25	800	37.5	57.58	24.62
10	0.25	0.30	2000	68.5	59.16	25.98
11	0.30	0.25	800	76.8	66.59	32.43
12	0.25	0.25	1400	55.9	74.02	38.17
13	0.25	0.25	1400	58.3	76.71	38.14
14	0.30	0.20	1400	86.7	44.82	26.55
15	0.20	0.20	1400	32.8	31.95	27.72
16	0.25	0.25	1400	55.7	76.45	37.39
17	0.25	0.25	1400	57.1	77.38	38.87

The variable C i.e. stirring speed also affects the particle size but up to a smaller extent which is not recognizable by the current design. As the stirring speed increased to 1400 rpm from 800 rpm, smaller NPs were obtained. However, further escalation in stirring speed does not result in size reduction of NPs, beside it an amplification in particle size was witnessed at 2000 rpm. The closeness of values obtained for actual experimental outcomes and values predicted by the Box-Behnken model (Table 5.8), evidently confirms the high precision for the projection of particle size of developed NNPs.

5.3.4.2 Effect of Independent Factors on EE (R2)

The BBD model anticipated the following polynomial equation for EE (R2):

$$R2 = 85.67 + 10.38A + 1.19B - 2.68C + 2.86AB + 6.96AC - 0.093BC - 18.72A^2 - 15.04B^2 - 0.73C^2$$

Where,

R2 is the EE of NNPs
A is the chitosan concentration (%)
B is the Na-TPP concentration (%)
C is the speed of stirring (rpm).

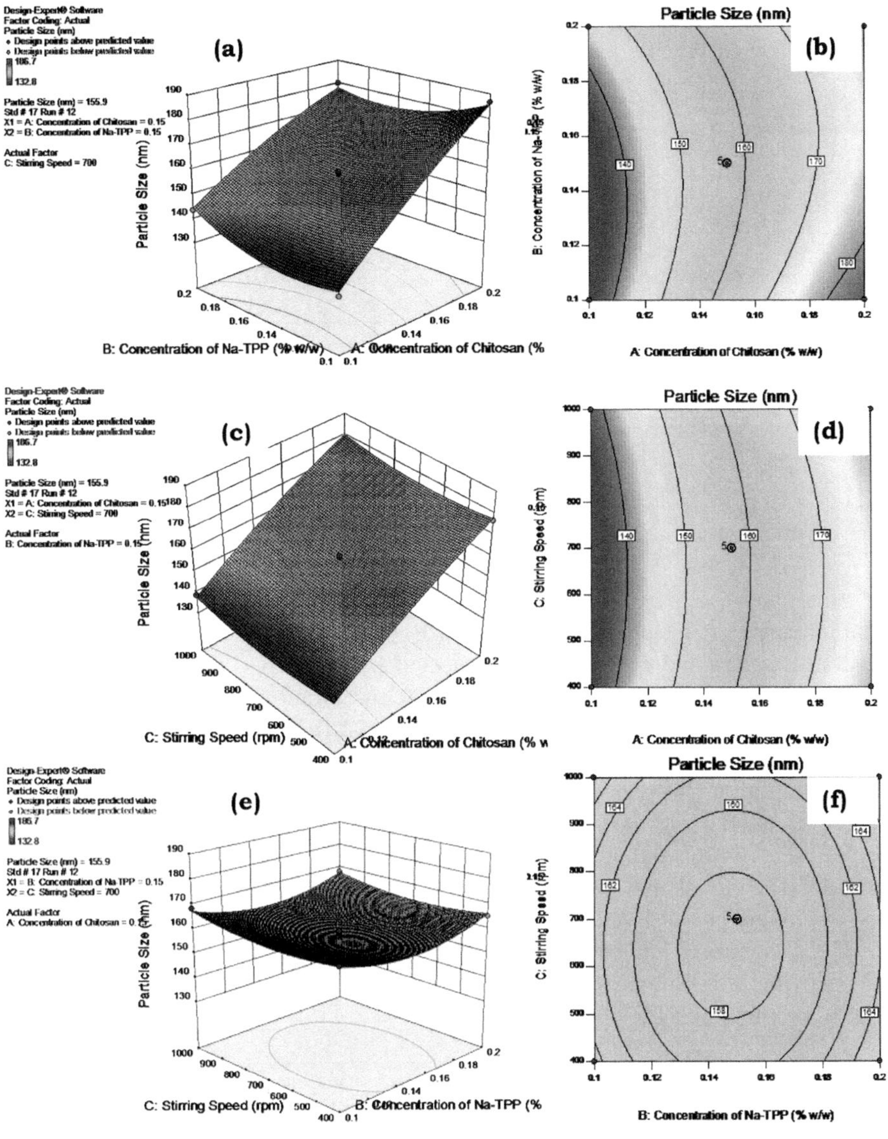

Fig. 5.7 Response surface and contour plots for particle size (R1) (**a**) Response surface plot of factor B vs. A against R1; (**b**) Contour plot of factor B vs. A against R1; (**c**) Response surface plot of factor C vs. A against R1; (**d**) Contour plot of factor C vs. A against R1; (**e**) Response surface plot of factor C vs. B against R1; (**f**) Contour plot of factor C vs. B against R1 [32]

Table 5.8 Actual and predicted values for particle size [32]

Run no.	Actual value	Predicted value
1	43.0	44.53
2	38.0	37.74
3	80.0	78.15
4	59.0	57.28
5	64.0	64.66
6	68.0	67.41
7	78.0	80.21
8	65.0	66.19
9	37.5	35.59
10	68.5	68.04
11	76.8	77.76
12	55.9	57.28
13	58.3	57.28
14	86.7	85.78
15	32.8	34.75
16	55.7	57.28
17	57.1	57.28

The influence of independent factors on EE was explored with the help of surface response curves and contour plots projected by the BBD model (Fig. 5.8). The prominent effect of independent variables A (concentration of chitosan) and B (concentration of Na-TPP) on dependent response R2 (EE) was clearly evident from different experimental runs. The entrapment of drug tends to increase when higher concentration of chitosan was incorporated. The higher extent of cross-linking ultimately resulted in elevated EE. The chitosan/Na-TPP ratio was also critical for better EE as it controls the utilization of available binding sites during the entrapment process. The model also reveals that variable C i.e. stirring speed does not have any significant effect on EE [32].

The closeness of values obtained for actual experimental results and values predicted by the Box-Behnken model (Table 5.9), evidently confirms the high precision for the prediction of EE of developed NNPs.

5.3.4.3 Effect of Independent Factors on Drug Loading (R3)

The BBD model anticipated the following polynomial equation for drug loading (R3):

$$R3 = 43.10 + 3.18A - 0.23B + 0.13C + 2.82AB + 0.22AC - 0.36BC - 4.49A^2 - 4.92B^2 - 4.90C^2$$

Where,

R3 is the drug loading

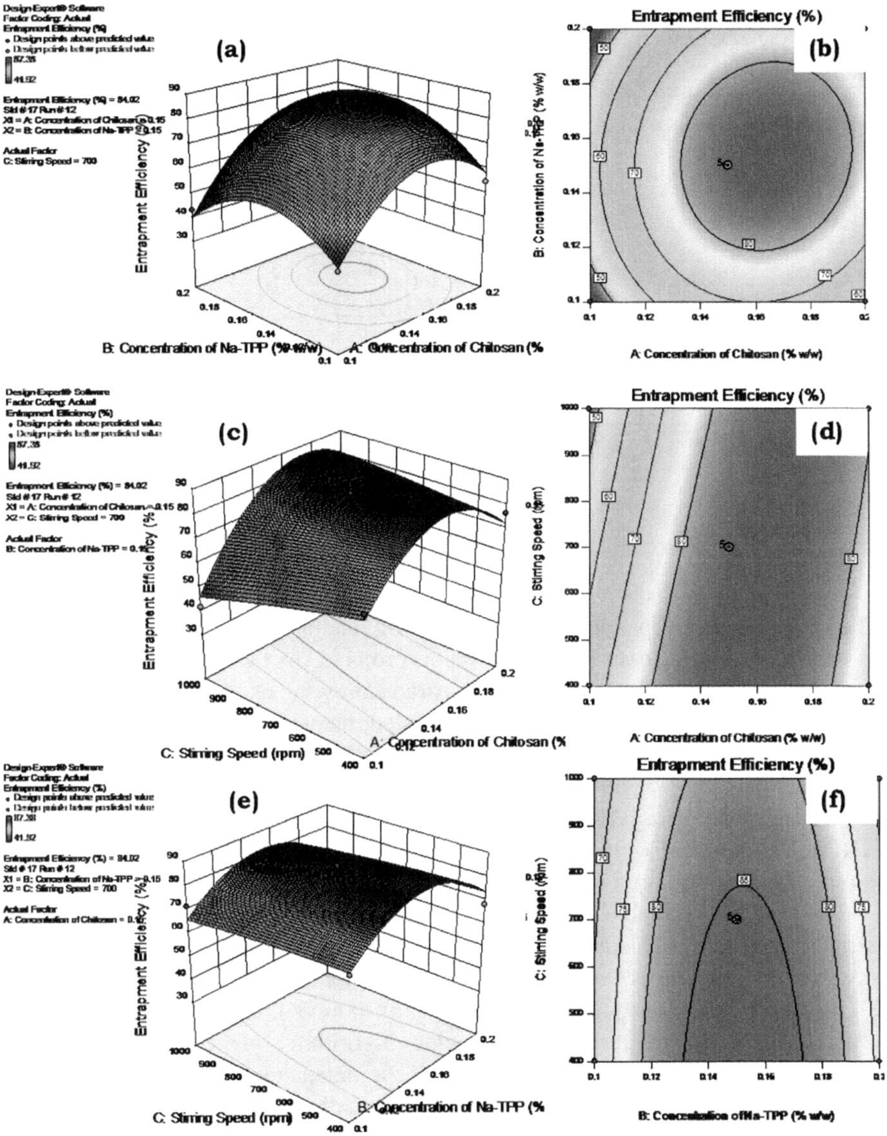

Fig. 5.8 Contour and response surface plots for EE (R2) (**a**) Response surface plot of factor B vs. A against R2; (**b**) Contour plot of factor B vs. A against R2; (**c**) Response surface plot of factor C vs. A against R2; (**d**) Contour plot of factor C vs. A against R2; (**e**) Response surface plot of factor C vs. B against R2; (**f**) Contour plot of factor C vs. B against R2 [32]

A is the chitosan concentration (%)
B is the Na-TPP concentration (%)
C is the speed of stirring (rpm).

Table 5.9 Actual and predicted values for EE [32]

Run no.	Actual value	Predicted value
1	33.27	29.85
2	31.92	36.19
3	57.58	56.33
4	73.79	75.67
5	60.44	61.30
6	61.63	56.11
7	68.77	70.87
8	58.34	63.86
9	57.58	55.48
10	59.16	58.30
11	66.59	62.32
12	74.02	75.67
13	76.71	75.67
14	44.82	48.24
15	31.95	33.20
16	76.45	75.67
17	77.38	75.67

The influence of independent variables on drug loading was explored with the help of surface response curves and contour plots projected by the BBD model (Fig. 5.9). With the results of different batches prepared following experimental design criteria, it was detected that the drug loading was augmented initially with a rise in chitosan concentration. But as the concentration was further increased, a downfall in drug loading was also observed. The chitosan/Na-TPP ratio was also critical for higher drug loading as it governs the effective cross-linking during the formation of NNPs. The model also reveals that variable C i.e. stirring speed does not have any significant effect on drug loading [32].

The closeness of values obtained for actual experimental results and values predicted by the Box-Behnken model (Table 5.10), evidently confirms the high precision for the prediction of drug loading of developed NNPs.

Based on the data acquired with Box-Behnken model, Design Expert software® utilizes ingrained numerical optimization algorithms for the optimization of nanoparticulate system. The impressions of the independent factors on the dependent responses were taken into account for the finalization of optimized formula to get the NNPs meeting the criteria selected for the formulation optimization. After defining the upper and lower limits for independent variables Design Expert software® consequentially yielded the optimized formula according to the goals lay down for the dependent responses. The suggested optimized formula was shown in Table 5.11.

The actual experimental results along with the predicted values for selected dependent responses for the NNPs prepared with optimized formula were summarized in Table 5.12.

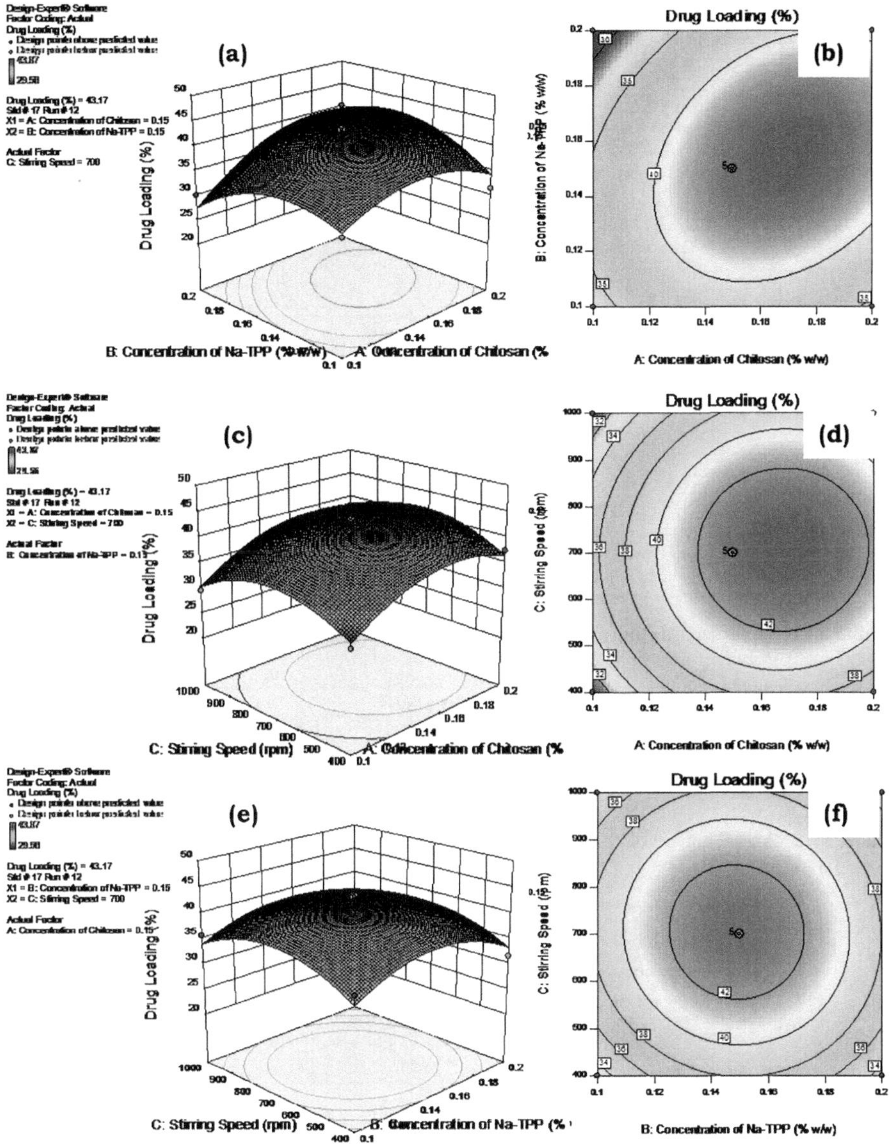

Fig. 5.9 Contour and response surface plots for drug loading (R3) (**a**) Response surface plot of factor B vs. A against R3; (**b**) Contour plot of factor B vs. A against R3; (**c**) Response surface plot of factor C vs. A against R3; (**d**) Contour plot of factor C vs. A against R3; (**e**) Response surface plot of factor C vs. B against R3; (**f**) Contour plot of factor C vs. B against R3 [32]

The software's ability to forecast accuracy was confirmed by low percentage prediction errors (<5%). These findings provide confidence in the use of statistical experimental designs for customizing pharmaceutical formulations with desired parameters during the development phase [32].

Table 5.10 Actual and predicted values for drug loading [32]

Run no.	Actual value	Predicted value
1	25.18	22.46
2	24.56	25.44
3	35.29	34.46
4	37.93	38.10
5	29.86	28.02
6	30.71	28.99
7	33.24	32.23
8	26.57	28.29
9	24.62	25.63
10	25.98	27.82
11	32.43	31.55
12	38.17	38.10
13	38.14	38.10
14	26.55	29.27
15	27.72	28.55
16	37.39	38.10
17	38.87	38.10

Table 5.11 Optimized formula for NNPs suggested by Design Expert software® [32]

S. no.	Factors	Suggested values
1	[A] Concentration of chitosan (%w/w)	0.243
2	[B] Concentration of Na-TPP (%w/w)	0.249
3	[C] Stirring speed (rpm)	1298.844

Table 5.12 Actual and predicted values of dependent responses for NNPs prepared with optimized formula [32]

S. no.	Dependent response	Predicted values	Actual values
1	[R1] Particle size (nm)	54.151	56.4
2	[R2] Entrapment efficiency (%)	74.425	75.64
3	[R3] Drug loading (%)	38.425	33.72

5.3.5 Control Strategy

The intent of control strategy is to assure the depicted quality of developed product without fail for all the batches. The components of the control strategy provide evidence for the ways that input material (drug substance and excipients), intermediate material (in-process materials), container closure system, and drug product controls, as well as in-process controls, affect the product quality. These controls should be based on product, formulation and process understanding and should include, at a minimum, control of the critical process parameters and material attributes [8, 31, 37].

A comprehensive development methodology for pharmaceutical products will churn out perception of the product and developmental process and point out areas of unpredictability. Variability sources that have the capability to affect the product quality should be recognized, properly understood, and then controlled. By comprehending the origins of variability and how they affect product quality, in-process materials, and downstream processes or processing, it is possible to move controls upstream and reduce the requirement for final-product analysis. Understanding the product and process, along with quality risk management, can help to control the process so that variability (such as that of raw materials) may be adjusted for in a flexible way to produce consistent product quality [20, 37].

An unconventional concept for manufacturing which encompasses the variety of input ingredients might be less meticulously restricted could be made possible by this process understanding. Alternatively, to guarantee consistent product quality, it would be conceivable to build an accustom process stride i.e., a phase that recognizes the interaction among the materials attributes and the right kind of process control.

When assessing if a material is achieving its quality attributes, alternate methods can be justified by an increased understanding of the product's performance. The usage of such substitutions may perhaps facilitate actual performance evaluation. These approaches minimize the requirement end product testing by adapting the actual performance evaluation, but still cannot take the place of the GMP-mandated review and quality control procedures before a batch is released.

A control strategy may consist of, but need not be restricted to, the following [13, 14, 20, 37]:

- Command over properties and quality of API, selected excipients, primary packaging materials;
- Product/process constraint(s);
- Command over all the unit operations protocols;
- Real-time or in-process testing of actual performance for depicted outcomes;
- A supervising schedule i.e., evaluation at frequent intervals for the verification of multivariate prediction models.

However, a control strategy may perhaps include assorted components i.e., some components might bank on final-product testing, while other may count on real-time performance evaluation.

5.3.6 Product Lifecycle Management and Continual Improvement

Health sector establishments can delve into innovative approaches that may improve the quality of product at any stage of product lifecycle. Process performance along with final product may possibly inspected to assure the consistent performance and dispensing the final product with quality attributes as anticipated by the designed experimental space. Such analysis could take account of trend analysis of the production procedure as supplementary experience is gathered thru the course of routine production. Perpetuation at regular time intervals is generally favourable for a number of design spaces that use mathematical/statistical models in order to endorse the models' performance. Model perpetuation can be exemplified as a pursuit whose control is decided by the internal quality management system of individual pharmaceutical establishments till the generated design space continues without any modification/variation. Gaining greater process knowledge may lead to the urge to spread out, cut down, or recharacterize the design space. Changes to design space are subject to regional requirements [8, 38, 39].

5.4 Future Perspectives

As we look to the future, the role of DoE is expected to expand and evolve, driven by advancements in technology, regulatory requirements, and the increasing complexity of pharmaceutical formulations. The future of DoE in pharmaceuticals will be heavily influenced by the amalgamation of cutting-edge tools e.g., artificial intelligence (AI), machine learning (ML), and high-throughput experimentation (HTE). These technologies will improve DoE's efficiency and accuracy by allowing it to analyze big datasets and identify complicated patterns that are difficult to detect using traditional approaches [1, 8, 16].

Artificial Intelligence and Machine Learning AI and ML can be used to predict the outcomes of experiments, optimize experimental designs, and analyze complex data sets. This will significantly reduce the time and cost associated with pharmaceutical development.

High-Throughput Experimentation HTE enables the quick screening of a wide range of experimental circumstances. When paired with DoE, HTE may accelerate formulation and process optimization, resulting in shorter development periods and more efficient resource utilization.

5.4.1 Individualized Medicine

The trend with respect to individualized medicine, where medications are custom-made as per the pathological condition and genetic makeup of individual patients, is strengthening the support for the utilization of DoE in pharmaceuticals. DoE will be crucial in developing personalized formulations and optimizing dosing regimens to ensure maximum efficacy and minimal side effects for individual patients.

Genomics and Proteomics The integration of genomics and proteomics data into DoE will enable the development of personalized medications that are custom-made to the genetic profile of specific patients.

Patient-Specific Formulations DoE can be used to optimize formulations for specific patient populations, taking into account factors such as age, weight, and comorbidities.

5.4.2 Regulatory Considerations

Different agencies responsible for the regulation of pharmaceuticals around the world are increasingly recognizing the value of DoE in pharmaceutical development. In the future, it is expected that the use of DoE will become a standard requirement for regulatory submissions, particularly for complex formulations and manufacturing processes [16, 24, 40].

Quality-by-Design (QbD) DoE is a vital component of the QbD framework, which put emphasis on product designing with assurance of envisaged quality from the outset. The adoption of QbD principles by regulatory agencies will drive the amplification in utilization of DoE in pharmaceutical development.

Regulatory Compliance The use of DoE will help pharmaceutical companies to demonstrate regulatory compliance by providing robust data on the optimization and control of their processes.

5.4.3 Sustainability and Cost Efficiency

The pharmaceutical industry is under increasing pressure to reduce its environmental footprint and improve cost efficiency. DoE can contribute to these goals by optimizing processes to reduce waste, energy consumption, and raw material usage.

Green Chemistry DoE can be used to develop environmentally friendly processes that minimize the use of hazardous chemicals and reduce waste generation.

Cost Reduction By optimizing processes and formulations, DoE can help pharmaceutical companies to reduce production costs and improve profitability.

5.4.4 Challenges and Opportunities

While the future of DoE in pharmaceutical optimization is promising, there are several challenges that need to be addressed:

Complexity of Biological Systems The complexity of biological systems can make it difficult to identify the optimal conditions for pharmaceutical products. Advanced statistical methods and computational tools will be needed to address this challenge.

Data Management The integration of large datasets from various sources (e.g., genomics, proteomics, clinical trials) will require robust data management systems and analytical tools.

Training and Expertise The successful implementation of DoE requires specialized knowledge and expertise. There will be a need for ongoing training and education to ensure that researchers and practitioners are conferred with the obligatory expertise.

5.5 Conclusion

DoE is a powerful statistical measure that has significantly impacted the pharmaceutical sector, particularly in the optimization of therapeutic products. DoE is now frequently employed at different stages of pharmaceutical development, including:

Formulation Development DoE helps in identifying the most favourable amalgamation of different excipients required and active pharmaceutical ingredients (APIs) to accomplish anticipated product characteristics such as stability, bioavailability, and release profile.

Process Optimization It is used to optimize manufacturing processes, ensuring consistent product quality and reducing production costs.

Analytical Method Development DoE aids in developing robust analytical methods that are critical for quality control and regulatory compliance.

The FDA embraced the Quality-by-Design methodology in 2004, as outlined in "Pharmaceutical cGMPs for 21st century - a risk-based approach". The "International Conference on Harmonization (ICH) Q8 Pharmaceutical Development, Q9 Quality Risk Assessment, and Q10 Pharmaceutical Quality System" outline specific prerequisites for the surety of product quality. QbD and DoE techniques facilitate the execution of ICH/Q8 and ICH/Q9. In view of the fact that the FDA acknowledged the QbD approach, DoE has been extensively used to stipulate a comprehensive perception of the product under consideration along with process adopted for production. The scientific literatures provide numerous examples of DoE applications for evaluating and optimizing pharmaceutical products and production processes [13, 14, 16, 20, 24].

The future of DoE with respect to optimization of pharmaceutical products seems promising, with various prospects for innovation and improvement. The integration of modern technologies, the trend toward customized medicine, regulatory considerations, and the emphasis on sustainability and cost efficiency will all contribute to the expanding usage of DoE in the pharmaceutical business. By tackling the obstacles and capitalizing on the opportunities, DoE will unrelentingly perform an imperative accountability in fostering the expansion of secure, efficacious, and superior-quality pharmaceutical products.

References

1. Aksu B, Paradkar A, de Matas M, Ozer O, Guneri T, York P. Quality by design approach: application of artificial intelligence techniques of tablets manufactured by direct compression. AAPS PharmSciTech. 2012;13(4):1138–46.
2. Gundogdu E, Demir ES, Ozgenc E, Yegen G, Aksu B. Applying quality by design principles in the development and prepaACS Omegaration of a new radiopharmaceutical: Technetium-99m-Imatinib mesylate. 2020;5(10):5297–305.
3. Huang J, Goolcharran C, Ghosh K. A Quality by Design approach to investigate tablet dissolution shift upon accelerated stability by multivariate methods. Eur J Pharm Biopharm. 2011;78(1):141–50.
4. Jiang M, Severson KA, Love JC, Madden H, Swann P, Zang L, et al. Opportunities and challenges of real-time release testing in biopharmaceutical manufacturing. Biotechnol Bioeng. 2017;114(11):2445–56.
5. Korakianiti E, Rekkas D. Statistical thinking and knowledge management for quality-driven design and manufacturing in pharmaceuticals. Pharm Res. 2011;28(7):1465–79.
6. Lebrun P, Giacoletti K, Scherder T, Rozet E, Boulanger B. A quality by design approach for longitudinal quality attributes. J Biopharm Stat. 2015;25(2):247–59.
7. Li K, Rogers G, Nashed-Samuel Y, Lee H, Mire-Sluis A, Cherney B, et al. Creating a Holistic Extractables and Leachables (E&L) program for biotechnology products. PDA J Pharm Sci Technol. 2015;69(5):590–619.
8. Miesle JE, Osei-Yeboah F, Pauli-Bruns A, Chen B, Manceva S, Wade JB, et al. Current approaches to design space development and regulatory applications for drug products: findings from the IQ utilization of design space for filings working group survey. Pharm Res. 2024;41(9):1775–86.
9. Lange R, Ter Heine R, van der Gronde T, Selles S, de Klerk J, Bloemendal H, et al. Applying quality by design principles to the small-scale preparation of the bone-targeting therapeutic radiopharmaceutical rhenium-188-HEDP. Eur J Pharm Sci. 2016;90:96–101.
10. Nistor I, Lebrun P, Ceccato A, Lecomte F, Slama I, Oprean R, et al. Implementation of a design space approach for enantiomeric separations in polar organic solvent chromatography. J Pharm Biomed Anal. 2013;74:273–83.
11. Sauri J, Millan D, Sune-Negre JM, Colom H, Tico JR, Minarro M, et al. Quality by Design approach to understand the physicochemical phenomena involved in controlled release of captopril SR matrix tablets. Int J Pharm. 2014;477(1–2):431–41.

12. Simoes A, Veiga F, Vitorino C. Question-based review for pharmaceutical development: an enhanced quality approach. Eur J Pharm Biopharm. 2024;195:114174.
13. Food, Drug Administration HHS. International conference on harmonisation; guidance on Q8(R1) pharmaceutical development; addition of annex; availability. Notice. Fed Regist. 2009;74(109):27325–6.
14. Food, Drug Administration HHS. International conference on harmonisation; guidance on Q10 pharmaceutical quality system; availability. Notice. Fed Regist. 2009;74(66):15990–1.
15. Mehta R, Teckoe J, Schoener C, Workentine S, Ferrizzi D, Rajabi-Siahboomi A. Investigation into the effect of ethylcellulose viscosity variation on the drug release of metoprolol tartrate and acetaminophen extended release multiparticulates-part I. AAPS PharmSciTech. 2016;17(6):1366–75.
16. Ter Horst JP, Turimella SL, Metsers F, Zwiers A. Implementation of Quality by Design (QbD) principles in regulatory dossiers of medicinal products in the European Union (EU) between 2014 and 2019. Ther Innov Regul Sci. 2021;55(3):583–90.
17. Wittkopp F, Welsh J, Todd R, Staby A, Roush D, Lyall J, et al. Current state of implementation of in silico tools in the biopharmaceutical industry-Proceedings of the 5th modeling workshop. Biotechnol Bioeng. 2024;121(9):2952–73.
18. Yekpe K, Abatzoglou N, Bataille B, Gosselin R, Sharkawi T, Simard JS, et al. Developing a quality by design approach to model tablet dissolution testing: an industrial case study. Pharm Dev Technol. 2018;23(6):646–54.
19. Page S, Coupe A, Barrett A. An industrial perspective on the design and development of medicines for older patients. Int J Pharm. 2016;512(2):352–4.
20. Food, Drug Administration HHS. International conference on harmonisation; guidance on Q8 pharmaceutical development; availability. Notice. Fed Regist. 2006;71(98):29344–5.
21. Aaberg C, Dahmen H, Davies C, Sandau PL, Srinivasan R. ISO 9001:2015 versus ICH Q10 - a comparison. PDA J Pharm Sci Technol. 2021;75(2):188–206.
22. Michaelis M, Leopold CS. Mixture design approach for early stage formulation development of a transdermal delivery system. Drug Dev Ind Pharm. 2015;41(9):1532–40.
23. Fonteyne M, Wickstrom H, Peeters E, Vercruysse J, Ehlers H, Peters BH, et al. Influence of raw material properties upon critical quality attributes of continuously produced granules and tablets. Eur J Pharm Biopharm. 2014;87(2):252–63.
24. McMahon ME, Abbott A, Babayan Y, Carhart J, Chen CW, Debie E, et al. Considerations for updates to ICH Q1 and Q5C stability guidelines: embracing current technology and risk assessment strategies. AAPS J. 2021;23(6):107.
25. Kumar A, Pandey AN, Jain SK. Nasal-nanotechnology: revolution for efficient therapeutics delivery. Drug Deliv. 2016;23(3):681–93.
26. Muteki K, Swaminathan V, Sekulic SS, Reid GL. De-risking pharmaceutical tablet manufacture through process understanding, latent variable modeling, and optimization technologies. AAPS PharmSciTech. 2011;12(4):1324–34.
27. Benchekroun YH, El Karbane M, Ihssane B, Haidara H, Azougagh M, Saffaj T. Application of design space, uncertainty, and risk profile strategies to the development and validation of UPLC method for the characterization of four authorized phosphodiesterase type 5 inhibitors to combat counterfeit drugs. J AOAC Int. 2020;103(3):715–24.
28. Jain S. Quality by design (QbD): A comprehensive understanding of implementation and challenges in pharmaceuticals development. Int J Pharm Pharm Sci. 2014;6(1):29–35.
29. Lawrence XYu. Pharmaceutical quality by design: product and process development, understanding, and control. Pharm Res. 2008;25(4):781–91.
30. Vora C, Patadia R, Mittal K, Mashru R. Risk based approach for design and optimization of stomach specific delivery of rifampicin. Int J Pharm. 2013;455(1–2):169–81.
31. Prajapati PB, Mistry KY, Shah SA. DoE-based Analytical Failure Modes Critical Effect Analysis (AFMCEA) to a Multipurpose-RP-HPLC method for the estimation of multiple FDC products of metformin hydrochloride using an analytical quality by design approach. J AOAC Int. 2022;105(4):986–98.

32. Kumar A. Development and evaluation of nanotechnology based dosage forms for intranasal delivery. Bilaspur: Guru Ghasidas Vishwavidyalaya; 2019.
33. Wong KY, Zeng D, Lin DY. Penalized regression for multiple types of many features with missing data. Stat Sin. 2023;33(2):633–62.
34. Lebrun P, Krier F, Mantanus J, Grohganz H, Yang M, Rozet E, et al. Design space approach in the optimization of the spray-drying process. Eur J Pharm Biopharm. 2012;80(1):226–34.
35. Harms J, Wang X, Kim T, Yang X, Rathore AS. Defining process design space for biotech products: case study of Pichia pastoris fermentation. Biotechnol Prog. 2008;24(3):655–62.
36. Peterson JJ. A Bayesian approach to the ICH Q8 definition of design space. J Biopharm Stat. 2008;18(5):959–75.
37. Kolhe P, Sunasara K, Lewis L, Bailey Piatchek M, Rodriguez R, Arch-Douglas K. Control strategy approach for a well-characterized vaccine drug product. PDA J Pharm Sci Technol. 2020;74(2):213–28.
38. Verch T, Campa C, Chery CC, Frenkel R, Graul T, Jaya N, et al. Analytical quality by design, life cycle management, and method control. AAPS J. 2022;24(1):34.
39. Schmidtsdorff S, Neumann J, Schmidt AH, Parr MK. Analytical lifecycle management for comprehensive and universal nitrosamine analysis in various pharmaceutical formulations by supercritical fluid chromatography. J Pharm Biomed Anal. 2021;197:113960.
40. Chatfield MJ, Borman PJ. Acceptance criteria for method equivalency assessments. Anal Chem. 2009;81(24):9841–8.

Chapter 6
Computational Modelling of Pharmaceutical Die Filling Processes

Naseeb Basha Shaik, Roja Rani Budha, Bikash Rajan Jena, Rajasekhar Reddy Alavala, and S. N. Koteswara Rao G.

Contents

N. B. Shaik (✉)
G. Pulla Reddy College of Pharmacy, Hyderabad, Telangana, India

R. R. Budha
H.K. College of Pharmacy, Mumbai, Maharashtra, India

B. R. Jena
Centurion University of Technology and Management (CUTM), R.Sitapur, Odisha, India

R. R. Alavala · S. N. K. Rao G.
Shobhaben Pratapbhai Patel School of Pharmacy and Technology Management, SVKM's NMIMS, Mumbai, Maharashtra, India

S. N. K. Rao G., R. R. Alavala (eds.), *Applications of Computational Tools in Drug Design and Development*, https://doi.org/10.1007/978-981-96-4154-3_6

Abstract Die filling is a critical stage in tablet compression, where the uniformity of powder distribution within the die cavity directly affects the tablet weight, hardness, content uniformity and complete quality. Traditional methods of optimizing this process can be time-consuming and prone to human error. Computational modeling of pharmaceutical die filling processes plays a crucial role in enhancing the efficiency, precision and scalability of solid dosage form manufacturing, particularly in tablet production. In this chapter computational modeling techniques were discussed, including discrete element modeling (DEM), computational fluid dynamics (CFD), finite element analysis (FEA) and CFD-DEM models provide advanced tools to simulate and predict the behavior of powders during die filling, to help in understanding the complex interactions between particles, the effects of powder flowability and the influence of machine parameters such as die geometry, fill height and compression forces. By leveraging these computational approaches, manufacturers can optimize die filling conditions, minimize variability in tablet weight, improve process control and reduce material waste. Artificial intelligence (AI)/Machine learning (ML) integration in pharmaceutical die filling optimization, by analyzing vast datasets from real-time sensors and historical production data for predictive modeling of new formulations enables faster development with minimal errors through AI models provides adaptive process control, adjusting parameters to improve consistency and product quality. Implementation of real-time process monitoring combined with closed-loop control systems to continuously adjust parameters during die filling. The computational modeling in die filling processes, emphasizes the design of more efficient, robust, and scalable tablet production systems, contributing to the development of high-quality pharmaceutical products.

Keywords Die filling · Computational modeling · Discrete element modeling (DEM) · Computational fluid dynamics (CFD) · Finite element analysis (FEA) · AI/ML in die filling

6.1 Introduction

6.1.1 Overview of Die Filling in Pharmaceutical Manufacturing

Powder compaction is extensively used in manufacturing of most commonly used in solid dosage forms in the pharmaceutical industry i.e. tablets. In the course of powder compaction, die filling stage is a powder is first deposited into a die and then compressed under high pressure to form a coherent powder compact. Die filling stage is a vital step in the tablet compression procedure within pharmaceutical tablet manufacturing. It directly impacts the consistency, quality and performance of the final tablet product. In this process, the pharmaceutical powder blend (which typically includes active pharmaceutical ingredients (APIs), excipients, binders and other additives) is filled into the die cavity of a tablet press. Uniformity of weight and content of the tablets in production process depend upon the die filling step ensures the performance of the formulations [1].

The proficiency of the die filling process is influenced by the powder properties (particle size, flowability, morphology and cohesion) and process conditions. Hence, it is important to understand the powder properties influence on the die filling process. The most significant characteristics in die filling are weight consistency and uniformity, which are greatly impacted by the powder quality principally flowability. This explains why powder flow characterization, powder properties, filling system and process prediction need attention in experimental and computational research have a major impact on flow behavior [2, 3].

6.1.2 Importance of Computational Modelling in Optimizing Die Filling Processes

A consistent and uniform die filling procedure is always preferred. Computational modelling is critical in optimizing die filling processes in the pharmaceutical industry, where precise and consistent die filling is required to produce high-quality goods. Inconsistent die filling can result in problems such as unequal weight, low mechanical strength, and reduced functionality. Manufacturers can use computer models to forecast, analyze and optimize die filling behavior without requiring significant physical testing, saving time, money and increasing process efficiency.

Uniformity and productivity of solid dosage forms can understand by die filling process provides helpful insight in optimizing the materials, equipment and processes. The Discrete Element Method (DEM), Computational Fluid Dynamics (CFD) and Finite Element Analysis (FEA) are the computational tools provides a detailed understanding of complex phenomena in the die filling process, allowing the identification of critical factors influencing powder behavior, flow, uniformity, and compaction. These simulations can predict powder behavior in the die cavity while accounting for issues like as particle interactions, powder flowability and die design [4, 5].

Computational modelling provides transient particle-scale information, such as velocities of individual particle, trajectories and forces interactions, which are typically difficult, if not impossible, to collect from physical experiments. Such particle-scale data is required to determine the primary elements influencing macroscopic granular flow behavior by assessing individual particle movement and interaction. Because of the ease of access, the analysis process is facilitated, and the experimental research is much easier to conduct parametric studies in computational simulations. Advances in computational modelling and optimization, together with a deeper understanding of powder behavior and processing dynamics, can considerably improve this process, resulting in more consistent tablet manufacture, less waste, and higher-quality pharmaceuticals [6].

6.1.3 Scope and Objectives of the Chapter

The scope and aims of computational modelling in pharmaceutical die filling operations are to improve the overall efficiency, quality, and consistency of tablet manufacturing. By simulating powder flow, die shape, compaction dynamics, and process parameters, computational models provide a thorough understanding of the factors that influence die filling. Optimizing key parameters not only avoids trial and error, but also improves product quality, reduces waste, and assures compliance with demanding regulatory criteria.

This chapter focusses on the computational modelling of the die filling process by simulating and optimizing powder behavior within the die cavity. Computational methods such as the Discrete Element Method (DEM), Finite Element Analysis (FEA), and Computational Fluid Dynamics (CFD) are presented, as well as the computational setup for die filling simulations. The use of computational tools in pharmaceutical manufacturing allows producers to meet the growing demand for high-quality, uniform tablets while also increasing production efficiency and lowering costs. As the industry shifts to more automated, data-driven techniques, computational modelling will continue to play a crucial role in designing the future of pharmaceutical production processes.

6.2 Fundamentals of Die Filling

6.2.1 *Definition and Significance of Die Filling in Tablet Manufacturing*

Die filling is a critical step in the tablet manufacturing process, specifically in the tablet compression stage, where a powder is transferred into the cavities of a tablet die. The primary goal of die filling is to fill each die cavity with the correct amount of powder blend, which will then be compacted into a solid tablet. During this phase precise and uniform powder distribution is crucial, as it directly influences the weight, density, hardness and other critical characteristics of the final tablet. The importance of die filling in the tablet production process ensures consistency, accuracy of tablet dosage, in prevention of manufacturing defects and optimization of production efficiency etc., To meet the required quality, it is necessary to ensure a complete, reproducible and consistent die filling (Figs. 6.1 and 6.2) [2].

6.2.2 *Key Parameters Influencing Die Filling Efficiency and Product Quality*

The die filling step of the tableting process is influenced by the characteristics of the powder and the process parameters. Powders flow when forces are applied, and how they flow depends on the bulk density, particle size, shape, distribution characteristics, temperature and humidity levels, and the powder's condition under

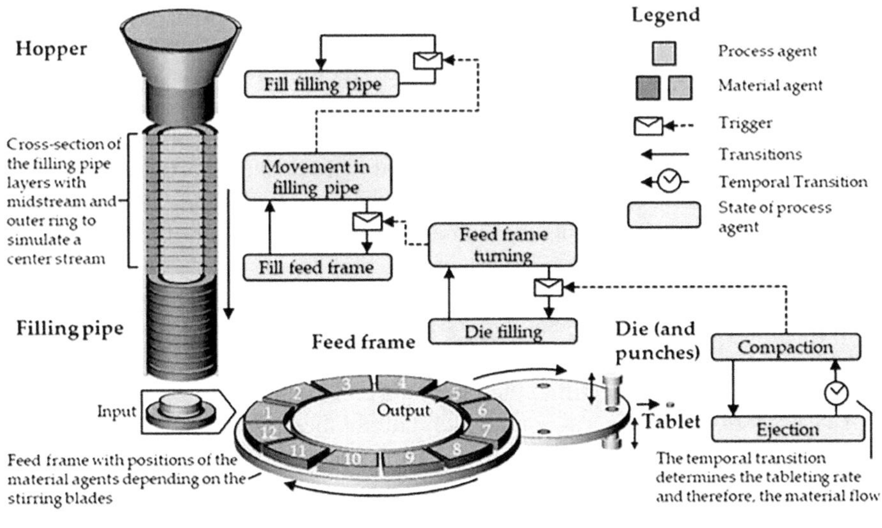

Fig. 6.1 Schematic representation of tableting process

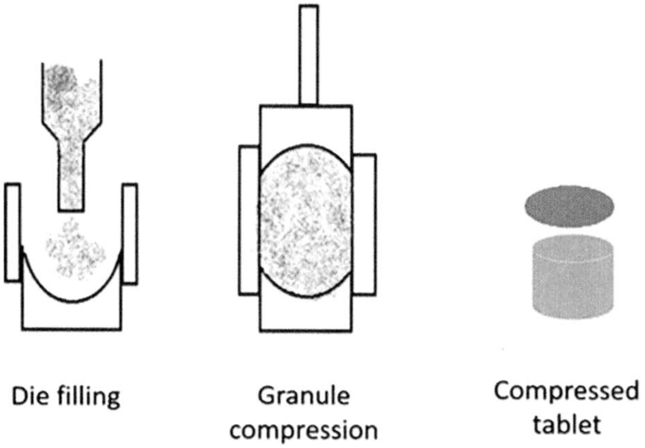

Fig. 6.2 Die filling process

processing circumstances. The parameters influencing the die filling efficiency are powder flow from a moving shoe, the filling velocity, escape of air from the die and escape of entrained air in the powder, influence the effectiveness of die filling, impacting factors like tablet weight uniformity, content consistency, hardness and overall production efficiency [2].

6.2.3 Traditional Methods and Challenges in Die Filling Optimization

In older manufacturing setups, die filling is frequently managed through manual changes, with workers managing the feeder system's flow rate and vibration settings, hopper geometry, and compact pressure. Traditional approaches for optimizing this process have evolved throughout time. Despite established procedures, achieving optimal die filling efficiency and product quality presents a number of problems, which might have an impact on total tablet manufacturing performance. Powder flow behavior is naturally varying due to a variety of characteristics such as particle size distribution, moisture content, particle shape, and the existence of cohesive forces.

Traditional approaches for die filling optimization have been used in the pharmaceutical industry for decades, but they face a number of obstacles that restrict their efficiency, particularly when tablet formulas and production settings evolve. Variability in powder flow, difficulties attaining consistency, resource intensiveness, and a lack of real-time process control all highlight these technologies' weaknesses. To overcome these issues, the industry is increasingly relying on advanced computational tools like DEM, CFD, automated monitoring systems, and real-time feedback loops to produce more precise, consistent, and efficient die filling. These new technologies provide a more robust answer to the challenges of die filling, increasing both process efficiency and product quality.

6.2.4 Role of Computational Modelling in Addressing These Challenges

For a better thoughtful of die filling, computational approaches DEM, CFD used to model to achieve promising results. Transient particle-scale information, such as individual particle velocities, trajectories, and interaction forces, can be obtained using computational modelling. This information is typically exceedingly difficult, if not impossible, to get through physical experiments. By analyzing the movement and interaction of individual particles, such particle-scale information is crucial for determining the main elements influencing the macroscopic granular flow behavior. Such information is easily accessible, which facilitates the analytical process as well. Another benefit of computational modelling over experimental research is that parametric studies are considerably simpler to conduct in computational simulations due to the freedom to define system parameters such as particle characteristics, die shape, shoe speed etc., Computational modeling addresses die filling challenges by enabling predictive analysis and optimization of flow behavior, thermal management and material properties. Computational modeling significantly contributes to improving the quality, efficiency and cost-effectiveness [7, 8].

6.3 Computational Approaches for Die Filling Modelling

A consistent and homogeneous die filling procedure is necessary to produce powder compacts of superior quality. Particle properties (size, density, and shape), shoe and die geometries, and operating conditions (shoe kinematics and powder level in the shoe) all have a substantial impact on the powder flow and packing, which define the results of die filling. The flow characterization should be carried out using a device that can capture the key characteristics of the actual die filling operations, such as the model die filling system, in order to quantify how well the material can flow in die filling.

The analysis and design of die filling systems can be greatly aided by computational modelling, which is based on the DEM and CFD. The simulations have yielded some significant findings. (1) The passage of fine particles into the die may be impeded by trapped air. (2) Suction filling can increase the filling rate and prevent air entrapment in the die; and (3) segregation in die filling can result from variations in particle size, density, and air sensitivity. The computational analysis, some useful suggestions helps to optimize the die filling process, makes the design process easier by providing particle flow and packing behavior in die filling and also helps to validate design by investigative the performance of a offered die filling system (Fig. 6.3) [4].

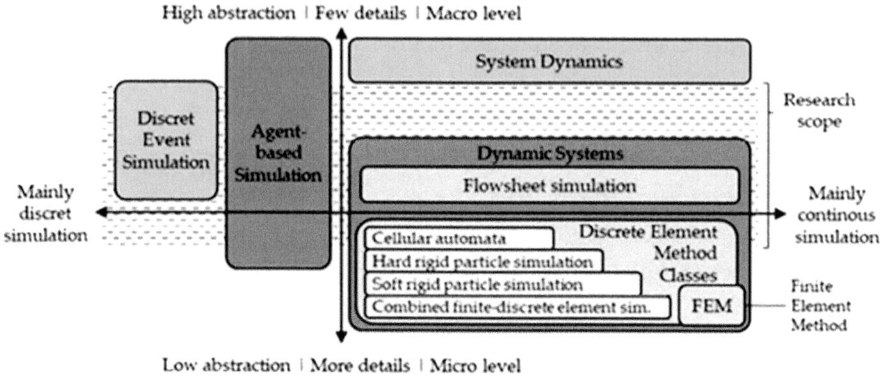

Fig. 6.3 Dynamic simulation methods for the tableting process

6.3.1 Overview of Different Computational Techniques (DEM, CFD, etc.)

In pharmaceutical die filling during tablet manufacturing, materials like powders or granules are compressed into uniform tablets. To simulate and optimize the die filling process, two main types of approaches are often used (a) Discrete-based approaches (DEM) and (b) Continuum-based approaches (CFD, FEM). Each method has distinct advantages and is suited to different features of the die filling process. DEM is particularly effective in simulating granular flow, whereas CFD is utilized to model cohesive or fluid-like particles. The Finite Element Method (FEM) is used to simulate the deformation and compaction of materials under pressure. DEM-FEM coupling enable more complex simulations of powder flow and material behavior under compression. The method used is determined by the characteristics of the die filling process as well as the level of information necessary for process optimization [9–19].

Discrete Element Method (DEM)

The DEM is increasingly being utilized to model and optimize the die filling process in tablet manufacture, particularly for powder compaction and granular material handling. Die filling is a vital step in tablet formulation in which powdered or granulated ingredients are pressed into dies to make homogeneous tablets. DEM gives information about the behavior of these materials, which helps to assure consistent tablet quality, reduce faults, and optimize manufacturing. The DEM is increasingly being utilized to model and optimize the die filling process in tablet manufacture, particularly for powder compaction and granular material handling. Die filling is a vital step in tablet formulation in which powdered or granulated ingredients are pressed into dies to make homogeneous tablets. DEM gives information about the behavior of these materials, which helps to assure consistent tablet quality, reduce faults, and optimize manufacturing [19].

DEM simulates the behavior of powders with varied particle sizes, shapes, and moisture concentrations in order to forecast how they will flow under various conditions. It contributes to the optimization of the powder feeding system, including hopper design and feeder mechanics, in order to maintain a continuous supply of powder for die filling. This decreases powder segregation and clumping. Before die filling, the powder is vibrated or subjected to external forces to aid in homogeneity and compaction. DEM can mimic these forces to better understand how they affect the flow and settling of particles in the die. This is especially beneficial in tablet press machines, which utilize vibration or tapping to settle powder into the die and ensure consistent filling. DEM also predicts particle compaction under pressure during the compression process. DEM can predict tablet density and strength after compression by predicting how particles interact as the die is filled and squeezed. This is critical for producing tablets that are both physically stable and capable of dissolving at the appropriate rate for optimum drug delivery.

DEM can be used to simulate the separation of powder particles depending on size, density, and shape. Segregation in pharmaceutical formulations can cause unequal distribution of active pharmaceutical ingredients (APIs), resulting in inconsistent doses in tablets. The mechanical qualities of materials, such as particle size, shape, surface roughness, and moisture content, might affect the flowability and behavior of the powder during die filling. DEM can simulate the impact of these material attributes on the overall die filling process and make recommendations for modifying the formulation or processing parameters to achieve optimal die filling [9, 20].

By simulating each powder particle as a single entity and examining the behavior of the particles in relation to one another, including contact interactions and particle deformation, by solving Newton's second law equations of motion, DEM takes into account the discrete nature of the powder particles. Numerous metrics, including the average number of connections, the volume proportion of particles, the orientation and distribution of contacts, the contact area, and the length of the links connecting the particle centroids, are used to define the powder behavior in the DEM approach. DEM was firstly developed by 'Cundall' for resolving soil mechanics problems [21], it has been expansively used to simulate the performance of powders flow pattern and discharge [22], filling, packing [23], shearing property [24], pile formation [25], fall mill bed layout and loading circumstances [26], during mixing and transportation [27, 28], and fluidized bed processes [29].

Computational Fluid Dynamics (CFD)

Die compaction, a popular powder processing method that involves three phases, is one of the unit activities in the tablet production process. Die filling involves transferring the powder from a hopper into a small space called a die. When the deposited powder is subjected to high compression pressure, compact formation occurs. The compact is then removed by ejection. It is always better to use a die filling process that is consistent and stable. In the end, non-homogeneous die filling may result in serious product flaws such distortion, poor strength, shrinking, and cracking. However, increasing production necessitates a speedy filling procedure.

Knowing the filling technique could offer crucial advice for maximizing the consistency and efficiency of goods, equipment, and processes (Fig. 6.4) [4, 8, 14, 15, 30].

The die fill is crucial since it limits tablet press productivity and establishes important tablet quality attributes like weight and content homogeneity [31]. A linked CFD-DEM simulation approach was used to model the die filling process under the effects of both air and vacuum [32, 33]. By using the DEM-CFD coupled approach [17] to calculate the viscous stress, pressure gradient and drag force, Guo et al. investigated the air and particle interaction that takes place during the die filling process. With the aid of air and vacuum models, the impact of air on powder flow was investigated.

Since pharmaceutical powders tend to be more cohesive, cohesive powders were also generated in order to examine the effects of cohesiveness. This study showed that air has a major impact on the powder flow in systems with lighter and lower particle sizes. The behaviour of the powder flow in the air was affected by density and particle size. One measure of flow behaviour that was found to remain constant during die filling under vacuum conditions was the dimensionless mass flow rate.

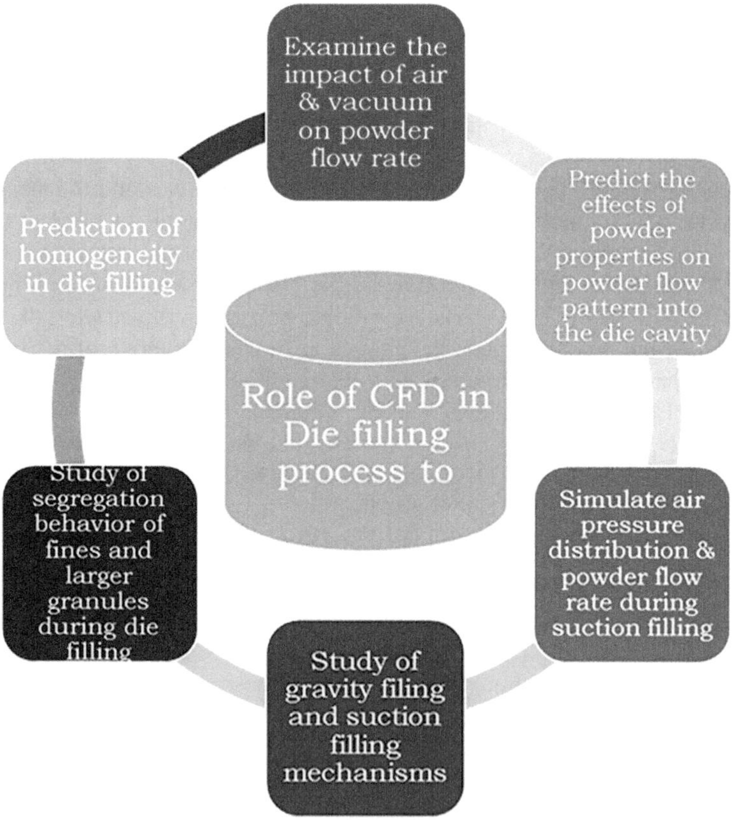

Fig. 6.4 Role of CFD in die filling process

Two mechanisms were implicated in the flow characteristics during die filling when air was present, and they were as follows: (1) The air had no influence on the system with larger and heavier particles, and the dimensionless mass flow rate was the same as in a vacuum (2). The effect of air was felt in systems with lighter and lower particles. As a result, as density and particle size increased, so did its dimensionless mass flow rate. Both a polydisperse and a monodisperse system under vacuum showed the same dimensionless mass flow rate. The polydisperse system under air conditioning showed a decreased dimensionless mass flow rate, suggesting that air has a major impact on powder flow.

Gravity filling and suction filling were the two types of die filling mechanisms that were observed [17, 33]. Wu et al. used a DEM-CFD coupled simulation approach to study the suction filling mechanism. To investigate how suction affects powder flow behavior, a comparison study between suction and gravity filling was conducted. This method made it easy to perform numerical simulations on the distribution of air pressure and powder flow rate, which were difficult to accomplish through physical trials. It also gave us a great deal of information about the suction mechanism in the die filling process. According to the results, a pressure gradient was produced over the powder bed during the suction filling process when the bottom punch was drawn downward, which increased the flow of powder from the hopper shoe into the die. Last but not least, by raising the shoe's powder flow rate and velocity, suction filling helped to increase die filling's productivity.

Wu et al. used a coupled DEM-CFD approach to investigate the segregation and flow behavior of binary mixtures made up of particles of varying sizes throughout the die filling process in both vacuum and air. The powder flow rate into the die during stationary shoe filling was lower in air conditioning than in vacuum, according to DEM-CFD simulations. Due to air entrapment in the area closer to the die's walls, the concentration of fine particles along the die's breadth at its centre was higher when air was present. In contrast, when hoover was present, the concentration of fine particles at the die's centre was lower. Because fines could pass through voids made by coarser particles, they settled in the bottom of the die in vacuum conditions. However, in air conditions, air prevented fines from settling vertically, indicating that segregation of fine particles along the die's height was forbidden. It was discovered that the particles were uniformly distributed throughout the air along die thickness. Because air readily escapes the die cavity before it is filled by the powder blend during the dominant nose flow in a moving shoe, there is very little effect of air on segregation and powder flow during this scenario. As a result, die filling results in the presence of vacuum and air remained unchanged. Segregation during die filling by moving shoe is influenced by both free surface segregation and the percolation of tiny particles through holes created by bigger particles. As a result, even though the near end of the die had a higher concentration of fines than the far end, the die was now filled with a uniform concentration of fines. In the examination of die filling, the coupled DEM-CFD simulation tool was beneficial [11]. It was only feasible to obtain tablets free of significant flaws such as low strength, breaking, shrinking, and deformation provided the powder blend was uniformly poured into the die without any particle segregation (Fig. 6.5).

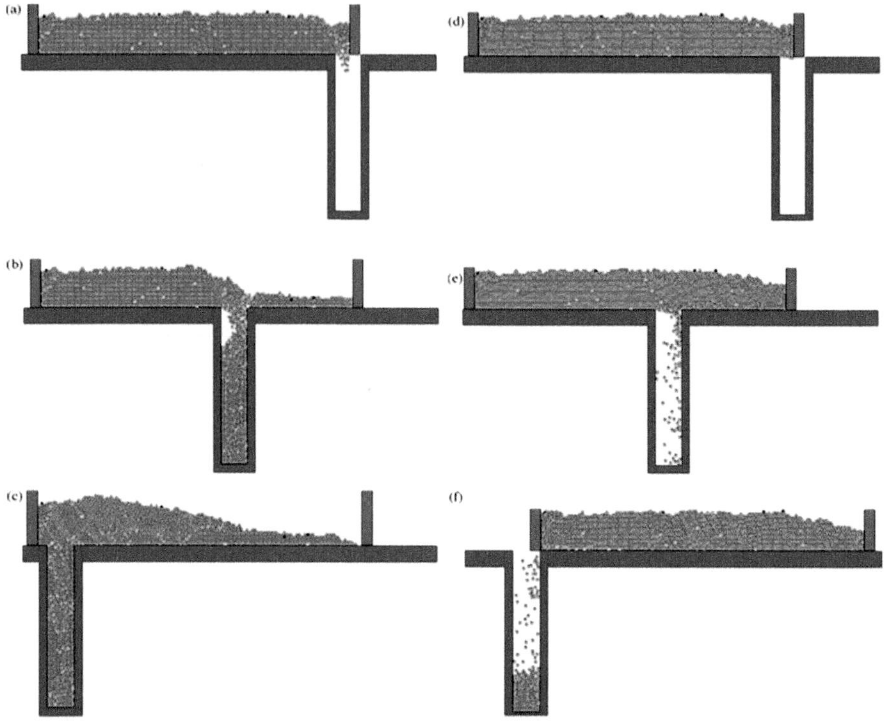

Fig. 6.5 Powder flow patterns at different filling speeds (**a–c**) 100 mm/s and (**d–f**) 400 mm/s during die filling [8]

Finite Element Method (FEM)

FEM is a powerful computational tool used to simulate and optimize various mechanical processes, including the die filling process in tablet manufacturing. It is typically employed in solid mechanics and structural analysis, plays a critical role in the optimization of the compaction and die filling stages, where powders or granules are filled into a die cavity to form tablets. It provides a detailed understanding of the stress and pressure distributions, material behavior under compaction and the interaction between powders and die tools.

FEM is commonly used to simulate the compaction process that occurs when powders or granules are deposited in a die cavity. During the filling and compression stages, forces are applied to the powder, and its reaction in terms of density, strength, and elasticity is important to the final tablet quality. It aids in simulating how powder or granular materials behave under compression, letting producers to anticipate how particles would interact with one another, the die walls, and the punches. Die geometry and tooling design are essential for producing consistent die filling and effective compaction. FEM simulates the stress and strain distributions within the die and powders, which aids in the optimization of die shapes, punch designs, and die wall roughness for improved powder flow and uniform filling. The die cavity must be designed to allow powder to flow smoothly into the cavity while minimizing air entrapment and segregation. FEM simulations can reveal flaws in die design, such as inefficient pressure distribution or powder flow.

Powders are subjected to high pressures during die filling and compression. FEM can model the stress and pressure distributions on the powder particles, which are critical for understanding how punches transmit force through the powder bed. This enables manufacturers to forecast how the material will compact and adjust the compaction force accordingly. The homogeneity of pressure distribution is critical to producing tablets with consistent hardness and density. FEM permits the detection of locations in the die where pressure may be unequal, resulting in variances in tablet quality. FEM enables manufacturers to understand how different particle sizes, shapes, and binder contents will react throughout the compaction and die filling processes. This is critical for optimizing the formulation and achieving consistent tablet characteristics. It can replicate the complete tablet formation process, including die filling and compaction. It predicts how powder will react when crushed and if it will attain the necessary density, hardness, and tablet quality. This involves modelling how powder will fill the die cavity, the influence of friction between particles and the die walls, and how the compaction force will affect particle rearrangement, bonding, or deformation during the compression process.

To find potential tablet problems that may occur during the die filling and compaction procedures. These flaws can include varying tablet weight, breaking, chipping, or weak tablets as a result of uneven compaction. By simulating alternative die filling and compaction scenarios, FEM helps discover the ideal operating conditions to limit the chance of failures and ensure that tablets are uniform in size, strength, and content [34].

Hybrid Methods (Coupled DEM-CFD or DEM-FEM)

Hybrid approaches that combine DEM with CFD or FEM are increasingly being employed in die filling simulations, particularly in the pharmaceutical industry. These methods combine the advantages of many simulation techniques to provide a more complete knowledge of the complicated processes that occur throughout the powder filling process. Hybrid approaches provide higher accuracy by combining granular dynamics (DEM) with fluid flow (CFD) or structural deformation (FEM), particularly when dealing with multiphase interactions or the mechanical behavior of powders under compaction stresses. This hybrid method combines DEM, which models granular material behavior, and CFD, which simulates fluid flow (including air and gas). The powder is regarded as a granular substance, and its interactions with the surrounding fluid are simulated simultaneously. DEM replicates the movement and interactions of individual powder particles, taking into account particle size, shape, friction, and cohesion. CFD simulates fluid flow, such as air or gas, via a die cavity or feeder system, and predicts how the fluid and powder interact. DEM models particle movement, settlement, and rearrangement during die filling.

In die filling, powders frequently interact with air or vacuum systems, altering particle flow. A DEM-CFD hybrid can simulate fluid-powder interactions, which is very beneficial for air-assisted filling. It enables simulation of how powders behave when exposed to air currents, which aids in the design of feeder systems and die cavity designs. DEM-CFD is useful for simulating how air aids in the flow of powder, preventing obstructions or separation. DEM-CFD is widely used in simulating powder feeders where air or gas is used to help transfer powder into the die cavity. By coupling DEM with CFD, helps in simulate how air flows through the feeder and interacts with the powder, ensuring that powders flow smoothly and uniformly, avoiding blockages, segregation or non-uniform filling. DEM-FEM simulations are particularly useful in optimizing die designs, allow the simulation of how powder fills the die cavity and how the powder is compacted under pressure. This hybrid technique improves the compaction process by ensuring that particles densify consistently and the final tablet has consistent hardness and density [4, 30, 35].

Lattice Boltzmann Method (LBM)
The Lattice Boltzmann Method (LBM) is a computational technique used to simulate fluid dynamics, particularly in complex flow systems such as powder flow, where the solid particles (powder) interact with a surrounding fluid (such as air or gas). In the die filling especially when air-assisted or vacuum-assisted powder filling systems are employed LBM can offer valuable insights into the fluid-powder interactions that influence the flow behavior and the final quality of tablets. The LBM simulates fluid flow using a lattice grid, with each lattice point representing a fluid particle's distribution function. These distribution functions describe the movement and interaction of fluid particles and effective for simulating complex fluid flow, especially in multiphase systems (fluid-solid or gas-powder interactions). In die filling, air is often used to assist the powder flow into the die cavity by simulating the behavior of both the powder particles and the air flow in a coupled manner, LBM helps to predict how the powder is transported, distributed and compacted in the die. LBM can simulate how the air interacts with the powder, helping optimize the air pressure, feeder design and die cavity shape and minimizing issues like segregation, blockages or non-uniform distribution [36].

6.3.2 Comparison of Discrete and Continuum-based Approaches [37]

The choice between discrete-based and continuum-based computational approaches depends on the nature of the material or system being modeled, the scale of analysis and the specific phenomena of interest. In the context of simulations for industrial processes such as die filling, casting and material flow, discrete and continuum-based approaches each have unique advantages and limitations. Discrete-based methods model systems as a collection of individual, distinct entities (e.g., particles, atoms, or discrete blocks) that interact with each other based on predefined rules or forces. DEM simulates the motion and interaction of particles of granular materials, powders etc., ideal for modeling systems where individual particle behavior is critical. Discrete methods are often used for modeling systems at the micro or particulate materials. Discrete methods model the direct interactions between individual particles, such as collisions, friction and adhesion, these interactions are typically governed by newtonian mechanics or specialized force models. Discrete methods can be computationally expensive because they require the calculation of interactions between large numbers of particles or discrete elements. As the number of elements increases, the computational requirements grow exponentially. Discrete methods provide a more accurate representation of systems with granular or particulate materials, where individual interactions are crucial for predicting behavior of the flow of powders.

Continuum—Based Approaches

Continuum-based methods treat the material or system as a continuous medium, where physical quantities like velocity, pressure, temperature and density vary smoothly across space. These approaches do not resolve individual particles but instead model the aggregate behavior of a large number of particles or molecules, typically using differential equations that describe mass, momentum and energy conservation. CFD and FEM models used for structural and thermal analysis, these models are often based on assumptions of elasticity, plasticity, or fluidity and they are ideal for simulating the bulk behavior of materials under stress, such as in casting solidification, fluid flow or heat conduction.

6.3.3 Advantages and Limitations of Each Method

Both discrete-based and continuum-based approaches are powerful tools for simulating and analyzing different material behaviors and physical phenomena, but each has its own advantages and limitations.

Advantages of discrete based approaches (DEM):

- Ideal for modeling systems where the behavior of individual particles matters. This includes granular materials, powders and systems where interactions between particles (collisions, friction, adhesion) significantly influence the overall behavior.
- Can accurately simulate micro-level phenomena like particle collisions, particle shape effects and the formation of particle clusters or jams.
- It excels at simulating non-linear interactions, such as contact forces, friction or elastic and plastic deformations at the particle level.
- Well-suited for systems that experience large deformations, local yield behavior or jamming, where continuum models may not handle accurately without sophisticated modifications.

- Discrete models can naturally represent materials that are inherently discontinuous, such as granular materials, powders or broken solids, where each particle behaves as an individual entity. This makes them useful for casting with particulate molds, bulk material handling and soil mechanics.
- Can handle complex, irregular particle shapes more naturally than continuum approaches. Ideal for simulating non-uniform material distributions or heterogeneous media.

Limitations of discrete-based approaches:

- As the number of particles increases, the computational cost of simulating their interactions can grow exponentially, this makes less feasible for very large systems, such as those involving thousands or millions of particles in industrial processes.
- The need to calculate interactions between each particle in the system leads to high memory and processing requirements. Simulating large, realistic systems often requires significant computational resources.
- Discrete models simulate individual particle behavior, which means that they are often used in microscale simulations. This can be a disadvantage when modeling large-scale phenomena, where the bulk behavior of the material is more important than the individual particle interactions.
- In some cases, the physical models used to describe particle interactions can be overly simplistic and may not capture more complex material behaviors accurately.
- Discrete models are not well-suited for simulating continuous or homogeneous materials like metals, liquids, or gases, especially at large scales.
- For complex fluid dynamics or multi-phase systems, discrete models may require hybrid approaches like coupling with CFD for fluid flow to make simulations more complex.

Advantages of continuum based approaches (CFD, FEM):

- Continuum-based methods are well-suited for large-scale simulations where the system can be approximated as a continuous medium like fluid flow in large systems. These approaches can efficiently model systems with millions of degrees of freedom, especially when the material properties are averaged over a large domain, computationally more efficient for simulating continuous materials.
- Handle a broad range of problems involving fluid dynamics, heat transfer, stress analysis and material deformation.
- Ideal for modeling the macroscopic behavior of fluids, solids and gases in industrial applications like die filling.

Limitations of continuum based approaches (CFD, FEM):

- Continuum-based models average over the properties of materials, which means that they cannot resolve fine-scale particle interactions or capture the heterogeneity of materials at the micro or particle scale.

- Cannot identify the granular effects, local particle rearrangements or micro-structural effects in powder compaction or granular flow in molds.
- Continuum models assume that the material is continuous and homogeneous on the scale of interest. This assumption may fail in cases where the material has discrete particles, interfaces or fractures, leading to a loss of accuracy in certain applications.
- Continuum methods can be challenging to apply directly, for problems involving free surfaces.
- Continuum models are generally effective for newtonian fluids, they may require specialized formulations or models to handle non-newtonian behaviors, where the material's flow properties change with shear rate.
- These methods require detailed constitutive models for material behavior to capture the correct material response, which can be challenging if the material behavior is poorly understood or exhibits complex, non-linear responses.

6.4 Computational Approaches in Die Filling Modelling

6.4.1 DEM Simulations for Granular Flow Analysis

The DEM is a numerical technique used to simulate and analyze the behavior of granular materials and particulate flows, which behave in complex ways due to interactions between discrete particles. DEM simulations are particularly useful in understanding phenomena like granular flow, compaction, jamming and segregation in systems that cannot be modeled effectively. DEM simulates the motion of a large number of discrete particles and the forces acting between particles. These particles are tracked as they move and interact over time, enabling the study of complex material behaviors at a particle level. The core of DEM lies in calculating the forces between particles, contact forces like compression and frictional forces, cohesion forces and collision forces. DEM can simulate configuration of granular flows in hoppers, chutes, conveyors and bins. The flow behavior is dominated by particle-particle interactions, which differ from the behavior of fluids or solids.

DEM granular flow simulations steps include interactions between particles and boundaries i.e. walls, chutes and boundary conditions are important for controlling flow behavior and preventing particles from passing through walls. The flow properties of the granular material velocity, density, shear stress, pressure are extracted and analyzed to understand the overall flow behavior and identify key phenomena like segregation, compaction or jamming.

6.4.2 *CFD in Modelling of Powder Flow and Compaction*

CFD is a powerful tool that can be applied to the modeling of powder flow and compaction used for modeling of granular flow and particle-level interactions can be used to model the fluid-like behavior of powders, especially in cases where powders behave as granular fluids or when powders are handled in systems involving significant airflow or liquid phases. In powder flow and compaction simulations, CFD often focuses on the fluid-particle interactions, where the powder is treated as a multiphase flow system consisting of particles and surrounding fluid. By coupling CFD with DEM or using other specialized methods, the overall behavior of the powder under various conditions can be effectively simulated.

Powders behave as a granular material and their flow can exhibit characteristics similar to both solids and fluids. The flow behavior is highly dependent on particle size, shape, moisture content and the presence of inter-particle forces. Unlike fluids, powders do not flow continuously, they exhibit complex rheological behavior, including cohesion, friction and shear. Modeling this behavior requires specialized techniques, often incorporating both CFD and DEM. During the compaction of powders, often resulting in significant changes in density, porosity and structure. The modeling of compaction involves understanding how the particles deform, the forces involved and the resulting structural changes. As powders are compacted, the bulk density increases and porosity decreases. CFD models can help to predict the final density distribution within a compacted powder mass.

CFD-DEM models for powder flow and compaction

CFD-DEM simulations for interactions between the particles (modeled by DEM) and the surrounding fluid (modeled by CFD). These interactions affect the flow of powders in systems like pneumatic conveyors or fluidized beds. DEM provides a way to simulate granular stress through contact forces between particles, which is necessary for studying compaction processes where the powder undergoes densification. In CFD-DEM simulations, defining particle-wall interactions, particle-particle collisions and the effect of external forces is critical to obtain realistic results for compaction or powder flow.

6.4.3 *DEM-CFD Coupled Approaches for Comprehensive Die Filling Simulations*

DEM- CFD coupled approaches provide a powerful tool for simulating complex granular flows in processes such as die filling, where granular materials are packed or compressed in dies. In the die filling process, the powder or granular material is often subjected to external forces (gravity, air pressure, or mechanical compaction) and its interaction with the die walls, as well as the fluid phase, can significantly affect the packing quality, compaction behavior, flowability and the formation of voids or segregation. DEM coupling with CFD allows for a more comprehensive understanding of the interactions between particles and fluids, capturing both particle-particle interactions and fluid-particle dynamics and flow patterns.

6.5 Modelling Powder Flow Behavior

6.5.1 Granular Properties and Rheological Models of Materials

Granular materials, such as powders and other particle compounds, have complex behavior that distinguishes them from fluids and solids. Understanding their characteristics and flow behavior is critical for processes such as powder handling, compaction, bulk materials transportation, and die filling. Granular materials frequently behave as a disordered collection of particles, impacted by interparticle interactions, external forces, and the presence of fluids or binders. To simulate the flow and deformation of granular materials, rheological models and granular material properties must be thoroughly understood and used. These models assist represent the flowability, compaction, shear strength, and viscoelasticity of granular materials under diverse situations, particularly when they transition from solid-like to fluid-like states.

Granular materials are characterized by a range of physical and mechanical properties that dictate their behavior in different processing environments, such as particle size distribution (PSD), particle shape, surface roughness, cohesion and inter particle forces, bulk density and void fraction, frictional and shear properties, powder compressibility, and so on. The rheological behavior of granular materials describes how they flow, deform, and respond to applied stresses. Depending on the flow conditions, granular materials can exhibit both viscoelastic and viscoelastic behavior. Several models have been constructed to characterize the flow properties and stress-strain relationships for granular materials, frequently by modifying models from fluid mechanics or solid mechanics.

Understanding the rheological characteristics of granular materials is critical for predicting and optimizing powder flow behavior into molds or dies. Inaccurate modelling of these features might result in segregation, non-uniform filling, or void development in the final product. Understanding the rheology of granular materials is critical for powder compaction because it influences how particles densify under pressure.

6.5.2 Particle-Particle and Particle-Wall Interactions

Granular materials have distinct flow characteristics due to interactions between individual particles and particles and container walls. These interactions are crucial for understanding and optimizing industrial processes such as die filling, powder compaction, bulk handling, and powder transportation. The two main interactions are particle-particle and particle-wall interactions. These include forces such as friction, cohesion, and adhesion, which have a direct impact on the flowability, compaction, segregation, and packing behavior of granular materials. This contributes to understanding the material's strength, flowability, and responsiveness to external stressors.

In fine powders or in the presence of moisture, cohesive forces between particles can have a considerable impact on material behavior. These forces are generated by capillary forces, electrostatic forces, chemical bonds, or vanderwaal forces. Cohesive forces can promote agglomeration, which occurs when particles clump together to form larger clusters, limiting the powder's ability to flow freely. These effects are especially relevant in powder compaction and mixing procedures. Granular materials display a variety of motions at the particle level, such as sliding and rolling. The combination of sliding, rolling and collisions contributes to the shear strength and flowability of the granular material. In dense packings, these behaviors play a crucial role in the material's overall rheological properties. In many applications, granular materials interact with container walls or boundaries, which affect how powders flow through hoppers, fill molds or behave in confined spaces. These interactions are important in processes such as powder transport, die filling and storage. The friction between the particle and the wall can affect the flowability of granular materials. Smooth surfaces result in lower friction, leading to easier flow but potentially higher segregation of particles (e.g., fine particles may segregate from coarse ones during flow). Rough surfaces increase the resistance to flow and can lead to blockages or arching in confined systems, where particles pile up and prevent further flow.

Modeling particle-particle and particle-wall interactions

To simulate and predict the behavior of granular materials, computational methods DEM used to simulate the motion of individual particles and calculates the forces between them based on the contact models described above. DEM models simulate particle-particle and particle-wall interactions, including friction, cohesion, contact deformation and energy dissipation allows for accurate predictions of material behavior in processes like powder flow, die filling and compaction. DEM simulations can model real-world phenomena, including jamming, segregation, flowability and density distribution, providing insight into material behavior in complex systems.

6.5.3 Validation of DEM Simulations with Experimental Data

Any simulation method must be assessed against experimental data to assure its correctness and dependability. This validation is critical for building confidence in DEM models and guaranteeing that they can be used effectively in real-world settings, such as die filling, powder compaction, bulk handling, and other industrial operations. Validation of DEM simulations is critical since they rely on models and assumptions regarding particle interactions, material properties, and boundary conditions. DEM parameters such as friction coefficients, particle stiffness, particle shape, and cohesion values are frequently defined using assumptions or idealizations. Validation against experimental data helps to calibrate these parameters so that they match observed behavior. Before utilizing DEM for process optimization, design, or scale-up, experimental validation ensures that the model appropriately predicts real-world results like as flow patterns, pressure distribution, or compaction behavior. Accurate DEM simulations can assist detect inefficiencies, bottlenecks, and important design parameters in systems like hoppers, molds, or compaction presses, leading to improved process control and product quality [38].

Methods for validating DEM simulations

There are numerous methods for validating DEM simulations, depending on the application and available experimental data. These methods are roughly classified into qualitative and quantitative validation. Qualitative validation involves comparing the visual and qualitative features of DEM simulations with experimental observations. In processes like die filling or hopper flow, DEM simulations can be compared with high-speed videos or optical imaging like particle tracking or digital imaging of real granular material flow. By visually comparing the flow patterns, segregation and particle distribution in the simulation and experiment, one can assess whether the simulation captures the general behavior of the material. In granular material handling, DEM simulations of arch formation or jamming can be compared with real-world experiments where arch formation leads to flow blockage. Observations of discharge rates and flow interruptions can serve as a qualitative benchmark. Whereas quantitative validation involves comparing specific measurable quantities between simulation results and experimental data. These comparisons allow for a more rigorous assessment of simulation accuracy.

Challenges in Validation

While DEM is a powerful tool, validation can be challenging for several reasons. Complexity of granular materials and their behavior can be influenced by many factors such as particle shape, size distribution, frictional properties, cohesion and external forces. Capturing these factors accurately in both experiments and simulations is often difficult. DEM simulations require a large number of parameters like friction coefficients, particle stiffness, damping coefficients etc., to be specified. These parameters must be calibrated against experimental data, and minor changes can cause considerable variances in simulation results. Particle-wall interactions, such as friction or adhesion, can vary based on wall roughness, material characteristics, and particle-wall contact geometry, making it challenging to precisely simulate.

Validation of DEM simulations with experimental data is an important step in confirming the model's correctness and applicability to real-world applications. By comparing flow rates, pressure distributions, force measurements, and other relevant characteristics across simulations and tests, the DEM model can be validated and its parameters refined to increase its prediction capabilities. While validation can be difficult due to the complexity of granular materials, applying a combination of qualitative comparisons, quantitative measures, and experimental procedures allows for a full assessment of DEM's predicted accuracy.

6.6 Simulation of Die Filling Dynamics [39]

6.6.1 Powder Feeding Mechanisms and Flow Patterns in the Die

The behavior of powders within the die cavity during tablet pressing is an important factor in making sure product quality and consistency. The way powder flows and fills a die has a considerable impact on the density distribution, compaction, homogeneity, and overall quality of the finished product. Powder feeding methods and subsequent flow patterns inside the die are controlled by a variety of parameters, including particle characteristics, powder handling processes, die design, and external pressures used throughout the process. Powder feeding, the technique of delivering powder into a mold or die cavity, is a typical stage in the compaction process. The method in which powders are fed into the die influences their distribution and packing density.

Several factors influence the flow of powder inside the die, including particle properties, the mechanism for feeding the powder, and die design. Accurate powder flow forecasts are critical for ensuring constant density distribution and meeting process compaction requirements. The first filling of the die cavity is crucial. The powder must be distributed evenly to ensure that the cavity is entirely filled and that the initial packing density is as uniform as feasible. When powders are put into a die, they frequently exhibit a non-uniform distribution, particularly when there is segregation or stacking. Fine particles may be more likely to separate from bigger particles, resulting in an uneven distribution in the die.

Die shape and size, as well as the geometry of the die cavity, influence powder filling. Narrower die cavities or cavities with complicated geometries might cause powder to flow less freely, resulting in non-uniform or partial filling. A conical or chambered die may encourage better powder flow, whereas deep, narrow cavities may trap air pockets, resulting in non-uniform compaction. After filling the die chamber with powder, compact it to reduce porosity and enhance density. The flow patterns during compaction are critical for assuring quality. The flowability of the powder during filling and compaction is principally determined by interparticle interactions and external forces applied. Particle size distribution and form are also important in establishing the flow pattern. The techniques like DEM simulations, it is possible to predict and optimize powder flow behavior to improve product consistency and process efficiency.

6.6.2 *Compression and Consolidation Processes During Die Filling*

In powder-based manufacturing processes such as tablet pressing and granulation, the compression and consolidation of powder materials during die filling are critical in defining the end product's quality, mechanical characteristics, and density distribution. Understanding the mechanisms of powder compaction, consolidation, and the pressures involved can aid in the optimization of the filling process, as well as improving overall manufacturing efficiency and product uniformity. Before compression and consolidation take place, the powder must be delivered into the die cavity. The filling process determines the initial distribution and arrangement of powder particles inside the die and it can be influenced by size, shape and cohesiveness of powder, feeding mechanisms like gravity feed, pneumatic feed or mechanical feed, die geometry like cylindrical, conical or complex shapes and external forces like vibration and compression. The compression process uses pressure to reduce the void space between particles and enhance density, whereas consolidation is the process by which a powder bed grows denser over time as particles redistribute and rearrange under applied pressure. Consolidation occurs during the die filling and compaction stages because to both inter-particle and mechanical strains.

To make robust, homogeneous tablets, the powder must consolidate during tablet pressing. The consolidation process is accomplished through pressure-induced compaction, in which the powder particles deform and join together to produce a solid, hard tablet. The consolidation is influenced by particle size and shape. Fine particles consolidate more easily than coarse particles, and irregularly shaped particles may have higher inter-particle friction, affecting consolidation. Lubricants can reduce friction during compression, improving the powder's flowability and uniform consolidation. Excess moisture can cause caking and poor consolidation, whereas insufficient moisture can result in inadequate particle bonding.

The compression and consolidation processes that occur during die filling are critical in defining the final quality and mechanical qualities of powder-based products. Efficient compaction results in a homogeneous density distribution, reduced porosity, and enhanced mechanical strength in the finished product. Understanding the parameters that drive compression and consolidation, such as powder qualities, die geometry, and applied pressures, allows producers to optimize these processes and ensure high-quality products with consistent properties. Advanced computational tools, like as DEM simulations, can provide useful insights into powder behavior during compression and consolidation, further improving process optimization.

6.6.3 Influence of Process Parameters on Tablet Uniformity and Integrity

The consistency and integrity of the tablets are crucial in the tablet manufacturing process to ensure the correct dosage, effectiveness, and quality of the finished product. The compaction and compression phases are critical in defining these characteristics, although various process parameters affect both tablet uniformity and tablet integrity. The PSD of the powder combination impacts the flowability and packing density. Powders with a narrow PSD pack more consistently, resulting in consistent tablet weight and content distribution. In contrast, a wide PSD may cause segregation and non-uniform filling, affecting weight and content consistency. A narrow PSD improves compaction behavior by ensuring that particles fill holes more efficiently, increasing the tablet's density and strength. However, a large PSD may result in insufficient particle bonding, resulting in low tablet hardness and friability. Irregularly formed particles may interlock, impacting flowability and homogeneity during powder loading. Spherical particles tend to flow and pack better, resulting in more constant tablet weight. Spherical particles compact more uniformly, increasing the tablet's mechanical strength. Irregular forms might cause weak spots or inconsistencies in bonding, resulting in tablets with poor mechanical qualities.

Excess moisture in the powder mixture may lead to caking or agglomeration, which results in poor powder flow and irregular die filling, affecting weight and content uniformity. Too much moisture may reduce tablet hardness due to the reduction in particle-particle bonding. Conversely, inadequate moisture may result in poor granulation and insufficient compaction. Binders are added to improve particle cohesion. The type and concentration of binder can affect tablet weight consistency. The binder plays a critical role in improving tablet hardness and friability by improving inter-particle bonding during compaction. Lubricants are used to reduce friction between powder particles and die walls. However, an excess of lubricant can lead to poor powder flow or segregation. Insufficient lubricant can cause sticking to the punch or die, resulting in non-uniform tablet weight. Lubricants must be used in the right proportion to ensure proper ejection from the die without compromising the tablet's strength. Overuse of lubricants can reduce the tablet's hardness, as they may reduce the bonding between particles, leading to poor mechanical properties.

Disintegrants help the tablet break apart upon ingestion, but they can affect uniformity by altering the flowability of the powder. Mechanical strength and disintegration rate are influenced by the type and amount of disintegrants used. Excessive disintegrants may weaken the tablet, resulting in friability, whereas insufficient disintegrants may cause sluggish or partial breakdown. The compression force exerted during the tablet forming process has a direct impact on tablet weight and uniformity. Insufficient compression force can result in non-dense tablets with poor weight uniformity, whilst too much force can cause tablet capping or lamination. Compression force is necessary to achieve the specified tablet hardness. Too little compression force results in weak tablets with limited mechanical strength. Excessive force can produce over-compression, producing brittle tablets prone to friability and cracking.

Tablet hardness is a reliable sign of uniform compaction. Inconsistent hardness across a batch could suggest uneven powder packing, poor binder dispersion, or incorrect compression settings. Tablet hardness is directly related to tablet integrity. A well-compacted tablet often has a homogeneous hardness, allowing it to tolerate the stresses of handling, storage, and transportation without breaking apart. Tablets with low hardness are more likely to break or chip. Dwell time is the amount of time the powder is compressed. If the dwell period is too short, the powder may not have enough time to pack and adhere properly, resulting in tablet irregularities. Longer dwell times often produce stronger tablets because the particles have more opportunity to undergo plastic deformation and bond formation. However, too extended dwell times might result in tablet over-compression and lower hardness due to over-bonding.

The ejection force is the force required to eject the produced tablet from the die. A high ejection force can induce tablet deformation or surface damage, reduce tablet strength, and cause the tablet to break or chip when removed from the die, jeopardizing its integrity. Powder flowability and compressibility can be affected by temperature and humidity levels. High humidity can cause powder clumping and poor flow, whereas excessive dryness can cause dusting and poor homogeneity, which can also affect tablet mechanical qualities. A number of process elements, including as powder properties, formulation ingredients, compression settings, and ambient conditions, have a significant impact on tablet uniformity and integrity. By optimizing these factors, we can ensure high-quality products that meet regulatory requirements.

6.7 Optimization of Die Filling Processes

6.7.1 *Design Optimization for Efficient Powder Flow and Uniform Die Filling*

Design optimization for efficient powder flow and uniform die filling is an important aspect of tablet manufacturing. To ensure optimal powder flow and uniform die filling, the design must consider both material characteristics such as particle size, shape, and distribution, as well as the physical conditions within the powder-handling and forming systems.

Improved flowability and less segregation during handling are guaranteed by a well-managed PSD. To prevent packing or clumping, a common strategy is to use a combination of fine and coarse particles. Larger particles improve flowability and lower friction, whereas finer particles fill in gaps. In general, spherical particles have better flow characteristics, less inter-particle friction, and are less likely to cluster or arch. It is possible to increase the packing density and achieve more consistent die filling by using powders with optimized forms. Because of electrostatic charges, wetness, or uneven surface texture, certain powders have a tendency to show cohesive qualities. Lubricants can enhance flow and decrease cohesiveness. Too much moisture can impede flow and cause powder to clump.

Feeder designs with a conical or funnel form promote gravitational flow and lessen the possibility of bridging or arching, which can happen in more flat or horizontal geometries. To reduce flow resistance, the die's transition from the powder hopper to the die cavity should be seamless. Steer clear of acute angles and corners since these can interfere with powder flow and result in inadequate die filling. Smoother filling and fewer places where powder could collect are two benefits of tapered walls. In order to preserve flow consistency, the entry point into the die cavity should be designed to minimize shear pressures on the powder. The powder in the die cavity can be better filled and compacted by using vibratory feeders.

6.7.2 *Sensitivity Analysis and Parameter Optimization Using Computational Models*

In order to improve the design and functionality of powder flow processes and guarantee consistent die filling, sensitivity analysis and parameter optimization are essential techniques. Computational models can be used to forecast the effects of process parameter changes on system behavior, resulting in more effective and optimized designs. Sensitivity analysis helps identify the relationship between input variables i.e. process parameters and the output response like powder flow characteristics, die filling uniformity, compaction quality assesses how sensitive the output is to changes in input parameters and helps prioritize which parameters are most critical to process performance [40, 41].

Sensitivity analysis for powder flow and die filling

Particle size, shape, density, cohesion, moisture content, die geometry, feeder geometry, venting systems, compaction pressure, vibration intensity, feed rate, compaction speed and hopper angle. Sensitivity analysis DEM or FEA simulation model used. Output variables like die filling uniformity, powder flowability, compaction density distribution. The final uniformity of powder density post-compaction and internal consistency, void formation and surface defects in the finished part. Local sensitivity analysis used to analyze how small changes in individual parameters affect the outcome. This is done by perturbing one parameter at a time while keeping others constant.

Global sensitivity analysis used to consider how variations in multiple parameters interact to affect the output. This is usually done using statistical methods such as monte-carlo simulations, variance-based methods or response surface methodology (RSM). Sensitivity analysis results analyzed by identifying the most influential parameters like size, shape or interactions between particles etc., Visualize the results through plots such as tornado plots, sensitivity indices to clearly communicate which parameters dominate the process. Heat maps, contour plots or scatter plots can visualize the effect of input parameters on the output response [40].

Parameter optimization

Parameter optimization involves finding the set of parameters that maximizes a desired outcome, such as improving powder flow, uniform die filling or part quality. This process can be iterative and often relies on computational models to search for the optimal set of conditions. The objective of optimization maximizing die filling uniformity, minimizing powder wastage, optimizing compaction efficiency and ensuring quality by minimizing defects like cracks or internal voids. The die's geometrical or physical limitations, material-specific limitations, and material or energy-related cost constraints are a few established constraints. Optimization algorithm several methods can be used to optimize the process parameters like gradient-based optimization like steepest descent, newton-raphson for continuous optimization tasks where the relationship between inputs and outputs is smooth. Genetic algorithms (GA) or simulated annealing (SA) for global optimization, especially when the parameter space is non-linear or discontinuous. Surrogate-based optimization uses response surface models like polynomial regression to approximate the process and reduce the computational cost. Use of computational models like DEM to model powder flow and interactions between particles under different parameter conditions. FEA used to simulate how compaction forces influence powder density and part integrity. Combine these models with optimization algorithms to perform a parametric search. The optimization algorithm iteratively adjusts process parameters based on the model's performance and feedback from the objective function.

The optimization process will typically yield a range of optimal solutions. The pareto front is a way to visualize trade-offs between competing objectives. The best combination of parameters that results in optimal flow behavior, uniform die filling and acceptable part quality. These parameters can then be validated through experiments or further simulation iterations. In the real power of combining sensitivity analysis and parameter optimization lies in their synergy. Sensitivity analysis identifies the key parameters that most affect the outcomes, while optimization systematically tunes these parameters to achieve the best performance. Perform the sensitivity analysis to identify the most important parameters, based on sensitivity analysis results, focus the optimization process on those critical parameters and use of DEM to explore different values of the parameters. Once optimized parameters are identified, validate the results through physical experimentation or higher-fidelity simulations to ensure that the optimization aligns with real-world performance. Reassess the process with new data or adjusted parameters to further refine the system for ongoing performance improvements.

Sensitivity analysis and parameter optimization using computational models are powerful tools in the design and improvement of powder flow processes and die filling. Sensitivity analysis helps identify which parameters most influence the outcome, while parameter optimization fine-tunes these parameters to achieve optimal performance. Combined, these methods allow for a more systematic and efficient approach for optimizing powder handling and compaction processes, ultimately improving the uniformity, quality and efficiency of the final product [42].

6.7.3 Case Studies Demonstrating Improved Process Efficiency Through Computational Optimization

Computational optimization, particularly using tools like DEM, FEA and CFD, has been effectively applied to pharma industry to enhance process efficiency, reduce material waste and improve product quality. Computational optimization has been successfully employed to improve powder flow and uniform die filling, here few case studies.

Case Study 6.1 Addressing air pockets and inconsistent fill in die filling for a sensitive formulation [43]

Die filling of a highly sensitive formulation used in extended-release tablets in B-Pharm company, the active ingredient in the tablet was prone to degradation in the presence of moisture and air, making precise control of the die filling process crucial. They observed air pockets and inconsistent fill weight, which led to unacceptable variation in tablet hardness and dissolution rates. During die filling, the sensitive formulation would trap air, leading to voids in the filled dies and resulting in variable tablet hardness. Variability in tablet weight caused by inconsistent die filling had an impact on both product quality and regulatory compliance. The formulation was susceptible to variations in flowability during the die filling process because to its sensitivity to moisture. The vacuum-assisted continuous powder feeding system minimizes air entrainment and guarantees a consistent, even flow into the die cavities. Throughout handling and storage, the formulation was maintained at regulated humidity levels to avoid alterations in the flowability and properties of the powder.

The air flow during the die filling process was simulated using CFD simulations. These models assisted in locating potential air traps in the die cavity and in refining the die design and feeder arrangement to reduce air entrapment. Additionally, the simulations assisted in determining the ideal powder flow rates for consistent die filling and forecasted the material flow behavior of the sensitive powder. An anti-static mechanism was incorporated into a screw feeder to enhance powder flow and avoid charge accumulation that might cause segregation. The die fill's uniformity was enhanced and compacted powder was broken up using vibration. Tablet hardness and homogeneity were improved by the vacuum-assisted feeder system and optimized die geometry, which greatly decreased air pockets and voids during die filling. This issue was lessened by 25%, and the API was dispersed equally among the tablets, improving content homogeneity. By removing air pockets, the dissolution profile variability seen in earlier batches was addressed and uniform tablet hardness was achieved. Powder flowability was maintained and no deterioration of the active ingredient was noted by controlling moisture levels during formulation processing and storage. Regulatory requirements for weight homogeneity and dissolve performance were satisfied by the reliable die filling and tablet quality.

Case Study 6.2 Improving die filling for high-dose powder formulations [43]
In die filling with a high-dose tablet formulation that comprised a lot of active pharmaceutical ingredients (APIs) in a matrix in G-Pharm. This resulted in a high dose because of incorrect powder flow. The high powder density led to issues with die fill uniformity, affecting tablet weight consistency and dissolution. In order to produce a more consistent powder blend with better flow characteristics, the team reformulated the product utilizing granulation processes. By adding binders that improved powder compaction, the excipient matrix was changed to increase powder cohesion and flowability.

The team was able to better understand the effects of powder characteristics, feeder settings, and die design by simulating the die filling of the high-dose powder using DEM. To guarantee steady powder feeding into the die, a loss-in-weight feeder was employed. To guarantee consistent filling without resulting in segregation, the feeder speed was meticulously adjusted. To guarantee constant die fill, the feeder rate was adjusted based on the weight of each tablet, which was continuously checked. Improved tablet weight consistency was the outcome of constant die filling with less segregation brought about by the feeder design and powder composition optimization. By reducing the weight fluctuation by 40%, the pills were guaranteed to meet weight uniformity regulations. Higher yields and less material waste resulted from a lower batch rejection rate brought on by enhanced die filling consistency.

Case Study 6.3 Powder flow optimization in pharmaceutical tablet compression [4, 8]
To increase tablet consistency and decrease production downtime, die filling and powder flow optimization are used in tablet compression. The powder flow during

the compression stage of tablet manufacturing influences the weight, hardness, and appearance of the finished product. Poor powder flow leads to inconsistencies in tablet properties, resulting in product rejection and the need for reprocessing. A DEM model was developed to simulate the behavior of pharmaceutical powder during the die filling and compression stages. The model incorporated factors such as powder cohesion, particle shape and compaction force. The analysis revealed that particle shape and moisture content had the greatest effect on flowability, while compaction speed and die geometry influenced tablet consistency and uniformity. A multi-parameter optimization was performed using particle shape modification and optimized die wall geometry. The goal was to maximize powder flow while minimizing variability in tablet weight and hardness. After optimization, the tablet weight variation decreased by 25% and hardness variation decreased by 15%, resulting in more consistent product quality. A 20% decrease in production downtime resulted from the flowability enhancements, which decreased the frequency of process stoppages brought on by powder bridging or clogging. By lowering the quantity of faulty pills, the optimized procedure increased tablet yield and raised total production efficiency by 10%.

Case Study 6.4 Modelling roll compaction process by combined DEM & FEM approach
The roll compaction process was effectively modelled using the DEM simulation. The two primary components of the intricate process of roll compaction are the conveying of powder using a screw feeder and the compaction of the powder between two counter-rotating rollers, where it experiences significant distortion. The behavior of granular material in roll compaction was examined in this paper using a combined DEM-FEM multi-scale technique. The flow of granular material into the compaction zone via the screw conveyor was modelled using DEM [44].

6.8 Quality by Design (QbD) in Die Filling [45]

6.8.1 Application of QbD Principles in Pharmaceutical Manufacturing

Instead of depending just on testing and inspection to guarantee quality, QbD is a methodical approach to pharmaceutical development that emphasizes integrating quality into the product and process from the beginning. Identifying and managing process variability is the aim of QbD in order to reliably generate a product with the required level of efficacy, safety, and quality. QbD offers a framework for die filling process design that reduces risks, maximizes efficiency, and guarantees constant product quality. A crucial stage in the manufacturing of tablets is die filling, which involves loading powder or grains into die cavities or molds prior to compression. Any deviation from this procedure may lead to problems like irregular tablet weight, hardness, or dissolution characteristics. As a result, using QbD principles in die filling is crucial to producing reliable, superior pharmaceuticals [46].

When QbD is used in die filling, the physical, chemical, biological, or microbiological characteristics that need to be managed to guarantee the intended product quality are known as CQAs. CQAs to guarantee that the amount of powder loaded into each die for tablet weight is consistent. For uniform content, the quantity of API in each tablet should be consistent. Uniform material binding and compaction for tablet hardness. Maintaining consistency in the tablet's rate of dissolving and disintegration. Die filling procedures can be optimized to reduce variability in these CQAs by using QbD, guaranteeing constant tablet quality.

The process factors that have an impact on the CQAs are called CPPs. The powder's cohesiveness, flowability, and particle size distribution all affect how evenly the die is filled during die filling. The uniformity of powder dispersion may be affected by the die's filling rate. The way powder fills the die and settles before compression can be influenced by the size and shape of the die cavities. The regularity of die filling can be affected by the type of powder feeder being used and its settings, such as filling speed and flow rate. The impact of each CPP on the CQAs must be thoroughly understood in order to apply the QbD strategy. To guarantee that the final product continuously satisfies the required requirements, process optimization entails determining the ideal ranges for each CPP.

DoE is a significant QbD technique that makes it possible to systematically investigate the connections between output factors like tablet weight and content uniformity and input variables like powder characteristics, die geometry, and feeder settings. It aids in determining the factors that have the biggest effects on die filling. The ideal operating circumstances to increase product consistency and reduce variability. Manufacturers can find the ideal die filling parameters by experimenting with various powder compositions, die designs, and process conditions using DoE. This aids in creating a reliable and repeatable filling die procedure. Identifying possible hazards to the die filling process, such as air entrapment, uneven flow, and powder segregation, and coming up with ways to reduce or control these risks are all part of risk assessment and control. Variability in particle size, shape, and flowability are common risk concerns in die filling that can impact the consistency of the material. Inconsistent tablet weight and inadequate compaction may result from trapped air during die filling. Inconsistencies in tablet composition may result from segregation during the filling process caused by variations in the size or density of powder particles.

Once the hazards associated with process variability in die filling have been recognized, manufacturers can design the process to reduce them with the use of QbD. In order to guarantee constant quality, the control approach entails setting up procedures and guidelines for keeping the process within a reasonable range. Real-time monitoring of crucial production parameters, including tablet weight, die fill volume, and powder flow rate, enables real-time die filling modifications. Based on real-time data, automated systems can modify variables like feed rate or compaction pressure to maintain the operation in ideal circumstances. A resilient process is one that functions reliably even when input materials, machinery, and ambient factors change. The die filling process can be greatly impacted by variations in powder characteristics, such as moisture content and particle size distribution. By making the process intrinsically robust, QbD seeks to reduce the effect of these variances on the quality of the finished product. By carefully choosing excipients or powders that flow more reliably or by creating dies that can handle a greater variety of powder qualities, QbD aims to reduce the die filling process' sensitivity to changes in material properties.

Large datasets from die filling simulations or experiments can be analyzed using a variety of QbD tool techniques, such as principal component analysis (PCA) or partial least squares regression (PLSR), which aid in determining the relationship between process variables and product quality attributes. During die filling, powder characteristics and process conditions can be tracked using instruments like NIR spectroscopy or X-ray powder diffraction, which can give real-time input for process modifications. Manufacturers are able to forecast process behavior and optimize die filling operations by using big data technologies to examine historical data from prior production runs.

In order to guarantee that die filling is continuously within predetermined quality limits and produces homogeneous tablet weight, hardness, and dissolve qualities, QbD is applied. By methodically locating and resolving sources of variability and risk, QbD lowers the chance of flaws including inconsistent active component distribution, poor compaction, and variance in tablet weight. Because of the strong process design and optimization made possible by QbD, development and scale-up procedures are streamlined, requiring less time-consuming troubleshooting and increasing production efficiency. QbD lowers expenses related to waste, poor quality, and non-compliance with regulations by decreasing variability and the need for rework or recalls [2, 47].

6.8.2 Role of Computational Modelling in QbD-Based Process Development

Modelling by computation in order to guarantee that the finished product continuously satisfies predetermined quality requirements, QbD-based process development is essential since it offers strong tools for predicting, optimizing, and controlling process behavior. Finding and comprehending the CPPs, forecasting how process modifications will impact the CQAs, and controlling the inherent variability in the manufacturing process all depend on computational modelling in QbD. It helps producers to reduce development time, costs, and dangers associated with traditional trial-and-error methods by allowing them to design and simulate different components of pharmaceutical operations prior to physical experimentation.

By providing insights into the linkages between processes and products, computational modelling supports data-driven decision-making. In particular, it aids in simulating process conditions and forecasting their impacts, characterizing process dynamics, and identifying crucial variables that affect CQAs. enhancing process robustness by anticipating how changes in equipment or raw materials may affect product quality, streamlining risk analysis, and enabling sensitivity analysis of process parameters in order to optimize the manufacturing process for consistent product quality. These skills fit in nicely with the objectives of QbD, which aims to comprehend, manage, and reduce manufacturing process variability in order to guarantee a reliable and superior end result.

The range of process conditions that reliably result in a product meeting the necessary quality standards can be defined as a design space using DoE-based models. Computational models can direct real-time process modifications to remain inside the design space once it has been defined. DEM is a technique used to simulate the motion and interaction of individual particles in granulation, powder flow, and compression processes, while CFD is used to predict fluid flow, heat transfer, and mixing behaviors. By forecasting how process variables like tablet strength and porosity will affect product quality, DEM can be utilized in QbD to plan and optimize the compaction process. Predictive models can be created using AI/ML approaches to assist forecast how modifications to raw materials or process conditions may affect the quality of the finished product.

Computational modelling simulates how changes in process circumstances impact product quality, which aids in risk evaluations. A design space, or the collection of process circumstances that guarantee constant quality, can be defined using QbD principles by combining different modelling approaches, such as DoE, CFD, and DEM. For every crucial process parameter, a range of acceptable operating conditions is defined and several scenarios are simulated using computational models. Once the design space is created, it allows for process modifications to be made with flexibility while preserving the quality of the final product. Lastly, real-time monitoring and control of crucial parameters during manufacturing is made possible using PAT tools combined with computational models [48].

In QbD-based process development, computational modelling aids in identifying critical process factors that impact product quality, improving comprehension of the process-product link. Models eliminate the need for costly and time-consuming experimental experiments by enabling the optimization of manufacturing parameters. Computational modelling speeds up the launch of new goods by enabling quick process design, simulation, and optimization. Predictive risk analysis is made possible by computational simulations, which guarantee that the process stays stable under various operating situations and help to detect any process faults before they happen. Computational modelling can assist in reducing development and manufacturing costs by reducing the need for trial-and-error experiments and optimizing the use of raw materials [45, 49].

6.8.3 Risk Assessment and Mitigation Strategies Using Predictive Modelling [50]

In order to assure constant product quality and process resilience, risk assessment is essential, particularly when using the QbD framework. Manufacturers can create plans to reduce risks and boost overall productivity by recognizing possible hazards early in the product development process. By employing data-driven simulations to predict how process variables, raw material variances, and other factors may affect the end product's quality, predictive modelling plays a crucial role in anticipating and resolving possible problems like CQA and CPPA potent method for simulating and forecasting the behavior of complex systems, including pharmaceutical production processes, is predictive modelling, which makes use of computational tools. Understanding the connections between raw materials, process parameters, and CQAs as well as how these factors change over time is the aim. Manufacturers can quantify the impact of uncertainties in process parameters, material attributes, and other factors on the quality of the final product, identify potential risks before they arise based on process and material variability, optimize the process design to reduce the likelihood of an out-of-spec product, and create a design space where the process remains robust and consistent under varying conditions by integrating predictive modelling into risk assessment.

A systematic experimental strategy that modifies important parameters in a structured manner to measure how they affect product quality. Multiple process variables and CQAs can be correlated using partial least squares (PLS) regression methods or principal component analysis (PCA). By modelling millions of potential process scenarios based on known inputs and their variability, Monte Carlo simulations shed light on risk and variability. Predictive models developed using AI and machine learning on massive datasets are able to spot patterns and connections that are difficult for conventional statistical methods to detect.

Manufacturers can prioritize risks and create mitigation plans based on the simulation results. For example, granulation time may be targeted for stricter control throughout production if it is shown to be a process parameter that is extremely sensitive to changes in dissolution rate. Predictive models can be used to identify risks and design control techniques to mitigate them. These could include feedback control systems, real-time process monitoring, or even modifications to raw material standards [51].

6.9 Challenges and Limitations of Computational Models [5, 52]

6.9.1 Accuracy and Validation of Computational Models

Computational models like CFD or DEM are frequently used in the die filling process of tablet manufacturing to model and optimize the flow of powders, granules, or other particle materials into die cavities. These models aid in forecasting the final tablet product's quality, compaction, and homogeneity. Nonetheless, these models' precision and dependability are essential for making wise choices while designing and expanding industrial processes. To obtain the required CQAs in die filling applications, such as constant tablet weight, content uniformity, and dissolving profiles, model accuracy and validation are crucial.

A number of variables can affect the precision of the computational models used in die filling. Die filling often involves interactions between powder particles and the gas phase. It need sophisticated multiphase models that can forecast how air movement affects powder behavior and vice versa to effectively model these interactions. The influence of vibration, feeding mechanism, die shape, and air entrainment, among other factors, Granular materials can vary greatly in their size distribution, bulk density, flowability, frictional characteristics, and particle shape, all of which can have an impact on die filling accuracy. For appropriate simulation inputs, these properties frequently require experimental determination. Boundary conditions such as inlet flow rates, die wall friction, hopper design, and die cavity shape have a significant impact on simulation accuracy. Even small variations in die geometry can significantly impact the powder flow and distribution in the die, which must be carefully modeled.

6.9.2 Computational Cost and Computational Resources Required for Simulations

Computational simulations, such as DEM, CFD and coupled DEM-CFD models, have become vital tools for optimizing the processes, from powder flow and tablet compression to granulation and coating. However, these simulations can be computationally intensive, depending on the level of detail required, the complexity of the system and the specific objectives of the simulation. Understanding the computational costs and resource requirements is essential for both the practical application and the scalability of these simulations.

The computational cost of simulations is influenced by several factors, which dictate the time and resources required to run simulations effectively. More detailed models require more computational power. High-fidelity DEM models that consider individual particle interactions, particle shape and material properties need a large number of particles and more complex calculations. Larger problems, such as simulating the behavior of a full-scale process require significantly more computational resources than smaller, lab-scale models. The number of degrees of freedom (DOF), grid size (in CFD) and particle count (in DEM) all scale with the size of the problem, directly influencing computational cost. Simulations that need to account for extended time periods, such as in powder compaction or tablet dissolution studies, require simulations to run for longer times and at higher resolutions, which increases computational expense.

For accurate simulations, CFD models require fine mesh grids to resolve gradients in velocity, pressure and temperature. Finer grids demand more memory and processing power. For DEM models, accurate representation of the system's behavior in small particle interactions requires the simulation of a large number of particles, which significantly increases the computational time. Smaller time steps are required for appropriate results in both DEM and CFD, particularly when simulating rapid events like fluid flow through narrow geometries or particle collisions. Computational costs rise when time steps are smaller because more iterations are needed to mimic a process. Processor power, such as Graphics Processing Units (GPUs), which speed up computations, memory, storage on high-speed solid state drives (SSDs), and time, are commonly used to quantify the computing resources needed to conduct simulations. Cloud-based computing platforms such as Google Cloud, Microsoft Azure, and Amazon Web Services (AWS) offer scalable resources for large-scale simulations. Depending on the particular requirements of the simulation, these platforms enable the rental of high-performance computing resources that may be scaled up or down.

The results of simulations can be utilized to lessen the requirement for time-consuming and expensive actual prototyping. Pharmaceutical companies can optimize process parameters, reduce risks, and enhance product quality by doing several simulations early in the development process. Over time, this can lead to significant time and cost savings. Depending on the model's complexity, scale, accuracy, and computing methodology, pharmaceutical manufacturing simulations might have a significant computational cost and resource requirements. However, it is now much more feasible to execute large-scale, high-fidelity simulations thanks to developments in parallel computing, GPU acceleration, and cloud-based infrastructure. Manufacturers can effectively use computer simulations to optimize operations and guarantee excellent product quality by carefully balancing the amount of information required with the computational resources available and by utilizing cost-effective techniques like cloud computing [53–55].

6.9.3 Integration of Complex Process Parameters and Material Properties in Simulations

Complex process factors and material qualities must be integrated for computational simulations to be accurate and relevant in pharmaceutical manufacture. The simulation's capacity to simulate actual circumstances and forecast the caliber and performance of the final product is directly impacted by these elements and characteristics. Combining process variables like temperature, humidity, mixing time, and compression force with material characteristics like powder flowability and compaction behavior in simulations enables a more thorough understanding of the procedure and aids in improving product design, efficiency, and quality.

For simulations to effectively depict the behavior of processes in the actual world, the process parameters must be carefully incorporated. The final granule size distribution and homogeneity are influenced by the rate of material mixing. The mixing speed has a direct impact on fluid-powder interactions, powder-powder interactions, and particle dynamics in both DEM and CFD simulations. The amount of time a substance spends in a mixer or granulator affects the size growth and agglomeration of particles. When granulating fluid flow interacts with particles in CFD simulations, this moment is crucial. The rate at which binder is added influences how moisture and binder are distributed across the powder bed in wet granulation. This is crucial for tracking granule size evolution and particle-particle interaction changes in simulations. Tablet hardness, friability, and dissolution rate are all influenced by the pressure used during compaction. Since the force distribution within the powder bed affects particle rearrangement and compaction behavior, this parameter needs to be precisely incorporated into simulations, especially in DEM models of tablet compression. For consistent tablet weight and content, the ease with which powder fills the die cavity is essential. Powder flowability is frequently modelled in DEM using die shape and particle characteristics, which are influenced by material qualities like inter-particle friction, cohesion, and particle size distribution.

The inherent qualities of the materials such as powders, granules, or tablets that are being processed are known as material attributes. For simulations to produce accurate and significant findings, these qualities must be accurately represented. The degree of particle cohesiveness affects powder flow, which is important for both granulation and powder filling, i.e., die filling. The flowability of powders and the packing behavior in dies can be predicted using simulations based on these qualities, which also affect how powders behave under compaction. Particles may experience elastic compaction or plastic deformation when compressed.

6.10 Future Directions and Emerging Technologies

6.10.1 Advances in Simulation Techniques for Die Filling Processes

As well known, the die filling process is a crucial stage in the production of pharmaceuticals, especially tablets, where precise dose, weight consistency, and content uniformity depend on the consistent, uniform distribution of powder in the die chamber. Reaching these results necessitates a thorough comprehension of die cavity interactions, granule properties, and powder flow behavior all of which are impacted by process variables including fill rate, vibration, and pressure as well as material attributes like flowability and compressibility.

The die filling process may now be predicted and optimized with greater accuracy thanks to recent developments in simulation techniques. These developments aid in resolving more conventional issues including inconsistent powder flow, low fill weight repeatability, and inefficient processes. One of the most popular simulation methods for comprehending how granular materials like powders and granules behave throughout the die filling process is the DEM. Based on their motion dynamics and physical characteristics, DEM models how individual particles interact with one another. Understanding how powders flow into the die requires an understanding of particle-particle interactions including friction, cohesion, and particle size distribution, which can now be more accurately represented thanks to recent developments in DEM for die filling. Recent developments in CFD for die filling allow for the simulation of multiphase flows, in which gases and powders interact inside the die. One of the biggest developments in simulating intricate die filling processes is the combined DEM-CFD. This coupled approach combines the strengths of both methods to provide a holistic model of the process, considering both particle mechanics and fluid dynamics.

While the current advancements in simulation techniques for die filling processes have significantly improved process understanding and optimization, the integration of machine learning (ML) and AI algorithms with DEM-CFD simulations is gaining momentum. These technologies can be used to optimize the process parameters more efficiently than traditional trial-and-error methods, predict material behavior under various conditions, learning from vast datasets of experimental and simulation data and enable adaptive simulations, where models improve their accuracy based on real-time data from the production line.

The development of more accurate material models that can dynamically capture the change in properties like powder compaction or changes in moisture content during the process will lead to more realistic simulations. Additionally, integrating multiscale modeling can further enhance the accuracy of material behavior predictions. With advancements in cloud computing, high-fidelity simulations, which once required significant computational infrastructure are now more accessible.

Real-time, collaborative simulation and optimization in production will be made possible by cloud-based platforms, which also lower simulation costs and facilitate quicker iteration and optimization. By combining PAT tools with simulation approaches, producers will be able to continuously monitor and modify the die filling process. In order to guarantee constant product quality, this integration will aid in the process' real-time optimization. Recent developments in die filling process simulation approaches have significantly improved the precision, effectiveness, and predictive power of pharmaceutical production simulations. The combination of DEM, CFD, and coupled DEM-CFD models has improved knowledge of fluid-powder interactions, powder dynamics, and process optimization. ML/AI and high-fidelity simulations have also been incorporated to enable more accurate and effective manufacturing processes that guarantee consistent product quality and lower production costs [56].

6.10.2 Integration of Machine Learning and Artificial Intelligence in Predictive Modelling

Pharmaceutical production is undergoing a radical change as a result of the incorporation of ML/AI into predictive modelling. These sophisticated computational tools can optimize intricate operations, improve forecasting capacities, and offer real-time insights into the dynamics of processes. Large data sets can be analyzed using ML/AI to find patterns and connections that traditional models would miss, improving process effectiveness, product quality, and speed to market. Model transparency is essential for guaranteeing adherence to good manufacturing standards (GMP) in the highly regulated pharmaceutical sector. By using explainable AI techniques, models' decision-making processes can be better understood, boosting confidence in predictions made by AI. Regulatory frameworks like the FDA's 21 CFR Part 11 and ICH Q10 for pharmaceutical quality systems must be followed when integrating ML/AI into pharmaceutical manufacturing. These frameworks demand that computational models and data be rigorously documented, validated, and traceable.

Predictive modelling for pharmaceutical manufacture using ML/AI has the potential to completely transform the way medications are made. The options are numerous, ranging from batch optimization and dynamic quality control to real-time process optimization and predictive maintenance. As the pharmaceutical sector continues to embrace digital transformation, the use of AI and ML will result in production processes that are more effective, economical, and of greater quality [57].

6.10.3 Potential for Virtual Process Optimization and Scale-Up Using Computational Tools

The pharmaceutical sector is under increasing pressure to create and manufacture high-quality medications more cheaply, effectively, and in accordance with strict regulations. Scaling up with computational tools and optimizing virtual processes are important tactics that have a lot of promise for overcoming these obstacles. Manufacturers can theoretically predict, optimize, and scale up complicated processes with computational tools including simulation techniques and ML/AI. This eliminates the need for trial-and-error testing, lowers risks, and speeds up time to market. In order to determine the ideal process parameters, design decisions, and material properties that result in the best production outcomes such as consistent product quality, increased efficiency, and lower costs virtual process optimization uses computational models and simulations. Manufacturers can optimize crucial parameters without doing significant physical experiments by forecasting the behavior of materials and processes [58].

In virtual process optimization process simulation by DEM and CFD are widely used to simulate the behavior of powders, fluids and granules in pharmaceutical manufacturing. These simulations allow manufacturers to understand how various parameters, such as filling speed, vibration during die filling, or binder concentration in granulation, affect the final product. With virtual process optimization can minimize material waste, reduce energy consumption and avoid the need for expensive trial-and-error experiments, to test various configurations and parameters in a fraction of the time it would take to run physical experiments, speeding up the development process, for more consistent products, improving CQAs and to quickly adapt to new materials or process changes without major reconfigurations or physical testing.

Scale-up of virtual process is a crucial step in pharmaceutical manufacturing, transitioning from small-scale laboratory production to large-scale commercial production. This process is typically fraught with challenges as small-scale experiments do not always directly correlate with large-scale behavior due to the nonlinear nature of scaling laws. Computational tools CFD and DEM simulations can be used to simulate processes like granulation, mixing and drying at different scales. These simulations help predict how process behavior changes as the scale increases and can suggest adjustments to process parameters that will ensure consistent product quality. The use of multi-scale modeling allows simulations to bridge the gap between laboratory-scale and industrial-scale processes.

Future directions in virtual process optimization and scale-up

The incorporation of artificial intelligence (AI), particularly reinforcement learning, with real-time process control systems will allow for continuous optimization during both the development and scale-up stages. The integration of multi-scale models will provide even more accurate predictions for process behavior, improving both optimization and scale-up capabilities. Cloud computing and digital twins will facilitate easier sharing of models and process data across teams and locations, enabling more agile and collaborative optimization and scale-up efforts. Virtual process optimization and scale-up using computational tools have the potential to significantly enhance pharmaceutical manufacturing by improving efficiency, reducing costs and ensuring product quality. By leveraging techniques such as process simulation, multi-scale modeling, ML/AI manufacturers can optimize their processes and scale up more confidently, ultimately reducing time-to-market and improving the sustainability of their operations [59].

As computational tools continue to evolve, they will play an increasingly important role in the development of robust, scalable and efficient pharmaceutical manufacturing processes. The future of virtual process optimization and scale-up is poised to further enhance the ability of pharmaceutical companies to meet the growing demand for high-quality medicines, delivered in a more cost-effective and timely manner.

6.11 Conclusion

Pharmaceutical die filling is a critical step in tablet manufacturing, where the uniformity and consistency of powder filling directly impact the quality attributes of the final product, such as tablet weight, content uniformity and mechanical strength. Because powders behave dynamically and are affected by a number of variables, such as die shape, moisture content, powder cohesion, and particle size distribution, achieving ideal die filling is a difficult task. By offering a virtual environment for studying and optimizing the die filling process, computational modelling is essential to overcome these obstacles. The detailed simulation of powder flow, interaction with die surfaces, and the ensuing effect on tablet properties is made possible by computational tools such as the DEM, CFD, and coupled DEM-CFD models. These tools also make it possible to optimize process parameters to improve die fill uniformity and decrease batch-to-batch variability. To minimize material waste and lower production costs by optimizing the usage of raw materials, as well as to predict and comprehend powder behavior under various processing circumstances without the need for significant trial-and-error testing.

By offering instruments to forecast product quality and manage process variability throughout manufacture, computational modelling also aids QbD and PAT methodologies, guaranteeing regulatory compliance and improving product consistency. The use of CFD and coupled DEM-CFD simulations in die filling process optimization has improved the accuracy of powder flow predictions under a range of process circumstances. By optimizing die cavity design, filling speeds, and vibration settings, these instruments have produced more consistent tablet properties and more uniform die fills. Predictive changes during die filling have been made possible by real-time process monitoring and adaptive process control utilizing ML/AI algorithms. This has decreased variability and guaranteed that the finished product satisfies specifications.

Advanced scale-up capabilities like virtual scale-up using computational models has become more accurate, allowing manufacturers to predict how changes in process parameters at the laboratory scale will scale up to larger production volumes. By simulating the entire manufacturing process, companies can minimize risks associated with scaling up and ensure that large-scale production remains consistent with small-scale tests. By virtually simulating the die filling process and optimizing conditions before physical experiments, computational tools have significantly reduced the time and cost required for process development and optimization. This is particularly important in high-volume production, where time-to-market and cost control are critical. While significant progress has been made in the application of computational modelling to die filling processes, there are still several areas where future research and innovation could enhance the efficiency and effectiveness of these techniques.

Future research could focus on the integration of multi-physics models, where various phenomena such as thermal effects, moisture dynamics and chemical interactions are coupled within the same simulation. This could provide a more comprehensive understanding of the interactions that influence powder behavior during die filling and compression. Combining DEM, CFD and FEA could help optimize not just the powder flow, but also the tablet integrity by simulating the entire compression cycle, including the material response to stress.

AI/ML integration to expand in pharmaceutical die filling optimization. By analyzing vast datasets from real-time sensors and historical production data. AI models can provide adaptive process control, adjusting parameters on the fly to improve consistency and product quality. Machine learning can also be applied to predictive modeling for new formulations, enabling faster development of new tablet types by predicting powder behavior, fill uniformity and final tablet properties before any physical testing. Implementation of real-time process monitoring combined with closed-loop control systems to continuously adjust parameters during die filling. This could lead to fully automated production lines that optimize conditions throughout the entire manufacturing process.

In conclusion, computational modelling has proven to be a transformative tool in optimizing the die filling process in pharmaceutical manufacturing. With advancements in simulation techniques, machine learning and real-time monitoring, the potential to improve product quality, increase process efficiency and reduce costs has never been greater. As research continues to advance and the industry moves towards more data-driven, adaptive production methods, computational tools will become indispensable in meeting the growing demand for high-quality, consistent pharmaceutical products.

References

1. Zakhvatayeva A, Zhong W, Makroo HA, Hare C, Wu CY. An experimental study of die filling of pharmaceutical powders using a rotary die filling system. Int J Pharm. 2018;553(1–2):84–96. https://doi.org/10.1016/j.ijpharm.2018.09.067.
2. Schomberg AK, Kwade A, Finke JH. The challenge of die filling in rotary presses—A systematic study of material properties and process parameters. Pharmaceutics. 2020;12(3):248. https://doi.org/10.3390/pharmaceutics12030248.
3. Xie X, Puri VM. Uniformity of powder die filling using a feed shoe: a review. Part Sci Technol. 2006;24(4):411–26. https://doi.org/10.1080/02726350600934663.

4. Guo Y, Wu CY. Computational modeling of pharmaceutical die filling processes. In: Predictive modeling of pharmaceutical unit operations. Woodhead Publishing; 2017. p. 253–71. https://doi.org/10.1016/B978-0-08-100154-7.00011-9.
5. Horner M, Joshi S, Waghmare Y. Predictive modeling of pharmaceutical unit operations; 2016. p. 383.
6. Sinka IC, Cocks AC. Evaluating the flow behaviour of powders for die fill performance. Powder Metall. 2009;52(1):8–11.
7. Wu CY, Cocks AC. Flow behaviour of powders during die filling. Powder Metall. 2004;47(2):127–36. https://doi.org/10.1179/003258904225015617.
8. Wu CY. DEM simulations of die filling during pharmaceutical tabletting. Particuology. 2008;6(6):412–8. https://doi.org/10.1016/j.partic.2008.07.008.
9. Bierwisch C, Kraft T, Riedel H, Moseler M. Die filling optimization using three-dimensional discrete element modeling. Powder Technol. 2009;196(2):169–79.
10. Bierwisch C, Kraft T, Riedel H, Moseler M. Three-dimensional discrete element models for the granular statics and dynamics of powders in cavity filling. J Mech Phys Solids. 2009;57(1):10–31.
11. Guo Y, Wu CY, Kafui KD, Thornton C. 3D DEM/CFD analysis of size-induced segregation during die filling. Powder Technol. 2011;206(1–2):177–88.
12. Guo Y, Wu CY, Thornton C. The effects of air and particle density difference on segregation of powder mixtures during die filling. Chem Eng Sci. 2011;66(4):661–73.
13. Nwose EN, Pei C, Wu CY. Modelling die filling with charged particles using DEM/CFD. Particuology. 2012;10(2):229–35.
14. Wu CY, Cocks AC. Numerical and experimental investigations of the flow of powder into a confined space. Mech Mater. 2006;38(4):304–24.
15. Ketterhagen WR. Simulation of powder flow in a lab-scale tablet press feed frame: effects of design and operating parameters on measures of tablet quality. Powder Technol. 2015;275:361–74.
16. Furukawa R, Shiosaka Y, Kadota K, Takagaki K, Noguchi T, Shimosaka A, Shirakawa Y. Size-induced segregation during pharmaceutical particle die filling assessed by response surface methodology using discrete element method. J Drug Deliv Sci Technol. 2016;35:284–93.
17. Wu CY, Guo Y. Numerical modelling of suction filling using DEM/CFD. Chem Eng Sci. 2012;73:231–8.
18. Hildebrandt C, Gopireddy SR, Scherließ R, Urbanetz NA. Investigation of powder flow within a pharmaceutical tablet press force feeder–A DEM approach. Powder Technol. 2019;345:616–32.
19. Tsunazawa Y, Shigeto Y, Tokoro C, Sakai M. Numerical simulation of industrial die filling using the discrete element method. Chem Eng Sci. 2015;138:791–809.
20. Chung YC, Ooi JY. Influence of discrete element model parameters on bulk behavior of a granular solid under confined compression. Part Sci Technol. 2007;26(1):83–96.
21. Cundall PA, Strack OD. A discrete numerical model for granular assemblies. Geotechnique. 1979;29(1):47–65.
22. Balevičius R, Kačianauskas R, Mroz Z, Sielamowicz I. Discrete element method applied to multiobjective optimization of discharge flow parameters in hoppers. Struct Multidiscip Optim. 2006;31:163–75.
23. Zhang ZP, Liu LF, Yuan YD, Yu AB. A simulation study of the effects of dynamic variables on the packing of spheres. Powder Technol. 2001;116(1):23–32.
24. Tykhoniuk R, Tomas J, Luding S, Kappl M, Heim L, Butt HJ. Ultrafine cohesive powders: from interparticle contacts to continuum behaviour. Chem Eng Sci. 2007;62(11):2843–64.
25. Matuttis HG, Luding S, Herrmann HJ. Discrete element simulations of dense packings and heaps made of spherical and non-spherical particles. Powder Technol. 2000;109(1–3):278–92.
26. Venugopal R, Rajamani RK. 3D simulation of charge motion in tumbling mills by the discrete element method. Powder Technol. 2001;115(2):157–66.

27. Gyenis J, Ulbert Z, Szépvölgyi J, Tsuji Y. Discrete particle simulation of flow regimes in bulk solids mixing and conveying. Powder Technol. 1999;104(3):248–57.
28. Kwapinska M, Saage G, Tsotsas E. Mixing of particles in rotary drums: a comparison of discrete element simulations with experimental results and penetration models for thermal processes. Powder Technol. 2006;161(1):69–78.
29. Link JM, Cuypers LA, Deen NG, Kuipers JA. Flow regimes in a spout–fluid bed: a combined experimental and simulation study. Chem Eng Sci. 2005;60(13):3425–42.
30. Hemamanjushree S, Tippavajhala VK. Simulation of unit operations in formulation development of tablets using computational fluid dynamics. AAPS PharmSciTech. 2020;21(3):103. https://doi.org/10.1208/s12249-020-1635-1.
31. Baserinia R, Sinka IC. Powder die filling under gravity and suction fill mechanisms. Int J Pharm. 2019;563:135–55. https://doi.org/10.1016/j.ijpharm.2019.01.068.
32. Mehta CH, Narayan R, Nayak UY. Computational modeling for formulation design. Drug Discov Today. 2019;24(3):781–8.
33. Tan XC, Liu QK, Zhu DQ, Yuan PQ, Cheng ZM, Yuan WK. Pyrolysis of heavy oil in the presence of supercritical water: the reaction kinetics in different phases. AICHE J. 2015;61(3):857–66.
34. Bayle JP, Reynaud V, Gobin F, Brenneis C, Tronche E, Ferry C, Royet V. Modeling of powder die compaction for press cycle optimization. In: TOP FUEL 2015-Reactor full performance 2015 Sep 14. https://doi.org/10.1051/epjn/2016018.
35. Martin NL, Schomberg AK, Finke JH, Abraham TG, Kwade A, Herrmann C. Process modeling and simulation of tableting—an agent-based simulation methodology for direct compression. Pharmaceutics. 2021;13(7):996.
36. Ramesh KK, Kumar GK, Swapna K, Datta D, Rajest SS. A review of medical image segmentation algorithms. EAI endorsed transactions on pervasive health and technology. 2021;7(27):e6.
37. Gómez-Silva F, Zaera R, Askes H. Discrete-continuum-discrete approach for the modeling of the dynamic behavior of 2D lattice systems. Thin-Walled Struct. 2024;203:112182. https://doi.org/10.1016/j.tws.2024.112182.
38. Bernhardt ML, Biscontin G. Experimental validation study of 3D direct simple shear DEM simulations. Soils Found. 2016;56(3):336–47.
39. Yang XS, Koziel S, Leifsson L. Computational optimization, modelling and simulation: past, present and future. Procedia Comput Sci. 2014;29:754–8.
40. Serban R. A parallel computational model for sensitivity analysis in optimization for robustness. Optim Methods Softw. 2009;24(1):105–21. https://doi.org/10.1080/10556780802394217.
41. Boukouvala F, Niotis V, Ramachandran R, Muzzio FJ, Ierapetritou MG. An integrated approach for dynamic flowsheet modeling and sensitivity analysis of a continuous tablet manufacturing process. Comput Chem Eng. 2012;42:30–47. https://doi.org/10.1016/j.compchemeng.2012.02.015.
42. Borgonovo E, Pangallo M, Rivkin J, Rizzo L, Siggelkow N. Sensitivity analysis of agent-based models: a new protocol. Comput Math Organ Theory. 2022;28(1):52–94.
43. Guo Y, Kafui KD, Wu CY, Thornton C, Seville JP. A coupled DEM/CFD analysis of the effect of air on powder flow during die filling. AICHE J. 2009;55(1):49–62. https://doi.org/10.1002/aic.11734.
44. Mazor A, Orefice L, Michrafy A, De Ryck A, Khinast JG. A combined DEM & FEM approach for modelling roll compaction process. Powder Technol. 2018;337:3–16. https://doi.org/10.1016/j.powtec.2017.04.053.
45. Adam S, Suzzi D, Radeke C, Khinast JG. An integrated Quality by Design (QbD) approach towards design space definition of a blending unit operation by Discrete Element Method (DEM) simulation. Eur J Pharm Sci. 2011;42(1–2):106–15. https://doi.org/10.1016/j.ejps.2010.10.013.
46. Jadhav JB, Girawale NN, Chaudhari RA. Quality by Design (QBD) approach used in development of pharmaceuticals. Int J Pure App Biosci. 2014;2(5):214–23.
47. Rogers AJ, Hashemi A, Ierapetritou MG. Modeling of particulate processes for the continuous manufacture of solid-based pharmaceutical dosage forms. PRO. 2013;1(2):67–127.

48. Hildebrandt C, Gopireddy SR, Scherließ R, Urbanetz NA. Assessment of material and process attributes' influence on tablet quality using a QbD and DEM combined approach. Powder Technol. 2019;345:390–404.
49. Mishra SM, Rohera BD. An integrated, quality by design (QbD) approach for design, development and optimization of orally disintegrating tablet formulation of carbamazepine. Pharm Dev Technol. 2017;22(7):889–903. https://doi.org/10.1080/10837450.2016.1199566.
50. Ojha VK, Schiano S, Wu CY, Snášel V, Abraham A. Predictive modeling of die filling of the pharmaceutical granules using the flexible neural tree. Neural Comput & Applic. 2018;29:467–81. https://doi.org/10.1007/s00521-016-2545-8.
51. Kelley B, Cromwell M, Jerkins J. Integration of QbD risk assessment tools and overall risk management. Biologicals. 2016;44(5):341–51.
52. Ketterhagen WR, Ende MT, Hancock BC. Process modeling in the pharmaceutical industry using the discrete element method. J Pharm Sci. 2009;98(2):442–70. https://doi.org/10.1002/jps.21466.
53. Carothers C, Ferscha A, Fujimoto R, Jefferson D, Loper M, Marathe M, Mosterman P, Taylor SJ, Vakilzadian H. Computational challenges in modeling and simulation. In: Research challenges in modeling and simulation for engineering complex systems. Springer; 2017. p. 45–74.
54. Buyya R, Ranjan R, Calheiros RN. Modeling and simulation of scalable Cloud computing environments and the CloudSim toolkit: challenges and opportunities. In: 2009 international conference on high performance computing & simulation. IEEE; 2009. p. 1–11.
55. Calheiros RN, Ranjan R, Beloglazov A, De Rose CA, Buyya R. CloudSim: a toolkit for modeling and simulation of cloud computing environments and evaluation of resource provisioning algorithms. Softw Pract Experience. 2011;41(1):23–50.
56. Yeom SB, Ha ES, Kim MS, Jeong SH, Hwang SJ, Choi DH. Application of the discrete element method for manufacturing process simulation in the pharmaceutical industry. Pharmaceutics. 2019;11(8):414. https://doi.org/10.3390/pharmaceutics11080414.
57. Bishnu SK, Alnouri SY, Al-Mohannadi DM. Computational applications using data driven modeling in process systems: a review. Digit Chem Eng. 2023;8:100111. https://doi.org/10.1016/j.dche.2023.100111.
58. Bokonda PL, Ouazzani-Touhami K, Souissi N. Predictive analysis using machine learning: review of trends and methods. In: 2020 International Symposium on Advanced Electrical and Communication Technologies (ISAECT). IEEE; 2020. p. 1–6.
59. Yang CC. Explainable artificial intelligence for predictive modeling in healthcare. J Healthc Info Res. 2022;6(2):228–39.

Chapter 7
Inhalation Aerosol Based Targeted Drug Delivery to Upper Airways: Computational Fluid Dynamics (CFD) Approach

Bikash Rajan Jena, Soumyashree Tripathy, Surya Kanta Swain, Roja Rani Budha, Naseeb Basha Shaik, Rajasekhar Reddy Alavala, S. N. Koteswara Rao G., Amiya Prusty, and Prasanna Parida

Contents

B. R. Jena (✉)
School of Pharmacy and Life Sciences, Centurion University of Technolgy and Management, Jatani, Odisha, India
e-mail: bikashranjan.jena@cutm.ac.in

S. Tripathy
Centurion University of Technology and Management, Khurda, Odisha, India

S. K. Swain
Amity Institute of Pharmacy, Amity University, Major Arterial Road, AA II, Newtown, Kadampukur, Kolkata, West Bengal, India

R. R. Budha
H.K. College of Pharmacy, Oshiwara, Jogeshwari (W), Mumbai, Maharashtra, India

N. B. Shaik
G. Pulla Reddy College of Pharmacy, Mehdipatnam, Hyderabad, Telangana, India

R. R. Alavala · S. N. K. Rao G.
Shobhaben Pratapbhai Patel School of Pharmacy and Technology Management, SVKM's NMIMS, Vile Parle (W), Mumbai, Maharashtra, India

A. Prusty
Faculty of Pharmacy, C.V Raman Global University, Bhubaneswar, Odisha, India

P. Parida
Institute of Pharmacy and Technology, Cuttack, India

S. N. K. Rao G., R. R. Alavala (eds.), *Applications of Computational Tools in Drug Design and Development*, https://doi.org/10.1007/978-981-96-4154-3_7

Abstract Inhalation therapy is utilized frequently to treat respiratory problems in children, including asthma. However, only a tiny fraction of inhaled drugs successfully reach the lungs, which reduces their effectiveness. Most pharmaceutical delivery systems and inhalation protocols are formulated and researched for adults, with doses adjusted for children based on their body weight. The relevance of physiological aspects, such as airway size and breathing strategies, as well as physical elements connected to particle mobility, such as aerosol dynamics, in influencing how particles are deposited in the body, has long been acknowledged. However, the use of this knowledge in understanding how these factors affect younger populations has been constrained. So computational fluid dynamics (CFD) is an initial method to investigate potential techniques for improving aerosol deposition in children. A computational model is utilised to recreate the movement of air and particle in the upper respiratory tract, with age groups ranging from 5 years old to adulthood. The model represents the anatomy of the upper airway in a basic but precise manner.

Current chapter signifies modern advancements, technologies associated for CFD and inhalation based targeted drug therapy for upper respiratory tracts and associated current status and future prospectives.

Keywords Aerosol deposition · Drug delivery · Inhalation · Lung · Dry power inhalers · Metered dose inhalers

7.1 Introduction

The pharmaceutical industry prioritises pulmonary medication delivery because it allows pharmaceuticals with limited bioavailability to be inhaled across a vast surface area of the lungs. Since the late 1950s, the pressurized metered-dose inhaler (pMDI) has been the most preferred therapeutic choice for asthma [1]. Regardless of recent progress in delivering drugs to respiratory tract, less than 30% of a dosage enters a patient's airways [2]. The remaining substance accumulates in the airway of the mouth and throat, as well as in the valve stem, nozzle, or valve holding chamber (if it exists) [3]. These regions can restrict airflow. Research results suggest that patients have difficulties coordinating the inhaler's activation and the following inhalation. The challenge might be linked to a restricted understanding of the device's physics and technical components [4, 5]. Effective treatment of respiratory disorders relies heavily on meticulous device design and precise medication composition, whether in suspension or solution form. Therapeutic medications targeted at the respiratory system (RS) may be delivered via aerosol particles, which are effective carriers. Historically, the act of respiration smoke produced by plant leaf combustion may relieve respiratory ailments [1, 2]. This method of delivering drugs has been acknowledged for an extensive duration. To ensure the accurate delivery of small aerosols into the lungs, the administration of drugs has to be more precisely targeted [8], since drug deposition is unevenly distributed inside the lungs [6, 7]. Aerosol distribution is a minimally intrusive technique that is very efficient at much lower concentrations compared to what is required for oral therapy. There are numerous therapeutic aerosol delivery devices available, including pressurised, MDIs, DPIs, medical nebulizers, solution mist inhalers, and nasal sprays. Aerosol medication offers a non-invasive approach to oral therapy that may achieve comparable efficacy with much-reduced dosing. Today, several techniques used for delivering therapeutic aerosols like pMDI, DPI, medical nebulizers, and nasal sprays. Inhalation therapy with pMDI is a widely employed and effective treatment for asthma.

This work aims to come up with a CFD model that accurately simulates the movement of aerosol particles from pMDI into a simulated mouth-throat structure [1–3]. Therapeutic aerosols are supplied via nasal inhalation and oral delivery, with inhalation therapy resulting in a minute amount of medicine reaching the appropriate target area. The management of respiratory conditions in children, particularly asthma, relies mostly on inhalation treatment [1–3]. Asthma is not only the most common long-term health problem worldwide [4], but recent research also indicates that the number of children affected by asthma has risen in the last 10 years [5–8]. It is acknowledged in this particular context that inhaled corticosteroids are efficacious drugs for diminishing airway inflammation [9]. It has been suggested that regularly using these treatments at a modest dose may decrease the likelihood of becoming sick or dying [10]. The advantage of inhalation treatment is that it quickly starts to have a therapeutic impact when the medicine is deposited in the tracheobronchial area [11]. Furthermore, it has been demonstrated that fewer adverse effects are linked to it when compared to taking it orally, and smaller amounts are needed. Table 7.1 embodies advantage and disadvantages of inhalation devices. Typically, medication for paediatric asthma is given by inhalation using dry powder inhalers, pressured metered-dose inhalers, or nebulizers [3, 12].

7.1.1 Advancements and Recompenses of Using the DPIs and MDIs for Diverse Devices

Although there are differences in the advantages and disadvantages of various kinds of devices used for delivering drugs to the respiratory system of children, nebulizers and MDIs are frequently suggested for younger paediatric patients, especially those

Table 7.1 Significance and shortcomings of diverse inhalation devices

Various devices containing inhalation	Significance	Shortcomings	References
pMDI	Small and densely packed. Moveable Multidose (around 200 doses) Closed atmosphere (It is important to ensure that stability is maintained and that there is no degradation observed)	Inhalation technique and patient co-ordination are essential. Maximum drug release in the mouth mg. The highest amount is 5 Partial range of drugs available	[11]
Nebulizers	No precise inhalation system It disperses medication solutions into aerosol form to the fullest extent. Carries huge dosages Appropriate for infants and persons too sickening	Prolonged period of time Massive Contents effortlessly pollute Relatively expensive and seems to be deprived delivery competence	[11]
DPI	Small and densely packed. Moveable Handy for management	The amount of respirable dosage is directly influenced by the rate at which air is inhaled. Humidity can lead to the clumping of powders and the softening of capsules. Most DPIs typically contain lactose.	[11]

between the ages of 5 and 6 [12–14]. Active devices, like nebulizers and MDIs, produce aerosols regardless of the patient's physical activity. On the other hand, dry powder inhalers (DPIs) are indicated for older children who can provide the necessary suction to break up the powdered medicine. Despite paediatric inhalation treatment is often used, a significant number of inhaled drugs enter the lungs too late to have their maximum therapeutic effects [15–17]. A significant percentage of the inhaled amount is usually wasted in anatomical areas outside the chest, such as the oral and pharyngeal cavities. Prior studies on living organisms have shown that the deposition efficiencies of nebulizers generally vary from 0.5% to 12% [16]. On the other hand, in the best conditions, pMDIand DPIs have shown efficiencies as high as 40% [16–18]. In order to address these drawbacks [19], further measures must be taken to improve, if not achieve ideal, therapeutic delivery in paediatric populations. The distinct behavioural and emotional characteristics of children, along with their tendency to ignore treatment suggestions, have an adverse effect on the management and treatment of illnesses. Furthermore, correct anatomical properties of children's lungs, such as airway size and breathing patterns, as well as aerodynamic parameters such as aerosol particle motion, are critical for assessing the efficiency of aerosol accumulation in their lungs. But, these essential physiological and physical traits are often disregarded. Typically, drug delivery devices and the instructions for using them by inhalation are created using research that focuses on adult adults [20–23]. However, thorough study on live organisms has long recognized the crucial significance of particle size in influencing how particles are deposited in the lungs and, therefore, how successful they are in deposition [24–27]. In paediatric patients, the dosage (measured in mg) and frequency of administration (e.g. every 12 h) of a particular medication (e.g. bronchodilators) are adjusted based on the child's body weight [28, 29]. Additional scrutiny and adjustment of this information for younger age cohorts is necessary. Simultaneously, it is important to consider the changes in lung structure that occur as one gets older. During the course of child growth, the respiratory passageways that transport air into the lungs undergo enlargement, enabling a greater amount of air to be inhaled and exhaled with each breath. Lung capacity continues to increase into adulthood [30–33]. Changes in respiratory rate and architecture have a significant impact on deposition results. These outcomes differ significantly between children and adults. Extensive research has been done in the lungs of people on the mechanics of aerosol accumulation. The efficacy of predicting the release of drugs using existing lung deposition models is limited because they underestimate deposition in the oropharynx. Recent studies on the placement of drug delivery systems in the oropharyngeal region have discovered additional factors that play a crucial role in defining these systems' main movement and placement. The digital fluid dynamics model is useful for studying particle movement and settling. The investigation of methods for applying nasal spray is also covered [34–36]. Table 7.2 represents the selected airway diameters as a function of age for illustrations [37]. By delivering pharmaceuticals to the upper respiratory tract via inhalation aerosols, a prevalent and extraordinarily effective method is achieved. This method enables the targeted delivery of pharmaceutical substances to particular areas of the respiratory system, such as the lungs or nasal passages. When it

Table 7.2 Particular airway diameters as a function of age

	Age in (yrs.)			
Diameter (mm)	5	10	25	Ref.
Parts of Mouth	11	14	19	[37]
Zones of Trachea	9	11.6	16	
Parts of Bronchi (Left & Right)	6.2	8	10.8	

comes to the treatment of various respiratory conditions, a targeted approach is crucial because it guarantees that the medication reaches the intended site of action. This is achieved by use of DPIs, pMDIs, and nebulizers, which deliver the aerosolized medication directly to the intended site. Using CFD to evaluate the distribution of pharmaceutical aerosols in the upper respiratory tract is an incredibly successful procedure. Using CFD to mimic the deposition patterns and movement of aerosol particles in order to analyse the effectiveness and efficiency of upper respiratory treatment could provide significant insights. By employing this approach, scientists and the pharmaceutical sector have the capability of improving the configuration of inhaler apparatus, developing drug delivery systems with greater accuracy, and assessing the impacts of numerous factors influencing aerosol deposition, including inhaling technique and particulate size. Airway diameter of different age groups is mentioned in Table 7.2 [37]. In simple terms, the purpose of CFD to the delivery of drugs via inhalation aerosols holds significant promise in augmenting our comprehension and streamlining the precise administration of drugs to the upper respiratory tract. This may ultimately result in improved outcomes for patients undergoing treatment for respiratory illnesses. By utilizing inhalation aerosols, CFD [39] can effectively evaluate the transport of medications to the upper airways. CFD models make substantial contributions to the knowledge of aerosol component movements in the upper respiratory tract, including deposition patterns and clearance mechanisms. The information shown above can be used to optimise inhaler configurations and drug administration strategies, ensuring that drugs reach their intended targets within the respiratory system. CFD offers substantial benefits in terms of cost and time efficiency in comparison to experimental methodologies. This facilitates the optimization of medication delivery systems for better efficiency [38–40].

7.2 Effectiveness of Drug Delivery via Aerosols and Its Key Factors

7.2.1 Inhalation Therapy Aided Common Features

Different physical processes impact particles that are inhaled, depending on their forms, densities, and sizes (including micro- and nanoparticles). These factors influence the extent to which the particles may reach diverse provinces of the respiratory system [57]. It is crucial to bear in mind that respiration is inherently connected to fluctuating aerodynamic circumstances that significantly influence the behavior of aerosol particles. Besides they have the ability to undergo changes in size over time, such as coagulation, condensation, or evaporation (in the case of liquid droplets). This is due to the influence of numerous patient- and technology-related variables. The movement of air or aerosol through the airways is influenced by specific characteristics of each patient, such as their inhalation dynamics and airway geometry. In contrast, inhalers utilize airflow passing through a specialized precursor to generate a medicinal aerosol. This methodology utilizes engineering techniques to enhance both the precursor and the inhalation device. While the patient's inhalation introduces the aerosol into the respiratory tract, many factors associated with both the patient and the technology contributes to the differential distribution of distinct components of the respiratory system (local deposition). Based on the provided information, it is evident that determining the precise dosage of an inhaled aerosol medication that effectively reaches the intended locations is not possible. This technique of drug delivery differs from several other ways such as injections, capsules, pills, and suppositories. Several factors connected to patients and technology, as shown in Fig. 7.1 which represents a diagram that visually depicts the procedure and variables that influence the efficiency of administering aerosol drugs. The movement of air or aerosol through the lungs is influenced by the patient's inhalation dynamics and airway morphology.

In point of fact, a medical aerosol is produced by employing an inhaler to force air out of a specific preparation. This procedure, which makes use of technology, enhances both the precursor and the breathing apparatus. The flow that is created by the patient is responsible for transporting aerosol into the respiratory system. It is at this point when patient-related and technical aspects come together, resulting in the medication being locally deposed in various regions of the RS. Therefore, the medicine that was given to the surface of the RS will be effective because it will be able to circumvent the barrier that is created by the fluids that cover the airways. Through the process of the settling particles interacting with the surface of the RS, the medication is ultimately guided to various areas of the respiratory system [15, 37–40].

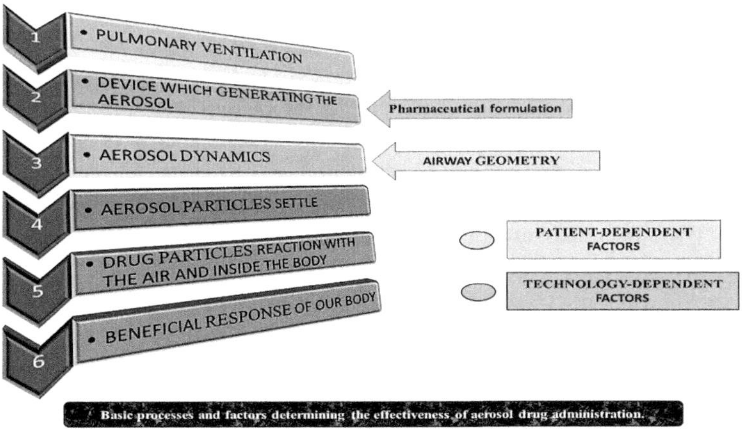

Fig. 7.1 Schematic representation illustrating process and factors determining the effectiveness of aerosol drug administration

7.2.2 *Deposition in the Respiratory Tract*

The mathematical techniques that are used to change and measure the features of potential therapeutic candidates are known as computational approaches for computer-aided drug design (CADD). A variety of software applications include these strategies into their operations. Software packages such as those described above include a variety of features that are available to the general public and can be purchased for commercial use. The following subgroup includes specific examples of crucial tools for Computer-Aided Drug Design (CADD), with a particular emphasis on those tools that are commonly utilized in our laboratory [14, 15]. There are a number of elements that can have an effect on the pattern of deposition: (1) the characteristics of the particles (2) the geometry of the airways, which varies depending on gender, age, and the occurrence of disease; and (3) the pattern of breathing, which includes frequency, tidal volume, and the amount of time without taking a breath. Inertial impaction, sedimentation, diffusion, interception, and electrostatic impacts are some of the key mechanisms that may be responsible for the deposition of aerosols [41–43].

7.2.2.1 Computational Fluid Dynamics in Medical Applications

A growing assortment of medical applications has adopted CFD, a potent instrument that is utilized to assess inhalation aerosol-based drug delivery to the upper airways. Through the simulation of aerosol particle flow and deposition patterns within the respiratory system, CFD enables a comprehensive evaluation of the efficiency and efficacy of drug delivery. In comparison to conventional experimental methods, this approach offers the potential for cost and time savings, the capability of simulating conditions that are challenging to replicate, and the capacity to generate an analysis with a significant degree of intricacy [44]. In general, CFD exhibits considerable potential when applied to the delivery of drugs to the upper airways via inhalation aerosols [45]. It can provide valuable insights into drug dissolution and clearance, and reduce the need for human subject testing, all of which contribute to the creation of more precise and individualized treatment options for respiratory diseases. The application of CFD has demonstrated significant promise in optimizing treatment strategies for respiratory diseases and enhancing drug delivery efficiency. Despite this, the complete implementation of numerical methodologies in the design of inhalers still faces obstacles. The comprehensive integration of coupled fluid dynamics and aerosol generation into inhaler design is hindered by their intricate characteristics [47]. In addition, constraints on computational capacity present obstacles to the complete implementation of CFD in the design of inhalers. In spite of these obstacles, developments in computational capacity and ongoing research bode well for the future of CFD in inhaler design. Table 7.3 represents CFD modelling approaches and applications.

Table 7.3 CFD modelling approaches and applications

S. no	CFD modelling approaches	Applications
1	Through the examination of variables such airway geometry, inhalation flow rate, and particle size.	Development of devices (MDIs, DPIs, and nebulizers)
2	CFD modelling assists in accurately forecasting the movement of particles and the patterns of their deposition within the respiratory tract of compositions.	Optimization of formulation (in order to optimize drug stability, bioavailability, and targeting efficacy, thereby enhancing therapeutic results).
3	CFD-based modelling enables to study aerosol deposition patterns in various respiratory diseases.	Disease-specific targeting (customizing drug delivery strategies according to disease-specific variables improves treatment effectiveness and reduces adverse effects).

7.2.2.2 Modelling Aerosol Dynamics in the Respiratory System

CFD models allowing for a better understanding of how different factors influences release of drugs to the upper airways.

CFD models helps researchers in better understanding of aerosol dynamics that includes simulating airflow patterns and particle deposition to analyze how various factors influence drug delivery to the upper airways. Moreover, CFD provides a platform to optimize drug delivery devices by virtually testing and enhancing their efficiency, mitigating the need for extensive physical experiments [38].

In particular, the utilization of CFD enables the evaluation of optimal particle size and inhalation techniques for effective drug delivery. This can further aid in customizing treatment plans for individual patients, resulting in enhanced therapeutic outcomes for respiratory diseases. Additionally, CFD offers spatially and temporally resolved predictions, revolutionizing the field of respiratory drug delivery by providing valuable insights into aerosol behavior and deposition patterns, ultimately leading to the development of more efficient and targeted drug delivery strategies [40]. With the advantages of time and cost savings, the ability to simulate difficult-to-study conditions, and the provision of a high level of detail, CFD has the potential to revolutionize respiratory drug delivery by enabling a deeper understanding of aerosol behavior, informing the development of more targeted and efficient drug delivery strategies, and improving treatment outcomes for respiratory diseases. This simulation technique is useful for studying aerosol dynamics in the respiratory system. It allows researchers to analyze airflow patterns and particle deposition, optimize drug delivery devices, personalize treatment plans, and develop more efficient drug delivery strategies.

Generally human respiratory system is a complex system comprising the extrathoracic, thoracic, and pulmonary parts which are considered to be the three primary sections of the respiratory system and out of which the extrathoracic airway or the head airway, serves as the primary defence mechanism against potentially hazardous compounds that are inhaled [45–47]. There are sixteen generations of branching airways that make up the tracheobronchial (TB) tree, which is also referred to as the conducting airway. The trachea is the only airway that is present in this configuration. In the pulmonary region (P), which is where gas exchange takes place, the alveolar sacs and ducts are the constituent parts. The majority of people breathe through their noses when they are at rest or when they are not exerting themselves too hard. Nevertheless, when they are exerting themselves vigorously, they switch to breathing through their mouths or their noses. Due to the fact that the resistance of the nasal airway is substantially higher than the resistance of the oral airway, people frequently switch to practicing oral/nasal breathing [40]. Certain individuals continue to breathe via their mouths even when they are at rest. It is possible for a person to inhale an airborne particle that is delivered close to them; the radius of the particle's aerodynamic diameter is directly proportional to its inhalability, which is another term for the particle's capacity to enter the head airway [40]. A portion of the airborne particles that are able to enter the human airway from aerosol wind tunnel is referred to as inhalability. Based on the findings of these wind tunnel tests [40, 45, 47–50], it has been discovered that the inhalability of particles smaller than 5 μm is close to 100%. The inhalability of particles decreases with increasing size and remains constant at 50% for particles exceeding approximately 50 μm [51–55].

Following its entry into the respiratory tract, the aerosol particle has the potential to be deposited at a number of different sites within the respiratory tract. In the course of respiration, the airflow goes through a number of different directions, including those that occur within the nasal/mouth, pharynx, and larynx regions, as well as the airway bifurcations [40, 45, 48]. Due to their inability to follow the flow of air, particles with a size higher than 0.5 μm have the potential to settle in these locations. This phenomenon is referred to as impaction, and it occurs when particles fall into these spaces. In point of fact, the deposition of particles in the region of the mouth and throat that occurs as a result of direct interaction continues to be responsible for a sizeable fraction of the quantity of medication that is expelled from pMDI and DPI. The process of sedimentation is the fundamental mechanism that is responsible for the accumulation of particles that are inhaled in the alveolar region and the tiny airways. On the other hand, particles that are smaller than 0.2 μm have the ability to be dispersed throughout the entire respiratory tract through the process of diffusion [40, 45, 47–49]. Diffusion deposition is the driving force behind the relevance of nanoparticles that are smaller than 100 nanometres. In situations where the length of the particle is comparable to the size of the airways in the lungs, the process of interception deposition plays a significant role in the process of catching elongated particles, which includes fibrous aerosols. As they are manufactured and delivered, pharmaceutical aerosols have the potential to acquire an electrical charge. This is especially true in the case of DPI and pMDI [40, 45, 48]. In the case of aerosols, the electrostatic charge status can be affected by a variety of factors, including the composition of the material, the additives, the formulation, and the structure of the device. The amount of electrostatic charges that are present on aerosols produced by various devices has been discovered to have an effect on the deposition of these fluids. Electrostatic phenomena suggest when the quantity of charges on particles exceeds a particular threshold it causes increase in particle deposition [40, 45, 47–50]. Some examples of these phenomena include image and space charge forces.

A series of experimental tests have been carried out within the human respiratory tract to ascertain particle deposition efficiency. These measurements were taken out on human volunteers as well as physical models of replicas of the airway. In order to accomplish this, spherical test particles that had been labelled with radiolabels or fluorescent compounds were utilized. Various independent organisation like ICRP and NCRP developed Lung deposition models, that are dependent on the anatomical structure of the airways as well as the usual breathing behaviours of adult males who are of average age [47, 49]. These models have been utilized extensively for the purpose of calculating the quantity of radiation that is acquired from radioactive aerosols, as well as for the purpose of determining the amount of hazardous particulate matter that is inhaled in both natural environments and occupational settings. In order to develop deposition equations for various portions of the airway, theoretical models make use of reduced airway geometry and particle deposition. These mechanisms include impaction, sedimentation, diffusion, and interception. Input data on particles with defined attributes and physiological parameters are used in the computation of both overall and localized aerosol depositions. This calculation is accomplished utilizing the input data received from the particles and validation was accomplished by the utilization of experimental data gathered both in vivo and in vitro at the same time. It has been demonstrated beyond a reasonable doubt that the ICRP, NCRP, and other lung deposition models are accurate through the examination of the data that is currently available [47, 49].

7.2.3 The Role of Aerosols in Upper Airways Treatment

Aerosols have a significant role in the treatment of upper airway disorders [48]. They allow for targeted drug delivery directly to the affected areas of the respiratory system, resulting in quicker and more effective relief of symptoms [48]. Inhalation aerosol-based drug delivery to the upper airways relies on the efficient generation of aerosol particles in the respiratory structure. By using CFD, researchers can simulate and analyze the behavior of aerosol particles in the respiratory system, providing valuable insights into drug deposition patterns and clearance mechanisms. This information is essential for optimizing drug delivery strategies and inhaler designs, ensuring that medications reach their intended target sites within the respiratory system. Furthermore, the use of aerosols in upper airways treatment can lead to more targeted and personalized treatment options for respiratory diseases, ultimately improving patient outcomes [50]. In addition to the treatment of respiratory diseases, aerosols role is significant in delivering medications to the nasal passages. This targeted approach allows for the in treatment of conditions such as allergic rhinitis and sinusitis. CFD is responsible into the deposition and clearance of aerosolized drugs in the nasal passages, facilitating the development of more effective nasal drug delivery strategies [51].

The use of CFD in inhalation aerosol-based drug delivery to the upper airways offers a promising avenue for enhancing drug delivery efficiency and optimizing treatment strategies for respiratory diseases [52]. While there are challenges in fully implementing numerical methods in inhaler design, ongoing research and advancements in computational power hold significant promise for the future of CFD in this field. With continued developments, CFD has the potential to revolutionize the design and optimization of inhalers, ultimately leading to improved patient care and outcomes in respiratory disease treatment. Overall, the integration of CFD in respiratory drug delivery research enables a deeper understanding of aerosol behavior and deposition patterns, leading to the development of more targeted and efficient drug delivery strategies [53].

7.2.4 Techniques Utilised in Airway Geometry

Scientists replicate the motion and deposition of therapeutic aerosols by using authentic inhalation patterns that closely resemble those produced by DPIs and nebulizers. The geometries provide a very precise and consistent representation of the upper airway in the human lung, which displays variations depending on age. It encompasses the anatomical components of the oral cavity and pharynx linked to the lungs' respiratory passages. The conductive bronchial airway tree accurately represents the physical features of a normal adult human (male) [47, 49–56]. To summarize, the airway tree comprises seven generations of airways divided into two branches and unevenly distributed. Figure 7.1 illustrates the whole configuration of the airway. Similarly Fig. 7.2 indicates that, in contrast to several other methods of drug delivery such as injections, capsules and tablets, and suppositories, it is not feasible to accurately determine the precise amount of an inhaled aerosol medicine that reaches the intended locations especially to the lungs.

It is essential to highlight that the respiratory system has 64 distal outputs, which are categorized according on their precise positions in the lungs. The configuration is graphically shown in Fig. 7.1a, with various hues used to differentiate the outputs for improved comprehension. To evaluate the attributes of the deposition mentioned, the entire structure is partitioned into three distinct regions: the oropharyngeal area, the trachea, and the other airways responsible for air conduction. The adult variation is scaled down to accurately mimic the airways present in children of a certain age. Previous studies have shown that the anatomical features of the upper respiratory airways in newborns and adults are similar [50]. Empirical investigations have shown that systematically reducing the size of airways in adults may provide relatively precise estimates of aerosol deposition in the airways of children [50, 57]. The aforementioned investigations identified inconsistencies in the scaling factors used to modify the dimensions of the mouth-throat and airways while transitioning from adults to children. It is crucial to comprehend that disparities between these

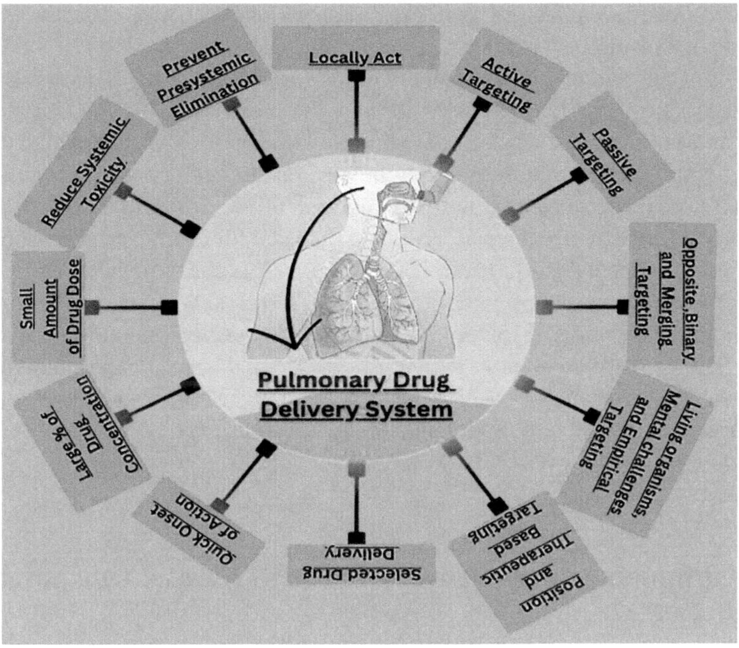

Fig. 7.2 Schematic representation of Pulmonary Drug delivery system

two measures may not consistently signify variances in the physiological maturation of separate lung sections as an individual age. Alternatively, these changes may be ascribed to the intrinsic variability within the same species since each research is first done using a distinct adult model. Table 7.1 represents a thorough summary of the anatomical characteristics of the airway, organised according to age groups [50, 57]. Therefore, it is essential to highlight this fact. Consequently, the airway shapes generated fall within the required range for proper bodily functioning. The purpose of this research is not to examine differences across other species, such as changes in patient traits or irregularities in the typical structure of the airway, such as airway synapsis [58].

7.2.5 Advancements in Inhalation Therapy for Respiratory Diseases

CFD has emerged as a valuable tool for optimizing inhaler design and improving drug delivery to the upper airways [49]. By simulating the flow of inhaled particles CFD can help researchers and clinicians better understand how different factors, such as inhaler design and patient-specific parameters, impact drug delivery to the upper airways. This knowledge can then be used to improve inhaler design, customize treatment plans for individual patients, and enhance overall therapeutic outcomes for respiratory diseases. In particular, CFD can provide insights into the optimal particle size and inhalation technique for effective drug delivery. By simulating the airflow and particle deposition in the nasal airway, CFD can aid in the design and optimization of drug delivery devices for the upper airways, ensuring efficient and targeted drug delivery to the desired sites of action. Furthermore, the use of CFD in respiratory drug delivery research allows for virtual testing and optimization of drug delivery devices, reducing the need for costly and time-consuming physical experiments [54]. Overall, CFD offers a cost-effective and efficient approach to studying drug delivery in the upper airways. With the ability to provide spatially and temporally resolved predictions, CFD has the potential to revolutionize the field of respiratory drug delivery [49]. By enabling a deeper understanding of aerosol behavior and deposition patterns, CFD can inform the development of more targeted and efficient drug delivery strategies, ultimately leading to improved treatment outcomes for respiratory diseases.

7.3 Optimizing Drug Delivery Through Aerosol Inhalation

Inhalation aerosol-based drug delivery to upper airways can be optimized through the use of CFD [48]. This technique allows for the simulation of airflow patterns and the deposition of inhaled particles, providing detailed insights into the effectiveness of different inhalation techniques.

The integration of CFD has opened avenues for developing more personalized and targeted treatment options for respiratory diseases. By simulating the flow of inhaled particles and their deposition patterns within the respiratory system, CFD offers a deeper understanding of aerosol behavior and informs the development of more efficient drug delivery strategies [51].

In the realm of personalized medicine, CFD has the potential to revolutionize the design and optimization of inhalers, ultimately leading to improved patient care and outcomes in respiratory disease treatment. This holds promise for enhancing drug delivery efficiency, optimizing treatment plans for individual patients, and improving therapeutic outcomes for respiratory diseases. The ability to analyze airflow patterns and particle deposition enables the customization of treatment plans, ensuring targeted and efficient drug delivery to the desired sites of action tailored to each patient's specific needs [51–54].

The ongoing research and advancements in computational power present significant promise for the future of CFD in enhancing drug delivery to the upper airways. With its ability to provide spatially and temporally resolved predictions, CFD offers a cost-effective and efficient approach to studying drug delivery in the upper airways.

In developing targeted drug delivery strategies, CFD is poised to play a pivotal role in personalized treatment plans and improved outcomes for individuals with respiratory diseases. So, the use of CFD in inhalation aerosol-based drug delivery to upper airways offers new opportunities for personalized and targeted treatment options that can enhance drug delivery efficiency and optimize treatment plans for respiratory diseases [47–49].

7.3.1 *Particle Size and Distribution Impact on Aerosol Therapy*

The impact of particle size and distribution in aerosol therapy is a crucial aspect that can be effectively studied and analyzed using CFD. By simulating the flow of respiratory aerosols, CFD can provide valuable insights into how particle size and distribution influence drug deposition within the upper airways, helping to optimize the design of inhaler devices and improve drug delivery efficiency [56]. These insights can inform the development of inhalation strategies that maximize drug delivery to target sites and minimize deposition in non-target areas, reducing potential side effects and improving therapeutic outcomes. Overall, using CFD in inhalation aerosol-based drug delivery to upper airways holds great potential for advancing personalized treatment options and optimization [49, 57]. The ability to analyze particle size and distribution through CFD provides crucial information for designing inhaler devices that can effectively target specific areas within the respiratory system. By optimizing the delivery of medication to the intended sites of action, the potential for reducing side effects and enhancing therapeutic outcomes is significantly increased.

Furthermore, the advancement of personalized medicine through the utilization of CFD in inhalation aerosol-based drug delivery to upper airways is a promising development. This approach allows for the customization of treatment plans based on individual patient needs, ensuring that drug delivery strategies are tailored to maximize efficacy and minimize potential risks or adverse effects.

In conclusion, the integration of CFD in aerosol therapy for respiratory diseases offers a paradigm shift in how drug delivery strategies are developed and optimized. By providing detailed insights into airflow patterns, particle deposition, and the impact of particle size and distribution, CFD enables the design of more targeted and efficient drug delivery strategies, ultimately leading to improved treatment outcomes for individuals with respiratory diseases [57].

7.3.2 *Simulation Techniques for Effective Aerosol Drug Administration*

Simulation techniques, such as CFD, play a crucial role for progress and optimization of aerosol drug administration methods [49]. By simulating the behavior of aerosol particles in the respiratory system, CFD can provide valuable insights into factors such as particle size, deposition patterns, and flow dynamics. These insights can then be used to design inhaler devices and drug formulations that optimize drug delivery to the target site within the respiratory system. This can ultimately lead to improved drug efficacy, reduced side effects, and better treatment outcomes for individuals with respiratory diseases. By accurately modeling the behaviour of the respiratory system aided aerosol units in using CFD, researchers and clinicians can gain a comprehensive understanding of how different variables, such as inhalation technique or device design, to the target site [58]. This knowledge can then inform the development of new drug delivery strategies that are optimized for specific patient populations and respiratory conditions. Overall, the integration of CFD in aerosol therapy for respiratory diseases holds great promise for improving drug delivery strategies and individualizing treatment plans. The combination of CFD and aerosol therapy for respiratory diseases has the potential to revolutionize drug delivery strategies by providing in-depth understanding of airflow patterns and aerosol behaviour in the respiratory system [48].

7.4 Addressing Computational Fluid Dynamics Model Validation

To advance aerosol inhalation treatment, it is crucial to address the need for further validation of CFD models. Integrating experimental data with simulation results can enhance the accuracy and reliability of CFD models, providing a more comprehensive understanding of aerosol behavior in the respiratory system. Collaborative efforts between researchers and clinicians to validate CFD models using empirical data from experimental studies can lead to more robust and clinically applicable simulation techniques. Furthermore, the development of more advanced and refined models is necessary to capture the complexities of aerosol behavior in the respiratory system by accurately predicting particles deposition and its transport in airway. The Use of *in Silico* Models for Optimal Drug Delivery, which are valuable tools for optimizing drug delivery in aerosol inhalation treatment [60].

Validating CFD models with empirical data from experimental studies is essential for ensuring the accuracy and reliability of simulation techniques in aerosol inhalation treatment. Collaborative efforts between researchers and clinicians can play a pivotal role in this validation process, as they can provide valuable insights and clinical applicability to CFD models [51]. By integrating experimental data with simulation results, helps researchers and clinicians to know the aerosol behavior. This validation process enhances the fidelity of CFD models, allowing for a more accurate representation of particle deposition, transport, and airflow dynamics within the respiratory system.

7.4.1 Refinement of Aerosol Inhaler Devices and Formulations

The optimization of inhaler devices and drug formulations for specific patient populations and respiratory conditions remains a challenge in aerosol inhalation treatment. Research initiatives should focus on developing more advanced and refined models to accurately capture the complexities of aerosol behavior, considering variables such as inhalation technique, lung geometry, and airflow patterns. By refining aerosol inhaler devices and formulations, tailored treatment options can be designed to maximize drug delivery efficiency while minimizing potential side effects [61].

7.5 Innovations in Aerosol Therapy Research and Development

7.5.1 Targeted Aerosol Therapies

Targeted aerosol therapies have seen several innovative developments aimed at enhancing drug delivery efficiency, minimizing side effects, and improving patient outcomes. Here are some notable innovations in this field:

7.5.1.1 Nanotechnology

Nanoparticles offer precise drug targeting and controlled release capabilities. Researchers are exploring various nanomaterials, such as liposomes, dendrimers, and polymeric nanoparticles, to encapsulate drugs for targeted delivery to specific cells or tissues within the respiratory tract. These nanoparticles can improve drug solubility, stability, and bioavailability while reducing systemic side effects.

7.5.1.2 Smart Inhalers

Smart inhalers incorporate sensors and connectivity features to monitor inhalation technique, medication adherence, and respiratory parameters in real-time. These devices can provide feedback to patients and healthcare providers, helping optimize treatment regimens and improve therapeutic outcomes [62].

7.5.1.3 Micro Fluidic Aerosol Devices

Micro fluidic technology enables precise control over droplet size, distribution, and velocity, facilitating targeted delivery of aerosolized drugs to specific regions of the lungs. Micro fluidic aerosol devices can generate monodisperse droplets with narrow size distributions, enhancing drug deposition efficiency and minimizing wastage.

7.5.1.4 Biological Targeting Ligands

Ligands such as antibodies, peptides, and aptamers can be conjugated to aerosolized drugs to achieve specific targeting to receptors or biomarkers over expressed in diseased cells or tissues. This approach enhances drug accumulation at the target site while reducing off-target effects on healthy tissues.

7.5.1.5 3D Printing of Inhalable Formulations

3D printing technology allows for the fabrication of customized inhalable formulations with complex geometries and tailored drug release profiles. Inhalable 3D-printed constructs can incorporate multiple drugs, excipients, and release-controlling polymers, offering personalized treatment options for patients with diverse respiratory conditions.

7.5.1.6 Gene Therapy Aerosols

Aerosol-mediated delivery of gene therapy vectors enables targeted gene transfer to the respiratory epithelium for the management of genetic lung diseases. Advances in viral and non-viral vector design, as well as gene editing technologies like CRISPR-Cas9, hold promise for more effective and durable gene correction in the lungs.

7.5.1.7 Ultrasound-Assisted Aerosol Delivery

Ultrasound can enhance aerosol deposition and penetration in the lungs by inducing transient changes in pulmonary mechanics and increasing airway permeability. Ultrasound-assisted aerosol delivery has shown potential for improving the distribution and efficacy of inhaled drugs, particularly in patients with compromised lung function or underlying respiratory conditions.

7.5.1.8 Personalized Inhalation Therapy

Integration of patient-specific factors such as lung morphology, physiology, and disease phenotype into inhalation therapy planning allows for personalized dosing regimens and treatment strategies. Computational modelling, imaging techniques, and biomarker profiling enable the customization of aerosol therapies to individual patient needs, optimizing therapeutic outcomes and minimizing adverse effects [62].

These innovations in targeted aerosol therapies hold promise for addressing medical needs for the treatment of various pulmonary diseases. Continued research and development efforts in this field are essential to harness the full potential of aerosol drug delivery for improved patient care.

7.6 The Impact of CFD on Inhalable Drug Formulations

CFD has significantly impacted the development and optimization of inhalable drug formulations by providing valuable insights into airflow patterns, particle deposition mechanisms, and drug delivery efficiency within the respiratory tract. Here are some key impacts of CFD on inhalable drug formulations:

7.6.1 Optimization of Inhaler Devices

CFD simulations allow for the assessment of airflow dynamics and particle behaviour within inhaler devices, including metered-dose inhalers (MDIs), DPIs, and nebulizers. By modelling device geometry, airflow rates, and particle properties, CFD can optimize inhaler design parameters such as mouthpiece geometry, an actuation mechanism, and flow rate to enhance drug delivery efficiency and consistency.

7.6.2 Prediction of Deposition Patterns

CFD models simulate the aerosolized drug particles deposition and transport within the respiratory tract, predicting deposition patterns in different lung regions (e.g., bronchi, bronchioles, alveoli). Understanding the factors influencing particle deposition, such as particle size, density, and airflow dynamics, enables the design of inhalable formulations tailored to target specific lung regions for optimal therapeutic effect.

7.6.3 Evaluation of Formulation Performance

CFD enables virtual testing of inhalable drug formulations by predicting factors such as lung deposition fraction, fine particle fraction (FPF), and regional deposition efficiency. This allows researchers to assess the performance of different formulation compositions, particle sizes, and excipients without the need for extensive in vitro or in vivo experiments, accelerating formulation optimization and reducing development costs [62].

7.6.4 Assessment of Inhalation Technique

CFD simulations can analyze the impact of patient-specific factors (e.g., inhalation flow rate, inspiratory profile) on drug delivery efficiency and deposition patterns. By integrating patient-specific data with CFD modelling, inhaler devices and drug formulations can be optimized to accommodate variations in inhalation techniques among different patient populations, improving treatment efficacy and patient adherence.

7.6.5 Investigation of Disease States

CFD models can simulate airflow patterns and particle deposition in diseased airways affected by conditions such as COPD, cystic fibrosis etc. By studying how disease-related factors (e.g., airway constriction, mucus accumulation) influence aerosol deposition, CFD facilitates the development of inhalable drug formulations tailored to specific disease states, optimizing therapeutic outcomes for patients [62].

7.6.6 Design of Targeted Drug Delivery Systems

CFD-guided design facilitates the development of targeted drug delivery systems that exploit physiological factors (e.g., airflow dynamics, mucociliary clearance) to enhance drug deposition at specific target sites within the lungs. By incorporating biological targeting ligands or controlled-release mechanisms into inhalable formulations, CFD-driven approaches enable precise delivery of therapeutics to diseased cells or tissues while minimizing systemic side effects [58, 59].

Overall, CFD plays a crucial role in advancing the development and optimization of inhalable drug formulations by providing detailed insights into airflow dynamics, particle behaviour, and deposition mechanisms within the respiratory tract. By integrating CFD with experimental studies and clinical data, researchers can accelerate the translation of innovative inhalable therapies from concept to clinical application, ultimately improving patient outcomes in respiratory medicine [62].

Advancements in CFD have significantly improved the understanding of particles behavior in from aerosol system. Through the integration of experimental data and the development of advanced CFD models, researchers and clinicians can now gain a more comprehensive understanding of the complexities involved in aerosol inhalation treatment [62]. These innovations have paved the way for the refinement of inhaler devices and drug formulations, allowing for the customization of treatment options to suit individual patient characteristics and respiratory conditions. Additionally, CFD models have shown promise in the development of novel approaches for targeted drug delivery [62].

7.7 Personalized Medicine Approaches in Aerosol Inhalation Treatment

Personalized medicine through computational aerosol targeting refers to the utilization of computational modelling techniques to tailor aerosol drug delivery systems to individual patients. This approach integrates patient-specific data such as lung anatomy, physiology, and disease characteristics **for** delivery of medications directly to the affected areas of the respiratory system [62]. Key components of this approach include:

7.7.1 Patient-Specific Data Collection

Gathering detailed information about the patient's respiratory system, including lung function tests, imaging scans (such as CT scans), and clinical history.

7.7.2 Computational Modelling

Using sophisticated computational algorithms to simulate airflow patterns, deposition mechanisms, and drug dispersion within the respiratory tract. These models consider factors such as airway geometry, particle size distribution, and inhalation dynamics.

7.7.3 Optimization Algorithms

Employing optimization techniques to identify the most effective aerosol delivery strategy for each individual patient. This may involve adjusting parameters such as inhaler device settings, particle characteristics, and dosing regimens to maximize drug deposition at the target site while minimizing systemic side effects [62].

7.7.4 Validation Studies

Conduct experimental studies to validate the predictions of the computational models and ensure the accuracy of the personalized drug delivery approach.

7.7.5 Clinical Implementation

Integrating personalized aerosol targeting into clinical practice, potentially through the development of specialized inhaler devices or inhalation therapies tailored to individual patients.

The benefits of personalized medicine through computational aerosol targeting may include improved treatment efficacy, reduced side effects, and better patient adherence to therapy. By delivering medications directly to the site of disease activity, this approach has the potential to enhance the effectiveness of respiratory treatments for conditions such as asthma, chronic obstructive pulmonary disease (COPD), cystic fibrosis, and respiratory infections. Additionally, it may enable more precise dosing adjustments based on individual patient responses, leading to optimized therapeutic outcomes [62].

The future of aerosol inhalation treatment lies in personalized medicine approaches, including patients lung geometry and airflow patterns. By leveraging CFD models and experimental data, researchers can tailor drug delivery strategies to specific patient populations, ultimately enhancing treatment efficacy and improving patient outcomes. Furthermore, the integration of personalized medicine approaches with innovative aerosol inhaler devices and formulations holds great promise in revolutionizing drug delivery strategies for respiratory diseases. Utilizing CFD in aerosol inhalation treatment has opened up possibilities for refining aerosol inhaler devices and drug formulations to maximize drug delivery efficiency and minimize potential side effects [62, 63]. The modern advantages and disadvantages of inhalation devices are outlined in Table 7.2 [64].

7.8 Personalized Medicine Approaches for Tailored Treatment Plans

The personalized medicine approaches consider individual patient characteristics, such as lung geometry and airflow patterns, can significantly enhance the effectiveness of aerosol inhalation treatment. By leveraging CFD techniques to simulate and analyze the behavior of aerosol drugs in the respiratory system, personalized treatment plans can be customized to maximize efficacy and improve patient outcomes. Collaborative research efforts aimed at integrating patient-specific data into CFD models can pave the way for personalized aerosol inhalation treatment tailored to the unique needs of individual patients.

7.9 Evaluating the Efficiency of Computational Models for Pulmonary Drug Delivery

One of the important aspects of computational models is their efficiency and accuracy. Evaluating the efficiency and accuracy of computational models for pulmonary drug delivery is essential for ensuring the reliability and applicability of these models in clinical practice. Collaborative research efforts focused on comparing computational predictions with empirical data from clinical studies can provide valuable insights into the performance of these models. By systematically evaluating the efficacy of computational models, researchers and clinicians can identify areas for improvement and refine the predictive capabilities of these models [62, 63].

Furthermore, the development of standardized validation protocols for computational models can ensure consistency and reliability across different research studies and clinical applications. This can facilitate the broader adoption of computational models, ultimately leading to more effective and personalized treatment strategies for respiratory diseases. Overall, CFD approaches used for refining aerosol inhaler devices and drug formulations to maximize drug delivery efficiency and efficacy.

7.10 Leveraging Advanced Computational Techniques for Drug Delivery Optimization

In addition to evaluating the efficiency of existing computational models, there is a growing interest in leveraging advanced computational techniques to optimize drug delivery in pulmonary drug delivery. By integrating advanced modeling approaches, such as machine learning and artificial intelligence, researchers can develop predictive models that account for individual patient variability and optimize drug delivery strategies based on personalized physiological and anatomical characteristics [56–59].

These advanced computational techniques have the potential to revolutionize drug delivery. Collaborative research endeavours that combine expertise in computational modeling, pharmacology, and clinical practice can drive the advancement of these techniques and their translation into clinical applications. By harnessing the power of advanced computational techniques not only helps in more personalized and effective approach to treatment but also ultimately improving patient outcomes and quality of life [62, 63]. Complete illustrations of CFD modelling approaches and its applications are demonstrated in Table 7.3 [35, 65–68].

7.11 Optimization Strategies Using CFD Approach

This novel technique used for optimizing inhalation aerosol-based targeted drug delivery to the upper airways. By simulating airflow patterns and aerosol deposition within the respiratory tract, CFD enables the design and evaluation of inhalation devices and formulations to improve drug delivery efficiency and therapeutic outcomes. So following are the different optimization strategies using CFD approach in inhalation aerosol-based targeted drug delivery to upper airways, highlighting its applications and potential benefits [65, 69–72].

7.11.1 Understanding Airflow Dynamics

CFD simulations provide insights into airflow dynamics within the upper airways, including the nasal cavity, pharynx, and larynx. By modeling the complex geometry of the respiratory tract and airflow patterns during inhalation, CFD allows researchers to identify regions of turbulent flow, stagnant zones, and areas of high particle deposition. Understanding airflow dynamics is crucial for optimizing inhalation device design and formulation to enhance drug delivery efficiency and minimize wastage.

7.11.2 Device Design Optimization

CFD simulations facilitate the optimization of inhalation device design to achieve optimal aerosol delivery to the upper airways. By modeling factors such as airflow velocity, particle size distribution, and device geometry, researchers can evaluate the effectiveness of MDIs, DPIs and nebulizers. CFD-guided device design optimization aims to enhance aerosolization efficiency, reduce device resistance, and ensure uniform drug deposition across the upper airways [73–77].

7.11.3 Formulation Development

CFD simulations play a crucial role in optimizing inhalation aerosol formulations for targeted drug delivery to the upper airways. By modeling aerosol particle behavior, deposition patterns, and drug release kinetics, researchers can evaluate the impact of formulation parameters, such as particle size, density, and surface morphology, on drug delivery efficiency. CFD-guided formulation development enables the design of inhalable aerosol formulations that maximize drug deposition in the upper airways while minimizing systemic absorption and side effects [72–78].

7.11.4 Patient-Specific Modelling

CFD-based patient-specific modeling allows for personalized optimization of inhalation aerosol-based drug delivery to the upper airways. By integrating patient-specific anatomical data, such as airway geometry, dimensions, and airflow dynamics, researchers can simulate drug deposition patterns in individual patients. Patient-specific CFD modeling enables the customization of inhalation device settings, dosing regimens, and drug formulations to optimize therapeutic outcomes and minimize variability in drug response among patients [73–77].

7.11.5 Validation and Verification

Validation and verification of CFD simulations are essential steps in ensuring the accuracy and reliability of optimization strategies for inhalation aerosol-based targeted drug delivery. Experimental studies, such as in vitro deposition studies and in vivo pharmacokinetic studies, are conducted to validate CFD predictions and verify the efficacy of optimized inhalation devices and formulations. Integration of experimental data with CFD simulations enhances the robustness and credibility of optimization strategies and facilitates translation into clinical practice [73–77].

Optimization strategies using CFD approach offer promising opportunities to enhance inhalation aerosol-based targeted drug delivery to the upper airways. By leveraging CFD simulations to optimize device design, formulation development, patient-specific modeling, and validation studies, researchers can overcome challenges associated with aerosol drug delivery and improve therapeutic outcomes for respiratory diseases [73–78].

CFD models have numerous advantages in comparison to semi-empirical and 1D whole-lung techniques. These models are developed by solving the fundamental equations that describe the movement of particles, allowing for the accurate analysis of issues such as changing flow patterns, turbulence, the spread of particles under turbulent conditions, the change in size of particles due to moisture absorption, and the interaction among particles and surfaces in complicated shapes. CFD simulations utilize realistic airway geometry to precisely measure deposition in intricate structures [73, 74]. CFD, a scientific tool, was first used over 20 years ago [75–79] and used realistic airway geometry [82].

CFD simulations are able to directly anticipate the impact of jet and spray momentum on pharmaceutical aerosols entering the mouth-throat and upper tracheobronchial airways. CFD models are computationally expensive, difficult to resolve flow dynamics in the wide bifurcating airways, and challenging to capture pharmaceutical aerosol formation and delivery physics. Due to this, these models have limitations in the respiratory tract [83–87]. Longest and Holbrook thoroughly review CFD models of extrathoracic, tracheobronchial, and alveolar airways [81, 84, 87].

In respiratory drug delivery, CFD simulates aerosol production, optimizes inhaler design and function, and model's airway deposition advancing respiratory drug delivery technology [88, 89]. Sommerfeld and Wong [90, 91] provide excellent aerosol generation simulation and multiscale modelling overviews.

7.12 Applications of CFD in Drug Delivery

CFD has arisen as a commanding tool for optimizing inhalation aerosol-based targeted drug delivery to the upper airways. By simulating airflow patterns and aerosol deposition within the respiratory tract, CFD can improve drug delivery efficiency and the therapeutic outcomes. So, to highlighting its applications and potential benefits in drug delivery, following are key benefit [35, 65–68].

7.12.1 Device Development

CFD technique optimizes inhaler device design, improve aerosol deposition patterns, and enhance drug delivery efficiency. By analyzing factors such as particle size, inhalation flow rate, and airway geometry, CFD simulations can help identify optimal inhaler formulations and delivery techniques for targeting specific regions of the respiratory tract. Additionally, CFD modeling enables researchers to evaluate the impact of patient-specific factors, such as age, gender, and respiratory conditions, on aerosol deposition and drug efficacy. It is feasible to develop and optimize inhalation devices, such as MDIs, DPIs, and nebulizers, with the use of CFD simulations. Through the examination of airflow patterns and drug dispersion within these devices, CFD contributes to the enhancement of efficient medication delivery and performance of the devices [35, 65–68].

7.12.2 Formulation Aided Optimization

CFD modelling assists in optimizing aerosol drug formulations by predicting deposition and their behavioural patterns in the respiratory tract. Researchers can use CFD simulations to enhance drug stability, bioavailability, and targeting efficiency, leading to improved therapeutic outcomes [35, 65, 91].

7.12.3 Disease-Specific Targeting

CFD modeling allows researchers to analyze the distribution of aerosol deposition in various respiratory conditions. Tailoring drug delivery strategies based on disease-specific characteristics enhances treatment efficacy and minimizes side effects [65, 66].

CFD plays a pivotal role in advancing the principles of inhalation aerosol-based targeted drug delivery. By simulating airflow dynamics, particle behavior, and deposition mechanisms, CFD facilitates optimization of inhalation devices and drug formulations, leading to improved therapeutic outcomes and personalized treatment strategies for respiratory diseases.

7.13 Current Challenges and Future Directions in Aerosol Inhalation Treatment

Further validation of CFD models, the integration of experimental data with simulation results, and the development of more advanced and refined models required to accurately capture the complexities of aerosol behavior for inhalation treatment. Additionally, the optimization of inhaler devices and drug formulations for specific patient populations and respiratory conditions remains a challenge [55–60]. To overcome these challenges, ongoing research efforts are focused on improving the accuracy and reliability of CFD models, integrating experimental data into model development and validation, and exploring new approaches to optimize drug delivery in aerosol inhalation treatment. Furthermore, the current Aerosol Inhalation Treatment consider individual patient characteristics, such as lung geometry and airflow patterns, which can further enhance the effectiveness of aerosol inhalation treatment and improve patient outcomes in the future. By accurately predicting aerosol particles property like its deposition and transport, CFD models can provide valuable insights in effectiveness of aerosol drug delivery to the upper airways [92–94]. Through the application of advanced numerical simulations, engineering principles, and medical knowledge, CFD offers unprecedented opportunities to enhance drug delivery efficiency, improve therapeutic outcomes, and advance patient care. By addressing key challenges, fostering interdisciplinary collaborations, and promoting ethical and regulatory compliance, CFD-based approaches have the potential to revolutionize respiratory medicine and enhance the quality of life for millions of patients worldwide [59–63].

7.14 Conclusion

The advancements in the field of aerosol inhalation treatment have shown great promise in improving drug release to the upper airways and optimizing treatment for respiratory diseases. By focusing on the refinement of CFD models, the optimization of inhaler devices and formulations, and the implementation of personalized medicine approaches, significant progress has been made toward personalized treatment strategies. The validation and integration of CFD models with experimental data from clinical studies have contributed to study realistic aerosol behavior in the respiratory system. This validation process has enhanced the precision and consistency of simulation techniques, leading to more robust and clinically applicable models for predicting aerosol transport and deposition. Furthermore, the refinement of aerosol inhaler devices and drug formulations has opened up possibilities for tailored treatment options designed to maximize drug delivery efficiency while minimizing potential side effects. These advancements, combined with the potential of personalized medicine approaches, shows potential future for revolutionizing drug delivery strategies for respiratory diseases. Continuous research and innovation in this field will undoubtedly lead to a paradigm shift in personalized treatment strategies, ultimately improving the quality of care for individuals with respiratory conditions. As the future unfolds, it is imperative to continue evaluating the efficiency and accuracy of computational models for pulmonary drug delivery, ensuring that these models useful for enhancing aerosol inhalation treatment.

References

1. Sanders M. Inhalation therapy: an historical review. Prim Care Respir J. 2007;16:71–81.
2. Stein SW, Thiel CG. The history of therapeutic aerosols: a chronological review. J Aerosol Med Pulm Drug Deliv. 2017;30:20–41.
3. Maccarone C, Masiero S. Spa therapy interventions for post respiratory rehabilitation in COVID-19 subjects: does the review of recent evidence suggest a role? Environ Sci Pollut Res. 2021;28:46063–6.
4. Kostrzon M, Sliwka A, Wloch T, Szpunar M, Ankowska D, Nowobilski R. Subterranean pulmonary rehabilitation in chronic obstructive pulmonary disease. Adv Exp Med Biol. 2019;1176:35–46.
5. Zając D, Russjan E, Kostrzon M, Kaczyńska K. Inhalations with brine solution from the 'Wieliczka' salt mine diminish airwayhyperreactivity and inflammation in a murine model of non-atopic asthma. Int J Mol Sci. 2020;21:4798.
6. Mohanty RR, Das S. Inhaled insulin—current direction of insulin research. J Clin Diagn Res. 2017;11:OE01–2.
7. Klonoff DC. Afrezza inhaled insulin: the fastest-acting FDA-approved insulin on the market has favorable properties. J Diabetes Sci Technol. 2014;8:1071–3.
8. Osowicka M, Janowiak P, Gorzewska A, Lichodziejewska-Niemierko M. Inhaled opioids for cancer pain relief: a narrative review. Medicine. 2022;101:e28921.

9. De Berardis D, Fornaro M, Orsolini L, Iasevoli F, Tomasetti C, De Bartolomeis A, Serroni N, Valchera A, Carano A, Vellante F. The role of inhaled loxapine in the treatment of acute agitation in patients with psychiatric disorders: a clinical review. Int J Mol Sci. 2017;18:349.
10. Glenardi G, Handayani T, Barus J, Mangkuliguna G. Inhaled levodopa (CVT-301) for the treatment of Parkinson disease. A systematic review and meta-analysis of randomized controlled trials. Neurol Clin Pract. 2022;12:139–48.
11. Miragoli M, Ceriotti P, Iafisco M, Vacchiano M, Salvarani N, Alogna A, Carullo P, Ramirez-Rodríguez GB, Patrício T, Degli Esposti L, et al. Inhalation of peptide-loaded nanoparticles improves heart failure. Sci Transl Med. 2018;10:eaan6205.
12. Cheng YS. Mechanisms of pharmaceutical aerosol deposition in the respiratory tract. AAPS Pharm Sci Tech. 2014;15(3):630–40. https://doi.org/10.1208/s12249-014-0092-0. Epub 2014 Feb 22. PMID: 24563174; PMCID: PMC4037474
13. Carrigy NB, Ruzycki CA, Golshahi L, Finlay WH. Paediatric in vitro and in silico models of deposition via oral and nasal inhalation. J Aerosol Med Pulm Drug Deliv. 2014;27(3):149–69. https://doi.org/10.1089/jamp.2013.1075.
14. Zhu B, Traini D, Young P. Aerosol particle generation from solution-based pressurized metered dose inhalers: a technical overview of parameters that influence respiratory deposition. Pharm Dev Technol. 2015;20:897–910.
15. Sosnowski TR. Towards more precise targeting of inhaled aerosols to different areas of the respiratory system. Pharmaceutics. 2024;16(1):97. https://doi.org/10.3390/pharmaceutics16010097. PMID: 38258107; PMCID: PMC10818612
16. Kostrzon M, Sliwka A, Wloch T, Szpunar M, Ankowska D, Nowobilski R. Subterranean pulmonary rehabilitation in chronic obstructive pulmonary disease. Adv Exp Med Biol. 2019;1176:35–46. https://doi.org/10.1007/5584_2019_354.
17. Newman SP. Drug delivery to the lungs: challenges and opportunities. Ther Deliv. 2017;8:647–61. https://doi.org/10.4155/tde-2017-0037.
18. Miragoli M, Ceriotti P, Iafisco M, Vacchiano M, Salvarani N, Alogna A, Carullo P, Ramirez-Rodríguez GB, Patrício T, DegliEsposti L, et al. Inhalation of peptide-loaded nanoparticles improves heart failure. Sci Transl Med. 2018;10:eaan6205. https://doi.org/10.1126/scitranslmed.aan6205.
19. Clark AR. Half a century of technological advances in pulmonary drug delivery: a personal perspective. Front Drug Deliv. 2022;2:871147. https://doi.org/10.3389/fddev.2022.871147.
20. Cheng YS, Zhou Y, Su WC. Deposition of particles in human mouth–throat replicas and a USP induction port. J Aerosol Med Pulm Drug Deliv. 2015;28:147–55.
21. Khajeh-Hosseini-Dalasm N, Longest PW. Deposition of particles in the alveolar airways: inhalation and breath-hold with pharmaceutical aerosols. J Aerosol Sci. 2015;79:15–30. https://doi.org/10.1016/j.jaerosci.2014.09.003.
22. Sosnowski TR, Moskal A, Gradoń L. Dynamics of oro-pharyngeal aerosol transport and deposition with the realistic flow pattern. Inhal Toxicol. 2006;18:773–80. https://doi.org/10.1080/08958370600748737.
23. Moskal A, Gradoń L. Temporary and spatial deposition of aerosol particles in the upper human airways during breathing cycle. J Aerosol Sci. 2002;33:1525–39. https://doi.org/10.1016/S0021-8502(02)00108-8.
24. Li Z, Kleinstreuer C, Zhang. Particle deposition in the human tracheobronchial airways due to transient inspiratory flow patterns. J Aerosol Sci. 2007;38:625–44. https://doi.org/10.1016/j.jaerosci.2007.03.010.
25. Haddrell AE, Lewis D, Church T, Vehring R, Murnane D, Reid JP. Pulmonary aerosol delivery and the importance of growth dynamics. Ther Deliv. 2017;8:1051–61. https://doi.org/10.4155/tde-2017-0093. PMC – PubMed
26. Tamadondar MR, Salehi K, Abrahamsson P, Rasmuson A. The role of fine excipient particles in adhesive mixtures for inhalation. AICHE J. 2021;67:e17150. https://doi.org/10.1002/aic.17150.

27. Cassidy JP, Amin N, Marino M, Gotfried M, Meyer T, Sommerer K, Baughman RA. Insulin lung deposition and clearance following Technosphere® insulin inhalation powder administration. Pharm Res. 2011;28:2157–64. https://doi.org/10.1007/s11095-011-0443-4.
28. Pritchard JN. Nebulized drug delivery in respiratory medicine: what does the future hold? Ther Deliv. 2017;8:391–9. https://doi.org/10.4155/tde-2017-0015.
29. Dorosz A, Urbankowski T, Zieliński K, Michnikowski M, Krenke R, Moskal A. Inhalation profiles through a dry powder inhaler: relation between inhalation technique and spirometric measures. J Aerosol Med Pulm Drug Deliv. 2021;34:346–57. https://doi.org/10.1089/jamp.2020.1663.
30. Beck-Broichsitter M. Aerosol production by vibrating membrane technology: analysis of the electrolyte effect on generated droplet size and nebulizer output rate. J Pharm Sci. 2017;106:2168–72. https://doi.org/10.1016/j.xphs.2017.04.062.
31. Saadat M, Manshadi MKD, Mohammadi M, Zare MJ, Zarei M, Kamali R, Sanati-Nezhad A. Magnetic particle targeting for diagnosis and therapy of lung cancers. J Control Release. 2020;328:776–91. https://doi.org/10.1016/j.jconrel.2020.09.017.
32. Ógáin ON, Li J, Tajber L, Corrigan OI, Healy AM. Particle engineering of materials for oral inhalation by dry powder inhalers. I—particles of sugar excipients (trehalose and raffinose) for protein delivery. Int J Pharm. 2011;405:23–35. https://doi.org/10.1016/j.ijpharm.2010.11.039.
33. Mortazavi H, Pakzad L. The hydrodynamics and mixing performance in a moving baffle oscillatory baffled reactor through computational fluid dynamics (CFD). Processes. 2020;8:1236.
34. Moskal A, Gradoń L. Temporary and spatial deposition of aerosol particles in the upper human airways during breathing cycle. J Aerosol Sci. 2002;33:1525–39.
35. Longest PW, Holbrook LT. In silico models of aerosol delivery to the respiratory tract—development and applications. Adv Drug Deliv Rev. 2012;64:296–311.
36. Crowe CT, Troutt TR, Chung J. Numerical models for two-phase turbulent flows. Annu Rev Fluid Mech. 1996;28:11–43.
37. Das P, Nof E, Amirav I, Kassinos SC, Sznitman J. Targeting inhaled aerosol delivery to upper airways in children: insight from computational fluid dynamics (CFD). PLoS One. 2018;13(11):e0207711. https://doi.org/10.1371/journal.pone.0207711.
38. Labiris NR, Dolovich M. Pulmonary drug delivery. Part II: the role of inhalant delivery devices and drug formulations in therapeutic effectiveness of aerosolized medications. Br J Clin Pharmacol. 2003;56(6):600–12. https://doi.org/10.1046/j.1365-2125.2003.01893.x.
39. Dutta R, Spence B, Wei X, Dhapare S, Hindle M, Longest PW. CFD guided optimization of nose-to-lung aerosol delivery in adults: effects of inhalation waveforms and synchronized aerosol delivery. Pharm Res. 2020;37(10):199. https://doi.org/10.1007/s11095-020-02923-8.
40. Ruzycki CA, Javaheri E, Finlay WH. The use of computational fluid dynamics in inhaler design. Expert Opin Drug Deliv. 2013;10(3):307–23. https://doi.org/10.1517/17425247.2013.753053.
41. Laube BL, Janssens HM, de Jongh FHC, Devadason SG, Dhand R, Diot P, Everard ML, Horvath I, Navalesi P, Voshaar T. What the pulmonary specialist should know about the new inhalation therapies. Eur Respir J. 2011;37:1308–417.
42. ICRP. Human respiratory tract model for radiological protection. London: Pergamon; 1994.
43. Anjilvel S, Asgharian B. A multiple-path model of particle deposition in the rat lung. Fund Appl Toxicol. 1995;28:41–50. https://doi.org/10.1006/faat.1995.1144.
44. Ali M. Pulmonary drug delivery. In: Kulkarni V, editor. Handbook of non-invasive drug delivery systems. New York: Elsevier; 2010. p. 209–46.
45. Kleinstreuer C, Zhang Z, Li Z, Roberts WL, Rojas C. A new methodology for targeting drug-aerosols in the human respiratory system. Int J Heat Mass Transf. 2008;51(3–24):5578–89. https://doi.org/10.1016/j.ijheatmasstransfer.2008.04.052.
46. Longest PW, et al. Use of computational fluid dynamics deposition modeling in respiratory drug delivery. Expert Opin Drug Deliv. 2019;16(1):7–26.
47. Kleven M, Melaaen MC, Djupesland PG. Computational fluid dynamics (CFD) applied in the drug delivery design process to the nasal passages: a review. J Mech Med Biol. 2012;12(01):1230002. https://doi.org/10.1142/s0219519411004526.

48. Longest PW, Bass K, Dutta R, Rani V, Thomas ML, El-Achwah A, Hindle M. Use of computational fluid dynamics deposition modeling in respiratory drug delivery. Expert Opin Drug Deliv. 2019;16(1):7–26. https://doi.org/10.1080/17425247.2019.1551875.
49. Jones N. The nose and paranasal sinuses physiology and anatomy. Adv Drug Deliv Rev. 2001;51(5):5–19. https://doi.org/10.1016/S0169-409X(01)00172-7.
50. Longest PW, Landon TH. In silico models of aerosol delivery to the respiratory tract—development and applications. Adv Drug Deliv Rev. 2012;64(4):296–311.
51. Larimi MM, et al. Numerical and experimental analysis of drug inhalation in realistic human upper airway model. Pharmaceuticals. 2023;16:406.
52. Berry M, Hargadon B, Morgan A, et al. Alveolar nitric oxide in adults with asthma: evidence of distal lung inflammation in refractory asthma. Eur Respir J. 2005;25(6):986–91.
53. Aleksandra V, et al. Coupled in silico platform: computational fluid dynamics (CFD) and physiologically-based pharmacokinetic (PBPK) modelling. Eur J Pharm Sci. 2018;113:171–84.
54. Walenga RL, Babiskin AH, Zhao L. In silico methods for development of generic drug-device combination orally inhaled drug products. CPT Pharmacometrics Syst Pharmacol. 2019;8(6):359–70. https://doi.org/10.1002/psp4.12413. Epub 2019 May 21. PMID: 31044532; PMCID: PMC6618094
55. Delvadia RR, Longest PW, Hindle M, Byron PR. In vitro tests for aerosol deposition. III: effect of inhaler insertion angle on aerosol deposition. J Aerosol Med Pulm Drug Deliv. 2013;26(3):145–56. https://doi.org/10.1089/jamp.2012.0989. Epub 2012 Oct 1
56. Zhu JH, How HS, Wang DY, Lee SJ, Lee HP. Evaluation of particle deposition efficiency in human oral passage: a CFD study. In: Goh J, editor. The 15th international conference on biomedical engineering. IFMBE proceedings, vol. 43. Cham: Springer; 2014. https://doi.org/10.1007/978-3-319-02913-9_62.
57. Ostrovski Y, Dorfman S, Mezhericher M, Kassinos KC, Sznitman J. Targeted drug delivery to upper airways using a pulsed aerosol bolus and inhaled volume tracking method. Flow Turbul Combust. 2018;102(1):73–87.
58. Tian G, Hindle M, Lee S, Longest PW. Validating CFD predictions of pharmaceutical aerosol deposition with in vivo data. Pharm Res. 2015;32(10):3170–87.
59. Longest PW, Tian G, Khajeh-Hosseini-Dalasm N, Hindle M. Validating whole-airway CFD predictions of DPI aerosol deposition at multiple flow rates. J Aerosol Med Pulm Drug Deliv. 2016;29(6):461–81. https://doi.org/10.1089/jamp.2015.1281. Epub 2016 Apr 15. PMID: 27082824; PMCID: PMC5220557
60. Li B, Feng Y. In silico study to enhance delivery efficiency of charged nanoscale nasal spray aerosols to the olfactory region using external magnetic fields. Bioengineering. 2022;9:40.
61. Feng Y, Chen X, Yang M, Dong K. Multiscale computational models for respiratory aerosol dynamics with medical applications. Comput Math Methods Med. 2019;2019:1–2.
62. Islam N, Cleary MJ. Developing an efficient and reliable dry powder inhaler for pulmonary drug delivery – a review for multidisciplinary researchers. Med Eng Phys. 2012;34(4):409–27.
63. Usmani OS, Singh D, Spinola M, et al. The prevalence of small airways disease in adult asthma: a systematic literature review. Respir Med. 2016;116:19–27.
64. Jose RJ, Oliver AN. The importance of inhaler adherence to prevent COPD exacerbations. Med Sci. 2019;7:54. https://doi.org/10.3390/medsci7040054.
65. Darquenne C. Aerosol deposition in health and disease. J Aerosol Med Pulm Drug Deliv. 2012;25(3):140–7.
66. Dolovich MB, Dhand R. Aerosol drug delivery: developments in device design and clinical use. Lancet. 2011;377(9770):1032–45.
67. Leung TF, Liu TC, Lam CW, Wong GW. Asthma control, inhaled corticosteroids adherence, and airway inflammation evaluation using electronic monitoring devices. J Asthma. 2013;50(6):652–9.
68. Ehrmann S, Roche-Campo F, Sferrazza Papa GF, Isabey D, Brochard L, Apiou-Sbirlea G. Aerosol therapy during mechanical ventilation: an international survey. Intensive Care Med. 2013;39(6):1048–56.

69. Longest PW, Tian G. Computations of aerosol deposition in the extrathoracic airways of children and adolescents. J Aerosol Sci. 2011;42(12):781–99.
70. Xi J, Kim J, Si XA, Zhou Y. Application of computational fluid dynamics in inhalation drug delivery: a critical review. Int J Comput Fluid Dyn. 2020;34(4):307–39.
71. Donovan MJ, Kim KJ, Johnson M, Laube BL. A simple technique to predict regional deposition of aerosols in the human nose. J Aerosol Med. 2005;18(3):294–302.
72. Kleinstreuer C, Zhang Z. Airflow and particle transport in the human respiratory system. Annu Rev Fluid Mech. 2010;42:301–34.
73. Tannehill JC, Anderson DA, Pletcher RH. Computational fluid mechanics and heat transfer. 2nd ed. Washington: Taylor and Francis; 1997. [Google Scholar]
74. Patankar S. Numerical heat transfer and fluid flow. CRC press; 1980. [Google Scholar]
75. Balashazy I, Hofmann W. Particle deposition in airway bifurcations-II. Expiratory flow. J Aerosol Sci. 1993;24:773–86. [Google Scholar]
76. Finlay W, Stapleton K, Yokota J. On the use of computational fluid dynamics for simulating flow and particle deposition in the human respiratory tract. J Aerosol Med. 1996;9(3):329–41. [Google Scholar]
77. Kimbell JS, Gross EA, Joyner DR, et al. Application of computational fluid dynamics regional dosimetry of inhaled chemicals in the upper respiratory tract of the rat. Toxicol Appl Pharmacol. 1993;121:253–63. [PubMed] [Google Scholar]
78. Comer JK, Kleinstreuer C, Hyun S, et al. Aerosol transport and deposition in sequentially bifurcating airways. J Biomech Eng. 2000;122(2):152–8. [PubMed] [Google Scholar]
79. Martonen TB, Guan XF. Effects of tumors on inhaled pharmacologic drugs II. Particle motion. Cell Biochem Biophys. 2001;35(3):245–53. [PubMed] [Google Scholar]
80. Xi J, Longest PW, Martonen TB. Effects of the laryngeal jet on nano- and microparticle transport and deposition in an approximate model of the upper tracheobronchial airways. J Appl Physiol. 2008;104(6):1761–77. [PubMed] [Google Scholar]
81. Longest PW, Vinchurkar S, Martonen TB. Transport and deposition of respiratory aerosols in models of childhood asthma. J Aerosol Sci. 2006;37:1234–57. [Google Scholar]
82. Sznitman J, Heimshch T, Wildhaber JH, et al. Respiratory flow phenomena and gravitational deposition in a three-dimensional space-filling model of the pulmonary acinar tree. J Biomech Eng. 2009;31:031010. [PubMed] [Google Scholar]
83. Lambert AR, O'Shaughnessy PT, Tawhai MH. Regional deposition of particles in an image-based airway model: large-eddy simulation and left-right lung ventilation asymmetry. Aerosol Sci Technol. 2011;45:11–25.
84. Inthavong K, Choi L-T, Tu J, et al. Micron particle deposition in a tracheobronchial airway model under different breathing conditions. Med Eng Phys. 2010;32:1198–212.
85. Lin C-L, Tawhai MH, McLennan G, et al. Characteristics of the turbulent laryngeal jet and its effect on airflow in the human intra-thoracic airways. Respir Physiol Neurobiol. 2007;157:295–309.
86. Choi J, Tawhai M, Hoffman EA, et al. On intra- and intersubject variability of airflow in the human lungs. Phys Fluids. 2009;21:101901.
87. Li Z, Kleinstreuer C, Zhang Z. Particle deposition in the human tracheobronchial airways due to transient inspiratory flow patterns. J Aerosol Sci. 2007;38:625–44. [Google Scholar]
88. Cui Y, Sommerfeld M. Forces on micron-sized particles randomly distributed on the surface of larger particles and possibility of detachment. Int J Multiphase Flow. 2015;72:39–52. [Google Scholar]
89. Cui Y, Sommerfeld M. Application of lattice-Boltzmann method for analysing detachment of micron-sized particles from carrier particles in turbulent flows. Flow Turbul Combust. 2018;100(1):271–97. [Google Scholar]
90. Sommerfeld M, Schmalfuß S. Numerical analysis of carrier particle motion in a dry powder inhaler. J Fluids Eng. 2016;138(4):041308. [Google Scholar]

91. Wong W, Fletcher DF, Traini D, et al. The use of computational approaches in inhaler development. Adv Drug Deliv Rev. 2012;64(4):312–22. [PubMed] [Google Scholar]
92. Anderson JD. Computational fluid dynamics: the basics with applications. McGraw-Hill Education; 1995.
93. Versteeg HK, Malalasekera W. An introduction to computational fluid dynamics: the finite volume method. 2nd ed. Pearson Education Limited; 2007.
94. Dutta R, Spence B, Wei X, Dhapare S, Hindle M, Longest PW. CFD guided optimization of nose-to-lung aerosol delivery in adults: effects of inhalation waveforms and synchronized aerosol delivery. Pharm Res. 2020;37:20. https://doi.org/10.1007/s11095-020-02923-8.

Chapter 8
Pharmaceutical Inhalation Compounds Development by Using In Silico Modeling Tools

Astha Patel, Sakshi Parab, Srishti Gupta, Pavan Kumar Chintamaneni, and S. N. Koteswara Rao G.

Contents

A. Patel · S. Parab · S. Gupta · S. N. K. Rao G. (✉)
Shobhaben Pratapbhai Patel School of Pharmacy and Technology Management, SVKM's NMIMS, Vile Parle (W), Mumbai, Maharashtra, India

P. K. Chintamaneni
GITAM Institute of Pharmacy, Rudraram, Hyderabad, Telangana, India

S. N. K. Rao G., R. R. Alavala (eds.), *Applications of Computational Tools in Drug Design and Development*, https://doi.org/10.1007/978-981-96-4154-3_8

Abstract In silico modeling has transformed the design of pharmaceutical inhalation compounds into a potent platform to simulate and optimize the delivery systems in drug design for inhalation therapy. These computational methods provide accurate predictions of drug behavior within the respiratory tract from particle deposition patterns to aerosol dynamics and pharmacokinetic-pharmacodynamic (PK-PD) profiles. This approach accelerates drug development by reducing reliance on extensive experimental trials, optimizing formulation and device efficiency, and ultimately enhancing the targeted delivery of therapies for respiratory infections and conditions. This chapter explores the major of inhalation therapy which includes optimization of particles size, innovation of various aerosolization techniques, and advancements in inhalation devices such as dry powder inhalers, and smart inhalers. It also focuses on the importance of computational fluid dynamics in modeling, aerosol behavior and its provision for an inclusion of patient-specific factors in drug deposition. It covers multifaceted challenges related to formulation stability, regulatory compliance, and device design which interleave between drug properties, mechanisms of delivery, and adherence to treatment. It goes further to discuss emerging trends in the use of artificial intelligence (AI), machine learning (ML), and hybrid in silico models to improve the accuracy of predictions and to individualize treatment approaches. The potential of in silico modeling to overcome classic hurdles of drug development to induce innovation, and ultimately facilitate the development of more effective, patient-centered inhalation therapies is important. This approach promises to redefine the scientific pharmaceutical landscape in terms of respiratory health through precision engineering and cutting-edge technology.

Keywords In silico modeling · Inhalation therapy · Simulation · Computational modeling · Drug development · Pulmonary

Abbreviations

ADME	Absorption, distribution, metabolism, and excretion
AI	Artificial intelligence
API	Active pharmaceutical ingredient
APSD	Aerodynamic particle size distribution
ARDS	Acute respiratory distress syndrome
CADD	Computer-aided drug design

CAGR	Compound annual growth rate
CF	Cystic fibrosis
CFD	Computational fluid dynamics
CFPD	Computational fluid-particle dynamics
CiPA	Comprehensive in vitro proarrhythmic assay
COPD	Chronic obstructive pulmonary disease
DPI	Dry powder inhalers
EGFR	Epidermal growth factor receptor
EMA	European Medicines Agency
European REACH legislation	European Union's Regulation on Registration, Evaluation, Authorization and Restriction of Chemicals
FDA	Food and drug administration
FEV1	Forced expiratory volume
FPF	Fine particle fraction
GMP	Good manufacturing practices
IPF	Idiopathic pulmonary fibrosis
IVIVCs	In vitro/in vivo correlations
JORRP	Juvenile onset recurrent respiratory papillomatosis
MDI	Metered-dose inhalers
MIC	Minimum inhibitory concentration
ML	Machine learning
MMAD	Mass median aerodynamic diameter
MPPD	Multiple-path particle dosimetry
NGI	Next-generation impactor
OIDPs	Orally inhaled drug products
PBPK	Physiologically based pharmacokinetic
PD	Pharmacodynamics
PK	Pharmacokinetics
PSD	Particle size distribution
QSP	Quantitative systems pharmacology
RNAs	Ribonucleic acid
TOBI	Tobramycin

8.1 Introduction

The integration of in silico modeling has significantly accelerated the advancement of inhaled pharmaceutical development. These computational methodologies have a powerful framework for simulating and predicting drug behaviour within the respiratory system. This approach is not only fast, but it also solves the complex issues related to inhaled therapy, like aerosol dynamics, particle deposition profiles, and pharmacokinetic parameters. The traditional experimental methods are lengthy and resource intensive. Consequently, in silico methodologies are gaining prominence to improve and accelerate the drug development paradigm [1]. Inhalation drug delivery systems are an advanced discipline of pharmaceutical technology that permits the direct delivery of drugs through the respiratory tract. The increasing incidence of such systems is because they have the potential to allow a rapid action of therapy and preferentially deliver medications to specific sites within the respiratory tract. This targeted approach presents many advantages in the treatment of respiratory diseases like asthma and chronic obstructive pulmonary disease (COPD). There is a wide range of inhalation drug delivery systems, including metered dose inhalers, dry powder inhalers, nebulizers, and others [2]. Inhaled pharmaceutical delivery in silico modeling refers to using computational methods to simulate and predict the behavior of inhaled drugs within the respiratory tract. This has been the most important approach for optimization of drug formulations and devices, thus resulting in the improvement of patient outcomes as well as reduction in development cost and time [3].

8.1.1 Importance of Inhalation Therapies

In the recent years there has been an increase in the importance of inhalation therapy in the pharmaceutical drug industry. The reason for this trend is the improvement in technology, an increased incidence rate of pulmonary diseases, and the need for correct drug delivery systems. Inhalation therapy offers significant advantages over traditional methods of administration, such as injection and oral drugs. As a result, many forms of respiratory diseases require this type of therapy. Inhalation therapy offers a targeted approach to drug delivery, directly to the lungs, minimizing systemic exposure and maximizing therapeutic effects within the respiratory system [4]. The growing incidence of chronic respiratory diseases is the major cause for the expansion of the inhalation therapies market. The latest available market research reports reflect an enormous global market for pulmonary drug delivery systems. Pulmonary drug delivery systems reached USD 53.24 billion in 2023. Reports have shown that the industry would continue to grow strongly and have a compound annual growth rate of 4.6% through the year 2030. This expanding market is a result not only of increased occurrences of respiratory disorders across the globe but also the increased needs for better targeted treatment models that can be delivered with the aid of inhalation therapy [5]. This route of administration has several key advantages: rapid onset of action, improved patient compliance, and fewer systemic side effects. The faster onset of action associated with inhaled drugs compared to oral or parenteral routes results from the direct delivery of the therapeutic agent to the pulmonary system. This avoids first-pass metabolism by the liver and gastrointestinal degradation, so it reaches the target site faster and the therapeutic effect occurs sooner. This advantage is particularly important in managing acute exacerbations of respiratory disorders such as asthma and COPD, where rapid relief of symptoms is important for maintaining the patient's well-being. These newer inhalation devices, for instance dry powder inhalers (DPIs) and metered-dose inhalers (MDIs), tend to emphasize friendliness towards patients and portability. Usage monitoring and reminders on medication would strengthen the patient's adherence to their prescribed schedules further by smart technology integrated in the device [6]. By focusing on the lungs directly, inhalation therapies have lower systemic side effects as associated with oral or injectable drugs. Because it targets one area in the body, the effect is usually seen through localized delivery with reduced dosage, minimizing potential side effects [7]. The COVID-19 pandemic has underscored critical need for effective inhalation therapies by hastening research and development in this area. Urgency to develop treatments that are administered via inhalation fuelled the research into aerosolized medications in managing the viral infections, particularly viral infections of the respiratory tract. For instance, inhaled nitric oxide and antiviral agents were some of the novel therapeutic interventions investigated for treating COVID-19, showing the versatility and importance of inhalation therapies in emerging global health challenges. Elevated levels of attention have therefore not only progressed our understanding of inhalation delivery of infectious diseases but also been a gateway to future application in pandemic preparedness and response [6]. Recent developments with inhalation technology have greatly pushed ahead the increasing relevance of the treatment. Dry powder inhalers have become extremely popular in the administration of inhaled drugs, especially due to their user-friendliness and their ability to provide exact drug dosing in a dry powder form, thus not requiring propellants like in the case of MDIs. This is helpful for patients who might not be able to coordinate actuation of an MDI with inhalation. Newer DPI technologies also include breath-actuated mechanisms. These mechanisms are designed to release medication only when the patient inhales. This ensures that a consistent dose is delivered independent of the patient's flow rate, overcoming a disadvantage of earlier DPI models which were more dependent on inspiratory effort. This improves the reliability and effectiveness of DPI for a greater number of patients, including those with weaker inspiratory capacity [7]. Smart inhalers with electronic sensors and connectivity enable real-time monitoring of the administration of medication and patterns of adherence in patients. This is a significant source of clinical value, particularly for those patients who suffer from chronic respiratory diseases and, therefore, have difficulties with consistent adherence to treatment. The pulmonary delivery of biologics stands for an emerging area of research and development. These complex therapeutic molecules, conventionally administered via parenteral routes, are now being formulated for inhalation, offering the potential for enhanced therapeutic efficacy within the respiratory system and a concomitant reduction in systemic adverse effects [8]. The growing importance of inhalation therapies in the pharmaceutical world is underscored by their distinct advantages in treating respiratory pathologies, concurrent technological innovation in delivery systems, and escalating demand driven by changing public health imperatives. Ongoing research and development in this field position inhalation therapies to assume an even more pivotal role in modern medical practice.

8.1.2 Challenges in the Development of Inhalation Products

The greatest hurdle in developing inhalation drug products lies in aligning the different expectations of the multiple stakeholders-including patients, healthcare providers, payers, and regulatory authorities-where each group has different expectations about efficacy, safety profile, usability, and cost. The formulation of a cohesive product specification that successfully reconciles such competing demands is essential but often complicated by the need for rigorous multidisciplinary collaboration involving operational, marketing, clinical research, and manufacturing teams [9]. Formulation stability is another important challenge, which requires the formulation to be stable with the integrity and efficacy of the drug over the shelf life of the product, besides being compatible with the chosen inhalation device. It demands careful choice of excipients and optimization of manufacturing processes to avoid degradation and ensure consistent delivery of the drug [10]. Device design also must focus on user-friendliness and ergonomic factors to maximize patient adherence and therapeutic outcomes. More complex devices are likely to lead to suboptimal use, especially in patient populations that have limited dexterity or suffer from cognitive impairment. Hence, intuitive device operation, clear feedback mechanisms such as visual or auditory cues, and robust design all come together to ensure correct and consistent drug delivery, which contributes toward improved patient outcomes [2]. Navigating the intricate regulatory landscape represents another major challenge in developing inhaled drug products. These drugs are held to strict safety and efficacy requirements by regulatory bodies like the FDA (U.S. Food and Drug Administration) and EMA (European Medicines Agency). Adherence to these guidelines requires extensive preclinical and clinical testing, coupled with rigorous validation processes, which can significantly prolong development timelines and add significantly to costs. Moreover, the scaling-up of laboratory-scale synthesis and formulation to compliant Good Manufacturing Practices (GMP) production is a delicate scaling task. This increases the complexity of cost management, process optimization, and operational efficiency because of the requirement for specialized equipment that can manage both small-scale clinical trial materials and large-scale commercial manufacturing. The scale-up process requires careful consideration of critical process parameters, rigorous process validation, and robust quality control measures to ensure consistent product quality and compliance with regulatory requirements throughout the product lifecycle [11].

The dosimetry of inhaled pharmaceuticals quantifying the dose delivered to the lungs and characterizing its later behavior in the respiratory system is fundamental to therapeutic efficacy. Measuring the fraction of drug that reaches the site of action and figuring out disposition within the complex anatomy of the respiratory tract requires a sophisticated in silico approach, integrated with robust experimental approaches both in vitro and in vivo. These methods try to illuminate the complex interplay of aerosol dynamics, mechanisms of particle deposition, and dissolution/absorption processes within the lung environment. Furthermore, inter-patient variability in inhalation technique can include factors such as inspiratory flow rate, tidal volume, and breath-holding duration that introduce considerable variability in the delivered dose and deposition pattern. This variability can lead to suboptimal therapeutic outcomes due to under-treatment or, conversely, increase the risk of adverse effects due to overexposure. Therefore, careful consideration of patient-specific factors and device design features that minimize the impact of variable inhalation technique are crucial for ensuring consistent and predictable drug delivery to the target site [10]. Patient adherence to prescribed inhalation therapies is critical to optimal therapeutic outcome; however, it remains a common clinical challenge. Suboptimal adherence is multifactorial, partly due to the inherent complexity of many inhalation devices. Effective use of a metered-dose inhaler, for example, requires precise coordination between actuation and inhalation, a skill that can be particularly difficult for elderly patients, children, and patients with physical or cognitive impairments. The resulting improper administration then minimizes pulmonary drug delivery with a subsequent decrease in therapy efficacy. Thus, all patients must be educated appropriately on the proper use of the device to ensure proper treatment benefits. Lack of training can lead to flawed administration techniques, including inhaling at an inappropriate rate, holding breath for the wrong amount of time or manipulating the device incorrectly, all which impact on drug delivery and clinical benefits. As a result, such ongoing education should include clear instructions, demonstrations, and regular follow-up assessments to take care of the difficulties a patient faces [3]. Although the inhalation pharmaceutical formulations present quite significant therapeutic benefits for treatment of respiratory diseases, it is encumbered with a complex array of diverse challenges, from meeting users' needs to satisfying strictly regulatory requirements. All such challenges need to be overcome using innovative device design, effective formulation strategies, and through comprehensive patient education programs that will help the continued advancement and success of inhalation therapies within the pharmaceutical market.

8.1.3 Role of In Silico Modeling in Drug Development

In silico modeling involves the use of computational simulations and algorithms to model biological processes, predict drug-target interactions, and optimize various stages of pharmaceutical development. This computational approach has gained widespread adoption within the pharmaceutical industry because it can reduce development costs, accelerate timelines, and enhance the precision of drug discovery and development processes. In silico methodologies include computer-aided drug design (CADD), molecular modeling that deals with the study of molecular structures and their interactions, and quantitative systems pharmacology (QSP) that combines physiological and pharmacological data to predict the behavior of drugs in complex biological systems. These methodologies are used to simulate biological systems in various levels of complexity: from molecular interactions to the physiology of whole organs, so that the behavior of preclinical and clinical drug candidates can be predicted. Further capability has been added to the capabilities of in silico modeling through the integration of paradigms in artificial intelligence (AI) and machine learning (ML), with their capability to analyse large complex datasets such as genomics, proteomics, and clinical data to develop sophisticated predictive models that inform decision-making in drug development from target identification and validation through lead optimization to preclinical testing, clinical trial design, and even post-market surveillance [12]. Figure 8.1 describes the role of this approach in improving drug development of inhalation compounds. In silico modeling is particularly useful at the early stages of drug discovery, by which means potential therapeutic candidates are identified based on predicted interactions with specific biological targets. Techniques such as virtual screening-a computational method based on molecular docking and scoring algorithms-allow the rapid screening of enormous chemical libraries, containing thousands or even millions of compounds. This high-throughput screening approach reduces the number of compounds that require subsequent experimental investigation, thereby streamlining the drug discovery process and reducing associated costs. In addition, in silico methods can be used to predict key pharmacological properties of candidate molecules, such as absorption, distribution, metabolism, and excretion (ADME) profiles, as well as potential toxicity liabilities. Early priority toward compounds with predicted favourable properties by in silico modeling allows researchers to focus available resources on the most promising candidates, making later steps in preclinical and clinical development more efficient and thus more likely to be successful. This early screening of the potential drug candidate also determines a potential off-target effect as well as provides information needed for designing more selective therapeutics that are safer [13, 14]. With these methodologies, comes the great prospect for enhancing the optimality of designs for clinical trials through patient specific models computational and virtual cohorts. Since these methods help in simulating patient response to the different treatment schedules, right dosing times may be decided, a best endpoint of the study could be identified, or an optimal population of enrolled patients. This predictive possibility would even allow the streamlining of inclusion/exclusion criteria toward a greater prospect of meeting significant effects of

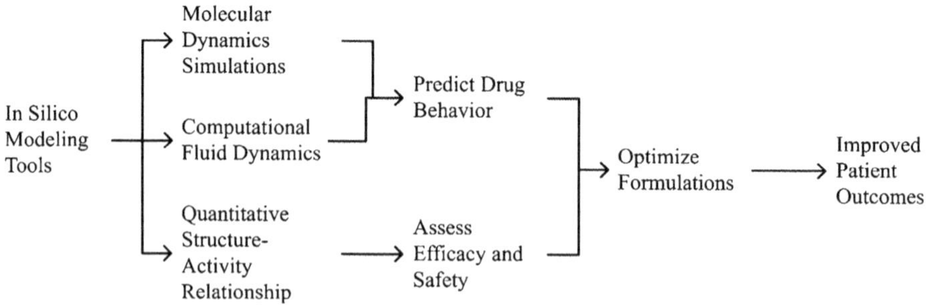

Fig. 8.1 In silico modeling in drug development

treatment through statistical power. Significantly, using virtual patients to create synthetic control arms can provide an attractive avenue for making a study more precise at the cost of minimizing the actual number of participants needed for a trial. This, in turn, reduces trial costs, speeds up timelines, and can mitigate concerns over ethical considerations associated with exposing patients to placebo or suboptimal treatments. Besides, in silico modeling can be used to detect possible sources of variability in patient response and can be used for designing more robust and efficient clinical trials [15].

In silico approaches make it easier to build predictive models for several pharmacokinetics (PK) and pharmacodynamic (PD) properties of drug candidates. By simulating ADME processes, researchers could get valuable information about the probable efficacy and safety profile of these compounds before entering resource-intensive clinical trials. These computational models predict such key PK parameters as bioavailability, clearance, volume of distribution, and half-life, but also important PD parameters regarding drug-target interactions and downstream effects. This predictive capability can provide informed decisions in lead optimization, dose selection, and trial design to increase the probability of success in the subsequent clinical development stages while reducing the risk of costly failures [16]. Ethical considerations in animal testing have pushed a large amount of shift towards the adoption of in silico approaches as useful alternatives or supplementary methods to the traditional preclinical studies. Through these methodologies, which make use of computational models to enlighten potential toxicological risks, such as off-target effects and drug-drug interactions, fewer animals need to be used for experimentation. This approach aligns with the principles of the 3Rs (Replacement, Reduction, and Refinement) in animal research, supporting more ethical and humane research practices while also providing an even more efficient and cost-effective approach to early safety assessment. In silico models can predict various toxicological endpoints, such as genotoxicity, cardiotoxicity, and hepatotoxicity, which allow researchers to prioritize safer drug candidates for further development and minimize reliance on animal models [17]. In silico modeling provides an effective drug repurposing approach, identifying existing drugs that are approved and may display therapeutic efficacy against new diseases based on their established mechanism of action and target interaction. This computational strategy will significantly accelerate development timelines versus de novo drug discovery as repurposed drugs already have been extensively preclinically and clinically evaluated in terms of safety, along with characterization of their pharmacokinetic and pharmacodynamic properties. By using computational tools like molecular docking, virtual screening, and network pharmacology, researchers can identify new potential indications for existing drugs. This will reduce the time and resources needed to get new treatments to patients. This approach has been most valuable in addressing emerging health crises and rare diseases, where rapid development of effective therapies is crucial [18].

8.2 Key Challenges in Pharmaceutical Inhalation Compound Development

The development of inhaled pharmaceutical formulations poses unique challenges within the pharma industry, mainly because this is an area where a complex interplay exists between drug formulation, device engineering, and regulatory constraints. The translational path from concept to market approval is inherently complex and potentially inhibitive to innovation as well as costly for developing.

8.2.1 Particle Size and Distribution for Pulmonary Deposition

The inherent complexity of inhaled drug products arises from the intricate interplay of formulation and delivery mechanisms, necessitating careful consideration of several critical factors during development. Particle size perhaps is one of the most significant physicochemical parameters, which impacts the pulmonary deposition. The particle size greatly affects their deposition within the respiratory tract. The nanoparticles with a size smaller than 1 μm mainly stay in the exhaled breath and have minimum lung penetration. The optimum lung deposition is shown by particles in the size range of 1–3 μm, where nearly 50–60% of particles reach the alveoli, which enables systemic absorption. Particles of sizes between 3 and 5 μm tend to be deposited in the smaller airways. Conversely, particles larger than 5 μm in size are primarily captured in the upper airways or oropharynx by inertial impaction, resulting in reduced therapeutic effectiveness and increased risk for toxic effects [19]. Optimal pulmonary delivery, according to much research, is achieved with particles that have a Mass Median Aerodynamic Diameter (MMAD) of 2.5–3 μm. This particle size range is favorable for efficient deposition in the peripheral lung regions while concurrently minimizing deposition in the oropharynx. Oropharyngeal deposition can lead to the development of localized adverse effects, such as oral thrush or voice changes, particularly when using inhaled corticosteroids. However, particles that are excessively oversized lead to impaction at high locations in the respiratory airways, limiting passage and deposition to target tissues while particles that are too small, likely will be exhaled rather than deposited, minimizing or impairing therapeutic activity [20]. This synergistic interplay of mechanisms for deposition underlines the critical importance to be placed on an optimized PSD. This optimization is therefore essential to achieve maximal pulmonary deposition while at the same time minimizing drug loss and the potential for systemic side effects [21]. Achieving a narrow particle size distribution (PSD) is a crucial element of successful development for inhalable formulations, although this comes with serious challenges. Processes of milling and spray drying used in manufacture lead to variations in the PSD of particles formed; it will eventually affect drug delivery efficiency. Besides, forces of inter particulates will lead to particle agglomeration that would make the effective size alter during dispersion through inhalation. Consequently, the design of inhalation devices, especially dry powder inhalers, needs to be taken into careful consideration to ensure that enough airflow is generated to de-agglomerate particles and ensure optimal delivery to the target lung regions [22].

8.2.2 Aerosolization and Formulation Optimization

Aerosolization may be defined as the processing of liquid or solid drug substances to a suspension of fine-particle size in an aerial medium, also known simply as an aerosol. The performance of inhalation therapies thus strongly depends on the tightly controlled regulation of both size and distribution within this aerosol. Effective pulmonary administration requires the formation of the particles in a size range, presumably between 1 and 5 μm. Sizes larger than 5 μm may have poor ability to penetrate the lung parenchyma, while sizes of less than 1 μm are likely to exit the lungs before reaching the part of the lung where effective delivery can take place. Ensuring the right particle size is therefore essential for optimized therapeutic effect [23]. The physical characteristics of the aerosolized particles, such as shape and surface properties, greatly influence their behavior during aerosolization and subsequent deposition within the respiratory tract. For example, irregularly shaped particles may be predisposed to aggregation, thereby compromising dispersion and potentially lowering overall delivery efficiency. There is a wide range of inhalation devices, including MDIs and DPIs, which work in separate ways to generate aerosols. MDIs generally use propellants to help generate the aerosol, whereas DPIs rely on the inspiratory airflow of the patient to deliver the powdered medication. Each of these delivery systems poses specific challenges to device design and the degree of patient compliance [24]. Optimizing formulations for inhalation delivery systems involves a complex process with considerations of several interacting factors. Such factors include the physicochemical properties of the API, such as solubility, stability, and interactions with excipients. Poorly soluble APIs often require unique formulation strategies to achieve the necessary lung deposition. The role of excipients in controlling formulation performance is crucial for controlling drug stability, aggregation behavior, and particle size distribution. The selection of proper excipients should be based on the compatibility of the chosen active pharmaceutical ingredient (API) and their effect on the aerosolization characteristics of the final product. Processes in manufacturing, such as spray drying and micronization, significantly affect the final product. These techniques need to be optimized to ensure consistent particle size and morphology while not compromising the stability of the API. Finally, compatibility of the formulation with the delivery device is critical. The design of an inhaler, such as the resistance of a dry powder inhaler, can affect delivery efficiency and usability by the patients. As such, knowing the interaction between formulation and device characteristics is essential when developing effective inhalation drugs [25].

8.2.3 *Deposition Patterns in the Lungs*

Deposition of inhaled particles within the lungs is a multifactorial process influenced by several physical mechanisms that mainly include sedimentation, inertial impaction, and diffusion. The interplay between particle size, shape, density, and the airflow dynamics in the respiratory tract determines these mechanisms. Sedimentation dominates the smaller airways, because at these locations, airway flow velocity is lower. At these locations, sedimentation of particles onto the surfaces of the airways, through gravity, occurs easily. Due to inertial impaction more substantial particles, with an aerodynamic diameter greater than 5 μm, tend to depart from the airstream due to inertia and impact the airway walls, especially where the airways bifurcate and in the larger airways. High concentration of particles, typically much smaller than 0.1 μm, move at random by diffusion, mostly caused by collision with surrounding gas molecules. This process supports particle deposition in the lungs' alveolar regions. Particles which have an aerodynamic diameter falling between 2 and 5 μm are considered ideal to be administered by inhalation; these can successfully reach every zone of the lungs-respiratory as well as conducting-allowing proper efficacy [26, 27]. Other factors that affect deposition of inhaled particles in the lungs include breathing patterns. The rate and depth of inhalation significantly affect the deposition of particles. A rapid inhalation favours deposition in the central airways due to increased inertial impaction. Slow, deep inhalation, on the other hand, promotes peripheral distribution through sedimentation. Breath-holding at the end of inspiration can also enhance deposition by allowing more time for sedimentation and diffusion to occur. Variations in the anatomy of the airway structure include different branching angles and diameters. Pathological conditions may also change the geometry of the airways and alter airflow dynamics due to bronchoconstriction, inflammation, and mucus hypersecretion, resulting in heterogenous distribution of drugs in the lungs. Some inhaled particles are hygroscopic; this means that they can absorb some of the moisture from their environment, the humid lung environment. Hygroscopic growth increases particle size, which affects the deposition mechanism and may increase deposition in certain lung areas. Temperature can also come into play since it influences particle size as well as on airflow characteristics [28].

The respiratory system has inherent defence mechanisms, including mucociliary clearance and immune responses, intended to remove foreign particles that have been inhaled. Such mechanisms can considerably hinder drug delivery by eliminating therapeutic agents prematurely from the lungs. Formulating inhalable drugs requires precise control over the particle size to ensure proper deposition in the desired lung region. Maintaining a particle size within the optimal range and simultaneously ensuring drug stability and efficacy is a significant technical challenge. Additionally, the formulation should be amenable to efficient aerosolization without compromising the integrity of the particles [29]. The design of inhalation devices directly affects the effective delivery of drugs. It must be well formulated and optimized for drug compositions and specific patient populations. Thus, variation in device performance tends to lead to drug delivery inconsistencies, making the inclusion of consideration of device characteristics within the scope of formulation development necessary. Patient use of prescribed inhalation technique is an important determinant for the success of treatment. Suboptimal use of inhalation technique can minimize lung delivery of drugs, regardless of formulation or device quality. It, therefore, identifies a role for patient education or training on appropriate use of devices [30]. Pulmonary deposition patterns are crucially important to the development of effective inhaled pharmaceuticals, regarding understanding complex patterns of airways and patient adherence. Improvement of effectiveness and clinical outcomes of inhalation therapy can be considerably enhanced through addressing the complexities and challenges involved in particle behavior, airway physiology, device performance, and patient adherence.

8.2.4 Drug Release Kinetics and Therapeutic Outcomes

Drug release kinetics means the rate at which a drug is liberated from its delivery system and becomes available to be absorbed and subsequently, for pharmacological action, at the target site. Formulations intended to be administered by inhalation require special consideration of the physicochemical properties of the drug, such as solubility, particle size distribution, and chemical stability. Controlled release strategies are often used to maximize therapeutic effectiveness by sustaining drug concentrations within the therapeutic window for as long as possible. Such strategies minimize fluctuation in drug levels, thus reducing dosing frequency and possibly improving compliance among patients. Various techniques are used to achieve controlled release, including incorporation of polymers for the modulation of drug release rates or designing formulations that have mucoadhesive properties, thus resisting mucociliary clearance and prolonging residence time in the lungs [31]. Mathematical modeling is an essential step for characterizing and predicting drug release kinetics from inhaled formulations. The use of several mathematical models, such as zero-order, first-order, Higuchi, and Korsmeyer-Peppas models, helps to describe the release profiles. The models are used to give insights into the mechanisms that govern drug release and influence formulation parameters. Zero-order release is associated with the constant rate of drug release with time, and first-order release is determined by a rate proportional to the concentration of the remaining drug within the formulation. The Higuchi model relates the drug release of the matrix system to be dependent on Fickian diffusion. The Korsmeyer-Peppas model is frequently applied in drug release analysis from polymeric systems and can indicate various mechanisms involved, including diffusion and erosion. By accurately modeling drug release kinetics, formulation scientists can optimize formulations toward achieving desired pharmacokinetic profiles, including specific rates and durations of release. This maximizes therapeutic efficacy and minimizes adverse effects [32]. The inherent anatomical and physiological characteristics of the respiratory system significantly influence drug deposition and absorption. The aerodynamic diameter of the inhaled particles is important in determining the place of deposition. Larger ones are mainly impacted in the upper respiratory tract, but smaller particles can reach to the deeper lung regions. Yet, these smaller particles are also subjected to rapid systemic absorption, which may lead to poor therapeutic concentrations at the desired site within the lung parenchyma [33].

8.3 In Silico Modeling Techniques in Inhalation Compound Development

In silico modeling techniques have significantly enhanced the development of inhalation drug delivery systems, which are critical for the treatment of respiratory diseases. Researchers can simulate various aspects of drug formulation and delivery using these techniques, optimize inhaler designs, and predict drug behavior in the lungs. This chapter discusses several key areas of in silico modeling relevant to inhalation compound development.

8.3.1 Formulation Modeling

Estimation of pulmonary drug delivery in experimental animals using in silico methods is very challenging because current technologies are unable to predict both the deposition and absorption of drugs in animal lungs. Some pharmacokinetic modeling tools, like GastroPlus™ and Mimetikos Preludium™, can estimate these parameters in human lungs but are not applicable to animal respiratory tracts, such as those of rats. Dosimetry models such as the Multiple-Path Particle Dosimetry model can estimate aerosol deposition in animals, including mice, rats, rabbits, rhesus monkeys, and pigs. However, deposition data from MPPD cannot be used straight away in software like GastroPlus™ to simulate pulmonary drug absorption because these two dosimetry models represent lung sub-regions differently. MPPD divides the deposition into pulmonary/alveolar and tracheobronchial regions. GastroPlus™ would require four regions: thoracic, extra-thoracic, bronchiolar, and alveolar. The presence of GastroPlus accounts for the probable swallowing of drugs deposited in the extra-thoracic region; that further complicates cross-software integration [34].

8.3.2 Computational Fluid Dynamics (CFD) for Aerosol Behavior

Computational Fluid Dynamics (CFD) is a modeling approach based on physics to predict fluid and particle behavior within realistic geometries. The method is especially useful in investigating aerosol dynamics in the human respiratory system during inhalation as shown in Fig. 8.2. CFD can forecast particle deposition and dispersion conditions for various flow conditions because it considers complicated interactions between particles and airflows.

The latest improvement has increased its ability to model size distribution and deposition patterns more accurately in different lung regions [35]. Computational Fluid Dynamics (CFD) provides an essential advantage in the design of generic orally inhaled drug products (OIDPs) by predicting the regional deposition fraction of the initial dose in various target regions. It considers the device-specific factors such as geometry, aerodynamic particle size distribution (APSD), spray angle, velocity, and orifice diameter. These predictions are preliminary screening values to assess the in vivo performance of drug-device combinations. However, prediction of both local and systemic drug concentrations needs a supplementary approach. The physiologically based pharmacokinetic, a compartmental technique, makes use of formulation and physiological parameters to predict pharmacokinetics. The integration of CFD with PBPK offers an overall prediction of an OIDP's in vivo performance through the combination of formulation and device characteristics [36]. Computational Fluid Dynamics (CFD) has become a more powerful and promising alternative approach to traditional methods of analysis for drug deposition patterns. However, CFD predicts accurately only when drug properties, respiratory geometry, and breathing conditions are modelled precisely. Development of an IVIVC that relies on the respiratory anatomy varies with inhalation patterns of the patient, the nature of the disease, as well as MDI dose distribution is a very trying task. Advanced CFD models are needed to capture the complicated interactions between devices and patient-specific factors. The commonly used Euler-Lagrange fluid-particle simulation approach in respiratory aerosol studies typically fails to consider the momentum exchange between discrete particles and the carrier fluid, which is an important consideration for enhancing model accuracy [37].

CFD Modeling ⟶ Predict Particle Behavior ⟶ Assess Device Factors ⟶ Regional Deposition Prediction ⟶ Integrate PBPK ⟶ Evaluate In Vivo Performance

Fig. 8.2 Workflow of computational fluid dynamics

8.3.3 Optimization of Inhaler Devices

Another significant application of in silico modeling is optimization of inhaler devices. The CFD can evaluate various design parameters, for example, nozzle geometry, airflow rates, and particle size distributions to enhance inhaler performance. Such an approach would allow fast prototyping and testing of new inhaler designs, reducing the number of trials and the development time without physical experiments, which helps save on costs [38].

8.3.4 Particle Size Distribution and Aerosol Performance

Particle size distribution, or PSD, is considered a critical factor that may influence the efficiency of delivering drugs to the lungs and hence determine aerosol performance in inhalation therapies. Particle size determines their deposition patterns in the respiratory tract. Optimal therapeutic outcomes require the inhaled particles to usually fall within the range of 1–5 μm. This size range can avoid the upper respiratory tract and to get into the peripheral lung areas where they can act with pharmacological effects. Particles greater than 5 μm tend to deposit in the upper airways or oropharynx. Conversely, particles less than 1 μm tend to be exhaled and often do not reach the desired site [39]. In silico modelling has been used to simulate effects of PSD variations on deposition profiles and therapeutic effects. By using these models, alterations in the size distribution of aerosolized particles change their behaviour during inhalation and subsequent deposition in different lung regions. Studies revealed that optimizing PSD improves deposition and bioavailability of lung drugs after inhalation. For example, studies have shown that particles having an aerodynamic diameter between 1 and 3 μm are most effective with reduced side effects since such sized particles are more likely to reach the alveoli for systemic absorption [39].

Other factors influencing aerosol performance include the inspiratory flow rate and the airway geometry. For instance, variations in the inspiratory flow rates impact the mass of fine particles as well as the general drug delivery efficiency from the inhalers. For example, increases in the flow rate of inhalation improve the delivered mass of drugs as well as the fraction of the fine particle emanating from dry powder inhalers (DPIs). This is important because a higher fine particle fraction correlates with better peripheral lung deposition. Further complications arise in the relationship between PSD and aerosol performance because of the physical properties of carrier particles used in formulations. Optimization of these properties, such as shape, surface roughness, and density, has been demonstrated to enhance aerosol performance considerably. For instance, smaller carrier particles have been correlated with improved aerosol performance due to their ability to generate a more favorable PSD for lung delivery.

8.3.5 Lung Deposition and Drug Delivery Modeling

Modeling lung deposition is therefore of paramount importance in explaining the effectiveness of a drug delivery to its target site in the respiratory system. The ability of aerosolized particles to navigate through the complex anatomy of the lung determines the effectiveness of inhalation therapy, wherein their size and velocity highly affect the deposition pattern. In silico models have emerged as powerful tools in this regard, enabling researchers to simulate the complex structures of the respiratory system and predict where aerosolized particles are likely to deposit. The lungs consist of a branching network of airways that can be modelled using various computational techniques, including computational fluid dynamics and stochastic models. Models developed with these techniques could be used to replicate anatomical features of the respiratory tract, such as the bifurcations and varied diameters of airways, which are critical to predicting particle behavior. Studies indicate that CFD simulations help in the elaboration of detailed insights regarding how variations in inhalation strategies will affect aerosol deposition into the airway tree and into the alveolar regions. With the modelling of airflow and particle dynamics, researchers may identify appropriate delivery strategies for various formulations of drugs with enhanced efficacy and reduced risk of undesirable side effects. One of the notable approaches is hybrid multiscale lung models that include detailed anatomical representations but simplified flow dynamics. This model can make efficient simulations at different scales, from large conducting airways down to the alveolar level. For instance, a numerical method was reported to predict particle deposition in a holistic model of the whole respiratory tract, which includes patient-specific upper airways and bifurcation units, which are adjustable to model different generations of airways. Such detail allows the researcher to study how the patient's anatomy impacts the outcome of drug delivery, particularly in the development of personal inhalation therapies [40].

8.3.6 Pharmacokinetic and Pharmacodynamic Modeling

In the in silico approaches in drug development with inhalation therapy, PK and PD modeling go to great depths to establish knowledge. They enable an outline to explain the drug performance of behavior after inhaling and which is essential in ensuring proper therapeutic efficacy coupled with safety. The ADME of inhaled drugs can be forecasted by pharmacokinetic modeling. The pharmacokinetics of inhaled drugs faces unique challenges since drugs should reach the lungs in an adequate concentration to achieve an effect. PK modeling gives the concentration of the drug at its site of action over time, which is essential for evaluating therapeutic efficacy. For instance, one study presented a simplified-geometry model that includes a mixed compartmental and spatially distributed representation of inhaled bronchodilators. Such a model enables researchers to compute airflow responses to treatment for virtual patients, producing results that align closely with observed clinical outcomes such as FEV1 profiles. This close alignment underlines the utility of this model in predicting how different dosages and administration regimens may influence drug action in a patient-specific manner. Pharmacodynamic modeling, on the other hand, deals with the biological effect of drugs on the body and how these effects change with time. PD models consider the connection between a drug's pharmacological response, including receptor binding and subsequent biological effects, and its concentration. PK-PD model integration enables researchers to explain both how much drug is brought to the lungs and to an extent, how well the drug interacts with its receptor sites to produce an appropriate therapeutic effect. This will be particularly important for drugs administered by inhalation: adequate unbound concentrations need to be achieved at the effect site for drugs used in this manner to function properly [41]. This integration of PK and PD modeling with formulation data and computational fluid dynamics (CFD) simulations provides a complete understanding of the behavior of a drug from inhalation to systemic circulation. For example, physiologically based pharmacokinetic (PBPK) models use anatomical and physiological characteristics to predict drug exposure in humans without depending on clinical data alone. These models consider the size of the organ, the blood flow rates, and the physicochemical properties of the drug inhaled. By simulating these interactions, researchers can predict variations in formulation or delivery method that can impact drug absorption and efficacy [42]. Furthermore, advances in mathematical modeling have enhanced our ability to analyze the complex interplay of pharmacokinetics and pharmacodynamics in inhaled treatments. For instance, very recently, it has been demonstrated that knowledge of kinetics of inhaled corticosteroids can improve efficacy by optimizing factors such as particle size and receptor binding affinity. This kind of knowledge is invaluable in tailoring therapies according to individual patient needs such that medications are both efficacious and safe [43].

8.4 Case Studies and Applications

8.4.1 Nebulization of Antibiotics for Cystic Fibrosis

Cystic fibrosis is a genetic disorder where the mucus produced is thick and sticky, clogging airways and leading to chronic lung infections. One of the biggest challenges in the management of cystic fibrosis is recurrent infection by Pseudomonas aeruginosa, a bacterium that thrives in the thick mucus and contributes to decline in lung function [44]. To counteract these infections, nebulized antibiotics have become a fundamental part of CF management as they deliver medications directly to the site of infection. Nebulization is one of the most effective ways of administering antibiotics in CF because it delivers drugs directly to the endobronchial sites where infections occur. This is more favourable as the local concentration of the antibiotics would be higher than the systemic exposure, reducing the likelihood of side effects from systemic administration. A study reports that for efficient action, the concentration of the antibiotics in bronchial secretions should at least be ten times more than the MIC to allow penetration through bacterial biofilms and result in bactericidal action. Among the antibiotics used for nebulization, tobramycin is one of the best studied and most widely used. Tobramycin solution for inhalation (TOBI) has been shown to significantly enhance lung function and reduce the frequency of pulmonary exacerbations in patients with chronic Pseudomonas aeruginosa infections. In a landmark study of 520 patients, patients receiving TOBI showed significant improvements in respiratory function as early as 2 weeks into treatment, with continued benefits through the course of the trial. Nebulized tobramycin also reduced hospitalization rates and decreased the need for intravenous antibiotics. Improvements in nebulization technology have further enhanced treatment results. Modern nebulizers are devised to follow and adapt the breathing pattern of patients. For example, intelligent nebulizers make aerosol output based on real-time feedback from inhalation flow rate, maximize drug delivery efficiency. Technological advancements that improve adherence during the treatment of a patient are when therapy becomes much more efficient and friendly [45].

8.4.2 Optimizing Stability and Performance Using In Silico Models

In silico modeling has now become an integral part of optimization of stability and performance in inhalation therapies, especially in nebulized formulations and DPIs. Such models make it easier to understand how formulation parameters impact drug delivery, stability, and therapeutic efficacy overall. The researchers can discover optimal formulations and delivery systems that help improve patient outcome by simulating different conditions and variables. In silico models allow the consideration of a wide range of formulation variables without the time-consuming and costly laboratory tests. For instance, through simulation, researchers can identify how the stability and performance of nebulized formulations depend on the excipients used, pH levels, or osmotic pressures. This predictive capability is particularly useful in designing lipid nanoparticles for drug delivery by nebulization. An example has shown that the optimization of composition, molar ratios, and structure of lipid nanoparticles can improve their performance significantly in delivering therapeutic RNAs to the lungs. The researchers found optimal formulations that enhanced the efficiency of drug delivery while maintaining stability by evaluating different clusters in chemical space [46]. The role of in silico modeling is also important in the assessment of aerosol performance. It has been demonstrated that, for example, droplet size and flow rate greatly affect the fine particle fraction (FPF) that is delivered to the lungs. Through simulation of various inhalation scenarios, one is able to predict the changes in drug deposition to various lung regions based on variations in these parameters. One study demonstrated how changes in flow rates during nebulization affected aerosol deposition patterns at different stages of a next-generation impactor (NGI), thus providing a more detailed understanding of how to optimize nebulizer performance. Modeling can also be used to identify potential adverse reactions associated with specific excipients or formulation components. This way, by simulating different physiological conditions, researchers can predict how changes in formulation might lead to issues such as bronchospasm or coughing during administration.

8.5 Challenges and Limitations of In Silico Modelling

One of the primary challenges in in silico modelling is the quality and the availability of the data. The accuracy of the in-silico models depends upon the fidelity of the input data. For example, in case of drug -induced nephrotoxicity, in order to improve the predictive models, the interdisciplinary collaboration and integration of multiple data sources is essential [47]. Similarly, the data possesses significant obstacles to the clinical translation of in silico models in the realm of cancer modelling which leads to limited amount of pertinent validation data [48]. Another significant challenge is balancing the model complexity with interpretability. Higher predictive power can be achieved by complex models, but they can be difficult to validate and interpret. In case of toxicology, this is considered to be relevant, where there occur no unique definition for a 'good model', and a balance between predictability and complexity is considered to be necessary [49]. Emphasizing the need for models which are both interpretable and accurate, the link between predictive capability and model complexity is highlighted, in case mechanical ventilation modelling [50]. The validation as well as verification of in silico models is considered to be a critical yet challenging task. To ensure their accuracy and reliability, the models must be validated against the clinical gold standards. This is considered to be a major hurdle in case of clinical application of in silico models, observed in the verification of study of the intracardiac blood flow models [51]. As noted in the prediction of splice-altering single nucleotide variants, the lack of large scale evaluation studies, leads to the further complication of the validation process [52]. Despite their potential value in the prediction of the chemical properties for the regulatory purposes, regulatory acceptance of in silico models is limited. For instance, the promotion of the use of alternative methods by the European REACH legislation, although the practical application remains constrained due to the acceptance issues and various stakeholders' influences [53]. Often in silico models requires significant time and computational resources, which can limit their practical applicant. To overcome the limitations, tailoring of the computational approaches to specific applications with relatively low complexity is necessary, as demonstrated in the study of left ventricular haemodynamic [54].

8.5.1 Complexities of Respiratory Biology and Lung Dynamics in In Silico Modeling

Due to the multiscale nature of the lung anatomy and physiology, in silico modelling of the respiratory system is complex. The intricate structure of the human lungs with vast airway networks leads to challenges in accurately stimulating aerosol deposition and airflow. The computational model must account for hierarchal architecture of the lungs, which spans for multiple scales from the microscopic cellular level to macroscopic organ level [55–57]. The complexities are further compounded by the need in order to incorporate the anatomical realism, which is seen in case of models that aim in order to attempt to capture the heterogenicity of the acinar environment [58]. By affecting the respiratory flow phenomenon, another layer of complexity arises from the variability of the inter-subject in the lung morphology. By supporting the advancements of the personalize medicines, the variability necessitates the development of the hybrid in silico models. Such models are considered crucial for understanding the disease dynamics such as chronic respiratory conditions and for optimize the inhalation therapies [56, 59, 60]. The respiratory system's dynamic nature, which is characterized by the interaction of gas and liquid phases and significant changes in the volume during breathing, possesses additional challenges. The in silico models must capture theses non-linear interactions in order to predict the emergent phenomenon which includes the dynamic force length relationship of the smooth muscles which are present in the airways and the spatially heterogenous ventilation during the bronchoconstriction. These models are considered to be essential in order to develop the therapeutic strategies as well as to understand the complex mechanisms of the diseases [56]. The in silico models are increasingly integrated with experimental data such as imaging and physiological measurements, for the enhancement of their accuracy and the predictive power. Validation of models against the clinical data and further supporting the development of the patient specific stimulations the allowances are provided by integration. For example, in case of acute respiratory distress syndrome (ARDS) the models which are tailored to the individual respiratory mechanics can be used to predict for the outcome in such situations [57, 60, 61].

8.5.2 Model Validation and Reliability

Ensure credibility and applicability in regulatory and clinical settings the model validation and reliability studies are needed in the in silico model which needs validation processes that includes, experimental data comparison, usage of various validation matrixes and adherence to the good modelling practices. Model validation is a critical process to ensure the accuracy and reliability of the computational models. Predict the outcomes with accuracy the validation involves the comparing of the model outputs with the experimental or real-world data to assess the model's ability.

For example, validation is performed in order to compare the model outputs to the data which is present in the literature, which can be either experimental or computational in order to ensure the model's reliability in for the prediction of the biomechanical behaviours in case of spinal facet joints [62]. Invitro tests are used in the stimulation results for comparing the development of the coronary stents, ensuring that the model is accurately replicated [63]. Reliability in in silico modelling refers to the consistency and dependency of the model's predictions across various datasets and scenarios. In order to assess the credibility of the model it is considered to be extremely crucial for models which are generally used in case of regulatory settings [64]. For the enhancement of reliability of the model, different validation matrixes and methods are used, as demonstrated in the validation of the mathematical model of EGFR+ lung adenocarcinoma, whereby a combination of statistical methods was used in order to ensure robust predictions [65]. Additionally, in order to assess the reliability of the models, the reliability estimation methods and hypothesis testing, are used which helps in order to understand the performance of the model under different conditions [66].

8.5.3 Data Availability and Quality

Data availability is an extremely critical factor in case of the development and application of the in-silico models. Particularly in case of ADME (absorption, distribution, metabolism, excretion), to build accurate and robust models it is essential to build high quality datasets. Impacting the utility for the model development the publicly available datasets, such as those which are used in case of the ADME properties, vary significantly in terms of the presentation and information content [67]. The availability of the experimental data is considered to be a major factor for the utilization of the in silico models, because it allows for the refinement and the validation of the predictive models [68]. For the enhancement of the quality of the in silico models both commercial as well as public databases are used in the integration of the data, although these databases often contains heterogenous data from different laboratories sources [69]. By influencing the predictive performance of the models the data quality which is used in case of the in silico modelling is considered to be paramount. Setting an upper limit as well as highlighting the importance of the reliable data certain types of quality assessments, experimental uncertainty needs to be managed [70]. The quality of the dataset, the use of algorithms as well as the endpoints which are available for the predictions of the accuracy of the in silico tools in case of the ADMET predictions are considered to be largely important [71]. For the enhancement of the model utility with emphasizing the development of the machine learning approaches in the in silico modelling the need for the high quality internal data as well as tailored descriptors is considered to be important [72].

8.5.4 Regulatory Acceptance and Future Adoption in Drug Development

With the help regulatory bodies such as United States Food and Drug Administration (USFDA) and European Medicines Agency (EMA) in silico modelling has gained increasing recognition in case of the drug development. In order to accept the evidence by these agencies which is obtained through stimulation and modelling as a part of the regulatory submission process, provides that the methods are considered to be deemed credible as well as "qualified" [64, 73]. A rigorous amount of validation, verification is required in order to assess the credibility of the in silico models, in order to enhance the drug's safety and efficacy [64, 73, 74]. The Comprehensive In Vitro Proarrhythmic assay (CiPA) was used in order to understand how the in silico models are being integrated into the regulatory framework in order to assess the proarrhythmic risk, which ultimately demonstrates the potential of these models which further influences the regulatory guidelines [75]. For the enhancement of the model's integration and credibility into the regulatory process the adoption in the in-silico modelling in the drug development is promising. Reducing the need for the extensive human and animal testing, can ultimately lead to the development of the virtual cohort in in silico trials and patient-specific models, thereby streamlining the drug development process [76, 77]. In order to improve the predictive capabilities of the in silico models, the advancement in case of the artificial intelligence (AI) and machine learning (ML) are expected, in order to make more reliable for the decision-making in case of the regulatory [78]. For the enhancement of the accuracy and reliability the expansion of databases, computational power are considered to be important in case of the in silico predictions, which eventually leads to the facilitation of the broader acceptance and their use in the drug development studies [79].

8.6 Future Directions and Conclusion

In silico modelling is a significant advancement across various domains. For enhancement in case of the reliability and predictability of the ADME/T models, the integration of the big data and system sciences are responsible, which are considered to be extremely crucial in case of the drug discovery and drug development [80]. Also for the prediction of the complex pharmacokinetics and pharmacodynamic behaviours, the use of artificial intelligence (AI) and machine learning (ML) is considered for the improvement in the accuracy of the models [79]. The development of hybrid-modelling framework with a combination of the agent-based and continuum-based methods is considered in order to provide comprehensive insights into the complex biological processes such as tumour growth and cancer development [81, 82]. In case of the real-world scenarios The gap between in vitro and in vivo stimulations, especially in the case of drug delivery systems, is considered to be another important area in order to focus, leading to the enhancement in the applicability of the in silico stimulations [83].

8.6.1 Emerging Trends in In Silico Modeling for Inhalation Therapy

For the stimulation of the respiratory airflow and in case of aerosol deposition, recent advancements in case of computational capabilities have enabled to develop a multiscale in silico drug model which handles the integration of the computational fluid-particle dynamics (CFPD). Now these models rather than focusing on the patient-based applications they are used to study the inter subject variability as well as for designing of the efficient inhalation therapies for larger population [55]. In silico modelling is also used to optimize the size and shape of the aerosol which helps in order to improve the drug delivery. For example, for the enhancement of the deposition efficiency fibre-shaped aerosols have been used in the respiratory tract, which consists of specific geometric concentrations showing promising results to help in targeting the different regions of the lungs. Now this approach helps to understand the design of the aerosols with largest payload capacity and maximum drug delivery efficiency [84]. High-resolution in silico models are used in developing the local aerosol deposition with unknown accuracy. These models can help in stimulating the entire lung system, which ultimately accounts for individual variations in case of the morphology of the lungs and in case of diseased states, such as idiopathic pulmonary fibrosis (IPF) [85]. For the enhancement in the development of the inhaled drug products the in-silico models are increasingly being integrated with the help of in vivo and in vitro data. This approach helps to identify the major factors such as drug absorption and deposition, which further leads to the facilitation of the design of the formulation following the optimized release as well as the rate of absorption [86, 87]. In order to understand the disease-specific changes in the lung structure and function, in silico models are also being used. For example, the models have been developed to stimulation of the effects of chronic obstructive pulmonary diseases (COPD) and asthma on the drug delivery, which leads to the understanding on how these conditions are responsible to alter the aerosol deposition patterns. This is extremely essential for developing effective therapies for the patients suffering from the respiratory diseases [88].

8.6.2 Integration of AI and ML in Drug Development

The integration of Artificial Intelligence (AI) and machine learning (ML) in the drug development for the inhalation therapy is a revolutionary field which is accounted by the enhancement in the personalization as well as precision in the treatment delivery. For the improvement in the targeting of the inhaled medications to a specific site within the respiratory system the optimization of the drug delivery systems with smart inhalers. The use of machine learning empowered smart inhaler was observed in case of the juvenile onset recurrent respiratory papillomatosis (JORRP) which was able in order to perform optimization of the drug delivery system by adjusting the position of the inhaler nozzle and diameter which was based on the patient specific data, thereby understanding the increased efficiency of the drug deposition into the targeted areas such as the parts which are larynx and glottis [89]. In context of the inhalation therapy, the AI-driven approaches are considered to be valuable in order to estimate the ability of the model and predict the pharmacokinetics and pharmacodynamics of the drug candidate, which is considered to be crucial for the optimization and delivery of the drug formulation [90]. The integration of AI in this field not only accelerates the development process but also reduces the cost and the need for extensive animal testing by providing more accurate predictions and drug behaviours in humans [91, 92].

8.6.3 In Silico Modelling in Personalized Medicine

The integration of in silico models with personalized medicine approaches and pharmacogenomics leads to the prediction of individual patient responses to the inhaled therapies. The pharmacokinetics and pharmacodynamics models of the drugs, helps to tailor the treatment to the environmental and generic factors which affect each patient [93]. The personalized approach aims at delivering the right drug at the right dose and at the right time, to enhance the treatment efficacy and minimization of side effects [94].

8.6.4 Future of Inhalation Compound Development

The future of pharmaceutical inhalation compound development is poised to benefit significantly from the advancement in the in-silico modelling tools. The use of modelling tools is being increased to stimulate various pharmacokinetic processes such as absorption, distribution, metabolism and excretion (ADME), which are crucial to predict the plasma levels of the drug which is inhaled as well as the bioavailability. In silico models such as GastroPlus, have shown promising results in the prediction of the pharmacokinetics profiles of the inhaled drug by the stimulation of the systemic absorption and lung deposition, which are critical for the development of the effective inhalation therapies [39, 95, 96]. Through the integration of multiscale biophysics into the in-silico pharmacology is expected for the enhancement of the precision of these models. This approach of stimulation allows the complex biological interactions at various stages from the molecular to the organ level. Such advancements could revolutionize the way in which the development of the inhalation compounds takes place, offering more personalized and therapeutic options [97]. Despite the advancements, the challenges remain in case of the accurate prediction of the pharmacokinetics and pharmacodynamics for the inhaled substances due to the complexities in the physiology of the lungs, variability in lung permeability and deposition [96]. However, the continuous involvement of the computational models, which includes the use of big data analytics and machine learning, is expected in order to overcome the hurdles, thereby providing more reliable predictions and facilitation of the drug development process [80, 98].

References

1. Ibrahim M, Verma R, Garcia-Contreras L. Inhalation drug delivery devices: technology update. Med Devices (Auckl) [Internet]. 2015 Feb 12 [cited 2024 Dec 24];8:131. Available from: https://pmc.ncbi.nlm.nih.gov/articles/PMC4334339/
2. Sou T, Bergström CAS. Contemporary formulation development for inhaled pharmaceuticals. J Pharm Sci [Internet]. 2021 Jan 1 [cited 2024 Dec 24];110(1):66–86. Available from: http://jpharmsci.org/article/S0022354920305001/fulltext
3. Bhattacharyya S, Sogali BS. Inhalation therapy – approaches and challenges. Asian J Pharm Clin Res [Internet]. 2018 Apr 1 [cited 2024 Dec 24];11(4):9–16. Available from: https://journals.innovareacademics.in/index.php/ajpcr/article/view/24117
4. Magramane S, Vlahović K, Gordon P, Kállai-Szabó N, Zelkó R, Antal I, et al. Inhalation dosage forms: a focus on dry powder inhalers and their advancements. Pharmaceuticals [Internet]. 2023 Nov 28 [cited 2024 Dec 24];16(12):1658. Available from: https://www.mdpi.com/1424-8247/16/12/1658/htm
5. Pulmonary Drug Delivery Systems Market Report Size, 2030 [Internet]. [cited 2024 Dec 24]. Available from: https://www.grandviewresearch.com/industry-analysis/pulmonary-drug-delivery-systems-market
6. Canto Mangana J, Schilder KA, Bretones-Pedrinaci JI, Blesa ARM, de Medina FS, Martínez-Augustin O, et al. A perspective current and past modes of inhalation therapy. Microb Biotechnol [Internet]. 2024 Feb 1 [cited 2024 Dec 24];17(2):e14419. Available from: https://pmc.ncbi.nlm.nih.gov/articles/PMC10883785/
7. Inhalable Drugs Market Size to Worth Around USD 65.51 Bn by 2034 – BioSpace [Internet]. [cited 2024 Dec 24]. Available from: https://www.biospace.com/press-releases/inhalable-drugs-market-size-to-worth-around-usd-65-51-bn-by-2034
8. Banat H, Csóka I, Paróczai D, Burian K, Farkas Á, Ambrus R. A novel combined dry powder inhaler comprising nanosized ketoprofen-embedded mannitol-coated microparticles for pulmonary inflammations: development, in vitro–in silico characterization, and cell line evaluation. Pharmaceuticals [Internet]. 2024 Jan 1 [cited 2024 Dec 24];17(1):75. Available from: https://www.mdpi.com/1424-8247/17/1/75/htm
9. Why are inhalation devices so difficult to develop? – ONdrugDelivery [Internet]. [cited 2024 Dec 25]. Available from: https://ondrugdelivery.com/why-are-inhalation-devices-so-difficult-to-develop/
10. Naik E. Challenges and opportunities in the development of inhalable formulations for pulmonary drug delivery. Asian J Pharm Technol Innov [Internet]. 2023 Feb 28 [cited 2024 Dec 25];11(40):1–2. Available from: https://www.asianpharmtech.com/articles/challenges-and-opportunities-in-the-development-of-inhalable-formulations-for-pulmonary-drug-delivery-97671.html
11. Hickey AJ. Emerging trends in inhaled drug delivery. Adv Drug Deliv Rev. 2020;157:63–70.

12. Al-Mohaya M, Mesut B, Kurt A, Çelik YS. In silico approaches which are used in pharmacy. J Appl Pharm Sci [Internet]. 2024 Apr 5 [cited 2024 Dec 25];14,(4):239–53. Available from: https://japsonline.com/abstract.php?article_id=4213&sts=2
13. In Silico Modeling: Accelerating drug development – Patheon pharma services [Internet]. [cited 2024 Dec 25]. Available from: https://www.patheon.com/us/en/insights-resources/blog/accelerating-drug-development-with-in-silico-modeling.html
14. Inhalation Aerosols. Inhalation aerosols 2019 Mar 21.
15. Piñero J, Furlong LI, Sanz F. In silico models in drug development: where we are. Curr Opin Pharmacol [Internet]. 2018 Oct 1 [cited 2024 Dec 25];42:111–21. Available from: https://pubmed.ncbi.nlm.nih.gov/30205360/
16. Chang Y, Hawkins BA, Du JJ, Groundwater PW, Hibbs DE, Lai F. A guide to in silico drug design. Pharmaceutics [Internet]. 2022 Jan 1 [cited 2024 Dec 25];15(1):49. Available from: https://pmc.ncbi.nlm.nih.gov/articles/PMC9867171/
17. Mamoshina P, Bueno-Orovio A, Rodriguez B. Dual transcriptomic and molecular machine learning predicts all major clinical forms of drug cardiotoxicity. Front Pharmacol [Internet]. 2020 May 21 [cited 2024 Dec 25];11:504317. Available from: https://clue.io/cmap
18. Brogi S, Ramalho TC, Kuca K, Medina-Franco JL, Valko M. Editorial: in silico methods for drug design and discovery. Front Chem. 2020;8:612.
19. Labiris NR, Dolovich MB. Pulmonary drug delivery. Part I: Physiological factors affecting therapeutic effectiveness of aerosolized medications. Br J Clin Pharmacol [Internet]. 2003 Dec 1 [cited 2024 Dec 26];56(6):588–99. Available from: https://onlinelibrary.wiley.com/doi/full/10.1046/j.1365-2125.2003.01892.x
20. Four Challenges for Pulmonary Drug Delivery – pharma excipients [Internet]. [cited 2024 Dec 25]. Available from: https://www.pharmaexcipients.com/news/four-challenges-for-pulmonary-drug-delivery/
21. Hinds WC. measuring particle size distribution in pharmaceutical aerosols. 2006 Oct 23;424.
22. Negi A, Nimbkar S, Moses JA. Engineering inhalable therapeutic particles: conventional and emerging approaches. Pharmaceutics [Internet]. 2023 Nov 30 [cited 2024 Dec 26];15(12):2706. Available from: https://www.mdpi.com/1999-4923/15/12/2706/htm
23. Labiris NR, Dolovich MB. Pulmonary drug delivery. Part II: The role of inhalant delivery devices and drug formulations in therapeutic effectiveness of aerosolized medications. Br J Clin Pharmacol [Internet]. 2003 Dec [cited 2024 Dec 26];56(6):600. Available from: https://pmc.ncbi.nlm.nih.gov/articles/PMC1884297/
24. Haddrell AE, Lewis D, Church T, Vehring R, Murnane D, Reid JP. Pulmonary aerosol delivery and the importance of growth dynamics. Ther Deliv [Internet]. 2017 Dec 1 [cited 2024 Dec 26];8(12):1051–61. Available from: https://www.tandfonline.com/doi/abs/10.4155/tde-2017-0093
25. Shaji J, Shaikh M. Current development in the evaluation methods of pulmonary drug delivery system. Indian J Pharm Sci [Internet]. 2016 May 26 [cited 2024 Dec 26];78(3):294–306. Available from: https://www.ijpsonline.com/articles/current-development-in-the-evaluation-methods-of-pulmonary-drug-delivery-system.html
26. Fernández Tena A, Casan Clarà P. Deposition of inhaled particles in the lungs. Arch Bronconeumol [Internet]. 2012 Jul 1 [cited 2024 Dec 26];48(7):240–6. Available from: https://www.archbronconeumol.org/en-deposition-inhaled-particles-in-lungs-articulo-S1579212912000845
27. Heyder J. Deposition of inhaled particles in the human respiratory tract and consequences for regional targeting in respiratory drug delivery. Proc Am Thorac Soc [Internet]. 2004 [cited 2024 Dec 26];1(4):315–20. Available from: www.atsjournals.org
28. Newman SP. Drug delivery to the lungs: challenges and opportunities. Ther Deliv [Internet]. 2017 Jul 1 [cited 2024 Dec 26];8(8):647–61. Available from: https://pubmed.ncbi.nlm.nih.gov/28730933/
29. Clarà PC, Jerez FR, Ramírez JB, González CM. Deposition and clinical impact of inhaled particles in the lung. Arch Bronconeumol. 2023;59(6):377–82.

30. Chrystyn H. Methods to identify drug deposition in the lungs following inhalation. Br J Clin Pharmacol [Internet]. 2001 [cited 2024 Dec 26];51(4):289. Available from: https://pmc.ncbi.nlm.nih.gov/articles/PMC2014454/
31. Hye T, Moinuddin SM, Sarkar T, Nguyen T, Saha D, Ahsan F. An evolving perspective on novel modified release drug delivery systems for inhalational therapy. Expert Opin Drug Deliv [Internet]. 2023 [cited 2024 Dec 26];20(3):335. Available from: https://pmc.ncbi.nlm.nih.gov/articles/PMC10699164/
32. Anderson S, Atkins P, Bäckman P, Cipolla D, Clark A, Daviskas E, et al. Inhaled medicines: past, present, and future. Pharmacol Rev [Internet]. 2022 Jan 1 [cited 2024 Dec 26];74(1):48–118. Available from: https://pharmrev.aspetjournals.org/content/74/1/48
33. Why Are Inhalation Devices So Difficult To Develop? – ONdrugDelivery [Internet]. [cited 2024 Dec 26]. Available from: https://ondrugdelivery.com/why-are-inhalation-devices-so-difficult-to-develop/
34. Shi C, Ignjatović J, Liu T, Han M, Cun D, Đuriš J, et al. In vitro – in vivo – in silico approach in the development of inhaled drug products: nanocrystal-based formulations with budesonide as a model drug. Asian J Pharm Sci [Internet]. 2021 May 1 [cited 2024 Dec 23];16(3):350. Available from: https://pmc.ncbi.nlm.nih.gov/articles/PMC8261257/
35. Ruzycki CA, Javaheri E, Finlay WH. The use of computational fluid dynamics in inhaler design. Expert Opin Drug Deliv [Internet]. 2013 Mar [cited 2024 Dec 23];10(3):307–23. Available from: https://pubmed.ncbi.nlm.nih.gov/23289401/
36. Walenga RL, Babiskin AH, Zhao L. In silico methods for development of generic drug–device combination orally inhaled drug products. CPT Pharmacometrics Syst Pharmacol [Internet]. 2019 Jun 1 [cited 2024 Dec 23];8(6):359. Available from: https://pmc.ncbi.nlm.nih.gov/articles/PMC6618094/
37. Talaat M. Revolutionizing respiratory medicine: an integrated computational and experimental approach to inhalation dosimetry and disease severity classification A dissertation presented. 2024.
38. Longest PW, Holbrook LT. In silico models of aerosol delivery to the respiratory tract – development and applications. Adv Drug Deliv Rev [Internet]. 2011 Mar 30 [cited 2024 Dec 23];64(4):296. Available from: https://pmc.ncbi.nlm.nih.gov/articles/PMC3258464/
39. Sharareh S-BSW. Role of in-silico modeling in drug development for inhalation treatment. J Mol Pharm Org Process Res. 2013;01(02):106.
40. Koullapis P, Ollson B, Kassinos SC, Sznitman J. Multiscale in silico lung modeling strategies for aerosol inhalation therapy and drug delivery. Curr Opin Biomed Eng [Internet]. 2019 Sep 1 [cited 2024 Dec 23];11:130. Available from: https://pmc.ncbi.nlm.nih.gov/articles/PMC7611802/
41. Matera MG, Calzetta L, Ora J, Rogliani P, Cazzola M. Pharmacokinetic/pharmacodynamic approaches to drug delivery design for inhalation drugs. Expert Opin Drug Deliv [Internet]. 2021 [cited 2024 Dec 23];18(7):891–906. Available from: https://pubmed.ncbi.nlm.nih.gov/33412922/
42. Derendorf H, Hochhaus G, Meibohm B, Möllmann H, Barth J. Pharmacokinetics and pharmacodynamics of inhaled corticosteroids. J Allergy Clin Immunol [Internet]. 1998 Apr 1 [cited 2024 Dec 23];101(4 SUPPL.):S440–6. Available from: http://www.jacionline.org/article/S0091674998701563/fulltext
43. Panunzi S, Gaz C, Cibella F, De Gaetano A. Validation of a simplified-geometry model of inhaled formoterol pharmacodynamics in asthmatic patients. Front Physiol. 2022;13:1018050.
44. Elborn JS, Blasi F, Burgel PR, Peckham D. Role of inhaled antibiotics in the era of highly effective CFTR modulators. Eur Respir Rev [Internet]. 2023 Jan 11 [cited 2024 Dec 23];32(167). Available from: https://publications.ersnet.org/content/errev/32/167/220154
45. Sermetgaudelus I, Lecocguic Y, Ferroni A, Clairicia M, Barthe J, Delaunay J, et al. Nebulized antibiotics in cystic fibrosis. Paediatr Drugs [Internet]. 2002 [cited 2024 Dec 23];4(7):455–67. Available from: https://pubmed.ncbi.nlm.nih.gov/12083973/

46. Lokugamage MP, Vanover D, Beyersdorf J, Hatit MZC, Rotolo L, Echeverri ES, et al. Optimization of lipid nanoparticles for the delivery of nebulized therapeutic mRNA to the lungs. Nat Biomed Eng [Internet]. 2021 Oct 6 [cited 2024 Dec 23];5(9):1059–68. Available from: https://www.nature.com/articles/s41551-021-00786-x
47. Li N, Shi J, Chen Z, Dong Z, Ma S, Li Y, Huang X, Li X. In silico prediction of drug-induced nephrotoxicity: current progress and pitfalls. Expert Opin Drug Metab Toxicol. 2024:1–13.
48. Hadjicharalambous M, Wijeratne PA, Vavourakis V. From tumour perfusion to drug delivery and clinical translation of in silico cancer models. Methods. 2021;185:82–93.
49. Gleeson M, Modi S, Bender AL, Marchese Robinson R, Kirchmair J, Promkatkaew M, Hannongbua S, Glen R. The challenges involved in modeling toxicity data in silico: a review. Curr Pharm Des. 2012;18(9):1266–91.
50. Hannon DM, Mistry S, Das A, Saffaran S, Laffey JG, Brook BS, Hardman JG, Bates DG. Modeling mechanical ventilation in silico-potential and pitfalls. Thieme Medical Publishers, Inc; 2022.
51. Obermeier L, Wiegand M, Hellmeier F, Manini C, Kuehne T, Goubergrits L, Vellguth K. Verification study of in silico computed intracardiac blood flow with 4D flow MRI. IEEE Trans Biomed Eng. 2024;71:2568–79.
52. Jian X, Boerwinkle E, Liu X. In silico prediction of splice-altering single nucleotide variants in the human genome. Nucleic Acids Res. 2014;42(22):13534–44.
53. Benfenati E, Diaza RG, Cassano A, Pardoe S, Gini G, Mays C, et al. The acceptance of in silico models for REACH: requirements, barriers, and perspectives. Chem Cent J. 2011;5:1–11.
54. Obermeier L, Wiegand M, Hellmeier F, Manini C, Kuehne T, Goubergrits L, et al. Verification study of in silico computed intracardiac blood flow with 4D flow MRI. IEEE Trans Biomed Eng. 2024;71(9):2568–79.
55. Koullapis P, Ollson B, Kassinos SC, Sznitman J. Multiscale in silico lung modeling strategies for aerosol inhalation therapy and drug delivery. Curr Opin Biomed Eng. 2019;11:130–6.
56. George SC. In silico modeling of respiratory structure, function, and disease. Drug Discov Today Dis Models. 2007;4(3):123–4.
57. Neelakantan S, Xin Y, Gaver DP, Cereda M, Rizi R, Smith BJ, et al. Computational lung modelling in respiratory medicine. J R Soc Interface. Royal Society Publishing. 2022;19:20220062.
58. Hofemeier P, Koshiyama K, Wada S, Sznitman J. One (sub-)acinus for all: fate of inhaled aerosols in heterogeneous pulmonary acinar structures. Eur J Pharm Sci. 2018;113:53–63.
59. Kassinos SC, JSznitman. Multiscale modeling of respiratory transport phenomena and intersubject variability. Annu Rev Fluid Mech. 2024;57:141–65.
60. Geitner CM, Becher T, Frerichs I, Weiler N, Bates JHT, Wall WA. An approach to study recruitment/derecruitment dynamics in a patient-specific computational model of an injured human lung. Int J Numer Method Biomed Eng. 2023;39(9):e3745.
61. Oakes JM, Mal GC, et al. Airflow and particle deposition simulations in health and emphysema: from in vivo to in silico animal experiments. Ann Biomed Eng. 2014;42:899–914.
62. Mengoni M. Biomechanical modelling of the facet joints: a review of methods and validation processes in finite element analysis. In: Biomechanics and modeling in mechanobiology, vol. 20. Springer Science and Business Media Deutschland GmbH; 2021. p. 389–401.
63. Antonini L, Mandelli L, Berti F, Pennati G, Petrini L. Validation of the computational model of a coronary stent: a fundamental step towards in silico trials. J Mech Behav Biomed Mater. 2021;122:104644.
64. Viceconti M, Pappalardo F, Rodriguez B, Horner M, Bischoff J, Musuamba TF. In silico trials: Verification, validation and uncertainty quantification of predictive models used in the regulatory evaluation of biomedical products. Methods. 2021;185:120–7.
65. Jacob E, Villain L, Ceres N, Palgen JL, L'Hostis A, Monteiro C, Kahoul R. The combination of statistical methods to compare observed and simulated data allowed to assess effectively the validity of mathematical model predictions in a context of a EGFR+ lung adenocarcinoma. Cancer Res. 2023;83:4287.

66. Rebba R, Mahadevan S. Computational methods for model reliability assessment. Reliab Eng Syst Saf. 2008;93(8):1197–207.
67. Przybylak, Covey-Crump P, Cronin. Characterisation of data resources for in silico modelling: benchmark datasets for ADME properties. Characterisation of data resources for in silico modelling: benchmark datasets for ADME properties. Expert Opin Drug Metab Toxicol. ISSN 1744-7607 [Internet]. Available from: http://researchonline.ljmu.ac.uk/
68. Oprisiu I, Winiwarter S. In silico ADME modeling. Systems medicine: integrative, qualitative and computational approaches. 1–3:208–22.
69. Anger L, WA, SK, SD, & RS. Generalized workflow for generating highly predictive in silico off-target activity models. J Chem Inf Model. 2014;54(9):2411–22.
70. Kramer C, Kalliokoski T, Gedeck P, Vulpetti A. The experimental uncertainty of heterogeneous public K(i) data. J Med Chem. 2012;55(11):5165–73.
71. Kar S, Leszczynski J. Open access in silico tools to predict the ADMET profiling of drug candidates. Expert Opin Drug Discov. 2020;15(12):1473–87.
72. Göller AH, Kuhnke L, Montanari F, Bonin A, Schneckener S, ter Laak A, et al. Bayer's in silico ADMET platform: a journey of machine learning over the past two decades. Drug Discov Today. 2020;25(9):1702–9.
73. Jose J, Mathew B, Parambi DG. In silico trial approach for biomedical products: a regulatory perspective. Comb Chem High Throughput Screen. 2022;25:1991–2000.
74. Musuamba FT, Skottheim Rusten I, Lesage R, Russo G, Bursi R, Emili L, et al. Scientific and regulatory evaluation of mechanistic in silico drug and disease models in drug development: Building model credibility. CPT Pharmacometrics Syst Pharmacol. 2021;10(8):804–25.
75. Li Z, Garnett C, Strauss DG. Quantitative systems pharmacology models for a new international cardiac safety regulatory paradigm: an overview of the comprehensive in vitro proarrhythmia assay in silico modeling Approach, CPT: pharmacometrics and systems pharmacology, vol. 8. American Society for Clinical Pharmacology and Therapeutics; 2019. p. 371–9.
76. Pappalardo F, Russo G, Tshinanu FM, Viceconti M. In silico clinical trials: concepts and early adoptions, vol. 20., Briefings in bioinformatics. Oxford University Press; 2019. p. 1699–708.
77. Gal J, Milano G, Ferrero JM, Saâda-Bouzid E, Viotti J, Chabaud S, et al. Optimizing drug development in oncology by clinical trial simulation: why and how? Brief Bioinform. 2017;19(6):1203–17.
78. Djuris J, Cvijic S, Djekic L. Model-informed drug development. In: Silico assessment of drug bioperformance following oral and percutaneous administration, vol. 17. Pharmaceuticals. Multidisciplinary Digital Publishing Institute (MDPI); 2024.
79. Alqahtani S. Improving on in-silico prediction of oral drug bioavailability. Expert Opin Drug Metab Toxicol. 2023;19:665–70.
80. Wang Y, Xing J, Xu Y, Zhou N, Peng J, Xiong Z, et al. In silico ADME/T modelling for rational drug design. Q Rev Biophys. Cambridge University Press. 2015;48:488–515.
81. de Montigny J, Iosif A, Breitwieser L, Manca M, Bauer R, Vavourakis V. An in silico hybrid continuum−/agent-based procedure to modelling cancer development: interrogating the interplay amongst glioma invasion, vascularity and necrosis. Methods. 2021;185:94–104.
82. Antonopoulos M, Dionysiou D, Stamatakos G, Uzunoglu N. Three-dimensional tumor growth in time-varying chemical fields: a modeling framework and theoretical study. BMC Bioinformatics. 2019;20(1):1–36.
83. Siepmann J. In-silico simulations of advanced drug delivery systems: what will the future offer? Int J Pharm. 2013;454(1):512–6.
84. Shachar-berman L, Bhardwaj S, Ostrovski Y, Das P, Koullapis P, Kassinos S, et al. In silico optimization of fiber-shaped aerosols in inhalation therapy for augmented targeting and deposition across the respiratory tract. Pharmaceutics. 2020;12(3):230.
85. Grill MJ, Biehler J, Wichmann KR, Rudlstorfer D, Rixner M, Brei M, et al. In silico high-resolution whole lung model to predict the locally delivered dose of inhaled drugs. 2023 Jul 7; Available from: http://arxiv.org/abs/2307.04757

86. Shi C, Ignjatović J, Liu T, Han M, Cun D, Đuriš J, et al. In vitro – in vivo – in silico approach in the development of inhaled drug products: nanocrystal-based formulations with budesonide as a model drug. Asian J Pharm Sci. 2021;16(3):350–62.
87. Bourlon M, Feng Y, Garcia-Contreras L. Designing aerosol therapies based on the integrated evaluation of in vitro, in vivo, and in silico data. Pharmaceutics. 2023;15(6):1695.
88. Martonen T, Fleming J, Schroeter J, Conway J, Hwang D. In silico modeling of asthma. Adv Drug Deliv Rev. 2003;55(7):829–49.
89. Islam M, Liu C, Cai C, Shah J, Feng Y. A user-centered smart inhaler algorithm for targeted drug delivery in juvenile onset recurrent respiratory papillomatosis treatment integrating computational fluid particle dynamics and machine learning, vol. 36. Physics of Fluids; 2024.
90. Bannigan P, Aldeghi M, Bao Z, Häse F, Aspuru-Guzik A, Allen C. Machine learning directed drug formulation development. Adv Drug Deliv Rev. 2021;175:113806.
91. Walters WP, BR. Critical assessment of AI in drug discovery. Expert Opin Drug Discov. 2021;16(9):937–47.
92. Zhong F, Xing J, Li X, Liu X, Fu Z, Xiong Z, Lu D, Wu X, Zhao J, Tan X, Li F. Artificial intelligence in drug design. Sci China Life Sci. 2018;61:1191–204.
93. Paliwal A, Jain S, Kumar S, Wal P, Khandai M, Khandige PS, Sadananda V, Anwer M, Gulati M, Behl T, Srivastava S. Predictive modelling in pharmacokinetics: from in-silico simulations to personalized medicine. Expert Opin Drug Metab Toxicol. 2024;20(4):181–95.
94. Marques L, Costa B, Pereira M, Silva A, Santos J, Saldanha L, et al. Advancing precision medicine: a review of innovative in silico approaches for drug development, clinical pharmacology and personalized healthcare. Pharmaceutics. Multidisciplinary Digital Publishing Institute (MDPI). 2024;16:332.
95. Salar-Behzadi S, Wu S, Mercuri A, Meindl C, Stranzinger S, Fröhlich E. Effect of the pulmonary deposition and in vitro permeability on the prediction of plasma levels of inhaled budesonide formulation. Int J Pharm. 2017;532(1):337–44.
96. Fröhlich E, Mercuri A, Wu S, Salar-Behzadi S. Measurements of deposition, lung surface area and lung fluid for simulation of inhaled compounds, vol. 7., Front Pharmacol. Frontiers Research Foundation; 2016.
97. Barros MT, Paci M, Tervonen A, Passini E, Koivumäki J, Hyttinen J, et al. From multiscale biophysics to digital twins of tissues and organs: future opportunities for in silico pharmacology. 2023 Jun 4; Available from: http://arxiv.org/abs/2306.02369
98. Komura H, Watanabe R, Mizuguchi K. The trends and future prospective of in silico models from the viewpoint of ADME evaluation in drug discovery. Pharmaceutics. Multidisciplinary Digital Publishing Institute (MDPI). 2023;15:2619.

Chapter 9
Computational Tools Employed in Chemoinformatics

Gouthami Thumma, Prabhakar Reddy Veerareddy, and Gangarapu Kiran

Contents

G. Thumma
University College of Pharmaceutical Sciences, Kakatiya University, Warangal, Telangana, India

P. R. Veerareddy
University College of Pharmaceutical Sciences, Palamuru University, Mahabubnagar, Telangana, India

G. Kiran (✉)
School of Pharmacy, Anurag University, Venkatapur (V), Ghatakser (M), Hyderabad, Telangana, India

S. N. K. Rao G., R. R. Alavala (eds.), *Applications of Computational Tools in Drug Design and Development*, https://doi.org/10.1007/978-981-96-4154-3_9

Abstract Chemoinformatics is the branch of chemistry field which develops the tools related to databases, calculation of descriptors, data visualization and analysis, and representation of molecules etc. The review described the computational tools used in chemoinformatics for the advancement in scientific research. Here, we presented various chemoinformatics tools related to databases both public and commercial, data curation, visualization and analysis of data. Open-source tools are highlighted and significance of each tool are mentioned in this chapter.

Keywords Computational tools · Chemoinformatics · Webservers · Docking · Auto dock

9.1 Introduction

Chemoinformatics is the branch of chemistry field which develops the tools related to databases, calculation of descriptors, data visualization and analysis, and representation of molecules etc. [1–3]. It mainly implies with design, organization, management, visualization and analysis of chemicals information. It is the combination of computational and theoretical chemistry which uses the graph theory, descriptor vectors and statistical learning. It is playing wide a role in drug design and discovery by identification of lead and helps in conversion of lead to drug discovery [4]. Cheminformatics has wide range of applications in the chemistry fields of analytical, inorganic, medicinal, physical and theoretical chemistry [2, 5].

9.2 History of Chemoinformatics

Chemoinformatics involves the development of tools related to chemistry by using the advanced technology tools. The application of informatics on chemicals has been known since inspection of computers in the field of science. Chemoinformatics is an emerging theoretical chemistry field with creation, development, storage, analysis, sharing, visualization, and organization.

Various scientist has coined the definition of chemoinformatics such as "Chemoinformatics is generic term that encompasses the design, creation, organization, management, retrieval, analysis, dissemination, visualization and use of chemical information" by G. Paris in 1998 [6].

"Cheminformatics is the mixing of those information resources to transform data into information and information into knowledge for the intended purpose of making better decisions faster in the area of drug lead identification and optimization" by Brown F.K. in 1998 [7].

"Chemoinformatics is the application of informatics methods to solve chemical problems" by J. Gasteiger in 2004 [8].

"Chemoinformatics is a field based on the representation of molecules as objects (graphs or vectors) in a chemical space". By Varnek A & Baskin I in 2011 [2].

Chemoinformatics is a new discipline combining several old fields' viz., chemical databases, structure activity modelling, drug design and computer-aided synthesis design. Many pioneer researchers have worked in the field of chemoinformatics and applied in wide area of scientific research. Some of researchers are Michael F. Lynch worked at University of Sheffield and developed automatic indexing procedure, and developed algorithms related to structure and sub-structure searching [9]. Prof Peter Willett from University of Sheffield has developed genetic algorithm for flexible docking, chemical similarity searching, and similarity and clustering in chemical information systems, Rascal, fingerprint based virtual screening methods, and open access mega journals [10]. Prof Corwin Hansch is a pioneer of QSAR and developed relationship between the physicochemical descriptors versus biological activity [11, 12]. Hansch is considered to be "father of QSAR" because of his work linking the physicochemical properties of compounds with their biological activity [13]. One of Hansch's most significant contributions was emphasizing the role of hydrophobicity (logP) in determining a compound's biological activity. Hansch equation relates biological activity to descriptors such as hydrophobicity (logP), electronic effects and steric factors. Hansch integrated statistical methods into chemistry enabling more accurate predictions of a molecule's biological activity. By using regression analysis which correlate biological activity with molecular properties, he made QSAR a powerful predictive tool in drug design [14, 15]. Johann Gasteiger is an emeritus professor in chemistry at the University of Erlangen-Nuremberg, Germany has made significant contributions in the field of chemoinformatics. Gasteiger was pioneer in the use of neural network in chemistry and co-authored textbook on chemoinformatics [16, 17]. He carried work on representation of chemical reactions, calculation of structure descriptors, Chemstable and computer representation of chemical compounds. Prof Irwin D. Kuntz, University of California has worked on structure based drug design and developed Dock software. Kuntz has published many research papers on docking algorithms, fingerprinting, protein folding, and identification of leads against bacteria, viruses and parasites [18].

9.3 Open Source: Open Tools

Open-source cheminformatics tools are software applications that provide functionalities for handling chemical information and data. These tools are widely used in fields such as drug design, discovery, materials science, and chemical engineering.

Here are some notable open-source cheminformatics tools:

1. **RDKit**: A collection of cheminformatics and machine learning tools written in C++ and Python. RDKit is widely used for tasks like molecular modeling, descriptor calculation, and molecular similarity searching.
2. **Open Babel**: A chemical file translation program that allows users to convert between different chemical formats. It also includes functionalities for molecular manipulation and descriptor calculations.
3. **ChemTools**: A Python library designed for the analysis and visualization of quantum chemistry data. It helps researchers interpret the results of electronic structure calculations.
4. **Cinfony**: A framework that integrates various cheminformatics libraries (like RDKit, Open Babel, and others) into a unified interface, making it easier to access and use multiple tools.
5. **Avogadro**: An advanced molecular editor and visualization tool designed for building and manipulating molecular structures. It supports a variety of file formats and includes computational chemistry capabilities.
6. **ChemPy**: A Python package for simulating and solving chemical kinetics and equilibrium problems. It's particularly useful for educational purposes and small-scale simulations.
7. **BioPython**: Although primarily focused on bioinformatics, it has cheminformatics functionalities that allow users to work with chemical data related to biological systems.
8. **Mordred**: A molecular descriptor calculator that can compute a wide variety of descriptors for molecules, useful for quantitative structure-activity relationship (QSAR) studies.

These tools facilitate research and development by allowing scientists to analyze chemical data, visualize molecular structures, and develop predictive models. The open-source nature means they are accessible to anyone and can be modified to suit specific needs. Table 9.1 Showed list of notable cheminformatics toolkits.

Table 9.1 List of notable cheminformatics toolkits

Tool	Applications	Websource page
Pymol	Molecular 3D graphics tool	https://www.pymol.org/
VMD (visual molecular dynamics)	3D molecular visualization tool used in molecular dynamics simulations	https://shorturl.at/COEJ9
Chimera	Molecular interactive visualization tool	https://shorturl.at/KIlJb
Discovery Studio Biovia	Visualization of molecules	https://discover.3ds.com/
Jmol	Chemical structures Java viewer	https://jmol.sourceforge.net/
DataWarrior	Chemical data visualization and analysis of data	https://openmolecules.org/datawarrior/index.html
Chemsketch (ACD Labs)	Chemical structures drawing	https://shorturl.at/EaGAe/
Chemdraw & Chemoffice	Drawing chemical structures	https://shorturl.at/IHjJK
ChemDoodle 2D	2D chemical structures drawing	https://www.chemdoodle.com/
Open Babel	For structures convertor and optimization	http://openbabel.org/
RDKit	Molecular manipulation, descriptor calculation, similarity searching, and visualization of chemical structures.	https://www.rdkit.org/
KNIME	Open source Chemtool for creating and analyzing databases	https://www.knime.com/
Jchem (Chemaxon)	Computing molecular descriptors and fingerprints for data analysis.	https://chemaxon.com/instant-jchem
CDK	Molecular modeling, descriptor calculation, chemical information retrieval, file format conversion, substructure searching, visualizations, and integration with other cheminformatics tools.	https://cdk.github.io/

Chemical space contains >10^{60} compounds and description of molecules which contains scaffolds, fingerprints, and whole molecule properties. The visualization methods are coordinate and network-based. The definition of chemical space is "an M-dimensional Cartesian space in which compounds are positioned based on a set of M physicochemical and/or cheminformatics descriptors." The major applications of chemical space by compound library design, compound selection, and compound classification by pharmacological effect and molecular target [19, 20].

9.4 Compound Databases Tools

Numerous cheminformatics compound databases have developed some are public, commercial, in-house, De novo, and virtual chemical libraries. Public databases are Zinc, National Cancer Institute, DrugBank, PHAROS, Pubchem, ChEMBL, Chemspider and Human Metabolome Database [21].

Zinc (http://zinc.docking.org/) contains purchasable molecules for virtual screening. The database contains about 230 million molecules in 3D formats and ready to dock. Zinc was developed by Irwin and Shoichet Laboratories at University of California, San Francisco. Zinc offers molecules in various formats like SMILES, SDF and 3D conformations. It contains search tools for structure, similarity, substructure and molecular properties searches. The database has the feature to apply filters like lead-like, scaffolds, fragment-like or drug-like [22].

National Cancer Institute (http://dtp.nci.nih.gov) contains screened compounds, DrugBank (http://www.drugbank.ca/) database can explore for Drug Knowledgebase, Interaction Checker, drug reactions, drug targets, drug classification and Pathways. The cases of drug discovery and repurposing, drug-protein relationships, contraindications, drug interactions, clinical trials, patents, pathways, and adverse effects and blackbox warnings can be explored [23].

PHAROS (https://pharos.nih.gov/) contains information of drugs which could be useful for drug repurposing. The primary mission of KMC is to create an integrated and comprehensive knowledge base for the Druggable Genome (DG), with a focus on shedding light on the less studied or poorly characterized areas of the DG. This effort specifically targets three major protein families commonly involved in drug development: Ion channels (ICs), G-protein-coupled receptors (GPCRs), and Kinases [24].

PubChem is a free chemical database managed by the NIH, providing extensive information on over 111 million compounds, including small molecules, mixtures, and bioactive compounds. It offers detailed chemical properties like molecular weight and solubility, along with biological activity data, making it a valuable resource for researchers. Users can perform advanced searches by structure, SMILES, InChI, molecular formula, or text (e.g., names or CAS numbers), with options for substructure and bioactivity searches. PubChem integrates bioassay data, supporting structure-activity relationships (SAR) and drug discovery. It is also connected to other NCBI resources, such as PubMed, GenBank, and external databases like ChEMBL, providing a comprehensive view of related biological and chemical information. The platform offers powerful tools, including a 3D structure viewer, property prediction features, and similarity search capabilities, along with access to chemical synthesis methods and supplier information. Free to use, PubChem allows bulk downloads and API access, enabling integration into user applications and workflows. It also includes chemical safety, toxicity data, and GHS classification. Educational resources and tutorials are available, making PubChem a crucial tool for researchers and students [25].

ChemSpider, a free chemical structure database maintained by the Royal Society of Chemistry (RSC), provides access to over 100 million compounds from various sources, offering detailed chemical properties like molecular weight, solubility, and spectra (NMR, IR, MS). It allows users to search by structure, SMILES, InChI, molecular formula, names, or CAS numbers, with substructure search capabilities. ChemSpider is both crowd sourced and expert-curated, ensuring reliable data, and it integrates with external databases such as PubChem and ChEMBL, as well as RSC literature. The platform includes interactive tools like a 3D viewer, property prediction features, and synthetic routes, supporting researchers in organic chemistry and drug discovery. It offers mobile access, an API for integration, links to chemical suppliers, and is freely accessible without registration, making it a valuable resource for chemists and researchers in drug discovery, bioinformatics, and toxicology [26].

9.5 Databases of Natural Products

Zinc database is having >250,000 natural products in catalogs category. BIOFACQUIM curated database for the natural products. The unique of the database containing natural compounds isolated and characterized in Mexico. COCONUT 2.0 database is the collection of natural isolated compounds. The database consists of collections of Herbal Ingrediant Targets (HIT), Piel Lab DB, CMNPD which has marine natural products, Herbal Ingredients in-vivo Metabolism (HIM), Traditional Chinese Medicine database, Analyticon discovery NP, African Natural Products (ANPDB), and NPACT. SuperNatural III is curated products obtained from living organism. It is user friendly provides information on toxicity class, physicochemical structure and vendor information.

Comprehensive Marine Natural Products Database (CMNPD) is a manually open access curated database containing the information of marine natural compounds. The features of the database are providing the information of molecular properities and PK properties, pharmacological activity data, taxonomy, and geographical distribution of marine organisms and citations. Other database related to marine natural products are Ocean Data Bank, CMNPD, and MarinLit. Table 9.2 represented Databases/Webservers of chemical data handling.

9.6 Tools for Data Curation of Chemicals

Chemical data curation is the technical process of integrating and organizing chemicals data from various sources. It involves publication, annotation, and presentation of the chemical data to maintain its value over time. The steps involved in the chemical data curation process are data identification, data cleansing, and data transformation. Data Curation of chemicals is essential for performing any task for

Table 9.2 Databases/webservers of chemical data handling

Pubchem	PubChem is an open chemistry database at the National Institutes of Health (NIH).	https://pubchem.ncbi.nlm.nih.gov/ [25]
ChEMBL	Its a manually curated database of bioactive molecules with drug-like properties	https://www.ebi.ac.uk/chembl/ [27]
Zinc [22]	Purchasable database of compounds for virtual screening.	https://zinc.docking.org/
DrugBank [28]	It contains drug data containing data distributed as categories, clinical trials, patents, pathways, and drug interactions	https://go.drugbank.com/
PHAROS [24]	It contains information of drugs which could be useful for drug repurposing	https://pharos.nih.gov/
National Cancer Institute [23]	NCI contains screened compounds	http://dtp.nci.nih.gov
Mcule [29]	Purchasable database with molecular modeling tools	https://mcule.com/
ChemSpider [30]	Chemical structure database maintained by the RSC, London	https://www.chemspider.com/
BIOFACQUIM [31]	It contains databases of natural compounds	http://zinc15.docking.org/catalogs/biofacquimnp/
COCONUT2.0 [32]	It is the collection of natural isolated compounds	https://coconut.naturalproducts.net/
SuperNatural III [33]	It is curated products obtained from living organism	https://bioinf-applied.charite.de/supernatural_3/
LOTUS [34]	It is an open source project for natural products	https://lotus.naturalproducts.net/
CHEMnetBASE [35]	It is collection of interactive database and libraries	https://www.chemnetbase.com/search/SimpleSearch.xhtml?dswid=-1695
Comprehensive marine natural products database (CMNPD) [36]	CMNPD is a manually open access curated database containing the information of marine natural compounds	https://www.cmnpd.org/
MarinLit [37]	It contains 42,183 marine compounds	https://marinlit.rsc.org/login?redirect=%2Fcompounds
Chemical Ocean Database [38]	Database contains chemical composition, material composition and distribution	https://www.odb.ntu.edu.tw/chem/

(continued)

Table 9.2 (continued)

BindingDB [39]	BindingDB contains 2.9 M data for 1.2 M compounds and 9.3 K targets.	https://www.bindingdb.org/rwd/bind/index.jsp
LEA3D [40]	LEA3D offers a facility to screen or to design small molecules	https://chemoinfo.ipmc.cnrs.fr/LEA3D/index.html
SenSaaS [41]	Rigid or flexible shape-based alignment of molecules	https://chemoinfo.ipmc.cnrs.fr/SENSAAS/index.html
Molinspiration [42]	Calculation of important molecular properties	https://www.molinspiration.com/

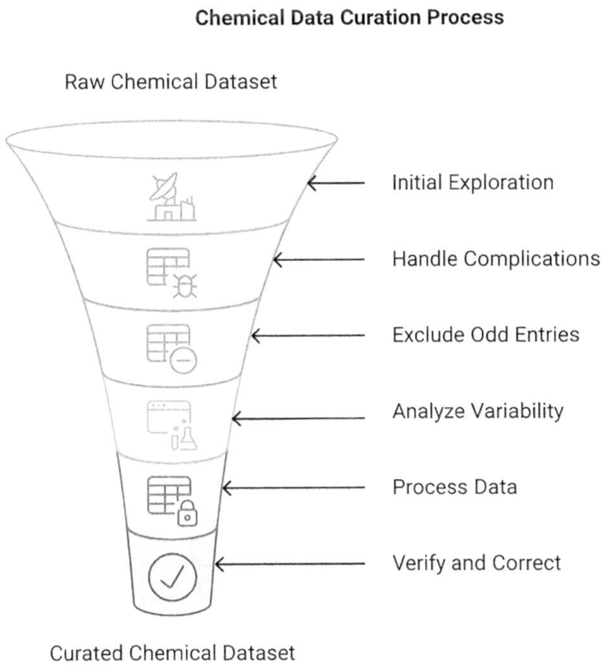

Fig. 9.1 Data curation process

chemoinformatics. Quality data set is very important for get a good machine learning model. If the data is full of mistakes, noisy, full of activity cliffs could not able to get a useful ML model. Data mining, data preprocessing, Data Extraction & curation are the steps involved for getting a useful data set which are indispensable procedure for any projects. Figure 9.1 showcased data curation process.

Some of cheminformatics tools used of Chemical data curation are mentioned below.

9.6.1 Datagrok (https://datagrok.ai/help/datagrok/)

Datagrok can load huge data set from ChEMBL and perform substructure searches, visualize, apply filters, and explore chemical space. The platform can clean, visualize, explore data and interpret models. Datagrok is very fast and powerful and can understand data and take action.

9.6.2 ChemEx (http://www.biotec.or.th/isl/ChemEx) [43]

ChemEx software developed by "Information Systems Laboratory, National Center for Genetic Engineering and Biotechnology (BIOTEC)", Thailand. ChemEx accelerates the curation of chemical data by efficiently extracting information about compounds, organisms, and assays from a vast array of publications. ChemEx tool is an chemical extraction system which extract both images and text in publications. The text annotator is designed to extract entities related to compounds, organisms, and assays from textual content. Additionally, the structure image recognition feature facilitates the conversion of chemical raster images into a machine-readable format. Users are provided with annotated text, along with summarized information regarding the identified compounds, the organisms that produce these compounds, and the corresponding assay tests.

9.6.3 Instant JChem (https://chemaxon.com/instant-jchem)

Instant JChem is developed by ChemAxon, a company that specializes in cheminformatics software. It provides a platform for data management and analysis in chemistry, enabling users to create and manage chemical structures, perform data analysis, and integrate with other data sources [3].

9.6.4 KNIME (https://www.knime.com/): [44]

KNIME (Konstanz Information Miner) is an open-source data analytics, reporting, and integration platform developed by the KNIME community, originally led by the University of Konstanz in Germany. It allows users to create data workflows by visually connecting nodes that represent various data processing tasks.

9.7 Key Features

- Visual Workflow Interface: Users can create workflows by simply dragging and dropping nodes, ensuring that the platform is user-friendly for both technical and non-technical individuals.
- Integration: Supports various data sources, including databases, files, and web services, and can integrate with other tools like Python, R, and Java.
- Extensibility: Users can add custom nodes and use extensions for specific tasks, such as text mining, machine learning, and bioinformatics.
- Data Visualization: Offers built-in visualization tools for analyzing and presenting data results.

9.7.1 Applications

1. Data Preprocessing: Cleaning and transforming data for analysis.
2. Machine Learning: Building predictive models using a range of algorithms.
3. Text Mining: Analyzing and extracting information from unstructured data.
4. Bioinformatics: Analyzing biological data, including genomics and proteomics.
5. Business Intelligence: Creating reports and dashboards for data-driven decision-making.
6. Pharmaceutical Research: Used in cheminformatics and drug discovery for analyzing chemical compounds.

KNIME's flexibility and extensive community support make it suitable for a wide range of applications across various industries.

9.8 Chemoinformatics Tools Used in Molecular Modeling

9.8.1 Molecular Docking

Docking is the interaction of protein and ligand into active site which could understand the mechanistic binding of ligand into the binding site.

Software's are available as open source and commercial, a part of these webservers have also developed for performing molecular docking. Several software tools are available to perform molecular docking some of them are listed below.

9.8.2 AutoDock Tools (http://ccsb.scripps.edu/): [45]

It is developed by The Scripps Research Institute (TSRI) and is part of the AutoDock suite, which is widely used for molecular docking studies. AutoDock Tools provides a graphical user interface for preparing input files for docking simulations and visualizing the results. It helps researchers set up ligand-receptor docking experiments and analyze the docking outcomes effectively. AutoDock and AutoDock Vina have become standard tools in computational biology and cheminformatics for studying molecular interactions, drug discovery, and design.

9.8.3 AutoDock Tools Dock (https://dock.compbio.ucsf.edu/): [46]

DOCK is developed by the University of California, San Francisco (UCSF). This molecular modeling software is mainly utilized to predict the binding interactions of small molecules, like potential drug candidates, with receptors that have a known 3D structure. DOCK has been a foundational tool in the field of CADD. DOCK is especially important in drug discovery processes, allowing researchers to pinpoint promising drug candidates and refine their interactions with biological targets.

9.8.4 AutoDock Tools Dock PyRx (https://pyrx.sourceforge.io/): [47]

Developed by The Scripps Research Institute, PyRx is a virtual screening tool tailored for computational drug discovery. It enables researchers to assess compound libraries against potential drug targets, offering medicinal chemists a comprehensive platform for virtual screening. PyRx supports the entire workflow, from preparing data and submitting jobs to analyzing results, making it a valuable resource in the drug discovery process.

9.8.5 AutoDock Tools Dock PyRx iGemDock (https://gemdock.life.nctu.edu.tw/)

A free and user-friendly docking software that offers flexible ligand docking and is suitable for academic research [48].

9.8.6 AutoDock Tools Dock PyRx iGemDock Smina (https://www.cheminformania.com/ligand-docking-with-smina/)

Smina command line tool for performing docking. Smina offers improved scoring and flexibility for docking small molecules to proteins [49].

9.8.7 AutoDock Tools Dock PyRx iGemDock R Dock (https://rdock.gitlab.io/)

R Dock is also command line tool with python feature. Designed for flexible ligand docking and protein-ligand complex refinement, particularly useful in drug design [50].

9.9 Webservers for Docking

9.9.1 SwissDock (https://www.swissdock.ch/)

An online docking service that allows users to upload their ligands and proteins, providing an easy-to-use web interface [51].

9.9.2 DockThor (https://dockthor.lncc.br/v2/)

A free website for docking for redocking, blind docking, virtual screening [52].

9.9.3 CB-Dock ("http://cao.labshare.cn/cb-dock/"): [53]

CB-Dock is a cutting-edge method for protein-ligand docking that automates the detection of binding sites. It identifies the center and dimensions of these sites, adjusts the docking box size according to the ligands being analyzed, and conducts molecular docking using AutoDock Vina. Comprehensive benchmarking shows that this cavity-centric approach significantly improves both the hit rate and accuracy compared to conventional blind docking techniques. By employing a curvature-based cavity detection method (CurPocket) to forecast binding sites on target proteins and using AutoDock Vina to establish the binding poses of ligands, CB-Dock optimizes the docking process and enhances precision.

9.9.4 SiBiOLEAD (https://sibiolead.com/)

It is a CRO company involved in CADD, the platform offers free automated ligand preparation and conformer search and degeneration and can dock upto 2 ligands with PLIP analysis.

9.9.5 Click Docking (https://mcule.com/apps/1-click-docking/)

It is developed by Mcule which is an online drug discovery platform which provides services related to compound database and molecular modeling tools.

9.9.6 DiffDock (https://310.ai/docs/docking/DiffDock)

It is a AI tool which performs deep docking and predicts the binding poses [54].

9.9.7 SeamDock ("https://bioserv.rpbs.univ-paris-diderot.fr/services/SeamDock/")

It is standard pipeline for performing protein ligand docking [55].

9.9.8 PatchDock ("https://bioinfo3d.cs.tau.ac.il/PatchDock/")

It is server for performing the receptor ligand docking [56].

9.10 Protein-Protein Docking

GRAMM Docking (https://gramm.compbio.ku.edu/): A web interface performs protein docking [57].

9.10.1 Zdock: (https://zdock.wenglab.org/)

It is developed by Zhiping Weng's Lab at University of Massachusetts for protein docking [58].

9.10.2 CABS-dock (https://biocomp.chem.uw.edu.pl/CABSdock)

A web interface used for docking of protein and peptide.

9.10.3 HDOCK Server (http://hdock.phys.hust.edu.cn/)

The HDOCK Server is significant for its ability to perform high-accuracy protein-protein and protein-ligand docking, enabling researchers to predict complex structures efficiently. It combines both template-based and free docking methods, enhancing the accuracy of structural predictions in drug discovery and biological research [59].

9.10.4 Molecular Dynamics Simulations

Molecular dynamic (MD) simulations is a computer simulations that analyze the movement of atoms and molecules over time.

Tools for MD simulations are available as webservers and software's.

9.10.5 GROMACS (https://www.gromacs.org/)

It is a open source and free software for MD simulations and analysis. Gromacs 2024 new features on automatic scaling of the target distribution, reduced artifacts, on pressure, from LJ pair interactions and high performance [60].

9.10.6 GENESIS (https://www.r-ccs.riken.jp/labs/cbrt/)

It has developed by Sugita group in RIKEN, Japan in 2009. It performs MD simulation with high performance for biomolecular systems. It has features to do MD simulations by atomistic, coarse-grained and QM/MM method [61].

9.10.7 LAMMPS (https://www.lammps.org/)

It is MD simulator performs classical MD simulation. It has applications to do MD simulation for solid-state materials (metals, semiconductors), coarse-grained and soft matter (biomolecules, polymers) [62].

9.10.8 Webgro (https://simlab.uams.edu/) [63]

WebGro is an automated online tool designed for conducting molecular dynamics simulations of macromolecules, such as proteins, either individually or in complex with ligands (small molecules). Leveraging the GROMACS simulation package, WebGro performs fully solvated molecular dynamics simulations. Users simply need to upload their protein file (in .pdb format), and WebGro will handle both the simulation and trajectory analysis. The submission page is pre-configured with default parameters based on established research related to GROMACS simulations. For a deeper understanding of each parameter, users are encouraged to consult relevant papers and tutorials.

9.10.9 MDWeb (https://mmb.irbbarcelona.org/MDWeb/) [64]

MDWeb is a personalized workspace that offers standardized protocols for preparing molecular structures, conducting molecular dynamics simulations, and analyzing simulation trajectories. Built upon renowned simulation software such as Amber, NAMD, and GROMACS, MDWeb integrates a range of preparation and analysis tools within a unified interface. While MDWeb does not grant free access to licensed software, it operates entirely on open-access modules and tools. The platform's internal engine is powered by MDMoby, a network of web services utilizing the BioMoby protocol, which can also be accessed programmatically via standard web service clients and APIs.

9.10.10 MDsrv (https://nglviewer.org/mdsrv/) [65]

MDsrv is a web-based platform designed to facilitate collaborative research by offering non-experts straightforward and rapid access to molecular dynamics (MD) simulations. Part of the nglviewer suite, MDsrv enables the visualization of MD trajectories via a web server, utilizing the advanced capabilities of the NGL viewer application.

9.10.11 PACKMOL (https://m3g.github.io/packmol/) [66]

PACKMOL generates initial configurations for MD simulations by efficiently arranging molecules within specified regions of space. This careful packing ensures that short-range repulsive interactions do not interfere with the simulations. With a wide range of spatial constraints applicable to molecules or their constituent atoms, PACKMOL simplifies the creation of organized systems, including lamellar, spherical, or tubular lipid structures.

9.11 Conclusion

In conclusion, computational tools in chemoinformatics play a crucial role in modern chemical research and drug discovery. By leveraging advanced algorithms and software, these tools facilitate the analysis, visualization, and interpretation of complex chemical data. They enhance the ability to predict molecular behavior, optimize compound structures, and streamline the drug development process. As technology continues to evolve, the integration of machine learning and artificial intelligence with chemoinformatics tools is expected to further accelerate discoveries and improve efficiencies in various fields of chemistry and biology. This chapter provided the chemoinformatics tools mostly open source tools with their applications.

References

1. Gasteiger J, Engel T. Chemoinformatics: a textbook. John Wiley & Sons; 2006.
2. Varnek A, Baskin II. Chemoinformatics as a theoretical chemistry discipline. Mol Inform. 2011;30(1):20–32.
3. Karthikeyan M, Vyas R, Karthikeyan M, Vyas R. Open-source tools, techniques, and data in chemoinformatics. In: Practical chemoinformatics. Springer India; 2014. p. 1–92.
4. Chen WL. Chemoinformatics: past, present, and future. J Chem Inf Model. 2006;46(6): 2230–55.

5. Chonde S, Kumara S, editors. Cheminformatics: an introductory review. IIE Annual conference proceedings. Institute of Industrial and Systems Engineers (IISE); 2014.
6. Engel T. Basic overview of chemoinformatics. J Chem Inf Model. 2006;46(6):2267–77.
7. Prakash N, Gareja D. Cheminformatics. J Proteomics Bioinform. 2010;3:249–52.
8. Bhalerao S, Verma D, Rohan L, Teli N, Vinodkumar S. Chemoinformatics: the application of informatics methods to solve chemical problems. Res J Pharm Biol Chem Sci. 2013;4:475.
9. Lynch MF. The Sheffield University generic chemical structures project. World Patent Inf. 1993;15(3):135–41.
10. Willett P. Searching techniques for databases of two-and three-dimensional chemical structures. J Med Chem. 2005;48(13):4183–99.
11. Hansch C. The physicochemical approach to drug design and discovery (QSAR). Drug Dev Res. 1981;1(4):267–309.
12. Gramatica P. A short history of QSAR evolution. [online][Accessed 26 Jan 2012] Available from: http://qsarworld com/Temp Fileupload/Shorthistoryofqsar pdf; 2011.
13. Mannhold R, Krogsgaard-Larsen P, Timmerman H. QSAR: Hansch analysis and related approaches. John Wiley & Sons; 2008.
14. Hansch C. Recent advances in biochemical QSAR. In: Correlation analysis in chemistry: recent advances. Springer US; 1978. p. 397–438.
15. Dastmalchi S, Hamzeh-Mivehroud M, Sokouti B. Quantitative structure–activity relationship: a practical approach. CRC Press; 2018.
16. Engel T, Gasteiger J. Applied chemoinformatics: achievements and future opportunities. John Wiley & Sons; 2018.
17. Cherkasov A, Muratov EN, Fourches D, Varnek A, Baskin II, Cronin M, et al. QSAR modeling: where have you been? Where are you going to? J Med Chem. 2014;57(12):4977–5010.
18. Pegg SC-H, Haresco JJ, Kuntz ID. A genetic algorithm for structure-based de novo design. J Comput Aided Mol Des. 2001;15:911–33.
19. Medina-Franco JL, Chávez-Hernández AL, López-López E, Saldívar-González FI. Chemical multiverse: an expanded view of chemical space. Mol Inform. 2022;41(11):2200116.
20. Medina-Franco JL, Martinez-Mayorga K, Meurice N. Balancing novelty with confined chemical space in modern drug discovery. Expert Opin Drug Discov. 2014;9(2):151–65.
21. Roy K. In silico drug design: repurposing techniques and methodologies. Academic Press; 2019.
22. Irwin JJ, Shoichet BK. ZINC− a free database of commercially available compounds for virtual screening. J Chem Inf Model. 2005;45(1):177–82.
23. Communications NCIOoC. National Cancer Institute: National Cancer Institute, Office of Cancer Communications; 1985.
24. Nguyen D-T, Mathias S, Bologa C, Brunak S, Fernandez N, Gaulton A, et al. Pharos: collating protein information to shed light on the druggable genome. Nucleic Acids Res. 2017;45(D1):D995–D1002.
25. Kim S, Chen J, Cheng T, Gindulyte A, He J, He S, et al. PubChem 2023 update. Nucleic Acids Res. 2023;51(D1):D1373–D80.
26. Ayers M. ChemSpider: the free chemical database. Ref Rev. 2012;26(7):45–6.
27. Gaulton A, Hersey A, Nowotka M, Bento AP, Chambers J, Mendez D, et al. The ChEMBL database in 2017. Nucleic Acids Res. 2017;45(D1):D945–D54.
28. Knox C, Wilson M, Klinger CM, Franklin M, Oler E, Wilson A, et al. DrugBank 6.0: the DrugBank knowledgebase for 2024. Nucleic Acids Res. 2024;52(D1):D1265–D75.
29. Kiss R, Sandor M, Szalai FA. http://Mcule. com: a public web service for drug discovery. J Chem. 2012;4(Suppl 1):P17.
30. Pence HE, Williams A. ChemSpider: an online chemical information resource. ACS Publications; 2010.
31. Pilón-Jiménez BA, Saldívar-González FI, Díaz-Eufracio BI, Medina-Franco JL. BIOFACQUIM: a Mexican compound database of natural products. Biomol Ther. 2019;9(1):31.

32. Nainala VC, Rajan K, Kanakam SRS, Sharma N, Weißenborn V, Schaub J, et al. COCONUT 2.0: a comprehensive overhaul and curation of the collection of open natural products database; 2024.
33. Dunkel M, Fullbeck M, Neumann S, Preissner R. SuperNatural: a searchable database of available natural compounds. Nucleic Acids Res. 2006;34(suppl_1):D678–D83.
34. Rutz A, Sorokina M, Galgonek J, Mietchen D, Willighagen E, Gaudry A, et al. The LOTUS initiative for open knowledge management in natural products research. elife. 2022;11:e70780.
35. Lafferty M. CHEMnetBASE. Issues Sci Technol Librariansh. 2007;50
36. Lyu C, Chen T, Qiang B, Liu N, Wang H, Zhang L, et al. CMNPD: a comprehensive marine natural products database towards facilitating drug discovery from the ocean. Nucleic Acids Res. 2021;49(D1):D509–D15.
37. Blunt JW, Munro MH. Is there an ideal database for natural products research? In: Natural products: natural products: discourse, diversity, and design. Wiley; 2014. p. 413–31.
38. Rho T. Efforts on improving comparability of ocean chemical data. Sea J Korean Soc Ocean. 2021;26(3):201–19.
39. Chen X, Liu M, Gilson MK. BindingDB: a web-accessible molecular recognition database. Comb Chem High Throughput Screen. 2001;4(8):719–25.
40. Douguet D, Munier-Lehmann H, Labesse G, Pochet S. LEA3D: a computer-aided ligand design for structure-based drug design. J Med Chem. 2005;48(7):2457–68.
41. Biyuzan H, Masrour M-A, Grandmougin L, Payan F, Douguet D. SENSAAS-flex: a joint optimization approach for aligning 3D shapes and exploring the molecular conformation space. Bioinformatics. 2024;40(3):btae105.
42. Mabkhot YN, Alatibi F, El-Sayed NNE, Al-Showiman S, Kheder NA, Wadood A, et al. Antimicrobial activity of some novel armed thiophene derivatives and petra/osiris/molinspiration (POM) analyses. Molecules. 2016;21(2):222.
43. Tharatipyakul A, Numnark S, Wichadakul D, Ingsriswang S, editors. ChemEx: information extraction system for chemical data curation. BMC Bioinformatics. 2012;13(Suppl 17):S9. Springer
44. Acito F. Predictive analytics with KNIME. In: Analytics for citizen data scientists Switzerland. Springer; 2023.
45. Huey R, Morris GM, Forli S. Using AutoDock 4 and AutoDock vina with AutoDockTools: a tutorial. The Scripps Research Institute Molecular Graphics Laboratory 2012;10550(92037):1000.
46. Knight IS, Mailhot O, Tang KG, Irwin JJ. DockOpt: a tool for automatic optimization of docking models. J Chem Inf Model. 2024;64(3):1004–16.
47. Dallakyan S, Olson AJ. Small-molecule library screening by docking with PyRx. In: Chemical biology: methods and protocols. Springer New York; 2015. p. 243–50.
48. Hsu K-C, Chen Y-F, Lin S-R, Yang J-M. iGEMDOCK: a graphical environment of enhancing GEMDOCK using pharmacological interactions and post-screening analysis. BMC Bioinformatics. 2011;12:1–11.
49. Minibaeva G, Ivanova A, Polishchuk P. EasyDock: customizable and scalable docking tool. J Chem. 2023;15(1):102.
50. Li L, Chen R, Weng Z. RDOCK: refinement of rigid-body protein docking predictions. Proteins: Struct Funct Bioinf. 2003;53(3):693–707.
51. Grosdidier A, Zoete V, Michielin O. SwissDock, a protein-small molecule docking web service based on EADock DSS. Nucleic Acids Res. 2011;39(suppl_2):W270–W7.
52. Santos KB, Guedes IA, Karl AL, Dardenne LE. Highly flexible ligand docking: benchmarking of the DockThor program on the LEADS-PEP protein–peptide data set. J Chem Inf Model. 2020;60(2):667–83.
53. Liu Y, Grimm M, Dai W-t, Hou M-c, Xiao Z-X, Cao Y. CB-dock: a web server for cavity detection-guided protein–ligand blind docking. Acta Pharmacol Sin. 2020;41(1):138–44.
54. Corso G, Stärk H, Jing B, Barzilay R, Jaakkola T. Diffdock: diffusion steps, twists, and turns for molecular docking. arXiv preprint arXiv:221001776; 2022.

55. Murail S, De Vries SJ, Rey J, Moroy G, Tufféry P. SeamDock: an interactive and collaborative online docking resource to assist small compound molecular docking. Front Mol Biosci. 2021;8:716466.
56. Schneidman-Duhovny D, Inbar Y, Nussinov R, Wolfson HJ. PatchDock and SymmDock: servers for rigid and symmetric docking. Nucleic Acids Res. 2005;33(suppl_2):W363–W7.
57. Singh A, Copeland MM, Kundrotas PJ, Vakser IA. GRAMM web server for protein docking. In: Computational drug discovery and design. Springer; 2023. p. 101–12.
58. Pierce BG, Wiehe K, Hwang H, Kim B-H, Vreven T, Weng Z. ZDOCK server: interactive docking prediction of protein–protein complexes and symmetric multimers. Bioinformatics. 2014;30(12):1771–3.
59. Yan Y, Tao H, He J, Huang S-Y. The HDOCK server for integrated protein–protein docking. Nat Protoc. 2020;15(5):1829–52.
60. Van Der Spoel D, Lindahl E, Hess B, Groenhof G, Mark AE, Berendsen HJ. GROMACS: fast, flexible, and free. J Comput Chem. 2005;26(16):1701–18.
61. Jung J, Mori T, Kobayashi C, Matsunaga Y, Yoda T, Feig M, et al. GENESIS: a hybrid-parallel and multi-scale molecular dynamics simulator with enhanced sampling algorithms for biomolecular and cellular simulations. Wiley Interdiscip Rev Comput Mol Sci. 2015;5(4):310–23.
62. FrantzDale B, Plimpton SJ, Shephard MS. Software components for parallel multiscale simulation: an example with LAMMPS. Eng Comput. 2010;26:205–11.
63. Paz JO, Batchelor WD, Pedersen P. WebGro: a web-based soybean management decision support system. Agron J. 2004;96(6):1771–9.
64. Hospital A, Andrio P, Fenollosa C, Cicin-Sain D, Orozco M, Gelpí JL. MDWeb and MDMoby: an integrated web-based platform for molecular dynamics simulations. Bioinformatics. 2012;28(9):1278–9.
65. Kampfrath M, Staritzbichler R, Hernández GP, Rose AS, Tiemann JK, Scheuermann G, et al. MDsrv: visual sharing and analysis of molecular dynamics simulations. Nucleic Acids Res. 2022;50(W1):W483–W9.
66. Martínez L, Andrade R, Birgin EG, Martínez JM. PACKMOL: a package for building initial configurations for molecular dynamics simulations. J Comput Chem. 2009;30(13):2157–64.

Chapter 10
Applications of Computational Tools in Formulation Development

Mansi Gandhi, Rohan Pai, and S. N. Koteswara Rao G.

Contents

Abstract The chapter explores how computational models help make pharmaceutical formulations. It talks about using molecular dynamics simulations, along with lab results and computer studies, to increase solubility, stability, and bioavailability of various formulations. These include cyclodextrin complexes, lipid nanoparticles, liposomes, self-nanoemulsifying drug delivery systems (SNEDDS), and other delivery systems. These computational approaches predict drug behavior, optimize formulations, and speed up drug development. The chapter highlights computational tools like AMBER software, GastroPlus, and molecular docking. These tools identify ideal complexation conditions, improving drug solubility and stability. The models help understand how drugs move through the body. They show how drugs get absorbed, distributed, metabolised and eliminated. Computational models allow researchers to design drug formulations that work better. Models help make more effective pharmaceutical products.

M. Gandhi · R. Pai (✉) · S. N. K. Rao G.
Shobhaben Pratapbhai Patel School of Pharmacy & Technology Management, SVKM's NMIMS, Mumbai, India
e-mail: rohan.pai@nmims.edu

S. N. K. Rao G., R. R. Alavala (eds.), *Applications of Computational Tools in Drug Design and Development*, https://doi.org/10.1007/978-981-96-4154-3_10

Keywords Computational models · Cyclodextrin complexes · Lipid nanoparticles · Liposomes · Self-nanoemulsifying drug delivery systems (SNEDDS) · Computational models · Molecular dynamics simulations

10.1 Introduction

Pharmaceutical development and drug delivery are a very critical process as it is a time consuming, costly, tedious method. Even after the final preparation of the formulation, it takes years to penetrate in the market because of the Regulatory constraints. Designing a formulation itself is an important phase in the development of the pharmaceutical product, as it requires selection of excipient, procedure, drug absorption, dissolution, encapsulation efficiency, solubility, etc. Formulation development in pharmaceuticals is a multifaceted process that involves the design and optimization of drug formulations to ensure their safety, efficacy, stability, and manufacturability. However, with the advent of computational tools, there has been a paradigm shift in the way formulations are designed, evaluated, and optimized.

A vast array of approaches and procedures known as computational tools make use of computers' capabilities to create, develop, simulate, and evaluate different aspects of formulation development as described in Table 10.1. With the aid of these instruments, scientists can expedite the search for the best formulations while gaining important insights into the physicochemical characteristics of medications and excipients. They can also forecast formulation behavior under various circumstances. The different computational tools that are currently in use are Molecular Dynamics (MD), Finite Element Method (FEM), Computational Fluid Dynamics (CFD), and Physiologically based Pharmacokinetics model (PBPK).

One of the main advantages of using these tools is being able to connect theoretical models and experimental data. The correlation between computational models and experimental data facilitates the creation of predictive models that precisely depict the behavior of intricate formulations. Through integration, iterative improvement is possible.

In MD simulations, a representation of the interactions between molecules in a formulation is simulated indicating their stability, solubility, and bioavailability. Moreover, molecular docking simulations can predict the binding affinity of drug molecules to target receptors hence accelerate design of drug delivery systems with increased therapeutic efficacy.

Table 10.1 Application of some of the widely used computational techniques

Computational tool	Application
PBPK models	Obtaining information about the ADME of the drug
Molecular dynamics	Identify interaction between different complex molecules
Computational fluid dynamics	Interaction of biological and physical components for drug production at different scales

The Physiologically based Pharmacokinetics (PBPK) model helps to combine the physicochemical characteristics of a drug and the Pharmaceutical formulation in order to forecast Absorption, Distribution, Metabolism and excretion (ADME). The powder during tablet compression is one of the physical phenomena that can be modelled by using the Finite Element method.

Additionally, computational tools have facilitated high-throughput screening of large chemical libraries for identification of potential drug candidates and excipients for formulation development. Virtual screening techniques expedite the identification of lead compounds with desirable pharmacokinetic and pharmacodynamic properties such as structure-based drug design and ligand-based. These methods help to promptly identify potential drugs through simulating how they bind to target receptors reducing the time and expenditure related to experimental screenings.

Artificial intelligence (AI) coupled with machine learning algorithms has also automated formulation development by implementing data analysis, pattern recognition, decision-making processes among others. This is very important because these algorithms can analyze extensive quantities of both experimental and computational data in order to detect correlations, patterns or trends that may not be obvious.

Therefore, it is possible to conclude that computational tools are essential to the development of modern formulations because they give researchers strong and adaptable tools for creating, assessing, and improving drug formulations. These tools enable researchers to overcome the difficulties associated with conventional experimental methods and expedite the discovery of safe, efficient, and commercially viable formulations. Examples of these tools include QSAR modeling, molecular dynamics simulations, and AI-driven approaches.

10.2 Molecular Dynamics

Molecular dynamics is a computational method employed to simulate intricate systems at the atomic scale by numerically solving equations of motion. This technique enables the study of complex biological molecules and their interactions, offering valuable insights into fundamental processes such as protein folding, ion transport, and enzyme catalysis, stability study, inclusion complex. Utilizing force fields to model atomic interactions, popular software tools like AMBER, NAMD, CHARMM, and GROMOS are commonly utilized for these simulations. When several computer cores are to be performed simultaneously, MD helps in reducing the time While molecular dynamics simulations are a powerful tool, challenges such as the refinement of force field accuracy and the computational intensity limiting simulation duration and system size remain areas of active research and development in the field [1].

10.3 PBPK Models

PBPK models are advanced physiologically based compartmental model as they are based on blood flows and tissue models which are interconnected in a physiological manner. These models help in providing information about the ADME of the drug as well as the formulation. These models require high computational, in vitro data, and the human data can be extrapolated from other species because of the physiological similarity. These models can even be used to simulate kinetics at different stages [2].

10.4 Applications of Computational Models in Formulation Development

Computational models can be used in various ways to obtain quicker and less tedious results which might enhance the formulation design and development. It can be used to study the solubility, the entrapment efficiency, PBPK, process design, excipient compatibility, etc.

10.4.1 Inclusion Complex

In a study, a combined approach of molecular dynamics (MD) simulation and in vitro experiments to investigate the enhancement of water solubility and stability of Emodin (ED) through inclusion complexation with β-cyclodextrins (BCDs). Emodin (ED), a natural compound with various pharmacological activities, often suffers from poor water solubility, limiting its bioavailability and therapeutic efficacy. Cyclodextrins (CDs) have been widely studied for their ability to improve the solubility and stability of hydrophobic compounds through inclusion complexation. MD simulation is conducted on the basis of the results obtained from the in vitro experiments to evaluate the stability of the EB-BCD complex. MD simulation was performed for 500 ns to evaluate the inclusion complex form by ED with native BCD. On the other hand, on heating the complex with HPBCD and SBEBCD the stability of these complexes increased. The data obtained, concluded that there are 2 orientations observed for the ED-BCD complex: R-form (resorcinol ring insertion) and C- form (cresol ring insertion. The R-form was predominant in the ED-BCD complex, while the C-form dominated in the ED-modified CD complex. The MD simulation, combined with water accessibility and atomic contact analysis, identified sulfobutylether-β-cyclodextrin (SBEBCD) as exhibiting superior encapsulation efficiency for ED compared to other CDs. The study concludes that MD simulations facilitated the identification of the optimal CD for ED complexation, thereby potentially improving its solubility and stability [3].

A research conducted using AMBER software for aimed to explore the optimal interaction between Rifampicin and native and derivatized cyclodextrins across different pH levels. The study immersed all Rifampicin-cyclodextrin complexes in water at constant temperature, varying pH levels (4, 7, 9), and low energy levels (around 5 kcal) to assess their impact on solubility and cyclodextrin type. Results indicated that solubility increased with rising pH levels, with greater stability observed in γ-cyclodextrin compared to β-cyclodextrin. When examining the complexation of Rifampicin with modified cyclodextrins (HPBCD and HPGCD), the most stable complexes involved Rifampicin being enclosed by the piperazine tail in the inner cavity of HPBCD and the bulky region in the surface cavity. In contrast, for HPGCD, the bulky region was enclosed in the inner cavity. The result obtained after the study were constant at pH 4 and 7, indicating that the stability and the solubility of Rifampicin has increased under the various pH conditions [4].

An inclusion complex was formed of epiisopiloturine with HPBCD and BCD through the in vitro and molecular dynamics. This study gave two different orientations: (1) Imidazole and benzene ring were opposite it to each other and (2) Imidazole and benzene rings are aligned parallel to each other. A higher affinity was observed towards HPBCD than BCD as per the MD simulation. This was because of the one extra hydrogen bonding present in HPBCD than BCD, which results in enhanced binding strength and increased affinity [5].

The study looked at how well a SOR:CD complex dissolved in water. They used computer simulations to do this. First, they minimized the energy of the water molecules. Then, they heated the complex to 298 K at normal air pressure. They examined it for 200 ns using special software called AMBER. They found two ways the complex could bind together (P-form and C-form). When they compared SBEBCD to other cyclodextrins, they saw it could wrap around the SOR molecule better. Fewer water molecules could get in between them. This means the SOR-SBEBCD complex will likely dissolve better and be more stable. Making it useful for medicines [6].

10.4.2 Liposomes

For the formulation of cordycepin liposomes to enhance anti colon cancer activity, in silico studies were used to study the interaction of cordycepin with lipid. This study helped in reducing the time of formulation as it causes less time than the traditional trial and error method. It was observed that phosphatidylcholine lipid was found to be most suitable for cordycepin absorption as it showed a permeation of the drug in the phospholipid layer with 1:20 ratio which proved that the encapsulation efficiency was high and the toxicity level was very low [7].

MD simulation on hypericin to predict the structure and stability of liposome is one more such example where MD has helped in formulation. A 10micro second simulation time was kept to evaluate different number of hypericin against constant lipid molecules in presence of water molecules. It was observed that as the no. of hypericin increased the no. of Course Grain particles also increased resulting in expansion of size of liposome. Also, no all the hypericin molecules were penetrated in the lipid bilayer some were remaining in the lipid water interface. This study gave a huge insight for formulation of liposomes and its drug delivery [8].

Parameter sensitivity assessment was carried for Rifampicin vesicular based liposomes through Gastroplus software to predict the drug dissolution, drug absorption in the various organs. Through this study it was observed that drug dissolution was independent of particle shape, size, solubility and pH of the media. Additionally, GastroPlus software was also used to predict the drug absorbed in the 9 segments of GIT (jejunum, caecum, colon, stomach, duodenum, etc.) and it was observed that approximately 51% of the drug was absorbed which concluded that it is better to deliver the drug through transdermal route instead of the traditional than the oral delivery. In this research GastroPlus helped in anticipating the *in vitro in vivo* behaviour with the pharmacokinetic property [9].

In the process of formulation development, the utilization of a hybrid in silico model has proven to be an invaluable tool in predicting the pharmacokinetic activity of temoporfin liposomes. This model was developed by constructing a multicompartmental framework to predict pharmacokinetics, followed by the simulation of in silico models based on rat and human pharmacokinetics. The model aided in the comparison of plasma profiles of commercially available temoporfin formulations and the evaluation of stability parameters, thereby enhancing the understanding of preclinical, in vivo, and in vitro data. STELLA software was used to perform this *in silico* studies [10].

10.4.3 Self-Emulsifying Drug Delivery Systems (SEDDS)

The loading of cyclosporin A (CyA) by lipid-based NPs made of poly (ethylene glycol) molecules and medium-chain fatty acids at different compositions has been studied using molecular docking (MD). The simulations demonstrated that the NP nanostructure and drug exposure to the aqueous environment were not significantly altered by an increase in drug loading. The NP surface is where the CyA molecules prefer to localize. However, a submerged localization of CyA inside the NP was seen if the particle was overloaded with CyA (increasing the number of drug molecules from 1 to 10). In order to assess the potential of MD simulations for identifying the nanostructure and localization of drug molecules within SEDDS droplets, a variety of lipid excipient systems common to SEDDS design were modeled in conjunction with the immunosuppressive drug molecule CyA, which is used in a variety of commercial applications. Molecule dynamics simulations were used to systematically show how the composition of the SEDDS excipient affects the droplet nanostructure and drug localization [11].

To improve the solubility and dissolution of meloxicam (MLX), stearyl amine (SA) was used in the development of self-nanoemulsifying drug delivery systems (SNEDDS). In the lipid environment (i.e., Labrafil M), MD simulations were used to study the function of SA and the relationship between SA and MLX. The key factors influencing solubility enhancement in the presence of SA were variations in pKa, the physiological charges of both molecules, and the ionic complexation-type interaction between SA and MLX. Liquid lipid solubility was found to be improved by the interactions between MLX and SA [12].

GastroPlus was used to evaluate the PK parameters of Isoniazid SNEDDS and Rifampicin SNEDDS. Three tabs were provided for the data obtained from in vitro studies or literature review to initiate the simulation process. A fasted rat with transit time was selected from the program to evaluate the plasma concentration and permeability. It was observed over the study that the permeability and solubility enhancement lead to increase in PK parameters [13].

10.4.4 Oral Formulations

The study utilized computational fluid dynamics (CFD) in conjunction with the finite element volume method to forecast the release of metformin from an extended-release formulation based on hydroxypropyl methyl cellulose. The CFD model created was employed to assess how the dissolution profile of the tablet is influenced by its geometry, specifically the tooling shape and size. Furthermore, a physiologically based pharmacokinetic (PBPK) model was established using GastroPlusTM software to predict the in vivo performance of a novel high-dose tablet (1000 mg) in comparison to the currently available low-dose (2 × 500 mg) marketed tablet, demonstrating comparable in vivo exposure. The outcomes of bioequivalence investigations on the high-dose tablet aligned with the simulation findings. Consequently, the developed model has the potential to expedite the product development process in this context [14].

To predict the individual plasma profile of the valproate enteric coated tablet formulation, an *in-silico* model was created on the basis of the in vitro dissolution obtained. This study was also supported by the data obtained that gastric condition does not affect drug release mechanism in small intestine. The valproate formulation was compared with 2 other Enteric coated formulation through in silico, which resulted in the result that the valproate formulation was most determinant. Therefore, this study helps in providing a better understanding towards the in vivo studies and individual or patient specific to assess patient in future studies [15].

10.4.5 Miscellaneous

10.4.5.1 Inhaled Drug Products

Budesonide's impact on inhaled drug delivery systems was investigated using the Gastroplus and MPPD models. The amount of budesonide in the rat lungs was examined using the MPPD model, and the information gathered from this software was then imported into Gastroplus to forecast the drug's absorption and intratracheal formulation's disposition. In an effort to save time and produce findings more quickly, this study was carried out in place of the in vivo investigation. For the purpose of the study, a rat that had already been fasted was chosen using the Gastroplus software's PKPlus add-in module in order to assess the pharmacokinetic characteristics of budesonide. In this study, two types of input data were used to test three different types of inhalation formulations: the first was the invivo values of particle deposition of the three formulations, and the second was the insilico predicted values based on particle size data from invitro aerodynamics. Additionally, the software's default Johnson dissolution equation was used to predict how the inhaled formulation would dissolve in the lungs. By the time the study was over, in silico modeling had supplied information about the absorption pattern and regional drug absorption in various lung segments, information that is required to operate an inhaled drug delivery system in cases where an in vivo study would not be able to provide [16].

10.4.5.2 In Gastroretentive Delivery System

A human GI tract model was created with Advance Compartmental Absorption and Transit (ACAT) which was linked with PBPK GastroPlus model to predict the absorption of Ranitidine gastric formulation This ACAT model had 9 compartments (i.e. stomach, duodenum, jejunum (2), ileum (3), caecum, ascending colon). These compartments were connected in a series in order to predict the drug transport, dissolution, absorption. It was assumed that ranitidine has a discontinuous absorption. The invitro data drug dissolution rates were used to predict the pharmacokinetic properties of Ranitidine. Mechanistic explanations for such unusual behaviour can be found in in silico PBPK models, which can capture specificities in drug absorption profile influenced by the combined effect of drug/formulation and physiological factors. As a result, the developed ranitidine specific PBPK model can be a useful tool in the development and assessment of additional oral dosage forms for this medication [17].

Molecular dynamics and PBPK modelling were used to accelerate the formulation of solid dispersion by inputting the invitro dissolution profile in order to obtain the invivo performance of solid dispersion of vemurafenib. AMBER software was used to simulate the development of solid dispersion, all the required excipients were added along with the optimum conditions for the development of SD, to obtain the best suited formulation process. PBPK model was used to predict the in vivo performance by comparing the results of the crystalline drug and the solid dispersion formulation, it was observed that the formulated SD of vemurafenib shows better bioavailability and dissolution rates than the crystalline VEM drug because of the amorphous characteristic of the formulation. Also, precipitation process can be examined by the PBPK model. Thus, the in vitro simulated to obtain the in-silico data can give a better understanding of the in vivo performance of the drug [18].

10.4.5.3 To Assess the Feasibility of Formulation Development

The impact of disintegration, dissolution, absorption, drug distribution and clearance factors influencing the oral bioavailability of atazanavir (ATZ) was investigated using biorelevant dissolution testing with PBPK modeling. The ATZ oral bioavailability was found to be sensitive to the volume of distribution rather than preabsorptive and absorptive factors, as indicated by the plasma profile obtained through simulation studies [19].

10.4.5.4 In Situ Gel Formulation

A MD based evaluation of excipients compatibility and the nanoemul in-situ gel formation of quercetin was conducted by Schrödinger tools. An elucidation of the transition from a sol to a gel phase in the formulation at 313 K was potentially explicable by alterations in the physical properties of the systems and the observed heightened interaction among components at two distinct temperatures, 275 K (1.85 °C) and 313 K (39.85 °C), discerned at a molecular scale. The study scrutinized the formulation's physical properties by assessing parameters such as density, the count of hydrogen bonds between the formulation and solvent, the radius of gyration of poloxamer-407 molecules, and the diffusion coefficients of the solvent and formulation constituents. These findings could significantly contribute to a more profound comprehension of the interactions between excipients and the drug in the formulation of pharmaceutical dosage forms [20].

10.4.5.5 Drug Delivery to Tumour

Development of an in-silico model to explore the effect of different drug regimes and changes in pharmacokinetic parameters on tumour exposure to the cytotoxic drug doxorubicin. The foundation of this computational model is mathematical modeling. The model has illustrated how drug levels in an aberrant vasculature of the tumor are influenced by the distance from supplying vessels. To simulate the growth of tumors, multi-scale three-dimensional models have been developed. By using computational methods to deliver drugs to tumors, it would be possible to gain a better understanding of the mechanisms underlying tumor growth and create more effective formulations [21].

10.4.5.6 Niosome Formulation

MD was used to prepare the bilayer for ifosofamide niosomes. All the excipients were mixed together and then ifosofamide was added on the top layer to interpret how they interacted with each other at 500 k temperature. Further, NVT was conducted to maintain the system temperature at 298 K as it is considered to be the optimum temperature for the preparation of niosomes. This study helps in saving a lot of time in preparation in lab through trial and error method as it gives accurate results about the drug -excipient binding and the stability of the bilayer thus formed [22].

10.5 Conclusion

Computational models not only help in drug discovery but also helps a lot in designing and formulating the final product as it reduces a lot of process time and chances of error also reduces as it helps in various process parameters. There are certain parameters which might not give accurate results which can be improved upon in the future.

References

1. Hospital A, Goñi JR, Orozco M, Gelpí JL. Molecular dynamics simulations: advances and applications. Vol. 8, Advances and applications in bioinformatics and chemistry. Dove Medical Press Ltd; 2015. p. 37–47.
2. Hack CE, Efremenko AY, Pendse SN, Ellison CA, Najjar A, Hewitt N, et al. Physiologically based pharmacokinetic modeling software. In: Physiologically based pharmacokinetic (PBPK) modeling: methods and applications in toxicology and risk assessment. Elsevier; 2020. p. 81–126.
3. Oo A. In silico and in vitro studies on inclusion complexation of anthraquinone derivatives with beta-cyclodextrin derivatives [Internet]. Available from: https://digital.car.chula.ac.th/chulaetd
4. Anjani QK, Domínguez-Robles J, Utomo E, Font M, Martínez-Ohárriz MC, Permana AD, et al. Inclusion complexes of rifampicin with native and derivatized cyclodextrins: in silico modeling, formulation, and characterization. Pharmaceuticals. 2022;15(1):20.
5. De Melo CM, Da Silva AL, De Melo KR, Da Silva PCD, de Souza ML, de Sousa ALMD, et al. In silico and in vitro study of epiisopiloturine/ hydroxypropyl-β-cyclodextrin inclusion complexes obtained by different methods. J Drug Deliv Sci Technol. 2021;65:102658.
6. Aman A, Ali S, Mahalapbutr P, Krusong K, Wolschann P, Rungrotmongkol T. Enhancing solubility and stability of sorafenib through cyclodextrin-based inclusion complexation: in silico and in vitro studies. RSC Adv. 2023;13(39):27244–54.
7. Khuntawee W, Amornloetwattana R, Vongsangnak W, Namdee K, Yata T, Karttunen M, et al. In silicoandin vitrodesign of cordycepin encapsulation in liposomes for colon cancer treatment. RSC Adv. 2021;11(15):8475–84.
8. Jämbeck JPM, Eriksson ESE, Laaksonen A, Lyubartsev AP, Eriksson LA. Molecular dynamics studies of liposomes as carriers for photosensitizing drugs: development, validation, and simulations with a coarse-grained model. J Chem Theory Comput. 2014;10(1):5–13.
9. Hussain A, Altamimi MA, Alshehri S, Imam SS, Singh SK. Vesicular elastic liposomes for transdermal delivery of rifampicin: in-vitro, in-vivo and in silico GastroPlus™ prediction studies. Eur J Pharm Sci. 2020;151:105411.
10. Jablonka L, Ashtikar M, Gao GF, Thurn M, Modh H, Wang JW, et al. Predicting human pharmacokinetics of liposomal temoporfin using a hybrid in silico model. Eur J Pharm Biopharm. 2020;149:121–34.

11. Benson SP, Pleiss J. Molecular dynamics simulations of self-emulsifying drug-delivery systems (SEDDS): influence of excipients on droplet nanostructure and drug localization. Langmuir. 2014;30(28):8471–80.
12. Parekh VJ, Desai ND, Shaikh MS, Shinde UA. Self nanoemulsifying granules (SNEGs) of meloxicam: preparation, characterization, molecular modeling and evaluation of in vivo anti-inflammatory activity. Drug Dev Ind Pharm. 2017;43(4):600–10.
13. Hussain A, Singh SK, Singh N, Prasad Verma PR. In vitro – in vivo - in silico simulation studies of anti-tubercular drugs doped with a self nanoemulsifying drug delivery system. RSC Adv. 2016;6(95):93147–61.
14. Chen W, Desai D, Good D, Crison J, Timmins P, Paruchuri S, et al. Mathematical model-based accelerated development of extended-release metformin hydrochloride tablet formulation. AAPS PharmSciTech. 2016;17(4):1007–13.
15. Karkossa F, Klein S. Individualized in vitro and in silico methods for predicting in vivo performance of enteric-coated tablets containing a narrow therapeutic index drug. Eur J Pharm Biopharm. 2019;135:13–24.
16. Shi C, Ignjatović J, Liu T, Han M, Cun D, Đuriš J, et al. In vitro – in vivo - in silico approach in the development of inhaled drug products: nanocrystal-based formulations with budesonide as a model drug. Asian J Pharm Sci. 2021;16(3):350–62.
17. Cvijic S, Ibric S, Parojcic J, Djuris J. An in vitro – in silico approach for the formulation and characterization of ranitidine gastroretentive delivery systems. J Drug Deliv Sci Technol. 2018;45:1–10.
18. Gao H, Wang W, Dong J, Ye Z, Ouyang D. An integrated computational methodology with data-driven machine learning, molecular modeling and PBPK modeling to accelerate solid dispersion formulation design. Eur J Pharm Biopharm. 2021;158:336–46.
19. Berlin M, Ruff A, Kesisoglou F, Xu W, Wang MH, Dressman JB. Advances and challenges in PBPK modeling – analysis of factors contributing to the oral absorption of atazanavir, a poorly soluble weak base. Eur J Pharm Biopharm. 2015;93:267–80.
20. Aithal GC, Nayak UY, Mehta C, Narayan R, Gopalkrishna P, Pandiyan S, et al. Localized in situ nanoemulgel drug delivery system of quercetin for periodontitis: development and computational simulations. Molecules. 2018;23(6):1363.
21. Hubbard ME, Jove M, Loadman PM, Phillips RM, Twelves CJ, Smye SW. Drug delivery in a tumour cord model: a computational simulation. R Soc Open Sci. 2017;4(5):170014.
22. Hajinezhad MR, Shahraki S, Nikfarjam Z, Davodabadi F, Mirinejad S, Rahdar A, et al. Development of a new vesicular formulation for delivery of Ifosfamide: evidence from in vitro, in vivo, and in silico experiments. Arab J Chem. 2023;16(9):105086.

Chapter 11
Computational Tools for Solubility Prediction

Sasmit Yewale, Rajasekhar Reddy Alavala, and Kunal Gokhale

Contents

S. Yewale · R. R. Alavala · K. Gokhale (✉)
Shobhaben Pratapbhai Patel School of Pharmacy and Technology Management, SVKM's NMIMS, Mumbai, India
e-mail: kunal.gokhale@nmims.edu

S. N. K. Rao G., R. R. Alavala (eds.), *Applications of Computational Tools in Drug Design and Development*, https://doi.org/10.1007/978-981-96-4154-3_11

Abstract Solubility prediction is a pivotal element in diverse fields such as drug discovery, material science, and environmental studies. This chapter delves into the computational tools and methodologies that facilitate accurate solubility predictions. It begins with classical methods like Raoult's Law, Henry's Law, and Hansen Solubility Parameters, providing foundational insights into solubility behavior. The discussion extends to advanced quantum mechanics-based approaches, including Density Functional Theory (DFT) and Molecular Dynamics (MD) simulations, which offer atomistic-level precision in modeling solvation energies and interactions.

Machine learning and data-driven approaches are emphasized for their ability to uncover complex patterns in chemical data using techniques such as neural networks, support vector machines, and random forests. These methods are bolstered by hybrid strategies that integrate quantum mechanics with machine learning, achieving improved accuracy and scalability.

Real-world applications highlight the significance of computational solubility prediction in optimizing drug bioavailability, designing innovative materials, and assessing environmental impacts of pollutants. Furthermore, case studies underscore the practical implications of combining experimental data with computational insights to resolve solubility challenges efficiently.

Emerging trends point towards the integration of multi-scale modeling, quantum-enhanced machine learning, and the use of open-access tools to advance predictive capabilities. The chapter concludes with a call for interdisciplinary collaboration and the development of standardized protocols to enhance the reliability and applicability of computational solubility prediction across scientific and industrial domains.

Keywords Solubility prediction · Computational tools · Machine learning · Quantum mechanics · Molecular dynamics · Drug design and material science

11.1 Introduction

Solubility is the ability of a substance to dissolve in a solvent to form a solution. The substance that dissolves is called the solute, and the solvent is usually a liquid. The process of dissolving is called solvation.

- **Importance of Solubility Prediction:**
- Predicting a substance's solubility is crucial as it helps in drug development, chemical synthesis, and industrial operations. For example, in drug development, medicinal chemists must choose which lead compounds to investigate further. Solubility prediction using software tools or computational models allows researchers to quickly identify compounds with desirable solubility properties. If Compound A is found to be more soluble in water than Compound B, it may have better bioavailability and pharmacokinetic attributes [1–3]. This helps the pharmaceutical company prioritize Compound A for further studies, saving time and money. Additionally, early solubility prediction allows researchers to address solubility challenges by exploring formulation techniques like salt formation or particle size reduction.
- **Solubility and its role in Pharmacokinetics:**

  1. **Absorption:** Solubility impacts a drug's ability to dissolve in digestive fluids and enter the bloodstream. Poor solubility can lead to low bioavailability [4].
  2. **Distribution:** Solubility determines how a drug spreads through the body. Lipid-soluble drugs permeate cell membranes and fatty tissues, while water-soluble drugs stay in the bloodstream and extracellular fluids.
  3. **Metabolism:** Solubility affects how quickly a drug is metabolized. Highly soluble drugs are metabolized faster, while those with low solubility may have longer half-lives.
  4. **Excretion:** Solubility influences drug elimination. Water-soluble drugs are excreted via urine, while lipid-soluble drugs may be metabolized into water-soluble forms or excreted through bile.

- **Benefits of Solubility Prediction:**

  - **Improved Absorption:** Poorly soluble medicines can be made more soluble to enhance absorption and bioavailability. This can be achieved through formulation optimization, solubilizing agents, or prodrugs that become active in the body.
  - **Controlled Release:** Solubility can be adjusted to control the drug's release rate. Sustained release formulations alter solubility over time, allowing for prolonged drug action and reduced dosing frequency.
  - **Targeting Particular Tissues:** Adjusting a drug's solubility allows for targeting specific tissues. Lipid-soluble drugs added, are ideal for central nervous system disorders, while water-soluble drugs may be better for systemic infections.

- **Application of Solubility Prediction:**

  Formulation Development: Solubility prediction aids in developing medication formulations with optimal solubility, enhancing bioavailability and effectiveness.

Drug Design: Predicting solubility is key in drug design, helping chemists select lead compounds with favourable solubility early in development, reducing the risk of solubility issues later.
ADME Optimization: Solubility prediction helps optimize a drug's absorption, distribution, metabolism, and excretion (ADME) by enabling the creation of compounds with desired physicochemical properties [5–7].
Lower Experimental Costs and Time: Solubility prediction models offer a cost-effective and efficient alternative to labour-intensive solubility experiments, quickly estimating the solubility of many compounds.
Risk Mitigation: By identifying potential solubility issues early, researchers can focus on compounds with better profiles or explore ways to enhance solubility, reducing risks during drug development.
Environmental Impact: Solubility prediction supports the development of environmentally friendly processes in the chemical industry by forecasting which chemicals will dissolve in eco-friendly solvents [8, 9].

Since solubility prediction has wide-ranging ramifications, it is extremely important in many fields. The following is a synopsis of its relevance in material science, environmental research, and drug discovery [10]**:**

1. Drug Discovery
   Medicinal chemists might choose lead compounds with advantageous pharmacokinetic properties early in the process by using solubility prediction.

   - Predicting solubility aids in optimizing the absorption, distribution, metabolism, and excretion (ADME) properties of drug candidates, improving therapeutic outcomes.
   - It directs formulation development to enhance medication solubility, bioavailability, and efficacy.

Anticipating issues linked to solubility can help researchers reduce risks and expedite the drug development process.

2. Material Science

   - It helps anticipate the solubility of polymers, nanoparticles, and other materials, guiding their formulation and processing.
   - It promotes the design and synthesis of innovative materials with specialized properties for specific applications.

Understanding phase behaviour, crystallization, and dissolution kinetics of materials is influenced by solubility prediction. It is essential for creating materials with desired mechanical, thermal, electrical, or optical properties.

3. Environmental Studies
   By comprehending the solubility behaviour of substances, researchers can assess their potential environmental impact and develop strategies for remediation and pollution control. Solubility prediction aids in predicting the solubility of

pollutants, contaminants, and pharmaceuticals in water, soil, and air. It also helps in assessing the fate and transport of chemicals in the environment.

By using mathematical models, algorithms, and simulations to estimate the solubility behaviour of chemical compounds, computational methods are essential in the solubility prediction process. These techniques provide useful information about the solubility characteristics of various chemicals and provide an economical and effective substitute for experimental methods. An overview of computational techniques and their function in solubility prediction is provided below [11]**:**

1. Quantitative Structure-Activity Relationships, or QSAR models, link a compound's chemical structure to its physical, chemical, or biological characteristics, such as solubility. These models measure the structural aspects related to solubility using molecular descriptors such molecular weight, hydrophobicity, and electrical properties. QSAR models are frequently created using machine learning algorithms, including linear regression, support vector machines (SVM), and artificial neural networks (ANN), using datasets of known solubility values.
2. Molecular Dynamics (MD) simulations: By simulating atom-to-molecule mobility and interactions over time, MD simulations shed light on how molecular-level chemical systems behave. MD simulations help clarify the thermodynamics and kinetics of solute-solvent interactions, such as solute dissolution, hydration, and solvation, in the context of solubility prediction.

   MD simulations can determine solubility-related parameters including solubility energy, solvation free energy, and solute-solvent interactions by examining the trajectories of molecules in solution.
3. Calculations using Quantum Mechanics (QM): QM computations provide highly accurate results for the electronic structure and energetics of molecules by solving the Schrödinger equation.

   By simulating intermolecular interactions, hydrogen bonding, and electrical factors that affect solute-solvent interactions, these computations can forecast solubility-related features.

   In QM calculations, solubility energy, solvation energy, and other solubility descriptors are frequently predicted using density functional theory (DFT) and ab initio techniques.
4. Statistical Thermodynamics: Using the concepts of thermodynamics, statistical thermodynamics techniques, like the Gibbs free energy calculation, forecast phase behaviour and solubility equilibrium.

   These techniques compute the solubility of a solute in a solvent under circumstances by considering variables like temperature, pressure, entropy, and enthalpy.

   Statistical thermodynamics methods shed light on the factors influencing solubility, such as entropy-driven and enthalpy-driven processes.

11.2 Classical Methods for Solubility Prediction [12]

The shake flask method [13] is a classical technique for figuring this out. It entails combining a sample of a solute with a buffer solution until saturation. After the saturated solution reaches equilibrium, the dissolved concentrations define the solution's ultimate solubility.

Let us understand **Shake Flask Method** in more detailed manner:

1. Get a Solution Ready: Transfer a known mass of the material whose solubility you need to ascertain into a dry, clean flask.
2. Add Solvent: Fill the flask with a precisely calculated volume of the solvent. The solvent should be used in an amount that will dissolve the substance to the greatest extent possible.
3. Shake: To guarantee that the solute and solvent are thoroughly mixed, seal the flask and give it a good shake. This phase is essential to achieving balance.
4. Permit Equilibration: Give the flask enough time to come to equilibrium by letting it sit in an undisturbed state. This could take a few hours or a full day, depending on how soluble the material is.
5. Filtration (if required): If the material is partially soluble in the solvent, strain the mixture to get rid of any solid particles that have not dissolved. Measuring only dissolved solute is ensured by this procedure.
6. Analyse: Find out how much solute is in the solution. This can be accomplished in several ways, including spectrophotometry, titration, and even just weighing a sample of the solution.
7. Repeat: To create a solubility curve or compare solubilities under various circumstances, repeat the procedure using different solute concentrations or solvents.
8. Finally, determine the substance's solubility under the specified circumstances by calculating the volume of the solvent and the solute's observed concentration in the solution.

Example- **Solubility of a Drug in a Solvent:** To ascertain a drug's solubility in a range of solvents, the pharmaceutical industry frequently use the shake flask method. Let us take the example of figuring out how soluble a novel medicinal molecule is in ethanol.

One would take a dry, clean flask and fill it with a known mass of the medicinal component. After that, you would fill the flask with a precise amount of ethanol. To dissolve the medication in the solvent, you would once more give the flask a good shake.

Following a predetermined amount of time for equilibration, you would filter the mixture if needed and use suitable analytical methods, such as high-performance liquid chromatography (HPLC), to determine the drug's concentration. You can ascertain the medication's solubility in ethanol under the specified circumstances by repeating this procedure with other drug masses and ethanol volumes. When creating dosage forms for the medication, such as tablets or capsules, this information is essential.

Highlighting the key points:

Prepare Solution: Place a known mass of the material in a clean flask.
Add Solvent: Add a calculated volume of solvent to dissolve the substance as much as possible.
Shake: Seal and shake the flask to mix thoroughly.
Equilibrate: Allow the mixture to sit undisturbed until equilibrium is reached.
Filter (if needed): Filter out undissolved solids to ensure only dissolved solute is measured.
Analyse: Determine the solute concentration using methods like spectrophotometry or titration.
Repeat: Conduct the process with different conditions to create a solubility curve.
Calculate Solubility: Use the solvent volume and solute concentration to calculate solubility under the given conditions.

We can also understand better with these 2 laws-.

11.2.1 Raoult's Law and Henry's Law: In This We Will Read About the Explanation of These Laws and Their Application in Predicting Solubility [14]

Henry's Law and Raoult's Law are crucial ideas for comprehending how solutions behave, especially when it comes to forecasting solubility.

1. **Raoult's Law**: This law states that the vapor pressure of a component in an ideal solution is directly proportional to its mole fraction in the solution. It's primarily applicable to solutions where the solute and solvent molecules are similar in size and nature. Raoult's Law can be used to predict the vapor pressure of a volatile component in a solution based on its mole fraction and vapour pressure of pure component.

The general form of Raoult's Law for an ideal solution is:

$$\boldsymbol{P\text{total}} = \boldsymbol{P\text{solvent}} \times \boldsymbol{x\text{solvent}} + \boldsymbol{P\text{solute}} \times \boldsymbol{x\text{solute}}$$

Where:

- Ptotal is the total vapor pressure of the solution,
- Psolvent is the vapor pressure of the pure solvent,
- Psolute is the vapor pressure of the pure solute (usually assumed to be negligible unless it is a volatile solute),
- $x$solvent is the mole fraction of the solvent in the solution, and
- $x$solute is the mole fraction of the solute in the solution.

For dilute solutions, where the solute concentration is low, $x$solute can be approximated to zero, simplifying Raoult's Law to:

$$\boldsymbol{P}\textbf{total} = \boldsymbol{P}\textbf{solvent} \times \boldsymbol{x}\textbf{solvent}$$

Let us understand with the help of diagrammatic representation (Fig. 11.1).

2. **Henry's Law**: This law states that the concentration of a gas in a solution is directly proportional to the partial pressure of that gas above the solution. Henry's Law is typically applied to dilute solutions and describes the solubility of gases in liquids. It is particularly useful in understanding gas solubility in liquids like water.

   $\boldsymbol{C} = \boldsymbol{k}\cdot\boldsymbol{P}$ Where:

- C is the concentration of the gas in the liquid (usually in mol/L or Molarity).
- P is the partial pressure of the gas above the liquid (usually in atmospheres, atm).
- k is Henry's Law constant, which is specific to the solute-solvent pair at a particular temperature.
- The value of k depends on factors such as the nature of the gas, the solvent, and the temperature.
- For ideal dilute solutions, the constant k is directly proportional to the solubility of the gas in the solvent. Therefore, you might also see Henry's Law expressed as:

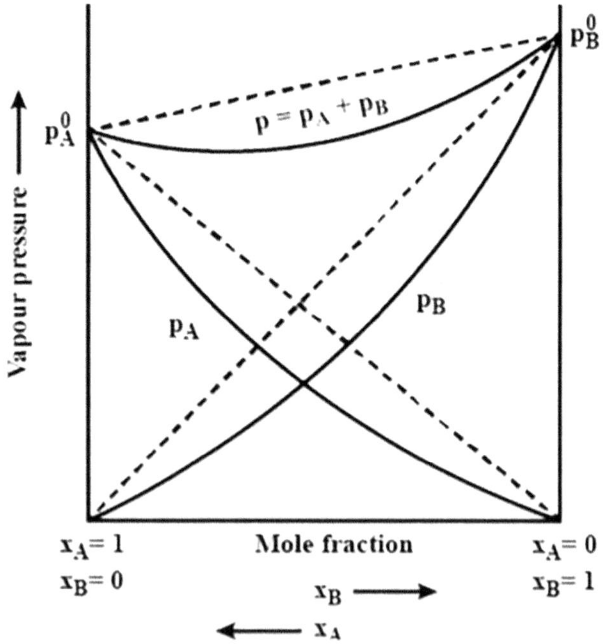

Fig. 11.1 Schematic representation of Raoult's Law

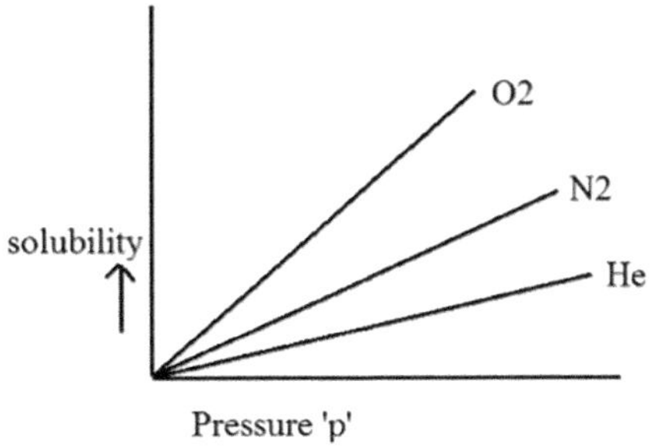

Fig. 11.2 Schematic representation of Henry's Law

$$C = H \cdot PC = H \cdot P$$

Where H is Henry's Law constant.

Let us understand with the help of a graph (Fig. 11.2):

When it comes to applying these rules to predict solubility: For instance, Raoult's Law applies more to liquid solutions of volatile solutes in volatile solvents. It can assist in forecasting the impact of a solute's presence on the solution's vapor pressure. Henry's Law applies more to the dissolution of gases in liquids. Based on the partial pressure of the gas above the solution, aids in the prediction of the solubility of gases in a liquid solvent. The activity coefficients of components in liquid mixtures can be predicted using group contribution techniques called UNIFAC (20) and UNIQUAC (21) which can be used to predict solubility.

11.2.2 Group Contribution Methods: Overview of Methods Such as UNIFAC and UNIQUAC [15]

1. UNIFAC: Based on the idea of breaking down a molecule into functional groups and considering their interactions, this system was created in the 1970s by Fredenslund, Rasmussen, and Prausnitz. Due to its accuracy and versatility, particularly for complex mixtures, it is widely used in the chemical engineering industry to predict activity coefficients by considering excess Gibbs free energy and combinatorial entropy

11.2.2.1 UNIFAC for Solubility

UNIFAC computes activity coefficients based on the interactions between these structural groups, considering the molecular structure of each component in the mixture. UNIFAC can be used to estimate liquid phase activity coefficients, which can subsequently be used to predict solubility in the context of solubility.

UNIFAC can be used, for instance, to estimate the activity coefficients of the solute and solvent in the solution if you wish to forecast the solubility of a solute in a solvent. The solute's solubility in the specified solvent is then estimated by combining these activity coefficients with the thermodynamic formula for solubility.

2. UNIQUAC: Developed in the 1970s as well by Maurer and Prausnitz. It is based on the division of molecules into functional groups, just like UNIFAC, but it considers more intricate relationships.

11.2.2.2 UNIQUAC for Solubility

UNIQUAC and UNIFAC both estimate activity coefficients in liquid mixtures using group contribution methods. It takes into consideration the chemical interactions between molecules as well as their size and form.

By estimating the activity coefficients of the solute and solvent in the solution, UNIQUAC can be used to predict the solubility of a solute in a solvent. After that, these activity coefficients are used to estimate solubility in equations like Raoult's law or Henry's law, which vary based on the solute's type and concentration range.

11.2.3 Hansen Solubility Parameters: Explanation of How These Parameters Are Used to Predict Solubility [16]

When evaluating phase equilibrium parameters, such as solubility, in a variety of industrial processes involving liquid mixtures, UNIFAC and UNIQUAC are both useful methods. But depending on the system and the range of conditions considered, their accuracy may differ. The three parameters known as Hansen Solubility Parameters (HSP) are based on the dispersion forces (δD), polar forces (δP), and hydrogen bonding forces (δH) of a material. They are used to describe the solubility behaviour of that substance. The HSP space is a three-dimensional representation of these parameters that is frequently used. To estimate solubility using Hansen Solubility Parameters, one usually takes the subsequent actions:

1. Ascertain the solvent and solutes' HSP: Determine the HSP values for the target solutes and the solvent. This entails calculating or measuring the contributions of each substance's hydrogen bonds, polarity, and dispersion.
2. Plot the HSP values of the solvent and solutes in the HSP space and compare them. The proximity of a solute's HSP values in this space can be used to forecast the solute's solubility in each solvent. Solutes should generally be more soluble in the solvent if their HSP values are comparable to that of the solvent.
3. Employ Distance Metrics: The compatibility between the HSP values of solvents and solutes can be quantitatively evaluated by using a variety of distance metrics, such as Hansen distance or Euclidean distance. The anticipated solubility increases with decreasing distance between the solvent and solute's HSP values.

4. Consider Practical Limitations: Although HSP can offer insightful information about solubility, practical applications must consider additional variables including temperature, pressure, and chemical interactions. Furthermore, especially in complex systems, HSP might not be able to adequately capture all facets of solubility behaviour.
5. Validate Experimentally: To verify the predictions generated by HSP, experimental validation is essential. Conducting solubility experiments under appropriate conditions is necessary to confirm the accuracy of the predictions. When evaluating solubility based on the forces of interaction between solvent and solute molecules, Hansen Solubility Parameters offer a systematic and intuitive approach. This technique aids in the selection of suitable solvents for a variety of applications in industries such as polymers, coatings, and medicines.

11.3 Quantum Mechanics-Based Approaches-

11.3.1 Density Functional Theory (DFT): How DFT Is Used to Calculate Solvation Energies and Predict Solubility [13, 17–19]

A computational quantum mechanical modelling technique called density functional theory (DFT) is used to comprehend and forecast the characteristics of molecules and materials. DFT may be used in several methods to predict solubility and calculate solvation energies:

1. Solvation Energy Calculation: The energy change related to the solute's dissolution in a solvent is called the solvation energy, and it may be determined using DFT. Modelling the interactions between the molecules of the solvent and the solute is required for this. The solvation energy can be found by computing the energy difference between the solvated and unsolvated states.
2. Prediction of Solvation Structure: DFT can also be used to predict the solvent molecules' arrangement surrounding the solute, as well as the structure of solvated molecules. Knowing how various solvents interact with solutes and how these interactions affect solvation energies is crucial.
3. Solubility Prediction: DFT can be used to predict solubility by integrating solvation energy calculations with thermodynamic models, such as the Gibbs free energy. The balance between the energy needed to break the solute-solute contact (lattice energy or cohesive energy) and the energy obtained from the solute-solvent interactions (solvation energy) determines solubility. Both of these energy contributions can be understood through the use of DFT.
4. Screening Solvents: Using their solvation energies as a guide, DFT can be used to screen possible solvents for a particular solute. One can forecast solubility

trends by calculating the solvation energies of a variety of solvents and determining which solvent is most likely to offer the best solvation for a given solute.**A real-life scenario to understand it better:**

- Design of Catalysts: DFT is utilized in the design and improvement of catalysts for chemical processes. DFT, for instance, was essential in the development of catalysts for the ammonia-producing Haber-Bosch process. DFT aids in the identification of the most effective catalysts for nitrogen fixation, enhancing the energy efficiency of the process by forecasting the electrical characteristics of various materials.
- Drug Design: Drug compounds' electrical structures and interactions with biological targets are modelled using DFT. In one instance, HIV protease inhibitor binding affinity was investigated using DFT, which aided in the development of more potent antiretroviral medications.
- Material Science: DFT is useful in the creation of cutting-edge materials like graphene. DFT is used by researchers to investigate the electronic characteristics of graphene and its derivatives. Table 11.1 showed a comparative table on the strengths and the limitations of DFT theory

Table 11.1 A comparative table on the strengths and the limitations of DFT theory

Aspect	Strengths	Limitations
Accuracy	Gives many systems a reasonable balance between computing expense and precision.	Possibly difficult to explain weak interactions and van der Waals forces precisely
Computational efficiency	More computationally efficient than MP2 or CCSD, two other quantum mechanical techniques.	Still potentially costly to compute for very big systems or lengthy simulations.
Versatility	Applicable to a wide range of systems including solids, liquids, and gases.	Accuracy highly depends on the choice of exchange-correlation functional.
Predictive power	Effective for predicting electronic properties, geometries, and reaction pathways.	May not always accurately predict excited states and transition states.
Scalability	Scales well with system size, making it suitable for studying larger molecules or periodic systems.	Limited in handling very large biological systems or complex environments without approximations.
User accessibility	Widely implemented in many computational chemistry software packages, making it accessible to researchers.	Requires expertise to choose the right functional and interpret results correctly.
Cost	Relatively cost-effective in terms of computational resources compared to other high-level methods.	Despite being efficient, still resource-intensive for extensive systems.

11.3.2 Molecular Dynamics (MD): Introduction to MD Simulations and Their Application in Solubility Prediction [20–24]

Computational methods called **Molecular Dynamics (MD)** simulations are used to examine how atoms and molecules move and interact over time. They are based on Newton's equations of motion, which track atoms' locations and velocities as they change over time in response to external forces. With the aid of MD simulations, scientists can better comprehend processes like protein folding, chemical interactions, and material qualities by gaining insight into the dynamic behavior of molecules.

MD simulations can be used to study the behavior of solute molecules in solvent environments in the context of solubility prediction. Researchers can estimate solubility qualities including solvation-free energies, solubility parameters, and solubility curves by modeling the interactions between solute and solvent molecules. These simulations help in medication design, material science, and environmental studies by predicting the solubility of a given molecule in several solvents under various conditions.

However, because of the intricacy of the molecular interactions involved, MD simulations for solubility prediction require precise force field parameters and adequate computer power. Furthermore, rigorous validation against experimental data is necessary to guarantee the accuracy of predictions. Despite these difficulties, molecular dynamics (MD) simulations provide insightful information about solubility processes that aid in the creation of novel materials.

A real-life scenario to understand it better:

- Protein Folding: The folding process of proteins is studied using molecular dynamics simulations. For instance, the application of MD in understanding the folding mechanics of the amyloid-beta peptide, which is linked to Alzheimer's disease, aids in the creation of treatment plans.
- Drug-Protein Interactions: MD simulations aid in the comprehension of the molecular interactions that occur between pharmaceuticals and their target proteins. For example, Gleevec (imatinib) binding to the BCR-ABL protein in chronic myeloid leukaemia was studied using MD, providing insights into mechanisms of medication resistance.
- Material Behaviour: The mechanical characteristics of materials under stress are studied using MD simulations. MD simulations are used in aerospace engineering to assist design stronger aircraft components by predicting the behaviour of materials under extreme pressure and temperature, such as carbon-fibre composites. Table 11.2 showed a comparative table on the strengths and the limitations of Molecular Theory

Table 11.2 A comparative table on the strengths and the limitations of Molecular Theory

Strengths	Limitations
Atomic-level detail	Computationally expensive
Provides detailed insights into the atomic and molecular interactions, allowing for an in-depth understanding of dynamic processes.	Requires significant computational resources, especially for large systems or long simulation times.
Time-dependent analysis	Limited by simulation time
Can simulate the time evolution of systems, capturing dynamic events such as folding, binding, and conformational changes.	Typically limited to nanoseconds to microseconds, which may not be sufficient to capture slower biological or chemical processes.
Versatility	Sensitivity to initial conditions
Applicable to a wide range of systems, including proteins, nucleic acids, membranes, and materials.	Results can be highly sensitive to the initial setup, such as the starting structure, force fields, and simulation parameters.
Force field-based	Approximations in force fields
Utilizes well-established force fields that can be customized for specific types of molecules or materials.	Force fields are based on approximations and may not always accurately represent all types of interactions, especially in complex systems.
Insight into mechanisms	Requires extensive validation
Helps elucidate mechanisms of molecular interactions, such as binding, diffusion, and conformational changes.	Results often need to be validated with experimental data or higher-level quantum mechanical methods.
Visualization of complex systems	Limited quantum mechanical accuracy
Allows for the visualization of molecular motions and interactions, aiding in understanding complex biological processes.	Classical MD does not account for quantum mechanical effects, which may be important in certain systems or reactions.
Customization and flexibility	Difficulty in modeling large systems
Can be tailored to specific research needs, including the choice of force fields, simulation conditions, and system size.	While MD can handle relatively large systems, scaling up to very large systems (e.g., entire cells) remains challenging.

11.3.3 Quantum Mechanics/Molecular Mechanics (QM/MM): Brief Overview of QM/MM Methods and Their Role in Solubility Prediction [25–29]

To examine complicated chemical systems, quantum mechanics/molecular mechanics (QM/MM) approaches combine the precision of QM with the effectiveness of MM. A zone treated with classical molecular mechanics (the surrounding environment) and a region treated with quantum mechanics (usually the active site or a selection of atoms) divides the system in QM/MM simulations.

1. QM Region: Chemical reactions, bond breaking/forming, and electronic structure computations all fall under the category of QM region interactions, which are the atoms and interactions that need to be accurately treated in quantum mechanics. Density functional theory (DFT) and ab initio methods are commonly used to characterize this region; these approaches properly explain electrical characteristics but are computationally demanding.

2. MM area: Force field models are used to treat the MM area, which encompasses the remainder of the system. Large molecular systems can be efficiently simulated thanks to the use of force fields in molecular mechanics, which approximate interatomic interactions based on empirical data.
3. Boundary Treatment: To maintain consistency between the two regions, coupling terms are inserted at the contact between the QM and MM sections. Information is transferred between the QM and MM areas by these coupling terms enabling the accurate description of interactions across the boundary.
4. Applications in Solubility Prediction: Because QM/MM techniques can record the interaction between solute and solvent at the electronic level, they are useful for researching solvation phenomena and forecasting solubility. QM/MM simulations are able to offer comprehensive insights into solute-solvent interactions, solvation energies, and solubility trends because they treat the solute quantum mechanically and the solvent/classical environment using molecular mechanics.
5. Calculating Solvation Energies: By taking into consideration the solvent's induced polarization effects as well as the solute's electronic structure, QM/MM techniques may precisely compute solvation energies. These computations aid in the prediction of solubility and offer numerical measurements of the interactions between the solute and the solvent.
6. Understanding of Solvation Mechanisms: QM/MM simulations help clarify the mechanisms behind solvation activities, such as hydrogen bonding interactions, solvent penetration, and hydration shell creation. Comprehending these mechanisms is essential for forecasting changes in solubility and creating compounds with the intended solubility characteristics.

Let us understand with help of diagram:

Enzyme Catalysis: QM/MM methods are used to study enzyme-catalysed reactions. For example, QM/MM was applied to investigate the catalytic mechanism of cytochrome P450 enzymes, which are involved in drug metabolism. This information is critical for understanding how drugs are processed in the body and for designing safer pharmaceuticals.

Photoreceptor Proteins: QM/MM simulations have been used to explore the photochemistry of rhodopsin, the proteins involved in vision. By modelling the light-induced reactions in these proteins, researchers gain insights into the molecular basis of vision and related disorders.

Chemical Reactions in Solution: QM/MM methods are used to study complex chemical reactions in solution, such as the hydrolysis of ATP in biological systems. This approach allows researchers to model the reaction mechanism with quantum precision while considering the influence of the surrounding water molecules, providing a more accurate understanding of biochemical processes.

All things considered; by combining precise quantum mechanical descriptions with effective molecular mechanics simulations, QM/MM approaches provide a potent way for researching solvation processes at the molecular level and forecasting solubility. Table 11.3 represented a comparative table on the strengths and the limitations of QM/MM.

Table 11.3 A comparative table on the strengths and the limitations of QM/MM

Strengths	Limitations
Accuracy in modelling complex systems: Combines the precision of quantum mechanics with the efficiency of molecular mechanics, making it ideal for studying large biomolecular systems.	**Computationally intensive**: The QM part requires significant computational resources, especially for larger systems or high-accuracy levels.
Flexible and scalable: Can be applied to a wide range of chemical and biological systems, including enzyme catalysis, protein-ligand interactions, and materials science.	**Complex setup**: Requires careful partitioning of the system into QM and MM regions, and errors can arise from inappropriate boundaries or interactions between the regions.
Captures electronic effects: Allows for accurate modeling of electronic structure changes, which is crucial for studying reactions and transition states.	**Limited by QM accuracy**: The accuracy of the QM region is dependent on the chosen level of theory, and higher accuracy can significantly increase computational cost.
Efficient for large systems: By treating only the reactive region with QM and the rest with MM, QM/MM can handle much larger systems than full QM methods.	**Parameter dependence**: The accuracy of the MM part is dependent on the quality of the force fields used, which may not always be well parameterized for the specific system studied.
Realistic environmental modeling: Can model reactions in realistic environments, such as enzymes or solvated systems, where full QM would be infeasible.	**Boundary effects**: The interface between QM and MM regions can sometimes lead to artifacts or inaccuracies in the simulation results.

- **"Sources and Methods for Gathering Solubility Data in Chemical Research":**
  1. Experimental Databases: Make use of experimental databases that provide information on the solubility of different chemicals in a range of solvents. ACD/Labs Solubility Database, PubChem, ChemSpider, and several literature sources are a few examples.
  2. Computational Databases: A few databases aggregate solubility data estimated by computational techniques like as molecular simulations or quantum chemistry. These databases can offer predictions for chemicals for which there is insufficient experimental evidence, or they can complement experimental data.
  3. Software for Predicting Properties: Based on a chemical compound's molecular structure, software programs such as ALOGPS, COSMOtherm, and several QSAR (Quantitative Structure-Activity Relationship) models predict the solubility values of the compound.
  4. Custom Data Collection: To obtain solubility data, you might need to run simulations or experiments if particular chemicals or solvents are of interest. This could entail predicting solubility values computationally or by performing solubility experiments in the lab.

- **"Key Descriptors and Factors Influencing Solubility Prediction in Chemical Systems":**

  1. Molecular Descriptors: Choose molecules that best represent the physicochemical characteristics of solvents and solutes that are important for solubility. Molecular weight, surface area, hydrogen bonding ability, lipophilicity, polarizability, and other characteristics might be some of these descriptors.
  2. Functional Groups: Choose certain chemical motifs or functional groups that are known to affect solubility and include them as characteristics. In polar liquids, for instance, hydrophilic groups such as hydroxyl (-OH) and carboxyl (-COOH) tend to promote solubility.
  3. Topological Indices: To describe the connection and structure of molecules, use topological indices. These indices include details on branching, symmetry, molecule size, and other structural characteristics that might influence solubility.
  4. 3D molecule Descriptors: Take into account adding three-dimensional (3D) descriptors that describe molecule forms and spatial configurations. These characteristics can shed light on how intermolecular interactions and steric factors affect solubility.
  5. Properties of the Solvent: Provide details on the solvent's polarity, dielectric constant, capacity to donate and receive hydrogen bonds, and accessibility. Features unique to each solvent can aid in capturing solvent-solute interactions that are essential for predicting solubility.
  6. Hydration Energy: Include characteristics pertaining to solvation-free energy, also known as hydration energy, which measure the energy shift brought about by a solute dissolving in a solvent. These characteristics give clear insight into the interactions between the solute and solvent that govern solubility.

11.4 Machine Learning and Data-Driven Approaches

11.4.1 Overview of Machine Learning Techniques Such as Neural Networks, Support Vector Machines, and Random Forests in Solubility Prediction [30–35]

Solubility prediction problems frequently make use of machine learning methods including neural networks, support vector machines (SVM), and random forests. An outline of each technique's operation in this situation is provided below:

1. Neural Networks: By identifying intricate patterns and correlations in the input data, neural networks, in particular deep learning models like convolutional neural networks (CNNs) or recurrent neural networks (RNNs), may be utilized for solubility prediction.

- Neural networks are capable of predicting the solubility values of substances by using their structural properties or chemical descriptors as input.
- When working with complicated chemical structures, deep learning models can be useful since they can automatically extract characteristics from raw data.

2. SVMs, or support vector machines:

   - For problems involving regression and classification, SVM is a supervised learning method.
   - Molecular descriptors as features and the related solubility values as labels can be used to train support vector machines (SVM) for solubility prediction.
   - In classification, SVM looks for the hyperplane that best divides data points into classes; in regression, it looks for the hyperplane that fits the data points with the greatest margin.

3. Random Forests: A group of decision trees make up the ensemble learning technique known as random forests.

   - Random forests are capable of handling non-linear correlations between features and solubility values in solubility prediction. In order to lessen overfitting and enhance generalization, every decision tree in the random forest is trained using a different random subset of the features and data.

Although very effective, data-driven methods for solubility prediction have many drawbacks and difficulties.

This work used a range of machine learning models to predict solubility, and it was published in Nature Communications. Support vector machines (SVM), Random Forest (RF), Multiple Linear Regression (MLR), Artificial Neural Networks (ANN), and other linear and non-linear models were compared by the researchers.

Important Results:

Comparatively speaking, non-linear models like Random Forest and Neural Networks outperformed linear models like MLR.

To evaluate the model's accuracy in predicting solubility, the study proposed new evaluation measures. Compared to more conventional metrics like R2 and RMSE, one such metric measured the proportion of predictions that fell within a specific error range from the actual experimental values.

According to the research, machine learning can accurately anticipate a wide variety of chemicals' solubility, which lessens the requirement for.

- **Addressing potential ethical considerations or biases that might arise from data-driven approaches in solubility prediction:**

  - **Accuracy and Quality of Data:**
  - Bias resulting from incomplete or faulty data Data-driven strategies mainly depend on the calibre of the incoming data. Predictions may be erroneous or deceptive if the data used to train the models is inadequate, biased, or of low

quality. This might result in false conclusions and potentially dangerous choices, particularly in domains such as medication development.

- Over-dependence on Past Information Solubility prediction models frequently rely on historical data, which could be biased or represent out-of-date procedures. For instance, previous data may be skewed in favor of particular chemical groups, which could result in predictions for new or underrepresented substances being less accurate.
- **Model Interpretability and Transparency:**
- **Opaque Models:** Certain data-driven strategies, like machine learning, might be "black boxes," which means it is difficult to understand how they make decisions internally. It may be difficult to comprehend the reasoning behind a certain prediction because to this lack of transparency, which could be problematic for researchers who need to validate the results or for regulatory approval.
- **Ethical Responsibilities:** It is ethically required to make sure that a solubility prediction model's predictions are dependable and understandable if it is to be utilized in crucial contexts such as drug development. This entails comprehending the model's constraints and offering justifications for forecasts.
- **Algorithm Design Bias:**
- Algorithmic Bias: It is possible that the solubility prediction algorithms unintentionally mirror the prejudices of the data they were trained on or the developers of the algorithms. For instance, the model could not function effectively for chemicals from different locations or sectors if the training data mostly consists of compounds from such industries.
- Impact on Innovation: By favoring some types of molecules over others, prediction model bias might hinder innovation. This can result in a lack of diversity among the chemicals being investigated, which could cause creative ideas that are not biased by the model to be overlooked.
- **Utilizing Predictions Ethically:**
- Making decisions: Without considering the larger picture or the inherent uncertainties in the projections, decision-makers run the risk of relying too heavily on predictions made by data-driven models. This could result in immoral choices being made, such developing a medication candidate without enough experimental validation based only on good solubility forecasts.
- Equity and Access: Data-driven technologies for predicting solubility should be available to a broad spectrum of researchers, not just those with substantial funding. Restrictions on access to these technologies have the potential to exacerbate disparities in scientific research and innovation by creating a divide between various research groups.
- **Data security and privacy:**

- Sensitive Data: Data security and privacy raise ethical questions when solubility prediction models are trained on confidential or sensitive data. It is critical to guarantee the security of sensitive data and prevent predictions from unintentionally disclosing private information.**Let us have a look at some**

recent case studies highlighting the successful implementation of these techniques:

1. **Predicting Water Solubility and Organic Solvents using Machine Learning Study Synopsis:** The goal of this work, which was published in Nature Communications, was to use multiple machine learning models to predict solubility. Multiple Linear Regression (MLR), Artificial Neural Networks (ANN), Support Vector Machines (SVM), and Random Forest (RF) were among the linear and non-linear models that the researchers compared.

   Principal Results:

   Compared to linear models like MLR, non-linear models—in particular, Random Forest and Neural Networks—performed noticeably better.

   The work improved model accuracy in solubility prediction by introducing new assessment measures. A more trustworthy evaluation than more conventional metrics like R2 and RMSE was offered by one such metric, which concentrated on the percentage of predictions that fell within a specific error range from the actual experimental data. The study showed that machine learning could predict a wide range of chemicals' solubility with accuracy, negating the necessity for in-depth experimental testing.

   Implications: According to the results, machine learning models—particularly non-linear ones—can be very successful at predicting solubility, an important property for a number of applications, including the discovery of new drugs.
2. **API Solubility Prediction Using Quantitative Structure-Property Relationship (QSPR) Models:** An Overview This case study examined the use of QSPR models to forecast the solubility of active pharmaceutical ingredients (APIs) in commercial crystallization procedures, and it was published in Frontiers of Chemical Science and Engineering.

   Principal Results:

   To enhance solubility predictions in binary solvent systems, the study integrated machine learning methods with conventional QSPR models.

   Because machine learning models can spot trends in vast datasets that traditional approaches would miss, they have been able to predict solubility with greater accuracy.

   By optimizing conditions for API solubility, the industrial crystallization process improved the efficiency and efficacy of pharmaceutical manufacture through the implementation of these models.

   Consequences: The study shown that combining machine learning with QSPR models can greatly enhance solubility prediction of API.
3. **Using Machine Learning to Predict Crystallization Propensity Study Overview:** Another study looked at the application of machine learning to predict APIs' propensity to crystallize, which is an important consideration in drug development.

   Key Findings: The parameters impacting API crystallization were analyzed using a combination of machine learning techniques in this research.

The work provided important insights for pharmaceutical formulation by identifying important molecular characteristics and environmental factors that influence crystallization.

Researchers were able to better comprehend and regulate this important part of drug development with the use of machine learning models that were able to predict crystallization outcomes with high accuracy.

Implications: By maximizing the solid-state characteristics of APIs, machine learning can be used to predict crystallization propensity, which can improve therapeutic formulations by increasing stability, bioavailability, and manufacturability.

11.4.2 Data Sources and Feature Selection for Machine Learning Models [19, 32, 33, 36–43]

Data Sources:

Use experimental databases to obtain information about the solubility of different compounds in different solvents. ACD/Labs Solubility Database, PubChem, ChemSpider, and several literature sources are a few examples. These databases provide insightful empirical information on solubility in many settings.

Computational Databases: A few databases compile solubility information derived from computational methods like quantum chemistry or molecular simulations. These databases can supplement experimental data or offer predictions for compounds for which there is insufficient experimental evidence or complement experimental data.

Data Collection: It can be required to use bespoke data collection techniques when particular compounds or solvents are of interest. Depending on the demands of the research, this may entail acquiring data from specialist sources, performing tests, or running simulations.

Feature Selection:

Select molecular descriptors that accurately convey the physicochemical properties of solutes and solvents that are pertinent to solubility. Molecular weight, surface area, hydrogen bonding capability, lipophilicity, polarizability, and other pertinent characteristics are a few examples of these descriptors. Include details on particular chemical motifs or functional groups that are known to affect solubility. For example, in polar liquids, hydrophilic groups such as hydroxyl (-OH) and carboxyl (-COOH) tend to enhance solubility.

Topological Indices: Use topological indices to explain the properties and structural connectivity of molecules. These indices include details on characteristics that can affect solubility, such as branching, symmetry, and molecule size.

3D Molecule Descriptors: Take into consideration adding three-dimensional descriptors that depict the form and spatial arrangement of molecules. These descriptors can shed light on the steric factors and intermolecular interactions that influence solubility.

Add characteristics that describe the characteristics of solvents, such as accessibility, hydrogen bond acceptance and donation ability, polarity, and dielectric constant. In solvent-solute interactions that are pertinent to solubility, solvent-specific properties are extremely important.

Include information about solvation-free energy or hydration energy, which measure the energy variations brought on by the solute dissolving in a solvent. These characteristics offer important insights into how solute-solvent interactions impact solubility.

11.4.3 Challenges and Limitations of Data-Driven Approaches in Solubility Prediction [44–59]

1. Data Quality and Availability: Prediction accuracy can be greatly impacted by the caliber and volume of data that is available to train machine learning models. Acquiring high-quality experimental data for a wide variety of chemicals can be difficult when predicting solubility. Furthermore, incomplete data for specific chemical classes or characteristic ranges could result in skewed models.
2. Feature Selection and Representation: It is essential to select pertinent molecular descriptors or features that accurately capture the characteristics and chemical structure of molecules. Finding the most informative aspects, however, can be difficult and call for subject-matter knowledge. Poor feature selection may result in less than ideal model performance.
3. Overfitting of the model: Overfitting happens when a model picks up noise or unimportant patterns from the training set, which makes it difficult for the model to generalize to new data. This is especially problematic for solubility prediction in cases where there are inadequate training data or sophisticated models. Cross-validation and regularization strategies can be used to reduce overfitting
4. Interpretability: Deep neural networks are one type of machine learning model that is difficult to interpret because of its intrinsic complexity. It can be difficult to comprehend how these models predict solubility, particularly in domains where scientific understanding and decision-making depend heavily on interpretability.
5. Domain Transferability: It's possible that models that were trained on solubility information for particular chemical classes or experimental setups won't translate well to other domains. Although they need to be carefully adjusted and validated, transfer learning techniques—which make use of information from related tasks or datasets—can aid in overcoming this constraint.
6. Uncertainty Estimation: Making decisions and evaluating risks depend heavily on estimating the degree of uncertainty surrounding model projections. It can be difficult to assess the accuracy of predictions made by many data-driven models

since they only offer point estimates rather than expressing uncertainty. This is particularly true in situations when there is little or no data extrapolation.

7. Ethical and Legal Considerations: Legal and ethical requirements, especially those pertaining to environmental effect and safety, must be followed by predictive models trained on chemical data. Predictions that are biased or erroneous may have serious repercussions, hence rigorous model validation and validation are required before implementation.

11.5 Hybrid Approaches

11.5.1 Combination of Quantum Mechanics and Machine Learning Methods for Improved Solubility Prediction [60–66]

By utilizing the precision of quantum mechanics computations and the scalability of machine learning techniques, the combination of these two fields of study can improve solubility prediction. While machine learning can effectively manage big datasets and capture intricate correlations between molecule characteristics and solubility, quantum mechanics can offer detailed insights into the molecular interactions influencing solubility. Molecular dynamics simulations and quantum machine learning are two promising approaches to improve the efficiency and accuracy of solubility prediction.

the prediction of solubility by combining machine learning methods with molecular dynamics simulations. Detailed information on molecular movements and interactions may be obtained via molecular dynamics simulations, and these insights can be utilized to create features for machine learning models that are trained to predict solubility.

11.5.2 Examples of Successful Hybrid Approaches and Their Applications [47, 48, 60, 61, 64, 65, 67–76]

- Combining Computational Models with Experimental Data: QSPR/QSAR Models Molecular descriptors obtained from chemical structures are integrated with experimental solubility data in Quantitative Structure-Property Relationship (QSPR) or Quantitative Structure-Activity Relationship (QSAR) models. These models allow for predictions for substances for which there are no experimental data by correlating characteristics of the chemical structure with solubility.
- Alphabets for machine learning: To enhance solubility predictions, hybrid models that integrate machine learning methods, including neural networks or random forests, can integrate molecular descriptors with experimental data. These

models are capable of handling non-linear correlations between solubility and chemical characteristics and can learn intricate patterns from the data.

- Bringing Together Experimental Data and Molecular Simulation:
- Simulations of Molecular Dynamics (MD): At the molecular level, interactions between solutes and solvents can be thoroughly understood by MD simulations. Researchers can improve solubility predictions and obtain a mechanistic knowledge of solvation mechanisms by merging MD simulations with experimental data.
- Calculating Free Energy: Accurate predictions of solvation free energies and solubility can be achieved by hybrid techniques that combine free energy calculations with experimental data. Solvation free energies for various chemical compounds can be computed using methods like thermodynamic integration (TI) or free energy perturbation (FEP).
- Consensus modeling and ensemble methods: ensemble learning: ensemble methods combine predictions from several different separate models to increase overall robustness and accuracy. Ensemble methods, including averaging results from various computational models or merging predictions from various QSPR/QSAR models, can improve prediction performance for solubility prediction.
- Consensus Modeling: To produce more accurate solubility estimates, consensus modeling combines predictions from multiple databases or computational techniques. Consensus models alleviate the shortcomings of individual approaches and yield more accurate forecasts by taking into account various sources of information.
- Machine Learning-Based Models Based on Physics:
- Combining Data-Driven and Physics-Based Models: Accurate solubility forecasts can be achieved by combining machine learning algorithms with physics-based models, such as thermodynamic or kinetic models, and taking advantage of both the adaptability of data-driven techniques and the interpretability of physics-based approaches.

- **Providing more detailed examples or hypothetical scenarios where hybrid approaches outperform traditional methods for better understanding:**

  1. **Drug Development and Discovery:**

     Conventional Method: Thousands of compounds are experimentally tested against a target in high-throughput screening (HTS), which is the traditional method of drug development. This procedure takes a lot of time, money, and resources.

     Hybrid Approach: The drug development process can be greatly accelerated by integrating Quantitative Structure-Activity Relationship (QSAR) modeling with Molecular Dynamics (MD) simulations. The dynamic behavior of possible therapeutic compounds and their interactions with target proteins are better understood thanks to MD simulations. The biological activity of novel compounds can then be predicted using QSAR models that have been improved by machine learning. This combination lessens the need for extensive experimental screening by enabling the quick discovery of interesting medication candidates.

Benefits: This hybrid strategy lowers expenses, shortens the timeframe for drug discovery, and raises the possibility of finding promising therapeutic candidates which give promising results.

2. **Designing Catalysts in Chemical Engineering:**

   Conventional Method: The conventional approach to catalyst design is a laborious and costly laboratory procedure that tests different catalyst formulations by trial and error.

   Hybrid Approach: Catalyst design can be significantly enhanced by combining machine learning (ML) models with density functional theory (DFT) computations. DFT predicts the activity and selectivity of possible catalysts by offering in-depth insights into their electronic structure. The effectiveness of novel catalyst compositions across a range of reactions can then be predicted by machine learning models trained on DFT-calculated parameters, negating the need for intensive laboratory testing.

   Benefit: By examining a wider variety of possible formulations than would otherwise be possible, this hybrid approach not only expedites the catalyst discovery process but also makes it possible to build more effective and selective catalysts.

3. **Nanotechnology and Material Science:**

   Conventional Approach: Conventional material design frequently entails the costly and time-consuming synthesis and testing of a large range of materials to find those with desired qualities.

   Hybrid Approach: Material design can be optimized by a hybrid approach that combines machine learning and quantum mechanics/molecular mechanics (QM/MM). With the use of QM/MM techniques, materials can be accurately modelled at the atomic level, accounting for both quantum and classical interactions. The design of materials with mechanical, thermal, or electrical properties can then be guided by machine learning's analysis of these interactions to predict the properties of novel materials or nanostructures.

   Benefit: By eliminating the need for costly and time-consuming experimental synthesis and testing, this method allows for the quick identification of materials with certain qualities.

4. Imaginary Situation: Improving Drug Solubility

   Consider a pharmaceutical business creating a novel medication that is poorly soluble. In the past, they would go through several iterations of trial-and-error formulation, experimenting with various excipients, solvents, and methods to improve solubility.

   Hybrid Method:

   To begin with, the business could make use of Molecular Dynamics (MD) simulations to comprehend the molecular interactions between the drug molecule and various solvents and excipients.

The most promising formulations could then be found by using Density Functional Theory (DFT) computations to forecast the energy changes related to these interactions.

Finally, the ideal formulation conditions for the medication could be predicted using a Machine Learning (ML) model trained on historical solubility enhancement data.

Result: With this strategy, the business could rapidly identify the best formulas, cutting down on the time and expense involved in trial and error process resulting in more effective bioavailability.

11.6 Case Studies and Applications

11.6.1 Case Studies Demonstrating the Application of Computational Tools for Solubility Prediction in Drug Design, Material Discovery, and Environmental Studies [20, 59, 77–95]

1. Drug Design:

   A Case Study Computational techniques are essential in drug discovery to forecast drug candidates' solubility and guarantee appropriate bioavailability. For instance, researchers can anticipate the solubility of medicinal compounds in various physiological conditions by using machine learning algorithms to analyze the electrical and structural characteristics of the molecules using quantum mechanics calculations. This method was used to successfully predict the water solubility of drug-like compounds in a study by combining machine-learning models with quantum-chemical descriptors. Researchers can rank drug candidates with the best solubility profiles for additional development by using computational methods.
2. Material Discovery:

   Case Study: Predicting a material's solubility is essential for a variety of applications, including catalysts, organic electronics, and energy storage devices. Computational techniques are a key component of this process. For example, in the field of organic electronics, scientists predict the solubility of organic semiconductors in various solvents using molecular dynamics simulations and quantum mechanics calculations. This helps with the design of solution-processable materials for flexible and effective electronic devices. Researchers can find innovative materials with specialized solubility qualities for particular technological applications more quickly by combining computational methods with experimental confirmation.
3. Environmental Study:

   Case Study: Predicting the solubility and fate of pollutants, contaminants, and environmentally significant compounds in aquatic and terrestrial ecosystems is a growing area of environmental studies that employ computational tools. For instance, scientists use machine learning methods and molecular dynamics simulations to forecast the solubility and transport characteristics of organic contaminants in soil and water systems. Computational methods can offer important insights into the environmental fate of contaminants and help guide risk assessment and remediation strategies by modeling the interactions between pollutants and environmental matrices. These investigations advance sustainable environmental management strategies and deepen our knowledge of the effects of chemicals on the environment.

- **In computational solubility prediction, potential applications and new directions include:**

  1. Integration of Multi-scale Modeling: To precisely capture solute-solvent interactions across various length and time scales, future developments are probably going to involve the integration of multi-scale modeling approaches, combining quantum mechanics, molecular dynamics simulations, and machine learning techniques. The construction of more thorough and accurate solubility models that can represent intricate molecular dynamics will be made possible by this combination.
  2. Quantum Machine Learning: By utilizing quantum computing capabilities to handle complex molecular systems and high-dimensional data more effectively, quantum machine learning techniques—such as quantum neural networks and quantum-enhanced algorithms—hold promise for increasing the accuracy and efficiency of solubility prediction.
  3. Data-driven Approaches: Deep learning and transfer learning are two examples of data-driven approaches that will continue to be important in solubility prediction due to the availability of vast datasets containing experimental solubility measurements and molecular characteristics. These methods are able to uncover important information from a variety of datasets and comprehend the intricate connections between solubility and molecular descriptors.
  4. Predictive Modeling for Complex Systems: Predicting solubility in mixtures, multi-component systems, and non-ideal circumstances are only a few of the problems that computational solubility prediction will increasingly concentrate on solving. We will create sophisticated modeling tools to appropriately account for the impacts of temperature, pressure, and solvent composition on solubility behavior. These techniques will include thermodynamic modeling, statistical mechanics, and machine learning.
  5. High-throughput Screening and Virtual Screening: To find novel compounds with desirable solubility properties for drug discovery, materials science, and other applications, computational tools will be used for high-throughput screening and virtual screening of large chemical libraries. In order to identify candidate compounds with the best solubility profiles for experimental validation, these screening techniques will make use of machine learning models trained on a variety of datasets.

6. Interdisciplinary Collaboration: To create novel modeling strategies, algorithms, and software tools suited to particular application domains, chemists, physicists, computer scientists, and engineers will collaborate across disciplinary boundaries in future computational solubility prediction research. This joint endeavor will propel the advancement of increasingly precise, effective, and adaptable computational methods for solubility forecasting in numerous scientific and industrial domains.

- **Let's have a view on the research articles** [96]:

1. The paper "Prediction of organic compound aqueous solubility using machine learning: a comparison study of descriptor-based and fingerprints-based models" by Arash Tayyebi et al. contrasts two methods for utilizing machine learning techniques to predict the aqueous solubility of organic compounds. Models based on descriptors and models based on fingerprints are two methods:

   Models based on descriptors: The molecular structure of the substances is represented by these models using a collection of chemical descriptors. Quantitative descriptions of several molecular characteristics, including size, shape, electronegativity, and hydrophobicity, are known as descriptors. Molecular weight, atom count, number of rotatable bonds, and topological indices are a few examples of descriptors. Choosing a selection of pertinent descriptors and using them as input features for machine learning algorithms like linear regression is a common practice for descriptor-based models.
   Fingerprint-based models: Binary bit strings called "molecular fingerprints" are used to represent the structural details of molecules. Certain structural characteristics (such as functional groups, atom kinds, and bond types) and their spatial configurations within the molecule are captured by these fingerprints, whether they are present or absent. Morgan fingerprints, MACCS keys, and Extended Connectivity Fingerprints (ECFP) are examples of common fingerprinting algorithms. These fingerprints are used as input features in fingerprint-based models, which help machine learning algorithms predict aqueous solubility.
   The study uses a dataset of organic compounds with known water solubility values to examine the effectiveness of these two modeling methodologies. The predicted accuracy and computing cost of each model are evaluated using metrics including mean absolute error (MAE), root mean square error (RMSE), coefficient of determination (R-squared), and computational efficiency.

2. A unified machine learning (ML) framework for predicting solubility across different organic solvents is introduced in the paper "A unified ML framework for solubility prediction across organic solvents" by Antony D. Vassileiou et al. In chemistry and related disciplines, predicting solubility is a crucial task that affects processes like medication formulation, chemical synthesis, and material design [97].

(a) **Goal**: The main objective of the research is to create a flexible machine learning framework that can reliably predict an organic compound's solubility in a variety of solvents
(b) **Unified Approach:** The authors suggest a unified framework that can manage predictions for a variety of solvents concurrently, in contrast to conventional solubility prediction models that are solvent-specific. The prediction process is streamlined by this unified methodology, which also enables wider applicability.
(c) **Machine Learning approaches**: To simulate the link between molecular characteristics and solubility across different solvents, the framework probably makes use of a variety of machine learning approaches, such as regression, classification, or deep learning. In the study, the precise algorithms and methods used would be covered in full.
(d) **Feature Engineering**: Choosing and designing pertinent features is a crucial step in the development of successful machine learning models. The authors probably investigate ways to derive useful information from molecular architectures that are positively correlated with solubility in various solvents.
(e) **Data**: Training and assessing the ML models need a large dataset that includes details regarding solubility in a variety of organic solvents. Molecular structures, experimental solubility measurements, and solvent environment data may be included in the dataset.
(f) Model Evaluation: A variety of metrics, including mean absolute error (MAE), root mean square error (RMSE), and coefficient of determination (R-squared), are probably used to evaluate the effectiveness of the machine learning framework. To guarantee the models' resilience and generalizability, cross-validation techniques may also be used.
(g) Uses: The created machine learning system may find use in drug development, syntheses, materials research, and other domains where precise solubility prediction is essential. Researchers and practitioners can gain useful insights from the framework, which delivers dependable predictions across a variety of solvents.

In conclusion, the paper presents a unique machine learning framework for solubility prediction that takes a holistic approach to solving the problem of solvent variety. This concept has ramifications for a number of fields of study and enterprises where solubility is important.

3. **Through dataset curation, Jintao Meng et al.'s study "Boosting the predictive performance with aqueous solubility dataset curation" focuses on enhancing the predictive performance of aqueous solubility models** [98].

   (a) **Dataset Curation**: When creating precise predictive models, the authors stress the significance of high-quality data. In order to guarantee that the data is relevant and reliable for the purpose of forecasting aqueous solubility, they go through a process known as dataset curation, which entails carefully choosing and preparing the data.

(b) **Improved Predictive Performance**: The authors want to improve the predictive capabilities of current models by carefully selecting the aqueous solubility dataset. This include cleaning up errors, eliminating noisy or unnecessary data points, and possibly adding more high-quality data to the dataset.
(c) **Methodology**: The article probably describes the precise steps taken to curate the dataset. These steps could involve feature engineering, data cleaning, outlier identification, and validation processes. These methods are essential to guaranteeing the curated dataset's efficacy and integrity.
(d) **Predictive Models**: The aqueous solubility predictive models are developed or improved upon using the curated information as a basis. To create precise models based on the carefully chosen data, the authors may use a variety of machine learning algorithms, regression techniques, or statistical approaches.
(e) **Evaluation Metrics:** Conventional evaluation metrics like mean absolute error (MAE), root mean square error (RMSE), coefficient of determination (R-squared), or cross-validation methods are used to evaluate the performance of the predictive models. These measures shed light on the models' dependability and accuracy.
(f) Impact: The paper talks about how enhanced predictive performance might affect a number of applications, including chemical engineering, environmental research, and medication development. Precise estimations of water solubility are necessary for process optimization, improving medication formulations, and creating safer and more potent compounds.
(g) Contributions: The approach for culling datasets and the study's capacity to enhance prediction accuracy in aqueous solubility modeling are its main points of contribution. The writers offer helpful insights for academics working in predictive modeling and related domains by highlighting the significance of data quality.

The paper offers a methodical strategy for increasing predictive performance through dataset curation overall, which has implications for developing better models for aqueous solubility prediction and other relevant fields of study and business.

4. **The use of machine learning techniques for solubility prediction is the main topic of the paper "Machine Learning for Solubility Prediction" by Tianyuan Zheng et al.** [99]

(a) **Interdisciplinary Collaboration**: The authors' expertise in chemistry and computer science emphasize an interdisciplinary approach to the solubility prediction problem. This partnership makes it possible to combine cutting-edge machine learning techniques with domain expertise.
(b) **Machine Learning Techniques**: It's possible that the article goes over different machine learning techniques and algorithms that are applied to solubility prediction. This could involve ensemble approaches, neural networks, regression techniques, or classification algorithms designed specifically for the solubility prediction job.

(c) **Feature Engineering**: Creating precise solubility prediction models requires the application of feature engineering. It is probable that the authors investigate techniques for identifying characteristic features from molecular structures or descriptors that capture pertinent attributes affecting solubility.
(d) **Sources of Data**: Numerous databases with details on molecular structures, experimental solubility tests, and related chemical properties may be used in the study. To clean, normalize, and standardize the data for training and evaluation, data preprocessing techniques can be used.
(e) **Model Evaluation**: Using common evaluation measures like mean absolute error (MAE), root mean square error (RMSE), or coefficient of determination (R-squared), the effectiveness of the machine learning models is evaluated. The models' generalizability may also be guaranteed by employing cross-validation procedures.
(f) **Validation and Benchmarking**: To show the predictive models' superiority and efficacy in solubility prediction, they are probably verified and benchmarked against current techniques or experimental data
(g) **Applications and Implications**: Materials science, agrochemicals, pharmaceuticals, and other industries are among those that stand to benefit from accurate solubility prediction. The potential uses of machine learning-based solubility prediction in medication development, formulation optimization, and chemical process development are covered in the article.
(h) **Future Directions**: The authors can suggest other lines of inquiry or difficulties in the area of solubility prediction, like adding more data sources, enhancing the interpretability of models, or tackling the models' applicability to other chemical classes and solvents

In conclusion, the paper provides a thorough analysis of machine learning methods for predicting solubility, emphasizing how they might spur new discoveries and advancements in science.

5. **A unique method for predicting the solubility of pharmacological compounds using fragment-pair representations and attention processes is presented in the paper "Fragment-pair based drug molecule solubility prediction through attention mechanism" by Jianping Liu et al.** [100]

(a) **Fragment-Pair Representation**: The authors suggest using sets of fragment pairs to represent medicinal compounds. By taking into account pairings of fragments inside the molecule, this representation is able to capture the structural properties of molecules.
(b) **Attention Mechanism**: During the prediction phase, the predictive model incorporates an attention mechanism that allows it to selectively focus on significant fragment pairs. Thanks to this technique, the model can dynamically balance the contributions of various fragment pairings according to how important they are for predicting solubility.
(c) **Machine Learning Model:** Based on the fragment-pair representations and attention mechanism, the article probably explains how to build a machine learning model. The model is capable of learning intricate correlations

between fragment pairs and solubility by utilizing methods like deep learning, neural networks, or attention-based architectures.

(d) **Training Data**: A dataset comprising medicinal compounds with known solubility values is used to train the algorithm. Molecular structures, experimental solubility measurements, and other pertinent chemical parameters might be included in this collection.

(e) **Validation and Evaluation**: Using common metrics like mean absolute error (MAE), root mean square error (RMSE), or coefficient of determination (R-squared), the predictive performance of the model is verified and assessed. The model's generalizability may be evaluated by using cross-validation procedures.

(f) Comparison with Current Methods: To show the usefulness of the suggested strategy and its potential for improvements over conventional procedures, it is likely compared to the performance of current solubility prediction methods

(g) Applications: Precise estimation of a medication's solubility is essential for drug development and pharmaceutical research. The suggested approach may be used to speed up drug discovery procedures, predict how drugs would behave in biological systems, and optimize medication formulations, according to the article.

(h) Future Research Directions: The authors may recommend examining new features, enhancing the interpretability of the model, or applying the method to other relevant computational chemistry applications

Overall, by utilizing fragment-pair representations and attention mechanisms, the work offers a novel strategy for drug molecule solubility prediction, with implications for improving computational techniques in pharmaceutical research and drug design.

6. [101] **An inventive method for predicting the solubility of drug-like molecules is shown in the paper "Accurate, reliable, and interpretable solubility prediction of drug-like molecules with attention pooling and Bayesian learning," written by Seongok Ryu and Sumin Lee.**

(a) **Goal**: The main goal of the research is to create a solubility prediction model with high levels of interpretability, accuracy, and dependability. Because solubility prediction affects medication formulation and bioavailability, it is an essential task in drug research and development.

(b) **Attention Pooling**: The prediction model's developers suggest including attention pooling methods. By concentrating on the most pertinent molecular properties for solubility prediction, the model is able to dynamically assess the significance of various molecular features through attention pooling.

(c) **Bayesian Learning**: By offering uncertainty estimates, Bayesian learning techniques are used to improve prediction dependability. In order to evaluate how reliable the model's predictions are, it is necessary to quantify prediction uncertainty, which is made possible by Bayesian approaches.

(d) **Machine Learning Model**: It appears that the paper explains how to build a machine learning model that combines Bayesian learning with attention pooling. The intricate connections between molecular characteristics and solubility are well captured by this model
(e) **Feature Representation**: A collection of features that capture the structural and physicochemical characteristics of drug-like compounds is used to represent them. It is possible to use feature engineering approaches to extract useful features from molecular structures.
(f) **Training and Evaluation:** A dataset of drug-like compounds with known solubility values is used to train and assess the predictive model. Standard evaluation metrics like mean absolute error (MAE) and root mean square error (RMSE) are used to evaluate the performance of the model.
(g) **Interpretability**: By offering predictions that are easy to grasp, the created model enables researchers to comprehend the underlying variables affecting solubility. This improves the model's usefulness in directing drug development decision-making procedures.
(h) **Applications**: There are many uses for the precise, dependable, and comprehensible solubility prediction model in pharmaceutical development and research. It can help with formulation optimization, pharmacokinetic modeling, and medication design considerations.
(i) **Contributions and Implications**: It is possible that the publication will address how the suggested method advances the subject of solubility prediction and how it might enhance drug discovery procedures while cutting experimental expenses

In conclusion, the paper offers a sophisticated method of solubility prediction that makes use of Bayesian learning and attention pooling approaches to produce predictions for drug-like compounds that are precise, dependable, and understandable.

Highlighting some common challenges faced in these case studies and how they were addressed:

1. **Data Quality and Availability Challenge:** Research like Tayyebi et al. (2018) and Meng et al. (2020) have shown that one of the main issues with solubility prediction is the quality and availability of data. The performance and generalizability of machine learning models can be severely impacted by the tiny, noisy, or incomplete datasets that are frequently used to train the models.

   Meng et al. (2020) curated a high-quality dataset on aqueous solubility as a solution to this problem. By meticulously removing inaccurate or inconsistent data points and standardizing the dataset, they improved the machine learning models' capacity for prediction. In a similar vein, additional research has stressed that to increase model resilience, thorough data preprocessing and the integration of several data sources are crucial.

2. **Feature Selection and Model Complexity Challenge**: Accurate modeling of molecular structures depends on choosing the appropriate features or descriptors. It was challenging for Vassileiou et al. (2019) and Zheng et al. (2020) to strike a balance between predicted accuracy and model complexity. Excessively intricate models may perform well when fitted to training data, but may overfit (fail to generalize to new data).

   Solution: In their comparison of fingerprint-based and descriptor-based models, Tayyebi et al. (2018) discovered that, depending on the model and dataset, certain features may be more effective. To improve model performance, they underlined the significance of feature engineering and selection. To deal with model complexity, Vassileiou et al. (2019) used machine learning algorithms that may automatically modify the model in order to prevent overfitting.
3. **Interpreting Model Predictions Difficulty:** Deep learning or sophisticated ensemble models can frequently be "black boxes," making it challenging to understand how they arrive at particular predictions. This posed a serious problem in research such as that of Ryu and Lee (2021), where comprehension of the model's decision-making procedure was essential for regulatory and scientific purposes.

   Solution: Ryu and Lee (2021) used Bayesian learning and attention methods to address this problem, revealing which chemical properties or fragments had the greatest influence on the model's predictions. By using this method, their solubility prediction models became more interpretable and valuable for real-world uses.
4. **Extrapolation Throughout Various Chemical Domains Challenge:** As demonstrated by the research conducted by Vassileiou et al. (2019) and Liu et al. (2021), models trained on a particular chemical space may not generalize well to different types of chemicals or solvents. This restricts the models' capacity to be applied in many contexts.

   Solution: To increase the model's adaptability, Vassileiou et al. (2019) created a unified machine learning framework that could predict solubility across a range of organic solvents. Fragment-pair based models were employed by Liu et al. (2021) to better represent the diversity of chemical structures and improve the model's cross-chemical generalization capacity.
5. **Reaching a Computational Efficiency and Accuracy Balance Problem:** Increasing computational complexity and time to achieve high prediction accuracy was a concern in the research by Meng et al. (2020) and Ryu and Lee (2021).

   Solution: Meng et al. (2020) reduced computing needs and maintained excellent accuracy in their models by streamlining the dataset and employing effective techniques. Ryu and Lee (2021) used attention pooling approaches, which lowered computing load and increased accuracy by concentrating computational resources on the most pertinent portions of the input data.

11.6.2 Prospects and Emerging Trends in Computational Solubility Prediction [102–109]

1. Increased Accuracy of Quantum Mechanics Calculations: To increase the accuracy of molecular property predictions related to solubility, such as solvation energies and solute-solvent interactions, more developments in quantum mechanics techniques, such as density functional theory (DFT) and wavefunction-based methods, are required.
2. Development of Novel Descriptors: Research ought to concentrate on finding and creating new molecular descriptors that accurately represent the physicochemical characteristics that affect solubility, such as intermolecular interactions, electronic structure, and molecular geometry. These descriptors ought to be applicable to various solute-solvent systems and computationally efficient.
3. Integration of Experimental Data: Computational models can improve forecast accuracy and dependability by including experimental solubility data. Research should concentrate on creating procedures for integrating experimental observations into computational frameworks, such as solubility databases and high-throughput experimental methods for solubility prediction.
4. Validation and Benchmarking: Carefully verifying the accuracy, dependability, and generalizability of computational solubility prediction techniques against experimental data is crucial for evaluating their suitability for use in a variety of chemical systems and environmental settings. To make comparison studies and computational tool performance evaluation easier, standardized datasets and evaluation procedures ought to be created.
5. Uncertainty Quantification: Research on computational solubility prediction should focus on quantifying uncertainty, particularly how mistakes spread from computational models and input data to predictions. Uncertainty estimating techniques, Bayesian inference, and probabilistic approaches can all offer important insights regarding the degree of confidence and dependability in solubility forecasts.
6. Open Access Software and Data Repositories: By creating open-access software tools and data repositories for computational solubility prediction, the scientific community will be encouraged to collaborate, be repeatable, and share information. By making computational methods, models, datasets, and validation protocols easily accessible, these resources should promote interdisciplinary research and innovation in solubility prediction.
7. Application to Complex Systems: Research should concentrate on applying computational solubility prediction techniques to multi-component systems, mixtures, polymers, and nanoparticles, among other complex systems. Comprehending the solubility behavior of these systems is essential for applications in environmental remediation, medication development, and materials design.
8. Interdisciplinary Collaboration: To advance computational techniques for solubility prediction, researchers from a variety of fields, including chemistry, phys-

ics, computer science, and engineering, must work together. The creation of novel techniques, algorithms, and applications that tackle the difficult problems of precisely and effectively forecasting solubility can be aided by interdisciplinary methods.

11.7 Conclusion

11.7.1 Importance of Computational Tools in Advancing Solubility Prediction and Its Impact on Various Industries

The use of computational methods, which offer quick and affordable ways to investigate the intricate relationships between solute and solvent molecules, is essential to the advancement of solubility prediction. The following are some salient features emphasizing their significance and influence on diverse industries:

1. Quickening the Process of Drug Discovery: Pharmaceutical researchers can search through sizable chemical databases for possible drug candidates with desired solubility profiles by using computational solubility prediction techniques. Drug discovery processes can be sped up by identifying molecules that are expected to have high solubility and bioavailability, which will result in the production of more potent treatments.
2. Formulation Design Optimization: Computational solubility prediction methods are useful for formulation design optimization in the food, cosmetics, and agrochemical industries. It is possible for producers to produce goods with enhanced stability, effectiveness, and sensory qualities by forecasting the interactions between various elements in a solution.
3. Improving Material Design: Computational solubility prediction in materials science allows scientists to create new materials with certain characteristics. Engineers may create new materials for everything from coatings and adhesives to medication delivery systems and energy storage devices by studying how solvents interact with polymers, nanoparticles, and other materials.
4. Environmental Impact Assessment: Chemicals like pesticides and pollutants are evaluated for their potential effects on the environment using computational methods for solubility prediction. Researchers can assess the possible hazards associated with these compounds and create plans for pollution control and prevention by simulating how they dissolve and spread in natural settings.
5. Cost and Time Savings: Labor-intensive, time-consuming, and costly are some of the drawbacks of using traditional experimental methods to measure solubility parameters. In lieu of extensive experimental testing, computational solubility prediction techniques provide a cost-effective alternative by enabling researchers to quickly screen a broad range of chemicals and solvent conditions.

6. Data-Driven Decision Making: Industries may make better judgments at every stage of the product development lifecycle by combining computational solubility prediction tools with experimental data and machine learning techniques. With the aid of these resources, researchers may spot patterns, make precise forecasts, and streamline procedures, which promotes creativity and increased productivity.

In general, computational techniques have transformed the field of solubility prediction by providing flexible and strong methods to tackle intricate issues in a range of industries, including environmental protection and drug discovery. These instruments will become more and more essential in promoting sustainability and innovation across a range of industries as they develop.

11.8 Suggestions for Further Research

To improve solubility predictions for complex systems, more research and development in this area should concentrate on the following areas:

- Integration of Multi-Scale Modeling: Create computational frameworks that integrate approaches to modeling solute-solvent interactions at different levels of detail, from atomistic to macroscopic scales. This will help to improve the accuracy of solubility predictions.
- Incorporation of Quantum Mechanical Methods: Investigate the use of quantum mechanical calculations, such as density functional theory (DFT) or correlated wave function methods, to precisely model solute-solvent interactions. Quantum mechanical approaches can offer comprehensive insights into solute-solvent interactions.
- Better Handling of Solvent Effects: Create sophisticated modeling techniques, such as explicitly handling solvent molecules in simulations and adding solvent polarization effects in quantum mechanical computations. Realistic solubility forecasts need accurate accounting for solvent effects, especially in polar or protic fluids.
- Methods for Big Data Analysis Based on Data: Make use of data-driven techniques, such deep learning and machine learning, to examine enormous databases of solubility data and identify important trends and correlations. Predictive model accuracy can be increased and hidden trends in solubility data can be found using advanced data analytics techniques.
- Improved Descriptor Creation: Make an investment in the creation of cutting-edge molecular descriptors that accurately represent the variety of physicochemical characteristics affecting solubility. This comprises descriptors designed for particular solvent classes or classes of chemicals for 3D structures and dynamic interactions.
- Conduct thorough validation and benchmarking experiments to evaluate the efficacy of computational solubility prediction techniques in a range of chemical

systems and experimental setups. Standardized criteria and datasets make it easier to compare various approaches fairly and point out areas that need improvement.

- Automation and High-Throughput Screening: Create high-throughput screening platforms and automated procedures for quick and economical solubility prediction. The identification of new solvents and solutes with desirable solubility qualities can be sped up by integration with experimental methods like combinatorial synthesis and high-throughput experimentation.
- Application to innovative Materials and Drug Discovery: Expand the use of computational solubility prediction techniques into new fields, including formulation development, drug discovery, and innovative material design. Innovation can be facilitated by modifying computational techniques to specifically handle certain fields' issues and accelerate the developmental efforts.

- **Future Directions and Emerging Fields**

1. Machine Learning and Quantum Chemistry Integrated Overview: A deeper integration of machine learning (ML) approaches with quantum chemistry methodologies is likely to be a feature of solubility prediction in the future. Researchers can produce more precise and understandable predictions by fusing the meticulous, first-principles calculations of quantum chemistry with the data-driven insights of machine learning. In situations when there is a dearth of experimental data, this hybrid approach may be particularly useful for forecasting the solubility of complicated or new chemicals.

   New Applications: For instance, by capturing both electronic structural features and statistical correlations in big datasets, combining density functional theory (DFT) computations with machine learning (ML) models can aid in the prediction of solubility across various chemical spaces.
2. Progress in Simulations of Molecular Dynamics (MD) Overview: A growing number of molecular studies are using MD simulations to examine a compound's solubility in different solvents. Static models have a hard time capturing the in-depth insights that these simulations offer on solvation processes, intermolecular interactions, and dynamic behaviors.

   New Applications: Improved solubility forecasts at varying pressure and temperature can be achieved by the application of improved sampling strategies and the combination of MD and ML algorithms. These developments may be very important for sectors such as pharmaceuticals, where accurate predictions of solubility are critical to the formulation of drugs.
3. Creating Hybrid Models Overview: Hybrid models are becoming more and more popular. Examples of these combinations include ML models and MD simulations or molecular docking. By utilizing the advantages of each technique, these models are able to produce solubility forecasts that are more precise and trustworthy.

New Applications: Hybrid techniques could be applied, for example, to investigate the impact of different excipients on medication solubility in complicated formulations or to predict solubility in mixed solvent systems. This would be very helpful in the drug-development process, as it's important to forecast solubility in various formulations.

4. High-Throughput Screening and AI-Driven Automation Overview: A new area of research that holds the potential to speed up the discovery and development of novel materials and medications is the application of AI-driven automation in solubility prediction. When AI algorithms are paired with automated high-throughput screening, they can quickly examine huge datasets, spot trends, and provide highly accurate solubility predictions.

   New Applications: In the pharmaceutical business, where vast libraries of chemicals must be swiftly screened, this strategy is especially helpful. Drug development can be accelerated by using AI-driven platforms to rank candidates according to their expected solubility and other important characteristics.

5. Data-Driven Approaches and Big Data Integration Overview: Combining big data analytics with solubility prediction tools will be crucial as more solubility data from diverse sources becomes available. By using huge and diverse datasets, these data-driven approaches can help reveal new insights, spot trends, and increase the forecast accuracy of models.

   New Applications: Global solubility datasets, which may be used to train more potent predictive models, may be developed as a result of the use of cloud-based platforms and cooperative data-sharing programs. Many industries, including materials research and pharmaceuticals, might gain from this.

6. Overview of Personalized Medicine and Solubility Prediction: Solubility prediction may find use in personalized medicine in the future. It might be able to customize medication formulations to optimize effectiveness and reduce negative effects by forecasting the solubility of pharmaceuticals in the unique biological settings of individuals.

   New Applications: To optimize drug therapy on a personalized basis, this strategy may involve merging patient-specific data, such as metabolic profiles and genetic information, with solubility prediction models.

7. Sustainability and Green Chemistry Overview: To help green chemistry activities, solubility prediction techniques are being developed in response to the increased emphasis on sustainability. With the use of these instruments, environmentally friendly solvents and procedures that reduce the usage of dangerous materials can be designed.

   New Applications: Predictive models can be used to find solvent substitutes with less of an adverse effect on the environment or to streamline procedures in the chemical production industry that use less energy and waste.

References

1. Solubility definition: Source: General chemistry or pharmacology textbooks often define solubility.
2. Dahan A, Miller JM. The solubility–permeability interplay and its implications in formulation design and development for poorly soluble drugs. AAPS J. 2012;14(2):244–51.
3. Lipinski CA, Lombardo F, Dominy BW, Feeney PJ. Experimental and computational approaches to estimate solubility and permeability in drug discovery and development settings. Adv Drug Deliv Rev. 2001;46(1–3):3–26.
4. Kerns EH, Di L. Drug-like properties: concepts, structure design and methods. 2nd ed. Boston: Academic; 2016.
5. Amidon GL, Lennernäs H, Shah VP, Crison JR. A theoretical basis for a biopharmaceutic drug classification: the correlation of in vitro drug product dissolution and in vivo bioavailability. Pharm Res. 1995;12(3):413–20.
6. Di L, Kerns EH. Drug-like properties: concepts, structure design and methods. 2nd ed. Boston: Academic; 2016.
7. Hou T, Xu X. Recent developments and application of virtual screening in drug discovery: an overview. Curr Pharm Des. 2004;10(9):1011–33.
8. Bajaj S, Singla D, Sakhuja N. Stability testing of pharmaceutical products. J Appl Pharm Sci. 2012;2(3):129–38.
9. Poliakoff M, Licence P. Green chemistry. Nature. 2007;450(7172):810–2.
10. Kwon GS. Polymeric drug delivery systems. Boca Raton: CRC Press; 2005. Reference: Schwarzenbach RP, Gschwend PM, Imboden DM. Environmental organic chemistry. 3rd ed. Hoboken: Wiley; 2016
11. Tropsha A. Best practices for QSAR model development, validation, and exploitation. Mol Inform. 2010;29(6–7):476–88.
12. Box KJ, Comer JE. The shake flask method for the determination of solubility. J Pharm Pharmacol. 2002;54(4):581–7.
13. Atkins PW, de Paula J. Atkins' physical chemistry. 10th ed. Oxford: Oxford University Press; 2014.
14. Fredenslund A, Jones RL, Prausnitz JM. Group-contribution estimation of activity coefficients in nonideal liquid mixtures. AICHE J. 1975 Feb;21(1):108–18.
15. Hansen CM. Hansen solubility parameters: a user's handbook. 2nd ed. Boca Raton: CRC Press; 2007.
16. Parr RG, Yang W. Density-functional theory of atoms and molecules. New York: Oxford University Press; 1989.
17. Kohn W, Sham LJ. Self-consistent equations including exchange and correlation effects. Phys Rev. 1965;140(4A):A1133–8.
18. Cramer CJ, Truhlar DG. Implicit solvation models: equilibria, structure, spectra, and dynamics. Chem Rev. 1999;99(8):2161–200.
19. Tomasi J, Mennucci B, Cammi R. Quantum mechanical continuum solvation models. Chem Rev. 2005;105(8):2999–3094.
20. Frenkel D, Smit B. Understanding molecular simulation: from algorithms to applications. 2nd ed. San Diego: Academic; 2001.
21. Karplus M, McCammon JA. Molecular dynamics simulations of biomolecules. Nat Struct Biol. 2002;9(9):646–52.
22. Jorgensen WL, Tirado-Rives J. Molecular modeling of organic and biomolecular systems using BOSS and MCPRO. J Comput Chem. 2005;26(16):1689–700.
23. Rapaport DC. The art of molecular dynamics simulation. 2nd ed. Cambridge: Cambridge University Press; 2004.
24. Allen MP, Tildesley DJ. Computer simulation of liquids. Oxford: Oxford University Press; 1987.
25. Friesner RA, Guallar V. Ab initio quantum chemical and mixed quantum mechanics/molecular mechanics (QM/MM) methods for studying enzymatic catalysis. Annu Rev Phys Chem. 2005;56:389–427.

26. Senn HM, Thiel W. QM/MM methods for biomolecular systems. Angew Chem Int Ed Engl. 2009;48(7):1198–229.
27. Warshel A, Levitt M. Theoretical studies of enzymic reactions: dielectric, electrostatic and steric stabilization of the carbonium ion in the reaction of lysozyme. J Mol Biol. 1976;103(2):227–49.
28. Lin H, Truhlar DG. QM/MM: what have we learned, where are we, and where do we go from here? Theor Chem Accounts. 2007;117(2):185–99.
29. Gao J, Truhlar DG. Quantum mechanical methods for enzyme kinetics. Annu Rev Phys Chem. 2002;53:467–505.
30. Lusci A, Pollastri G, Baldi P. Deep architectures and deep learning in chemoinformatics: the prediction of aqueous solubility for drug-like molecules. J Chem Inf Model. 2013;53(7):1563–75.
31. Yuan W, Son N, Wang S. Predicting aqueous solubility of compounds using deep learning models. In: 2019 IEEE International Conference on Bioinformatics and Biomedicine (BIBM); 2019. p. 2646–53.
32. Tetko IV, Tanchuk VY, Villa AE. Prediction of n-octanol/water partition coefficients from PHYSPROP database using artificial neural networks and atom-based systematic fragmentation. J Chem Inf Comput Sci. 2001;41(5):1407–21.
33. Hou T, Xu X. ADME evaluation in drug discovery. 5. Correlation of Caco-2 permeation with simple molecular properties. J Chem Inf Comput Sci. 2003;43(5):1710–7.
34. Svetnik V, Liaw A, Tong C, Wang T, Sheridan RP, Song Q. Random forest: a classification and regression tool for compound classification and QSAR modeling. J Chem Inf Comput Sci. 2003;43(6):1947–58.
35. Sheridan RP, Wang WM, Liaw A, Ma J, Gifford EM. Extreme gradient boosting as a method for quantitative structure-activity relationships. J Chem Inf Model 2016;56(12):2353–2360. The 3 case studies- Machine Learning in Organic Solvent and Water Solubility Prediction: Reference: Xiong, Z., Yang, Z., Hou, Y., & Zhu, R. (2024). Machine learning-assisted solubility prediction for organic solvents and water. *Nature Communications*. This study compared various machine learning models to predict solubility and introduced new evaluation metrics for accuracy assessment. Quantitative Structure-Property Relationship (QSPR) Models in API Solubility Prediction: Reference: Chinta, S., & Rengaswamy, R. (2019). Machine learning derived QSPR models to predict drug solubility in binary solvent systems. *Industrial & Engineering Chemistry Research*. The study combined QSPR models with machine learning to enhance solubility predictions in pharmaceutical manufacturing. Machine Learning in Crystallization Propensity Prediction: Reference: Ghosh, A., et al. (2019). Assessment of machine learning approaches for predicting the crystallization propensity of active pharmaceutical ingredients. *Cryst Eng Comm*. This research applied machine learning to predict the crystallization of APIs, aiding in better drug formulation.
36. ACD/Labs Solubility Database [Internet]. Advanced Chemistry Development, Inc. [cited 2024 Jun 2]. Available from: https://www.acdlabs.com/products/pc_admet/solubility/
37. Kim S, Chen J, Cheng T, Gindulyte A, He J, He S, et al. PubChem in 2021: new data content and improved web interfaces. Nucleic Acids Res. 2021;49(D1):D1388–95. Pence HE, Williams A. ChemSpider: an online chemical information resource. J Chem Educ 2010;87(11):1123–4
38. Schütt O, Gedeck P, Rogers-Evans M, Naumann T, Engkvist O. Computational approaches to small molecule solubility. Med Chem Commun. 2018;9:125–44.
39. Liu J, Wang X, Hu Y, Chen L, Feng S. Prediction of solubility of organic compounds in supercritical carbon dioxide using quantum chemistry and statistical learning methods. J Chem Eng Data. 2016;61(3):1259–69. Wishart DS. DrugBank: a comprehensive resource for in silico drug discovery and exploration. Nucleic Acids Res. 2006;34(suppl_1)
40. Delaney JS. ESOL: estimating aqueous solubility directly from molecular structure. J Chem Inf Comput Sci. 2004;44(3):1000–5. Todeschini R, Consonni V. Molecular Descriptors for Chemoinformatics. 2nd ed. Weinheim: Wiley-VCH; 2009

41. Todeschini R, Consonni V, Pavan M, Manganaro A. Challenges in the calculation of topological indices: the case of self-returning walks and multiple edges. J Chem Inf Comput Sci. 2003;43(6):1580–91.
42. Feher M, Schmidt JM. Property distributions: differences between drugs, natural products, and molecules from combinatorial chemistry. J Chem Inf Comput Sci. 2003;43(1):218–27.
43. Abraham MH, Ibrahim A, Zissimos AM. Determination of sets of solute descriptors from chromatographic measurements. J Chromatogr A. 2004;1037(1–2):29–47.
44. Sliwoski G, Kothiwale S, Meiler J, Lowe EW Jr. Computational methods in drug discovery. Pharmacol Rev. 2014;66(1):334–95.
45. Zhang L, Tan J, Han D, Zhu L. Prediction of drug solubility and permeability in drug discovery and development. Comb Chem High Throughput Screen. 2012;15(4):333–9.
46. Williams HD, Trevaskis NL, Charman SA, Shanker RM, Charman WN, Pouton CW, et al. Strategies to address low drug solubility in discovery and development. Pharmacol Rev. 2013;65(1):315–499.
47. Todeschini R, Consonni V. Molecular descriptors for chemoinformatics. 2nd ed. Weinheim: Wiley-VCH; 2009.
48. Cherkasov A, Muratov EN, Fourches D, Varnek A, Baskin II, Cronin M, et al. QSAR modeling: where have you been? Where are you going to? J Med Chem. 2014;57(12):4977–5010.
49. Heikamp K, Bajorath J. Interpreting features and feature weights of support vector machines in virtual screening: understanding the whole is greater than the sum of its parts principle. J Comput Aided Mol Des. 2013;27(9):831–45.
50. Hawkins DM. The problem of overfitting. J Chem Inf Comput Sci. 2004;44(1):1–12.
51. Cawley GC, Talbot NL. On over-fitting in model selection and subsequent selection bias in performance evaluation. J Mach Learn Res. 2010;11:2079–107.
52. Montavon G, Samek W, Müller KR. Methods for interpreting and understanding deep neural networks. Digit Signal Process. 2018;73:1–15.
53. Lipton ZC. The mythos of model interpretability. Commun ACM. 2018;61(10):36–43.
54. Weiss K, Khoshgoftaar TM, Wang D. A survey of transfer learning. J Big Data. 2016;3(1):9.
55. Pan SJ, Yang Q. A survey on transfer learning. IEEE Trans Knowl Data Eng. 2010;22(10):1345–59.
56. Gal Y, Ghahramani Z. Dropout as a Bayesian approximation: representing model uncertainty in deep learning. In: Proceedings of the 33rd International Conference on Machine Learning (ICML-16); 2016. p. 1050–9.
57. Kendall A, Gal Y. What uncertainties do we need in Bayesian deep learning for computer vision? In: Advances in Neural Information Processing Systems (NIPS); 2017. p. 5574–84.
58. Mullard A. AI in drug design: artificial intelligence is changing the way drugs are discovered. Nature. 2017;548(7666)
59. LeCun Y, Bengio Y, Hinton G. Deep learning. Nature. 2015;521(7553):436–44.
60. Rupp M, Tkatchenko A, Müller KR, von Lilienfeld OA. Fast and accurate modeling of molecular atomization energies with machine learning. Phys Rev Lett. 2012;108(5):058301.
61. Behler J, Parrinello M. Generalized neural-network representation of high-dimensional potential-energy surfaces. Phys Rev Lett. 2007;98(14):146401.
62. Gilmer J, Schoenholz SS, Riley PF, Vinyals O, Dahl GE. Neural message passing for quantum chemistry. In: Proceedings of the 34th International Conference on Machine Learning (ICML-17); 2017. p. 1263–72.
63. Huang B, von Lilienfeld OA. Communication: Understanding molecular representations in machine learning: The role of uniqueness and target similarity. J Chem Phys. 2016;145(16):161102.
64. Noé F, Tkatchenko A, Müller KR, Clementi C. Machine learning for molecular simulation. Annu Rev Phys Chem. 2020;71:361–90.
65. Schütt KT, Sauceda HE, Kindermans PJ, Tkatchenko A, Müller KR. SchNet – A deep learning architecture for molecules and materials. J Chem Phys. 2018;148(24):241722.
66. Unke OT, Meuwly M. A reactive, scalable, and transferable model for molecular energies from a neural network approach based on local information. J Chem Phys. 2019;150(15):154102.

67. Hansch C, Fujita T, ρ-σ-π analysis. A method for the correlation of biological activity and chemical structure. J Am Chem Soc. 1964;86(8):1616–26.
68. Dror RO, Dirks RM, Grossman JP, Xu H, Shaw DE. Biomolecular simulation: a computational microscope for molecular biology. Annu Rev Biophys. 2012;41:429–52.
69. Shirts MR, Mobley DL, Chodera JD, Pande VS. Accurate and efficient corrections for missing dispersion interactions in molecular simulations. J Phys Chem B. 2007;111(45):13052–63.
70. Chodera JD, Mobley DL. Entropy-enthalpy compensation: role and ramifications in biomolecular ligand recognition and design. Annu Rev Biophys. 2013;42:121–42.
71. Breiman L. Bagging predictors. Mach Learn. 1996;24(2):123–40.
72. Freund Y, Schapire RE. Experiments with a new boosting algorithm. In: Proceedings of the 13th International Conference on Machine Learning (ICML-96); 1996. p. 148–56.
73. Papadopoulos G. Inductive, transductive, and consensus modeling for QSAR/QSPR (PhD thesis). ETH Zurich: Zurich; 2008.
74. Lind P, Maltseva T, Gaspar HA, Grisoni F, Loureiro A, Pasikanti KK, et al. Consensus modeling for HTS assays. J Chem Inf Model. 2020;60(9):4324–35.
75. Pérez González J, Godoy AL, Vázquez GE, Fernández MP. Predictive models of aqueous solubility based on molecular topology and the extended Wiener index. J Chem Inf Comput Sci. 2002;42(4):730–7.
76. Jorgensen WL, Maxwell DS, Tirado-Rives J. Development and testing of the OPLS all-atom force field on conformational energetics and properties of organic liquids. J Am Chem Soc. 1996;118(45):11225–36.
77. He J, Zhang J, Li J, Wang X, Tang Y. Predicting drug solubility using machine learning models combined with molecular dynamics simulations. J Chem Inf Model. 2019;59(11):5075–85.
78. Li Y, Yang L, Du H, Yang F, Song W, Wei Y. QSPR models for predicting solubility of drug-like molecules: rigorous model development and validation. J Chem Inf Model. 2018;58(8):1570–81.
79. Llinàs A, Glen RC, Goodman JM. Solubility challenge: can you predict solubilities of 32 molecules using a database of 100 reliable measurements? J Chem Inf Model. 2008;48(7):1289–303.
80. Kwon S, Park Y, Kim H, Kim D. Computational prediction of solubility of organic semiconductors for organic electronic devices. J Phys Chem C. 2017;121(49):27444–53.
81. Lee H, Park S, Cho JH, Lee JK, Kim S, Kim K. Machine learning-based prediction of solubility parameters for polymers. Sci Rep. 2020;10(1):15741.
82. Gorai P, Stevanović V, Toberer ES. Computationally guided discovery of thermoelectric materials. Nat Rev Mater. 2017;2(9):17053.
83. Wang J, Liu G, Zhang J, Li W, Zhou W, Liu H, et al. Prediction of solubility of persistent organic pollutants in water using machine learning models. Environ Sci Technol. 2018;52(3):1516–25.
84. Arnold RJ, Hess TF, Ward TJ, Tummala SR, Barton CA. Solubility predictions of organic compounds using a multivariate quantitative structure-activity relationship model. J Chem Inf Model. 2016;56(10):1972–82.
85. Zhang X, Liang Y, Zhu H, Zhang H, Yang Z. Prediction of solubility of agrochemicals using QSPR models combined with experimental data. Environ Sci Pollut Res Int. 2020;27(15):18294–305.
86. Warshel A, Levitt M. Theoretical studies of enzymatic reactions: dielectric, electrostatic and steric stabilization of the carbonium ion in the reaction of lysozyme. J Mol Biol. 1976;103(2):227–49.
87. Roux B, Simonson T. Implicit solvent models. Biophys Chem. 1999;78(1–2):1–20.
88. Cao Y, Guerreschi GG, Aspuru-Guzik A. Quantum Neuron: an elementary building block for machine learning on quantum computers. arXiv preprint arXiv:1711.11240. 2017 Nov 30.
89. Biamonte J, Wittek P, Pancotti N, Rebentrost P, Wiebe N, Lloyd S. Quantum machine learning. Nature. 2017;549(7671):195–202.

90. Tan C, Sun F, Kong T, Zhang W, Yang C, Liu C. A survey on deep transfer learning. In: Proceedings of the 27th International Conference on Artificial Neural Networks (ICANN 2018); 2018 Oct 4. p. 270–9.
91. Kirkpatrick S, Gelatt CD, Vecchi MP. Optimization by simulated annealing. Science. 1983;220(4598):671–80.
92. Ma XH, Zhu F,Liu X,Shi Z,Zhang JX, Yang SY et al., Virtual screening methods as a tool for drug discovery from large chemical libraries, Curr. Med. Chem. 2012;19(32):5562–71
93. Brown N, McKay B, Gilardoni F, Gasteiger J. A graph-based genetic algorithm and its application to the multiobjective evolution of median molecules. J Chem Inf Comput Sci. 2004;44(3):1079–87.
94. Tabor DP, Roch LM, Saikin SK, Kreisbeck C, Sheberla D, Montoya JH, et al. Accelerating the discovery of materials for clean energy in the era of smart automation. Nat Rev Mater. 2018;3(5):5–20.
95. Aspuru-Guzik A, Persson K. Materials acceleration platform: accelerating advanced energy materials discovery by integrating high-throughput methods and artificial intelligence. Mater Horiz. 2018;5(4):500–4.
96. Tayyebi A, Safavi A, Khosravian A, Rostami S, Rahimi F, Samadi KH. Prediction of organic compound aqueous solubility using machine learning: a comparison study of descriptor-based and fingerprints-based models. J Mol Graph Model. 2018;79:1–10.
97. Vassileiou AD, Zhang S, Kaxiras E. A unified ML framework for solubility prediction across organic solvents. J Chem Phys. 2019;151(8):084106.
98. Meng J, Wang Y, Chen X, Liu X, Wang H, Wu D, et al. Boosting the predictive performance with aqueous solubility dataset curation. Mol Pharm. 2020;17(11):4102–14.
99. Zheng T, Zhang W, Li J, Lv X, Zhang L, Wang R. Machine Learning for Solubility Prediction. Mol Pharm. 2020;17(9):3292–303.
100. Liu J, Zhang Y, Yu Y, Shi Y, Li J. Fragment-pair based drug molecule solubility prediction through attention mechanism. J Cheminform. 2021;13(1):33.
101. Ryu S, Lee S. Accurate, reliable, and interpretable solubility prediction of drug-like molecules with attention pooling and Bayesian learning. J Chem Inf Model. 2021;61(5):2327–36.
102. Increased Accuracy of Quantum Mechanics Calculations: Reference: Bartók AP, Payne MC, Kondor R, Csányi G. Gaussian approximation potentials: the accuracy of quantum mechanics, without the electrons. Phys Rev Lett. 2010;104(13):136403.
103. Development of Novel Descriptors: Reference: Ramakrishnan R, Dral PO, Rupp M, von Lilienfeld OA. Quantum chemistry structures and properties of 134 kilo molecules. Sci Data 2014;1:140022.
104. Integration of Experimental Data: Reference: Huuskonen J. Estimation of aqueous solubility for a diverse set of organic compounds based on molecular topology. J Chem Inf Comput Sci 2000;40(3):773.
105. Validation and Benchmarking: Reference: Klimisch HJ, Andreae M, Tillmann U. A systematic approach for evaluating the quality of experimental toxicological and ecotoxicological data. Regul Toxicol Pharmacol 1997;25(1):1–5.
106. Uncertainty Quantification: Reference: Chiang SY, Lin HH, Lin JJ. An overview of model assessment, model uncertainty, and model selection for practitioner consideration. J Hydroinf 2008;10(1):1–11.
107. Open Access Software and Data Repositories: Reference: O'Boyle NM, Morley C, Hutchison GR. Pybel: A Python wrapper for the OpenBabel cheminformatics toolkit. Chem Cent J. 2008;2(1):5.
108. Application to Complex Systems: Reference: Sun Q, Cai J, Zhao G, Wang J, Zhang X, Zhang H. Prediction of the solubility of 2,4,6-trinitrotoluene in organic solvents using group contribution methods. Ind Eng Chem Res 2011;50(14):8563–8571.
109. Gasteiger J. From atoms and bonds to three-dimensional molecular space: molecular descriptors applied to QSAR. Top Curr Chem. 1987;175:1–38.

Chapter 12
Advances in the Computational Prediction of Absorption Prediction of Pharmaceuticals

Saba Wahid A. M. Khan, Aditi Pandey, Somesh Mishra, Roja Rani Budha, and S. N. Koteswara Rao G.

Contents

S. W. A. M. Khan (✉) · A. Pandey · S. Mishra · R. R. Budha
H.K. College of Pharmacy, Oshiwara, Jogeshwari (W), Mumbai, Maharashtra, India

S. N. K. Rao G.
Shobhaben Pratapbhai Patel School of Pharmacy and Technology Management, SVKM's NMIMS, Vile Parle (W), Mumbai, Maharashtra, India

S. N. K. Rao G., R. R. Alavala (eds.), *Applications of Computational Tools in Drug Design and Development*, https://doi.org/10.1007/978-981-96-4154-3_12

Abstract Discovery of drugs has already been advanced with AI and ML, that have made substantial advances in a few scientific domains, including CADD. DL is excellent at predicting medication qualities, analysing chemical structures, and improving molecular design, particularly when combined with CNNs. Drug interactions are predicted by AI-driven models, which also offer information on possible side effects or synergy. AI models find novel therapeutic applications for current medications by examining large datasets of pharmacological characteristics and clinical results. ML is useful for QSAR models, which uses molecule's chemical structure for predicting biological activity of the drug. AI speeds up screening large chemical libraries at high throughput, cutting down on the duration of time required to locate possible therapeutic volunteers. AI also helps in creating new therapeutic compounds from the ground up, guaranteeing the necessary qualities while reducing adverse effects. Even with bright futures, issues like interpretability, overfitting, data quality, and computational resource requirements still exist. In conclusion, AI and ML are revolutionizing drug research and advancing the goal of safer, more individualized, and more effective medications.

Keywords In silico · Software · Optimization · Pharmacokinetics · Validation techniques · QSAR model · AI QSAR pharmacokinetic databases

Abbreviations

ABC	ATP-Binding Cassette Transporters
ADM	Applicability Domain Measure
ADME	Absorption, Distribution, Metabolism, and Excretion
ADMET	Absorption, Distribution, Metabolism, Excretion, and Toxicity
AI	Artificial Intelligence
aiQSAR	Artificial Intelligence-Driven Quantitative Structure-Activity Relationship
ANN	Artificial Neural Network
API	Active Pharmaceutical Ingredient
AUC	Area Under Curve
AUC	Area under curve
BA	Oral Bioavailability
BBB	Blood-Brain Barrier
BCS	Biopharmaceutics Classification System
CADD	Computer-Aided Drug Discovery (Pharmaceutical sciences) Combined Annotation Dependent Depletion (genomics) (4.2.3 case studies)
CART	Classification and Regression Tree
ChEMBL	Chemical Database of Bioactive Molecules
CL	Clearance

CLint	Intrinsic Clearance
CLr	Renal Clearance
C_{max}	Maximum Concentration
COVID-19	Coronavirus Disease 2019
Cpds	Compounds
CV	Coefficient of Variation
CYP	Cytochrome P450 Enzymes
DDIs	Drug-Drug Interactions
DL	Deep Learning
EMA	European Medicines Agency
ES Compounds	External Set Compounds
F	Bioavailability
F_a	Fraction of Absorbed Dose
F_{abs}	Absorption rate
FDA	Food and Drug Administration
GBM	Gradient Boosting Machine
GCN	Graph Convolutional Network
GCNN	Graph Convolutional Neural Network
GEMDOCK	Generic Evolutionary Method for Molecular Docking
GIT/GI	Gastrointestinal
GLIDA	GPCR–Ligand Database
GPOPT	Conjugate Gradient Optimization
GPR	Gaussian Process Regression
GROMACS	Groningen Machine for Chemical Simulations
HBD	Hydrogen Bond Donors
HIA	Human Intestinal Absorption
HPPB	Human Plasma Protein Binding
HSPiP	Hansen Solubility Parameters in Practice
HSVR	Hierarchical Support Vector Regression
IPM	Isopropyl Myristate
IVIVC	In Vitro–In Vivo Correlation
K_a	Absorption Rate Constant
KNN	K-Nearest Neighbour
K_p	Brain-to-Plasma Concentration Ratio
$K_{p,uu,brain}$	Unbound Brain-to-Plasma Partition Coefficient
LBDD	Ligand-Based Drug Design
LMT	Logistic Model Trees
Log BB	Logarithmic Form of Brain-to-Plasma Concentration Ratio
LogitBoost	Boosted Logistic Regression
ML	Machine Learning
MLR	Multiple Linear Regression
MLRs	Multiple Linear Regressions
MR	Magnetic Resonance
MRPs	Multi drug Resistance associated Proteins
MRPs	Multidrug Resistance-Associated Proteins

MW	Molecular Weight
NAMD	Nanoscale Molecular Dynamics
NCE	New Chemical Entity
NCI	National Cancer Institute
NearZeroVar	Near-Zero-Variance
NMP	N-Methyl-2-Pyrrolidone
OATPs	Organic Anion Transporting Polypeptides
OCTs	Organic Cation Transporters
PBPK	Physiologically Based Pharmacokinetic
PCR	Principal Component Regression
P_{eff}	Effective Permeability
pH	Power of Hydrogen
PK	Pharmacokinetics
PK/PD	Pharmacokinetics/Pharmacodynamics
pK_a	Acid Dissociation Constant
PK-DB	Pharmacokinetic Databases
PLANTS	Protein-Ligand Ant System for Molecular Docking
PLSR	Partial Least Squares Regression
PLSRs	Partial Least Squares Regressions
PSA	Polar Surface Area
QSAR	Quantitative Structure-Activity Relationship
REST API	Representational State Transfer Application Programming Interface
RF	Random Forest
RIF	Rifampicin
SARS-CoV-2	Severe Acute Respiratory Syndrome Coronavirus
SBDD	Structure-Based Drug Design
sdwd	Sparse Distance Weighted Discrimination
SLC	Solute Carriers
sub-Q	Subcutaneous
SVM	Support Vector Machine
svmLinearWeights	Linear Support Vector Machines with Class Weights
SYBYL-X	Molecular Modelling and Simulation Software
$t_{1/2}$	Half-Life
T_{max}	Time to Reach Maximum Concentration
TS Compounds	Test Set Compounds
USFDA	United States Food and Drug Administration
V_d	Volume of Distribution
XGB	Extreme Gradient Boosting
ZINC	ZINC Is Not Commercial (free pharmacophore search software for screening the purchasable subset of the ZINC database

12.1 Introduction

The process of getting a medicine approved by the FDA has grown more time-consuming and costly in recent decades. A successful medication development process typically costs between $600 and $800 million and takes 10–15 years. Pharmacokinetic (PK) problems and human toxicities account for 50% of the medications that do not make it to market, while efficacy accounts for 29%. Since difficulties including poor oral absorption, unfavourable distribution, fast metabolism or excretion, and drug-induced toxicity can show up clinically as a lack of efficacy, the true failure rate due to PK and toxicity issues may be substantially greater [1]. The four key parameters of the PK process are (The ADME process). Absorption is defined as the process in which a drug reaches to the main circulatory system i.e. the systemic circulation on its administration. The movement of medication molecules from the bloodstream to different tissues and organs is known as distribution. A drug undergoes metabolism when it is digested by the enzymes into metabolites, which are then expelled from the body. The process by which intact drug molecules are passively or actively eliminated from the body by bile or urine is known as excretion [2]. The primary biopharmaceutical parameters influencing the fraction of the administered dose which is absorbed (F_{abs}) and its absorption rate are the API's solubility, intestinal transit time, the luminal stability (chemical and enzymatic), intestinal permeability, and dissolution. Drug products frequently need pharmaceutical research to create a profile of plasma concentration over time that guarantees an ideal pharmacological response with tolerable negative effects to achieve high systemic bioavailability. For modified release drugs, which are intended to enhance pharmacological response, this is especially important. With coefficients of variation (CV) around 60%–120%, oral products with low bioavailability ($F < 25\%–35\%$) typically show higher intra- and interindividual variability in plasma exposure [3]. The percentage of an oral medication that enters the bloodstream in its unaltered molecular form is known as bioavailability, and it is a crucial pharmacokinetic parameter [4]:

$$\mathbf{F} = \mathbf{F}_{\mathbf{abs}} \times (\mathbf{1} - \mathbf{E}_{\mathbf{G}}) \times (\mathbf{1} - \mathbf{E}_{\mathbf{H}})$$

Before we can say that these processes are completely understood, we must first obtain a far deeper understanding of the interactions between sophisticated oral dose forms and the intricate, dynamic gastrointestinal (GI) physiology across all age groups, from infants to the elderly [4–6]. Gaining this knowledge is necessary to incorporate it into sophisticated software that may be used in development of drug and regulatory decision-making processes.

The overall profile of drug absorption in the digestive tract is influenced by motility, transit length, solubility, dissolution, precipitation, and stability, while regional intestinal effective permeability (P_{eff}) is a significant biopharmaceutical parameter. A beneficial tool to understand and forecast GI drug absorption and bioavailability, which are related to the intestinal absorption potential, which is the BCS, classes of [4, 7] (Fig. 12.1).

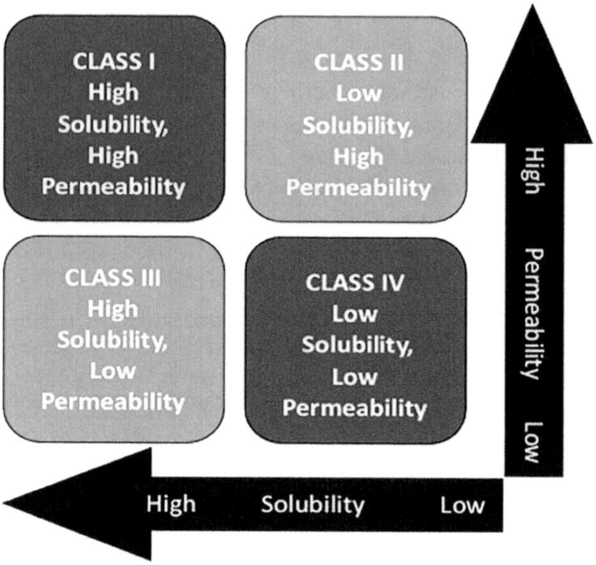

Fig. 12.1 Biopharmaceutics Classification System (BCS) [7]

Computer prediction of pharmacological absorption significantly impacts the success rates of new therapeutics; the former is also an important step in the drug research. A study developed model correlated absorption rate constant (K_a) using molecular descriptors for 22 passively absorbed drugs that achieved strong predictive accuracy for fraction of dose absorbed (F_a) accessing parameters like PSA and HBD [21]. SVM yields a prediction 96.85% accuracy for oral bioavailability that indicates effectiveness of ML. This was another approach that employed SVM on a dataset of 511 compounds [22]. There were disadvantages in traditional methods found regarding accuracy in forecasting behaviour of drugs in biological systems that lead to high attrition rates during clinical trials. Improvisation in predicting accuracy related to drug absorption, permeability and solubility were made by enhancing the computational models that leverage molecular descriptors. This review aims to explore these innovative methodologies, their applications, and the future directions for improving predictive models in pharmacokinetics.

12.2 Theoretical Foundation

12.2.1 Pharmacokinetics (PK) Basics

12.2.1.1 What Is Pharmacokinetics?

PK is an investigation of how drugs are absorbed, distributed, metabolized, and excreted (ADME) in the body over time (more on absorption is covered in Sect. 12.2). It examines how drugs are absorbed into the bloodstream, how they are circulating throughout tissue, how they are chemically altered (metabolized), and how they are eventually excreted. A primary objective of pharmacokinetics is to comprehend concentration of drug at the target location and its impact on therapeutic effects. Although measuring drug concentration at specific receptor sites is not feasible (as these sites are often deep within tissues like the myocardium), pharmacokinetic principles help us estimate drug levels in the body by analysing accessible fluids such as blood, plasma, or urine. Overall, pharmacokinetics provides crucial insights into how a drug behaves in the body [8].

12.2.1.2 Bioavailability

Bioavailability, the portion of a medication that enters the bloodstream in its unmodified for, is a critical element in pharmacokinetics. It simply refers to the amount of a medicine that is chemically intact following the absorption process when it is taken orally. Hence, higher bioavailability ensures better therapeutic effects, as the drug reaches the bloodstream in its active form. When a drug is administered orally or through intramuscular injection, it must be absorbed into the body before it can take effect. If assumed that the medicine is 100% absorbed in some situations, like this one-compartment model, which means that its bioavailability (F) is 1. This shows that the medication enters the bloodstream unaltered and in its entirety. The absorption rate (K_a) is often significantly faster than the rate of elimination. Using a procedure known as the method of residuals, we can determine K_a by examining the drug's absorption. This entails distinguishing between the elimination and absorption processes and calculating the plasma concentration that would result from instantaneous absorption. The absorption rate can be inferred from the discrepancy between the estimated and real concentrations. Back-extrapolation is a crucial method for determining K_a because it enables us to separate the effects of absorption and excretion. A new line, known as the residual line, is created by charting the actual measured concentrations versus the extrapolated values. The absorption rate K_a is indicated by the slope of the line; greater slopes signify faster absorption. Predicting plasma drug concentrations following an oral dose is made easier by knowing K (elimination rate) and K_a. The rate at which a drug is made available in the bloodstream is crucial for evaluating different drug formulations and brands [8, 9].

12.2.1.3 Volume of Distribution (V_d), Clearance (CL) and Half Life ($t_{1/2}$) and Their Relation

V_d and CL are two key pharmacokinetic features that influence drug dosing, efficacy, and elimination. A theoretical volume called V_d connects the quantity of medications in the body to their plasma concentration. It shows the apparent extent to which a medication is dispersed across bodily tissues. Whereas a high V_d shows widespread tissue distribution, a low V_d indicates the medication is mostly confined to the plasma. The loading dosage, or the quantity of medication required to reach a therapeutic plasma concentration, is determined in part by V_d [10]. The quantity of plasma that a medication is removed from in a certain amount of time is referred to as clearance, providing insight into the efficiency of elimination processes, whether through renal excretion, liver metabolism. Unlike half-life, which depends on both clearance and volume of distribution (V_d), clearance is a direct measure of elimination capacity [10, 11]. The drug's half-life ($t_{1/2}$), a crucial factor in determining how long it remains in the system, is influenced by both V_d and CL. $t_{1/2}$ is the amount of time needed for a drug's plasma concentration to drop to half of what it was before. This parameter is essential for understanding how quickly a drug is eliminated from the body, which in turn aids in determining dosing regimens and predicting drug behaviour over time. Most drugs exhibit 'first-order kinetics', meaning their elimination rate is proportional to the drug concentration in the plasma. In this instance, after each half-life, almost 50% of the medication is eliminated from the body. For example, 50% of a drug that has a half-life of 4 hours will still be present after this time; 25% will still be present after another 4 hours; and so on. However, some compounds show "zero-order kinetics," where the drug is eliminated at a constant rate regardless of concentration. This means that the amount cleared is not dependent on the amount of drug present, resulting in a linear fall as opposed to an exponential one. The half-life is given by the formula [10]:

$$\mathbf{t_{1/2} = 0.693 \times V_d / CL}$$

This relationship shows that for a given clearance, a drug with a larger V_d will have a longer half-life. Conversely, if clearance increases (e.g., through enhanced renal function), the drug is eliminated faster, reducing its half-life [10, 12].

12.2.1.4 First-Pass Effect

The First Pass Effect, also known as the First Pass Metabolism, is another significant PK phenomena that affects the bioavailability of drugs given orally. After being absorbed from the gastrointestinal tract and passing via the portal vein to the liver, a drug reaches the systemic circulation. During this journey, a substantial portion of the drug may undergo metabolic conversion, resulting in reduced systemic availability. Current pharmacokinetic models typically assume that drugs administered both intravenously and orally enter the same vascular pool. The AUC of plasma concentration-time after oral delivery, however, can be significantly lower than that obtained after intravenous treatment, according to published research, even if a medication is completely absorbed. This discrepancy is mostly caused by the first-pass effect, particularly in the liver, where drug-metabolizing enzymes can significantly reduce the drug's concentration before it enters the bloodstream. Along with liver metabolism, the first pass effect may also impact the gut that has drug metabolising enzymes, but in smaller amounts than liver. Oral absorption, including the unstirred water layer, membrane constraints, and flow limitations of the drug is affected by the pK_a, solubility, and structural similarity to endogenous substrates, influence factors that are physicochemical properties of the drug. It is important to understand these properties to analyse the process various drug's work in the GI system, thus, affecting its overall absorption and bioavailability. The outcomes of the first-pass effect are essential for the design and clinical evaluation of new drugs intended for oral use [13].

12.2.1.5 C_{max} and T_{max}

C_{max} and T_{max} are critical data that characterize a drug's distribution and absorption inside the body following administration. C_{max} is the maximum amount of a drug in circulation that is peak plasma concentration obtained after delivery because higher concentrations can be related with improved pharmacological activity and adverse reactions, T_{max}, the time needed to attain this peak concentration, is an essential factor for evaluating the drug's instant effects and potential toxicity. It gives detailed information on the drug's absorption rate and is affected by various factors, including the drug's formulation, the way it is administered, and the individual characteristics of each patient, including GI motility and metabolic rates. Both Cmax and Tmax are obtained by calculating the AUC-time curve, which shows the total portion of drug available for a specific time period. They are evaluated using descriptive statistics, such as arithmetic means, across individuals in clinical studies to summarize the PK behaviour of a drug compound. Logarithmic modifications may theoretically give a more accurate examination of PK data, even though PK factors are commonly established on the underlying premise of an exponential half-life. However, the arithmetic means of C_{max} and T_{max} are commonly reported because they provide straightforward interpretations of drug exposure and can facilitate comparisons between different formulations or studies [14].

12.2.1.6 Drug Transporters

The pharmacokinetic mechanisms of ADME depend heavily on drug transporters, which are unique membrane proteins that help medicines pass through cellular membranes. These transporters can be divided into two main categories: SLC and ABC transporters. Transporters aid in the absorption of medications from the gastrointestinal system into the circulation. For instance, the intestinal epithelium contains SLC transporters that facilitate the absorption of several drugs and nutrients include OCTs and OATPs. By facilitating the medications' entry into organs including the liver, kidneys, and CNS, including the BBB, they can also affect how drugs are distributed. Once inside the cells, drugs may undergo metabolic transformations primarily in the liver, where SLC and ABC transporters help medicines be excreted by promoting their efflux into bile or blood. The two phases of metabolism i.e., modification and conjugation reactions make up the metabolic process itself, and transporters are crucial to the ultimate removal of medications. Drug metabolites, for example, can be exported from hepatocytes via transporters including MRPs, which affects the body's overall drug clearance. Additionally, drug transporters are essential in DDIs. When multiple drugs are administered, they can compete for the same transporters, potentially leading to altered pharmacokinetics and unexpected therapeutic outcomes. Additionally, genetic polymorphisms in transporter genes can result in variations in transporter function among individuals, affecting drug absorption and elimination. This variability is critical in personalized medicine, as it underscores the importance of considering genetic factors when predicting a patient's response to therapy [15, 16].

12.2.1.7 Lipid Solubility and Permeability

Particularly for medications that are sparingly hydrophilic, lipids are essential for improving the solubility and bioavailability of pharmaceuticals. Lipid-based drug formulations help in keeping the drug in a dissolution phase within gastrointestinal tract, thereby improving absorption. They offer advantages such as minimizing food effects, reducing inter-individual variability, and being relatively easy to prepare with common excipients. However, the main challenge lies in understanding the complex pharmacokinetic and biopharmaceutical aspects, such as how these lipid formulations are processed after oral administration and predicting their in vivo behaviour from in vitro tests. When drugs enter the body, they must cross cell membranes to reach their target sites. The cell membrane consists of a lipid bilayer which is selectively permeable, allowing lipophilic drugs to diffuse more easily than polar or larger molecules. Factors like a molecular size of drug, degree of ionization, and lipophilicity are key to its ADME. For large lipophilic drugs, passage across the membrane occurs through direct diffusion, while smaller, water-soluble substances may travel through aqueous channels or paracellular pathways. It is one of the important parameters to predict drug behaviour and ensure optimal therapeutic outcomes [17, 18].

12.2.2 Absorption

Absorption is a fundamental process in pharmacokinetics, where a medication enters the bloodstream after being administered. It may be passive or active, depending on characteristics of medication and surroundings. To analyze a drug's bioavailability, the % amount a drug reaches in bloodstream and the potential it must produce therapeutic effects, efficiency and absorption rate are imorprtant factors. The GI membrane's integrity is a physiological component that may have therapeutic effect. Some physiological factors like pH, surface area, gastrointestinal membrane integrity, and transport mechanisms also have a substantial impact, but the permeability and solubility are the primary parameters that govern medication absorption. While permeability is a drug's capability of passage across cell membranes, solubility refers to a medication's capacity of dissolution in GI fluids. Fluids.

Fa has a direct bearing on its therapeutic effectiveness and requires consideration of both parameters to predict the fraction of the medicine that is absorbed. in-vitro and in-silico techniques along with other models are used to predict absorption using many models. It assesses how medications dissolve inside the GI tract and pass through membrane. For passively absorbed and highly soluble drugs, certain models like the 2/4/A1 rat intestinal model have shown good predictability. Predicting absorption is vital for understanding drug dosing, potential interactions, and optimizing drug formulations. It helps ensure that drugs achieve the desired therapeutic concentration in the bloodstream, guiding the design of safe and effective medications [19, 20].

12.2.2.1 Absorption Mechanisms

The drug's physicochemical characteristics, formulation, and biological milieu all affect the absorption process, which differs based on the mode of administration (oral, i.v., topical, etc.). For instance, solid dosage forms like tablets must disintegrate and dissolve before it gets absorbed in the blood, making the rate of dissolution a critical factor. The small intestine is typically a primary location for absorption of drug because of its extensive surface area along with permeability. Additionally, factors such as pH, ionization state of the drug, and gastrointestinal motility can significantly affect absorption efficiency. The primary mechanisms of absorption include passive diffusion, where lipid soluble drugs transverse cell membranes along concentration gradients; facilitated diffusion, which involves specific transport proteins that aid in drug movement; and active transport, a process that requires energy to move drugs against their concentration gradients [23–25]. A schematic representation of various stages involved in oral absorption is presented in Fig. 12.2 [26].

12.2.2.2 Physiological Barriers to Absorption

Numerous physiological obstacles that affect the absorption of microparticles from the gastrointestinal (GI) tract have a major effect on their bioavailability. First, the efficacy of microparticles can be diminished by chemical degradation caused by changes in pH within the GI system, from the stomach's acidic environment to the intestines' alkaline settings. Enzymatic degradation by digestive enzymes such as proteases and lipases further break down microparticles, limiting their absorption. Additionally, microparticles may bind to food or other substances in the GI lumen, forming complexes that reduce their free concentration and absorption potential. As a physical barrier, the unstirred water layer, a stagnant fluid close to the intestinal epithelium, slows down the passage of particles to the absorptive surface. Particles may be trapped by the mucus layer once it is close to the epithelial cells, preventing them from reaching the surface of the epithelium. While membrane permeability controls the possibility that microparticles will pass through by passive diffusion,

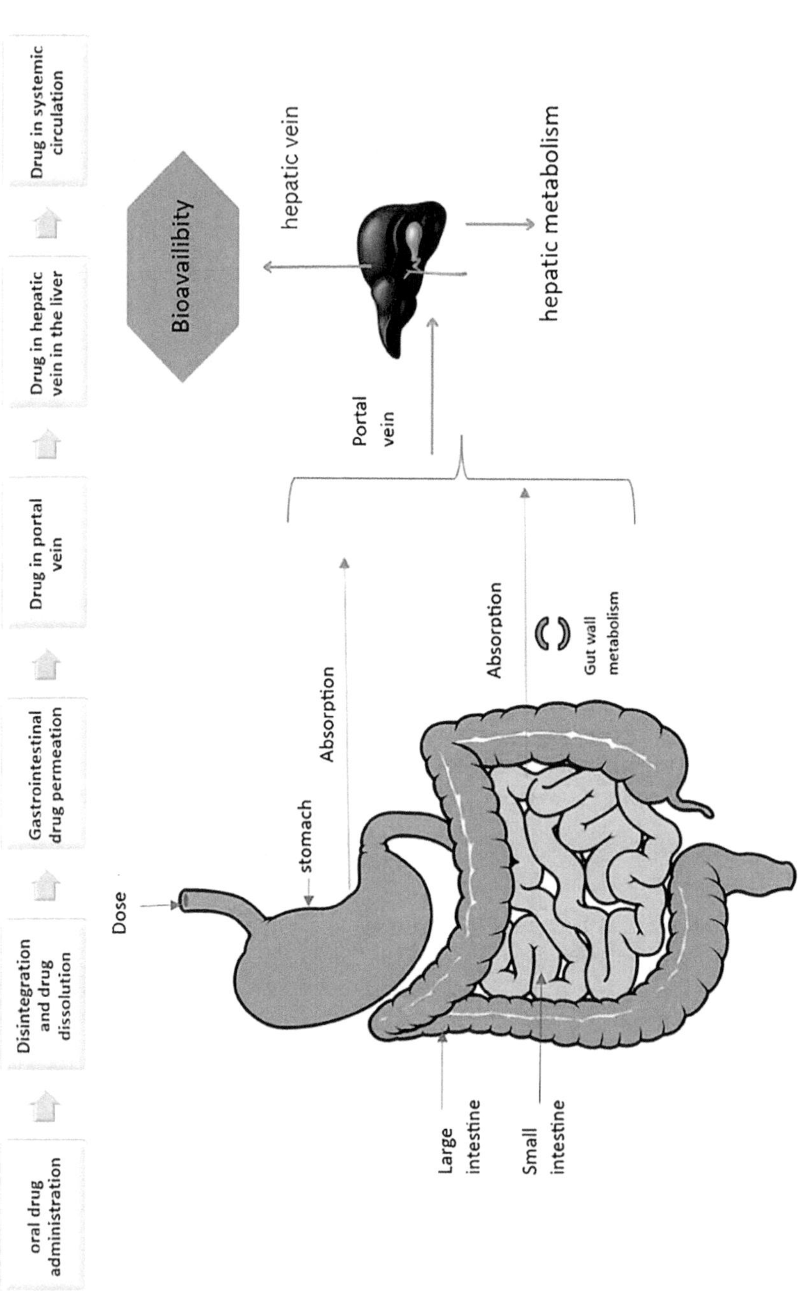

Fig 12.2 Schematic diagram of oral absorption [26]

active transport, or endocytosis, tight junctions between epithelial cells inhibit the paracellular transit of bigger particles. After absorption, pre-systemic metabolism presents further challenges, with the gut wall metabolizing microparticles before they reach systemic circulation, followed by hepatic metabolism. Together, these barriers along with factors such as particle size, surface tension, and ζ-potential-contribute to the generally low and variable absorption of microparticles in oral drug delivery systems [32–34].

12.2.3 *Factors Influencing Drug Absorption*

The way a medicine is administered, its chemical makeup, and the physiological circumstances of each individual patient all affect how well the drug is absorbed. Passive diffusion, the most popular method of drug transport across cell membranes, is controlled by several variables, including molecular size, concentration gradient, lipid solubility, degree of ionization, and protein binding. Smaller molecules diffuse more easily, and an increased concentration gradient accelerates diffusion across membranes. Lipid-soluble drugs cross cell membranes more readily because the cell membrane is primarily made of phospholipids, allowing non-polar substances to dissolve and pass through more efficiently. The degree of ionization also plays a crucial role, as only un-ionized forms of the drug can freely cross membranes, and the ionization level depends on the environment's pH and the drug's pK_a. Changes in plasma protein levels, particularly albumin, have a major impact on medication availability because protein binding can restrict absorption because only the free, unbound fraction of the drug can pass membranes. Other mechanisms of absorption include pinocytosis, which is the engulfing of large molecules like insulin into cells, and filtration through aqueous channels between epithelial cells, although this is more common in tissues with 'leaky' epithelia such as the gut and renal tubules. In contrast, tight barriers like the blood-brain barrier restricts such transport type. Additionally, factors such as local blood flow at the site of injection influence the absorption rate, with heat or massage enhancing blood flow and thereby increasing absorption [34, 36].

12.3 Computational Methods for Absorption Prediction

12.3.1 *QSAR Models*

For chemical data analysis, QSAR modelling is a well-known computer technique. Such models are made by empirical, linear, or nonlinear correlation with chemical descriptors that are derived from molecular structures with the experimentally found characteristics or bioactivities of such molecules. Then, they're used to forecast or create novel compounds that has particular desired characteristics. AI-driven QSAR modelling revolutionised drug's development that predicts biological activities from molecular structures. Choosing molecular descriptors and numerical values is a step in this procedure which indicate a molecule's physical characteristics. Pre-processing of data is done to ensure accuracy, and statistical methods, like regression analysis, are employed to link these descriptors to biological activities. Advanced ML algorithms further enhance these predictions. QSAR models identifies potential drug candidates much faster than traditional methods by enabling rapid virtual screening of vast compound libraries. They also provide insights into how molecular structure affects biological activity because it supports rational drug design. Recent advancements in AI and machine ML have significantly improved accuracy and versatility in QSAR models' sector, making them essential tools for modern drug discovery. The integration of AI not only speeds up the identification of promising drug candidates but also reduces the need for extensive experimental testing, paving the way for more efficient and innovative pharmaceutical development [35]. The aiQSAR method can be applied in two distinct ways; cross validating (or re-testing) the training set and predicting the endpoint values for compounds in evaluation values for compounds in the endpoint values for compounds in the evaluation set [37]. The workflow of aiQSAR is presented in Fig. 12.3.

Step-by-Step Process: Step 1—Calculating Descriptors: Using all of the provided 1D and 2D descriptors, compute 3839 descriptors for all TS and ES compounds using the Dragon 7 software. It is also possible to import a numerical matrix of descriptors calculated for each TS and ES compound.

Step 2: Fingerprint Calculation: Each TS and ES compound's fingerprint can be found using the R package "rcdk." The "pubchem" and "extended" fingerprint types can be obtained using the "get.fingerprint" function. Extended fingerprints employ a defined fingerprinting algorithm that folds the sequence to 1024 and assesses routes up to six chemical bonds in depth, whereas Pubchem fingerprints are structural keys created by the Pubchem database.

Step 3: Determine Fingerprint Similarities: Utilizing the R package "fingerprint," determine the Tanimoto distance, which quantifies the fingerprint similarities between the target chemical and all TS compounds. Choosing the Local Group in Step Four: Based on Tanimoto distances, use an algorithm to choose 20 to 50 TS compounds that are most similar to the target chemical. Choose every chemical that satisfies the first set of requirements. Refine the selection process using the average ranks of two similarity measures if the necessary number is not reached.

Step 4a: Training Set: Select the local group of similar compounds from the remaining TS after removing a target chemical.

Step 4b: Evaluation Set: For a target chemical in the ES, select the local group of similar chemicals from the entire TS.

Step 5: Filtering descriptors: Descriptors with missing values should be discarded for any compound in the target compound or the local group. With the help of the R package "caret," use the "nearZeroVar" function to remove the zero-variance and near-zero-variance traits. Since this method is autonomous for each target molecule, different descriptors can be selected each time.

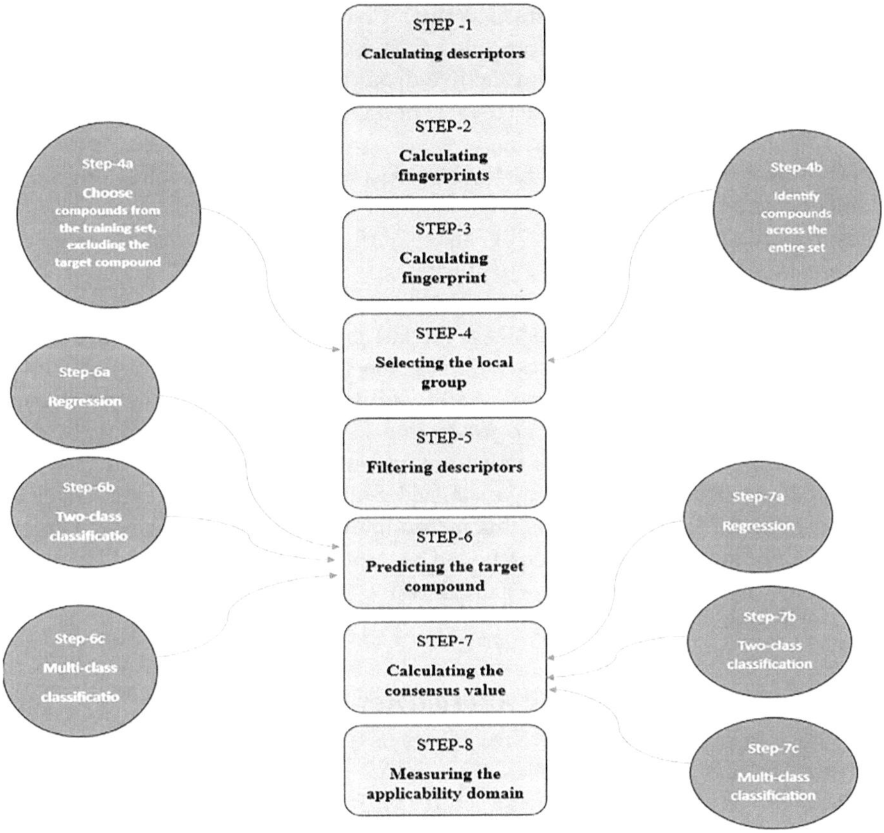

Fig. 12.3 Flow chart of the QSAR model development [37]

Step 6: Target Compound Prediction: Create several mathematical models from the local group using the R package "caret," then use each one to predict the endpoint value of the target chemical. Different methods are used depending on the endpoint type and are selected based on their performance in an independent evaluation of a large number of datasets.

Phase 6a: Regression: Use methods like rf (Random Forest), extratrees (Random Forest by Randomization), glmboost (Boosted Generalized Linear Model), gaussprPoly (Gaussian Process with Polynomial Kernel), and glmnet (Lasso and ElasticNet Regularized Generalized Linear Models) for regression endpoints.

Step 6b: Binary Classification: To make use of methods like LMT (Logistic Model Trees), sdwd (Sparse Distance Weighted Discrimination), glmboost, rf, extra-Trees, LogitBoost (Boosted Logistic Regression), and svmLinearWeights (Linear Support Vector Machines with Class Weights).

Step 6c: Multi-Class Classification: Apply methods to multi-class endpoints, including rf, extraTrees, LogitBoost, LMT, and sdwd.

Step 7: Determine the Consensus Value: Combine the predictions from each method to create a single output value. The processes employed vary according to the type of endpoint. Any of the five distinct models that are more than 10% higher than the overall highest TS value or more than 10% lower than the lowest TS value should be eliminated in.

Step 7a: Regression. Calculate the final forecast by averaging the remaining values predicted by each model.

Step 7b: Binary Classification: Take the majority vote from each of the seven different models to determine the final forecast.

Step 7c: Multi-Class Classification: Select the most likely class to be predicted from the five distinct two models as the last prediction. If there is a tie, choose the class with the lowest representation in the TS.

Step 8: Evaluating the Domain of Applicability: By contrasting the target compound with others in its local group, you can find its Applicability Domain Measure (ADM). Assign the target compound a rank between 1 to 5, with a higher average similarity being associated with a higher ADM rating. Based on the intended ADM cutoff, use this metric to evaluate the coverage of chemical space pertinent to the target compound's class and decide if the final forecast is trustworthy or ought to be disregarded [37].

12.3.2 Machine Learning Approaches

Various ML algorithms are employed by QSAR modelling to reveal relationships between properties and molecular structure. MLR is widely used because it is easy to interpret and has computational efficiency, though it is prone to overfitting and sensitivity to outliers [31]. PLSR and PCR (lower dimensionality) are converting correlated features into uncorrelated principal components, but these only capture linear relationships. More advanced algorithms are popular for their ability to handle complex, high-dimensional data like ANN, RF, and SVM. ANN, despite its strong fitting capability, needs expert handling to avoid overfitting, but recent advancements in DL have renewed interest in its application. SVM and RF are favoured because they're stable and easy to implement. Also, algorithms such as GPR, GBM, and XGBoost have shown modelling. Simpler algorithms like KNN and CART were employed early on but have become less popular with the advent of more powerful methods [32].

12.3.3 Physiologically Based Pharmacokinetic (PBPK) Modeling

PBPK models provide a mechanistic framework to integrate compound and system data for prediction of drug exposure in humans [41]. While they're scientifically well-founded, they can significantly reduce the uncertainty and increase confidence in extrapolations outside of populations or scenarios that have been evaluated. The use of PBPK models in preclinical and clinical development phases is expanding. By clarifying the mechanisms underlying drug exposure, they aid in the selection and decision-making of prospective drugs during preclinical development. PBPK models support regulatory submissions and internal decision-making during clinical development. Recent proposed guidelines from the USFDA and the EMA highlight the increasing importance of PBPK models. Despite their similar advantages, it is still difficult to gain trust in PBPK models for organ impairment, pediatric dose selection, and non-drug interaction (DDI) applications. This difficulty arises from poor in vitro–in vivo correlation (IVIVC), especially for transporter-dependent or noncytochrome P450 (CYP)-mediated elimination pathways, and the lack of sufficient clinical datasets to resolve parameter non-identifiability [40, 42, 43]. The unique benefit of PBPK modeling and simulations is its mechanistic foundation. Making the most of this strength requires building confidence in a mechanistically sound model that has been authorized for a specific use. This calls for a deep comprehension of the mechanisms pertinent to the current inquiry. Depending on the application, different criteria must be met in order to build trust in these mechanisms.

Figure 12.4 illustrates these requirements for the three main PBPK application areas, which comprise the bulk of regulatory submissions: drug–drug interactions (DDI), extrapolation from a base population to pediatrics or special populations, and absorption and formulation-related applications and formulation-related applications.

PBPK modelling for absorption and formulation-related applications involves predicting oral drug absorption by integrating factors like permeability, dissolution, particle size, and release rates, as well as selecting formulations based on these predictions. In order to support bio waivers, it also entails establishing bioequivalence by IVIVC and forecasting how food or proton pump inhibitors may affect drug absorption. The product of the fraction absorbed (F_{abs}) and the fraction escaping intestinal loss (F_g), with intestinal loss resulting from metabolism or efflux, is known as gut bioavailability (F_g). High confidence in absorption predictions is achieved when F_{abs} is reliably assessed, especially for lipophilic small molecules. Understanding mechanisms like efflux and gut metabolism is crucial. For CYP3A substrates, in vitro data can aid quantification, but for non-CYP metabolism, intravenous PK data is essential. Without such data, characterizing mechanisms for poorly soluble compounds is challenging [39, 42].

In general, for all three keys PBPK applications, fewer processes influencing drug exposure (such as drug dissolution, metabolism, and elimination routes) lead to fewer related parameters, lower total uncertainty, and higher model prediction confidence. For instance, medications classified as BCS class I have the fewest parameters and, as a result, the least uncertainty, giving forecasts the highest level of confidence. These pharmaceuticals are primarily eliminated through single or large CYP-mediated metabolism. The sole requirements for a BCS I drug are those related to drug clearance and enzyme inhibition or induction activity because these compounds are frequently not affected by transporters or restricted by absorption. For BCS Class I drugs, biowaivers are commonly given, removing the need for

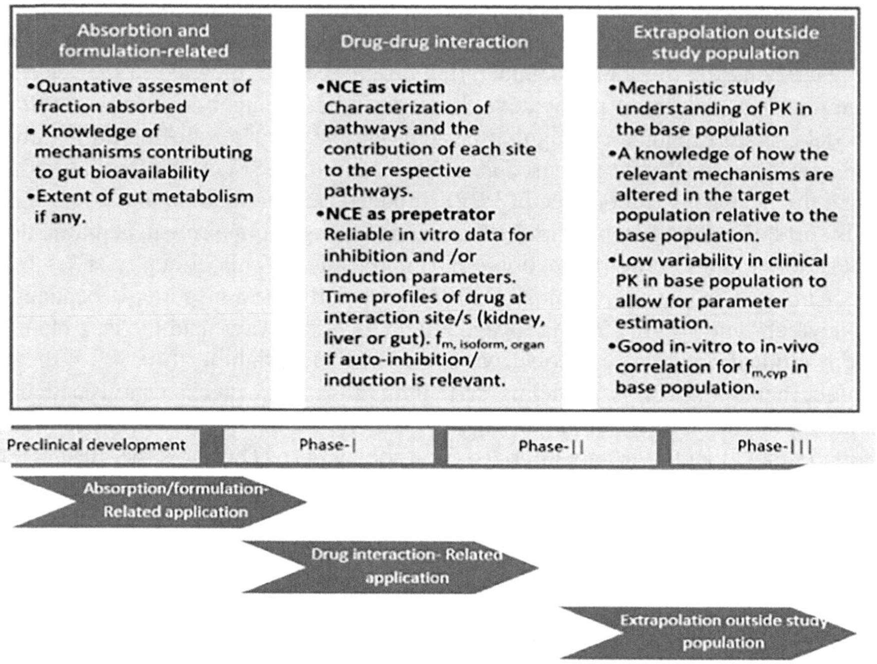

To achieve a high level of confidence in PBPK predictions across the three broad application categories, certain requirements must be met. These categories are also positioned along the value chain. The greater the variability and the smaller the cohort size, the wider the range of the estimated parameter. If this range approaches the entire spectrum of plausible values, the parameter estimation exercise becomes less meaningful. Key terms include PK (pharmacokinetics), NCE (new chemical entity), DDI (drug–drug interaction), and PBPK (physiologically based pharmacokinetics) [43].

Fig. 12.4 Important PBPK application areas [42]

modeling. However, PBPK modeling can help with the solubility-and/or permeability-limited, transporter-dependent exposure, face absorption, and drug-drug interaction (DDI) problems with BCS Class II–IV medications. Applications for these classes include predicting absorption [44, 45], the effects of proton pump inhibitors (PPIs) [46], food effects, bioequivalence assessment through in vitro-in vivo correlation (IVIVC) for formulation bridging, and DDI assessment [42]. A comprehensive list of these applications is detailed by Shebley et al [47]. The unique feature of PBPK models is their capacity to integrate thorough chemical and system knowledge, allowing for precise predictions of drug exposure, drug-drug interactions (DDI), and absorption in untested situations or unstudied populations. But only when the processes pertinent to the particular application are highly trusted can their full potential be achieved. Hypothesis testing can greatly increase confidence in PBPK model predictions when there aren't enough clinical datasets to address parameter non-identifiability. Expanding the uses of 1PBPK models requires efforts to enhance in vitro assays, comprehend the variables affecting drug exposure and its variability across various persons and populations, and increase confidence in system parameters. Future regulatory agency adoption of these models is anticipated to grow as they develop and gain from the scientific community's cumulative experience.

12.3.4 *In Silico Tools and Software*

Significant progress has been achieved in computational chemistry, especially in the development of in silico tools and software for modeling and simulating absorption processes. These computational methods are crucial for understanding how substances interact with various media, which is vital in applications ranging from drug development to environmental science. By utilizing advanced algorithms and high-performance computing, researchers can accurately predict absorption characteristics, minimizing the need for extensive experimental trials. This accelerates the research and development process and improves our ability to design more effective materials and pharmaceuticals. As the demand for efficient and sustainable solutions grows, integrating computational methods into absorption studies is at the forefront of scientific innovation, providing new insights into molecular interactions and transport phenomena.

12.3.4.1 In Silico

Predicting ADME with in silico models has become critical. The fundamental issue in in silico ADME screening is to develop strong models with high predictability. Recent studies on model construction have primarily used AI and ML technologies, such as ANN, RF, SVM, and tree-based approaches [48, 49]. When investigating the relationship between physicochemical qualities and ADME, the essential properties to evaluate are lipophilicity, solubility, ionization, topology, and molecular weight (MW). Solubility and membrane permeability are crucial for oral absorption, as highlighted by the Biopharmaceutics Drug Classification System, which identifies these parameters as key factors in absorption. The importance of solubility and permeability is clear because the maximum absorbable dose is substantially determined by water solubility and absorption rate constants, given a certain intestinal water volume. Poor water solubility and membrane permeability pose significant challenges in selecting appropriate solvents for achieving sufficient exposure in toxicokinetic studies and in developing formulations for clinical trials, often leading to higher failure rates [50]. Numerous researchers have conducted in vivo brain exposure studies, resulting in total brain-to-plasma concentration ratios, known as K_p, or its logarithmic form, log BB. These parameters indicate the extent to which compounds cross the blood-brain barrier (BBB), which is formed by the endothelial cells of brain capillaries. Due to the labour-intensive and time-consuming nature of studies using mice or rats, many researchers have developed qualitative or quantitative computational models (such as classification or QSAR models). They have also transitioned from using methods like MLRs and PLSRs to more advanced techniques like SVMs and ANNs, leveraging new technological advancements (Table 12.1).

12.3.4.2 Software

Computational methods are integrated into computer-aided drug discovery (CADD) to improve the effectiveness of drug development. To find and improve possible drug candidates, it uses techniques including machine learning, virtual screening, and molecular modeling. CADD significantly reduces time and costs associated with traditional drug discovery, which can take over a decade and cost approximately $800 million [65, 66]. Recent advancements in data availability and computational power have expanded the searchable chemical space, enabling the identification of novel compounds [67, 68]. CADD uses a systematic, stepwise process so that it finds and enhances possible medication options, binding site identification, where computational methods analyse protein structures to find cavities favourable for ligand binding. Following this, molecular docking simulates the way small molecules react with the target by predicting their conformations and binding affinities. It involves generation of various ligand poses within the binding site and uses scoring functions to assess their stability and interaction energy. Lead optimisation is the next important step, where promising compounds are refined based on experimental data and computational feedback. Techniques like molecular dynamics simulations provides insights into the altering behaviour of ligand-target interactions, that allows for further adjustments to enhance efficacy. Pharmacophore modelling complements this by evaluating essential features of the ligand that are important for activity, guiding the design of new compounds that maintain these characteristics while improving overall drug-like properties. This iterative process of prediction, testing, and refinement accelerates the development of effective therapeutics tailored to specific biological targets [67, 70] (Fig. 12.5).

Several tools and software packages are widely used for computer-aided drug discovery (CADD), each offering unique functionalities for various drug development process. These tools collectively enhance the efficiency and accuracy of drug discovery processes by integrating computational method with experimental data [71]. Table 12.2 presents tools and software used in CADD.

In the realm of computational design of drug, several software programs and tools are used to support the various phases of the drug discovery process. Sites of binding or interaction prediction tools, such as Cavitator, and ConCavity, are crucial for identifying potential binding sites on protein targets. These tools analyse the protein's surface and cavities to predict where a drug molecule might interact most effectively. By simulating the interaction between a drug candidate and its target protein, docking programs such as AutoDock, GEMDOCK, and idock aid researchers in understanding the drug's binding affinity and orientation inside the binding

Table 12.1 represents a summary of current in silico predictive models for screening ADME

Assay	Dataset			Type of model	Algorithms	Predictive perfomance	Ref.
	No. of cpds	source	parameters				
Solubility	11,780–22,209	In-house assay data	Kinetic solubility	Two-classification model	RF, GCNN	Specifity = 0.90–0.91, senstivity = 0.63–0.71, AUC-ROC = 0.87–0.90	[51]
	12,674	Two public databases	Aqueous solubility	Combination of two-classification and regression models	Gradient boosting and recursive RF	R^2 = 0.87 for censensus model senstivity = 0.80, specificity = 0.96	[52]
Permeability	130	Literature	Caco-2 cells	QSAR	Multivariant linear regression	R = 0.76	[53]
	386	Marketed drugs	MDCK cells	QSAR	PLS, SVM	AUC = 0.81, for SVM, 0,84 for SVM	[54]
HIA	141	ChEMBL, research and serotonin database	Fa	Two-classifier model, QSAR	Hierarchical combination of classification and regression	Accuracy = 0.765, precision = 0.782 R^2 = 0.379	[55]
HPPB	117	Data from literature	Unbound fraction	QSAR	PLS	Q^2 = 0.69	[59]
BBB	7162	7 studies	Two-classification model	……	Light gradient boosting machine	Selectivity = 0.90, specificity = 0.94	[60]
$K_{p,uu,brain}$	640	In-house	QSAR	…….	Conjugate gradient optimization (GPOPT)	R^2 = 0.536 for GPOPT	[61]
Human liver microsome	28,138	Proprietary cpds	CL_{int}	Two-classification model	RF, XGB, GCN	RF, XGB, GCN	[62, 63]
CLr	636	Drugs and drug-like compounds	….	Global model	MLR and RF	R^2 = 0.21–0.36	[64]

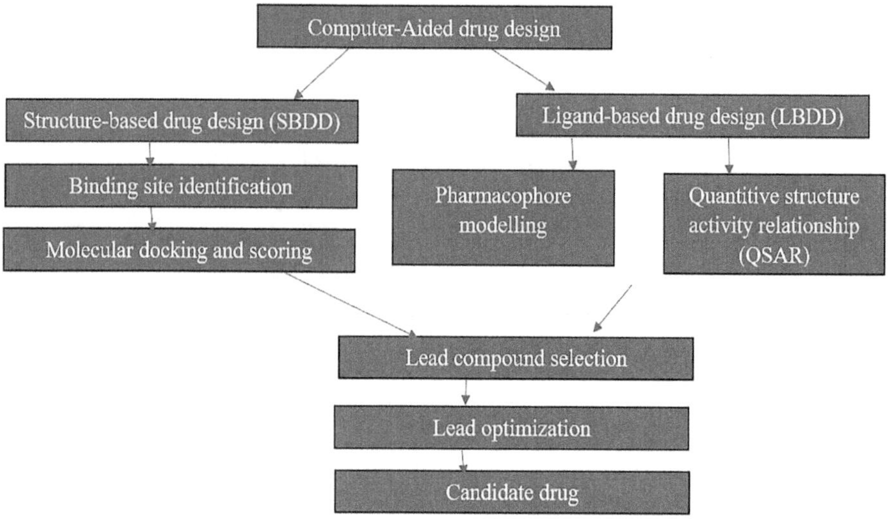

Fig. 12.5 overview of typical CADD overflow [69]

site. To anticipate the three-dimensional structure of proteins based on the known structures of homologous proteins, homology modeling tools such as EsyPred3D and Modeller are crucial. When a target protein's experimental structure is not available, this is very helpful. PLANTS to find the ones that have the highest likelihood of binding to the target protein which are virtual screening tools used to evaluate large libraries of compounds. It helps reducing the number of possible drug candidates by this method, which improves the efficiency of the drug development process. Each of these tools plays an important role in the intricate and multifaceted process of computer-aided drug design (CADD), that contributes to the development of new and effective therapeutic agents [69].

12.4 Data Sources and Validation Techniques: Overview of Computational Prediction in Pharmacokinetics

In last few years, challenges have persisted for enhancing drug absorption and BA, yet these methods are often constraint in addressing issues related to having poorly solubility and impermeability. Rising experimental techniques and computational models, however, offer promising solutions. Among these advancements, in silico methods are gaining significant traction in overcoming bioavailability hurdles by providing predictive insights into drug absorption and pharmacokinetic behaviours. Tools like GastroPlus™ and SimCyp® are being used to predict gastrointestinal (GIT) absorption, passive intestinal membrane permeability, and pharmacokinetic processes, allowing for a more streamlined drug development process. These

Table 12.2 tool and software used in various computational drug design [69]

Function	Program/ server	Free/commercial	Description
Prediction of interaction sites/ binding	ConSurf	Free	Surface-mapping
	ConCavity	Free	Employs evolutionary sequencing and 3D structural models.
	Cavitator	Free	Grid-based geometric analysis for predicting binding pockets.
	ProBis	Free	Local structure alignment
Modelling and molecular dynamics packages	Amber	Free	Molecular dynamics simulation packages
	Discovery studio	Commercial	A molecular mechanics simulation program utilizing the CHARMM force field.
	GROMACS	Free	Molecular dynamics simulation package
	Desmond	Free/commercial	Molecular dynamics simulation program
	NAMD	Free	Molecular dynamics simulation package
	SYBYL-X	Commercial	Molecular modeling and simulation package
Databases	DrugBank	Free	Comphrensive database of target and drug
	GLIDA	Free	GPCR ligand database
	PubChem	Free	Database of small compounds
	ZINC	Free	A library of commercially available compounds for virtual screening
	ChemSpider	Free/ commercial	Database of chemical structures
	ChEMBL	Free/ commercial	Manually curated database of bioactive compounds
	NCI	Free/ commercial	Large database of curated compounds
Docking	AutoDock	Free	Flexible side chain
	cDocker	Commercial	Incorporates side chain flexibility at the atomic scale using grid-based docking
	Idock	Free	Flexible ligand docking
	PLANTS	Free for academic	Uses ant colony optimization (ACO) method
	GEMDOCK	Free	Docking using generic evolutionary method
	PatchDock	Free	Uses shape complementarity method of rigid docking

(continued)

Table 12.2 (continued)

Function	Program/ server	Free/commercial	Description
Homology modeling	RaptorX	Free for non-commercial use	Protein folding prediction based on distance metrics using deep learning techniques.
	Biskit	Free	A suite of python libraries for structural bioinformatics.
	Phyre2	Free	A web server utilizing HMM for structure prediction, functional analysis, and mutation assessment
	EsyPred3D	Free	MODELLER based automated web server
	Modeller	Free	A homology/comparative modeling tool that employs spatial restraints and de novo loop modeling.

computational models not only optimize time and resources but also aid in virtual screening for solubility-related processes such as solubilisation, dissolution, and precipitation. With a growing reliance on in silico methods, the fusion of computational and experimental approaches is accelerating drug development, providing a more robust framework for understanding and enhancing drug bioavailability [38].

12.4.1 Experimental Validation Methods

12.4.1.1 In-Vitro vs In-Vivo Validation

In validation methods of, the focus is on comparing in vitro models, like cell-based assays, with in in vivo or human studies to assess and refine absorption predictions. Cell lines, especially the Caco-2 model, are widely used to estimate human intestinal passability due to their ability to mimics the resistant nature of intestine. They offer a controlled environment that can indicate the permeability of a compound without needing animal or human testing. These assays, however, have limitations, as they may not fully capture the complexities of the human gastrointestinal (GI) system, such as GI motility and regional differences in absorption [4, 56]. Animal models allow for a more comprehensive assessment of drug absorption by capturing physiological conditions. While useful, they can sometimes have limited translatability to humans due to species differences. Advanced models integrate in vivo findings with computational tools to enhance the accuracy of human predictions [56, 57]. Modern computational tools combine both in-vitro and in-vivo data to improvise predictions. These tools allow for the simulation of plasma concentration versus time profiles by combining in vitro permeability data with in vivo pharmacokinetic parameters, providing a balanced approach between efficiency and accuracy. examples are Simcyp, GastroPlus, and GI-Sim [58] (Table 12.3).

Table 12.3 Features of invitro and in vivo validation methods [4, 56–58]

Features	In-vivo validation	In-vitro validation
Primary model	Animal models	Cell lines ($CaCO_2$)
Predictive power	Effective for human translation	Effective for screening
Environment	Complex (biological)	Controlled
Limitations	Species to species variance (results may vary)	Lacks GI complexity

12.4.1.2 Bioavailability Studies

Evaluating a drug's absorption and bioavailability requires both in vitro and in vivo methods. Early screening of drug permeability and passive absorption is made possible by in vitro models, such as Caco-2 cell lines, which offer controlled systems that replicate gastrointestinal (GI) barriers. Although these techniques are fundamental for evaluating a compound's early metabolic characteristics, they do not fully capture the intricacies of human gastrointestinal physiology. On the other hand, in vivo investigations, which are mostly carried out using animal models, allow scientists to assess systemic exposure and bioavailability in more biologically accurate settings. Although species-specific metabolic variations might affect the prediction accuracy of these models, they take into consideration a variety of physiological parameters [4, 56, 72]. The estimating power of in-vitro and in-vivo procedures is improved by in silico techniques in addition to laboratory models. Computational models, like quantitative structure-activity relationships (QSAR), leverage molecular profile and machine learning to predict a drug's bioavailability, which can reduce the need for animal studies. However, each approach has inherent limitations: in vitro models do not replicate the full biological complexity of the human GI system, and in vivo outcomes couldn't possibly be fully correlating with human outcomes due to interspecies variations. Therefore, validation against high quality human data remains essential [72, 73]. Advancements in bioavailability studies emphasize refining these models, especially through the development of information-rich databases that can improve validation accuracy. These evolving techniques shape drug discovery and may reduce reliance on extensive clinical trials by enabling robust early-phase bioavailability assessments [73].

12.4.1.3 Permeability and Solubility Studies

In computational absorption speculations of pharmaceuticals, solvability and penetrability studies represents key validation techniques. These studies leverage both experiments and computer-based simulation methods to better understand and predict pharmacokinetics, especially focusing on gastrointestinal and transdermal absorption. For example: solubility of rifampicin (RIF) was explored in various permeation enhancers across temperature and pressure ranges, validated through models like Van't Hoff and Apelblat, and evaluated using HSPiP software to identify solvents enhancing RIF solubility, such as limonene and IPM, particularly favourable for transdermal or oral administration [74, 75]. Complementary to these solubility assessments, permeability predictions are often developed using quantitative structure activity relationship (QSAR) models, as shown in studies where the hierarchical support vector regression (HSVR) model demonstrated high accuracy in predicting intestinal permeability based on drug descriptors and transport routes [76]. Furthermore, solvent studies for RIF indicated that solvents like N-methyl-2-pyrrolidone (NMP) could be highly suitable for subcutaneous (sub-Q) delivery, a finding further validated with GastroPlus software to predict pharmacokinetics responses, supporting its potential use in treating skin tuberculosis via sub-Q administration [77].

12.4.1.4 Clinical Trials as Ultimate Validations

In pharmaceutical research, clinical trials serve as the final verification of computer absorption predictions. Despite the remarkable precision with which computer models can describe medication absorption pathways, clinical studies are necessary to verify that these predictions are applicable in real-world settings. Scientists assess medication safety, effectiveness, and bioavailability in the intricate milieu of the human body; clinical trials supply crucial human data that validates and enhances computer models. By contrasting anticipated PK profiles with real-world clinical outcomes, this procedure provides information that enhances model dependability and directs the creation of safer and more efficient medications [78]. Along with this, preclinical research is important to bridge the gap between clinical application and theoretical modeling. These investigations allow researchers to assess important pharmacokinetic and toxicological characteristics through both in vitro and in vivo evaluations before human trials. Computer models are improvised with help of preclinical research by replicating human biology and drug interactions in lab settings. This enhances the prediction accuracy of these models with respect to the ADME and toxicity of drugs. The reliability of computer models can be greatly enhanced by including preclinical study outcomes. For instance, data from several preclinical trials informs model parameters and assumptions, that lead to more complex predictions. Also, positive findings in later clinical trials highlights the significance of these investigations as a validation stage by commonly correlating with good results in preclinical trials. The data obtained from preclinical research are essential for refining treatment tactics and reducing late-stage clinical failures, since the pharmaceutical industry increasingly depends upon the computational tools to accelerate drug development [79].

12.4.1.5 Cross Validation with Pharmacokinetic Databases

PK-DB is an open-access pharmacokinetic database through which researchers can now efficiently cross-validate pharmacokinetic data across clinical trials. PK-DB addresses long-standing accessibility challenges by providing curated information on various parameters, such as patient demographics (age, body weight, genetic variants), intervention details (dosing, substance, route), and pharmacokinetic metrics (clearance, half-life, AUC). Through its standardized data and strong validation protocols, PK-DB enables robust cross-study meta-analyses and data integration with computational models like PBPK, PK/PD, and population pharmacokinetics. These features facilitate individualized and stratified modelling, making PK-DB a pivotal tool for advancing pharmacokinetic research through cross-validation [80]. The flowchart below represents the PK-DB curation process that includes extracting and importing data, curating study details, normalizing units, validating, calculating parameters, and enabling access via REST API or web interface (Fig. 12.6):

12.4.2 Case Studies of Computational Predictions

Case studies of computational prediction highlight the diverse applications and methodologies used in various fields as follows:

12.4.2.1 Computational Prediction of Phage-Bacterium Interactions

In the realm of microbiology, understanding the interactions between bacteriophages (viruses that infect bacteria) and their bacterial hosts is crucial for developing phage therapy as an alternative to antibiotics. A well-known case study used machine learning techniques to accurately anticipate these interactions. Like KN, RF, SVM, and ANN algorithm are used by the researcher to analyze a dataset including diverse phage and bacterial features. The study emphasized the need of parameter tuning—adjusting algorithm parameters to enhance performance—and thorough dataset selection for improving prediction accuracy. The researchers observed a 93% F1-score after considerable model modification and rigorous cross-validation. This high degree of accuracy illustrates the computational models' ability to anticipate phage-host interactions, opening the path for more focused and successful phage therapeutics in clinical settings [80, 81].

12.4.2.2 Forecasting Models During the COVID-19 Pandemic

Global public health experts faced previously unheard-of difficulties because of the COVID-19 outbreak, necessitating rapid and accurate forecasting models to inform decision-making. During the early stages of the epidemic in Sweden, a systematic

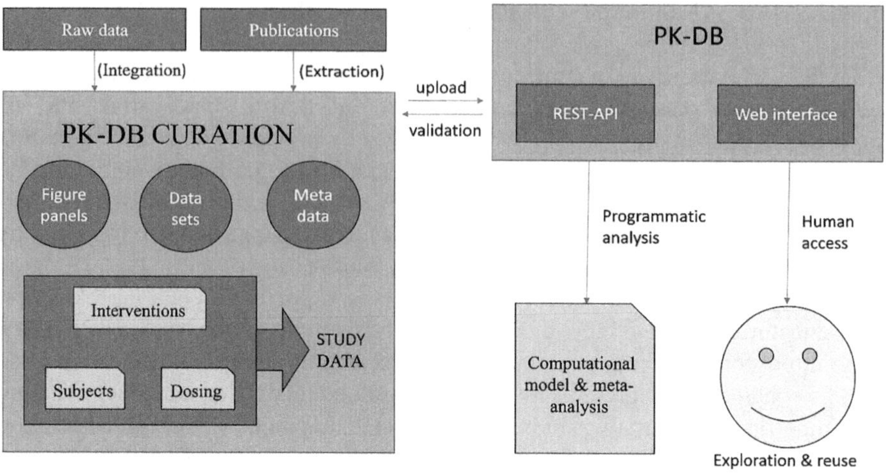

Fig. 12.6 Validation protocol of pharmacokinetics [80]

evaluation was conducted to assess the precision of several computer models developed. This evaluation identified 52 papers that used a variety of modeling methodologies, including statistical analysis, machine learning, and epidemiological simulations. Only four of these studies had good prediction accuracy when compared to actual infection rates and consequences. The findings underlined the difficulty that early models faced as their understanding of SARS-CoV-2, transmission patterns, and public health initiatives evolved. This case study demonstrates both the potential and limitations of computational predictions in critical situations, emphasizing the significance of continuous model modifications as new information becomes available [82].

12.4.2.3 Combined Annotation Dependent Depletion (CADD) Framework

The CADD framework provides a systematic technique for assessing the potential effect of genetic variations within the human genome and shows a significant advancement in computational genomics. It differentiates between proxy-neutral variations, which are unlikely to affect function, and proxy-deleterious variants, which are more likely to cause harm by using ML approaches. CADD assigns ratings to individual variations based on their expected effects on gene function after training on a diverse set of genomic characteristics. This grading method found to be helpful for researchers prioritize for the future investigation in the clinical genetics and genomic medicine. A great deal of studies has been demonstrated CADD's utilization in searching pathogenic variations related to illnesses like cancer and unusual genetic disorders. Computational predictions can transform our comprehension of genetic variation and its consequences for health and disease [30]. The groundbreaking ability of computational prediction in a variety of domains has been showcased by these case studies that also includes microbiology, public health, toxicology, and genetics. They show how modern computational tools may increase our knowledge of complex biological systems, affecting public health initiatives, assessing the adverse effects of chemicals, and interpreting genetic data, eventually leading to better outcomes in research and healthcare.

12.5 Applications in Drug Development

Computer-aided drug discovery (CADD) applications are broad and mainly improves the effectiveness and efficiency of drug development in the pharmaceutical industry. An important application is SBDD, that generates new drugs by using the 3D structures of target proteins. This method allows increasing the likelihood of therapeutic success and researchers to formulate compounds that fit precisely into the active site of the target. There's also an important application is ligand-guided drug development, which uses existing knowledge of active substances to identify new candidates with similar pharmacological properties. By understanding the structural features and biological activities of known ligands, scientists can refine new compounds to obtain improved safety and efficacy profiles. Along with these, virtual screening is a robust technique that enables for the quick evaluation of extensive chemical libraries against biological targets. This method majorly cuts down both time and costs in the early stages of drug development by swiftly discovering promising candidates for further study. A crucial aspect of CADD, molecular docking, simulating the interactions between small molecules and target macromolecules. This method forecasts binding affinities and helps in lead optimization by providing data into how modifications to a compound can improve its interaction with the target. Moreover, QSAR modelling applies mathematical methods to link molecular structure with biological activity, helping in the design of compounds that demonstrate required therapeutic effects. Pharmacophore modelling further refines drug design by discovering essential features required for biological activity, providing guidance to researchers in formulating new compounds that maintain these vital characteristics. ADMET prediction asses the relevant properties of drug candidates in the early phases of development. By evaluating these factors at the outset, researchers can improve the safety profiles of drug candidates and reduce the threats of failures in later stages [30, 66, 77]. QSAR modelling have essential application into several fields, particularly in drug innovation and development. Here represent some key applications:

12.5.1 Drug Discovery

QSAR models make it possible to anticipate a compound's biochemical effect based on its chemical entities, they are essential to the drug discovery process. Researchers can quickly and easily identify promising drug candidates due to these predictive abilities, which fosters the drug development timeline. Scientists can computationally analyze extensive chemical libraries, narrowing down the options to those most likely to exhibit the desired biological effects by employing QSAR models. This method not only speeds up the process of discovering potential drugs but also majorly decreases the costs related to research and development testing. Thus, QSAR modeling has become a crucial tool in contemporary pharmaceutical research, allowing the swift and economical discovery of new therapeutic agents [27, 28].

12.5.2 Virtual Screening

QSAR models significantly improves the efficiency of the drug development process and allow for the virtual analysis of extensive compound libraries. Without the need for immediate physical synthesis or biological testing, this computational approach allows researchers to assess hundreds of potential chemicals based on their expected biological activity. QSAR models the hit-to-lead optimization stage, where initial "hits" from screening are refined into viable "lead" compounds for further development by discovering the most promising candidates early on. Not only this prioritization conserves time and resources, but also it enhances the percentage of discovering effective drug candidates, eventually accelerating the modifications from initial discovery to clinical use [27, 29].

12.5.3 Pharmacokinetics

QSAR model predictions the pharmacokinetic properties of substances, including drug absorption, distribution, metabolism, and excretion (ADME). These properties must be understood for evaluating a drug's safety and effectiveness. The compound's molecular structure are evaluated by QSAR models to determine how well a drug will be absorbed into the bloodstream, circulate through the body, be metabolized by enzymes, and be excreted. Potential issues in the drug development process, such as low bioavailability or toxic metabolites, which could affect the drug's efficacy or safety are identified early in the drug development process through these predictive abilities. As a result, QSAR models play a significant role in optimizing the development of new medications by offering important insights into their pharmacokinetic behaviour, ultimately enhancing the efficiency and success of drug discovery and development [28].

12.5.4 *Integration with Machine Learning*

The analysis of Quantitative Structure-Activity Relationship (QSAR) models has been greatly enhanced by the integration of learning techniques. Vast and intricate datasets are processed and analyzed by machine learning algorithms, uncovering complex patterns and relationships that traditional methods often fail to detect. More accurate predictions of a compound's bioactivity are enabled by this capability, which is based on its chemical structure. Additionally, their predictive accuracy is continuously refined, making them increasingly reliable, as more data is processed by these models. These advancements are regarded as critical in drug discovery, as they help expedite the process, reduce expenses, and increase the probability of discovering new, effective medications. By leveraging machine learning, QSAR modeling has been developed into a robust tool for therapeutic agent design and optimization [28].

12.6 Challenges and Future Directions

12.6.1 *Limitations of Current Models*

Numerous advantages, such as speeding up and enhancing the drug development process, are offered by Computer-aided Drug Design (CADD) but it also presents some challenges. In the CADD field, a critical issue is the shortage of AI and ML experts. To tackle this, strategies like focused recruitment and specialized training programs must be adopt by the organization. Bridging the gap organization like in-silico medicine are taking lead by developing a skilled workforce capable of utilizing advanced computational modeling for drug development. By addressing these challenges, better strategies can be create, and more efficient drug discovery processes can be achieved [87].

12.6.1.1 Accuracy of Predictive Models

A significant challenge present in enhancing the accuracy of computational models in computer-aided drug design (CADD). The theoretical models which are relied by Molecular dynamics simulations, docking scores, and machine learning that may not adequately represent the complexities of biological systems. A deep understanding of scoring techniques is important for improving precision. For predicting how well drugs will bind to their targets, these plays a crucial role. Scoring parameters need to be meticulously calibrated, various chemical descriptors should be utilized, and ongoing validation with experimental data is crucial to improve the accuracy. For instance, prediction precision is increased thoroughly by comparing docking scores against established binding affinities. By refining the equilibrium between sensitivity and specificity, researchers can trust in scoring algorithms is bolstered, thereby minimizing errors in drug discovery determination [86, 88].

12.6.1.2 Data Quality and Quantity

The quality and quantity about data utilized for training to determine the efficacy of CADD tools. Incorrect projections might be result from poor quality or inadequate data. A recurring issue, particularly in the context of machine learning in drug development are the absence of curated, high-quality datasets. Removing outliers and establishing consistent data formatting helps optimize molecular interaction datasets, reducing errors and increasing the trustworthiness of computer models. Additionally, implementing standardized experimental protocols, such as consistent assay conditions and endpoint measurements, further improves data quality in CADD, ensuring robust and dependable results [89].

12.6.1.3 Unbalanced Reliance on Computational Estimation

Although CADD is a strong tool, excessive dependence on its projections in absence of further analytical or experimental confirmation might result in misplaced attempts. Balancing computational predictions with experimental evidence is crucial for successful drug discovery [90].

12.6.1.4 Time and Computational Cost

Modern CADD approaches, particularly those requiring large molecular dynamics demonstration or complicated machine learning models, need substantial computer assets. The associated costs, both in terms of time and infrastructure, can be prohibitive for some research groups [91].

12.6.1.5 Depicting Molecular Flexibility

Bioactive molecule, which are featuring potential therapeutic chemicals and interacting proteins, are very adaptable. Accurately representing this flexibility, particularly in techniques like molecular docking, is challenging and can significantly impact the results of CADD studies [92].

12.6.1.6 Interpretability of AI Models

Machine learning and artificial intelligence models get more complicated, making predictions more difficult to comprehend. The 'black box' nature of AI models can make it difficult to comprehend why a specific molecule is anticipated to be active or how its structure could be modified [93]. As Computer Aided Drug Design (CADD) incorporates AI developments, the drug profession is poised for big alterations. The future of therapeutic advances will be dependent on the appropriate and successful implementation of these technologies.

12.6.2 Integration with Other Computational Approaches

AI and ML have greatly affected scientific domains, including Computer-Aided Drug Design (CADD). This junction occurs transforming drug discovery methods. Deep learning, particularly convolutional neural networks (CNNs), is effective in predicting pharmacological characteristics, analyzing chemical structures, and optimizing molecular design [85]. AI-powered algorithms may be used to estimate how different drugs might interact with each other, revealing possible synergies or bad responses when numerous medications are administered concurrently. AI has played an important role in uncovering novel therapeutic utilization for current medications by analyzing large extensive data of pharmacological characteristics and clinical outcomes. QSAR models, improved through machine learning, provide more precise estimation of a molecular biological activity using its chemical structure. Artificial intelligence accelerates rapid screening of extensive chemical archives, considerably lowering the time required compared to traditional approaches. AI also helps to create innovative medicinal compounds from scratch, ensuring that they have the necessary qualities while reducing unwanted negative effects. Despite the exciting promises for AI and machine learning in CADD, problems persist. Data integrity, model over-optimization, model transparency, and the requirement for considerable computing assets are all issues that must be addressed. Overall, AI and machine learning are revolutionizing drug discovery, bringing the creation of more effective, safer, and personalized medications closer to reality [86].

12.6.3 Emerging Technologies

Artificial intelligence (AI) and machine learning (ML) have recently made significant advancements across various scientific fields, including Computer-Aided Drug Design (CADD), revolutionizing drug discovery processes [86]. Predicting the pharmacological properties, examining chemical structures, and molecular design enhancing exceptionally effective through deep learning by specifically convolutional neural networks (CNNs). Drug-drug interactions can estimate by AI powered models, insights into possible synergies or adverse effects are offered by AI models when multiple drugs are administered together. Additionally, new therapeutic uses for existing drugs have been identified using AI models by sifting through extensive databases of pharmacological properties and clinical data. Machine learning found to be useful for the Quantitative structure-activity relationship (QSAR), as they are capable of more precise predictions regarding a molecule's biological activity provided based on its chemical structure. Artificial intelligence can accelerate the high-throughput screening of expansive chemical libraries can accelerate by artificial intelligence, significantly reduces the time needed to find new drug candidates compared to traditional methods. Furthermore, the development of new therapeutic agents from the ground up can assisted by AI, the necessary properties are ensured while minimizing potential side effects. However, in computer-aided drug design (CADD) despite the considerable promise by Artificial Intelligence and ML, challenges like data quality, overfitting, interpretability, and the demand for substantial computational resources still need to address [84, 85]. Nonetheless, drug research is being revolutionizing by AI and ML, developing the creation of more effective, safer, and personalized treatments.

12.7 Conclusion

In computer-aided drug design (CADD), the combination of machine learning (ML) and artificial intelligence (AI) can used to accelerate the process of drug discovery. The precision of predictions regarding drug properties, streamline molecular design, and increase the efficiency of high-throughput screening, these technologies significantly reducing both time and materials needed. AI-powered algorithms can predict potential drug interactions, offering insights into possible benefits or side effects while numerous bioactive compounds are used together. In identifying new therapeutic applications for existing drugs by examining extensive databases of pharmacological data and clinical results the AI model plays a crucial role. Advances of quantitative structure-activity relationship (QSAR) models are refined by machine learning (ML), based on its chemical makeup enables more precise estimation of a molecule's biological effect. Furthermore, reducing the risk of negative side effects while AI can create new use of existing therapeutic compounds from the ground up, tailoring them to exhibit the desired features. Although, by these advancements brought their own set of challenges. The quantity and quality of the training data crucially depends on the advancement of AI and machine learning models, which poses a major obstacle.Unreliable results and inaccurate predictions can be produced by faulty or low-quality data. Another issue is overfitting, where models excel with training data but struggle to apply their learning to new, unseen data. Additionally, understanding how AI models make their predictions is a key concern, as complex systems like deep learning networks often function as "black boxes," complicating the interpretation of their decision-making processes. Furthermore, the considerable computational resources needed for training and deploying AI models can be a significant barrier, especially for smaller research organizations or companies with tight budgets. In conclusion, while AI and ML are reshaping drug discovery to be more effective and efficient, tackling these challenges is essential to fully harness their potential in computer-aided drug design (CADD). The collaboration of AI, ML, and CADD offers exciting opportunities for developing more efficient, safer, and personalized medications, bringing us closer to significant advancements in medical science.

References

1. Lin J, Sahakian DC, De Morais SM, Xu JJ, Polzer RJ, Winter SM. The role of absorption, distribution, metabolism, excretion and toxicity in drug discovery. Curr Top Med Chem. 2003;3(10):1125–54.
2. Jang GR, Harris RZ, Lau DT. Pharmacokinetics and its role in small molecule drug discovery research. Med Res Rev. 2001;21(5):382–96.
3. Hellriegel ET, Bjornsson TD, Hauck WW. Interpatient variability in bioavailability is related to the extent of absorption: implications for bioavailability and bioequivalence studies. Clin Pharmacol Ther. 1996;60(6):601–7.
4. Dahlgren D, Lennernäs H. Intestinal permeability and drug absorption: predictive experimental, computational and in vivo approaches. Pharmaceutics. 2019;11(8):411.
5. Wang K, Amidon GL, Smith DE. Physiological dynamics in the upper gastrointestinal tract and the development of gastrointestinal absorption models for the immediate-release oral dosage forms in healthy adult human. Pharm Res. 2023;40(11):2607–26.
6. Astolfi ML, Protano C, Schiavi E, Marconi E, Capobianco D, Massimi L, Ristorini M, Baldassarre ME, Laforgia N, Vitali M, Canepari S. A prophylactic multi-strain probiotic treatment to reduce the absorption of toxic elements: in-vitro study and biomonitoring of breast milk and infant stools. Environ Int. 2019;130:104818.
7. Yang Y, Zhao Y, Yu A, Sun D, Yu LX. Oral drug absorption: evaluation and prediction. In: Developing solid oral dosage forms 2017 Jan 1. Academic Press. p. 331–54.
8. DiPiro JT. Concepts in clinical pharmacokinetics. ASHP; 2010.
9. Rowe P. Pharmacokinetics. Bookboon; 2012.
10. Mason RW, McQueen EG, Keary PJ, James NM. Pharmacokinetics of lithium: elimination half-time, renal clearance and apparent volume of distribution in schizophrenia. Clin Pharmacokinet. 1978;3:241–6.
11. Rowland M, Benet LZ, Graham GG. Clearance concepts in pharmacokinetics. J Pharmacokinet Biopharm. 1973;1(2):123–36.
12. Hallare J, Gerriets V. Half-life. Treasure Island: StatPearls Publishing; 2023.
13. Doherty MM, Pang KS. First-pass effect: significance of the intestine for absorption and metabolism. Drug Chem Toxicol. 1997;20(4):329–44.
14. Julious SA, Debarnot CA. Why are pharmacokinetic data summarized by arithmetic means? J Biopharm Stat. 2000;10(1):55–71.
15. Zhang Y. Overview of transporters in pharmacokinetics and drug discovery. Curr Protoc Pharmacol. 2018;82(1):e46.
16. Marquez B, Van Bambeke F. ABC multidrug transporters: target for modulation of drug pharmacokinetics and drug-drug interactions. Curr Drug Targets. 2011;12(5):600–20.
17. Benet LZ, Kroetz D, Sheiner L, Hardman J, Limbird L. Pharmacokinetics: the dynamics of drug absorption, distribution, metabolism, and elimination. In: Goodman and Gilman's the pharmacological basis of therapeutics, vol. 3; 1996. p. e27.
18. Yang J, Li K, He D, Gu J, Xu J, Xie J, Zhang M, Liu Y, Tan Q, Zhang J. Toward a better understanding of metabolic and pharmacokinetic characteristics of low-solubility, low-permeability natural medicines. Drug Metab Rev. 2020;52(1):19–43.
19. Loucks J, Yost S, Kaplan B. An introduction to basic pharmacokinetics. Transplantation. 2015;99(5):903–7.
20. Riviere JE. Absorption, distribution, metabolism, and elimination. In: Veterinary pharmacology and therapeutics. 9th ed; 2009.
21. Linnankoski J, Mäkelä JM, Ranta VP, Urtti A, Yliperttula M. Computational prediction of oral drug absorption based on absorption rate constants in humans. J Med Chem. 2006;49(12):3674–81.
22. Kumar R, Sharma A, Varadwaj PK. A prediction model for oral bioavailability of drugs using physicochemical properties by support vector machine. J Nat Sci Biol Med. 2011;2(2):168–73.
23. Grogan S, Preuss CV. Definition/Introduction.

24. Gu CH, Kuldipkumar A, Chauhan H. Factors affecting drug absorption and disposition. In: Pharmaceutics 2024 Jan 1. Academic Press. p. 499–516.
25. Blumer JL. Principles of drug disposition in the critically ill child. In: Pediatric critical care 2011 Jan 1. Mosby. p. 1538–52.
26. Abuhelwa AY, Williams DB, Upton RN, Foster DJ. Food, gastrointestinal pH, and models of oral drug absorption. Eur J Pharm Biopharm. 2017;112:234–48.
27. Neves BJ, Braga RC, Melo-Filho CC, Moreira-Filho JT, Muratov EN, Andrade CH. QSAR-based virtual screening: advances and applications in drug discovery. Front Pharmacol. 2018;9:1275.
28. Tropsha A, Isayev O, Varnek A, Schneider G, Cherkasov A. Integrating QSAR modelling and deep learning in drug discovery: the emergence of deep QSAR. Nat Rev Drug Discov. 2024;23(2):141–55.
29. Ancuceanu R, Tamba B, Stoicescu CS, Dinu M. Use of QSAR global models and molecular docking for developing new inhibitors of c-src tyrosine kinase. Int J Mol Sci. 2019;21(1):19.
30. Schubach M, Maass T, Nazaretyan L, Röner S, Kircher M. CADD v1. 7: using protein language models, regulatory CNNs and other nucleotide-level scores to improve genome-wide variant predictions. Nucleic Acids Res. 2024;52(D1):D1143–54.
31. Piir G, Kahn I, García-Sosa AT, Sild S, Ahte P, Maran U. Best practices for QSAR model reporting: physical and chemical properties, ecotoxicity, environmental fate, human health, and toxicokinetics endpoints. Environ Health Perspect. 2018;126(12):126001.
32. Wu Z, Zhu M, Kang Y, Leung EL, Lei T, Shen C, Jiang D, Wang Z, Cao D, Hou T. Do we need different machine learning algorithms for QSAR modeling? A comprehensive assessment of 16 machine learning algorithms on 14 QSAR data sets. Brief Bioinform. 2021;22(4):bbaa321.
33. Ashford M. Gastrointestinal tract–physiology and drug absorption. Aulton's pharmaceutics e-book: the design and manufacture of medicines 2017;300.
34. Norris DA, Puri N, Sinko PJ. The effect of physical barriers and properties on the oral absorption of particulates. Adv Drug Deliv Rev. 1998;34(2–3):135–54.
35. Muhammad U, Uzairu A, Ebuka Arthur D. Review on quantitative structure activity relationship (QSAR) modeling. J Anal Pharm Res. 2018;7(2):240–2.
36. Chillistone S, Hardman JG. Factors affecting drug absorption and distribution. Anaesth Intensive Care Med. 2017;18(7):335–9.
37. Vukovic K, Gadaleta D, Benfenati E. Methodology of aiQSAR: a group-specific approach to QSAR modelling. J Chem. 2019;11:1–9.
38. Soliman ME, Adewumi AT, Akawa OB, Subair TI, Okunlola FO, Akinsuku OE, Khan S. Simulation models for prediction of bioavailability of medicinal drugs—the interface between experiment and computation. AAPS PharmSciTech. 2022;23(3):86.
39. Deepika D, Kumar V. The role of "physiologically based pharmacokinetic model (PBPK)" new approach methodology (NAM) in pharmaceuticals and environmental chemical risk assessment. Int J Environ Res Public Health. 2023;20(4):3473.
40. Parasrampuria DA, Benet LZ, Sharma A. Why drugs fail in late stages of development: case study analyses from the last decade and recommendations. AAPS J. 2018;20:1–6.
41. Jones HM, et al. Physiologically based pharmacokinetic modeling in drug discovery and development: a pharmaceutical industry perspective. Clin Pharmacol Ther. 2015;97:247–62.
42. Peters SA, Dolgos H. Requirements to establishing confidence in physiologically based pharmacokinetic (PBPK) models and overcoming some of the challenges to meeting them. Clin Pharmacokinet. 2019;58(11):1355–71.
43. Mehrotra N, Bhattaram A, Earp JC, Florian J, Krudys K, Lee JE, Lee JY, Liu J, Mulugeta Y, Yu J, Zhao P. Role of quantitative clinical pharmacology in pediatric approval and labeling. Drug Metab Dispos. 2016;44(7):924–33.
44. Johnson TN, Zhou D, Bui KH. Development of physiologically based pharmacokinetic model to evaluate the relative systemic exposure to quetiapine after administration of IR and XR formulations to adults, children and adolescents. Biopharm Drug Dispos. 2014;35(6):341–52.

45. Hansmann S, Darwich A, Margolskee A, Aarons L, Dressman J. Forecasting oral absorption across biopharmaceutics classification system classes with physiologically based pharmacokinetic models. J Pharm Pharmacol. 2016;68(12):1501–15.
46. Dodd S, Kollipara S, Sanchez-Felix M, Kim H, Meng Q, Beato S, Heimbach T. Prediction of ARA/PPI drug-drug interactions at the drug discovery and development interface. J Pharm Sci. 2019;108(1):87–101.
47. Shebley M, Sandhu P, Emami Riedmaier A, Jamei M, Narayanan R, Patel A, Peters SA, Reddy VP, Zheng M, de Zwart L, Beneton M. Physiologically based pharmacokinetic model qualification and reporting procedures for regulatory submissions: a consortium perspective. Clin Pharmacol Ther. 2018;104(1):88–110.
48. de Souza Neto LR, Moreira-Filho JT, Neves BJ, Maidana RL, Guimarães AC, Furnham N, Andrade CH, Silva FP Jr. In silico strategies to support fragment-to-lead optimization in drug discovery. Front Chem. 2020;8:93.
49. Ferreira LL, Andricopulo AD. ADMET modeling approaches in drug discovery. Drug Discov Today. 2019;24(5):1157–65.
50. Komura H, Watanabe R, Mizuguchi K. The trends and future prospective of in silico models from the viewpoint of ADME evaluation in drug discovery. Pharmaceutics. 2023;15(11):2619.
51. Siramshetty V, Williams J, Nguyễn ĐT, Neyra J, Southall N, Mathé E, Xu X, Shah P. Validating ADME QSAR models using marketed drugs. SLAS Discov. 2021;26(10):1326–36.
52. Falcón-Cano G, Molina C, Cabrera-Pérez MÁ. ADME prediction with KNIME: in silico aqueous solubility consensus model based on supervised recursive random forest approaches. ADMET DMPK. 2020;8(3):251–73.
53. Kamiya Y, Omura A, Hayasaka R, Saito R, Sano I, Handa K, Ohori J, Kitajima M, Shono F, Funatsu K, Yamazaki H. Prediction of permeability across intestinal cell monolayers for 219 disparate chemicals using in vitro experimental coefficients in a pH gradient system and in silico analyses by trivariate linear regressions and machine learning. Biochem Pharmacol. 2021;192:114749.
54. Broccatelli F, Salphati L, Plise E, Cheong J, Gobbi A, Lee ML, Aliagas I. Predicting passive permeability of drug-like molecules from chemical structure: where are we? Mol Pharm. 2016;13(12):4199–208.
55. Czub N, Szlęk J, Pacławski A, Klimonczyk K, Puccetti M, Mendyk A. Artificial intelligence-based quantitative structure–property relationship model for predicting human intestinal absorption of compounds with serotonergic activity. Mol Pharm. 2023;20(5):2545–55.
56. Price E, Kalvass JC, DeGoey D, Hosmane B, Doktor S, Desino K. Global analysis of models for predicting human absorption: QSAR, in vitro, and preclinical models. J Med Chem. 2021;64(13):9389–403.
57. Sjögren E, Thorn H, Tannergren C. In silico modeling of gastrointestinal drug absorption: predictive performance of three physiologically based absorption models. Mol Pharm. 2016;13(6):1763–78.
58. Stenberg P, Norinder U, Luthman K, Artursson P. Experimental and computational screening models for the prediction of intestinal drug absorption. J Med Chem. 2001;44(12):1927–37.
59. Fagerholm U, Hellberg S, Alvarsson J, Arvidsson McShane S, Spjuth O. In silico prediction of volume of distribution of drugs in man using conformal prediction performs on par with animal data-based models. Xenobiotica. 2021;51(12):1366–71.
60. Shaker B, Yu MS, Song JS, Ahn S, Ryu JY, Oh KS, Na D. LightBBB: computational prediction model of blood–brain-barrier penetration based on LightGBM. Bioinformatics. 2021;37(8):1135–9.
61. Kosugi Y, Mizuno K, Santos C, Sato S, Hosea N, Zientek M. Direct comparison of the prediction of the unbound brain-to-plasma partitioning utilizing machine learning approach and mechanistic neuropharmacokinetic model. AAPS J. 2021;23:72.
62. Sasahara K, Shibata M, Sasabe H, Suzuki T, Takeuchi K, Umehara K, Kashiyama E. Predicting drug metabolism and pharmacokinetics features of in-house compounds by a hybrid machine-learning model. Drug Metab Pharmacokinet. 2021;39:100395.

63. Sasahara K, Shibata M, Sasabe H, Suzuki T, Takeuchi K, Umehara K, Kashiyama E. Feature importance of machine learning prediction models shows structurally active part and important physicochemical features in drug design. Drug Metab Pharmacokinet. 2021;39:100401.
64. Chen J, Yang H, Zhu L, Wu Z, Li W, Tang Y, Liu G. In silico prediction of human renal clearance of compounds using quantitative structure-pharmacokinetic relationship models. Chem Res Toxicol. 2020;33(2):640–50.
65. Medina-Franco JL. Grand challenges of computer-aided drug design: the road ahead. Front Drug Discov. 2021;1:728551.
66. Baig MH, Ahmad K, Rabbani G, Danishuddin M, Choi I. Computer aided drug design and its application to the development of potential drugs for neurodegenerative disorders. Curr Neuropharmacol. 2018;16(6):740–8.
67. Ahire SM, Jadhav SP, Shewale VV, Pawar PS, Kokande AM, Sonawane DP, Patil DM. A review on computer-aided drug design and discovery. J ReAttach Ther Develop Diversities. 2023;6(10s (2)):1573–82.
68. Chandershekar A, Bhaskar A, Mekkanti MR, Rinku M. A review on Computer Aided Drug Design (CAAD) and it's implications in drug discovery and development process. Int J Health Care Biol Sci. 2020;1(1):27–33.
69. Rajkishan T, Rachana A, Shruti S, Bhumi P, Patel D. Computer-aided drug designing. In: Advances in bioinformatics. Springer; 2021. p. 151–82.
70. Batool M, Ahmad B, Choi S. A structure-based drug discovery paradigm. Int J Mol Sci. 2019;20(11):2783.
71. Yu W, MacKerell AD. Computer-aided drug design methods. In: Antibiotics: methods and protocols. New York: Humana Press; 2017. p. 85–106.
72. Pelkonen O, Boobis AR, Gundert-Remy U. In vitro prediction of gastrointestinal absorption and bioavailability: an experts' meeting report. Eur J Clin Pharmacol. 2001;57:621–9.
73. Torres Silva F, HG Trossini G. The survey of the use of QSAR methods to determine intestinal absorption and oral bioavailability during drug design. Med Chem. 2014;10(5):441–8.
74. Hussain A, Altamimi MA, Alshehri S, Imam SS. Assessment of solubility and Hansen solubility parameters of rifampicin in various permeation enhancers: experimental and computational approach. J Mol Liq. 2021;328:115432.
75. Lee MH, Ta GH, Weng CF, Leong MK. In silico prediction of intestinal permeability by hierarchical support vector regression. Int J Mol Sci. 2020;21(10):3582.
76. Mahdi WA, Hussain A, Altamimi MA, Alshehri S, Bukhari SI, Ahsan MN. Experimental solubility, thermodynamic/computational validations, and GastroPlus-based in silico prediction for subcutaneous delivery of rifampicin. AAPS PharmSciTech. 2021;22:1–4.
77. Rentzsch P, Witten D, Cooper GM, Shendure J, Kircher M. CADD: predicting the deleteriousness of variants throughout the human genome. Nucleic Acids Res. 2019;47(D1):D886–94.
78. Collin CB, Gebhardt T, Golebiewski M, Karaderi T, Hillemanns M, Khan FM, Salehzadeh-Yazdi A, Kirschner M, Krobitsch S, EU-STANDS4PM consortium, Kuepfer L. Computational models for clinical applications in personalized medicine—guidelines and recommendations for data integration and model validation. J Pers Med. 2022;12(2):166.
79. Wu F, Zhou Y, Li L, Shen X, Chen G, Wang X, Liang X, Tan M, Huang Z. Computational approaches in preclinical studies on drug discovery and development. Front Chem. 2020;8:726.
80. Grzegorzewski J, Brandhorst J, Green K, Eleftheriadou D, Duport Y, Barthorscht F, Köller A, Ke DY, De Angelis S, König M. PK-DB: pharmacokinetics database for individualized and stratified computational modeling. Nucleic Acids Res. 2021;49(D1):D1358–64.
81. Leite DM, Brochet X, Resch G, Que YA, Neves A, Peña-Reyes C. Computational prediction of inter-species relationships through omics data analysis and machine learning. BMC Bioinformatics. 2018;19:151–9.
82. Gerlee P, Jöud A, Spreco A, Timpka T. Computational models predicting the early development of the COVID-19 pandemic in Sweden: systematic review, data synthesis, and secondary validation of accuracy. Sci Rep. 2022;12(1):13256.

83. Nakagomi H, Mochizuki H, Inoue M, Hirotsu Y, Amemiya K, Sakamoto I, Nakagomi S, Kubota T, Omata M. Combined annotationdependent depletion score for BRCA1/2 variants in patients with breast and/or ovarian cancer. Cancer science. 2018 Feb;109(2):453–61.
84. Gómez-Bombarelli R, Wei JN, Duvenaud D, Hernández-Lobato JM, Sánchez-Lengeling B, Sheberla D, Aguilera-Iparraguirre J, Hirzel TD, Adams RP, Aspuru-Guzik A. Automatic chemical design using a data-driven continuous representation of molecules. ACS Central Sci. 2018;4(2):268–76.
85. Schneider P, Walters WP, Plowright AT, Sieroka N, Listgarten J, Goodnow RA Jr, Fisher J, Jansen JM, Duca JS, Rush TS, Zentgraf M. Rethinking drug design in the artificial intelligence era. Nat Rev Drug Discov. 2020;19(5):353–64.
86. Niazi SK, Mariam Z. Computer-aided drug design and drug discovery: a prospective analysis. Pharmaceuticals. 2023;17(1):22.
87. Jorgensen WL. The many roles of computation in drug discovery. Science. 2004;303(5665):1813–8.
88. Warren GL, Andrews CW, Capelli AM, Clarke B, LaLonde J, Lambert MH, Lindvall M, Nevins N, Semus SF, Senger S, Tedesco G. A critical assessment of docking programs and scoring functions. J Med Chem. 2006;49(20):5912–31.
89. Walters WP, Murcko MA. Prediction of 'drug-likeness'. Adv Drug Deliv Rev. 2002;54(3):255–71.
90. Kitchen DB, Decornez H, Furr JR, Bajorath J. Docking and scoring in virtual screening for drug discovery: methods and applications. Nat Rev Drug Discov. 2004;3(11):935–49.
91. Dror RO, Dirks RM, Grossman JP, Xu H, Shaw DE. Biomolecular simulation: a computational microscope for molecular biology. Annu Rev Biophys. 2012;41(1):429–52.
92. Teague SJ. Implications of protein flexibility for drug discovery. Nat Rev Drug Discov. 2003;2(7):527–41.
93. Ching T, Himmelstein DS, Beaulieu-Jones BK, Kalinin AA, Do BT, Way GP, Ferrero E, Agapow PM, Zietz M, Hoffman MM, Xie W. Opportunities and obstacles for deep learning in biology and medicine. J R Soc Interface. 2018;15(141):20170387.

GPSR Compliance

The European Union's (EU) General Product Safety Regulation (GPSR) is a set of rules that requires consumer products to be safe and our obligations to ensure this.

If you have any concerns about our products, you can contact us on ProductSafety@springernature.com

In case Publisher is established outside the EU, the EU authorized representative is:

Springer Nature Customer Service Center GmbH
Europaplatz 3
69115 Heidelberg, Germany

Batch number: 08468046

Printed by Printforce, the Netherlands